PROCEEDINGS OF THE 14th WORLD CONGRESS
International Federation of Automatic Control

Beijing, P.R. China, 5 - 9 July 1999

(in 18 volumes)

Edited by

Han-Fu CHEN
Chinese Academy of Sciences
Beijing, P.R.China

Dai-Zhan CHENG
Chinese Academy of Sciences
Beijing, P.R.China

and

Ji-Feng ZHANG
Chinese Academy of Sciences
Beijing, P.R.China

VOLUME E
ROBUST CONTROL II

Volume Editors

A. ISIDORI (Italy) and T. GLAD (Sweden)

Published for the

INTERNATIONAL FEDERATION OF AUTOMATIC CONTROL

by

PERGAMON
An Imprint of Elsevier Science

UK	Elsevier Science Ltd, The Boulevard, Langford Lane, Kidlington, Oxford, OX5 1GB, UK
USA	Elsevier Science Inc., 660 White Plains Road, Tarrytown, New York 10591-5153, USA
JAPAN	Elsevier Science Japan, Tsunashima Building Annex, 3-20-12 Yushima, Bunkyo-ku, Tokyo 113, Japan

First edition 1999

Library of Congress Cataloging in Publication Data

A catalogue record for this book is available from the Library of Congress

British Library Cataloguing in Publication Data

A catalogue record for this book is available from the British Library

ISBN 0-08-043216 6

Transferred to Digital Printing 2009

14th Triennial World Congress of IFAC,
the International Federation of Automatic Control

Hosted by
The Chinese Association of Automation

International Program Committee
Han-Fu Chen (China), *Chairman*
Luis Basanez (Spain), *Vice-Chair*
Mogens Blanke (Denmark)
Xiren Cao (Hong Kong, China), *Vice-Chair*
Tian-You Chai (China)
Jose B. Cruz, Jr. (USA), *Vice-Chair*
Manfred Deistler (Austria)
Tyrone Duncan (USA)
Lei Guo (China)
Yasushi Hashimoto (Japan)
Alberto Isidori (Italy)
Laszlo Keviczky (Hungary)
Vladimir Kucera (Czech Republic)
Lennart Ljung (Sweden)
Lena Martensson (Sweden)
Micheal K. Masten (USA)
Thomas McAvoy (USA)
Umit Ozguner (USA)
Juan de la Puente (Spain)
Tzyh-Jong Tarn (USA)

National Organizing Committee
Yong-Xiang Lu (China), *Chairman*
Han-Fu Chen (China), *Co-Chair*
Ru-Wei Dai (China), *Co-Chair*
Bo-Lin Sun (China), *Secretary General*
Gong-Qing Zhang (China), *Treasurer*
Shu-Hui Deng (China)
Yan-Da Li (China)
Song-De Ma (China)
K. L. Mak (Hong Kong, China)
Yu-Geng Xi (China)
Jing-Heng Yang (China)
Kan Zhang (China)

National Congress Advisory Board
Jian-Song (China), *Chairman*
Yu-Chi Ho (USA)
Qi-Heng Hu (China)
Tzyh-Jong Tarn (USA)
Jia-Chi Yang (China)
Guang-Ya Zhu (China)

The Chinese Association of Automation
Ru-Wei Dai, *Co-President, (1997-2002)*
Han-Fu Chen, *Co-President, (1997-2002)*

IFAC President (1996-1999)
Yong-Zai Lu (China)

CONTENTS OF VOLUMES

PREFACE

As the last triennial World Congress of the International Federation of Automatic Control (IFAC) of this century, the 14th IFAC Congress is recognized as a preeminent event for the worldwide community of control systems science and engineering. It is a great pleasure and honor for Chinese control researchers and engineers to host the Congress in Beijing.

Beijing is one of the oldest cities in the world. Its civilization can be traced back to 1000 BC when it was a small town. Since the Jin Dynasty (AD 1115), and during the Yuan, Ming and Qing Dynasties, Beijing has been the capital. It is famous partly for its many historical sites, including the Great Wall, the Forbidden City, the Summer Palace, the Ming Tombs, etc. Beijing is also famous for the development of automatic machines over many centuries, including the armillary sphere for simulating the Universe and the seismograph for detecting earthquakes, making the city one of great world centres for automation.

Beijing is also a young city; every day it shows a new face to the world. The fast pace of change in Beijing is fuelled by the rapid growth of the economy and the development of science and technology, including automation. China has vast potential for the development of automatic control. Beijing cordially welcomes you!

In the light of the outstanding efforts of our predecessors, our goal in organizing this Congress is to continue to increase the quality of the papers and of the Congress activities. We hope this Congress will be remembered as a showcase for the fruitful results on control theory and application achieved since the beginning of this century. In addition, we expect that several of the results and ideas demonstrated in this Congress will have lasting influence in the next century on the directions taken by the field of control systems engineering.

The response from the international control community to the call for papers was strong. The International Program Committee (IPC) received 2169 submitted papers, 53 invited session proposals and eight panel discussion proposals from 68 countries and regions. We were impressed by the high quality of the submissions. The technical program for the Congress covers a broad range of topics, with sessions arranged into nine congress tracks and subdivided into 45 technical areas. Reviews were conducted by all of the IFAC Technical Committees, with the help of hundreds of worldwide experts. Through this rigorous review procedure 1556 papers, one keynote speech, five plenary lectures, 47 invited session proposals and seven panel discussion proposals were accepted. They came from 59 countries and regions. The IPC is confident that the program that has been selected will provide significant contributions to each of the subject areas of the congress tracks.

There are five plenary lectures scheduled, one at the start of each of the five days of the Congress. The plenary speakers include some of the world's most outstanding authorities in various control fields. The topics of the plenary lectures include surveys, prospects and introductions to advanced control theory and applications. In selecting the plenary speakers, we have been guided by a desire to provide perspectives on theoretical topics, important applications, and other issues of common interest. It was a difficult task to select five plenary speakers among so many outstanding candidates who were nominated by our colleagues.

Following the plenary addresses, the congress program is composed of more than 1500 papers, to be presented in 214 lecture-type sessions and 47 poster-type sessions, the latter arranged in three time clusters. The paper sessions are supplemented by seven panel discussions. No distinction is made between poster papers and lecture papers in the Proceedings.

The Proceedings of the Congress are published in two forms, printed and CD-ROM. The Proceedings in its printed form consists of 18 volumes: 17 subject volumes, which contain all lecture and poster papers and a plenary volume. The plenary volume contains the keynote speech, plenary papers, panel discussions, and the table of contents, author index and subject index for the full set of subject volumes.

The CD-ROM, containing all the material in the 18 volumes, is provided to each registered participant as part of the registration material. Both the printed Proceedings and the CD-ROM will be marketed after the Congress, as the official Congress Proceedings, by the IFAC Publisher, Elsevier Science Ltd.

We expect that the reader, whether a participant in the Congress, or a researcher or practising engineer studying the material afterwards, will find the Congress publications useful, informative and helpful in their related creative works.

The Congress organizers have made every effort to create a high quality environment for the Congress, but China, as a developing country, still may seem unsatisfactory in many respects, including the language barrier. You may feel inconvenienced and you may need to be patient in communicating your needs. Imagination, plus a little spirit of adventure, may help you to have some unexpected and unforgettable experiences in this part of the world, which, in the end, makes life colorful.

Finally, we would like to thank all IPC members, all IFAC Technical Committee chairs and all reviewers for their valuable contribution to the Congress. We are also indebted to all our friends from different parts of the world who have given support, help and useful suggestions.

Han-Fu CHEN, Daizhan CHENG and Jifeng ZHANG

CONTENTS OF VOLUME E

ROBUST CONTROL II
(Technical Area 2c-II)

14th Triennial World Congress, Beijing, P.R. China

OPTIMAL CONTROL FOR TIMED AUTOMATA [1]

Eugene Asarin * Oded Maler **

* *Institute for Information Transmission Problems, 19 Bol. Karetnyi per. 101447 Moscow, Russia,* asarin@iitp.ru
** Verimag, *2, av. de Vignate, 38610 Gières, France,* Oded.Maler@imag.fr [2]

Abstract: Timed automata enrich discrete event systems with quantitative timing information, incorporating information about delays and durations into discrete models. In this paper our previous work on controller synthesis for timed automata, which dealt with qualitative properties such as avoidance of forbidden states, is extended to the following problem: given a timed game automaton, find a strategy (=controller) which brings it to a desired set of states as fast as possible.

Keywords: Timed Automata, Optimal Control, Dynamic Programming

1. INTRODUCTION

Timed automata (TA) of Alur and Dill (1994) are a system model where quantitative timing constraints are added to discrete transition systems. So far most of the work on verification and synthesis for TA concentrated on qualitative all-or-nothing properties and the only (important) role of the timing information was to constrain the set of possible behaviors.

In order to treat the control synthesis problem our previously-introduced model of *Timed Game Automaton* (TGA) from Asarin et al. (1995, 1998) is used. It is the usual TA where the actions are partitioned into those of the controller and those of the uncontrolled environment. On this model various controller synthesis problems for *qualitative* properties can be formulated and solved. One example is the eventuality problem: find the set of "winning" states from which the controller can enforce the TGA into a set F, and compute the "strategy" for these states. In this work the following natural *quantitative* generalization is solved: find for each state the *minimal* time in which the controller can enforce the automaton into F, regardless of uncontrolled events, and construct a controller which achieves this optimum for each state.

As in Asarin et al. (1998) the solution is obtained by a backward fixed-point calculation on the state-space of the TGA. The main difference is that the iteration is performed on a function from the TGA state-space into $\mathbb{R}$ which denotes roughly the min-max of the temporal distance from the state to F (this is similar to value/policy iterations used in dynamic programming, e.g. Bertsekas (1995)). Since the state-space of a TGA is non-countable, these functions cannot be tabulated. Fortunately, all the value functions encountered in the iteration of our synthesis algorithm belong to a special well-founded class of functions admitting simple linear-algebraic representation. This guarantees the termination of the algorithm.

The rest of the paper is organized as follows: In section 2 the TGA model is re-introduced. In section 3 the controller synthesis problem is for-

[1] This work was partially supported by the European Community Esprit-LTR Project 26270 VHS (Verification of Hybrid systems), the French-Israeli collaboration project 970MAEFUT5 (Hybrid Models of Industrial Plants) and by Research Grants 97-01-00692 and 96-15-96048 of Russian Foundation of Basic Research.
[2] Verimag is a joint laboratory of CNRS, UJF, and INPG

mulated and our synthesis algorithm is described. In section 4 correctness and termination of the algorithm are proved. An example and some remarks conclude the paper.

2. GAMES ON TIMED SYSTEMS

2.1 *The Context*

Timed games are extensions of discrete games (or equivalently of plant models used in the theory of supervisory control for discrete event systems, Ramadge and Wonham (1987)) where the players' actions may take place anywhere on the physical time axis, subject to certain timing constraints. For such games the model of TA of Alur and Dill (1994) is used, in which automata are equipped with auxiliary continuous variables called *clocks* which grow uniformly when the automaton is in some state. The clocks interact with the transitions by participating in pre-conditions (guards) for certain transitions and they are possibly reset when some transitions are taken.

In this continuous-time setting, a player might choose at a given moment to wait some time t and *then* take a transition. In this case, it should consider not only what the adversary can do *after* this action but also the possibility that the latter *will not wait for t time*, and perform an action at some $t' < t$.

For a more elaborate description of the game-theoretic approach to untimed and timed synthesis, the reader is encouraged to look at the lengthy introductions of Asarin et al. (1995, 1998) as well as Tomlin et al. (1998).

2.2 *Timed Game Automata*

Let Q be a finite set of *states* and let $X = \mathbb{R}_+^d$ for some integer d be the *clock space.* Denote elements of X as $\boldsymbol{x} = (x_1, \ldots, x_d)$ and use $\boldsymbol{x} + t$ for $\boldsymbol{x} + (t, t, \ldots, t)$. Elements of $Q \times X$ are called *configurations.* Let $K \in \mathbb{N}$ be a constant fixed throughout the paper — it is the largest constant in the definition of the TGA. A subset of X is called a *zone* if it can be obtained as a boolean combination of inequalities of the form $x_i \leq c$, $x_i < c$, $x_i - x_j \leq c$, where $c \in \{0, 1, \ldots, K\}$. The set of zones is denoted by $\mathcal{Z}(X)$. A zone Z is *right-open* if for any $\boldsymbol{x} \in Z$ there is a $\tau > 0$ such that the interval $(\boldsymbol{x}, \boldsymbol{x} + \tau)$ is also included in Z. These properties of subsets of X extend naturally to subsets of $Q \times X$, e.g. $P \subseteq Q \times X$ is called a zone if it can be written as finite union of sets of the form $\{q_i\} \times P_i$ such that all the P_i's are zones. A function $\rho : X \to X$ is a *reset* function if it sets some of the coordinates of its argument to 0 and leaves the others intact. The set of all such functions is denoted by $\mathcal{J}(X)$. Assume two finite sets of actions A and B, a special empty action ε and let $A^\varepsilon = A \cup \{\varepsilon\}$ and $B^\varepsilon = B \cup \{\varepsilon\}$. The set A represents our actions (i.e. possible actions of the controller), while B stands for uncontrollable actions of the environment. The action ε stands for "wait and see".

Definition 1. (Timed Game Automaton). A TGA is a tuple $\mathcal{A} = (Z, A, B, T^A, T^B, \delta, \rho)$ where $Z \subseteq Q \times X$ is a zone, Q and X are the state and clock spaces, A and B are two distinct action alphabets, $T^A \subseteq Q \times X \times A^\varepsilon$ and $T^B \subseteq Q \times X \times B^\varepsilon$ are timing constraints for the two types of actions, the functions $\delta : Q \times A^\varepsilon \times B^\varepsilon \to Q$ and $\rho : Q \times A^\varepsilon \times B^\varepsilon \to \mathcal{J}(X)$ indicate which state is reached when performing a (possibly joint) action and which clocks are reset in that occasion.

Further requirements are the following: for every state q and action $a \in A^\varepsilon$, the set $T^A(q, a) = \{\boldsymbol{x} : (q, \boldsymbol{x}, a) \in T^A\}$ is a zone. Similar requirements hold for T^B. Assume that $\delta(q, \varepsilon, \varepsilon) = q$ and that $\rho(q, \varepsilon, \varepsilon)$ is the identity function (if both sides refrain from action nothing happens). The automaton is required to be *strongly non-Zeno,* that is, in every cycle in the transition graph of the automaton (induced by δ), there is at least one transition which resets a clock variable x_i to zero, and at least one transition which can be taken only if $x_i \geq 1$. This is a very important condition as it prevents the controller and the environment to achieve their goals using unrealistic tricks that stop time.

The last requirement is that the zone $T^A(q, \varepsilon)$ is right-open for any q. The meaning of this requirement is that the A player can wait at time t_0 only if he or she can wait a little more.

Intuitively, when the automaton is at a configuration $(q, \boldsymbol{x})$, time can progress as long as both players agree, that is, $(q, \boldsymbol{x}, \varepsilon) \in T^B \cap T^A$. As soon as one of them can take an action, i.e. $(q, \boldsymbol{x}, a) \in T^A$ for some $a \in A$ or $(q, \boldsymbol{x}, b) \in T^B$ for some $b \in B$, or both, a transition can be taken. This can be formalized as follows:

Definition 2. (Steps and Runs). A *joint step* of a TGA $\mathcal{A}$ is $(q, \boldsymbol{x}) \longrightarrow (q', \boldsymbol{x}')$ which is either

(1) a time step (of duration t):.
$(q, \boldsymbol{x}) \xrightarrow{t} (q, \boldsymbol{x} + t)$, such that $t > 0$, and for every $t' < t$,
$(q, \boldsymbol{x} + t', \varepsilon) \in T^A \cap T^B$.

(2) a discrete step $(q, \boldsymbol{x}) \xrightarrow{(a,b)} (q', \boldsymbol{x}')$ such that $(a, b) \neq (\varepsilon, \varepsilon)$, $(q, \boldsymbol{x}, a) \in T^A$, $(q, \boldsymbol{x}, b) \in T^B$, $q' = \delta(q, a, b)$, and $\boldsymbol{x}' = \rho(q, a, b)(\boldsymbol{x})$.

A *run* of a TGA $\mathcal{A}$ starting from $(q_0, \boldsymbol{x}_0)$ is a sequence of joint steps
$\xi : (q_0, \boldsymbol{x}_0) \longrightarrow (q_1, \boldsymbol{x}_1) \longrightarrow \dots$

Note that $(q, \boldsymbol{x}, \varepsilon) \in T^A \cap T^B$ means that both A and B agree to let time progress by a *positive* amount. On the other hand it is possible to reach a state where ε is not permitted by one or more of the two players: in this case, the only thing that can happen then is a discrete step.

For ease of notation introduce the total transition function $\bar{\delta} : Q \times X \times A^\varepsilon \times B^\varepsilon \to Q \times X \cup \{\top, \bot\}$ defined by $\bar{\delta}(q, \boldsymbol{x}, a, b) = (\delta(q, a, b), \rho(q, a, b)(\boldsymbol{x}))$ in the "normal" case when $(a, b) \neq (\varepsilon, \varepsilon)$, $(q, \boldsymbol{x}, a) \in T^A$, and $(q, \boldsymbol{x}, b) \in T^B$. For completeness put it equal to a special "bad" state $\bot$ when $(q, \boldsymbol{x}, a) \notin T^A$ and a "good" state $\top$ when $(q, \boldsymbol{x}, a) \in T^A, (q, \boldsymbol{x}, b) \notin T^B$. Finally, put $\bar{\delta}(q, \boldsymbol{x}, \varepsilon, \varepsilon) = (q, \boldsymbol{x})$ when $(q, \boldsymbol{x}, \varepsilon) \in T^A \cap T^B$.

Clearly, replacing T^A by any subset of it, will restrict the range of action that the player A can chose at certain configurations, and hence decrease the set of behaviors of the TGA. Controller synthesis problems can be formulated as finding appropriate restrictions of T^A, referred to as strategies. These strategies are not necessary deterministic as they might allow A more than one action at a given configuration.

3. THE PROBLEM AND THE ALGORITHM

Definition 3. (Optimal Control Problem). Given a TGA $\mathcal{A}$ and a set $F \subset Q \times X$ find a strategy $T^A_* \subseteq T^A$ for player A which allows him or her to reach the target set F as fast as possible whatever player B does.

The following algorithm is a generalization of the synthesis method for eventuality games on timed automata. The algorithm iterates over the value function which gives an upper estimate of arrival times to the set F. When a value function $f : Q \times X \cup \{\top, \bot\} \to \mathbb{R}_+ \cup \{0, \infty\}$ at any step of the iteration is found, it means that a controller which allows to reach F from $(q, \boldsymbol{x})$ in no more than $f(q, \boldsymbol{x})$ time is also found. The algorithm for finding the value function has the following form:

Algorithm 1. (Value Iteration for TGA).
{Initialization}

$$f_0(q, \boldsymbol{x}) = \begin{cases} 0 \text{ if } (q, \boldsymbol{x}) \in F \cup \{\top\} \\ \infty \quad \text{otherwise;} \end{cases}$$

$n := 0;$

{Iteration}

repeat

$n := n + 1;$
$f_n := \pi(f_{n-1});$

until $f_n = f_{n-1}$;

{Strategy extraction}
$f_* := f(n);$
$T^A_* = \alpha(f_*).$

The operators used in the algorithm are as follows:

$$\pi(f) = \min\{f, \pi_{\mathrm{A}}(f), \pi_{\mathrm{I}}(f)\},$$

where

$$\pi_{\mathrm{A}}(f)(q, \boldsymbol{x}) = \min_{a \in A} \max_{b \in B^\varepsilon} f(\bar{\delta}(q, \boldsymbol{x}, a, b))$$

and

$$\pi_{\mathrm{I}}(f)(q, \boldsymbol{x}) = \inf_{t \in \mathbb{R}_+} v(q, \boldsymbol{x}, t) \tag{1}$$

where

$$v(q, \boldsymbol{x}, t) = \max(\sup_{\tau < t} g(q, \boldsymbol{x}, \tau), t + f(q, \boldsymbol{x} + t)) \tag{2}$$

and

$$g(q, \boldsymbol{x}, \tau) = \max_{b \in B}(\tau + f(\bar{\delta}(q, \boldsymbol{x} + \tau, \varepsilon, b))) \tag{3}$$

Intuitively $g(q, \boldsymbol{x}, \tau)$ specifies the worst (largest) f of the configuration in which the game might be after B has performed an action at τ while A was idling. Similarly $v(q, \boldsymbol{x}, t)$ is the worst thing that can happen to A after deciding to idle for t time at $(q, \boldsymbol{x})$: it includes all the possible outcomes of B's actions at $\tau < t$. The best waiting time for A is the t which minimizes $v(q, \boldsymbol{x}, t)$ and its outcome π_{I} is compared with π_{A} which denotes the best outcome of an immediate action.

The strategy extraction operator for a configuration $z = (q, \boldsymbol{x})$ is given by the formula: $\alpha(f)(z) =$

$$\begin{cases} \alpha_A(f)(z) \text{ if } \pi_{\mathrm{A}}(f)(z) < \pi_{\mathrm{I}}(f)(z) \\ \{\varepsilon\} \text{ if } \pi_{\mathrm{I}}(f)(z) < \pi_{\mathrm{A}}(f)(z) \\ \alpha_A(f)(z) \cup \{\varepsilon\} \text{ if } \pi_{\mathrm{A}}(f)(z) = \pi_{\mathrm{I}}(f)(z) < \infty \\ \emptyset \text{ if } \pi_{\mathrm{A}}(f)(z) = \pi_{\mathrm{I}}(f)(z) = \infty \end{cases}$$

where $\alpha_A(f)(z) =$

$$\{a \in T^A(z) | \max_{b \in B^\varepsilon} f(\bar{\delta}(z, a, b) = \pi_{\mathrm{A}}(f(z))\}.$$

Now a subtlety of real-valued time should be addressed. The bad news about the strategy extracted by α is that it can violate the right-openness condition. This is due to the fact of the life that sometimes player A has to take a transition the sooner the better but after time t_0. This is reflected in the strategy $\alpha(f)$ as follows: player A has action ε enabled until t_0 *including* t_0 and a "good" discrete transition enabled after t_0. But formally speaking, this strategy is blocking at time t_0. This is the reason to introduce its non-blocking approximations $\alpha_\varsigma(f)$ for small $\varsigma > 0$, which will satisfy the right-openness condition (but use non-integers in the guards). It will be shown in the next section, that taking this strategy for ς small enough player a can reach the aim f in time arbitrarily close to $f_*(q,x)$.

The ς-relaxed strategy is obtained from $\alpha(f)$ by enabling the ε action on a narrow stripe: $\alpha_\varsigma(f) = \alpha(f) \cup \{(q,x,\varepsilon) : (q,x) \in S_\varsigma(f)\}$, where

$$S_\varsigma(f) = \mathcal{I}\{(q,x)|\pi_{\mathrm{I}}(f)(q,x) < \pi_{\mathrm{A}}(f)(q,x) + \varsigma\},$$

and $\mathcal{I}$ stands for interior of a set.

4. CORRECTNESS PROOF

4.1 *Partial correctness*

Our first aim is to prove that if the algorithm converges then it gives the optimal least-restrictive solution of the problem. The next two lemmata can be proved by a straightforward induction over n. Let $\lambda > 0$ be any given (small) positive number.

Lemma 4. If player A uses the strategy $T^A_{n,\varsigma} = \alpha_\varsigma(f_{n-1}))$ from an initial position in (q,x) with $v = f_n(q,x) < \infty$, then the TGA reaches the target set F in no more then $v + n\varsigma$ time with no more then n transitions whatever the adversary does.

Corollary 5. Suppose that algorithm 1 converges in N iterations. Let $\varsigma = \lambda/N$ If player A uses the strategy $T^A_{*,\varsigma} = \alpha_\varsigma(f_*)$ from an initial position (q,x) such that $v = f_*(q,x) < \infty$ then the TGA reaches the target set F in no more then $v + \lambda$ time whatever the adversary does.

Lemma 6. For every (q,x) and n, player B can prevent the TGA from reaching F during at least n steps or $f_n(q,x) - \lambda$ time, whatever player A does.

Corollary 7. Whatever player A does from an initial position in (q,x) player B can prevent the TGA from reaching the target set F during at least $f_*(q,x) - \lambda$ time.

Corollary 8. If from an initial state (q,x) player A makes a transition not in T^A_*, then the adversary B can prevent the TGA from reaching the target set F during strictly more than $f_*(q,x)$ time.

These corollaries imply the partial correctness of the algorithm.

Claim 9. Suppose the algorithm 1 converges. Then

(1) The function $f_*(q,x)$ is the infimal time necessary for A to drive the TGA from (q,x) to F whatever B does;

(2) The strategy $T^A_{*,\varsigma} = \alpha_\varsigma(f_*)$ guarantees arrival to F in time $f_*(q,x) + \lambda$;

(3) It is least restrictive: any strategy which guarantees arrival to F in time $f_*(q,x)$ is a sub-strategy of T^A_*.

4.2 *Simple functions*

Now two things should be stated. The first one is effectiveness. The problem is that the algorithm works over functions from $Q \times \mathbb{R}^n$ to $\mathbb{R}$ and a finite representation of these functions is needed in order to implement the algorithm. The second issue is to prove that the algorithm converges. Effectiveness of the algorithm and its finite convergence follow from the fact that all the functions used in the algorithm belong to the following class which is closed under the operator π.

Definition 10. A function $f : X \rightarrow \mathbb{R}_+ \cup \{0,\infty\}$ is called *simple* iff it can be represented as

$$f(x) = \begin{cases} c_i & \text{if } x \in D_i \\ d_j - x_{l_j} & \text{if } x \in E_j \end{cases}$$

where $D_i, i = 1,\ldots,M$ and $E_j, j = 1,\ldots,N$ are zones, $E_j \subseteq \{x|x_{l_j} \leq K\}$ and $c_i, d_j \in \mathbb{N} \cup \{\infty\}$. A function $f : Q \times X \rightarrow \mathbb{R}_+ \cup \{0,\infty\}$ is simple if $f(q,\cdot)$ is simple for any fixed $q \in Q$

The additional boundedness condition on the E_j's means that the value of x_{l_j} can influence f only when $x_{l_j} < K$, the largest constant in the definition of the TGA. Clearly a simple functions admits a finite representation as a list of zones D_i and E_j and of constants c_i and d_j. It is easy to see that equality of simple functions is decidable, and that they satisfy some closure properties:

Lemma 11. If $f(x)$ and $g(x)$ are simple and $\rho(x)$ a reset function, then $\max(f,g)$, $\min(f,g)$ and $f(\rho(x))$ are also simple. The set $\{x|f(x) < g(x)\}$ is a zone.

All the closure results proved in this subsection are effective in the sense that given a representation of original functions as simple, representations of resulting functions can be found algorithmically.

Another property of simple functions will be crucial for the proof of convergence of the algorithm.

Lemma 12. (Well-foundedness). Any decreasing sequence of simple functions is finite.

4.3 *Effectiveness and convergence*

The key step in the proof of convergence and effectiveness of the algorithm 1 is the closure of simple function under operators used in the algorithm. Let $\Delta(\boldsymbol{x}, Z)$ be the largest backward distance from $\boldsymbol{x}$ to a zone Z, i.e. $\Delta(\boldsymbol{x}, Z) = \sup\{t \geq 0 | \boldsymbol{x} + t \in Z\}$. It is easy to see that Δ is simple in the first argument.

Lemma 13. (Closure of simple functions under π). If $f(q, \boldsymbol{x})$ is simple, then $\pi_{\mathrm{A}}(f)$, $\pi_{\mathrm{I}}(f)$ and $\pi(f)$ are simple. Their representations can be found effectively.

PROOF. For $\pi_{\mathrm{A}}(f)$ it follows immediately from lemma 11 and the fact that for any fixed a, b and q the function $\bar{\delta}(q, \cdot, a, b)$ is a reset function (restricted to a zone).

The case of $\pi_{\mathrm{I}}(f)$ is more difficult. Looking at equations (1), (2), and (3) one can see that $\pi_{\mathrm{I}}(f)$ is obtained by elimination (via inf and sup) of the t variable from functions whose domain includes state-space and time. For example $g(q, \boldsymbol{x}, \tau)$ of (3) is equal to

$$g(q, \boldsymbol{x}, \tau) = \tau + \max_{b \in B}(\bar{\delta}(q, \boldsymbol{x} + \tau, \varepsilon, b))$$

and from the closure of simple function under reset, one can conclude that $g(q, \boldsymbol{x}, t)$ is of the form $t + f'(q, \boldsymbol{x} + t)$ for a simple function f'. This motivates the definition of the following class of functions:

Definition 14. (Bizones and Nice Functions). A *bizone* is a union of sets of the form $\{(\boldsymbol{x}, t) | \boldsymbol{x} \in C \wedge \boldsymbol{x} + t \in D\}$ where C and D are zones. A function $f : X \times \mathbb{R}_{+} \to \mathbb{R}_{+} \cup \{0, \infty\}$ is referred to as *nice* iff it can be represented as

$$g(\boldsymbol{x}, t) = \begin{cases} c_i + t \text{ if } (\boldsymbol{x}, t) \in D_i \\ d_j - x_{l_j} \text{ if } (\boldsymbol{x}, t) \in E_j, \end{cases}$$

where $D_i, i = 1, \ldots, M$ and $E_j, j = 1, \ldots, N$ are bizones, $E_j \subseteq \{\boldsymbol{x} | x_{l_j} \leq K\}$ and $c_i, d_j \in \mathbb{N} \cup \{\infty\}$.

Sublemma 1. (Properties of Nice Functions).

(1) If $f(\boldsymbol{x})$ is simple, then $t + f(\boldsymbol{x} + t)$ is nice;
(2) if g and h are nice, then so are $\min\{g, h\}$ and $\max\{g, h\}$;
(3) if g is nice, then $h(\boldsymbol{x}, t) = \sup_{\tau < t} g(\boldsymbol{x}, \tau)$ is nice;
(4) if g is nice, then $\inf_{t \in \mathbb{R}_{+}} g(\boldsymbol{x}, t)$ is simple.

PROOF. (1) Immediate from definitions.

(2) Let

$$g(\boldsymbol{x}, t) = \begin{cases} c_i + t \text{ if } (\boldsymbol{x}, t) \in D_i \\ d_i - x_{l_i} \text{ if } (\boldsymbol{x}, t) \in E_i, \end{cases}$$

$$h(\boldsymbol{x}, t) = \begin{cases} r_m + t \text{ if } (\boldsymbol{x}, t) \in P_m \\ s_m - x_{l_n} \text{ if } (\boldsymbol{x}, t) \in Q_m, \end{cases}$$

and let

$$u(\boldsymbol{x}, t) = \max\{g(\boldsymbol{x}, t), h(\boldsymbol{x}, t)\} = \\ = \begin{cases} g(\boldsymbol{x}, t) \text{ if } g(\boldsymbol{x}, t) > h(\boldsymbol{x}, t) \\ h(\boldsymbol{x}, t) \text{ if } g(\boldsymbol{x}, t) \leq h(\boldsymbol{x}, t), \end{cases}$$

To obtain a representation for $u(\boldsymbol{x}, t)$ each line of the formula for g is combined with each line of the formula for h and it is to verify that the following types of sets, which appear when comparing lines of these formulae, are bizones:

$\{(\boldsymbol{x}, t) | r + t \leq c + t\}$. In fact it is either $\mathbb{R}_{+}^{n+1}$ or $\emptyset$, hence a bizone.
$\{(\boldsymbol{x}, t) | s - x_i \leq c + t\}$. It can be written in the form $\{(\boldsymbol{x}, t) | x_i + t \geq s - c\}$, which is a bizone.
$\{(\boldsymbol{x}, t) | s - x_i \leq d - x_i\}$. It is either $\mathbb{R}_{+}^{n+1}$ or $\emptyset$, hence a bizone.
$\{(\boldsymbol{x}, t) | s - x_i \leq d - x_j\}$. It can be written in the form $\{(\boldsymbol{x}, t) | x_i - x_j \geq s - d\}$, which is a bizone.

When intersected with $D_i \cap P_m$, $D_i \cap Q_m$, $E_i \cap P_m$, and $E_i \cap Q_m$ these sets produce bizones participating in the definition of the function $u(\boldsymbol{x}, t)$. The boundedness condition is inherited from the bizones E_i and Q_m. The reasoning for min is identical.

(3) Let

$$g(\boldsymbol{x}, \tau) = \begin{cases} c_i + \tau \text{ if } (\boldsymbol{x}, \tau) \in D_i \\ d_j - x_{l_j} \text{ if } (\boldsymbol{x}, \tau) \in E_j, \end{cases}$$

and $h(\boldsymbol{x}, t) = \sup_{\tau < t} g(\boldsymbol{x}, \tau)$.

Observe that the set $[\boldsymbol{x}, \boldsymbol{x} + t] = \{\boldsymbol{x} + \tau | t \in [0, t]\}$ intersects finitely many D_i and E_j sets from the definition of g. Associate with them the following (partial) functions:

$u_j(\boldsymbol{x}, t) = d_j - x_{l_j}$ if $[\boldsymbol{x}, \boldsymbol{x} + t] \cap E_j \neq \emptyset$;
$v_i(\boldsymbol{x}, t) = c_i - t$ if $\boldsymbol{x} + t \in D_i$
and $w_i(\boldsymbol{x}, t) = c_i + \Delta(\boldsymbol{x}, D_i)$
if $[\boldsymbol{x}, \boldsymbol{x} + t] \cap D_i \neq \emptyset \ \wedge \ \boldsymbol{x} + t \notin D_i$

All the defining conditions are bizones, the functions are nice in $\boldsymbol{x}$ and t and so is h which is their max.

(4) Similar to statement 3.

This concludes the proof of the sublemma. □

It follows immediately from the sublemma that $\pi_{\mathrm{I}}(f)$ is simple. The case of $\pi(f)$ is immediate from two previous cases and Lemma 11. □

Corollary 15. The algorithm 1 is effective and always converges.

PROOF. Effectiveness follows immediately from lemma 13.

The sequence f_n is a decreasing sequence of simple functions. By virtue of Lemma 12 it stabilizes. □

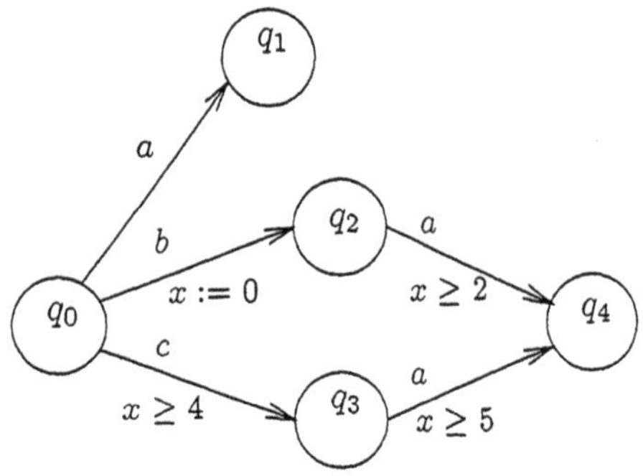

Fig. 1. A TGA with a trivial adversary. Missing guards and invariants are *true*.

Together with Claim 9 this finishes the proof of correctness of the algorithm.

Theorem 16. (Main Result). The algorithm 1 always converges and produces the less-restrictive optimal strategy for the optimal control problem for Timed Automata.

5. EXAMPLE

Consider the 1-clock TGA of figure 1 where the adversary B is trivial and the target set F is $\{q_4\}$. From q_0 player A can choose between waiting, going to q_1 (a losing sink), to q_2 (while resetting the clock) or to q_3. Intuitively, for smaller values of x it might be better to do b and go to q_2 because A does not lose much by resetting and can benefit from the smaller transition guard. The value function and strategy obtained by our algorithm are depicted below.

q	x	f	T_*^A
q_4	$[0,\infty)$	0	$\{\varepsilon\}$
q_1	$[0,\infty)$	∞	$\emptyset$
q_2	$[0,2)$	$2-x$	$\{\varepsilon\}$
	$[2,\infty)$	0	$\{a\}$
q_3	$[0,5)$	$5-x$	$\{\varepsilon\}$
	$[5,\infty)$	0	$\{a\}$
q_0	$[0,3)$	2	$\{b\}$
	$[3,3]$	2	$\{\varepsilon,b\}$
	$(3,4)$	$5-x$	$\{\varepsilon\}$
	$[4,5]$	$5-x$	$\{\varepsilon,c\}$
	$(5,\infty)$	0	$\{c\}$

As one can see, q_4 is a winning state right from the start while from q_1 you can never reach q_4. The clock spaces of q_2 and q_3 are partitioned into two intervals: one after the clock values reach their respective guards, where the value of f is 0, and the interval before that, where f measures the time until the satisfaction of the guards. Finally $f(q_0,x)$ is obtained as $\min\{f(q_1,x), f(q_2,0), f(q_3,x)\}$ and one can observe that $x=3$ is the breakpoint after which it is better to take the c transition to q_3. In the absence of an adversary, not all the complexities of the algorithm are demonstrated in this example.

6. CONCLUDING REMARKS

This work constitutes yet another step in the process of lifting classical results from automata to timed automata, and the main result can be rephrased as finding (min-max) shortest paths in non-countable (but well-behaving) graphs. The algorithm can be easily extended to models where integer costs are associated with transitions. However, assigning different costs of time to different states will transform the model into a more general hybrid system (such as the ones whose synthesis problem is treated by Wong-Toi (1997)), and will probably make the problem harder, if not undecidable.

From the application point of view, many problems related to digital circuit design, or scheduling problems in manufacturing and telecommunication, can be formulated as optimal control problems in the framework introduced here. The authors believe that unifying the qualitative "property-based" approach, which is dominant in verification, with optimization-oriented approaches used elsewhere is important for the development of hybrid systems research.

References

R. Alur and D. L. Dill. A theory of timed automata. *Theoretical Computer Science*, 126: 183–235, 1994.

E. Asarin, O. Maler, and A. Pnueli. Symbolic controller synthesis for discrete and timed automata. In P. Antsaklis, W. Kohn, A. Nerode, and S. Sastry, editors, *Hybrid Systems II*, volume 999 of *LNCS*, pages 1–20. Springer, 1995.

E. Asarin, O. Maler, A. Pnueli, and J. Sifakis. Controller synthesis for timed automata. In *Proc. IFAC Conference on System Structure and Control*, pages 469–474. IRCYN, 1998.

D.P. Bertsekas. *Dynamic Programming and Optimal Control.* Athena Scientific, 1995.

P.J. Ramadge and W.M. Wonham. Supervisory control of a class of discrete event processes. *SIAM Journal of Control and Optimization*, 25: 206–230, 1987.

C. Tomlin, J. Lygeros, and S. Sastry. Synthesizing controllers for nonlinear hybrid systems. In T. Henzinger and S. Sastry, editors, *Hybrid Systems: Computation and Control*, volume 1386 of *Lecture Notes in Computer Science*, pages 360–373. Springer, 1998.

H. Wong-Toi. The synthesis of controllers for linear hybrid automata. In *Proc. CDC'97*, pages 4607–4612. IEEE, 1997.

REACHABILITY COMPUTATION FOR LINEAR HYBRID SYSTEMS

Gerardo Lafferriere * George J. Pappas ** Sergio Yovine ***

* *Department of Mathematical Sciences*
Portland State University
Portland, OR 97207, USA
gerardo@mth.pdx.edu
** *Department of Electrical Engineering and Computer Sciences*
University of California at Berkeley
Berkeley, CA 94720, USA
gpappas@eecs.berkeley.edu
*** *VERIMAG*
Centre Equation, 2 Avenue de Vignate
38610 Gieres, France
Sergio.Yovine@imag.fr

Abstract: Linear hybrid systems are finite state machines with linear vector fields of the form $\dot{x} = Ax$ in each discrete location. Very recently, the reachability problem for classes of linear hybrid systems was shown to be decidable. In this paper, the decidability result is extended to capture classes of linear hybrid systems where in each location the dynamics are of the form $\dot{x} = Ax + Bu$, for various types of inputs.

Keywords: Reachability, linear systems, finite state machines.

1. INTRODUCTION

Hybrid systems combine discrete event systems with differential equations. One of the most important problems for hybrid systems is the *reachability problem* which asks whether some unsafe region is reachable from an initial region. The reachability problem for hybrid systems is known to be difficult. In order to tackle the complexity of the reachability problem, we adopt a computational point of view.

The important issue for computational algorithms for systems with infinite state spaces is *decidability*. The known decidable classes include timed automata (Alur and Dill, 1994), multirate automata (Alur *et al.*, 1995), and rectangular hybrid automata (Puri and Varaiya, 1994; Henzinger *et al.*, 1995). Undecidable classes of hybrid systems are described by Henzinger *et al.* (1995) and the references therein. Recently, it was shown that *the reachability problem is decidable for a class of linear hybrid systems*, that is hybrid systems with linear vector fields of the form $\dot{x} = Ax$ at each discrete location (Lafferriere *et al.*, 1998*b*). This result is significant given the wide applicability of linear systems in control theory.

This decidability result is obtained using new mathematical and computational techniques. In particular, the notion of *o-minimality* from *model theory* is used to define a class of hybrid systems, called *o-minimal hybrid systems.* In (Lafferriere *et al.*, 1998*a*), it is shown that all o-minimal hybrid systems admit finite bisimulations, the first step in showing decidability. For decidability, one needs to make the bisimulation algorithm computationally feasible. This is achieved using *math-*

ematical logic as the main tool to symbolically represent sets with logic formulas. The main computational tool for set manipulation is *quantifier elimination.* Since quantifier elimination exists for the theory of reals with addition and multiplication (Tarski, 1951), one either finds or transforms into subclasses of o-minimal hybrid system which are definable in this decidable theory. This leads to a decidable class of linear hybrid systems.

In this paper, the above decidability result is extended to computationally solve the reachability problem for hybrid systems with continuous dynamics of the form $\dot{x} = Ax + Bu$ in each discrete location. This requires investigation of not only classes of linear vector fields, but also classes of inputs for which this problem is decidable. In addition, conditions on the interactions between the input and the natural dynamics of the system are also important for this extension.

2. HYBRID SYSTEMS

The class of hybrid systems to be considered is defined as follows.

Definition 1. A *hybrid system* is a tuple $H = (X, X_0, X_F, F, E, I, G, R)$ where

- $X = X_D \times X_C$ is the state space with $X_D = \{q_1, \ldots, q_n\}$ and X_C a manifold.
- $X_0 \subseteq X$ is the set of initial states
- $X_F \subseteq X$ is the set of final states
- $F : X \longrightarrow TX_C$ assigns to each discrete location $q \in X_D$ a vector field $F(q, \cdot)$
- $E \subseteq X_D \times X_D$ is the set of discrete transitions
- $I : X_D \longrightarrow 2^{X_C}$ assigns to each location a set $I(q) \subseteq X_C$ called the invariant.
- $G : E \longrightarrow X_D \times 2^{X_C}$ assigns to $e = (q_1, q_2) \in E$ a guard of the form $\{q_1\} \times U$, $U \subseteq I(q_1)$.
- $R : E \longrightarrow X_D \times 2^{X_C}$ assigns to $e = (q_1, q_2) \in E$ a reset of the form $\{q_2\} \times V$, $V \subseteq I(q_2)$.

The hybrid system H has two types of transitions $\xrightarrow{e}$ and $\xrightarrow{\tau}$ where $(q, x) \xrightarrow{e} (q', x')$ for $e \in E$ iff $(q, x) \in G(e)$ and $(q', x') \in R(e)$, and $(q_1, x_1) \xrightarrow{\tau} (q_2, x_2)$ iff $q_1 = q_2$ and there exists $\delta \geq 0$ and a curve $x : [0, \delta] \longrightarrow M$ with $x(0) = x_1$, $x(\delta) = x_2$ and for all $t \in [0, \delta]$ it satisfies $dx/dt = F(q_1, x(t))$ and $x(t) \in I(q_1)$. A trajectory of H is a finite concatenation of transitions. Having defined $\xrightarrow{\tau}$ and $\xrightarrow{e}$ allows us to formally define $Pre_\tau(P)$ and $Pre_e(P)$ for $e \in E$ and any region $P \subseteq X$ as

$$Pre_\sigma(P) = \{q \in Q \mid \exists p \in P \text{ and } q \xrightarrow{\sigma} p\} \quad (1)$$

where σ is either τ or any $e \in E$. The following standard bisimulation algorithm, constructs a reachability preserving, finite state quotient of the hybrid system (provided the algorithm terminates). Because of the initialized structure of the discrete transitions, the initial partition $\mathcal{S}_q$ is a partition which is compatible with all relevant sets of the hybrid system associated with each discrete location q.

Algorithm (Bisimulation Algorithm for Hybrid Systems)
Set $X/\sim \; = \bigcup_q \mathcal{S}_q$
for $q \in X_D$
  while $\exists$ $P, P' \in \mathcal{S}_q$ such that $\emptyset \neq P \cap Pre_\tau(P') \neq P$
    Set $P_1 = P \cap Pre_\tau(P')$; $P_2 = P \setminus Pre_\tau(P')$
    refine $\mathcal{S}_q = (\mathcal{S}_q \setminus \{P\}) \cup \{P_1, P_2\}$
  end while
end for

There are two issues that need to be resolved. The first is to find classes of hybrid systems for which this algorithm terminates. The second is to find classes of systems for which the algorithm is computable. The termination issue is tackled by the notion of o-minimality from model theory.

3. MATHEMATICAL LOGIC AND MODEL THEORY

This section provides a brief introduction to mathematical logic and model theory. The reader is referred to Chang and Keisler (1977) for more details.

3.1 *Languages and formulas*

A *language* is a set of symbols separated into three groups: relations, functions and constants. The sets $\mathcal{L}_0 = \{<, +, -, 0, 1\}$, $\mathcal{L}_R = \{<, +, -, \cdot, 0, 1\}$, and $\mathcal{L}_{\exp} = \{<, +, -, \cdot, 0, 1, \exp\}$ are examples of languages where $<$ (less than) is the relation, $+$ (plus), $-$ (minus), $\cdot$ (product) and exp (exponentiation) are the functions, and 0 (zero) and 1 (one) are the constants.

Let $\mathcal{V} = \{x, y, z, x_0, x_1, \ldots\}$ be a countable set of *variables.* The set of *terms* of a language is inductively defined as follows. A term θ is a variable, a constant, or $F(\theta_1, \ldots, \theta_m)$, where F is a m-ary function and θ_i, $i = 1, \ldots, m$ are terms. For instance, $x - 2y + 3$ and $x + yz^2 - 1$ are terms of $\mathcal{L}_0$ and $\mathcal{L}_R$, respectively. In other words, terms of $\mathcal{L}_0$ are linear expressions and terms of $\mathcal{L}_R$ are polynomials with integer coefficients. Notice that integers are the only numbers allowed in expressions (they can be obtained by adding up the constant 1).

The *atomic formulas* of a language are of the form $\theta_1 = \theta_2$, or $R(\theta_1, \ldots, \theta_n)$, where θ_i, $i = 1, \ldots, n$

are terms and R is an n-ary relation. For example, $xy > 0$ and $x^2 + 1 = 0$ are terms of $\mathcal{L}_R$.

The set of *(first-order) formulas* is recursively defined as follows. A formula ϕ is an atomic formula, $\phi_1 \wedge \phi_2$, $\neg\phi_1$, $\forall x : \phi_1$ or $\exists x : \phi_1$, where ϕ_1 and ϕ_2 are formulas, x is a variable, $\wedge$ (conjunction) and $\neg$ (negation) are the boolean connectives, and $\forall$ (for all) and $\exists$ (there exists) are the quantifiers.

Examples of $\mathcal{L}_R$-formulas are $\forall x\ \forall y : xy > 0$, $\exists x : x^2 - 2 = 0$, and $\exists w : xw^2 + yw + z = 0$. The occurrence of a variable in a formula is *free* if it is not inside the scope of a quantifier; otherwise, it is *bound*. For example, x, y, and z are free and w is bound in the last example. We often write $\phi(x_1, \ldots, x_n)$ to indicate that $x_1, \ldots, x_n$ are the free variables of the formula ϕ. A *sentence* of $\mathcal{L}_R$ is a formula with no free variables. The first two examples are sentences.

3.2 *Models*

A *model* of a language consists of a non-empty set S and an interpretation of the relations, functions and constants. For example, $(\mathbb{R}, <, +, -, \cdot, 0, 1)$ and $(\mathbb{Q}, <, +, -, \cdot, 0, 1)$, are *models* of $\mathcal{L}_R$ with the usual meaning of the symbols.

A set $Y \subseteq S^n$ is *definable* in a language if there exists a formula $\phi(x_1, \ldots, x_n)$ such that $Y = \{(a_1, \ldots, a_n) \in S^n \mid \phi(a_1, \ldots, a_n)\}$. For example, over $\mathbb{R}$, the formula $x^2 - 2 = 0$ defines the set $\{\sqrt{2}, -\sqrt{2}\}$. Two formulas $\phi(x_1, \ldots, x_n)$ and $\psi(x_1, \ldots, x_n)$ are *equivalent* in a model, denoted by $\phi \equiv \psi$, if for every assignment $(a_1, \ldots, a_n)$ of $(x_1, \ldots, x_n)$, $\phi(a_1, \ldots, a_n)$ is true if and only if $\psi(a_1, \ldots, a_n)$ is true. Equivalent formulas define the same set.

3.3 *Theories*

A *theory* is a subset of sentences. Any model of a language defines a theory: *the set of all sentences which hold in the model.* By abuse of notation the theory $(\mathbb{R}, <, +, -, \cdot, 0, 1)$ will refer to the collection of formulas of $\mathcal{L}_R$ which are true in the model (and similarly for other languages). Only models over $\mathbb{R}$ will be considered.

Definition 2. The theory of $\mathcal{L}$ is *o-minimal* ("order minimal") if every definable subset of $\mathbb{R}$ is a finite union of points and intervals (possibly unbounded).

Examples of o-minimal theories include

(1) $\mathcal{R} = (\mathbb{R}, <, +, -, \cdot, 0, 1)$,

(2) $\mathcal{R}_{\text{sin}} = (\mathbb{R}, <, +, -, \cdot, \sin|_{[-\pi,\pi]}, 0, 1)$ (van den Dries, 1986),

(3) $\mathcal{R}_{\text{exp}} = (\mathbb{R}, <, +, -, \cdot, \exp, 0, 1)$ (Wilkie, 1996), and

(4) $\mathcal{R}_{\text{sin,exp}} = (\mathbb{R}, <, +, -, \cdot, \sin|_{[-\pi,\pi]}, \exp, 0, 1)$ (van den Dries and Miller, 1994).

Based on this notion a new class of hybrid systems is defined.

Definition 3. A hybrid system $H = (X, X_0, X_F, F, E, I, G, R)$ is said to be *o-minimal* if

- $X_C = \mathbb{R}^n$
- for each $q \in X_D$ the flow of F_q is complete (flow is definable for all time)
- for each $q \in X_D$ all the relevant sets (guards, invariants, initial conditions etc) and the flow of F_q are definable in an o-minimal extension of $(\mathbb{R}, <, +, -, \cdot, 0, 1)$.

Termination of the bisimulation algorithm is guaranteed by the following theorem.

Theorem 4. Every o-minimal hybrid system admits a finite bisimulation. In particular, the bisimulation algorithm, terminates for o-minimal hybrid systems.

PROOF. See (Lafferriere *et al.*, 1998*a*).

4. DECIDABLE LINEAR HYBRID SYSTEMS

Based on the previous result, if one can find or transform into classes of o-minimal hybrid systems for which each step of the bisimulation algorithm is decidable then one obtains a class of hybrid systems for which the following problem is decidable.

Problem 5. (Reachability Problem) Is there a trajectory of H which starts in X_0 and ends in X_F?

This is achieved for a class of linear hybrid systems. For this class, each step of the algorithm can be posed as a quantifier elimination problem in the theory $(\mathbb{R}, +, -, \cdot, <, 0, 1)$ which is known to be decidable (Tarski, 1951).

Theorem 6. Consider an o-minimal hybrid system H where $X_C = \mathbb{R}^n$, all relevant sets are semialgebraic and for all $q \in X_D$, $F(q, x) = Ax$ where A is a matrix with rational entries and A belongs in one of the following classes

- A is nilpotent
- A is diagonalizable and has real, rational eigenvalues

- A has purely imaginary eigenvalues of the form $i\omega$, $\omega \in \mathbb{Q}$ and its real Jordan form is block diagonal with 2×2 blocks.

Then the reachability problem for the linear hybrid system H is decidable.

PROOF. See (Lafferriere *et al.*, 1998*b*).

5. EXTENSIONS

The goal of this paper is to extend Theorem 6 to include linear hybrid systems where in each discrete locations the dynamics are of the form

$$\dot{x} = Ax + Bu$$

for various types of inputs u. This requires an extension of the definition of hybrid systems to allow time-varying dynamics.

Definition 7. A hybrid system $H = (X, X_0, X_F, F, E, I, G, R)$ with X, X_0, X_F, E, I, G, and R as in Definition 1 is said to have *time-varying dynamics* if

$$F : X \times \mathbb{R} \to TX_C,$$

that is, for each $q \in X_D$, $F(q, \cdot, \cdot)$ defines a time-varying vector field on X_C.

The main result is the following.

Theorem 8. Consider a hybrid system H with time-varying dynamics where all relevant sets are semialgebraic and for all $q \in X_D$ the dynamics are of the form $\dot{x} = Ax + Bu$ where A,B are matrices with rational entries and in addition one of the following holds:

- A is nilpotent and each entry of u is a polynomial in t
- A is diagonalizable, has real rational eigenvalues, and each entry of u is of the form $e^{\mu t}$ with $\mu \in \mathbb{Q}$ not an eigenvalue of A.
- A has purely imaginary eigenvalues of the form $i\omega$, $\omega \in \mathbb{Q}$, and the entries in the input u are of the form $\sin(\alpha t)$ or $\cos(\alpha t)$ with $\alpha \in \mathbb{Q}$ and $\alpha \neq \pm\omega$.

Then the reachability problem for the hybrid system H is decidable.

PROOF. First an associated (time-invariant) hybrid system as in Definition 1 is constructed, which has an equivalent reachability problem. This is done by introducing time as a new variable x_{n+1}. More precisely, we define the new hybrid system $\widetilde{H} = (\widetilde{X}, \widetilde{X}_0, \widetilde{X}_F, \widetilde{F}, \widetilde{E}, \widetilde{I}, \widetilde{G}, \widetilde{R})$ is defined as follows: $\widetilde{X} = X_D \times \mathbb{R}^{n+1}$, $\widetilde{X}_0 = X_0 \times \mathbb{R}$, $\widetilde{X}_F = X_F \times \mathbb{R}$, $\widetilde{E} = E$, $\widetilde{I} : X_D \to 2^{\mathbb{R}^{n+1}}$, $\widetilde{G} : \widetilde{E} \to X_D \times 2^{\mathbb{R}^{n+1}}$, $\widetilde{R} : \widetilde{E} \to X_D \times 2^{\mathbb{R}^{n+1}}$, and, for $\tilde{x} = (x, x_{n+1})^T \in \mathbb{R}^{n+1}$,

$$\widetilde{F}(q, \tilde{x}) = \begin{pmatrix} A_q x + B_q u_q(x_{n+1}) \\ 1 \end{pmatrix}.$$

Let

$$\phi_q(s, t, x_0) = e^{A_q(t-s)} x_0 + \int_s^t e^{A_q(t-\tau)} B_q u_q(\tau)\, d\tau.$$

Then the flow of the vector field $\widetilde{F}_q = \widetilde{F}(q, \cdot)$ is given by

$$\phi_{\widetilde{F}_q}(t, (x_0, s_0)) = (\phi_q(s_0, t, x_0), t + s_0)^T.$$

This shows that if ϕ_q is definable in an o-minimal theory then so is the flow of the corresponding vector field $\widetilde{F}_q$. Moreover, the reachability problem for H is decidable if and only if the reachability problem for $\widetilde{H}$ is decidable. Since the original sets are semialgebraic, they are definable in the o-minimal theory $\mathcal{R}$. The following two facts will suffice to conclude the proof:

(1) the flows of the vectors fields $\widetilde{F}_q$ are definable in an o-minimal theory which extends $\mathcal{R}$.
(2) for any two semialgebraic sets V, W, the formula

$$x \in V \wedge \exists y \in W, \exists t, \exists s : t > s \wedge \phi_q(s, t, x) = y \qquad (2)$$

is equivalent to a formula in $\mathcal{R}$.

The first fact together with Theorem 4 shows that the bisimulation algorithm terminates. The second fact shows that the statement $x \in V \cap Pre_\tau(W)$ can be effectively decided. Therefore the two facts complete the proof.

In order to prove both facts Formula 2 must be analyzed for each type of possible control. From now on the subscript q will be dropped to simplify notation.

If A is nilpotent then each entry of e^{At} is a polynomial in t. Therefore, each entry of $\phi(s, t, x)$ is a polynomial in t and s and hence definable in $\mathcal{R}$.

If A is diagonalizable with real eigenvalues then each entry of e^{At} is a linear combination of the functions $e^{\lambda_i t}$ with λ_i an eigenvalue of A. Since each entry of u is of the form $e^{\mu_j t}$, the entries of $e^{-A\tau} Bu(\tau)$ are linear combinations of $e^{(\mu_j - \lambda_i)t}$. By the assumption $\mu_j - \lambda_i \neq 0$ it results (after integration) that the entries of $\phi(s, t, x)$ are also linear combinations of the functions $e^{\lambda_i t}$, $e^{(\mu_j - \lambda_i)t}$, $e^{\lambda_i s}$, and $e^{(\mu_j - \lambda_i)s}$, and therefore definable in $\mathcal{R}_{\exp}$.

Finally, in the case of imaginary eigenvalues as described, the entries of e^{At} are linear combinations of $\cos(\omega_k t)$, $\sin(\omega_k t)$ with $i\omega_k$ eigenvalues of A.

Because of the form of the input u in this case, the entries of $e^{-A\tau}Bu(\tau)$ are linear combinations of the products $\cos(\alpha_j t)\cos(\omega_k t)$, $\cos(\alpha_j t)\sin(\omega_k t)$, $\sin(\alpha_j t)\cos(\omega_k t)$, and $\sin(\alpha_j t)\sin(\omega_k t)$. Using standard product formulas these can be rewritten as linear combinations of $\sin((\alpha_j \pm \omega_k)t)$ and $\cos((\alpha_j \pm \omega_k)t)$. Since $\alpha_j \neq \pm\omega_k$, integration again results in linear combinations of $\sin((\alpha_j \pm \omega_k)t)$ and $\cos((\alpha_j \pm \omega_k)t)$ and therefore $\phi(s,t,x)$ is definable in $\mathcal{R}_{\text{sin}}$.

The analysis above implies that all relevant sets and relevant flows of the hybrid system are definable in the o-minimal theory $\mathcal{R}_{\text{sin,exp}}$. This concludes the proof of the first fact.

To prove the second fact it suffices to show that the formula $\eta(x,y) \equiv \exists s,t,t > s : \phi(s,t,x) = y$ is equivalent to a formula in $\mathcal{R}$.

In the case of nilpotent A and polynomial inputs, the formula already is in $\mathcal{R}$. Assume now that A is diagonal and the entries of u are exponential functions as indicated above. The formula $\eta(x,y)$ can be written more explicitly (entry by entry) as

$$y_i = \sum_k \psi_{ik}(x)e^{\mu_k t}e^{-\mu_k s} + \sum_l a_{il}(e^{\nu_l t} - e^{\nu_l s})$$

for some linear functions ψ_{ik} of x and rational constants a_{il}, μ_k, and ν_l.

Let d_k denote the least common denominator of the μ_k and c_l the least common denominator of the ν_l (both assumed positive). Set $d_0 = \prod d_k$, $c_0 = \prod c_l$, $p_k = \mu_k d_0$, and $q_l = \nu_l c_0$. Then $\eta(x,y)$ is equivalent to

$$\exists\tilde{s}\,\exists\tilde{t} : d_0\tilde{t} > c_0\tilde{s} \;\wedge$$
$$y_i = \sum_k \psi_{ik}(x)e^{p_k\tilde{t}}e^{-p_k\tilde{s}} + \sum_l a_{il}(e^{q_l\tilde{t}} - e^{q_l\tilde{s}})$$

Using the substitutions $z = e^{\tilde{t}}$, $w = e^{-\tilde{s}}$, the previous formula is equivalent to

$$\exists z\,\exists w : z > 0, w > 0 \;\wedge\; w^{c_0}z^{d_0} > 1$$
$$y_i = \sum_k \psi_{ik}(x)z^{p_k}w^{p_k} + \sum_l a_{il}(z^{q_l} - w^{-q_l})$$

which can be rewritten in $\mathcal{R}$ by multiplying through by $\prod w^{q_l}$.

In the case of purely imaginary eigenvalues the formula for $\phi(s,t,x)$ can be rewritten in terms of polynomials in $\cos(t)$, $\sin(t)$, $\cos(s)$, and $\sin(s)$ with rational coefficients. The substitutions $z = \cos(t)$, $w = \sin(t)$ and the identity $z^2 + w^2 = 1$ (and similarly for the s variable) lead to the equivalent formula in $\mathcal{R}$. This concludes the proof of the theorem.

Remark 9. The above results remain valid if the inputs are allowed to be linear combinations of the functions of the corresponding type: exponentials in case of real eigenvalues, and sinusoidal in the case of imaginary eigenvalues. In all cases the same "resonance" restrictions apply on the parameters μ and α.

6. CONCLUSIONS

This paper extends the decidability results for linear hybrid systems to include systems with inputs of a special type. The techniques used involved the notion of a bisimulation and elimination of quantifiers.

Areas for future research include extensions of Theorem 8 to differential inclusions of the form

$$\dot{x} = Ax + Bu \qquad u \in U$$

where the input set is a rectangular set. This requires certain monotonicity conditions on the solution integral, which will be valid on a smaller class of linear hybrid systems.

Acknowledgment: This research is supported by the Army Research Office under grants DAAH 04-95-1-0588 and DAAH 04-96-1-0341.

7. REFERENCES

Alur, R. and D.L. Dill (1994). A theory of timed automata. *Theoretical Computer Science* **126**, 183–235.

Alur, R., C. Coucoubetis, N. Halbwachs, T.A. Henzinger, P.H. Ho, X. Nicolin, A. Olivero, J. Sifakis and S. Yovine (1995). The algorithmic analysis of hybrid systems. *Theoretical Computer Science* **138**, 3–34.

Chang, C. C. and H. J. Keisler (1977). *Model Theory.* second ed.. North-Holland.

Henzinger, T.A., P.W. Kopke, A. Puri and P. Varaiya (1995). What's decidable about hybrid automata?. In: *Proceedings of the 27th Annual Symposium on Theory of Computing.* ACM Press. pp. 373–382.

Lafferriere, G., G.J. Pappas and S. Sastry (1998*a*). O-minimal hybrid systems. Technical Report UCB/ERL M98/29. University of California at Berkeley. Berkeley, CA. Submitted to *Mathematics of Control, Signals, and Systems.*

Lafferriere, G., G.J. Pappas and S. Yovine (1998*b*). Decidable hybrid systems. Technical report. University of California at Berkeley. Berkeley, CA. To appear in *Hybrid Systems: Computation and Control, 1999.*

Puri, A. and P. Varaiya (1994). Decidability of hybrid systems with rectangular differential inclusions. In: *Computer Aided Verification.* pp. 95–104.

Tarski, A. (1951). *A decision method for elementary algebra and geometry.* second ed.. University of California Press.

van den Dries, L. (1986). A generalization of the tarski-siedenberg theorem, and some nondefinability results. *Bulletin of the American Mathematical Society* **15**(2), 189–193.

van den Dries, L. and C. Miller (1994). On the real exponential field with restricted analytic functions. *Israel Journal of Mathematics* **85**, 19–56.

Wilkie, A. J. (1996). Model completeness results for expansions of the ordered field of real numbers by restricted pfaffian functions and the exponential function. *Journal of the American Mathematical Society* **9**(4), 1051–1094.

USING DYNAMICAL INVARIANTS IN THE ANALYSIS OF HYBRID DYNAMICAL SYSTEMS

Kevin X. He and Michael D. Lemmon [1]

Department of Electrical Engineering
University of Notre Dame
Notre Dame, IN 46556, USA
fax:(219)-631-4393
Email: xhe, lemmon@maddog.ee.nd.edu

Abstract: This paper examines subsets of a hybrid system's state space which are invariant with respect to cycles of discrete events. It is well known that such dynamical invariants play an important role in the stability and supervision of such systems. The primary contribution of this paper is to show how such dynamical invariants may be identified for hybrid systems whose switching logics are characterized by Petri networks. *Copyright* ©1999 IFAC

Keywords: Hybrid System, Petri Nets, Dynamical Invariants, Branching Processes.

1 Introduction

This paper examines subsets of a hybrid system's state space which are invariant with respect to cycles of discrete events. It is well known (Branicky, 1996) (Hou et al., 1996) (He et al. 1998) (Deshpande et al., 1995) that such dynamical invariants play an important role in the stability of hybrid automata (Alur et al., 1996). This paper examines the role of such dynamical invariants in hybrid systems whose switching logics are characterized by ordinary Petri nets.

In spite of their recent popularity, hybrid automata suffer from significant limitations. Verifying system specifications posed as predicates in a temporal logic (Henzinger et al, 1995) are known to be undecidable for most hybrid automata (Puri et al, 1993). It is only for the class of linear rectangular hybrid automata (Puri et al, 1993) and o-minimal systems (Lafferriere et al, 1998) that decidability can be established. Part of the reason for this undecidability rests with the fact that verification attempts to characterize hybrid system invariants in a necessary and sufficient manner. Other work (Johansson et al, 1998) (Petterson et al, 1996) has provided sufficient tests for hybrid system stability which provide computationally tractable approximations to these dynamical invariants.

Another limitation of hybrid automata is that they capture concurrency in an unnatural manner. The graph structure underlying the hybrid automata cannot directly represent systems requiring synchronization of discrete processes. In order to model process synchronization, the hybrid automata needs to embed synchronization controls in the predicates guarding the firing of an arc. As a result, an arc's guard predicates mix discrete and continuous valued states of the hybrid system in a way which greatly complicates the identification of discrete invariants in the hybrid system. Since the identification of such invariants is a necessary prerequisite for studying the stability of hybrid systems (He et al., 1998), the hybrid automaton may not provide the best framework in which to study hybrid system stability and performance.

This paper studies hybrid system stability and performance using a modeling framework in which the sys-

[1]We gratefully acknowledge the partial financial support of the Army Research Office (DAAH04-96-10285 and DAAG5-98-1-0199) and the National Science Foundation (NSF-ECS95-31485)

tem's discrete switching behavior is represented by an ordinary Petri net. The model in this paper is a refinement of the programmable timed Petri net (Lemmon et al., 1998). It is not related to hybrid Petri nets (Flaus et al., 1997) using continuous-valued tokens. Petri nets are used because they provide a natural means of capturing process synchronization. As a result the discrete behavior of the system is isolated to the graph structure of the network, thereby leaving guard predicates on network arcs to only be dependent on the continuous-valued states of the hybrid system. By confining the discrete dynamics to the network's graph, we can use partial order methods (Vogler et al., 1992) to easily identify cycles of discrete events and deduce the causal relationships between these various cycles. Therefore the use of Petri nets in the modeling of hybrid systems appears to provide an attractive framework for the analysis and subsequent synthesis of hybrid systems.

The remainder of this paper is organized as follows. Section 2 presents the hybrid Petri net model used in this paper. Section 3 formally introduces the dynamical invariants of this system. Section 4 summarizes the principal contribution of this paper.

2 Hybrid Petri Networks

The hybrid Petri net presented in this section is closely related to the hybrid automaton (Alur et al., 1996) and more general versions of this structure were referred to as the programmable timed Petri net (PTPN) (Lemmon et al., 1998). A hybrid Petri net may be represented by the triple (N, Δ, L) where N is an ordinary Petri net, Δ is a finite set of Lipschitz continuous vector fields over $\Re^n$, and L is a mapping from the transitions and arcs of N onto predicates in a propositional logic whose truth values are interpreted with respect to the continuous states of the system. On a more heuristic basis, we can view N as capturing the discrete part of the system, Δ as capturing the continuous part of the system, and L as capturing the interaction between the continuous and discrete parts of the system.

The discrete part of the system is represented by the Petri network, N. The network is represented by the triple (S, T, F) where S is a set of places, T is a set of transitions, and $F \subset (S \times T) \cup (T \times S)$ is a set of input and output arcs. A transition or place is often referred to as a *node* The *preset* of a node $x \in S \cup T$ (also denoted as $\bullet x$) is the set of all nodes $y \in S \cup T$ such that $(y, x) \in F$. Similarly, the *postet* of a node $x \in S \cup T$ (also denoted as $x\bullet$) is the set of all nodes y such that $(x, y) \in F$. The discrete state of the network at time τ is represented by the *marking* vector $\mu \in Z^{|S|}$ of non-negative integers and the *firing* vector $q \in \{0,1\}^{|T|}$.

Following the traditional interpretation from the Petri net literature, μ identifies the number of tokens that are marking the places in the network while q identifies which transitions are *firing*. In particular, we say that place $s_i \in S$ is marked under μ if the ith component of μ is positive (i.e., $\mu_i > 0$). Similarly, we say that transition $t_i \in T$ is firing under q if $q_i = 1$.

The continuous part of the system is represented by the collection $\Delta = \{f_1, \ldots, f_N\}$ of Lipschitz continuous vector fields over $\Re^n$. The continuous state of the system at time τ is characterized by the 4-tuple, (f, τ_0, x_0, x) where $f \in \Delta$, $\tau_0 \in \Re$, $x_0 \in \Re^n$, and $x : (\tau_0, \tau] \to \Re^n$. This four-tuple, therefore defines an initial value problem

$$\frac{dx(\tau)}{d\tau} = f(x(\tau)) \quad (1)$$

$$x(\tau_0) = x_0 \quad (2)$$

which is satisfied by the absolutely continuous function x in the function space $\mathcal{C}(-\infty, \infty)$. The value that x takes at time τ is denoted as $x(\tau)$. Combining this 4-tuple, with the discrete system state, the hybrid system's state is represented by the 6-tuple $z = (\mu, q, \dot{x}, x, x_0, \tau_0)$. z takes values in the set $Z^{|S|} \times \{0,1\}^{|T|} \times \Delta \times \mathcal{C}(-\infty, \infty) \times \Re^n \times \Re$. This rather complex object represents the hybrid state for the hybrid Petri net.

We now define a propositional logic, $\mathcal{P}$ whose semantics are defined with respect to states in $\mathcal{H}$. The atomic formulae in $\mathcal{P}$ are either

- *invariant predicates* of the form $[\dot{x} = f_j]$, $[x_0 = c]$, or $[\tau_0 = \tau]$
- or *guard predicates* of the form $[x'(\tau) b R c]$ where R is a binary relation, $<$ or $>$ and $b, c \in \Re^n$.

The logic, $\mathcal{P}$, consists of all well formed formulae (wffs). A wff in $\mathcal{P}$ is defined recursively in the usual manner. In other words any atomic predicate in $\mathcal{P}$ is a wff. Furthermore, the logical conjunction, disjunction, or not of a wff is a wff. The semantics of this logic are defined with respect to the hybrid state z. In other words a predicate in $\mathcal{P}$ evaluates to true with respect to hybrid state z if z satisfies the conditions in the predicate. A predicate $p \in \mathcal{P}$ will be called a *guard* or *invariant* predicate if its atomic formulae are all either guard or invariant predicates.

A function $z : \Re \to \mathcal{H}$ will be called a hybrid trajectory. The value that function z takes at time $\tau \in \Re$ is denoted as

$$z^{(\tau)} = (\mu^{(\tau)}, q^{(\tau)}, \dot{x}^{(\tau)}, x^{(\tau)}, x_0^{(\tau)}, \tau_0^{(\tau)}) \quad (3)$$

The superscripts denote the dependence on the time τ. We use superscripts here since several of the substates (i.e. $\dot{x}$ and x) are also functions of $\Re$. Note that the state trajectory $x^{(\tau)}$ will be an absolutely continuous function over $(-\infty, \tau]$ while $x^{(\tau)}(\tau_1)$ denotes the value in $\Re^n$ that this function takes at time τ_1. We can, without loss of generality, drop the superscript here and simply denote $x^{(\tau)}$ as x. The truth values of the predicates in $\mathcal{P}$ are defined with respect to a hybrid trajectory z. There are two types of predicates to consider; those obtained from the transition labels $L(t)$ and those obtained from labeling arcs in F. A transition $t_i \in T$ is mapped onto an invariant predicate $L(t_i) \in \mathcal{P}$. The hybrid state $z^{(\tau)} \in \mathcal{H}$ at time τ is said to satisfy $L(t_i)$ (also denoted as $z^{(\tau)} \models L(t_i)$) if $q_i^{(\tau)} = 1$ and the substates $\dot{x}^{(\tau)}$, $x_0^{(\tau)}$, and $\tau_0^{(\tau)}$ satisfy $L(t_i)$. An input arc, (s_i, t) is mapped onto a guard predicate $L((s_i, t))$. The hybrid state $z^{(\tau)} \in \mathcal{H}$ satisfies $L((s_i, t))$ (denoted as $z^{(\tau)} \models L((s_i, t))$) if $\mu_i^{(\tau^-)} > 0$ and $x(\tau^-)$ satisfies the guard condition $L((s_i, t))$. An output arc (t_i, s) is mapped onto a guard predicate $L((t_i, s))$. The hybrid state $z^{(\tau)} \in \mathcal{H}$ satisfies $L((t_i, s))$ (denoted as $z^{(\tau)} \models L((t_i, s))$) if $q_i^{(\tau^-)} > 0$ and $x(\tau^-)$ satisfies the guard predicate. In order to ensure consistency, we will require that $L((t, s_1)) = L((t, s_2))$ for all $t \in T$. In other words all output arcs of a transition t are labeled with the same guard. On a heuristic level, we interpret the invariant predicates on the transition as reset conditions for the timer. The guard conditions are interpreted as conditions on the continuous state which allow the enabling or completed firing of a transition.

The dynamics of the hybrid Petri net are defined as follows. A switching time τ is called a *rising-edge* for event i if $q_i^{(\tau^+)} = q_i^{(\tau^-)} + 1$ and is called a *falling-edge* for event i if when $q_i^{(\tau^+)} = q_i^{(\tau^-)} - 1$ for some i. The hybrid trajectory z is said to satisfy the hybrid system (N, Δ, L) if

- for any $\tau \in (\tau_1, \tau_2]$ where τ is not a rising edge event and τ_1 is a rising edge for event i. then $z^{(\tau)} \models L(t_i)$.

- for any τ which is a rising edge for event i then $z^{(\tau^-)} \models L((s_j, t_i))$, $q_i^{(\tau^+)} = q_i^{(\tau^-)} + 1$, and

$$\mu_j^{(\tau^+)} = \begin{cases} \mu_j^{(\tau^-)} - 1 & \forall\ s_j \in \bullet t_i \\ \mu_j^{(\tau^-)} & \text{othwerwise} \end{cases} \tag{4}$$

- for any τ which is a falling edge for event i then $z^{(\tau^-)} \models L((t_i, s_j))$, $q_i^{(\tau^+)} = q_i^{(\tau^-)} - 1$, and

$$\mu_j^{(\tau^+)} = \begin{cases} \mu_j^{(\tau^-)} + 1 & \forall\ s_j \in t_i \bullet \\ \mu_j^{(\tau^-)} & \text{otherwise} \end{cases} \tag{5}$$

In the first bullet, time is partitioned into disjoint intervals separated by rising edge events. This first bullet says that the continuous subsystem that is active during this interval is determined by the firing transition that generated the rising edge event. During this interval, the continuous dynamics of the system satisfy the initial value problem associated with the selected timer. The discrete state only changes on a rising or falling edge event. The second bullet states that rising edge events reset the firing vector and remove tokens from the preset of the firing transition when the continuous state satisfies the guard conditions. The third bullet refers to falling edge events and says that these events are triggered when a transition completes firing. The action of a completed transition is to add tokens to the transition's postset when the guard condition on the output arcs are satisfied.

It should also be remarked that the trajectories satisfying the hybrid Petri net automatically exclude chattering solutions associated with "Zeno" type systems. In other words, the interval $(\tau_1, \tau_2]$ is always assumed to be a compact set of nonzero measure so that infinitely fast switching between subsystems is not allowed.

As can be seen this definition of a hybrid Petri net is similar to the hybrid automaton with respect to the use of logical predicates in conditioning the firing of transitions. What is different here, however, is that the synchronization of discrete switchings is modeled by the Petri network. In this way the discrete and continuous parts of the system are completely separated and this separation will allow us to use discrete methods to determine discrete invariants for the hybrid system.

3 Dynamical Invariants of Hybrid Petri Nets

It has been recognized that cycles of discrete events are important in the study of hybrid system stability and performance. Viability analysis (Deshpande et al., 1995), fixed point analyses used by model checking methods (Alur et al., 1996) (Henzinger et al., 1995) and the stability analyses in (Hou et al, 1996) (He et al., 1998) all rely on the existence of subsets of hybrid states that are invariant under a cycle's action. In light of this prior work a necessary first step in the analysis of hybrid system stability and performance involves identifying cycles within the discrete-event switching system.

For hybrid automata containing no explicit synchronization mechanisms, the fundamental cycles of the system can be readily identified from the model's graph. Finite state machines, however, do not allow the modeling of process synchronization unless synchronization controls are placed in labels that condition the firing of an arc. These synchronization controls, however, introduce additional cycles into the system which cannot be readily identified from the finite state machine's graph structure. It is for this reason that the hybrid Petri net structure of section 2 was introduced. The network, N, used in section 2 allows us to represent synchronization controls directly in the graph of the system. In this section, we show how partial order methods (Vogler et al., 1992) can be used to identify cycles within the hyrid Petri net's graph.

Consider a trajectory z satisfying the hybrid Petri net, (N, Δ, L). Associated with this trajectory is a sequence of transitions (also called the *occurrence sequence*) $\sigma = t_{i_1}, t_{i_2}, \cdots, t_{i_j}, \cdots$ representing the sequence of transitions executed by the trajectory. This sequence is easily obtained from the rising-edge generating events i_j (for $j = 1, \ldots, \infty$). According to our hybrid Petri net, the continuous subsystem activated by a rising edge event is active until the next rising edge event occurs. Therefore this sequence of transitions represents the sequence of systems that the trajectory z switches between.

Now consider a subsequence $\sigma_s = t_{i_k}, t_{i_{k+1}}, \cdots, t_{i_{k+N}}$ which starts at the kth rising-edge event and terminates at tne $k+N$th rising edge event. Let τ_{k+N} be the time when the $k+N$th rising edge event occurred. The subsequence is said to be a *cycle* if there exists a hybrid trajectory z_e satisfying (N, Δ, L) such that $z_e^{(\tau)} = z^{(\tau)}$ for $\tau < \tau_{k+N}$ and such that the occurrence sequence σ_{k+N} generated by rising edge events for $\tau > \tau_{k+N}$ has the form $\sigma_{k+N} = \sigma_s^*$ This definition says that there exists a hybrid trajectory which can repeatedly execute the subsequence σ_s. The cycle will be said to be *fundamental* if it contains no other cycle of the network. Because of the way the hybrid Petri net was defined, the occurrence sequence σ lies in the P-type Petri net language generated by the Petri net N. By the Petri net pumping lemma (Jantzen, 1987), we know that any such sequence can be written as the concatenation of three subsequences, xyz, such that xy^iz is also an element of the network's P-type language for all $i \geq 1$. The subsequence y is, of course, the cycle defined above for our hybrid Petri net and hence the pumping lemma allows us to assert that all cycles of the hybrid Petri net will be contained within the cycles of the network N for that system.

Fundamental cycles of a Petri net can be identified using the network's reachability tree. However, the size of the reachability tree grows in an exponential manner with the network's size. Futhermore, except for bounded Petri nets, the reachability tree identifies a P-type language which can be significantly larger than that of the original network. Rather than using the reachability tree, this paper proposes using a prefix of the network's maximal branching process (Engelfriet, 1991). The construction of a maximal branching process of a network is a partial order method (Vogler et al., 1992). Partial order methods such as unfoldings can provide a very efficient means of enumerating the reachable markings of a network (McMillan, 1992). In our case, we construct a branching process which can be used to enumerate a fundamental set of finite duration occurrence sequences (including fundamental cycles) from which the network's P-type language can be generated.

The details concerning the construction of the occurrence network and the extraction of *fundamental runs* from this network are provided in the appendix. In the remainder of this section, a simple example is provided illustrating the method. Consider a hybrid Petri net whose network, N, is shown in figure 1. Using the cutset defined in the appendix, the branching process (N', h') of N is constructed. The configurations extracted from the occurrence net N' of the branching process are shown in figure 2. Within each configuration we can identify three fundamental cycles: cycle $\sigma_1 = (t_1, t_3, t_7)$, $\sigma_2 = (t_3, t_5)$, and $\sigma_3 = (t_2, t_4, t_6)$.

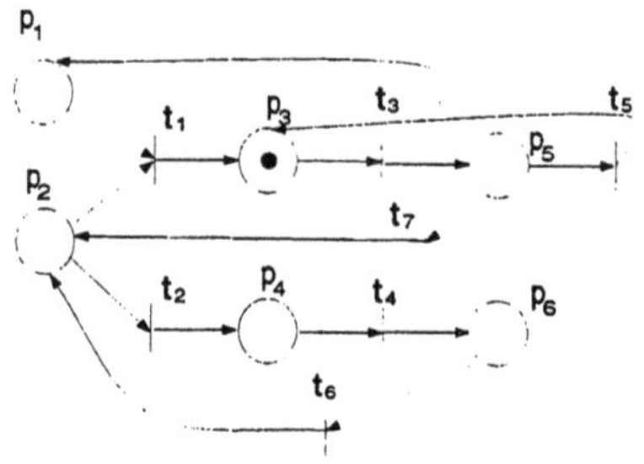

Figure 1: Hybrid System's Petri Net

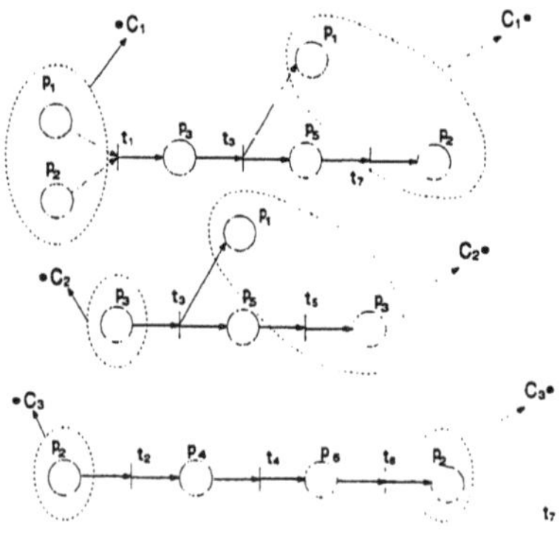

Figure 2: Hybrid System Cycles

The cycles identified in the above example represent a set of possible behaviors that the hybrid Petri net can execute repeatedly. These cycles, however, can only be executed if the enabling conditions embodied in the guard predicates for the arcs connecting transitions in the cycle hold true. Clearly, these guard predicates will not hold for every initial set of continuous states. What we want to do is find a subset of $\Re^n$ for which these enabling conditions are repeatedly satisfied. In essence, this means we are looking for a subset of $\Re^n$

which is invariant under the action of the cycle. If such an invariant can be found for any cycle of the network, then we know that it is possible to start and supervise the system so that it does not stop (i.e. is deadlock free) and so that the continuous states of the system are bounded (i.e. the resulting switched system is stable in the sense of Lyapunov).

The determination of such invariants for fundamental cycles is a "hard" problem. Necessary and sufficient conditions characterizing these sets lead to the undecidable nature of the hybrid automata verification problems (Puri et al., 1993). The virtue of the hybrid Petri net lies in its ability to more easily model a very rich set of hybrid behaviors, but there is no reason to expect that the identification of invariants within hybrid Petri nets to be any easier than for the hybrid automaton. In the face of these decidability issues, one approach involves determining tight sufficient conditions that approximate the network invariants. Such methods were established in (Petterson et al., 1996) and (Johansson et al., 1998) for switching systems in which the switching sets form conic sectors and the fundamental systems are linear time-invariant plants. For these problems, the existence of the invariant set can be posed as a linear matrix inequality (LMI) feasiblity problem. In (He et al., 1998), this approach was used with the fundamental cycles generated by hybrid automata to study hybrid system stability. These LMI methods can also be used to establish the existence of invariants for the fundamental cycles in hybrid Petri nets.

4 Summary

This paper has presented a hybrid system model in which Petri nets are used to model the discrete switching behavior of the system. It was shown how partial order methods could be used to extract fundamental cycles of the network's P-type Petri net language as a first step in identifying subsets of the system's state space which are invariant under the cycle's action. The results in this paper represent preliminary results attempting to provide a systematic framework for the analysis and synthesis of hybrid systems.

5 Appendix: Branching Processes

Given two nodes y and z in net N, we say that y precedes z ($y < z$) if and only if there exists a sequence of arcs (a *path*) of the form $\sigma = (x_1, x_2), (x_2, x_3), \cdots (x_{N-1}, x_N)$ such that $x_1 = y$ and $x_N = z$. Two nodes x_1 and x_2 are in *conflict* in net N if there exists distinct transitions $t_1, t_2 \in T$ such that $\bullet t_1 \cap \bullet t_2 \neq \emptyset$ and $t_i < x_i$ for $i = 1, 2$. We say that two nodes x_1 an x_2 are concurrent if they are neither in precedence or conflict.

Given a network N, denote $Min(N)$ as the set of places where $\bullet p = \emptyset$. An *occurrence network* is a finitary acyclic network $N = (P, T, I, O)$ with initial marking μ_0 such that $\| \bullet p \| \leq 1$ for all $p \in P$, no transition $t \in T$ is in self-conflict and $\mu_0(p) = 1$ if and only if $p \in Min(N)$. Let $N_1 = (P_1, T_1, I_1, O_1)$ and $N_2 = (P_2, T_2, I_2, O_2)$ be two nets with initial markings μ_{01} and μ_{02}, respectively. A net homomorphism $h : P_1 \cup T_1 \rightarrow P_2 \cup T_2$ is a mapping between nodes of N_1 and N_2 such that

1. $h(P_1) \subseteq P_2$, $h(T_1) \subseteq T_2$
2. for all $t \in T_1$, the restriction of h to $\bullet t$ is a bijection between $\bullet t$ (in N_1) and $\bullet h(t)$ (in N_2). Similarly for $t\bullet$ and $h(t)\bullet$.
3. and the restriction of h to $\text{Min}(N_1)$ is a bijection between $\text{Min}(N_1)$ and $\text{Min}(N_2)$.

A *branching process* of network N with initial marking μ_0 is a pair (N', h) such that

1. $N' = (P', T', I', O')$ is an occurrence network
2. h is a homomorphism from N' to N such that if $\bullet t_1 = \bullet t_2$ and $h(t_1) = h(t_2)$ then $t_1 = t_2$.

An unfolding of N is a maximal branching process of network N (Engelfriet, 1991).

The unfolding may be infinite in size. We need to identify a finite prefix of the unfolding that allows us to identify all minimal cycles of the network. Well known prefixes for unfoldings allow us to enumerate all reachable markings of a Petri net (McMillan, 1992) (Esparza et al., 1996) In our analysis, however, we are interested in enumerating all fundamental cycles of a network. This may be accomplished by using the following prefix of the unfolding N'. Define the cutset of N' as the set of all places, $p \in P'$ such that $p\bullet = \emptyset$ or there exists $p_1 \in N'$ and $p_2 \in N$ such that $p_1 < p$ and $h(p_1) = h(p) = p_2$. Essentially, the first condition identifies all markings from which there are no output arcs. The other condition identifies what is sometimes refered to as the dupset of the process. The dupset consists of places that have appeared previously in the branching process and hence can be used to identify cycles within the network.

Given a branching process (N', h) of network N, we define a *configuration* or *fundamental run* as a set, C, of transitions in the occurrence net N' such that for all $t \in C$, if there exists t' such that $t' < t$, then $t' \in$

C and t' and t are not in conflict for all t and t' in C. Configurations represent fundamental runs of the system which we can use to identify cycles within the network as shown in the paper's example.

References

Alur, R., Henzinger, T and Ho, P-H (1996). "Automatic symbolic verification of embedded systems", In: *IEEE Transactions on Software Engineering*, 22:181-201, 1996.

Branicky, M.S. (1998), "Multiple Lyapunov functions and other analysis tools for switched and hybrid systems", *IEEE Trans. on Automatic Control*, 43(4), April 1998. Deshpande A., and Varaiya, P. (1995). "Viable control of hybrid systems", In: *Hybrid Systems II*, A. Nerode (ed.), LNCS Vol. 999, Springer-Verlag, Berlin, 1995.

Engelfriet, J. (1991) "Branching processes of Petri nets". In: *Acta Informatica 28*, 575–591, 1991.

Esparza, J., Romer S. and Vogler, W. (1996). "An improvement of McMillan's unfolding algorithm". In: *Proceedings of Tools and Algorithms for the construction and Analysis of Systems* (T. Margaria and B. Steffen, Eds.). Vol. 1055 of *Lecture Notes in Computer Science*, Springer-Verlag, Berlin, 1996.

Flaus, J.-M. and Alla, H. (1997), "Structural analysis of hybrid systems modeled by hybrid flow nets", in *Proceedings of the European Control Conference 97*, Brussels, Belgium, 1997. Jantzen, M. (1987), "Language Theory of Petri Nets", in ",*Petri Nets: central models and their properties, Advanced in Petri Nets 1986*, W. Brauer, W. Reisig, and G. Rosenberg (eds.), Lecture Notes in Computer Science Vol 254-I, pp. 397-412, Springer-Verlag, 1987.

He, K.X. and Lemmon, M.D. (1998), "Lyapunov stability of continuous-valued systems under the supervisions of discrete-event transition sys- tems", In: *Proceeding of the Berkeley 1998 Workshop on Hybrid systems.*, pp: 175-189, Berkeley, CA, April 13-15, 1998.

Henzinger, T., Ho, P.-H., and Wong-Toi, H. (1995), "A user's guide to HyTech", in *First Workshop on Tools and Algorithms for the Construction and Analysis of Systems: TACAS94*, Lecture Notes in Computer Science, Vol. 1019, Springer-Verlag, pp. 41-71, 1995.

Hou, L., Michel, A.N., and Ye, H. (1996), "Stability analysis of switched systems", in *Proceedings of the 35th Conference on Decision and Control*, Kobe Japan, 1996.

Johansson, M. and Rantzer, A. (1998), "Computation of piecewise quadratic Lyapunov functions for hybrid systems", *IEEE Transactions on Automatic Control*, Vol. 44:3, 1998.

Lafferriere, G., Pappas, G.J., and Sastry, S. (1998), "o-minimal hybrid systems", *Technical report UCB/ERL M98/29*, University of California Berkeley, May 1998.

Lemmon, M.D., He, K.X. and Bett, C.J. (1998) "Modeling Hybrid Control Systems using Programmable Timed Petri Nets". In: *Proceeding of L'Automatisation des Processus Mixtes (ADPM'98)*, Rheims France, March 19-20, 1998, to appear in *European* Journal of Automatic Control (JESA), 1999.

McMillan, K. (1992) "Using unfoldings to avoid the state explosion problem in the verification of asynchronous circuits". In: *Computer Aided Verification Fourth International Workshop, CAV'92* (B.V. Bochmann and D.K. Probst, Eds.). Vol. 663 of *Lecture Notes in Computer Science*, Springer Verlag, Berlin, pp. 164–177, 1992.

Petterson, S. and Lennartson, B. (1996), "Stability and robustness of hybrid systems", in *Proceedings of the 35th Conference on Decision and Control*, Kobe Japan, 1996.

Puri, A. and Varaiya, P. (1994), "Decidability of hybrid systems with rectangular differential inclusions", in *Computer Aided Verification: CAV'94*, D. Dill (ed.), Lecture Notes in Computer Science, Vol. 818, Springer-Verlag, pp. 81-84, 1994.

Vogler, W. (1992), *Modular Construction and Partial ORder Semantics of Petri Nets*, Lecture Notes in Computer Science, Vol. 625, Springer-Verlag, 1992.

14th Triennial World Congress, Beijing, P.R. China

E-2c-05-4

COMPUTATION OF CONTROLLERS FOR NONLINEAR HYBRID SYSTEMS [1]

C. Tomlin* J. Lygeros S. Sastry****

* *Department of Aeronautics and Astronautics*
Stanford University, Stanford, CA 94306, USA
tomlin@leland.stanford.edu
** *Department of Electrical Engineering, U.C. Berkeley, CA 94720, USA*
lygeros,sastry@eecs.berkeley.edu

Abstract. A *hybrid system* is one which involves the interaction of discrete and continuous dynamics; in a *nonlinear hybrid system*, the continuous dynamics may be nonlinear and therefore may model quite general physical systems. Designing controllers for hybrid systems involves generating both the continuous nonlinear control law as well as the logic for the *controllable transitions* between discrete states. In this paper, we present our technique for synthesizing controllers for nonlinear hybrid systems, a method which is based on solving a system of coupled partial differential equations. Then, we discuss a numerical algorithm based on *level set methods* for solving this system of equations. *Copyright © 1999 IFAC*

Keywords. Hybrid systems, partial differential equations, computational methods, air traffic control.

1. INTRODUCTION

For about the past six years, researchers in the traditionally distinct fields of control theory and computer science verification have proposed models, verification, and controller synthesis techniques for complex, safety critical systems. The area of *hybrid systems* is loosely defined as the study of systems which involve the interaction of discrete event and continuous time dynamics, with the purpose of proving properties such as reachability and stability. The discrete event models naturally accommodate linguistic and qualitative information, and are used to model modes of operation of the system, and the continuous dynamics model the physical processes themselves.

One class of approaches to modeling and analysis of hybrid systems has been to extend techniques for finite state automata to include systems with simple continuous dynamics. Models and decidability results have been obtained for timed automata (Alur and Dill, 1994), linear hybrid automata (Alur *et al.*, 1993), and hybrid input/output automata (Lynch *et al.*, 1996). A second class of models and analysis techniques for hybrid systems has developed out of research in continuous state space and continuous time dynamical systems and control. The emphasis here has been on extending the standard modeling, reachability and stability analyses, and controller design techniques to capture the interaction between the continuous and discrete dynamics (Branicky, 1994; Deshpande, 1994; Lygeros, 1996). The idea of posing the controller synthesis problem as a discrete game between the system and its environment is attributed to Church (Church, 1962) and solved by Büchi and Landweber (Büchi and Landweber, 1969). Controller synthesis on timed automata was first developed in (Maler

[1] Research supported by a Stanford University Terman Fellowship, by NASA under grant NAG 2-1039, and by ONR under grant N0014-97-1-0946.

et al., 1995), and an algorithm for controller synthesis on linear hybrid automata is presented in (Wong-Toi, 1997). One area in which results have been hard to come by is the efficient *computation* of reachable sets for hybrid systems whose dynamics are nonlinear. Only recently, some attempts to directly approach this problem have been reported in the literature (Dang and Maler, 1998; Greenstreet and Mitchell, 1998).

Our approach to hybrid systems modeling and controller synthesis incorporates accurate, nonlinear models of the continuous dynamics with models for discrete event dynamics. We include continuous and discrete input variables to model both parameters that the designer may control as well as disturbance parameters that the designer must control against. Using analysis based on traditional discrete and continuous optimal control techniques, and on two-person zero-sum game theory for automata and continuous dynamical systems, we derive the Hamilton-Jacobi partial differential equations whose solutions describe *exactly* the boundaries of reachable sets. Only then do we approximate: we use a clever numerical technique to solve this equation. These equations are the heart of our general controller synthesis technique for hybrid systems, in which we calculate feedback control laws for the continuous and discrete variables which guarantee that the hybrid system remains in the "safe subset" of the reachable set. In this paper we overview this methodology for computing reachable sets of hybrid systems. Our hybrid system model is developed from that of (Lygeros, 1996), and our controller synthesis technique is presented in (Tomlin, 1998), which builds on the work of (Tomlin *et al.*, 1998*c*; Lygeros *et al.*, 1998; Lygeros *et al.*, 1999). Our motivation for this work arose out of attempting to verify the safety of a class of conflict resolution maneuvers for aircraft, in (Tomlin *et al.*, 1998*a*), and of flight management systems for aircraft, in (Tomlin *et al.*, 1998*b*). Several worked examples are presented in (Tomlin, 1998).

2. NONLINEAR HYBRID SYSTEMS

Our model assumes that there exist both controllable and uncontrollable discrete actions, as well as continuous control inputs and continuous disturbance inputs.

A *hybrid automaton* is a tuple: $H = ((Q \times X), (U \times D), (\Sigma_1 \times \Sigma_2), f, \delta, Inv, I, Y, h, \Omega)$ where Q is a finite set of discrete states, $X = \mathbb{R}^n$, $U \subseteq \mathbb{R}^u$ is the set of continuous control inputs, $D \subseteq \mathbb{R}^d$ is the set of continuous environment inputs, known as disturbances, $\Sigma = \Sigma_1 \times \Sigma_2$ is a finite set of actions, where Σ_1 denotes the set of discrete control inputs, and Σ_2 the set of discrete disturbance inputs, $f : Q \times X \times U \times D \to \mathbb{R}^n$, $\delta : Q \times X \times \Sigma_1 \times \Sigma_2 \to 2^{Q \times X}$ is the discrete transition function, $Inv \subseteq Q \times X$ is the invariant associated to each discrete state, $I \subseteq Q \times X$ is a subset of initial states, Y is the set of continuous outputs, $h : Q \times X \to 2^Y$ is the output map, and Ω is an acceptance condition (here $\Omega = (\Box F)$, meaning that the state of the system must remain within a set $F \subseteq Q \times X$). The space of acceptable control and disturbance trajectories are denoted by $\mathcal{U} = \{u(\cdot) \in PC^0 \mid u(t) \in U \; \forall t \in \mathbb{R}\}$, $\mathcal{D} = \{d(\cdot) \in PC^0 \mid d(t) \in D \; \forall t \in \mathbb{R}\}$, where PC^0 denotes the set of piecewise continuous functions of $\mathbb{R}$. We assume that the actions of the controller are modeled by the continuous variable u as well as the discrete action σ_1, and that the actions of the disturbance, or the environment, are modeled by both d and σ_2. The goal of the controller is to force the system to remain inside F, and the goal of the disturbance is to force the system to leave this same set.

A **hybrid time trajectory** is a sequence of intervals of the real line $\tau = [0, \tau_0'][\tau_1, \tau_1'][\tau_2, \tau_2'] \ldots [\tau_k, \tau_k')$ such that $\tau_{i-1}' = \tau_i$ and $\tau_i \leq \tau_i'$. The index k may be finite or infinite. For $t \in \mathbb{R}$, we use $t \in \tau$ to represent $t \in [\tau_i, \tau_i']$ for some i. Denote by $\mathcal{T}$ the set of all hybrid time trajectories. A **hybrid system trajectory** is defined as $(\tau, q[\cdot], x(\cdot), u(\cdot), d(\cdot), \sigma_1[\cdot], \sigma_2[\cdot])$ where $\tau \in \mathcal{T}$, $q[\cdot] \in Q^\omega$, $x(\cdot) : \tau \to X$, $u(\cdot) : \tau \to U$, $d(\cdot) : \tau \to D$, $\sigma_1[\cdot] \in \Sigma_1^\omega$, and $\sigma_2[\cdot] \in \Sigma_2^\omega$. The initial condition satisfies $(q[0], x(0)) \in I$; the discrete evolution satisfies $(q[i+1], x(\tau_{i+1})) \in \delta(q[i], x(\tau_i'), \sigma_1[i], \sigma_2[i])$ for all i and τ; the continuous evolution satisfies $(q[i], x(t)) \in Inv$ and $\dot{x}(t) = f(q[i], x(t), u(t), d(t))$ for $t \in [\tau_i, \tau_i']$; and the output evolution satisfies $y(t) \in h(q[i], x(t))$, for $t \in [\tau_i, \tau_i']$. The state trajectory is the $(q[\cdot], x(\cdot))$ such that the above conditions are satisfied. To ensure that the laws for continuous evolution are meaningful we assume that the function f is *globally Lipschitz* in x and continuous in u and d.

3. CONTROLLER SYNTHESIS FOR NONLINEAR HYBRID SYSTEMS

Consider the problem of synthesizing controls $(u(\cdot), \sigma_1[\cdot])$, in the presence of environmental disturbances $(d(\cdot), \sigma_2[\cdot])$, for the nonlinear hybrid automaton H.

In (Tomlin, 1998), we showed that for purely discrete systems the computation of the largest set of states for which the control can guarantee that the acceptance condition Ω is met despite the action of the disturbance can be computed via the *discrete Hamilton-Jacobi equation*:

$$J(q, i-1) - J(q, i) = \min\left\{0, \max_{\sigma_1}\min_{\sigma_2}\right.$$

$$[\min_{q' \in \delta(q,\sigma_1,\sigma_2)} J(q', i) - J(q, i)]\}$$

where $J(q, i) : Q \times \mathbb{Z}_- \rightarrow \{0, 1\}$, with $J(q, 0) = 1$ if $q \in F$ and $J(q, 0) = 0$ otherwise. It is easy to show that $J(q, i) = 1$ if the control has a sequence of actions which will ensure that the system remains in F for at least i steps, for all possible actions of the disturbance. A fixed point of (1), $J(q, i-1) = J(q, i) \triangleq J^*(q)$ is reached after a finite number of iterations, for all finite Q. The "safe set of states", i.e., the set of initial states for which the control can guarantee that the system remains in F for ever, is then $\{q \in Q | J^*(q) = 1\}$.

Similarly, in (Tomlin, 1998) it was shown that for continuous systems, $\dot{x}(t) = f(x(t), u(t), d(t))$, this computation can be performed via the following *Hamilton-Jacobi partial differential equation*:

$$-\frac{\partial J^*(x,t)}{\partial t} = \begin{cases} H^*(x, \frac{\partial J^*(x,t)}{\partial x}) \\ \quad \text{for } \{x \in X \mid J^*(x,t) > 0\} \\ \min\{0, H^*(x, \frac{\partial J^*(x,t)}{\partial x})\} \\ \quad \text{for } \{x \in X \mid J^*(x,t) \leq 0\} \end{cases} \tag{1}$$

where $J(x, u(\cdot), d(\cdot), t) : \mathbb{R}^n \times \mathcal{U} \times \mathcal{D} \times \mathbb{R}_- \rightarrow \mathbb{R}$ is defined so that $J(x, u(\cdot), d(\cdot), t) = l(x(0))$, for some $l(x)$ with $l(x) > 0$ for x inside F, $l(x) = 0$ for x on the boundary of F and $l(x) < 0$ outside F. $J^*(x, t)$ is defined as $J^*(x, t) = \max_{u \in \mathcal{U}} \min_{d \in \mathcal{D}} J(x, u(\cdot), d(\cdot), t)$, and $H^*(\cdot)$ is the familiar *Hamiltonian* of optimal control: $H^*(x, p) = \max_{u \in U} \min_{d \in D} p^T f(x, u, d)$ for costate p. It can be shown that if (1) reaches a fixed point at time t^*, then the safe set of states is $\{x | J^*(x, t^*) \geq 0\}$, otherwise it is $\{x | J^*(x, -\infty) \geq 0\}$.

For nonlinear hybrid systems, we again seek to construct the largest set of states for which the control can guarantee that the acceptance condition Ω is met despite the action of the disturbance. For any set $K \subseteq Q \times X$ define the controllable and uncontrollable predecessors of K by:

$$\begin{aligned} Pre_1(K) =& \{(q, x) \in Q \times X | \exists \sigma_1 \in \Sigma_1 \ \forall \sigma_2 \in \Sigma_2 \\ & \delta(q, x, \sigma_1, \sigma_2) \subseteq K\} \cap K \\ Pre_2(K) =& \{(q, x) \in Q \times X | \forall \sigma_1 \in \Sigma_1 \ \exists \sigma_2 \in \Sigma_2 \\ & \delta(q, x, \sigma_1, \sigma_2) \cap K^c \neq \emptyset\} \cup K^c \end{aligned} \tag{2}$$

In other words, the controllable predecessor of K, $Pre_1(K)$, contains all states in K for which the controllable actions can force the state to remain in K for at least one step in the discrete evolution. The uncontrollable predecessor $Pre_2(K)$, on the other hand, contains all states in K^c as well as all states from which the uncontrollable actions may be able to force the state outside K. We also define the "reach-avoid" operator:

Definition 1 (Reach-Avoid) *Consider two subsets $G \subseteq Q \times X$ and $E \subseteq Q \times X$ such that $G \cap E = \emptyset$. The reach-avoid operator is defined as*

$$\begin{aligned} &Reach(G, E) = \{(q, x) \in Q \times X \mid \\ &\forall u \in \mathcal{U} \ \exists d \in \mathcal{D} \text{ and } t \geq 0 \text{ such that} \\ &(q(t), x(t)) \in G \text{ and } (q(s), x(s)) \notin E \text{ for } s \in [0, t]\} \end{aligned} \tag{3}$$

where $(q(s), x(s))$ is the continuous state trajectory of $\dot{x}(s) = f(q(s), x(s), u(s), d(s))$ starting at (q, x). The set $Reach(G, E)$ describes those states from which, for all $u(\cdot) \in \mathcal{U}$, there exists a $d(\cdot) \in \mathcal{D}$, such that the state trajectory $(q(s), x(s))$ can be driven to G while avoiding an "escape" set E.

Consider the following algorithm (motivated by the algorithm proposed in (Wong-Toi, 1997) for linear hybrid automata):

Let $W^0 = F, W^{-1} = \emptyset, i = 0$.
While $W^i \neq W^{i-1}$ do
  $W^{i-1} = W^i \setminus Reach(Pre_2(W^i), Pre_1(W^i)))$ (4)
  $i = i - 1$
end

Since at each step, $W^{i-1} \subseteq W^i$, the set W^i decreases monotonically as i decreases. If the algorithm terminates, we denote the fixed point as W^*. The calculation of *Reach* can be carried out by appropriately modifying the Hamilton-Jacobi construction of (1), as we describe below. Let $l_G : X \rightarrow \mathbb{R}$ and $l_E : X \rightarrow \mathbb{R}$ be differentiable functions such that $G \triangleq \{x \in X : l_G(x) \leq 0\}$ and $E \triangleq \{x \in X : l_E(x) \leq 0\}$ (in general G and E may be expressed as the intersection of a set of differentiable functions). Consider the following system of interconnected Hamilton-Jacobi equations:

$$-\frac{\partial J_G^*(x,t)}{\partial t} = \begin{cases} H_G^*(x, \frac{\partial J_G^*(x,t)}{\partial x}) \\ \quad \text{for } \{x \in X \mid J_G^*(x,t) > 0\} \\ \min\{0, H_G^*(x, \frac{\partial J_G^*(x,t)}{\partial x})\} \\ \quad \text{for } \{x \in X \mid J_G^*(x,t) \leq 0\} \end{cases} \tag{5}$$

and

$$-\frac{\partial J_E^*(x,t)}{\partial t} = \begin{cases} H_E^*(x, \frac{\partial J_E^*(x,t)}{\partial x}) \\ \quad \text{for } \{x \in X \mid J_E^*(x,t) > 0\} \\ \min\{0, H_E^*(x, \frac{\partial J_E^*(x,t)}{\partial x})\} \\ \quad \text{for } \{x \in X \mid J_E^*(x,t) \leq 0\} \end{cases} \tag{6}$$

where $J_G(x, u(\cdot), d(\cdot), t) = l_G(x(0))$ and
$J_E(x, u(\cdot), d(\cdot), t) = l_E(x(0))$, and

$$H_G^*(x, \frac{\partial J_G^*}{\partial x}) = \begin{cases} 0, \text{for } \{x \in X \mid J_E^*(x,t) \le 0\} \\ \max_{u \in U} \min_{d \in D} \frac{\partial J_G^*}{\partial x} f(x,u,d) \text{ otherwise} \end{cases} \quad (7)$$

$$H_E^*(x, \frac{\partial J_E^*}{\partial x}) = \begin{cases} 0, \text{for } \{x \in X \mid J_G^*(x,t) \le 0\} \\ \min_{u \in U} \max_{d \in D} \frac{\partial J_E^*}{\partial x} f(x,u,d) \text{ otherwise} \end{cases} \quad (8)$$

Theorem 1 (Characterization of Reach-Avoid) *If $J_G^*(x,t)$ ($J_E^*(x,t)$ respectively) satisfies the Hamilton-Jacobi equation (5) ((6) respectively), and if it converges uniformly in x as $t \to -\infty$ to a function $J_G^*(x)$ ($J_E^*(x)$ respectively). Then,*

$$Reach(G, E) = \{x \in X \mid J_G^*(x) < 0\} \quad (9)$$

Proof: See (Tomlin, 1998). ■

4. COMPUTATION OF CONTROLLERS

In practice, the usefulness of the algorithm for hybrid controller synthesis depends on our ability to efficiently compute the optimal control and disturbance trajectories $(u^*(\cdot), d^*(\cdot))$, as well as solutions to the Hamilton-Jacobi partial differential equation (1). Numerical solutions are potentially complicated by the facts that the right hand side of (1) is non-smooth and that the initial data F may have non-smooth boundary, that $(u^*(\cdot), d^*(\cdot))$ may be discontinuous, and that the solution $J^*(x,t)$ may develop shocks over time. In this section, we discuss a numerical technique developed by Osher and Sethian which computes the viscosity solution to the Hamilton-Jacobi equation, ensuring that discontinuities are preserved. We conclude with a discussion of approximation techniques to approximate non-smooth sets with smooth sets.

4.1 *Level Set Method for Evolving Curves*

Consider the Hamilton-Jacobi equation (1). We define the *viscosity solution* (Crandall *et al.*, 1984) to (1) as the solution as $\epsilon \to 0$ of the regularized partial differential equation:

$$-\frac{\partial J_\epsilon^*(x,t)}{\partial t} = \begin{cases} H^*(x, \frac{\partial J_\epsilon^*(x,t)}{\partial x}) + \epsilon \Delta J_\epsilon^*(x,t) \\ \quad \text{for } \{x \in X \mid J_\epsilon^*(x,t) > 0\} \\ \min\{0, H^*(x, \frac{\partial J_\epsilon^*(x,t)}{\partial x})\} + \epsilon \Delta J_\epsilon^*(x,t) \\ \quad \text{for } \{x \in X \mid J_\epsilon^*(x,t) \le 0\} \end{cases}$$

with boundary condition $J_\epsilon^*(x,0) = l_\epsilon(x)$. The level set methods of Osher and Sethian (Osher and Sethian, 1988) are computational schemes for propagating interfaces in which the speed of propagation is governed by a partial differential equation. These methods choose the solution of (1) (with zero viscosity) to be the one obtained from the regularized system as the viscosity coefficient ϵ vanishes. We present an outline of the method below for a two-dimensional example.

In order for the numerical scheme to closely approximate the gradient $\frac{\partial J^*(x,t)}{\partial x}$, especially at points of discontinuity, an appropriate approximation to the spatial derivative must be used. Consider an example in two dimensions, with X discretized into a grid with spacing Δx_1 and Δx_2. The *forward difference operator* D^{+x_i} at $x = (x_1, x_2)$ is defined as (for x_1, similarly for x_2):

$$D^{+x_1} J^*(x,t) = \frac{J^*((x_1 + \Delta x_1, x_2), t) - J^*(x,t)}{\Delta x_1} \quad (10)$$

The *backward difference operator* D^{-x_i} is defined as (for x_1, similarly for x_2):

$$D^{-x_1} J^*(x,t) = \frac{J^*(x,t) - J^*((x_1 - \Delta x_1, x_2), t)}{\Delta x_1} \quad (11)$$

Similarly, the *central difference operator* D^{0x_i} is defined as (for x_1, similarly for x_2):

$$D^{0x_1} J^*(x,t) = \frac{D^{+x_1} J^*(x,t) + D^{-x_1} J^*(x,t)}{2} \quad (12)$$

At each grid point $x = (x_1, x_2)$, the partial derivatives $\frac{\partial J^*(x,t)}{\partial x_1}$ and $\frac{\partial J^*(x,t)}{\partial x_2}$ may be approximated to first order using either the forward, backward, or central difference operators. The correct choice of operator depends on the direction of $f(x, u^*, d^*)$ (in our case it depends on $-f(x, u^*, d^*)$ since we compute backwards in time). If $-f(x, u^*, d^*)$ flows from left to right (from smaller to larger values of x_1), then D^{-x_1} should be used to approximate $\frac{\partial J^*(x,t)}{\partial x_1}$ (and vice versa); and if $-f(x, u^*, d^*)$ flows from bottom to top (from smaller to larger values of x_2), then D^{-x_2} should be used to approximate $\frac{\partial J^*(x,t)}{\partial x_2}$ (and vice versa). Such an approximation is called an *upwind* scheme, since it uses information upwind of the direction that information propagates.

The algorithm for the two dimensional example proceeds as follows. Choose a domain of interest in X and discretize the domain with a grid of spacing $\Delta x_1, \Delta x_2$. Let x_{ij} represent the grid point $(i\Delta x_1, j\Delta x_2)$ and let $\tilde{J}^*(x_{ij}, t)$ represent the numerical approximation of $J^*(x_{ij}, t)$. Using the boundary condition $J^*(x, 0) = l(x)$, compute $\tilde{J}^*(x_{ij}, 0)$ for each x_{ij}.

Let $t = 0$.

While $\tilde{J}^*(x_{ij}, t) \neq \tilde{J}^*(x_{ij}, t-\Delta t)$ perform the following steps:

(1) Compute

$$u^*(x_{ij}, \frac{\partial \tilde{J}^*(x_{ij},t)}{\partial x_1}, \frac{\partial \tilde{J}^*(x_{ij},t)}{\partial x_2}) \qquad (13)$$

$$d^*(x_{ij}, \frac{\partial \tilde{J}^*(x_{ij},t)}{\partial x_1}, \frac{\partial \tilde{J}^*(x_{ij},t)}{\partial x_2}) \qquad (14)$$

using the initial approximations to the derivatives

$$\frac{\partial \tilde{J}^*(x_{ij},t)}{\partial x_1} = D^{0x_1}, \quad \frac{\partial \tilde{J}^*(x_{ij},t)}{\partial x_2} = D^{0x_2} \; (15)$$

(2) Calculate $f(x_{ij}, u^*, d^*)$

(3) If $-f(x_{ij}, u^*, d^*)$ flows from larger to smaller values of x_1, let

$$\frac{\partial \tilde{J}^*(x_{ij},t)}{\partial x_1} = D^{+x_1} \qquad (16)$$

otherwise let

$$\frac{\partial \tilde{J}^*(x_{ij},t)}{\partial x_1} = D^{-x_1} \qquad (17)$$

If $-f(x_{ij}, u^*, d^*)$ flows from larger to smaller values of x_2, let

$$\frac{\partial \tilde{J}^*(x_{ij},t)}{\partial x_2} = D^{+x_2} \qquad (18)$$

otherwise let

$$\frac{\partial \tilde{J}^*(x_{ij},t)}{\partial x_2} = D^{-x_2} \qquad (19)$$

(4) Compute $\tilde{J}^*(x_{ij}, t-\Delta t)$:
For x_{ij} such that $J^*(\tilde{x}_{ij}, t) > 0$,

$$\tilde{J}^*(x_{ij}, t-\Delta t) = \tilde{J}^*(x_{ij}, t) + \Delta t \frac{\partial \tilde{J}^*(x_{ij},t)}{\partial x} f(x_{ij}, u^*, d^*)$$

For x_{ij} such that $J^*(\tilde{x}_{ij}, t) \leq 0$,

$$(x_{ij}, t-\Delta t) = \begin{cases} \tilde{J}^*(x_{ij}, t) + \Delta t \frac{\partial \tilde{J}^*(x_{ij},t)}{\partial x} f(x_{ij}, u^*, d^*) \\ \quad \text{if } \frac{\partial \tilde{J}^*(x_{ij},t)}{\partial x} f(x_{ij}, u^*, d^*) < 0 \\ \tilde{J}^*(x_{ij}, t) \text{ otherwise} \end{cases} \qquad (20)$$

This method is applied successfully to the computation of a reach set in aircraft conflict resolution, in (Tomlin, 1998).

4.2 *Approximation of* $J^*(\cdot)$

A different approach to solving the Hamilton-Jacobi PDE relies on approximating the value function J^* with function chosen from a class parameterized by a finite number of parameters. These parameters are then tuned to improve the approximation. For certain classes of discrete problems it can be shown that if the class of approximating functions is rich enough and the parameter tuning is done properly, this method can produce very good solutions for very complex problems (Bertsekas and Tsitsiklis, 1996). We seek to extend this approach to the continuous domain. As an illustration, consider a class of approximating functions for which the parameters enter linearly. In other words, at each time t assume that the value function $J^*(x,t)$ is approximated by:

$$\tilde{J}(x,t) = \sum_{i=1}^{N} r_i(t) b_i(x) = B(x)^T R(t)$$

where b_i is a finite collection of basis functions. For example, b_i could be polynomials of degree up to M, i.e. for $x \in \mathbb{R}^n$, b_i is the collection of functions $1, x_1, \ldots, x_n, x_1^2, x_1 x_2, \ldots, x_n^M$. It is a well known fact that this class of functions can (for sufficiently large M) approximate a continuous function over a compact set arbitrarily closely. Starting with an initial approximation $\tilde{J}(x,0) = B(x)^T R(0)$ of $J^*(x,0)$, the Hamilton-Jacobi equation can be approximated by:

$$-B(x)^T \dot{R}(t) = \begin{cases} H^*(x, \frac{\partial B}{\partial x}(x) R(t)) \\ \quad \text{for } \{x \in X \mid B(x)^T R(t) > 0\} \\ \min\{0, H^*(x, \frac{\partial B}{\partial x}(x) R(t))\} \\ \quad \text{for } \{x \in X \mid B(x)^T R(t) \leq 0\} \end{cases} \qquad (21)$$

Overall, the Hamilton-Jacobi PDE can be approximated by an ODE of the form:

$$B(x)^T \dot{R}(t) = B(x)^T \tilde{H}(R(t))$$

where $\tilde{H}$ denotes the linearly parameterized approximation for the right hand side. Equating the coefficient for each b_i and solving the approximating ODE for R produces and estimate $\tilde{J}(x,t) = B(x)^T R(t)$ for J^*.

5. CONCLUDING REMARKS

It is essential that a bound on the error due to approximation be known at each step of the algorithm, in order to guarantee that the computed safe set is a conservative approximation to the actual safe set. Also, even if all of the issues associated with computing solutions to the continuous Hamilton-Jacobi equation are resolved, there are more computational challenges arising from the discrete dynamics. The first is *undecidability*: the class of hybrid systems for which algorithms like the one presented here are guaranteed to terminate is known to be restricted (Henzinger *et al.*, 1995). Another problem is the requirement that the controller resulting from our algorithm is *non-Zeno*, i.e. does not enforce the safety requirement by preventing time from diverging. The algorithm proposed here has no way of preventing such behavior. Adding this requirement to our controllers is likely to be a major challenge, as it involves reasoning with infinite executions and liveness specifications. This is the subject of on-going research.

6. REFERENCES

Alur, R. and D. Dill (1994). A theory of timed automata. *Theoretical Computer Science* **126**, 183–235.

Alur, R., C. Courcoubetis, T. A. Henzinger and P.-H. Ho (1993). Hybrid automata: An algorithmic approach to the specification and verification of hybrid systems. In: *Hybrid Systems* (Robert L. Grossman, Anil Nerode, Anders P. Ravn and Hans Rischel, Eds.). pp. 366–392. Springer Verlag. New York.

Bertsekas, Dimitri P. and John N. Tsitsiklis (1996). *Neuro-dynamic Programming.* Athena Scientific. Belmont, Massachusetts.

Branicky, M. S. (1994). Control of Hybrid Systems. PhD thesis. Department of Electrical Engineering and Computer Sciences, Massachusetts Institute of Technology.

Büchi, J. R. and L. H. Landweber (1969). Solving sequential conditions by finite-state operators. In: *Proceedings of the American Mathematical Society.* pp. 295–311.

Church, A. (1962). Logic, arithmetic, and automata. In: *Proceedings of the International Congress of Mathematicians.* pp. 23–35.

Crandall, M. G., L. C. Evans and P.-L. Lions (1984). Some properties of viscosity solutions of Hamilton-Jacobi equations. *Transactions of the American Mathematical Society* **282**(2), 487–502.

Dang, T. and O. Maler (1998). Reachability analysis via face lifting. In: *Hybrid Systems: Computation and Control* (S. Sastry and T.A. Henzinger, Eds.). pp. 96–109. Number 1386 In: *LNCS.* Springer Verlag.

Deshpande, A. (1994). Control of Hybrid Systems. PhD thesis. Department of Electrical Engineering and Computer Sciences, University of California at Berkeley.

Greenstreet, M.R. and I. Mitchell (1998). Integrating projections. In: *Hybrid Systems: Computation and Control* (S. Sastry and T.A. Henzinger, Eds.). pp. 159–174. Number 1386 In: *LNCS.* Springer Verlag.

Henzinger, T. A., P. W. Kopke, A. Puri and P. Varaiya (1995). What's decidable about hybrid automata. In: *Proceedings of the 27th Annual ACM Symposium on Theory of Computing.*

Lygeros, J. (1996). Hierarchical, Hybrid Control of Large Scale Systems. PhD thesis. Department of Electrical Engineering and Computer Sciences, University of California at Berkeley.

Lygeros, J., C. Tomlin and S. Sastry (1998). On controller synthesis for nonlinear hybrid systems. In: *Proceedings of the IEEE Conference on Decision and Control.* Tampa, FL. pp. 2101–2106.

Lygeros, J., C. Tomlin and S. Sastry (1999). Controllers for reachability specifications for hybrid systems. *Automatica.* To appear.

Lynch, N., R. Segala, F. Vaandrager and H.B. Weinberg (1996). Hybrid I/O automata. In: *Hybrid Systems III.* pp. 496–510. Lecture Notes in Computer Science 1066. Springer Verlag.

Maler, O., A. Pnueli and J. Sifakis (1995). On the synthesis of discrete controllers for timed systems. In: *STACS 95: Theoretical Aspects of Computer Science* (Ernst W. Mayr and Claude Puech, Eds.). pp. 229–242. Lecture Notes in Computer Science 900. Springer Verlag. Munich.

Osher, S. and J. A. Sethian (1988). Fronts propagating with curvature-dependent speed: Algorithms based on Hamilton-Jacobi formulations. *Journal of Computational Physics* **79**, 12–49.

Tomlin, C., G. J. Pappas and S. Sastry (1998*a*). Conflict resolution for air traffic management: A case study in multi-agent hybrid systems. *IEEE Transactions on Automatic Control* **43**(4), 509–521.

Tomlin, C., J. Lygeros and S. Sastry (1998*b*). Aerodynamic envelope protection using hybrid control. In: *Proceedings of the American Control Conference.* Phildelphia, PA. pp. 1793–1796.

Tomlin, C., J. Lygeros and S. Sastry (1998*c*). Synthesizing controllers for nonlinear hybrid systems. In: *Hybrid Systems: Computation and Control* (T. Henzinger and S. Sastry, Eds.). pp. 360–373. Lecture Notes in Computer Science 1386. Springer Verlag. New York.

Tomlin, Claire J. (1998). Hybrid Control of Air Traffic Management Systems. PhD thesis. Department of Electrical Engineering, University of California, Berkeley.

Wong-Toi, H. (1997). The synthesis of controllers for linear hybrid automata. In: *Proceedings of the IEEE Conference on Decision and Control.* San Diego, CA.

14th Triennial World Congress, Beijing, P.R. China

E-2c-05-5

A HYBRID DYNAMICAL SYSTEM OF ORDER N WITH ALL TRAJECTORIES CONVERGING TO (N-1)! LIMIT CYCLES

Andrey V. Savkin*

* Department of Electrical and Electronic Engineering, University of Western Australia, Nedlands, WA 6907, Australia, email: savkin@ee.uwa.edu.au, fax: +(618) 93801065

Abstract. The paper analyzes an example of the discrete control of a continuous variable system. We prove that the switched server system with one server and an arbitrary number of buffers has a finite set of periodic trajectories which attracts all other trajectories of the system. *Copyright © 1999 IFAC*

Keywords. Limit Cycles, Stability, Dynamic Systems, Networks, Switching

1. INTRODUCTION

Hybrid dynamical systems have attracted considerable attention in recent years. In general, hybrid systems are those that combine continuous and discrete behavior and involve, thereby, both continuous and discrete state variables.

One important type of hybrid dynamical systems is the class of discretely controlled continuous-time systems. Two interesting examples of such systems were inspired by the models for flexible manufacturing systems (Chase *et al.* 1993). They were called "the switched arrival system" and "the switched server system". These dynamical systems are of interest on their own right but have also been used to model certain aspects of flexible manufacturing systems. These examples can also be interpreted as models for simple dynamically routed closed queueing networks. It is well known that the switched arrival system exhibits a chaotic behavior whereas, under certain assumptions, the dynamics of the switched server system is eventually periodic (Chase et al. 1993). However, only the case of systems with three buffers was considered. The systems with three buffers can be reduced to planar systems, which makes their analysis a much easier task.

In the current paper, we give a complete qualitative analysis of the switched server system consisting of one server and n buffers with a simple and quite natural server switching feedback strategy. This switching strategy is more realistic for manufacturing applications and very similar to switching policies for flexible manufacturing systems. This system has no equilibrium points. Hence the simplest possible attractors are limit cycles. We prove that the state space of this system can be partitioned into a collection of $(n-1)!$ unbounded regions such that each of them is invariant and contains one limit cycle. Moreover, all trajectories with initial conditions from a fixed region converge to the corresponding limit cycle. Hence any trajectory of this system is asymptotically periodic and the switched server system always exhibits a regular stable predictable behavior. This conclusion is very important for applications.

Our example shows that it is typical even for very simple hybrid systems to exibit a qualitative behavior which is quite unusual for either ordinary differential equations or discrete event systems.

2. THE SWITCHED SERVER SYSTEM

Consider the following single-machine flexible manufacturing system or switched server system (Perkins and Kumar 1989).

This system consists of n buffers, with work arriving to the buffer j at a constant rate $p_j > 0$ where $j = 1, 2, \dots, n$. Also, the system contains one machine or server that removes work from any selected buffer at a constant rate $p > 0$. Furthermore, whenever the server switches from one buffer to another, a set-up time $\delta > 0$ is required.

We refer to the contents of buffers as "work"; it will be convinient to to think of work as a fluid, and a buffer as a tank. In applications, work can represent a continuous approximation to the discrete flow of parts in manufacturing systems (as in (Perkins and Kumar 1989)), or jobs in a computer system, etc.

The example can also be thought of as a simple instance of the switched controller problem.

Let $x_j(t)$ be the amount of work in the buffer j at time t. Then $x_j(t)$ is a continuous variable of this system. The location of the server is a control variable. This variable is a discrete one. Any trajectory of the switched server system is defined by our switching feedback policy and initial condition

$$x_j(0) = x_j^0 \quad \forall j = 1, 2, \ldots, n \tag{1}$$

where $x_j^0 \geq 0$.

We assume that the following assumption holds (Perkins and Kumar 1989):

Assumption 1 $p > p_1 + p_2 + \ldots + p_n$.

It is obvious that if Assumption 1 does not hold then the system is unstable in the following sense: its trajectories are not bounded on $[0, \infty)$.

Switching strategy Introduce the set

$$K_0 \triangleq \left\{ \begin{array}{c} (x_1, \ldots, x_n) \in \mathbf{R}^n : \\ x_1 \geq 0, \ldots, x_n \geq 0 \end{array} \right\}.$$

Furthermore, introduce a map $\mathcal{I}$ from the set K_0 to the set of indeces $\{1, 2, \ldots, n\}$ as follows:

$$\mathcal{I}(x_1, x_2, \ldots, x_n) = \min_{j=1,\ldots,n} \left\{ j : \frac{x_j}{p_j} = \max\left[\frac{x_1}{p_1}, \ldots, \frac{x_n}{p_n}\right] \right\}.$$

In other words, $\mathcal{I}(x_1, x_2, \ldots, x_n)$ is the index j at which the maximum of $\frac{x_j}{p_j}$ is achieved, and if the maximum is achieved at several j, we take the minimum among them.

Here, we propose the following simple switching strategy:

P1: The server starts with the buffer j such that $j = \mathcal{I}(x_1^0, x_2^0, \ldots, x_n^0)$.

P2: The server removes work from the current buffer until it is empty.

P3: Whenever, the server has emptied one buffer at time t, it switches to the buffer j such that $j = \mathcal{I}(x_1(t), x_2(t), \ldots, x_n(t))$.

This control feedback policy is quite natural and very similar to the control policies for manufacturing systems considered in (Perkins and Kumar 1989).

Now we show that this system can be described by a system of logic-differential equations. Indeed, introduce a set of discrete variables

$$Q \triangleq \{q_0, q_1, \ldots, q_{n-1}, q_n\}.$$

Furthermore, introduce the following vectors

$$a(q_0) \triangleq \begin{pmatrix} p_1 \\ p_2 \\ \vdots \\ p_{n-1} \\ p_n \end{pmatrix}, \quad a(q_1) \triangleq \begin{pmatrix} p_1 - p \\ p_2 \\ \vdots \\ p_{n-1} \\ p_n \end{pmatrix}, \quad a(q_2) \triangleq \begin{pmatrix} p_1 \\ p_2 - p \\ \vdots \\ p_{n-1} \\ p_n \end{pmatrix}, \ldots, \quad a(q_n) \triangleq \begin{pmatrix} p_1 \\ p_2 \\ \vdots \\ p_{n-1} \\ p_n - p \end{pmatrix}. \tag{2}$$

In this system, $x_j(t)$ is the amount of work in buffer j at time t, the discrete state q_j corresponds to the case when the server is removing work from the buffer j, the discrete state q_0 corresponds to the case when the server is switching from one buffer to another.

Moreover, introduce map $\mathcal{I}_Q$ from the set K_0 to the set Q as follows:

$$\mathcal{I}_Q(x_1, x_2, \ldots, x_n) = q_j \quad \text{if} \quad \mathcal{I}(x_1, x_2, \ldots, x_n) = j.$$

Notation Let $\nu(t)$ be a function of time. Then $\nu(t+0) \triangleq \lim_{\epsilon>0,\ \epsilon\to 0} \nu(t+\epsilon)$.

Then the above switched server system can be described by the following equations:

$$\textbf{if} \quad q(t) = q_j \quad \textbf{then} \quad \dot{x}(t) = a(q_j). \tag{3}$$

Furthermore, our switching rule **P1, P2, P3** can

be described as

$$\begin{array}{ll} & (\textbf{if } q(t) = q_j \;\; j = 1, 2, \ldots, n \\ & \qquad\qquad \textbf{and } x_j(t) = 0) \\ \textbf{then} & \begin{pmatrix} q(\tau) := q_0 \quad \forall \tau \in (t, t+\delta], \\ q(t+\delta+0) := \\ \mathcal{I}_Q(x_1(t), x_2(t), \ldots, x_n(t)). \end{pmatrix}. \end{array} \quad (4)$$

It should be pointed out that, for any solution $[x(t), q(t)]$ of the system (3), (4), (1), $x(t)$ is continuous and $q(t)$ is piecewise-constant and left-continuous. Note also that the solution apparently exists, is unique, and can be defined on $[0, +\infty)$. Furthermore, it can be easily seen, that for any solution $[x(t), q(t)]$ of the system (3), (4), (1), there exists a sequence $\{t_k\} |_{k=0}^{\infty}$ such that $t_0 = 0$, $t_1 \geq t_0$, $t_{k+1} > t_k$ for all $k = 1, 2, \ldots$, $\lim t_k = +\infty$ as $k \to +\infty$, $q(t_k) \neq q(t_k + 0)$, and

$$q(t) = const \quad \forall t \in (t_k, t_{k+1}] \quad \forall k = 0, 1, 2, \ldots. \quad (5)$$

Definition 1 The sequence t_k is called the switching time sequence of the solution $[x(t), q(t)]$.

Definition 2 A solution $[x(t), q(t)]$ of the system (3), (4), (1) is said to be a limit cycle if there exists a time $T > 0$ such that

$$x(t+T) = x(t), \quad q(t+T) = q(t) \quad \forall t \geq 0.$$

Definition 3 Let $[x(t), q(t)]$ be a limit cycle of the system (3), (4), (1). Furthermore, let $T > 0$ be its mimimal period, and let t_k be its switching time sequence. An integer number s is said to be the order of this limit cycle if

$$t_s \leq T < t_{s+1}.$$

Definition 4 Let $[x(t), q(t)]$ be a limit cycle of the system (3), (4), (1), and let t_k be the switching time sequence of $[x(t), q(t)]$, and s be its order. Furthermore, let $[\hat{x}(t), \hat{q}(t)]$ be any other solution of this system and let $\hat{t}_k$ be its switching sequence. Then $[\hat{x}(t), \hat{q}(t)]$ is said to converge to $[x(t), q(t)]$ if the following condition holds:

There exists a number $N > 0$ such that

$$q(t_k) = \hat{q}(\hat{t}_{k+N}) \quad \forall k = 0, 1, 2, 3, \ldots$$

and

$$\lim_{i \to +\infty} \hat{x}(\hat{t}_{is+N+j}) = x(t_j) \quad \forall j = 1, 2, \ldots, s. \quad (6)$$

Remark It can be easily seen that Definition 4 implies the following property:

Let $[x(t), q(t)]$ be a limit cycle of the system (3), (4), (1), and let $[\hat{x}(t), \hat{q}(t)]$ be any other trajectory, which converges to $[x(t), q(t)]$. Then, it follows from condition (6) that

$$\lim_{t_* \to +\infty} \sup_{\hat{t} \geq t_*} \inf_{t \geq t_*} \|\hat{x}(\hat{t}) - x(t)\| = 0. \quad (7)$$

The condition (7) is the standard definition of convergence to a limit cycle from the classical qualitative theory of ordinary differential equations (see e.g. (Petrovski 1966)).

3. THE MAIN RESULT

In this section, we present a complete qualitative analysis of the hybrid dynamical system under consideration.

We will use the following notation

$$(n-1)! \triangleq 1 \times 2 \times 3 \times \cdots \times (n-1).$$

Now we are in a position to present the main result of this paper.

Theorem 1 Consider the switched server system (3), (4), (1) where $p > 0, p_1 > 0, p_2 > 0, \ldots, p_n > 0$ are any parameters such that condition Assumption 1 holds. Then this system has $(n-1)!$ limit cycles. Furthermore, any trajectory of the system converges to one of them.

The proof of this result will be given in the full version of the paper.

REFERENCES

Chase, C., J. Serrano and P. Ramadge (1993). 'Periodicity and chaos from switched flow systems: Contrasting examples of discretely controlled continuous systems'. IEEE Transactions on Automatic Control **38**(1), 70–83.

Perkins, J. and P.R. Kumar (1989). 'Stable, distributed, real-time scheduling of flexible manufacturing/assembly/disassembly systems'. IEEE Transactions on Automatic Control **34**(2), 139–148.

Petrovski, I. (1966). Ordinary Differential Equations. Pover Publications Inc.. New York.

Copyright © 1999 IFAC
14th Triennial World Congress, Beijing, P.R. China

HIERARCHICAL HYBRID SYSTEMS: GEOMETRY, CONTROLLABILITY AND APPLICATIONS TO AIR TRAFFIC CONTROL ⋆

Peter E. Caines *,[1] **Ekaterina S. Lemch** *

* *Department of Electrical and Computer Engineering, McGill University, 3480 University Street, Montreal, Quebec, Canada, H3A 2A7, ⟨peterc,lemch⟩@cim.mcgill.edu*

Abstract: In this paper the notions of finite analytic partition, dynamical consistency, and partition machines (originally developed in (Wei, 1995; Caines and Wei, 1998) for dynamical systems on differentiable manifolds) is generalized to hybrid systems. The theory is illustrated via a highly simplified air traffic management example which employs new conditions for the global controllability of nonlinear systems (Caines and Lemch, 1998; Lemch and Caines, 1998).

Keywords: Hybrid models, Nonlinear systems, Controllability, Mechanical systems, Air traffic control.

1. INTRODUCTION

Many complex control systems in engineering, such as air traffic management systems (Tomlin *et al.*, 1998), motor drives (Balluchi *et al.*, 1998) and intelligent highway systems (Lygeros *et al.*, 1998), have a hybrid nature, in the sense that (i) at the lowest level they can be characterized by continuous differential equations, (ii) at the highest level by a discrete mechanism, and (iii) the evolution of the overall system is described by the interaction of all system levels.

In this paper a hybrid (base) system is modelled as a quintuple consisting of a state space (which is the direct product of a set of discrete states and an n-dimensional manifold), sets of admissible continuous and discrete controls, a family of controlled autonomous vector fields assigned to each discrete state, and a (partially defined) map of discrete transitions. Next, generalizing (Caines and Wei, 1998), the notion of a finite analytic partition Π of a state space of a hybrid system is defined. Then the notion of dynamical consistency is generalized to that of *hybrid dynamical consistency*. Based on these notions, the partition machine H^{Π} of a hybrid system H is defined in such a way that, in the class of *in-block controllable partitions*, the controllability of the high level (described by the partition machine, which is a discrete finite state machine) is equivalent (under some technical conditions) to the controllability of the low level (described by differential equations). This approach promises to be useful in the analysis of complex dynamical systems in that it permits the design of hierarchical hybrid controllers.

An important problem within this framework is that of finding conditions for a partition to satisfy the in-block controllability hypothesis. In Section 3, a form of local accessibility for nonlinear control systems is introduced called the *continuous fountain condition.* To verify the in-block controllability hypothesis using the theory of this paper it is sufficient to establish that (i) the con-

⋆ Work partially supported by NSERC grant number OGP 0001329 and NASA-Ames Research Center grant number NAG-2-1040

[1] Also affiliated with the Canadian Institute for Advanced Research

tinuous fountain condition and (ii) a recurrence condition hold. Furthermore, for the so-called energy slice partitions of Hamiltonian systems (and, in fact, partitions of more general systems), the dense recurrence condition under a distinguished control is an inherent property which does not need explicit verification (whenever each slice is precompact).

In Section 4, an application of the theory to a highly simplified air traffic management system is presented.

2. HYBRID PARTITION MACHINES FOR HYBRID SYSTEMS

Consider a *hybrid system* H which, in this paper, is taken to be the quintuple

$$H = \{L \triangleq Q \times M, U, \Sigma, f, \Gamma\}, \qquad (1)$$

where
$Q = \{q_1, \cdots, q_k\}$ is a set of discrete states (which are called *control locations*);
$M \subset \Re^n$ is an n-dimensional manifold;
L is called the *state space of* H;
$U \subset \Re^u$ is a set of admissible controls;
Σ is a finite set of transition labels;
$f : Q \times M \times U \to TM$ is an autonomous vector field assigned to each control location;
$\Gamma : L \times \Sigma \to L$ is an untimed (partially defined) map of discrete transitions. □

Similar models of hybrid systems are employed by Tomlin *et al.* (1998), Broucke (1998), Lygeros *et al.* (1998), Branicky *et al.* (1994), and others.

Henceforth in this paper the notation

$$\Gamma((q_1, y_1); \sigma(t^*)) \equiv \Gamma((q_1, y_1); \sigma) = (q_2, y_2)$$

(where $q_1, q_2 \in Q, y_1, y_2 \in M, \sigma \in \Sigma$) means that for any specified instant $t^* \in \Re$ and control σ,

(i) $\lim_{t \to t^{*-}} q_t = q_1, \lim_{t \to t^{*+}} q_t = q_{t^*} = q_2$;
(ii) $\lim_{t \to t^{*-}} y(t) = y_1, \lim_{t \to t^{*+}} y(t) = y(t^*) = y_2$.

According to the description above, the continuous and discrete state trajectories are continuous from the right. In fact the discrete state is piecewise constant from the right.

The symbol Σ shall denote the original set Σ extended with the identity element $\{id\}$.

It shall be assumed that the system H is *non-zeno* in the sense that the class of inputs $\{\sigma\}$ is restricted so that in finite time only a finite number of discrete transitions may be generated.

The continuous dynamics of the hybrid system in each control location $q \in Q$ are described by a differential equation of the form

$$S^q: \quad \dot{x} = f(q, x, u), \qquad x \in M, u \in U, \qquad (2)$$

for which the assumptions are

(i) U is the sets of all bounded piecewise $C^k(\Re^1)$ ($k \geq 1$) functions of time with limits from the right (i.e. functions which are k times continuously differentiable, except possibly at a finite number of points in any compact interval, and which are bounded and have limits from the right);
(ii) for each $q \in Q$, $f(q, \cdot, \cdot)$ is continuously differentiable with respect to its arguments $x \in M$ and $u \in U$;
(iii) M is a non-empty, open, connected n-dimensional manifold.

The symbol $\phi_q(t, x, u)$ shall denote the transition function of the vector field $f(q, x, u)$ with respect to the initial condition $x \in M$ and control $u \in U$.

Generalizing (Wei, 1995; Caines and Wei, 1998) for the case $card(Q) = 1$, we formulate

Definition 2.1. A *finite analytic partition of the state space* $L = Q \times M$ of a hybrid system H to be to be a collection $\Pi = \{\pi_q; q \in Q\}$ where each π_q, $q \in Q$, is a finite analytic partition of M. By definition a finite analytic partition π_q is a pairwise disjoint collection of subsets $\pi_q = \{X_1^q, X_2^q, \cdots, X_{k_q}^q\}$ such that every X_i^q is non-empty, open, path-connected, and is such that

$$\bar{M} = \bigcup_{i=1}^{k_q} (X_i^q \cup \partial X_i^q),$$

where, further, the boundary ∂X_i^q of every block X_i^q is a locally finite union of connected components of $n - p$ dimensional, $p \geq 1$, analytic manifolds (possibly with boundary). □

Definition 2.2. Let $p = (q, x) \in L$, $p' = (q', x') \in L$ and let $R \subset L$. p' *is accessible from* p *with respect to* R if there exists a finite sequence of control pairs

$$\{u(t), t \in [T_0, T_1); \sigma(T_1)\},$$
$$\cdots$$
$$\{u(t), t \in [T_{k-1}, T_k); \sigma(T_k)\},$$

such that the state trajectory $p(\cdot)$ defined as follows

$$\begin{aligned} p(T_0) &= p; \\ p(t) &= (q_{T_i}, \phi_{q(T_i)}(t, x(T_i), u)), \\ &\quad t \in (T_i; T_{i+1}), 0 \leq i \leq k-1; \\ p(T_{i+1}) &= \Gamma(\lim_{t \to T_{i+1}^-} p(t); \sigma(T_{i+1})) \end{aligned}$$

satisfies (i) $p(t) \in R$, for all $t \in [T_0; T_k]$; and (ii) $p(T_k) = p'$.

The set of all accessible from p (with respect to $R \subset L$) states shall be denoted by $A^R(p)$.

The state space L of H is said to be *controllable* if $A^L(p) = L$, for all $p \in L$. □

Definition 2.3. Let $q_1, q_2 \in Q$ and let $X^{q_1} \in \pi_{q_1} \in \Pi$, $Y^{q_2} \in \pi_{q_2} \in \Pi$, where Π is a finite analytic partition of the state space of H. $\langle X^{q_1}, Y^{q_2}\rangle$ is said to be a *hybrid dynamical consistent pair* (HDC) if and only if for any $x \in X^{q_1}$ there exist a state $x' \in X^{q_1}$ such that (q_1, x') is accessible from (q_1, x) with respect to $q_1 \times X^{q_1} \subset L$ and at least one of the following conditions holds:

(a) there exists a discrete control $\sigma \in \Sigma$ such that $\Gamma((q_1, x'); \sigma) = (q_2, y)$, for some $y \in Y^{q_2}$

OR

(b) $q_2 = q_1$, i.e. $X^{q_1}, Y^{q_2} \in \pi_{q_1}$, and there exists a control $u \in U$ defined on a finite time interval $[0, T]$ such that
   (i) $\phi_{q_1}(t, x', u) \in X^{q_1}$, for all $t \in [0; t^*)$, $0 < t^* < T$;
   (ii) $\phi_{q_1}(t^*, x', u)$ is a facial boundary state of $\partial X^{q_1} \cap \partial Y^{q_2}$, i.e. it lies in the relative interior of $n-1$ dimensional connected components of the boundaries ∂X^{q_1} and ∂Y^{q_2}; and
   (iii) $\phi_{q_1}(t, x', u) \in Y^{q_2}$, for all $t \in (t^*, T]$. □

The notion of hybrid dynamical consistency for hybrid systems is evidently a generalization of the notion of *dynamical consistency (DC)* introduced in (Caines and Wei, 1998) for control systems on differentiable manifolds.

Definition 2.4. Let H be a hybrid system

$$H = \{L \triangleq Q \times M, U, \Sigma, f, \Gamma\}$$

and $\Pi = \{\pi_q; q \in Q\} = \{\{X_1^q, \cdots, X_{k_q}^q\}; q \in Q\}$ be a finite analytic partitions of L. Consider the system $H^\Pi = \{L^\Pi, \bar{E}, \Phi^\Pi\}$, where
L^Π, which is called the *state space of* H^Π, is defined as

$$L^\Pi \triangleq \{(q, i); q \in Q, i \in \{1, \cdots, k_q\}\};$$

$\bar{E} = \{E_{q_1,i}^{q_2,j}; q_1, q_2 \in Q, 1 \le i \le k_{q_1}, 1 \le j \le k_{q_2}\}$ is a finite set of transition labels;
$\Phi^\Pi : L^\Pi \times \bar{E} \to L^\Pi$ is a (partially defined) map such that $\Phi^\Pi((q_1, i); E_{q_1,i}^{q_2,j}) = (q_2, j)$ if and only if $\langle X_i^{q_1}, X_j^{q_2}\rangle$ is a hybrid dynamical consistent pair. Otherwise $\Phi^\Pi((q_1, i); E_{q_1,i}^{q_2,j})$ is not defined.

The finite state machine H^Π is called the *hybrid partition machine* of the hybrid system H. □

Definition 2.5. Let Π be a finite analytic partition of the state space L of H. Π is called *hybrid in-block controllable* if for any $X^q \in \pi_q \in \Pi$, $q \in Q$, and any $x \in X^q$, the set of accessible from (q, x) with respect to $q \times X^q \subset L$ states is equal to $q \times X^q$. □

Theorem 2.1. Let Π be a hybrid in-block controllable partition of L. Then
H^Π is controllable (as a finite state machine)

$$\Longleftrightarrow$$

$L^* \triangleq \bigcup_{q \in Q} q \times (M - \partial\pi_q)$ is controllable with respect to $\tilde{L} \triangleq \bigcup_{q \in Q} q \times (M - NF(\pi_q))$, where $NF(\pi_q)$, $q \in Q$, is the set of all non-facial states of the partition π_q. □

3. FOUNTAINS, RECURRENCE AND CONTROLLABILITY

This section of the paper concerns the global controllability of nonlinear systems of the form (2). There is an extensive literature on various forms of the accesibility and controllability problem for nonlinear systems. In particular, in (Jurdjevic, 1997; Kunita, 1979; Lobry, 1974), it has been shown that control affine systems satsfying a dense recurrence condition and a full rank Lie algebra condition are (globally) controllable. In (Manikonda and Krishnaprasad, 1997), sufficient conditions for controllability of affine nonlinear control systems (where the drift vector field is a Lie-Poisson reduced Hamiltonian vector field) are presented.

A state $x \in M$ is said to be a *continuous fountain* if (i) there exists an open ball neighbourhood $B_\rho(x) \subset M$ such that for all open ball neighbourhoods $B_\delta(x)$, where $\delta < \rho$, the accessibility and co-accessibility sets from x relative to $B_\delta(x)$ are open when the state x is deleted; and (ii) whenever (i) holds throughout an open neighborhood $N(x)$ of x, the limit supremum of the radii ρ of the open ball neighbourhoods of $y \in N(x)$ for which (i) holds is continuous at x.

Example 3.1. Consider the following system

$$\begin{aligned} \dot{x} &= u, \; x(0) = x_0; \\ \dot{y} &= x^2, \; y(0) = y_0; \end{aligned} \tag{3}$$

where M is taken to be $\Re^2$.
Then for any initial condition $p = (x_0, y_0)$ and any $\delta > 0$, the accessible set $A^\delta(p) = (\{(x, y);\ y > y_0\} \cup \{p\}) \cap B_\delta(p)$ and the co-accessible set $CA^\delta(p) = (\{(x, y);\ y < y_0\} \cup \{p\}) \cap B_\delta(p)$ are such that $(A^\delta(p) - \{p\})$ and $(CA^\delta(p) - \{p\})$ are open. Hence any $p \in M$ is a continuous fountain. □

A state $x \in M$ is called *control recurrent* if it lies in a non-trivial positive limit set under some control $u \in \mathcal{U}$, i.e. x is not an equilibrium point under u and $x = \lim_{n \to \infty} \phi(t_n, x, u)$, for some sequence $\{t_n; n = 1, 2, \cdots\}$ such that $\lim_{n \to \infty} t_n = \infty$.

Theorem 3.1. Assume that the system S on the open connected state space M is such that every state $x \in M$ is a continuous fountain and through each x in M there exists a non-trivial orbit. Then M is controllable. □

Theorem 3.2. Assume that the system S on the open connected state space M is such that every state $x \in M$ is a continuous fountain and for every x in M there exists a state dependent control $u_x(\cdot) \in \mathcal{U}(\Re^n; \Re^m)$ such that x is control recurrent under u_x. Then M is controllable. □

3.1 *The Fountain Condition for Control Affine Systems*

Consider a system S of the *input-linear* or *control affine* class of nonlinear time-invariant control systems

$$S: \quad \dot{x} = f(x) + \sum_{i=1}^{m} g_i(x)u_i, \tag{4}$$

where $f, g_1, g_2, \cdots, g_m$ are smooth mappings from $\Re^n$ into $\Re^n$.

It can be shown that the fountain condition for control affine systems does not imply the Lie Algebra Rank Condition.

Example 3.2. Let $h_1(x,y)$, $h_2(x,y)$ be $C^\infty(\Re \times \Re)$ functions such that

(i) $h_1(x,y) \neq 0$, if $(x,y) \in S_1$, and $h_1(x,y) = 0$, if $(x,y) \in \Re^2 - S_1$,
(ii) $h_2(x,y) \neq 0$, if $(x,y) \in S_2$, and $h_2(x,y) = 0$, if $(x,y) \in \Re^2 - S_2$,

where $S_1 \subset \Re^2$ is the open unit disk with the center at (0,1) and $S_2 \subset \Re^2$ is the open unit disk with the center at (0,-1).
Consider the following control affine system

$$\begin{aligned} \dot{x} &= (h_1(x,y) + h_2(x,y))u, \\ \dot{y} &= 1. \end{aligned} \tag{5}$$

Then the accessible set from $p = (0,0)$ and co-accessible set to p are as shown in Figure 1. These sets are such that $A^\delta(p) - \{p\}$ and $CA^\delta(p) - \{p\}$ are open, for any $\delta > 0$. Hence p is a fountain. On the other hand, LARC fails at p since the function $g \triangleq (h_1(x,y) + h_2(x,y))$ and all its partial derivatives are equal to zero at p and thus the dimension of the associated Lie algebra of the system (5) at p is equal to 1. □

Conversely, the Lie Algebra Rank Condition does not imply the fountain condition (Caines and Lemch, 1998).

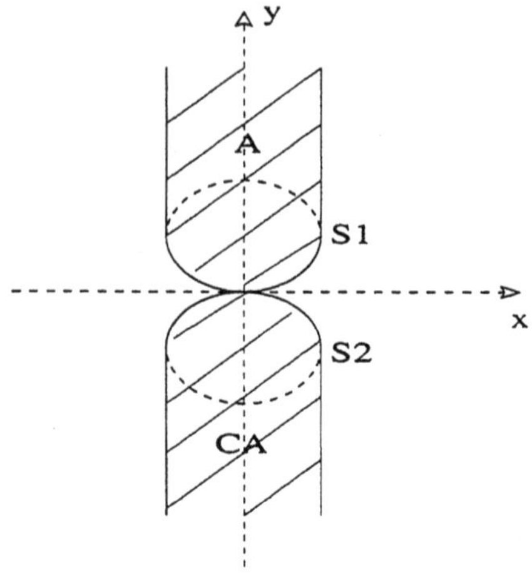

Fig. 1. Fountain condition does not imply LARC.

3.2 *Hamiltonian Control Systems*

In this section an application of the results of Section 3 to *affine Hamiltonian control systems* is considered. Such systems have a smooth Hamiltonian of the form

$$H(q,p,u) = H_0(q,p) - \sum_{j=1}^{m} H_j(q,p)u_j,$$

where $H_0(q,p)$ is the *internal Hamiltonian (energy)* and H_j, $j = 1, 2, \cdots, m$, are the *interaction* or *coupling* Hamiltonians.

Definition 3.1. A u_0 energy slice $ES(H^-, H^+)$ of a Hamiltonian control system is the set of states for which the value of H, under the state dependent control $u_0 \in \mathcal{U}(\Re^n; \Re^m)$, lies between some fixed values H^- and H^+

$$ES(H^-, H^+) \triangleq \{(p,q);\ (p,q) \in E \text{ and } H^- < H(p,q,u)_{|u=u_0} < H^+\}.$$

□

Theorem 3.3. A Hamiltonian control system for which all states are continuous fountains and all equilibrium points under some constant control $u_0 \in \mathcal{U}$ are isolated is such that any precompact connected component of a u_0 energy slice is controllable. □

Note that arguments analogous to those used in the proof of Theorem (3.3) can be applied to any dynamical system for which there is a measure preserving flow and precompact slices based on any (not necessarily energy) first integral of the system motion. The theorem is stated for affine Hamiltonian systems and energy based slices because of their great importance.

4. APPLICATION TO A PLANAR AIRCRAFT SYSTEM

Air traffic control is an example of a hybrid control system which consists of differential equations coupled with a discrete event system. There is an extensive literature on this subject, see, for example, (Tomlin *et al.*, 1998; Perry, 1997), and the reference therein. Using the approach developed in this paper we propose a decomposition of the state space of the system, which models the relative motion of two planar aircraft.

The motion of an aircraft is described by the left invariant vector field

$$\dot{g} = gX, \qquad (6)$$

where

$$g = \begin{bmatrix} \cos\phi & -\sin\phi & x \\ \sin\phi & \cos\phi & y \\ 0 & 0 & 1 \end{bmatrix},$$

$$X = \begin{bmatrix} 0 & -\omega & v \\ \omega & 0 & 0 \\ 0 & 0 & 0 \end{bmatrix};$$

x, y, ϕ represent the planar position and orientation of the aircraft; v, ω represent the translational and rotational velocities.

Let $g_1(x_1, y_1, \phi_1)$, $g_2(x_2, y_2, \phi_2)$, and $g(x, y, \phi)$ denote the configuration of aircraft 1, the configuration of aircraft 2, and the relative configuration of aircraft 2 with respect to aircraft 1, respectively. Then

$$\dot{g}_1 = g_1 X_1, \quad \dot{g}_2 = g_2 X_2, \quad \text{and} \quad g_2 = g_1 g,$$

where X_1 and X_2 are given by the velocity (v_1, ω_1) of aircraft 1 and the velocity (v_2, ω_2) of aircraft 2. The relative motion of two aircraft is described by

$$S_r \quad \begin{cases} \dot{x} = v_2 \cos\phi + \omega_1 y - v_1 \\ \dot{y} = v_2 \sin\phi - \omega_1 x \\ \dot{\phi} = \omega_2 - \omega_1. \end{cases} \qquad (7)$$

The state space for the system (7) is taken to be

$$M \triangleq \Re^1 \times \Re^1 \times [0; 2\pi).$$

For any given $u \triangleq (v_1, \omega_1, v_2, \omega_2)$ such that $\omega_1 \neq 0, \omega_2 \neq 0$, the function

$$F(x, y, \phi; u) \triangleq (y - \frac{v_1}{\omega_1} + \frac{v_2}{\omega_2}\cos\phi)^2 + (x - \frac{v_2}{\omega_2}\sin\phi)^2$$

stays constant along the relative (with respect to aircraft 1) trajectory of the aircraft 2. In other words, $F(x, y, \phi; u)$ is a first integral of (7).

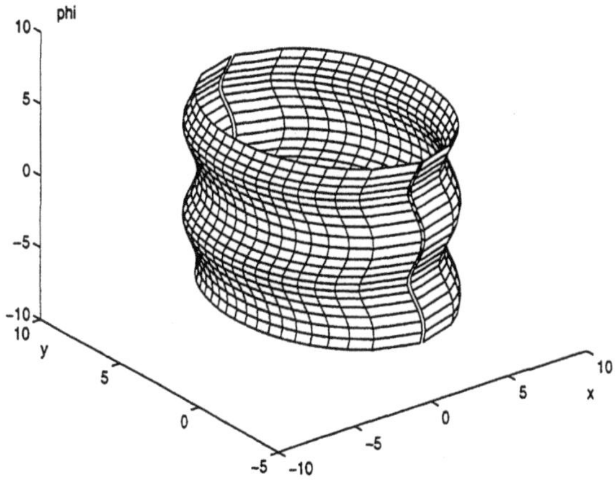

Fig. 2. An F-slice of the system (7).

Definition 4.1. Let $F(x, y, \phi; u)$, where $u = (v_1, \omega_1, v_2, \omega_2)$, be a first integral of the system (7). A u_0 *F-slice* $S(A^-, A^+)$ *of the control system (7)* is the set of states for which the value of F under given controls $u_0 = (v_1^0, \omega_1^0, v_2^0, \omega_2^0)$ lies between some fixed values A^- and A^+

$$S(A^-, A^+) \triangleq \{ (x, y, \phi);\ (x, y, \phi) \in M \text{ and } A^- < F(x, y, \phi; u)_{|u=u_0} < A^+ \}.$$

□

Lemma 4.1. (i) For any $0 \leq A^- < A^+ < \infty$, $S(A^-, A^+)$ is non-empty, precompact, and path-connected.

(ii) Consider the system (7) on some finite time interval $[T_0; T_1]$. Assume that the velocity $\bar{v}_1, \bar{\omega}_1$ of the aircraft 1 are constant on $[T_0; T_1]$ and $\bar{\omega}_1 \neq 0$. Each state $(x, y, \phi) \in M$ is a continuous fountain with respect to controls (v_2, ω_2).

(iii) The flow of the system (7) is measure preserving. □

Applying the methodology described in Section 3 it can be shown that each u_0 F-slice (with $\omega_1 \neq 0, \omega_2 \neq 0$) constitutes a controllable subset of the system (7), where (v_1, ω_1) are treated as constant parameters and (v_2, ω_2) are treated as control inputs.

5. CONCLUSION

The air traffic system described in Section 4 can be modelled as a hybrid system (Caines and Lemch, 1999), in which each control location $q_i \in Q$ is characterized by the velocity (v_1^i, ω_1^i) of aircraft 1. Then, as has been shown in the previous section, it is possible to construct a partition π_{q_i} in such a way that each block $X^{q_i} \in \pi_{q_i}$ is a u_0 F-slice of the system (7), for some $u_0 =$

$(v_1^i, \omega_1^i, v_2^o, \omega_2^o)$. Hence $\Pi = \{\pi_q;\ q_i \in Q\}$ is in-block controllable.

Current air traffic control management is completely centralized and provides instructions to all aircraft regardless of the specific objectives of an individual aircraft (Perry, 1997). The approach described in this paper would allow a certain degree of decentralization. Within the *hybrid partition machine* framework, a discrete controller supervises its continuous subsystems via the feedback relations; furthermore, each continuous subsystem is itself (internally) subject to feedback control. The resulting hierarchical control structure is such that the high level controller makes supervisory decisions, while the low level controllers will actually generate the required state trajectories.

Given a partition π, where each block $X \in \pi$ is a connected component of an energy slice, one can generate finer partitions using (i) $u = constant \neq 0$ energy slices, (ii) different integrals of motion, or (iii) controllable transversal manifolds, in particular *transversal foliations* (Broucke, 1998). The resulting blocks may constitute neighbourhoods of system states of possible interest and hence form candidate targets for high level trajectory control. Even though, in general, the hybrid in-block controllability property is not preserved under the transversal decomposition of an energy slice block, the associated hybrid partition machine stays in the class of hybrid between-block controllable machines (Caines and Lemch, 1999).

In conclusion, the notions of state aggregation described in this paper promise to facilitate the analysis and design of complex control systems; in addition to the traffic management example introduced above, we also mention in that context mechanical systems with many degrees of freedom and large space structures (Goh and Caughey, 1985*a*; Goh and Caughey, 1985*b*).

6. REFERENCES

Balluchi, A., M. Di Benedetto, C. Pinello, C. Rossi and A. Sangiovanni-Vincentelli (1998). Hybrid control for automotive engine management: The cut-off case. In: *Proceedings of the first international workshop, Hybrid Systems: Computation and Control.* Springer-Verlag. Berkeley, CA. pp. 13–32.

Broucke, M. (1998). A geometric approach to bisimulation and approxamte verification of hybrid systems. In: *Proceedings of the 37th IEEE Control Systems Society Conference on Decision and Control.* Tampa, FL.

Caines, P.E. and E.S. Lemch (1998). On the global controllability of hamiltonian and other nonlinear systems: fountain and recurrence. In: *Proceedings of the 37th IEEE Control Systems Society Conference on Decision and Control.* Tampa, FL. pp. 3575–3580.

Caines, P.E. and E.S. Lemch (1999). Partition machines for hybrid systems with applications to air traffic management systems. In: *Proceedings of AAAI Spring Symposium on Hybrid System.*

Caines, P.E. and Y-J. Wei (1998). Hierarchical hybrid control systems: A lattice theoretic formulation. *IEEE Transactions on Automatic Control* **37**(4), 501–508.

Goh, C.J. and T.K. Caughey (1985*a*). On the stability problem caused by finite actuator dynamics in the collocated control of large space structures. *Int. Journal of Control* **41**(3), 787–802.

Goh, C.J. and T.K. Caughey (1985*b*). A quasi-linear vibration suppression technique for large space structures via stiffness modification. *Int. Journal of Control* **41**(3), 803–811.

Jurdjevic, V. (1997). *Geometric Control Theory.* Cambridge University Press.

Kunita, H. (1979). On the controllability of nonlinear systems with applications to polynomial systems. *Applied Mathematics and Optimization* **5**, 89–99.

Lemch, E. and P. E. Caines (1998). Hierarchical hybrid systems: Partition deformations and application to the acrobot system. In: *Proceedings of the first international workshop, Hybrid Systems: Computation and Control.* Springer-Verlag. Berkeley, CA. pp. 237–252.

Lobry, C. (1974). Controllability of nonlinear systems on compact manifolds. *SIAM J. Control* **12**, 1–4.

Lygeros, J., D.N. Godbole and S. Sastry (1998). Verified hybrid controllers for automated vehicles. *IEEE Transactions on Automatic Control* **43**(4), 509–521.

Manikonda, V. and P.S. Krishnaprasad (1997). Controllability of lie-poisson reduced dynamics. In: *American Control Conference.* Albuquerque, New Mexico. pp. 2203–2207.

Perry, T.S. (1997). In search of the future of air traffic control. *IEEE Spectrum,* August, pp. 19–35.

Tomlin, C., G.J. Pappas and S. Sastry (1998). Conflict resolution for air traffic management: A study in multiagent hybrid systems. *IEEE Transactions on Automatic Control* **43**(4), 509–521.

Wei, Y-J (1995). Logic Control: Markovian Fragments, Hierarchy and Hybrid Systems. PhD thesis. Department of Electrical and Computer Engineering, McGill University, Montreal, Canada.

PASSIVITY-BASED SPEED CONTROL (PBC) OF INDUCTION MOTOR

Chen, Feng Xu, Wenli

Department of Automation, Tsinghua University, Beijing, P.R.China, 100084
Telephone: 8610-62782422
Fax: 8610-62781993
E-mail: chenfeng@mail.cic.tsinghua.edu.cn

Abstract: Passivity-based control (PBC) is a new strategy for induction motor control. In this paper, a speed controller is designed based on the PBC torque controller in (Ortega and Espinosa, 1993). At the same time the field orientation can be attained asymptotically. This controller is simple and simulation results show that it has good static and dynamic performance. *Copyright © 1999 IFAC*

Key words: passive speed control induction motor

1. INTRODUCTION

A lot of nonlinear control strategies of electrical machines have been developed since vector control. They can be roughly classified into four groups: feedback linearizing schemes, Lyapunov-based methods, sliding mode control and passivity-based control (PBC). All of them have pluses and minuses. Though PBC isn't a new idea in robot control, it was introduced in motor control only in recent years.

Like the feedback linearizing scheme, PBC is also a nonlinear feedback control strategy, but they are different greatly. Though linearizing schemes theoretically realized accurate decoupling control, they have some disadvantages: 1) the necessary and sufficient condition for linearization can't be held all the time; 2) singular points exist; 3) the controller is complex and the computation load is heavy. By contrast, PBC does not decoupling the system, but it has an outstanding advantage – simplicity, because it does not cancel all the nonlinearities (neglecting the so-called "workless forces"). As the result it doesn't have any preliminary requirement or singular point. It also has less computation work and better robustness.

When PBC was introduced in motor control firstly (Ortega and Espinosa, 1993), it only realized torque control and had a very simple torque controller. In recent literature (Ortega, et al.,1996), (Levent, et al., 1997), some types of speed controllers were also developed. Besides the common properties of PBC, they possess a new virtue that the rotor-current feedback is not needed. This gives PBC methods a good future for practical application. But they lose the simplicity of the torque controller to some degree and need the acceleration of the rotor to be measured.

In this paper we develop a new speed controller for induction motors using the idea of the torque controller presented in (Ortega and Espinosa, 1993). We know that the induction motor system can be divided into two subsystems: external speed loop and internal current loop. Because of the coupling property of induction motor, the current loop can't be entirely included in the speed loop. Compared with electrical subsystem the mechanical part of induction motor is much slower. So we presents a new idea of designing the speed controller. First, design a torque controller as described in (Ortega and Espinosa, 1993), considering the speed is a slowly changing parameter, which can be handled as a constant. Then the feedback from speed error to torque reference is established to realize the speed control. Though this speed controller requires the rotor-current feedback, it is very simple and has good robustness. The simulation results are very close to those given in (Ortega, 1996).

However, the hypothesis that the speed is a slowly changing parameter is not correct all the time. It is influenced by the choice of parameters of the speed controller. Because it is difficult to analyze theoretically, a discussion based on simulation results will be given in section 4.

2. INTRODUCTION TO PBC AND TORQUE CONTROLLER DESIGN

Romeo Ortega and other researchers have made great contribution to establish PBC. They, by carefully analyzing the motor system, discovered that the state-space representation couldn't clearly indicate its energy relation. Therefore, by using Euler-Lagrange equation, they described the motor and found the so-called "workless force", which does not affect the energy balance. Because of the known tight relationship between energy balance and system stability, we can easily conclude that the workless force does not influence the system's stability. If we don't take the workless force into account in controller design, we can get a simpler controller. This is the reason for PBC controller's simplicity.

Now we begin to design a PBC torque controller for induction motors. To do so the most important thing is to define the workless force. The induction motor in the d-q reference frame can be described :

$$A\begin{bmatrix}\dot{q}_1\\ \dot{q}_2\\ \dot{q}_3\\ \dot{q}_4\end{bmatrix}=\begin{bmatrix}u_1\\ u_2\\ 0\\ 0\end{bmatrix} \tag{1a}$$

$$A=\begin{bmatrix} R_s+L_sp & -u_3L_s & L_{sr}p & -u_3L_{sr}\\ u_3L_s & R_s+L_sp & u_3L_{sr} & L_{sr}p\\ L_{sr}p & -(u_3-\nu\dot{q}_5)L_{sr} & R_r+L_rp & -(u_3-\nu\dot{q}_5)L_r\\ (u_3-\nu\dot{q}_5)L_{sr} & L_{sr}p & (u_3-\nu\dot{q}_5)L_r & R_r+L_rp \end{bmatrix}$$

$$J\ddot{q}_5+f\dot{q}_5=y-y_L \tag{1b}$$

In equation (1a), R_s, R_r, L_s, L_r are resistance and inductance of stator and rotor, respectively, L_{sr} is the mutual inductance, ν is the number of pole pair, and p is the differential symbol. State variables:

$$[\dot{q}_1 \quad \dot{q}_2 \quad \dot{q}_3 \quad \dot{q}_4] = [i_{sd} \quad i_{sq} \quad i_{rd} \quad i_{rq}]$$

are stator and rotor currents on d-q axes. Control variables $[u_1 \quad u_2 \quad u_3]$ = $[u_{sd} \quad u_{sq} \quad \omega_s]$ are stator voltage on d-q axes and frequency, $\dot{q}_5$ is the rotor speed.. In equation (1b) J is the rotor inertia, f is the damp coefficient, y and y_L are generated and laden torque respectively,

Now using Euler-Lagrange equation, the induction motor can be described by

$$D\ddot{q}+b(\dot{q},u_3)\dot{q}+R\dot{q}+W=M\begin{bmatrix}u_1\\ u_2\end{bmatrix} \tag{2}$$

the generated torque :

$$y = L_{sr}(\dot{q}_2\dot{q}_3 - \dot{q}_1\dot{q}_4) \quad (3)$$

the coefficient matrixes are given by:

$$D = \begin{bmatrix} L_s I_2 & L_{sr} I_2 \\ L_{sr} I_2 & L_r I_2 \end{bmatrix}, R = \begin{bmatrix} R_s I_2 & 0 \\ 0 & R_r I_2 \end{bmatrix}$$

$$b(\dot{q}, u_3) = \begin{bmatrix} L_s J_2 & L_{sr} J_2 \\ L_{sr} J_2 & L_r J_2 \end{bmatrix} (u_3 - v\dot{q}_5)$$

$$M = [I_2 \quad 0]^T$$

$$W = [-v\dot{q}_5(L_s\dot{q}_2 + L_{sr}\dot{q}_4) \quad v\dot{q}_5(L_s\dot{q}_1 + L_{sr}\dot{q}_3) \quad 0 \quad 0]^T$$

I_2 is 2×2 identity matrix, J_2 is 2×2 anti-symmetric matrix, i.e. $J_2 = \begin{bmatrix} 0 & -1 \\ 1 & 0 \end{bmatrix} = -J_2^T$

Written in the form of "Newton second law", equation (2) becomes:

$$D\ddot{q} = -[b(\dot{q}, u_3)\dot{q} + R\dot{q} + W] + Mu \quad (4)$$

If we choose the total energy as $H = \frac{1}{2}\dot{q}^T D\dot{q}$, then its derivative is:

$$\dot{H} = \dot{q}^T D\ddot{q} = -\dot{q}^T b\dot{q} + \dot{q}^T(-R\dot{q} + Mu - W)$$

Notice that b is skew-symmetric, i.e. $b = -b^T$, therefore $\dot{q}^T b\dot{q} \equiv 0$. So "$b\dot{q}$" is a workless force. It doesn't play any role in energy balance and system's stability. As the result, it needn't to be canceled by the control law.

Define state variables' references as

$$\dot{q}_d = [\dot{q}_{d1} \quad \dot{q}_{d2} \quad \dot{q}_{d3} \quad \dot{q}_{d4}]$$

and the corresponding torque references as

$$y_d = L_{sr}(\dot{q}_{d2}\dot{q}_{d3} - \dot{q}_{d1}\dot{q}_{d4}).$$

If error vector is $\dot{e} = \dot{q} - \dot{q}_d$, then the error equation is:

$$D\ddot{e} + [b(\dot{q}, u_3) + R]\dot{e} = \psi$$

where ψ is disturbance.

$$\psi = M\begin{bmatrix} u_1 \\ u_2 \end{bmatrix} - \{D\ddot{q}_d + [b(\dot{q}, u_3) + R]\dot{q}_d + W\} \quad (5)$$

Choose Lyapunov function as $H_d = 1/2\dot{e}^T D\dot{e}$.

Thus, its derivative is $\dot{H}_d = -\dot{e}^T R\dot{e} + \dot{e}^T\psi$

Apparently, if $\psi \equiv 0$, $\Rightarrow \dot{H}_d < 0 \Rightarrow \dot{e} \to 0$ (by Lyapunov theorem). So in the final state we get the desired torque y_d. In the next chapter we will introduce a new PBC speed controller, based on this torque controller.

3. DESIGN METHODOLOGY OF PBC SPEED CONTROLLER

The basic idea of designing the new speed controller is: first, design a torque controller, considering speed as a constant; then build feedback from the speed error to torque reference to control the speed.

3.1. Design of torque controller.

Before designing torque controller, we should choose references for state variables in order to acquire the desired torque in the stable state.

As in field orientation, define state variables' references as:

$$\dot{q}_d = [\dot{q}_{d1} \quad \dot{q}_{d2} \quad \dot{q}_{d3} \quad \dot{q}_{d4}],$$

$$\dot{q}_{d3} = 0, \dot{q}_{d4} = \beta = const.$$

Let $\psi \equiv 0$, solving the equation (5), we can get $\dot{q}_{d1}, \dot{q}_{d2}$ and the control law of u:

$$\dot{q}_{d1} = -\frac{y_d}{L_{sr}\beta}, \dot{q}_{d2} = -\frac{L_r\beta}{L_{sr}};$$

$$u_1 = -(L_s\dot{q}_2 + L_{sr}\dot{q}_4)v\dot{q}_5 + \gamma_1$$

$$u_2 = (L_s\dot{q}_1 + L_{sr}\dot{q}_3)v\dot{q}_5 + \gamma_2$$

$$u_3 = v\dot{q}_5 + \gamma_3$$

with

$$\gamma_1 = -\frac{R_s y_d}{L_{sr}\beta} - \frac{R_r\beta^3}{y_d}(L_{sr} - \frac{L_s L_r}{L_{sr}})$$

$$\gamma_2 = -\frac{R_s L_r \beta}{L_{sr}} - \frac{R_r L_s \beta}{L_r}, \qquad \gamma_3 = \frac{R_r\beta^2}{y_d}$$

$\gamma_1, \gamma_2, \gamma_3$ are constants.

Rotor currents can be obtained through observer omitted in this paper, see (Ortega and Espinosa, 1993).

3.2. Design of the speed controller.

To control the speed, we could not use a constant torque reference any more. We should build feedback from the speed error. The form of speed controllers can be various. If we use proportion controller, then the torque reference is:

$$y_d = -k_1 e_\omega = -k_1(\dot{q}_5 - \dot{q}_{d5})$$

The controller parameter k_1 greatly affects the system stability, because large k_1 can make the system's dynamic violent. As a sequence, the speed cannot be thought as constant. But because of the nonlinearity of the induction motor, it is difficult to analyze the effect of k_1 theoretically. So we use simulation methods instead.

To accelerate the system's dynamic response, a stator current feedback (with control parameter k_2=15) and stator frequency acceleration term (with control parameter k_3=0.2) are added. So the final control law is:

$$u_1 = -(L_s\dot{q}_2 + L_{sr}\dot{q}_4)v\dot{q}_5 + \gamma_1 - k_2(\dot{q}_1 - \dot{q}_{d1})$$
$$u_2 = (L_s\dot{q}_1 + L_{sr}\dot{q}_3)v\dot{q}_5 + \gamma_2 - k_2(\dot{q}_2 - \dot{q}_{d2})$$
$$u_3 = v\dot{q}_5 + \gamma_3 - k_3(\dot{q}_5 - \dot{q}_{d5})$$

4. SIMULATION

The numerical values of the four-pole squirrel-cage induction motor used in (Ortega and Espinosa,1993), were chosen, that is R_s=0.687 Ω ,R_r =0.842 Ω ,L_s=84mH, L_r=85.2mH, L_{sr}=81.3mH, J=0.03kgm², v=2, start states $[\dot{q}_{10} \quad \dot{q}_{20} \quad \dot{q}_{30} \quad \dot{q}_{40}]$=[-21.87 −14.8 0 14.1]A, start speed: 100rad/s, the desired speed: 146rad/s. Simulation results are shown in Fig. 1-4:

Fig. 1a and 1b show that the system has good static and dynamic performance when the parameter of speed controller k_1=0.2, and the stator current feedback and stator frequency acceleration term are added. If k_1=1(Fig.2a), there will be a long time oscillation in system transition to the final stable state. When k_1=3(Fig.2b), an undamped vibration appears. From these simulation results we know that the parameter adjustment through simulation or experiment is needed for designing speed controllers in practice.

Fig.3 and Fig.4 show that system's dynamic response is slowed, if the stator current feedback and the stator frequency acceleration term are deleted. In Fig.3 even an oscillation occurs.

5. CONCLUSION

In essence, PBC is an undecoupling nonlinear feedback control. Because not all of the nonlinearities are canceled, it has a simple control law. Based on the existing torque controller, we presented a new speed controller, which preserves the advantage of PBC controller - simplicity. Simulation results show that the new speed controller has a simple control law, good static and dynamic property, and robustness. The main drawback is that the rotor-current feedback is needed.

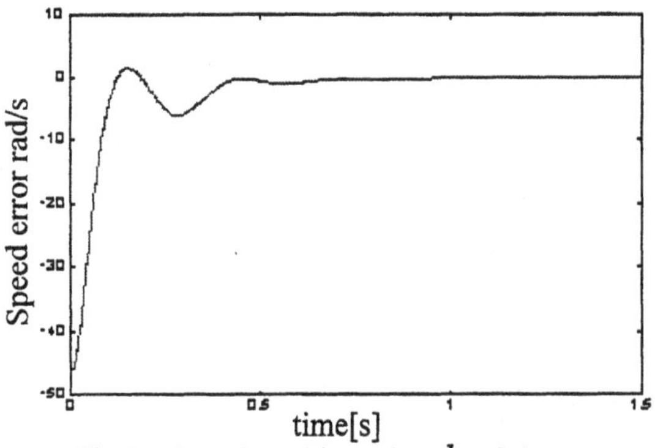

Fig.1a. Speed tracking when k_1=0.2.

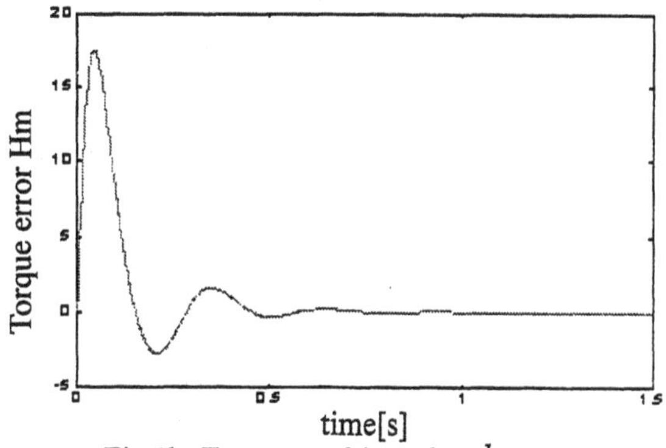

Fig.1b. Torque tracking when k_1=0.2.

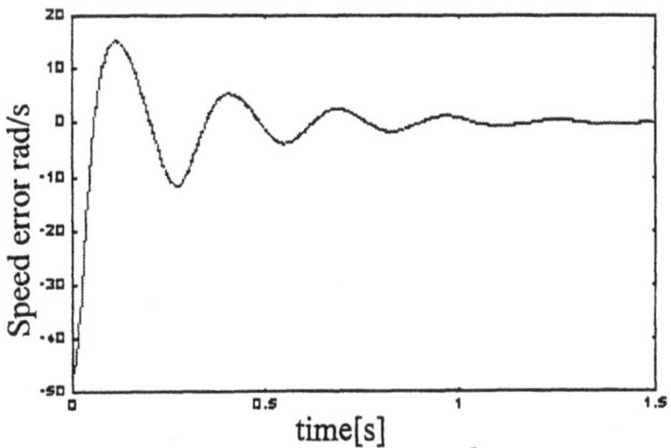

Fig.2a. Speed tracking when k_1=1.

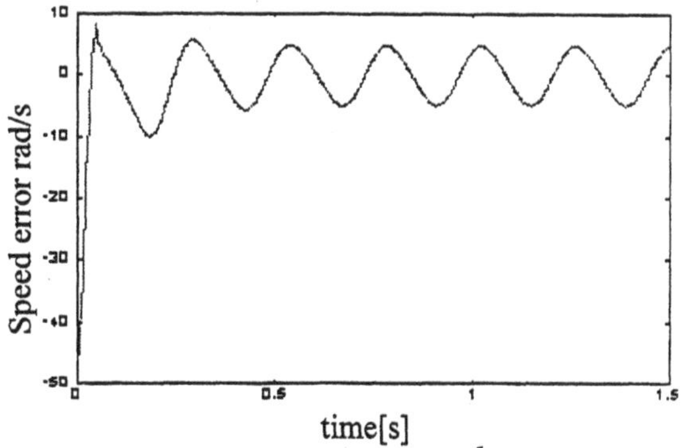

Fig.2b. Speed tracking when k_1=3.

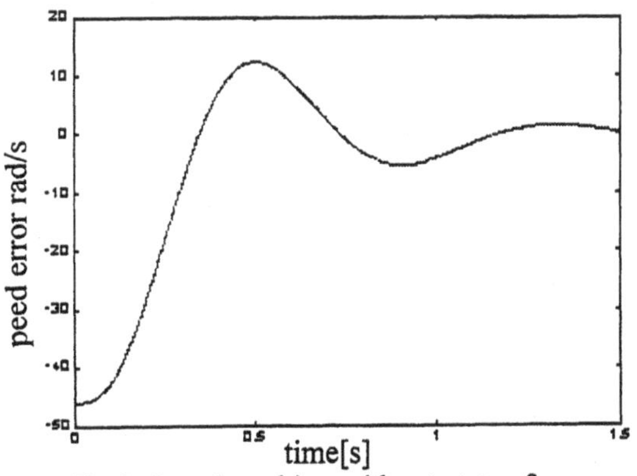

Fig.3. Speed tracking without stator frequency acceleration term (k_1=0.2).

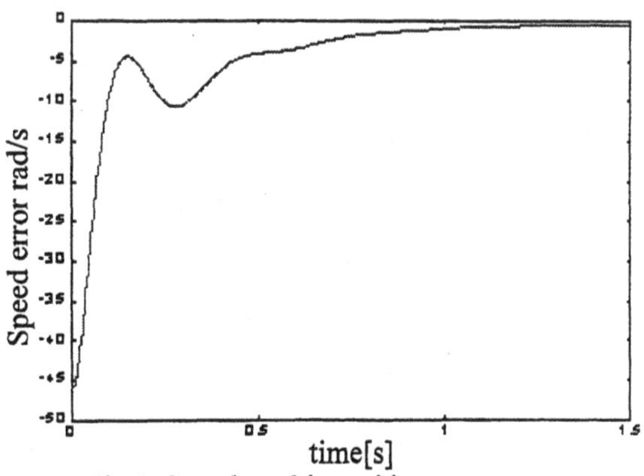

Fig.4. Speed tracking without stator current feedback (k_1=0.2).

REFERENCE

Espinosa G., Ortega R. and Nicklasson P. (1997). Torque and flux tracking of induction motors. Int. J. Robust Nonlinear Cont., Vol.7. No. 1. pp. 1-10.

Kim Ki-Chul, Romeo Ortega and Ali Charara (1997). Theoretical and experimental comparison of two nonlinear controllers for current-fed induction motors. Vol. 5. No. 3. pp. 338-348.

Levent U. Gokdere and Marwan A. Simaan (1997). A passivity-based method for induction motor control. IEEE Trans. on industrial electronics. Vol. 44. No. 5. pp. 688-695.

Ortega R. and Espinosa G. (1993). Torque regulation of induction motors. Automatica Vol. 29. No. 3. pp. 621-633.

Ortega R., Per J. Nicklasson and Espinosa G. (1996). On speed control of induction motors. Automatica Vol. 32. No. 3. pp. 455-460.

CONTROLLABILITY AND SINGULARITIES IN THE N-TRAILER SYSTEM WITH KINGPIN HITCHING

Claudio Altafini

Optimization and Systems Theory
Royal Institute of Technology, Stockholm, Sweden
altafini@math.kth.se

Abstract: When one or more trailers of the so-called n-trailer mobile robot present kingpin hitching i.e. they are not hooked directly on the midpoint of the axle of the preceding trailer, the system has a more complicated dynamics than in the standard case. These off-axle hitches can be substituted by a passive steering wheel with the steering input calculated via nonlinear feedback from the configuration state. The new singularities associated with this general n-trailer system can be physically interpreted in terms of the virtual steering wheels. Furthermore, in the regular points controllability can be deducted from the existing results for the standard n-trailer system. *Copyright © 1999 IFAC*

Keywords: Mobile robots, Nonlinear Systems, Controllability, Singularity.

1. INTRODUCTION

The n-trailer system, i.e. a mobile robot pulling n trailers is a typical example of system in which the nonholonomic kinematic constraints, due to the assumption of rolling without sleeping of the wheels, play a determinant role. The configuration of the system is given by two position coordinates and $n + 1$ angles and, normally, two are the inputs available, representing the steering and translational actuators of the pulling cart, even though systems with more steering inputs have been studied in (Bushnell *et al.*, 1993) and (Tilbury *et al.*, 1995). The number of trailers can be increased at will: adding a trailer simply means adding a nonholonomic constraint to the system or, equivalently, an orientation angle to the configuration space.

The usual model for the pulling cart and for the trailers consists of a single axle with two wheels that are assumed parallel, neglecting, in the steering pair, the difference due to the fact that rotations must occur around a single center of rotation, see (Alexander and Maddox, 1988). Each axle is hitched to the preceding trailer by means of a rigid bar. In the case extensively treated in the literature, starting from Laumond (Laumond, 1990), every axle is hitched exactly in the middle of the axle in front of it, i.e. only axle-to-axle connection are present. This is called the *standard* n-trailer problem.

In many practical cases, the standard n-trailer is not a realistic situation: in fact, to improve manoeuvrability or due to the cumbersome shape of the trailer body, it frequently happens that a trailer cannot be attached exactly in the middle of the preceding axle but at a positive distance from it. When one or more trailers present kingpin hitching, the system is named *general* n-trailer. Examples of such systems are the truck and trailers that we normally see on our highways or a real car pulling a trailer, or special articulated vehicles, like the so-called LHD (Load-Haul-Dump) a two-axle truck used for mining applications.

A rich literature has grown for the standard n-trailer problem dealing with controllability (Laumond, 1990), open loop motion generation (Murray and

Sastry, 1993; Rouchon *et al.*, 1993*a*) closed loop control strategies (Bloch *et al.*, 1992; Sørdalen, 1993) for different control problems like point stabilization, point tracking or path following, see (Kolmanovsky and McClamroch, 1995) for a survey.

On the other hand, the general n-trailer problem has not attracted much attention, although several studies deal with specific configurations, like for example the above mentioned two-unit articulated vehicle (DeSanctis, 1997; Hemami and Polotski, 1997). A couple of significant contributions, however, are due to Laumond, who proved controllability for a standard n-trailer system driven by a two-unit articulated vehicle (Laumond, 1990), and to Rouchon *et al.* in the context of differential flatness: it is shown in (Rouchon *et al.*, 1993*b*) that a system with an off-hitched trailer is not flat if other trailers are attached behind it. Indeed kingpin hitching "breaks the chain" that constitute the "backbone" of the n-trailer system so that the two-input chained form is lost (and therefore flatness). We will show, however, that this breaking points (the off-hitching joints) can be replaced by "virtual" passive steering wheels whose steering angle is uniquely determined by the configuration state by means of nonlinear feedback.

The dynamic system for the general n-trailer has a more complex structure than the standard one, but it can be described as an extra number of terms to be added to the basic dynamic equations. This suggests that some of the techniques for the standard n-trailer can be generalized to our system. In particular, results on controllability obtained by (Laumond, 1990; Murray and Sastry, 1993) can be extended quite directly. The more complex nature of the system reflects in an higher number of singularities when compared to the standard case, see (Jean, 1996). We give a physical interpretation of these extra singularities in terms of the steering angles of the above mentioned virtual wheels.

The paper proceeds as follow: we first review the notion of nonlinear controllability applied to driftless systems (Section 2), then calculate the kinematic equations of the general n-trailer in Section 3 and show how the virtual wheels pop up in Section 4. The comparison, in Section 5, between the equations of the standard and general n-trailer systems leads to the considerations about the singularities of the general problem in Section 6. Finally we state the result on controllability in Section 7.

2. NONLINEAR CONTROLLABILITY

In this section we review standard material on nonlinear controllability taken from (Hermann and Krener, 1977) and already used by several authors in the same context of proving controllability for drift-free nonholonomic systems (Laumond, 1990; Murray and Sastry, 1993).

For a given nonlinear system, controllability means the existence of an open connected component of the configuration space connecting two given postures. Given a drift-free system of the form

$$\dot{x} = \sum_{k=1}^{r} g_k(x) u_k$$

where x is defined on a manifold $M \subseteq \mathbb{R}^n$, u on a subspace $\mathcal{U} \subseteq \mathbb{R}^r$ and g_i are independent vector fields on M, the system is said *controllable* if for any $x_0, x_1 \in M$ there exist $T > 0$ and $u[0,T] \to \mathcal{U}$ s.t. $x(0) = x_0$ and $x(T) = x_1$.

The system is *small-time controllable* (*locally controllable*) at x_0 if we can reach nearby points in arbitrarily small amounts of time remaining in a neighborhood of x_0.

A useful concept to check controllability is the rank of the Control Lie Algebra. From the distribution $\Delta = span\{g_1, \ldots, g_r\}$ the Control Lie Algebra can be defined as the smallest distribution Δ_{LA} that contains Δ and is closed under Lie bracket operation. If at a point x_p of the configuration space the $rank \Delta_{LA} = n$, the system is locally controllable at x_p. In other words it means that there exists a finite family of vector fields that generates the tangent space $T_p M$ at x_p. The system is controllable if it is locally controllable at any point $x_p \in M$. The rank condition can be checked building a filtration for the system. For this scope different families of vector fields have been proposed in the literature (Laumond, 1993*a*; Laumond, 1993*b*; Murray and Sastry, 1993).

In particular, for the n-trailer the main problem is to verify the rank condition for a generic value of n, which means the need of having an iterative procedure like the one proposed by Laumond (Laumond, 1993*a*), and to cope with singularities. The index at which the filtration stops at a point x_p is called the *degree of nonholonomy* of the system at x_p. When x_p satisfies the Lie Algebra rank condition then the system is said *maximally nonholonomic* at x_p. Singular points for the standard n-trailer system have the physical meaning of orthogonal angles between consecutive trailers and they lead to a different degree of nonholonomy, see (Jean, 1996) for a complete discussion.

3. KINEMATIC MODEL FOR THE GENERAL N-TRAILER

Suppose we have a generalized n-trailer system with m ($m \leq n$) of the trailers not directly attached at the center of the previous axle but at a positive distance M_i from this point. Under

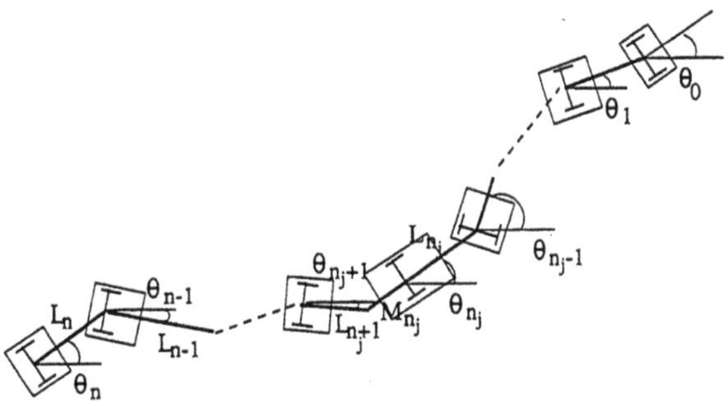

Fig. 1. The general n-trailer system.

the assumption of rolling without slipping, we can use the nonholonomic constraints on the $i-th$ and $i+1-th$ axles of the robot to obtain a recursive dynamic equation for the orientation angle. Let θ_i be the orientation angle of the $i-th$ axle and v_i its translational velocity, for $i \in \{0, 1, \dots, n-1\}$. From the geometry of the vehicle, see Fig. 1, we get:

$$\begin{cases} x_{i+1} = x_i - L_{i+1}\cos\theta_{i+1} - M_i\cos\theta_i \\ y_{i+1} = y_i - L_{i+1}\sin\theta_{i+1} - M_i\sin\theta_i \end{cases}$$

where the L_i is the distance between the i-th axle and the hitching point of the same trailer and M_i is the distance between the i-th axle and the kingpin hitching point of the following trailer. Moreover, from the nonholonomic constraint on (x_i, y_i) we know that

$$\begin{cases} \dot{x}_i = v_i\cos\theta_i \\ \dot{y}_i = v_i\sin\theta_i \end{cases}$$

Inserting into the one-form

$$\dot{x}_{i+1}\sin\theta_{i+1} - \dot{y}_{i+1}\cos\theta_{i+1} = 0$$

we get the recursive equation for the orientation angle θ_{i+1} as a function of θ_i and v_i, $i \in \{1, \dots, n-1\}$:

$$\dot{\theta}_{i+1} = \frac{v_i\sin(\theta_i - \theta_{i+1})}{L_{i+1}} - \frac{M_i\cos(\theta_i - \theta_{i+1})\dot{\theta}_i}{L_{i+1}} \qquad (1)$$

Also the calculation of the velocity of the axle $i+1$ is slightly more complicated than in the standard n-trailer problem:

$$v_{i+1} = v_i\cos(\theta_i - \theta_{i+1}) + M_i\sin(\theta_i - \theta_{i+1})\dot{\theta}_i \quad (2)$$

This accounts for the intuitive phenomenon that, in presence of off-hitching, a trailer can have a (small) positive velocity and the following trailer a negative one, when the angle between the two trailers $\theta_i - \theta_{i+1}$ is changing with high rate. Obviously, both eq. (1) and eq. (2) reduce to the well-known equations for the standard n-trailer problem when no off-hitching is present, namely when $M_i = 0$.

The n-trailer system has two physical inputs, corresponding to translational and steering actions of the car pulling the trailers. Calling $\beta_1 \triangleq \theta_0 - \theta_1$, at the kinematic level we can consider these two inputs to be the steering speed $\omega_0 = \dot{\beta}_1 = \dot{\theta}_0 - \dot{\theta}_1$ and the translational speed v_0 of the driving cart.

To complete the state space model of the general n-trailer system, we need the cartesian coordinates of one of the middle points of the axles: for the purposes of this paper it is convenient to choose (x_0, y_0) of the driving cart. The complete kinematic model for the general n-trailer problem is then:

$$\dot{x}_0 = v_0\cos\theta_0 \qquad (3)$$

$$\dot{y}_0 = v_0\sin\theta_0 \qquad (4)$$

$$\dot{\theta}_0 = \omega_0 \qquad (5)$$

$$\dot{\theta}_{i+1} = \frac{v_i\sin(\theta_i - \theta_{i+1})}{L_{i+1}} - \frac{M_i\cos(\theta_i - \theta_{i+1})\dot{\theta}_i}{L_{i+1}} \qquad (6)$$

$i \in \{1, \dots, n-1\}$, with the relation (2) for the recursive calculation of the velocities v_i, and $n-m$ of the M_is equal to 0.

4. THE VIRTUAL STEERING WHEELS

The basic idea is that the n-trailer system with m off-hitching joints can be converted into an $n+m$-trailer system with $m+1$ steering axles, adding a steerable wheel at each of the aforementioned joints not directly hitched on the preceding axle. These m *virtual* steering wheels are *passive*, in the sense that their steering angles are (uniquely) determined by the configuration and by the dynamic equations of the system. This is equivalent to say that their inputs are obtained by means of feedback from the configuration of the system and only the driving unit has exogenous input.

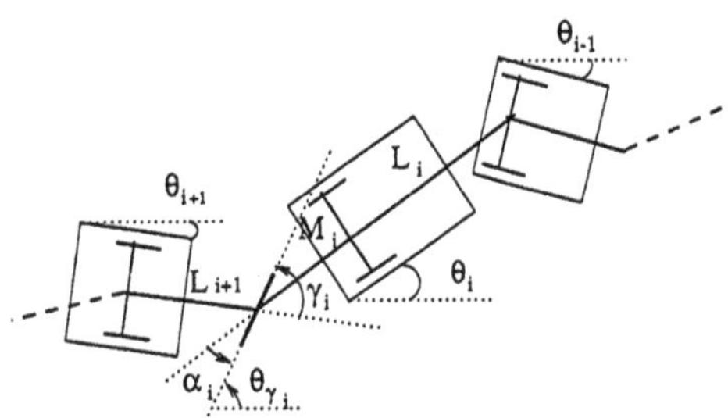

Fig. 2. The virtual steering wheel.

First we prove that a passive steering wheel is indeed admissible by the system and then, in

Section 6, we show that these virtual steering wheels provide physical insight into the extra singularities of the system due to the kingpin hitching.

Proposition 1. Consider the 2-trailer off-hitching connection between the trailers i and $i+1$. This subsystem is equivalent to a standard 3-trailer system with a steering wheel in the middle (Fig. 2). If θ_{γ_i} is the orientation angle of the steering wheel and its steering angle is defined as $\gamma_i \triangleq \theta_{\gamma_i} - \theta_{i+1}$, then it must be:

$$\gamma_i = \theta_i - \theta_{i+1} + \arctan\left(-\frac{M_i}{v_i}\dot{\theta}_i\right) \tag{7}$$

Proof. Consider the sketch in Fig. 2. If such a virtual wheel exist, then, due to the rigidity of the connection between M_i and L_i, the dynamic equation for the i-th orientation angle θ_i, obtained in general from eq. (1), must also be expressible as a function of the angle α_i (representing the steering angle of the virtual wheel with respect to the preceding trailer):

$$\dot{\theta}_i = -\frac{v_i}{M_i}\tan\alpha_i$$

where the minus sign reflects the fact that the wheel is following and not preceding the axle. At the same time, the steering angle of the virtual wheel with respect to the following trailer $i+1$, γ_i, must be such that

$$\dot{\theta}_{i+1} = \frac{v_{i+1}}{L_{i+1}}\tan\gamma_i \tag{8}$$

Therefore, a physical solution exists (i.e. the wheel is admissible) iff

$$\gamma_i = \theta_i - \theta_{i+1} + \alpha_i$$

Substituting into eq. (1), we get:

$$\frac{v_{i+1}\tan\gamma_i}{L_{i+1}} = \frac{v_{i+1}\tan(\theta_i - \theta_{i+1})}{L_{i+1}} + \frac{M_i v_i \tan\alpha_i}{M_i L_{i+1}\cos(\theta_i - \theta_{i+1})}$$

using eq. (2):

$$\left[v_i\cos(\theta_i - \theta_{i+1}) + M_i\sin(\theta_i - \theta_{i+1})\left(-\frac{v_i}{M_i}\tan\alpha_i\right)\right] \cdot (\tan\gamma_i - \tan(\theta_i - \theta_{i+1})) = \frac{v_i\tan\alpha_i}{\cos(\theta_i - \theta_{i+1})}$$

i.e.

$$\tan\gamma_i = \frac{\tan(\theta_i - \theta_{i+1}) + \tan\alpha_i}{1 - \tan(\theta_i - \theta_{i+1})\tan\alpha_i} = \tan(\theta_i - \theta_{i+1} + \alpha_i)$$

□

5. COMPARISON BETWEEN STANDARD AND GENERAL N-TRAILER SYSTEMS

In what follows it is convenient to use the following notation: we call $n_1, \ldots, n_m$, $n_j < n_{j+1}$, $n_m < n$ the indices of the axles having nonnull off-hitching ($M_{n_j} \neq 0$). We can group together the axles between two consecutive steering wheels: $\{0, 1, \ldots, n_1\}, \ldots, \{n_{j-1}+1, \ldots, n_j - 1, n_j\}, \ldots, \{n_m + 1, \ldots, n-1, n\}$. Each of the groups of axles, together with the steering wheel in front of it, will constitute a steering train. In case of two consecutive axles having off-hitching, then a steering train is reduced to a single axle. We call this a *degenerate* steering train. With this notation we can rewrite the dynamic equations of the orientation angles of the general n-trailer problem as:

$$\dot{\theta}_{n_j+1} = \frac{v_{n_j+1}\tan(\theta_{n_j} - \theta_{n_j+1})}{L_{n_j+1}} - \frac{M_{n_j}\dot{\theta}_{n_j}}{L_{n_j+1}\cos(\theta_{n_j} - \theta_{n_j+1})}$$

$$v_{n_j+1} = v_{n_j}\cos(\theta_{n_j} - \theta_{n_j+1}) + M_{n_j}\sin(\theta_{n_j} - \theta_{n_j+1})\dot{\theta}_{n_j}$$

$$j \in \{1, \ldots, m\}$$

$$\dot{\theta}_{n_j-i} = \frac{v_{n_j-i}\tan(\theta_{n_j-i-1} - \theta_{n_j-i})}{L_{n_j-i}}$$

$$v_{n_j-i} = v_{n_j-i-1}\cos(\theta_{n_j-i-1} - \theta_{n_j-i})$$

with $j \in \{1, \ldots, m+1\}$, $i \in \{0, 1, \ldots, n_j - n_{j-1} - 2\}$ and $n_{m+1} = n$.

We assume, in what follows, that the steering trains are not degenerate.

In order to show the similarities of the dynamic eq. (3)-(6) with those of the standard n-trailer it is convenient to rewrite them in terms of the relative orientation angles $\beta_i \triangleq \theta_{i-1} - \theta_i$, $i \in \{1, \ldots, n\}$, expressing all v_i as function of v_0. The system becomes:

$$\dot{x}_0 = v_0\cos\theta_0 \tag{9}$$

$$\dot{y}_0 = v_0\sin\theta_0 \tag{10}$$

$$\dot{\theta}_0 = \omega_0 \tag{11}$$

$$\dot{\beta}_{n_j+1} = v_0\left(\prod_{k=1}^{n_j}\cos\beta_k\right)\prod_{k=1}^{j-1}\left(1 + \frac{M_{n_k}}{L_{n_k}}\tan\beta_{n_k}\tan\beta_{n_k+1}\right) \cdot \left(\frac{\tan\beta_{n_j}}{L_{n_j}} - \frac{\sin\beta_{n_j+1}}{L_{n_j+1}} + \frac{M_{n_j}}{L_{n_j}L_{n_j+1}}\tan\beta_{n_j}\cos\beta_{n_j+1}\right) \tag{12}$$

$$\dot{\beta}_{n_j+i} = v_0\left(\prod_{k=1}^{n_j+i-1}\cos\beta_k\right)\prod_{k=1}^{j}\left(1 + \frac{M_{n_k}}{L_{n_k}}\tan\beta_{n_k}\tan\beta_{n_k+1}\right) \cdot \left(\frac{\tan\beta_{n_j+i-1}}{L_{n_j+i-1}} - \frac{\sin\beta_{n_j+i}}{L_{n_j+i}}\right) \tag{13}$$

$j \in \{1, \ldots, m\}$, $i \in \{2, \ldots, n_{j+1} - n_j\}$.

Clearly, the components due to the kingpin hitching enter linearly into the system as extra terms added to the basic dynamic equations of the standard n-trailer, which are reobtained choosing $M_{n_j} = 0$:

$$\begin{aligned}
\dot{x}_0 &= v_0 \cos\theta_0 \\
\dot{y}_0 &= v_0 \sin\theta_0 \\
\dot{\theta}_0 &= \omega_0 \\
\dot{\beta}_{i+1} &= v_0 \left(\prod_{k=1}^{i} \cos\beta_k\right)\left(\frac{\tan\beta_i}{L_i} - \frac{\sin\beta_{i+1}}{L_{i+1}}\right)
\end{aligned}$$

$i \in \{1, \ldots, n\}$.

6. SINGULARITIES

Proposition 1 affirms that a virtual steering wheel can be placed on the kingpin hitching nodes and gives a value for its steering angle. Furthermore, from eq. (8) we see that the tangent of the virtual steering angle enters into the dynamic equation of the heading of the following trailer. Therefore one can ask whether the "virtual" singularity introduced by $\tan\gamma_j$ must be considered in the analysis of the controllability of the model or less. The corresponding singular points can be rewritten using the formula:

$$\alpha_j = \arctan\left(-\frac{M_{n_j}}{v_{n_j}}\dot{\theta}_{n_j}\right) = \gamma_j - \beta_{n_j+1}$$

as

$$\dot{\theta}_{n_j} = -\frac{v_{n_j}}{M_{n_j}}\tan(\gamma_j - \beta_{n_j+1}) = \frac{v_{n_j}}{L_{n_j}}\tan\beta_{n_j}$$

or, after some manipulations,

$$\tan\gamma_j = \frac{\tan\beta_{n_j+1} - \frac{M_{n_j}}{L_{n_j}}\tan\beta_{n_j}}{1 + \frac{M_{n_j}}{L_{n_j}}\tan\beta_{n_j+1}\tan\beta_{n_j}} \tag{14}$$

The singular point

$$\gamma_j \rightarrow \frac{\pi}{2} \; mod \; \pi$$

is then equivalent to

$$1 + \frac{M_{n_j}}{L_{n_j}}\tan\beta_{n_j+1}\tan\beta_{n_j} \rightarrow 0. \tag{15}$$

With eq. (15), the question posed above can now easily be answered looking at eq. (12)-(13): the "virtual" singularities are indeed singular points of the Control Lie Algebra since $\left(1 + \frac{M_{n_j}}{L_{n_j}}\tan\beta_{n_j+1}\tan\beta_{n_j}\right) = 0$ implies that the vector field associated with the input v_0 has the last $n - n_j - 1$ components that are null, so it cannot generate a full rank distribution only by means of Lie bracketing. These extra singularities, which are not present in the standard n-trailer system, have then the "nice" physical interpretation of one of the virtual wheels being orthogonal to the velocity vector of the following trailer.

To summarize: in the standard n-trailer we have that the singular locus is given by (see (Jean, 1996))

$$\mathcal{S}_S = \left\{\beta_i = \frac{\pi}{2} \; mod \; \pi\right\}; \; i \in \{1, \ldots, n\}$$

whereas, in the general n-trailer, it is larger:

$$\mathcal{S}_G = \left\{\beta_i = \frac{\pi}{2} \; mod \; \pi\right\} \bigcup \left\{\gamma_j = \frac{\pi}{2} \; mod \; \pi\right\} \tag{16}$$

$i \in \{1, \ldots, n\}$, $j \in \{1, \ldots, m\}$.

7. CONTROLLABILITY OF THE GENERAL N-TRAILER

Theorem 2. The general n-trailer system (9)-(13) is controllable out of the singular locus $\mathcal{S}_G$.

Proof. We give only an informal argument as proof. The rigorous mathematical formulation requires a lot of bookkeeping without adding anything significant to the discussion.

The proof is a direct consequence of the bilinearity of the Lie bracket operation:

$$[X_1 + X_2,\, Y] = [X_1,\, Y] + [X_2,\, Y]$$

In fact, from eq. (9)-(13) we have that the vector fields that form the distribution for the general n-trailer $\Delta_G = span\{g_{1_G}, g_{2_G}\}$ can be written as the sum of the corresponding g_{1_S}, g_{2_S} of the standard n-trailer plus an extra component on the vector field associated with the tangential speed input v_0:

$$\begin{aligned}
g_{1_G} &= g_{1_S} + g_{1_H} \\
g_{2_G} &= g_{2_S}
\end{aligned}$$

The first $n_1 + 3$ components of g_{1_H} are equal to 0. If we call:

$$p_j \triangleq \frac{M_{n_j}}{L_{n_j}}\tan\beta_{n_j}\tan\beta_{n_j+1} \qquad j \in \{1, \ldots, m\}$$

then the generic $n_j + i + 3$ element of the vector g_{1_H} is:

$$g_{1_H}[n_j + i + 3] = \left(\prod_{k=1}^{n_j+i-1} \cos\beta_k\right) \cdot \sum_{k=1}^{j}\left(\sum_{(k)\in C(k,j)} (p_{l(1)} \cdot \ldots \cdot p_{l(k)})\right)\left(\frac{\tan\beta_{n_j+i-1}}{L_{n_j+i-1}} - \frac{\sin\beta_{n_j+i}}{L_{n_j+i}}\right)$$

where $C(k, j)$ is the set of possible combinations of k numbers in the first j integers. The component $n_j + 4$, representing the equation of the axle immediately after the kingpin hitching, has a slightly different expression:

$$g_{1_H}[n_j + 4] = \left(\prod_{k=1}^{n_j} \cos\beta_k\right) \frac{M_{n_j} \tan\beta_{n_j} \cos\beta_{n_j+1}}{L_{n_j} L_{n_j+1}}$$
$$+ \left(\prod_{k=1}^{n_j} \cos\beta_k\right) \sum_{k=1}^{j-1} \left(\sum_{l(k)\in C(k)} \left(p_{l(1)} \cdot \dots \cdot p_{l(k)}\right) \right) \cdot$$
$$\cdot \left(\frac{\tan\beta_{n_j}}{L_{n_j}} - \frac{\sin\beta_{n_j+1}}{L_{n_j+1}} + \frac{M_{n_j} \tan\beta_{n_j} \cos\beta_{n_j+1}}{L_{n_j} L_{n_j+1}} \right)$$

Now, we can use the result of Laumond for the standard n-trailer. In (Laumond, 1993*a*) a family of vector fields is proposed that generates the Control Lie Algebra for the standard n-trailer. The family is composed of $n + 3$ vector fields for any value of n. Out of the singular locus $\mathcal{S}_G$, the result still holds also for our Δ_G in the sense that the vector fields of the family remain independent as in the standard case. The rank condition being satisfied, the system is locally controllable in any regular point. □

Obviously, the verification of controllability in a global sense requires more effort and a technique like the one proposed by (Jean, 1996) should be exploited.

8. CONCLUSION

This paper generalizes the results on controllability to an n-trailer system with one or more kingpin hitching connections between its trailers. To the extra singularities introduced by these off-axle connections a physical meaning can be associated if virtual passive steering wheels are placed on the kingpin hitching joints.

9. REFERENCES

Alexander, J.C. and J.H. Maddox (1988). On the maneuvering of vehicles. *SIAM Journal of Applied Mathematics* **48**, 38–51.

Bloch, A., M. Reyhanoglu and N.H. McClamroch (1992). Control and stabilization of nonholonomic dynamical systems. *Trans. on Automatic Control* **37**(11), 1746–1757.

Bushnell, L.G., D. Tilbury and S. Sastry (1993). Steering three-input chained form nonholonomic sysymes using sinusoids: the fire truck example. In: *Proc. of the 2nd European Control Conference*. Groningen, The Netherlands. pp. 1432–1437.

DeSanctis, R.M. (1997). Modeling and path-tracking for a Load-Haul-Dump mining vehicle. *ASME Journal of Dynamic System Meas. and Control* **119**, 40–47.

Hemami, A. and V. Polotski (1997). Dynamics of a dual-unit articulated vehicle for path tracking control problem formulation. In: *Proc. 1997 IEEE Conf. on Control Applications*. Hartford, CT. pp. 173–176.

Hermann, R. and A.J. Krener (1977). Nonlinear controllability and observability. *IEEE Trans. on Automatic Control* **22**(5), 728–740.

Jean, F. (1996). The car with n trailers: characterization of the singular configuration.. *EISAM: control optimization and calculus of variations* **1**, 241–266.

Kolmanovsky, I. and N.H. McClamroch (1995). Developments in nonholonomic control problems. *IEEE Control System Magazine* pp. 20–36.

Laumond, J-P. (1993*a*). Controllability of a multibody mobile robot. *IEEE Trans. on Robotics and Automation* **9**, 755–763.

Laumond, J.P. (1990). Nonholonomic motion planning versus controllability via the multibody car system example. Technical Report STAN-CS-90-1345. Dep. of Computer Science, Stanford University.

Laumond, J.P. (1993*b*). Singularities and topological aspects in nonholonomic motion planning. In: *Nonholonomic Motion Planning* (Z. Li and J.F. Canny, Eds.). Kluwer Academic. Boston, MA.

Murray, R. and S. Sastry (1993). Nonholonomic motion planning: steering with sinusoids. *IEEE Trans. on Automatic Control* **38**, 700–716.

Rouchon, P., M. Fliess, J. Levine and P. Martin (1993*a*). Flatness, and motion planning: the car with n trailers. In: *Proc. 2nd European Control Conference*. Groningen, The Netherlands. pp. 1518–1522.

Rouchon, P., M. Fliess, J. Levine and P. Martin (1993*b*). Flatness, motion planning and trailer systems. In: *Proc. 32nd IEEE Conf. on Decision and Control*. San Antonio, Texas. pp. 2700–2075.

Sørdalen, O.J. (1993). Exponential stabilization of a car with n trailers. In: *Proc. IEEE 32nd Conf. on Decision and Control*. San Antonio, Texas. pp. 978–983.

Tilbury, D., O. Sørdalen, L. Bushnell and S. Sastry (1995). A multisteering trailer system: conversion into chained form using dynamic feedback. *IEEE Trans. on Robotics and Automation* **11**(6), 807–818.

APPLICATION OF ITERATIVE LEARNING FOR CONSTANT TORQUE CONTROL OF SWITCHED RELUCTANCE MOTORS

N. C. SAHOO, J. X. XU, and S. K. PANDA

Department of Electrical Engineering,
National University of Singapore, Singapore – 119260
E-mail: elexujx@nus.edu.sg

Abstract: The paper deals with the investigations on an iterative learning approach to determine the desired current waveforms for switched reluctance motors, which give rise to ripple-free torque. The current waveforms are generated by repeated corrections from iteration to iteration starting from the conventional rectangular pulse profile as the initial waveform. The scheme requires much less *a priori* knowledge of the magnetic characteristics of the motor. The algorithms have been formulated for both one-phase-on and two-phase-on schemes, for a four-phase switched reluctance motor, in the light of the principles behind iterative learning. Based on the observations from the test results of these schemes, a modified scheme has been proposed by incorporating a suitable commutation process, often called as torque sharing functions, in order to generate reasonably smooth current waveforms for the ease of tracking by the stator circuit of the motor. The performances of all the proposed schemes have been verified by computer simulation. *Copyright ©1999 IFAC*

Keywords: Iterative learning, switched reluctance motor, torque ripple, current waveform.

1. INTRODUCTION

The switched reluctance motor (SRM) has a lot of advantages due to its low cost and rugged construction. The SRM is operated in deep magnetic saturation to increase the output power density. Due to the saturation effect and the variation of magnetic reluctance, all pertinent characteristics of the machine model, i.e., flux-linkage, inductance, torque etc., are nonlinear functions of both rotor position and current. Because of these nonlinearities, the generated torque contains significant ripples when the motor is excited by the conventional rectangular pulse excitation.

The problem of torque ripple minimization has been addressed by a number of researchers (Filicori, *et al.*, 1993; Taylor, 1991). The techniques suggested in these works are mainly based on exhaustive measurement of magnetic characteristics of the SRM. All these methods address one fundamental issue, i.e., the determination of a suitable current/flux waveform.

In the present work, the problem from has been tackled from a control engineering point of view. The current is treated as input to the SRM and output is the torque. There are a couple of reasonably accurate models of the torque (Taylor, *et al.*, 1988; Ilic-Spong, *et al.*, 1987) taking the effect of magnetic saturation into account. In spite of the availability of good models for the torque, it is not possible to derive an inverse expression characterizing the amount of phase currents required to produce a given torque at some specified rotor position, without resorting to iterative numerical techniques.

This paper describes an efficient approach for finding the solution of the nonlinear function, i.e., current as a function of torque and rotor position, in a tabular form by iterative learning (Arimoto, *et al.*, 1984). There are a number of strategies one can adopt, i.e., for example, one-phase-on scheme and two-phase-on scheme. Both the schemes have been addressed in their simplest forms. Further, a modified two-phase-on scheme incorporating a set of novel pre-specified torque sharing functions (TSFs) has been proposed. This scheme generates reasonably smooth current profiles.

The paper is organized as follows. Section 2 briefly describes the flux and torque models of the SRM used in this study. In Section 3, the proposed iterative learning schemes, for both one-phase-on and two-phase-on strategies, have been detailed. Section 4 concludes the paper.

2. SRM TORQUE MODEL

The SRM has a salient pole stator with concentrated windings and also a salient pole rotor with no magnets or coils. The basic principles of SRM are described in detail in (Lawrenson, *et al.*, 1980). Here, a four-phase SRM has been used. All the phases are assumed to be symmetrical. Throughout the paper, reference to a generic phase j is reflected as a subscript in the variables. Because of the double saliency of the motor and the intentional operation of the motor under magnetic saturation, the inductance of phase $j(j=1,\cdots,4)$, L_j, is a function of both rotor position θ and the current I_j. However, under the assumption of linear magnetics, inductance vs. rotor position profile can be approximated in a trapezoidal manner over one rotor pole pitch as shown in Fig. 1 (solid line). Without loss of generality, we assume that phase 1 is characterized by this profile. The significance of the sinusoidal broken line curve is explained at a later stage in this section. The various parameters of the profile are defined as:
β_s = stator pole arc, β_r = rotor pole arc ($\beta_s < \beta_r$), α_r = rotor pole pitch ($\beta_s + \beta_r < \alpha_r$), L_a = phase inductance at the aligned position, and L_u = phase inductance at the unaligned position.

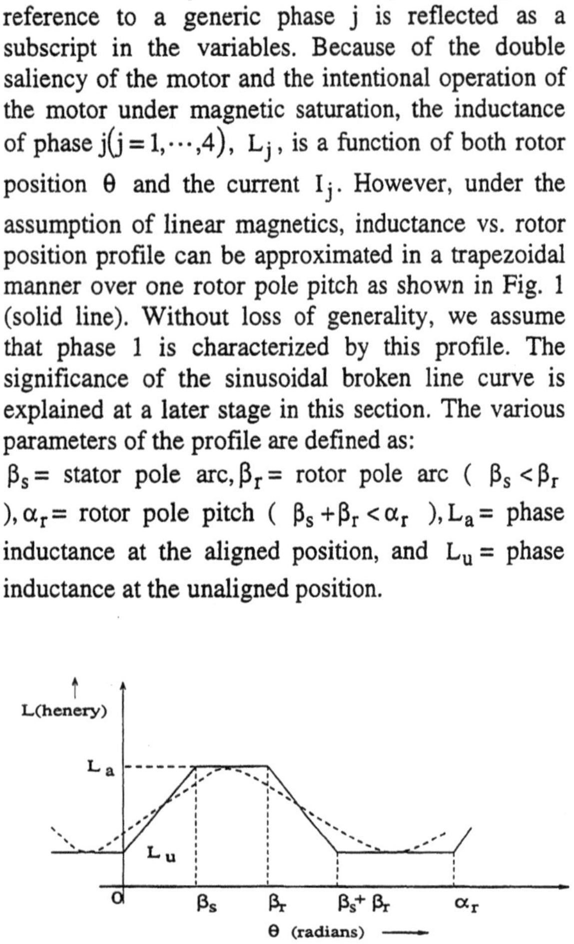

Fig. 1. Approximated inductance vs. rotor position for phase # 1.

This inductance profile can be mathematically described by

$$L_1(\theta)=\begin{cases} L_u+K\theta & , \quad 0\le\theta\le\beta_s \\ L_a & , \quad \beta_s\le\theta\le\beta_r \\ L_u-K(\theta-\beta_s-\beta_r) & , \quad \beta_r\le\theta\le\beta_s+\beta_r \\ L_u & , \quad \beta_s+\beta_r\le\theta\le\alpha_r \end{cases} \tag{1}$$

and that of phase j is,

$$L_j(\theta)=L_1\left(\theta-\frac{\alpha_r}{4}(j-1)\right) \tag{2}$$

where K is the slope of the profile in the zone of increasing inductance, given by equation (3). The parameters of the inductance profile are taken from (Buja, *et al.*, 1993) :

$$K=\frac{L_a-L_u}{\beta_s}. \tag{3}$$

Output power = 7.5 kW, Number of phases = 4, Number of rotor poles (N_r) = 6, Number of stator poles (N_s) = 8, Rated speed = 1900 rpm, L_u = 10 mH, L_a = 110 mH, α_r = 1.05 rad, β_s = 0.35 rad, β_r = 0.42 rad, ϕ_s = 1.2 Wb.

When any one phase of the SRM is energized, the torque developed, T_j, is expressed as

$$T_j(\theta_e,I_j)=N_r\frac{\partial}{\partial\theta_e}W_c(\theta_e,I_j) \tag{4}$$

where θ_e and N_r are respectively the electrical angle and the number of rotor poles and $\theta_e=N_r\theta$. The co-energy, W_c, is defined as

$$W_c(\theta_e,I_j)=\int_0^{I_j}\phi_j(\theta_e,i_j)di_j \tag{5}$$

where $\phi_j(\theta_e,I_j)$ is the flux-linkage and, under linear magnetics of the SRM, is expressed as:

$$\phi_j(\theta_e,I_j)=L_j(\theta_e)I_j(\theta_e). \tag{6}$$

Hence, the final expression for the torque can be derived as

$$T_j(\theta_e,I_j)=\frac{N_r}{2}\frac{\partial L_j(\theta_e)}{\partial\theta_e}I_j^2(\theta_e). \tag{7}$$

The total torque is the algebraic sum of the individual phase torques.

A more accurate expression of the torque can be obtained by taking a flux-linkage model that incorporates the effect of magnetic saturation. Such an efficient model has been proposed in (Ilic-Spong, *et al.*, 1987), which is also used here. This nonlinear flux-linkage model is expressed by

$$\phi_j(\theta_e,I_j)=\phi_s\{1-\exp(-I_jf_j(\theta_e))\}, \quad I_j\ge 0 \tag{8}$$

where ϕ_s is the saturated flux-linkage and $f_j(\theta_e)$ is expressed as a strictly positive Fourier series expansion. However, for the degree of accuracy for the present investigation, the first two terms are included, as suggested in (Ilic-Spong, *et al.*, 1987), of the Fourier series. Here, we have evaluated $f_j(\theta_e)$ by a sinusoidal approximation (Panda and Dash, 1996) of the inductance profile. Thus, the torque produced by phase j can be obtained as

$$T_j(\theta_e,I_j)=\frac{N_r\phi_s}{f_j^2(\theta_e)}\frac{df_j(\theta_e)}{d\theta_e}\begin{Bmatrix}1-[1+I_jf_j(\theta_e)] \\ \exp(-I_jf_j(\theta_e))\end{Bmatrix}. \tag{9}$$

3. BASIC ILC SCHEME FOR SRM

In this section, the iterative learning control (ILC) schemes for learning the desired current waveform for some specified desired torque is described. Both one- and two-phase-on strategies are addressed. The

SRM is assumed to be rotating with constant speed and the desired current profile is instantaneously established in the phase windings. The designs of the learning system are done using electrical angles (The length of an electrical cycle is equal to one rotor pole pitch.)

3.1 One-Phase-On Scheme

In this scheme, only one phase is conducting at any instant. The complete span of an electrical cycle is allotted equally amongst all the four phases. Thus each phase gets a conduction angle of $\pi/2$ radians (elect.). A phase is selected for conduction only when it is passing through a region of increasing inductance The phases are switched sequentially after successive intervals of $\pi/2$ radians (elect.). Thus, in the context of ILC, the interval of repetition of the process output is $\pi/2$ radians (elect.). The system does not change because of this particular mode of phase switching. Thus, in each conduction interval, an incoming phase starts with the current profile learnt during the conduction interval of the immediate outgoing phase, modifies it based on the torque error during its own conduction and passes it to the next phase at the end of its conduction. This mechanism is iterated again and again.

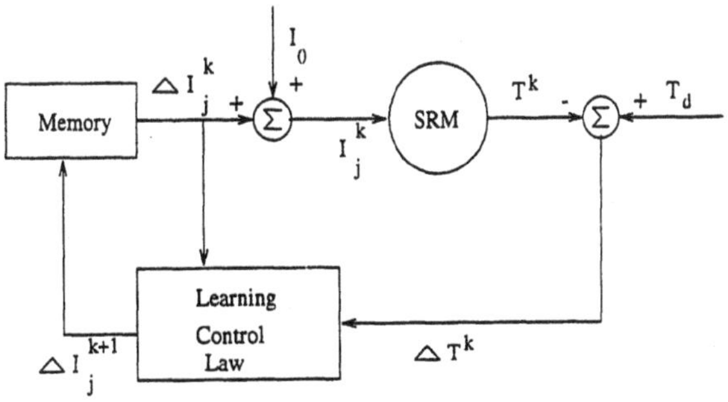

Fig. 2 Iterative learning control scheme for SRM.

The iterative corrections to the current profile are based on the torque error. In the conventional ILC method, as there is no correction done in the very first cycle, the performance of the system, in this first cycle, greatly depends upon the initial current profile chosen. Here, a fairly judicious choice of initial current is done from the linear torque model (7). To be more specific, for a desired torque T_d, the initial current, I_0, is computed from equation (10). Again, to make it as simple as possible, the derivative in the equation (10), $dL_j/d\theta_e$, is taken to be the constant slope of the inductance profile in the increasing inductance region evaluated from the trapezoidal approximation of the inductance profile (Fig. 1). Finally, the learning laws are stated formally in equations (11) and (12) without any reference to any particular phase. In these equations, $i(=1,\cdots,N)$ is the sample number within the iteration itself, k is the iteration index, $\Delta T(=T_d - T)$ is the torque error, and G is the learning gain. The idea is illustrated in Fig.2.

$$T_d = \frac{N_r}{2}\frac{dL_j}{d\theta_e} I_0^2 . \qquad (10)$$

$$\Delta I^{k+1}(i) = \Delta I^k(i) + G\,\Delta T^k(i) . \qquad (11)$$

$$I^{k+1}(i) = I_0 + \Delta I^{k+1}(i) . \qquad (12)$$

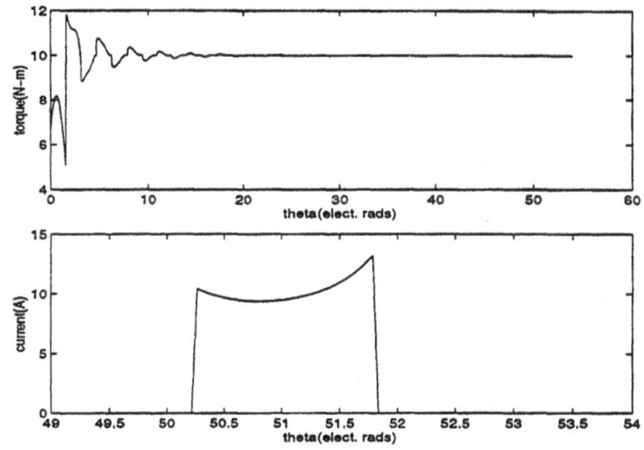

Fig. 3 The iterative learning of torque profile and the current profile for phase # 1 (after completion of learning) for a desired torque of 10 N-m (G = 0.9).

The number of samples for each phase (N) is chosen to be 32. The motor torque T is computed from its nonlinear expression (9). Fig. 3 shows the iterative control of the torque profile and the correspondingly learnt current profile for a desired torque of 10 N-m. The simulation tests for different levels of the desired torque are carried out. The results indicate that the learning gains should be different for different torque levels for convergence as well as the fastness of convergence. From these results, it is clearly observed that the convergence (in terms of peak/average error) is achieved on the iteration axis. These results clearly indicate the efficacy of the iterative learning control. However, the need to make this learning gain adaptive arises.

Adaptive learning gain: For this purpose, the expression for torque is expanded in a Taylor's series form, and neglecting higher order terms, as:

$$T = T_j = T_0 + \left.\frac{\partial T_j}{\partial I_j}\right|_{\theta_e} \Delta I_j . \qquad (13)$$

It is assumed that the current is updated at an instant when the rotor position is held constant. Thus, the other first order term involving the derivative of torque with respect to rotor position is neglected. T_0 may be regarded to be equivalent to the torque obtained from a linear model.

Now, the direct transmission term (Arimoto, *et al.*, 1984) D_s^{eq} for the SRM problem is:

$$D_s^{eq} = \left.\frac{\partial T_j}{\partial I_j}\right|_{\theta_e} = N_r \phi_s \frac{df_j(\theta_e)}{d\theta_e} I_j \exp\left(-I_j f_j(\theta_e)\right) \qquad (14)$$

The learning gain G is made adaptive by the following the principles of ILC by:

$$G = \frac{1}{D_s^{eq}} . \tag{15}$$

Here, the conduction interval must be selected such that $D_s^{eq} \neq 0$. Such an undesirable situation will arise in any one of the following situations.

(a) Current is zero. **(b)** $df_j(\theta_e)/d\theta_e$ is zero. **(c)** Current is infinitely high (a physically impossible case!).

The third condition is of theoretical interest only. However, its numerical possibility is avoided by the initial approximation to the solution. With regard to the first two conditions, this is automatically eliminated due to the following reasons. A nonzero current is assumed in the very first interval and a phase is excited only when it is passing through a positive torque producing region.

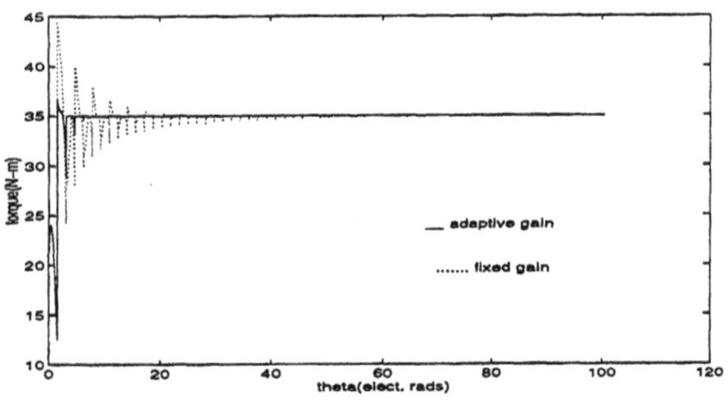

Fig. 4 Comparison of the iterative learning of torque profile for adaptive and fixed learning gains (desired torque = 35 N-m, fixed gain = 0.6).

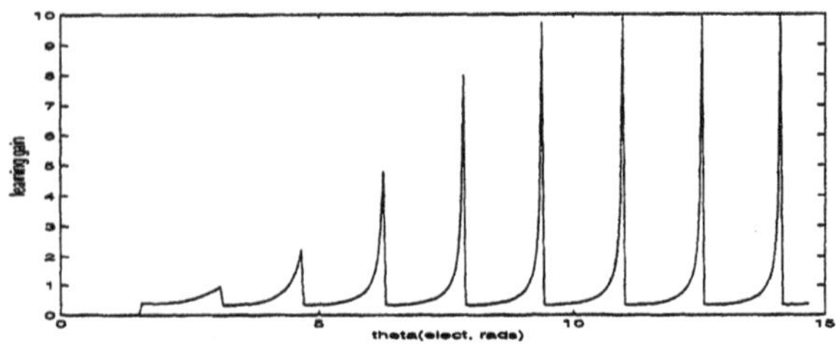

Fig. 5 The variations of the adaptive learning gain for a desired torque of 35 N-m.

The adaptive learning law is simulated for a wide range of desired torque levels. Fig. 4 shows a comparison of the learning of torque profile for a desired torque of 35 N-m. Clearly, the superiority of the adaptive learning gain over fixed gain is reflected. The finally learnt current profiles are same in both the cases. The variations of the adaptive gain is shown in Fig. 5. As to the nature of the current waveform thus obtained, it has the following features.

1. It has got sharp rising and falling edges. This is expected because a phase is instantly asked to increase its torque output from zero to the desired value at the rising edge (The opposite is the case at the falling edge).

2. The amount of rise/fall at the respective edges are not necessarily equal. Particularly, as the desired torque level rises, the amount of fall is much more than the amount of rise at the respective edges. It can be clearly seen Fig. 3.

It is worth mentioning here that such high values of current will result in very high resistive loss in the stator circuit. Further, it may also be very difficult to inject/extinguish such high values of currents in a very short time. Thus, it is necessary to seek for a scheme where the magnitudes of currents can be lessened. The immediate remedy is to go for a scheme where the total torque can be shared with another suitable phase during the period when the current has a very high value. This modified strategy is addressed as two-phase-on scheme and is described below.

3.2 *Two-Phase-On Scheme:*

Here, at any instant, two phases are allowed for conduction at the most There are typically two types of conduction intervals following each other, i.e., an interval where only one phase is allowed to conduct and the other where two phases are allowed to conduct. In the interval where two phases are conducting, the phase which has been conducting, alone, immediately before goes out and a new phase is introduced. A phase is selected for conduction only when it is passing through its increasing inductance region. The interval during which two phases are conducting is called as an overlap period and the other one is regarded as non-overlap period.

There are two different learning systems operating independently of each other, i.e., one for the overlap period and another for the non-overlap period. Again, in the overlap period, there are two learning subsystems, i.e., one for the incoming phase and another for the outgoing phase. Thus, three different learning gains are: G_{on} for the non-overlap period; G_{in} and G_{out} for the incoming and the outgoing phases, respectively, in the overlap period. The currents are defined accordingly, i. e., I_{on}, I_{in}, and I_{out} respectively. G_{on} is computed according to equation (15), as it is nothing but the one-phase-on scheme in this interval. For the computation of the other two gains, the following steps are derived.

For the overlap period, the following inequality must be satisfied for the convergence of the learning algorithm,.

$$\left| 1 - \left(D_s^{in} G_{in} + D_s^{out} G_{out} \right) \right| < 1 . \tag{16}$$

By setting the L.H.S. of this inequality to zero, as done before, $D_s^{in} G_{in} + D_s^{out} G_{out} = 1$. (17)

Now, G_{in} and G_{out} are selected as (with $K_1 + K_2 = 1$): $G_{in} = \frac{K_1}{D_s^{in}}$, $G_{out} = \frac{K_2}{D_s^{out}}$. (18)

In our study, $K_1 = K_2 = 0.5$. The learning laws for the complete current waveform are defined as below.

Non-overlap period:.

$$\Delta I_{on}^{k+1}(i) = \Delta I_{on}^{k}(i) + G_{on}\,\Delta T^{k}(i) \tag{19}$$

$$I_{on}^{k+1}(i) = I_0 + \Delta I_{on}^{k+1}(i). \tag{20}$$

Overlap period (incoming phase):

$$\Delta I_{in}^{k+1}(i) = \Delta I_{in}^{k}(i) + G_{in}\,\Delta T^{k}(i). \tag{21}$$

$$I_{in}^{k+1}(i) = I_0 + \Delta I_{in}^{k+1}(i). \tag{22}$$

Overlap period (outgoing phase):

$$\Delta I_{out}^{k+1}(i) = \Delta I_{out}^{k}(i) + G_{out}\,\Delta T^{k}(i). \tag{23}$$

$$I_{out}^{k+1}(i) = I_0 + \Delta I_{out}^{k+1}(i). \tag{24}$$

In the computer simulations, the lengths of the overlap and non-overlap intervals are chosen to be 60^0 (elect.) and 30^0 (elect.) respectively. Fig. 6 shows the learning of torque. The current profiles are shown in Fig. 7. Clearly, the current levels are reduced. It is really interesting to observe how a torque sharing is automatically achieved during the overlap interval

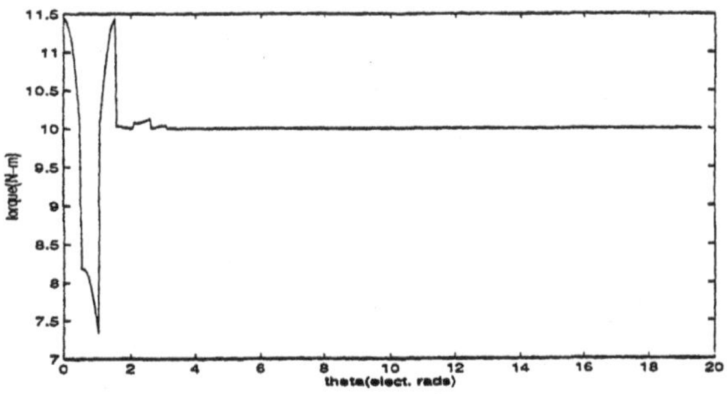

Fig. 6. Iterative learning of torque in two-phase-on scheme (T_d = 10 N-m).

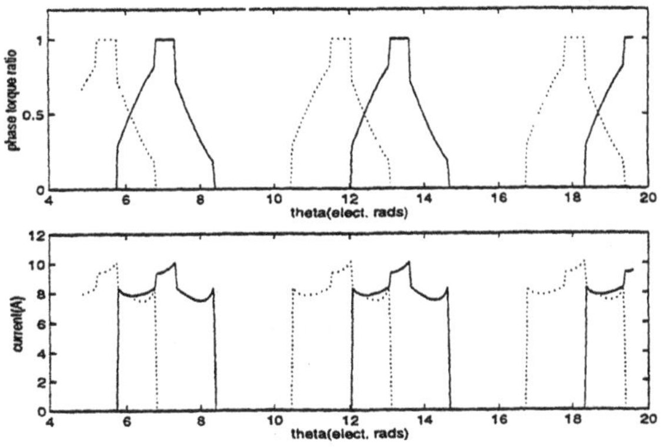

Fig. 7. Torque sharing between two consecutive phases (-phase1, ... phase 4) and the correspondingly learnt current profiles (T_d = 10 N-m).

Thus, the two-phase-on scheme has a clear edge over the one-phase-on scheme in terms of peak value of the required current. However, the current profile still has sudden rising/falling segments. This happens because of the very structure of the scheme. Because, the tractability of the desired current profile mainly rests upon its frequency content, the bandwidth of the stator phase circuit, we need to examine these issues. For a faithful tracking of the current profile in the stator circuit, one viable approach will be to modify the desired current profile, if necessary, so that most of its dominant frequency contents lie within the bandwidth of the stator circuit. In essence, a smoother current profile is desired. Because, the sharing of torque between the phases determines the current profile, it is necessary to assign smoother torque sharing functions (TSFs).

3.3 *ILC Method with Torque Sharing Functions:*

The objective is to coordinate the production of torque by the individual phases so that the total torque at any instant is equal to the desired torque. A TSF is defined for a phase only in the region where it can produce positive torque. For rest of the electrical cycle, the TSF for that phase is zero. There are a number of ways one can define these TSFs. An arbitrary choice of the TSF will not be suitable for our purpose. The simplest trajectory that can be used for this purpose is the cubic segment (CS). The cubic segment, here, is a cubic polynomial of θ_e. This function is characterized by four arbitrary parameters. With the two-phase-on scheme, each TSF will have a rising segment followed by one constant segment and a falling segment. Fig. 8 shows two sets of TSFs for a desired torque of 5 N-m. In the top figure, the overlap angle is $\pi/3$ (elect. rads). In the bottom figure, the overlap angle is $\pi/2$ (elect. rads.). The functions are reasonably smooth in both the cases.

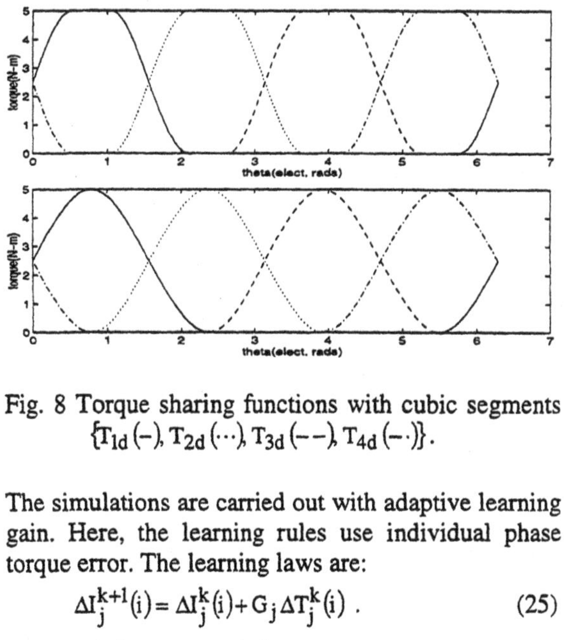

Fig. 8 Torque sharing functions with cubic segments $\{T_{1d}(-), T_{2d}(\cdots), T_{3d}(--), T_{4d}(-\cdot)\}$.

The simulations are carried out with adaptive learning gain. Here, the learning rules use individual phase torque error. The learning laws are:

$$\Delta I_j^{k+1}(i) = \Delta I_j^{k}(i) + G_j\,\Delta T_j^{k}(i)\ . \tag{25}$$

$$I_j^{k+1}(i) = I_0 + \Delta I_j^{k+1}(i). \tag{26}$$

The learning gain for each individual phase must satisfy the convergence condition. Hence, the equation for the gain computation is same as equation (15) ($G_j D_{sj} = 1$ and D_{sj} is same as D_s^{eq}). However, as the learning law for each phase operates for the complete electrical cycle (D_{sj} goes through the zero and negative values), the learning gain G_j is modified as (ε = 0.001):

$$G_j = \begin{cases} \dfrac{1}{D_{sj}+\varepsilon} & , \text{if } D_{sj} \geq 0 \\ \dfrac{-1}{|D_{sj}|+\varepsilon} & , \text{if } D_{sj} < 0 \end{cases} \tag{27}$$

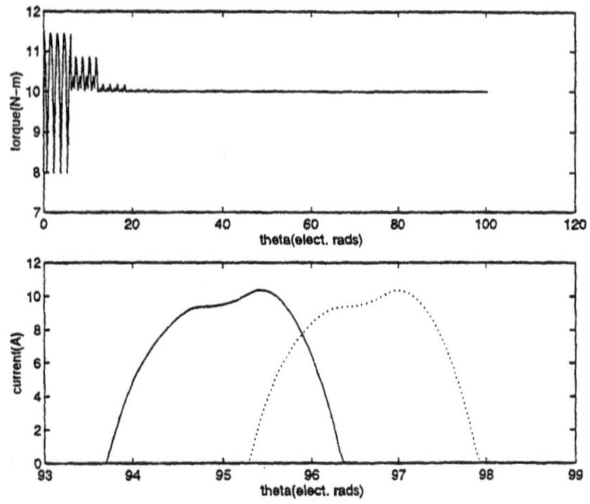

Fig. 9 The iterative learning of torque and the finally learnt current profiles for two consecutive phases for a desired torque of 10 N-m {phase 1 (–), phase 2 (..)}.

As the negative torque producing region of any phase is of no interest and, further, in order to avoid possible undesired numerical outcome near the point when a phase is just near the boundary of positive and negative torque producing regions, the following precaution can be incorporated (tolerating very negligible torque error). The constant η is a very small positive fraction (much less than 1).If $\frac{df_j(\theta_e)}{d\theta_e} < \eta$, then $I_j(i) = 0$. The number of samples is again same as before. For an overlap angle of $\pi/3$ (elect. rads), the learning of torque and the final current profiles are shown in Fig. 9. The learnt current waveforms have no sharp rising/falling segments. From the results, it is clear that the torque-sharing function approach for obtaining a smoother current profile is fruitful.

4. CONCLUSION

Determination of a suitable current profile for minimizing the torque ripples in the switched reluctance motor is a burning problem. We have presented a sequence of investigative studies on the use of iterative learning control method for solving this problem. As it turns out, there are a number of possible schemes to meet this goal, i.e., starting from the simplest one-phase-on scheme to a two-phase-on scheme incorporating the specification of torque-sharing functions for the individual phases. All the three different schemes reported yield us different solutions to our goal. Each solution is the best possible that can be achieved under that specific scheme. However, from a high performance application point of view, the two-phase-on scheme with the specifications of torque-sharing functions is a viable approach considering the limitations of the stator circuit.

In the implementations of all the three schemes, estimation or measurement of electromagnetic torque of the motor is essential. Barring this, the algorithms are quite simple. The beauty of the ILC approach to this problem lies in the fact that it is able to learn the desired current waveform that is highly nonlinear by iterative corrections, which is difficult to obtain by any conventional means. Because of the very structure of the ILC method used here, the performance is normally poor in the very first cycle unless special care is taken (as done in this work).

The present work is expected to serve as a stepping stone to the applicability of ILC method to the SRM problem. A lot of improvements on the present state of the methods can be done, i.e., for example, search for an optimal torque sharing function, incorporating the effect of speed of the motor into the learning process etc.

REFERENCES

Arimoto, S., S. Kawamura, and F. Miyasaki (1984). Bettering the operation of robots by learning. *Journal of Robotic Systems,* **1**(2), 123-140.

Buja, G. S., R. Menis, and M. I. Valla (1993). Variable structure control of an SRM drive. *IEEE Trans. Ind. Electronics,* **IE-40**, (1), 56-63.

Filicori, F., C. G. L. Bianco, and A. Tonielli (1993). Modeling and control strategies for a variable reluctance direct drive motor. *IEEE Trans. Ind. Electronics*, **40** (1), 105-115.

Ilic-Spong, M., R. Marino, S. M. Peresada, and D. G. Taylor (1987). Feedback linearizing control of switched reluctance motors. *IEEE Trans. Automatic Control,* **32**(5), 371-379.

Lawrenson, P. J., J. M. Stephenson, P. T. Blenkinsop, J. Corda, and N. N. Fulton (1980). Variable-speed switched reluctance motors. *IEE Proc. Pt. B.*, **127**(4), 253-265.

Panda, S. K., and P. K. Dash (1996). Application of nonlinear control to switched reluctance motors: a feedback linearisation approach. *IEE Proc. Pt. B.*, **143**(5), 371-379.

Taylor, D. G., M. J. Woolley, and M. Ilic-Spong (1988). Design and implementation of a linearizing and decoupling feedback transformation for switched reluctance motors. *Proc. 17th Symp. Incr. Motion contr. sys. and devices*, Champaign, 173-184.

Taylor, D. G. (1991). An experimental study on composite control of switched reluctance motors. *IEEE Control Syst. Mag.*, **11**(2), 31-36.

14th Triennial World Congress, Beijing, P.R. China

E-2c-06-4

COMPARISON OF NONLINEAR CONTROL DESIGNS FOR A DUCTED FAN MODEL

Jie Yu, Ali Jadbabaie, James Primbs, and Yun Huang

Control and Dynamical Systems, Mail Code 107-81, California Institute of Technology, Pasadena, CA 91125, USA
jyu, alij, jprimbs, yunhuang@cds.caltech.edu

Abstract: In this paper we compare different nonlinear control design methods by applying them to the planar model of a ducted fan engine. Those design can be divided into two steps. The first step requires the derivation of a Control Lyapunov Function (CLF), while the second involves using an existing CLF to generate a controller. The main premise of this paper is that by combining the best of these two phases, it is possible to find controllers that achieve superior performance when compared to those that apply each phase independently. All of the results are compared to the optimal solution which is approximated by solving a trajectory optimization problem with a sufficiently large time horizon. *Copyright © 1999 IFAC*

Keywords: Nonlinear control design, Optimal Control, LPV Methods, Model Predictive Control, Ducted Fan, Control Lyapunov Functions

1. INTRODUCTION

For more than three decades, nonlinear stabilization has been a major part of control engineering research.Despite the fact that nonlinear control design is much harder than its linear counterpart, several state of the art techniques have been developed to tackle the problem of nonlinear stabilization. They can be roughly divided into two major groups. The first group is comprised of those methods that generate a control Lyapunov Function (CLF), while the second group takes an existing CLF and uses it to generate a stabilizing controller.

Motivated by the notion of combining the advantages of these two viewpoints into the design methodology, a workshop was organized at the American Control Conference in June 1997. Several nonlinear design techniques were applied to simple toy examples generated by the converse HJB method Doyle *et al.* (1996), and they were compared to the optimal solution which was known in advance. The purpose of this paper is to continue this comparison further, and add some other design methods to the existing ensemble. The methods are applied to a ducted fan, a research setup developed at Caltech as a testbed for nonlinear control design. In this paper we report the simulation results performed using various nonlinear methods on a simplified model of the fan, and experimental results will be presented in our future papers.

This paper is organized as follows: In section 2, we describe the state space model for the ducted fan engine. Section 3 deals with the methods that obtain a control Lyapunov function. In section 4, we describe methods that utilize the CLF to generate a stabilizing controller. A comparison of different methods is performed in section 5, and finally our conclusions are presented in section 6.

2. CALTECH DUCTED FAN MODEL

The Caltech Ducted Fan is a small flight control experiment whose dynamics are representative of either a Harier in hover mode or a thrust vectored aircraft such as F18-HARV or X-31 in forward flight Murray (1998). This system has been used for a number of studies and papers. In particular, a comparison of several linear and nonlinear controllers was performed in Kantner *et al.* (1995); Bodenheimer *et al.* (1996); Nieuwstadt and R.M.Murray (1996). In this section we describe the simple planar model of the fan shown in Figure 1 which ignores the stand dynamics.

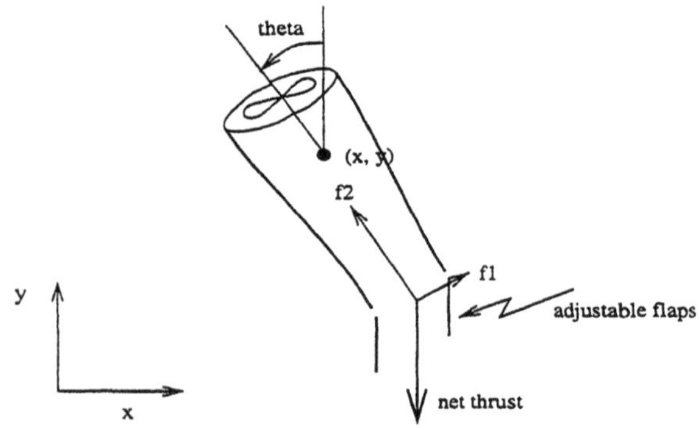

Fig. 1. Planar ducted fan model

Let (x, y, θ) denote the position and orientation of a point on the main axis of the fan that is distance l from the center of mass. We assume that the forces acting on the fan consist of a force f_1 perpendicular to the axis of the fan acting at a distance r and a force f_2 parallel to the axis of the fan. Assuming m, J, and g to be the mass of the fan, the moment of inertia, and the gravitational constant respectively, the equations of motion can be written as follows:

$$\begin{aligned} m\ddot{x} &= -d\dot{x} + f_1 \cos\theta - f_2 \sin\theta \\ m\ddot{y} &= -d\dot{y} + f_1 \sin\theta + f_2 \cos\theta - mg \\ J\ddot{\theta} &= r f_1 \end{aligned} \quad (1)$$

where the drag terms are modeled as viscous friction with d being the viscous friction coefficient. The detailed data as : $m = 4.0kg$,$g = 9.8m/sec^2$,$J = 0.0475kgm^2$,$r = 0.26m$,$l = 0.00213m$,$d = 0.1Nsec$. It is convenient to redefine the inputs so that the origin is an equilibrium point of the system with zero input. If we let $u_1 = f_1$ and $u_2 = f_2 - mg$.

$$\begin{aligned} m\ddot{x} &= -mg\sin\theta - d\dot{x} + u_1 \cos\theta - u_2 \sin\theta \\ m\ddot{y} &= mg(\cos\theta - 1) - d\dot{y} + u_1 \sin\theta + u_2 \cos\theta \\ J\ddot{\theta} &= r u_1 \end{aligned} \quad (2)$$

These equations are referred to as the *planar ducted fan equations*. The following quadratic cost function was used for comparison of different design techniques:

$$\mathcal{J} = \int_0^\infty (\bar{x}^T(t) Q \bar{x}(t) + u^T(t) R u(t)) dt$$

where $\bar{x} = [x\ \dot{x}\ y\ \dot{y}\ \theta\ \dot{\theta}]^T$, $R = I$, and Q is chosen to be a diagonal matrix with the following diagonal terms:

$$Q = diag([\,10\ 1\ 10\ 1\ 1\ 1\,]) \quad (3)$$

We chose the following initial condition:

$$[\,x\ \dot{x}\ y\ \dot{y}\ \theta\ \dot{\theta}\,] = [\,5\ 5\ 5\ 0\ -0.9\pi/2\ 0\,] \quad (4)$$

3. GENERATION OF CLFS

Consider the following nonlinear system:

$$\dot{x} = f(x) + g(x)u \quad (5)$$

where $x \in \mathbb{R}^n$, $u \in \mathbb{R}^m$. A *control Lyapunov function* (CLF) is a $\mathbf{C}^1$, proper, positive definite function $V : \mathbb{R}^n \to \mathbb{R}_+$ such that:

$$\inf_u [V_x(x)f(x) + V_x(x)g(x)u] < 0 \quad (6)$$

for all $x \neq 0$ Artstein (1983); Sontag (1983, 1989). It can be shown that the existence of a CLF for the system (5) is equivalent to the existence of a globally asymptotically stabilizing control law $u = k(x)$ which is smooth everywhere except possibly at $x = 0$ Artstein (1983).

In what follows, some of the widely used methods in nonlinear control design are presented , and interpreted in the context of the search for a control Lyapunov function.

3.1 *Jacobian Linearization*

Perhaps the simplest method of deriving a CLF is to use the Jacobian linearization of the system and generate a CLF by solving an LQR problem. In the case of the nonlinear system

$$\dot{x} = f(x) + g(x)u$$

A and B are assumed to be :

$$A = \frac{\partial f(x)}{\partial x}|_{x=0} \quad B = \frac{\partial g(x)}{\partial x}|_{x=0}$$

Obviously the obtained CLF $V(x) = x^T P x$ by solving corresponding Riccati Equation will be valid only in a region around the equilibrium. Therefore, we should not expect good performance from initial conditions far from the origin. This is indeed the case as simulation results show

3.2 *Global Linearization*

The basic idea behind this approach is to model a nonlinear system as a Polytopic Linear Differential Inclusion (PLDI) Boyd *et al.* (1994). The

PLDI describing the planar ducted fan model can be written as:

$$\dot{x} = \sum_{i=1}^{2} \alpha_i(t)(A_i x + B_i u)$$
$$u = -Kx \tag{7}$$

The problem of quadratic stability of the obtained PLDI, can be recast as an LMI feasibility problem which can be solved very efficiently using interior point methods.

Minimize

$$tr(Z)$$

Subject to:

$$Y > 0$$
$$\begin{bmatrix} N & YQ^{1/2} & X^T R^{1/2} \\ Q^{1/2}Y & -I & 0 \\ R^{1/2}X & 0 & -I \end{bmatrix} < 0$$
$$\begin{bmatrix} Z & x_0^T \\ x_0 & Y \end{bmatrix} > 0$$
$$i = 1,2$$

where $N = YA_i^T + A_iY - B_iX - X^TB_i^T$, $Y = P^{-1}$, and $X = KY$ are the change of variables made to recast the matrix inequalities as LMIs Boyd *et al.* (1994). Q and R are chosen as before, and A_1, B_1 are obtained by linearization of the ducted fan model at the origin and A_2 and B_2 are chosen such that the dynamics lie in the convex hull described by (7). Using the LMI formulation of the LQR problem for PLDIs Boyd *et al.* (1994), we can find a CLF (given by $V(x) = x^TPx$) for the ducted fan model for positive values of θ. However, a global constant quadratic CLF does not exist.

3.3 *Frozen Riccati Equation (FRE) Method*

This method was first introduced by Cloutier *et. al.* in Cloutier *et al.* (1996). The basic idea behind this method, sometimes called *State Dependent Riccati Equations*, is to solve the Riccati equations online, at each time step. Although results are often promising, there are no rigorous justifications for even maintaining mere stability. Nevertheless, the simplicity of the implementation makes the FRE approach a plausible alternative in some applications. To apply this method, the planar ducted fan model is written as:

$$\dot{x} = A(x)x + g(x)u \tag{8}$$

At each *frozen state* the Riccati equation is solved, and then the resulting state feedback controller is feedback to the system, that is, the state feedback nonlinear control law is obtained by solving the following:

$$0 = A(x)^T P(x) + P(x)A(x) - P(x)g(x)g^T(x)P(x) + Q$$
$$u = -g^T(x)P(x)x \tag{9}$$

The quantity $V(x) = x^TP(x)x$ generated by this technique is in general only a local CLF. Furthermore, one of the major drawbacks of this method is the lack of a systematic procedure for selecting, among the infinite possibilities, a single parameterization for $f(x)$ (in the form of equation (8)) which achieves stability and acceptable performance. However, in the case of the ducted fan model, the *obvious parameterization* of $f(x)$ appears to work in simulation studies.

3.4 *Linear Parameter Varying (LPV) Methods*

In this technique, the following so-called quasi-LPV representation of a nonlinear input-affine system is used to design a state feedback controller

$$\dot{x} = A(\rho(x))x + B(\rho(x))u \tag{10}$$

Assume the underlying parameter ρ varies in the allowable set

$$\mathcal{F}_{\mathcal{P}}^{\nu} := \{\rho \in C^1(\mathbf{R}^+, \mathbf{R}^m) : \rho \in \mathcal{P},\ \underline{\nu}_i \leq \dot{\rho}_i \leq \overline{\nu}_i,\}$$

where $\mathcal{P} \subset \mathbf{R}^m$ is a compact set. If there exists a positive definite $X(\rho)$ such that the following inequality is satisfied

$$\begin{bmatrix} -\sum_{i=1}^{m} \underline{\overline{\nu}}_i \frac{\partial X}{\partial \rho_i} + L & XC^T \\ CX & -I \end{bmatrix} < 0 \tag{11}$$

for all $\rho \in \mathcal{P}$ where $L = AX + XA^T - BB^T$, $C(\rho) = Q^{\frac{1}{2}}(\rho(x))$, then the closed-loop system is stable with the state feedback

$$u(x) = -B^T(\rho(x))X^{-1}(\rho(x))x.$$

Moreover, an upper bound on the optimal value function $V^*(x)$ (which also serves as a CLF) is given by:

$$V(x) = x^TX^{-1}(\rho(x))x \geq V^*(x).$$

The notation $\sum_{i=1}^{m} \underline{\overline{\nu}}_i(\rho)$ in (11) means that every combination of $\overline{\nu}_i(\rho)$ and $\underline{\nu}_i(\rho)$ should be included in the inequality. For instance, when $m = 2$, $\overline{\nu}_1(\rho) + \overline{\nu}_2(\rho)$, $\overline{\nu}_1(\rho) + \underline{\nu}_2(\rho)$, $\underline{\nu}_1(\rho) + \overline{\nu}_2(\rho)$ and $\underline{\nu}_1(\rho) + \underline{\nu}_2(\rho)$ should be checked individually. In other words, (11) actually represents 2^m inequalities. Additionally, solving (11) involves gridding the parameter space $\mathcal{P}$ and choosing a finite set of basis for $X(\rho)$ (see Wu *et al.* (1996) for details).

For the ducted fan, $\rho = \theta$ is chosen as the varying parameter, and the operation range $\mathcal{P} = [-\frac{\pi}{2}, \frac{\pi}{2}]$.

The bound on the rate variation on θ is set to 10, i.e., $|\dot{\theta}| \leq 10$. Both $A(\rho)$ and $B(\rho)$ are the same as in the model used for the FRE method. A set of basis functions is chosen to compute $X(\rho)$, i.e., $X(\rho) = \sum_{i=1}^{5} f_i(\rho)X_i$ where X_i's are symmetric coefficient matrices and $\{f_i(\rho)\}$ are fifth order Legendre polynomials on $\mathcal{P}$:
$\{f_i(\rho)\} = \{1, \frac{2}{\pi}\theta, (3(\frac{2}{\pi}\theta)^2 - 1)/2, (5(\frac{2}{\pi}\theta)^3 - 3(\frac{2}{\pi}\theta))/2, (35(\frac{2}{\pi}\theta)^4 - 30(\frac{2}{\pi}\theta)^2 + 3)/2\}$

4. CLF BASED CONTROL SCHEMES

4.1 *Sontag's formula*

Perhaps the most important formula for producing a stabilizing controller based on the existence of a CLF was introduced in Sontag (1989) and has come to be known as Sontag's formula. Below, we present a slight variation of Sontag's original formula:

$$u_s = \begin{cases} -\left[\dfrac{V_x f + \sqrt{M}}{V_x g g^T V_x^T}\right] g^T V_x^T & V_x g \neq 0 \\ 0 & V_x g = 0 \end{cases} \tag{12}$$

where $M = (V_x f)^2 + q(x)(V_x g g^T V_x^T)$, $V_x = \frac{\partial V}{\partial x}$. While this formula has obtained recognition for possessing many desirable properties (particularly related to the continuity of the control law Sontag (1989)), it also has an interpretation in terms of optimal control. Sontag's formula, in essence, uses the directional information given by the CLF, V, and scales it properly to solve the Hamilton-Jacobi-Bellman (HJB) equation. That is, Sontag's formula can be "derived" by assuming the control action to be of the form:

$$u = -\frac{\lambda(x)}{2} g^T V_x^T$$

and determining λ by solving the HJB equation pointwise with $\lambda(x)V_x$ substituting for the gradient of the value function. In particular, if V has level curves that agree with those of the value function, then Sontag's formula produces the optimal controller. On the other hand, when a CLF does not closely resemble the value function, poor performance may result Freeman and Primbs (1996). Results from Sontag's formula with the CLF from LPV can be found in the comparisons section.

4.2 *MPC+CLF*

This section deals with combining the CLF obtained using methods described in the previous section with a Model Predictive Control (MPC) strategy. First, however we review some results on MPC.

4.2.1. *Model Predictive Control (MPC)* In Model Predictive Control (cf. García *et al.* (1989); Mayne and Michalska (1990)), the current control at state x and time t is obtained by determining on-line the optimal control $\hat{u}$ over the interval $[t, t+T]$ respecting the following objective:

$$\inf_{u(\cdot)} \int_{t}^{t+T_h} (q(x(\tau)) + u^T(\tau)u(\tau))d\tau + \Phi(x(t+T))$$
$$s.t.\ \dot{x} = f(x) + g(x)u$$

The resulting control trajectory is implemented until a new state update occurs, usually at pre-specified sampling intervals of time T_s. Repeating these calculations yields a feedback control law. This type of implementation is commonly referred to as receding or moving horizon. As is evident from this sort of control scheme, obtaining a reduced value of the performance index is of utmost importance.

Model predictive control has traditionally been plagued by stability questions. For nonlinear systems, these have been addressed in various ways, including the use of end constraints (which require that the state be identically zero at the end of the horizon T), and by various other methods Mayne and Michalska (1990); Michalska and Mayne (1993) Unfortunately, these do not always result in satisfactory solutions when considered in the framework of practical implementation. Often such additional constraints introduce a number of problems related to the feasibility of the on-line optimization. Hence, while receding horizon control exploits the ability to compute on-line, this also introduces many difficulties ranging from stability to feasibility of the on-line optimization. Nevertheless, as computer power increases, receding horizon control and on-line computation become increasingly attractive.

4.2.2. *MPC extensions of CLF formulas* In Freeman and Kokotović (1995), it was proposed to use a CLF in conjunction with the following optimization to determine a control law:

Pointwise Min-Norm

$$minimize\ u^T u \tag{13}$$
$$subject\ to\ V_x(f + gu) \leq -\sigma(x) \tag{14}$$

where σ is some positive definite function and the optimization is solved pointwise (i.e. for each x). This formula pointwise minimizes the control energy used while requiring that V be a Lyapunov function for the closed loop system and decrease by at least σ at every point. Sontag's formula (12) results by choosing:

$$\sigma_s = \sqrt{(V_x f)^2 + q(x)(V_x g g^T V_x^T)}$$

In Primbs *et al.* (1998) the following MPC extension of pointwise min-norm controllers was proposed:

MPC+CLF

$$minimize \int_t^{t+T_h} (q(x) + u^T u)dt \quad (15)$$

$$s.t. \; V_x(f + gu(t)) \leq -\epsilon\sigma(x) \quad (16)$$

$$V(x(t+T_h)) \leq V(x_\sigma(t+T_h)) \quad (17)$$

where $1 \geq \epsilon > 0$ and x_σ represents the state trajectory from the pointwise min-norm controller with parameter $\sigma(x)$. The constraint (16) requires that V is a Lyapunov function for the closed loop system and is applied over the sampling interval $[t, t+T_s]$ in which the optimizing control trajectory is implemented. The second constraint (17) obviates the need for a terminal weight, using the information from the CLF instead.

This formulation of MPC avoids the stability problems of standard MPC by incorporating the stability properties of the CLF. This allows the horizon to be varied on-line and the optimization to be preempted before the optimizing solution is found, all without losing stability. When no horizon is present, it reduces to the pointwise min-norm problem, which also serves as a feasible trajectory for the MPC problem when a horizon is present. In order to test this hybrid strategy on the planar model of the ducted fan, we used the CLF obtained by the Quasi-LPV method described in section 3.4. Simulations were performed for two different time horizons, $T_h = 0.1$ and $T_h = 0.3$ are resulted in trajectories quite similar to the optimal trajectory. Note that as we increase the time horizon, the response improves as expected.

5. COMPARISONS

As was mentioned earlier, by choosing a large time horizon we found the optimal cost for the quadratic cost $\mathcal{J}$ from the chosen initial condition by solving a single trajectory optimization. Values of the cost function for all of the methods described in this paper are given in the following table.

A review of Table 1 leads to some interesting observations. At the top is MPC without the use

Table 1. Values of the cost function $\mathcal{J}$ using different methods for the ducted fan example.

Method	Cost
MPC w/o CLF ($T_h = 0.3, T_s = 0.05$)	unstable
Jacobian Linearization+LQR	unstable
FRE	2801
Global Linearization+LQR using LMIs	2423
Quasi-LPV	1795
CLF from LPV+Sontag	1761
CLF from LPV+MPC ($T_h = 0.1s, T_s = 0.05$)	1572
CLF from LPV+MPC($T_h = 0.3s, T_s = 0.05$)	1451
'Optimal'	1362

of a terminal weight or a CLF. Despite the fact that it exploits on-line computation to directly address the performance aspect of the optimal control problem, it lacks a stability guarantee, and in this case is actually unstable. Not so surprising a result is that JL+LQR is found to be unstable. This illustrates the true nonlinear nature of the problem and indicates that nonlinear techniques are needed. Only slightly more sophisticated than JL+LQR is the FRE technique. Even though it also lacks global stability guarantees, it is stabilizing from the chosen initial condition, although with a rather poor cost.

Next, we find that the Global Linearization techniques provide a guarantee of stability and reduce the cost even further. When even more off-line computation is thrown at the problem as in LPV techniques, the cost reduces even further. Note that in LPV and FRE techniques, a design choice is involved in the selection of a state dependent representation. Although no systematic procedure was used (nor does one exist), the results obtained here were the best of the state dependent representations that were tested.

The results from LPV on down were all qualitatively similar to the optimal trajectory. Applying Sontag's formula with the aid of the CLF from LPV results in a cost and trajectory very similar to that obtained from the standard LPV implementation. When this approach was extended to an on-line MPC implementation, a significant drop in the cost was observed. As the horizon was increased from $T_h = 0.1$ to $T_h = 0.3$, the cost steadily decreased, providing the lowest cost observed in any of the simulations.

To summarize, in general the following trends were observed. While not uniformly true, the more detailed and sophisticated techniques, which generally involve extensive off-line analysis, tended to outperform the simpler, less theoretically sound techniques. Extensive computation was also found to be extremely beneficial, especially when employed in an on-line manner, but only when used under the guidance of a solid theoretical framework. This accounts for the difference between the near optimal MPC+CLF controllers and the unstable standard MPC approach. A question of increasing importance will be whether on-line techniques such as MPC+CLF can realistically be implemented in a real time environment such as the Caltech ducted fan experiment. This will be a topic of ongoing interest as this research continues.

Note that since the dynamics of the planar fan model are not static feedback linearizable, we did not include that here. Also, there are a variety of other methods that generate a CLF such as backstepping and forwarding which were not explored because these methods require the nonlinear system to be transformed to a special structure.

6. CONCLUSIONS

The purpose of this paper was to perform a comparison between several existing nonlinear design methods on the planar model of a ducted fan. The design schemes were divided into two groups. The first dealt with different methods of generating Control Lyapunov Functions (CLFs). Despite the fact that the CLF generating methods have their standard method of generating a controller, it was shown via simulation that if the CLF methodology is combined with an MPC framework, the hybrid method works better than any of the individual ones. Simulation results were performed for a specific initial condition which incorporated the true nonlinear nature of the problem. Further research in this direction is under progress, and our goal is to do a similar comparison with more sophisticated models and eventually with the Caltech ducted fan experiment.

References

Artstein, Z. (1983). Stabilization with relaxed controls. *Nonlinear Anal.* **7**(11), 1163–1173.

Bodenheimer, B., P. Bendotti and M. Kantner (1996). Linear parameter-varying control of a ducted fan engine. *Int.J.Robust and Nonlinear Control* **6**, 1023–1044.

Boyd, S., L. El Ghaoui, E. Feron and V. Balakrishnan (1994). *Linear Matrix Inequalities in System and Control Theory*. Vol. 15 of *Studies in Applied Mathematics*. SIAM. Philadelphia, PA.

Cloutier, J.R., C.N. D'Souza and C.P. Mracek (1996). Nonlinear regulation and nonlinear H_∞ control via the state-dependent riccati equation technique. In: *Proc. 1st Internat. Conf. on Nonl. Problems in Aviation and Aeraspace.* Daytona Beach, FL.

Doyle, J. C., J. Primbs, B. Shapiro and V. Nevistić (1996). Nonlinear games: examples and counterexamples. In: *Proceedings of the* 35th *IEEE Conference on Decision and Control* . Kobe, Japan. pp. 3915–3920.

Freeman, R. and J. Primbs (1996). Control lyapunov functions: New ideas from an old source. In: *Proceedings of the* 35th *IEEE Conference on Decision and Control* . Kobe, Japan. pp. 3926–3931.

Freeman, R. and P. Kokotović (1995). Optimal nonlinear controllers for feedback linearizable systems. In: *Proceedings of the 1995 American Control Conference.* Seattle, Washington. pp. 2722–2726.

García, C. E., D. M. Prett and M. Morari (1989). Model predictive control: Theory and practice - A survey. *Automatica* **25**(3), 335–348.

Kantner, M., B. Bodenheimer, P.Bendotti and R. M. Murray (1995). An experimental comparison of controllers for a vectored thrust,ducted fan engine. In: *Proceedings of the 1995 American Control Conference.* Seattle. pp. 1956–1961.

Mayne, D. Q. and H. Michalska (1990). Receding horizon control of nonlinear systems. *IEEE Trans. Aut. Control* **35**(7), 814–824.

Michalska, H. and D.Q. Mayne (1993). Robust receding horizon control of constrained nonlinear systems. *IEEE Trans. Aut. Control* **38**(11), 1623–1633.

Murray, R.M. (1998). Modeling of the caltech ducted fan, class notes for cds 111. Technical report. California Institute of Technology. Control and Dynamical Systems 107–81, Pasadena, CA 91125.

Nieuwstadt, M.van and R.M.Murray (1996). Real time trjectory generation for differentially flat systems. In: *Proceedings of the 1996 IFAC World Congress.*

Primbs, J. A., V. Nevistić and John C. Doyle (1998). On Control Lyapunov Functions and Receding Horizon Control: Part II a receding horizon extension of pointwise min-norm controllers. *Submitted to IEEE TAC.*

Sontag, E.D. (1983). A lyapunov-like characterization of asymptotic controllability. *SIAM J. Contr. Optimiz.* **21**(3), 462–471.

Sontag, E.D. (1989). A 'universal' construction of Artstein's theorem on nonlinear stabilization. *Syst. Contr. Lett.* **13**(2), 117–123.

Wu, Fen, Hua Xin Yang, Andy Packard and Greg Becker (1996). Induced L_2-Norm control for lpv systems with bounded parameter variation. *Int. Journal of Nolinear and Robust Control* **vol.6**, 983–998.

ADAPTIVE OUTPUT FEEDBACK STABILIZATION OF A CLASS OF UNCERTAIN CONTINUOUS STIRRED TANK REACTORS

R. Antonelli and A. Astolfi

Centre for Process Systems Engineering,
Imperial College of Science, Technology and Medicine,
London SW7 2BY, UK
FAX:+44-171-5946606, E-mail:{r.antonelli,a.astolfi}@ic.ac.uk

Abstract: The problem of global temperature control for a general class of chemical reactors is addressed and solved using Lyapunov techniques. The main results of the paper show that, under reasonable and easy to test assumptions, global temperature control can be achieved with a bounded output feedback control law and with an adaptive output feedback control law. The control bounds can be very general and tight, and the only condition to be fulfilled is a simple and natural feasibility condition, namely that the steady state control is admissible. The theory is illustrated on a simple simulation example.

Keywords: Chemical Reactors, Lyapunov Function, Global Asymptotic Stability, Temperature Regulation.

1. INTRODUCTION

Feedback stabilization of chemical reactors represents a challenging control problem because of the inherent nonlinearity of the process. During the past years many researchers have widely studied the behavior of the Continuous Stirred Tank Reactors (CSTRs) and analyzed the problem of temperature and concentration regulation for this class of process. Departing from the early work of Aris and Amundson (1958), which determined the stability of a reactor using the linearization method and analyzed the possibility of control by various means, important studies have been made by Alvarez and Gonzales (1989). Their approach consisted of using nonlinear geometric control tools to stabilize unstable steady states of a CSTR by analyzing the performance of proportional and proportional-integral controls. Recently, Alvarez *et al.* (1991) studied the behavior of nonlinear controllers for the specific case of an exothermic, first order, irreversible reactor. Global boundedness of trajectories and practical stabilization of a chemical reactor driven by an input-output linearizing type feedback were presented by Alvarez-Ramirez (1994). Finally Alvarez-Ramirez *et al.* (1996) proved that a CSTR, in the presence of reaction rate uncertainty, can be stabilized by applying a high-gain feedback law. Viel *et al.* (1997*b*) applied a state feedback controller to stabilize the temperature of a chemical reactor in spite of uncertainties on the reaction kinetics and proposed a robust state observer to overcome the problem of partial state measurement. Then Viel *et al.* (1997*a*) extended these results to a general class of exothermic reactor. A different approach is shown by Fradkov *et al.* (1997) who proposed a semi-adaptive state feedback controller to overcome the problem of parameter uncertainty.

It must be noted that most of the nonlinear controller techniques assume that all state variables are measured or estimated. In industrial practice, the estimation of unmeasured state variables and unknown parameters is not a trivial problem. Although advanced methods for accurate measurement of the concentrations in a chemical reactor are available, these have rarely been used in industrial plants because of very high operating costs. The problem of state variable measurement or estimation can be overcome by introducing an output feedback controller able to regulate and globally stabilize the reactor temperature at its set point without requiring any information about concentrations.

This paper is organized as follows. In Section 2 a general class of CSTRs is proposed and analyzed. Since this model takes into account the jacket dynamics, it is more complex than that examined by Viel *et al.* (1997*b*) and better represents the real process. Then, in Section 3, some important assumptions that will be used in demonstrating the main theorems are presented. A bounded output feedback control law and an adaptive control law are proposed in Section 4 and Section 5, respectively. Finally, Section 6 presents a specific reactor example and numerical simulations to illustrate the performance of the bounded controllers.

2. CHEMICAL REACTOR DYNAMICS

Consider a CSTR in which m reactions take place between n chemical species and where a jacket is used to supply or remove heat from the system. The heat provision or removing process is regulated by manipulation of the heating or cooling fluid flow rate. Reactor and chemical reactions are modeled using the following assumptions.

- The reactor and jacket are perfectly stirred.
- The reactor and jacket volumes are constant.
- The kinetics are "separable" kinetics, *i.e.* $r(x,T) = k(T)\phi(x)$.
- The thermal exchanges between the reactor and the jacket are expressed by constant heat transfer parameters q_i and the amount of heat retained in the reactor walls is considered negligible.
- The reaction enthalpies ΔH_r are supposed constant.

Under the above assumptions, the mass and energy balances applied to the jacket reactor give the following mathematical model (see *e.g.* Gibon-Fargeot *et al.* (1994)):

$$\begin{cases} \dot{x} = Cr(x,T) + d(x^{in} - x) \\ \dot{T} = B(x,T) + q_1(T_j - T) + d(T^{in} - T) \\ \dot{T}_j = -q_2(T_j - T) + u \end{cases} \quad (1)$$

where:

- $x = \text{col}(x_r, x_p) \in R^r \times R^p$ is the vector of the $r+p = n$ concentrations of reactants and products;
- T is the reactor temperature;
- T_j is the jacket temperature;
- $r(x,T) = k(T)\varphi(x) \in R^m$ is the vector of reaction kinetics. The vector $k(T) \in R^m$ is composed of positive bounded functions of the temperature and the functions $(\varphi_1(x), \cdots, \varphi_m(x)) \in R^m$ are non-negative functions of the reactants concentrations;
- $B(x,T) = br(x,T) \in R$ where $b^T \in R^m$ are coefficients that are positive constants if the reaction is exothermic and negative if the reaction is endothermic;
- $C \in R^{n \times m}$ is the matrix of the stoichiometric coefficients;
- $x^{in} \in R^n$ is the vector of non-negative and constant feed concentrations ;
- $d \in R$ is the dilution rate;
- $q_1 \in R$ is the side reactor heat transfer parameter;
- $q_2 \in R$ is the side jacket heat transfer parameter ;
- T^{in} is the stream feed temperature;
- u is the control input.

The above variables can be expressed in terms of the physical parameters listed in the Appendix.

3. CONTROL PROBLEM FORMULATION

The control objective is to guarantee that, under certain general assumptions, the reactor and jacket temperatures can be stabilized at a given set point (T^*, T_j^*) by means of an output feedback control law that feeds back only the temperature values. It makes sense to consider as equilibrium point only the temperature set as the concentrations achieve their set point when the temperature is at its steady state. The following assumptions are considered.

A1) For $T = T^*$ the system

$$\dot{x} = Cr(x,T^*) + d(x^{in} - x)$$

has a single equilibrium point $x = x^*$, which is globally asymptotically stable (GAS).

A2) There exist a positive definite and proper function $W^*(x - x^*)$ and two positive definite matrices P and Q such that

$$\dot{W}^* = \left[\frac{\partial W^*}{\partial x}\right]^T \left(Cr(x,T^*) + d(x^{in} - x)\right) \leq -\underline{\lambda}(P)\phi^2(x - x^*) \leq 0$$

and

$$\left\|\frac{\partial W^*}{\partial x}\right\|^2 \leq \overline{\lambda}^2(Q)\phi^2(x - x^*)$$

with $\underline{\lambda}(P)$ and $\overline{\lambda}(Q)$ the minimum and the maximum eigenvalue of matrix P and Q.

A3) The functions $k(T) \in R^m$ satisfy the Lipschitz condition at T^*. This implies that

$$||Cr(x,T) - Cr(x,T^*)|| \le M|T - T^*|.$$

Let x^*, T^*, T_j^*, u^* be the steady state values of the state and input variables respectively. Defining the "deviation" variables $\psi = x - x^*$, $\zeta = T - T^*$, $\eta = T_j - T_j^*$ and $\delta = u - u^*$, the system (1) can be written as

$$\begin{cases} \dot{\psi} = G(\zeta,\psi)\psi + Ck(\zeta + T^*)\varphi(x^*) + d(x^{in} - x^*) \\ \dot{\zeta} = -(q_1 + d)\zeta + q_1\eta + p_1 \\ \dot{\eta} = q_2\zeta - q_2\eta + \delta u \end{cases}$$

where

$$\delta u = k_1(\cdot)\zeta + k_2(\cdot)\eta$$

$$G(\zeta,\psi) = \Big(Ck(\zeta + T^*)\tilde{\varphi}(\psi, x^*) - d\Big)$$

$$p_1 = q_1(T_j^* - T^*) + d(T^{in} - T^*) + B(x,T)$$

$$B(x,T) = B_0 + B_1\psi + B_2\zeta$$

with

$$B_0 = B_0(x^*,T^*) = \sum_{i=1}^{m} k_{0i}b_i\varphi_i(x_i^*)\exp\left(\frac{-\lambda_{0i}}{T^*}\right),$$

$$B_1 = \left[k_{0i}b_i \exp\left[\frac{-\lambda_{0i}}{T^*}\right]\left[\frac{\varphi_i(x_i) - \varphi_i(x_i^*)}{x_i - x_i^*}\right], \dots\right.$$

$$\left. \dots, k_{0m}b_m \exp\left[\frac{-\lambda_{0m}}{T^*}\right]\left[\frac{\varphi_m(x_m) - \varphi_m(x_m^*)}{x_m - x_m^*}\right]\right],$$

$$B_2 = \sum_{i=1}^{m} k_{0i}b_i\varphi_i(x_i)\left[\frac{\exp(\frac{-\lambda_{0i}}{T}) - \exp(\frac{-\lambda_{0i}}{T^*})}{T - T^*}\right].$$

Finally, $\tilde{\varphi}(\psi, x^*)$ is a computable function and $k_1(\cdot)$ and $k_2(\cdot)$ are functions to be determined.

Remark 1. In general, the structure of the function $\varphi_i(x_i)$ differs from reaction to reaction and is proportional to the product of the concentrations of the reactants, where the concentration of each reactant is raised to the power given by the stoichiometric coefficient of that component.

4. CONTROL DESIGN 1

In this section an output feedback control law that feeds back T and T_j and relies on the knowledge of the temperatures set point (T^*, T_j^*), the control set point u^* and a lower bound $\underline{q_2}$ on the thermal coefficient q_2 is presented.

Theorem 1. *Consider the CSTR described by equations (1). Assume that* ***A1, A2, A3*** *hold. Assume moreover that*

B1) *$u \in [\underline{u}, \overline{u}]$ where $\underline{u}$ and $\overline{u}$ are the lower and the upper bound on the input u, respectively;*

B2) *the set point values (T^*, T_j^*, u^*) and a lower bound $\underline{q_2}$ on the thermal coefficient q_2 are known;*

B3) *the set point control u^* is admissible, i.e. $u^* \in [\underline{u}, \overline{u}]$;*

B4) *B_1 and B_2 are such that*

$$B_1\frac{\psi}{\phi} < \frac{\underline{\lambda}(P)}{M\overline{\lambda}(Q)}(d - |B_2|_{max}) \qquad (2)$$

$$B_2 < d. \qquad (3)$$

Moreover let

$$u = u^* + k_1(\cdot)\zeta + k_2(\cdot)\eta. \qquad (4)$$

Then, for any $k_1(\cdot)$ and $k_2(\cdot)$ such that $u \in [\underline{u}, \overline{u}]$ and

$$k_2(\cdot) < \underline{q}_2 \qquad (5)$$

$$\Big(k_1(\cdot) + k_2(\cdot)\Big) + \frac{k_1^2(\cdot)}{4\underline{q}_2} \le 0 \qquad (6)$$

the closed loop (1)-(4) is GAS at the equilibrium point (x^*, T^*, T_j^*).

Proof 1. By assumption **A2** and **A3** , the following relations hold

$$\dot{W}^*(x) = \left[\frac{\partial W^*}{\partial x}\right]^T \Big[Cr(x,T) + d(x^{in} - x)\Big]$$

$$\le \left[\frac{\partial W^*}{\partial x}\right]^T \Big[Cr(x,T^*) + d(x^{in} - x)\Big]$$

$$+ \left\|\frac{\partial W^*}{\partial x}\right\| \cdot ||Cr(x,T) - Cr(x,T^*)||.$$

$$\dot{W}^*(x) \le -\alpha_1\phi^2(x - x^*) + \alpha_2\phi(x - x^*)|T - T^*|.$$

Consider the positive definite and proper function

$$W^*(x) + \lambda V(\zeta, \eta).$$

where λ is a positive real number and

$$V(\zeta,\eta) = q_2\zeta^2 + q_1\eta^2.$$

After some manipulation, the following relation holds

$$\dot{W}^* + \lambda\dot{V} \le -\left[\sqrt{\alpha_1}\phi - \frac{\|\pi(\psi,\phi)\|^2}{2\sqrt{\alpha_1}}\zeta\right]^2$$

$$- \left[\sqrt{2\lambda q_1(q_2 - k_2)}\eta - \frac{\lambda q_1(2q_2 + k_1)}{\sqrt{2\lambda q_1(q_2 - k_2)}}\zeta\right]^2$$

$$-\lambda\left[2q_2(q_1+d-B_2) - \frac{\|\pi(\psi,\phi)\|^2}{4\alpha_1\lambda} - \frac{q_1(2q_2+k_1)^2}{2(q_2 - k_2)}\right]\zeta^2$$

with $\pi(\psi,\phi) = \alpha_2 + 2\lambda q_2 B_1\psi/\phi$. The first two terms are negative for any non-zero value of the variables ϕ, ζ and η, while the third is indefinite. However, let λ be defined as

$$\lambda = \lambda^* = \frac{\alpha_2}{2q_2}\left|\frac{\phi}{B_1\psi}\right|_{max}.$$

Then, it is possible to show that, for any non-zero value of ϕ, ζ and η, and for any choice of k_1 and k_2 fulfilling the conditions (5)-(6),

$$\dot{W}^* + \lambda^*\dot{V} < 0.$$

As a result, the closed loop system (1)-(4) is GAS at the equilibrium point (x^*, T^*, T_j^*).

Remark 2. It is worth noting that all considerations made above are valid both in the exothermic and endothermic case. Our control structure is capable of performing the global temperature stabilization of a CSTR whatever type of reaction takes place, provided that the conditions (2) and (3) are satisfied.

5. CONTROL DESIGN 2

In this section a new output feedback control law is proposed that feeds back T_j and requires only the knowledge of the jacket temperature set point T_j^*.

Theorem 2. *Consider the CSTR described by equations (1). Assume that **A1**, **A2**, **A3** hold. Assume moreover that*

C1) *$u \in [\underline{u}, \overline{u}]$;*
C2) *the set point value T_j^* is known;*
C3) *$u^* \in [\underline{u}, \overline{u}]$;*
C4) *B_1 and B_2 are such that the conditions (2) and (3) are satisfied.*

Then the dynamic output feedback control law

$$\begin{cases} u = \theta \\ \dot{\theta} = -\dfrac{(T_j^* - T_j)(\theta - \overline{u})(\theta - \underline{u})}{(\overline{u} - \underline{u})} \end{cases} \tag{7}$$

with $\theta(0) \in (\underline{u}, \overline{u})$ is such that:

1) *(x, T, T_j) converge to the steady state values;*
2) *θ is bounded.*

Proof 2. Consider the positive definite and proper function

$$W^*(x) + \lambda^* V(\zeta, \eta) + \omega(\theta) \tag{8}$$

where

$$\omega(\theta) = 2\lambda q_1 \omega^*(\theta) = 2\lambda q_1 \log\left|\frac{(\theta - \overline{u})^\tau}{(\theta - \underline{u})^{\tau+1}}\right|$$

with $\tau = \dfrac{u^* - \overline{u}}{\overline{u} - \underline{u}} < 0$. Observe that

$$\frac{\partial \omega^*}{\partial \theta}(\theta - \overline{u})(\theta - \underline{u}) + \theta = u^*.$$

Then, it is possible to show that, for any non-zero value of ϕ, ζ and η, and for any choice of k_1 and k_2 fulfilling the conditions (5)-(6),

$$\dot{W}^* + \lambda^*\dot{V} + \dot{\omega} < 0.$$

As a result, the closed loop system (1)-(7) is GAS at the equilibrium point (x^*, T^*, T_j^*).

6. APPLICATION EXAMPLE

To illustrate the above results, a classical example of continuous stirred jacket reactor, in which the first order, irreversible and endothermic, reaction $A \rightarrow B$ takes place, is presented. Let λ_0 be the activation energy and k_0 be a physical constant. The chemical reaction rate is assumed to follow an Arrhenius type kinetic

$$k(T) = k_0 \exp\left(-\frac{\lambda_0}{T}\right)$$

and the reactor is modeled in accordance with the Hypotheses ***Hp1*** to ***Hp5***. In this special case the system (1) is

$$\begin{cases} \dot{x}_A = -k(T)x_A + d(x_A^{in} - x_A) \\ \dot{x}_B = k(T)x_A - dx_B \\ \dot{T} = bk(T)x_A + q_1(T_j - T) + d(T^{in} - T) \\ \dot{T}_j = -q_2(T_j - T) + u \end{cases} \tag{9}$$

and simple considerations show that Assumptions **A1** to **A3** hold.

Case 1. (Non adaptive control) Consider the bounded output feedback control law

$$u = \overline{m}_1 + \overline{m}_2 \tag{10}$$

with

$$\overline{m}_1 = \max\Big(0, \beta_1 \tanh(T^* - T)\Big)$$

$$\overline{m}_2 = \max\Big(\varepsilon_2, \beta_2 \tanh\big(T^* + \alpha(T_j^* - T^*) - T_j\big)\Big)$$

$\beta_1 > 0$, $\beta_2 > u^*$, $0 < \varepsilon_2 \leq u^*$ and

$$\alpha = 1 + \frac{1}{(T_j^* - T^*)}\text{atanh}\left(\frac{u^*}{\beta_2}\right).$$

The control law (10) is constrained and its lower and upper bounds are, respectively

$$\underline{u} = \varepsilon_2 \leq u \leq (\beta_1 + \beta_2) = \overline{u}. \tag{11}$$

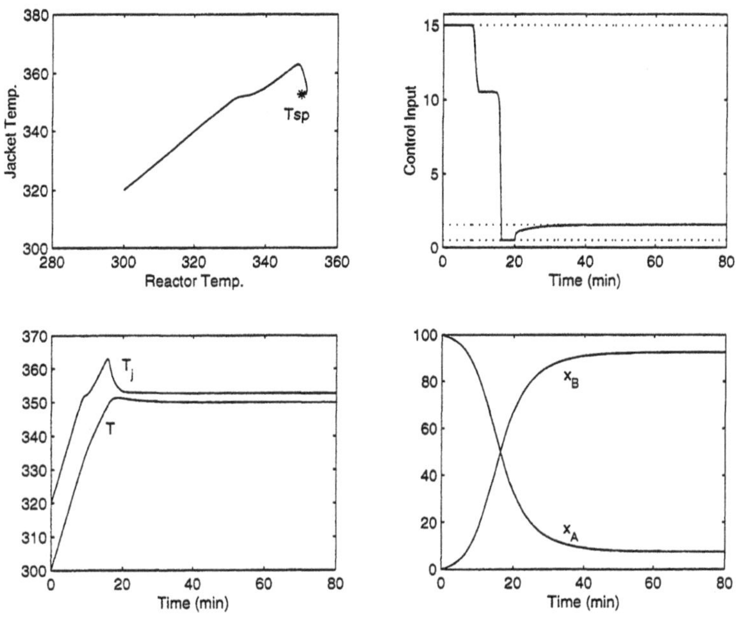

Fig. 1. Simulation results for an endothermic reactor with the bounded output feedback control law (10).

Simple calculations lead to the form

$$\delta u = u - u^* = k_1(\cdot)(T - T^*) + k_2(\cdot)(T_j - T_j^*)$$

with $k_1(\cdot) = \dfrac{\overline{m}_1}{T - T^*}$ and $k_2(\cdot) = \dfrac{\overline{m}_2 - u^*}{T_j - T_j^*}$.

For any values of β_1, β_2 and ε_2 satisfying the relations (5) and (6), the reactor described by equations (9) is stabilizable by means of the bounded output feedback control law (10) that does not require the estimation of the concentrations of the chemical species, but only the knowledge of the following information

- set point reactor temperature T^*;
- set point jacket temperature T_j^*;
- set point control input u^*;
- a lower bound on the thermal coefficient q_2.

Case 2. (Adaptive control) Consider the adaptive control law

$$\begin{cases} u = \theta \\ \dot{\theta} = -\dfrac{(T_j^* - T_j)\Big(\theta - (\beta_1 + \beta_2)\Big)(\theta - \varepsilon_2)}{\Big((\beta_1 + \beta_2) - \varepsilon_2\Big)} \end{cases} \tag{12}$$

with $\theta(0) \in (\varepsilon_2, \beta_1 + \beta_2)$ and β_1, β_2, ε_2 as in equation (10). Notice that the only knowledge of the jacket temperature set point T_j^* is required to stabilize the reactor represented by system (9).

The following initial conditions and parameter values were used in the simulation example for both cases: $x_A(0) = 100\ mol \cdot m^{-3}$, $x_B(0) = 0$, $T(0) = 300K$, $T_j(0) = 320K$, $\theta(0) = 13.5K \cdot min^{-1}$, $k_0 = 1.3677 \times 10^8\ min^{-1}$, $\lambda_0 = 7290\ K$, $d = 0.01\ min^{-1}$, $x_A^{in} = 100\ mol \cdot m^{-3}$, $T^{in} = 300K$, $b = -0.01196\ K \cdot m^3 \cdot mol^{-1}$, $q_1 = 0.191\ min^{-1}$, $q_2 = 0.574\ min^{-1}$, $T^* = 350\ K$, $T_j^* = 352.6\ K$, $x_A^* = 7.5\ mol \cdot m^{-3}$, $\beta_1 = 10$, $\beta_2 = 5$, $\varepsilon_2 = 0.5$. Figures 1 and 2 show the jacket and reactor temperatures of the controlled system, as well as the control action and the trajectory of the concentrations, when the bounded output feedback law in equation (10) and the adaptive control law in equation (12) are applied respectively.

7. CONCLUSION

The difficulty in the measurement or the estimation of all state variables motivated the search for output feedback control laws, *i.e.* feedback control laws which generate the control signals on the basis of the temperature measurements only. In the present work a general class of CSTRs has been studied and the problem of global temperature stabilization for both endothermic and exothermic reactions has been dealt with. The dynamic model proposed for this process is more complex than the one studied in previous papers because of the inclusion of the equation describing the jacket dynamics (see Viel *et al.* (1997*b*) and Viel *et al.* (1997*a*)). Consequently, the robust stabilization by output feedback of a general class of reactors has been achieved. First, an output feedback control law relying on the knowledge of the temperatures set point and control set point generally available is presented. Then, an adaptive control law based only on the jacket temperature set point value is examined. The global stability of both closed loop systems has been proved despite a partial lack of knowledge of the kinetic parameters. Extensions of the presented methods are directed

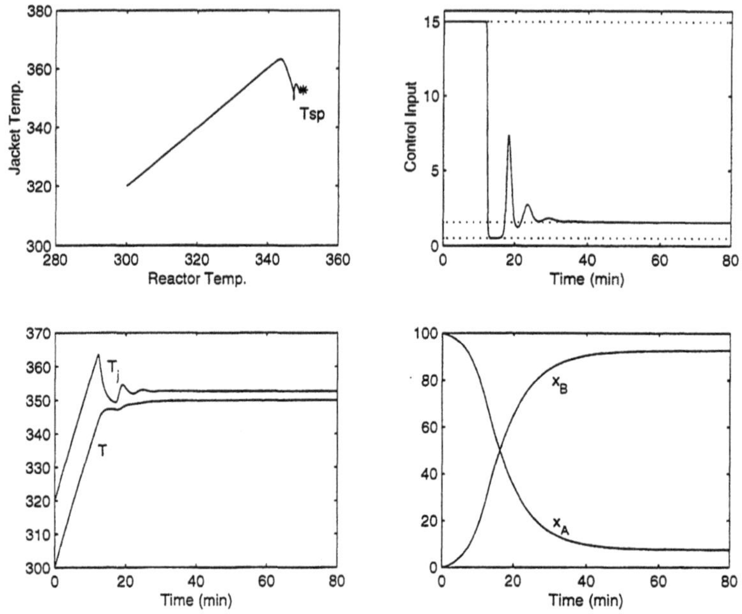

Fig. 2. Simulation results for an endothermic reactor with the adaptive control law (12).

toward a larger class of systems, including reactors characterized by recycle streams.

8. REFERENCES

Alvarez, J. and E. Gonzales (1989). Global nonlinear control of a continuous stirred reactor. *Chemical Engineering Science* **44**(5), 1147–1160.

Alvarez, J., J. Alvarez and R. Suarez (1991). Nonlinear bounded control for a class of continuous agitated tank reactors. *Chemical Engineering Science* **46**(12), 3235–3249.

Alvarez-Ramirez, J. (1994). Stability of a class of uncertain continuous stirred chemical reactors with a nonlinear feedback. *Chemical Engineering Science* **49**(11), 1743–1748.

Alvarez-Ramirez, J., R. Suarez and R. Femat (1996). Control of continuous-stirred tank reactors: stabilization with unknown reaction rates. *Chemical Engineering Science* **51**(17), 4183–4188.

Aris, R. and N. Amundson (1958). An analysis of chemical reactor stability and control. *Chemical Engineering Science* **7**(8), 121–131.

Fradkov, A., R. Ortega and G. Bastin (1997). Semi-adaptive control of convexly parametrized systems with application to temperature regulation of chemical reactors. *European Control Conference.*

Gibon-Fargeot, A.M., H. Hammouri and F. Celle (1994). Nonlinear observers for chemical reactors. *Chemical Engineering Science* **49**(14), 2287–2300.

Viel, F., F. Jadot and G. Bastin (1997*a*). Global stabilization of exothermic chemical reactor under input constraints. *Automatica* **33**(8), 1437–1448.

Viel, F., F. Jadot and G. Bastin (1997*b*). Robust feedback stabilization of chemical reactors. *IEEE Trans. Autom. Control* **42**(4), 473–481.

9. APPENDIX

A_t	heat exchange area, m^2;
V_1	volume of the reacting mixture in the tank, m^3
V_2	volume of the double shell reactor, m^3
C_1	reactor contents heat capacity, $kJ \cdot kg^{-1} \cdot K^{-1}$
C_2	heat transfer fluid heat capacity, $kJ \cdot kg^{-1} \cdot K^{-1}$
ρ_1	reactor contents density, $kg \cdot m^{-3}$
ρ_2	heat transfer fluid density, $kg \cdot m^{-3}$
Q	volumetric flow rate of the feed stream $m^3 \cdot min^{-1}$
U	global heat exchange , $kJ \cdot m^{-2} \cdot K^{-1} \cdot min^{-1}$
ΔH_{ri}	heat of the ith reaction at temperature T, $kJ \cdot mol^{-1}$
$d = Q/V_1$	dilution rate, min^{-1}
$q_1 = \dfrac{UA_t}{\rho_1 C_1 V_1}$	side reactor heat transfer coefficient
$q_2 = \dfrac{UA_t}{\rho_2 C_2 V_2}$	side jacket heat transfer coefficient
$b_i = \dfrac{(-\Delta H_{ri})}{\rho_1 C_1}$	heat of the ith reaction.

SWINGING UP OF THE SPHERICAL PENDULUM

A.S. Shiriaev* H. Ludvigsen* O. Egeland*

* *Department of Engineering Cybernetics*
Norwegian University of Science and Technology
N-7034 Trondheim, NORWAY.
Anton.Shiriaev@itk.ntnu.no
Halgeir.Ludvigsen@itk.ntnu.no
Olav.Egeland@itk.ntnu.no

Abstract: The stabilization problem for the upright position of the controlled spherical pendulum is solved in the sense that all solutions tend to the upright position. The suggested stabilizing regulator is a passive controller derived using the speed-gradient method in combination with a switching conditions. It is shown that a straightforward energy-based passive controller of the speed-gradient type does not even solve the stabilization problem for the compact invariant set containing the upright position.

Keywords: passivity, V-detectability, spherical pendulum, attractability of hyperbolic equilibrium, control of first integrals

1. INTRODUCTION

Passive nonlinear systems have attracted much attention from the nonlinear control community, and such interest has been motivated by both theoretical and practical reasons. Passive nonlinear systems are encountered in many applications, for instance, mechanical and electro-mechanical systems. At present, due to the efforts of many authors, in particular, due to Byrnes *et al.* (1991), the complete solution of the stabilization problem for passive nonlinear systems with a *unique* equilibrium point stabilization problem has been obtained.

Another important problem concerning the passive nonlinear systems is the stabilization of a prescribed invariant set, which is possibly noncompact. Some results in this direction can be found in (Shiriaev and Fradkov, 1998; Shiriaev, 1998*a*; Shiriaev, 1998*c*).

In this paper we restrict our attention to a detailed analysis of the controlled spherical pendulum motions with the objective to render the upper equilibrium point to be an *unique* attractive point of the closed loop system (and further to stabilize this point). The controlled spherical pendulum is a passive nonlinear system with two equilibrium points. In these points, in contrast to the equilibriums of the pendulum in a plane, the vector field of the unforced spherical pendulum does not have any linear approximations. The last property does not depend on the parametrization but originates from the conservation of the angular momentum around the vertical axis.

The unforced spherical pendulum has two independent first integrals (the total energy H_0 and the generalized moment p_0, corresponding to rotation in the horizontal plane) and is a completely integrable nonlinear system. The upright position belongs to the invariant subset of the phase space defined by $H_0 = 2mgl$ and $p_0 = 0$, where m, l are mass and length of the pendulum; g is the acceleration of gravity. By proposition 1 this subset is a stable manifold of the upright position. We therefore propose to use the energy-based speed-gradient algorithm (Fradkov, 1996; Åström and Furuta, 1996; Fradkov *et al.*, 1997; Shiriaev, 1998*c*) to stabilize a compact subset of the cylindrical state space defined by $H_0 = 2mgl$ and $p_0 = 0$.

As it is shown in theorem 6 the straightforward passive controller based on energy considerations may not provide the stabilizability of the prescribed goal set. The solution of the closed loop system may tend to a cycle, which does not belong to the goal set. Even if the closed loop system solution tends to the stable manifold of the upright position this feedback control does not provide the attractability of this point and the closed loop system solution may perform infinite rotations. This reflects the fact that in general the upright position will be non-unique ω-limit point of the closed-loop system.

It turned out, and this is the main contribution of the paper, that a simple modification of the speed-gradient-energy method by introducing switching leads to global attractability of the upright position of the spherical pendulum.

The paper is organized as follows. Section 2 contains the problem statement and the important auxiliary results concerning the detailed analysis of the unforced spherical pendulum motions. The main results of the paper is presented in section 3. The computer simulations are collected in section 4 and conclusions in section 5.

2. PRELIMINARIES AND PROBLEM STATEMENT

The motions of the controlled spherical pendulum are described by the Euler-Lagrange equations

$$\frac{d}{dt}\frac{\partial \mathcal{L}(q,\dot{q})}{\partial \dot{q}_1} - \frac{\partial \mathcal{L}(q,\dot{q})}{\partial q_1} = g_1(q,\dot{q})u_1, \quad (1)$$

$$\frac{d}{dt}\frac{\partial \mathcal{L}(q,\dot{q})}{\partial \dot{q}_2} = g_2(q,\dot{q})u_2, \quad (2)$$

where $q = (q_1, q_2)$, $\dot{q} = (\dot{q}_1, \dot{q}_2)$ are the coordinates and velocities of the spherical pendulum; $\mathcal{L}(q,\dot{q})$ is the Lagrangian function

$$\mathcal{L}(q,\dot{q}) = \frac{1}{2}ml^2\left[(\dot{q}_1)^2 + (\dot{q}_2)^2\sin^2 q_1\right] - \\ -mgl(1-\cos q_1); \quad (3)$$

m, l are mass and length of the pendulum, g is the acceleration of gravity; $g_1(q,\dot{q})$, $g_2(q,\dot{q})$ are scalar smooth functions; u_1, u_2 are control functions; see figure 1.

We will understand stabilization of the set S as rendering S attractive, i. e. providing that all trajectories of closed loop system tend to S as t tends to $+\infty$. This paper is devoted to the problem: *define a state feedback regulator which stabilizes the upright position of the spherical pendulum*, more precisely, rendering the upright position as

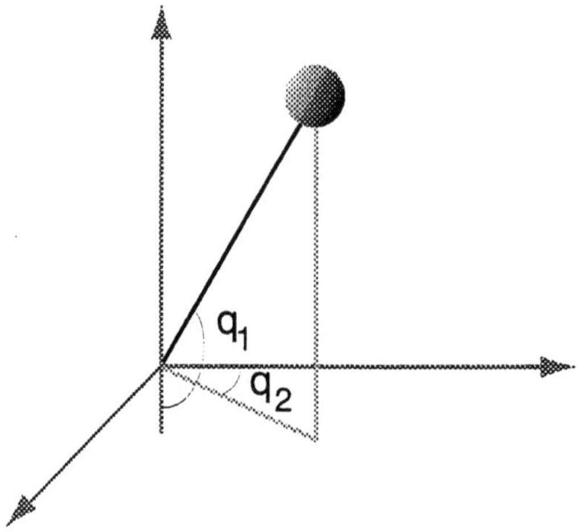

Fig. 1. The spherical pendulum

an unique attractive point for the closed loop system.

It is convenient to choose as a state space for the spherical pendulum the cylindrical space with a sphere of radius l in the base, i. e. $0 \le q_1 \le \pi$, $0 \le q_2 < 2\pi$. The unforced system (1), (2) is completely integrable. Indeed, the functions

$$H_0(q,\dot{q}) = \frac{1}{2}ml^2\left[(\dot{q}_1)^2 + (\dot{q}_2)^2\sin^2 q_1\right] + \\ +mgl(1-\cos q_1) \quad (4)$$

$$p_0(q,\dot{q}) = ml^2(\dot{q}_2)\sin^2 q_1 \quad (5)$$

are the independent first integrals of the unforced spherical pendulum with $H_0(q,\dot{q}) \ge 0$, $p_0(q,\dot{q}) \in \mathbf{R}^1$. Take the constants $H^* \ge 0$, $p^* \in \mathbf{R}^1$ and introduce the scalar function V and the dummy output $y = (y_1, y_2)^T$ of the system (1), (2) as follows

$$V(q,\dot{q}) = \frac{r_1}{2}\left[H_0(q,\dot{q}) - H^*\right]^2 + \frac{r_2}{2}[p_0 - p^*]^2 . \quad (6)$$

$$y_1 = r_1 \cdot \dot{q}_1 \cdot g_1(q,\dot{q}) \cdot [H_0(q,\dot{q}) - H^*], \\ y_2 = g_2(q,\dot{q}) \cdot (r_1 \cdot \dot{q}_2 \cdot [H_0(q,\dot{q}) - H^*] + \\ +r_2 \cdot [p_0(q,\dot{q}) - p^*]), \quad (7)$$

where r_1, r_2 are some positive constants. The functions V, y possess the important property: the controlled spherical pendulum (1), (2) with output function y is a passive system with nonnegative storage function V, see (Willems, 1972). In particular, the derivative of the function V along the solution of the system (1), (2) takes the form

$$\frac{d}{dt}V(q,\dot{q}) = y_1u_1 + y_2u_2.$$

The solution of the upright position stabilization problem will take advantage of the passivity, and the feedback regulator will be formed by the modified speed-gradient algorithm, see (Fradkov, 1990).

The analysis of the qualitative behaviour of the closed loop system solutions with the feedback regulator formed by the speed-gradient algorithm is essentially based on the detailed investigation of the motions of the unforced spherical pendulum. The rest of this section is devoted to such important auxiliary results.

Proposition 1. Denote Λ_p as the maximal set of the trajectories $(q, \dot{q})$ of the unforced spherical pendulum such that the point p of the cylindrical phase space is an ω-limit point of any motion from Λ_p. Then

1). the set $\Lambda_{(0,0,0,0)} = \{(0,0,0,0)\}$, i. e. there is no other trajectory except the point $(0,0,0,0)$, for which this equilibrium is its ω-limit point;

2). the set $\Lambda_{(\pi,0,0,0)}$ has the form

$$\Lambda_{(\pi,0,0,0)} = \left\{ (\hat{q}_1(t), q_2, \dot{\hat{q}}_1(t), 0) \right\},$$

where q_2 is any constant from $[0, 2\pi)$; the pair $(\hat{q}_1(t), \dot{\hat{q}}_1(t))$ is a solution of the equation

$$ml^2 \cdot \ddot{\hat{q}}_1(t) = -mgl \cdot \sin \hat{q}_1(t)$$

subjected to the constraint

$$\frac{1}{2} ml^2 (\dot{\hat{q}}_1(t))^2 + mgl(1 - \cos \hat{q}_1(t)) = 2mgl, \quad \forall t \geq 0.$$

Moreover, the upright position is unique ω-limit point of any trajectory from $\Lambda_{(\pi,0,0,0)}$. □

Consider the motion $(q, \dot{q})$ of the unforced spherical pendulum with $\dot{q}_2 \neq 0$. By proposition 1 this trajectory does not belong to the sets $\Lambda_{(0,0,0,0)}$, $\Lambda_{(\pi,0,0,0)}$. Therefore the smooth function $q_1(t)$ is strictly bounded away from zero and π. By this reason the trajectory $(q, \dot{q})$ satisfies the equations

$$\ddot{q}_1 = \sin q_1 \cdot \cos q_1 \cdot (\dot{q}_2)^2 - \frac{g}{l} \cdot \sin q_1, \qquad (8)$$

$$\ddot{q}_2 = -2 \cdot \frac{\cos q_1}{\sin q_1} \cdot \dot{q}_1 \cdot \dot{q}_2, \qquad (9)$$

which is well defined for all $t \geq 0$.

Theorem 2. Suppose that the functions g_1, g_2, are not identically equal to zero along any solution of the unforced spherical pendulum. Let $(q, \dot{q})$ be any motion of the unforced spherical pendulum consistent with $y(t) = 0$ for all $t \geq 0$, where y is defined by (7). Then only the following cases are possible:

1). $H_0(q(t), \dot{q}(t)) = H^*$, $p_0(q(t), \dot{q}(t)) = p^*$;

2). $(q(t), \dot{q}(t)) = (\pi, 0, 0, 0)$ or $(q(t), \dot{q}(t)) = (0, 0, 0, 0)$;

3). $(\hat{q}(t), \dot{\hat{q}}(t)) = (q_1, q_2 \cdot t, 0, q_2)$, where the constants q_1, q_2 satisfy the equations

$$\begin{cases} r_1 q_2 [H_0(\hat{q}, \dot{\hat{q}}) - H_*] + r_2 [p_0(\hat{q}, \dot{\hat{q}}) - p^*] = 0 \\ q_2^2 \cos q_1 = \frac{g}{l} \end{cases} \qquad (10)$$

Deciphering of the equation (10) for the different values of the constants H^*, p^* leads to the following corollaries of theorem 2.

Corollary 3. Given the prescribed constants $H^* \geq 0$, $p^* \in \mathbf{R}^1$ and the set

$$\mathcal{D}_{\alpha,\beta} = \{(q, \dot{q}) : |H_0(q, \dot{q}) - H^*| \geq \alpha, \\ |p_0(q, \dot{q}) - p^*| \leq \beta\},$$

suppose that the assumptions of theorem 2 are valid. Then for any $\alpha > 0$, $\beta > 0$ there exist positive constants r_1, r_2 such that for any solution of the unforced spherical pendulum with initial conditions from $\mathcal{D}_{\alpha,\beta}$ consistent with $y(t) = 0$ for all $t \geq 0$ only case 2) of theorem 2 is possible. □

Corollary 4. Let $H^* = 0$, $p^* = 0$ and suppose that the assumptions of theorem 2 are valid. Let r_1, r_2 be any positive constants. Then for an arbitrary motion of the unforced spherical pendulum consistent with $y(t) = 0$ for all $t \geq 0$ only case 2) of theorem 2 is possible. □

Remark 5. Theorem 2 essentially describes the motions of the unforced spherical pendulum which are V-detectable, see (Shiriaev, 1998*c*). It is worth also to mention that Corollary 4 states the local zero-state detectability of the spherical pendulum with respect to the downward equilibrium and provides the local asymptotic stabilizability of this equilibrium, see also (Shiriaev, 1998*b*). □

3. MAIN RESULTS

We will investigate the feedback regulator suggested by the speed-gradient method to solve the upright position stabilization problem. This allows for taking advantage of the passivity properties of the system.

Taking the values $H^* = 2mgl$, $p^* = 0$, we obtain that the set V_0 of zeros of the function V, introduced in (6), corresponds the invariant subset of the spherical pendulum state space which contains the upright position. So the first step is to consider the stabilizing problem of the set V_0.

Let $\phi_1(z)$, $\phi_2(z)$ be any scalar smooth functions satisfying the inequalities

$$\varepsilon_1 |z|^2 \leq z\phi_1(z), \; z\phi_2(z) \leq \varepsilon_2 |z|^2, \quad \forall z \in \mathbf{R}^1, (11)$$

where $\varepsilon_2 \geq \varepsilon_1 > 0$, and take the regulator of the form

$$u_1 = -\phi_1(y_1), \quad u_2 = -\phi_2(y_2). \qquad (12)$$

The following Theorem reflects the qualitative behaviour of the solutions of the closed loop system

(1), (2), (12) with the objective to make all solutions converge to the stable manifold of the upright position of the unforced spherical pendulum, which by Proposition 1 coincides with the compact set $V_0 = \{[q, \dot{q}]; H_0(q,) = 2mgl, p_0(q, \dot{q}) = 0\}$.

Theorem 6. Let $H^* = 2mgl$, $p^* = 0$ and r_1, r_2 be any positive constants, and the functions g_1, g_2 not identically equal to zero along any solution of the unforced spherical pendulum. Let ϕ_1, ϕ_2 be any scalar smooth functions satisfying the inequalities (11). Define the regulator by relation (12). Then

a). for any solution $(q, \dot{q})$ of the closed loop system, which does not coincide with the equilibrium points $(0,0,0,0)$, $(\pi, 0, 0, 0)$, only the following alternatives are possible:
 1). $\lim_{t\to+\infty} H_0(q(t), \dot{q}(t)) = H^*$,
 $\lim_{t\to+\infty} p_0(q(t), \dot{q}(t)) = 0$;
 2). as $t \to +\infty$, the trajectory $(q(t), \dot{q}(t))$ tends to the unique cycle Γ of the unforced pendulum defined by Theorem 2;

b). for any solution $(q(t), \dot{q}(t))$ of the closed loop system (1), (2), (12), which corresponds to the case a1), $\forall \varepsilon > 0$ there exists the sequence of moments of time $\{T_n\}_{n=1}^{+\infty}$, $T_n \to +\infty$, such that $|q_1(T_n) - \pi| < \varepsilon$, $|\dot{q}_1(T_n)| < \varepsilon$. □

Remark 7. Theorem 6 does not guarantee that even in the best case, when the solution $(q, \dot{q})$ of the closed loop system (1), (2), (12) tends to the set $V_0 = \{(q, \dot{q}) : H_0(q, \dot{q}) = 2mgl,\ p_0(q, \dot{q}) = 0\}$, the solution $(q, \dot{q})$ will be close to the upper equilibrium point for all sufficiently large moments of time. But this trajectory will be an infinite number of times in any neighbourhood of the upright position, see also the results of the computer simulations. □

Thus the detailed analysis of the closed loop system (1), (2) with the feedback control (12) derived from the standard speed-gradient algorithm, shows that even with the weakening goal of making the set V_0 globally asymptotically stable, is not achieved.

Below we describe the algorithm which solves the problem of stabilizing of the upright point. The solution is based on a modification of the speed-gradient method and consists of two steps: first, construct a feedback control such that along the closed loop system trajectory $(q, \dot{q})$ the variable $\dot{q}_2(t)$ achieves the zero value in a finite time; second, define the feedback control with $u_2 \equiv 0$ providing the stabilization of the pendulum upright position. Both steps will use the passivity of the controlled spherical pendulum, but in contrast to (12) the suggested feedback control will contain several points of switching and at the first step provides the stabilization of the zero value of p_0 in finite time. This makes it possible to reduce the problem of *the spherical pendulum* upright point stabilization to the problem of the upright position of *the pendulum*. The last problem as shown in (Shiriaev *et al.*, 1998) can be successfully solved.

Theorem 8. Let $\delta > 0$ and $\phi_2(z)$ be any smooth scalar function satisfying the inequality (11) and the smooth function g_2 not identically equal to zero along any solution of the unforced spherical pendulum. Take the regulator

$$u_1(t) \equiv 0$$

$$u_2(t) = \begin{cases} -\phi_2\left(g_2(q, \dot{q}) \cdot [p_0(q, \dot{q}) - \delta]\right), & \text{if } \dot{q}_2 < 0 \\ -\phi_2\left(g_2(q, \dot{q}) \cdot [p_0(q, \dot{q}) + \delta]\right), & \text{if } \dot{q}_2 > 0 \\ 0, & \text{if } \dot{q}_2 = 0 \end{cases} \tag{13}$$

Then any solution $(q, \dot{q})$ of the closed loop system (1), (2), (13) achieves for finite time $T > 0$ the point $(q(T)), \dot{q}(T))$ with $\dot{q}_2(T) = 0$. □

Thus any solution $(q, \dot{q})$ of the closed loop system (1), (2), (13) comes in finite time $T = T(q(0), \dot{q}(0))$ to the point $(q_1(T)), q_2^*, \dot{q}_1(T), 0)$, which differs from the origin, where q_2^* is some constant from $[0, 2\pi)$. In this point the feedback control is switched such that along the closed loop system solution for $t \geq T$ the important relation $p_0(q(t), \dot{q}(t)) = 0$ (or equally $\dot{q}_2(t) = 0$) is preserved. Then the motion of the controlled spherical pendulum is reduced to

$$(q(t), \dot{q}(t)) = (\hat{q}_1(t), q_2^*, \dot{\hat{q}}_1(t), 0),$$

where $(\hat{q}_1, \dot{\hat{q}}_1)$ satisfies the equation of the controlled pendulum

$$ml^2 \cdot \ddot{\hat{q}}_1(t) + mgl \cdot \sin \hat{q}_1(t) = g_1(q, \dot{q})u_1. \tag{14}$$

The following theorem, extracted from (Shiriaev *et al.*, 1998), provides the opportunity to the stabilization of the upright position of pendulum.

Theorem 9. Let $2mgl > \varepsilon > 0$ and $\phi_1(z)$ be any smooth scalar function satisfying the inequality (11) and the smooth scalar function g_1 not identically equal to zero along any solution of the unforced pendulum. Then the regulator

$$u_1 = \begin{cases} u^*, & \text{if } q_1 = 0,\ \dot{q}_1 = 0 \\ 0, & \text{if } H_0(q, \dot{q}) = 2mgl \\ -\phi_1\left(\dot{q}_1 g_1(q, \dot{q})[H_0(q, \dot{q}) - 2mgl + \varepsilon]\right), & \text{if } H_0(q, \dot{q}) > 2mgl \\ -\phi_1\left(\dot{q}_1 g_1(q, \dot{q})[H_0(q, \dot{q}) - 2mgl - \varepsilon]\right), & \text{if } H_0(q, \dot{q}) < 2mgl \end{cases} \tag{15}$$

where u^* is any nonzero number, globally makes the unstable equilibrium point $[q, \dot{q}] = [\pi, 0]$ of the controlled pendulum (14) to be a unique ω-limit point of the closed-loop system. □

Remark 10. The closed loop system (14), (15) is a nonlinear system with discontinuous right side in the switching points of the control function. Thus the question of the definiteness of the solution of the system can arise. It follows from the proof of theorem 9 that for any initial condition the solution of the system (14), (15) is an absolutely continuous function, which is well defined on $[0, +\infty)$, and the derivative of this function is a piece-wise smooth function, which can have no more than one point of discontinuity. □

Taking advantage of the theorem 8 and theorem 9 we can write the final form of the upright position stabilizing regulator.

Theorem 11. Take positive constants δ, ε with $2mgl > \varepsilon$, any nonzero constant u^* and the scalar smooth functions $\phi_1(z)$, $\phi_2(z)$ satisfying (11). Then the regulator

$$u_1 = \begin{cases} 0, \text{ if } \dot{q}_2 \neq 0 \text{ or } H_0(q,\dot{q}) = 2mgl \\ u^*, \text{ if } q_1 = 0, \ \dot{q}_1 = 0 \\ -\phi_1\left(\dot{q}_1 g_1(q,\dot{q})[H_0(q,\dot{q}) - 2mgl + \varepsilon]\right), \\ \qquad \text{if } (H_0(q,\dot{q}) > 2mgl) \,\&\, (\dot{q}_2 = 0) \\ -\phi_1\left(\dot{q}_1 g_1(q,\dot{q})[H_0(q,\dot{q}) - 2mgl - \varepsilon]\right), \\ \qquad \text{if } (H_0(q,\dot{q}) < 2mgl) \,\&\, (\dot{q}_2 = 0) \end{cases}$$

$$u_2 = \begin{cases} 0, & \text{if } \dot{q}_2 = 0 \\ -\phi_2\left(g_2(q,\dot{q}) \cdot [p_0(q,\dot{q}) - \delta]\right), & \text{if } \dot{q}_2 < 0 \\ -\phi_2\left(g_2(q,\dot{q}) \cdot [p_0(q,\dot{q}) + \delta]\right), & \text{if } \dot{q}_2 > 0 \end{cases}$$

globally makes the upright position of the controlled spherical pendulum (1), (2) to be a unique ω-limit point of the closed loop system. □

Remark 12. Since the submission of this paper the authors became aware of the interesting work of Bloch *et al.* (1998), where using the different arguments the local asymptotic stabilizability of the upright position of the spherical pendulum on a cart was established. To emphasize the difference between the presented results and (Bloch *et al.*, 1998) it is necessary to mention that our results are global. Due to this advantage we can establish only the attractability of the upright position. □

4. COMPUTER SIMULATIONS

In all the simulations the following parameters have been used: $m = 1kg$, $l = 1m$, $g = 9.81m/s^2$, $q_1(0) = 1rad$, $\dot{q}_1(0) = 1rad/s$, $q_2(0) = \pi/2rad$, $\dot{q}_2(0) = 1rad/s$.

Figure 2 shows the motion of the unforced spherical pendulum

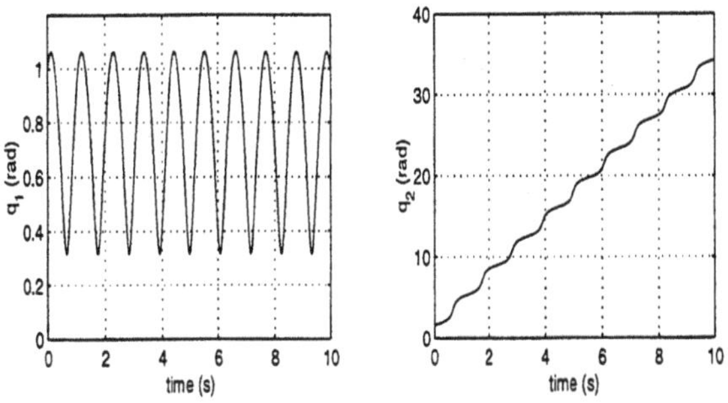

Fig. 2. The motion of the unforced spherical pendulum

By Theorem 6 the upright position is a relative (i. e. with respect to only the part coordinate q_1, $\dot{q}_1$ of the state vector) ω-limit point of any solution of the closed loop system along which the functions H_0, p_0 tend to the prescribed values $2mgl$ and 0. But it is possible that there exist other ω-limit points. The last means that the closed loop system solution may be rotating. Figure 3 presents such a motion of the controlled pendulum with $u_1 = -y_1$, $u_2 = -y_2$ as defined in (7).

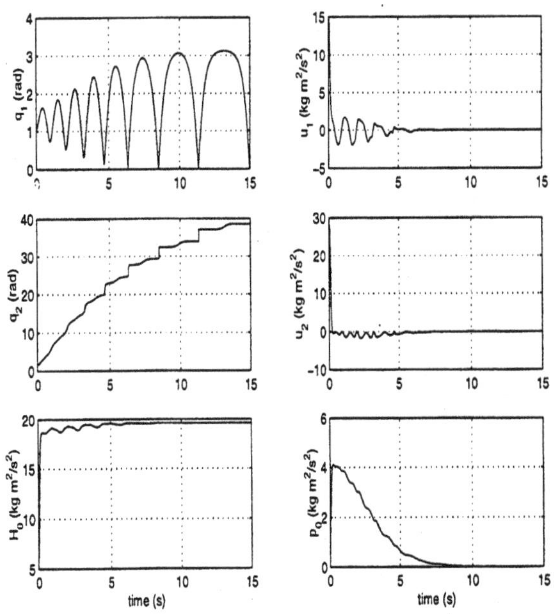

Fig. 3. The oscillations of the spherical pendulum

Figure 4 shows the motion with the regulator defined in Theorem 11, using $\phi_1(z) = \phi_2(z) = z$

5. CONCLUSIONS

The spherical pendulum is a nonlinear system with two equilibrium points. This paper is devoted to the solution of stabilizing the upright position. The main idea of the solution is based on the facts that, first, the spherical pendulum is a passive system with respect to some natural

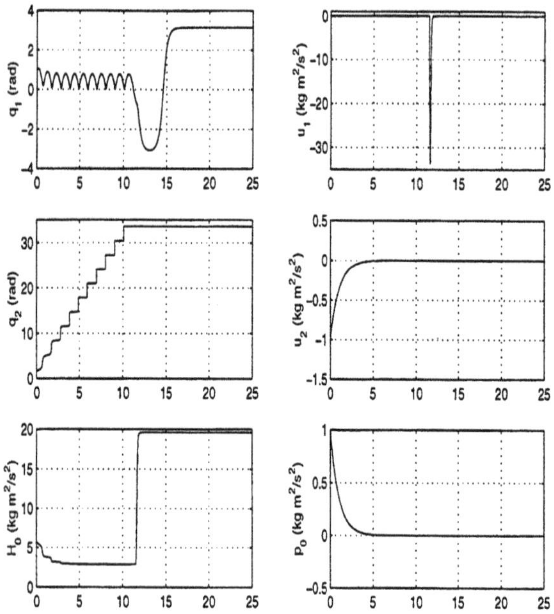

Fig. 4. The stabilization of the spherical pendulum upright position

defined storage function V and, second, the zero level V_0 of this storage function V coincides with the stable manifold of the upright position.

So by stabilizing the compact set V_0 one can hope that the upright position will be an unique attractive point of the closed loop system. To this end it seems naturally to apply the speed-gradient method with the objective to stabilize the set V_0. However, it was shown that for any passive controller the union of ω-limit sets of all closed loop system trajectories will consist of three connected invariant sets: the downward position of the spherical pendulum, some cycle and the set V_0. This means that any passive controller does not provide the global asymptotic stability of V_0. Therefore, it does not provide the attractability of the upright position.

The main contribution of the paper is the modification of the speed-gradient method. It was shown that, consecutively stabilizing the prescribed values of the first integrals of the unforced spherical pendulum and using the switching between the regulators, the upright position becomes a unique attractive point of the closed loop system. The obtained rigorous results are successfully tested by the simulations.

6. REFERENCES

Bloch, A.M., N.E. Leonard and J.E. Marsden (1998). Matching and stabilization by the method of controlled Lagrangians. In: *Proceedings of the 37th CDC.* IEEE. Tampa, USA. pp. 1446–1451.

Byrnes, C.I., A. Isidori and J.C. Willems (1991). Passivity, feedback equivalence, and the global stabilization of minimum phase nonlinear systems. *IEEE Transactions on Automatic Control* **36**, 1228–1240.

Fradkov, A.L. (1990). *Adaptive control of complex systems.* Moscow: Nauka. (in Russian).

Fradkov, A.L. (1996). Swinging control of nonlinear oscillations. *International Journal of Control* **64**(6), 1189–1202.

Fradkov, A.L., I.A. Makarov, A.S. Shiriaev and O.P. Tompchina (1997). Control of oscillations in Hamiltonian systems. In: *Proceedings of the 4th European Control Conference.* Brussels.

Åström, K.J. and K. Furuta (1996). Swinging up a pendulum by energy control. In: *Proceedings of the 13th IFAC World Congress.* Vol. E. San Francisco. pp. 37–42.

Shiriaev, A.S. (1998*a*). Control of oscillations in affine nonlinear systems. In: *Proceedings of the IFAC Conference on System Structure and Control.* Vol. 3. Nantes, France. pp. 789–794.

Shiriaev, A.S. (1998*b*). New tests of zero-state detectability. In: *Proceedings of the 4th IFAC Symposium on Nonlinear Control Systems Design.* Vol. 2. Enschede, The Netherlands. pp. 533–538.

Shiriaev, A.S. (1998*c*). The notion of V-detectability and stabilization of invariant sets of nonlinear systems. In: *Proceedings of the 37th CDC.* Tampa. pp. 2509–2514.

Shiriaev, A.S. and A.L. Fradkov (1998). Stabilization of invariant set of nonlinear systems. In: *Nonlinear dynamical systems.* St. Petersburg State University. St. Petersburg, Russia. (in Russian).

Shiriaev, A.S., O. Egeland and H. Ludvigsen (1998). Global stabilization of unstable equilibrium point of pendulum. In: *Proceedings of 37th CDC.* IEEE. Tampa. pp. 4584–4585.

Willems, J.C. (1972). Dissipative dynamic systems. Part 1. General Theory. *Arch. Rat. Mech. Anal.* **45**(5), 321–351.

STABILITY IN CELLULAR NEURAL NETWORKS WITH DELAYS

Jinde Cao

Adult Education College, Yunnan University, Kunming 650091, P. R. China
E-mail: jdcao@ynu.edu.cn

Abstract: In this paper, the author analyses further problems of global stability for a class of cellular neural networks with delays (DCNN) by means of Lyapunov functional method and some new analysis technique, some new criteria of stability are obtained under more general conditions. These criteria can be applied to design globally stable networks and thus have important significance in signal processing, especially in moving image treatment. *Copyright © 1999 IFAC*

Keywords: cellular neural networks, Lyapunov functional, stability, signal processing, delay.

1. INTRODUCTION

It is well-known that the cellular neural networks (CNN), which have been used to solve many problems which is not easily solvable by means of tradition methods, have been widely applied in many fields. we know that the CNN is formed by many units called cells, the structure of the CNN is similar to that found in cellular automata, namely, any cell in a cellular neural network is connected only to its neighbor cells. A cell contains linear and nonlinear circuit elements, which typically are linear capacitors, linear resistors, linear and nonlinear controlled sources, and independent sources. The circuit diagram and connection pattern implementing for the CNN can find in Chua and Yang (1988). The CNN can be applied in signal processing, moreover, it can be used to solve some image processing and pattern recognition problems, however, processing of moving image and pattern recognition require introduction of delay in the signals transmitted among the cells, i.e. require to consider the CNN with delays (DCNN) which can be described by delayed differential equations (namely, functional differential equations). The studies of stability on the DCNN and CNN are of theoretical and applicable important significance in the design of networks and signal processing, especially in moving image treatment. There exist some results of stability for CNN and DCNN, see (Cao, 1998; Cao and Zhou, 1998; Chua and Yang, 1988a, 1988b; Civalleri, *et al.*, 1993; Liao, 1994; Lu, 1997; Roska, *et al.*, 1992). The purpose of this paper is to derive some new criteria for global asymptotic stability of the DCNN, which are independent of delays, by using of Lyapunov functional method (Cao, 1998; Cao and Wan, 1997; Cao and Zhou, 1998) and analysis technique of new inequality

$$3abc \le a^3 + b^3 + c^3 (a,b,c \ge 0).$$

Thus some results related references (Cao, 1998; Cao and Zhou, 1998; Chua, 1988a, 1988b; Civalleri, *et al.*, 1993; Liao, 1994; Lu, 1997; Roska, *et al.*, 1992) and the references cited therein are extended and improved, and some corresponding results become actually some special cases of our results.

In the following, consider the system of DCNN described by functional differential equations with delays

$$x_i'(t) = -c_i x_i(t) + \sum_{j=1}^{n} a_{ij} f_j(x_j(t))$$
$$+\sum_{j=1}^{n} b_{ij} f_j(x_j(t-\tau_j)) + I_i \ (c_i > 0, i = 1,2,\cdots n) \quad (1)$$

in which n corresponds a neural network with n units; $x_i(t)$ corresponds to the state vector of the i-th unit at time t; $f_j(x_j(t))$ denotes the output of the j-th unit at time $t; a_{ij}, b_{ij}, I_i, c_i$ are constant, a_{ij} denotes the strength of the j-th unit on the i-th unit at time t, b_{ij} denotes the strength of the j-th unit on the i-th unit at time $t-\tau_j, I_i$ denotes the external bias on the i-th unit, τ_j corresponds to the transmission delay along the axon of the j-th unit and is not negative constant, c_i represents the rate with which the i-th unit will reset its potential to the resting state in isolation when disconnected from the network and external inputs.

In the following, assume that each of the relation between the output of the cell $f_i(i=1,2,\cdots,n)$ and the state of the cell have following properties

(H_1) $f_i(i=1,2,\cdots,n)$ is bounded on R;

(H_2) There exists a number μ_i such that

$$|f_i(u)-f_i(v)| \le \mu_i |u-v|$$

for any $u,v \in R$.

It is easy to find from (H_2) that f_i is a continuous function on R. In particular, if the relation between the output of the cell and the state of the cell is described by a piecewise-linear function

$$f_i(x) = (|x+1|-|x-1|)/2,$$

then it is easy to see that the function f_i clearly satisfy the hypotheses (H_1) and (H_2) above, and $\mu_i \equiv 1 (i=1,2,\cdots,n)$. The circuit implementing of the equations (1) can be referred to references (Chua and Yang, 1988; Roska and Chua, 1992).

2. STABILITY ANALYSIS OF DCNN

Lemma 1 Assume that the output of the cell function $f_i(i=1,2,\cdots,n)$ satisfy the hypotheses (H_1) and (H_2) above. Then there exists an equilibrium for the DCNN (1).

Proof If let $x^* = (x_1^*, x_2^*, \cdots, x_n^*)^T$ denote an equilibrium of the DCNN(1), then x^* satisfies the nonlinear algebraic equations

$$x_i^* = c_i^{-1}[\sum_{j=1}^{n} a_{ij} f_j(x_j^*) + \sum_{j=1}^{n} b_{ij} f_j(x_j^*) + I_i]$$

$$= \sum_{j=1}^{n} [c_i^{-1}(a_{ij}+b_{ij})] f_j(x_j^*) + c_i^{-1} I_i \qquad (2)$$

Let

$$B = [c_i^{-1}(a_{ij}+b_{ij})]_{n\times n},$$

$$I = (c_1^{-1}I_1, c_2^{-1}I_2, \cdots, c_n^{-1}I_n)^T,$$

$$f(x^*) = (f_1(x_1^*), f_2(x_2^*), \cdots, f_n(x_n^*))^T,$$

then the system (2) can be written in vector-matrix notation in the form

$$x^* = F(x^*) = Bf(x^*) + I \qquad (3)$$

thus x^* is a fixed point of the map $F: R^n \to R^n$. The existence of a fixed point of the map F can be shown by the well known Brouwer's fixed point theorem. In fact, the i-th component of $F(x)$ satisfy the following form

$$|(F(x))_i| = \sum_{j=1}^{n} [c_i^{-1}(a_{ij}+b_{ij})] f_j(x_j) + c_i^{-1} I_i$$

$$\le \sum_{j=1}^{n} |c_i^{-1}(a_{ij}+b_{ij})| |f_j(x_j)| + c_i^{-1}|I_i|$$

$$\le \sum_{j=1}^{n} |c_i^{-1}(a_{ij}+b_{ij})| M + c_i^{-1}|I_i|$$

where $x = (x_1, x_2, \cdots, x_n)^T$, $M = \max_{1\le i\le n} \sup_s |f_i(s)|$.

Let

$$K = \max_{1\le i\le n}(\sum_{j=1}^{n} |c_i^{-1}(a_{ij}+b_{ij})| M + c_i^{-1}|I_i|),$$

then

$$F(R^n) \subset \Omega$$

in which

$$\Omega = \left\{(x_1, x_2, \cdots, x_n) \in R^n \,|\, |x_i| \le K, i=1,2,\cdots,n\right\}.$$

It is easy to see that the map F is continuous. Let map F be restricted on Ω, i.e. $F|_\Omega : \Omega \to \Omega$. Obviously $F|_\Omega$ maps the bounded closed and convex set Ω on R^n into itself; hence by the well known Brouwer's theorem, the map F has at least one fixed point say $x^* = (x_1^*, x_2^*, \cdots, x_n^*)^T$. This completes the proof.

Remark: Note that Brouwer's theorem does not guarantee the uniqueness of the fixed point. However, in this paper, some sufficient criteria on the system (1), which will guarantee not only the uniqueness of the equilibrium but also its recall by its global asymptotic stability, is derived. The uniqueness of the equilibrium will follow from the global asymptotic stability to be established below.

Lemma 2 For the DCNN (1), suppose that the output of the cell $f_i(i=1,2,\cdots,n)$ satisfy the hypotheses (H_1) and (H_2) above. Then all solutions of the DCNN (1) remain bounded for $[0,+\infty)$.

Proof It is easy to observe that all solutions of the DCNN (1) satisfy differential inequalities of the form

$$-c_i x_i(t) - \lambda_i \le x_i'(t) \le -c_i x_i(t) + \lambda_i \qquad (4)$$

where $\lambda_i = \sum_{j=1}^{n}(|a_{ij}|+|b_{ij}|)\sup_{s\in R}|f_j(s)| + |I_i|$. Using (4) above, one can easily prove that solutions of the

DCNN (1) remain bounded on $[0,+\infty)$. This completes the proof.

Theorem For the DCNN (1), suppose that the output of the cell $f_i\,(i=1,2,\cdots,n)$ satisfies the hypotheses (H_1) and (H_2) above. Assume furthermore that the system parameters $a_{ij}, b_{ij}\,(i,j=1,2,\cdots,n)$ satisfy the following conditions:

$$\sum_{j=1}^{n}[(|a_{ij}||\mu_j|^{3\alpha_j^*}+|a_{ij}||\mu_j|^{3\beta_j^*}+|a_{ji}||\mu_i|^{3\gamma_i^*}+$$
$$(|b_{ij}||\mu_j|^{3\alpha_j}+|b_{ij}||\mu_j|^{3\beta_j}+|b_{ji}||\mu_i|^{3\gamma_i})]<3c_i$$

in which $\mu_j\,(j=1,2,\cdots,n)$ is constant numbers of the hypotheses $(H)_2$ above, and $\alpha_j,\beta_j,\gamma_j,\alpha_j^*,\beta_j^*,\gamma_j^*$ are any real numbers with $\alpha_j+\beta_j+\gamma_j=1$ and $\alpha_j^*+\beta_j^*+\gamma_j^*=1\,(j=1,2,\cdots,n)$. Then the equilibrium x^* of the DCNN (1) is globally asymptotically stable independent of delays.

Proof If $x^*=(x_1^*,x_2^*,\cdots,x_n^*)^T$ is an equilibrium of the DCNN (1), one can derive from (1) that the deviations $y_i(t)=x_i(t)-x_i^*\,(i=1,2,\cdots,n)$ satisfy

$$y_i'(t)=-c_iy_i(t)+\sum_{j=1}^{n}a_{ij}(f_j(x_j^*+y_j(t))-f_j(x_j^*))$$
$$+\sum_{j=1}^{n}b_{ij}(f_j(x_j^*+y_j(t-\tau_j))-f_j(x_j^*)) \qquad (5)$$

Obviously, $(0,0,\cdots,0)^T$ is an equilibrium of (5). To prove the global asymptotic stability of x^* of the DCNN (1), it is sufficient to prove the global asymptotic stability of the trivial solution of (5). The local existence of solutions of (1) and (5) follows by the method of steps, the existence on $[0,+\infty)$ will be a consequence of our analysis below. Now consider the Lyapunov functional defined by

$$V(t)=V(y)(t)=\frac{1}{3}\sum_{i=1}^{n}(|y_i(t)|^3$$
$$+\sum_{j=1}^{n}|b_{ij}||\mu_j|^{3\gamma_j}\int_{t-\tau_j}^{t}|y_j(s)|^3ds) \qquad (6)$$

where $\gamma_j\,(j=1,2,\cdots,n)$ are any real numbers.

Calculating the upper right derivative D^+V of V along the solution of (5), It can follow that

$$D^+V\le\sum_{i=1}^{n}[|y_i(t)|^2(-c_i|y_i(t)|+\sum_{j=1}^{n}|a_{ij}||\mu_j||y_j(t)|$$
$$+\sum_{j=1}^{n}|b_{ij}||\mu_j||y_j(t-\tau_j)|$$
$$+\frac{1}{3}\sum_{j=1}^{n}|b_{ij}||\mu_j|^{3\gamma_j}(|y_j(t)|^3-|y_j(t-\tau_j)|^3)]$$

$$=\sum_{i=1}^{n}[(-c_i|y_i(t)|^3+\sum_{j=1}^{n}|a_{ij}||\mu_j||y_i(t)|^2|y_j(t)|$$
$$+\sum_{j=1}^{n}|b_{ij}||\mu_j||y_i(t)|^2|y_j(t-\tau_j)|$$
$$+\frac{1}{3}\sum_{j=1}^{n}|b_{ij}||\mu_j|^{3\gamma_j}(|y_j(t)|^3-|y_j(t-\tau_j)|^3)]$$
$$=\sum_{i=1}^{n}[(-c_i|y_i(t)|^3$$
$$+\sum_{j=1}^{n}|a_{ij}|(|\mu_j|^{\alpha_j^*}|y_i(t)|)(|\mu_j|^{\beta_j^*}|y_i(t)|)(|\mu_j|^{\gamma_j^*}|y_j(t)|)$$
$$+\sum_{j=1}^{n}|b_{ij}|(|\mu_j|^{\alpha_j}|y_i(t)|)(|\mu_j|^{\beta_j}|y_i(t)|)(|\mu_j|^{\gamma_j}|y_j(t-\tau_j)|)$$
$$+\frac{1}{3}\sum_{j=1}^{n}|b_{ij}||\mu_j|^{3\gamma_j}(|y_j(t)|^3-|y_j(t-\tau_j)|^3)] \qquad (7)$$

where $\alpha_j,\beta_j,\gamma_j,\alpha_j^*,\beta_j^*,\gamma_j^*$ are any real numbers with $\alpha_j+\beta_j+\gamma_j=1$ and $\alpha_j^*+\beta_j^*+\gamma_j^*=1$. Now estimating the right side of (7) by using of the inequality $3abc\le a^3+b^3+c^3\,(a,b,c\ge 0)$, It easily follows that

$$D^+V\le\sum_{i=1}^{n}[(-c_i|y_i(t)|^3+\sum_{j=1}^{n}|a_{ij}|\frac{1}{3}((|\mu_j|^{\alpha_j^*}|y_i(t)|)^3$$
$$+(|\mu_j|^{\beta_j^*}|y_i(t)|)^3+(|\mu_j|^{\gamma_j^*}|y_j(t)|)^3)$$
$$+\sum_{j=1}^{n}|b_{ij}|\frac{1}{3}((|\mu_j|^{\alpha_j}|y_i(t)|)^3$$
$$+(|\mu_j|^{\beta_j}|y_i(t)|)^3+(|\mu_j|^{\gamma_j}|y_j(t-\tau_j)|)^3)$$
$$+\frac{1}{3}\sum_{j=1}^{n}|b_{ij}||\mu_j|^{3\gamma_j}(|y_j(t)|^3-|y_j(t-\tau_j)|^3)]$$
$$=\sum_{i=1}^{n}[(-c_i|y_i(t)|^3+\sum_{j=1}^{n}|a_{ij}|\frac{1}{3}((|\mu_j|^{\alpha_j^*}|y_i(t)|)^3$$
$$+(|\mu_j|^{\beta_j^*}|y_i(t)|)^3+(|\mu_j|^{\gamma_j^*}|y_j(t)|)^3)$$
$$+\sum_{j=1}^{n}|b_{ij}|\frac{1}{3}((|\mu_j|^{\alpha_j}|y_i(t)|)^3+(|\mu_j|^{\beta_j}|y_i(t)|)^3)$$
$$+\frac{1}{3}\sum_{j=1}^{n}|b_{ij}||\mu_j|^{3\gamma_j}|y_j(t)|^3]$$
$$=\sum_{i=1}^{n}[(-c_i|y_i(t)|^3+\sum_{j=1}^{n}\frac{1}{3}|a_{ij}|(|\mu_j|^{3\alpha_j^*}+|\mu_j|^{3\beta_j^*})|y_i(t)|^3$$
$$+\sum_{j=1}^{n}\frac{1}{3}|a_{ji}||\mu_i|^{3\gamma_i^*}|y_i(t)|^3$$
$$+\sum_{j=1}^{n}\frac{1}{3}|b_{ij}|(|\mu_j|^{3\alpha_j}+|\mu_j|^{3\beta_j})|y_i(t)|^3$$
$$+\frac{1}{3}\sum_{j=1}^{n}|b_{ji}||\mu_i|^{3\gamma_i}|y_i(t)|^3]$$
$$=\sum_{i=1}^{n}[-(c_i-\frac{1}{3}\sum_{j=1}^{n}((|a_{ij}||\mu_j|^{3\alpha_j^*}+|a_{ij}||\mu_j|^{3\beta_j^*}+|a_{ji}||\mu_i|^{3\gamma_i^*})$$
$$+(|b_{ij}||\mu_j|^{3\alpha_j}+|b_{ij}||\mu_j|^{3\beta_j}+|b_{ji}||\mu_i|^{3\gamma_i}))]|y_i(t)|^3$$

$$\leq -r\sum_{i=1}^{n}|y_i(t)|^3 \quad (8)$$

where

$$r = \min_{1\leq i\leq n}[-(c_i - \frac{1}{3}\sum_{j=1}^{n}((|a_{ij}|\mu_j^{3\alpha_j^*} + |a_{ij}|\mu_j^{3\beta_j^*} + |a_{ji}|\mu_i^{3\gamma_i^*}) + (|b_{ij}|\mu_j^{3\alpha_j} + |b_{ij}|\mu_j^{3\beta_j} + |b_{ji}|\mu_i^{3\gamma_i}))] > 0$$

A consequence of (8) is that

$$V(y)(t) + r\int_0^t \sum_{i=1}^{n}|y_i(s)|^3 ds \leq V(y)(0) \quad (9)$$

It follows from (9) that

$$\int_0^{+\infty}\sum_{i=1}^{n}|y_i(t)|^3 dt < +\infty \quad (10)$$

According to Lemma 2, $x_i(t)$ is bounded on $(0,+\infty)$. This implies boundedness of $y_i(t), y_i'(t)$ on $(0,+\infty)$;hence $|y_i(t)|, |y_i(t)|^3$ is uniformly continuous on $(0,+\infty)$. By Barbalat's Lemma (Gopalsamy, 1992), it follows that

$$\lim_{t\to+\infty}\sum_{i=1}^{n}|y_i(t)|^3 = 0 \quad (11)$$

It follows from (11) that zero solution of the equation (5) is globally asymptotically stable for any delays, thus the equilibrium of the DCNN (1) is also globally asymptotically stable. This completes the proof.

3. CONCLUSION

In this paper, analysis technique of new inequality

$$3abc \leq a^3 + b^3 + c^3 (a,b,c \geq 0)$$

is firstly applied and some new sufficient criteria, which have a wider adaptive range and can be applied to design unconditional globally stable cellular neural networks with delays(DCNN), is obtained. These possess highly important significance in some applied fields, for instance, the global optimization problems. Moreover, the results in this paper are easily verifiable and the each of the output of the cell $f_j (j = 1,2,\cdots,n)$ is quite general, and it has very wide adjustable leeway because our criteria possess many adjustable real parameters

$$\mu_j > 0, \alpha_j, \beta_j, \gamma_j, \alpha_j^*, \beta_j^*, \gamma_j^* \in R$$

with $\alpha_j + \beta_j + \gamma_j = 1$ and $\alpha_j^* + \beta_j^* + \gamma_j^* = 1$ $(j = 1, 2,\cdots,n)$, these are of important significance in the design of DCNN and signal processing, especially in moving image treatment. In addition, one did not also assume the symmetry of the connection matrix $(a_{ij})_{n\times n}, (b_{ij})_{n\times n}$ and one only assume the output of the cell (i.e. the nonlinear properties of the cell) with the hypotheses (H_1) and (H_2) above, and not requiring differentiable or strictly monotonously increasing.

REFERENCES

Cao, J. D. (1998). Stability of delayed cellular neural network. *Information and Control*, to be published, (in Chinese).

Cao, J.D. and D.M.Zhou (1998). Stability analysis of delayed cellular neural network. *Neural Networks*, in press.

Cao, J.D. and S.D. Wan (1997). The global asymptotic stability of Hopfield neural network with delay. *J. of Biomath.*, **12**, 60-63 (in Chinese).

Chua, L. O. and L. Yang (1988). Cellular neural networks: Theory. *IEEE Trans. on CAS*, **35**, 1257-1272.

Chua, L. O. and L.Yang (1988). Cellular neural networks: Applications. *IEEE Trans. on CAS*, **35**, 1273-1290.

Civalleri, P. P. and M.Gilli (1993). On stability of cellular neural networks with delay. *IEEE Trans. on CAS*, **40**, 157-165.

Gopalsamy, K. (1992). *Stability and Oscillation in Delay Differential Equations of Population Dynamics*. Kluwer Academic Publishers, Dordrecht.

Liao, X.X. (1994). Mathematical theory (II) of cellular neural networks. *Science in China* (Series A), **24**, 1037-1046 (in Chinese).

Lu, H.T. and Z.Y.He (1997). Unconditioned stability of cellular neural networks with delays. *Acta Electronica Sinica*, **25**, 1-4 (in Chinese).

Roska T. and L.O. Chua (1992). Cellular neural networks with nonlinear and delay-type template. *Int. J. Circuit Theory Appl.*, **20**, 469-481.

ROBUST H_∞-ALMOST DISTURBANCE DECOUPLING PROBLEM WITH STABILITY FOR A CLASS OF NONLINEAR TIME-DELAY SYSTEMS [1]

Guo-ping Lu * Yu-fan Zheng ** Daniel W. C. Ho ***

* *Department of Automatic Control,*
Nantong Institute of Technology, Jiangsu 226007, P.R. China
** *Institute of Systems Science,*
East China Normal University, Shanghai 200062, P.R. China
*** *Department of Mathematics,*
City University of Hong Kong, Hong Kong
Email: madaniel@cityu.edu.hk

Abstract: This paper deals with the robust H_∞-almost disturbance decoupling problem with stability (RADDPS) for a class of nonlinear time-delay systems with uncertainty. The paper presents a sufficient condition for solvability of RADDPS, and a design of state feedback controller if it is solvable. Illustrative example is given to demonstrate the applicability of the proposed approach.

Keywords: Nonlinear system, Time-delay, Robust H_∞-control, Almost disturbance decoupling.

1. INTRODUCTION

Disturbance decoupling problems with stability (DDPS) have been discussed for linear and nonlinear systems intensively in literature (Wonham 1985, Isidori 1995). It is known that *almost disturbance decoupling problem* (Weiland, *et al.* 1989) is also an interesting and practical topic in control theory for many years. The almost disturbance decoupling problem is to find an feedback control law such that the closed-loop system achieves an arbitrary small level of disturbance attenuation in the sense that, for example, the L_2-norm ratio between disturbance and output functions is less than an arbitrarily small number $\gamma > 0$ while maintaining internal stability (Weiland, *et al.* 1989).

For linear systems the above mentioned problem has been completely solved. For nonlinear systems relevant results are available only for single-input and single-output systems (Isidori 1996). In a different formulation, (Lin, *et al.* 1996, Lin 1997) discuss the H_∞- almost disturbance decoupling problems with internal stability of linear systems subject to input saturation and input-additive disturbance.

In this paper we assume that: (1) state feedback is applied; and (2) the output variables are state variables. Under the above two assumptions, *the almost disturbance decoupling problem with stability (ADDPS)* for a class of nonlinear systems is developed. We shall further extend ADDPS to the robust H_∞ almost disturbance decoupling problem with stability (RADDPS) in this work.

It is known to us that time delay is commonly encountered in various engineering systems, such as chemical processes, hydraulic and rolling mill systems. As time delay usually results in unsat-

[1] This work was supported by Natural Science Foundation of China, Doctoral Education Foundation of China, the Australia Research Council, Hong Kong Research Grant Council #9040285.

isfactory performances and is frequently a source of instability, many researchers have paid serious attention to the problems caused by time delay. Recently several authors have used different approaches such as quadratic Lyapunov function, robust H_∞ control, linear matrix inequalities to study the linear systems with time delay and uncertainty (Lee, *et al.* 1994, Trinh, *et al.* 1994).

In this paper we study the RADDPS for a class of nonlinear systems with time delay and uncertain ties. The class of nonlinear systems with time delays and parameter uncertainties considered in this paper are described as follows:

$$\dot{x}(t) = \sum_{i=0}^{r}[A_i + \Delta A_i(t)]x(t-d_i) + [A_{r+1} + \Delta A_{r+1}]f[x(t), x(t-d_1), \cdots, x(t-d_r)] + [B + \Delta B(t)]u(t) + Dw(t) \tag{1}$$

where $x(t) \in \mathbf{R}^n$, $u(t) \in \mathbf{R}^m$ and $w(t) \in \mathbf{R}^p$ are respectively the state, the control input and the disturbance input; $d_0 = 0$ and $d_i \geq 0$ $(i = 1, 2, \cdots, r)$ are the time delays; and A_i, B, D are known constant matrices with appropriate dimensions $(i = 0, 1, \cdots, r+1)$; $\Delta B(t)$, $\Delta A_i(t)$ are appropriately dimensional matrices representing time-varying parameter uncertainties, which are sometimes denoted as ΔB and ΔA_i for $i = 0, 1, \cdots, r+1$; $f[x(t), x(t-d_1), \cdots, x(t-d_r)]$ is known vector-valued continuous function. It is also a vector-valued function of time t. For convenience, we denote $x_{d_i} = x(t-d_i)$, $(i = 0, 1, \cdots, r)$, x_{d_0} as x. The following assumption is important for our analysis in our paper:

Assumption 1.1: *The uncertainty parameters $\Delta A_i(t); i = 0, 1, \cdots, r+1$, and $\Delta B(t)$ satisfy the following condition*

$$\Delta A_i(t) = H_i F_i(t) N_i, \quad \Delta B(t) = H_B N F_b(t) N_b \tag{2}$$

where H_i, N_i, H_b, N_b are known constant matrices with appropriate dimensions. The matrix-valued functions $F_i(t)$, $F_b(t)$ $(i = 0, 1, 2, \cdots, r+1)$ are time-varying Lebesgue integrable functions, which describe system parameter uncertainties, satisfying that

$$F_i(t)F_i'(t) \leq I, \quad F_b(t)F_b'(t) \leq I \tag{3}$$

where $i = 0, 1, 2, \cdots, r+1$.

If the $\Delta A_i(t)$ and $\Delta B(t)$ satisfy Assumption 1.1, then they are called *allowable uncertainty parameters.*

Assumption 1.2: *Assume the parameter matrices D, B, H_b and the nonlinear vector f satisfy the following conditions*

(i) there exist constant matrices D_0, H_{b_0} with appropriate dimensions such that

$$D = BD_0, \qquad H_b = BH_{b_0} \tag{4}$$

(ii) there exist a constant $c_i > 0$ and constant matrices E_i $(i = 0, 1, 2, \cdots, r)$ with appropriate dimensions such that

$$f'f \leq \sum_{i=0}^{r} c_i x_{d_i}' E_i' E_i x_{d_i} \tag{5}$$

where $f = f[x(t), x(t-d_1), \cdots, x(t-d_r)]$.

The precise description of RADDPS for a class of nonlinear time-delay systems will be given in next section. Throughout this paper, the following notations will be used.

X' denotes the transpose of matrix $X \in \mathbf{R}^{n\times m}$, $\|X\|$ denotes Euclidean norm of X. I denotes an identity matrix with appropriate dimensions. Given two matrices X and Y with the same dimensions, the notation $X > Y$ $(X \geq Y)$ means that $X - Y$ is positive definite (semidefinite). For $x \in \mathbf{R}^n$, $\|x\|$ denotes the Euclidean norm of x. L_2^n denotes the set of all measurable functions $x(t) \in \mathbf{R}^n, 0 \leq t < \infty$ such that $\int_0^\infty \|x(t)\|^2 dt < \infty$. The L_2-norm of $x(t) \in L_2^n$ is defined as $\|x\|_{L_2} = \left(\int_0^\infty \|x(t)\|^2 dt\right)^{1/2}$.

The rest of this paper is organized as follows: In Section 2, we give a precise formulation for RADDPS, which is considered in this paper. Section 3 presents the main results of the paper, *i.e.*, the solvability conditions for the proposed RADDPS. An illustrative example is presented in Section 4 and the conclusion remarks are drawn in Section 5.

2. PROBLEM STATEMENT OF RADDPS

Consider the following general nonlinear system with time delays and parameter uncertainties:

$$\dot{x}(t) = f[x(t), x(t-d_1), \cdots, x(t-d_r), \Delta_1(t), \cdots, \Delta_q(t), w(t)] \tag{6}$$

where $x(t) \in \mathbf{R}^n$ and $w(t) \in \mathbf{R}^p$ for $t \in \mathbf{R}$ are the state and the disturbance input, respectively. Let Δ_j^* be a set of piece-wise-continuous matrix-valued functions of time t, $j = 1, 2, \cdots, q$. By $\Delta_j(t) \in \Delta_j^*$, we describe the time-varying parameter uncertainties,

Now we give two definitions for the system (6):

Definition 2.1: *(Robust L_2-Stability) The system (6) is of robust L_2-stability if the solution $x(t); t \geq 0$ of system (6) with initial condition $x(t) = 0$ $(t \leq 0)$ satisfies $x(t) \in L_2^n$ subject that $w(t) \in L_2^p$ and $\Delta_j(t) \in \Delta_j^*$ $(j = 1, 2, \cdots, q)$.*

Definition 2.2: *(Robust L_2-γ-Gain Performance) Given $\gamma > 0$ system (6) is of robust L_2-γ-gain performance if with initial condition $x(t) = 0$ $(t \leq 0)$*

$$||x||_{L_2} \leq \gamma ||w||_{L_2}$$

for any $w(t) \in L_2^p$ and $\Delta_j(t) \in \Delta_j^*$ $(j = 1, 2, \cdots, q)$.

Based on the above definitions, we present the definition of the RADDPS for the system (1) as follows.

RADDPS (Robust H_∞ Almost Disturbance Decoupling Problem with Internal Stability) For a *priori* given $\gamma > 0$, to find, if possible, a linear state feedback law, under which the resulting closed loop system of (1) has the following properties.

(1) In the absence of the disturbance $w(t)$, $x = 0$ is globally asymptotically stable equilibrium.

(2) It is of robust L_2-stability and robust L_2-γ-gain performance subject to all allowable parameter uncertainties conditions.

Our main result is shown in the following Lemma.

Lemma 2.1: *(Wang, et al. 1992) Assume that A, H, N are real constant matrices with appropriate dimensions and $F(t)$ is a matrix-valued function with appropriate dimension, then the following matrix-inequalities hold.*

(a) For any $\epsilon > 0$

$$HF(t)N + N'F'(t)H' \leq \epsilon^{-1}HH' + \epsilon N'N \quad (7)$$

if $F'(t)F(t) \leq I$.

(b)

$$[A + HF(t)N][A + HF(t)N]' \leq A(I - \epsilon N'N)^{-1}A' + \epsilon^{-1}HH'$$

with $\epsilon > 0$ satisfying $\epsilon N'N < I$ and $F'(t)F(t) \leq I$.

3. MAIN RESULTS

In what follows we assume that there exists an $\epsilon > 0$ such that

$$R_\epsilon := 2I - \epsilon^{-1}H_{b_0}H_{b_0}' - \epsilon N_b'N_b > 0. \quad (8)$$

and it is always possible to find a sufficiently small $\epsilon_i > 0$ such that $\epsilon_i N_i' N_i < I$ $(i = 1, 2, \cdots, r+1)$.

Now let

$$T_i := A_i(I - \epsilon_i N_i' N_i)^{-1} A_i' + \epsilon_i^{-1} H_i H_i' \quad (9)$$

and establish a matrix Riccati equation as follows.

$$PA_0 + A_0'P - P(BR_\epsilon B' - \sum_{i=1}^{r+1} T_i - \epsilon_0 H_0 H_0')P + \sum_{i=1}^{r} c_i E_i' E_i + \epsilon_0^{-1} N_0' N_0 + rI + Q = 0 \quad (10)$$

where Q and ϵ_0 are properly chosen constant positive definite matrix and positive constant, respectively.

If there exists a positive definite solution P to the algebraic Riccati equation (10), then we can find a state feedback law as follows:

$$u(t) = -(1 + \eta)B'Px(t) \quad \eta \geq 0 \quad (11)$$

The following theorem is the main result of this paper.

Theorem 3.1: *Assume that system (1) satisfies the conditions (2)-(5), then its RADDPS is solvable if there exists a positive definite solution P for algebraic Riccati equation (10). Furthermore, the feedback law (11) gives a solution to the RADDPS for system (1).*

Proof: Choose a Lyapunov functional candidate

$$V = x'(t)Px(t) + \sum_{i=1}^{r} \int_{t-d_i}^{t} x'(s)(I + c_i E_i' E_i)x(s)ds, \quad (12)$$

Thus, the derivative of V along any trajectory $x(t); t \in \mathbf{R}$ of the closed-loop system (1) under (11) is

$$\dot{V} = 2x'P\sum_{i=0}^{r}(A_i + \Delta A_i)x_{d_i} + 2x'P(A_{r+1} + \Delta A_{r+1})f + 2x'P(B + \Delta B)u \quad (13)$$

$$+2x'PDw + x'\sum_{i=1}^{r}(I + c_i E_i' E_i)x - \sum_{i=1}^{r} x_{d_i}'(I + c_i E_i' E_i)x_{d_i} \quad (14)$$

By (4) and Lemma 2.1 we have

$$2x'PDw = 2x'PBR_\epsilon^{1/2}R_\epsilon^{-1/2}D_0 w \leq \eta x'PBR_\epsilon B'Px + \eta^{-1}w'D_0'R_\epsilon^{-1}D_0 w \quad (15)$$

Using (5), (11) and (15), for some $\eta > 0$, we have

$$\dot{V} \le 2x'P\sum_{i=0}^{r}(A_i + \Delta A_i)x_{d_i} + 2x'P(A_{r+1} + \Delta A_{r+1})f - 2(1+\eta)x'P(B+\Delta B)B'Px + x'\sum_{i=1}^{r}(I + c_iE_i'E_i)x - \sum_{i=1}^{r}x_{d_i}'(I + c_iE_i'E_i)x_{d_i} + \sum_{i=1}^{r}c_ix_{d_i}'E_i'E_ix_{d_i} - f'f + \eta x'PBR_\epsilon B'Px + \eta^{-1}w'D_0'R_\epsilon^{-1}D_0w \quad (16)$$

Let

$$\Omega := \begin{pmatrix} M & \Omega_0 \\ \Omega_0' & -I \end{pmatrix} \quad (17)$$

where

$$M := P(A_0 + \Delta A_0) + (A_0 + \Delta A_0)'P - (1+\eta)P[(B+\Delta B)B' + B(B+\Delta B)']P + \sum_{i=1}^{r}(I + c_iE_i'E_i) + Q + \eta PBR_\epsilon B'P \quad (18)$$

$$\Omega_0 := (\,P(A_1 + \Delta A_1) \quad \cdots \quad P(A_{r+1} + \Delta A_{r+1})\,),$$

then

$$\dot{V} \le -x'Qx + \eta^{-1}w'D_0'R_\epsilon^{-1}D_0w + X\Omega X' \quad (19)$$

where

$$X := (\,x' \quad x_{d_1}' \quad x_{d_2}' \quad \cdots \quad x_{d_r}' \quad f'\,)$$

Now we show that

$$\Omega \le 0 \quad (20)$$

For any $\epsilon_0, \epsilon > 0$, by Lemma 2.1, we have

$$P\Delta A_0 + \Delta A_0'P \le \epsilon_0 PH_0H_0'P + \epsilon_0^{-1}N_0'N_0,$$

$$-(1+\eta)P[(B+\Delta B)B' + B(B+\Delta B)']P = -2(1+\eta)PBB'P - (1+\eta)PB(H_{b_0}F_bN_b + N_b'F_b'H_{b_0}')B'P \le -PBR_\epsilon B'P - \eta PBR_\epsilon B'P \quad (21)$$

By (18) and (21), we obtain

$$M \le PA_0 + A_0'P - P(BR_\epsilon B' - \epsilon_0 H_0H_0')P + \epsilon_0^{-1}N_0'N_0 + \sum_{i=1}^{r}c_iE_i'E_i + rI + Q \quad (22)$$

It follows from (10) that (22) is equivalent to

$$M \le -P\sum_{i=1}^{r+1}T_iP \quad (23)$$

By Lemma 2.1, for any $i = 1, 2, \cdots, r+1$, we have

$$P(A_i + \Delta A_i)(A_i + \Delta A_i)'P \le PT_iP \quad (24)$$

Furthermore, we have

$$M + P\sum_{i=1}^{r+1}P(A_i + \Delta A_i)(A_i + \Delta A_i)'P \le 0$$

By means of Schur Complement Lemma, (20) holds, see *e.g.*, (Iwasaki, *et al.* 1994). Therefore, (16) implies

$$\dot{V} \le -x'Qx + \eta^{-1}w'D_0'R_\epsilon^{-1}D_0w \quad (25)$$

If $w = 0$, then for any allowable parameter uncertainties $F_i(t), F_b(t)$, $(i = 0, 1, 2, \cdots, r+1)$

$$\dot{V} \le -x'Qx \quad (26)$$

Therefore, we conclude that the resulting closed loop system is asymptotically stable.

Let $x(t)$ be the trajectories of the closed-loop system of (1) under (12) with initial condition $x(t) = 0$ $(t \le 0)$. Integrating both sides of (25) from 0 to t, we have

$$\lambda_{\min}(Q)\int_0^t \|x\|^2 ds \le \eta^{-1}\|D_0'R_\epsilon^{-1}D_0\|\int_0^t \|w\|^2 ds \quad (27)$$

Obviously, it follows from $w(t) \in L_2^p$ that $x(t) \in L_2^n$ and for any $w(t) \in L_2^p$ (27) implies

$$\|x\|_{L_2} \le \sqrt{\frac{\|D_0'R_\epsilon^{-1}D_0\|}{\eta\lambda_{\min}(Q)}}\|w\|_{L_2} \quad (28)$$

i.e., for any $\gamma > 0$, we can choose

$$\eta^*(\gamma) = \frac{\|D_0'R_\epsilon^{-1}D_0\|}{\gamma^2\lambda_{\min}(Q)}.$$

If $\eta \ge \eta^*(\gamma)$, then we have

$$\|x\|_{L_2} \le \gamma\|w\|_{L_2}$$

This, therefore, completes the proof of this theorem. □

Remarks:

(1) If (A_0, B) is stabilizable and there exist constant matrices $H_0 = BH_{00}$, A_{0i}, H_{0i} and $\epsilon_i > 0$ such that $H_i = BH_{0i}$, $A_i = BA_{0i}$ $(i = 1, 2, \cdots, r+1)$ and

$$\epsilon_i N_i'N_i < I,$$

$$R_\epsilon - \sum_{i=1}^{r+1}\left[A_{0i}(I - \epsilon_i N_i'N_i)^{-1}A_{0i}' + \epsilon_i^{-1}H_{0i}H_{0i}'\right] > 0$$

then, let

$$R(\epsilon_0) := R_\epsilon - \sum_{i=1}^{r+1}[A_{0i}(I - \epsilon_i N_i' N_i)^{-1} A_{0i}' + \epsilon_i^{-1} H_{0i} H_{0i}'] - \epsilon_0 H_{00} H_{00}', \qquad (29)$$

the matrix-valued function $R(\epsilon_0)$ is continuous function of ϵ_0. Therefore, there exists a sufficiently small $\epsilon_0 > 0$ such that $R(\epsilon_0) > 0$. It further implies that the algebraic Riccati equation (10) has a unique positive definite matrix solution P, see *e.g.*, (Knobloch, *et al.* 1993). By Theorem 4.1, the RADDPS for the system (1) is solvable.

(2) The parameters ϵ, ϵ_i introduced in the algebraic Riccati equation (10) is helpful to find a solvable solution. This fact can be shown in the illustrative example in Section 4.

(3) When $r = 0$, system (1) describes a class of nonlinear systems without time delay. In this case, Theorem 4 also gives a sufficient condition of RADDPS, and a design of state feedback law for a class of nonlinear systems without time delay. □

4. NUMERICAL EXAMPLE

Consider the second order nonlinear system with time delay described as follows.

$$\frac{d^2y(t)}{dt^2} - (4 + sint)\frac{dy(t)}{dt} + (3 + sint)\frac{dy(t-1)}{dt}$$

$$-costy(t-1) + (1 - |sint|)$$

$$\times \sqrt{y^2(t) + (\frac{dy(t)}{dt})^2 + y^2(t-1) + (\frac{dy(t-1)}{dt})^2}$$

$$= (4 + |cost|)u(t) + w(t) \qquad (30)$$

Let

$$\frac{dy(t)}{dt} = x_2(t), \quad y(t) = x_1(t),$$
$$x(t) = (\,x_1(t) \quad x_2(t)\,)',$$
$$g = \sqrt{x_1^2(t) + x_2^2(t) + x_1^2(t-1) + x_2^2(t-1)},$$

then the system (30) can be transformed into system (1), where

$$r = 1, \ F_0(t) = sint, \ F_1(t) = cost, \ F_2(t) = |sint|,$$
$$F_b(t) = |cost|, \ A_0 = \begin{pmatrix} 0 & 1 \\ -3 & 4 \end{pmatrix}, \ A_1 = \begin{pmatrix} 0 & 0 \\ 0 & -1 \end{pmatrix},$$
$$A_2 = \begin{pmatrix} 0 \\ -1 \end{pmatrix}, \ B = \begin{pmatrix} 0 \\ 4 \end{pmatrix}, \ D = \begin{pmatrix} 0 \\ 1 \end{pmatrix},$$
$$D_0 = H_{b0} = \frac{1}{4}, \ C_0 = C_1 = 1, \ E_0 = E_1 = I,$$
$$N_0 = (\,-1 \quad 1\,), \ N_1 = (\,1 \quad -1\,), \ N_2 = N_b = 1,$$
$$H_0 = H_1 = H_2 = H_b = (\,0 \quad 1\,)'.$$

If we choose

$$\epsilon_0 = 1, \ \epsilon = \epsilon_1 = \frac{1}{4}, \ \epsilon_2 = \frac{1}{2}, \ Q = 0.1I,$$

then the algebraic Riccati equation (11) has a unique positive definite solution as follows by MATLAB.

$$P = \begin{pmatrix} 7.4010 & 0.3810 \\ 0.3810 & 1.0210 \end{pmatrix}$$

Therefore, the state feedback controller

$$u(t) = -(1 + \eta)(1.5239x_1(t) + 4.0839x_2(t)) \qquad (31)$$

solves the RADDPS for the system (30). That is, if we choose

$$\eta^*(\gamma) = \frac{5}{12\gamma^2},$$

then the control law (31) solves the RADDPS.

The simulation results are given in Figures 1-3, where the initial condition is taken by $x_1(t) = 1, x_2(t) = 1$ for $t \leq 0$ and the $\eta = 0.5$. In this case the $\gamma < 1$.

5. CONCLUSION

This short paper highlights some new ideas of implementing a robust controller for RADDPS for a special class of nonlinear time-delay systems. Under the strong assumptions in (4) and (5), a state feedback controller is found via solving an algebraic Riccati equation. Further analysis are carried out in future study to relax these strong assumptions on RADDPS to a wider class of nonlinear systems.

6. REFERENCES

Isidori, A., (1995) *Nonlinear control systems, 3rd Edition, Spinger-Verlag, Berlin*

Isidori, A., (1996) A note on almost disturbance decoupling for nonlinear minimum phase system, *Systems & Control Letters*, Vol. 27, pp. 191-194

Isidori, A., (1996) Global almost disturbance decoupling with stability for non minimum-phase single-input single-output nonlinear systems, *Systems & Control Letters*, Vol. 27, pp. 115-122

Iwasaki, T. and R. E. Skelton, (1994) All controllers for the general H_∞ control problem: LMI existence condition and state space formulas, *Automatica*, Vol. 30, pp. 1307-1317

Knobloch, H. W., A. Isidori and D. Flockerzi, (1993) *Topics in Control Theory, Basel, Boston, Berlin, Birkhäuser, Verlag*

Lee, J. H., S. W. Kim and W. H. Kwon, (1994) Memoryless H_∞ controllers for state delayed systems, *IEEE Trans. Automat. Contr.*, Vol. 39, pp. 784-788

Lin, Z., A. Saberi and A. R. Teel, (1996) The almost disturbance decoupling problem with internal stability for linear systems subject to input saturation-state feedback case, *Automatica*, Vol. 32, pp. 619-624

Lin, Z., (1997) H_∞-almost disturbance decoupling problem with internal stability for linear systems subject to input saturation, *IEEE Trans. Automat. Contr.*, Vol. 42, pp. 992-995

Saberi, A. and P. Sannuti, (1988) Global stabilization with almost disturbance decoupling of a class of uncertain nonlinear systems, *Int. J. Contr.*, Vol. 41, pp. 1655-1704

Trinh, H and M. Aldeen, (1994) Stabilization of uncertain dynamic delay systems by memoryless feedback controllers, *Int. J. Control*, Vol. 59, pp. 1525-1542

Wang, Y., L. Xie and C. E. de Souza, (1992) Robust control of a class of uncertain nonlinear systems, *Systems & Control Letters*, Vol. 19, pp. 139-149

Weiland, S. and J. C. Willems, (1989) Almost disturbance decoupling, *IEEE Trans. Automat. Contr.*, Vol. 34, pp. 277-286

Wonham, W. M., (1985) *Linear Multivariable Control: a Geometric Approach, Springer Verlag*

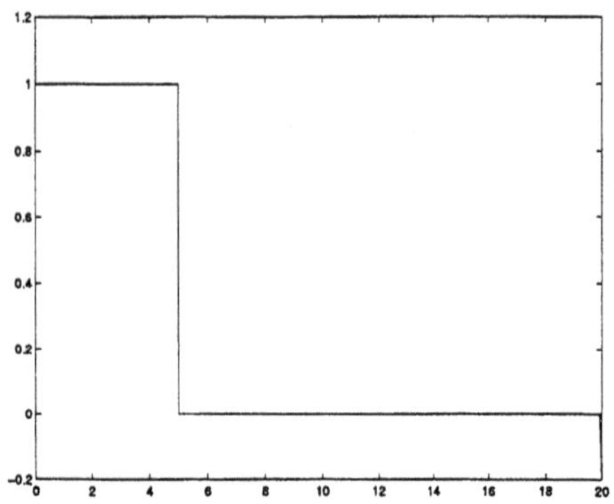

Fig. 1.1. $w(t)$

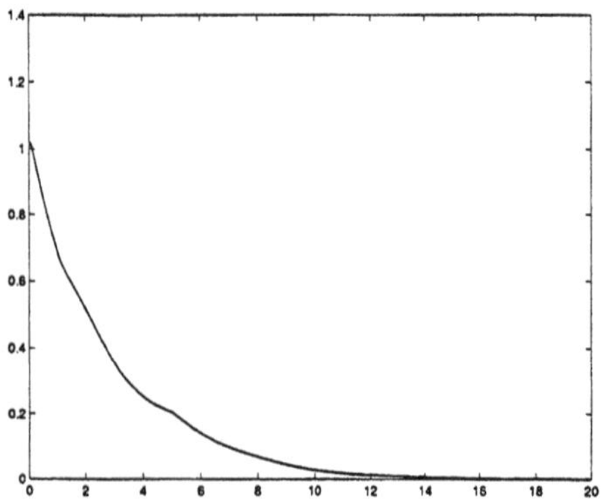

Fig. 1.2. $y(t)$

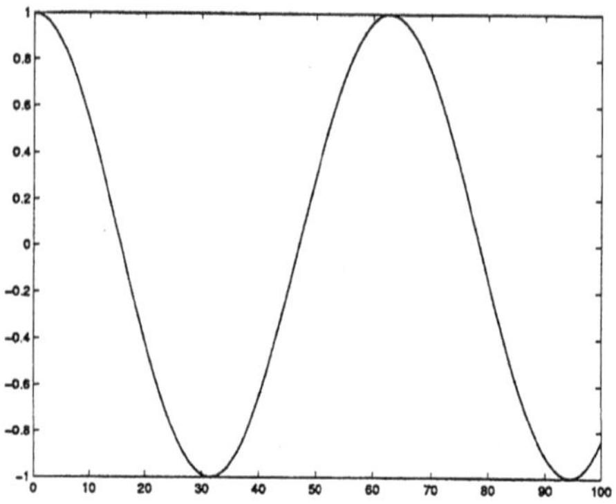

Fig. 2.1. $w(t)$

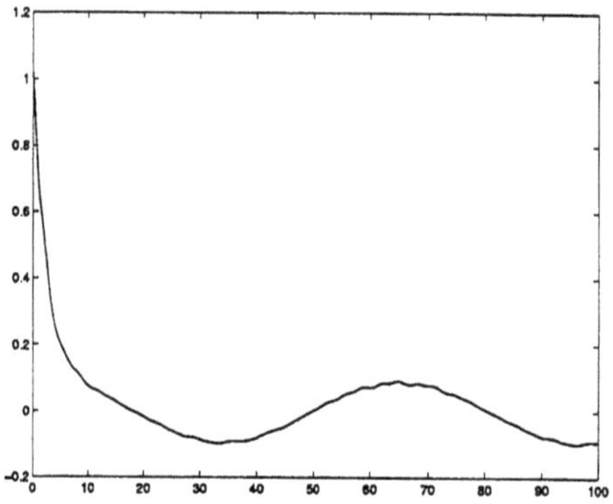

Fig. 2.2. $y(t)$

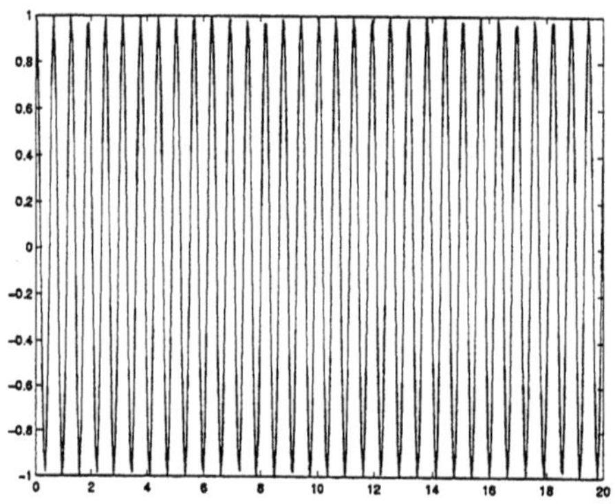

Fig. 3.1. $w(t)$

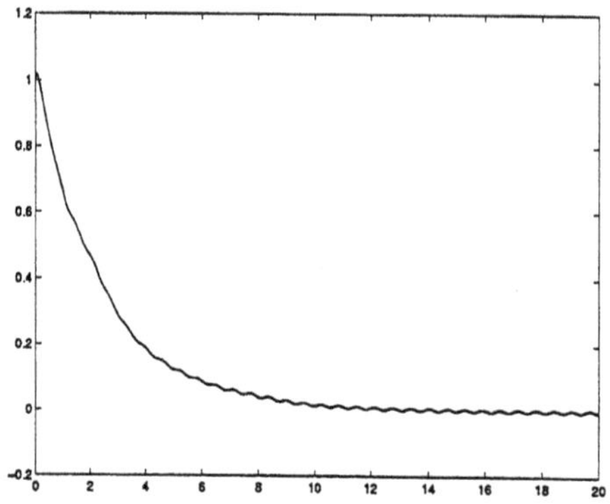

Fig. 3.2. $y(t)$

ADAPTIVE ROBUST CONTROL OF UNCERTAIN SYSTEMS WITH TIME-VARYING STATE DELAY

Jian-Xin Xu * Qing-Wei Jia * Tong-Heng Lee *

* *Department of Electrical Engineering*
National University of Singapore
10 Kent Ridge Crescent
Singapore 119260

Abstract: In this paper, a new adaptive robust control scheme is developed for a class of uncertain dynamical systems with time-varying state delay, unknown parameters and disturbances. By incorporating adaptive techniques into the robust control method, we propose a continuous adaptive robust controller which guarantees the uniform boundedness of the system and at the same time, the regulating error enters an arbitrarily designated zone in a finite time. The proposed controller is independent of the time-delay, hence it is applicable to a class of dynamical systems with uncertain time delays. The paper includes simulation studies demonstrating the performance of the proposed control scheme. *Copyright © 1999 IFAC*

Keywords: Adaptive robust control, time-delay system, uncertain system

1. INTRODUCTION

Time-varying delay has been often encountered in various engineering systems. Since the existence of such delay is frequently a source of instability, considerable attention has been paid to the study of systems with delay (Cheres *et al.*, 1989; Wu and Mizukami, 1996; Shyu and Yan, 1994; Hale, 1977; Phoojaruenchanachai and Furuta, 1992).

Cheres *et al.* (Cheres *et al.*, 1989) proposed mini-max controllers for time-varying dynamical systems with uncertain parameters, disturbance, and a bounded time-varying state delay. It is shown that the proposed min-max controllers can guarantee the uniform asymptotic stability of this class of uncertain time-delay systems. Because the so-called min-max controller is discontinuous, in (Cheres *et al.*, 1989) a continuous controller is employed to take the place of the min-max controller in the practical implementation. Thus, the asymptotic stability result can no longer be guaranteed practically. Continuous linear and nonlinear state feedback controllers are proposed in (Wu and Mizukami, 1996), where uncertain delay systems similar to (Cheres *et al.*, 1989) are considered. Uniform ultimate boundedness and uniform asymptotic stability can be guaranteed respectively. Note that the control methods proposed both in (Cheres *et al.*, 1989) and (Wu and Mizukami, 1996) have to solve the Riccati equation, whose analytical solutions are generally difficult to find and numerical solutions are time-consuming to obtain. It is also worth noting that all the controllers in (Cheres *et al.*, 1989; Wu and Mizukami, 1996) are designed based on the bounds of the unknown parameters, system disturbances and time-delay states. Hence the controller design may be conservative and may incur large control action.

In this paper, an adaptive robust control scheme is proposed for a class of uncertain dynamical systems with time-varying state delay, which includes the systems considered in (Cheres *et al.*, 1989; Wu and Mizukami, 1996; Shyu and Yan, 1994) as its subsets. To reduce the robust control efforts, we incorporate adaptive techniques to deal with the unknown parameters as well as the unknown gains in the bound of the delayed states. Here the uncertainties are supposed to be composed of two different categories: the first can be separated and expressed as the product of a set of known functions of states and a set of unknown constants, and the other category is not separable but with partially known bounding functions. The first category of uncertainties as well as the upper bound of the delayed states is dealt with by means of the well used adaptive control method. Meanwhile an adaptive robust method is applied to deal with the second category of uncertainties, where the unknown parameters in the upper bounding function are estimated with adaptation.

Since the time delay in the system states is assumed to be a non-negative continuous function with unknown bound, memory-based robust control methods fail to work. Analogous to (Cheres *et al.*, 1989) and (Wu and Mizukami, 1996), the Razumikhin-type theorem (Hale, 1977) is used to address this problem, but more general system uncertainties, nonlinearities as well as adaptive mechanisms are taken into account. As a consequence, the proposed controller is completely independent of the time-delay. It is shown that the proposed controller can guarantee the uniform boundedness of the system and at the same time, the regulating error enters an arbitrarily designated zone in a finite time.

This paper is organized as follows. Section 2 describes the class of nonlinear uncertain dynamical systems with time-varying state delay to be controlled. Section 3 gives the design procedure of the adaptive robust controller. Section 4 gives the stability analysis. Simulation-based studies on the proposed method are given in Section 5.

2. PROBLEM FORMULATION

Consider a class of uncertain dynamical system described by

$$\begin{aligned}\dot{\mathbf{x}}(t) = \mathbf{f}(\mathbf{x},t) + A_h(\mathbf{p},t)\mathbf{f}_h(\mathbf{x}(t-h(t)),t)\\ +B_0(t)\{[I+E(\mathbf{x},\mathbf{p},t)]\mathbf{u}(t)+\mathbf{d}_1(\mathbf{x},\mathbf{p},t)\\ +\mathbf{d}_2(\mathbf{x},\mathbf{p},t)\}\end{aligned} \quad (1)$$

where $\mathbf{x} = [x_1, x_2, \cdots, x_n]^\top \in \mathcal{R}^n$ is the measurable state vector of the system; $\mathbf{u}(t) \in \mathcal{R}^m$ is the control input vector of the system; $B_0(t) \in \mathcal{R}^{n\times m}$ is the input distribution matrix with full column rank; $\mathbf{f}(\mathbf{x},t) \in \mathcal{R}^n$ is a nonlinear function; $\mathbf{p} \in \mathcal{P}$ is an unknown system parameter vector and $\mathcal{P}$ is the set of admissible system parameters; $E(\mathbf{x},\mathbf{p},t)$ represents all possible perturbations of the input distribution matrix; $\mathbf{d}_1(\mathbf{x},\mathbf{p},t)$, $\mathbf{d}_2(\mathbf{x},\mathbf{p},t) \in \mathcal{R}^m$ are system uncertainties which may be nonlinear functions of the state $\mathbf{x}$, unknown parameter $\mathbf{p}$ and time t; the time delay $h(t)$ is a bounded and continuous function, i.e., $0 \le h(t) \le h^*$, where h^* is a nonnegative constant which may be unknown. The initial condition for (1) is given by

$$x(t) = \phi(t), \qquad t \in [t_0 - h^*, t_0] \quad (2)$$

where $\phi(t)$ is a continuous function on $[t_0 - h^*, t_0]$.

In this paper we make the following assumptions:

First, a continuous function vector $\mathbf{g}(\mathbf{x},t) \in \mathcal{R}^n$, $\mathbf{g}(0,t) = 0$ is chosen such that *Assumption A1* and *Assumption A2* are satisfied (Chen, 1987).

Assumption A1: For the chosen function vector $\mathbf{g}(\mathbf{x},t)$, there exists a C^1 Lyapunov function $V_1(\mathbf{x},t)$: $\mathcal{R}^n \times \mathcal{R}^+ \Longrightarrow \mathcal{R}^+$ and $\mathcal{K}$-class functions φ_i $(i = 1,2,3)$, such that

$$\varphi_1(||\mathbf{x}||) \le V_1(\mathbf{x},t) \le \varphi_2(||\mathbf{x}||)$$
$$\partial V_1/\partial t + (\partial V_1/\partial \mathbf{x})\mathbf{g}(\mathbf{x},t) \le -\varphi_3(||\mathbf{x}||),$$

where $||\cdot||$ is the Euclidean norm for vectors and the spectral norm for matrices. Note that this assumption assures that the zero state of the system described by

$$\dot{\mathbf{x}} = \mathbf{g}(\mathbf{x},t) \quad (3)$$

is uniformly asymptotical stable.

Assumption A2: There exists a function vector $v(\mathbf{x},t) \in \mathcal{R}^m$ such that

$$\mathbf{f}(\mathbf{x},t) - \mathbf{g}(\mathbf{x},t) = B_0 v(\mathbf{x},t). \quad (4)$$

Remark 1: *Assumption A2* restricts the structure of $\mathbf{g}(\mathbf{x},t)$. It should be noted that this assumption can always be satisfied if the system is with enough inputs (Chen, 1987). When the system (1) is linear, then the matching condition can be interpreted in a more precise way. Suppose $\mathbf{f}(\mathbf{x},t) = A\mathbf{x}$ and the pair (A, B_0) is stabilizable. Then one may choose $\mathbf{g}(\mathbf{x},t) = (A + B_0K)\mathbf{x}$, where K is a gain matrix which is chosen such that all the eigenvalues of $A + B_0K$ are with negative real parts. In such a case, it is easily to verify that *Assumption A2* is guaranteed with $v = -K\mathbf{x}$.

Assumption A3: For $E \in \mathcal{R}^{m\times m}$, there exist known bounding functions $r_{max}(\mathbf{x},t)$ and $r_{min}(\mathbf{x},t)$ such that

$$\forall t \in [0,\infty) \qquad \forall \mathbf{x} \in \mathcal{D} \qquad \forall \mathbf{p} \in \mathcal{P}$$
$$r_{max} \ \ge\ \lambda(\frac{1}{2}E + \frac{1}{2}E^\top) \ \ge\ r_{min} \ >\ -1 \quad (5)$$

where $\lambda(\cdot)$ indicates the eigenvalues of "$\cdot$".

Assumption A4: The following matching condition holds

$$A_h(\mathbf{p},t) = B_0(t)H_h(\mathbf{p},t)$$

where $||H_h(\mathbf{p},t)|| \le \rho_h(t)$ with $\rho_h(t) > 0$ a known function. Moreover

$$||\mathbf{f}_h(\mathbf{x}(t-h(t),t)|| \le \eta(\mathbf{x},t)||\mathbf{x}(t-h(t))||$$

where $\eta(\mathbf{x},t)$ is a positive function.

Assumption A5: The structured uncertainty $\mathbf{d}_1(\mathbf{x},\mathbf{p},t) = [d_{11}(\mathbf{x},\mathbf{p},t),\cdots,d_{1m}(\mathbf{x},\mathbf{p},t)]^\top \in \mathcal{R}^m$ can be expressed as

$$\begin{aligned}\mathbf{d}_1(\mathbf{x},\mathbf{p},t) &= \Theta(\mathbf{p})\xi(\mathbf{x},t)\\ \Theta &= diag(\theta_1^\top,\cdots,\theta_n^\top)\\ \xi &= [\xi_1,\xi_2,\cdots,\xi_n]^\top \end{aligned} \quad (6)$$

where $\theta_i^\top = [\theta_i^1,\cdots,\theta_i^{p_i}]$ is an unknown parameter vector and $\xi_i^\top = [\xi_i^1,\cdots,\xi_i^{p_i}]$ is a known function vector. The non-structured uncertainty $\mathbf{d}_2(\mathbf{x},\mathbf{p},t) = [d_{21}(\mathbf{x},\mathbf{p},t),\ \cdots\ ,d_{2n}(\mathbf{x},\mathbf{p},t)]^\top$ is bounded such that

$$\begin{aligned}&\forall t\in\mathcal{R}^+ \quad \forall\mathbf{x}\in\mathcal{D} \quad \forall\mathbf{p}\in\mathcal{P}\\ &||\mathbf{d}_2(\mathbf{x},\mathbf{p},t)|| \le \rho_d(\mathbf{x},\mathbf{q},t)\end{aligned} \quad (7)$$

where $\mathcal{D}$ is a compact subset of $\mathcal{R}^n$ in which the solution of (1) uniquely exists. $\rho_d(\mathbf{x},\mathbf{q},t)$ is an upper bounding function with unknown parameters (gains) $\mathbf{q}\in\mathcal{P}$. Here $\rho_d(\mathbf{x},\mathbf{q},t)$ is differentiable and convex to $\mathbf{q}$, that is

$$\rho_d(\mathbf{x},\mathbf{q}_2,t) - \rho_d(\mathbf{x},\mathbf{q}_1,t) \le (\mathbf{q}_2-\mathbf{q}_1)^\top \left.\frac{\partial\rho_d}{\partial\mathbf{q}}\right|_{\mathbf{q}_1} \quad (8)$$

Remark 2: (Cheres *et al.*, 1989) and (Wu and Mizukami, 1996) only consider the system uncertainty $\mathbf{d}_2$ together with a known upper bounding function, which is a particular case of (7) in the sense that $\mathbf{q}$ is known.

For simplicity, in this paper some time we use the abbreviation M in stead of $M(\cdot)$, where $M(\cdot)$ represents an arbitrary function or matrix used in this paper, and $\mathbf{x}_h$ instead of $\mathbf{x}(t-h(t))$.

3. THE ADAPTIVE ROBUST CONTROLLER

The adaptive robust technique is used in this section to develop a controller which guarantees the global boundedness of the system.

Under the *Assumption A2-5*, the system (1) can be transformed into

$$\begin{aligned}\dot{\mathbf{x}} &= \mathbf{f} + B_0[(I+E)\mathbf{u} + H_h\mathbf{f}_h + \mathbf{d}_1 + \mathbf{d}_2]\\ &= \mathbf{g} + B_0[(I+E)\mathbf{u} + v + H_h\mathbf{f}_h + \mathbf{d}_1 + \mathbf{d}_2]\end{aligned} \quad (9)$$

Define the parameter error matrices as

$$\tilde{\mathbf{q}} = \mathbf{q} - \hat{\mathbf{q}} \quad (10)$$

$$\tilde{\Theta} = \Theta - \hat{\Theta} \quad (11)$$

$$\tilde{\omega} = \omega - \hat{\omega} \quad (12)$$

where $\hat{\mathbf{q}}$, $\hat{\Theta}$ and $\hat{\omega}$ are the estimates of $\mathbf{q}$, Θ and ω respectively, and ω is defined in (24).

The control law $\mathbf{u}$ and the corresponding adaptive law are chosen to be

$$\mathbf{u} = \mathbf{u}_c + \mathbf{u}_v \quad (13)$$

$$\mathbf{u}_c = -v - \hat{\Theta}\xi - \mathbf{v}_d \quad (14)$$

$$\mathbf{u}_v = -\frac{r_{max}^2||\mathbf{u}_c||^2\alpha}{(1+r_{min})(r_{max}||\alpha^\top\mathbf{u}_c||+\varepsilon_1)} \quad (15)$$

$$\mathbf{v}_d = \frac{\hat{\rho}_d^2}{\hat{\rho}_d||\alpha||+\varepsilon_2}\alpha + \frac{k^2\hat{\omega}^2{\rho'_h}^2}{k\hat{\omega}\rho'_h||\alpha||+\varepsilon_3}\alpha + \frac{\delta_1{\rho'_h}^2\varrho^2}{\delta_2\rho'_h\varrho||\alpha||+\varepsilon_4}\alpha \quad (16)$$

$$\dot{\hat{\Theta}} = \Gamma_1(\alpha\xi^\top - \mu_1\hat{\Theta}) \quad (17)$$

$$\dot{\hat{\mathbf{q}}} = \Gamma_2(||\alpha||\left.\frac{\partial\rho_d}{\partial\mathbf{q}}\right|_{\hat{\mathbf{q}}} - \mu_2\hat{\mathbf{q}}) \quad (18)$$

$$\dot{\hat{\omega}} = \gamma_3(k\rho'_h||\alpha|| - \mu_3\hat{\omega}) \quad (19)$$

where $\varepsilon_1,\varepsilon_2,\varepsilon_3$, ε_4, δ_1, δ_2 and k are positive constants; and

$$\hat{\rho}_d \triangleq \rho_d(\mathbf{x},\hat{\mathbf{q}},t) \quad (20)$$

$$\alpha \triangleq B_0^\top\partial V_1/\partial\mathbf{x} \quad (21)$$

$$\rho'_h \triangleq \eta\rho_h \quad (22)$$

$$\varrho \triangleq ||\mathbf{x}|| + \sqrt{\lambda_{max}(\Gamma_1^{-1})}||\hat{\Theta}|| + \sqrt{\lambda_{max}(\Gamma_2^{-1})}||\hat{\mathbf{q}}|| + \sqrt{\gamma_3^{-1}}||\hat{\omega}|| \quad (23)$$

$$\omega \triangleq \sqrt{\lambda_{max}(\Gamma_1^{-1})}||\Theta|| + \sqrt{\lambda_{max}(\Gamma_2^{-1})}||\mathbf{q}|| \,. \quad (24)$$

γ_3 is a positive constant; Γ_i, $i=1,2$ are positive definite matrices chosen to be

$$\Gamma_i = diag(\gamma_{i1},\gamma_{i2},\cdots,\gamma_{in}). \quad (25)$$

μ_i, $i=1,2,3$, which constitute the μ-modification scheme, are defined as

$$\mu_i = \begin{cases} k_{ei}(\varepsilon_0 - ||\mathbf{x}||) & \mathbf{x}\in E_0\\ 0 & elsewhere\end{cases} \quad (26)$$

$$(27)$$

where k_{ei}, $i=1,2,3$ are positive constants.

$$E_0 \triangleq \{\mathbf{x} : ||\mathbf{x}|| < \varepsilon_0\} \quad (28)$$

where ε_0 is a positive constant specifying the desired regulating error bound.

4. CONVERGENCE ANALYSIS

For the above mentioned adaptive robust controller (13)-(19), we have the following theorem.

Theorem 1: By properly choosing the function vector $\mathbf{g}(\mathbf{x},t)$ such that $\varphi_3(||\mathbf{x}||) \geq k_0 V_1(\mathbf{x},t)$, where k_0 is a positive constant to be defined later, the proposed adaptive robust control law ensures that the system trajectory enters the set E_0 in a finite time. Moreover, the parameter estimation errors are bounded by the set

$$D = \{\tilde{\Theta}, \tilde{\mathbf{q}}, \tilde{\omega} : \quad trace\{\tilde{\Theta}^\top \tilde{\Theta}\} + \tilde{\mathbf{q}}^\top \tilde{\mathbf{q}} + \tilde{\omega}^2 < k'[\frac{1}{2}k_{e1}\varepsilon_0\, trace\{\Theta^\top \Theta\} + \frac{1}{2}k_{e2}\varepsilon_0\, \mathbf{q}^\top \mathbf{q} + \frac{1}{2}k_{e3}\varepsilon_0\, \omega^2 + \varepsilon]\} \tag{29}$$

where k' is defined to be

$$k' = \frac{1}{k''\ min\{\lambda_{min}(\Gamma_1^{-1}), \lambda_{min}(\Gamma_2^{-1}), \gamma_3^{-1}\}}$$

$$k'' = \frac{2min\{k_0, k_{e1}\delta, k_{e2}\delta, k_{e3}\delta\}}{max\{1, \lambda_{max}(\Gamma_1^{-1}), \lambda_{max}(\Gamma_2^{-1}), \gamma_3^{-1}\}}\ .$$

$\lambda_{max}(A)$ and $\lambda_{min}(A)$ indicates the maximum and minimum eigenvalues of the matrix A respectively. δ and k_0 are positive values to be defined later. □□

Proof:

The following positive definite function V is selected

$$V = V_1(\mathbf{x},t) + \frac{1}{2}trace\{\tilde{\Theta}^\top \Gamma_1^{-1}\tilde{\Theta}\} + \frac{1}{2}\tilde{\mathbf{q}}^\top \Gamma_2^{-1}\tilde{\mathbf{q}} + \frac{1}{2}\gamma_3^{-1}\tilde{\omega}^2 \tag{30}$$

Taking the derivative of V along the trajectory of the dynamic system (1) with the control law (13)-(19), and using (9) we have

$$\begin{aligned}
\dot{V} &= \partial V_1/\partial t + (\partial V_1/\partial \mathbf{x})\{\mathbf{g} + B_0[(I+E)\mathbf{u} + v \\
&\quad + H_h\mathbf{f}_h + \mathbf{d}_1 + \mathbf{d}_2]\} - trace\{\tilde{\Theta}^\top \Gamma_1^{-1}\dot{\hat{\Theta}}\} \\
&\quad - \tilde{\mathbf{q}}^\top \Gamma_2^{-1}\dot{\hat{\mathbf{q}}} - \gamma_3^{-1}\tilde{\omega}\dot{\hat{\omega}} \\
&= \partial V_1/\partial t + (\partial V_1/\partial \mathbf{x})\mathbf{g} + \alpha^\top[(I+E)\mathbf{u}_v + E\mathbf{u}_c \\
&\quad - \mathbf{v}_d + H_h\mathbf{f}_h + \tilde{\Theta}\xi + \mathbf{d}_2] - \gamma_3^{-1}\tilde{\omega}\dot{\hat{\omega}} \\
&\quad - trace\{\tilde{\Theta}^\top(\alpha\xi^\top - \mu_1\hat{\Theta})\} - \tilde{\mathbf{q}}^\top \Gamma_2^{-1}\dot{\hat{\mathbf{q}}} \\
&= \partial V_1/\partial t + (\partial V_1/\partial \mathbf{x})\mathbf{g} + \alpha^\top[(I+E)\mathbf{u}_v + E\mathbf{u}_c \\
&\quad - \mathbf{v}_d + H_h\mathbf{f}_h + \tilde{\Theta}\xi + \mathbf{d}_2)] - \gamma_3^{-1}\tilde{\omega}\dot{\hat{\omega}} \\
&\quad - \alpha^\top\tilde{\Theta}\xi + \mu_1 trace\{\tilde{\Theta}\hat{\Theta}^\top\} - \tilde{\mathbf{q}}^\top \Gamma_2^{-1}\dot{\hat{\mathbf{q}}} \\
&\leq \partial V_1/\partial t + (\partial V_1/\partial \mathbf{x})\mathbf{g} - \alpha^\top \mathbf{v}_d + r_{max}||\alpha^\top \mathbf{u}_c|| \\
&\quad - \alpha^\top(I+E)\frac{r_{max}^2||\mathbf{u}_c||^2\alpha}{(1+r_{min})(r_{max}||\alpha^\top \mathbf{u}_c|| + \varepsilon_1)} \\
&\quad + ||\alpha|| \cdot ||H_h|| \cdot ||\mathbf{f}_h|| + ||\alpha|| \cdot ||\mathbf{d}_2|| \\
&\quad + \mu_1 trace\{\tilde{\Theta}\hat{\Theta}^\top\} - \tilde{\mathbf{q}}^\top \Gamma_2^{-1}\dot{\hat{\mathbf{q}}} - \gamma_3^{-1}\tilde{\omega}\dot{\hat{\omega}}
\end{aligned} \tag{31}$$

In above derivation the following property of *trace* is used:

$$trace\{Q^\top \mathbf{v}\mathbf{w}^\top\} = \mathbf{v}^\top Q\mathbf{w} \tag{32}$$

where $\mathbf{v} \in \mathcal{R}^{n\times 1}$, $\mathbf{w} \in \mathcal{R}^{l\times 1}$, and $Q \in \mathcal{R}^{n\times l}$.

From *Assumption A1*, we have

$$\partial V_1/\partial t + (\partial V_1/\partial \mathbf{x})\mathbf{g} \leq -\varphi_3(||\mathbf{x}||)\ .$$

Following the Razumikhin-type theorem (Hale, 1977), we further assume that for a positive number $\vartheta > 1$, the following inequality holds

$$V(\mathbf{x}(\tau), \tilde{\Theta}(\tau), \tilde{\mathbf{q}}(\tau), \tilde{\omega}(\tau), \tau) < \vartheta^2 V(\mathbf{x}(t), \tilde{\Theta}(t), \tilde{\mathbf{q}}(t), \tilde{\omega}(t), t) \tag{33}$$

where $t - h^* \leq \tau \leq t$. Then we can easily have (see Appendix A)

$$||\mathbf{x}(t - h(t))|| \leq \vartheta\varrho + k\omega \tag{34}$$

with ϱ and ω defined in (23) and (24) respectively. It can be seen that this condition is more general than (Cheres *et al.*, 1989) and (Wu and Mizukami, 1996) because the upper bound in the right hand side of (34) may not be zero even when $\mathbf{x}(t)$=0.

Using the fact that

$$\alpha^\top(I+E)\alpha = \alpha^\top(I + \frac{E+E^\top}{2})\alpha$$

$$\frac{I + \frac{1}{2}E + \frac{1}{2}E^\top}{1 + r_{min}} \geq 1\ ,$$

it follows that

$$\begin{aligned}
\dot{V} &\leq -\varphi_3(||\mathbf{x}||) - \alpha^\top \mathbf{v}_d + \eta\rho_h||\alpha|| \cdot ||\mathbf{x}_t|| \\
&\quad + ||\alpha|| \cdot ||\mathbf{d}_2|| + \mu_1 trace\{\tilde{\Theta}\hat{\Theta}^\top\} - \tilde{\mathbf{q}}^\top \Gamma_2^{-1}\dot{\hat{\mathbf{q}}} \\
&\quad - \gamma_3^{-1}\tilde{\omega}\dot{\hat{\omega}} + \varepsilon_1 \\
&\leq -\varphi_3(||\mathbf{x}||) - \alpha^\top \mathbf{v}_d + \rho_h'||\alpha|| \cdot (\vartheta\varrho + k\omega) \\
&\quad + ||\alpha|| \cdot ||\mathbf{d}_2|| + \mu_1 trace\{\tilde{\Theta}\hat{\Theta}^\top\} - \tilde{\mathbf{q}}^\top \Gamma_2^{-1}\dot{\hat{\mathbf{q}}} \\
&\quad - \gamma_3^{-1}\tilde{\omega}\dot{\hat{\omega}} + \varepsilon_1 \\
&\leq -\varphi_3(||\mathbf{x}||) - \frac{\hat{\rho}_d^2||\alpha||^2}{\hat{\rho}_d||\alpha|| + \varepsilon_2} - \frac{k^2\hat{\omega}^2\rho_h'^2||\alpha||^2}{k\hat{\omega}\rho_h'||\alpha|| + \varepsilon_3} \\
&\quad - \frac{\delta_1\rho_h'^2\varrho^2||\alpha||^2}{\delta_2\rho_h'\varrho||\alpha|| + \varepsilon_4} + \vartheta\rho_h'\varrho||\alpha|| + k\omega\rho_h'||\alpha|| \\
&\quad + \rho_d||\alpha|| - ||\alpha||(\mathbf{q} - \hat{\mathbf{q}})^\top \left.\frac{\partial\rho_d}{\partial\mathbf{q}}\right|_{\hat{\mathbf{q}}} - k\tilde{\omega}\rho_h'||\alpha|| \\
&\quad + \mu_1 trace\{\tilde{\Theta}\hat{\Theta}^\top\} + \mu_2\tilde{\mathbf{q}}^\top\hat{\mathbf{q}} + \mu_3\tilde{\omega}\hat{\omega} + \varepsilon_1
\end{aligned}$$

It can be easily verified that

$$-\frac{k^2\hat{\omega}^2\rho_h'^2||\alpha||^2}{k\hat{\omega}\rho_h'||\alpha|| + \varepsilon_3} + k\omega\rho_h'||\alpha|| - k\tilde{\omega}\rho_h'||\alpha|| \leq \varepsilon_3\ .$$

If choosing δ_1/δ_2 sufficiently large, we also have

$$-\frac{\delta_1 {\rho_h'}^2 \varrho^2 ||\alpha||^2}{\delta_2 \rho_h' \varrho ||\alpha|| + \varepsilon_4} + \vartheta \rho_h' \varrho ||\alpha||$$
$$= \frac{(-\delta_1 + \delta_2 \vartheta){\rho_h'}^2 \varrho^2 ||\alpha||^2 + \varepsilon_4 \vartheta \rho_h' \varrho ||\alpha||}{\delta_2 \rho_h' \varrho ||\alpha|| + \varepsilon_4}$$
$$\leq \frac{\varepsilon_4 \vartheta \rho_h' \varrho ||\alpha||}{\delta_2 \rho_h' \varrho ||\alpha|| + \varepsilon_4}$$
$$\leq \varepsilon_4 .$$

It then follows that

$$\begin{aligned}
\dot{V} \leq & -\varphi_3(||\mathbf{x}||) - \frac{\hat{\rho}_d^2 ||\alpha||^2}{\hat{\rho}_d ||\alpha|| + \varepsilon_2} + \rho_d ||\alpha|| \\
& -||\alpha||(\rho_d - \hat{\rho}_d) + \mu_1 trace\{\tilde{\Theta}^\top \hat{\Theta}\} \\
& +\mu_2 \tilde{\mathbf{q}}^\top \hat{\mathbf{q}} + \mu_3 \tilde{\omega}\hat{\omega} + \varepsilon_1 + \varepsilon_3 + \varepsilon_4 \\
\leq & -\varphi_3(||\mathbf{x}||) + \mu_1 trace\{\tilde{\Theta}^\top(\Theta - \tilde{\Theta})\} \\
& +\mu_2 \tilde{\mathbf{q}}^\top(\mathbf{q} - \tilde{\mathbf{q}}) + \mu_3 \tilde{\omega}(\omega - \tilde{\omega}) + \varepsilon \\
\leq & -\varphi_3(||\mathbf{x}||) - \frac{1}{2}\mu_1 trace\{\tilde{\Theta}^\top \tilde{\Theta}\} - \frac{1}{2}\mu_2 \tilde{\mathbf{q}}^\top \tilde{\mathbf{q}} \\
& -\frac{1}{2}\mu_3 \tilde{\omega}^2 + \frac{1}{2}\mu_1 trace\{\Theta^\top \Theta\} + \frac{1}{2}\mu_2 \mathbf{q}^\top \mathbf{q} \\
& +\frac{1}{2}\mu_3 \omega^2 + \varepsilon. \qquad (35)
\end{aligned}$$

where $\varepsilon = \varepsilon_1 + \varepsilon_2 + \varepsilon_3 + \varepsilon_4$.

By choosing $\mathbf{g}(\mathbf{x}, t)$ such that

$$\varphi_3(||\mathbf{x}||) \geq k_0 V_1(\mathbf{x}, t) \qquad (36)$$
$$k_0 = \frac{\varepsilon + c_1}{\varphi_1(\varepsilon_0)} \qquad (37)$$

where c_1 is an arbitrary positive constant, then from (26) we have

$$\dot{V} \leq -\varphi_3(||\mathbf{x}||) + \varepsilon \leq -c_1 \ , \forall\, \mathbf{x} \in \mathcal{R}^n - E_0, \quad (38)$$

which implies that there exists a constant $0 < \varepsilon_0' < \varepsilon_0$ such that

$$\forall \mathbf{x} \in \mathcal{R} - E_0' \qquad \dot{V} < 0 \qquad (39)$$

where $E_0' \triangleq \{\mathbf{x} : \ ||\mathbf{x}|| < \varepsilon_0'\}$ is a subset of E_0. This indicates that the system will enter the set E_0 in a finite time.

When $\mathbf{x} \in E_0'$, it is obvious that

$$k_{ei}(\varepsilon_0 - \varepsilon_0') \leq \mu_i \leq k_{ei}\varepsilon_0, \quad i = 1, 2, 3 \ . \quad (40)$$

Define $\delta = \varepsilon_0 - \varepsilon_0' > 0$, then from (26), (35) and (40) we obtain

$$\begin{aligned}
\dot{V} \leq & -\varphi_3(||\mathbf{x}||) - \frac{1}{2}k_{e1}\delta\ trace\{\tilde{\Theta}^\top \tilde{\Theta}\} - \frac{1}{2}k_{e2}\delta\ \tilde{\mathbf{q}}^\top \tilde{\mathbf{q}} \\
& -\frac{1}{2}k_{e3}\delta\ \tilde{\omega}^2 + \frac{1}{2}k_{e1}\varepsilon_0\ trace\{\Theta^\top \Theta\} \\
& +\frac{1}{2}k_{e2}\varepsilon_0\ \mathbf{q}^\top \mathbf{q} + \frac{1}{2}k_{e3}\varepsilon_0\ \omega^2 + \varepsilon \\
\leq & -k''V + \frac{1}{2}k_{e1}\varepsilon_0\ trace\{\Theta^\top \Theta\} + \frac{1}{2}k_{e2}\varepsilon_0\ \mathbf{q}^\top \mathbf{q} \\
& +\frac{1}{2}k_{e3}\varepsilon_0\ \omega^2 + \varepsilon \qquad (41)
\end{aligned}$$

where k'' is defined in (29).

By solving (41) we can establish that

$$\begin{aligned}
V(t) \leq & e^{-k''t}V(t=0) + \frac{1}{k''}[\frac{1}{2}k_{e1}\varepsilon_0\ trace\{\Theta^\top \Theta\} \\
& +\frac{1}{2}k_{e2}\varepsilon_0\ \mathbf{q}^\top \mathbf{q} + \frac{1}{2}k_{e3}\varepsilon_0\ \omega^2 + \varepsilon]
\end{aligned}$$

which implies that the parameter estimation errors $\tilde{\Theta}$, $\tilde{\mathbf{q}}$ and $\tilde{\omega}$ converge exponentially to the residual set (29). □□

Remark 3: It is easy to find an appropriate scalar function V_1 and a vector $\mathbf{g}$ to meet the condition (36). For instance, we can simply choose $V_1 = \frac{1}{2}\mathbf{x}^\top\mathbf{x}$ and $\mathbf{g} = -\frac{\varepsilon + c_1}{2\varphi_1(\varepsilon_0)}\mathbf{x}$, then

$$\varphi_3(||\mathbf{x}||) = \frac{\varepsilon + c_1}{\varphi_1(\varepsilon_0)} V_1.$$

5. SIMULATION STUDY

In this section, a simulation study is given to demonstrate the effectiveness of the proposed control scheme.

Consider a nonlinear uncertain time-varying system described by (1) associated with

$$\mathbf{f} = \begin{bmatrix} -1.5x_1(1 + x_2^2) \\ sin(2t)x_1 + 0.3x_1x_2 \end{bmatrix}, \qquad B_0 = \begin{bmatrix} 0.0 \\ 1.0 \end{bmatrix},$$

$$H_h = [cos(x_1) \ \ sin(x_1x_2)] \ , \ \ \mathbf{f}_h = 0.9\mathbf{x}(t - h(t)) \ ,$$

$$d_1 = [\theta_1 \ \ \theta_2] \begin{bmatrix} sin(x_1) \\ x_1x_2 \end{bmatrix},$$

$$d_2 = \sqrt{0.7sin^2(8x_1) + 0.8cos^2(5t)x_2^2} \ .$$

The real values of the unknown parameter θ_1 and θ_2 are 0.3 and -0.4 respectively. The delay function $h(t)$ is given in Fig.1. The upper bound of d_2 is $\rho_d = \sqrt{q_1 + q_2x_2^2}$ with unknown constants q_1 and q_2. We choose the function vector $\mathbf{g}(\mathbf{x}, t)$ as

$$\mathbf{g} = \begin{bmatrix} -1.5x_1(1 + x_2^2) \\ -5x_2 \end{bmatrix},$$

then obviously $v = [sin(2t)x_1 + 0.3x_1x_2 + 5x_2]$.

The control and adapting parameters are chosen to be

$$\varepsilon_1 = \varepsilon_2 = \varepsilon_3 = \varepsilon_4 = 0.1, \quad k_{e1} = k_{e2} = k_{e3} = 1.0,$$
$$\Gamma_1 = \Gamma_2 = diag(1, 1), \qquad \gamma_3 = 1 \ .$$

The initial state condition is $\mathbf{x}(t) = [-0.7cos(t) \quad -0.4cos(t)]^\top$, $t \in [-h^*, 0]$, and the initial values of the parameter estimates $\hat{\Theta}$, $\hat{\mathbf{q}}$ and $\hat{\omega}$ are zeros.

The simulation results are given in Fig.2. It can be seen that the control scheme has achieved a good performance, and the control signal is quite smooth.

6. CONCLUSION

In this paper, the control problem for a class of dynamical systems with time-varying state delay, unknown parameters and disturbances is considered. By combining adaptive control and robust control methods, a new continuous adaptive robust controller is developed which guarantees the uniform boundedness of the system and assure that the regulating error enter the arbitrarily designated zone in a finite time. The effectiveness of the control schemes is verified by both theoretical analysis and simulation studies.

7. REFERENCES

Chen, Y.H. (1987). Adaptive robust model-following control and application to robot manipulators. *Journal of Dynamic Systems, Measurement, and Control* **109**, 209–215.

Cheres, E., S. Gutman and Z.J. Palmor (1989). Stabilization of uncertain dynamic systems including state delay. *IEEE Trans. Automatic Control* **34**, 1199–1203.

Hale, J.K. (1977). *Theory of functional differential equations.* New York: Springer-Verlag.

Phoojaruenchanachai, S. and K. Furuta (1992). Memoryless stabilization of uncertain linear systems including time-varying state delays. *IEEE trans. on automatic control* **37**, 1022–1026.

Shyu, K.K. and J.J. Yan (1994). Variable structure model following adaptive control for systems with time-varying delay. *Control Theory and Advanced Technology* **10**, 513–521.

Wu, Hansheng and Koichi Mizukami (1996). Linear and nonlinear stabilizing continuous controllers of uncertain dynamical systems including state delay. *IEEE trans. on automatic control* **1**, 116–121.

Xu, Jian-Xin, Tong-Heng Lee and Qing-Wei Jia (1997). An adaptive robust control scheme for a class of nonlinear uncertain systems. *International Journal of Systems Science* **28**, 429–434.

APPENDIX A: The Bound of $\mathbf{x}(t - h(t))$

Denote $\Sigma = trace\{\tilde{\Theta}^\top \Gamma_1^{-1}\tilde{\Theta}\} + \tilde{\mathbf{q}}^\top \Gamma_2^{-1}\tilde{\mathbf{q}} + \gamma_3^{-1}\tilde{\omega}^2$, then for $t - h^* \leq \tau \leq$, from (33) we have

$$\frac{1}{2}\mathbf{x}^\top(\tau)\mathbf{x}(\tau) + \frac{1}{2}\Sigma(\tau) \leq \frac{\vartheta^2}{2}[\mathbf{x}^\top(t)\mathbf{x}(t) + \Sigma(t)].$$

It follows that

$$\begin{aligned}\mathbf{x}^\top(\tau)\mathbf{x}(\tau) &\leq \vartheta^2(\mathbf{x}^\top(t)\mathbf{x}(t) + \Sigma(t)) \\ &\leq \vartheta^2[\mathbf{x}^\top(t)\mathbf{x}(t) + \lambda_{max}(\Gamma_1^{-1})||\tilde{\Theta}||^2 \\ &\quad + \lambda_{max}(\Gamma_2)^{-1}||\tilde{\mathbf{q}}||^2 + \gamma_3^{-1}\tilde{\omega}^2] \\ &\leq \vartheta^2(||\mathbf{x}|| + \sqrt{\lambda_{max}(\Gamma_1^{-1})}||\tilde{\Theta}|| \\ &\quad + \sqrt{\lambda_{max}(\Gamma_2^{-1})}||\tilde{\mathbf{q}}|| + \sqrt{\gamma_3^{-1}}||\tilde{\omega}||)^2 .\end{aligned}$$

Then we get

$$\begin{aligned}||\mathbf{x}(\tau)|| &\leq \vartheta(||\mathbf{x}|| + \sqrt{\lambda_{max}(\Gamma_1^{-1})}||\tilde{\Theta}|| \\ &\quad + \sqrt{\lambda_{max}(\Gamma_2^{-1})}||\tilde{\mathbf{q}}|| + \sqrt{\gamma_3^{-1}}||\tilde{\omega}|| \\ &\leq \vartheta(||\mathbf{x}|| + \sqrt{\lambda_{max}(\Gamma_1^{-1})}||\hat{\Theta}|| + \sqrt{\gamma_3^{-1}}\omega \\ &\quad + \sqrt{\lambda_{max}(\Gamma_2^{-1})}||\hat{\mathbf{q}}|| + \sqrt{\gamma_3^{-1}}||\hat{\omega}|| \\ &\quad + \sqrt{\lambda_{max}(\Gamma_1^{-1})}||\Theta|| + \sqrt{\lambda_{max}(\Gamma_2^{-1})}||\mathbf{q}||) \\ &\leq \vartheta\varrho + (1 + \sqrt{\gamma_3^{-1}})\vartheta\omega \\ &\leq \vartheta\varrho + k\omega \qquad (42)\end{aligned}$$

where k is a positive value which is sufficiently large such that $k > (1 + \sqrt{\gamma_3^{-1}})\vartheta$; ϱ and ω are defined in (23) and (24) respectively. □□

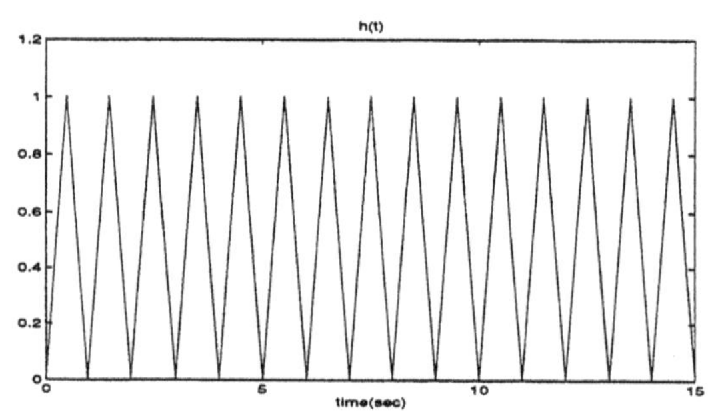

Fig. 1. Time-varying delay h(t).

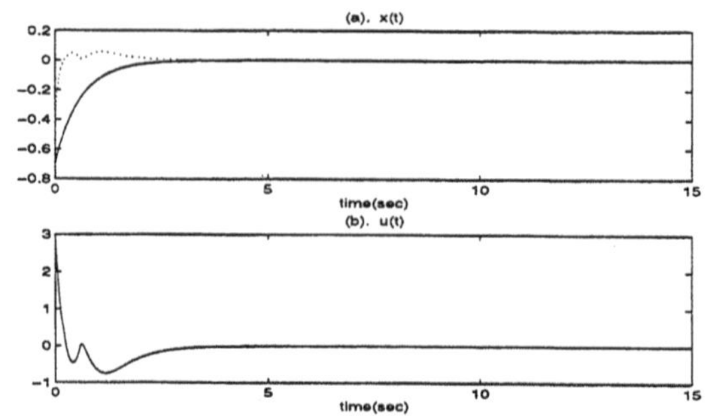

Fig. 2. (a). —: $x_1(t)$; ···: $x_2(t)$; (b). Evolution of input voltage u.

E-2c-07-4
14th Triennial World Congress, Beijing, P.R. China

CONSTRAINED FEEDBACK CONTROL OF A CLASS OF ADDITIVELY PERTURBED NONLINEAR NEUTRAL TIME-DELAY SYSTEMS

I.D. Clarkson[†] and D.P. Goodall[‡]

Control Theory & Applications Centre, Coventry University, Priory Street, Coventry, CV1 5FB, UK.

[†]*Fax.: +44 (0) 1203 838052* *Email: I.Clarkson@coventry.ac.uk*
[‡]*Fax.: +44 (0) 1203 838080* *Email: D.Goodall@coventry.ac.uk*

Abstract: Adopting a deterministic methodology based on Lyapunov stability theory and Lyapunov-Krasovskiǐ functionals, *delay-independent* stability criteria are proposed which are sufficient to ensure that all global closed-loop trajectories for a class of perturbed nonlinear delay systems of the neutral-type, subject to constraints on the control input, uniformly asymptotically converge to the state origin. An application of the results to a specific sub-class is provided and a class of robust stabilizing controls, satisfying prescribed constraints, is presented.

Keywords: Control constraints; asymptotic stability; uncertain nonlinear neutral delay systems; feedback stabilization; Lyapunov-Krasovskiǐ functionals.

1. INTRODUCTION

Almost all industrial processes exhibit inherent delay properties which arise due to the fact that intervals of time between the application of a stimuli and the system's response can occur. In addition, as a reflection of nature's variability and our 'limited' understanding, 'imperfections' frequently occur. The combined influence of these two phenomenon can have a significant affect on the dynamic behaviour of the system; in particular, these influences may have an adverse effect on the stability properties of the system. Consequently, there has been recent interest in stability analysis of perturbed time-delay systems; for example, (Amemiya and Leitmann, 1994; Choi and Chung, 1995; Clarkson and Goodall, 1997; Goodall, 1998; Verriest and Aggoune, 1998; Wu and Mizukami, 1996) utilized deterministic methodologies.
For this investigation, the uncertainty in the system is modelled as a class of nonlinear time-varying additive perturbations influencing a known idealized (nominal) model, and is assumed to be time, state, delayed-state and/or input dependent. Within this class of perturbations, the uncertainty can be divided into two distinct categories: *matched* and *residual* (or *unmatched*); the difference being that the controller can directly influence the former but only indirectly influence the latter. Additionally, *a priori* bounding knowledge of the system's uncertainty is assumed, which assures the robustness of any stabilizing feedback controller.
The main feature of such controllers is that the gains depend explicitly on the upper bounds of the uncertainty as a consequence of which 'high' gains may be encountered. This can be undesirable as the controllers may not satisfy any constraints imposed on the system and so, subsequently, there is need to synthesize controllers that can operate

within an acceptable threshold level. This situation has been addressed for delay free systems, with and without the presence of uncertainty, by, for example (Glazos and Żak, 1995; Lu, 1997) and for delay systems by (Clarkson and Goodall, 1998; Goodall, 1997; Hennet and Tarbouriech, 1997).
For this stabilizability problem, the known nominal time-delay system is assumed to be time-varying, nonlinear and of the neutral-type; that is time-delays arise in the state velocity vector. Such systems, without the presence of uncertainty, have been investigated by (Hui and Hu, 1997; Tchangani, *et al.*, 1998), but, to the knowledge of the authors, there is very little literature on the stabilization of uncertain neutral systems subject to control constraints. For uncertain linear systems of the neutral type, see (Goodall, 1997).
In this paper, under an appropriate feasibility condition, a class of stabilizing feedback controls, with memory, satisfying a prescribed set of constraints is presented. The controls have a similar structure to those in (Wu and Mizukami, 1996) which enable *global asymptotic* stability results to be attained. Utilizing Lyapunov stability theory and Lyapunov-Krasovskii functionals, *delay-independent* stability criteria are derived that are sufficient to ensure the stabilization of a general class, and a specific sub-class, of nonlinear uncertain neutral-type time-delay systems, under a constrained control hypothesis.

2. NOMENCLATURE

Let $\mathbb{R} := (-\infty,\ \infty)$, $\mathbb{R}_0^+ := [0,\ \infty)$ and define the Euclidean inner product on the vector space $\mathbb{R}^n$ and the induced norm by $\langle\cdot,\cdot\rangle$ and $\|\cdot\|$, respectively. If $\mathcal{X} \subset \mathbb{R}^p$ and $\mathcal{Y} \subset \mathbb{R}^q$, $C(\mathcal{X};\mathcal{Y})$ denotes the space of all continuous functions mapping $\mathbb{R}^p \to \mathbb{R}^q$, whereas $C^1(\mathcal{X};\mathcal{Y}) \subset C(\mathcal{X};\mathcal{Y})$ denotes the space of continuous functions which have continuous first order partial derivatives. Additionally, introduce the notation $x_t = x_t(r) := x(t+r)$ ($r \in [-\tau, 0]$, $\tau > 0$) to denote the restriction of $x(\cdot)$ to the interval $[t-\tau, t]$ and note that if $x_t \in C([-\tau,\ 0];\mathbb{R}^n)$, then $\|x_t\|_\tau := \sup_{r\in[-\tau,0]} \|x(t+r)\|$. Let a set of bounded functions in $C([-\tau,\ 0];\mathbb{R}^n)$ be defined by $\mathcal{Q}_A := \{q \in C([-\tau,\ 0];\mathbb{R}^n) : \|q\|_\tau < A,\ 0 < A < \infty\}$.

3. A CLASS OF NEUTRAL FUNCTIONAL DIFFERENTIAL EQUATIONS

The primary aim of this investigation is to synthesize robust continuous *bounded* feedback controllers to ensure global uniform asymptotic stability, (see (Kolmanovskii and Myshkis, 1992) for a definition) of a general class, and then a specific sub-class, of imperfectly known nonlinear time-delay systems, which are perturbations to a known nominal delay system of the neutral type.

The uncertain systems, to be described in subsequent sections, form a sub-class of a more general class of neutral functional differential equations with the following structure

$$\frac{\mathrm{d}}{\mathrm{d}t}\Big(x(t) - \tilde{G}(t, x_t)\Big) = F(t, x_t),\ t \geq t_0, \tag{1}$$

$$x_{t_0}(\theta) = \phi(\theta), \quad \theta \in [-\tau, 0], \quad \phi(0) = x^0, \tag{2}$$

where $x(t) \in \mathbb{R}^n$, F, $\tilde{G} : [t_0,\ \infty) \times \mathcal{Q}_A \to \mathbb{R}^n$ are continuous maps with $F(t,0) \equiv \tilde{G}(t,0) \equiv 0$, and $\phi \in C([-\tau,\ 0];\mathbb{R}^n)$. Also, it is assumed that there exists $l > 0$ such that $\|F(t,\vartheta)\| < l$ for all $t \geq t_0$, $\vartheta \in \mathcal{Q}_A$.

Remark. Clearly (1)-(2) reduces to a retarded functional differential equation if $\tilde{G}(t,\vartheta) \equiv 0$.

To achieve the main objective, it will be necessary to invoke the following lemma, associated with system (1)-(2), which can be found in Theorem 1.4, Chapter 4 in (Kolmanovskii and Myshkis, 1992). The conditions imposed on $\tilde{G}$, $F : \mathbb{R} \times \mathcal{Q}_A \to \mathbb{R}^n$ are sufficiently smooth to ensure existence and indefinite continuation of solutions.

Lemma 1 *Let $w(t,\vartheta) := \vartheta(0) - \tilde{G}(t,\vartheta)$ and define Ω to be the set of scalar non-decreasing functions $\ell \in C(\mathbb{R};\mathbb{R})$, such that $\ell(r) > 0$ for $r > 0$ and $\ell(0) = 0$. Consider system* (1)-(2) *and suppose there exists $\ell_1 \in \Omega$ and real $0 < \xi < 1$ such that*

$$\left\|\tilde{G}(t+s,\vartheta) - \tilde{G}(t,\chi)\right\|_\tau \leq \ell_1(s) + \xi\,\|\vartheta - \chi\|_\tau, \tag{3}$$

for all ϑ, $\chi \in \mathcal{Q}_A$, $t \geq t_0$ and $s \geq 0$. In addition, if there exists $\ell_i \in \Omega$, $i = 2,3,4$, with $\ell_2(\zeta) \to \infty$ as $\zeta \to \infty$, and a C^1 functional $V : [t_0,\ \infty) \times \mathcal{Q}_A \times \mathbb{R}^n \to \mathbb{R}_0^+$, such that

(a) *for all $(t,\vartheta) \in \mathbb{R} \times \mathcal{Q}_A$,*

$$\ell_2(\|w(t,\vartheta)\|) \leq V(t,\vartheta,w(t,\vartheta)) \leq \ell_3(\vartheta),$$

(b) *along solutions to* (1)-(2),

$$\dot{V}(t,x_t,w(t,x_t)) + \ell_4(\|x_t(0)\|) \leq 0,$$

*the equilibrium, $x = 0$, is **globally uniformly asymptotically stable**.*

4. HYPOTHESES AND PROBLEM FORMULATION

4.1 A General Class of Additively Perturbed Neutral Time-Delay Systems

For a general class of systems under consideration, the nominal (known) time delay system, considered here, is a nonlinear system of the neutral-type described by

$$\frac{\mathrm{d}}{\mathrm{d}t}\Big(x(t) - H(t, x(t-\tau))\Big) = f(t, x(t), x(t-\tau)) + \sum_{i\in\mathcal{I}} u_i(t) g_i(x(t)),\quad \tau > 0,\ t \in (t_0,\ \infty), \tag{4}$$

$$x_{t_0}(\theta) = \phi(\theta), \quad \theta \in [-\tau,\ 0], \quad \phi(0) = x^0, \tag{5}$$

where $x(t) \in \mathbb{R}^n$ is the *state* vector, $u(t) \in \mathbb{R}^m$ $(1 \leq m \leq n)$ is the *control* (or *input*) vector and $\mathcal{I} := \{1, 2, \ldots, m\}$. In addition, τ is a *bounded* delay which affects the system and with $x^0 \in \mathbb{R}^n$ specified, $\phi(\cdot)$ is a continuous vector-valued function satisfying the initial condition $x(t_0) = \phi(0) = x^0$. It is supposed that $g_i \in C(\mathbb{R}^n; \mathbb{R}^n)$ is locally uniformly bounded with respect to the state, and $f \in C(\mathbb{R} \times \mathbb{R}^n \times \mathbb{R}^n; \mathbb{R}^n)$ and $H \in C(\mathbb{R} \times \mathbb{R}^n; \mathbb{R}^n)$, satisfying $f(t, 0, 0) = H(t, 0) = 0$ for all t, are known. Moreover, for any $0 < A < \infty$, there exists $f_A > 0$ such that $\|f(t, \zeta, \psi)\| < f_A$ for all $t \geq t_0$ and $\zeta,\ \psi \in \mathcal{Q}_A$.

For this nominal time-delay system the following hypotheses are assumed to hold.

Hypothesis 1

(a) Let $z(t, \vartheta) := \vartheta(0) - H(t, \vartheta(-\tau))$. Suppose there exists $\omega_i \in \Omega$, $i = 1, 2, \ldots, 5$, with $\omega_1(\zeta) \to \infty$ as $\zeta \to \infty$, a C^1 functional $V_1 : \mathbb{R}^n \to \mathbb{R}_0^+$, a known continuous function $\rho : [t_0, \infty) \to [0, \bar{\rho}]$, with $\bar{\rho} \geq 0$, and known real constants a, b, $c > 0$, such that the following inequalities hold

  (i) $V_1(0) = 0$.

  (ii) $\omega_1(\|z(t, \vartheta)\|) \leq V_1(z(t, \vartheta)) \leq \omega_2(\vartheta)$.

  (iii) $\|\nabla V_1(z(t, \vartheta))\| \leq$
  $a\,\{\omega_3(\|\vartheta(0)\|)\}^{\frac{1}{2}} + b\,\{\omega_3(\|\vartheta(-\tau)\|)\}^{\frac{1}{2}}$.

  (iv) $\langle \nabla V_1(z(t, \vartheta)), f(t, \vartheta(0), \vartheta(-\tau)) \rangle \leq$
  $\rho(t)\,\{\omega_3(\|\vartheta(0)\|)\}^{\frac{1}{2}}\{\omega_3(\|\vartheta(-\tau)\|)\}^{\frac{1}{2}}$
  $+ c\,\omega_5(\|\vartheta(-\tau)\|) - c\,\omega_4(\|\vartheta(0)\|)$,

  where $\omega_4(r) \geq \omega_3(r) + \omega_5(r)$ for all $r \in \mathbb{R}$.

(b) There exists $\omega_6 \in \Omega$ and real constant $0 < \xi < 1$ such that, for all $\zeta,\ \psi \in \mathcal{Q}_A$, $t \geq t_0$ and $s \geq 0$,

$$\|H(t+s, \zeta) - H(t, \psi)\|_\tau \leq \omega_6(s) + \xi\,\|\zeta - \psi\|_\tau.$$

The class of systems to be investigated consists of the nominal nonlinear system, (4), perturbed by unknown functions $p_i : \mathbb{R} \times \mathbb{R}^n \times \mathbb{R}^n \times \mathbb{R}^m \to \mathbb{R}^m$ and $q : \mathbb{R} \times \mathbb{R}^n \times \mathbb{R}^n \times \mathbb{R}^m \to \mathbb{R}^n$ which are assumed to be continuous and belong to a known non-empty class which comprise all possible system uncertainty, including any known time-dependent and/or nonlinear elements. In particular, the class of systems to be studied has the form

$$\frac{\mathrm{d}}{\mathrm{d}t}\Big(x(t) - H(t, x(t-\tau))\Big) = f(t, x(t), x(t-\tau)) + \sum_{i \in \mathcal{I}} \big(u_i(t) + p_i(t, x(t), x(t-\tau), u(t))\big) g_i(x(t)) + q(t, x(t), x(t-\tau), u(t)), \quad (6)$$

$$x_{t_0}(\theta) = \phi(\theta), \quad \theta \in [-\tau,\ 0], \quad \phi(0) = x^0, \quad (7)$$

where the uncertainty functions are assumed to satisfy the following boundedness assumptions.

Hypothesis 2 For all $(t, y, \bar{y}, u) \in \mathbb{R} \times \mathbb{R}^n \times \mathbb{R}^n \times \mathbb{R}^m$, there exists known continuous functions $\kappa_i : [t_0,\ \infty) \to [0,\ \bar{\kappa}_i]$, with $0 \leq \bar{\kappa}_i < 1$ for all i, and known real constants α_i, β_i, γ_i, β, $\gamma \geq 0$, such that

$$|p_i(t, y, \bar{y}, u)| \leq \alpha_i + \beta_i\{\omega_3(\|y\|)\}^{\frac{1}{2}} + \gamma_i\{\omega_3(\|\bar{y}\|)\}^{\frac{1}{2}} + \kappa_i(t)|u_i|,$$

$$\|q(t, y, \bar{y}, u)\| \leq \beta\,\{\omega_3(\|y\|)\}^{\frac{1}{2}} + \gamma\,\{\omega_3(\|\bar{y}\|)\}^{\frac{1}{2}}.$$

To ensure that the class of feedback controls satisfies the constraints, of the form $|u_i(t)| \leq \rho_i$, $\rho_i > 0$ for all i with ρ_i prescribed, the following feasibility hypothesis is required.

Hypothesis 3 $\rho_i - (1 - \bar{\kappa}_i)^{-1}\alpha_i > 0$, for all i.

Finally, global stability can be achieved provided an additional hypothesis is satisfied.

Hypothesis 4 There exists real constants δ_1, $\delta_2 > 0$ such that, for all i,

$$|\langle \nabla V_1(z(t, \vartheta)), g_i(\vartheta(0)) \rangle| \leq \delta_1\,\{\omega_3(\|\vartheta(0)\|)\}^{\frac{1}{2}} + \delta_2\,\{\omega_3(\|\vartheta(-\tau)\|)\}^{\frac{1}{2}}.$$

Remark. Hypotheses 1(a)(iii) and 4 impose severe restrictions on the nonlinearity of the functions g_i.

The following subsection describes a second class of systems to be investigated, which is a sub-class of those systems described by (6)-(7).

4.2 A Specific Sub-Class of Perturbed Neutral Time-Delay Systems

Before defining the specific sub-class of systems of (6)-(7), some additional notation is introduced. Let $\mathbb{R}^{p \times p}$ represent the set of real square matrices of order $p \times p$ and if $P \in \mathbb{R}^{p \times p}$, $P > 0$ denotes that P is positive definite. For P, $Q \in \mathbb{R}^{p \times p}$, the statement $P > Q$ denotes that $P - Q$ is positive definite. Finally, let $\sigma_{\min(\max)}(\cdot)$ denote the minimum (maximum) eigenvalue of a symmetric positive definite matrix.

For the specific sub-class of systems, to be investigated in Section 6, it is supposed that the nominal system has the structure

$$\frac{\mathrm{d}}{\mathrm{d}t}\Big(x(t) - Gx(t-\tau)\Big) = f_1(x(t)) + f_2(x(t-\tau)) + \sum_{i \in \mathcal{I}} u_i(t) g_i(x(t)), \quad \tau > 0\ t \in (t_0,\ \infty), \quad (8)$$

$$x_{t_0}(\theta) = \phi(\theta), \quad \theta \in [-\tau,\ 0], \quad \phi(0) = x^0, \quad (9)$$

where $G \in \mathbb{R}^{n \times n}$, $f_1 \in C(\mathbb{R}^n; \mathbb{R}^n)$, $f_2 \in C(\mathbb{R}^n; \mathbb{R}^n)$ and, for all i, $g_i \in C(\mathbb{R}^n; \mathbb{R}^n)$ is locally uniformly bounded with respect to the state.

It is assumed that system (8)-(9) is subject to the

following hypotheses.

Hypothesis 5

(a) Given any real positive definite symmetric K_1, $K_2 \in \mathbb{R}^{n\times n}$, and any real continuous symmetric matrix-valued function $x \mapsto Q(x) \in \mathbb{R}^{n\times n}$, with $Q(x) > 0$ for all x, and, if it exists, $\liminf_{\|x\|\to\infty} Q(x) > 0$, there exists a unique symmetric $P > 0$ which satisfies the following algebraic Ricatti-type matrix equation

$$(\mathrm{D}f_1)^{\mathrm{T}}(x)P+P(\mathrm{D}f_1)(x)+K_2+Q(x)+R(x) \\ +(\mathrm{D}f_1)^{\mathrm{T}}(x)PK_1^{-1}P(\mathrm{D}f_1)(x) = O, \quad (10)$$

for which

$$R(x) := (\mathrm{D}f_2)^{\mathrm{T}}(x)PK_2^{-1}P(\mathrm{D}f_2)(x)+G^{\mathrm{T}}K_1G \\ -(\mathrm{D}f_2)^{\mathrm{T}}(x)PG-G^{\mathrm{T}}P(\mathrm{D}f_2)(x) \geq 0 \text{ for all } x,$$

where $(\mathrm{D}f)(x)$ denotes the Fréchet derivative of f at x; namely the Jacobian matrix of f.

(b) $0 < \|G\| < 1$.

Remark. Condition (10) is a fairly severe restriction on the nonlinearity of the functions f_1 and f_2.

For this specific case, the class of uncertain time-delay systems consists of the nominal system, described by (8)-(9), with analogous augmentation as for the more general system (6)-(7); namely

$$\frac{\mathrm{d}}{\mathrm{d}t}\Big(x(t) - Gx(t-\tau)\Big) = f_1(x(t)) + f_2(x(t-\tau)) \\ +\sum_{i\in\mathcal{I}} \big(u_i(t) + p_i(t, x(t), x(t-\tau), u(t))\big)g_i(x(t)) \\ + q(t, x(t), x(t-\tau), u(t)), \quad (11)$$

$$x_{t_0}(\theta) = \phi(\theta), \quad \theta \in [-\tau,\ 0], \quad \phi(0) = x^0, \quad (12)$$

whilst the uncertainty functions are assumed to satisfy Hypothesis 6, below.

Hypothesis 6 For all $(t, y, \bar{y}, u) \in \mathbb{R} \times \mathbb{R}^n \times \mathbb{R}^n \times \mathbb{R}^m$, there exists real known constants $\hat{\alpha}_i\ \hat{\beta}_i\ \hat{\gamma}_i\ \tilde{\beta},\ \tilde{\gamma} \geq 0$ and known continuous functions $\kappa_i : [t_0, \infty) \to [0,\ \bar{\kappa}_i]$, with $0 \leq \bar{\kappa}_i < 1$ for all i, such that

$$|p_i(t, y, \bar{y}, u)| \leq \hat{\alpha}_i + \hat{\beta}_i \|y\| + \hat{\gamma}_i \|\bar{y}\| + \kappa_i(t)|u_i|,$$
$$\|q(t, y, \bar{y}, u)\| \leq \tilde{\beta}\|y\| + \tilde{\gamma}\|\bar{y}\|.$$

Remark. Observe that for Hypotheses 2 and 6, p_i is the *matched* uncertainty whereas q is the *residual* uncertainty.

Similar to Hypothesis 3, the following feasibility hypothesis is required if the the class of feedback controls is to satisfy the control constraints.

Hypothesis 7 $\rho_i - (1-\bar{\kappa}_i)^{-1}\hat{\alpha}_i > 0$, for all i.

This paper considers the problem of stabilizing systems (6)-(7) and (11)-(12) when the components of the control input satisfy constraints of the form $|u_i| \leq \rho_i$, $\rho_i > 0$ for all i with ρ_i prescribed. Given initial states x^0 and ϕ each satisfying (7) and (12), the principal objective is to design a class of state-feedbacks, $\mathcal{F}$, subject to the imposed constraints and dependent on some design parameters, which will globally stabilize systems (6)-(7) and (11)-(12).

5. STABILIZATION OF THE GENERAL CLASS OF NEUTRAL SYSTEMS

In this section, a sufficient condition for the stabilization of system (6)-(7) is presented, assuming Hypotheses 1 - 4 hold. The class of constrained feedback controls, with memory, consist of nonlinear functions, $\tilde{f}(t, x_t) = [\tilde{f}_1(t, x_t), \ldots, \tilde{f}_m(t, x_t)]^{\mathrm{T}} \in \mathcal{F}$ which are continuous in $\mathbb{R}^n \setminus \{0\}$ and depend upon some design parameters. Each component of the control, $\tilde{f}_i(t, x_t)$, consists of two parts; a function used to stabilize the nominal nonlinear system, modelled by (4)-(5), and, loosely speaking, a multiplicative gain component designed to counteract any uncertainty in the system. The components of the controller, $\tilde{f}_i(t, x_t)$, are of the same structure as those used in (Wu and Mizukami, 1996), and are given by

$$(t, x_t) \mapsto \tilde{f}_i(t, x_t) := \\ -\frac{(1-\bar{\kappa}_i)^{-1}\alpha_i^2\,\langle\nabla V_1(z(t, x_t)), g_i(x)\rangle}{\alpha_i|\langle\nabla V_1(z(t, x_t)), g_i(x)\rangle| + \mu_i\,\omega_3(\|x\|)}, \quad (13)$$

where, for all i, the design parameters $\mu_i > 0$ are real.

Remark. Hypothesis 3 ensures that the controller, defined by (13), satisfies the prescribed constraints.

A Lyapunov-Krasovskiĭ functional candidate of the form

$$V(x_t, z(t, x_t)) := V_1(z(t, x_t)) \\ + \int_{t-\tau}^{t} \lambda\,\omega_4(\|x(\theta)\|_\tau) + (c-\lambda)\,\omega_5(\|x(\theta)\|_\tau)\,\mathrm{d}\theta, \quad (14)$$

where $0 \leq \lambda < c$, is utilized for the stability analysis. Invoking Lemma 1 and utilizing $u(t) = \tilde{f}(t, x_t)$, defined in (13), the following result can be deduced.

Define $\bar{\mu} := \sum_{i\in\mathcal{I}} \mu_i$, $\bar{\beta} := \sum_{i\in\mathcal{I}} \beta_i$, $\bar{\gamma} := \sum_{i\in\mathcal{I}} \gamma_i$ and $l := c - \bar{\rho} - (a+b)(\beta+\gamma) - (\delta_1+\delta_2)(\tilde{\beta}+\tilde{\gamma})$.

Suppose $l > 0$, then μ_i can be chosen to ensure

$$\bar{l} := l - \bar{\mu} > 0. \quad (15)$$

Theorem 1 *Assume Hypotheses* 1 - 4 *hold. If*

$$l > 0 \quad (16)$$

then, with design parameters μ_i chosen to satisfy (15), *the state origin is globally uniformly asymptotically stable for the class of systems modelled by* (6)-(7) *with constrained feedback control $u(t) = \tilde{f}(t, x_t)$, with $\tilde{f}$ given by* (13).

Proof The proof of Theorem 1 requires the existence and continuation of solutions, and satisfaction of the conditions of Lemma 1. The previously stated conditions on f, g_i, p_i, q, V_1 and ω_3 are sufficient for the existence and indefinite continuation of solutions, see (Kolmanovskii and Myshkis, 1992). Identifying system (6)-(7) with that of (1)-(2), it is evident $H(t,\vartheta) \equiv \tilde{G}(t,\vartheta)$ and in view of the conditions imposed on $H(t,\vartheta)$, condition (3) is satisfied. To confirm the remainder of Lemma 1, an analysis is performed using the Lyapunov-Krasovskiĭ functional candidate, given by (14). Invoking Hypothesis 1(a)(ii), straightforward calculations confirm that $V(x_t, z(t,x_t))$ satisfies

$$\omega_1(\|z(t,x_t)\|) \le V(x_t, z(t,x_t)) \le \omega_2(x_t) + \tau\left\{\lambda\,\omega_4(\|x_t\|_\tau) + (c-\lambda)\omega_5(\|x_t\|_\tau)\right\}.$$

Thus, property (a) of Lemma 1 is verified, provided $0 \le \lambda < c$. The parameter λ is introduced for analytical purposes and in the ensuing analysis it is shown that an appropriate value of λ can be chosen which satisfies $0 \le \lambda < c$. It now remains to verify condition (b) of Lemma 1.

As a consequence of Hypotheses 1(a), 2 and (13) it can be shown that along solutions to (6)-(7), for almost all t,

$$\begin{aligned}\dot V(x_t, z(t,x_t)) &\le -(c-\lambda-a\,\beta-\bar\mu)\,\omega_3(\|x(t)\|)\\ &+ (\bar\rho + a\,\gamma + b\,\beta)\,\{\omega_3(\|x(t)\|)\}^{\frac12}\,\{\omega_3(\|x_t(-\tau)\|)\}^{\frac12}\\ &+ \sum_{i\in\mathcal{I}} |\langle\nabla V_1(z(t,x_t)), g_i(x(t))\rangle|\\ &\times\left\{\{\omega_3(\|x(t)\|)\}^{\frac12}\beta_i + \{\omega_3(\|x_t(-\tau)\|)\}^{\frac12}\gamma_i\right\}\\ &- (\lambda - b\gamma)\,\omega_3(\|x_t(-\tau)\|). \qquad (17)\end{aligned}$$

For global stability, Hypothesis 4 is required. Thus, almost everywhere, inequality (17) simplifies to

$$\begin{aligned}\dot V(x_t, z(t,x_t)) &\le -(c-\lambda-a\beta-\delta_1\bar\beta-\bar\mu)\,\omega_3(\|x(t)\|)\\ &+(\bar\rho + a\,\gamma + b\,\beta + \delta_2\,\bar\beta + \delta_1\,\bar\gamma)\\ &\times\{\omega_3(\|x(t)\|)\}^{\frac12}\,\{\omega_3(\|x_t(-\tau)\|)\}^{\frac12}\\ &-(\lambda - b\,\gamma - \delta_2\,\bar\gamma)\,\omega_3(\|x_t(-\tau)\|),\\ &= -\langle\omega(x_t), M_\lambda\,\omega(x_t)\rangle,\end{aligned}$$

where $\omega(x_t) := [\{\omega_3(\|x(t)\|)\}^{\frac12}\ \{\omega_3(\|x_t(-\tau)\|)\}^{\frac12}]^{\mathrm{T}}$,

$$M_\lambda = \begin{bmatrix} c-\lambda-k_1 & -\frac12 k_2\\ -\frac12 k_2 & \lambda - k_3\end{bmatrix},$$

with $k_1 := a\,\beta + \delta_1\,\bar\beta + \bar\mu$, $k_2 := \bar\rho + a\,\gamma + b\,\beta + \delta_2\,\bar\beta + \delta_1\,\bar\gamma$ and $k_3 := b\,\gamma + \delta_2\,\bar\gamma$.

It remains to show that there exists $\lambda = \lambda^*$, satisfying $0 \le \lambda^* < c$, such that $M_{\lambda^*} > 0$.

Now, $\det(M_\lambda) = -\lambda^2 + (c-k_1+k_3)\lambda + k_1\,k_3 - k_3\,c - \frac14 k_2^2$ and it can be shown that the function $\lambda \mapsto -\lambda^2 + (c-k_1+k_3)\lambda$ has a local maximum, which occurs when $\lambda = \lambda^* := \frac12(c-k_1+k_3)$.

Subsequently, a straightforward calculation shows that $\det(M_{\lambda^*}) = \frac14\,\bar l\,(\bar l + 2\,k_2)$ and, in view of (15)-(16), M_{λ^*} is positive definite. It is also easily verified that $0 \le \lambda^* < c$.

Hence, along solutions to (6)-(7) and for almost all t,

$$\dot V(x_t, z(t,x_t)) \le -\sigma_{\min}(M_\lambda^*)\,\omega_3(\|x(t)\|).$$

Finally, invoking Lemma 1, the result readily follows.

□

6. STABILIZATION OF THE SPECIFIC SUB-CLASS OF NEUTRAL SYSTEMS

This final section investigates global uniform asymptotic stability for the class of systems described by (11)-(12), which are a sub-class of (6)-(7).

The proposed feedback controller for this system is

$$x_t \mapsto \tilde f_i(x_t) := -\frac{2(1-\bar\kappa_i)^{-1}\hat\alpha_i^2\,\langle g_i(x), P\tilde z(x_t)\rangle}{2\hat\alpha_i|\langle g_i(x), P\tilde z(x_t)\rangle| + \eta_i s^2(x)}, \quad (18)$$

where $s^2(x) := \int_0^1 \langle x, Q(\zeta x)x\rangle\,\mathrm{d}\zeta$, the design parameters $\eta_i > 0$, for all i, are real and $\tilde z(\vartheta) := \vartheta(0) - G\vartheta(-\tau)$.

Remark. Hypothesis 7 ensures that the controller, (18), satisfies the prescribed constraints.

Define $\bar\eta := \sum_{i\in\mathcal{I}}\eta_i$, $\hat\beta := \sum_{i\in\mathcal{I}}\hat\beta_i$, $\hat\gamma := \sum_{i\in\mathcal{I}}\hat\gamma_i$ and

$$\begin{aligned}k &:= 1 - \{\sigma_{\min}(\tilde Q)\}^{-\frac12}\,(\delta_1+\delta_2)\,(\hat\beta+\hat\gamma)\\ &\quad - 2\,\{\sigma_{\min}(\tilde Q)\}^{-1}\,(\|P\| + \|PG\|)\,(\tilde\beta + \tilde\gamma).\end{aligned}$$

Suppose $k > 0$, then η_i can be chosen to ensure

$$\bar k := k - \bar\eta > 0. \quad (19)$$

The following corollary may now be deduced from Theorem 1.

Corollary 1 *Assume Hypotheses* 5 - 7 *hold. If*

$$\{\sigma_{\min}(\tilde Q)\}^{\frac12} > \frac12\left\{(\delta_1+\delta_2)\,(\hat\beta+\hat\gamma) + \sqrt{\left((\delta_1+\delta_2)(\hat\beta+\hat\gamma)\right)^2 + 8(\tilde\beta+\tilde\gamma)(\|P\|+\|PG\|)}\right\}, \quad (20)$$

then utilizing the constrained feedback controller defined by (18) *and with* η_i *satisfying* (19), *the state origin of system* (11), *subject to the initial condition* (12), *is globally uniformly asymptotically stable.*

Proof To prove this result, the approach adopted is that of identifying Hypotheses 5 - 7 with those of Hypotheses 1-3, respectively, and then invoking Theorem 1 to satisfy the conditions of Lemma 1. Consider first the problem of identifying Hypothesis 5 with Hypothesis 1. Let $x_t \mapsto V_1(\tilde{z}(x_t)) := \langle \tilde{z}(x_t), P\tilde{z}(x_t)\rangle$, then Hypothesis 1(a)(i)-(ii) holds with $\omega_1(\zeta) = \sigma_{\min}(P)\zeta^2$, $\omega_2(\zeta) = \sigma_{\max}(P)\,\|\tilde{z}(\zeta)\|_\tau^2$.

As a consequence of Hypothesis 5(a) there exists a constant positive definite symmetric matrix $\tilde{Q} \in \mathbb{R}^{n\times n}$ such that $Q(x) > \tilde{Q}$ for all x. Consequently,

$$\|x\| \leq \{\sigma_{\min}(\tilde{Q})\}^{-\frac{1}{2}} s(x), \qquad \text{for all } x, \tag{21}$$

and utilizing this result, Hypothesis 1(a)(iii) holds with $\omega_3(\|x\|) = s^2(x)$, $a = 2\,\|P\|\,\{\sigma_{\min}(\tilde{Q})\}^{-\frac{1}{2}}$ and $b = 2\,\|PG\|\,\{\sigma_{\min}(\tilde{Q})\}^{-\frac{1}{2}}$.

In view of (21), the two inequalities

$$\langle y, My\rangle + 2\langle y, z\rangle + \langle z, M^{-1}z\rangle \equiv \langle y + M^{-1}z, M[y + M^{-1}z]\rangle \geq 0,$$
$$\text{and } 2\,|\langle My, z\rangle| \leq \langle z, Nz\rangle + \langle My, N^{-1}My\rangle,$$

for positive definite symmetric $M,\ N \in \mathbb{R}^{n\times n}$ and $y,\ z \in \mathbb{R}^n$,

$$\langle \nabla V_1(\tilde{z}(t, x_t)), f_1(x(t)) + f_2(x_t(-\tau))\rangle \leq -s^2(x(t)) - \tilde{s}^2(x(t)) + \tilde{s}^2(x_t(-\tau)),$$

where $\tilde{s}^2(x) := \int_0^1 \langle x, R(\zeta x)x\rangle\, d\zeta$. Therefore, Hypothesis 1(a)(iv) is valid with $c \equiv 1$, $\omega_4(\|x\|) = s^2(x) + \tilde{s}^2(x)$, $\omega_5(\|x\|) = \tilde{s}^2(x)$ and $\rho(t) \equiv 0$. Additionally, Hypothesis 5(b) ensures Hypothesis 1(b) (and hence (3)) holds. Note that the inequality $\omega_4(r) \geq \omega_3(r) + \omega_5(r)$, for all $r \in \mathbb{R}$, holds.

Identifying Hypothesis 6 with Hypothesis 2 and Hypothesis 3 with Hypothesis 7, invoking (21), $\alpha_i = \hat{\alpha}_i$, $\beta_i = \hat{\beta}_i\{\sigma_{\min}(\tilde{Q})\}^{-\frac{1}{2}}$, $\gamma_i = \hat{\gamma}_i\{\sigma_{\min}(\tilde{Q})\}^{-\frac{1}{2}}$, $\beta = \tilde{\beta}\{\sigma_{\min}(\tilde{Q})\}^{-\frac{1}{2}}$ and $\gamma = \tilde{\gamma}\{\sigma_{\min}(\tilde{Q})\}^{-\frac{1}{2}}$.

The result then follows from Theorem 1.

□

Remarks

(i) The stability criteria, (16) and (20), are *independent* of the delay. Note that such stability criteria tend to be more conservative than their *delay-dependent* counterparts.

(ii) The main drawback of using a controller with memory is that it requires the delay to be known precisely for implementation purposes; information which may not be readily available.

(iii) For any persistently acting residual uncertainty, global stability can still be achieved, but the result is now defined in terms of a compact set, containing the state origin. This situation has been addressed for a retarded functional differential equation by (Clarkson and Goodall, 1998).

REFERENCES

Amemiya, T. and G. Leitmann (1994). A method for designing a stabilizing control for a class of uncertain linear delay systems. *Dyn. and Con.*, **4**, 147-167.

Choi, H.H and M.J. Chung (1995). Memoryless stabilization of uncertain dynamic systems with time-varying delayed states and controls. *Automatica*, **31**, 1349-1351.

Clarkson, I.D. and D.P. Goodall (1997). Memoryless feedback controllers for imperfectly known nonlinear systems with time-varying delay. *Sys. Sci.*, **23**, 59-73.

Clarkson, I.D. and D.P. Goodall (1998). Global asymptotic stabilization of uncertain nonlinear delay systems via constrained feedback control. *Proc. 37th IEEE Conf. Decision and Control, Tampa, Florida, USA.*

Glazos, M.P. and S.H. Żak (1995). Practical stabilization of nonlinear/uncertain dynamic systems with bounded controllers. *Int. J. Con.*, **62**, 153-171.

Goodall, D.P. (1997). Constrained feedback control of imperfectly known, linear, time-delay systems of neutral-type. *Proc. RAS/IFAC 3rd Int. Workshop on Singular Solutions and Perturbations in Control Systems: SSPCS'97, Pereslavl-Zalessky, Russia*, 109-114.

Goodall, D.P. (1998). Stability criteria for feedback controlled, imperfectly known, bilinear systems with time-varying delay. *Math. & Comp. in Sim.*, **45**, 279-290.

Hennet, J.-C. and S. Tarbouriech (1997). Stability and stabilization of delay differential systems. *Automatica*, **33**, 347-354.

Hui, G.-D. and G.-D. Hu (1997). Simple criteria for stability of neutral systems with multiple delays. *Int. J. Sys. Sci.*, **28**, 1325-1328.

Kolmanovskii, V.B. and A. Myshkis (1992). *Applied Theory Of Functional Differential Equations.* Kluwer Academic Publishers, Boston, USA.

Lu, P. (1997). Tracking control of nonlinear systems with bounded controls and control rates. *Automatica*, **33**, 1199-1202.

Tchangani, A.P., M. Dambrine and J.P Richard (1998). Stability, attraction domains, and ultimate boundedness for nonlinear neutral systems. *Math. & Comp. in Sim.*, **45**, 291-298.

Verriest, E.I. and W. Aggoune (1998). Stability of nonlinear differential delay systems, *Math. & Comp. in Sim.*, **45**, 257-267.

Wu, H. and K. Mizukami (1996). Linear and nonlinear stabilizing continuous controllers of uncertain dynamical systems including state delay. *IEEE. Trans. Auto. Con.*, **41**, 116-121.

ON SOME MODEL TRANSFORMATIONS FOR STABILITY REGIONS ANALYSIS OF LINEAR SYSTEMS WITH MULTIPLE DELAYS

Silviu-Iulian Niculescu

HEUDIASYC (UMR CNRS 6599), Université de Technologie de Compiègne,
Centre Recherche Royallieu, BP 20529, 60205, Compiègne, France; E-mail: silviu@hds.utc.fr

Abstract. This note focuses on the analysis of some stability regions of a class of linear delay-differential systems involving several delays. Sufficient conditions for characterizing *bounded* and *unbounded* stability regions in the delays parameter space are given in terms of matrix inequalities. The novelty of the proposed approach lies in the unitary analysis of *delays-independent*, *delays-dependent* or *mixed delay-independent / delay-dependent* stability by using a *larger class* of *parametrized model transformations* of the initial system.

Keywords. Delays; model transformation; LMIs.

1 INTRODUCTION

It is well known that the stability of (dynamical) physical or biological systems including delays is a problem of recuring interest since the existence of a delay may induce instability or poor performances (Kolmanovskii and Nosov, 1986).

Real systems cannot always be described using single delay models. For example, the evolution of two distinct and conflicting populations is often described by a two dimensional nonlinear system involving two delays (Stépán, 1989; Kuang, 1993). Other biological examples and further comments are given in MacDonald (1989), Campbell and Bélair (1992). For the complexity of behaviour analysis, see, for instance, an der Heiden (1992).

For linear systems with multiple delays, one of the ideas used in the stability analysis was to extend the *delay-independent* (Kamen, 1982) and *delay-dependent* (Hertz *et al.* 1984) notions from the single delay case: thus, we may encounter *delays-independent* (Hale *et al.* 1985; Feron *et al.* 1992), *mixed delay-independent/delay-dependent* (Niculescu *et al.*, 1998, Kolmanovskii *et al.* 1999) or *delays-dependent* stability (Niculescu *et al.* 1996), function if the corresponding result involves information or not on each (some) delay(s). Inspired by Hale and Huang (1993), we can also use *bounded* or *unbounded* stability regions in the *delays parameter space*. The relationship between such notions can be found in Niculescu *et al.* (1997b) (and the references therein; see also the next section). The scalar case involving two delays τ_1 and τ_2 as parameters was completely treated in Hale and Huang (1933), and the particular $\tau_2 = 2\tau_1$ was considered in Schoen and Geering (1993). Some delays-independent criteria can be found in Boese (1995), Chen and Latchman (1995) (frequency-domain).

In this paper, we consider a class of linear systems described by delayed differential equations involving several (not *commensurate*) delays. To the best author's knowledge, the complete characterization of the geometry of the stable regions in the delays parameter space is still an open problem (Diekmann *et al.*, 1995); remarks on its complexity ($\mathcal{NP}$-hard) can be found in Toker and Ozbay (1996).

We are interested on characterizing some stability regions in a time-domain framework. The *boundedness* of such regions will be analyzed using appropriate (parametrized) *model transformations*. This way allows to overcome some of the problems presented in Kolmanovskii *et al.* (1999) or Niculescu *et al.* (1998) on the choice of the delay sets corresponding to *delay-dependent* or *delay-independent* type properties. To the best author's knowledge, there are no results in the literature avoiding such stability analysis in a time-domain framework.

Indeed, in the Liapunov's second method applied to multiple delay systems, we generally use the *original model*, or if not, *one* obtained by *integration*. Each model is related to some *delay-independent* and *delay-dependent* oriented methodologies, respectively (Niculescu, 1997). The novelty of the proposed approach lies in the way to construct a *'general'* model transformation (similar to the one used for LMI based *state feedback* construction; see Boyd *et al.* 1994).
Thus, the stability problem will be 'transformed' into an optimization problem (see also Niculescu, 1997; Li and de Souza, 1997), using some supplementary degree-of-freedom given by the *model transformation class*. In this way, we shall compute the stability regions, by an appropriate choice of the *model* in the corresponding *class*. Note that the proposed results recover or improve some of existing stability results (Section 3). The notations are standard (Hale and Lunel, 1993).

2 PROBLEM STATEMENT AND MAIN RESULTS

Consider the following system:

$$\dot{x}(t) = Ax(t) + \sum_{i=1}^{r} A_i x(t - \tau_i) \quad (1)$$

with the initial condition

$$x(t_0 + \theta) = \phi(\theta), \quad \forall\, \theta \in [-\bar{\tau}, 0];$$
$$\bar{\tau} = \max_{i=\overline{1,r}}\{\tau_i\}, \quad (t_0, \phi) \in \mathbf{R}_+ \times \mathcal{C}^{v}_{n,\tau} \quad (2)$$

where $x(t) \in \mathbf{R}^n$, τ_i $(i = \overline{1,r})$ are (constant) delays, etc. Following Hale and Huang (1993), the *stable region* is defined as a maximal connected set $\mathcal{D}$ included in $\mathbf{R}^r_+$ which contains the origin $0 \in \mathbf{R}^r$ of the delays-parameter space $(\tau_1, \ldots \tau_r)$ such that for each r-uplet $(\tau_1, \ldots, \tau_r) \in \mathcal{D}$ the considered system is asymptotically stable. Thus, when the stability region is bounded then the corresponding stability criteria is always of *delay-dependent* type in *each* delay. Note also that an unbounded stability region does *not guarantee* that the *delay-independent* property holds in *each* delay. Hale and Huang (1993) (two-delay scalar system) have shown, using a frequency-based approach, that the stability can be *delay-independent* in one delay, but *dependent* of the size of the other delay. Such case was called *mixed delay-independent / delay-dependent* stability, see, e.g. Niculescu *et al.* (1997b, 1998), Kolmanovskii *et al.* (1999). One of the standard *stability problem* to be considered is:

find if the stability region of the system (1)-(2) in the delays parameter space is bounded or not. If not, find corresponding sets of delays for which boundedness of the stability region is still preserved.

The 'limiting' case will correspond to *delay-independent* stability guaranteed for *each* delay. In fact, if the stability region is unbounded, we have *two problems* to handle: firstly, finding the *set of delays* ensuring *boundedness*, and, secondly, *maximizing* the *corresponding* stability *regions*. We shall see later what is the exact meaning of each problem.

Let us focus now on a *time-domain* approach: Liapunov's second method and comparison principle based techniques. It is clear that for such problem, a natural *assumption* is the *asymptotic stability* of:

$$\dot{x}(t) = \left(A + \sum_{i=1}^{r} A_i\right) x(t). \quad (3)$$

The schemes used in the literature can be summarized as follows:

Bounded stability regions (delays-dependent case). We can interpret (1) defined on $\mathcal{C}_{n,\bar{\tau}}$ as an equation on $\mathcal{C}_{n,2\bar{\tau}}$ (by integration over each delay τ_i interval):

$$\dot{x}(t) = \left(A + \sum_{i=1}^{r} A_i\right) x(t)$$
$$+ \sum_{i=1}^{r} A_i \int_{t-\tau_i}^{t} \left(Ax(\theta) + \sum_{i=1}^{r} A_i x(\theta - \tau)\right) d\theta. \quad (4)$$

The asymptotic stability of (4) guarantees the asymptotic stability of (1) (see the proof in Niculescu *et al.* 1994).
The idea of such transformation is to express the original system (1) as a differential equation involving "stable" terms $A + \sum_{i=1}^{r} A_i$, and some integral terms (interpreted as "uncertainty") describing the "delay presence". Some stability results can be found in Goubet *et al.* (1997) (comparison principle), or in Li and de Souza (1997), Niculescu *et al.* (1997a) (Liapunov's second method).

Unbounded stability regions with *delay-dependence* property for *some delays* (*mixed* case): the idea is to use the corresponding transformation only for the *delays* for which we have *delay-dependent* property. Assume a decomposition of the delay vector:

$$\tau = [\tau_1 \quad \tau_2 \quad \ldots \quad \tau_r]$$

of the form $\tau = [\tau_{dep} \quad \tau_{ind}]$, where τ_{dep} contains, without loss of generality (via an appropriate index permutation, etc.), the first k delays $\{\tau_1, \ldots, \tau_r\}$, and:

$$\dot{x}(t) = \left(A + \sum_{i=1}^{k} A_i\right) x(t) + \sum_{i=k+1}^{r} A_i x(t-\tau_i)$$
$$+\sum_{i=1}^{r} A_i \int_{t-\tau_i}^{t} \left(Ax(\theta) + \sum_{i=1}^{k} A_i x(\theta-\tau_i)\right) d\theta. \quad (5)$$

The problem can be handled similarly to the previous case, with the difference that the stability regions are *bounded* only in *some subspace* of the delays-parameter space (For further discussions, see, e.g., Kolmanovskii *et al.* 1998) . One of the problems difficult to solve is to find the *corresponding* set τ_{dep}. In fact, a natural question arises: How to construct correctly such set? We shall see later a way to overcome such problem.

Unbounded stability regions with *delay-independence* in *each delay* (*delays-independent* case): we work directly on (1) and we assume the stability of:

$$\dot{x}(t) = Ax(t),$$

which is a *necessary condition* (Hale *et al.* 1985) for the delays-independent stability, but is not sufficient (Niculescu, 1997). The analysis method is similar to the previous ones, etc.

In conclusion, in the time-domain framework, the computation of the considered stability regions is reduced to analyze the stability of the original system or of some model transformations with respect to the Hurwitz stability of some particular linear systems. To the best author's knowledge, all the existing results in the control literature do not allow to recover both *bounded* and *unbounded* type regions by an *unique* model transformation.

2.1 Model transformation and related results

An *idea* not sufficiently exploited in the literature (see, Goubet *et al.* 1997) is to "find" some decomposition of the each $A_i x(t-\tau_i)$ in order to improve the delay bounds. The proposed technique allow to *optimize* such decomposition. Thus, we introduce the following *class* of *model transformations* (with an appropriate ϕ defined on $\mathcal{C}_{n,2\bar{\tau}}$, etc.):

$$\mathcal{M}_C = \left\{x(t_0,\phi) \ : \ \dot{x}(t) = \left(A + \sum_{i=1}^{r} C_i\right) x(t) \right.$$
$$+\sum_{i=1}^{r}(A_i - C_i)x(t-\tau_i)$$
$$\left. +\sum_{i=1}^{r} C_i \int_{t-\tau_i}^{t} \left(Ax(\theta) + \sum_{i=1}^{r} A_i x(\theta-\tau_i)\right) d\theta \right\}, \quad (6)$$

where the 'parameter' C is given by:

$$C = [C_1 \ldots C_r].$$

It is clear that (6) includes the original system ($C_i = 0$) and respectively (4) or (5), if $C_i = A_i$ for all i or for some indices i in τ_{dep}.

Using similar arguments to Niculescu *et al.* (1994), for a given C matrix, the asymptotic stability of the corresponding model in the class $\mathcal{M}_C$ guarantees the same property for the original system. Due to the form of the system, the term "$C_i x(t)$" can be seen as a *stabilizing* term (it improves the stability of $\dot{x}(t) = Ax(t)$, etc.) Using a Liapunov-Krasovskii functional approach, we shall have the following *model representation* dependent result:

Theorem 1: *The linear system (1)-(2) is asymptotically stable for all the delays τ_i, $i = \overline{1,r}$ such that:*

$$0 \le \tau_i \le \tau_i^0, \quad (7)$$

if there exist $(r+1)^2$ matrices $P, S_{ij} > 0$ ($i = \overline{0,r}$, $j = \overline{1,r}$) and $S_h > 0$ ($h = \overline{1,r}$), and r matrices W_l ($l = \overline{1,r}$) such that:

$$\left[\begin{array}{c|c} \mathcal{Z}_{11} & \mathcal{Z}_{12} \\ \hline \mathcal{Z}_{12}^T & \mathcal{Z}_{22} \end{array}\right] > 0, \quad (8)$$

where:

$$\mathcal{Z}_{11} = \begin{bmatrix} \mathcal{X}_1(P,W) & \tau_1^0 PW_1M & \ldots & \tau_r^0 PW_rM \\ \tau_1^0 M^T W_1^T P & \tau_1^0 \mathcal{R}_1 & & 0 \\ \vdots & & \ddots & \\ \tau_r^0 M^T W_r^T P & 0 & & \tau_r^0 \mathcal{R}_r \end{bmatrix},$$

$$\mathcal{Z}_{12} = \begin{bmatrix} PA_1 - W_1 & \ldots & PA_r - W_r \\ 0 & & 0 \\ \vdots & & \vdots \\ 0 & & 0 \end{bmatrix},$$

$\mathcal{Z}_{22} = diag(S_1, \ldots, S_r)$

and

$$\mathcal{X}_1(P,W) = -(A^TP + PA) + \sum_{i=1}^{r}(W_i + W_i^T)$$
$$-\sum_{i=0}^{r} S_i - \sum_{i=1}^{r} \tau_i^0 \left(S_{i0} + \sum_{j=1}^{r} S_{ij}\right),$$
$$M = [A\, A_1 \ldots A_r] \in \mathbf{R}^{n\times n(r+1)},$$
$$\mathcal{R}_i = diag(S_{i0}, S_{i1}, \ldots, S_{ir}), i = \overline{1,k}.$$

The corresponding model in $\mathcal{M}_C$ is given by:

$$C_i = P^{-1}W_i, \quad i = \overline{1,r}.$$

The Liapunov-Krasovskii functional considered here is similar to the one proposed in Kolmanovskii *et al.* (1998) (see the full version of the paper):

$$V(x_t) \;=\; V_1(x_t) + V_2(x_t) + V_3(x_t), \qquad (9)$$

where:

$$V_1(x_t) = x(t)^T P x(t) + \sum_{i=1}^{r} \int_{-\tau_i}^{0} x(t+\theta)^T S_i x(t+\theta) d\theta,$$

$$V_2(x_t) = \sum_{i=1}^{r} \int_{-\tau_i}^{0} \left(\int_{t+\xi}^{t} x(\xi)^T S_{i0} x(\xi) d\xi \right) d\theta,$$

$$V_3(x_t) = \sum_{i=1}^{r} \left(\sum_{j=1}^{r} \int_{-\tau_i-\tau_j}^{-\tau_j} \int_{t+\xi}^{t} x(\xi)^T S_{ij} x(\xi) d\xi \right) d\theta.$$

2.2 On the computation of τ_{dep} and τ_{ind}

One of the problem not completely handled in Kolmanovskii *et al.* (1999) is the computation of the delay *partition* $\tau = [\tau_{dep}\ \tau_{ind}]$. Their idea was to find a maximal set τ_{ind} by imposing some delays from the delay vector τ to be 0. Such searching technique was related to the Hurwitz stability of some matrices associated to the original system.

The approach presented here allow to overcome such problem. Assume that there exist a delay vector $\tau^0 = \tau^*$ for which the matrix inequality (8) (now an LMI) has a solution. Note that due to the Hurwitz stability of the matrix $A + \sum_{i=1}^{r} A_i$, there *always exist* some 'small' τ_i^0, for which the LMI (8) is feasible (Niculescu, 1997).

In order to define τ_{dep} and τ_{ind} starting with $\tau^0 = \tau^*$, we shall use the following *idea*: one can simply "detect" a *delay-independent* property with respect to a given i-th delay (τ_i^0). Indeed, if $\tau_i^0 \to +\infty$ is allowed in (8), for some $i \in \{1, \ldots, r\}$ (with all τ_j^0, $j \neq i$, fixed), then the norms should satisfy $||W_i|| \to 0$ and $||S_{ik}|| \to 0$ ($k = \overline{0,r}$) and (8) is still feasible. The *corresponding algorithm* can be described as follows:

• First step: the problem τ_i^0 is an element of τ_{ind} is reduced to solve the following *optimization* problem

$$\begin{cases} \textbf{maximize} \;\; \tau_i^0 \;\; s.t. \\ (8) \;\; \text{satisfied} \end{cases},$$

for all *fixed* $\tau_j^0 = \tau_j^*$ ($j \neq i$), which is a standard LMI problem. It is clear that if $\tau_i^0 = \infty$, than the i-th delay is an *element* of the *set* τ_{ind}, if not it is an element of τ_{dep}. If $\tau_i^0 = \infty$ we shall set the corresponding $W_i \equiv 0$ and $\tau_i^0 S_{ik} \equiv 0$ ($k = \overline{0,r}$) and (8) will be reduced conformly, etc.

• Next step: Restart the algorithm with a different index i (for example $i \to i+1$; see the comments given below).

In this manner, we may easily compute the sets τ_{ind} and τ_{dep} with respect to the initial *delays set* $\tau^0 = \tau^*$. However, various situations may appear: for example, delay-independent stability in τ_i, and delay-dependent in τ_j, $i \neq j$, but also the case delay-independent with respect to τ_j and delay-dependent with respect to τ_i for the *same* delay system (see, for instance, Hale and Huang 1993). In such situations, we may define *various* sets τ_{ind} and τ_{dep}, depending on the "direction" firstly considered. In conclusion, a *simple* $i \to i+1$ 'step-by-step' algorithm is not sufficient for covering all the possible partition of the set τ in τ_{ind} and τ_{dep}.

2.3 Optimizing stability regions

If the computed set τ_{dep} is empty, we have the *delays-independent* case, and the *unbounded* set is the maximal allowable.

If τ_{dep} is not empty, then we may have *2 situations*:
• If τ_{dep} has only *one element*, the maximal allowable delay is given by via a *generalized eigenvalue optimization problem* with an LMI constraint (see also the first step of the algorithm described in the previous section, etc.) Such case is also similar to the *delay-dependent* problem for the *single delay case* (one dierction search).
• If τ_{dep} has more elements, then we may use the *delays-dependent* type *optimization algorithms* proposed in Niculescu *et al.* (1996, 1998) for the subset τ_{dep} in the delays parameter space, etc. For the sake of simplicity, we do not detail such cases here. Further remarks can be found in Niculescu *et al.* (1997b).

3 COROLLARIES

Let us concentrate to the limiting "delays-independent" and "delays-dependent" stability cases, which can be easily obtained by an appropriate choice of the corresponding Liapunov-Krasovskii functional. Thus, we shall have the following:

Corollary 1 (Delays-independent) *The linear system (1)-(2) is asymptotically stable if there ex-*

ist r + 1 matrices $P, S_i > 0$ $(i = \overline{1,r})$ *such that:*

$$\begin{bmatrix} -A^TP - PA - \sum_{i=1}^{r} S_i & PA_1 & \dots & PA_r \\ A_1^TP & S_1 & & 0 \\ \vdots & & \ddots & \\ A_r^TP & 0 & & S_r \end{bmatrix} > 0.$$

In this case, $V(x_t)$ becomes $V(x_t) = V_1(x_t)$ (see, e.g., Feron *et al.*, 1992; Boyd *et al.*, 1994).

Corollary 2 (Delays-dependent)
The linear system (1)-(2) is asymptotically stable for all the delays τ_i, $i = \overline{1,r}$ *such that:*

$$0 \leq \tau_i \leq \tau_i^0, \quad i = \overline{1,r} \tag{10}$$

if there exist $(k+1)r+1$ *matrices* $P, S_{ij} > 0$ $(i = \overline{0,k},\ j = \overline{1,r})$ *such that:*

$$\begin{bmatrix} \mathcal{X}_2(P) & \tau_1^0 PA_1M & \dots & \tau_r^0 PA_rM \\ \tau_1^0 M^TA_1^TP & \tau_1^0 \mathcal{R}_1 & & 0 \\ \vdots & & \ddots & \\ \tau_k^0 M^TA_r^TP & 0 & & \tau_k^0 \mathcal{R}_r \end{bmatrix} > 0 \tag{11}$$

where

$$\mathcal{X}_2(P) = -\left[\left(A + \sum_{i=1}^{r} A_i\right)^T P + P\left(A + \sum_{i=1}^{r} A_i\right)\right] - \sum_{i=1}^{r} \tau_i^0 \left(S_{i0} + \sum_{j=1}^{r} S_{ij}\right),$$

$$M = [A\ A_1 \dots A_r] \in \mathbf{R}^{n \times n(r+1)},$$

$$\mathcal{R}_i = diag(S_{i0}, S_{i1}, \dots, S_{ir}), i = \overline{1,k}.$$

In this case,

$$V(x_t) = x(t)^TPx(t) + V_2(x_t) + V_3(x_t).$$

The corresponding matrix inequality can be seen as the first block of (8) with $S_i \equiv 0$ $(i = \overline{(k+1),r})$.

This corollary allows to *improve* the delays-dependent type conditions proposed in Niculescu *et al.* (1997a) via a Liapunov-Razumikhin function:

$$V(x(t)) = x(t)^TPx(t),$$

but with some particular constraints on P (by introducing 'other' positive-definite matrices S_{ij}, such that $S_{ij} \leq P$). Indeed, in the matrix inequality (11) there are no constraints on the matrices S_{ij} with respect to the matrix P. A similar form (with respect to some matrix transformations by expressing the matrix inequality in terms of $X = P^{-1}$ instead of P, etc.) has been proposed in Li and de Souza (1997), Niculescu *et al.* (1998). Note that the result given in Li and de Souza (1997) can be extended to multiple delays by constraints of the form $(i = \overline{1,k}, j = \overline{1,r})$

$$\begin{bmatrix} X & M(X, A, A_j) \\ M(X, A, A_j)^T & N(S_{i0}, S_{ij}) \end{bmatrix} \geq 0, \tag{12}$$

with $X > 0$, $M(X, A, A_j) = XA^T$ or $= XA_j^T$ (function of the corresponding term in $\int_{-\tau_i}^{0} \dot{x}(t+\theta)d\theta$), and $N(S_{i0}, S_{ij}) = S_{i0}$ or $= S_{ij}$. In our frame, X corresponds to P^{-1}, etc. Niculescu *et al.* (1998) have proven that we may have alternative (12), i.e. "majorants" instead of "minorants" on X, but with completely equivalent results, etc.
The BMI based results proposed in Niculescu *et al.* (1996) imply the use some products $\beta_{ij}P$, with β_{ij} positive scalars instead of (positive-definite) matrices S_{ij}.

A different technique was considered in Niculescu *et al.* (1994), where a Liapunov-Krasovskii functional of the form:

$$V(x_t) = \sup_{\theta \in [-\tau, 0]} e^{\delta\theta} x(t+\theta)^T Px(t+\theta),$$

(for some real δ) is used. In fact, such result turns out to be similar to the Liapunov-Razumikhin based one (Niculescu, 1997), and this functional reflects the *limits* of such approach for the linear (time-invariant) delay system stability problem.

In conclusion, all these results are *analytically* more conservative than the proposed one, etc.

4 CONCLUSION

In this paper we have analyzed some *stability regions* in the delays parameter space for a class of linear systems involving several delays. We have used a Liapunov-Krasovskii functional approach combined with some *model transformation* constructing techniques. The results presented allow to recover ot to improve some of *delays-independent*, *delays-dependent* or *'mixed' delay-independent / delay-dependent* ones from the control literature. Note that the proposed criteria can be easily extended to structured uncertain linear systems with several delays (see also Li and de Souza, 1997) and also to the delay-independent or delay-dependent (Niculescu, 1997) stabilization problem (from the closed-loop stability point of view).

REFERENCES

Boese, F. G., 1995, Stability criteria for second-order dynamical systems involving several time-delays, *SIAM J. Math. Anal.*, **26**, 1306-1330.

Boyd, S., El Ghaoui, L., Feron, E. and Balakrishnan, V., 1994, *Linear matrix inequalities in system and control theory*. SIAM, Philadelphia.

Campbell, S. A. and Bélair, J., 1992, Multiple-delayed differential equations as models for biological control systems. *Proc. World Congress of Nonlinear Analysts'92*, 3110-3117.

Chen, J. and Latchman, H. A., 1995, Frequency sweeping tests for stability independent of delay. *IEEE Trans. Automat. Contr.*, **40**, 1640-1645.

Diekmann, O.*et al.* 1995, *Delay equations, Functional-, Complex and Nonlinear Analysis*, Springer-Verlag, New York.

Feron, E., Balakrishnan, V. and Boyd, S., 1992, A design of stabilizing state feedback for delay systems via convex optimization. *Proc. 31st IEEE Conf. Dec. Contr.*, Tucson, Arizona, US, 147-148.

Goubet-Bartholoméüs, A., Dambrine, M. and Richard, J. -P., 1997, Stability of perturbed systems with time-varying delays. *Syst. & Contr. Lett.*, **31**, 155-163

Hale, J. K. and Verduyn Lunel, S. M., 1993, *Introduction to Functional Differential Equations*. Applied Math. Sciences, **99**, Springer-Verlag, New York.

Hale, J. K., Infante, E. F. and Tsen, F. S. P., 1985 Stability in linear delay equations. *J. Math. Anal. Appl.,* **105**, 533-555.

Hale, J. K. and Huang, W., 1993, Global geometry of the stable regions for two delay differential equations. *J. Math. Anal. Appl.,* **178**, 344-362.

an der Heiden, U., 1992, Oscillations and chaos in nonlinear delay differential equations with applications to physiology. *Proc. World Congress of Nonlinear Analysts'92*, 3095-3107.

Hertz, D., Jury, E. I. and Zeheb, E., 1984, Stability independent and dependent of delay for delay differential systems. *J. Franklin Inst.*, **318**, 143-150.

Kamen, E. W., 1982, Linear systems with commensurate time delays: Stability and stabilization independent of delay. *IEEE Trans. Automat. Contr.*, **AC-27**, 367-375, 1982; corrections in *IEEE Trans. Automat. Contr.*, **AC-28**, 248-249, 1983.

Kolmanovskii, V. B. and Nosov, V. R., 1986, *Stability of Functional Differential Equations*, Academic Press, New York.

Kolmanovskii, V. B., S. I. Niculescu and J. P. Richard, 1999, On the Liapunov-Krasovskii Functionals for Stability Analysis of Linear Delay Systems, *Int. J. Contr.* (to appear).

Kuang, Y., 1993, *Delay differential equations with applications in population dynamics*. Academic Press, Boston.

Li, X. and de Souza, C. E., 1997, Criteria for robust stability and stabilization of uncertain linear systems with state delay. *Automatica*, **33**, 1657-1662.

MacDonald, N., 1986, Two delays may not destabilize although either delay can. *Math Biosciences*, **82**, 127-140.

MacDonald, N., 1989, *Biological delay systems: linear stability theory*. Cambridge Univ. Press, Cambridge.

Niculescu, S. -I., *et al.*, 1994, Robust stability and stabilization for uncertain linear systems with state delay: Multiple delays case (II). *Proc. IFAC Workshop on Robust Control Design*, Rio de Janeiro, Brazil, 475-480.

Niculescu, S. -I., Dion, J. -M. and Dugard, L., 1996, Delays-dependent stability for linear systems with several delays: An LMI approach. *Proc. 13th IFAC World 1996*, San Francisco, US, 1996.

Niculescu, S. -I., 1997, *Systèmes à retard. Aspects qualitatifs sur la stabilité et la stabilisation*, Diderot, Paris, Nouveaux Essais.

Niculescu, S. I., 1998, On some model transformations for stability regions analysis of linear systems with multiple delays (full version of the paper). *Internal Note HeuDiaSyC'98.*

Niculescu, S. -I., *et al.*, 1997a, Stability of linear systems with delayed state: An L.M.I. approach. in *JESA*, special issue on "Analysis and control of time-delay systems", **31**, 955-970.

Niculescu, S. -I., *et al.* 1997b, Stability and robust stability of time-delay systems: A guided tour. in *Stability and Control of Time-Delay Systems* (L. Dugard and E. I. Verriest, Eds.), LNCIS, Springer-Verlag, London, **228**, 1-71.

Niculescu, S. -I., J. -M. Dion and L. Dugard, 1998, Mixed delay-independent / delay-dependent stability of linear systems with delayed state: An LMI approach. in *Proc. IFAC Workshop on Linear Time-Delay Systems*, Grenoble, France.

Schoen, G. M. and Geering, H. M., 1993, Stability condition for a delay differential system. *Int. J. Contr.,* **58**, 247-252.

Toker, O. and Ozbay, H., 1996, Complexity issues in robust stability of linear delay-differential systems. *Math., Contr., Signals, Syst.*, **9**, 386-400.

Wang, S. S., Chen, B. S. and Lin, T. P., 1987, Robust stability of uncertain time-delay systems. *Int. J. Contr.*, **46**, 101-107.

14th Triennial World Congress, Beijing, P.R. China

FUNDAMENTAL PRINCIPLE AND SYNTHESIS OF VARIABLE STRUCTURE CONTROL FOR NONLINEAR NEUTRAL CONTROL SYSTEMS WITH TIME-DELAYS *

Cun-Chen Gao†, Zhen-Guang Yang†, Xin-Zheng Zhang‡ and Yong-Qing Liu‡

†Department of Mathematics and Computer Science
Yantai Teachers University, Yantai, 264025, P.R.China
‡Department of Automatic Control Engineering
South China University of Technology, Guangzhou, 510641, P.R.China

† *Email ccgao@mail.ytnc.edu.cn*

Abstract: This paper investigates the practicality problem of the equivalent control method for nonlinear neutral type variable structure control system (VSCS) with time-delays. Existence of the sliding mode for the system is given, and the conditions under which the sliding process has nothing to do with the external disturbance and parameter perturbation are obtained. At last, the synthetical method of the VSCS is provided and illustrated by an example of chemical process. *Copyright © 1999 IFAC*

Keywords: nonlinear, neutral, variable structure control, delays, sliding-mode control.

1. INTRODUCTION

In general, many practical problems can be described by nonlinear neutral type control systems with time-delay, i.e time-delay not only exists in state variables but also exists in the rate of change of the state variable. Owing to the system is affected by many disturbance factors besides time-delays, studying for the problems of robustness and adaptive for the neutral type control systems with time-delays is paid great attention (D.A. O'ccnnor; C.C. Gao). This paper investigates the practicality problem of the equivalent control method for nonlinear neutral type control systems with time-delays by employing the theory of variable structure control(VSC). In section 2, existence of the sliding mode for the system is given. In section3, conditions under which the sliding process has nothing to do with the external disturbance and parameter perturbation are obtained. In section 4, the synthetical method of the VSC is provided. In section 5, an illustrating example of chemical process is given. for the example, we know, the VSC of the system has some best dynamic quality.

*The Project is Supported by National Natural Science Fund of P.R.China, No. 69874005

2. SLIDING MODE EQUATION

We consider the nonlinear neutral type control systems with time-delay

$$\dot{x}(t) = f(t, x_t, \dot{x}_t) + Bu(t), \tag{1}$$

where $x \in R^n$ is a state vector, $u \in R^m$ is a control vector, $B \in R^{n\times m}$ is a constant matrix, $x_t \in C^1, f$ is continuous with regard to the variable t, and it is a bounded linear or nonlinear functional vector for x_t, here, linear space

$$C^1 = C^1([-\tau, 0], R^n), (\tau > 0 \text{ is a constant})$$

is composed of the whole differentiable functions by $[-\tau, 0] \longmapsto R^n, x_t$ and $\dot{x}_t$ are defined, respectively, by

$$x_t(\theta) = x(t+\theta) \text{ and } \dot{x}_t(\theta) = \frac{d}{dt}x(t+\theta)(-\tau \leq \theta \leq 0).$$

It is quite evident that the system (1) include the linear time invariant neutral type control systems with the multi-groups multi-delays as follows:

$$\begin{aligned}\dot{x}_i(t) &= \sum_{j=1}^{n} a_{0ij}^{(1)} x_j(t) + \sum_{k=1}^{N_1}\sum_{j=1}^{n} a_{kij}^{(1)} x_j(t-\tau_{kij}^{(1)}) \\ &+ \sum_{s=1}^{N_2}\sum_{j=1}^{n} a_{sij}^{(2)} \dot{x}_j(t-\tau_{sij}^{(2)}) + \sum_{j=1}^{m} b_{ij} u_j(t) \\ &(i=1,\cdots,n),\end{aligned} \quad (2)$$

where $a_{kij}^{(1)}(k = 0,1,\cdots,N_1), a_{sij}^{(2)}(s = 1,\cdots,N_2)$ and $b_{ij}(i,j = 1,\cdots,n)$ are all constants, $0 < \tau_{ki1}^{(1)} < \cdots < \tau_{kin}^{(1)}(k = 1,\cdots,N_1), 0 < \tau_{si1}^{(2)} < \cdots < \tau_{sin}^{(2)}(s = 1,\cdots,N_2), \tau = \max_{1\le i\le n}\{\tau_{kin}^{(1)}, \tau_{sin}^{(2)}; k = 1,\cdots,N_1, s = 1,\cdots,N_2\}$. Let $\phi(\theta) \in C^1$, its norm is defined by

$$\| \phi \| = \sup_{\theta\in[-\tau,0]} \| \phi(\theta) \| . \quad (3)$$

For every initial vector function $\phi \in C^1$ and control $u(t)$, let $x(t) = x(t, t_0, \phi, u)$ is the solution of (1) which satisfied the initial conditions $x_{t_0} = \phi$ and $\dot{x}_{t_0} = \dot{\phi}$.
Suppose that the switching vector function $S(t)$ have the form as

$$S(t) \overset{\text{def.}}{=} S(x(t)) = Cx(t), \quad (4)$$

where $C \in R^{m\times n}$ is a constant matrix in awaiting for determined, the control vector function $u(t)$ in (1) is discontinuous on the hyperplane $S(t) = 0$. Now, we need to design switching surface $S(t) = Cx(t) = 0$ (that is, we need to determine the matrix C) and VSC law $u(t)$, such that the solution of (1) tend towards the equilibrium state $x = 0$, and the solution have the certain adaptive and robustness.

From (1), (4) and $S(t) = 0$, the derivative of the switching surface $S(x(t))$, can be obtained

$$\dot{S}(x(t)) = C\dot{x}(t) = C[f(t, x_t, \dot{x}_t) + Bu(t)] = 0. \quad (5)$$

Suppose that CB is a invertible matrix, the equivalent control vector u_{eq} can be solved by (5)

$$u_{eq}(t) = -(CB)^{-1}Cf(t, x_t, \dot{x}_t). \quad (6)$$

Substituting (6) into (1) we can get an ideal sliding mode equation of the control system (1) in the following

$$\begin{cases} \dot{x}(t) &= [I - B(CB)^{-1}C]f(t, x_t, \dot{x}_t) \\ Cx(t) &= 0. \end{cases} \quad (7)$$

In the actual system, because of the existence of every non-ideal factor is inevitable, the real sliding motion don't precisely occur on switching surface $S(t) = 0$, which occur in some $\sigma(\sigma > 0)$ neighborhood

$$U(0,\sigma) = \{x \mid \| Cx(t) \| < \sigma\}$$

(that is the boundary layer) of the switching surface $S(t) = 0$.

Now, we will prove that the real sliding mode $\hat{x}(t)$ uniformly converges into the ideal sliding mode $x(t)$ in arbitrary finite time when the thickness of the boundary layer σ satisfies: $\sigma \to 0$.

In the boundary layer $\| S(t) \| = \| Cx(t) \| < \sigma$, using the regularizing method, we can obtained the differential equation of the real sliding motion for control system (1) as follows:

$$\dot{\hat{x}}(t) = f(t, x_t, \dot{x}_t) + B\hat{u}(t), \quad (8)$$

where $\hat{u}(t)$ is a control vector after the regularization, which is continuous on the switching surface $S(t) = 0$ and have considered some non-ideal factors.

Now, we will establish a relation between the real sliding mode $\hat{x}(t)$ and the ideal sliding mode $x(t)$.
Theorem 2.1: In the interval $[t_0, T]$, suppose that the solutions of the real motion equation (8) are all in the boundary layer $\| S(t) \| \le \sigma(\sigma > 0), f(t, x_t, \dot{x}_t)$ is continuous with respect to the variable t, and satisfies Lipschitz condition for x_t. Then for any initial states ϕ and ψ which satisfy the inequality $\| \phi - \psi \| \le \alpha\sigma(\alpha > 0; \phi, \psi \in C^1)$, there is a positive number b such that the real sliding mode $\hat{x}(t) = x(t, t_0, \psi, \hat{u})$ and the ideal sliding mode $x(t) = x(t, t_0, \phi, u_{eq})$ satisfy the following relation of inequality

$$\| \hat{x}(t) - x(t) \| \le b\sigma, t \in [t_0, T]. \quad (9)$$

Proof: In the boundary layer $\| S(t) \| \le \sigma$, we have

$$\dot{S}(t) = C\dot{x}(t) = Cf(t, x_t, \dot{x}_t) + CB\hat{u}(t),$$

thus

$$\hat{u}(t) = (CB)^{-1}\dot{S}(t) - (CB)^{-1}Cf(t, x_t, \dot{x}_t), \quad (10)$$

substituting (10) into (8) we can yield a real sliding equation of (1) in the following

$$\dot{x}(t) = B(CB)^{-1}\dot{S}(t) + [I - B(CB)^{-1}C]f(t, x_t, \dot{x}_t). \quad (11)$$

Therefore, using the separating step method, we can indicate solutions of (7) and (11) in interval $I_i = [t_0 + i\tau, t_0 + (i+1)\tau](i = 1, 2, \cdots)$, respectively, as the following

$$x(t) = x(t_0 + i\tau) + [(I - B(CB)^{-1}C] \int_{t_0+i\tau}^{t} f(t, x_\rho, \dot{x}_\rho)d\rho \quad (12)$$

and

$$\begin{aligned}\hat{x}(t) &= \hat{x}(t_0 + i\tau) + \int_{t_0+i\tau}^{t} B(CB)^{-1}\tfrac{dS}{d\theta}d\theta \\ &+ [(I - B(CB)^{-1}C] \int_{t_0+i\tau}^{t} f(\rho, \hat{x}_\rho, \dot{\hat{x}}_\rho)d\rho \\ &= \hat{x}(t_0 + i\tau) + B(CB)^{-1}[S(\hat{x}(t)) - S(\hat{x}(t_0 + i\tau))] \\ &+ [I - B(CB)^{-1}C] \int_{t_0+i\tau}^{t} f(\rho, \hat{x}_\rho, \dot{\hat{x}}_\rho)d\rho,\end{aligned} \quad (13)$$

in view of the conditions of Theorem 2.1, there exist a positive constant m_1, such that

$$\| B(CB)^{-1}[S(x(t)) - S(x(t_0 + i\tau))] \| \leq m_1\sigma. \quad (14)$$

From (12) minus (13), and consider that $f(t, x_t, \dot{x}_t)$ satisfies Lipschitz condition for x_t, using (14) we can obtain that

$$\| \widehat{x}(t) - x(t) \| \leq \| \widehat{x}(t_0 + i\tau) - x(t_0 + i\tau) \| + m_1\sigma$$
$$+ \| I - B(CB)^{-1}C \| \int_{t_0+i\tau}^{t} L \| \widehat{x}_\rho - x_\rho \| d\rho, \quad (15)$$

where L is a Lipschitz constant. By (15), we have

$$\| \widehat{x}_t - x_t \| \leq \| \widehat{x}_{t_0+i\tau} - x_{t_0+i\tau} \| + m_1\sigma$$
$$+ \| I - B(CB)^{-1}C \| \int_{t_0+i\tau}^{t} L \| \widehat{x}_\rho - x_\rho \| d\rho. \quad (16)$$

From spreading Gronwall inequality we immediately known

$$\| \widehat{x}_t - x_t \| \leq (\| \widehat{x}_{t_0+i\tau} - x_{t_0+i\tau} \| + m_1\sigma)\beta, t \in I_i$$
$$(i = 0, 1, \cdots), \quad (17)$$

where

$$\beta = \exp(\| I - B(CB)^{-1}C \| L\tau \sup_{\rho\in[t_0,\tau]} \| \widehat{x}_\rho - x_\rho \|). \quad (18)$$

Due to $\widehat{x}_{t_0} = \phi, x_{t_0} = \psi$ and $\| \phi - \psi \| \leq \alpha\sigma$ in interval I_0, from (17), there is a positive number $b_1 = (m_1 + \alpha)\beta$ such that

$$\| \widehat{x}_{t_0} - x_{t_0} \| \leq b_1\sigma, t_0 \in I_0. \quad (19)$$

Using (17) sufficiently, and it is similar to the estimation of (19), there is a positive number $b_{i+1} = (m_1 + b_i)\beta$ in interval $I_i (i = 1, 2, \cdots)$ such that

$$\| \widehat{x}_t - x_t \| \leq b_{i+1}\sigma, t \in I_i (i = 1, 2, \cdots). \quad (20)$$

For any finite interval $[t_0, T]$, there is a positive integer N such that $t_0 + N\tau < T \leq t_0 + (N+1)\tau$, that is

$$[t_0, T] \subset \bigcup_{i=0}^{N} I_i = [t_0, t_0 + (N+1)\tau]. \quad (21)$$

Therefore, taking $b = \max\{\alpha, b_1, \cdots, b_{N+1}\}$, we immediately known the result of Theorem 2.1 is right. The proof of Theorem 2.1 is completed.

Theorem 2.1 indicate that, using equivalent control method, the sliding equation of the neutral type control system with time-delay can be obtained, and unique. The result has extremely important signification. So long as the control vector $u(T)$ can ensure that the phase-trajectory of the system gets into the sliding mode, we can investigate the variable trend of the state vector after the system (1) goes to the sliding mode by the ideal sliding equation (4). Especially, when $\tau = 0$, Theorem 2.1 is a result of (1) with respect to the linear concentrating parameter system (V.I. Utkin). If $\dot{x}(t - \tau) = 0$, then Theorem 2.1 is exactly the result of retarded type control system (F.S. Yuan).

3. INVARIANCE OF DISTRUBANCE AND PERTURBATION ON SLIDING MODE

Besides affecting of the time-delays, the system (1) are affected by the external disturbance and parameter perturbation. From this, we will discuss the invariance for the two cases on the sliding mode, respectively.

A. Disturbed System

State equation of a disturbed system is defined by

$$\dot{x}(t) = f(t, x_t, \dot{x}_t) + Bu(t) + Dg, \quad (22)$$

where Dg is the external disturbance of system (1), $D \in R^{n\times l}$ is a constant matrix, $g = g(t, x_t, \dot{x}_t, p) \in R^l$ is a vector function, p is an uncertain amount. The conventional switching function $S(t)$ can be defined by

$$S(x(t)) = Cx(t), S \in R^m. \quad (23)$$

The sliding mode is all situated in the subspace $S_0 = \mathrm{Ker}C$ of $S(t) = 0$. Let $\dot{S}(t) = 0$ we can obtain that

$$Cf(t, x_t, \dot{x}_t) + CBu(t) + CDg = 0. \quad (24)$$

Suppose that $CB \in R^{n\times m}$ is a invertible matrix, from (24), the equivalent control can be determined by

$$u_{eq}(t) = -(CB)^{-1}C[f(t, x_t, \dot{x}_t) + Dg]. \quad (25)$$

Substituting (25) into (22), the sliding mode equation can be yield as follows:

$$\dot{x}(t) = [I - B(CB)^{-1}C]f(t, x_t, \dot{x}_t) + [I - B(CB)^{-1}C]Dg. \quad (26)$$

Therefore, the sliding motion is invariance for the disturbance g, if and only if

$$B(CB)^{-1}CDg(t, x_t, \dot{x}_t, p) = Dg(t, x_t, \dot{x}_t, p). \quad (27)$$

i.e have

Theorem 3.1: The sliding mode of the system (22) is not affected by the disturbance g, if and only if

$$\mathrm{rank}(B, D) = \mathrm{rank}(B) \quad (28)$$

or

$$D = B\widetilde{D}, \widetilde{D} \in R^{m\times l}. \quad (29)$$

The conditions (28) or (29) is a invariance condition of the system (22) for the disturbance, or to say that a entirely matching condition of the system with the disturbances.

B. System of Parameter Perturbation

State equation of the system of parameter perturbation is defined by

$$\dot{x}(t) = f(t, x_t, \dot{x}_t) + Bu(t) + \Delta f(t, x_t, \dot{x}_t, p) + \Delta Bu(t), \quad (30)$$

where p is an uncertain amount, Δf and ΔB indecate corresponding perturbation amount, respectively. Now, we study the irrelevant condition of the sliding mode with Δf and ΔB.

Theorem 3.2: The sliding mode of the system (30) has nothing to do with the parameter perturbations Δf and ΔB, if and only if

$$\text{rank}(B, \Delta f + \Delta Bu) = \text{rank}(B). \quad (31)$$

Proof: The ideal sliding mode motion equation of the system (30) is defined as follows:

$$\begin{cases} \dot{x}(t) & = f(t, x_t, \dot{x}_t) + Bu(t) + \Delta f + \Delta Bu(t) \\ Cx(t) & = 0. \end{cases} \quad (32)$$

From $\dot{S}(t) = 0$, we have

$$Cf + CBu(t) + C\Delta f + C\Delta Bu(t) = 0.$$

When CB invertible, can get the equivalent control in the following

$$u(t) = -(CB)^{-1}Cf - (CB)^{-1}C(\Delta f + \Delta Bu(t)), \quad (33)$$

substituting (33) into (30) can be obtained the sliding mode equation

$$\dot{x}(t) = [I - B(CB)^{-1}C]f + [I - B(CB)^{-1}C](\Delta f + \Delta Bu(t)). \quad (34)$$

If

$$B(CB)^{-1}C(\Delta f + \Delta Bu(t)) = \Delta f + \Delta Bu(t), \quad (35)$$

then $\Delta f + \Delta Bu(t)$ will not effect the sliding mode. The condition (35) is equivalent to

$$\text{rank}(B, \Delta f + \Delta Bu) = \text{rank}(B). \quad (36)$$

or

$$\Delta f + \Delta Bu = Bz, z \in R^m. \quad (37)$$

The proof of Theorem 3.2 is completed.

Let $L(B)$ is a linear space extended by column vector of the matrix B, then (37) is equivalent to $\Delta f + \Delta Bu \in L(B)$. The conditions (35), (36) or (37) are that the sliding mode has nothing to do with parameter perturbations $\Delta f + \Delta Bu$.

Remark 1: Geometric Significance of the System Invariance for the Disturbance g. From W. B. Gao, and pay attention to (27), we known that $g \in \mathbf{N}(I - B(CB)^{-1}C)$, $x \in \mathbf{R}(I - B(CB)^{-1}C)$, that is, the disturbance g is in the zero-space of operator $I - B(CB)^{-1}C$, and the sliding mode motion is in the image space of operator $I - B(CB)^{-1}C$. The two spaces is mutually a orthogonal supplemented subspace, that is

$$\mathbf{N}(I - B(CB)^{-1}C) \bigoplus \mathbf{R}(I - B(CB)^{-1}C) = X. \quad (38)$$

From this know that the sliding mode has the invariance and the completely adaptive for the disturbation g.

4. SYNTHESIS OF VARIABLE STRUCTURE CONTROL FOR SYSTEM

Now, we consider the synthesis problem of the VSC for the neutral type control system with time-delays in the external disturbance and parameter perturbation. Its synthesis steps is that

Step 1: Designing a VSC law $u(t)$ which make the sliding mode is realized.

Step 2: Designing the switching manifold $S(x(t)) = Cx(t) = 0$ such that the state of the system on the sliding manifold has some best dynamic quality.

We consider general neutral type control system with time-delays

$$\dot{x}(t) = f(t, x_t, \dot{x}_t) + \Delta f(t, x_t, \dot{x}_t, p) + Bu(t) + \Delta Bu(t) + Dg, \quad (39)$$

where p is an uncertain amount, Δf and ΔB is parameter perturbation, respectively, Dg is an external disturbance, $D \in R^{n \times l}$, $g = g(t, x_t, \dot{x}_t, p) \in R^n$ is a vector function.

Theorem 4.1: Suppose that the internal and external disturbance of the neutral type VSCS with time-delays (39) satisfies the conditions in the following:

$$(H_1) \qquad \| \Delta f(t, x_t, \dot{x}_t, p) \| \le f_0$$

$$(H_2) \qquad \| \Delta B \| \le \frac{b_0}{\| (CB)^{-1}C \|}$$

$$(H_3) \qquad \| Dg \| \le g_0$$

where f_0, b_0 and g_0 are all positive constants, and $b_0 < 1$. Then the phase-trajectory of the system (39) achieve the sliding motion in finite time under affected by VSC law $u(t)$:

$$u(t) = -(CB)^{-1}[Cf(t, x_t, \dot{x}_t) + KS + \varepsilon \text{sign}(S)], \quad (40)$$

where $K = \text{diag}(k_i)_{m \times m}, \varepsilon = \text{diag}(\varepsilon_i)_{m \times m}$ and $k_i > 0, \varepsilon_i > 0$, satisfy some inequality as follows:

$$\begin{cases} k_i > b_0 \\ (1 - b_0)\varepsilon_i > \varepsilon_{i0} + \| C \| (f_0 + g_0 + b_0) \| f(t, x_t, \dot{x}_t) \| \\ \varepsilon_{i0} > 0, i = 1, 2, \cdots, m. \end{cases} \quad (41)$$

Proof: Taking $V(x(t)) = \frac{1}{2}S^\top S$ as Liapunov function, we have

$$\begin{aligned} & \frac{d}{dt}V(x(t)) |_{(39)} = S^\top \dot{S} = S^\top C\dot{x} \\ & = -S^\top[K - C\Delta B(CB)^{-1}K]S \\ & + S^\top(C\Delta f - C\Delta B(CB)^{-1}Cf) + S^\top[CDg - \\ & \varepsilon\text{sign}(S) - C\Delta B(CB)^{-1}\varepsilon\text{sign}(S) \\ & \le -\sum_{i=1}^{m} k_i(1 - b_0)S_i^2 - \sum_{i=1}^{m}[\varepsilon_i - b_0\varepsilon_i \\ & - \| C \| (f_0 + g_0 + B_0 \| f \|)] \| S_i \| \\ & < -\sum_{i=1}^{m} k_i(1 - b_0)S_i^2 - \sum_{i=1}^{m} \varepsilon_{i0} \| S_i \|, \end{aligned} \quad (42)$$

where S_i is the i-th subvariable. In view of the comparison principle of the solution and (42), for any $S(x(t_0)) = Cx(t_0) \neq 0$, there is a number $T : T > t_0$ such that $S(x(T)) = Cx(T) = 0$. That is, the state of the system (39) realized the sliding mode motion. The proof of Theorem 4.1 is completed.

Remark 2: Theorem 4.1 indicate that the determining of control vector $u(t)$ needn't measure the uncertain amount p of the system, the gain coefficients k_i and $\varepsilon_i (i = 1, 2, \cdots, m)$ can be determined when we known only the variable range of p.

Remark 3: When input $u(t)$ of the system exist pure time-delay τ, the input $u(t)$ of the system can be immediately obtained by (40) as follows:

$$\begin{aligned} u(t) = & -(CB)^{-1}[Cf(t+\tau, x_{t+\tau}, \dot{x}_{t+\tau}) \\ & +KS(t+\tau)\varepsilon \text{sign}(S(t+\tau)), \end{aligned} \tag{43}$$

where $S(t+\tau) = Cx(t+\tau)$.

Remark 4: The above mentioned Theorem 2.1, Theorem 3.1, Theorem 3.2, and Theorem 4.1 are also hold for even more widespread nonlinear neutral control system with time-delays, which include that

$$\dot{x}(t) = f(t, x(t), x(t-\tau), \dot{x}(t-\tau)) + Bu(t), \tag{44}$$

$$\begin{aligned} \dot{x}(t) = & \int_{-\tau}^{0} \eta(t,\theta)x(t+\theta)d\theta \\ & +f(t, x(t), x(t-\tau), \dot{x}(t-\tau)) + Bu(t), \end{aligned} \tag{45}$$

and the cases of these systems has the external disturbance and parameter perturbation

$$\begin{aligned} \dot{x}(t) = & f(t, x(t), x(t-\tau), \dot{x}(t-\tau)) \\ & +\Delta f(t, x(t), x(t-\tau, \dot{x}(t-\tau), p) \\ & +(B+\Delta B)u(t) + Dg, \end{aligned} \tag{46}$$

$$\begin{aligned} \dot{x}(t) = & \int_{-\tau}^{0} [\eta(t,\theta) + \Delta\eta(t,\theta,p)]x(t+\theta)d\theta \\ & +f(t, x(t), x(t-\tau), \dot{x}(t-\tau)) \\ & +\Delta f(t, x(t), x(t-\tau), \dot{x}_(t-\tau), p) \\ & +(B+\Delta B)u(t) + Dg. \end{aligned} \tag{47}$$

They include every cases without neutral term (F. S. Yuan).

Remark 5: From the above four theorems, we can obtain the corresponding theorems (F. S. Yuan) of Retarded Type control system

$$\dot{x}(t) = f(t, x_t) + \Delta f(t, x_t, p) + (B+\Delta B)u(t) + Dg. \tag{48}$$

From Theorem 4.1, we known that $x_1 = -C_1^{-1}C_2x_2$ can be obtained by $S(t) = Cx(t) = 0$ after the state of the neutral type control system with time-delay enter the sliding mode motion, where $C = (C_1, C_2)$, $x = (x_1^\top, x_1^\top)^\top, C_1 \in R^{m\times m}$ and $C_2 \in R^{m\times(n-m)}$ are all inversable matrices, $x_1 \in R^m, x_2 \in R^{n-m}$. Therefore, the stability problem of the sliding mode (7) for the system (1) will change into the stability problem with respect to the state subvariable x_2 of $n-m$ dimision on the sliding surface $S(t) = 0$. This property is a big advantage for the synthesis of neutral type control system with time-delays. Because of reducing for dimision, which will enormously decrease difficult to determine the place of some characteristic values after the linearization for the neutral type VSCS with time-delays, thus we can take corresponding matrix C such that the solution of the sliding equation is asymptotical stable.

5. ILLUSTRATE EXAMPLE

At first, we introduce a Lemma (Y. Q. Liu; J. Hale) to need in the following example.

Lemma 1 Considering neutral type functional differential equation

$$\dot{x}(t) = ax(t) + bx(t-\tau) + c\dot{x}(t-\tau). \tag{49}$$

The zero-solution of (49) is asymptotical stable, if and only if the following conditions are satisfied:

$$\begin{cases} a+b<0 \\ |c|<1 \end{cases} \tag{50}$$

where

$$0 < \tau < \frac{2(1-c^2)^{\frac{1}{2}}}{(b^2-a^2)^{\frac{1}{2}}} \arctan \frac{[(1-c^2)(b^2-a^2)]^{\frac{1}{2}}}{(1+c)(a-b)}. \tag{51}$$

Example 1: Let us consider the series Dissolved Trough (DT) in Figure 1. The solute in a hopper is conveyed into the material mouth by conveyer belt. If we increase some deliver material amount, then the solution concentration of the DT1 is changed when the solute of increasing is conveyed into the material mouth from the hopper after which is fallen in DT1, that is, the alteration of the solution concentration in the DT1 is lain behind a conveyed time τ_1 from the hopper to the material mouth than the changed to increase amount of material. Postscript, it is also needed a short time τ_2 when the solution in DT1 want to flow into DT2.

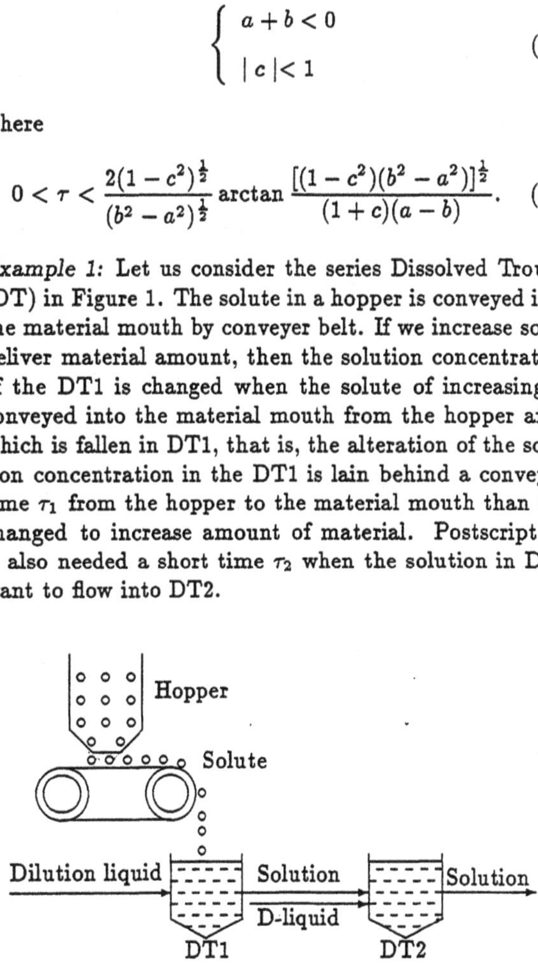

Fig.1 The series dissolved trough

From paper of C. H. Zhou, we known that the concentration of dissolving in DT $i (i = 1, 2)$ satisfies the following control system

$$\begin{cases} T_1\dot{x}_1(t) + x_1(t) & = K_1 u(t-\tau_1) \\ T_2\dot{x}_2(t) + x_2(t) & = K_2 x_1(t-\tau_2) + R_2\dot{x}_1(t-\tau_2), \end{cases} \tag{52}$$

where $T_i, K_i > 0 (i = 1, 2), R_2 > 0$ are all constants, u is a deliver material amount which make the solution concentration x_i in DT i tend toward the equivalent state $x_{i0} (i = 1, 2)$. There is no harm in supposing $x_{i0} = 0 (i = 1, 2)$, otherwise making a transform $\bar{x}_{i0} = x_i - x_{i0}$ can turn into $\bar{x}_{i0} = 0$. The term $R_2\dot{x}_1(t-\tau_2)$ at the right of the second form in (52) indicate that the direct proportion relation between the rate of change for concentration $x_2(t)$ and which for concentration $x_1(t-\tau_2)$.

Choosing the switching surface is that $S(t) = x_1 + cx_2 = 0$, then the sliding equation of (52) can be obtained by the equivalent control method as follows:

$$\dot{x}_2(t) = -T_2^{-1}x_2(t) - T_2^{-1}K_2 c x_2(t-\tau_2) + R_2\dot{x}_1(t-\tau_2). \tag{53}$$

In view of Lemma 1, the phase-trajectory of the system (53) asymptotically tend towards the equilibrium state $x_{i0} = 0$ along the sliding surface $S(t) = 0$ when

$$-\max\{-\frac{1}{K_2}, \frac{T_2}{R_2}\} < c < \frac{T_2}{R_2}.$$

This moment τ_2 should satisfies the following condition:

$$\begin{cases} 0 < \tau_2 < \dfrac{2(T_2^2 - R_2^2c^2)^{\frac{1}{2}}}{(K_2^2c^2-1)^{\frac{1}{2}}} \\ \arctan \dfrac{[(T_2^2 - R_2^2c^2)(K_2^2c^2-1)]^{\frac{1}{2}}}{(1-R_2c)(K_2c-1)} \\ T_2K_2 > R_2. \end{cases} \tag{54}$$

Therefore, from (53) and (54), the design of the switching manifold $S(t) = Cx(t) = 0$ can be finished so long as we take c to be satisfied in the following

$$\frac{1}{K_2} < c < \frac{T_2}{R_2}. \tag{55}$$

Next, the design the VSC law $u(t)$ such that the phase-trajectory of system (52) enter the sliding mode motion in the finite time. For this, taking

$$\begin{aligned} u(t) = \; & K_1^{-1}T_1(T_1^{-1}x_1(t+\tau_1) + cT_2^{-1}x_2(t+\tau_1) \\ & -cT_2^{-1}K_2x_1(t+\tau_1-\tau_2) \\ & -cT_2^{-1}R_2\dot{x}_1(t+\tau_1-\tau_2)) \\ & -K_1^{-1}T_1[kS(x(t+\tau_1)) + \varepsilon \mathrm{siag}(S(x(t+\tau_1)))] \end{aligned} \tag{56}$$

where $k, \varepsilon > 0$. Then

$$\dot{S}(t)\mid_{(52)} = -kS(t) - \varepsilon\mathrm{siag}(S(t)). \tag{57}$$

Therefore, from (57) the control law (56) can ensure that the state of the system (52) enter the sliding surface $S(t) = 0$ in the finite time, thus it has accomplished the sliding motion and has ensured that the system (52) is asymptotical stable.

6. CONCLUSION

In this paper, we study the practicality problem of the equivalent control method for nonlinear neutral variable structure control system with time delays. Existence of the sliding mode for the system is given, and the conditions under which the sliding process has nothing to do with the external disturbance and parameter perturbation are obtained. At last, the synthetical method of the VSCS is provided and illustrated by an example of chemical process.

For choosing method of C in (4), we shall discuss it in another paper.

REFERENCES

O'ccnnor,D.A., and T. J. Tarn (1983). On stabilization by state feedback for neutral differential difference equations. *IEEE Trans. Automat. Contr.*, **28**, 615-618.

Gao, C.C., and Y. Q. Liu (1995). Stability of variable structure control systems. *Proc. 1995 American Control Conf.*, 1538-1539, Washington.

Utkin, V.I (1981). Sliding Mode in Optimization and Control. Chapter 1-4, Moscow, Nauka.

Yuan, F. S. (1993). VSC of systems with time-delays. *Ph. D. Paper of South China Uni. of Tech.*, 1-80, Guangzhou.

Gao, W. B. (1990). Theoretical Basis of Variable Structure Control. Chapter 2, Chinese Science and Technology Press, Beijing.

Yuan, J., Z. X. Huang and G. Z. Gao (1984). The problem of the stability for the neutral type equivalent differential difference equation and ordinary differential equation. *Acta Mathematica Sinica*, **27**, 716-720.

Hale, J. (1977). Theory of Functional Differential Equations. Chapter 1; 12, Springer-Verlag, New Youk.

Liu, Y. Q., C. C. Gao and F. S. Yuan (1998). Variable Structure Control of Delays Compositive Systems. Chapter 2, Press in South-China University of Technology, Guangzhou.

Zhou, C. H. (1980). Control Principle of Chemical Process. Chapter 1; 2, Chemical Industry Press, Beijing.

14th Triennial World Congress, Beijing, P.R. China

THE EFFECT OF MULTIPLICATIVE INPUT UNCERTAINTY ON INPUT-OUTPUT FEEDBACK LINEARIZATION

Lalitha S. Balasubramhanya ** **Francis J. Doyle III** *,1

* *Dept. of Chem. Eng., Univ. of Delaware, Newark, DE 19716, USA*
** *School of Chemical Engineering, Purdue University, USA*

Abstract: Input-output linearization is a powerful method for the design of controllers for nonlinear systems. However, the exact linearization fails when there is plant-model mismatch. An important question is whether the structure of the uncertainty in conjunction with the linearization can be exploited to circumvent this problem. In this paper, multiplicative input uncertainty in MIMO nonlinear systems in considered. It is demonstrated that the effect of the uncertainty is more pronounced in systems with ill-conditioned characteristic matrices as opposed to systems with ill-conditioned gain matrices. Appropriate choice of the linearizing parameters in the input-output linearizing scheme offsets the effect of an ill-conditioned characteristic matrix and enables the design of a controller with significantly improved closed-loop performance.

Keywords: Multivariable Control, Nonlinear Systems, Feedback Linearization, Uncertainty, Robust Control.

1. INTRODUCTION

Multiple-input, multiple-output (MIMO) systems with a high condition number, such as high-purity distillation columns, are inherently difficult to control (Skogestad *et al.*, 1988). Most controllers designed for acceptable nominal performance tend to invert the plant (either explicitly or implicitly) resulting in controllers with high condition numbers as well. Uncertainty in the model requires that the controller must be detuned considerably to guarantee stability (Morari and Zafiriou, 1989). Hence, from a control perspective, it is important to understand the impact of ill-conditioning on robust controller design.

In recent years, input-output linearization (IOL) has gained increasing acceptance as an effective nonlinear control strategy (see (Allgöwer and Doyle III, 1996) for representative discussion of application to process systems). As no model can perfectly capture the physical characteristics of a system, it is important to analyze the effect of uncertainty in this approach. Several investigators have proposed nonlinear robust controller designs when faced with uncertainty (for representative references, see (Calvet, 1989; Kravaris and Palanki, 1988)). Most of these approaches have focussed on calculating the input move to guarantee stability based on a Lyapunov stability criterion for some bounded uncertainty. The effect of uncertainty is most pronounced for ill-conditioned systems and Kravaris and Soroush (1990) state that decoupling using input-output linearization for an ill-conditioned system leads to performance degradation. In this work, the emphasis is on an analysis of the effect of uncertainty when an IOL control strategy is used. In particular, the relevance of the condition number of the gain matrix and the characteristic matrix in the control design for a model with uncertainty will be demonstrated.

[1] Corresponding author (fdoyle@udel.edu)

2. BACKGROUND

Input-output linearization involves a nonlinear coordinate transformation that exactly converts a nonlinear system into a linear system in the transformed coordinates (Isidori, 1995). This is motivated by the fact that the resultant linear system can be controlled using an appropriate linear controller. The controller schematic is sketched in Figure 1. Ψ is the nonlinear coordinate transformation that converts the nonlinear input-output $(\boldsymbol{u}-\boldsymbol{y})$ plant into a linear input-output $(\boldsymbol{v}-\boldsymbol{y})$ plant. A suitable tracking controller can be designed using linear internal model control (IMC). The IMC controller consists of a suitable inverse of the linear plant and a filter to tune the closed-loop time constant.

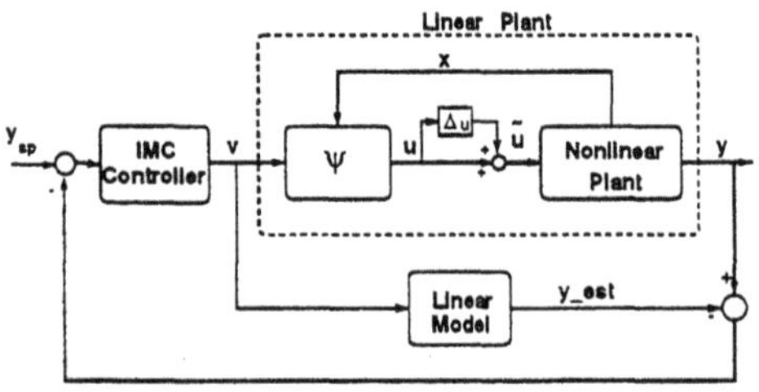

Fig. 1. Multiplicative input uncertainty in an IOL control framework.

In this work, we consider a nonlinear, multivariable, control-affine systems of the following form:

$$\dot{\boldsymbol{x}} = \boldsymbol{f}(\boldsymbol{x}) + \boldsymbol{g}(\boldsymbol{x})\boldsymbol{u}$$
$$\boldsymbol{y} = \boldsymbol{h}(\boldsymbol{x}) \tag{1}$$

A few definitions are introduced before stating the nonlinear control design.

Definition 1. Given a MIMO nonlinear system (Equation 1), the ith output, y_i, has a relative order r_i at a point $\boldsymbol{x}^0$ if in a neighborhood U of $\boldsymbol{x}^0$

$$L_{\boldsymbol{g}_j} L_{\boldsymbol{f}}^k h_i(\boldsymbol{x}) = 0 \qquad k = 1, \cdots, r_i - 2 \qquad \forall \boldsymbol{x} \in U$$

and the row vector

$$L_{\boldsymbol{g}} L_{\boldsymbol{f}}^{r_i-1} h_i(\boldsymbol{x}^0) = [L_{\boldsymbol{g}_1} L_{\boldsymbol{f}}^{r_i-1} h_i(\boldsymbol{x}^0) \cdots L_{\boldsymbol{g}_m} L_{\boldsymbol{f}}^{r_i-1} h_i(\boldsymbol{x}^0)]$$

is nonzero.

Definition 2. If each output y_i possesses a relative degree r_i, the *characteristic matrix* can be written as follows:

$$\boldsymbol{C}_h(\boldsymbol{x}) = \begin{bmatrix} L_{\boldsymbol{g}_1} L_{\boldsymbol{f}}^{r_1-1} h_1(\boldsymbol{x}) & \cdots & L_{\boldsymbol{g}_m} L_{\boldsymbol{f}}^{r_1-1} h_1(\boldsymbol{x}) \\ \vdots & \ddots & \vdots \\ L_{\boldsymbol{g}_1} L_{\boldsymbol{f}}^{r_m-1} h_m(\boldsymbol{x}) & \cdots & L_{\boldsymbol{g}_m} L_{\boldsymbol{f}}^{r_m-1} h_m(\boldsymbol{x}) \end{bmatrix}$$

Now we describe the input-output feedback linearizing control law. Consider an arbitrary set of constants $\alpha_{i,k} \in \Re (i = 1, \ldots, m; k = 0, \ldots, r_i)$ satisfying the following conditions:

$$(I) \qquad \det[(\sum_{k=0}^{r_1} \alpha_{1,k} s^k) \cdots (\sum_{k=0}^{r_m} \alpha_{m,k} s^k)] \neq 0$$

$$(II) \qquad \det[\alpha_{1,r_1} \cdots \alpha_{m,r_m}] \neq 0$$

If each of the outputs in the original system have a well defined relative degree and the corresponding characteristic matrix, $\boldsymbol{C}_h(\boldsymbol{x})$, is non-singular, the state feedback

$$\boldsymbol{u} = \left([\alpha_{1,r_1} \cdots \alpha_{m,r_m}][L_{\boldsymbol{g}} L_{\boldsymbol{f}}^{r_1-1} h_1 \cdots L_{\boldsymbol{g}} L_{\boldsymbol{f}}^{r_m-1} h_m]^T \right)^{-1} \cdot \left(v - \sum_{i=1}^{m} \sum_{k=0}^{r_i} \alpha_{i,k} L_{\boldsymbol{f}}^k h_i \right)$$

induces the closed-loop response

$$\sum_{k=0}^{r_i} \alpha_{i,k} \frac{d^k y_i}{dt^k} = v_i \qquad i = 1, \ldots, m \tag{2}$$

$\alpha_{i,k}$ can be chosen to ensure the stability and control the dynamics of the linear system in the transformed coordinates.

For ease of representation, we introduce the technical assumption that $r_i = r$, $i = 1, \cdots, m$ (this can be relaxed). Hence, one can introduce the following matrix:

$$\alpha^{(k)} = \begin{bmatrix} \alpha_{1,k} & 0 & \cdots & & 0 \\ 0 & \alpha_{2,k} & \cdots & & 0 \\ & & \ddots & \ddots & \\ 0 & \cdots & & \alpha_{m-1,k} & 0 \\ 0 & \cdots & & 0 & \alpha_{m,k} \end{bmatrix} \qquad k = 0, 1, \ldots, r-1$$

and the input-output linearizing control law can be represented as

$$\boldsymbol{u} = \left[\alpha^{(r)} \boldsymbol{C}_h(\boldsymbol{x})\right]^{-1} \left[v - \alpha^{(r)} L_{\boldsymbol{f}}^r \boldsymbol{h}(\boldsymbol{x}) - \alpha^{(r-1)} L_{\boldsymbol{f}}^{r-1} \boldsymbol{h}(\boldsymbol{x}) \cdots - \alpha^{(0)} \boldsymbol{h}(\boldsymbol{x})\right] \tag{3}$$

Returning to the discussion in the introduction, Equation 3 represents the relationship between the input vector $\boldsymbol{u}$ and the new input $\boldsymbol{v}$ (depicted as Ψ in Figure 1) such that there is a decoupled, linear relationship in each input-output channel. The resultant linear system between $\boldsymbol{v}$ and $\boldsymbol{y}$ is represented as G_l. Linear single input single output (SISO) IMC controllers can be designed for each individual channel to yield the closed-loop transfer function given by

$$y_i(s) = \frac{y_{sp_i}}{(\lambda s + 1)^r} \tag{4}$$

where y_{sp_i} is the setpoint for the ith channel. The filter time constant, λ, is equal to the closed-loop time constant of the system. By decreasing (increasing) λ, the system can be tuned to be more aggressive (sluggish).

To better understand the role of the coordinate transformation, consider a linear state-space system model given by $[A, B, C, D]$. The Lie derivatives for this system can be represented in the Laplace domain via:

$$\boldsymbol{y}(s) = C(sI - A)^{-1}B\boldsymbol{u}(s)$$

$$\vdots$$

$$s^r \boldsymbol{y}(s) = CA^r(sI - A)^{-1}B\boldsymbol{u}(s) + CA^{r-1}B\boldsymbol{u}(s)$$

The coordinate transformation and pole placement of the transformed system using the $\alpha^{(i)}$ matrices results in the corresponding linear version of Ψ:

$$\boldsymbol{v}(s) = \alpha^{(r)}CA^{r-1}B\boldsymbol{u}(s) + \alpha^{(r)}CA^r(sI - A)^{-1}B\boldsymbol{u}(s) + \cdots + \alpha^{(0)}C(sI - A)^{-1}B\boldsymbol{u}(s)$$

Finally, the transformation, Ψ, relating $\boldsymbol{u}$ and $\boldsymbol{v}$ is given by:

$$\begin{aligned} \boldsymbol{u}(s) &= [\alpha^{(r)}CA^r(sI - A)^{-1}B + \alpha^{(r)}CA^rB + \\ &\quad \alpha^{(r-1)}CA^{r-1}(sI - A)^{-1}B + \cdots \\ &\quad + \alpha^{(0)}C(sI - A)^{-1}B]^{-1}\boldsymbol{v}(s) \\ &= [G_l^{-1}(s)G(s)]^{-1}\boldsymbol{v}(s) = \Psi\boldsymbol{v}(s) \end{aligned} \quad (5)$$

It is straightforward to show that the coordinate transformation using IOL and subsequent IMC control is equivalent to a direct IMC control design.

3. IMPACT OF MULTIPLICATIVE INPUT UNCERTAINTY

It is difficult to capture the behavior of a complex physical system with equations; invariably there is plant-model mismatch or uncertainty associated with the model. Linear model uncertainty can be described in a number of ways including additive uncertainty, input multiplicative uncertainty and output multiplicative uncertainty (Morari and Zafiriou, 1989). In this paper, we focus our attention on linear multiplicative input uncertainty. To obtain a mathematical representation of the input multiplicative uncertainty, let G be the nominal linear model of the plant while the actual plant $G_p \in \Pi$ where Π is the set of possible plants. Then the real plant can be expressed in term of the nominal model as $G_p = G(I + L_I)$, where all the uncertainty associated with the input can be lumped into the term L_I (Morari and Zafiriou, 1989). To represent all the plants in the set of plants Π, bounds on L_I are considered. A bound can be defined as $\overline{\sigma}(L_I) \leq \Delta_u(\omega)\ \forall\omega$ where $\Delta_u = \max_{G_p \in \Pi} \overline{\sigma}(L_I)$ is the maximum uncertainty associated with all the plants (G_p) in the set Π.

For a nonlinear system, robustness analysis is much more complicated as the uncertainty description can also be defined using a nonlinear function. In the input-output linearization framework, one approach requires that certain matching conditions be satisfied by the uncertainty (Corless and Leitmann, 1981). Actuator uncertainty in a linear system is often modeled using an input multiplicative uncertainty description. Since the actuator uncertainty affects a nonlinear system in a similar manner, a linear multiplicative uncertainty description in combination with a nonlinear system is studied in this work. Figure 1 shows how such uncertainty can be represented using a Δ_u block. By its very definition, linear multiplicative input uncertainty satisfies the matching conditions described in (Corless and Leitmann, 1981).

The nonlinear transformation block Ψ represents the control law by Equation 3. However, the actual input to the nonlinear plant is assumed to be corrupted by linear uncertainty:

$$\begin{aligned} \tilde{\boldsymbol{u}} &= (I + \Delta_u)\boldsymbol{u} = \boldsymbol{u} + \Delta_u \left[\alpha^{(r)}\boldsymbol{C}_h\right]^{-1} \\ &\quad \cdot \left[\boldsymbol{v} - \alpha^{(r)}L_{\boldsymbol{f}}^r\boldsymbol{h} - \alpha^{(r-1)}L_{\boldsymbol{f}}^{r-1}\boldsymbol{h} \cdots - \alpha^{(0)}\boldsymbol{h}\right] \end{aligned} \quad (6)$$

By shifting the uncertainty through the transformation block Ψ, the linear nominal model can be preserved and a nonlinear uncertainty block can be added to the structure. That is, the problem can now be treated as a linear plant with a state dependent multiplicative input uncertainty, Δ_v. A schematic representation is shown in Figure 2.

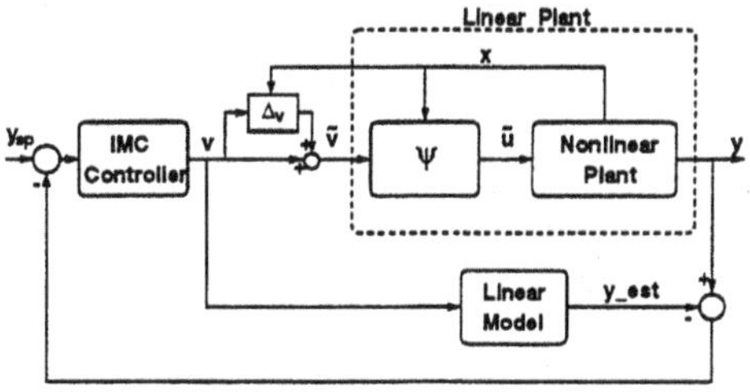

Fig. 2. Transformation of the multiplicative input uncertainty.

The uncertainty (Δ_u) in the input ($\boldsymbol{u}$) to the nonlinear plant is transformed into the uncertainty (Δ_v) in the input ($\boldsymbol{v}$) to the linear plant in the transformed coordinates. While constructing the transformation, however, the relationship between $\boldsymbol{v}$, the output of the linear IMC controller, and $\tilde{\boldsymbol{u}}$, the input to the nonlinear plant, has to be preserved. Equation 6 describes the relationship between $\tilde{u}$ and v using the uncertainty description Δ_u. The relationship between $\tilde{u}$ and v using the uncertainty description Δ_v is given by:

$$\begin{aligned}\tilde{u} &= [\alpha^{(r)}C_h(x)]^{-1}[(I+\Delta_v)v - \alpha^{(r)}L_f^r h(x) - \\ &\quad \alpha^{(r-1)}L_f^{r-1}h(x)\cdots - \alpha^{(0)}h(x)] \\ &= u + [\alpha^{(r)}C_h(x)]^{-1}\Delta_v v \end{aligned} \tag{7}$$

The uncertainty is transformed through Ψ and the "linear plant" and "linear model" are the same. This idea is depicted in Figure 2.

Equation 6 and Equation 7 can be equated and rearranged to obtain an expression for the nonlinear operator Δ_v.

$$\begin{aligned}\Delta_v v &= [\alpha^{(r)}C_h(x)]\Delta_u[\alpha^{(r)}C_h(x)]^{-1}[v - \alpha^{(r)}L_f^r h(x) - \\ &\quad \alpha^{(r-1)}L_f^{r-1}h(x)\cdots - \alpha^{(0)}h(x)]\end{aligned}$$

Thus, with this uncertainty description, we effectively have a linear system described by a linear model with nonlinear state dependent uncertainty. To further analyze the dynamics of this nonlinear, state dependent uncertainty block, an open-loop observer is constructed to evaluate the states, (Figure 3). $\Delta_v v$ can then be represented by the following nonlinear state-space model:

$$\begin{aligned}\dot{w} &= [f(w) - g(w)(I+\Delta_u)(\alpha^{(r)}C_h)^{-1}\alpha^{(r)}L_f^r h(w)\dots \\ &\quad -g(w)(I+\Delta_u)(\alpha^{(r)}C_h)^{-1}\alpha^{(0)}h(w)] \\ &\quad +g(w)(I+\Delta_u)(\alpha^{(r)}C_h)^{-1}v \\ &= f_v(w) + g_v(w)v \\ \Delta_v v &= -(\alpha^{(r)}C_h)\Delta_u(\alpha^{(r)}C_h)^{-1}(\alpha^{(r)}L_f^r h(w) + \dots \\ &\quad +\alpha^{(0)}h(w)) + (\alpha^{(r)}C_h)\Delta_u(\alpha^{(r)}C_h)^{-1}v \\ &= h_v(w) + l_v(w)v\end{aligned} \tag{8}$$

where w are the states of the observer.

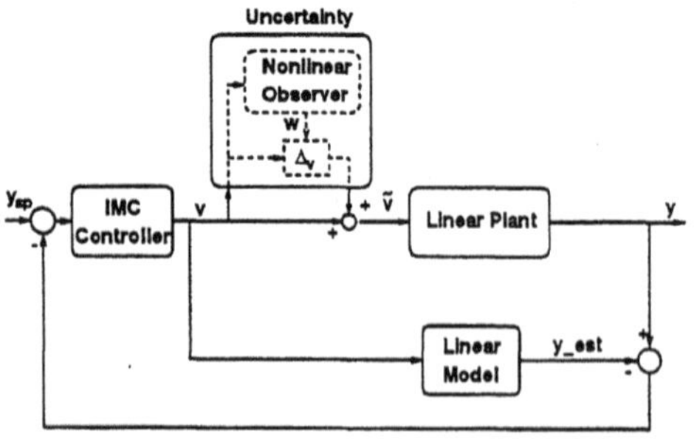

Fig. 3. Feedback linearized plant with a linear IMC controller and a nonlinear, state dependent uncertainty.

From Equation 8, it is seen that when $\Delta_u = 0$, $\Delta_v = 0$ as well. However, when $\Delta_u \neq 0$, Δ_v depends on the choice of $\alpha^{(r)}\dots\alpha^{(0)}$. Smaller values of $\alpha^{(r)}\dots\alpha^{(0)}$ result in faster dynamics of the observer and lower gains of Δ_v.

The parameter $\alpha^{(r)}$ defines the dynamics of the linear system G_l and the transformation Ψ. Small values of this parameter lead to faster dynamics in G_l. This results in faster feedback of the state information to the transformation block and, hence, the mismatch is reduced.

4. CONDITION NUMBERS OF THE GAIN AND THE CHARACTERISTIC MATRICES

In this section various case studies to demonstrate the effect of the condition numbers (κ) of the characteristic matrix and the gain matrix will be considered. *Note*: A system with high condition number, above 70, is generally considered ill-conditioned. The four cases considered are described in Table 2.

Table 1. Case study parameters

	$\kappa(G)$	$\kappa(C_h)$
Case Study 1	154.99	65.98
Case Study 2	6.13	65.98
Case Study 3	1.89	1.43
Case Study 4	159.11	2.88

Simulations show that the effect of uncertainty is amplified for systems with ill-conditioned characteristic matrix irrespective of the condition number of the gain matrix. The choice of $\alpha^{(r)}$ also plays an important role in the analysis of the robustness of the IOL controller.

4.1 *Case Study 1*

Consider a nonlinear system given by

$$\dot{x} = \begin{bmatrix}\dot{x}_1 \\ \dot{x}_2\end{bmatrix} = \begin{bmatrix}-0.01x_1 - 0.01x_2 - 0.05x_2^2 \\ -0.01x_2\end{bmatrix} + \begin{bmatrix}0.4 & -0.5 \\ -0.3 & 0.4\end{bmatrix}\begin{bmatrix}u_1 \\ u_2\end{bmatrix}$$

$$y = \begin{bmatrix}1 & 0 \\ 0 & 1\end{bmatrix}x$$

The condition of the gain matrix is 154.9935 which indicates the system is ill-conditioned and the condition number for the characteristic matrix is 65.9848 which also borders on ill-conditioning.

For this example, we select $\lambda = 20$ and $\alpha^{(0)} = I$, while $\alpha^{(1)}$ is varied. The choice of $\alpha^{(1)}$, which defines the dynamic behavior of the linear system, does not affect the closed-loop response as long as λ is unchanged. The various values of $\alpha^{(1)}$ considered are: $\alpha^{(1)} = \text{diag}(0.05, 0.05)$, $\text{diag}(0.5, 0.5)$, and $\text{diag}(5, 5)$.

The response of the closed-loop system to a setpoint change of $y_{sp} = [0.8\ 0]^T$ is shown in Figure 4. For higher values of $\alpha^{(1)}$, *i.e.*, larger time constants of the linear system, both the outputs deviate considerably from the reference trajectory. However, as can be observed, for lower values of $\alpha^{(1)}$ the system trajectory is close to the reference trajectory. This follows directly from the theory in section 3. In the calculation of Δ_v (Equation 8), $\alpha^{(1)}$ plays a crucial part. Lower values of $\alpha^{(1)}$ result in lower uncertainty in the linear system being controlled by the linear IMC controller.

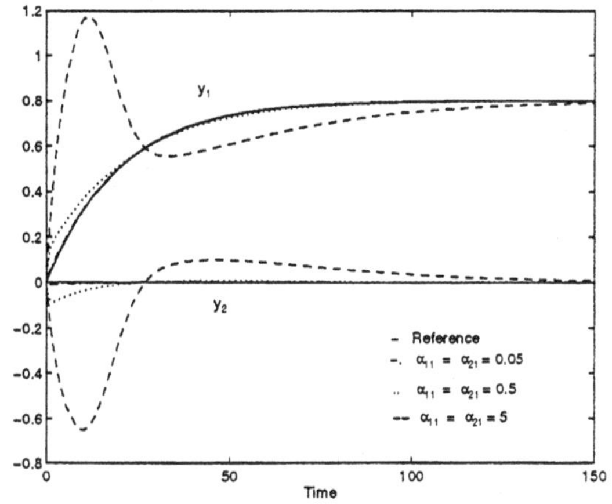

Fig. 4. Case study 1 – step setpoint change of $\boldsymbol{y}_{sp} = [0.8\ 0]^T$

4.2 *Case Study 2*

The nonlinear system considered in this section has an ill-conditioned characteristic matrix and a well conditioned gain matrix. The nonlinear system can be represented as:

$$\dot{\boldsymbol{x}} = \begin{bmatrix} -0.015x_1 + 0.02x_2 + 0.1x_2^2 \\ 0.015x_1 - 0.014x_2 \end{bmatrix} + \begin{bmatrix} 0.4 & -0.5 \\ -0.3 & 0.4 \end{bmatrix} \boldsymbol{u}$$

$$\boldsymbol{y} = \begin{bmatrix} 1 & 0 \\ 0 & 1 \end{bmatrix} \boldsymbol{x}$$

The condition number of the gain matrix is equal to 6.1256, while the condition number of the characteristic matrix is 65.9848, making it ill-conditioned.

A 20% uncertainty in the input streams is considered as in the previous case study. The closed-loop time constant, λ, is chosen to be 20. $\alpha^{(1)}$ is varied as per the previous case study while $\alpha^{(0)}$ is a 2×2 identity matrix.

The response of the system to a setpoint change of $\boldsymbol{y}_{sp} = [0.8\ 0]^T$ for various values of $\alpha^{(1)}$ is shown in Figure 5. The responses of the systems in both case studies 1 and 2 to the setpoint change are extremely close. This is explained when it is noted that the condition number for the characteristic matrix for both the cases is equal to 65.9848. Thus, even though the system in case study 2 has a well behaved gain matrix, its ill-conditioned characteristic matrix makes it sensitive to uncertainty and the choice of $\alpha^{(1)}$.

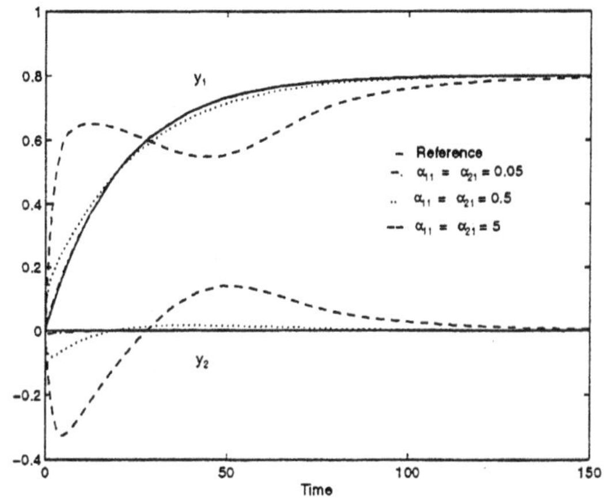

Fig. 5. Case study 2 – values of $\alpha^{(1)}$ for a step setpoint change of $\boldsymbol{y}_{sp} = [0.8\ 0]^T$

4.3 *Case Study 3*

The third case study involves the following nonlinear system:

$$\dot{\boldsymbol{x}} = \begin{bmatrix} -0.01x_1 - 0.01x_2 - 0.05x2^2 \\ -0.01x_2 \end{bmatrix} + \begin{bmatrix} 0.4 & 0.5 \\ -0.3 & 0.4 \end{bmatrix} \boldsymbol{u}$$

$$\boldsymbol{y} = \begin{bmatrix} 1 & 0 \\ 0 & 1 \end{bmatrix} \boldsymbol{x}$$

The condition number of the gain matrix is 1.8909 and the condition number of the characteristic matrix is 1.4295. This is relatively uninteresting from a robust control point of view, as the system is very well behaved. As in the previous sections, 20% input multiplicative uncertainty in the two inputs is considered.

Given the space limitations, no results are plotted, but it suffices to say that there are no robustness problems for this very benign system.

4.4 *Case Study 4*

The final case study deals with a system which has a well behaved characteristic matrix but an ill-conditioned gain matrix. Traditionally, it is inadvisable to decouple this system as the inverse based controller then becomes ill-conditioned as well (Kravaris and Soroush, 1990). However, from the studies conducted it can be shown that as long as the characteristic matrix is well behaved, the nonlinear system can be input-output linearized and decoupled without significant loss in performance of the controller.

The nonlinear system is defined by:

$$\dot{x} = \begin{bmatrix} -0.01x_1 - 0.001x_2 - 0.005x_2^2 \\ 0.2x_1 - 0.01x_2 \end{bmatrix} + \begin{bmatrix} 0.4 & -0.1 \\ -0.3 & 0.4 \end{bmatrix} u$$

$$y = \begin{bmatrix} 1 & 0 \\ 0 & 1 \end{bmatrix} x$$

The condition number of the gain matrix is 159.1078 while that of the characteristic matrix is 2.8840. Δ_u, $\alpha^{(1)}$ and $\alpha^{(0)}$ are defined as before. The closed-loop time constant, λ, is equal to 20.

A setpoint change of $\boldsymbol{y}_{sp} = [0.8\ 0]^T$ is considered. The response is shown in Figure 6. As long as the characteristic matrix is well behaved, the ill-conditioned gain matrix does not play a crucial part. Hence it can be observed that for different values of $\alpha^{(1)}$, the system tracks the reference trajectory fairly closely. The response of the system in case study 4 deviates more from the reference trajectory when $\alpha_{11} = 5$ than in case study 3 (not plotted). This can be attributed to higher condition number of the gain matrix (and also a small increase in the condition number of the characteristic matrix) for the case study 4. On the other hand, the deviation is not as pronounced as in the case study 2, Figure 5, where the condition number of the characteristic matrix is very high.

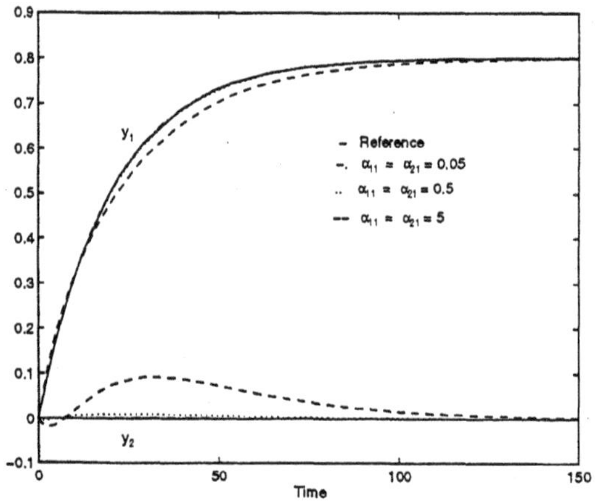

Fig. 6. Case study 4 – step setpoint change of $\boldsymbol{y}_{sp} = [0.8\ 0]^T$

5. CONCLUSIONS

Input-output linearization is an important tool in nonlinear control synthesis. Hence, it is important to understand the factors that play a role in the design and tuning of this controller.

Robustness of any controller when faced with plant-model mismatch is extremely important. From linear robust control theory it is known that an inverse based controller for a plant with an ill-conditioned gain matrix is ill-conditioned as well and has poor robustness properties. However, for nonlinear controllers developed using input-output linearizing scheme, the robustness properties of the controller depend on the scaled characteristic matrix. It is assumed that the nonlinear plant is corrupted by linear multiplicity uncertainty which can arise due to actuator measurement uncertainty. The different case studies considered are outlined in Table 1.

From the results, we can conclude that the matrix which determines the robustness of the controller is the characteristic matrix. Case studies 2 and 4 demonstrate good robustness properties while case studies 1 and 3 do not. However, for case studies 1 and 3, the choice of the time constants of the linear plant plays an important role. Systems with lower time constants have better control action compared to the linear systems with higher time constants as the uncertainty present in the outer linear loop of the IOL controller is reduced. However, more integrator computational effort is required when the nonlinear system is linearized to a system with a smaller time constant. The results discussed here hold as long as the zero dynamics of the system without uncertainty are stable.

6. REFERENCES

Allgöwer, F. and F.J. Doyle III (1996). Nonlinear Process Control - Which Way to the Promised Land?. In: *Proceedings of the Fifth International Conference on Chemical Process Control.* Tahoe City, CA.

Calvet, J. P. (1989). A Differential Geometric Approach for the Nominal and Robust Control of Nonlinear Chemical Processes. PhD thesis. Georgia Institute of Technology, Atlanta, GA. Chemical Engineering.

Corless, M. and G. Leitmann (1981). Continuous State Feedback Guaranteeing Uniform Ultimate Boundedness for Uncertain Dynamic Systems. *IEEE Trans. Autom. Control* **26**(5), 1139–1144.

Isidori, A. (1995). *Nonlinear Control Systems.* third ed.. Springer-Verlag, New York, NY.

Kravaris, C. and M. Soroush (1990). Synthesis of multivariable nonlinear controllers by input/output linearization. *AIChE J.* **36**(2), 249–264.

Kravaris, C. and S. Palanki (1988). Robust nonlinear state feedback under structured uncertainty. *AIChE J.* **34**(7), 1119–1127.

Morari, M. and E. Zafiriou (1989). *Robust Process Control.* Prentice-Hall, Englewood Cliffs, NJ.

Skogestad, S., M. Morari and J. C. Doyle (1988). Robust Control Of Ill-Conditioned Plants: High-Purity Distillation. *IEEE Trans. Autom. Control* **33**, 1092–1105.

LOCAL NONLINEAR PERFORMANCE ASSESSMENT FOR SINGLE-CONTROLLER DESIGN

Alexander J. Stack ** **Francis J. Doyle III** *,[1]

* *Dept. of Chem. Eng. Univ. of Delaware, Newark, DE 19716*
** *School of Chemical Engineering, Purdue University*

Abstract: A novel approach to operation-oriented nonlinearity is presented that utilizes a coherence analysis of multiple local excitations to investigate the nonlinearity of a process model in the neighborhood of a given process operation curve. By using the IMC structure with perfect control and a first order filter, a control structure is designed that exhibits a range of nonlinearity, controlled by a single parameter. This structure can then be assessed for performance and stability. Examples are given for a SISO Hammerstein model using a simplified μ-analysis. *Copyright © 1999 IFAC*

Keywords: Nonlinearity, Nonlinear Analysis, Model-based Control, Process Control.

1. INTRODUCTION

Improvements in process control have been identified as perhaps the most important source of increased productivity within the process industries, and would also have significant environmental and safety benefits. A particular bottleneck for improvements in plant-wide control is the lack of effective interaction between process design and process control.

As is emphasized in (Allgöwer and Doyle III, 1996), there is a need to develop system analysis tools, including ones that "capture the influence that nonlinearities have with respect to achieving certain performance goals in the closed-loop". This issue has been surveyed by Poolla et al. (1990), who focus on the effect of different uncertainties (parametric and dynamic) on the performance of linear time-invariant versus nonlinear time-varying control. In (Stack and Doyle III, 1997*b*) a methodology was introduced that advanced towards this goal by facilitating an analysis of the influence of economic (through the performance objective) and plant nonlinearity on the nonlinearity of the optimal structure of the corresponding control system. In this paper, an approach that utilizes some of the tools from the earlier work is introduced as a first step towards answering a deeper question: given a particular plant, what performance can be expected from nonlinear as opposed to linear control?

2. COHERENCE ANALYSIS

Magnitude-squared coherence (or simply "coherence") is a frequency dependent measure of the linear relationship between two stationary time signals (Carter, 1993). It has been utilized for discrete and continuous chemical engineering systems in (Pearson and Ogunnaike, 1994) and (Stack and Doyle III, 1997*a*; Stack and Doyle III, 1997*b*), respectively. The following is a brief description of the detailed discussion given in the latter paper.

For two random processes $u(t)$ and $x(t)$, the magnitude squared coherence is given by

$$\gamma_{ux}^2(f) = \frac{| S_{ux}(f) |^2}{S_{uu}(f)S_{xx}(f)} \qquad (1)$$

where $S(f)$ is the appropriate two-sided cross-spectral density function. Coherence values are

[1] Corresponding author (fdoyle@udel.edu)

bounded between zero and one, where a linear structure accounts for the entire relationship between $u(t)$ and $x(t)$ at that frequency, and zero, where there is no evidence of a linear relationship. Generally, coherence suppression can also occur due to noise in the input and output measurements, unmeasured disturbances, and errors from the coherence estimation process. In this work, a preliminary model representing a potential plant design is assumed, so exact values for the input and output will be known, avoiding noise and disturbance complications. This leaves a nonlinear relationship between $u(t)$ and $x(t)$, as well as coherence estimation errors, as the only source of coherence values less than one.

Following the weighted overlapped segment averaging (WOSA) method (Carter, 1993), the input and output data are divided into segments, with a 50 percent overlap between adjacent segments, which are detrended. Each segment is then multiplied by a Hanning weighting function and a fast Fourier transform (FFT) is performed on each segment. The magnitude squared values are time averaged over each frequency bin to obtain the power spectral densities of the input and output ($S_{uu}(f)$ and $S_{xx}(f)$), while the products are averaged to form the cross-spectral density $S_{ux}(f)$. N, t_s, and h are used to denote the total number of points analyzed, sampling time, and integration time respectively. For the work presented in this paper, the following values were employed: (N_i = 16384, $\sigma_x^2 = 1$, $t_s = 0.0005$, h = 0.0001, S = 256).

Additional technical details regarding the treatment of nonstationary signals and the analysis of variance of the coherence estimates are detailed in (Stack and Doyle III, 1997*a*; Stack and Doyle III, 1997*b*).

2.1 *Multiple Local-Excitation Coherence Analysis*

Dynamic control of an operating plant as it moves between desired setpoints is often the major challenge for process control, as opposed to control of a system at an operating point. The coherence analysis procedure briefly outlined above involves analyzing the data from one simulation run with a white noise input centred on one operating point. While it is possible to expand the examined region of operation (by increasing the variance on the white signal) to include more than one operating point, this approach to assessing dynamic behaviour is unsatisfactory. When moving between operating points the control engineer causes the plant to move along a desired control trajectory. By taking the above approach, a space of trajectories is searched that may stray far from any reasonable control trajectory. As only a portion of this data will correspond to the relevant control trajectory, the resulting coherence spectrum will be a conservative estimate of the model nonlinearity.

This paper proposes a method to constrain our data collection to points in a region that follows a desired trajectory by concatenating input and output data from simulations run at a number of points along the trajectory. The white noise inputs should be small enough to ensure that the data collected is "local" compared to the density of the operating points. These data records will then be detrended, concatenated, and then analyzed using the WOSA procedure described above. This method, which may be described as multiple local-excitation coherence analysis, will test how well one linear relationship accounts for the observed inputs and outputs in the neighborhood of the control trajectory.

For example, suppose that a nonlinear system is approximated at five points by a first-order linear transfer function with a time constant of one and gains of 1,2,4,10 and 20 respectively. The concatenated input and output data records and resulting coherence spectrum are shown as Figure 1. Although each individual data set is for a linear system, and if analyzed gives a uniform coherence of one, the coherence of the concatenated data shows coherence suppression representing the nonlinearity of the overall system over those five operating points.

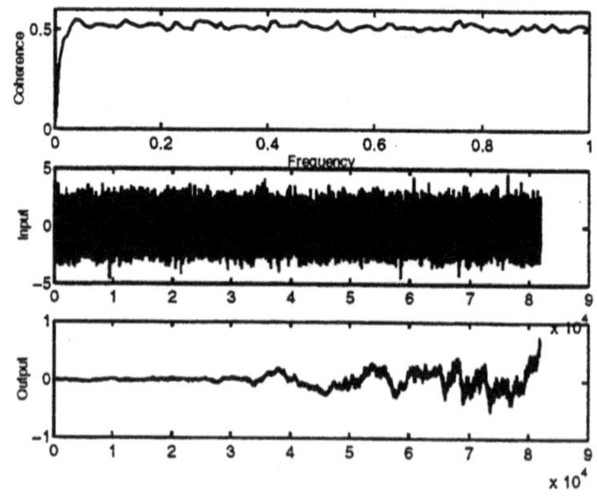

Fig. 1. Coherence analysis for the multiple linear model system.

However, the concatenation cannot be treated as an algorithmic exercise. Large differences between the final point of one data set and the first point of the next introduce discontinuities in the data, leading to extra, unwanted coherence suppression. For example, rearranging the data runs used in Figure 1 results in Figure 2. A much larger effect can be seen for models that are multiple local-excitation linear. A possible solution to this is to eliminate some points from the preceding data set to reduce this discontinuity. However, a large truncation will be undesirable, as it is important to represent each operating point with similar numbers of data points. It is probable that the more strongly stationary the data is, the less likely that coherence suppression will occur from this source.

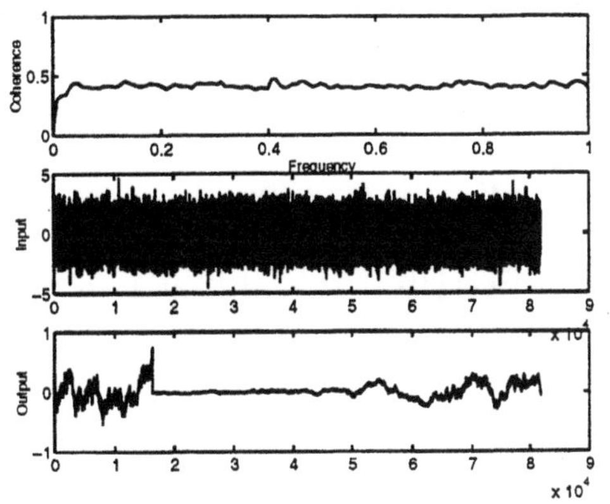

Fig. 2. Coherence analysis for the multiple linear model system, changed order of concatenation.

3. INTERNAL MODEL CONTROL STRUCTURE

For purposes of this analysis, a control structure is desired that i) is simply tunable over a wide range of nonlinearity, ideally through the adjustment of one parameter; and ii) represents a "good" control formulation for all values of the parameter. Internal model control (IMC) (Morari and Zafiriou, 1989) theory incorporating perfect control and a first-order filter provides such a framework.

In this framework, $\tilde{p}$ is a model of the plant p and q is a controller. By choosing the model inverse as the controller q ($q = \frac{1}{\tilde{p}}$) "perfect control" is implied, where the error $e = y - r$ equals zero. In reality, such a controller is infeasible (Morari and Zafiriou, 1989), but it is useful to investigate how plant characteristics limit the achievable performance independent of particular controller formulations. In particular, if $\tilde{p} = p$, then the perfect controller has the best performance achievable with any controller.

It is tempting to assume that q will have the same nonlinearity as $\tilde{p}$ in the case of perfect control. However, consider the coherence analysis of Wiener and Hammerstein models: these are exact inverses, but the coherence plots of Wiener plots decrease at higher frequencies while those of Hammerstein models are flat (Pearson and Ogunnaike, 1994). This implies that exact inverses can have different nonlinearities at at least some frequencies. However, it is difficult to make any general observations beyond this.

If the plant p is proper (for example, $p(s) = (s+1)^{-1}$), then the resulting q will be improper ($q(s) = (s+1)$), and may be realized by pre-multiplying with a first order filter $F = \frac{1}{\lambda s+1}$ to create

$$Q(s,\lambda) = \frac{s+1}{\lambda s+1} \tag{2}$$

which will approach q as λ approaches zero. Note that the nonlinearity of Q will also approach that of q as λ approaches zero.

The IMC model may be put into the classic feedback form by letting:

$$C(s,\lambda) = \frac{Q(s,\lambda)}{1-\tilde{p}(s)Q(s,\lambda)} \tag{3}$$

The filter parameter λ can be increased to slow the response of the closed-loop controller C, achieving superior control by effectively making the relationship between $\tilde{r}$ and u more linear, in general. Thus, tuning λ has the effect of tuning the nonlinearity of the perfect IMC control structure C, from its maximum nonlinearity as λ approaches zero to a more linear relationship as λ is increased (in general).

For ease of analysis, a SISO model with a known exact inverse will be analyzed. In general, models with exact inverses will not be applicable. As discussed in (Doyle *et al.*, 1995), the exact nonlinear inverse may be replaced with a relaxed inverse, where the norm of a residual signal is minimized, requiring only an inverse for the linear portion of the nonlinear model.

Hammerstein systems consist of a linear dynamic plant with a static nonlinearity, such as valve nonlinearity, on the input. These have been used to model a wide range of process systems (Eskinat *et al.*, 1991; Ouarti and Edgar, 1993). The Hammerstein system analyzed in this paper is:

$$\dot{x} = \frac{-x}{50} - 50\left[\frac{-3u}{10+8u}\right] \tag{4}$$

$$y = x \tag{5}$$

with a time delay of 0.1 on the linear subsystem. Note that the linear dynamics correspond to a transfer function with a gain of -50 and time constant of 50, and that u, x and y refer to the input, state and output respectively. The relaxed inverse for this Hammerstein system is

$$\dot{x} = \frac{-x}{\lambda} + \frac{50-\lambda}{50\lambda^2}u$$

$$y = \frac{-3-8(x-\frac{u}{\lambda})}{10(x-\frac{u}{\lambda})} \tag{6}$$

with a time delay of -0.1. This was analyzed using the multiple local-excitation coherence procedure detailed above. Simulations were run for inputs of 1,2,3 ... 18 with a white noise input of variance of 0.25. The simulations were run on an euler integrator in MATLAB with an integration time of 0.0001 and a sampling time of 0.0005. 16794 data points were collected at each point. To delete any dynamics arising from moving to new operating points, the first 410 data points in each record were truncated.

The resulting coherence spectra for each value of λ appear flat, except for a roll off at frequency zero that arises from the coherence estimation procedure. There appear to be no discontinuities in the concatenated output that would suppress the coherence values, and the outputs passed the run tests for windows of length 256 points. A plot of mean coherence versus λ is shown in Figure 3. As can be seen, they show the general trends espoused above: the system becomes linear as λ is increased. In this case, the coherence of the perfect inverse equals zero, and the realization of the IMC inverse has a mean coherence of zero for values of λ less than 56.

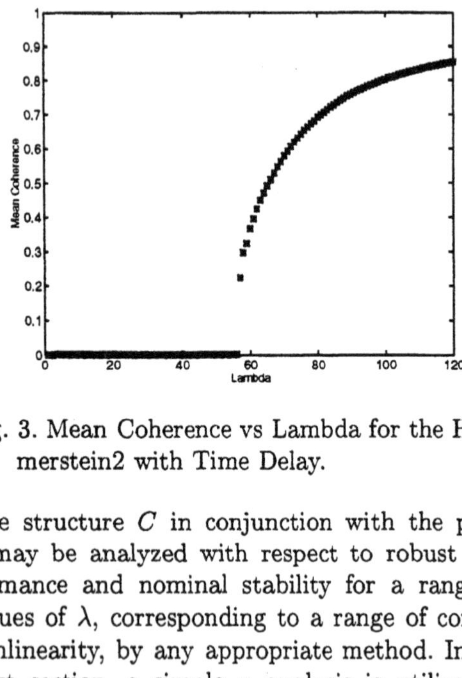

Fig. 3. Mean Coherence vs Lambda for the Hammerstein2 with Time Delay.

The structure C in conjunction with the plant p may be analyzed with respect to robust performance and nominal stability for a range of values of λ, corresponding to a range of control nonlinearity, by any appropriate method. In the next section, a simple μ analysis is utilized to demonstrate the potential results of the proposed approach.

4. PERFORMANCE ANALYSIS

Following the approach in (Doyle *et al.*, 1992), weights W_e, W_u, W_d and W_n are assigned to e, u, d, n in Figure 4. These weights may be regarded as economic penalties on variation in the respective channels. It is desired to keep the matrix:

$$\begin{bmatrix} W_e & 0 \\ 0 & W_e \end{bmatrix} \begin{bmatrix} PS & S \\ T & CS \end{bmatrix} \begin{bmatrix} W_d & 0 \\ 0 & W_n \end{bmatrix} \quad (7)$$

small in some sense, where S and T denote the sensitivity and complementary sensitivity functions. Doyle et al. (1992) use the specification

$$\| \left(|W_1 S|^2 + |W_2 T|^2\right)^{\frac{1}{2}} \|_\infty < 1 \quad (8)$$

$$|W_1| = |W_e| \left(|W_d P|^2 + |W_n|^2\right)^{\frac{1}{2}}$$

$$|W_2| = |W_u| \left(|W_n|^2 |P|^{-2} + |W_d|^2\right)^{\frac{1}{2}}$$

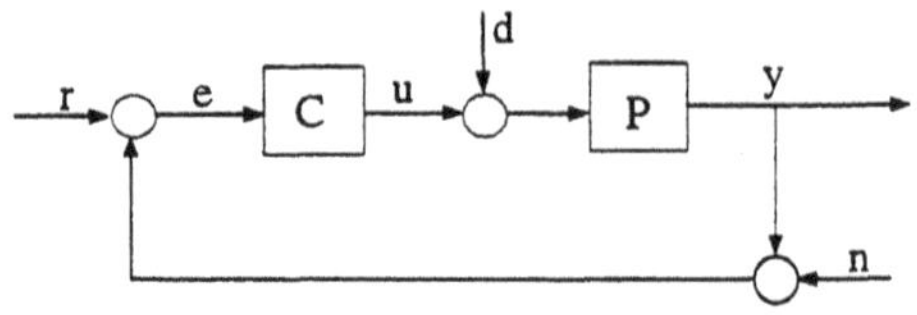

Fig. 4. Set-up and nomenclature for the performance weights.

In this paper, W_d and W_n are set equal to a positive constant K, and the value of W_e is found such that

$$\| \, |W_e| \left(|KP|^2 + K^2\right)^{\frac{1}{2}} |S| \, \|_\infty = 1 \quad (9)$$

corresponding to robust performance, and the value if W_u is found to make

$$\| \, |W_u| \left(K^2 |P|^{-2} + K^2\right)^{\frac{1}{2}} |T| \, \|_\infty = 1 \quad (10)$$

corresponding to nominal stability. As this is an analysis for linear models $P(s)$ and $C(s)$, the tests are run at a number of operating points, and the most restrictive weights found are reported. By comparing values of W_e and W_u found by the above procedure to various values of λ, graphs of robust performance and nominal stability vs the nonlinearity of C may be created. Note that this test should also apply for a region surrounding a control trajectory: similar to the ideas in extended linearization (Baumann and Rugh, 1986) a linearized model should also provide useful information for points in some region surrounding the equilibrium point.

These tests were performed at K equal to one. Sample graphs of these weights with W_e and W_u equal to one and 0.1 and a plant $P(s) = 1/s$ are given in Figure 5.

4.1 *Analysis of the Hammerstein System*

The robust performance and nominal stability analyses for the Hammerstein system for $K = 1$ may be found in Figure 6. The general trends are

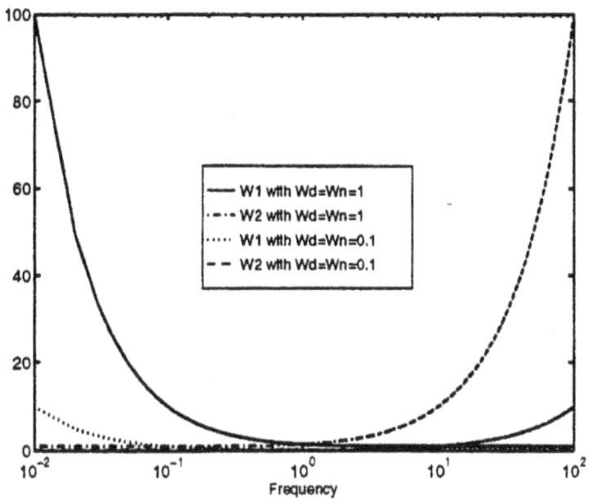

Fig. 5. Sample Weights

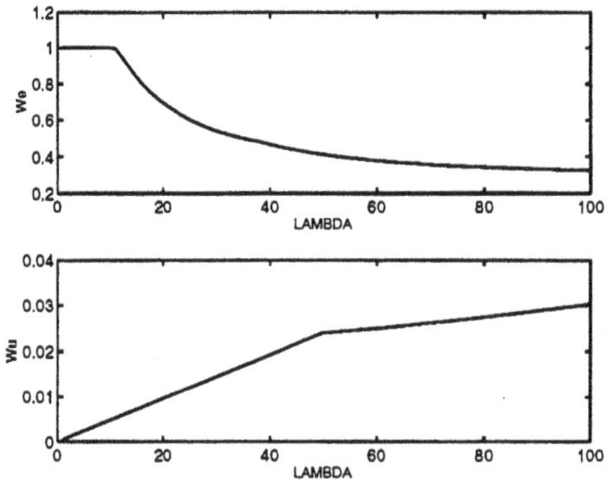

(a) Sorted by λ

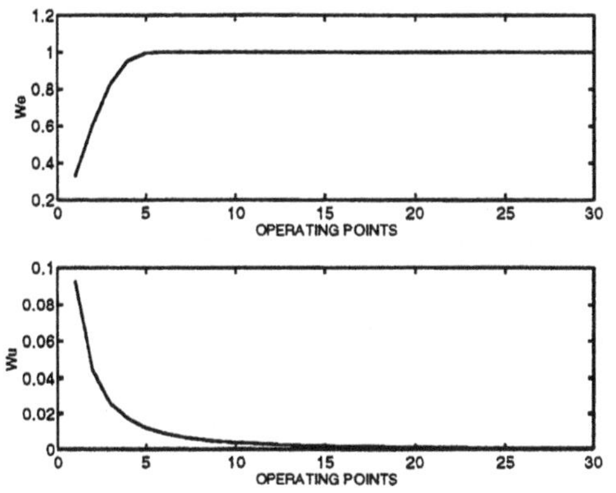

(b) Sorted by u steady state value

Fig. 6. μ-analysis for the Hammerstein model.

as expected, with a performance falloff as λ is increased and the control action becomes more linear. However, the nominal stability becomes possible for higher values of W_u as λ is increased.

In the plot of W_u versus λ, there is a noticeable change where the value of λ crosses the value of the time constant from equation 5. This is not unexpected, as the value of λ corresponds to the speed of the IMC controller: for smaller value of λ, the controller is attempting to move faster than the actual system to be controlled, leading to instability. As λ approaches zero, the performance metric is at its maximum while the stability metric falls to a minimum: this is also expected, as $\lambda = 0$ represents the perfect inverse. The plot of W_e vs λ contains a plateau for values of λ of ten or less: the cause of this is unknown.

In the plots sorted by λ, the smallest weights satisfying equations 9 and 10 over all the operating points for each λ are reported. By contrast, in the plots sorted by operating point, the smallest weights over all λ's for each operating point are reported. Thus, the plots in Figure 6b reveal the worst performance and stability shown by the realized IMC controller over the range of λ's considered. These can show that for operating points in the range $u = 5$ to 30, the performance metric W_e can be guaranteed to be at least one. However, in regions surrounding operating points with u less than five, this guarantee is lost. Analogous observations may also be made with respect to nominal stability and the values of W_u. This can be related to the asymptotic behaviour of the steady state locus of the original system.

5. DISCUSSION

Although still at a preliminary stage, these results suggest the possibility of quantitative insights into the tradeoff between controller nonlinearity, performance and stability. However, this work in many ways raises more questions than it answers. The following discussion attempts to bring together the above results, and more importantly to elucidate goals and directions for future research.

A desire to narrow the analysis of nonlinearity to a neighborhood of a control trajectory motivates the definition of multiple local-excitation coherence analysis. The analysis of the realized IMC controller for the Hammerstein system yields results predicted by IMC theory. However, it should be noted that the procedure followed in this paper strictly defines the "neighborhood" of the control trajectory investigated for the input only. The output is unconstrained, and may vary widely. The procedure may be refined in future work by focusing on the number and density of operating points to be tested, and the variance of the input signals used. It may be speculated that as the number of points investigated increases and the input variance decreases, the procedure may converge towards a "true" multiple local-excitation coherence spectrum. However, advances to be made in this area call for a more rigorous analysis. Alternative, less empirical approaches to nonlinearity analysis, such as the methods outlined in (Allgöwer, 1995), may also be productively extended to analyzing models around a trajectory.

The realization of the IMC inverse is shown to exhibit a range of nonlinearity as the first-order filter parameter is tuned. Furthermore, the structure contains information from the plant model, and thus contains a "good" controller in the sense that it will track setpoint changes for all values of λ. However, except for the perfect controller when λ equals zero, it is unclear what relationship the realized IMC controller bears to the best performance a single-controller of that nonlinearity could provide. Ideally, a structure should be designed that exhibits a wide range of nonlinearity, is tunable in a simple manner, and gives the best practical or theoretical performance for a single controller of that nonlinearity.

Note that the above discussion has been in the context of a single feedback control system. It has not addressed the possible gains and complications to be made by considering multiple control schemes, time-varying controllers, and other complications upon this simple control approach.

Although the μ-analysis used is simplistic, it demonstrates the potential quantifiable insights to be gained from tools of this type. From Figure 6a, it may be decided that a value of $\lambda = 50$ is an appropriate trade off between stability and performance. From Figure 3, this is seen to be a very nonlinear control scheme. However, Figure 6a also reveals that for this model, the performance asymptotes for larger values of λ. If it is desired to increase stability (accepting the performance degradation), we can then use Figure 3 to see the corresponding controller nonlinearity. While it is difficult to make sweeping statements, the realized IMC controller could be used as a standard against which to benchmark the performance and stability of various single-control schemes of an equivalent nonlinearity.

6. CONCLUSIONS

The results in this paper should be viewed as a preliminary step towards developing analysis tools to explore the link between economics, plant and controller design, and operational performance. Multiple local-excitation coherence analysis is a relatively straightforward method to perform a frequency-dependent nonlinearity analysis for a neighborhood of a state trajectory. This is an improvement over other suggested nonlinearity measures which have not considered this refinement, and broadly speaking consider nonlinearity over some elliptical region.

The realized IMC inverse control scheme is shown to exhibit a wide range of nonlinearity. The nonlinearity of the control structure is tunable through manipulation of the first-order filter parameter λ. By performing a μ-analysis incorporating economic weights, the relationship between economics, control scheme nonlinearity, robust performance, and nominal stability can be quantified.

7. ACKNOWLEDGMENTS

The authors gratefully acknowledge funding from the NSF (NYI Award CTS 9257059).

8. REFERENCES

Allgöwer, F. and F.J. Doyle III (1996). Nonlinear Process Control - Which Way to the Promised Land?. In: *Proceedings of the Fifth International Conference on Chemical Process Control.* Tahoe City, CA.

Allgöwer, F.A. (1995). Definition and computation of a nonlinearity measure and application to approximate I/O-linearization. Technical Report 95-1. Stuttgart University.

Baumann, W.T. and W.J. Rugh (1986). Feedback control of nonlinear systems by extended linearization. *IEEE Trans. Autom. Control* **AC-31**, 40–46.

Carter, G. C., Ed.) (1993). *Coherence and Time Delay Estimation.* IEEE Press, New York, NY.

Doyle, F.J., B.A. Ogunnaike and R.K. Pearson (1995). Nonlinear Model-Based Control Using Second Order Volterra Models. *Automatica* (31), 697–714.

Doyle, J. C., B. Francis and A. Tannenbaum (1992). *Feedback Control Theory.* Macmillan Publishing Co.

Eskinat, E., S. H. Johnson and W. L. Luyben (1991). Use of Hammerstein models in identification of nonlinear systems. *AIChE J.* **37**(2), 255–268.

Morari, M. and E. Zafiriou (1989). *Robust Process Control.* Prentice-Hall, Englewood Cliffs, NJ.

Ouarti, H. and T.F. Edgar (1993). The use of approximate models and exact linearization for control of nonlinear processes. In: *Proc. American Control Conf.*. San Francisco, CA. pp. 2268–2273.

Pearson, R. K. and B. A. Ogunnaike (1994). Detection of unmodelled disturbance effects by coherence analysis. In: *IFAC Symposium on Advanced Control of Chemical Processes.* Kyoto, Japan. pp. 458–463.

Poolla, K., J. Shamma and K. Wise (1990). Linear and nonlinear controllers for robust stabilization problems: A survey. In: *Proceedings of IFAC World Congress.* pp. 176–183.

Stack, A. J. and F. J. Doyle III (1997*a*). Application of a control-law nonlinearity measure to a chemical reactor. *AIChE J.* **43**, 425–447.

Stack, A.J. and F.J. Doyle III (1997*b*). The optimal control structure approach to measuring control-relevant nonlinearity. *Comput. Chem. Engng.* **21**, 998–1009.

I/O LINEARIZATION BASED μ SYNTHESIS FOR NONLINEAR ROBUST CONTROL SYSTEMS

Zhihong Peng , Min Wu , Zixing Cai

College of Information Engineering, Central South University of Technology, Changsha, Hunan, 410083, P. R. China, Tel: +86 731 8879628, E-mail: iice@csut.edu.cn

Abstract: This paper deals with μ synthesis for nonlinear robust control systems based on I/O linearization. By nonlinear state feedback, nonlinear systems can be linearized as controllable linear systems, such that uncertain nonlinear systems can be expressed as linear fractional transformations (LFTs) on the generalized linearized plants and uncertainty. By H_∞ optimization based μ synthesis technique, controllers can be obtained for the LFTs. Then the nonlinear robust control laws can be constructed by combining the controllers and the nonlinear state feedback. An example is given to illustrate the design. *Copyright © 1999 IFAC*

Keywords: nonlinear systems; linearization; uncertainty; robust control; H-infinity

1. INTRODUCTION

In recent years, nonlinear control systems[1] have become an important area in the field of automatic control and have been studied extensively (Fujita, et al., 1995; Katebi and Zhang, 1995; Wei, 1995; Wu, et al., 1996). A number of techniques have been developed for the analysis and synthesis for nonlinear control systems. One such technique (Fujita, et al., 1995) is using the Taylor series expansion to approximate the nonlinear system. Another technique (Katebi and Zhang, 1995) is to replace the nonlinear element by its sinusoidal input describing function to develop a quasi-linear nominal model for control analysis and design. Using these two techniques, the designed optimal controller for the linear approximate system may only be sub-optimal for the nonlinear system or not at all. A recently developed technique (Isidori, 1996) exact linearization of nonlinear systems with complete relative degree provides an useful method for designing controllers and has been implemented in aeronautics, robotics and other fields. Since this method has the restriction of complete relative degree, I/O linearization introduced by Kravaris and Chung (1987), which can also exactly linearize nonlinear systems even without complete relative degree, where nonlinear systems with complete relative degree are special cases, with the virtue of implementing independent, decoupled control loops, seems much appealing (Thomas and Nazmul, 1994).

Feedback control is indispensable for most physical systems. To synthesis a feedback control system, a precise mathematical model for the plant is required. It is known, however, that a design model can not always express the behavior of the real physical plant, resulting in an ideal mathematical model has various uncertainties. Therefore, a fundamental problem in control engineering is the design of feedback controllers that are insensitive to errors in the control design model. Zames (1981) introduced

[1]A project supported by the National Natural Science Foundation of China

the H_∞ control theory based on the small gain theorem, which gives the necessary and sufficient conditions for the stability of a system with uncertainty. However, the small gain theorem considers only one-block uncertainty, which will be conservative when the uncertainty is block-diagonal. This motivated the development of the structured singular value μ (Doyle, 1982), which deals with block-diagonal uncertainty and can consider the robust stability and robust performance simultaneously, that is, robust performance problem is equivalent to a robust stability problem with augmented uncertainty. Two basic methods for μ synthesis are *D-K* iteration (Doyle, 1985) and μ-*K* iteration (Lin, et al., 1993).

With the development of robust control, its applications to nonlinear systems have also been developed. H_∞ method based nonlinear robust control are developed in two aspects. One is to use H_∞ control to the linearized systems, which is inevitably conservative. Another is to use H_∞ method directly to the nonlinear systems, which involves solving two Hamilton-Jacobi inequalities which are nonlinear partial differential inequalities and the solutions are difficult to seek. However, μ method based nonlinear robust control has not been developed completely. In this paper, a H_∞ based μ synthesis method for robust control of affine nonlinear systems, which involves the description of nonlinear robust control problem, the I/O linearization of nonlinear systems, and μ synthesis for the linearized systems, is obtained. An illustrative example is given in this paper.

2. NONLINEAR ROBUST CONTROL PROBLEM

For a given nonlinear system (here we just consider single-input single-output system) with output multiplicative uncertainty, Δ is the normalized model error satisfying $\|\Delta\|_\infty < 1$, W(s) is the weighting function to the model error, here we just simply cover the real parametric uncertainty with norm bounded dynamical uncertainty.

Assume the nominal state space model of the uncertain nonlinear system is described as follows

$$\dot{\mathbf{x}} = \mathbf{f}(\mathbf{x}) + \mathbf{g}(\mathbf{x})u \qquad (1a)$$

$$y = h(\mathbf{x}) \qquad (1b)$$

where $\mathbf{x} = [x_1 \ x_2 \ \cdots \ x_n]^T$, u and y are state vector, input and output, respectively, and $\mathbf{f}(\mathbf{x})$, $\mathbf{g}(\mathbf{x})$ and $h(\mathbf{x})$ are smooth vector fields. It can be exactly linearized by the input/output linearization by means of the nonlinear state feedback

$$u = \alpha(\mathbf{x}) + \beta(\mathbf{x})v \qquad (2)$$

with v as an external reference input of the linearized system, and linear vector differential operators of the form

$$\hbar_r = \sum_{k=0}^{r} \beta_k \frac{d^k}{dt^k} \qquad (3)$$

satisfy $\beta_r \neq 0$ and $\sum_{k=0}^{r} \beta_k s^k \neq 0$ such that

$$\hbar_r y = v \qquad (4)$$

Then the linearized nonlinear uncertain system diagram block is described in Fig. 1, which can be transformed as Fig. 2 (Zhou, et al., 1996), Δ is normalized uncertainty with

$$\|\Delta\|_\infty < 1 \qquad (5)$$

G(s) is generalized plant including linearized nominal model and weighting function, which can be parted as

$$\mathbf{G}(s) = \begin{bmatrix} \mathbf{G}_{11} & \mathbf{G}_{12} \\ \mathbf{G}_{21} & \mathbf{G}_{22} \end{bmatrix} \qquad (6)$$

In Fig. 2, the following control law

$$v = Ky \qquad (7)$$

is used. Then the closed-loop control system can be transformed as Fig. 3, where **M** is the lower linear fractional transformation (LFT) on **G**(s) and controller *K*, i.e.,

$$\mathbf{M} = F_l(\mathbf{G}, K) = \mathbf{G}_{11} + \mathbf{G}_{12}K(I - \mathbf{G}_{22}K)^{-1}\mathbf{G}_{21} \qquad (8)$$

The structured singular value of the matrix **M**,

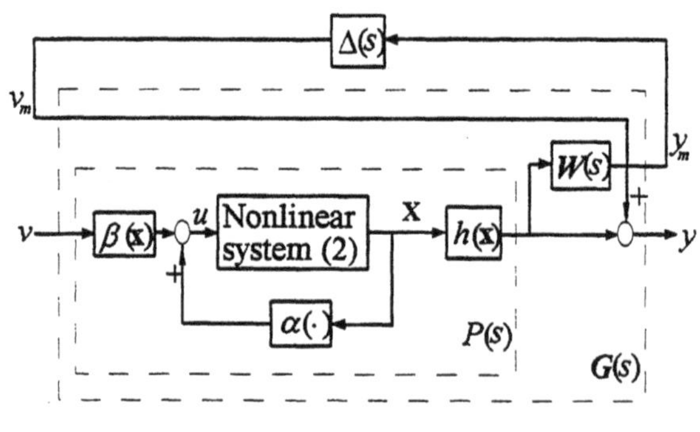

Fig. 1. The structure of the linearized system with output multiplicative uncertainty.

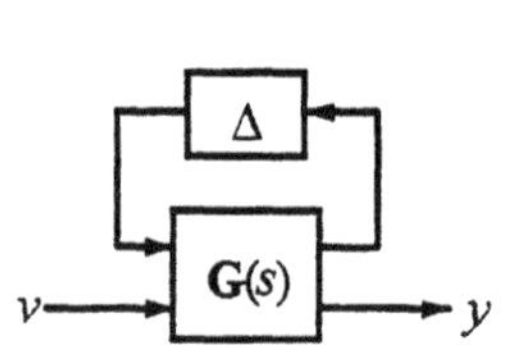

Fig. 2. The linearized system.

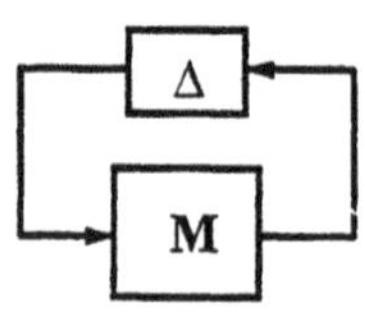

Fig. 3. The closed-loop control system.

denoted $\mu_\Delta(\mathbf{M})$, is defined by

$$\frac{1}{\mu_\Delta(\mathbf{M})} = \inf\{\sigma_{\max}(\Delta):\det(I-\Delta\mathbf{M})=0\} \quad (9)$$

and if no Δ exists to satisfy $\det(I-\Delta\mathbf{M})=0$, let $\mu_\Delta(\mathbf{M})=0$ (Zhou, et al., 1996), where $\sigma_{\max}(\cdot)$ denotes maximum singular value. The control objective is to find a stabilizing controller K minimizing H_∞ norm of $\mu_\Delta(\mathbf{M})$, i.e.,

$$\min_{K \text{ stabilizing } G} \left\| \mu_\Delta[F_l(\mathbf{G}, K)] \right\|_\infty \quad (10)$$

In this paper, the following nonlinear robust control problem is considered.

Nonlinear Robust Control Problem: Find a pair of nonlinear state feedback functions $\alpha(\mathbf{x})$ and $\beta(\mathbf{x})$, linear vector differential operators $\hbar_r$ and a stabilizing controller K satisfying the control objective (10).

3. LINEARIZATION OF UNCERTAIN NONLINEAR SYSTEMS

Lemma 1 Assume the nominal nonlinear system (1) has relative degree r at a point $\mathbf{x}^0$. The assumption means that

(i) $L_\mathbf{g}L_\mathbf{f}^k h(\mathbf{x})=0$ for all $\mathbf{x}$ in a neighborhood U of $\mathbf{x}^0$ and all $k<r-1$

(ii) $L_\mathbf{g}L_\mathbf{f}^{r-1}h(\mathbf{x}^0)\neq 0$

where $L_\mathbf{f}h(\mathbf{x})$ denotes the Lie derivative of $h(\mathbf{x})$ along $\mathbf{f}(\mathbf{x})$ and $L_\mathbf{f}^k h(\mathbf{x}) = L_\mathbf{f}L_\mathbf{f}^{k-1}h(\mathbf{x})$ with $L_\mathbf{f}^0 h(\mathbf{x}) = h(\mathbf{x})$.

The following result (Kravaris and Chung, 1987) gives the necessary and sufficient conditions for the I/O linearization of the nonlinear system (1).

Theorem 1 The I/O linearization of the nonlinear system (1) is possible if and only if the output y possesses a relative degree r and $L_\mathbf{g}L_\mathbf{f}^{r-1}h(\mathbf{x}^0)\neq 0$ for all $\mathbf{x}^0$.

If the above conditions hold and $\beta_r \neq 0$ then the nonlinear state feedback law is

$$u = [\beta_r L_\mathbf{g}L_\mathbf{f}^{r-1}h(\mathbf{x})]^{-1}[v - \sum_{k=0}^{r}\beta_k L_\mathbf{f}^k h(\mathbf{x})] \quad (11)$$

Therefore, the procedure leading to the construction of a nonlinear state feedback (2) and of linear vector differential operators (3) linearizing the nonlinear system (1) consists of the following steps.

Linearization Procedure:

(i) From $\mathbf{f}(\mathbf{x})$, $\mathbf{g}(\mathbf{x})$ and $h(\mathbf{x})$, check the conditions in Lemma 1.

(ii) If both are satisfied, set the accordingly β_k, for $0 \le k \le r$, then compute $\beta_r L_\mathbf{g}L_\mathbf{f}^{r-1}h(\mathbf{x})$ and $\sum_{k=0}^{r}\beta_k L_\mathbf{f}^k h(\mathbf{x})$.

(iii) Set

$$\alpha(\mathbf{x}) = -\frac{\sum_{k=0}^{r}\beta_k L_\mathbf{f}^k h(\mathbf{x})}{[\beta_r L_\mathbf{g}L_\mathbf{f}^{r-1}h(\mathbf{x})]}, \quad \beta(\mathbf{x}) = \frac{1}{[\beta_r L_\mathbf{g}L_\mathbf{f}^{r-1}h(\mathbf{x})]} \quad (12)$$

By the above linearization procedure, a linear and controllable system is obtained, where

$$y(s) = (\sum_{k=0}^{r}\beta_k s^k)^{-1} v(s) \quad (13)$$

For an uncertain nonlinear system, assume the linearized system can be described by the structure shown in Fig. 1, and the uncertainty can be expressed by a weighting function $\mathbf{W}(s)$ and a normalized uncertainty $\Delta(s)$ satisfying $\|\Delta(s)\|_\infty < 1$. Then the closed-loop transfer function from input v to output y, is denoted $T_{yv}(s)$, can be expressed as follows

$$T_{yv}(s) = [1+\Delta(s)\mathbf{W}(s)]P(s) \quad (14a)$$

where

$$P(s) = \frac{1}{\sum_{k=0}^{r}\beta_k s^k} \quad (14b)$$

is the transfer function of the linearized system. And we have

$$\begin{bmatrix} y_m \\ y \end{bmatrix} = \mathbf{G}(s)\begin{bmatrix} v_m \\ v \end{bmatrix} \quad (15)$$

$$\mathbf{G}(s) = \begin{bmatrix} \mathbf{G}_{11} & \mathbf{G}_{12} \\ \mathbf{G}_{21} & \mathbf{G}_{22} \end{bmatrix} = \begin{bmatrix} 0 & \mathbf{W}(s)P(s) \\ 1 & P(s) \end{bmatrix} \quad (16)$$

Then, $T_{yv}(s)$ can be expressed by the upper LFT on G(s) and $\Delta(s)$, i.e.,

$$T_{yv}(s) = F_u(\mathbf{G},\Delta) = \mathbf{G}_{22} + \mathbf{G}_{21}\Delta(I-\mathbf{G}_{11}\Delta)^{-1}\mathbf{G}_{12} \quad (17)$$

4. μ SYNTHESIS OF NONLINEAR ROBUST CONTROLLER

In Fig. 2, $\Delta(s)$ is model uncertainty and assume the performance requirement on the system is the ability of rejecting output disturbance, which is not affected by linearization. Then the linearized system can be added a fictitious weighting function $W_f(s)$, a fictitious normalized uncertainty $\Delta_f(s)$ satisfying $\|\Delta_f(s)\|_\infty < 1$ and a stabilizing controller K to construct the closed-loop control system shown in Fig. 4, where

$$\begin{bmatrix} y_m \\ y_f \end{bmatrix} = \widetilde{\mathbf{M}} \begin{bmatrix} v_m \\ v_f \end{bmatrix} \tag{18}$$

$$\widetilde{\mathbf{M}} = \begin{bmatrix} \widetilde{\mathbf{M}}_{11} & \widetilde{\mathbf{M}}_{12} \\ \widetilde{\mathbf{M}}_{21} & \widetilde{\mathbf{M}}_{22} \end{bmatrix} = \begin{bmatrix} \mathbf{W}T & \mathbf{W}T \\ W_f S & W_f S \end{bmatrix} \tag{19}$$

where

$$T = PK(I-PK)^{-1}, \quad S = (I-PK)^{-1} \tag{20}$$

are the nominal complementary sensitivity function and the nominal sensitivity function, respectively.

Assume further that the performance requirement is the maximum singular value of the frequency response function from v_f to y_f should be less than 1 at all frequencies. It is clear that $W_f(s)$ and $\Delta_f(s)$ are used to incorporate a fictitious performance requirement on the closed-loop control system. Let $T_f(s)$ denotes the transfer function from v_f to y_f, then

$$T_f(s) = F_u(\widetilde{\mathbf{M}}, \ \Delta) = \widetilde{\mathbf{M}}_{22} + \widetilde{\mathbf{M}}_{21}\Delta(I - \widetilde{\mathbf{M}}_{11}\Delta)^{-1}\widetilde{\mathbf{M}}_{12} \tag{21}$$

and the performance requirement on the closed-loop control system can be described by

$$\|F_u(\widetilde{\mathbf{M}}, \ \Delta)\|_\infty < 1 \tag{22}$$

The equivalent structure of Fig. 4 is shown in Fig. 5, where

$$\widetilde{\Delta}(s) = \begin{bmatrix} \Delta(s) & 0 \\ 0 & \Delta_f(s) \end{bmatrix} \tag{23}$$

$$\widetilde{\mathbf{G}}(s) = \begin{bmatrix} \widetilde{\mathbf{G}}_{11} & \widetilde{\mathbf{G}}_{12} \\ \widetilde{\mathbf{G}}_{21} & \widetilde{\mathbf{G}}_{22} \end{bmatrix} = \begin{bmatrix} 0 & 0 & \mathbf{W}(s)P(s) \\ W_f(s) & W_f(s) & W_f(s)P(s) \\ 1 & 1 & P(s) \end{bmatrix} \tag{24}$$

$$\widetilde{\mathbf{M}} = F_l(\widetilde{\mathbf{G}}, \ K) = \widetilde{\mathbf{G}}_{11} + \widetilde{\mathbf{G}}_{12}K(I - \widetilde{\mathbf{G}}_{22}K)^{-1}\widetilde{\mathbf{G}}_{21} \tag{25}$$

The following two results are important for analysis of robust stability and robust performance of the closed-loop control system. The first result is called robust stability theorem and the second is called robust performance theorem.

Theorem 2 The closed-loop control system is robust stable, if and only if

$$\mu_\Delta(\widetilde{\mathbf{M}}_{11}) = \mu_\Delta[\mathbf{W}T] = \mu_\Delta[\mathbf{W}PK(I-PK)^{-1}] < 1 \tag{26}$$

at all frequencies.

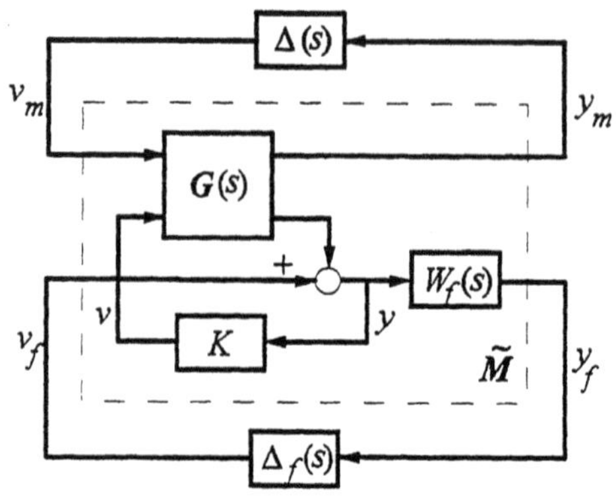

Fig. 4. The structure using a fictitious uncertainty.

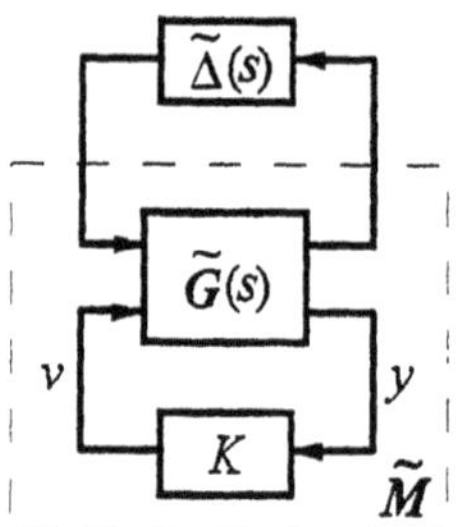

Fig. 5. The equivalent structure of Fig. 4.

Theorem 3 The closed-loop control system has robust performance, i.e., the expression (22) is satisfied, if and only if

$$\mu_{\widetilde{\Delta}}(\widetilde{\mathbf{M}}) < 1 \tag{27}$$

at all frequencies.

From the above two theorems, it is shown that the robust performance analysis can be replaced by an equivalent robust stability analysis using the fictitious weighting function $W_f(s)$ and the fictitious normalized uncertainty $\Delta_f(s)$. Therefore, robust stability of the closed-loop control system shown in Fig. 4 is equivalent to robust performance of the original closed-loop control system shown in Fig. 2.

From theorem 3, the performance requirement on the closed-loop control system can be described by

$$\left\|\mu_{\widetilde{\Delta}}\left[F_l(\widetilde{\mathbf{G}}, \ K)\right]\right\|_\infty < 1 \tag{28}$$

This is a μ synthesis problem. In this paper, consider the optimization problem of the expression (28), i.e., find a controller K satisfying

$$\min_{K \text{ stabilizing } \widetilde{G}} \left\|\mu_{\widetilde{\Delta}}\left[F_l(\widetilde{\mathbf{G}}, \ K)\right]\right\|_\infty \tag{29}$$

This optimization problem has been proved difficult to solve and now no solution is available. However, an approximate solution has been given by (Doyle, 1985) which is called D-K iteration. An alternative algorithm is the μ-K iteration introduced by Lin, et al. (1993) which can also be used to obtain a robust controller.

On the basis of the previous discussion, the procedure of obtaining a nonlinear robust controller consists of the following steps.

Procedure of Obtaining Nonlinear Robust Controller:
(i) From a given uncertain nonlinear system, find its nominal nonlinear system (1).
(ii) Based on the linearization procedure in section 3, linearize the nominal nonlinear system (1) as a linear and controllable system.
(iii) According to the uncertainty of the nonlinear system, describe the linearized system by the structure shown in Fig. 2.
(iv) For the linearized system, design a linear robust controller K by using D-K or μ-K iteration.
(v) By the expressions (2), (7)and (12), obtain the corresponding nonlinear robust controller.

5. PARADIGM

In uncertain nonlinear systems, consider the case of

$$\begin{bmatrix}\dot{x}_1\\ \dot{x}_2\\ \dot{x}_3\end{bmatrix}=\begin{bmatrix}0\\ x_1+x_2^2\\ x_1-x_2\end{bmatrix}+\begin{bmatrix}\exp(x_2)\\ \exp(x_2)\\ 0\end{bmatrix}u \qquad (30a)$$

$$y=p(x_1-x_2),\quad p\in(0.8,\ 1.2) \qquad (30b)$$

Choose $\mathbf{W}(s)$=0.2 and

$$\begin{bmatrix}\dot{x}_1\\ \dot{x}_2\\ \dot{x}_3\end{bmatrix}=\begin{bmatrix}0\\ x_1+x_2^2\\ x_1-x_2\end{bmatrix}+\begin{bmatrix}\exp(x_2)\\ \exp(x_2)\\ 0\end{bmatrix}u \qquad (31a)$$

$$y=x_1-x_2 \qquad (31b)$$

as the nominal nonlinear model of the system, which has a relative degree of 2, less than the dimension of the state space, i.e., 3, and then set $\beta_0=\beta_1=0,\quad \beta_2=1$, which results in

$$P(s)=\frac{1}{s^2} \qquad (32)$$

Therefore, the nonlinear state feedback can be obtained by the expression (12).

$$u=\frac{-2x_2(x_1+x_2^2)-v}{(1+2x_2)\exp(x_2)} \qquad (33)$$

Then the system (30) can be transformed as the uncertain linear system shown in Fig. 2, where

$$\mathbf{G}(s)=\begin{bmatrix}0 & \frac{0.2}{s^2}\\ 1 & \frac{1}{s^2}\end{bmatrix} \qquad (34)$$

follows from the expression (16).The performance weighting function $W_f(s)=\dfrac{20}{s+0.1}$ is chosen to reject output disturbances at low frequency band, then by using D-K iteration, a controller

$$K(s)=\frac{-2919057s^2-47.9286s-3.5111}{s^3+176.3922s^2+365.4534s+34.7824} \qquad (35)$$

is obtained. $\widetilde{\mathbf{M}}$ can be computed by expression (19). The μ curves of $\widetilde{\mathbf{M}}_{11}$ and $\widetilde{\mathbf{M}}$ are shown in Fig. 6 and Fig. 7. It is shown that the robust stability and robust performance requirements are met. The output y is shown in Fig. 8 under the output disturbance d=0.2(t-10) when the stable output is 1. It is clear that the nonlinear robust controller consists of the expressions (2), (33) and (35).

6. CONCLUSION

By I/O linearization, nonlinear systems can be linearized no matter the relative degree of the systems are complete or not. Then nonlinear robust control problem can be expressed as μ

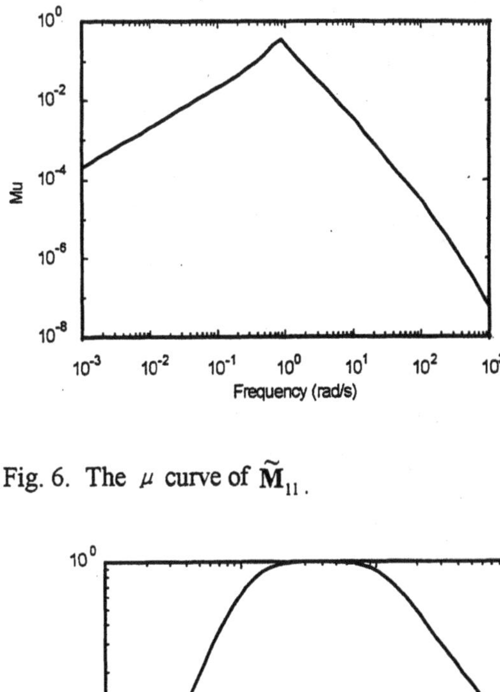

Fig. 6. The μ curve of $\widetilde{\mathbf{M}}_{11}$.

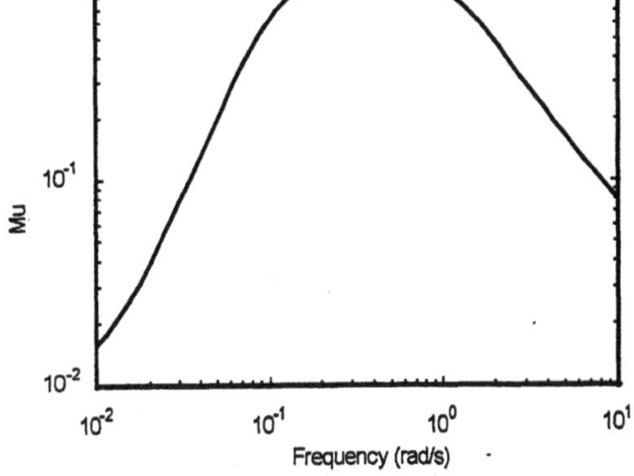

Fig. 7. The μ curve of $\widetilde{\mathbf{M}}$.

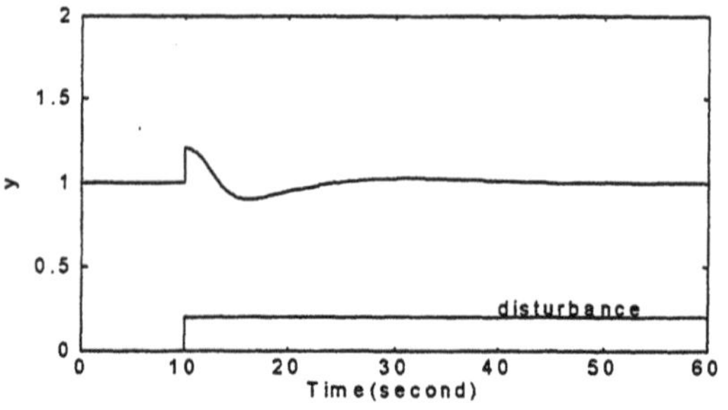

(a) The output response (p=0.81)

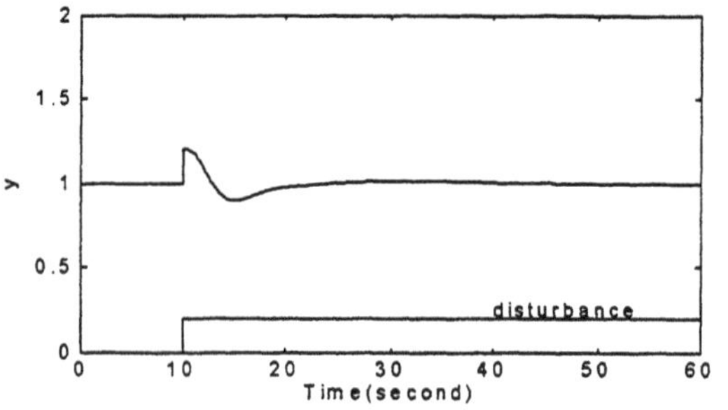

(b) The output response (p=1)

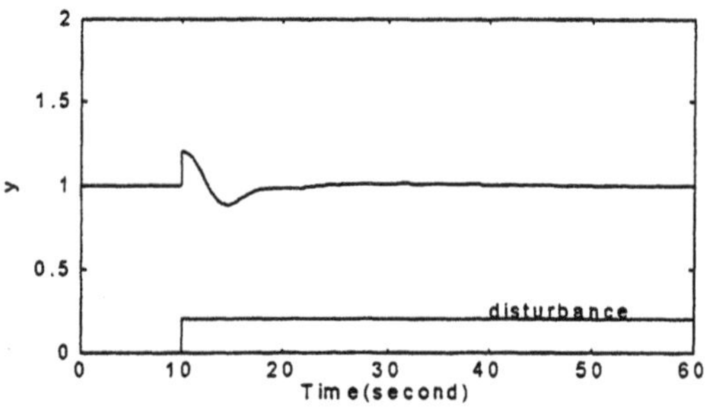

(c) The output response (p=1.19)

Fig. 8. The disturbance rejection output response.

synthesis problem of the linearized systems which are LFTs on the generalized linearized plant and uncertainty. Based on the LFTs, the linear robust controllers can be obtained by μ synthesis technique. The nonlinear robust control laws can then be constructed by the nonlinear state feedback and the linear robust controllers. Simulation result shows that this method has high stability robustness and performance robustness.

REFERENCES

Doyle, J. C. (1982). Analysis of feedback systems with structured uncertainties. *Proc IEE, Part-D*, **129**, 242-250.

Doyle, J. C. (1985). Structure uncertainty in control system design. *Proc. 24th IEEE Conf. on Decision and Control*, 260-265.

Fujita, M., T. Namerikawa, F. Matsumura and K. Uchida (1995). μ -synthesis of an electromagnetic suspension system. *IEEE Trans. Auto. Control*, **40**, 530-536.

Isidori, A. (1996). *Nonlinear Control Systems*. Springer-Verlag, Berlin.

Katebi, M. R. and Y. Zhang (1995). H_∞ control analysis and design for nonlinear systems. *INT. J. Control*, **61**, 459-474.

Kravaris, C., and C. B. Chung (1987). Nonlinear state feedback synthesis by global input/output linearization. *AIChE Journal*, **33**, 592-604.

Lin, J. L., I. Postlethwaite and D. W. Gu (1993). μ -K iteration: a new algorithm for μ synthesis. *Automatic*, **29**, 219-224.

Thomas, P. and M. K. Nazmul (1994). Nonlinear control of a bioreactor model using exact and I/O linearization. *INT. J. Control*, **60**, 499-519.

Wu, M., W. H. Gui, J. Xu and Z. X. Cai (1996). State-space approach for nonlinear H_∞ control (in Chinese). *Computing Technology and Automation*, **15**, 1-7.

Wei, K. (1995). Nonlinear H_∞ control and its application to rigid spacecraft. *IEEE Trans. Auto. Control*, **40**, 1281-1285.

Zames, G. (1981). Feedback and optimal sensitivity: model reference transformations, multiplicative seminorms, and approximate inverse. *IEEE Trans. Auto. Control*, **26**, 301-320.

Zhou, K. M., J. C. Doyle and K. Glover (1996). *Robust and Optimal Control*. Prentice-Hall Inc, New Jersey.

14th Triennial World Congress, Beijing, P.R. China

E-2c-08-4

ROBUST STABILITY ANALYSIS OF MIMO SYSTEMS WITH ASYMMETRIC NONLINEARITIES

Francisco Gordillo * Javier Aracil * Anibal Ollero *

* *Escuela Superior de Ingenieros*
Camino de los Descubrimientos s/n,
Sevilla-41092 Spain
E-mail: [gordillo,aracil,aollero]@cartuja.us.es

Abstract: The aim of this paper is to study the existence of limit cycles and undesirable equilibrium points in nonlinear, MIMO control systems in the presence of uncertainties. This study is important from the engineering point of view because such limit cycles and equilibria may change the global behaviour of the system, e.g. global stability may be lost. The analysis is performed using the describing function method. This paper extend the results of previous work to the case of asymmetric nonlinearities with the use of the dual describing function. The method is illustrated by means of an example. *Copyright © 1999 IFAC*

Keywords: Nonlinear Systems, Stability Analysis, Robustness, Singular Values, Harmonic Balance.

1. INTRODUCTION

There are several approaches to analyze the stability of a nonlinear control system. The problem is so difficult, and the variety of systems is so large that it appears that a single method is not (and probably will not be) able to cope with all the problems. Some of the methods such as the Lyapunov direct method and the conicity criterion (Vidyasagar, 1993) are quite rigurous but can give very conservative results.

In this field, the classical, approximate method of the describing function is still useful. This method compensates its lack of correctness with a less conservative nature. In this way it can complement the other methods.

In recent years, some results of the modern theory of robust linear control have been applied to the quasi-linear system obtained when it is applied the describing function to the nonlinear part of the system (Ferreres and Fromion, 1998; Gordillo et al., 1998; Katebi and Zhang, 1995). With this extension some guarantees of the unexistence of limit cycles can be given in the presence of uncertainties. All these extensions start with the nonlinear system decomposed in its linear and non-linear parts. The nonlinear part is assumed to have odd symmetry. Nevertheless, this is a hard assumption and will be unlikely fulfilled.

This paper presents the way to perform robust stability analysis of MIMO systems with asymmetric nonlinearities. In fact, the solution can be again found in the classical theory of the describing function with its variant the dual describing function (Atherton, 1975; ?; Genesio et al., 1995) which can cope with asymmetric nonlinearities. Furthermore, the method is also suitable for looking for equilibrium points. Therefore, assuming that no strange attractors will emerge, the robust global stability of the system is analyzed with the proposed method.

The paper is organized as follows. In the next section one of the existing methods to analyze the robust stability of MIMO systems with symmetric nonlinearities is recalled. In Sect. 3 the dual describing function method for the analysis of limit cycles and equilibrium points in MIMO systems is expounded. In Sect. 4 the extension of the result of Sect. 2 to the case of asymmetric nonlinearities is presented. Section 5 deals ith the practical issues about the application of the method. Section 6 shows an example of application. The paper closes with a Section of conclusions.

2. ROBUST ANALYSIS OF LIMIT CYCLES USING SINGULAR VALUES: SYMMETRIC NONLINEARITIES

Consider the system of Fig. 1 in which a nonlinear, MIMO system has been decomposed in a linear block represented by its transfer matrix $G(s)$ and a nonlinear block which is assumed to be memoryless. This is a very conventional way to represent nonlinear control systems, which can be used as a basis to apply the describing function method (Khalil, 1996; Vidyasagar, 1993). $\Delta(s)$ represents the uncertainties which can be considered model errors (this is usually called multiplicative error model). Other interesting approach is to include in $\Delta(s)$ the approximation errors of the describing function (Katebi and Zhang, 1995). With this approach the method acquires correctness but it is more conservative.

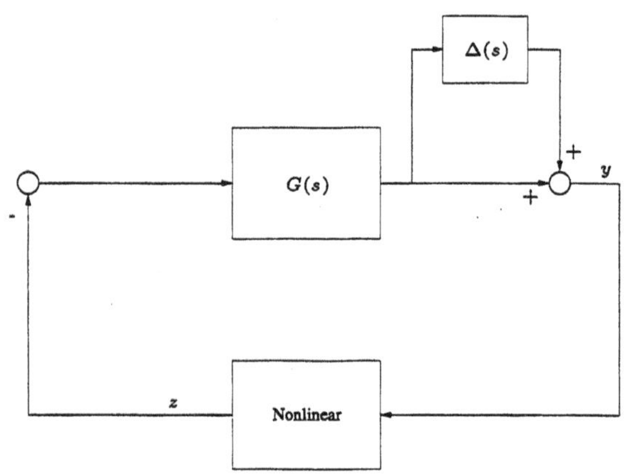

Fig. 1. Nonlinear Control Loop with Multiplicative Error Model

When the nonlinear block has odd-symmetry it can be approximated by its describing function matrix N. As the nonlineartity is assumed to be memoryless N will not depend on ω and therefore it will be written $N(a)$ where a is a vector of amplitudes.

The first-order harmonic balance provides the following equation to predict the existence of limit cycles:

$$\det(I + G(j\omega)N(a)) = 0 \tag{1}$$

Theorem 1. In the feedback loop of Fig. 1 with a symmetric nonlinear block, no solution of the harmonic balance equation (1) will exist if the following conditions are satisfied:

(1) There is no solution of Eq. (1) for the nominal closed loop ($\Delta = 0$);
(2) The model error fulfils

$$\bar{\sigma}(\Delta(j\omega)) < \frac{1}{\Gamma(a,\omega)} \qquad \forall a, \forall \omega \tag{2}$$

with

$$\Gamma(a,\omega) = \bar{\sigma}\left(G(j\omega)N(a)(I + G(j\omega)N(a))^{-1}\right)$$

Furthermore, there exists a rational transfer function matrix Δ that satisfies

$$\bar{\sigma}(\Delta(j\omega)) \leq \frac{1}{\Gamma(a,\omega)} \qquad \forall a, \forall \omega$$

such that a solution of Eq. (1) exists.

The proof can be found in (Gordillo et al., 1998). Similar results can be obtained for other types of uncertainty models.

In the next sections this result is extended to the case of asymmetric nonlinearities.

3. THE DUAL DESCRIBING FUNCTION METHOD

In the dual describing function method, a first order harmonic balance is performed, so the limit cycles are assumed to be of the form:

$$y_k(t) = a_k^0 + \mathrm{Re}(a_k^1 e^{j\omega t}) \quad \omega > 0, k = 1, \ldots, n \tag{3}$$

Where y_k with $k = 1, \ldots, n$, are the outputs of the linear part, $a_k^0, a_k^1 \in \mathbb{R}$ and $a_k^1 \geq 0$. Notice that while the amplitude of the limit cycles is not necessarily the same for all the outputs, their frequency ω is assumed to be the same. Since $a_k^1 \in \mathbb{R}$ it is assumed that there is not phase shifts between the limit cycles. In a formal analysis these phase shifts should be considered but it can be shown that they do not affect to the final result of this section (Aracil et al., 1999).

This expresion covers the case of symmetric limit cycles ($a_k^0 = 0$) and the equilibrium points ($a_k^1 = 0$).

Let the signals $y_k(t)$ of equation (3) enter in the nonlinear part, and consider its steady state periodic output signals $z_i(t)$. These signals can be expanded in series form as:

$$z_i(t) = \sum_{l=1}^{n} \left(N_{il}^0 a_l^0 + \mathrm{Re}(N_{il}^1 a_l^1 e^{j\omega t}) + \ldots\right)$$

Where N_{il}^0 and N_{il}^1 are, respectively, the bias gain and the first order harmonic gain from the *l-th* input of the nonlinear part to its *i-th* output. They can be computed with the usual expressions of the Fourier coefficients as:

$$N_{il}^0 = N_{il}^0(a_l^0, a_l^1, \omega) = \frac{1}{2\pi a_l^0} \int_{-\pi}^{\pi} z_i(t) d(\omega t) \quad (4)$$

$$N_{il}^1 = N_{il}^1(a_l^0, a_l^1, \omega) = \frac{1}{\pi a_l^1} \int_{-\pi}^{\pi} z_i(t) e^{-j\omega t} d(\omega t) \quad (5)$$

As the nonlinear part is assumed to be memoryless N_{il}^0 and N_{il}^1 will not depend on ω. In order to fulfil the first-order harmonic balance, the zero and first order terms of $y_k(t)$ must be equal to the corresponding terms of the output of the linear part when its input is $z_i(t)$ (neglecting the higher harmonics). In matrix form:

$$a^0 = -G(0)N^0(a^0, a^1)a^0 \quad (6)$$

$$a^1 = -G(j\omega)N^1(a^0, a^1)a^1 \quad (7)$$

Where N^0 and N^1 are respectively the matrices $[N_{il}^0]$ and $[N_{il}^1]$, $a^0 = (a_1^0, a_2^0, \ldots, a_n^0)^T$ and $a^1 = (a_1^1, a_2^1, \ldots, a_n^1)^T$. Equations (6)–(7) can be rewritten as:

$$(I + G(0)N^0(a^0, a^1))a^0 = 0$$

$$(I + G(j\omega)N^1(a^0, a^1))a^1 = 0$$

In order these equations not to have a trivial solution in a^0, a^1, the following conditions must be fulfilled:

$$\det(I + G(0)N^0(a^0, a^1)) = 0 \quad (8)$$

$$\det(I + G(j\omega)N^1(a^0, a^1)) = 0 \quad (9)$$

4. ROBUST STBILITY ANALYSIS USING SINGULAR VALUES: ASYMMETRIC NONLINEARITIES

Consider again the system of Fig. 1. Now the nonlinear block is allowed to be asymmetric.

Theorem 2. In the feedback loop of Fig. 1 with an asymmetric nonlinear block, no solution of the harmonic balance equations (8–9) will exist if the following conditions are satisfied:

(1) There is no solution of Eqs. (8) and (9) for the nominal closed loop ($\Delta = 0$);
(2) The model error satisfies one of the following conditions:

$$\bar{\sigma}(\Delta(0)) < \frac{1}{\Gamma^0(a^0, a^1)} \qquad \forall a^0, a^1 \in \mathbb{R}^n \quad (10)$$

$$\bar{\sigma}(\Delta(j\omega)) < \frac{1}{\Gamma^1(a^0, a^1, \omega)} \qquad \forall a^0, a^1 \in \mathbb{R}^n, \forall \omega > 0 \quad (11)$$

with

$$\Gamma^0 = \bar{\sigma}\left(G(0)N^0(I + G(0)N^0)^{-1}\right)$$

$$\Gamma^1 = \bar{\sigma}\left(G(j\omega)N^1(I + G(j\omega)N^1)^{-1}\right)$$

Furthermore, there exists a rational transfer function matrix Δ that satisfies

$$\bar{\sigma}(\Delta(0)) \leq \frac{1}{\Gamma^0(a^0, a^1)} \qquad \forall a^0, \forall a^1$$

$$\bar{\sigma}(\Delta(j\omega)) \leq \frac{1}{\Gamma^1(a^0, a^1, \omega)} \qquad \forall a^0, \forall a^1, \forall \omega$$

such that a solution of Eqs. (8–9) exists.

Proof. The proof is similar to the one of internal stability for linear systems, see for example (Green and Limebeer, 1995).

Equations (8–9) are of the form:

$$\det(I + F^\beta(\gamma)) = 0 \quad (12)$$

where $F^\beta = GN^0$ and $\gamma = (a^0, a^1)^T$ for Eq. (8) and $F^\beta = GN^1$ and $\gamma = (\omega, a^0, a^1)^T$ for Eq. (9). In the following, the necessary condition for Eq. (12) not to be satisfied is obtained.

Condition 1 implies that

$$\det(I + F^\beta(\gamma)) \neq 0 \qquad \forall \gamma \quad (13)$$

When $\Delta \neq 0$ Eqs. (8–9) must be applied with $(I + \Delta)G$ instead of G. Suppose that Δ brings the perturbed system to the limit where a solution of Eqs. (8–9) appears, so for a vector of parameters $\gamma = \gamma_0$

$$\begin{aligned} 0 &= \det\left(I + (I + \Delta)F^\beta(\gamma_0)\right) \\ &= \det\left(I + F^\beta(\gamma_0) + \Delta F^\beta(\gamma_0)\right) \\ &= \det\left[\left(I + \Delta F^\beta(\gamma_0)(I + F^\beta(\gamma_0))^{-1}\right) (I + F^\beta(\gamma_0))\right] \\ &= \det\left(I + \Delta F^\beta(\gamma_0)(I + F^\beta(\gamma_0))^{-1}\right) \det\left(I + F^\beta(\gamma_0)\right) \end{aligned}$$

Which with Eq. (13) implies that

$$\det\left(I + \Delta F^\beta(\gamma_0)(I + F^\beta(\gamma_0))^{-1}\right) = 0$$

This situation cannot arise if

$$\bar{\sigma}\left(\Delta F^\beta(\gamma)(I + F^\beta(\gamma))^{-1}\right) < 1 \qquad \forall \gamma$$

Which is fulfilled if

$$\bar{\sigma}(\Delta) < \frac{1}{\bar{\sigma}\left(F^\beta(\gamma)(I + F^\beta(\gamma))^{-1}\right)} \qquad \forall \gamma$$

Substituting γ and F^{β} for their respective functions in Eqs. (8–9) the first part of the theorem is obtained. The proof of the second part (which is a constructive proof) is identical to the corresponding one of linear, internal stability (see (Green and Limebeer, 1995, pg. 42)).

□

Similar theorems can be written for other types of model errors.

In this way a measure of robustness against limit cycles is obtained. Condition 1 can be verified with the easy extension of the methods of (Ollero et al., 1998) or checking if $\underline{\sigma}(I + G(0)N^0(a^0, a^1))) \neq 0$ and $\underline{\sigma}(I + G(j\omega)N^1(a^0, a^1))) \neq 0$ which assures that the Eqs. represented by (13) are fulfilled.

5. PRACTICAL ASPECTS

In this section some issues concerning the practical application of Theorem 2 are exposed. The first question than can be posed is how to calculate the describing functions. The analytical description of describing functions can only be obtained in some simple cases. This is perhaps the main reason (apart from its approximate nature) why the method has not obtained a great success lately. Nevertheless, with the great power of modern computers the computation can be performed numerically. To do this the experiment which define the describing function can be simulated and integrals (4–5) can be obtained numerically. In this way the method can be applied to complex nonlinearities for which other rigourous stability method will probably fail. With this approach stability analysis of fuzzy control systems has been performed in (Aracil et al., 1999; Gordillo et al., 1998).

A second point is how to apply Theorem 2. The theorem states two sufficient conditions on the error model in order to predict the inexistence of limit cycles. Therefore, if one of the conditions is fulfilled no limit cycles will be predicted. The best way is to check condition (10) in the first place and then if it is not fulfilled to check condition (11).

The verification of condition (10) can be performed calculating

$$\max_{a^0, a^1} \bar{\sigma} \left(G(0)N^0(a^0, a^1)(I + G(0)N^0(a^0, a^1))^{-1}\right)$$

It must be noticed that a^0 and a^1 are vectors of dimension equal to the number of outputs of the linear part but this optimization can be performed numerically. The same procedure can be applied to the verification of condition (11) but, in this case, for several values of ω (a grid must be defined for ω). The result can be plotted in a graph in which the x-axis is ω (or $\log \omega$ as is usual in frequency plots).

All this procedure involves a great amount of computations but it is affordable with modern computers. It should be taken into account that the problem at stake (the stability analysis of nonlinear, MIMO systems) is a very difficult problem.

6. APPLICATION EXAMPLE

In order to illustrate the presented method consider the following example. Let the linear part of Fig. 1 be represented by the following transfer matrix:

$$G(s) = \begin{pmatrix} \dfrac{2}{(s+1)^2} & \dfrac{1}{(s+1)^2} \\ \dfrac{1}{(s+1)^2} & \dfrac{1}{(s+1)^2} \end{pmatrix}$$

while the nonlinearities are presented in Fig. 2. The form of these nonlinearities are taken from (Gray and Brown, 1984) with slight modifications in order to make the nonlinearities being asymmetric. The linear part is different from the one of that reference to avoid the existence of limit cycles in the nominal case.

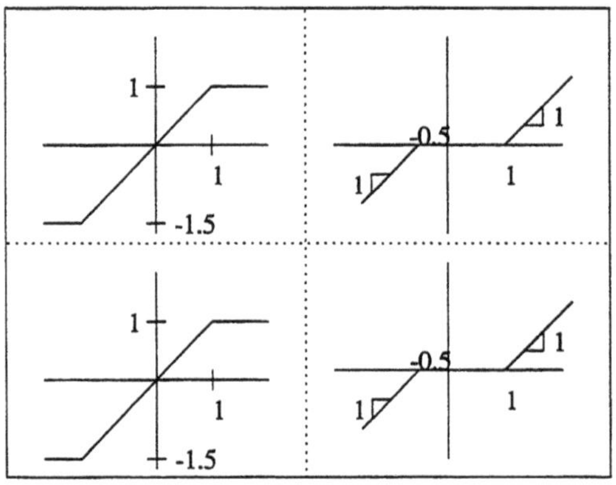

Fig. 2. Nonlinear Block of the Example

Performing the optimization procedure exposed in the previous section it can be verified that

$$\min_{a^0, a^1} \underline{\sigma}(I + G(0)N^0(a^0, a^1))) = 0.89 \neq 0$$

$$\min_{a^0, a^1, \omega} \underline{\sigma}(I + G(j\omega)N^1(a^0, a^1))) = 0.618 \neq 0$$

so the first condition of Theorem 2 is fulfilled. This means that the local behaviour of the system around the equilibrium point is, in fact, global. It can be easily verified that this behaviour is stable.

Repeating the procedure to compute the upper bound of $\Delta(0)$,

$$\max_{a^0, a^1} \bar{\sigma} \left(G(0)N^0(a^0, a^1)(I + G(0)N^0(a^0, a^1))^{-1}\right) =$$

$$11363 = 1/8.8 \times 10^{-5}$$

Therefore if $\bar{\sigma}(\Delta(0)) < 8.8 \times 10^{-5}$ the system will continue being gobally stable. As this number is quite small it will probably be necessary to check the second condition, that is, the upper bound on $\Delta(\omega)$. Performing the optimization procedure for 12 values of ω between 10^{-3} and 100 the graph shown in Fig. 3 is obtained. In spite of being $\bar{\sigma}(\Delta(0)) \geq 8.8 \times 10^{-5}$ the global stability of the system is predicted provided that $\bar{\sigma}(\Delta(\omega))$ is below this graph.

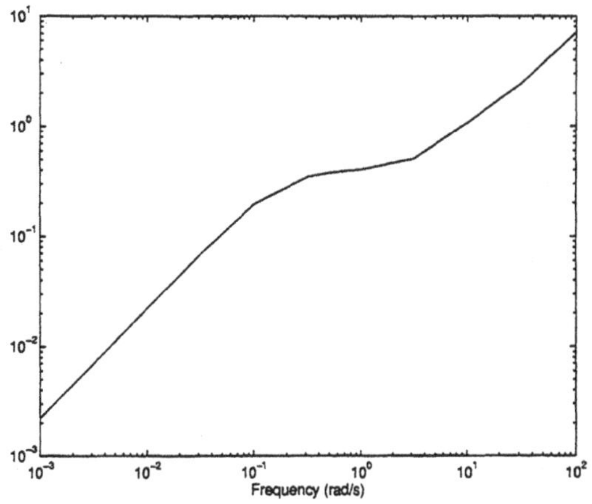

Fig. 3. Upper bound of $\bar{\sigma}(\Delta(\omega))$

7. CONCLUSIONS

A method to test the global stability of a MIMO system including asymmetrical nonlinearities in the presence of uncertainties has been presented. The method is approximate since it is based on the firt harmonic balance but it can completely exact if the error of the approximation is included in the uncertainties. In this case the method will be more conservative.

The method is applicable to a wide variety of nonlinearities as it has been shown by means of an example.

ACKNOWLEDGEMENTS

This work has been done in the framework of the project FAMIMO funded by the European Commission in the ESPRIT program (LTR Project 21911). The authors would also like to thank the Spanish Agency CICYT for partially supporting this work under grants TAP96–1184–C04–01, TAP97–0553 and TAP98–0541.

References

Aracil, J., F. Gordillo, F. Cuesta and A. Ollero. Global Stability Analysis of Takagi-Sugeno Fuzzy Control Systems. In: *Stability Issues in Fuzzy Control* (J. Aracil Ed.). Springer-Verlag. In press.

Atherton, D. P. (1975) *Nonlinear Control Engineering. Describing Function Analysis and Design*, Van Nostrand Reinhold.

Ferreres G. and V. Fromion (1998). Nonlinear analysis in the presence of parametric uncertainties. *Int. J. Control*, **69:5**, 695–716.

Genesio, R., M. Basso and A. Tesi (1995). On the prediction of limit cycle bifurcations in feedback systems. *Proc. of the 3rd European Control Conference*, Rome, Italy.

Gordillo, F., J. Aracil and A. Ollero, (1998). Frequency Domain Analysis of Multivariable Fuzzy Control Systems. In: *Application of Multi-Variable System Techniques* (Whalley, R. and M. Ebrahimi, Eds.), pp. 243–252. Professional Engineering Publishing Ltd.

Gray, J.O. and L.S. Brown (1984). A computer-aided design suite for the analysis and design of nonlinear control systems. *Trans. Inst. M. C.* **6:1**, 13–25.

Green, M. and D.J.N. Limebeer (1995). *Linear Robust Control*, Prentice-Hall.

Katebi, M.R. and Y. Zhang (1995) H_∞ control analysis and design for nonlinear systems, *Int. J. Control*, **61:2**, 459–474.

Khalil, H.K. (1996), *Nonlinear Systems*, Prentice Hall.

Ollero, A., J. Aracil and F. Gordillo, (1998), Stability Analysis of MIMO Fuzzy Control Systems in the Frequency Domain. *FUZZ-IEEE'98*, Anchorage, USA, pp. 49–54.

Vidyasagar, M., *Nonlinear Systems Analysis*, (1993), Prentice-Hall.

ROBUSTNESS OF EQUILIBRIA IN NONLINEAR SYSTEMS [1]

Lennart Andersson and Anders Rantzer

Department of Automatic Control, Lund Institute of Technology
Box 118, S-221 00 Lund, Sweden
fax: +46 46 13 81 18, email: lennart,rantzer@control.lth.se

Abstract: Robust stability of equilibria in nonlinear systems is considered. The nonlinear system is assumed to have a parametric uncertainty. Stability is investigated using the uncertain linearized system. Particular attention is paid to the fact that the location of the equilibria depends on the uncertain parameters. It is shown that structured singular values in several cases can be used to analyze also the effects of moving equilibria. *Copyright © 1999 IFAC.*

Keywords: Robust stability, Robust equilibria, Nonlinear systems, Uncertain dynamic systems, Linearization

1. INTRODUCTION

There are many methods for investigation of stability in nonlinear control systems, see for example Khalil (1996). Methods for global stability analysis of control systems include the describing function method, Lyapunov methods and methods related to integral quadratic constraints such as for example the Popov criteria and the small gain theorem. Local stability in a neighborhood of an equilibrium is usually shown based on Lyapunov methods. One of the results of Lyapunov theory is that local exponential stability of an equilibrium of a nonlinear system can be analyzed by studying the linearized system.

Robustness to disturbances is an important property of a control system. The methods referred to above as global, have shown to be well suited for analyzing robust stability. A lot of attention has also been paid to robust stability analysis of uncertain linear systems, see for example Zhou et al. (1996). However, less attention has been paid to systems with moving equilibria. Some work considering unstructured perturbations was reported in the paper Michel and Wang (1993). In this paper we consider constant parametric uncertainties.

The paper is organized as follows. In Section 2 properties of nonlinear systems are discussed resulting in a useful problem formulation. The problem is solved in Section 3 using a change of variables, and illustrated by some examples. This method is not always applicable. In Section 4 approximation methods are discussed as a possible resort. The paper is concluded in Section 5.

2. PROBLEM FORMULATION

In this section we first consider some properties of uncertain nonlinear systems; stability and moving equilibria. Based on these properties we state the problem.

Consider the nonlinear system

$$\frac{dx}{dt} = f(x,\delta), \tag{1}$$

where $f : (D,\Omega) \to R^n$ is continuously differentiable, and $\delta \in \Omega \subset R^r$ is an uncertain

[1] This work has been supported by the Swedish Research Council for Engineering Sciences under grant 95-759.

parameter vector. The equilibria, x_0, are given by

$$f(x^0,\delta)=0. \tag{2}$$

Some of these equilibria may move as δ changes. The rate of change with respect to changes in the parameter δ is given by

$$\frac{dx^0}{d\delta}=-\left[\frac{\partial f}{\partial x^0}\right]^{-1}\frac{\partial f}{\partial \delta}. \tag{3}$$

At points where

$$\det\left(\frac{\partial f}{\partial x^0}(x^0,\delta)\right)=0$$

it may happen that the number of equilibria is changed.

Example 1. Consider the simple nonlinear system

$$\dot{x}=\delta-x^2.$$

This system has no equilibria when $\delta<0$, one equilibrium at the origin when $\delta=0$ and two equilibria $x^0=\pm\sqrt{\delta}$ when $\delta>0$. The location of the equilibria are different for different values of $\delta>0$, and the number of equilibria changes at $\delta=0$.

Exponential stability of an equilibrium, x^0, of the nonlinear system may be analyzed by considering stability of the linearized system,

$$\frac{d(\Delta x)}{dt}=\frac{\partial f}{\partial x}(x^0,\delta)\Delta x, \tag{4}$$

where $\Delta x=x-x^0$.

Theorem 1. (Lyapunov's indirect method). Consider the nonlinear system (1) for a given value of $\delta\in\Omega$. Assume that x^0 is an equilibrium, that is $f(x^0,\delta)=0$. Let

$$A=\frac{\partial f}{\partial x}(x^0,\delta)$$

then

- x^0 is asymptotically stable if $\mathrm{Re}(\lambda_i)<0$ for all eigenvalues, λ_i, of A
- x^0 is unstable if $\mathrm{Re}(\lambda_i)>0$ for one or more eigenvalues, λ_i, of A

Proof See for example Khalil (1996).

Example 2. Consider the nonlinear system

$$\frac{dx}{dt}=f(x,\delta)=\delta(\delta^4-x^2)=\delta(\delta^2+x)(\delta^2-x).$$

It holds that

$$A=\frac{\partial f}{\partial x}(x^0,\delta)=-2\delta x^0.$$

The equilibrium $x^0=\delta^2>0$ is stable when $\delta>0$ and unstable when $\delta<0$, see Figure 1. This shows that the stability properties of this equilibrium changes at $\delta=0$. Similarly, the equilibrium $x^0=-\delta^2<0$ is unstable when $\delta>0$ and stable when $\delta<0$. This means that there for the same value of δ are both stable and unstable equilibria; *e.g.* for $\delta=1$ the equilibrium $x^0=1$ is stable and $x^0=-1$ is unstable. Note also that the same equilibrium location may be stable for one value of δ and unstable for an other value; for example the point $x=1$ is a stable equilibrium for $\delta=1$ and an unstable equilibrium for $\delta=-1$.

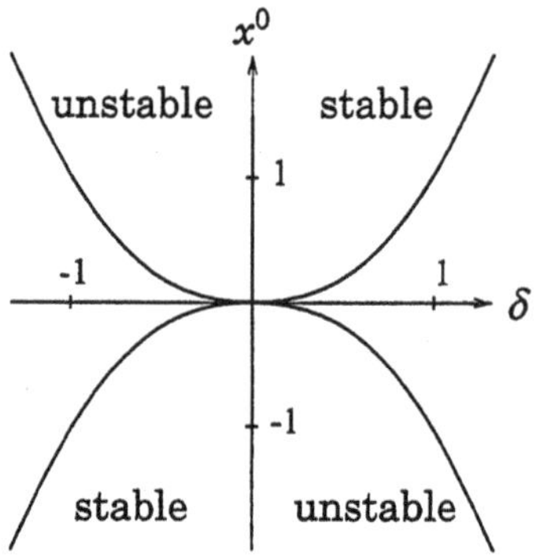

Fig. 1. The location of the equilibria x^0 vary with the parameter δ. It is seen that there are two equilibria for each value of $\delta\neq 0$, and that the same location x^0 is an equilibrium for more than one value of δ.

2.1 *A Robust Stability Problem*

The examples above point out two important considerations for investigation of robust local stability. The first is to take into account that x^0 changes if δ is changed. The second is to define, for each value of δ, which of the equilibria we are interested in. Both these considerations are captured by assuming that the equilibrium of interest is a function of the uncertain parameter

$$x^0=g(\delta),\qquad \delta\in\Omega. \tag{5}$$

The robust stability problem can then be formulated in the following way.

Problem 1. Determine if for all $\delta\in\Omega$ the eigenvalues of

$$J(\delta)=\frac{\partial f}{\partial x}(g(\delta),\delta) \tag{6}$$

stay in the open left half plane.

If g is continuous and $f\in C^1$ then an equivalent problem formulation is as follows.

Problem 2. Assume that there exists $\delta_{nom} \in \Omega$ such that the eigenvalues of

$$\frac{\partial f}{\partial x}(g(\delta_{nom}), \delta_{nom})$$

are in the open left half plane. Determine if

$$\det\left(i\omega I - \frac{\partial f}{\partial x}(g(\delta), \delta)\right) \neq 0$$

for all $\delta \in \Omega$ and all $\omega \in R$.

2.2 *Problem Reformulation*

The robust stability problem can be solved in different ways. One approach is to make a grid over possible values of the vector δ and for each of these values calculate the equilibrium of interest and the corresponding eigenvalues. This will result in long computation times if the number of uncertainties is high.

A similar problem also occurs for uncertain linear systems. For such systems the method of structured singular values (μ-analysis) have been developed, see for example Zhou et al. (1996) and Balas et al. (1993). These methods can be used to determine stability of an uncertain linear system

$$\frac{dz}{dt} = A(\delta)z$$

where $A(\delta)$ is rational in the elements of the uncertain parameter vector δ. These methods can therefore be directly applied to solve our problem if the Jacobian in Equation (6) is rational in the elements of δ. The remaining part of the paper will therefore focus on finding rational Jacobians.

2.3 *Heat Exchanger Example*

The example in this section illustrates the importance of taking into account that the location of the equilibrium depends on the uncertain parameter. Also when doing so, the resulting Jacobian is shown to be rational, which makes stability analysis easy using available methods.

Consider the heat exchanger in Figure 2. Two liquid flows q_1 and q_2 with temperatures $T_{1,in}$ and $T_{2,in}$, respectively, are entering the heat exchanger. We assume that the liquids in the two containers are well mixed, and that the outflows are as large as the inflows. The temperature difference between the two containers, $T_2 - T_1$, results in a heat transfer.

We will consider the flow q_2 as the control signal, T_1 as the measured output signal and the flow q_1 as well and the inflow temperatures $T_{1,in}$ and $T_{2,in}$ as disturbances.

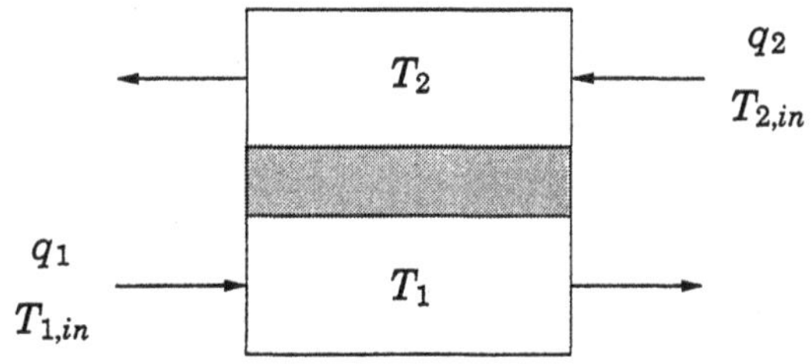

Fig. 2. Heat exchanger

A very simple nonlinear model for the heat exchanger is

$$\frac{dT_1}{dt} = \frac{q_1}{V_1}(T_{1,in} - T_1) - \frac{\alpha_1}{V_1}(T_1 - T_2)$$
$$\frac{dT_2}{dt} = \frac{q_2}{V_2}(T_{2,in} - T_2) - \frac{\alpha_2}{V_2}(T_2 - T_1),$$

where $\{V_i\}$ are the container volumes and

$$\alpha_1 = \frac{kA}{\rho_1 C_{p_1}}, \qquad \alpha_2 = \frac{kA}{\rho_2 C_{p_2}},$$

where A is the heat transfer area, $\{k_i\}$ heat transfer constants, $\{C_{p_i}\}$ heat capacities and $\{\rho_i\}$ the liquid densities. The system is nonlinear because q_1 and q_2 enter multiplicatively.

The nonlinear system has one equilibrium described by

$$q_1^0 = \alpha_1 \frac{T_1^0 - T_2^0}{T_{1,in}^0 - T_1^0}, \qquad q_2^0 = \alpha_2 \frac{T_2^0 - T_1^0}{T_{2,in}^0 - T_2^0}.$$

Linearization around this equilibrium gives the linear system

$$\frac{dx}{dt} = \begin{bmatrix} -\frac{q_1^0 + \alpha_1}{V_1} & \frac{\alpha_1}{V_1} \\ \frac{\alpha_2}{V_2} & -\frac{q_2^0 + \alpha_2}{V_2} \end{bmatrix} x + \begin{bmatrix} 0 \\ \frac{T_{2,in}^0 - T_2^0}{V_2} \end{bmatrix} u + w$$
$$y = \begin{bmatrix} 1 & 0 \end{bmatrix} x \qquad (7)$$

where the states, control signal and output of the linearized system are

$$x = \begin{bmatrix} T_1 - T_1^0 \\ T_2 - T_2^0 \end{bmatrix}, \quad u = q_2 - q_2^0, \quad y = T_1 - T_1^0$$

and the disturbances

$$w = \begin{bmatrix} \frac{T_{1,in}^0 - T_1^0}{V_1} & \frac{q_1^0}{V_1} & 0 \\ 0 & 0 & \frac{q_2^0}{V_2} \end{bmatrix} \begin{bmatrix} q_1 - q_1^0 \\ T_{1,in} - T_{1,in}^0 \\ T_{2,in} - T_{2,in}^0 \end{bmatrix}.$$

We assume that the temperature T_1 is controlled using a PID-controller,

$$G_{pid}(s) = K(1 + \frac{1}{sT_i} + sT_d),$$

which manipulates the flow q_2, see Figure 3.

The integral action in the controller implies that the temperature T_1, for a stable closed loop

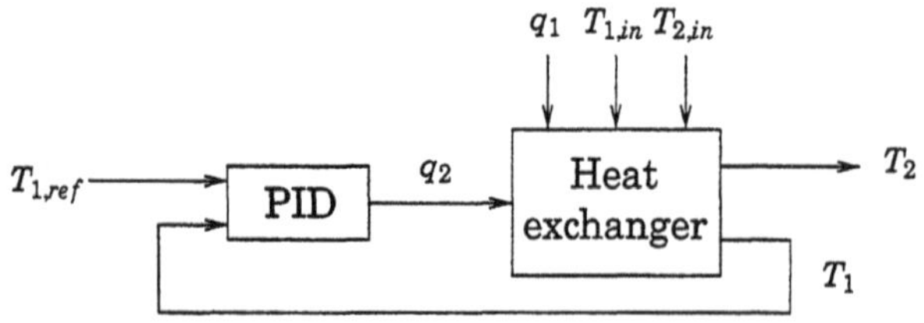

Fig. 3. PID-control of the heat exchanger

system, will be correct in stationarity. Assuming that all the inputs to the closed loop system are constant we obtain the following equations specifying the equilibrium:

$$\begin{aligned} T_1^0 &= T_{1,ref}^0 \\ T_2^0 &= T_1^0 + \frac{q_1^0}{\alpha_1}(T_1^0 - T_{1,in}^0) \\ q_2^0 &= \frac{q_1^0\alpha_2(T_1^0 - T_{1,in}^0)}{q_1^0(T_{1,in}^0 - T_1^0) + \alpha_1(T_{2,in}^0 - T_1^0)}. \end{aligned} \tag{8}$$

These equations show in a direct way how the location of the equilibrium is changed when, one of the stationary inputs to the closed loop system is changed.

Assume that the controller parameters are chosen as $K = 2$, $T_i = 0.2$ and $T_d = 0.05$ and that the heat exchanger parameters are $\alpha_1 = \alpha_2 = 1$ and $V_1 = V_2 = 1$. Also assume that the stationary input temperatures are

$$T_{1,in}^0 = 0, \quad T_{2,in}^0 = 30, \quad T_{1,ref}^0 = 10.$$

If the stationary input flow $q_1^0 = 1$, then

$$q_2^0 = 1, \quad T_1^0 = 10, \quad T_2^0 = 20,$$

and the linearized system (7) becomes

$$\begin{aligned} \frac{dx}{dt} &= \begin{bmatrix} -2 & 1 \\ 1 & -2 \end{bmatrix} x + \begin{bmatrix} 0 \\ 10 \end{bmatrix} u + w \\ y &= \begin{bmatrix} 1 & 0 \end{bmatrix} x. \end{aligned}$$

The eigenvalues of the closed loop system are -4.7 and $-0.17 \pm 4.6i$. The PID-design is on purpose very poor.

Now assume that the flow q_1^0 is uncertain. For what values of q_1^0 is the closed loop system stable? To answer this question we must take into account that the other variables, in stationarity, are affected by q_1^0, as described by the equations (8), and illustrated in Figure 4. In this example it is easy to express the equilibrium variables in the uncertain parameter q_1^0. The uncertain linearized system is

$$\begin{aligned} \frac{dx}{dt} &= \begin{bmatrix} -1-q_1^0 & 1 \\ 1 & \dfrac{-2}{2-q_1^0} \end{bmatrix} x + \begin{bmatrix} 0 \\ 10(2-q_1^0) \end{bmatrix} u + w \\ y &= \begin{bmatrix} 1 & 0 \end{bmatrix} x. \end{aligned}$$

The elements of the linearized system are rational functions of q_1^0 and the model can therefore be rewritten in a way suitable for robustness analysis. The closed loop uncertain linear system is stable for $0.78 < q_1^0 < 2.0$. We then also draw the conclusion that the equilibrium of the uncertain nonlinear system is stable for $0.78 < q_1^0 < 2.0$.

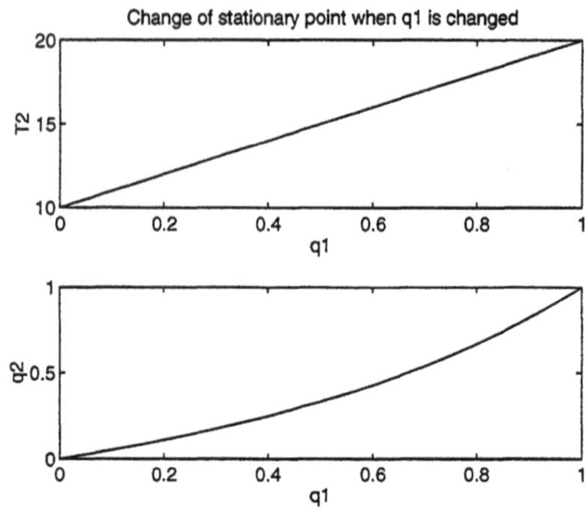

Fig. 4. The stationary point moves as q_1 is changed.

It is important to take into account that the location of the equilibrium is changed when the uncertain flow q_1^0 is changed. If we do not, we would obtain the *incorrect* uncertain linear system

$$\begin{aligned} \frac{dx}{dt} &= \begin{bmatrix} -1-q_1^0 & 1 \\ 1 & -2 \end{bmatrix} x + \begin{bmatrix} 0 \\ 10 \end{bmatrix} u + w \\ y &= \begin{bmatrix} 1 & 0 \end{bmatrix} x, \end{aligned}$$

which would lead us to believe that the closed loop system is stable for $0.53 < q_1^0$.

3. CHANGE OF VARIABLES

Linearizing the heat exchanger, with an uncertain flow, resulted in a linear system where the state-space matrices were rational functions of the uncertain parameter. Stability analysis of such systems can be done using available methods. If, however, the state-space matrices are not rational functions of the uncertain parameters these methods do not work. One way to cope with this problem is to use a change of variables.

Assume that the Jacobian given in Equation (6) is not rational in the elements of δ, then it might be possible to use a change of variables

$$\delta = T(\varepsilon)$$

such that

$$\frac{\partial f}{\partial x}(g(T(\varepsilon)), T(\varepsilon))$$

is rational in the elements of the new uncertain variable ε.

A common situation is that the Jacobian is a rational function of the elements in x^0 and δ. The main difficulty is then the function $x^0 = g(\delta)$. This function is usually not rational in the elements of δ and it may also be difficult to find an analytical expression. Ideas of how these problems can be solved in special cases are given using two examples.

3.1 *Double Integrator Example*

In this example the function $g(\delta)$ is easy to find, but it is not rational. A change of variables is used to cope with this problem.

Consider the process

$$\frac{dx_1}{dt} = x_2$$
$$\frac{dx_2}{dt} = f(u) + v,$$

which is a double integrator with an input saturation. The saturated input signal $f(u)$ together with a load disturbance v, enters the double integrator. The process is controlled by a PID-controller. The process and its connection to the PID-controller is shown in Figure 5.

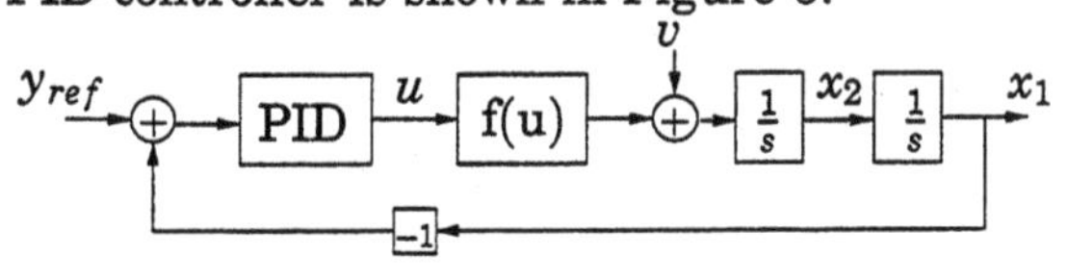

Fig. 5. PID-control of a double integrator with an input saturation.

The nonlinear closed loop system has one equilibrium. The stationary value of the output $x_1 = x_1^0$ is determined by the constant reference value $y_{ref} = y_{ref}^0$, as the controller contains an integrator. The input to the process integrators must be zero in stationarity; $v = v^0$ must therefore be compensated by a stationary control signal. The location of the equilibrium for different values of y_{ref}^0 and small enough v^0 is therefore

$$x_1^0 = y_{ref}^0$$
$$x_2^0 = 0$$
$$u^0 = f^{-1}(-v^0).$$

Linearizing the process model at the equilibrium results in the linearized model

$$\frac{d(\Delta x_1)}{dt} = \Delta x_2$$
$$\frac{d(\Delta x_2)}{dt} = \frac{\partial f}{\partial u}(u^0)\Delta u + \Delta v,$$

where we have introduced $\Delta x_1 = x_1 - x_1^0$, $\Delta x_2 = x_2 - x_2^0$ and $\Delta v = v - v^0$.

Now assume that we would like to investigate for what constant values of the disturbance $|v^0| < \pi/2$ the closed-loop system is stable when

$$f(u) = \arctan(u).$$

The matrix

$$\frac{\partial f}{\partial u}(u^0) = \frac{1}{1+(u^0)^2} = \frac{1}{1+\tan^2(v^0)} = \cos^2(v^0)$$

is not rational in v^0. We would therefore like to introduce a new uncertain parameter, ε, instead of v^0, in such a way that the Jacobian is rational in ε. There are different possibilities to do this.

The most immediate idea would be to let $\varepsilon = u^0$ so that

$$\frac{\partial f}{\partial u}(u^0) = \frac{1}{1+\varepsilon^2},$$

which is rational in ε. When we have analyzed the values of ε for which the closed-loop system is stable we find the corresponding values of v^0 using the equation

$$v^0 = -\arctan(\varepsilon).$$

Another possible change of uncertain variable is to introduce

$$\varepsilon = \cos^2(v^0)$$

so that

$$\frac{\partial f}{\partial u}(u^0) = \varepsilon.$$

The best choice among different possibilities depends on how difficult the relation between ε and δ (in this case v^0) will be, as well as on the complexity of the obtained stability problem.

3.2 *Another Example*

In this example it is illustrated how the need for explicit equilibrium expressions, that is explicit expressions for $g(\delta)$, can be avoided using a change of variables.

Consider the closed-loop nonlinear system

$$\frac{dx}{dt} = -x + (\delta + x)^2,$$

where δ is an uncertain parameter. The equilibria are the solutions to the equality

$$x^0 = (\delta + x^0)^2.$$

Let $\Delta x = x - x^0$, then the linearized system becomes

$$\frac{d(\Delta x)}{dt} = [-1 + 2(\delta + x^0)]\Delta x.$$

If we use the same approach as in the previous example we replace x^0 by

$$x^0 = \frac{1}{2} - \delta \pm \sqrt{\frac{1}{4} - \delta}, \quad \delta \leq \frac{1}{4}$$

resulting in the linearized system

$$\frac{d(\Delta x)}{dt} = \pm\sqrt{1-4\delta}\cdot\Delta x.$$

This expression is not rational in the uncertainty δ and we may proceed with a change of variables as in the previous example to see that the equilibrium

$$x^0 = \frac{1}{2} - \delta - \sqrt{\frac{1}{4} - \delta}, \qquad \delta < \frac{1}{4},$$

is stable while the equilibrium

$$x^0 = \frac{1}{2} - \delta + \sqrt{\frac{1}{4} - \delta}, \qquad \delta < \frac{1}{4},$$

is unstable.

A disadvantage with this approach is that we have to find an analytical expression for the equilibrium of interest, which may be difficult or impossible. This can sometimes be avoided if we use a change of variables without explicit expressions for $x^0 = g(\delta)$.

Consider the change of variables

$$\varepsilon = \delta + x^0$$

then

$$x^0 = (\delta + x^0)^2 = \varepsilon^2$$
$$\delta = \varepsilon - x^0 = \varepsilon(1-\varepsilon).$$

The interpretation of this change of variables is that we instead of considering δ as uncertain consider $\sqrt{x^0}$ as the uncertain parameter. The resulting linearized system is

$$\frac{d(\Delta x)}{dt} = (2\varepsilon - 1)\Delta x,$$

which is rational in the uncertain parameter ε. The uncertain system is stable for $\varepsilon < \frac{1}{2}$.

As a result the stable equilibria and the corresponding values on δ are parameterized as

$$x^0 = \varepsilon^2, \quad \delta = \varepsilon(1-\varepsilon), \qquad \varepsilon < \frac{1}{2},$$

and the unstable as

$$x^0 = \varepsilon^2, \quad \delta = \varepsilon(1-\varepsilon), \qquad \varepsilon > \frac{1}{2}.$$

To obtain the location and type of equilibria as a function of δ numerical methods may be used.

4. APPROXIMATION

The idea so far has been to write the Jacobian as a rational function of some uncertain parameters

$$J(\delta) = \frac{\partial f}{\partial x}(g(\delta), \delta) = Q(\varepsilon),$$

where Q is a matrix rational in the elements of ε, $\delta = T_\delta(\varepsilon)$, and $x^0 = T_x(\varepsilon)$.

Not all systems have a Jacobian that can be rewritten as a rational function of the uncertainties. In such cases a possible approach would be to approximate it by such a matrix function and capture the approximation error by an additional uncertainty. The ideas can be outlined as follows.

Assume that the Jacobian $\frac{\partial f}{\partial x}(x^0, \delta)$ is a rational function of the scalars x^0 and $|\delta| \leq 1$, while $x^0 = g(\delta)$ is not. Consider Maclaurins formula

$$x^0 = g(\delta) = g(0) + \frac{dg}{d\delta}(0)\delta + \underbrace{\frac{1}{2}\frac{d^2g}{d\delta^2}(\theta\delta)\cdot\delta^2}_{\varepsilon},$$

where $0 \leq \theta \leq 1$. Replacing the last term by an uncertainty, ε, results in a Jacobian rational in δ and ε. The first and second derivatives of $g(\delta)$ can be obtained by differentiating the expression $f(g(\delta), \delta) = 0$ two times. A bound on ε is then obtained using the supremum

$$|\varepsilon| \leq \sup_{|\delta|\leq 1, 0\leq\theta\leq 1} \frac{1}{2}\frac{d^2g}{d\delta^2}(\theta\delta)\cdot\delta^2.$$

5. CONCLUSIONS

In this paper we investigated robust stability of equilibria in nonlinear systems having parametric uncertainty. The main difficulty was that the location of the equilibria depends on the uncertain parameters. An example illustrated the importance of taking this into account. A problem formulation was given to cope with moving equilibria. Examples were given in which the problem could be solved based on a change of variables. When this is possible needs to be further investigated. At the end of the paper approximation methods were discussed as a possible resort when a change of variables does not work.

REFERENCES

BALAS, G., J. DOYLE, K. GLOVER, A. PACKARD, and R. SMITH (1993): *Mu-Analysis and Synthesis Toolbox*. The Math Works Inc.

KHALIL, H. (1996): *Nonlinear Systems*, second edition. Prentice Hall.

MICHEL, A. N. and K. WANG (1993): "Robust stability: perturbed systems with perturbed equilibria." *System & Control Letters*, **No 21**, pp. 155–162.

ZHOU, K., J. DOYLE, and K. GLOVER (1996): *Robust and Optimal Control*. Prentice-Hall.

THE SELF-STABLE REGION APPROACH FOR SECOND ORDER NONLINEAR UNCERTAIN SYSTEMS [1]

Huang Yi* Han Jingqing*

* *Institute of Systems Science, Chinese Academy of Sciences, Beijing, 100080, P.R. China*
Email: yao@bamboo.iss.ac.cn
Fax: 86-10-62561748

Abstract: In this paper, an inherent property that second order systems, which can be stable, have the Self-Stable Regions (SSRs) is proposed. Based on this property, a new synthesis method–the Self-Stable Region approach is suggested for a class of second-order nonlinear systems. The global stability of the closed-loop systems is proved. At last, the advantages of the SSR approach are analized. *Copyright ©1999 IFAC*

Keywords: Uncertain dynamic systems, second-order systems, control system synthesis, the Self-Stable Region (SSR)

1. INTRODUCTION

The control problem of nonlinear uncertain systems has received lots of extensive researches for many years. Many methods have been proposed and some progress has been made. Han,et al.(1997) pointed out that some second order systems have an inherent property that there exists certain region in which all trajectories of the system remaining in it after certain time T will converge to the origin. Based on this property, a new approach has been put forward to construct a continuous feedback law such that all system's trajectories move toward the region and eventually remain in it. In Huang and Han (1997,1998), a new concept, "the Self-Stable Region (SSR)" was deduced from this inherent property and a systematic recursive constructive method, "the SSR approach", was given to construct SSRs for the system

$$x^{(n)} = f(x, \dot{x}, \cdots, x^{(n-1)}) + u,$$

where $f(x, \dot{x}, \cdots, x^{n-1})$ is uncertain, $f(0, \cdots, 0) = 0$.

In this paper, we will first discuss the existence of the SSR for general second order systems and then extend this method to more general nonlinear control systems. Furthermore, we will give the global stabilization result of the SSR approach, while Han, et al., (1997), Huang and Han (1997,1998) only considered the global convergence case. At last, we will analize the advantages the SSR approach.

In Section 2, the concept of the SSR is introduced. In Section 3, we will discuss the existence of the SSR for system

$$\begin{cases} \dot{x}_1 = f_1(x_1, x_2), \\ \dot{x}_2 = f_2(x_1, x_2). \end{cases} \qquad (1)$$

In Section 4, the SSR approach and the global stabilization proof will be given . Some simulation results are shown in Section 5 and the properties of the SSR approach are concluded in Section 6.

[1] Project supported by the national natural science foundation of China, the youth foundation of Chinese Academy of Sciences and the national key project foundation of China

2. THE CONCEPT OF SSR

In this section, we will introduce the concept of the Self-Stable Region which has been given in Huang and Han (1997,1998). The following notations will be used: $\mathbf{C}$ denotes the phase plane, $\mathbf{C}^+$ and $\mathbf{C}^-$ denote, respectively, the right and the left half planes and $\overline{Q}$ denotes the closure of a region Q.

Let's first study an inherent property of the system

$$\sum_2 : \quad \begin{cases} \dot{x}_1 = x_2, \\ \dot{x}_2 = f(x_1, x_2). \end{cases}$$

where $f(x_1, x_2)$ is uncertain.

It's clear that all convergent trajectories can't converge to the origin along any direction in the first and the third quadrants. Assuming that G is any "cone" region, which lies in the second and the forth quadrants with the vertex $(0,0)$ and that $\overline{G}$ excludes the axes except the origin (as shown in Fig.1), it's obviously that for any point $(x_1, x_2) \in G$, $x_1 x_2 = \frac{1}{2}\frac{dx_1^2}{dt} < -\varepsilon < 0 (\forall |x_1| > \delta > 0)$. So, if there is a trajectory $(x_1(t), x_2(t))$ staying in G after certain time T, $(x_1(t))^2$ will be monotonically decreasing in G so long as $|x_1(t)| \geq \delta > 0$. Hence, $x_1(t) \longrightarrow 0 (t \longrightarrow \infty)$. Since $\overline{G}$ excludes x_2-axis except $(0,0)$, this trajectory can only converge to the origin. This property is an inherent property which is independent on the function $f(x_1, x_2)$. In this sense, we call G a Self-Stable Region for the system $\sum_2$.

Definition 1 (Huang and Han, 1998). Assuming G is a region in the state space, where the origin is its vertex, if it satisfies the condition that any system's trajectory, which remains in it after certain time, will eventually converge to the origin, then G is called the Self-Stable Region (SSR) of the system.

It's clear that, in the variable structure control method, all stable sliding modes for the system $\sum_2$ must lie in the second and the forth quadrants. So SSR concept reflects the essence of the sliding plane. But the restriction that the sliding mode is a superplane is relaxed by replacing the sliding mode with the SSR concept and the chattering can be easily avoided.

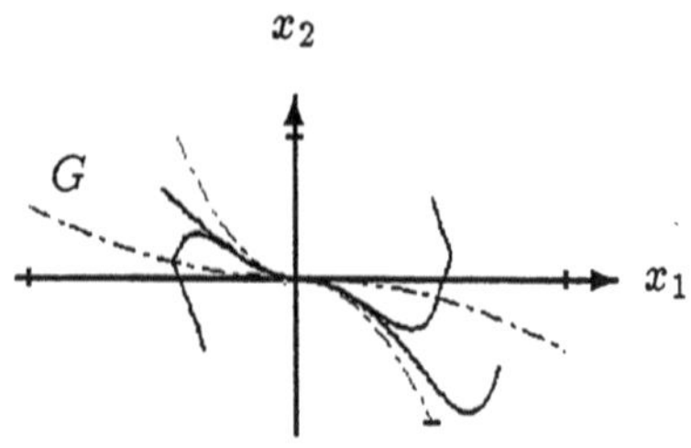

Fig. 1. The SSR for $\sum_2$

As what Han et al.,(1997), Huang and Han (1997,1998) have shown, the significance of SSR concept is that the Self-Stable property suggests a new synthesis method for uncertain systems: to determine a Self-Stable Region for the given system first, and then construct a feedback law such that all state trajectories will be attracted by the chosen SSR. Therefore, by using this synthesis method, we only need to pay attention to a certain large sector with the origin being its vertex. However, most synthesis method paid attention only to the origin or a superplane. Hence, the SSR approach will greatly reduce the complexity of the synthesis problem of uncertain systems.

In the next section, we will study the existence of the SSR for the system (1)

3. THE EXISTENCE OF THE SSR

Consider the system (1) again. Based on the sign of the function $f_1(x_1, x_2)$, the phase plane can be divided into three parts: P^+, P^- and P^0.

$$\begin{aligned} P^+ &= \{(x_1, x_2) : \quad f_1(x_1, x_2) > 0\}, \\ P^- &= \{(x_1, x_2) : \quad f_1(x_1, x_2) < 0\}, \\ P^0 &= \{(x_1, x_2) : \quad \mathbf{C} - (P^+ \cup P^-)\}. \end{aligned}$$

Lemma 1. The system (1) can be stable only if

(A) (i) The projection of $Q^+ = P^- \cap \mathbf{C}^+$ on the x_1-axis is $\mathbf{R}^+$;
The projection of $Q^- = P^+ \cap \mathbf{C}^-$ on the x_1-axis is $\mathbf{R}^-$;
(ii) $(0,0) \in \overline{Q^+} \cap \overline{Q^-}$.

Proof. We prove Lemma 1 by contradiction. It's clear that all converging trajectories must converge to the origin through the region Q^+ or Q^-.

(1)Without loss of generality, we suppose that the projection of Q^+ on the x_1-axis is a proper subset of $\mathbf{R}^+$, i.e. $\exists [\overline{x}_1, \overline{\overline{x}}_1] > 0$ and the region $\tilde{Q}$

$$\tilde{Q} = \{(x_1, x_2) | f_1(x_1, x_2) \geq 0, \forall \overline{x}_1 \leq x_1 \leq \overline{\overline{x}}_1, \forall x_2\}.$$

Then all trajectories $(x_1(t), x_2(t))$ starting from $x_1(0) \geq \overline{x}_1$ cannot cross over $\tilde{Q}$ to converge to the origin. So the system can't be stable.

(2)Suppose that $(0,0) \notin \overline{Q^+} \cap \overline{Q^-}$. Then the origin of the closed-loop system will be unstable and the system can't be stable. □

Example 1.

(1). $f_1(x_1, x_2) = -x_2^2 + 4x_1$. Q^+ is shown in Fig.2. Since there doesn't exist the appropriate region Q^-, any trajectory starting from $\mathbf{C}^-$ can't converge to the origin. So the system can't be stable.

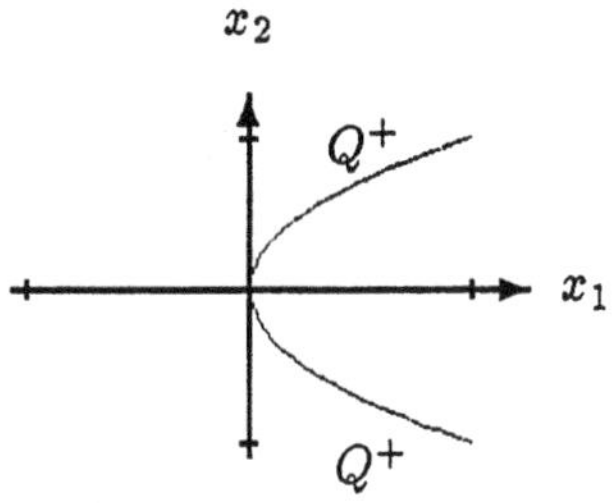

Fig. 2. Q^+, Q^- for $f_1 = -x_2^2 + 4x_1$

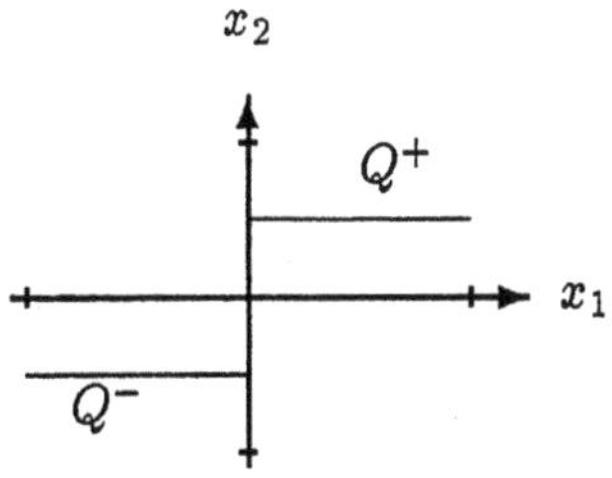

Fig. 3. Q^+, Q^- for $f_1 = sign(x_1) - x_2$

(2). $f_1(x_1, x_2) = sign(x_1) - x_2$. Q^+ and Q^- are shown in Fig.3. Since $(0,0) \notin \overline{Q^+} \cap \overline{Q^-}$, any trajectory can't converge to the origin and the system can't be stable.

Next, it's assumed that the system (1) satisfies the condition (A). Based on Definition 1 and Lemma 1, it's easy to know that the system (1) has SSRs in the region $Q^+ \cup Q^-$.

Theorem 1. If G is a region in the state space, where the origin is its vertex, $G \subset Q^+ \cup Q^-$ and $\overline{G}$ excludes the x_2-axis except the origin, then G is a SSR of the system (1).

Proof. Suppose that there exists $(x_1(t), x_2(t)) \in G$, $\forall t > T$. It's obvious that $\forall (x_1(t), x_2(t)) \in Q^+$, $x_1(t) > 0$, $\dot{x}_1(t) < 0$ and $\forall (x_1(t), x_2(t)) \in Q^-$, $x_1(t) < 0$, $\dot{x}_1(t) > 0$. So $\forall \varepsilon > 0$, $\exists \delta > 0$, s.t. $\forall (x_1(t), x_2(t))(|x_1(t)| > \delta > 0) \in G$, $x_1(t)\dot{x}_1(t) < -\varepsilon < 0$. Therefore $(x_1(t))^2$ is monotonically decreasing in G so long as $|x_1(t)| > \delta$ and will approach a limit as $t \to \infty$. Hence, $x_1(t) \to 0 (t \to \infty)$. Since $\overline{G}$ excludes the x_2-axis except $(0,0)$, $x_2(t) \to 0 (t \to \infty)$. Hence G is a SSR of the system (1). □

Example 2.

(1). $f_1(x_1, x_2) = x_2$. Q^+ and Q^- are the forth and the second quadrants, respectively. As what Fig.1 has shown, in $Q^+ \cup Q^-$, any "cone" region with vertex $(0,0)$ and its closure excludes the axes is a SSR of the system.

(2). $f_1(x_1, x_2) = sin(x_2)$. Q^+ and Q^- are shown in Fig.4. In the oblique line region, any "cone" region with vertex $(0,0)$ is a SSR of the system.

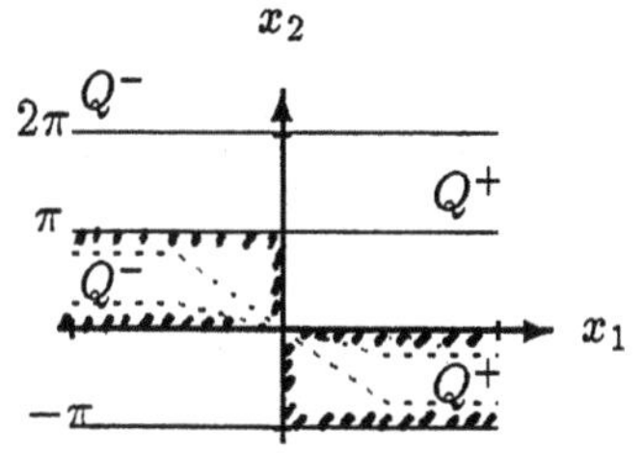

Fig. 4. The SSR for the system (1) in the case $f_1 = sin(x_2)$

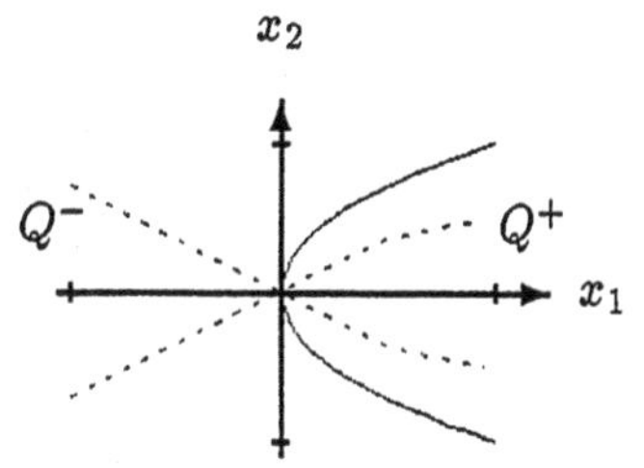

Fig. 5. The SSR for the system (1) in the case $f_1 = x_2^2 - 4x_1$

(3). $f_1(x_1, x_2) = x_2^2 - 4x_1$. Q^+ and Q^- are shown in Fig.5 in which any "cone" region with vertex $(0,0)$ is a SSR of the system.

Although the stability of the system (1) must be determined by the functions $f_1(x_1, x_2)$ and $f_2(x_1, x_2)$, Lemma 1 gives the necessary condition for a stable system, which is determined only by the function $f_1(x_1, x_2)$. Therefore, the existence of SSR is an inherent property of the systems that can be stable.

4. THE SSR APPROACH

Consider the nonlinear control system

$$\begin{cases} \dot{x}_1 = f_1(x_1, x_2), \\ \dot{x}_2 = f_2(x_1, x_2) + u, \end{cases} \tag{2}$$

where $f_1(x_1, x_2)$ and $f_2(x_1, x_2)$ are functions contain uncertainties. It's assumed that the system (2) can be stabilized. From Lemma 1 and Theorem 1, it must satisfies the condition (A) and has SSRs. Otherwise, it can't be stabilized.

Based on Definition 1 and Theorem 1, we want to choose a SSR and to synthesize a continuous feedback law such that the closed-loop system's trajectories can eventually remain in the chosen SSR.

Theorem 2. If there exist continuous and piecewise smooth functions $h_+(x_1)$, $h_-(x_1)$, a continuous function $h_0(x_1)$ and a continuous positive definite function $g(x_1, x_2) > 0, g(0,0) = 0$ satisfying

(B) (i) $h_+(x_1) > h_-(x_1), \quad \forall x_1 \neq 0;$
$h_+(0) = 0, \quad h_-(0) = 0;$
the points on $x_2 = h_+(x_1)$ and
$x_2 = h_-(x_1)$ are interior points of
$Q^+ \cup Q^-$ except $(0,0)$;
$h_+(x_1) > h_0(x_1) > h_-(x_1), \forall x_1 \neq 0;$
$h_0(0) = 0;$ (3)

(ii) $g(x_1, x_2) > q(x_1)|f_1(x_1, x_2)| + |f_2(x_1, x_2)|,$
$\forall |\lambda(x_1, x_2)| \geq 1,$ (4)

where

$$q(x_1) \geq \max\{|\frac{d(h_+(x_1))}{dx_1}|, |\frac{d(h_-(x_1))}{dx_1}|\},$$

$$\lambda(x_1, x_2) = \begin{cases} \dfrac{e(x_1, x_2)}{h_+(x_1) - h_0(x_1)}, & e > 0, \\ 0, & e = 0, \\ \dfrac{e(x_1, x_2)}{h_0(x_1) - h_-(x_1)}, & e < 0, \end{cases} \quad (5)$$

and

$$e(x_1, x_2) = x_2 - h_0(x_1),$$

then the system's trajectories will eventually converge to the origin under the continuous feedback law

$$u(x_1, x_2) = -g(x_1, x_2) \cdot sat(\lambda(x_1, x_2)). \quad (6)$$

Proof. Define

$$G = \{(x_1, x_2)|h_+(x_1) \geq x_2 \geq h_-(x_1), \forall x_1\}.$$

Since $G \subset Q^+ \cup Q^-$ and $\overline{G}$ excludes the x_2-axis except the origin, according to Theorem 1, G is a SSR of the system (2).

Define $e^{\pm}(x_1, x_2) = \pm(x_2 - h_{\pm}(x_1))$. It's clear that $\forall (x_1, x_2) \in G$, $e^+(x_1, x_2) < 0$ and $e^-(x_1, x_2) < 0$ (except the origin). From (4) and (6),

$$\begin{aligned} \dot{e}^+ |_{e^+ \geq 0} &= -(g(x_1, x_2) + \frac{dh_+(x_1)}{dx_1} f_1(x_1, x_2) \\ &\quad - f_2(x_1, x_2)) < 0, \\ &\quad \text{when } (x_1, x_2) \neq (0, 0), \\ \dot{e}^- |_{e^- \geq 0} &= -(g(x_1, x_2) - \frac{dh_-(x_1)}{dx_1} f_1(x_1, x_2) \\ &\quad + f_2(x_1, x_2)) < 0, \\ &\quad \text{when } (x_1, x_2) \neq (0, 0). \end{aligned}$$

Therefore, any trajectory $(x_1(t), x_2(t))$ has the following property:

1) If $(x_1(t), x_2(t))$ is outside G, it will go into it in a finite time, i.e. it will arrive at the boundary of G in a finite time if it doesn't converge to the origin directly and on the boundary of G, the gradient of the trajectory points to its interior.

2) If $(x_1(t), x_2(t))$ is in G, it will remain in it.

From the definition of G

$$(x_1(t), x_2(t)) \longrightarrow 0.$$

□

Theorem 2 has generalized the approach proposed in Han, et al., (1997) to a more common class of nonlinear systems. We call this method the Self-Stable Region approach.

It's well known that the selection of a suitable Lyapunov function for an arbitrarily given system is still unsolved. So the Lyapunov method has limitations in solving practical problem even though it seems to have certain advantages in analysis. From Theorem 2, the key in using the SSR approach is the proper selection of a SSR and the continuous positive definite function $g(x_1, x_2)$ according to the condition (B). It's more easier than the selection of a Lyapunov function. So the SSR approach avoids the difficulty of choicing the Lyapunov function.

In Theorem 2, the closed-loop system is globally convergent. Next, we will study how to choice SSRs such that the system (2) can be globally stabilized by the feedback (6).

Theorem 3. If $h_+(x_1) - h_0(x_1) = h_0(x_1) - h_-(x_1) \triangleq g_1(x_1)$ and

$$sign(\frac{dg_1^2(x_1)}{dt}) = sign(\frac{dx_1^2}{dt}), \forall (x_1, x_2) \in G$$

then the feedback law (6) will globally stabilize the closed-loop system (2).

Proof. Define a Lyapunov function

$$V(t) = \begin{cases} e^2(x_1, x_2), & \forall |\lambda(x_1, x_2)| > 1, \\ g_1^2(x_1), & \forall |\lambda(x_1, x_2)| \leq 1. \end{cases}$$

$V(t)$ is continuous on the boundary of G. Since $h_+(x_1) - h_0(x_1) = h_0(x_1) - h_-(x_1) = g_1(x_1)$, the formula (4) is equivalent to

$$\begin{aligned} g(x_1, x_2) > & |\frac{d(h_0(x_1))}{dx_1} f_1(x_1, x_2)| + |f_2(x_1, x_2)| \\ & + |\frac{d(g_1(x_1))}{dx_1} f_1(x_1, x_2)|, \\ & \forall |\lambda(x_1, x_2)| \geq 1. \end{aligned}$$

Then

$$
\begin{aligned}
\frac{1}{2}\frac{de^2}{dt} &= e\dot{e} \\
&= e(\dot{x}_2 - \frac{dh_0(x_1)}{dx_1}\dot{x}_1) \\
&= e(f_2(x_1,x_2) - \frac{dh_0(x_1)}{dx_1}f_1(x_1,x_2) \\
&\quad -g(x_1,x_2)sat(\lambda(x_1,x_2))) \\
&= \begin{cases} -e(g(x_1,x_2) + \frac{dh_0(x_1)}{dx_1}f_1(x_1,x_2) \\ -f_2(x_1,x_2)), \forall e \geq g_1(x_1) \\ \\ e(g(x_1,x_2) - \frac{dh_0(x_1)}{dx_1}f_1(x_1,x_2) \\ +f_2(x_1,x_2)), \forall e \leq -g_1(x_1) \end{cases} \\
&< -g_1(x_1)|\frac{dg_1(x_1)}{dx_1}f_1(x_1,x_2)| \\
&\leq 0.
\end{aligned}
$$

From the Proofs of Theorem 1,

$$
sign(\frac{dg_1^2(x_1)}{dt}) = sign(\frac{dx_1^2}{dt}),
$$
$$
\forall(x_1,x_2) \in G \text{ and } (x_1,x_2) \neq (0,0).
$$

So

$$
\frac{dV(t)}{dt} = \begin{cases} \frac{de^2}{dt} < 0, \ \forall(x_1,x_2) \notin G, \\ \\ \frac{g_1^2}{dt} < 0, \quad \forall(x_1,x_2) \in G. \end{cases}
$$

Hence, the closed-loop system is globally stable.□

5. SIMULATION RESULTS

Example 3. Given nonlinear system

$$
\begin{cases} \dot{x}_1 = sin(x_2), \\ \dot{x}_2 = x_1 + u. \end{cases} \tag{7}
$$

From the system's Q^+ and Q^- shown in Fig.4 and the condition (3), we construct a SSR as

$$
\begin{aligned}
h_0 &= -\frac{\pi}{2} \cdot sat(x_1(t)), \\
h_+ &= h_0 + \frac{\pi}{4} \cdot |sat(x_1(t))|, \\
h_- &= h_0 - \frac{\pi}{4} \cdot |sat(x_1(t))|.
\end{aligned}
$$

By the condition (4), we choose

$$
g(x_1,x_2) = \pi|sin(x_2)| + 2|x_1|.
$$

Fig.6 shows the closed-loop system's trjectories starting from initial points $(-1.5, 10.0)$ and $(-0.1, 10.0)$. They all move into the SSR G and smoothly converge to the origin under the control law.

Example 4. Given uncertain system

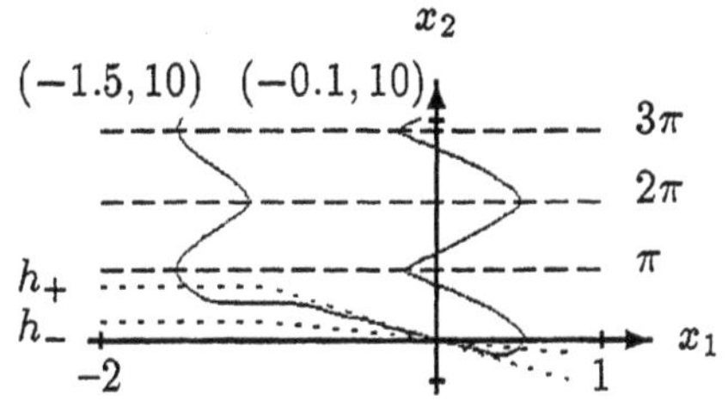

Fig. 6. Trajectories start from different points

$$
\begin{cases}
\dot{x}_1 = \overline{f}_1(t,x_1,x_2) \\
\quad = a \cdot sat(c_1 + t) \cdot (x_2^2 - 4x_1), \\
\dot{x}_2 = \overline{f}_2(t,x_1,x_2) + u \\
\quad = b \cdot x_1 \cdot sat(c_2 + cos(x_1)) \cdot \\
\quad\quad sat(c_3 + sin(x_2)) \cdot \\
\quad\quad sat(c_4 + sin(t)) + u, \\
\quad\quad 0 < a < 1, \quad 0 < b < 2, \quad c_i \geq 0.
\end{cases} \tag{8}
$$

The system's Q^+ and Q^- are shown in Fig.5. From the condition (3), we construct a SSR

$$
\begin{aligned}
h_0(x_1(t)) &= 0.0, \\
h_+(x_1(t)) &= \begin{cases} \sqrt{x_1}, & x_1 \geq 1, \\ x_1, & 0 \leq x_1 < 1, \\ -x_1, & x_1 < 0. \end{cases} \\
h_-(x_1(t)) &= -h_+(x_1(t)).
\end{aligned}
$$

Then by the condition (4), we choose

$$
q(x_1) = \begin{cases} \frac{2.0}{\sqrt{x_1}}, & x_1 \geq 1, \\ 2.0, & x_1 < 1, \end{cases}
$$

and

$$
g(x_1,x_2) = q(x_1)|x_2^2 - 4x_1| + 2|x_1|.
$$

Fig.7 shows that the trajectories of the different systems with the uncertain parameters (a,b,c_1,c_2,c_3,c_4) being $(1, 0.5, 0.5, 1, 1, 1)$, $(1, 1.5, 1, 0, 1, 1)$ and $(0.4, 1, 0, 1, 0, 0)$ all smoothly converge to the origin under the same control law.

Furthermore, since $g_1^2(x_1) = h_+^2(x_1)$ and

$$
\frac{dg_1^2(t)}{dt} = \begin{cases} \dot{x}_1, & x_1 \geq 1, \\ 2x_1\dot{x}_1, & x_1 < 1, \end{cases}
$$

$$
sign(\frac{dg_1^2(x_1)}{dt}) = sign(\frac{dx_1^2}{dt}), \quad \forall(x_1,x_2) \in G.
$$

Therefore, the closed-loop system is globally stable.

6. CONCLUSIONS

The paper has generalized the SSR approach to a common class of second order nonlinear uncertain systems.

The basic idea of the SSR approach is to utilize an inherent property of systems that can be stable – the existence of SSR. In the sliding mode control, discontinuous control law has been used to make the sliding mode be reachable. Although some

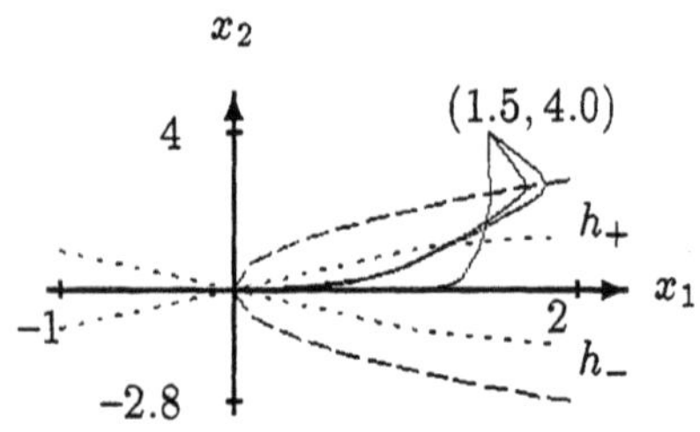

Fig. 7. Trajectories of different systems under the same control

ways have been suggested to suppress chatterings (Slotine, 1984, Burton and Zinober, 1986), the basic idea of these ways is to soften the switching. By introducing the SSR concept, the restriction that the sliding mode be a superplane is relaxed and the continuous feedback law can be adopted to make the state trajectories converge to the designed SSR. So the chatterings can be easily avoided.

The SSR approach is totally constructive. According to the condition (A) and (B), the geometric property of the SSRs is clear and the feedback law (6) is easily to be designed. Consequently, the construction procedure is much easier than construction a Lyapunov function.

The smoothness or continuity of $f_1(x_1, x_2)$ and $f_2(x_1, x_2)$ is not needed in the SSR approach. This approach only needs the basic informations of the model. First, from the condition (A), only the information about the sign of $f_1(x_1, x_2)$ is needed while determining the SSR G. So plants whose P^+ and P^- are the same may have a same SSR. For example, a same G can be chosen for funtions $f_1(x_1, x_2) = x_2^m$(m is any odd number). Furthermore, if the information about the region P^+ and P^- is enough to satisfy the condition (A), the SSR can be determined. Second, from the condition (4), the continuous positive definite function $g(x_1, x_2)$ is determined only by the dominant estimation of the functions $f_1(x_1, x_2)$ and $f_2(x_1, x_2)$ when the SSR G is chosen. The exact information about the systems' models is not needed. Hence, this approach doesn't place strict restrictions such as smoothness or matching conditions on the nonlinear uncertain models. These advantages can greatly lighten the burdens on identification and modeling. So this method will easily be used. On the other hand, for the given SSR G and a continuous positive definite function $g(x_1, x_2)$

$$g(x_1, x_2) > q(x_1)|f_1(x_1, x_2)| + |f_2(x_1, x_2)|,$$

uncertain system

$$\begin{cases} \dot{x}_1 = \overline{f}_1(t, x_1, x_2), \\ \qquad \overline{f}_1(t, 0, 0) = 0, \\ \dot{x}_2 = \overline{f}_2(t, x_1, x_2) + u(x_1, x_2), \\ \qquad \overline{f}_2(t, 0, 0) = 0, \end{cases} \tag{9}$$

can be robustly stabilized by the same control law (6) if it satisfies

$$sign(\overline{f}_1(t, x_1, x_2)) = sign(f_1(x_1, x_2)), \\ \forall (x_1(t), x_2(t)) \in G,$$

and

$$\begin{matrix} |\overline{f}_1(t, x_1, x_2)| \leq |f_1(x_1, x_2)|, \\ |\overline{f}_2(t, x_1, x_2)| \leq |f_2(x_1, x_2)|, \end{matrix} \forall (x_1(t), x_2(t)) \notin G,$$

no matter the uncertainties are linear, nonlinear or time-varying. So the bounds of the uncertain system family which can be stabilized by the constructed feedback law is easily to be determined.

REFERENCE

Burton, J.A. and A.S.I. Zinober (1986). Continuous approximation of variable structure control, *Int. J. System Sci*, **17**, 875-885.

Han, J.Q., W. Wang and Y. Huang (1997). A synthesis method for two-order uncertain systems–auto-stable region approach. *Systems Science and Mathmatical Sciences*, **10**, 152-159.

Huang, Y. and J.Q. Han (1997). A new synthesis method for high order nonlinear uncertain systems – Self-Stable Region approach, *Proceedings of 97' ASCC*, **II**, 391-394.

Huang, Y. and J.Q. Han (1998). A new synthesis method for uncertain systems-the Self-Stable Region Approach, *Int. J. Systems Sci*, to appear.

Slotine, J. E.(1984). Sliding controller design for nonlinear systems, *Int.J. Control*, **40**, 421-431.

14th Triennial World Congress, Beijing, P.R. China

A ROBUST DISTRIBUTED CONTROLLER FOR A SINGLE-LINK SCARA/CARTESIAN SMART MATERIALS ROBOT

S. S. Ge*, T. H. Lee*, J. Q. Gong* and G. Zhu***

* *Department of Electrical Engineering, National University of Singapore, Singapore 119260. Fax: (+65) 779-1103, E-mail: elegesz@nus.edu.sg*

*** *CieMed, National University of Singapore, Scicence Park II, Singapore 117684*

Abstract. In this paper, dynamic modelling and controller design are presented for a SCARA/Cartesian smart materials robot, a flexible SCARA/Cartesian robot bonded with piezoelectric actuators and sensors for better control performance. A novel distributed controller is developed directly based on the models described by partial differential equations, which guarantees the globally exponential stability of the closed-loop system. For practical implementation, finte dimensional controllers are also presented which can stablise the system with arbitrary any number of flexible modes. *Copyright © 1999 IFAC*

Key Words. Robust Distributed control, Smart materials Flexible Manipulator

1. Introduction

The SCARA/Cartesian robots are widely used in industry, especially in automatic manufacturing assembly (Lou *et al.*, 1995). For a flexible robot to carry a large payload and move at high speed, the effects of link flexibility will be non-negligible. The undesired residual vibrations make flexible robots difficult to control with high precision. For a traditional single-link flexible robot, there is only one motor acting as an actuator at the base. However, the flexible robot system is governed by PDEs and the system is of infinite dimensionality, complicated controllers have to be used to achieve control task precisely (Bayo, 1987) (Ge *et al.*, 1996; Ge *et al.*, 1997; Ge *et al.*, 1998)(Siciliano and Book, 1988; Zhu and Ge, 1998).

As an alternative, the concept of smart materials flexible robots emerges, where the flexible links are built with materials embedded or bonded with networks of distributed actuators, sensors, and processors (Hagood *et al.*, 1990)(Ge *et al.*, 1998; Ge *et al.*, 1999). While it retains all the advantages of a conventional flexible beam, such as light weight, high operational speed, it has additional sensing and control capability. With the application of smart materials, better control performance is expected to be obtainable easily.

In this paper, a single link smart materials flexible SCARA/Cartesian robot is discussed. Based on the Hamilton's principle, a set of PDEs and boundary conditions (BCs) are obtained first. A robust distributed controller is developed directly based on the PDE model rather than a finite element method (FEM) model or assumed mode method (AMM) model. For the ease of implementation, finte dimensional controllers are also presented which can stablise the system with arbitrary any number of flexible modes.

2. System Description and Dynamic Modelling

The system discussed in this paper is a single-link flexible SCARA/Cartesian robot with piezoelectric actuators and sensors distributed along its length, and a point mass payload at the tip. The schematic diagram of the smart materials beam is shown in Figure 1, the geometry of the system is shown in Figure 2, and the notations are defined in (Ge *et al.*, 1998; Ge *et al.*, 1999). The robot is in the horizontal plane. Thus, the effects of gravity are neglected which is always true for space applications. The base of the robot can only move along the y axis of the plane.

The Hamilton principle states that $\delta \int_{t_0}^{t_1}(E_k - E_p + W)dt = 0$ where $\delta(\ )$ denotes the variational operator, E_k is the system kinetic energy, E_p is the system potential energy, and W is the virtual work done by external forces. The fundamental relations of piezoelectric materials are given as follows (Ikeda, 1990)

$$\mathbf{F} = \mathbf{cS} - \mathbf{hD}, \ \mathbf{E} = -\mathbf{h}^T\mathbf{S} + \beta\mathbf{D} \tag{1}$$

where all the notations are defined in (Ge *et al.*,

1998; Ge *et al.*, 1999).

After complicated derivation, the following PDEs is obtained

$$m_s\ddot{s}+\rho_L\int_0^L\ddot{w}(x,t)dx+m_3[\ddot{s}+\ddot{w}(L,t)]+d_1(t)=f \quad (2)$$

$$-\rho_L[\ddot{s}+\ddot{w}(x,t)]+d_2(x,t)=c_L\frac{\partial^4 w(x,t)}{\partial x^4}+h_{aL}\frac{\partial^2 D_{ay}(x,t)}{\partial x^2}+h_{sL}\frac{\partial^2 D_{sy}(x,t)}{\partial x^2} \quad (3)$$

$$\beta_{aL}D_{ay}(x,t)+h_{aL}\frac{\partial^2 w(x,t)}{\partial x^2}+d_3(x,t)=bv \quad (4)$$

$$\beta_{sL}D_{sy}(x,t)+h_{sL}\frac{\partial^2 w(x,t)}{\partial x^2}+d_4(x,t)=0 \quad (5)$$

where $m_s = m_b + \rho_L L$, and BCs are

$$w(0,t)=0, \quad \frac{\partial w(0,t)}{\partial x}=0 \quad (6)$$

$$m_3[\ddot{s}+\ddot{w}(L,t)]=c_L\frac{\partial^3 w(L,t)}{\partial x^3}+h_{aL}\frac{\partial D_{ay}(L,t)}{\partial x}+h_{sL}\frac{\partial D_{sy}(L,t)}{\partial x} \quad (7)$$

$$c_L\frac{\partial^2 w(L,t)}{\partial x^2}+h_{aL}D_{ay}(L,t)+h_{sL}D_{sy}(L,t)=0 \quad (8)$$

where all the constants are defined in (Ge *et al.*, 1998; Ge *et al.*, 1999). It should be noted that d_i, $i=1,2,3,4$ are arbitrarily added to model uncertainties and disturbances.

Assume that d_1, d_2, d_3, and d_4 satisfy

$$|d_1|\le\epsilon_1, |d_2|\le\epsilon_2, |d_3|\le\epsilon_3, |d_4|\le\epsilon_4, \dot{d}_4=0 \quad (9)$$

$$|\frac{\partial d_3}{\partial x}|\le\sigma_1, |\frac{\partial d_4}{\partial x}|\le\sigma_2, |\frac{\partial^2 d_3}{\partial x^2}|\le\sigma_3, |\frac{\partial^2 d_4}{\partial x^2}|\le\sigma_4 \quad (10)$$

where ϵ_i, $i=1,2,\cdots,4$ are known positive constants; and σ_1, σ_2, σ_3, and σ_4 are known positive constants.

Although the above equations (2)-(5) and (6)-(8) explicitly illustrate the physical principles of the system, they can be further written into more user-friendly forms for controller design.

Substituting equations (3)-(5), (7) and (8) into (2), and integrating by parts, we arrive at

$$m_b\ddot{s}+[c_0\frac{\partial^3 w(0,t)}{\partial x^3}+\frac{bh_{aL}}{\beta_{aL}}\frac{\partial v(0,t)}{\partial x}]+d_5=f \quad (11)$$

where $c_0 = c_L - \frac{h_{aL}^2}{\beta_{aL}} - \frac{h_{sL}^2}{\beta_{sL}}$ and $d_5 = d_1(t) + \int_0^L x d_2(x,t)dx + \frac{h_{aL}}{\beta aL}ad_3(0,t) + \frac{h_{sL}}{\beta_{sL}}d_4(0,t) - \frac{h_{aL}}{\beta aL}d_3(L,t) - \frac{h_{sL}}{\beta_{sL}}d_4(L,t) + L[\frac{h_{aL}}{\beta aL}\frac{\partial d_3(L,t)}{\partial x} + \frac{h_{sL}}{\beta_{sL}}\frac{\partial d_4(L,t)}{\partial x}]$.

From inequalities (9) and (10), we know that d_5 is bounded by $|d_5|\le\epsilon_5$ with

$$\epsilon_5=\epsilon_1+\frac{1}{2}L^2\epsilon_2+|\frac{2h_{aL}}{\beta aL}|\epsilon_3+|\frac{2h_{sL}}{\beta_{sL}}|\epsilon_4+|\frac{Lh_{aL}}{\beta_{aL}}|\sigma_1+|\frac{Lh_{sL}}{\beta_{sL}}|\sigma_2 \quad (12)$$

Substituting equations (4) and (5) into equation (3), we arrive at

$$-\rho_L[\ddot{s}+\ddot{w}(x,t)]=c_0\frac{\partial^4 w(x,t)}{\partial x^4}+\frac{bh_{aL}}{\beta_{aL}}\frac{\partial^2 v(x,t)}{\partial x^2}-\frac{h_{aL}}{\beta_{aL}}\frac{\partial^2 d_3}{\partial x^2}-\frac{h_{sL}}{\beta_{sL}}\frac{\partial^2 d_4}{\partial x^2}-d_2 \quad (13)$$

3. Robust Distributed Controller Design

Because the PDE model derived in the last section is infinite dimensional, it is straightforward to develop a distributed controller which is also infinite dimensional. Under the assumption that the electrical field can be controlled precisely, controllers of infinite dimensionality is implementable. Therefore, we shall discuss it theoretically at first, and the implementation issues of the corresponding finite dimensional controller will be investigated in Section 4 because currently only finite dimensional controller can be implemented.

Let the desired translational trajectory $s_d \in C^2$ and define $e = s_d - s$.

Theorem 3.1 Under assumptions (9) and (10), the system governed by PDEs (11), (13), and constrained by BCs (6)-(8) can be stabilized by controller

$$\mathbf{u}=\begin{bmatrix} f \\ v \end{bmatrix}=\begin{bmatrix} f_1+f_r \\ v_1+v_r \end{bmatrix} \quad (14)$$

where

$$f_1 = -k_{pe}m_b e - k_{de}m_b\dot{e} + m_b\ddot{s}_d(t) \quad (15)$$

$$v_1 = \frac{\beta_{aL}}{bh_{aL}}\Big[-c_0\frac{\partial^2 w(x,t)}{\partial x^2}-\frac{1}{2}\rho_L\ddot{s}x^2+\rho_L k_{pw}\int_0^x\int_0^x w(x,t)dxdx+\rho_L k_{dw}\int_0^x\int_0^x \dot{w}(x,t)dxdx\Big] \quad (16)$$

$$f_r = -sgn(\dot{e})\epsilon_5 \quad (17)$$

$$v_r = \int_0^x\int_0^x sgn(\dot{w}(x,t))\epsilon_6 dxdx \quad (18)$$

with k_{pe}, k_{de}, k_{pw}, $k_{dw} > 0$,

$$\epsilon_6=\frac{\beta_{aL}}{bh_{aL}}\Big[\epsilon_2+\Big|\frac{h_{aL}}{\beta_{aL}}\Big|\sigma_3+\Big|\frac{h_{sL}}{\beta_{sL}}\Big|\sigma_4+\Big|\frac{k_{pw}}{2h_{sL}}\Big|\epsilon_4 x^2\Big] \quad (19)$$

and ϵ_5 is given in (12) with $x\in[0,L]$.

In addition, all the signals in the closed-loop are bounded.

Proof. Consider the following Lyapunov function candidate

$$V=\frac{1}{2}[k_{pw}\int_0^L w^2(x,t)dx+\int_0^L \dot{w}^2(x,t)dx+k_{pe}e^2+\dot{e}^2] \quad (20)$$

It is obvious that equation (20) is positive definite.

Taking the time derivative, applying equations (11), (13) and control law (14)-(19), we have

$$\dot{V} \leq -\int_0^L k_{dw}\dot{w}^2 dx - k_{de}\dot{e}^2 \quad (21)$$

which is negative semi-definite.

Actually, under the control law (14), the closed-loop system can be described by

$$\ddot{e} + k_{de}\dot{e}k_{pe}e + \frac{1}{m_b}[sgn(\dot{e})\epsilon_5 + d_5] = 0 \quad (22)$$

$$\begin{aligned}&\ddot{w}(x,t) + k_{dw}\dot{w}(x,t) + k_{pw}w(x,t)\\ &+\frac{1}{\rho_L}\Big\{[sgn(\dot{w})\epsilon_2-d_2]+\left|\frac{h_{aL}}{\beta_{aL}}\right|\left[sgn(\dot{w})\sigma_3-\frac{\partial^2 d_3}{\partial x^2}\right]\\ &+\left|\frac{h_{sL}}{\beta_{sL}}\right|\left[sgn(\dot{w})\sigma_4-\frac{\partial^2 d_4}{\partial x^2}\right]+\left|\frac{k_{pw}}{h_{sL}}\right|\Big[\frac{1}{2}sgn(\dot{w})\epsilon_4 x^2\\ &+\int_0^x\int_0^x d_4 dx dx\Big]\Big\} = 0 \quad (23)\end{aligned}$$

It means that the infinite dimensional system is totally decoupled under the infinite dimensional controller.

Because k_{pe}, k_{de}, k_{pw}, and $k_{dw} > 0$, from equations (22) and (23), it can be found that both the tracking error e and the deflection $w(x,t)$ approaches 0 asymptotically.

From equation (21), it can be seen that V is bounded. Consequently, $e(t)$, $\dot{e}(t)$, $w(x,t)$, $\dot{w}(x,t)$ are bounded. Under the assumption of small deflections, $\frac{\partial w^2(x,t)}{\partial x^2}$ is also bounded. Therefore, $v_1(x,t)$ and f_1 are bounded. Because of the assumptions of bounded disturbances (9) and (10), it can be found that f and v are bounded. Subsequently, it can be shown that all the signals in the closed-loop are bounded.

□

4. Finite Dimensional Controller Design

In practice, only a finite number of piezoelectric actuators can be bonded along the beam. Without losing generality, we assume that the i-th actuator is located between the interval $[a_i, b_i]$ with $0 \leq a_i, b_i \leq L$, $a_1 = 0$, $a_i < b_i$ $(i = 1, \cdots, n)$, $b_i \leq a_{i+1}$ $(i = 1, \cdots, n-1)$, and n being the number of the actuators. Therefore, piezoelectric voltages are only applied to the intervals $\cup_{i=1}^{i=n}[a_i, b_i]$ and $v(x,t) \equiv 0$, $\forall x \notin \cup_{i=1}^{i=n}[a_i, b_i]$. At the same time, because the commonly used AMM can capture the most dominant features of the system, it will be employed to describe the system.

According to the formulation of AMM, it is assumed that

$$w(x,t) = \sum_{i=1}^{i=N} \psi_i(x)q_i(t) \quad (24)$$

$$D_{aL}(x,t) = \sum_{i=1}^{i=N} -\frac{h_{aL}}{\beta_{aL}}\psi_i''(x)q_i(t) \quad (25)$$

$$D_{sL}(x,t) = \sum_{i=1}^{i=N} -\frac{h_{sL}}{\beta_{sL}}\psi_i''(x)q_i(t) \quad (26)$$

where $q_i(t)$'s are the generalized coordinates, N is the number of modes to be considered, and ψ_i's are the modes shape functions which are defined as (Ge *et al.*, 1998)

$$\begin{aligned}\psi_i(x)=&A_i[\cosh\frac{\nu_i x}{L}-\cos\frac{\nu_i x}{L}\\ &-\gamma_i(\sinh\frac{\nu_i x}{L}-\sin\frac{\nu_i x}{L})]=A_i\bar{\psi}_i(x) \quad (27)\end{aligned}$$

with ν_i, γ_i and A_i satisfying

$$\begin{aligned}&1+\cosh\nu_i\cos\nu_i+\frac{m_3\nu_i}{\rho_L L}(\sinh\nu_i\cos\nu_i\\ &-\cosh\nu_i\sin\nu_i) = 0 \quad (28)\end{aligned}$$

$$\gamma_i = (\cosh\nu_i+\cos\nu_i)/(\sinh\nu_i+\sin\nu_i) \quad (29)$$

$$A_i = [\int_0^L \bar{\psi}_i^2(x)dx + \frac{m_3}{\rho_L}\bar{\psi}_i^2(L)]^{-1/2} \quad (30)$$

Applying Lagrange-Euler formulation, the following AMM model of system is obtained

$$\mathbf{M}_A\ddot{\mathbf{Q}} + \mathbf{K}_A\mathbf{Q} + \mathbf{d} = \mathbf{F}_A \quad (31)$$

where

$$\mathbf{Q} := [\ s\ \ q_1\ \ q_2\ \ \cdots\ \ q_N\ ]^T \in R^{N+1} \quad (32)$$

is the generalized coordinates vector, $\mathbf{M}_A$, $\mathbf{K}_A$, $\mathbf{F}_A$ are defined in (Ge *et al.*, 1998; Ge *et al.*, 1999), and $\mathbf{d} \in R^{N+1}$ is the disturbance vector which is bounded. It should be noted that the modes shape function $\psi_i(x)$'s satisfy the following orthogonal conditions:

$$\rho_L\int_0^L \psi_i\psi_j dx+m_3\psi_i(L)\psi_j(L)=\begin{cases}0 & i\neq j\\ \rho_L & i=j\end{cases} \quad (33)$$

$$[c_L-\frac{h_{aL}^2}{\beta_{aL}}-\frac{h_{sL}^2}{\beta_{sL}}]\int_0^L \psi_i''\psi_j''dx=\begin{cases}0 & i\neq j\\ \omega_i^2\rho_L & i=j\end{cases} \quad (34)$$

For the AMM model, by substituting equation

(24) into equation(16) and solving for some multiple integrations, v_1 becomes

$$v_1 = \sum_{i=1}^{i=N}[k_1 q_i(t) + k_2\dot{q}_i(t)]\psi_i''(x) + k_3 x^2\ddot{s} \quad (35)$$

where

$$\begin{aligned} k_1 &= \frac{\beta_{aL}}{bh_{aL}}[\frac{h_0 h_{sL}}{\beta_{sL}} + k_{pw}\rho_L(\frac{L}{\nu_i})^4] \\ k_2 &= k_{dw}\frac{\rho_L\beta_{aL}}{bh_{aL}}(\frac{L}{\nu_i})^4 \\ k_3 &= \frac{\rho_L\beta_{aL}}{2bh_{aL}} \end{aligned}$$

Note that $q_i(t)$ and $\dot{q}_i(t)$ as

$$q_i(t) = -\frac{h_{sL}}{\beta_{sL}}\frac{L^4}{\nu_i^4}\int_0^L D_{sy}(x,t)\psi_i''(x)dx \quad (36)$$

$$\dot{q}_i(t) = -\frac{h_{sL}}{\beta_{sL}}\frac{L^4}{\nu_i^4}\int_0^L \dot{D}_{sy}(x,t)\psi_i''(x)dx \quad (37)$$

with both D_{sy} and $\dot{D}_{sy}$ being measurable.

Similarly, v_r in equation (18) can be implemented as

$$\begin{aligned} v_r(x,t) &= \int_0^x\int_0^x sgn(\dot{w}(x,t))\epsilon_6 dx dx \\ &= \int_0^x\int_0^x sgn(\sum_{i=1}^{i=N}\psi_i(x)\dot{q}_i(t))\epsilon_6 dx dx \quad (38) \end{aligned}$$

while the force controls f_1 (15) and f_r (17) remain the same.

Applying equation (24) to system (11)-(13), considering that only a finite number of piezoelectric actuators are utilized, we have the corresponding finite dimensional dynamic model:

$$m_b\ddot{s} + [c_0\sum_{i=1}^{i=N}\psi_i'''(0)q_i(t) + \frac{bh_{aL}}{\beta_{aL}}\frac{\partial v(0,t)}{\partial x}] + d_5 = f \quad (39)$$

$$\begin{aligned} -\rho_L[\ddot{s} + \sum_{i=1}^{i=N}\psi_i(x)\ddot{q}_i(t)] &= c_0\sum_{i=1}^{i=N}\psi_i''''(x)q_i(t) \\ &+ \frac{bh_{aL}}{\beta_{aL}}\frac{\partial^2 v(x,t)}{\partial x^2} - \frac{h_{aL}}{\beta_{aL}}\frac{\partial^2 d_3}{\partial x^2} \\ &- \frac{h_{sL}}{\beta_{sL}}\frac{\partial^2 d_4}{\partial x^2} - d_2, \ \forall x \in \cup_{i=1}^{i=n}[a_i, b_i] \quad (40) \end{aligned}$$

$$\begin{aligned} -\rho_L[\ddot{s} + \sum_{i=1}^{i=N}\psi_i(x)\ddot{q}_i(t)] &= c_0\sum_{i=1}^{i=N}\psi_i''''(x)q_i(t) \\ &- \frac{h_{aL}}{\beta_{aL}}\frac{\partial^2 d_3}{\partial x^2} - \frac{h_{sL}}{\beta_{sL}}\frac{\partial^2 d_4}{\partial x^2} - d_2, \ \forall x \notin \cup_{i=1}^{i=n}[a_i, b_i] \quad (41) \end{aligned}$$

Theorem 4.1 Finite dimensional system (39)-(41) can be asymptotically stabilized by spatially discretized controller $\mathbf{u} = [\ f_1 + f_r, \ \ v_1 + v_r\ ]^T$ with f_1 and f_r given by equations (15) and (17), v_1 and v_r taking the forms of (35) and (38), $\forall x \in \cup_{i=1}^{i=n}[a_i, b_i]$, v_1 and $v_r \equiv 0$, $\forall x \notin \cup_{i=1}^{i=n}[a_i, b_i]$.

Proof. Substituting f_1 in (15), f_r in (17), v_1 in (35) and v_r in (38) into system equations (39)-(41), we will in turn obtain the following closed-loop system equations

$$\ddot{e} + k_{de}\dot{e} + k_{pe}e + \frac{1}{m_b}[sgn(\dot{e})\epsilon_5 + d_5] = 0 \quad (42)$$

$$\begin{aligned} &\sum_{i=1}^{i=N}\psi_i(x)\ddot{q}_i(t) + k_{dw}\sum_{i=1}^{i=N}\psi_i(x)\dot{q}_i(t) + k_{pw}\sum_{i=1}^{i=N}\psi_i(x)q_i(t) \\ &+ \frac{1}{\rho_L}\Big\{\Big[sgn(\sum_{i=1}^{i=N}\psi_i(x)\dot{q}_i(t))\epsilon_2 - d_2\Big] \\ &+ \left|\frac{h_{aL}}{\beta_{aL}}\right|\Big[sgn(\sum_{i=1}^{i=N}\psi_i(x)\dot{q}_i(t))\sigma_3 - \frac{\partial^2 d_3}{\partial x^2}\Big] \\ &+ \left|\frac{h_{sL}}{\beta_{sL}}\right|\Big[sgn(\sum_{i=1}^{i=N}\psi_i(x)\dot{q}_i(t))\sigma_4 - \frac{\partial^2 d_4}{\partial x^2}\Big] \\ &+ \left|\frac{k_{pw}}{h_{sL}}\right|\Big[\frac{1}{2}sgn(\sum_{i=1}^{i=N}\psi_i(x)\dot{q}_i(t))\epsilon_4 x^2 \\ &+ \int_0^x\int_0^x d_4 dx dx\Big]\Big\} = 0, \forall x \in \cup_{i=1}^{i=n}[a_i, b_i] \quad (43) \end{aligned}$$

$$\begin{aligned} -\rho_L[\ddot{s} + \sum_{i=1}^{i=N}\psi_i(x)\ddot{q}_i(t)] &= c_0\sum_{i=1}^{i=N}\psi_i''''(x)q_i(t) \\ &- \frac{h_{aL}}{\beta_{aL}}\frac{\partial^2 d_3}{\partial x^2} - \frac{h_{sL}}{\beta_{sL}}\frac{\partial^2 d_4}{\partial x^2} - d_2, \forall x \notin \cup_{i=1}^{i=n}[a_i, b_i] \quad (44) \end{aligned}$$

Consider the following Lyapunov function candidate

$$V = \frac{1}{2}[k_{pw}\int_0^L w^2(x,t)dx + \int_0^L \dot{w}^2(x,t)dx + k_{pe}e^2 + \dot{e}^2]$$

Using the closed-loop equations (42) and (43), the time derivative of (45) can be obtained

$$\begin{aligned} \dot{V} &= -k_{dw}(\sum_{i=1}^{i=N}\psi_{(}x)\dot{q}_i(t))^2 - k_{de}\dot{e}^2 \\ &- \frac{1}{m_b}\dot{e}[sgn(\dot{e})\epsilon_5 + d_5(0,t)] \\ &- \frac{1}{\rho_L}\sum_{i=1}^{i=N}\psi_{(}x)\dot{q}_i(t)\Big\{\Big[sgn(\sum_{i=1}^{i=N}\psi_{(}x)\dot{q}_i(t))\epsilon_2 - d_2\Big] \\ &+ \left|\frac{h_{aL}}{\beta_{aL}}\right|\Big[sgn(\sum_{i=1}^{i=N}\psi_{(}x)\dot{q}_i(t))\sigma_3 - \frac{\partial^2 d_3}{\partial x^2}\Big] \\ &+ \left|\frac{h_{sL}}{\beta_{sL}}\right|\Big[sgn(\sum_{i=1}^{i=N}\psi_{(}x)\dot{q}_i(t))\sigma_4 - \frac{\partial^2 d_4}{\partial x^2}\Big] \\ &+ \left|\frac{k_{pw}}{h_{sL}}\right|\Big[\frac{1}{2}sgn(\sum_{i=1}^{i=N}\psi_{(}x)\dot{q}_i(t))\epsilon_4 x^2 \end{aligned}$$

$$+\int_0^x\int_0^x d_4 dx dx\Big]\Big\}$$

$$\leq -k_{dw}(\sum_{i=1}^{i=N}\psi_(x)\dot{q}_i(t))^2dx - k_{de}\dot{e}^2 \quad (45)$$

which is negative semi-definite. Therefore, $e \to 0$ and $\sum_{i=1}^{i=N}\psi_(x)q_i(t) \to 0$ asymptotically $\forall x \in \cup_{i=1}^{i=n}[a_i, b_i]$, while no conclusion can be drawn for $\sum_{i=1}^{i=N}\psi_i(x)q_i(t)$, $\forall x \notin \cup_{i=1}^{i=n}[a_i, b_i]$. On the other hand, ψ_i's are linearly independent, from the statement that $\sum_{i=1}^{i=N}\psi_i(x)q_i(t) \to 0$, $\forall x \in \cup_{i=1}^{i=n}[a_i, b_i]$ asymptotically, it can be deduced that $q_i(t) \to 0$ asymptotically, which is independent of x. Furthermore, $\psi_i(x)$ is bounded $\forall x \in [0, L]$, we can conclude that $\sum_{i=1}^{i=N}\psi_i(x)q_i(t) \to 0$ asymptotically, $\forall x \in [0, L]$ from equation (24).

□

Remarks:

1. As for the sensors, it is also true that only a finite number of piezoelectric sensors can be utilized in a practical system. Equations (36) and (37) give a method on the ways to obtain $q_i(t)$ and $\dot{q}_i(t)$, which requires spatially continuously distributed measurements of $D_{sy}(x,t)$, $\dot{D}_{sy}(x,t)$ and is not within the state of the art (Baruh and Choe, 1990). However, with a finite number of sensor outputs, $q_i(t)$ and $\dot{q}_i(t)$ can be approximated by interpolation or extrapolation of the sensor measurements as discussed in (Baruh and Choe, 1990). For brevity, we will not repeat here.

2. Because ψ_i's are known functions of x, ψ_i'''s are ready to be derived off-line and $v(x,t)$ can be constructed as long as q_i's and $\dot{q}_i$'s can be obtained as shown in equations (36).

5. Experimental Results

We shall experimentally evaluate the effectiveness of the proposed controller in vibration suppression by using the piezoelectric actuators and sensors. For simplicity, the base of the link is fixed in the experiment setup, which means that there is no force control. This, however, does not restrict the validity for testing the performance of the controllers proposed because the key point of controller design here is the suppression of the residual vibration using the piezoelectric voltage control rather than the force control.

The simple experiment setup is shown in Figure 3, which consists of a clamped flexible beam with a pair of piezoelectric actuator and sensor embedded near its base (the distance from the base to the center of sensor/actuator is 0.025m). The lengths of the beam, actuator, and sensor are 0.35m, 0.05m, and 0.05m, respectively. As stated earlier, since the base is clamped, there is no force control in the experiment. The input is the voltage v applied to the piezoelectric actuator and the output of the system is the sensor voltage v_o, which is proportional to the time derivative of the deflection w.

In the experiment, there is only one pair of piezoelectric actuator and sensor, the control law (14) is much simplified. In order to implement the integral, numerical summation is employed. Sensor voltage v_o is used to replace $\dot{w}$ and to calculate w. Subsequently, (14) is transferred into the following form

$$\begin{aligned}v(x,t) &= v(0.05, kT)\\ &= -217768\int_0^{.05}\int_0^x\int_0^{kT} v_o(\tau)d\tau dx dx\\ &\quad -362948\int_0^{.05}\int_0^x v_o(kT)dx dx\\ &\approx -544.42\sum_{i=0}^{i=k} v_o(iT) - 907.37v_o(kT) \quad (46)\end{aligned}$$

where $T = 0.005$ seconds is the sample interval. The first approximation in equation (46) is due to that there is only one sensor and all the voltages for $x \in [0, 0.05)$ are assumed to be v_o. The second approximation changes the time integral into the numerical summation.

Figure 4 show the open-loop response of the sensor voltage. Figures 5 and 6 give the sensor voltages in the closed-loop system and the corresponding control voltages. It can be seen that the residual vibrations can be effectively suppressed with the proposed controller.

6. Conclusion

In this paper, a new flexible SCARA/Cartesian robot system has been proposed, which fully combines the advantages of flexible robots and piezoelectric materials. Directly based on the PDE model, a novel distributed controller has been developed, which can guarantee the globally stability of the closed-loop system. For practical implementation, finte dimensional controllers have been developed, which can stablise the system with arbitrary any number of flexible modes. .

7. REFERENCES

Baruh, H. and K. Choe (1990). Sensor Placement in Structural Control. *Journal of Guidance, Control, and Dynamics*, **13**(3), 524-533.

Bayo, E. (1987). A Finite-Element Approach to Control the End-Point Motion of a Single-Link Flexible Robot. *Int. J. Robotics Research*, **4**(1), 63–75.

Blevins, R. D. (1979). *Formulas for Natural Frequency and Mode Shape.* Van Nostrand Reinhold Company, New York.

Cannon, R. H. and E. Schmitz (1984). Initial Experiments on the End-Point Control of a Flexible One-Link Robot. *Int. J. Robotics Research,* **3**(3), 62–75.

Ge, S. S., T.H. Lee and G. Zhu (1996). Genetic algorithm tuning of Lyapunov-based controllers: An application to a single-linke flexible robot system. *IEEE Transactions on Industrial Electronics,* **43**(5), 567-674.

Ge, S. S., T.H. Lee and G. Zhu (1997). A nonlinear feedback controller for a single-link flexible manipulator based on a finite element model. *Journal of Robotic Systems,* **14**(3), 165-178.

Ge, S. S., T.H. Lee and G. Zhu (1998). Asymptotically stable end-point regulation of a flexible SCARA/Cartesian Robot. *IEEE/ASME Trans. on Mechatronics,* **3**(2), 138-144.

Ge, S. S., T.H. Lee and J. Q. Gong (1998). Dynamic Modelling of a Smart Materials Robot. *AIAA Journal,* **36**(8), 1466-1478.

Ge, S. S., T.H. Lee and J. Q. Gong (1999). A Robust Distributed Controller of a Single-link SCARA/Cartesian Smart Materials Robot. To Appear in *Mechatronics.*

Hagood, N. W., W. H.Chung and A.V. Flotow (1990). Modelling of Piezoelectric Actuator Dynamics for Active Structural Control. *Journal of Intelligent Systems and Structures,* **1**, July, 327-355.

Ikeda, T. (1979). *Oxford Science Publications.* Oxford.

Luo, Z. H., N. Kitamura and B.Z. Guo (1995). Shear Force Feedback Control of Flexible Robot Arms. *IEEE Trans. Robotics and Automation,* **11**(5), 760-765.

Siciliano, B. and W.J. Book (1988). A Singular Perturbation Approach to Control Lightweight Flexible Manipulators. *Int. J. Robotics Research,* **7**(4), 79–90.

Zhu G. and S.S. Ge (1998). A quasi-tracking approach for finite time control of a mass beam system. To Appear in *Automatica.*

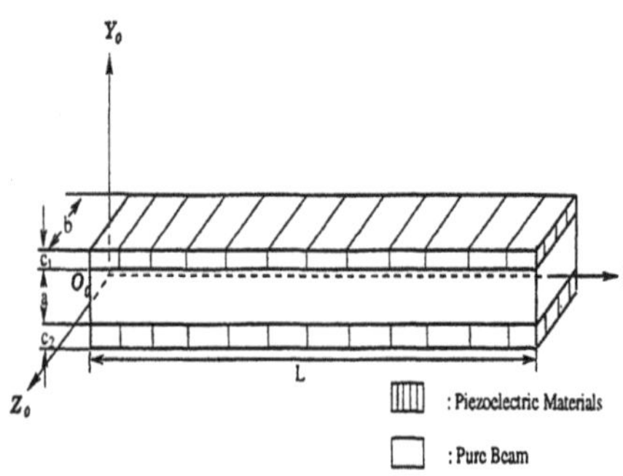

Fig. 1. Schematic of the smart materials beam

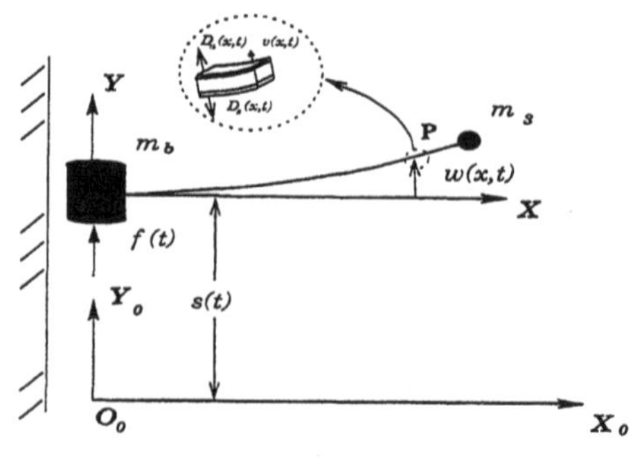

Fig. 2. Geometry of the system

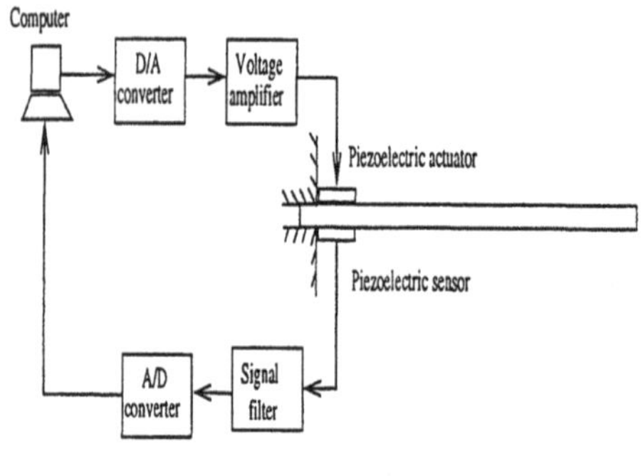

Fig. 3. Schematics of the experiment setup

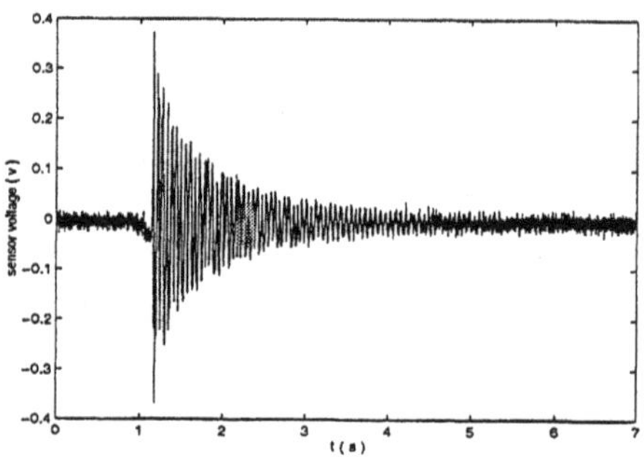

Fig. 4. Sensor voltages under free vibration

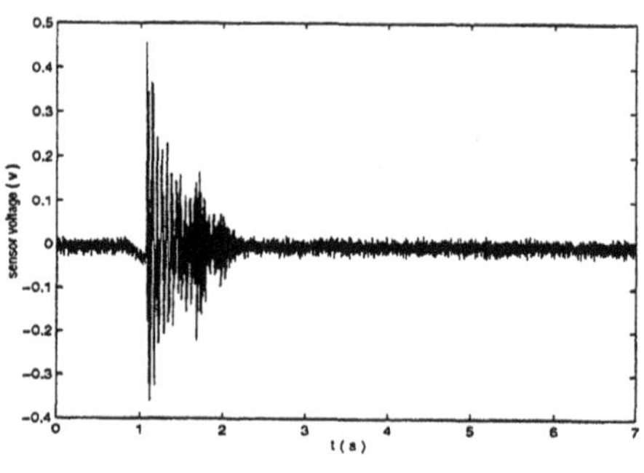

Fig. 5. Sensor voltages in the closed-loop

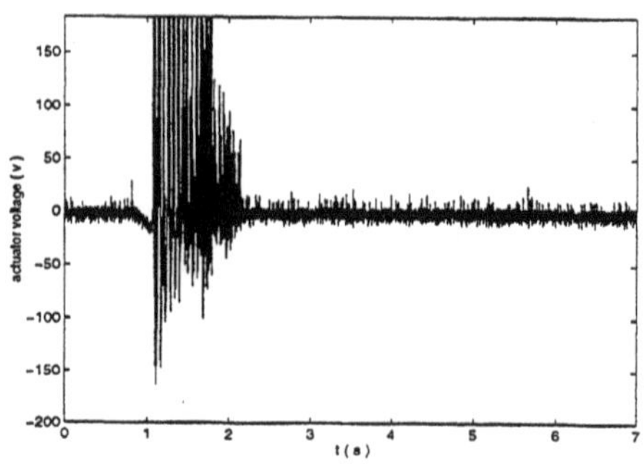

Fig. 6. Voltage control signals

Proportional and Integral Regulation of Irrigation canal Systems governed by the St Venant Equation

Cheng-Zhong XU and Gauthier SALLET

INRIA & ESA CNRS 7035, I.S.G.M.P.
Bât. A, Université de Metz, Ile du saulcy, F-57045 - Metz Cedex 01, France
Tel: (33) 3 87 54 72 80 — Fax: (33) 3 87 54 72 77. e-mail: xu@saulcy.loria.fr

Abstract : The purpose of our paper is to study control and regulation of irrigation canal systems using the St Venant equation. As the linearized models of these systems around equilibrium points are described by symmetric hyperbolic partial differential equations of two independent variables, we consider a class of symmetric hyperbolic systems with boundary observations. We present the exact observability and exponential stability results of Rauch and Taylor in the general case. These results are made explicit by providing estimates of necessary observation length for observability and estimates of exponential decay rate for stability in function of structure parameters. Then, we apply these results to show that for an one-reach irrigation canal system governed by the St Venant equation, the linearized model around an equilibrium point is exponentially stable for the open loop system. To guarantee a robust regulation and a robust stability of the closed loop system we propose proportional and integral output feedback controllers (PI-Controllers) which assure exponential stabilization of the linearized model and suppress (known or unknown) constant disturbances in the system. The work that we present here brings out a systematic synthesis method for designing boundary stabilizing PI-Controllers for irrigation systems based on linearized models of the St Venant equation.

Key words symmetric hyperbolic systems, exact observability, robust stabilization, exponential stability, irrigation canals, boundary PI-controllers.

1 Introduction

In the paper we consider an one-reach irrigation canal system governed by the St Venant equation

$$\left\{\begin{array}{l}\dfrac{\partial S(x,t)}{\partial t}+\dfrac{\partial Q(x,t)}{\partial x}=0,\\ \dfrac{\partial Q(x,t)}{\partial t}+\dfrac{\partial}{\partial x}\left[\dfrac{Q^2(x,t)}{S(x,t)}\right]+gS(x,t)\dfrac{\partial Z(x,t)}{\partial x}\\ +g\eta^2R^{-4/3}\left[\dfrac{Q^2(x,t)}{S(x,t)}\right]=0,\\ Q(0,t)=Q_{in}(t),\\ Z(L,t)=Z_{out}(t),\\ y(t)=\begin{pmatrix}Z(0,t)\\ Q(L,t)\end{pmatrix}.\end{array}\right. \tag{1}$$

The model is used by hydraulic engineers to describe dynamic behaviour of the unsteady flow in an open canal for shallow water [10]. In the equation (1), constant L is the length of the canal and $x \in [0, L]$. The other variables are specified in the following : $S(x,t)$ denotes the cross sectional area, $Z(x,t)$ the vertical position of water surface, $Q(x,t)$ the water flow, g the gravity acceleration constant, η the manning coefficient, P the wetted perimeter, R the hydraulic radius ($=S/P$), $Q_{in}(t)$ the upstream flow rate in the canal and $Z_{out}(t)$ the downstream level in the canal.

Our taking $Q_{in}(t)$ and $Z_{out}(t)$ as control variables and $y(t)$ as output observation leads to a nonlinear hyperbolic system with boundary control and boundary observation. The purpose of our paper is to study stabilization and regulation of the system using boundary PI-controllers. As a first approach we propose designing robust boundary PI-controllers for the system based on linearized models around equilibrium points. On the one hand, linearized models of the system (1) are governed by linear symmetric hyperbolic equations. On the other hand, a practical irrigation system is rather than a network constructed from a few models like (1) and appropriate joint boundary conditions. Motivated by these observations our objective is to develop a systematic method for designing PI-controllers in the general context of symmetric hyperbolic systems.

The symmetric hyperbolic system to be considered is as follows :

$$\left\{\begin{array}{l}\dfrac{\partial R(x,t)}{\partial t}=A(x)\dfrac{\partial R(x,t)}{\partial x}+B(x)R(x,t),\\ R^-(0,t)=D_0R^+(0,t)\\ R^+(1,t)=D_1R^-(1,t)\\ R(x,0)=R_0(x),\\ y(t)=\begin{pmatrix}R^-(1,t)\\ R^+(0,t)\end{pmatrix}\end{array}\right. \tag{2}$$

where $(x,t) \in [0,1] \times \mathbb{R}^+$, $R(x,t)$ is a $n \times 1$ vector function for $(x,t) \in [0,1] \times \mathbb{R}^+$, $B(x)$ is a real $n \times n$ matrix function, D_0 and D_1 are real constant matrices, $A(x)$ is a diagonal matrix function partitioned as

$$A(x)=\begin{pmatrix}A^-(x) & 0\\ 0 & A^+(x)\end{pmatrix}$$

with

$$A^{-}(x) = \mathrm{diag}(\lambda_1(x), \lambda_2(x), ..., \lambda_p(x))$$

and

$$A^{+}(x) = \mathrm{diag}(\lambda_{p+1}(x), \lambda_{p+2}(x), ..., \lambda_{p+q}(x))$$

$(p + q = n)$, and correspondingly

$$R(x,t) = \begin{pmatrix} R^{-}(x,t) \\ R^{+}(x,t) \end{pmatrix}.$$

In (2), $y(t) \in \mathbb{R}^n$ is our output observation. Hereafter, we denote by $B^*(x)$ the transposed matrix of $B(x)$, and $A_x(x)$ the derivative matrix of $A(x)$. Suppose the following conditions to be satisfied for the system (2).

H.1. (*regularity*) : $B(\cdot) \in C^0([0,1]; \mathbb{R}^{n\times n})$ and $A(\cdot) \in C^1([0,1]; \mathbb{R}^{n\times n})$.

H.2. (*strictly hyperbolic*) : $\lambda_i(x) < 0$, $i = 1, 2, ..., p$ for any $x \in [0,1]$ and $\lambda_{p+i} > 0$, $i = 1, 2, ..., q$ for any $x \in [0,1]$.

H.3. (*strict dissipativity*) : There is some $r > 0$ such that for any $w = \begin{pmatrix} w^- \\ w^+ \end{pmatrix} \in \mathbb{R}^p \times \mathbb{R}^q$,

$$w^* \left[B(x) + B^*(x) - A_x(x)\right] w \leq 0, \quad (3)$$

$$(w^-)^*(A^-(1) + D_1^* A^+(1) D_1) w^- \leq -r\|w^-\|^2_{\mathbb{R}^p} \quad (4)$$

and

$$(w^+)^*(A^+(0) + D_0^* A^-(0) D_0) w^+ \geq 0. \quad (5)$$

Using the method of characteristics developed in [5] and [6] we show that under H.1–H.3, the observation system (2) is exactly observable. Actually, we prove, as the reader will see, that the system (2) has exact observability with fewer outputs than those given in (2). As in [5] and [6] we prove exponential stability of the system from the observability result and standard semigroup property. Exponential stability and exact observability have been first established for the general case (2) in [5] and [6]. These results are made explicit in our paper by providing estimates of exponential decay rate in function of structure parameters, which is essential for our subsequent design of PI-controllers. Since the linearized model of the irrigation system (1) around an equilibrium point under consideration can be put into the form (2) but with the boundary control, using the general result of [5] and [6] we show that for the boundary control equal to zero, the linearized model is exponentially stable. Then, we give robust boundary PI-controllers which assure exponential stabilization of the linearized model and suppress asymptotically constant disturbances. The robustness and practical interest of PI-controllers have been shown in [1] and [12]. Our design of stabilizing PI-controllers is carried out by adapting to our case the theoretical results presented in [1], [12] and [13].

Multivariable robust PI-controller theory has been introduced by Davison in [1] for finite dimensional linear systems. The generalization of the theory to infinite dimensional systems has been established in [12] for analytic semigroups. This generalization has been proved to be true in [13] for strongly continuous semigroups. Essentially, in [12] [13] either the control operator or the observation operator is assumed bounded in order to have the PI-controlled closed loop system to be well posed. However, in our linearized irrigation system both of them are unbounded. Moreover, the semigroup underlying the linearized model is not analytic. *A priori*, it is not clear that the closed loop system obtained from a proportional and integral output feedback is well posed. Theoretical answers in this direction can be found for the class of regular systems in the recent work [2]. On the other hand, the method of spectrum assignment developed in [12] based on the perturbation theory cannot be directly applied to our case. Here, for the irrigation system we get the well-posedness of the PI-controlled system using directly the structure of symmetric hyperbolic systems. We propose also necessary transformations which make possible to apply the perturbation theory for doing spectrum assignment. However, we show that the tuning of I-controllers can be still performed by measuring the outputs of the process corresponding to step signals as described in [12]. The necessary and sufficient condition for existence of I-controllers is shown to be true for the linearized St Venant model.

The integral control of the irrigation system has been studied in [3] based on an abstract semigroup approach. However, no exponential stability has been proved in [3] for the system. The proportional control which is also important (see [12]) has not been addressed by [3]. The main contributions of our paper are

- providing estimates of exponential decay rate for a class of symmetric hyperbolic systems;
- applying the theory of symmetric hyperbolic systems to study stability of linearized St Venant models;
- proposing a design method for stabilizing boundary PI-controllers for linearized irrigation canal systems.

Our paper is organized as follows. In section 2 we present exact observability and exponential stability results for the general case (2). There, explicit estimates are provided of necessary observation length for observability and of exponential decay rate for stability in function of structure parameters. In section 3, based on a linearized St Venant model we present a method of synthesis for designing stabilizing boundary PI-controllers. Section 4 contains our discussions on future research directions.

2 Observability and stability of symmetric hyperbolic systems

We denote by H and Y respectively the state space and the output space for the system (2), where $H = [L^2(0,1)]^n$ and $Y = \mathbb{R}^n$ with

$$< f, g >_H = \sum_{i=1}^{n} \int_0^1 f_i(x) g_i(x) dx$$

and

$$< x, y >_Y = \sum_{i=1}^{n} x_i y_i .$$

Define the unbounded operator $\mathcal{A}$ associated with the system under consideration as follows :

$$\mathcal{D}(\mathcal{A}) = \left\{ f = \begin{pmatrix} f^- \\ f^+ \end{pmatrix} \in H; \begin{pmatrix} f_x^- \\ f_x^+ \end{pmatrix} \in H, \right.$$

$$\left. \begin{array}{l} f^-(0) = D_0 f^+(0) \\ f^+(1) = D_1 f^-(1) \end{array} \right\} \tag{6}$$

and for any $f \in \mathcal{D}(\mathcal{A})$,

$$\mathcal{A} f(x) = A(x) f_x(x) + B(x) f(x). \tag{7}$$

The output operator C is a linear mapping defined by $C : \mathcal{D}(\mathcal{A}) \to Y = \mathbb{R}^n$ such that

$$Cf = (f^-(1), \sqrt{A^+(0) + D_0^* A^-(0) D_0} f^+(0)), \tag{8}$$

where (and hereafter) $\sqrt{Q}$ is the uniquely defined positive square root operator for a positive operator Q. Now, setting

$$\phi(t) = \begin{bmatrix} R^-(\cdot, t) \\ R^+(\cdot, t) \end{bmatrix},$$

we rewrite the system (2) as evolution equation on H,

$$\left\{ \begin{array}{l} \dot{\phi}(t) = \mathcal{A}\phi(t) \\ \phi(0) = R_0 \\ y(t) = C\phi(t). \end{array} \right. \tag{9}$$

The following definition can be found in (c.f. [9]).

Definition 1 *Let $\mathcal{A}$ be the generator of a C_0-semigroup (noted as $e^{t\mathcal{A}}$) on H and C be a linear operator from $\mathcal{D}(\mathcal{A})$ to Y. The operator C is said $\mathcal{A}$-admissible if for some $T > 0$, the observation mapping $Ce^{\cdot\mathcal{A}} : \mathcal{D}(\mathcal{A}) \to L^2([0,T], Y)$ is continuous. The system (9) is called exactly observable if for some $T > 0$, there is a $K_T > 0$ such that*

$$K_T \|f\|_H^2 \le \int_0^T \|Ce^{t\mathcal{A}} f\|_Y^2 dt \le K_T^{-1} \|f\|_H^2 \tag{10}$$

holds for any $f \in \mathcal{D}(\mathcal{A})$. The semigroup $e^{t\mathcal{A}}$ is said exponentially stable if

$$\|e^{t\mathcal{A}}\|_{\mathcal{L}(H)} \le Me^{-\omega t} \tag{11}$$

holds for some $M > 0$ and $\omega > 0$.

Theorem 1 *1. The operator $\mathcal{A}$ defined in (6) and (7) generates a C_0-semigroup of contraction on H.*

2. The output operator C in (8) is $\mathcal{A}$-admissible.

3. For any initial condition $R_0 \in \mathcal{D}(\mathcal{A})$, the symmetric hyperbolic system (2) has a unique solution

$$R(\cdot,\cdot) \in C([0,1] \times [0,T]) \bigcap H^1([0,1] \times [0,T])$$

for any $T > 0$.

In the system (2), the characteristic curves are $\dot{x}(t) = -\lambda_i(x(t))$, $i = 1, 2, ..., n$. For the positive direction we define

$$\rho_1(x) = \max_{1 \le m \le p} (\lambda_m(x)) < 0 \tag{12}$$

and for the negative direction,

$$\rho_2(x) = \min_{p+1 \le m \le n} (\lambda_m(x)) > 0. \tag{13}$$

If each $\lambda_i(x)$, $i = 1, 2, ..., n$ is Lipschitz continuous, the function $\rho_i(x)$, $i = 1, 2$ are also Lipschitz continuous. First, let us consider the differential equation

$$\begin{array}{l} \dot{x}_2(t) = -\rho_2(x_2(t)) \\ x_2(0) = 1. \end{array} \tag{14}$$

The vector field $\rho_2(x)$ being Lipschitz continuous and positive, the unique solution $x_2(t)$ of (14) is a C^1-diffeomorphisme. We denote by $\vartheta_2(x)$ the inverse function such that $x_2(\vartheta_2(x)) = x$ for $x \in [0,1]$. The same property is true for the positive direction which is described by

$$\begin{array}{l} \dot{x}_1(t) = -\rho_1(x_1(t)) \\ x_1(\vartheta_2(0)) = 0. \end{array} \tag{15}$$

The minimum time required for any point in $[0,1]$ to be back along the characteristics of the system is determined by

$$T_O = \int_0^1 \left[\frac{1}{\rho_2(x)} - \frac{1}{\rho_1(x)} \right] dx. \tag{16}$$

Let $\sigma_{max}(\cdot)$ denote the the largest eigenvalue of a positive matrix $(\cdot)$. Using the method of characteristics developed in [6] we obtain the following estimates in function of the structure parameters (see also [16] [4]).

Theorem 2 *1.The system (2) is exactly observable on any time intervale of length $\geq T_O$: for any $R_0 \in H$,*

$$\left(\frac{1}{K_O}\right) \int_0^{T_O} \{r||R^-(1,t)||^2_{\mathbb{R}^p} + ||\sqrt{A^+(0) + D_0^* A^-(0) D_0} R^+(0,t)||^2_{\mathbb{R}^q}\} \, dt \geq ||R_0||^2_H, \quad (17)$$

where T_O and r were defined respectively in (16) and (4), and

$$K_O = r \left[\max_{x \in [0,1], 1 \leq k \leq p} (-\lambda_k(x)) \right]^{-1} \times$$

$$\times \sigma^{-1}_{max}(I + D_1^* D_1) exp(-2k_1 - T_O k_2), \quad (18)$$

$$k_1 = \max_{x \in [0,1]} \sqrt{\sigma_{max}(B^*(x)(A^{-1}(x))^2 B(x))} \quad (19)$$

and

$$k_2 = \max_{x \in [0,1]} \sigma_{max}(A_x(x) - B^*(x) - B(x)). \quad (20)$$

2. The semigroup e^{tA} associated with the system (2) is exponentially stable :

$$||e^{tA}||_{\mathcal{L}(H)} \leq \sqrt{\frac{1}{1 - K_O}} exp\left(-t \frac{1}{2T_O} \ln(\frac{1}{1 - K_O})\right), \quad (21)$$

where T_O and K_O were defined respectively in (16) and (18).

Remark 1 All the above results obtained for the space domain $[0, 1]$ remain still true for any interval $[0, L]$. It is sufficient to normalize the space variable x to get useful estimates.

3 Stabilizing boundary PI-controllers for the linearized St Venant equation

We assume that in (1) there is an algebraic relation between $S(x,t)$ and $Z(x,t)$ once the cross-section geometry is fixed (*e.g.*, trapeze section) :

$$S(x,t) = \mathcal{S}(Z(x,t)). \quad (22)$$

Choose $Z(x,t)$ and $Q(x,t)$ as state variables, and $Q_{in}(t)$ and $Z_{out}(t)$ as control variables. The upstream level and the downstream flow rate are the output observations :

$$y(t) = (Z(0,t), Q(l,t)). \quad (23)$$

The one-reach canal system described by (1) and (22) is a nonlinear multivariable system with boundary control and boundary observation. Our synthesis of stabilizing boundary PI-controllers will be carried out based on the linearized St Venant model around an equilibrium point which has been used in [3].

Proposition 1 *Consider the equilibrium point for which η and R are positive constants in (1), $Q_{in}(t) = Q^0_{in}$, $Z_{out}(t) = Z^0_{out}$,*

$$\frac{dZ^0(x)}{dx} = -\alpha \text{ and } \frac{dS^0(x)}{dx} = 0. \quad (24)$$

and v is a positive constant such that

$$\left.\frac{d\mathcal{S}(z)}{dz}\right|_{z=Z^0(x)} = v. \quad (25)$$

Then, the governing equations of the linearized system of (1) around the equilibrium point are :

$$\left\{ \begin{array}{l} \frac{\partial}{\partial t}\begin{pmatrix} z(x,t) \\ q(x,t) \end{pmatrix} = \begin{pmatrix} 0 & -a_1 \\ -a_2 & -a_3 \end{pmatrix} \frac{\partial}{\partial x}\begin{pmatrix} z(x,t) \\ q(x,t) \end{pmatrix} \\ + \begin{pmatrix} 0 & 0 \\ a_4 & -a_5 \end{pmatrix} \begin{pmatrix} z(x,t) \\ q(x,t) \end{pmatrix} \\ q(0,t) = u_1(t) \\ z(L,t) = u_2(t) \\ y(t) = (z(0,t), q(L,t)), \end{array} \right. \quad (26)$$

where

$$z(x,t) = Z(x,t) - Z^0(x), \quad q(x,t) = Q(x,t) - Q^0(x) \quad (27)$$

$$u_1(t) = Q_{in}(t) - Q^0_{in}, \quad u_2(t) = Z_{out}(t) - Z^0_{out}$$

$$a_1 = \frac{1}{v}, \quad a_2 = gS^0 - v\left(\frac{Q^0}{S^0}\right)^2, \quad a_3 = 2\left(\frac{Q^0}{S^0}\right) \quad (28)$$

$$a_4 = gv\left[\alpha + \left(\frac{\eta Q^0}{S^0 R^{2/3}}\right)^2\right]$$

$$a_5 = 2g\eta^2 Q^0 (S^0)^{-2} R^{-4/3}. \quad (29)$$

Remark 2 It is obvious that $a_i > 0$ for any $i \neq 2$ in (26). Moreover, from the hydraulic point of view, $a_2 > 0$ in (28) is a necessary condition for the model (1) to be valid.

The general form of boundary PI-controllers for (26) is

$$\left\{ \begin{array}{l} \begin{pmatrix} u_1(t) \\ u_2(t) \end{pmatrix} = K_p(y(t) - y^r) + k_I K_I \xi \\ \dot{\xi} = y(t) - y^r, \end{array} \right. \quad (30)$$

where $K_P, K_I \in \mathbb{R}^{2\times2}$ and $k_I \in \mathbb{R}$, and y^r is the constant reference signals to track. The role of the proportional part K_P is to speed up the exponential decay rate for stability if necessary. Introducing the integral part is to make the output track the constant reference signals and reject constant disturbances. However, even for an exponentially stable system, if the integral gains K_I and k_I are not properly choosed, the closed loop system may have eigenvalues in the right half complex plane, which results in unstability of the controlled system. So, it is not trivial to assign the spectrum in the open left half plane by means of an appropriate choice of K_I. We propose a design method for proportional matrix K_P, integral matrix K_I and integral gain k_I such that a) the system (26) controlled by the dynamic output feedback (30) is well-posed and the associated semigroup is exponentially stable; b) the output regulation is guaranteed :

$$\lim_{t\to+\infty} \|y(t) - y^r\|_{\mathbb{R}^2} = 0$$

independently of any constant known or unknown perturbation.

As proved in [12] condition a) implies b). Therefore, it is sufficient to choose K_P, K_I and k_I such that the closed loop system results in an exponentially stable semigroup. Denote by λ_1 the negative eigenvalue of the first matrix in (26) and by λ_2 the positive one :

$$\begin{cases} \lambda_1 = -\frac{1}{2}\left(a_3 + \sqrt{a_3^2 + 4a_1a_2}\right) \\ \lambda_2 = \frac{1}{2}\left(-a_3 + \sqrt{a_3^2 + 4a_1a_2}\right). \end{cases} \tag{31}$$

Our main result is the following.

Theorem 3 *1. The open loop system (26) with $u_1 = 0$ and $u_2 = 0$ is exponentially stable. 2. There exists a $k_I^* > 0$ such that for any $k_I \in]0, k_I^*]$, the PI-controller (30) stabilizes exponentially the linearized St-Venant system (26) provided that in (30),*

$$K_P = \begin{pmatrix} -\frac{\lambda_2}{a_1} & 0 \\ 0 & -\frac{a_1}{\lambda_1} \end{pmatrix} \tag{32}$$

and

$$K_I(1,1) = -\frac{\lambda_2}{a_1},\ K_I(1,2) = -1$$

$$K_I(2,1) = -\exp\left[\frac{K^2 l(\lambda_1+\lambda_2)}{\lambda_1\lambda_2}\right]$$

$$K_I(2,2) = -\frac{a_1\left(\lambda_2 - \lambda_1 + 2\lambda_1 \exp\left[\frac{K^2 l(\lambda_1+\lambda_2)}{\lambda_1\lambda_2}\right]\right)}{\lambda_1(\lambda_1+\lambda_2)}, \tag{33}$$

where

$$K^2 = \frac{g\eta^2 Q^0}{S^0 R^{4/3}} \tag{34}$$

Several remarks are in order :

1. On diagonalizing the first matrix in (26) we find that the linearized system satisfies all the hypothesis H.1–H.3. For $u_1 = u_2 = 0$, exponential stability of (26) follows from 2. of Theorem 2. Our choice K_P in (32) has been made such that the corresponding matrices D_0 and D_1 in (2) are equal to zero. The PI-controlled system (26) and (30) is governed by the semigroup whose generator is defined as follows :

$$\mathcal{D}(A_c) = \left\{ \begin{pmatrix} f_1 \\ f_2 \\ f_3 \end{pmatrix} \in (H^1[0,L])^2 \times \mathbb{R}^2; \right.$$

$$\left. \begin{pmatrix} f_1(0) \\ f_2(L) \end{pmatrix} = \begin{pmatrix} \frac{a_1}{\lambda_2-\lambda_1} & 0 \\ 0 & \frac{\lambda_1}{\lambda_1-\lambda_2} \end{pmatrix} k_I K_I f_3 \right\} \tag{35}$$

and for any $f \in \mathcal{D}(A_c)$, $g(x) = A_c f(x)$, where

$$g_1(x) = \begin{pmatrix} \lambda_1 & 0 \\ 0 & \lambda_2 \end{pmatrix} \frac{\partial}{\partial x} \begin{pmatrix} f_1(x) \\ f_2(x) \end{pmatrix}$$

$$+K^2 \begin{pmatrix} -1 & 1 \\ 1 & -1 \end{pmatrix} \begin{pmatrix} f_1(x) \\ f_2(x) \end{pmatrix},$$

$$g_2(x) = \begin{pmatrix} \frac{a_1}{\lambda_2 - \lambda_1} & 0 \\ 0 & -\frac{\lambda_1\lambda_2}{a_1(\lambda_1-\lambda_2)} \end{pmatrix} k_I K_I f_3$$

$$+ \begin{pmatrix} 1 & 0 \\ 0 & -\frac{\lambda_1}{a_1} \end{pmatrix} \begin{pmatrix} f_2(0) \\ f_1(L) \end{pmatrix}. \tag{36}$$

The question is how to choose K_I in order to assign the spectrum. The difficulty is that the domain $\mathcal{D}(A_c)$ depends on K_I. We propose some bijective transformation which allows to remove the difficulty. Then, we can apply the theory developed in [12] to choose K_I. In our case, the necessary and sufficient condition for existence of I-controllers is satisfied :

$$0 \neq \det(\tilde{C}_2 - \tilde{C}_1\hat{A}^{-1}\tilde{B}_1) =$$

$$= \frac{\lambda_1^2 - \lambda_2^2}{a_1\left(\lambda_1 \exp\left[\frac{K^2 L(\lambda_1+\lambda_2)}{\lambda_1\lambda_2}\right] + \lambda_2\right)} < 0,$$

where

$$\mathcal{D}(\tilde{A}) = \{(f_1, f_2) \in (H^1[0,L])^2;\ f_1(0) = f_2(L) = 0\} \tag{37}$$

and for any $f \in \mathcal{D}(\tilde{A})$,

$$\tilde{A}f(x) = \begin{pmatrix} \lambda_1 & 0 \\ 0 & \lambda_2 \end{pmatrix} f_x(x) + K^2 \begin{pmatrix} -1 & 1 \\ 1 & -1 \end{pmatrix} f(x), \tag{38}$$

$\tilde{B}_1 : \mathbb{R}^2 \to (L^2[0,L])^2$ such that

$$\tilde{B}_1 f = \begin{pmatrix} \frac{K^2(x-L) - \lambda_1}{L} & \frac{K^2 x}{L} \\ \frac{K^2(L-x)}{L} & \frac{\lambda_2 - K^2 x}{L} \end{pmatrix} f, \tag{39}$$

$\tilde{C}_1 : \mathcal{D}(\tilde{A}) \to \mathbb{R}^2$ such that

$$\tilde{C}_1 f = \begin{pmatrix} 1 & 0 \\ 0 & -\frac{\lambda_1}{a_1} \end{pmatrix} \begin{pmatrix} f_2(0) \\ f_1(L) \end{pmatrix} \tag{40}$$

and $\tilde{C}_2 : \mathbb{R}^2 \to \mathbb{R}^2$ such that

$$\tilde{C}_2 f = \begin{pmatrix} 1 & 0 \\ 0 & -\frac{\lambda_2}{a_1} \end{pmatrix} f. \tag{41}$$

Using Mathematica we find

$$K_I = -\tilde{B}_3^{-1}(\tilde{C}_2 - \tilde{C}_1\tilde{A}^{-1}\tilde{B}_1)^{-1} \quad (42)$$

as indicated in (33), where $\tilde{B}_3 : \mathbb{R}^2 \to \mathbb{R}^2$ such that

$$\tilde{B}_3 f = \begin{pmatrix} \frac{a_1}{\lambda_2 - \lambda_1} & 0 \\ 0 & \frac{\lambda_1}{\lambda_1 - \lambda_2} \end{pmatrix} f. \quad (43)$$

This choice of K_I assignes the spectrum to the open left half complex plane for the closed loop system. Finally, Huang's theorem is applied for proving exponential stability of the designed PI-controllers ([11], [13]).

2. It is very important to note that the matrix $\tilde{C}_2 - \tilde{C}_1\tilde{A}^{-1}\tilde{B}_1$ is just the step response of the system obtained following the tuning procedure proposed in [12]. As a consequence, it is not necessary to know exactly the system operators.

4 Conclusions

In this paper, we have presented exact observability and stability of a class of linear symmetric hyperbolic systems. Exact observability has allowed us to provide an estimate of exponential decay rate for stability of the systems. That result has been exploited fully for proposing a systematic design method for boundary PI-controllers for the linearized St Venant equation. The St Venant equation is used to describe dynamic behaviour of an one-reach shallow water canal system. Both the controls and the observations are boundary with the underlying semigroup being non-analytic. Necessary transformations have been proposed so that the robust PI-controllers design theory in [1], [12] and [13] is applicable to our case. Our future research directions are to test the capacity of the designed PI-controllers for controlling the nonlinear St Venant equation and to extend our approach for a network of multi-reach canal systems [10].

References

[1] E.J. Davison, Multivariable tuning regulators : the feedforward and robust control of a general servomechanism problem, *IEEE Transaction Autom. Contr.*, Vol. AC-21, No.1, pp.35-47, 1976.

[2] G. Weiss, Regular linear systems with feedback, *Math. Control Signals Systems* (1994) 7: 23-57.

[3] H. Bounit, H. Hammouri and J. sau, Regulation of an irrigation canal system through the semigroup approach, pp.261-267, *RIC'97 Proceedings* [10].

[4] C.Z. Xu, Exact observability and exponential stability of infinite dimensional bilinear systems, *Math. Control Signals Systems*, 9, pp.73-93 (1996).

[5] J. Rauch and M. Taylor, Exponential decay of solutions to hyperbolic equations in bounded domain, *Indiana University Mathematics Journal*, vol.24, No.1, pp.79-86 (1974).

[6] D.L. Russell, Controllability and stabilizability theory for linear partial differential equations : Recent progress and open questions, *Siam Review*, Vol.20, No.4, pp.639-739 (1978).

[7] C.Z. Xu, J.P. Gauthier and I. Kupka, Exponential stability of the heat exchanger equation, *The Proceedings of the European Control Conference*, Groningen, The Netherlands, pp.303-307 (1993).

[8] P.D. Lax and R.S. Phillips, Local boundary conditions for dissipative symmetric linear differential operators, *Communications on Pure and Applied Mathematics*, vol.XIII, pp.427-455, (1960).

[9] D.L. Russell and G. Weiss, A general necessary condition for exact observability, *SIAM J. Control and Optimization*, vol.32, no.1, pp.1-23 (1994).

[10] RIC'97 Proceeding, *Regulation of Irrigation Canals : State of the Art of Research and Applications*, International Workshop, April 22-24 1997, Marrakech, Morocco.

[11] F.L. Huang, Characteristic conditions for exponential stability of linear dynamical systems in Hilbert spaces, *Ann. of Diff. Equations*, 1(1),1985, pp.43-56.

[12] S.A. Pohjolainen, Robust multivariable PI-controllers for infinite dimensional systems, *IEEE Transaction Automatic Control*, vol.27, no.1, pp.17-30 (1985).

[13] C.Z. Xu and H. Jerbi, A robust PI-controller for infinite dimensional systems, *Int. J. Control*, vol.61, no.1, pp.33-45 (1995).

[14] T. Kato, *Perturbation theory for linear operators*, Springer-Verlag, 1976.

[15] I. Lasiecka and R. Triggiani, Finite rank, relative bounded perturbation of semigroups—Part I : well-posedness and boundary feedback hyperbolic dynamics, *Annali Scuola Normale Superior-Pisa*, Classe di Scienze, Serie IV, XII (4), pp.641-668 (1985).

[16] C.Z. Xu, Exponential stability of a class of infinite dimensional time-varying linear systems, *Proceedings of the International Conference on Control and Information*, 5-9 June 1995. Hong Kong.

14th Triennial World Congress, Beijing, P.R. China

ADAPTIVE CONTROL OF COORDINATED OPERATION OF TWO ROBOTIC MANIPULATORS

S. S. Ge* L. Huang T. H. Lee***

* *Dept of Electrical Engineering, National University of Singapore, Singapore 119260, E-mail: elegesz@nus.edu.sg*
** *Electronics Communication Eng Dept, Singapore Polytechnic, Singapore 139651*

Abstract: In this paper, the adaptive control of coordinated operation of two robotic manipulators are studied by considering the relative motion between the manipulated object and the end effector of a manipulator. The controller guarantees the convergence of the position of the constrained object and the manipulators, and the constraint force to their desired values. Simulation results are included to verify the effectiveness of the approach . *Copyright © 1999 IFAC*

Keywords: Robotics, Force Control, Adaptive Control, Constraints, Dynamics

1. INTRODUCTION

The dynamics of constrained robotic system is highly nonlinear and coupled. Through a nonlinear transformation, it can be transformed to a reduced differential equation describing an unconstrained robot motion, and an algebraic equation describing the constraint force (McClamroch and Wang, 1988). The concept of pseudo-velocity was also introduced to reduce the order of the system (Kankaanranta and Kovio, 1988).The techniques of both the nonlinear transformation and the order reduction were further utilized in the control of constrained multiple manipulators (Ramadorai *et al.*, 1992) and (Yao *et al.*, 1994). Based on these models, many attractive control strategies were developed, such as nonlinear feedback control (McClamroch and Wang, 1988), variable structure control (Yao *et al.*, 1994), adaptive control (Yuan, 1996) and (Slotine and Li, 1989) and neural network control (Ge *et al.*, 1998), to name a few.

Normally the constrained object is assumed to be motionless in the case of constrained single robot. Though the motion of the object is considered in the modeling of constrained multiple manipulators, it is assumed to be held tightly by all the end effectors of manipulators. These assumptions are not applicable to some applications in which both the movement of the object and its relative motion with respect to the end effector of a manipulator are required (Ge *et al.*, 1997). The common feature of these kinds of applications is that one robotic arm (manipulator 2) holds a work piece tightly, and another robotic arm (manipulator 1) doing the assembly or machining on the work piece.

The modeling and control of the system described above will be discussed in this paper. In Section 2, the dynamic model of the system is derived. In Section 3, an adaptive control strategy is proposed to control the robots so that the positions of the constraint object and the end-effectors, and the constraint forces to converge to their desired values. In Section 4, the simulation studies are used to show the effectiveness of the controller.

2. DYNAMIC MODEL

The system under study is schematically shown in Figure 1. The closed kinematic relationships in the system are given as follows

$$x_c = x_o + R_o x_{co} \quad x_h = x_o + R_o x_{ho}$$
$$R_c = R_o R_{co} \quad R_h = R_o R_{ho} = R_o$$

where x_c and x_h are the position vector of the contact points c and h respectively; R_c and R_h are the rotation matrices of the contact frames; and R_{co} and R_{ho} are their descriptions in the object coordinate.

Differentiating the above equations with respect to time t yields

$$v_c = Av_o + R_A v_{co}, \; v_h = Bv_o \qquad (1)$$

where $v_c = [\dot{x}_c^T \; \omega_c^T]^T$, $v_h = [\dot{x}_h^T \; \omega_h^T]^T$, $v_{co} = [\dot{x}_{co}^T \; \omega_{co}^T]^T$, $v_{ho} = [\dot{x}_{ho}^T \; \omega_{ho}^T]^T$, $v_o = [\dot{x}_o^T \; \omega_o^T]^T$ and

$$R_A = \begin{bmatrix} R_o & 0 \\ 0 & R_o \end{bmatrix} \quad A = \begin{bmatrix} I^{3\times3} & -S(R_o x_{co}) \\ 0 & I^{3\times3} \end{bmatrix}$$
$$B = \begin{bmatrix} I^{3\times3} & -S(R_o x_{ho}) \\ 0 & I^{3\times3} \end{bmatrix}$$

with $S(R_o x_{ho})$ being a skew-symmetric matrix and $u, \; v \in R^3$.

Assume that the contact between the end-effector of manipulator 1 and the constraint object is a point contact at c, and it is required to follow the following trajectory in the object coordinate

$$\Phi(r_{co}) = 0$$

The contact force f_c and the resulting equivalent force f_o on the object are given by

$$f_c = n_c\lambda, \; f_o = -A^T n_c \lambda$$

where λ is a Lagrange multiplier related to the magnitude of the force, and n_c is the unit normal vector at c.

Combining the dynamics of the manipulators and the object give rise to the dynamics of the whole system

$$M(q)\ddot{q} + C(q,\dot{q})\dot{q} + G(q) = \tau + J^T(q) n_c \lambda \quad (2)$$

where

$$M(q) = \begin{bmatrix} M_1(q_1) & 0 \\ 0 & M_2(q_2) \end{bmatrix}$$
$$C(q, \; \dot{q}) = \begin{bmatrix} C_1(q_1 \; \dot{q}_1) & 0 \\ 0 & C_2(q_2 \; \dot{q}_2) \end{bmatrix}$$
$$G(q) = \begin{bmatrix} G_1(q_1) \\ G_2(q_2) \end{bmatrix}, \quad J(q) = [J_1(q_1) \quad -AJ_2(q_2)]$$
$$q = [q_1^T \; q_2^T]^T \; \tau = [\tau_1^T \; \tau_2^T]^T$$

$M_i(q_i)$ is the inertia matrices, $C_i(q_i, \dot{q}_i)$ is the coriolis and centrifugal force matrices, $G_i(q_i)$ is the gravitational forces, τ_i is the joint torques and $J_i(q_i)$ is the Jacobian matrices ($i = 1, 2$).

Due to the holonomic constraint (1), a set of *independent* n coordinates $q^1 = [q_1^1 \ldots q_n^1]^T$ can be chosen such that

$$q = q(q^1) \qquad (3)$$

Differentiating equation (3) with respect to time t twice leads to

$$\dot{q} = L(q^1)\dot{q}^1 \qquad (4)$$
$$\ddot{q} = L(q^1)\ddot{q}^1 + \dot{L}(q^1)\dot{q}^1 \qquad (5)$$

where $L(q^1) = \partial q / \partial q^1$.

Substituting equations (4) and (5) into equation (2), the reduced order dynamic model of the system is obtained as follows

$$M^1(q^1)\ddot{q}^1 + C^1(q^1, \dot{q}^1)\dot{q}^1 + G^1(q^1)$$
$$= \tau + J^{1T}(q^1) n_c \lambda \qquad (6)$$

where

$$M^1(q^1) = M(q^1)L(q^1)$$
$$C^1(q^1, \, dotq^1) = M(q^1)\dot{L}(q^1) + C(q^1, \; \dot{q}^1)L(q^1)$$
$$G^1(q^1) = G(q(q^1)), \quad J^1(q^1) = J(q(q^1))$$

It can be easily proved that the dynamic model (6) has the following main properties

Property 1 The terms $L(q^1)$, $J^1(q^1)$ and n_c satisfy the relationship: $L^T(q^1)J^{1T}(q^1)n_c = 0$

Property 2 $M_L(q^1) \triangleq L^T(q^1)M^1(q^1)$ is symmetric positive definite ($s.p.d$), and bounded upper and below.

Property 3 Define $C_L(q^1, \dot{q}^1) = L^T(q^1)C^1(q^1, \dot{q}^1)$, then $N_L = \dot{M}_L(q^1) - 2C_L(q^1, \; \dot{q}^1)$ is skew-symmetric if $C_i(q_i, \; \dot{q}_i)(i = 1, 2)$ is in the Christoffel form, i.e., $x^T N_L x = 0, \; \forall x \in R^n$.

Property 4 The dynamics described by equation (6) are linear in the parameters, i.e.,

$$M^1(q^1)\ddot{q}_r^1 + C^1(q^1, \; \dot{q}^1)\dot{q}_r^1 + G^1(q^1)$$
$$= \Psi(q^1, \dot{q}^1, \dot{q}_r^1, \ddot{q}_r^1)P \qquad (7)$$

where $P \in R^l$ is the vector of the parameters of interest; $\Psi(*) \in R^{n\times l}$ is the regressor matrix of a known function; and $\dot{q}_r^1, \; \ddot{q}_r^1 \in R^n$.

3. CONTROLLER

Let $r_d(t)$ be the desired trajectory of the object, $r_{cod}(t)$ and $\lambda_d(t)$ be the desired trajectory on the object and the desired constraint force. One of the control objectives is to drive the manipulators so that $r(t)$ and $r_{co}(t)$ track $r_d(t)$ and $r_{cod}(t)$ respectively. Since $q^1(t)$ completely determines $r(t)$ and $r_{co}(t)$, it is equivalent to make $q^1(t)$ track the desired joint position $q_d^1(t)$. The second objective is to make $\lambda(t)$ track $\lambda_d(t)$.

Define

$$e^1 = q_d^1 - q^1, \quad e_\lambda = \lambda_d - \lambda \tag{8}$$

$$r^1 = \dot{e}^1 + K_e e^1, \ \dot{q}_r^1 = \dot{q}_d^1 + K_e e^1 \tag{9}$$

where $K_e > 0$.

It is obvious that

$$r^1 = \dot{q}_r^1 - \dot{q}^1 \tag{10}$$

In practice, P is usually unknown. Let $\hat{P}$ be its estimate and and $\tilde{P} = P - \hat{P}$.

Consider the following controller

$$\tau = \Psi(q^1, \dot{q}^1, \dot{q}_r^1, \ddot{q}_r^1)\hat{P} - J^{1T}(q^1)n_c\lambda_d - J^{1T}(q^1)k_\lambda \int_0^t e_\lambda(\tau)d\tau + L^{-T}(q^1)Kr^1 \tag{11}$$

where $k_\lambda > 0$ and $K > 0$.

Applying the control law (11) to the dynamic system (6), the following closed-loop system dynamics is obtained,

$$\begin{aligned} &M^1(q^1)\ddot{q}^1 + C^1(q^1, \dot{q}^1)\dot{q}^1 + G^1(q^1) \\ &= \Psi(q^1, \dot{q}^1, \dot{q}_r^1, \ddot{q}_r^1)\hat{P} - J^{1T}(q^1)n_c e_\lambda \\ &-J^{1T}(q^1)n_c k_\lambda \int_0^t e_\lambda(\tau)d\tau + L^{-T}(q^1)Kr^1 \end{aligned} \tag{12}$$

Pre-multiplying both sides of the above equation by $L^T(q^1)$ and using **Property 1**, it follows that

$$\begin{aligned} &M_L(q^1)\ddot{q}^1 + C_L(q^1, \dot{q}^1)\dot{q}^1 + G_L(q^1) \\ &= L^T(q^1)\Psi(q^1, \dot{q}^1, \dot{q}_r^1, \ddot{q}_r^1)\hat{P} + Kr^1 \end{aligned} \tag{13}$$

Pre-multiplying equation (7) by $L^T(q^1)$, it follows that

$$\begin{aligned} &M_L(q^1)\ddot{q}_r^1 + C_L(q^1, \dot{q}^1)\dot{q}_r^1 + G_L(q^1) \\ &= L^T(q^1)\Psi(q^1, \dot{q}^1, \dot{q}_r^1, \ddot{q}_r^1)P \end{aligned} \tag{14}$$

By subtracting equation (13) from equation (14) and using equation (10), it yields

$$\begin{aligned} &M_L(q^1)\dot{r}^1 + C_L(q^1, \dot{q}^1)r^1 + Kr^1 \\ &= L^T(q^1)\Psi(q^1, \dot{q}^1, \dot{q}_r^1, \ddot{q}_r^1)\tilde{P} \end{aligned} \tag{15}$$

From equations (7) and (12), it is obvious that

$$\begin{aligned} &J^{1T}(q^1)n_c(e_\lambda + k_\lambda \int_0^t e_\lambda(\tau)d\tau) \\ &= L^{-T}(q^1)Kr^1 - \Psi(q^1, \dot{q}^1, \dot{q}_r^1, \ddot{q}_r^1)\tilde{P} \\ &+ (\Psi(q^1, \dot{q}^1, \dot{q}_r^1, \ddot{q}_r^1) - \Psi(q^1, \dot{q}^1, \dot{q}^1, \ddot{q}^1))P \end{aligned} \tag{16}$$

For the convergence of the tacking errors e^1 and e_λ, the following theorem is presented.

Theorem 1. For the closed-loop dynamic system (13), if the parameters are updated by

$$\dot{\hat{P}} = \Gamma\Psi^T(q^1, \dot{q}^1, \dot{q}_r^1, \ddot{q}_r^1)L(q^1)r^1 \tag{17}$$

where Γ is a constant positive definite matrix, then $e^1 \to 0$ and e_λ is bounded as $t \to \infty$, and all the closed loop signals are bounded.

Proof

Consider the Lyapunov function,

$$V = \frac{1}{2}r^{1T}M_L(q^1)r^1 + \frac{1}{2}\tilde{P}^T\Gamma^{-1}\tilde{P} \tag{18}$$

Differentiate equation (18) with respect to time t,

$$\begin{aligned} \dot{V} &= \tilde{P}^T\Gamma^{-1}(\Gamma\Psi^T(q^1, \dot{q}^1, \dot{q}_r^1, \ddot{q}_r^1)L(q^1)r^1 - \dot{\hat{P}}) \\ &- r^{1T}Kr^1 \end{aligned} \tag{19}$$

where **Property 3** and equation (15) were used to derive the above equation,

Substituting the adaptation law (17) into the above equation leads to

$$\dot{V} = -r^{1T}Kr^1 \leq 0 \tag{20}$$

As $K > 0$, $r^1 \in L_2^n$. From the definition of r^1 in equation (9), $e^1 \to 0$, $q^1(t) \to q_d^1(t)$ as $t \to \infty$, and $\dot{e}^1 \in L_2^n$. From the closed kinematics (3), $q \to q_d$ when $t \to \infty$. Obviously $\tilde{P}$ is bounded in the sense of Lyapunov stability.

Because $e^1 \to 0$, $\dot{e}^1 \in L_2^n$ and $\tilde{P}$ is bounded, from equation (15), $\dot{r}^1 \in L_\infty^n$. As $r^1 \in L_2^n$, $r^1 \to 0$ as $t \to \infty$. From the definition of r^1 , $\dot{e}^1 \to 0$, $e^1 \to 0$ as $t \to \infty$.

Because $r^1 \to 0$, $e^1 \to 0$, $\dot{e}^1 \to 0$ as $t \to \infty$ and $\tilde{P}$ is bounded, it is obvious that the right hand side of equation (16) is bounded, and thus e_λ is bounded.

Q.E.D.

Before discussing the convergence of e_λ, the following lemma (Sadegh and Horowitz, 1996) is reproduced here for the completeness of presentation.

Lemma 1. Let $f(t): R^+ \rightarrow R^n$ be a uniformly continuous almost everywhere (u.c.a.e) function. Then for any $p_0 > 0$, $\lim_{t\to\infty} f(t) = 0$,

$$iff \lim_{t\to\infty} \int_t^{t+p} f(\tau)d\tau = 0 \text{ for all } 0 < p < p_0$$

Thus the following theorem about the convergence of e_λ is readily available. The proof can be similarly completed as the one in (Sadegh and Horowitz, 1996).

Theorem 2. For the closed loop system consisting of dynamic model (6), control law (11) and adaptation law (17), if

(1) $\ddot{q}_d^1$ is uniformally continuous almost everywhere (u.c.a.e.), and
(2) $\Psi_L(q_d^1, \dot{q}_d^1, \dot{q}_d^1, \ddot{q}_d^1) = L^T(q_d^1)\Psi(q_d^1, \dot{q}_d^1, \dot{q}_d^1, \ddot{q}_d^1)$ is persistently exciting,

then $e_\lambda \rightarrow 0$ as $t \rightarrow \infty$.

Proof

Denote $\Psi_L(q^1, \dot{q}^1, \dot{q}_r^1, \ddot{q}_r^1) = L^T(q^1)\Psi(q^1, \dot{q}^1, \dot{q}_r^1, \ddot{q}_r^1)$. The tracking errors equation (15) is re-written as follows

$$\begin{aligned} &M_L(q^1)\dot{r}^1 + C_L(q^1,\ \dot{q}^1)r^1 + Kr^1 \\ &= \Psi_L(q^1, \dot{q}^1, \dot{q}_r^1, \ddot{q}_r^1)\tilde{P} \end{aligned} \quad (21)$$

Define

$$\begin{aligned} f(t) &= \Psi_L(q^1, \dot{q}^1, \dot{q}_r^1, \ddot{q}_r^1)\tilde{P}(t) \\ f_1(t) &= M_L(q^1)\dot{r}^1(t) \\ f_2(t) &= C_L(q^1,\ \dot{q}^1)r^1(t) + Kr^1(t) \end{aligned}$$

Then equation (21) can be rewritten as

$$f_1(t) + f_2(t) = f(t) \quad (22)$$

Integrating both sides of equation (22) in the interval $[t, t+p]$ $(\leq p \leq p_0)$ yields

$$\int_t^{t+p} f_1(\tau)d\tau + \int_t^{t+p} f_2(\tau)d\tau = \int_t^{t+p} f(\tau)d\tau \quad (23)$$

and

$$\int_t^{t+p} f_1(\tau)d\tau = \int_t^{t+p} M_L(q^1)(\tau)\dot{r}^1(\tau)d\tau$$

$$\int_t^{t+p} f_2(\tau)d\tau = \int_t^{t+p} [C_L(q^1, \dot{q}^1)r^1(\tau) + Kr^1(\tau)]d\tau$$

By expanding the integral $\int_t^{t+p} f_1(\tau)d\tau$, it follows that

$$\begin{aligned} \int_t^{t+p} f_1(\tau)d\tau &= M_L(q^1)(t+p)r^1(t+p) \\ &- M_L(q^1)(t)r^1(t) - \int_t^{t+p} \dot{M}_L(q^1)(\tau)r^1(\tau)d\tau \end{aligned}$$

Thus

$$\begin{aligned} \left\|\int_t^{t+p} f_1(\tau)d\tau\right\| &\leq \sup_{q^1} \|M_L(q^1)\|(\|r^1(t+p)\| + \|r^1(t)\|) \\ &+ p_0(\sup_{t\leq\tau\leq t+p})\|\dot{M}_L(q^1)(\tau)\|\|r^1(\tau)\|) \end{aligned} \quad (24)$$

Notice that $\dot{M}_L(q^1)(t)$ can be written as

$$\dot{M}_L(q^1)(\tau) = \sum_{i=1}^{n} \frac{\partial M_L(q_i^1)}{\partial q_i^1}(\tau)\dot{q}_i^1 \quad (25)$$

As proved in **Theorem 1**, $r^1 \rightarrow 0$, $e^1 \rightarrow 0$ and $\dot{e}^1 \rightarrow 0$ as $t \rightarrow \infty$, thus $\dot{M}_L(q^1)(\tau)$ is bounded. In addition, $\sup_{q^1} \|M_L(q^1)\|$ is bounded from **Property 1**. Therefore

$$\lim_{t\to\infty} \int_t^{t+p} f_1(\tau)d\tau = 0$$

From the fact that $r^1 \rightarrow 0$ as $t \rightarrow \infty$, it is also obvious that

$$\lim_{t\to\infty} \int_t^{t+p} f_2(\tau)d\tau = 0, \ \lim_{t\to\infty} \int_t^{t+p} f(\tau)d\tau = 0$$

From **Lemma 1**,

$$\lim_{t\to\infty} f(t) = \lim_{t\to\infty} \Psi_L(q^1, \dot{q}^1, \dot{q}_r^1, \ddot{q}_r^1)\tilde{P}(t) = 0 \quad (26)$$

Consider the following inequality

$$\begin{aligned} &\|\Psi_L(q_d^1, \dot{q}_d^1, \dot{q}_d^1, \ddot{q}_d^1)\tilde{P}\| \\ &\leq \|\Psi_L(q^1, \dot{q}^1, \dot{q}_r^1, \ddot{q}_r^1) - \Psi_L(q_d^1, \dot{q}_d^1, \dot{q}_d^1, \ddot{q}_d^1)\|\|\tilde{P}\| \\ &+ \|\Psi_L(q^1, \dot{q}^1, \dot{q}_r^1, \ddot{q}_r^1)\tilde{P}\| \end{aligned} \quad (27)$$

Because $r^1 \rightarrow 0$ as $t \rightarrow \infty$, thus

$$\lim_{t\to\infty} \|\Psi_L(q^1, \dot{q}^1, \dot{q}_r^1, \ddot{q}_r^1) - \Psi_L(q_d^1, \dot{q}_d^1, \dot{q}_d^1, \ddot{q}_d^1)\| = 0$$

Based on the above equation, equation (26) and inequality (27),

$$\Psi_L(q_d^1, \dot{q}_d^1, \dot{q}_d^1, \ddot{q}_d^1)\tilde{P} \rightarrow 0 \text{ as } t \rightarrow \infty$$

Let

$$Q(t,t+\delta) = \int_t^{t+\delta} \Psi_L^T(q_d^1,\dot q_d^1,\dot q_d^1,\ddot q_d^1)(\tau)\Psi_L(q_d^1,\dot q_d^1,\dot q_d^1,\ddot q_d^1)(\tau)d\tau$$

Since $\Psi_L(q_d^1,\dot q_d^1,\dot q_d^1,\ddot q_d^1)$ is persistent exciting, then for some $\delta > 0$ and all t,

$$Q(t,t+\delta) \geq \alpha I > 0 \tag{28}$$

From integration by parts and the adaptation law (17), it follows that

$$\tilde P^T(t)Q(t,t+\delta)\tilde P(t) = 2\int_t^{t+\delta}\tilde P(\tau)Q(\tau,\tau+\delta)\Gamma\Psi_L(q^1,\dot q^1,\dot q_r^1,\ddot q_r^1)r^1(\tau)d\tau + \int_t^{t+\delta}[\tilde P^T\Psi_L^T(q_d^1,\dot q_d^1,\dot q_d^1,\ddot q_d^1)\Psi_L(q_d^1,\dot q_d^1,\dot q_d^1,\ddot q_d^1)](\tau)\tilde P d\tau$$

From equation (28) and the fact that $r^1 \to 0$ as $t \to \infty$ proven in **Theorem 1**, it can be seen that the right side of the above equation converges to zero when $t \to \infty$. Since $Q(t,t+\delta) \geq \alpha I > 0$, then it can be concluded that $\tilde P \to 0$ when $t \to \infty$.

It has been proven that $r^1 \to 0$, $e^1 \to 0$ and $\dot e^1 \to 0$ as $t \to \infty$ in **Theorem 1**. With $\tilde P \to 0$, the conclusion can be made that $\Psi(q^1, \dot q^1, \dot q_r^1\ \ddot q_r^1) \to \Psi(q^1,\dot q^1,\ddot q^1))$ when $t \to \infty$. Thus, from equation (16),

$$J^{1T}(q^1)n_c(e_\lambda + k_\lambda \int_0^t e_\lambda(\tau)d\tau) \to 0$$

Therefore

$$e_\lambda + k_\lambda \int_0^t e_\lambda(\tau)d\tau \to 0$$

because of the choice of $K_\lambda > 0$, $e_\lambda(\tau) \to 0$ as $t \to \infty$.

Q.E.D.

Remark The condition for the convergence of force is more stringent than those for the convergence of position. It requires that the trajectory q_d^1 be planned so that $\ddot q_d^1$, $L^T(q_d^1)\Psi(q_d^1,\dot q_d^1,\dot q_d^1,\ddot q_d^1)$ meet the conditions in **Theorem 2**. Otherwise only the boundedness of the force error can be guaranteed.

4. SIMULATION

The system used for simulation is schematically shown in Figure 2. The rectangular object is rigidly attached to manipulator 2 which moves along the horizontal plane; the end effector of manipulator 1 of two degrees of freedom is to follow a straight line describe by

$$x_{co} - y_{co} = 0 \tag{29}$$

The object coordinate $X_oO_oY_o$ is at the object mass center O_o. The length, mass and moment of inertia of each link of manipulator 1 are represented by d_i, m_i and I_i $(i = 1,2)$ respectively. Let l_i $(i = 1,2)$ represent the distance of the mass center of each link from the respective joint. The mass of manipulator 2 together with the object is M_2. The joints variable for two manipulators are $q_1 = [\theta_1\ \theta_2]^T$ and $q_2 = x$ respectively. Note that x is actually a linear displacement of the object along the horizontal plane. The gravitational acceleration is $g = 9.8m/s^2$.

The geometric parameters are $d_1 = d_2 = 0.3m$, $l_1 = l_2 = 0.15m$, $a = 0.2$ and $b = 0.5m$. The true values of the mass and inertia parameters are assumed to be $m_1 = m_2 = 0.1kg$,$M_2 = 0.2kg$, $I_1 = I_2 = 0.3kgm^2$, which are unknown for controller design. The true parameters P is $[0.6\ 0.04\ 0.3\ 0.44\ 0.15\ 0.06\ 0.06]^T$, while the initial estimated values of parameter $P(0)$ is $\hat P(0) = [0.2\ 0.01\ 0.4\ 0.4\ 0.1\ 0.2\ 0.2]^T$. The controller gain matrices are chosen as $K_e = 20I^{2\times2}$, $K = 15I^{2\times2}$ and $k_\lambda = 15$.

The desired trajectory on the object and the trajectory of the object are given by

$$x_{co} = y_{co} = -\frac{1}{12}cos(0.5t+1)$$
$$x_o = \frac{1}{10}(1-\sin(2t+6))$$

and desired force is $\lambda_d = 4N$.

The adaptation gain matrix Γ in adaptation law (17) is chosen as $\Gamma = 15I^{7\times7}$. The position tracking performances of the object and the force tracking performances are plotted in Figures 3 and 4 respectively. The control torques for the manipulators are given in Figure 5. It can be seen that under the proposed controller and the adaptation law, the positions and the forces converge to their desired values, and the torques are in the reasonable ranges.

5. CONCLUSION

In this paper, the coordinated dynamic control of two manipulators manipulating a constraint object is discussed by considering the relative motion of the object with respect to the manipulators. The dynamic model of such a system is established and an adaptive control law is then developed which can guarantees the asymptotic convergence of positions, and boundedness of the constraint force. The condition for the convergence of the constraint force is also presented. Simulation results showed that the proposed controller works quite well.

REFERENCES

Ge, S.S., X.Q. Chen, S. Xie, and D.L. Gu (1997). Motion and force control of a Cartesian arm and a rotary table. *Proc. of the IEEE Singapore International Symposium on Control Theory and Applications (SISCTA'97)*, pp. 286-289.

Ge, S.S., T.H. Lee and C.J. Harris (1998). *Adaptive neural network control of robotic manipulators.* World Scientific Publishing Co.

McClamroch, N.H. and D.W.Wang (1988). Feedback stabilization and tracking of constrained robots.*IEEE Trans. on Auto. Contr.*, **33**, pp. 419–426.

Kankaanranta, R.K. and H.N.Kovio (1988). Dynamics and simulation of compliant motion of a manipulator. *IEEE J. of Robotics and Auto.*, **4**, pp. 163-173.

Ramadorai, A. K., T. J. Tarn and A. K. Bejczy (1992). *Task definition, decoupling and redundancy resolution by nonlinear feedback in multi-robot object handling, Robotic Lab. Report, SSM-RL-92-16*, Washington University.

Sadegh, N. and R. Horowitz (1990). Stability and robustness analysis of a class of adaptive controller for robotic manipulators. *Int. J. of Robotics Research*, **9**, pp. 74-92.

Slotine, J-J. and W. Li (1987). On the adaptive control of robot manipulators. *Int. J. of Robotics Research*, **6**, pp. 147-157.

Slotine, J-J., and W. Li (1989).Composite adaptive control of robot manipulators. *Automatica*, **25**, pp. 509-519.

Yao, B., S. P. Chan and D. Wang (1994). Unified formulation of variable structure control schemes for robot manipulators. *IEEE Trans. on Auto. Contr.*, **39**,pp. 371–376.

Yuan, J. (1996). Composite adaptive control of constrained robots. *IEEE Trans. on Robotics and Auto.*, **12**, pp. 640-645.

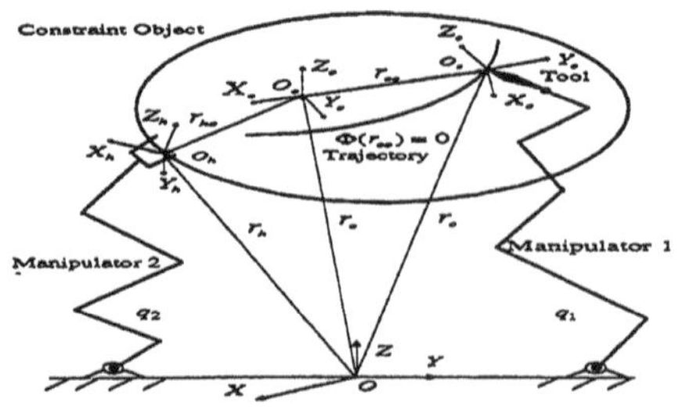

Fig. 1. Coordinated Operation of Two Robots

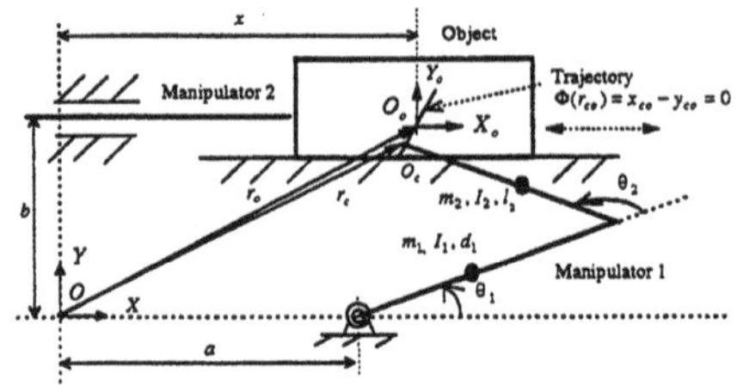

Fig. 2. Schematic of Simulation Example

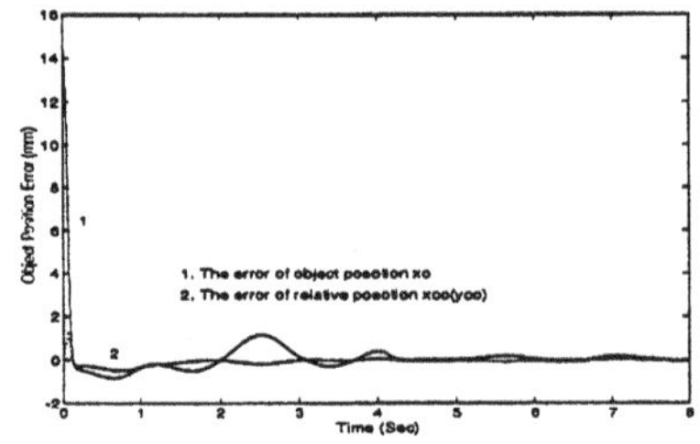

Fig. 3. Position Tracking Errors

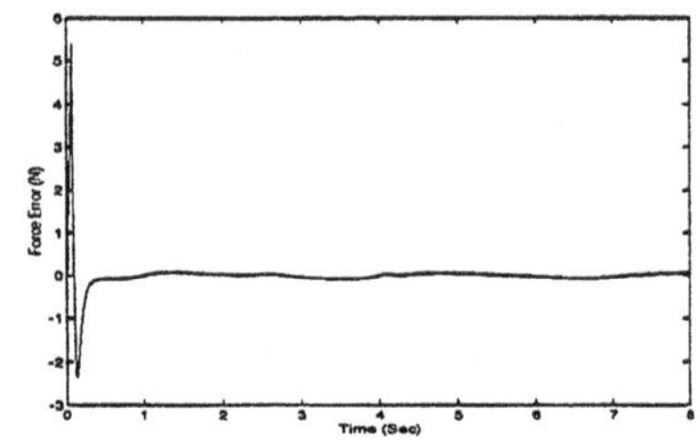

Fig. 4. Constraint Force Error

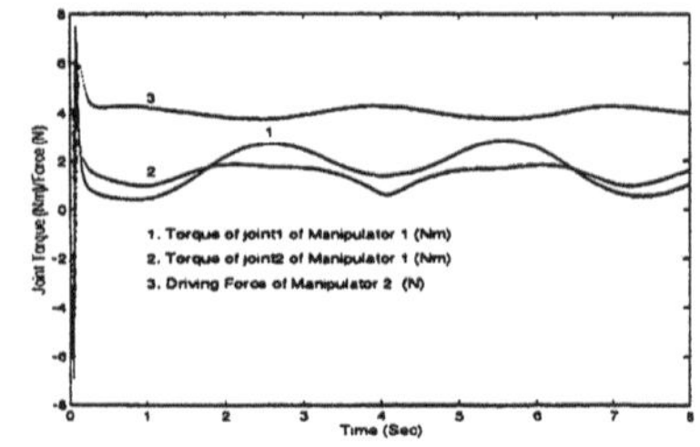

Fig. 5. Torques/Forces of Manipulators

NON LINEAR FEEDBACK STABILIZATION OF A ROTATING BODY-BEAM SYSTEM

B. Chentouf and J.-F. Couchouron

INRIA-Lorraine (CONGE project) & University of Metz
ISGMP, Bât A, Ile du Saulcy 57045 - METZ (FRANCE)
e-mail: {chentouf, couchour}@loria.fr

Abstract: This paper deals with **nonlinear** feedback stabilization of a flexible beam clamped at a rigid body and free at the other end. We assume that there is no damping and the feedback law proposed here consists of a nonlinear control torque applied on the rigid body and either a nonlinear boundary control moment or a nonlinear boundary control force or both of them applied at the free end of the beam. This **nonlinear** feedback, which insures the exponential decay of the beam vibrations, extends the **linear** case studied in (Laousy *et al.*, 1996) to a more general class of controls. This new class of controls is in particular of the interest to be robust. *Copyright © 1999 IFAC*

Keywords: Feedback stabilization, Nonlinear systems, Perturbation, Lyapunov stability.

1. INTRODUCTION

The purpose of this paper is to study the nonlinear feedback stabilization of the system presented in the figure 1.

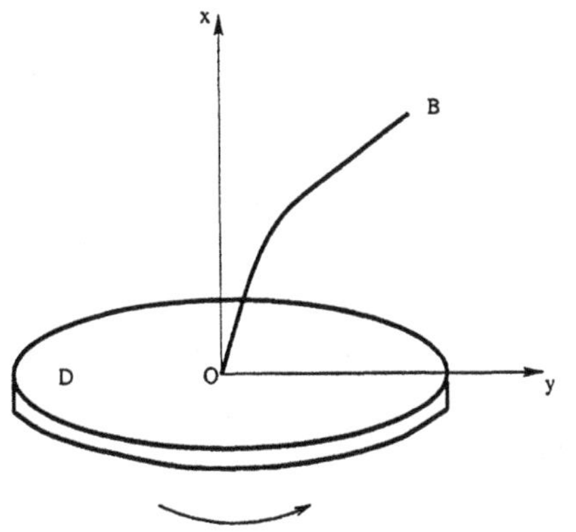

Fig. 1. The body-beam system.

This system has been introduced by Baillieul and Levi (Baillieul and Levi, 1987). It consists of a disk (D) with a beam (B) attached to its center and perpendicular to the disk plane. The disk (D) rotates freely about its axis and the motion of the beam (B) is confined to a plane perpendicular to the disk (see fig. 1). In this paper, we take a nonlinear control torque exerted on the disk (D) while a nonlinear boundary control moment and/or a nonlinear boundary control force are applied at the free end of the beam (B).
As in the linear case (Laousy *et al.*, 1996), we will show that in our case, the system is still stabilizable.

The stabilization problem of the body-beam system has been extensively studied in the literature, see (Baillieul and Levi, 1987), (Bloch and Titi, 1990), (Xu and Baillieul, 1993), (Xu and Sallet, 1992), (Morgül, 1991), (Morgül, 1993), (Morgül, 1994) and (Coron and d'andréa-Novel, 1998). In (Baillieul and Levi, 1987), the authors showed that with structural damping and without control, the body-beam system has a finite number of rotating equilibrium states. Later, Bloch and Titi (Bloch and Titi, 1990) showed that for the more difficult case of viscous damping, a linear inertial manifold exists for the body-beam system. By taking into account the effect of damping, it was shown in (Xu and Baillieul, 1993) that for any constant angular velocity smaller than the critical one, an exponentially stabilizing feedback torque control law was proved. Recently, exponential stabilization of the body-beam system was established in (Laousy *et al.*, 1996) as soon as one of the two **linear** boundary controls (force or moment) is present at the free end of the beam with the control torque of the disk. The contribution of this paper consists of extending the class of controls proposed in (Laousy *et al.*, 1996). In fact, our work is motivated by two arguments: First, the interest of such an extension is to highlight the robustness of controls. Actually, a minimum degree of robustness is needed in applications since no sensors neither actuators can provide an infinitely precise linear control as required in (Laousy *et al.*, 1996). Second, a wide class of controls is provided. For instance, this helps us to avoid eventually some saturation phenomena in the actuators.

The paper is organized as follows: in section 2, we deal with notations and main result. The section 3 is devoted to a sketch of proof of this main result. We postpone in an Appendix the nonlinear basic theoretic tools used for the body-beam system.

The body-beam system is modeled by

$$\begin{cases} \rho y_{tt} + EI y_{xxxx} = \rho\omega^2(t)y, \\ y(0,t) = y_x(0,t) = 0, \\ y_{xx}(1,t) = \Gamma_1(t), \\ y_{xxx}(1,t) = \Gamma_2(t), \\ \dot{\omega}(t) = \dfrac{\Gamma_3(t) - 2\rho\omega(t)\displaystyle\int_0^1 y(t)y_t(t)dx}{I_d + \rho\displaystyle\int_0^1 y(t)^2 dx}, \end{cases} \tag{1}$$

where the positive constants EI, ρ and I_d are respectively the flexural rigidity, the mass per unit length of the beam, and the disk's moment of inertia; $\omega(t)$ is the angular velocity of the disk at time t and the length of the beam is chosen as $L = 1$. Moreover, $\Gamma_1(t), \Gamma_2(t)$ and $\Gamma_3(t)$ are respectively the moment control, force control and the torque control to be determined so that the solution's energy of the resulting closed-loop system decays to zero in some functional space.

In this paper, we propose a stabilizing **nonlinear** controls. More precisely, given $\omega^* \in \mathbb{R}$ our candidate for feedback control is

$$\begin{aligned} \Gamma_1(t) &= -f(y_{tx}(1,t)), \\ \Gamma_2(t) &= g(y_t(1,t)), \\ \Gamma_3(t) &= -\gamma(\omega - \omega^*), \end{aligned} \tag{2}$$

but this time contrary to (Laousy *et al.*, 1996), f, g and γ are nonlinear real functions satisfying the following assumptions F1), F2) and F3)

F1) $f + g \not\equiv 0$ and $f, g \in \{h \in C^0(\mathbb{R}); h$ is increasing and $h(0) = 0\}$.

F2) There exist nonnegative constants L_1, L_2 and L_3 such that $L_1 + L_2 > 0$ and for all $x \in \mathbb{R}$,

$$L_1 \mid x \mid \leq \mid f(x) \mid \leq L_3 \mid x \mid,$$
$$L_2 \mid x \mid \leq \mid g(x) \mid \leq L_3 \mid x \mid .$$

F3) The function γ is Lipschitz on each bounded subset of $\mathbb{R}$ and for some $L_4 > 0$,

$$\gamma(x)x \geq 0, \quad \mid \gamma(x) \mid \geq L_4 \mid x \mid, \ \forall x \in \mathbb{R}.$$

The conditions F1)-F2) ensure in particulier the presence of at least one control at the flexible beam. For the phase space of the system, we take the real Hilbert space

$$X = \mathcal{H} \times \mathbb{R} = H_0^2 \times L^2(0,1) \times \mathbb{R},$$

equipped with the inner product

$$\langle (u, v, \xi), (\tilde{u}, \tilde{v}, \tilde{\xi}) \rangle_X = \langle (u, v), (\tilde{u}, \tilde{v}) \rangle_{\mathcal{H}} + \xi\tilde{\xi},$$

where for $n \in \mathbb{N}$,

$$H_0^n = \{u \in H^n(0,1);\ u(0) = u_x(0) = 0\}$$

and the space $\mathcal{H}$ is endoweded with the inner product

$$\langle (u, v), (\tilde{u}, \tilde{v}) \rangle = \int_0^1 [EI u_{xx}\tilde{u}_{xx} - \rho{\omega^*}^2 u\tilde{u} + \rho v\tilde{v}]dx.$$

Note that the norm induced by this scalar product is equivalent to the usual one of the Hilbert space $H^2(0,1) \times L^2(0,1)$ provided that $\mid \omega^* \mid < \sqrt{9EI/\rho}$; see (Laousy *et al.*, 1996) for details.

Set $\Phi = (y, z)$. Then, the closed loop system (1)-(2) can be written as follows

$$\begin{cases} \dot{\Phi}(t) + \tilde{A}\Phi(t) \\ \qquad + \left(0, ({\omega^*}^2 - \omega^2(t))y(t)\right) = 0, \\ \dot{\omega}(t) = \dfrac{-\gamma(\omega(t) - \omega^*)}{I_d + \rho\|y(t)\|^2_{L^2(0,1)}} \\ \qquad - \dfrac{2\rho\,\omega(t) < y(t), z(t) >_{L^2(0,1)}}{I_d + \rho\|y(t)\|^2_{L^2(0,1)}}, \end{cases} \tag{3}$$

where $\tilde{A}$ is the **nonlinear** operator defined by

$$\begin{aligned} \mathcal{D}(\tilde{A}) = \{&(y, z) \in H_0^4 \times H_0^2; \\ &y_{xx}(1) = -f(z_x(1)),\ y_{xxx}(1) = g(z(1))\}, \end{aligned} \tag{4}$$

and

$$\tilde{A}(y, z) = \left(-z, \frac{EI}{\rho}y_{xxxx} - {\omega^*}^2 y\right). \tag{5}$$

We note that the domain $\mathcal{D}(\tilde{A})$ is not a linear space. Now, let us introduce the subsystem

$$\begin{cases} \dot{\Phi}(t) + \tilde{A}\Phi(t) = 0, \\ \Phi(0) = \Phi_0. \end{cases} \quad (6)$$

In return, it is easy to verify that the system (3) is equivalent to the following one

$$\left(\dot{\Phi}(t), \dot{\omega}(t)\right) + (A + B)\left(\Phi(t), \omega(t)\right) = 0, \quad (7)$$

where

$$\begin{cases} A(\Phi, \omega) = \left(\tilde{A}\Phi, 0\right), \\ B(\Phi, \omega) = \left(0, (\omega^{*2} - \omega^2)y, \right. \\ \left. \dfrac{-\gamma(\omega - \omega^*) - 2\rho\,\omega(t) < y, z >_{L^2}}{I_d + \rho\|y\|^2_{L^2}}\right), \end{cases} \quad (8)$$

with $\mathcal{D}(A) = \mathcal{D}(\tilde{A}) \times \mathbb{R}$.

Unfortunately, it is not clear that the system (7) is well posed since the classical theory of perturbation can not be applied in our framework. This is due to the **nonlinearity** of the operator A and the **lack of accretivity** of the operator B. On the other hand, for our system, we see that the nonlinearity of the operator A occurs in its domain which is not a linear space. Really, this is a very bad case of nonlinearity since in view of the previous features, the following tools fail: first, the classical Duhamel's formula as well as the nonlinear Duhamel's variation of constant formula obtained in (Banks, 1988); second, the usual arguments of (Brezis, 1973) and (Pazy, 1983) about the density of the domain. In return, the known results of M. Pierre (Pierre, 1976) ensure the existence of local solutions of (7) and all the problem is to show that these solutions are global. We postpone in Appendix some indications on the mathematical tools needed to treat this situation.

The main result of our paper is the exponential stability of the global system (7). Of course as in (Laousy *et al.*, 1996), the decay rate, although exponential, is not uniform. The following theorem generalises the results of (Laousy *et al.*, 1996) in the sense that if the controls are linear in (2), we find again the result of (Laousy *et al.*, 1996). Our main result is

Theorem 1. *Suppose that* $| \omega^* | < \sqrt{9EI/\rho}$. *Then under the hypotheses F1)-F3), for each initial data* $(\Phi_0, \omega_0) \in \mathcal{D}(A)$ *the solution* $(\Phi(t), \omega(t))$ *of the closed-loop system (7) tends exponentially to* $(0, \omega^*)$ *in* X *as* $t \to +\infty$.

2. PROOF OF THE MAIN RESULT

2.1 *The subsystem (6)*

We start with the following proposition

Proposition 1. *Assume that* $| \omega^* | < \sqrt{9EI/\rho}$ *and that the hypothesis F1)-F2) are satisfied.*
1) The nonlinear operator $\tilde{A}$ *defined by (4)-(5) is m-accretive in* $\mathcal{H}$ *with dense domain. Thus* $-\tilde{A}$ *generates a contractive semigroup* $e^{-\tilde{A}t}$.
2) The semigroup $e^{-\tilde{A}t}$ *is asymptotically stable in* $\mathcal{H}$.
3) Let Φ *be the solution of (6) stemmed from* $\Phi_0 \in \mathcal{H}$. *Then there exist constants* $\tilde{M}$, $\tilde{\mu} > 0$ *such that*

$$\|\Phi(t))\| \leq \tilde{M}e^{-\tilde{\mu}t}\|\Phi_0\|, \quad \forall t \geq 0.$$

Sketch of the proof of Proposition 1.
1) One can easily check that, under the hypothesis F1), $\tilde{A}$ is m-accretive. We point out that the density of $\mathcal{D}(\tilde{A})$ in $\mathcal{H}$ is not obvious since $\mathcal{D}(\tilde{A})$ is not a linear space. So the proof of this density is not standard and needs the condition F2). Then, the rest of the first assertion 1) is a consequence of the method of Brezis (Brezis, 1973).
2) Since the nonlinear semigroup $e^{-\tilde{A}t}$ is contractive and since $\mathcal{D}(\tilde{A})$ is dense in $\mathcal{H}$, it suffices to prove the stability for any initial data $\Phi_0 = (y_0, z_0) \in \mathcal{D}(\tilde{A})$. This can be obtained by using the LaSalle's invariance principle (Jurdjevic and J. P. Quin, 1978)-(Dafermos and Slemrod, 1973) and an adaptation of computations of (Conrad and Pierre, 1990) .
3) As in the proof of 2), we may assume that $\Phi_0 = (y_0, z_0) \in \mathcal{D}(\tilde{A})$. In order to use the multiplier method, let us define a functional

$$\xi(t) = 2\int_0^1 xy_t y_x dx.$$

Next, given $\epsilon > 0$ we introduce as in (Conrad and Rao, 1993) and (Rao, 1993) the perturbed energy by

$$E^\epsilon(t) = E_1(t) + \epsilon\xi(t), \quad \forall t \geq 0,$$

where

$$E_1(t) = \frac{1}{2}\|(y(t), z(t)))\|^2_{\mathcal{H}}.$$

Using F1)-F2), we show after a careful computation that there is a positive constant $K > 0$ such that

$$\dot{E}^\epsilon(t) \leq -2M^{-\frac{1}{2}}KE^\epsilon(t).$$

Solving this differential inequality, we get the required exponential decay with

$$\mu = 2M^{-\frac{1}{2}}K.$$

□

We are going to indicate, in the next subsections, a sketch of proof of Theorem 1. For this, we need the hypotheses and the results of Appendix.

2.2 *Well posedness of the global system (7)*

First, we have shown in Proposition 1 that the operator $\tilde{A}$ defined by (4)-(5) is m-accretive in $\mathcal{H}$ with dense domain $\mathcal{D}(\tilde{A})$. We deduce thus, from (8), that the operator A is also m-accretive with dense domain $\mathcal{D}(A)$ in X and thus the hypothesis HI) cited in the Appendix is satisfied. Second, thanks to F3) , the operator B defined in (8) is Lipschitz on bounded subsets of X and therefore B satisfies the hypothesis HII) of the Appendix. Third, the Lyapunov function V given in (9) (see the next paragraph) ensures that the hypothesis HIII) of the Appendix holds. We deduce from Theorem 2 given in the Appendix, that for any $(\Phi_0, \omega_0) \in X$, there is a unique solution $(\Phi(t), \omega(t)) \in \mathcal{C}([0, +\infty]; X)$ of the system (7).

Remark 1. *This solution $(\Phi(t), \omega(t))$ is a strong one if $(\Phi_0, \omega_0) \in \mathcal{D}(A)$.*

2.3 *Exponential stability of the global system (7)*

As in (Xu and Baillieul, 1993) and (Laousy *et al.*, 1996), consider the function $V : X \to \mathbb{R}^+$ defined by

$$V(\Phi, \omega) = \frac{1}{2}\{I_d\,(\omega - \omega^*)^2 - \omega^{*2}\int_0^1 \rho y^2 dx + \int_0^1 (\rho z^2 + EIy_{xx}^2)dx + (\omega - \omega^*)^2 \int_0^1 \rho y^2 dx\}. \tag{9}$$

We claim that this function is a reasonable choice of Lyapunov function . Indeed, it is easy to check that

$$V(\Phi, \omega) \geq K||(\Phi, \omega)||_X,$$

for some positive constant K. On the other hand, using the Remark 1 and the assumption (F), a straightforward computation shows that for any initial condition $(\Phi_0, \omega_0) \in \mathcal{D}(A)$,

$$\begin{aligned}\frac{dV}{dt}(\Phi, \omega)(t) &= -EIy_{xt}(1,t)\; f(y_{xt}(1,t)) \\ &+EIy_t(1,t)\; g(y_t(1,t)) \\ &-(\omega(t) - \omega^*)\; \gamma(\omega(t) - \omega^*) \leq 0,\end{aligned} \tag{10}$$

for a.e. $t \geq 0$. Thus, the function $t \longmapsto V(\Phi, \omega)(t)$ is non increasing on $[0, +\infty[$.

Now, one will restrict oneself with the system

$$\dot{\Phi}(t) + (\tilde{A} + \tilde{B})\Phi(t),$$

where $\Phi = (y, z)$ and

$$\tilde{B}\Phi = (0, (\omega^{*2} - \omega^2)y).$$

It is obvious that the operator $(\omega^{*2} - \omega^2)\tilde{B}$ has the property

$$||(\omega^{*2} - \omega^2)\tilde{B}\Phi(t)||_{\mathcal{H}} \leq |\omega^{*2} - \omega^2|||\Phi(t)||_{\mathcal{H}}.$$

From (9) and (10), it results that $\omega^* - \omega \in L^2([0, \infty[; \mathbb{R}) \cap L^\infty([0, \infty[; \mathbb{R})$ and thus the condition (HV) of the Appendix holds with $\mu(t) = |\omega^{*2} - \omega(t)^2|$. (Really, we have the stronger condition $\lim_{t\to+\infty} \omega(t) = \omega^*$). Therefore, as we have shown, all the assumptions of Theorem 3 of the Appendix are satisfied. Thus, there exist positive constants M and κ such that

$$||\Phi(t)||_{\mathcal{H}} \leq Me^{-\kappa t}, \ \forall t \geq 0.$$

Finally, returning to the second equation of (7) and using F3), we prove analogously to (Xu and Baillieul, 1993) and (Laousy *et al.*, 1996) that $\omega^* - \omega$ tends also exponentially to zero . This ends the proof of Theorem 1. □

APPENDIX

Let X be a real Banach space and $T \in \mathbb{R}_+^*$. This part deals with evolution inclusions of the following type,

$$CP(x^0, A, B, T) = \begin{cases} \dot{x}(t) + Ax(t) + B(t, x(t)) \ni 0, \ t \in [0, T], \\ x(0) = x^0 \in \overline{\mathcal{D}(A)} \subset X, \end{cases}$$

In general when the unbounded (multivalued) operator A is **nonlinear** and the everywhere defined operator B is **non accretive**, it is not clear that $CP(x^0, A, B, T)$ is well-posed. Indeed, the known results in the literature in such a situation ask additional assumptions as for instance the equicontinuity of the semigroup e^{-tA} generated by the operator $-A$ (see (Bothe,), (Evans, 1977), (Gutman, 1983), (Vrabie, 1987)) or more general topological assumptions (Couchouron and Kamenski, 1998). But these kinds of assumptions on the semigroup e^{-tA} did not hold for the studied body-beam system. In order to overcome these difficulties, we had to use a result due to M. Pierre (Pierre, 1976) and which is based on the fixed point method.

We will introduce in this paragraph the following hypotheses

(HI) A is a multivalued unbounded nonlinear m-accretive operator with domain $\mathcal{D}(A)$.

HII) B is locally Lipschitz on $[0,T] \times X$ and bounded on the bounded subsets of $[0,T] \times X$.

HIII) For every $x^0 \in \overline{\mathcal{D}(A)}$, there is a constant $C(T)$ such that for all $T_0 \in [0,T[$, each solution x of $CP(x^0, A, B, T_0)$ satisfies $\|x(t)\| \le C(T)$, $\forall t \in [0, T_0]$.

The solutions of $CP(x^0, A, B, T)$ will be understood in the following "weak" sense:

Definition 1. *The continuous function x_* is a solution of $CP(x^0, A, B, T)$ if it is the mild solution of the quasi-autonomous problem*

$$\begin{cases} \dot{x}(t) + Ax(t) \ni -B(t, x_*(t)), \ t \in [0,T], \\ x(0) = x^0 \in \overline{\mathcal{D}(A)} \subseteq X. \end{cases}$$

The **Theorem 2** below extends the well-known Picard-Lindeloff-Lipschitz Theorem.

Theorem 2. *Assume that (HI), (HII) and (HIII) hold. Then, for $x^0 \in \overline{\mathcal{D}(A)}$, the problem $CP(x^0, A, B, T)$ has a unique solution $S(t)x^0$ and the function $x^0 \mapsto S(t)x^0$ is continuous on $\overline{\mathcal{D}(A)}$, for all $t \in [0,T]$.*

The second result of the Appendix is the exponential decay of $\|x(t)\|$ where x is the solution of $CP(x^0, A, B, +\infty)$. For this, we introduce the two following conditions:

(HIV) The problem $CP(x^0, A, B, +\infty)$ has a unique solution x (that is $CP(x^0, A, B, T)$ has a unique solution for all $T > 0$).

(HV) There is a locally integrable positive real valued function μ on $[0, +\infty[$ satisfying

$$\begin{cases} \|B(t, x(t))\| \le \mu(t)\|x(t)\|, \\ \lim_{\xi \to +\infty} \int_{\xi}^{\xi+\delta} \mu(\tau)d\tau = 0, \ \forall \delta > 0. \end{cases}$$

Theorem 3. *Assume that the* **nonlinear** *semigroup e^{-At} generated by the operator $-A$ is exponentially (uniformly) stable. Then, under the hypotheses (HI), (HIV) and (HV), there exist positive constants M and κ such that*

$$\|x(t)\| \le Me^{-\kappa t}, \ \forall t \ge 0.$$

Comments on Theorem 2. First, we know that there exists a unique local solution $x(\cdot)$ of $CP(x^0, A, B, T)$ (see pp. 126 of (Pierre, 1976)). Then following the same computations, we get thanks to (HIII) and Gronwall lemma the global solution of $CP(x^0, A, B, T)$.

Comments on Theorem 2. We show thanks to Bénilan's integral inequalities (see (Bénilan, 1972) or (Bénilan *et al.*,)) and Gronwall's Lemma that for any $t \ge t_0$, we have

$$\|x(t)\| \le M_1 e^{-\omega_0 \frac{t}{\delta}} e^{\omega_0 (2 + \frac{t_0}{\delta})} \|x(t_0)\|,$$

where the constants M_1, ω_0, δ and t_0 are chosen as follows:

$$\begin{cases} \|e^{-At}\| \le M_1 e^{-\omega_1 t}, \\ \omega_0 > \max(1, \omega_1), \\ \delta > \dfrac{1}{\omega_1}(\ln M_1 + 2\omega_0), \\ \int_{\xi}^{\xi+\delta} \mu(\tau)d\tau \le \dfrac{\omega_0}{\delta e^{\omega_0 \delta}}, \ \forall \xi \ge t_0. \end{cases}$$

3. REFERENCES

Baillieul, J. and M. Levi (1987). Rotational elastic dynamics. *Physica* **27 D**, 43-62.

Banks, S. (1988). *Mathematical theories of nonlinear systems.* Prentice Hall. New York, Toronto, Sydney, Tokyo.

Bénilan, P. (1972). *Equations d'évolutions dans un espace de Banach quelconque et applications.* Thèse Paris XI. Orsay.

Bénilan, P., M. G. Crandal and A. Pazy. *Nonlinear evolution equations in Banach spaces.* Monograph in preparation.

Bloch, A. M. and E. S. Titi (1990). On the dynamics of rotating elastic beams. In: *Proc. Conf. New Trends Syst. theory* Genoa, Italy, July 9-11. (Conte, Perdon, and Wyman, Eds.). MA: Birkhäuser, Cambridge.

Bothe, T. Multivalued perturbations of m-accretive differential inclusions. To appear in *Israël J. Math.*

Brezis, H. (1973). *Opérateurs maximaux monotones et semi-groupes de contractions dans les espaces de Hilbert.* North-Holland, Amsterdam, London, (1973).

Conrad, F. and M. Pierre (1990). Stabilization of Euler-Bernoulli beam by nonlinear boundary feedback. *Rapport de recherche INRIA*, **1235**.

Conrad, F. and B. Rao (1993). Decay of solutions of the wave equation in a star-shaped domain with nonlinear boundary feedback. *Asymptotic Analysis*, **7**, 159-177.

Coron, J. M. and B. d'andréa-Novel (1998). Stabilization of a rotating body-beam without damping. To appear in *IEEE Trans. Automat. Contr.*

Couchouron J.-F. and M. Kamenski (1998). A unified topological point view for integro-differential inclusions. *Lecture Notes in Nonlinear Analysis*, **2**.

Dafermos, C. M. and M. Slemrod (1973). Asymptotic behavior of non linear contractions semigroups. *J. Func. Anal* , **14**, pp. 97-106.

Evans, L. C. (1977). Nonlinear evolution equations in an arbitrary Banach space. *Israel Journal of Mathematics*, **26**, no. I.

Gutman, S. (1983). Evolutions governed by m-accretive plus compact operators. *J. Nonlinear Analysis, TMA*, **7**, no. 7, 707-715.

Jurdjevic, V. and J. P. Quin (1978). Controllability and stability. *J. Diff. Equations*, **28**, no. 3, 381-389.

Laousy H., C. Z. Xu and G. Sallet (1996). Boundary feedback stabilization of a rotating body-beam system. *IEEE Trans. Automat. Contr.*, **41**, **2**, 241-245.

Morgül, O. (1991). Orientation and stabilization of a flexible beam attached to a rigid body: Planar motion. *IEEE Trans. Automat. Contr.*, **36**, no. 5, 953-963.

Morgül, O. (1993). Constant angular velocity control of a rotating flexible structure. *in Proc. 2nd Conf., ECC'93., Groningen, Netherlands*, 299-302.

Morgül, O. (1994). Control of a rotating flexible structure. *IEEE Trans. Automat. Contr.*, **39**, no. 52, 351-356.

Pazy, A. (1983). *Semigroups of linear operators and applications to partial differential equations*. Springer Verlag. New york.

Pierre, M. (1976). Perturbations localement lipschitziennes et continues d'opérateurs *m*-accretifs. *Proceeding of the American Mathematical Society*, **58**, 124-128.

Rao, B. (1993). Decay estimate of solution for hybrid system of flexible structures. *Euro. J. Appl. Math*, **4**, 303-319.

Vrabie, I. I. (1987). *Compactness methods for nonlinear evolutions*. Pitman Monographs and surveys in Pure and Applied Mathematics, John Wiley & Sons, Inc., New York.

Xu, C. Z and J. Baillieul (1993). Stabilizability and stabilization of a rotating body-beam system with torque control. *IEEE Trans. Automat. Contr.*, **38**, no. 12, 1754-1765.

Xu, C. Z and G. Sallet (1992). Boundary stabilization of a rotating flexible system. *in Lecture Notes in Control and Information Sciences*. (R. F. Curtain, A. Bensoussan, and J. L. Lions, Eds.). vol. 185, pp. 347-36. New York.

E-2c-10-5
14th Triennial World Congress, Beijing, P.R. China

COMPARATIVE STUDIES OF P AND PI TYPE ITERATIVE LEARNING CONTROL ALGORITHMS FOR A CLASS OF PROCESS CONTROL PROBLEMS

Jian-Xin Xu, Kok-Hwa Liew and Chang-Chieh Hang

National University of Singapore Department of Electrical Engineering
10 Kent Ridge Crescent Singapore 119260
elexujx@nus.edu.sg

Abstract: This paper is concerned with the application of Iterative Learning Control (ILC) to a class of process systems that can be approximated as a first order plus dead-time model. The P-type ILC with Previous Cycle Feedback (PCF) and Current Cycle Feedback (CCF) are investigated. It is shown that ILC with PCF may not be suitable for the above mentioned system due to the large initial tracking error, which in turn results in control signal saturation, and thus degrades the control performance. On the other hand, ILC with CCF performs much better but the appropriate gain has to be searched by trial and error. To overcome this difficulty, P and PI-type ILC which are based on well established auto-tuning schemes are proposed to set the learning control gains. Experimental studies on the level control of a coupled-tank system verify the effectiveness and implementability of the proposed ILC algorithms. *Copyright © 1999 IFAC*

Keywords: Iterative Learning Control, Process Control, Auto-tuning.

1. INTRODUCTION

Controlling a nonlinear system in real time is a challenging task, which usually involves extensive mathematical formulation and intensive computation. Learning control however offers an easy and effective method to implement controllers in practice. The main strategy of the ILC is to improve the performance of the system being controlled iteratively by using the information obtained from the previous trials, and eventually to obtain an optimal control input that gives desired output with acceptable accuracy. Owing to the effectiveness of ILC for tracking control tasks in the presence of system uncertainties, increased attention has been given and a great number of ILC schemes have been developed hitherto (Lee, et al., 1997). Up to date, ILC has been mainly applied to and verified by motion control problems such as robotics control, servo- mechanisms, etc. (Longman, et al., 1997).

In this study, we demonstrated the design of real-time learning controllers for a class of process control problems that can be approximated as a first order plus dead-time model. It is well known that a large number of process control problems, though nonlinear in nature, can be approximated and effectively modeled as:

$$G(s) = \frac{Ke^{-\tau s}}{1 + sT_a} \qquad (1)$$

where K is the plant gain, τ is the apparent dead-time and T_a is the apparent time constant.

In comparison with the optimal or other control techniques, the ILC controller requires less prior knowledge regarding system parameters and thus is characterized by its simplicity. In this paper, we apply three types of the simplest ILC schemes to a coupled-tank system to investigate the effectiveness. Through substantial experiments that are conducted on the coupled-tank system, it was evident that ILC with PCF may not be suitable for the above mentioned system due to the large initial tracking error, which in turn results in control signal saturation. On the other hand, ILC with CCF performs much better but the appropriate gain has to be searched by trial and error. To overcome the difficulty, PI auto-tuning schemes is used to set the learning control gains. Meanwhile, learning techniques are applied to further improve the control performance.

In Section 2, we discuss the architecture of the learning controller and the three ILC schemes that to be investigated. Section 3 gives the overview of the coupled-tank system and the experimental setup. Comparative studies are conducted in Section 4 in terms of experimental results.

2. THE LEARNING CONTROLLERS

The architecture of the learning controllers that are implemented here takes the form of Figure 1. It consists of two major parts: a feedforward compensation, u^{ff} for the new control iteration and a linear feedback component. As noted, only the measurable sensor output voltage is used as the feedback signal, namely the P-type ILC. The selection of P-type ILC instead of D-type is justified below: (a) D-type ILC required the derivative of the error signal $\dot{e}(t)$ which is not directly available, (b) Although numerical methods can be used to approximate $\dot{e}(t)$, it is very sensitive to measurement noise and may destroy the system stability.

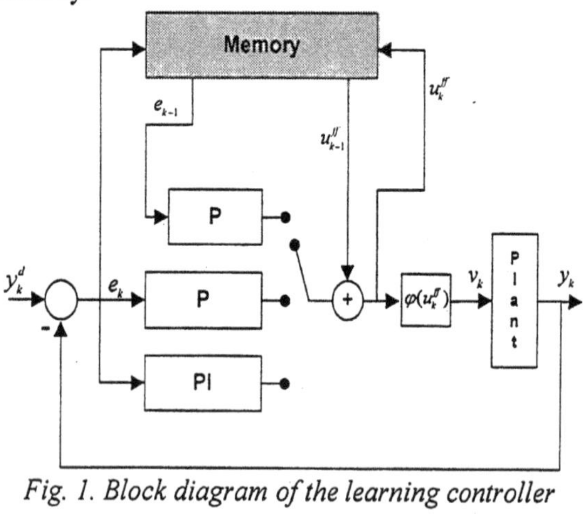

Fig. 1. Block diagram of the learning controller

A real system can only generate limited control signals. Thus an input saturation function is introduced and defined as:

$$\phi(u^{ff}(ih)) = \begin{cases} U_{upper} & if \ u^{ff}(ih) > U_{upper} \\ u^{ff}(ih) & if \ U_{lower} \le u^{ff}(ih) \le U_{upper} \\ U_{lower} & if \ u^{ff}(ih) < U_{lower} \end{cases} \quad (2)$$

Input update law of the learning control is given as:

$$u_{k+1}^{ff}(ih) = u_k^{ff}(ih) + l[e_k(ih), e_{k+1}(ih)] \qquad ih \in [0,T] \quad (3)$$

where $l[.]$ is the learning gain, k is the trial number, i is the time index, h is the sampling interval and $e_k(ih) = y_k^d(ih) - y_k(ih)$ is the error signal at k-th trial.

The objective of ILC is to ensure the asymptotic convergence with respect to the iteration within a finite tracking period. In this paper, only three ILC schemes will be investigated, namely the P-type ILC with Previous Cycle Feedback (PCF), P-type ILC with Current Cycle Feedback (CCF) and PI-type ILC. The formulation for these schemes are given as below:

<u>*P-type ILC with PCF:*</u>
In this case, the update of the control signal of the whole tracking period $[0,T]$ is based on the error profile obtained from the previous trial. It is purely a feed-forward control in time horizon:

$$u_{k+1}^{ff}(ih) = u_k^{ff}(ih) + Le_k(ih) \quad (4)$$

<u>*P-type ILC with CCF:*</u>
This is the combination of feed-forward and feedback control. The update law requires the current error signal as feedback to contribute to the overall control action:

$$u_{k+1}^{ff}(ih) = u_k^{ff}(ih) + Le_{k+1}(ih) \quad (5)$$

<u>*PI-type ILC:*</u>
If the feedback controller in Figure 1 is replaced with a Proportional-Integral controller, we have the PI-type ILC as formulated below:

$$u_{k+1}^{ff}(ih) = u_k^{ff}(ih) + w_{k+1}(ih) \quad (6)$$

$w^{k+1}(ih)$ is the PI controller output and is given as:

$$\begin{aligned} w_{k+1}(ih) &= P_{k+1}(ih) + I_{k+1}(ih) \\ P_{k+1}(ih) &= K_c e_{k+1}(ih) \\ I_{k+1}(ih) &= I_{k+1}(ih-h) + \frac{K_c h}{T_i} e_{k+1}(ih) \end{aligned} \quad (7)$$

where K_c and T_i are the learning gains that are selected based on Ziegler-Nichols auto-tuning.

3. THE COUPLED-TANK SYSTEM

Experiments are carried out to evaluate the performance of the learning control schemes that has been discussed in Section 2. The experiments utilized a coupled-tank system which can be approximated as a first order plus dead-time plant. The main objective here is to control the water level in the tank. This control problem is commonly found in process industries whereby it is required to supply certain

A = Cross-sectional area of both tanks

a_1 = Cross-sectional area of orifice 1

a_2 = Cross-sectional area of orifice 2

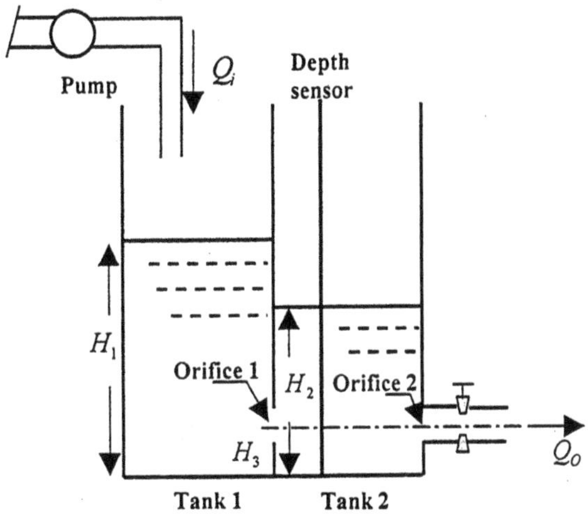

Fig. 2. Schematic diagram of the coupled-tank system

fluid to a chemical reactor at a constant rate.

Normally, a "hold-up" tank or reservoir is used to filter out any variations in the upstream supply flow and at the same time to ensure a temporary supply of reactant in case of process failure in supplying upstream.

3.1 The Plant Model

The plant employed in the experiment is a coupled-tank system. The schematic diagram of the system is shown in Figure 2. It consists of two "hold-up" tanks, a pump and a depth sensor.

The mathematical model of the coupled-tank system can be described as:

$$\begin{bmatrix} \dot{H}_1 \\ \dot{H}_2 \end{bmatrix} = \begin{bmatrix} -k_1/A & k_1/A \\ k_1/A & -(k_1+k_2)/A \end{bmatrix} \begin{bmatrix} H_1 \\ H_2 \end{bmatrix} + \begin{bmatrix} 1/A \\ 0 \end{bmatrix} Q_i \quad (8)$$

where $k_1 = \dfrac{C_{d1} a_1 \sqrt{2g}}{2\sqrt{H_1 - H_2}}$,

$k_2 = \dfrac{C_{d2} a_2 \sqrt{2g}}{2\sqrt{H_2 - H_3}}$,

g = Gravitational constant,

C_{d1}, C_{d2} = discharge coefficient, approximate 0.6,

$H_1, H_2, H_3, A, a_1, a_2, Q_i$ are defined as in Figure 2.

At steady state, the inflow rate (Q_i) and outflow rate (Q_0) are equal. From equation (8), it is clear that the outflow rate is a function of square root of the fluid level; the system essentially is nonlinear.

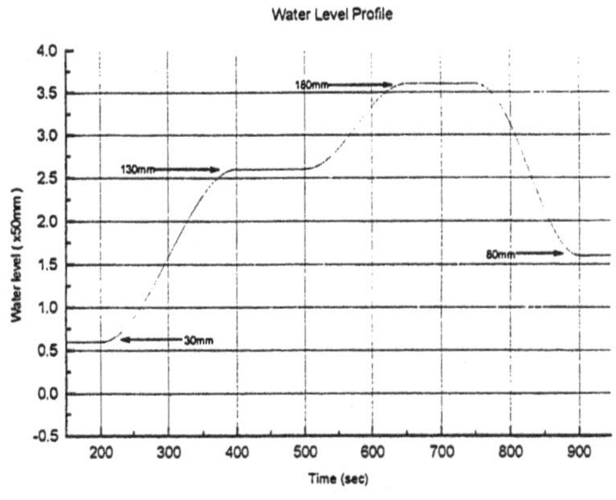

Fig. 3. The tracking profile of fluid level in tank 2

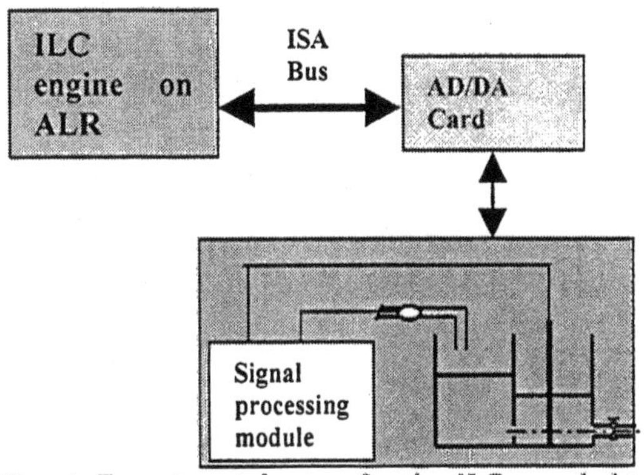

Fig. 4. Experimental setup for the ILC-coupled-tank system

Nevertheless, it is possible to simplify the plant transfer function to a first order plus a dead-time structure. The simplified plant model is estimated by obtaining the setpoint changing response and is given as:

$$G(s) = \frac{Ke^{-\tau s}}{1+sT} = \frac{0.72e^{-27s}}{1+246s} \quad (9)$$

The basic control problem here is to control the water level in tank 2 (H_2) by varying the speed of the pump in such a way that H_2 follows a pre-defined water level profile as given in Figure 3.

3.2 Experimental Setup

Figure 4 shows the overall setup of the experiment. It consists of three major sub-units:

(a) The coupled-tank apparatus which consists of a transparent plexi-glass container with dimension of 200mm x 100mm x 300mm and is divided into two tanks through a partition. Flow between the tanks is by means of a series of holes drilled at the base of the partition. These holes constitute "orifice 1" in Figure 2. Water is pumped from a reservoir into the first tank by a variable speed pump, which is driven by a D.C. motor. The depth of the fluid is measured using a parallel track depth sensor, which are attached in tank 1 and 2. Change in the water level is detected by means of the change in electrical resistance measured. This provides an electrical signal that is proportional to the height of the water.

(b) The Iterative Learning Control engine which was written in C++ and implemented on ALR computer with 16MB RAM and running at 90MHz. Besides, an in-house software, FlexControl™ is loaded to provide the real-time data capturing and viewing.

(c) A signal-processing module is installed in the system setup as well. The main function of this module is to provide a translation (signal conditioning) of the voltage levels between the AD/DA card and the actual plant to be controlled. It also provides the current drive to the D.C motor of the pump. The data acquisition card that is used in this experiment is from Data Translation (DT2801).

3.3 Filtering and Non-linearity Compensation

In this sub-section, we would like to highlight two important problems that are encountered during the implementation stage.

Filtering: While implementing the above control laws, it was realized that filtering is an important practical issue. Due to the nature of learning control, high frequency components(measurement noise) may be accumulated in subsequent trials. With a filter preceding the previous cycle control input, the high frequency components are suppressed. For example, the update control law for equation (4) should be modified to include the filter as:

$$u_{k+1}^{ff}(ih) = f\left[u_k^{ff}(ih)\right] + Le_k(ih) \qquad (10)$$

where $f[.]$ stands for a low pass filter with the cut-off frequency of ω_0.

The selection of the filter cut-off frequency ω_0 is important. If ω_0 is too small, the system will not learn properly as the desired frequency components are suppressed as well. On the other hand, if ω_0 is too large, the tracking output will be noisy. As a guide, it is desirable to select the filter cut-off frequency about 5 times the system bandwidth and it should be 10 times below the sampling frequency. In this experiment, a simple first order low pass filter $\frac{a_0}{s+\omega_0}$ is used with the cut-off frequency $\omega_0 = 0.15 rads^{-1}$ and d.c gain $\frac{a_0}{\omega_0} = 1$. The sampling frequency is set to 1 Hz.

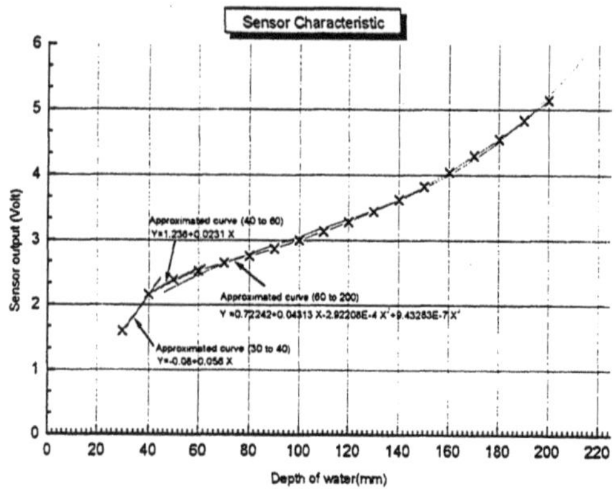

Fig. 5. Sensor characteristic (obtained through experiment)

Nonlinear Depth Sensor: The degree of non-linearity of the whole system is increased as the depth sensor itself is nonlinear. The sensor characteristic is plotted in Figure 5. This non-linearity is taken into account in the implementation where the graph is approximated by a third order polynomial equation and two linear equations as shown in the figure.

4. EXPERIMENT AND COMPARATIVE STUDIES

This section provides the experimental results for the systems that are discussed in Section 2 and 3. The graphs plotted here are the real time data captured while the system is running. The ILC engine is implemented in a PC and the aim is to control the water level in the coupled-tank system such that it follows the prescribed profile as shown in Figure 3. Root Mean Square (RMS) error as defined in equation (11) is used as the performance index for the comparison among the schemes given above.

$$RMS = \sqrt{\frac{\sum_{i=1}^{N} |e(i)|^2}{N}} \qquad (11)$$

N = Number of points within the tracking period $[0, T]$

4.1 P-type ILC with PCF

Figure 6 shows the tracking performance of this ILC scheme at Trial 2 when the learning gain is selected to be 3. The water overflowed after 420 seconds due to the constantly large control input.

This is due to the zero control input at Trial 1, the error value stored in the memory is large (setpoint

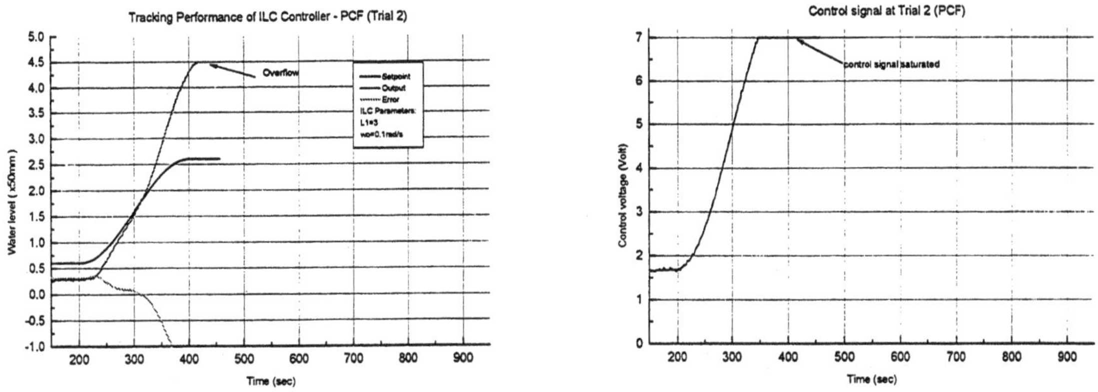

Fig. 6. Tracking Performance of P-type ILC with PCF at Trial 2

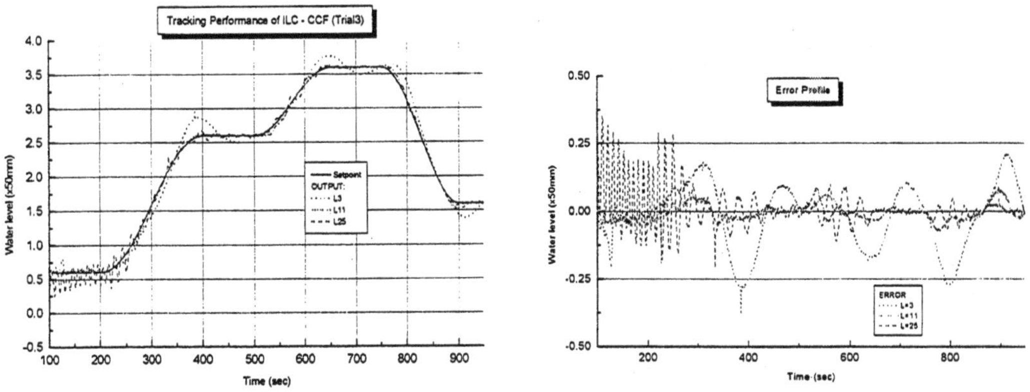

Fig. 7. Tracking performance & error profile of ILC with CCF

itself). From Figure 6, we can see that these large errors caused the control input at the second trial to saturate at maximum level. Further experiments conducted on the same system with lower learning gains (between 0.5 to 0.8) shows that although the control input did not saturate, the system takes infinite iterations to reach a satisfactory performance due to the low convergence rate. Thus, the P-type ILC with PCF is not suitable for controlling this coupled-tank system.

4.2 P-type ILC with CCF

In this case, the learning control scheme as formulated in equation (5) is implemented. The performance of three learning gains which range from low to high are presented in this paper. Figure 7 shows the tracking performance of this ILC scheme with L=3, 11 and 25 at Trial 3.

From the tracking performance that presented, we conclude that P-type ILC with CCF is a good candidate for the coupled-tank system. The RMS error profile for L=3, 9, 11, and 25 is shown in Figure 8. With this learning scheme, the RMS error can be reduced by about 10 times (i.e. RMS error for L=11.0 reduced from 0.2704 to 0.0297). Also, the high gain learning is not desirable as small oscillation appeared in Trial 3 (Figure 7). On the other hand, too

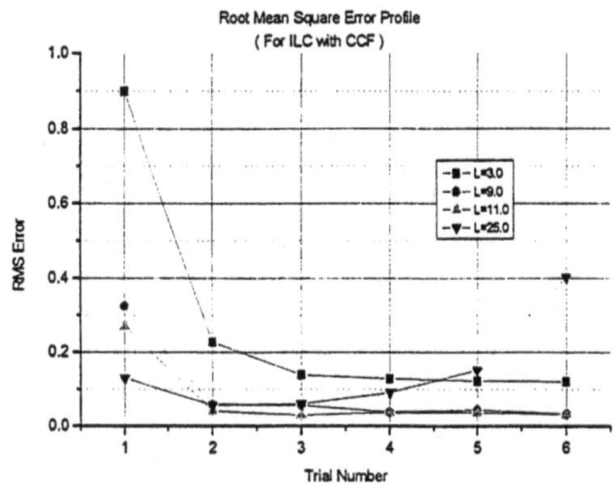

Fig. 8. RMS error profile for ILC with CCF

small the gain may defeat the purpose of learning as shown in Figure 8.

Although P-type ILC with CCF gives a good tracking performance, it depends very much on the learning gain that is selected. It needs a systematic way to tune the learning gain. In the last subsection, we explore the possibility of using the well-known PI auto-tuning to set the appropriate learning gain.

4.3 PI-type ILC

There exist many tuning or auto-tuning algorithms such as the gain/phase margin and the ITAE index for the widely used conventional PID controllers (ÅstrÖm, et al., 1988). The main idea of the PI-type

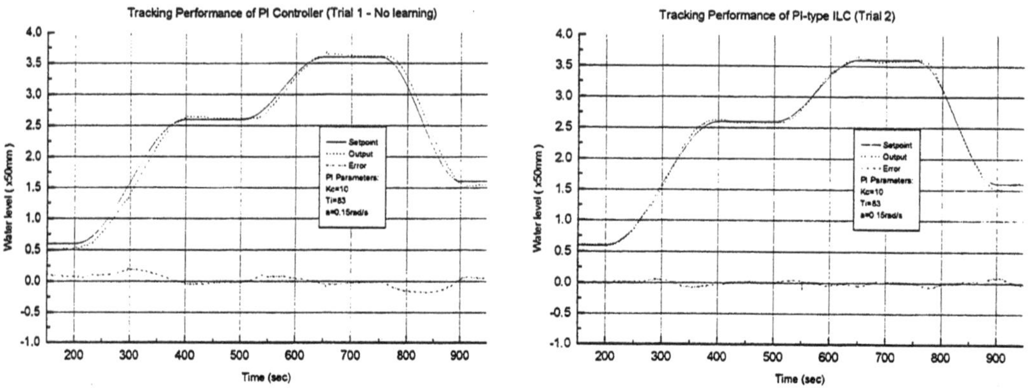

Fig. 9. Tracking performance of PI-type ILC

ILC is to utilize these auto-tuning methods to set the learning gains. For the coupled-tank experiment, the Ziegler-Nichols auto-tuning rule is used to set the learning gains due to its simplicity. The ultimate gain and period are obtained experimentally as $K_u = 22.5$ and $T_u = 99.2$ respectively. If Proportional(P) control is used to control the plant, the Ziegler-Nichols tuning rule suggested that the P-control gain is set as $\frac{1}{2}K_u = 11.25$. In section 4.2, it is shown that the CCF scheme with $L = 11.0$ achieves excellent tracking performance. Thus, the proportional gain as suggested can indeed be used to set the learning gain for the CCF scheme.

To further reduce the steady state error, we used the Proportional-Integral (PI) control to set the learning gains. The tuning rule gives the learning gains as $K_c \approx 10$ and $T_i \approx 83$. Figure 9 shows the tracking performance of this scheme for Trial 1 and Trial 2. It can be seen that tracking error is further reduced from Trial 1 to Trial 2.

Figure 10 gives the RMS error plots for all the learning schemes that have been discussed. Overall, the learning schemes perform better than the conventional PI controller. The achievable RMS error is as low as 0.027. For the conventional PI control, the RMS error is 0.095 and is about 70% higher. The PI-type ILC is preferred as it remains as the conventional PI-control in the first trial and achieves better performance in the second trial.

Consider the fact that the auto-tuned learning controller can converge very fast, and in practice there always exists unpredictable disturbance and measurement noise, it is recommended to cease the learning process after 2 or 3 trials. This is experimentally verified in Figure 10.

5. CONCLUSIONS

Iterative learning control is proposed for the tracking control of the finite-time operation on a first order

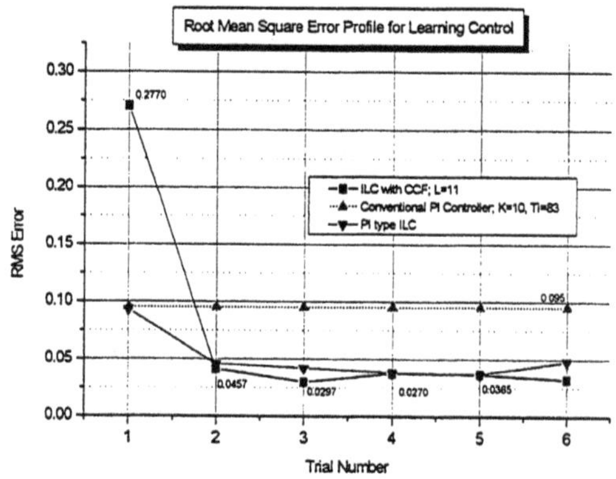

Fig. 10. RMS Error profile for learning control

plus dead-time plant. Experiments conducted on the coupled-tank system show that P-type ILC with PCF scheme is not suitable for controlling the system while ILC with CCF scheme gives good tracking performance. The use of auto-tuning method to set the learning gains has effectively improved the tracking performance after the first trial. The implementation of the ILC is relatively easy as it requires only a number of memory components plus a simple P or PI feedback controller.

REFERENCES

H.S. Lee and Z. Bien (1997)."A note on convergence property of iterative learning controller with respect to sup norm," Automatica, **Vol. 33(4)**, pp. 1591-1593.

J.X. Xu (1997). "Analysis of iterative learning control for a class of nonlinear discrete-time systems," Automatica, **Vol. 33(10)** pp. 1905-1907.

K.J. ÅstrÖm, and T. Hägglund (1988). "Automatic Tuning of PID Controllers," Instrument Society of America.

R.W. Longman and Chun-Ping Lo (1997). "Generalized holds, ropple attenuation, and tracking additional outputs in learning control," Journal of Guidance, Control, and Dynamics, **Vol. 20(6)**, pp. 1207-1214.

14th Triennial World Congress, Beijing, P.R. China

ADAPTIVE H_∞-CONTROL FOR ROBUST VISUAL FEEDBACK SYSTEMS

Akira Maruyama * and **Masayuki Fujita** **

* *Nachi-Fujikoshi Corp., Fujikoshi-Honmachi, Toyama, 930–8511, Japan*
** *Faculty of Engineering, Kanazawa University, Kodatsuno, Kanazawa, 920-8667, Japan*
E-mail: fujita@t.kanazawa-u.ac.jp

Abstract: This paper deals with an adaptive H_∞-control for the Lagrangian systems with visual feedback in presence of unknown parametric uncertainties. Based on the adaptive manipulator control, the proposed controller consists of a visual PD feedback part and the full dynamics feedforward part with the update law of the parametric uncertainty. The L_2-gain performance analysis of the proposed controller is carried out and its asymptotic property is examined. *Copyright © 1999 IFAC*

Keywords: Nonlinear control, H_∞ control, Visual motion, Disturbance rejection, Adaptive control

1. INTRODUCTION

A lot of researches have been investigated the control of the manipulator systems described by the Euler-Lagrange equations, so-called *Lyapunov based method.* The Lyapunov based method tackled the control problem by exploiting the physical structure of the manipulator(Arimoto, 1996). Recently the robust manipulator control against the various uncertainties has been studied, e.g, the adaptive control(Ortega, *et al.*, 1989), the performance analysis of the disturbance attenuation properties in the nonlinear H_∞ setting (Scherpen, *et al.*, 1997) and so on.

In recent years, the use of the visual information in the feedback loop, called *visual servo* or *visual feedback*, has also attracted the attention(Hutchinson, *et al.*, 1996). More recently, the Lyapunov based method has been successfully applied to the visual feedback problem (Kelly, 1996, Mauryama, *el al.*, 1997). Furthermore, the robust visual feedback control against the torque disturbances and the target motions in L_2 gain sense has been derived (Maruyama, *el al.*, 1998). These works have been studied under the assumption of the exact knowledge of the manipulator model parameters. However the exact manipulator parameters can not be generally obtained since the uncertainties inevitably exist in the manipulator model. The adaptive H_∞ control problem in presence of both the parametric uncertainty and the external disturbance has been investigated for the nonlinear systems (Lu, *et al.*, 1997). The design of the adaptive H_∞ controller requires solving a nonlinear Hamilton-Jacobi inequality. In general, this solution is difficult to compute.

The present paper proposed an adaptive visual feedback control algorithm for the manipulator systems with the parametric uncertainty. Futhermore, the visual feedback controller guarantees the robustness to the external disturbances, such as the manipulator torque disturbances and/or the unknown motion of the image target object. The robustness is measured by the L_2 gain of nonlinear systems, see (van der Schaft, 1996). The storage function is constructed by the adaptive H_∞ technique introduced in (Lu, *et al.*, 1997). The image feature parameter potential

(Maruyama, it et al., 1997, 1998) is the important role of the L_2 gain performance analysis for the visual feedback system.

Throughout this paper, the Euclidean norm of vector $x \in \mathcal{R}^n$ is defined as $\|x\| := \sqrt{x^T x}$. For all vector functions $f : [0, \infty) \to \mathcal{R}^n$, L_2 is the Banach space with respect to the L_2 norm $\|f\|_2^2 := \int_0^\infty \|f(t)\|^2 dt$.

This paper is organized as follows. Section 2 shows the visual feedback system model with both the parametric uncertainty and the external disturbances. Further the visual feedback control problem in the adaptive H_∞ setting is formulated. In Section 3, it is shown that the L_2 gain of the closed loop system can be made sufficiency small with a suitable selection of the feedback gains. Then the stability and asymptotic property is shown in order to confirm the internal stability. In Section 4, a simulation example is made for the visual feedback system with the proposed controller. The conclusions are finally given in Section 5.

2. SYSTEM MODEL AND PROBLEM STATEMENT

2.1 *Camera and Manipulator Model*

Consider a non-redundant planar manipulator with the world frame $\Sigma_w = \{X_w\ Y_w\ Z_w\}$ depicted in Figure 1. Suppose a camera is mounted on the manipulator end-effector attached a frame $\Sigma_c = \{X_c\ Y_c\ Z_c\}$. A frame $\Sigma_i = \{X_i\ Y_i\}$ is defined in the camera image plane and its origin is the intersection of the optical axis with the image plane. Here it is assumed that the axes X_i and Y_i parallel the axes X_c and Y_c respectively, and the planes X_c–Y_c and X_w–Y_w are separated by the focal length $l > 0$. The object point is located at $[x_p\ y_p\ z_p]^T$ with respect to the frame Σ_w. It is assumed that $z_p > l$ and the object point motion is restricted on the $X_w - Y_w$ plane, i.e. $\dot{z}_p = 0$.

Under the above assumptions, the projective geometry of the camera is modeled by a scaled orthographic projection as follows:

$$f(p, c(q), \theta_c(q)) = se^{-\hat{\theta}_c(q)}(b - c(q)), \quad (1)$$

where $q \in \mathcal{R}^2$ is the joint angle vector of the non-redundant manipulator, $c(q) \in \mathcal{R}^2$ and $\theta_c(q) \in \mathcal{R}$ are the camera position and its orientation with respect to Σ_w, respectively, and given by the manipulator direct kinematics. $b := [x_p\ y_p]^T$ is the object position and $s > 0$ is a constant scaling factor which consists of the focal length l and the depth z_p. The form $\hat{a} \in so(2), a \in \mathcal{R}$ can be represented by a following 2×2 skew-symmetric matrix:

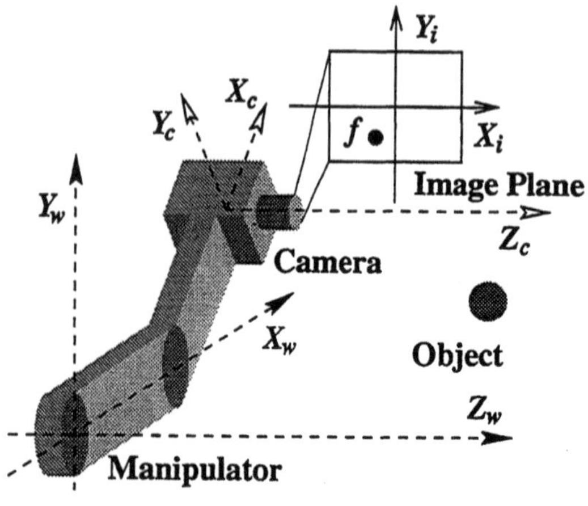

Fig. 1. Visual Feedback Configuration

$$\hat{a} := \begin{bmatrix} 0 & -a \\ a & 0 \end{bmatrix}.$$

The time derivative of (1) gives us the following equation,

$$\dot{f} = -se^{-\hat{\theta}_c} J_p(q)\dot{q} - \hat{\dot{\theta}}_c f + se^{-\hat{\theta}_c}\dot{b}, \quad (2)$$

where $\dot{b}$ is the unknown velocity vector of the target and $J_p(q) \in \mathcal{R}^{2\times2}$ is the manipulator Jacobian which is used in the differential kinematics equation, $\dot{c} = J_p(q)\dot{q}$.

A standard method for deriving the dynamic equations of the manipulator is via the so-called Euler-Lagrange equations. It is common to write the Euler-Lagrange equations in matrix form as

$$M(q)\ddot{q} + C(q, \dot{q})\dot{q} + g(q) + w = \tau, \quad (3)$$

where τ is the vector of control input torques, w is the external torque disturbances vector, $M(q)$ is the inertia matrix, $C(q, \dot{q})\dot{q}$ is the Coriolis and centrifugal torques vector and $g(q)$ is the gravitational torques vector. It is well known that the inertia matrix $M(q)$ is positive definite and the matrix $\dot{M}(q) - 2C(q, \dot{q})$ is the skew–symmetric. The dynamic equations (3) are linear in the parameters, i.e.

$$M(q)a + C(q, \dot{q})v + g(q) = Y(q, \dot{q}, v, a)\rho, \quad (4)$$

with ρ is the vector of the manipulator's constant parameters and $Y(\cdot)$ is a matrix known as the *regressor*.

2.2 *Parametric Uncertainties of Manipulator*

This paper focuses only on the parametric uncertainty of the manipulator dynamics (3). Needless to say, it is recognized that the camera model (1) has the parametric uncertainty of the scaling factor s. However it should be remarked that the proposed controller is expected to have a robustness against the uncertainty of the scaling factor s, since no scaling factor is used.

With respect only to the parametric uncertainty, the nominal manipulator model is defined as

$$M_0(q)a + C_0(q,\dot{q})v + g_0(q) = Y(q,\dot{q},v,a)\rho_0, \tag{5}$$

where M_0, C_0, g_0 and ρ_0 represent nominal values vis-a-vis to parametric uncertainty of M, C, g and ρ. Therefore the parametric uncertainty is expressed by the following equations :

$$\begin{aligned}(M(q) - M_0(q))a + (C(q,\dot{q}) - C_0(q,\dot{q}))v \\ + (g(q) - g_0(q)) = Y(q,\dot{q},v,a)\tilde{\rho},\end{aligned} \tag{6}$$

where $\tilde{\rho} := \rho - \rho_0$. Note that the time derivative of the uncertainty $\tilde{\rho}$ equals to 0, because ρ and ρ_0 are the constant values.

2.3 *Visual Feedback Control Problem*

The visual feedback control objective is to design a control input τ such that the camera position c coincides with the object position b. $c = b$ is equivalent to $f = 0$, because the scaling factor s is positive and $e^{-\hat{\theta}_c}$ is the nonsingular matrix in the camera model (1).

The following assumptions are needed:

A1 There is a manipulator joint configuration q_0 for which $f = 0$. It should be noted that q_0 have no need to be known or measured.
A2 The unknown desired position q_0 is an isolated solution of $f(q_0) = 0$.
A3 The Jacobian J_p is the nonsingular matrix.

A1 assures that the visual feedback problem is solvable and A2 and A3 are necessary conditions to examine the asymptotic property.

With the parametric uncertainty $\tilde{\rho}$ in mind, a nominal visual feedback controller is defined as

$$\begin{aligned}\tau &= u + M_0(\lambda\dot{\eta}) + C_0(q,\dot{q})(\lambda\eta) + g_0(q), \\ &= u + Y(q,\dot{q},\lambda\eta,\lambda\dot{\eta})\rho_0,\end{aligned} \tag{7}$$

where $\eta := J_p^T(q)e^{\hat{\theta}_c}f$, $\lambda \in \mathcal{R}$ is positive and u is the control input that will be designed to achieve the visual feedback control objective. The definition of the nominal controller is same as the visual feedback control law proposed in (Maruyama, et al., 1998).

This paper deals with the adaptive H_∞ control design for the visual feedback system. The objective of the design is to attenuate the impact of the exogenous disturbances and the error induced by the initial guess of the parameters. Here the disturbance torque w and the unknown target velocity $\dot{b}$ are regarded as the exogenous disturbances denoted by $r := [w^T\ \dot{b}^T]^T$. The impact is measured by the L_2-gain from the disturbance r to the controlled output,

$$z = [\xi^T\ f^T]^T. \tag{8}$$

where $\xi := \dot{q} - \lambda\eta$. Substituting the control laws (7) into (3) gives us the following closed loop system:

$$\begin{aligned}M(q)\dot{\xi} &= -C(q,\dot{q})\xi + u + w - Y(q,\dot{q},\lambda\eta,\lambda\dot{\eta})\tilde{\rho}, \\ \dot{f} &= -se^{-\hat{\theta}_c}J_p(q)\dot{q} - \dot{\hat{\theta}}_c f + se^{-\hat{\theta}_c}\dot{b}, \\ z &= [\xi^T\ f^T]^T.\end{aligned} \tag{9}$$

The precise statement of the visual feedback control problem is given by

Visual feedback control problem in the L_2 gain sense : *Suppose $\varepsilon > 0$ and $\gamma > 0$ are given. Design a controller u such that the closed-loop system (9) with $[q^T(0)\ \dot{q}^T(0)]^T = [q_0^T\ 0]^T$ satisfies*

$$\int_0^T \|z\|^2 dt \le \gamma^2 \int_0^T \|r\|^2 dt + \varepsilon, \tag{10}$$

for all $T \in \mathcal{R}_+, r \in L_2$ and $\tilde{\rho}$.

The ε term represents the effect due to the mismatching between the initial guess and the true value of the uncertainty. In order to guarantee the H_∞-performance (10), it is required to show that the closed-loop system (9) is dissipative with respect to supply rate $\frac{1}{2}(\gamma^2\|r\|^2 - \|z\|^2)$. It is enough to find a storage function V for each $\tilde{\rho}$, s.t. the following differential dissipation inequality holds:

$$\dot{V} \le \frac{1}{2}(\gamma^2\|r\|^2 - \|z\|^2). \tag{11}$$

2.4 *The H_∞ Visual Feedback Controller for the nominal model*

This subsection describes the case where the parametric uncertainty $\tilde{\rho} = 0$. This issue was discussed in (Mauryama, *et al.*, 1998) and the following result was obtained.

Theorem 1. Let $\tilde{\rho} = 0$. Suppose a visual feedback control law is given by

$$u = -K_1\xi + k_2\eta, \tag{12}$$

which has the gains K_1 and k_2 such that the following inequality holds

$$-\xi^T K_1 \xi - k_2 \lambda f^T e^{-\hat{\theta}_c} J_p J_p^T e^{\hat{\theta}_c} f + \frac{1}{2}\xi^T\xi + \frac{1}{2} f^T f + \frac{1}{2\gamma^2}\xi^T \xi + \frac{k_2^2}{2\gamma^2} f^T f \leq 0, \tag{13}$$

Then the closed-loop system (9) and (12) satisfies the H_∞ criterion (10).

In the previous work (Maruyama, *et al.*, 1998), it is shown that the inequality (13) is a sufficient condition which a storage function,

$$V = \frac{1}{2}\xi^T M(q)\xi + \frac{k_2}{2s}\|f(q)\|^2, \tag{14}$$

is satisfied the differential dissipation inequality as follows:

$$\dot{V} + \frac{1}{2}(\|z\|^2 - \|r\|^2) \leq 0, \tag{15}$$

for $r \in L_2$. The storage function (14) is defined as the sum of the manipulator's kinetic energy and the *image feature parameter potential function.*

3. ADAPTIVE H_∞ VISUAL FEEDBACK CONTROL WITH PARAMETRIC UNCERTAINTY

3.1 L_2 *gain performance analysis*

In this subsection, consider the solutions to the visual feedback control problem in the adaptive H_∞-setting is considered. When the uncertainty $\tilde{p}$ is known, the function V is a solution to the differential dissipation inequality. However, in the case where the uncertainty $\tilde{p}$ is unknown, it is required that the storage function reflects the parameter estimated value p.

The following approach is motivated by the adaptive H_∞ control introduced in (Lu, *et al.*, 1997). Based on the well-known adaptive manipulator control (Ortega, *et al.*, 1989), the control input u is defined as

$$\dot{p} = -\Gamma^{-1} Y^T(q, \dot{q}, \lambda\eta, \lambda\dot{\eta})\xi, \tag{16}$$

$$u = Y(q, \dot{q}, \lambda\eta, \lambda\dot{\eta})p - K_1\xi + k_2\eta, \tag{17}$$

where p is the estimated vector of the parametric uncertainties, Γ is the positive definite and the equation (16) is the parameter update law with the initial condition $p(0)$. Substituting the control laws (16) and (17) into (9), the following equation is obtained after some algebra.

$$\begin{aligned} M(q)\dot{\xi} &= -C(q,\dot{q})\xi + u + w + Y(p - \tilde{p}), \\ \dot{f} &= -se^{-\hat{\theta}_c} J_p(q)\dot{q} - \dot{\hat{\theta}}_c f + se^{-\hat{\theta}_c}\dot{b}, \\ \dot{p} &= -\Gamma^{-1} Y^T \xi. \end{aligned} \tag{18}$$

The storage function candidate is the aforementioned storage function V with an additional quadratic p-term:

$$W = V + \frac{1}{2}(p - \tilde{p})^T \Gamma (p - \tilde{p}). \tag{19}$$

Calculating the time derivative of the positive definite function W yields

$$\begin{aligned} \dot{W} &= \xi^T(-C\xi - K_1\xi + k_2\eta - w + Y(p - \tilde{p})) \\ &\quad + \frac{k_2}{s} f^T(-se^{-\hat{\theta}_c} J_p \dot{q} - \dot{\hat{\theta}}_c f + se^{-\hat{\theta}_c}\dot{b}) \\ &\quad + \frac{1}{2}\xi^T \dot{M}\xi - (p - \tilde{p})^T \Gamma(\Gamma^{-1} Y^T \xi), \\ &= \frac{1}{2}\xi^T(\dot{M} - 2C)\xi - \frac{k_2}{s} f^T \dot{\hat{\theta}}_c f \\ &\quad - \xi^T K_1 \xi - k_2 \lambda \eta^T \eta - \xi^T w + k_2 f^T e^{-\hat{\theta}_c}\dot{b}. \end{aligned}$$

The skew-symmetry of the matrices $\dot{M} - 2C$ and $\dot{\hat{\theta}}_c$ derives the following equation:

$$\dot{W} = -\xi^T K_1 \xi - k_2 \lambda \eta^T \eta - \xi^T w + k_2 f^T e^{-\hat{\theta}_c}\dot{b}.$$

After completing-the-square, the following inequality holds:

$$\begin{aligned} &\dot{W} + \frac{1}{2}\|z\|^2 - \frac{\gamma^2}{2}\|r\|^2 \\ &= -\frac{\gamma^2}{2}\left\|r - \frac{1}{\gamma^2}\begin{bmatrix} -\xi \\ k_2 f \end{bmatrix}\right\|^2 \\ &\quad - \xi^T K_1 \xi - k_2 \lambda f^T e^{-\hat{\theta}_c} J_p J_p^T e^{\hat{\theta}_c} f \\ &\quad + \frac{1}{2}\xi^T \xi + \frac{1}{2} f^T f + \frac{1}{2\gamma^2}\xi^T\xi + \frac{k_2^2}{2\gamma^2} f^T f. \end{aligned}$$

Note that the worst case disturbance is given by

$$r = \frac{1}{\gamma^2}\begin{bmatrix} -\xi \\ k_2 f \end{bmatrix}.$$

Hence it can be verified that (11) holds if the following inequality is satisfied:

$$-\xi^T K_1 \xi - k_2 \lambda f^T e^{-\hat{\theta}_c} J_p J_p^T e^{\hat{\theta}_c} f + \frac{1}{2}\xi^T\xi + \frac{1}{2} f^T f + \frac{1}{2\gamma^2}\xi^T \xi + \frac{k_2^2}{2\gamma^2} f^T f \leq 0. \tag{20}$$

The above discussion is summarized in the following theorem.

Theorem 2. Given a positive scalar ε and choose the visual feedback controller with the gains K_1, k_2, λ and Γ such that the inequality (20) and

$$(p(0) - \tilde{p})^T \Gamma (p(0) - \tilde{p}) \leq \varepsilon, \tag{21}$$

hold. Then the closed-loop system (18) satisfies (10).

Proof. The equation (20) gives us the following equation:

$$\dot{W} \le -\frac{1}{2}(\|z\|^2 - \gamma^2\|r\|^2). \qquad (22)$$

By integrating both sides of (22) from 0 to T and noticing $W(T) \ge 0$, $q(0) = q_0$ and $\dot{q}(0) = 0$, it is obtained that

$$\int_0^T \|z\|^2 dt \le \gamma^2 \int_0^T \|r\|^2 dt + (p(0) - \bar{p})^T \Gamma (p(0) - \bar{p}),$$

which implies

$$\int_0^T \|z\|^2 dt \le \gamma^2 \int_0^T \|r\|^2 dt + \varepsilon, \qquad (23)$$

for all $T \ge 0$. □

The inequality (20) is written as the following quadratic form:

$$-\begin{bmatrix}\xi^T & f^T\end{bmatrix}\begin{bmatrix}P_{11} & 0\\ 0 & P_{22}\end{bmatrix}\begin{bmatrix}\xi\\ f\end{bmatrix} \le 0. \qquad (24)$$

where

$$P_{11} = K_1 - \frac{1}{2}I - \frac{1}{2\gamma^2}I,$$
$$P_{22} = k_2\lambda e^{-\hat{\theta}_c} J_p J_p^T e^{\hat{\theta}_c} - \frac{1}{2}I - \frac{k_2^2}{2\gamma^2}I,$$

The positive semi-definiteness of the matrix

$$P := \begin{bmatrix}P_{11} & 0\\ 0 & P_{22}\end{bmatrix}$$

is a necessary and sufficient condition which the inequality (20) is satisfied.

Remark 1. The proposed control law (16) and (17) including the gain condition (20) does not require the camera parameter s which consists of the focal length and so on. Therefore the proposed control design is expected to have a robustness against the uncertainties of the camera parameters. It should be noted that there are the following reasons why the our controller need not use the camera parameters:

- By exploiting the image feature parameter potential, the storage function (14) is constructed, and
- The L_2 gain performance analysis makes use of the skew-symmetric property of $\hat{\hat{\theta}}$.

3.2 *Stability analysis*

This subsection deals with the internal stability of the closed-loop visual feedback system, i.e., the asymptotic property of the closed-loop system without the external disturbances. The following theorem shows the asymptotic stability and the parameter convergence.

Theorem 3. Under the assumption of A1, A2 and A3, consider the closed-loop system (18). Suppose the each gains satisfying the inequality (20) If $r = 0$, then the closed-loop system is Lyapunov stable. In addition, the states of the system satisfy that $q \to q_0$ and $\dot{q} \to 0$ as $t \to \infty$. Moreover, the update law satisfies that $\dot{p} \to 0$ as $t \to \infty$.

Proof. This result follows from the fact that the storage function W becomes the Lyapunov function. □

4. A DESIGN EXAMPLE

This section shows the simulation results to confirm the effectiveness of the proposed visual feedback control design.

4.1 *Simulation Setup and Parameters*

Suppose a 2 DOF planar manipulator with a CCD camera. The planar manipulator has the following regressor (Fujita, *et al.*, 1997):

$$Y(q,\dot{q},v,a) = \begin{bmatrix} a_1 & a_1+a_2 & Y_{13}(q,\dot{q},v,a) & \cos q_1 & \cos q_1 & \cos(q_1+q_2)\\ 0 & a_1+a_2 & Y_{23}(q,\dot{q},v,a) & 0 & 0 & \cos(q_1+q_2)\end{bmatrix}$$

where $v = [v_1\ v_2]^T \in \mathcal{R}^2$, $a = [a_1\ a_2]^T \in \mathcal{R}^2$ and

$$Y_{13}(q,\dot{q},v,a) = (2a_1 + a_2)\cos q_2 - (\dot{q}_2 v_1 + \dot{q}_1 v_2 + \dot{q}_2 v_2)\sin q_2,$$
$$Y_{23}(q,\dot{q},v,a) = a_1 \cos q_2 + \dot{q}_1 v_1 \sin q_2.$$

Let us assume that the nominal inertia parameters ρ_0 of unloaded and the parameters ρ with payload (15.0 kg) are given by

$$\rho_0 = \begin{bmatrix}0.25 & 0.028 & 0.033 & 7.57 & 4.08 & 1.63\end{bmatrix},$$
$$\rho = \begin{bmatrix}0.85 & 0.20 & 0.27 & 7.57 & 33.5 & 13.4\end{bmatrix}.$$

It is assumed that the camera parameter s is equal to 8.53×10^{-2} and the camera has a 256×256 pixel array with interpixel spacing 0.05 (pixel/m). Further, it is considered that the target motion is circular whose radius is 4.00×10^{-2} (m) and its angular velocity is 0.8π (rad/s).

To achieve the performance with $\gamma = 4.0$, the gains are selected as follows:

$$K_1 = \text{diag}(2.0, 2.0),$$
$$k_2 = 10.0,$$
$$\lambda = 0.1,$$
$$\Gamma = \text{diag}(1.0, 1.0, 1.0, 0.05, 0.05, 0.05).$$

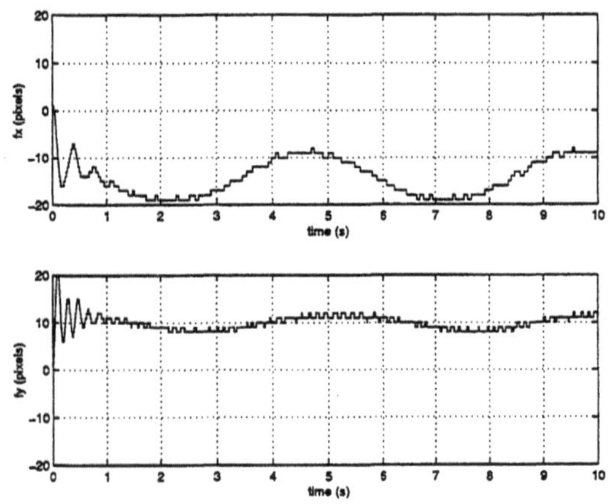

Fig. 2. Responses using Algorithm (12)

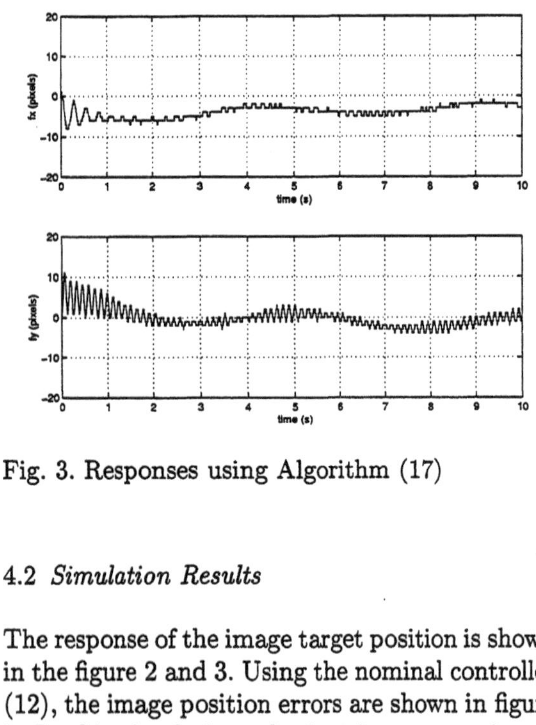

Fig. 3. Responses using Algorithm (17)

4.2 *Simulation Results*

The response of the image target position is shown in the figure 2 and 3. Using the nominal controller (12), the image position errors are shown in figure 2. In this simulation, the inertia parametric uncertainties $\tilde{\rho}$ cause the steady state errors which are around 10 pixels.

The adaptive H_∞ controller (17) results in the performance shown in figure 3. In this case, the steady-state errors are smaller than the result of nominal controller. These simulation results show that the adaptive H_∞ visual feedback control has a robustness against the parametric uncertainties. However the responses of the adaptive H_∞ control are more oscillatory than the nominal controller. It is seemed that this oscillation is caused by unmodeled dynamics such as the quantization of the image data since adaptive control technique performs poorly in presence of unmodeled dynamics.

5. CONCLUSIONS

This paper shows that the adaptive H_∞ control for the visual feedback system is designed via the PD type controller plus dynamics feedforward and the adaptation laws. First, the manipulator model with camera and its parametric uncertainty are introduced. Next, a robust visual feedback controller under the parameter perturbation and the external disturbance is derived. The proposed control is motivated by the adaptive H_∞ theory for nonlinear systems. Finally, it is confirmed that the stability and the asymptotic property of the visual feedback controller.

REFERENCES

Arimoto, S. (1996). *Control Theory of Non-Linear Mechanical Systems, A Passivity-Based and Circuit-Theoretic Approach,* Oxford Univ. Press.

Fujita, M. and N. Tanaka, (1997). A New Type of Robust Tracking Control of Robot Manipulators Based on Generalized SP-D Control Scheme, *Proc. ICRA,* Albuquerque, NM, 2371–2376.

Hutchinson, S., G. Hager and P. Corke (1996). A Tutorial on Visual Servo Control, *IEEE Trans. Robotics and Automation,* **12**, 651–670.

Kelly, R. (1996). Robust Asymptotically Stable Visual Servoing of Planar Robots, *IEEE Trans. Robotics and Automation,* **12**, 759–766.

Lu, W. and A. Packard (1997). Adaptive H_∞-Control for Nonlinear Systems, *Proc. ACC* Albuquerque, NM, 1208–1212.

Maruyama, A. and M. Fujita (1997). Robust Visual Servo Control for Planar Manipulators with the Eye-in-Hand Configuration, *Proc. of the 36th CDC,* San Diego, CA, 2551–2552.

Maruyama, A. and M. Fujita (1998). L_2-Gain Performance Analysis for Nonlinear Robust Visual Servo Control, *Proc. ACC,* Philadelphia, PA. 2932–2936.

Ortega, R. and M. Spong (1989). Adaptive Motion Control of Rigid Robots: a Tutorial, *Automatica,* **25**, 877–888.

van der Schaft, A. (1996). *L_2-Gain and Passivity Techniques in Nonlinear Control,* Lecture Notes in Control and Information Sciences, 218, Springer-Verlag.

Scherpen, J. and R. Ortega (1997). On Nonlinear Control of Euler–Lagrange Systems: Disturbance Attenuation Properties, *System & Control Letters,* **30**, 49–56.

Copyright © 1999 IFAC
14th Triennial World Congress, Beijing, P.R. China

NON-TIME REFERENCED TRACKING CONTROL WITH APPLICATION IN UNMANNED VEHICLE

Wei Kang * **Ning Xi** **,[1]

* *Mathematics Department, Naval Postgraduate School, Monterey, CA 93943, USA*
** *Department of Electrical Engineering, Michigan State University, East Lansing, MI 48824-1226, USA*

Abstract: A new design method for non-time based tracking controller is presented. The key step is the introduction of a motion reference variable based on the desired and measurable system output. It enables the construction of control systems with integrated planning capability, in which planning becomes real-time closed-loop process. The control system is capable of coping with unexpected events and uncertainties. The new design method converts a controller designed by traditional time-based approach to a new controller with respect to a non-time based action reference. The design method is exemplified by a unmanned vehicle tracking control problem. *Copyright © 1999 IFAC*

Keywords: Feedback control, Autonomous vehicles

1. INTRODUCTION

The objective of tracking control is to ensure the output of a system to track a given reference input. A common feature of many path tracking feedback design is a state trajectory following approach. The reference input or desired path is usually described as a function of time. Therefore, most of systems are modeled as a time-based dynamic system. However, non-time based controllers are often preferred in applications of robotics. Developing non-time based controller has attracted the attention of researchers in different fields. In (Xi, 1993a), the event-based controller design was first introduced. Since then, it has been successfully applied to robot motion control (Tarn, 1993), multi-robot coordination (Xi, 93b), force and impact control (Wu, 1995), robotic teleoperation (Tarn, 1995) and manufacturing automation (Xi, 1997).

[1] Research supported in part by NSF IRI 9796287 and NSF IRI 9796300.

In the event-based planning and control scheme, the function of the Action Reference block in Figure 1, (b) is to compute the action reference parameter s on-line, based on sensory measurements. The planner generates the desired value to the system, according to the on-line computed action reference parameter s. In other words, the action plan is adjusted on line which enables the planner to handle unexpected events or uncertainties. For example, the event-based planning and control scheme has been successfully applied to deal with unexpected obstacle in a robot motion (Xi, 1993a). If the robot is blocked by the obstacle, the action reference also stops evolving. As a result the desired input will be constant and the error will remain unchanged. In this case, if the conventional time-based motion plan were implemented, the system error would keep increasing as the planner generates the desired input according to the original fixed plan. As a result, the system would become unstable.

Non-time based controller has also been applied to mobile robot tracking control. DeSantis (1994) and Sampei (1991) provide controllers to track lines and circles. In (Kaminer, 1998), tracking of a trimming trajectory based on the distance to the trajectory and the arc length of the trajectory is derived and implemented for unmanned air vehicles.

In this paper, we adopt a view point which is different from all the non-time based approaches mentioned above. We focus on the problem of finding controllers for the tracking of a desired path described as a function of non-time based action reference. Instead of developing new design algorithms based on the non-time based action reference, we use the existing design method for time-based controller. Then, the time variable in the feedback is substituted by a special transformation so called state-to-reference projection. The resulting feedback becomes non-time based. The new approach has the following four advantages. First, the model of the system is a general nonlinear dynamical control system. The method is applicable to a wide range of tracking control problems. Second, a transformation is provided to transform a time dependent controller into a non-time based controller. It bridges the traditional state trajectory tracking with the non-time based design. The results enable us to design non-time based feedback using the existing, well known methods of feedback design in time domain such as LQR and H_∞ control. The third advantage of our design algorithm is the flexibility in choosing action references. The arc length of the desired path is often used as the motion reference. However, there are many curves for which the formula of computing arc length is complicated. Parameters other than arc length can be used as action references in the design to simplify the controller. The fourth advantage of our method is that the transformation from state space to the action reference is not necessarily the orthogonal projection. For many curves, orthogonal projection requires on-line numerical solution of minimization. Our method allows the designer to use any mapping satisfying a projection condition. A good choice for the projection mapping can dramatically reduce the on-line computational time.

2. MOTION REFERENCE PROJECTION METHOD

The dynamics of mechanical or electrical systems is usually modeled by differential equations in which the free variable is the time variable t. A desired trajectory is often described as a function of time, denoted by $x_d(t)$. x represents the state of the system. In a typical tracking control problem, a controller can be designed so that the trajectory of the plan system $x(t)$ asymptotically approaches the desired trajectory $x_d(t)$. There are many design methods in the literature for tracking control. Since the design is based on the model driven by t, the controller is time-based. However, in the control of autonomous vehicles, the desired path can be easily described as a function of non-time based action reference s. In this section, we prove the result which transforms a time-based feedback controller into a non-time based controller for a desired path driven by motion reference s.

A system is defined by the equation

$$\dot{x} = f(x, u), \ y = h(x), \tag{1}$$

where $u \in I\!R^m$ is the control input, and $x \in I\!R^n$ is the state of the system. The desired path is usually given in a output space. Suppose that $y = h(x)$ represents the output of the system. The desired path is given by a non-time based motion plan $y = y_d(s)$, where s is the motion reference.

The first step in the controller design is to obtain a corresponding time-based motion plan based on the given velocity. If s is the non-time based motion reference, it requires that s is an strictly increasing function of time t, i.e. $s = v(t)$. How to pick the function $v(t)$ depends on the desired speed. For example, if the problem require that the system is operated so that s is increasing at a constant speed v_0, then $v(t) = v_0 t$.

The second step is to find a time-based feedback, $\tilde{u}(x,t)$, to track the path $y_d(v(t))$. There are many well known design algorithms can be found in the literature of linear and nonlinear control theory. The feedback satisfies

$$\lim_{t\to\infty} (h(x(t)) - y_d(v(t))) = 0.$$

Furthermore, if the initial position is on the desired path, then the trajectory of the controlled system follows the path. More specifically, there exists an initial condition of the system x_0 such that the trajectory satisfies $h(x(t)) = y_d(t)$. Denote this path by $x_d(s)$ or $x_d(v(t))$.

The third step is to find a suitable transformation, $s = \gamma(x)$, from the state space to the reference s. The transformation satisfies

$$\gamma(x_d(s)) = s \tag{2}$$

For example, given any state x_0, let $x_d(s_0)$ be the orthogonal projection from x_0 to $x_d(s)$. If we define $\gamma(x_0) = s_0$, then it satisfies (2). The transformation satisfying (2) is called a state-to-reference projection. However, orthogonal projection is not the only way to define γ. In section 3, the projection to x-axis is adopted as γ, which is a simpler formula than orthogonal projection.

The last step is to construct the non-time based feedback. Let

$$u(x) = \tilde{u}(x, v^{-1}(\gamma(x))) \quad (3)$$

where $\tilde{u}(x,t)$ is the feedback found in the second step, $\gamma(x)$ is a state-to-reference projection. The closed-loop system is

$$\dot{x} = f(x, u(x)). \quad (4)$$

Given any state x and any time t, the distance from x to the point $x_d(t)$ in the desired path is denoted by $d(x,t)$. So, $d(x,t) = \|x - x_d(t)\|$ is a function from $\mathbb{R}^n \times \mathbb{R}$ to $\mathbb{R}$. The open set $U(x_d, r)$ consists of $(x,t) \in \mathbb{R}^n \times \mathbb{R}$ such that $d(x,t)$ is less than r, i.e.

$$U(x_d, r) = \{(x,t) \in \mathbb{R}^n \times \mathbb{R} \mid d(x,t) < r\}.$$

A δ neighborhood of x_d is a subset of $\mathbb{R}^n$ defined by

$$W(\delta) = \{x \in \mathbb{R}^n \mid d(x, x_d) < \delta\}.$$

In this paper, we assume that the vector field $F(x,u)$, the feedback $\tilde{u}(x,t)$, and their derivatives are bounded. Assume that the derivatives of $v(t)$ and $v^{-1}(s)$ are bounded. The closed loop system with feedback $\tilde{u}(x,t)$ is

$$\dot{x} = f(x, \tilde{u}(x,t)) \quad (5)$$

The non-time based feedback is

$$u(x) = \tilde{u}(x, v^{-1}(\gamma(x))) \quad (6)$$

The closed-loop system with non-time based feedback is

$$\dot{x} = f(x, u(x)). \quad (7)$$

Let $\tilde{\phi}(t, t_0, x_0)$ be the trajectory of (5) with the initial time t_0 and initial condition $x(t_0) = x_0$. We assume that there exists a neighborhood $U(x_d, r)$ so that $(x_0, t_0) \in U(x_d, r)$ implies that $\tilde{\phi}(t, t_0, x_0)$ approaches x_d exponentially. Furthermore, the trajectory satisfies

$$\begin{aligned} me^{-\alpha_1(t-t_0)}d(x_0,t_0) &\le d(\tilde{\phi}(t,t_0,x_0),t) \\ &\le Me^{-\alpha_2(t-t_0)}d(x_0,t_0) \end{aligned} \quad (8)$$

for some $m > 0$, $M > 0$, $\alpha_1 > 0$ and $\alpha_2 > 0$. The following function is used as our Lyapunov function. It is defined by

$$V(x_0, t_0) = \int_{t_0}^{\infty} d(\tilde{\phi}(t, t_0, x_0), t)dt. \quad (9)$$

We assume that all derivatives of $V(x,t)$ of order less than or equal to three are bounded in a neighborhood $U(x_d, r)$. From (8), it is easy to show that

$$md(x,t) \le V(x,t) \le Md(x,t) \quad (10)$$

for some $m > 0$ and $M > 0$.

Theorem 2.1. There exists a neighborhood $W(\delta)$ of the desired trajectory $x_d(s)$ such that the trajectory $x(t)$ of (7) satisfies

$$lim_{t\to\infty} d(\ x(t), v^{-1}(\gamma(x(t)))\ ) = 0.$$

provided $x(0)$ is in $W(\delta)$. This implies that the trajectory of the closed-loop system with a time-invariant feedback asymptotically approaches the desired path.

Proof. The trajectory of (5) with initial time t_0 and initial state $x(t_0) = x_0$ is denoted by $\tilde{\phi}(t, t_0, x_0)$. Similarly, a trajectory of (7) is denoted by $\phi(t, t_0, x_0)$. By the definition of $V(x,t)$, it is easy to check that the derivative of $V(x,t)$ along $\tilde{\phi}(t, t_0, x_0)$, a trajectory of (5), satisfies $\frac{d}{dt}V(x(t),t) = -d(x(t),t)$. Therefore,

$$\begin{aligned} &\frac{\partial V(x,t)}{\partial x} f(x, \tilde{u}(x,t)) + \frac{\partial V(x,t)}{\partial t} \\ &= -d(x,t). \end{aligned} \quad (11)$$

The desired trajectory $x_d(v(t))$ is a solution to the equation (5). Therefore, $\dot{x}_d(v(t)) = f(x_d, \tilde{u}(x_d, t))$. Since $s = \gamma(x_d(s))$, we have $t = v^{-1}(\gamma(x_d(v(t))))$. The derivative of this equation with respect to t yields

$$\frac{\partial v^{-1}}{\partial s}\frac{\partial \gamma}{\partial x} f(x_d, \tilde{u}(x_d, t)) = 1. \quad (12)$$

Now, let's consider the closed-loop system (7) with the feedback (6). The derivative of $V(x, v^{-1}(\gamma(x)))$ in the direction of (7) is

$$\begin{aligned} \dot{V}(x,, v^{-1}(\gamma(x))) &= \frac{\partial V}{\partial x} f(x, \tilde{u}(x, v^{-1}(\gamma(x)))) \\ &+ \frac{\partial V}{\partial t}(x, v^{-1}(\gamma(x))) + \left(\frac{\partial v^{-1}}{\partial s}\frac{\partial \gamma}{\partial x} f - 1\right)\frac{\partial V}{\partial t}. \end{aligned}$$

From (11), we have

$$\begin{aligned} &\dot{V}(x,, v^{-1}(\gamma(x))) = -d(x,, v^{-1}(\gamma(x))) \\ &+ \left(\frac{\partial v^{-1}}{\partial s}\frac{\partial \gamma}{\partial x} f - 1\right)\frac{\partial V}{\partial t}(x, v^{-1}(\gamma(x))). \end{aligned} \quad (13)$$

By the definition of $V(x,t)$, it is easy to check that $V(x,t) > 0$ for all x in a neighborhood of $x_d(t)$ and $V(x_d(v(t)), t) = 0$. Therefore, $V(x,t)$ has a minimum value at $x_d(t)$. It can be proved that

$$V(x,t) = (x - x_d(t))^T Q(x,t)(x - x_d(t))$$

for some positive definite matrix $Q(x,t)$. Therefore, $\frac{\partial V}{\partial t}(x,t) = P(x,t)(x - x_d(t))$ for some row vector $P(x,t)$. In a neighborhood $U(x_d, r)$, $P(x,t)$ is bounded. So,

$$\left|\frac{\partial V(x,t)}{\partial t}\right| \le M_1 d(x,t) \quad (14)$$

in a neighborhood $U(x_d, r)$ for some $M_1 > 0$. Meanwhile, (12) implies

$$\frac{\partial v^{-1}}{\partial s}\frac{\partial \gamma}{\partial x}f(x, u(x)) - 1 = 0$$

if x is a point on $x_d(s)$. Therefore,

$$\left|\frac{\partial v^{-1}}{\partial s}\frac{\partial \gamma}{\partial x}f(x, u(x)) - 1\right| < \frac{1}{2M_1} \quad (15)$$

in a neighborhood $W(\delta)$. From (13), (14) and (15), we have

$$\dot{V}(x,, v^{-1}(\gamma(x))) < -\frac{d(x, v^{-1}(\gamma(x)))}{2}.$$

From (10),

$$\dot{V}(x,, v^{-1}(\gamma(x))) < -\frac{1}{2M}V(x,, v^{-1}(\gamma(x)))$$

in a neighborhood of x_d. By Gronwall's inequality, $V(x,, v^{-1}(\gamma(x)))$ approaches zero exponentially along a trajectory $\phi(t, t_0, x_0)$ if x_0 is in a neighborhood of x_d. From (10),

$$md(x, t) \leq V(x, t).$$

Therefore, $d(x, v^{-1}(\gamma(x)))$ approaches zero exponentially along a trajectory $\phi(t, t_0, x_0)$ if x_0 is in a neighborhood, $W(\delta)$, of $x_d(s)$. $\triangleleft$

The Theorem 1 has indicated that following the above design procedure, a non-time based feedback can be obtained to stabilize the tracking control. Furthermore, all design methods for traditional time-based controller can still be used. This significantly simplifies the design procedure for a non-time based controller.

3. PATH TRACKING FOR A UNMANNED VEHICLE

In this section, tracking control of a unmanned vehicle using state feedback is discussed. Consider a kinematic model of the vehicle with front and rear wheels, as shown Figure 2. The rear wheels are aligned with the vehicle while the front wheels are allowed to spin about the vertical axes. The constraints on the system arise by allowing the wheels to roll and spin, but not slip. Let (x, y, ϕ, θ) denote the configuration of the car, parameterized by the location of the rear wheels. The control system model is (Murry, 1993)

$$\begin{aligned} \dot{x} &= u_1 \cos\theta, \ \dot{y} = u_1 \sin\theta \\ \dot{\theta} &= \frac{u_1}{l} tan\phi, \ \dot{\phi} = u_2, \end{aligned} \quad (1)$$

where u_1 corresponds to the forward velocity of the rear wheels of the car and u_2 corresponds to the velocity of the steering wheel, the angle of the car body with respect to the horizontal is θ, the steering angle with respect to the car body is ϕ, (x, y) is the location of the rear wheels, l is the length between the front and the rear wheels. In the following, we will first derive a time-based feedback for an arbitrary path given by $y = f(x)$. The design does not depend on its motion reference. A motion reference, then is defined for a given path (a sine curve is adopted as our example), and the state-to-reference projection will be found to convert the time-based feedback to a non-time based feedback. Simulations are carried out to show the stability of the tracking control.

We defined the following change of coordinates to achieve linearization of the last three equations in (1),

$$\begin{aligned} z_1 &= y, \ z_2 = \sin\theta, \ z_3 = \frac{1}{l}\cos\theta\tan\phi \\ \alpha_1 &= -\frac{u_1}{l^2}\sin\theta(\tan\phi)^2, \ \beta_1 = \frac{\cos\theta}{l\cos^2\phi} \\ v &= \alpha_1 + \beta_1 u_2 \end{aligned} \quad (2)$$

The last three equations in (1) are transformed into

$$\dot{z}_1 = u_1 z_2, \ \dot{z}_2 = u_1 z_3, \ \dot{z}_3 = v \quad (3)$$

Suppose that the desired path is the following function of time

$$x = x_d(t), \ y = y_d(t).$$

Let's first find a feedback for u_1. Since $\frac{d}{dt}(x - x_d) = u_1\cos\theta - \dot{x}_d$, define

$$u_1 = \frac{1}{\cos\theta}(\dot{x}_d - (x - x_d)). \quad (4)$$

Under (4), the first equation of (1) becomes

$$\frac{d}{dt}(x - x_d) = -(x - x_d).$$

Its solution asymptotically approaches zero as $t \to \infty$. The feedback for u_2 comes from (3). Define

$$\begin{aligned} e_1 &= z_1 - y_d, \ e_2 = z_2 - \frac{\dot{y}_d}{u_1}, \\ e_3 &= z_3 - \frac{1}{u_1}\frac{d}{dt}(\frac{\dot{y}_d}{u_1}) \end{aligned} \quad (5)$$

Then,

$$\begin{aligned} \dot{e}_1 &= u_1 e_2, \ \dot{e}_2 = u_1 e_3 \\ \dot{e}_3 &= v - \frac{d}{dt}(\frac{1}{u_1}\frac{d}{dt}(\frac{\dot{y}_d}{u_1})) \end{aligned} \quad (6)$$

The right side of the last equation in (6) satisfies

$$\begin{aligned} &\frac{d}{dt}(\frac{1}{u_1}\frac{d}{dt}(\frac{\dot{y}_d}{u_1})) \\ &= \frac{\dddot{y}_d}{u_1^2} - \frac{3\ddot{y}_d\dot{u}_1}{u_1^3} + \frac{3\dot{y}_d\dot{u}_1^2}{u_1^4} - \frac{\dot{y}_d\ddot{u}_1}{u_1^3} \end{aligned} \quad (7)$$

The derivatives of u_1 can be computed from (4). Notice that $\ddot{u}_1$ contains the term $\dot{\phi}$, which depends on u_2. Therefore, the expression(7) has the form

$$\alpha_2(x,y,\theta,\phi,t)+\beta_2(x,y,\theta,\phi,t)u_2 \quad (8)$$

To stabilize (6), we define

$$\begin{aligned} v = &\alpha_2(x,y,\theta,\phi,t)+\beta_2(x,y,\theta,\phi,t)u_2 \\ &+u_1(a_1e_1+a_2e_2+a_3e_3) \end{aligned} \quad (9)$$

Then, (6) is equivalent to

$$\begin{aligned} &\dot{e}_1 = u_1e_2, \quad \dot{e}_2 = u_1e_3 \\ &\dot{e}_3 = u_1(a_1e_1+a_2e_2+a_3e_3) \end{aligned} \quad (10)$$

Select the values of $a_i, i=1,2,3$, so that all the eigenvalues of

$$A = \begin{bmatrix} 0 & 1 & 0 \\ 0 & 0 & 1 \\ a_1 & a_2 & a_3 \end{bmatrix}$$

are on the left half plane. Assume that

$$-\frac{\pi}{2} < \theta(t) < \frac{\pi}{2}. \quad (11)$$

The solution of (10) is $e(t) = e_0\exp(A\int_0^t u_1(\tau)d\tau)$. From (4), $\int_0^t u_1(\tau)d\tau$ increases as x_d increases. Since A is stable, we know that $e(t)$ approaches zero exponentially as x_d increases. Therefore, $y - y_d(t) \to 0$ as x_d increases. To find a formula for u_2, we solve for u_2 from (9) and the last equation in (2). The feedback is

$$u_2 = \frac{\alpha_2-\alpha_1}{\beta_1-\beta_2}+\frac{u_1}{\beta_1-\beta_2}(a_1e_1+a_2e_2+a_3e_3).$$

To summarize, define

$$\begin{aligned} &\alpha_1(\theta,\phi) = -\frac{1}{l^2}\sin\theta(\tan\phi)^2, \\ &\beta_1(\theta,\phi) = \frac{\cos\theta}{l\cos^2\phi} \\ &\alpha_2(x,y,\theta,\phi,t)+\beta_2(x,y,\theta,\phi,t)u_2 \\ &= \frac{\dddot{y}_d}{u_1^2}-\frac{3\ddot{y}_d\dot{u}_1}{u_1^3}+\frac{3\dot{y}_d\dot{u}_1^2}{u_1^4}-\frac{\dot{y}_d\ddot{u}_1}{u_1^3}. \end{aligned} \quad (12)$$

The feedback is

$$\begin{aligned} u_1 &= \frac{1}{\cos\theta}(\dot{x}_d-(x-x_d)) \\ u_2 &= -\frac{\alpha_2(x,y,\theta,\phi,t)-\alpha_1(\theta,\phi)}{\beta_2(x,y,\theta,\phi,t)-\beta_1(\theta,\phi)} \\ &+u_1(a_1e_1+a_2e_2+a_3e_3). \end{aligned} \quad (13)$$

Theorem 3.1. Under the feedback (13), $(x(t),y(t))$ of (1) exponentially approaches a desired path $(x_d(t),y_d(t))$ as x_d increases, provided that (11) is satisfied.

Remark. The graph of any differentiable function $y=f(x)$ satisfies the condition (11). For instance, the curve $y=\sin x$ and the lines $y=ax+b$ satisfy the condition. However, a circle does not satisfy (11). Feedbacks to track circles can be found in the literature (see, for instance, (DeSantis, 1994)). ◁

The feedback (13) depends on time. However, it can be converted into a non-time based feedback using state-to-reference projection as proved in § 2. In the following, we use the method developed in § 2 and the feedback (13) to design non-time based feedbacks for the tracking of the curve $y=\sin x$. The key step is to find a motion reference for the task. The parametric equation for this curve using arc length as the parameter is not simple. Furthermore, finding the orthogonal projection from a point (x,y) to the curve involves solving nonlinear equations. Different from circles and lines, arc length and orthogonal projection are not the best choice in this case for real time computation. As pointed out in § 2, the design method proposed in this paper is not based on a specific motion reference. It works for arbitrary choice of the reference. For the *sin* curve, a simple choice of motion reference is $s=x$. Our design algorithm does not require orthogonal projection neither. We use the following γ which simply project any (x,y) onto the point in the curve with the same first coordinate,

$$\gamma(x,y)=x. \quad (14)$$

A time-based feedback is computed using the formula (13) to track the curve

$$x=t, \quad y=\sin t.$$

For the reason of simplicity, the function $s=v(t)$ is $v(t)=1$. This implies that the speed of x on the desired curve is 1. However, it is not necessary to assume constant velocity. Our design method is applicable even if the speed is not a constant. The feedback is calculated by using (13). The time variable t in the feedback is substituted by the state-to-reference projection $\gamma(x,y)=x$ to achieve a time-invariant state feedback.

In the simulation, the length of the vehicle is $l=1$. The coefficients in (13) are $a_1=-5$, $a_2=-9$ and $a_3=-5$. The initial position of the vehicles are $(0,2)$ and $(3,-1.8589)$. The simulation results are shown in Figure 3 and Figure 4. The dotted curves in Figure 3a and 4a are the paths of the vehicle in xy-plane. Figure 3b and 4b shows the value $y-y_d$, the difference between the actual path of the vehicle and the desired path. Although the stablity of the system is local, the simulation shows that the system is stable for large variations of the initial conditions $x(0)$ and $y(0)$ under the non-time based feedback. Actually, the initial condition $x(0)$ can be anywhere.

4. CONCLUSION

A new design method for non-time based tracking controller has been developed. It can convert a controller designed by traditional time-based approach to a new non-time based controller. It significantly simplifies the design procedure. The method has been applied to design a tracking controller of a unmanned vehicle. The simulation results have clearly demonstrated the advantage of non-time based controller. More importantly, the results have provided a efficient and systematic approach to design non-time based tracking controller for a general nonlinear dynamic system. It is an important step in the advancement of developing a theoretical foundation for sensor-referenced intelligent control.

REFERENCES

DeSantis, R. M. (1994). Path-tracking for a tractor-trailer-like robot. The International Journal of Robotics Research, Vol. 13, No. 6, 533-544.

Kaminer, I., Pascoal, A., Hallberg, E. and Silvestre, C. (1998). Trajectory tracking for autonomous vehicles: an integrated approach to guidance and control. J. of Guidance, Control and Dynamics, Vol. 21, pp. 29-38.

Murry, R. M. and Sastry, S. S. (1993). Nonholonomic motion planning: steering using sinusoids. IEEE Trans. Automat. Contr., Vol. 38, pp. 700-716.

Sampei, M., Tamura, T., Itoh, T. and Nakamichi, M. (1991). Path tracking control of trailer-like mobile robot. IEEE/RSJ International Workshop on Intelligent Robots and Systems IROS'91, pp. 193-198, Osaka, Japan.

Tarn, T.J., Xi, N., Guo, C. and Bejczy, A.K. (1995). Fusion of Human and Machine Intelligence for Telerobotic Systems. The Proceedings of the 1995 IEEE International Conference on Robotics and Automation, Nagoya, Japan.

Tarn, T. J., Bejczy, A.K. and Xi, N. (1993). Intelligent Planning and Control for Robot Arms. Proceedings of the IFAC 1993 World Congress, Sydney, Australia, 18-23.

Wu, Y., Tarn, T.J. and Xi, N. (1995). Force and Transition Control with Environmental Uncertainties. The Proceedings of the 1995 IEEE International Conference on Robotics and Automation, Nagoya, Japan.

Xi, N. and Tarn, T.J. (1997). Integrated Task Scheduling and Action Planning/Control for Robotic Systems Based on a Max-Plus Algebra Model. Proceedings of IEEE/RSJ International Conference on Intelligent Robotics and Systems, Grenoble, France.

Xi, N. (1993a). Event-Based Planning and Control for Robotic Systems. Doctoral Dissertation, Washington University.

Xi, N., Tarn, T.J. and Bejczy, A.K. (1993b). Event-Based Planning and Control for Multi-Robot Coordination. Proceedings of the 1993 IEEE International Conference on Robotics and Automation, Atlanta, Georgia, pp. 251-258.

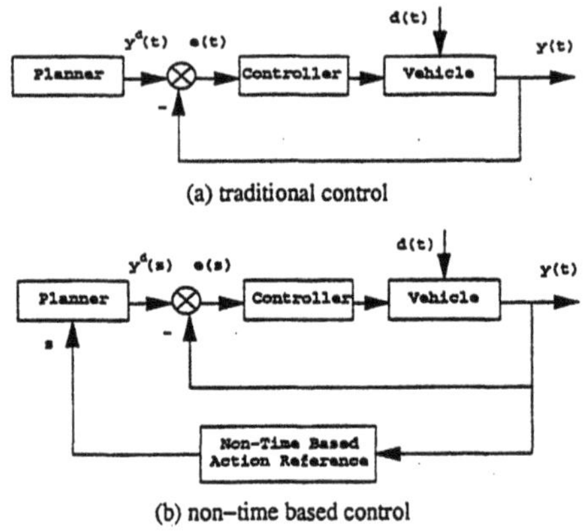

Figure 1: Comparison of non-time based and time based control

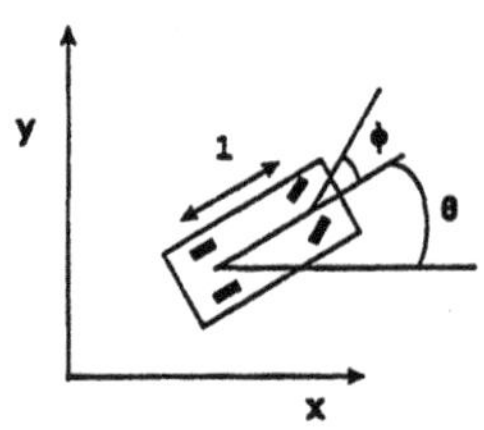

Figure 2: The configuration of the car is determined by the location of the back wheels (x, y), and the angles θ and ϕ.

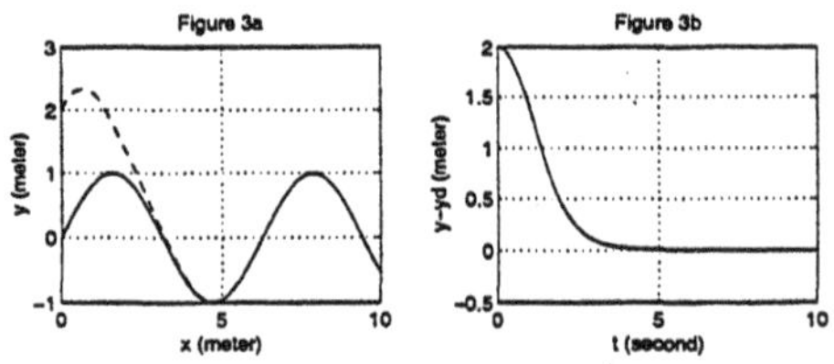

Figure 3: Tracking a sine curve

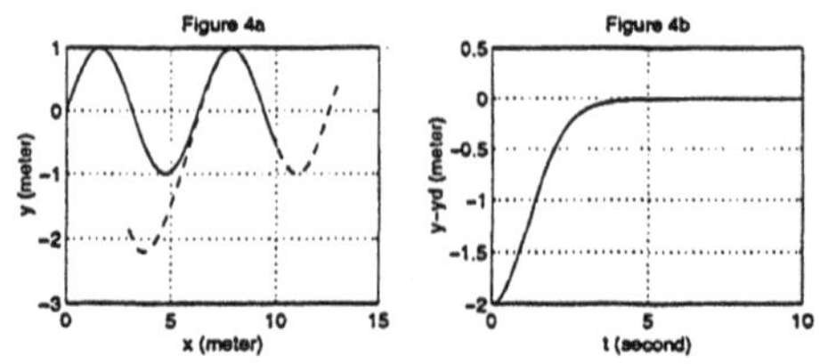

Figure 4: Tracking a sine curve

14th Triennial World Congress, Beijing, P.R. China

E-2c-11-2

ROTATING STALL CONTROL OF MG3 COMPRESSOR MODELS GOVERNED BY PARTIAL DIFFERENTIAL EQUATIONS

MingQing Xiao *,[1] **Tamer Başar** **,[2]

* *University of California at Davis, CA 95616, USA.*
Email: xiao@math.ucdavis.edu. Fax: 530-752-6635.
** *University of Illinois at Urbana-Champaign, IL 61801, USA.*
Email: tbasar@control.csl.uiuc.edu. Fax: 217-244-1653.

Abstract: In a recent paper, Moore and Greitzer developed a full compressor model, governed by a set of three simultaneous nonlinear partial differential equations for pressure rise, and average and disturbed values of the flow coefficient, as functions of time and angle around the compressor. Here we consider feedback stabilization of this model with general cubic compressor characteristics, and show that the controller designed eliminates rotating stall (a fundamental aerodynamic instability in axial flow compressors – a complex nonlinear phenomenon that limits their performance) as the pressure rise reaches to the maximum. *Copyright © 1999 IFAC*

Keywords: Moore-Greitzer Model, Rotating stall, Stability, Evolution equation, Nonlinear feedback control

1. INTRODUCTION

In general, the performance of a compression system is limited by two dynamic instabilities called *rotating stall* and *surge*, which occur in the flow through the compressor. Problems of compressor instability have been of concern to aircraft engine designers for two decades because the engine performance is effectively reduced by rotating stall and surge. Rotating stall, which corresponds to a traveling wave of gas around the annulus of the compressor, occurs when a nonaxisymmetric flow disturbance develops (around the annulus of the rotor) and causes severe reduction in the performance of the compressor. As rotating stall develops, axisymmetric oscillations across the compression system, known as surge, occur. Surge is a low-frequency, large-amplitude oscillation of the mean flow rate in the compressor which induces high blade, causing stress levels and possible reverse flow which affects flow conditions throughout the entire compression system. These instability phenomena must be avoided during the flight operations.

1.1 *Physical Model and General Features of Rotating Stall in Axial Flow Compression Systems*

The basic compression system is shown in Fig. 1. It consists of an inlet duct, a compressor, a downstream duct, a plenum and a throttle. The compressor operates between the inlet duct and the downstream duct. The flow enters the compressor from the inlet duct and exits into the plenum through the downstream duct. The throttle controls the flow through the system at the plenum exit in order to model the turbine. The compressor geometry is shown in Fig. 2. Variable ϕ in Fig. 1 represents the local, unsteady axial velocity at the compressor face, which depends on both angle θ

[1] Research supported in part by AFOSR under F49620-95-1-0409
[2] Research supported in part by the U.S. Department of Energy under Grant DE-FG-02-97-ER-13939

around the wheel and (dimensionless) time ξ. Ψ is the pressure in the plenum, which is defined as

$$\Psi := \frac{\text{exit static pressure - inlet total pressure}}{\text{density} \times \text{mean rotar speed}^2}$$

The annulus-averaged axial flow coefficient Φ of ϕ around the wheel is defined as

$$\Phi(\xi) := \frac{1}{2\pi}\int_0^{2\pi} \phi(\theta,\xi)d\theta$$

In order to understand the rotating stall, it is necessary to describe the observed behavior through experimental investigations. For a detailed experimental set-up as well as experimental facility, we refer the reader to (Day *et al.* 1979). Here we summarize the rotating stall behavior in the $\Phi\Psi$-plane in Fig. 3, where it can be seen that the important feature for rotating stall is the *discontinuous drop* in the pressure-rise characteristic as the throttle is closed. The mass flow jumps to significantly reduced values (pressure rise usually drops to less than half the prestall value), occurring from point A to B in the figure. If one opens the throttle, the mass flow leaves the rotating stall regime at point C, which is different from that at which it enters. Thus there is an associated effect of hysteresis between the onset and the cessation of stall. In some cases, such hysteresis is substantial and can not be tolerated during engine operation.

1.2 *Mathematical Model*

The full model compression system (Moore-Greitzer model) can be described by the following simultaneous nonlinear partial differential equations (see (Moore and Greitzer 1990)):

$$l_c\frac{d\Phi}{d\xi} = -\Psi(\xi) + \frac{1}{2\pi}\int_0^{2\pi} \psi_c(\Phi + \phi'_\eta\big|_{\eta=0})d\theta \quad (1)$$

$$l_c\frac{d\Psi}{d\xi} = \frac{1}{4B^2}(\Phi(\xi) - \Phi_T(u)) \quad (2)$$

where the upstream disturbance velocity potential $\phi'(\xi,\theta,\eta)$ satisfies Laplace's equation

$$\phi'_{\theta\theta} + \phi'_{\eta\eta} = 0, \quad (\theta,\eta) \in [0,2\pi] \times (-\infty,0) \quad (3)$$

The boundary conditions are periodic in θ,

$$\frac{\partial}{\partial\xi}(m\phi' + \frac{1}{a}\phi'_\eta) - \Big(\Psi_c(\Phi + \phi'_\eta) - \frac{1}{2\pi} \times \int_0^{2\pi} \Psi_c(\Phi + \phi'_\eta)d\theta - \frac{1}{2a}\phi'_{\theta\eta}\Big) = 0 \text{ at } \eta = 0 \quad (4)$$

$$\phi' = 0 \text{ at } \eta = -\infty \quad (5)$$

where the subscripts ξ, θ, η denote partial differentiation. Φ is the annulus-averaged axial flow coefficient, Ψ is the annulus-averaged total-to-static pressure rise coefficient, l_c is the effective length of the compression system, Ψ_c is called a general characteristic function of compressors, B is the Greitzer B-parameter which determines whether a given compressor is more likely to enter rotating stall or surge, m is the duct parameter, a is the internal compressor lag, ξ is the dimensionless time, θ is the circumferential angle around the compressor annulus, and Φ_T is the control function of the throttle opening.

Moore and Greitzer (Moore and Greitzer 1990) developed the model (1)-(4) to explain the observed behavior. One of the attractive features of this three states model is that it captures the main dynamical behavior of compressor systems, including both rotating stall and surge. Usually, high-pressure-rise is needed in order to increase performance and efficiency of the compressor systems. But this is at the price of a significantly reduced stability margin. Hence it is necessary to use active control to suppress the rotating stall as well as surge in order to extend the stable operating range and to improve engine performance. So far most results for analysis and control of the stall and surge were based on simplified models obtained by a Galerkin projection of the PDE describing the stall dynamics onto its first Fourier models (for example, see (Liaw and Abed 1996), (McCaughan 1990)). After the projection, the MG3 PDE model becomes a model described by three ODEs, and thus a bifurcation-theoretic approach becomes possible. Although the single harmonic Galerkin method is a useful approximation for the general transients and gives a correct picture of the nonlinear effect of rotating stall on performances, it cannot however accurately represent wave of the relaxation type of transition which arise in fully developed rotating stall, as pointed out by Moore and Greitzer(Moore and Greitzer 1990). Certain controller designs based on the first mode projection fail to work properly (for example see (Henderson and Sparks 1996), (Leonessa *et al.* 1997)). For the control of full model compression systems, a back-stepping technique was introduced in (Banaszuk *et al.* 1997). However, the back-stepping controller introduced requires the information of the derivative of the disturbance velocity potential, which actually is quite difficult to be measured in real applications. Another recent work on stabilization of the full model compression system is (Xiao 1998), which has focussed on the case when the compressor characteristic is not a cubic polynomial.

In this paper, we study the stabilization of rotating stall of the Moore-Greitzer model with a general cubic compressor characteristics. Since only throttle opening control will result in undesirable jump and hysteresis behavior as shown in Fig. 3, an active controller has to be introduced in order to improve the system performance. Rotating stalls appear mainly because of the non-nominal equilibrium points of the compression systems. We propose here a feedback controller that eliminates all but the nominal equilibrium point as the annulus-averaged total-to-static pressure rise coefficient reaches its maximum value. The controller uses only information on the average of the disturbance velocity potential and the pressure rise coefficient, both of which are easily accessible. We also demonstrate that the regularity of the initial condition is an important factor in controlling the full model system.

2. ABSTRACT MATHEMATICAL SETTING

Let $L^2[0,2\pi]$ be the space of all square integrable 2π-periodic functions. For any $\phi \in L^2[0,2\pi]$, its Fourier series is

$$\phi(\theta) = \sum_{n=0}^{\infty}(a_n \cos n\theta + b_n \sin n\theta).$$

Define linear operator $K : L^2[0,2\pi] \to L^2[0,2\pi]$ by

$$K(\phi) = ama_0 + \sum_{n=1}^{\infty}\{1+\frac{am}{n}\} \times (a_n \cos n\theta + b_n \sin n\theta)$$

for any ϕ in its Fourier series as above. Clearly, K is a positive definite, self-adjoint linear operator on $L^2[0,2\pi]$. Thus $<\phi,\psi>_K := <\phi, K\psi>_{L^2[0,2\pi]}$ defines an equivalent inner product on $L^2[0,2\pi]$, and we denote by $L_K^2[0,2\pi]$ the space which consists of elements of $L^2[0,2\pi]$ with inner product $<\cdot,\cdot>_K$. The solution of Laplace's equation (3) has the form

$$\phi'(\xi,\theta,\eta) = \sum_{n=1}^{\infty} e^{n\eta}(a_n(\xi)\cos n\theta + b_n(\xi)\sin n\theta).$$

Let $g := \phi'_\eta|_{\eta=0}$, which represents the disturbance velocity at the entrance of the compressor. By using the operator K defined above, (4) is equivalent to

$$\frac{\partial g}{\partial \xi} = -\frac{1}{2}K^{-1}\frac{\partial g(\xi,\theta)}{\partial\theta} + aK^{-1}\left(\psi_c(\Phi(\xi)+g(\xi,\theta)) - \overline{\psi}_c\right)$$

where

$$\overline{\psi}_c(\xi) := \frac{1}{2\pi}\int_0^{2\pi}\psi_c(\Phi(\xi)+g(\xi,\theta))d\theta. \quad (6)$$

Let $X = L_K^2[0,2\pi] \times \mathbf{R} \times \mathbf{R}$. Throughout this paper, $\|\cdot\|$ denotes the norm of X and $<\cdot,\cdot>$ denotes the inner product of X, which are defined as

$$<x_1,x_2> = a^{-1} <g_1,g_2>_{L_K^2[0,2\pi]} + l_c\Phi_1\Phi_2 + (4l_cB^2)\Psi_1\Psi_2$$

$$\|x\| = \sqrt{<x,x>},\ x_i = (g_i,\Phi_i,\Psi_i)$$

where $x, x_i \in X, i = 1,2$. Thus, equations (1)-(4) can be transformed into an evolution equation on Hilbert space X:

$$\frac{dx}{d\xi} = Ax + f(x), \quad x(0) = x_0 \quad (7)$$

where

$$x = \begin{bmatrix} g \\ \Phi \\ \Psi \end{bmatrix}, \quad A = \begin{bmatrix} -\frac{1}{2}K^{-1}\frac{\partial}{\partial\theta} & 0 & 0 \\ 0 & 0 & 0 \\ 0 & 0 & 0 \end{bmatrix} \quad (8)$$

with $D(A) = \left\{x \in X \mid \frac{\partial g}{\partial\theta} \in L_K^2[0,2\pi] \text{ and } g(0) = g(2\pi)\right\}$ and

$$f(g,\Phi,\Psi) = \begin{bmatrix} aK^{-1}(\psi_c(\Phi+g) - \overline{\psi}_c) \\ \frac{1}{l_c}(\overline{\psi}_c - \Psi) \\ \frac{1}{4l_cB^2}(\Phi - \Phi_T) \end{bmatrix} \quad (9)$$

A function $x \in C([0,t_f];X)$ is a weak (mild) solution of (7) on $[0,t_f]$ if $x(0) = x_0 \in X$, $f(x) \in L^1(0,t_f;X)$ and if for each $w \in D(A^*)$ the function $<x(\xi),w>$ is absolutely continuous on $[0,t_f]$ and satisfies

$$\frac{d}{d\xi} <x(\xi),w> = <x(\xi),A^*w> + <f(x),w>$$

for almost all $\xi \in [0,t_f]$. A function $x : [0,t_f) \to X$ is a classical solution of (7) on $[0,t_f)$ if x is continuous on $[0,t_f)$, continuously differentiable on $(0,t_f)$, $x(\xi) \in D(A)$ for $0 < \xi < t_f$ and (7) is satisfied on $[0,t_f)$.

3. EXISTENCE OF A GLOBAL BOUNDED SOLUTION

Henceforth we will assume that the compressor characteristics has the form

$$\psi_c(\Phi) = a_0 + a_1\Phi + a_2\Phi^2 + a_3\Phi^3 \quad (10)$$

where $a_0 > 0, a_3 < 0$. In this section and the next one, we also assume that the throttle characteristic has the form

$$\Phi_T(\Psi) = \gamma\sqrt{\Psi} \tag{11}$$

where γ is proportional to the cross-sectional area of the throttle, which is used to control the exit mass flow of the compressor.

Theorem 1. The evolution equation (7) admits a bounded global weak solution with the compressor characteristics (10) and the throttle characteristic (11) if the initial condition $x_0 = (g_0, \Phi_0, \Psi_0) \in X$, $\Psi_0 > 0$.

Outline of the Proof: If (7) admits an unbounded solution, according to (Chan *et al.* 1992), there exists $t_{\max} > 0$ such that (7) admits a local solution on the interval $[0, t_{\max})$ and $\lim_{\xi \to t_{\max}} ||x(\xi)|| = +\infty$. We first claim that $t_{max} = +\infty$. Since $a_3 < 0$ we have that

$$\int_0^{2\pi} \left[a_1 + a_2\Phi(2 + \frac{1}{2\pi}) + 3a_3\Phi^2(1 + \frac{1}{2\pi})\right]$$

$$\times g^2(s,\theta)d\theta + a_3 \int_0^{2\pi} g^4(s,\theta)d\theta \le C_1$$

$$\psi_c(\Phi)\Phi = (a_0 + a_1\Phi + a_2\Phi^2 + a_3\Phi^3)\Phi \le C_2$$

where C_1, C_2 are constants. Thus we have

$$||x(\xi, x_0)||^2 \le (C_1 + C_2)\xi + ||x_0||^2$$

which implies that $||x(\xi, x_0)||$ is unbounded over $[0, t_{\max})$ only if $t_{\max} = +\infty$.

Now assume that $||x(\xi)||^2 = a^{-1}||g(\xi)||^2_{L^2_K[0,2\pi]} + l_c\Phi^2(\xi) + (4l_cB^2)\Psi^2(\xi) \to +\infty$ as $\xi \to \infty$. Note that $||g(\xi)||^2_{L^2_K[0,2\pi]}$ has to be bounded over the interval $[0, \infty)$ since $a_3 < 0$. Hence $||x(\xi)||^2 \to \infty$ would imply that $\Phi^2(\xi) + \Psi^2(\xi) \to \infty$ as $\xi \to \infty$. However, a direct calculation shows that

$$\sup_{[0,+\infty)} \{\Phi^2(\xi) + \Psi^2(\xi)\} < +\infty$$

Therefore, we have

$$\sup_{[0,\infty)} ||x(\xi)|| \le C \quad \text{for some constant } C$$

That is, (7) admits a bounded global weak solution. □

Remark: For simplicity, we can assume that $a_2 = 0$ because the term Φ^2 can be eliminated by an appropriate translation of the variable. Furthermore we also assume that $a_1 > 0$, because otherwise the system is always stable.

4. ASYMPTOTIC BEHAVIOR: FULL-DEVELOPED ROTATING STALL

A sequence $\{f_n\}_{n=1}^{\infty} \subset L^2[0, 2\pi]$ converges weakly to $f \in L^2[0, 2\pi]$ provided that

$$\int_0^{2\pi} f_n(\theta)g(\theta)d\theta \to \int_0^{2\pi} f(\theta)g(\theta)d\theta \quad \text{as } n \to \infty$$

for each $g \in L^2[0, 2\pi]$. For a given admissible feedback $\Phi_T = \Phi_T(x)$ for system (7), let $S(t)x_0 := x(t, x_0)$; we know that operator $S(t)$ is a (nonlinear) semi-group on X, that is,

$$\begin{cases} S(t+s) = S(t) \cdot S(s), & \forall s, t \ge 0, \\ S(0) = I & \text{(Identity in } X). \end{cases}$$

For each $x_0 \in X, \vartheta^+(x_0) = \cup_{t \in \mathbf{R}^+} S(t)x_0$ defines the positive orbit. The ω-limit set of x_0 is the set given by $\omega(x_0) = \{y_0 \in X :$ there exists a sequence $t_n \to +\infty$ as $n \to +\infty$ such that $S(t_n)x_0 \to y_0$ on $X\}$, and the weak ω-limit set of x_0 is the set given by $\omega_w(x_0) = \{y_0 \in X :$ there exists a sequence $t_n \to +\infty$ as $n \to +\infty$ such that $S(t_n)x_0 \to y_0$ weakly *on* $X\}$. A stationary point, or an equilibrium point of (7) is a point $x_0 \in X$ such that

$$S(t)x_0 = x_0, \qquad \forall t \ge 0$$

A subset $\mathcal{A}$ of X is a (functional) invariant set for the semi-group $S(t)$ if

$$S(t)\mathcal{A} = \mathcal{A}, \qquad \forall t \ge 0$$

S is said to be sequentially weakly continuous on X at t if x_n converges weakly to x in X implies that $S(t)x_n$ converges weakly to $S(t)x$ in X. From the mathematical point of view, rotating stall can be described as follows: When the plenum pressure and the angle-averaged through flow are independent of time ξ, that is $\Phi = constant, \Psi = constant$, and the upstream disturbance velocity potential g is not identical to zero, such a behavior of the system (7) is called rotating stall. According to Bessel inequality, the rotating stall of (7) can be expressed in the form given by the next theorem.

Theorem 2. The rotating stall $(g_0^\infty, \Phi_0^\infty, \Psi_0^\infty)$ of (7) is given by

$$g_0^\infty(\theta) = \sum_{n=1}^{\infty} \alpha_n \cos n\theta + \beta_n \sin n\theta$$

where

$$\sum_{n=1}^{\infty} \alpha_n^2 + \beta_n^2 \le -\frac{2\pi}{3a_3}\psi_c'(\Phi_0^\infty) \tag{12}$$

The rotating stall characteristic (i.e. the steady performance Ψ_0^∞) has a lower bound given by

$$\Psi_0^\infty \geq c_0 - 2a_1\Phi_0^\infty - 8a_3(\Phi_0^\infty)^3 \qquad (13)$$

Theorem 3. Let $x_0 = \{g_0, \Phi_0, \Psi_0\}$. Assume that $g_0 \in L^2[0, 2\pi]$. Let $\{g_0^\infty, \Phi_0^\infty, \Psi_0^\infty\}$ represent the the behavior of full-developed rotating stall, then we have

$$\int_0^{2\pi} \left[\psi_c'\left(\Phi_0^\infty\right) + a_3[g_0^\infty(\theta)]^2\right][g_0^\infty(\theta)]^2 d\theta = 0 \qquad (14)$$

$$\Psi_0^\infty - \psi_c(\Phi_0^\infty) - \frac{3a_3\Phi_0^\infty}{2\pi}\int_0^{2\pi}[g_0^\infty(\theta)]^2 d\theta = 0 \qquad (15)$$

$$\Phi_0^\infty - \gamma\sqrt{\Psi_0^\infty} = 0 \qquad (16)$$

5. A NONLINEAR STABILIZING CONTROLLER

The goal for this section is to design a feedback controller $\Phi_T(u, \Psi) = (\gamma + u)\sqrt{\Psi}$ in order to suppress the rotating stall (such that $g \to 0$ in an appropriate norm as $\xi \to \infty$). This brings us to the following:

Theorem 4. Consider the nonlinear feedback controller

$$u = \sqrt{\frac{q^2 \int_0^{2\pi} g^2(\xi, \theta)d\theta + a_1}{(-3a_3)\Psi}} \qquad (17)$$

where q is an arbitrary nonzero number. Let

$$\gamma_c := \sqrt{\frac{a_1}{-3a_3}}\left(1/(a_0 + \frac{2}{3}a_1\sqrt{\frac{a_1}{-3a_3}}) - 1\right)$$

and assume that $\gamma > \gamma_c \geq 0$.

(1) If $g_0 \in L^2[0, 2\pi], \Psi_0 > 0$, then (7) is weakly stable under the feedback control $\Phi_T(u, \Psi)$, that is

$$\int_0^{2\pi} g(\xi, \theta)w(\theta)d\theta \to 0 \quad \text{as } \xi \to \infty \qquad (18)$$

for any $w \in L^2[0, 2\pi]$.

(2) If $g_0 \in H^1[0, 2\pi], \Psi_0 > 0$, then (7) is strongly stable, under the feedback control $\Psi(u, \Psi)$, that is

$$||g(\xi)||_{L_K^2[0.2\pi]} \to 0 \quad \text{as } \xi \to \infty. \qquad (19)$$

Outline of the Proof: As in the proof of Theorem 1, we can show that the feedback system admits a global bounded weak (classical) solution if $g(0) \in L^2[0, 2\pi]$ ($g(0) \in H^1[0, 2\pi]$). Thus for any $\{g_0^\infty, \Phi_0^\infty, \Psi_0^\infty\} \in \omega_w(x_0)$ ($\{g_0^\infty, \Phi_0^\infty, \Psi_0^\infty\} \in \omega_(x_0)$), it satisfies equations (14)-(16) with $\gamma\sqrt{\Psi}$ replaced by Φ_T. Using (17) and (14), it yields

$$-q^2\left[\int_0^{2\pi}[g_0^\infty]^2 d\theta\right]^2 + a_3\int_0^{2\pi}[g_0^\infty]^4 d\theta \geq 0 \qquad (20)$$

Thus $g_0^\infty = 0$ a.e. on $[0, 2\pi]$ because of $a_3 < 0$. According to equations (15) and (16), we have $\Psi_0^\infty = \psi_c\left(\Phi_0^\infty\right)$ and $\Phi_0^\infty = (\gamma + \sqrt{\frac{a_1}{(-3a_3)}})\sqrt{\Psi_0^\infty}$. The proof of second part has been employed the fact that the embedding $H^1[0, 2\pi] \hookrightarrow L^2[0, 2\pi]$ is compact. □

Remark: An important feature of the feedback control (17) is that we just need to measure the average of the upstream disturbance in addition to the pressure rise, even though it basically is still a distributed controller. If we project the system and (17) onto its first Fourier modes, (17) becomes

$$u = \sqrt{\frac{q^2\pi^2 A^2 + a_1}{(-3a_3)\Psi}} \qquad (21)$$

where A is the amplitude of the first-term disturbance velocity. The controller (21) is now finite dimensional and can stabilize the Galerkin projection of the full MG3 model onto its first modes. It actually changes the bifurcation from subcritical bifurcation to supercritical bifurcation of first mode (see (Liaw and Abed 1996) for more details). In the simulations we have used the cubic compressor characteristic from Moore and Greitzer (Moore and Greitzer 1990):

$$\psi_c(\Phi) = \phi_0 + H\left[1 + \frac{3}{2}\left(\frac{\Phi}{W} - 1\right) - \frac{1}{2}\left(\frac{\Phi}{W} - 1\right)^3\right]$$

where ϕ_0, H, and W are parameters which represent, respectively, the shut-off pressure rise, semi-height, and semi-width of the characteristic. The values of above parameters are $a = 1/3, l_c = 6.0, m = 2.0, H = 0.32, W = 0.18, \phi_0 = 0.23$ and $B = 0.1$. Figure 4 exhibits the pressure drop when the parabolic throttle characteristic is applied. Figure 5 – Figure 7 show the system response when the proposed feedback controller is used.

6. CONCLUDING REMARKS

We have studied the asymptotic behavior and stabilization of the full model compression system with rotating stall control. A stabilizing controller with the average of the upstream disturbance velocity and pressure rise as output measurement has been proposed, which is able to suppress the rotating stall as the pressure reaches its maximum. The effect of initial conditions on the full MG3 model system has been illustrated in the paper, which is more accurate than that obtainable by any finite-dimensional projections.

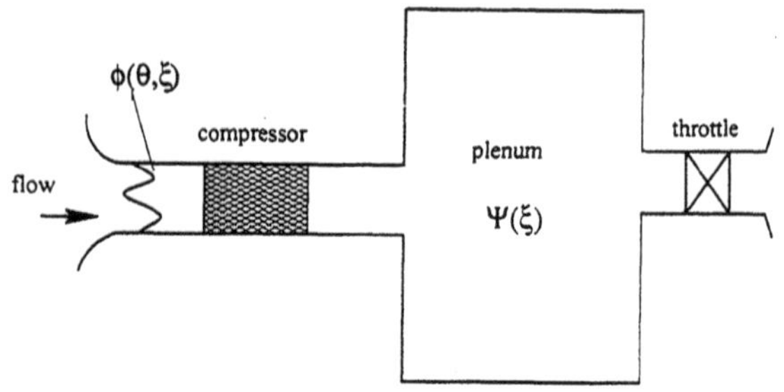

Fig. 1. Axial flow compression system

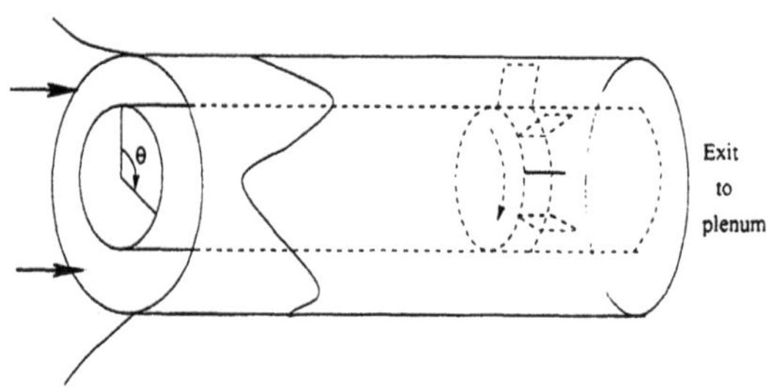

Fig. 2. The compressor geometry

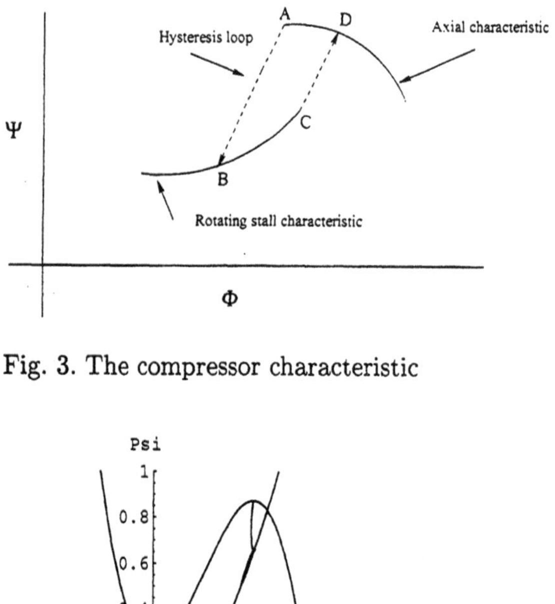

Fig. 3. The compressor characteristic

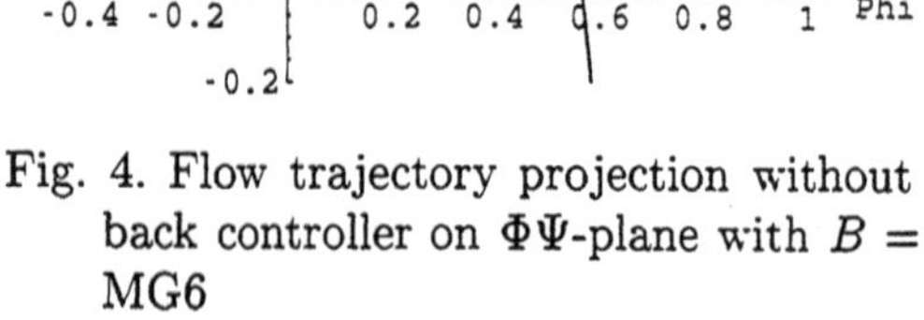

Fig. 4. Flow trajectory projection without feedback controller on $\Phi\Psi$-plane with $B = 0.1$ - MG6

7. REFERENCES

Banaszuk, A., H. Hauksson and I. Mezic (1997). A backstepping controller for a nonlinear partial differential equation model of compression system instabilities. *Preprint.*

Chan, W., M. Xiao and Y. Zhao (1992). Feedback stabilization of a class of multivalued nonlinear distributed parameter systems. *Nonlinear Analysis: Theory, Methods and Applications* **6**, 911-921.

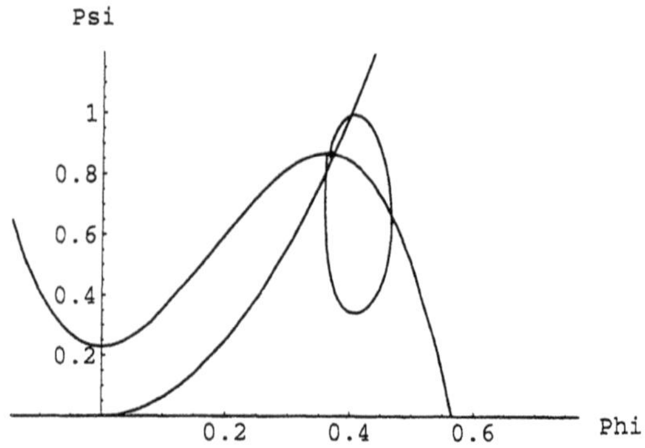

Fig. 5. Flow trajectory projection on $\Phi\Psi$-plane with proposed controller and $B = 0.1$ - MG6

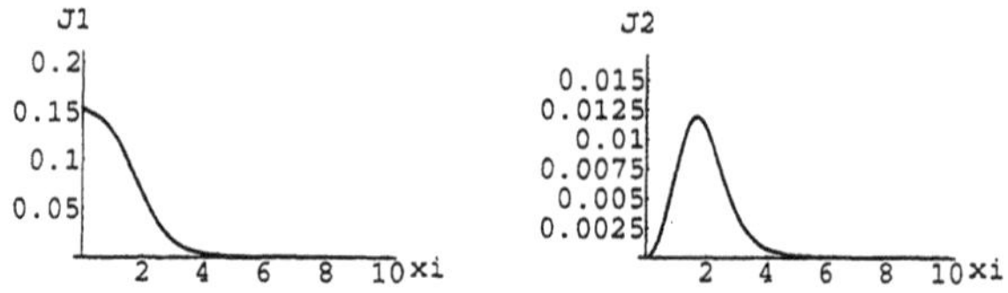

Fig. 6. Two-mode stall call evolution with proposed controller and $B = 0.1$ - MG6

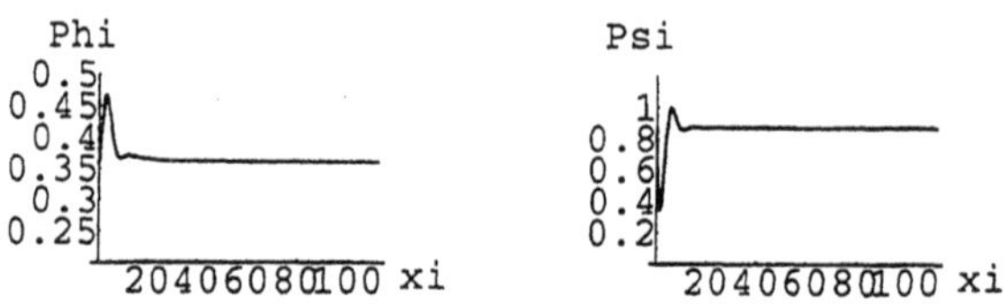

Fig. 7. Flow and Pressure rise responses vs dimensionless time ξ- MG6

Day, I.J., E.M. Greitzer and N.A. Cumpsty (1979). Prediction of compressor performance in rotating stall. *Journal of Engineering for Power* **100**, 1–14.

Henderson, E. and A. Sparks (1996). On the suitability of bifurcation stabilization control laws for high order moore-greitzer compressor models. *Preprint.*

Leonessa, A., V.S. Chellaboina and W.M. Haddad (1997). Globally stabilizing controllers for multi-mode axial flow compressors via equilibria-dependent lyapunov functions. *Proc. of American Control Conference* pp. 993–997.

Liaw, D.C. and E.H. Abed (1996). Active control of compressor stall inception: A bifurcation theoretic approach. *Automatica* **32**, 109–115.

McCaughan, F. (1990). Bifurcation analysis of axial flow compressor stability. *SIAM J. of Appl. Math.* **50**, 1232–1253.

Moore, F.K. and E.M. Greitzer (1990). A theory of post-stall transients in axial compression systems: Part i–development of equations. *ASME Journal Engineering for Gas Turbines and Power* **108**, 68–76.

Xiao, M. (1998). Stabilization of the full model compression system. *Proc. of 37th IEEE CDC* pp. 2575–2580.

MODELLING AND CONTROL OF A HIGH-PRECISION POSITIONNING SYSTEM ACTUATED BY A LINEAR SYNCHRONOUS MOTOR *

J. Boichot * E. Delaleau ** N. V. Diep * J. Lévine **** E. Parzy *****

* *Université Paris-Sud, Centre Scientifique d'Orsay, Laboratoire des Signaux et Systèmes, CNRS – Supélec, Plateau de Moulon, 91192 Gif-sur-Yvette, France. Presently with General Electric Medical Systems Europe (Dept Générateurs), 283 rue de la Minière, BP 34, BUC Cedex, France.*
** *Université Paris-Sud, Centre Scientifique d'Orsay, Laboratoire des Signaux et Systèmes, CNRS – Supélec, Plateau de Moulon, 91192 Gif-sur-Yvette, France. E-mail:* delaleau@lss.supelec.fr
*** *Micro-Contrôle, Z. I. de Beaune-la-Rolande, 45340 Beaune-la-Rolande, France.*
**** *Centre Automatique et Systèmes, École Nationale Supérieure des Mines de Paris, 35, rue Saint-Honoré, 77305 Fontainebleau Cedex, France. E-mail:* levine@cas.ensmp.fr

Abstract: We present a hierarchical controller for a linear synchronous motor with permanent magnets for a high-precision positionning system developed by Newport-France. This system is supposed to realize relatively fast linear displacements from 1 count (4nm) to 400mm. A displacement of 25mm must be achieved in less than 0.2s with maximum acceleration of 1 g and current and voltage limitations, and with an error in the final position less than 1 count. A nonlinear model of the plant is obtained. The motion planning and tracking design are done thanks to its flatness property. The corresponding tracking controller is then realized in a hierarchical way by a high level force control loop and a low level current loop. The control performances are presented on real experiments. *Copyright ©1999 IFAC*

Keywords: nonlinear control, flatness, hierarchical control, linear synchronous motor, position control.

1. INTRODUCTION

In this paper, we are concerned with the control of a linear synchronous drive with permanent magnets designed to realize high-accuracy positionning and relatively fast displacements for the inspection of micro-circuits. Clearly, the accuracy must be in accordance with the size of the circuits, generally of order less than 0.1μm, and fast displacements are needed to increase the productivity of the machine. In this perspective, linear synchronous motors have many advantages such as low frictions, simple design, and high accelerations compared to its low weight and relatively small size. This relatively new topic has already received

* Work supported by Micro-Contrôle-Newport.

some attention (see (Futami and Furutani, March 1989)). The purpose of this study is to increase our mastering of this kind of motor. It has been done in a collaboration between the company Micro-Contrôle/Newport France and two Research Laboratories.

Though the real machine is designed for displacements in 3 dimensions (see figure 1), we focus attention here on 1 dimensional aspects for simplicity's sake.

The machine is made up with a strip of permanent magnets, 2 rails and an optical scale allowing position measurements with a resolution of 4nm. The motor (see figure 2) is made up with a stator (the permanent magnet strip) fixed on the basis, a power supply and a moving body with a pair of coils whose currents are measured. The platform is attached to this moving body. The controller is realized by a real-time computer designed on purpose by Micro-Contrôle.

According to various implementation constraints, a hierarchical structure of the controller is desired and we show how the natural structure of the system is particularly adapted to such a design.

The paper is organized as follows: we first present a model of the plant. Then, we show that the system is flat and the consequences on motion planning and tracking. We further deduce a hierarchical controller and we finally present some real experiments.

Fig. 1. The plant.

2. MODELLING

Linear induction motors have received a great deal of attention in the recent literature, and several textbooks such as (Yamamura, 1972; Poloujadoff, 1980; Gieras, 1994) are devoted to this subject. Surprisingly, studies on linear synchronous motors are less frequent. One may refer to (Xiong and Nasar, 1989; Mizumu and Yamada, 1992; Yoshimura *et al.*, 1995; Jung *et al.*, 1997) and to the Proceedings (LDIA, 1995) [1] for some aspects of their modelling.

2.1 *Notations and assumptions*

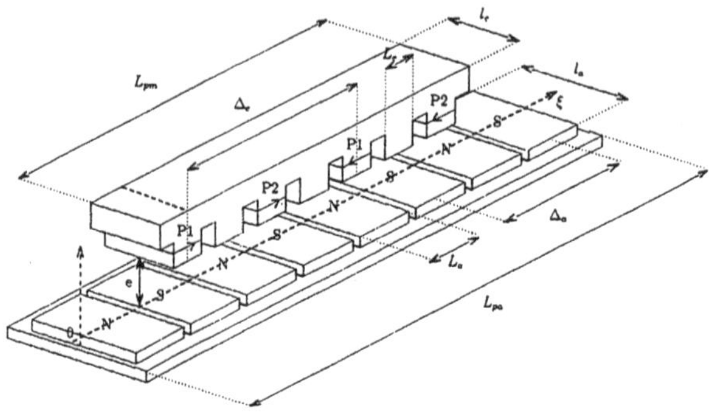

Fig. 2. Permanent magnets linear synchronous motor with two phases.

2.1.0.1. *Notations:*

- u_i: voltage imposed to phase i, $i = 1, 2$;
- i_i: current across phase i;
- ψ_i : total flux of phase i;
- R_i: resistance of phase i;
- L_i: coefficient of self-induction of phase i;
- M: mutual inductance between the two phases;
- K_{E_i}: coefficient of electromotive force of phase i.

2.1.0.2. *Modelling assumptions:*

- The magnetic circuits are not saturated; thus the fluxes are linear functions of the currents. Moreover, magnetic circuits are perfectly laminated, and currents only flow inside coils.
- The current density can be assumed to be constant in the section of the conductors (absence of skin effect).
- The mobile part of the motor is assumed to be at constant altitude over the magnetic path where the magnets are joined. Hence the air-gap is constant (contrarily to most of the rotating machines). The only possible movement of the mobile part of the motor is an horizontal translation along the rails.
- Mechanical losses are neglected and the whole stored magnetic energy is converted in mechanical energy.
- The vertical component of the magnetic field is assumed periodic with respect to space (regularity of the magnets) and only the first harmonic is considered. This is the standard "sinusoidal" hypothesis.

[1] Reference kindly communicated by M. Poloujadoff, University Paris VI.

- The characteristics of the two circuits are assumed to be identical, i.e. $R_1 = R_2 = R$, $L_1 = L_2 = L$, $K_{E_1} = K_{E_2} = K_E$. Moreover, $M \approx 0$ and is neglected.

2.2 *The system*

Recall that we only consider here the dynamics of a single motor moving along a fixed axis. We denote by (ξ, v) its position and velocity along this axis. By Newton's law, and denoting by F_{ext} the sum of the exterior forces acting on the platform, we immediately get that $m\dot{v} = F_{\text{ext}} = F_{\text{m}} + F_{\text{fr}} + F_{\text{l}}$ where F_{m} is the force produced by the motor, F_{fr} is the friction force, and F_{l} the resistive force of the load, considered as a perturbation. The frictional force depends on the position and velocity and may also depend on time (see e.g., (Bo and Pavelescu, 1982; Karnopp, 1985; Haessig and Friedland, 1991)). Its precise form is not needed here and is omitted.

The force F_m generated by the motor is the sum of two terms F_{u} and F_{det}, the effective and detent forces. F_{u} is given by

$$F_{\text{u}} = K_F\left(i_1\cos(\alpha\xi) + i_2\sin(\alpha\xi)\right) \tag{1}$$

where $\frac{2\pi}{\alpha}$ is the length between two successive south poles in the magnet strip and K_F is the force constant of the motor. F_{det} is given by

$$F_{\text{det}} = F_d\sin(2\alpha(\xi + \Delta_d)) \tag{2}$$

where Δ_d is the distance between the centers of the two phases and F_d a constant (see (Yoshimura *et al.*, 1995) or (Jung *et al.*, 1997)[2] for more details on the detent force).

The fluxes being assumed linear with respect to the currents i_1 and i_2 of the two phases, the electromotive force is, in the first phase, given by $K_E v\cos(\alpha\xi)$ and in the second phase by $K_E v\sin(\alpha\xi)$. Thus, by Kirchhoff's law, we get

$$u_1 = L\frac{di_1}{dt} + Ri_1 + K_E v\cos(\alpha\xi)$$
$$u_2 = L\frac{di_2}{dt} + Ri_2 + K_E v\sin(\alpha\xi)\ .$$

Note that if there are no mechanical losses, $K_E = K_F$.

The model reads:

$$\dot{\xi} = v \tag{3a}$$
$$m\dot{v} = K_F\left(i_1\cos(\alpha\xi) + i_2\sin(\alpha\xi)\right) + F_d\sin(2\alpha(\xi+\Delta_d)) + F_{\text{fr}}(\xi, v, t) + F_{\text{l}} \tag{3b}$$
$$L\frac{di_1}{dt} = -Ri_1 - K_E v\cos(\alpha\xi) + u_1 \tag{3c}$$
$$L\frac{di_2}{dt} = -Ri_2 - K_E v\sin(\alpha\xi) + u_2 \tag{3d}$$

[2] Reference kindly communicated by C. Marchand, Laboratoire de génie électrique de Paris.

2.3 *Direct and quadrature current coordinates*

The above model was expressed in the coordinates (ξ, v, i_1, i_2) with (u_1, u_2) the input voltages. A simpler form can be obtained by changing the coordinates, in the spirit of the well-known Park transformation, into (ξ, v, i_d, i_q) where i_d and i_q are the direct and quadrature currents:

$$\begin{aligned} i_d &= \ \ i_1\cos(\alpha\xi) + i_2\sin(\alpha\xi) \\ i_q &= -i_1\sin(\alpha\xi) + i_2\cos(\alpha\xi) \end{aligned} \tag{4}$$

and by replacing the input voltages by the equivalent pair (u_d, u_q) of direct and quadrature voltages:

$$\begin{aligned} u_d &= \ \ u_1\cos(\alpha\xi) + u_2\sin(\alpha\xi) \\ u_q &= -u_1\sin(\alpha\xi) + u_2\cos(\alpha\xi) \end{aligned} \tag{5}$$

This change of coordinates is indeed suggested by the fact that F_{u}, the efficient part of the force delivered by the motor, (see (1)), is directly generated by the current i_d, and is independent of i_q. The inverse transformations are obtained by elementary trigonometric computations and are omitted.

The system, in these new coordinates, reads:

$$\dot{\xi} = v \tag{6a}$$
$$m\dot{v} = K_F i_d + F_d\sin(2\alpha(\xi+\Delta_d)) + F_{\text{fr}}(\xi, v, t) + F_{\text{l}} \tag{6b}$$
$$L\frac{di_d}{dt} = -Ri_d + L\alpha v i_q - K_E v + u_d \tag{6c}$$
$$L\frac{di_q}{dt} = -Ri_q - L\alpha v i_d + u_q\ . \tag{6d}$$

3. FLATNESS

We now establish that the model (3), or equivalently (6), is flat (see (Fliess *et al.*, 1995; Fliess *et al.*, 1999) for an introduction to flatness).

Roughly speaking, a control system with m input variables is flat if there exists a set of m independent variables, called a flat output or linearizing output, such that every variable of the system can be expressed in terms of the flat output and a finite number of its derivatives. The main applications of this property are motion planning and trajectory tracking (Fliess *et al.*, 1995; Fliess *et al.*, 1999).

Here, the input variables are u_1 and u_2, or equivalently u_d and u_q. In this section, the perturbative resistive force F_l is assumed to be known. Let us show that $y = (y_1, y_2)$, with

$$y_1 = \xi \ , \quad y_2 = i_q \tag{7}$$

is a flat output. Clearly, $v = \dot{y}_1$ and, by (6b), we have

$$\begin{aligned} i_d = \frac{1}{K_F}\{m\ddot{y}_1 - F_d \sin(2\alpha(y_1 + \Delta_d)) \\ -F_{fr}(y_1, \dot{y}_1, t) - F_l\} \stackrel{\text{def}}{=} I_d(y_1, \dot{y}_1, \ddot{y}_1) \ . \end{aligned} \tag{8}$$

Differentiating (8) and using (6c), we get that u_d is a function of y_1 and derivatives up to the 3rd order and, by (6d), that u_q is a function of y_1 and derivatives up to the 3rd order and of y_2 and $\dot{y}_2$. Thus, we have proved that all the state and input variables $(\xi, v, i_d, i_q, u_d, u_q)$ can be expressed in terms of $(y_1, \dot{y}_1, \ddot{y}_1, y_1^{(3)}, y_2, \dot{y}_2)$, namely in terms of the components of $y = (y_1, y_2)$ and a finite number of their time derivatives, which proves that y is a flat output and that the system (3), or equivalently (6), is flat.

Remark. Notice that flatness only depends on the structure of the model and not on the particular electromagnetic characteristics of the process. More precisely, it would still hold true if we considered, in place of the present model, a "generic" model of the form

$$\begin{aligned} & m\ddot{\xi} = F(\xi, v, i_1, i_2, t) \\ & \frac{d(L_1 i_1)}{dt} + \frac{d(M i_2)}{dt} + R_1 i_1 + K_{E_1} v \cos(\alpha\xi) \\ & \qquad = u_1 \\ & \frac{d(M i_1)}{dt} + \frac{d(L_2 i_2)}{dt} + R_2 i_2 + K_{E_2} v \sin(\alpha\xi) \\ & \qquad = u_2 \end{aligned} \tag{9}$$

with L_1, L_2 and the coefficient of mutual inductance M possibly depending on (ξ, v), provided that $\frac{\partial F}{\partial i_1}$ or $\frac{\partial F}{\partial i_2}$ is non zero.

3.1 *Interpretation*

The interpretation of the first component of the flat output $y_1 = \xi$, the position of the motor, is immediate. The second component $y_2 = i_q$, since the dissipated energy in the circuits by Joule effect is

$$E = R(i_1^2 + i_2^2) = Ri_d^2 + Ri_q^2$$

describes the losses. It can be minimized by choosing $i_q = 0$, which implies that the currents i_1 and i_2 satisfy the well-known relation:

$$i_1 \sin(\alpha\xi) = i_2 \cos(\alpha\xi) \ . \tag{10}$$

The previous analysis shows that the motor can only follow trajectories whose jerk (third order derivative) is everywhere defined and bounded. Moreover, i_q's trajectories must be at least differentiable, which is obviously the case if $i_q(t) \equiv 0$.

3.2 *Application to motion planning*

Recall that a typical reference trajectory of the position must start from a prescribed idle position ξ_0 (with $v_0 = 0$) at time t_0, and arrive at another idle prescribed position ξ_1 (with $v_1 = 0$) at time $t_1 > t_0$. Moreover, the acceleration must not exceed 1 g, the amplitude of the maximum currents i_1 and i_2 must remain bounded by $I_{\max}$, the current i_q must be 0 all along the path, and the amplitude of the maximum voltages u_1 and u_2 must remain bounded by $U_{\max}$.

According to (7), (8), the problem consists, in a first step, in designing a 3 times differentiable curve for ξ, with $i_q = 0$, such that the initial and final constraints are met. The second step consists in choosing the minimum duration $t_1 - t_0$ such that the acceleration, current and voltage contraints are satisfied.

Considering the 3rd degree polynomial with respect to time:

$$\xi^*(t) = \xi_0 + (\xi_1 - \xi_0)\left(\frac{t - t_0}{t_1 - t_0}\right)^2 \left(3 - 2\left(\frac{t - t_0}{t_1 - t_0}\right)\right)$$

it satisfies $\xi^*(t_0) = \xi_0$, $\dot{\xi}^*(t_0) = v(t_0) = 0$, $\xi^*(t_1) = \xi_1$, $\dot{\xi}^*(t_1) = v^*(t_1) = 0$. Then, the references of the direct current i_d^* and voltage u_d^* are deduced as above, with $i_q^*(t) \equiv 0$, which implies that $u_q^*(t) = L\alpha\dot{\xi}^*(t) i_d^*(t)$. The original currents and voltages are then given by

$$i_1^* = i_d^* \cos(\alpha\xi^*) \ , \quad i_2^* = i_d^* \sin(\alpha\xi^*) \tag{11}$$

$$\begin{aligned} u_1^* = u_d^* \cos(\alpha\xi^*) - L\alpha\dot{\xi}^* i_d^* \sin(\alpha\xi^*) \\ u_2^* = u_d^* \sin(\alpha\xi^*) + L\alpha\dot{\xi}^* i_d^* \cos(\alpha\xi^*) \ . \end{aligned} \tag{12}$$

Finally, let us sketch how the final time t_1 may be tuned to satisfy the constraints. The reference velocity and acceleration are given by

$$\begin{aligned} \dot{\xi}^*(t) &= 6\left(\frac{\xi_1 - \xi_0}{t_1 - t_0}\right)\left(\frac{t - t_0}{t_1 - t_0}\right)\left(1 - \left(\frac{t - t_0}{t_1 - t_0}\right)\right) \\ \ddot{\xi}^*(t) &= 6\frac{\xi_1 - \xi_0}{(t_1 - t_0)^2}\left(1 - 2\left(\frac{t - t_0}{t_1 - t_0}\right)\right) \ . \end{aligned}$$

The maximum velocity is reached at $t = \frac{t_1 + t_0}{2}$ and is equal to

$$v_{\max} = \frac{3}{2}\left(\frac{|\xi_1 - \xi_0|}{t_1 - t_0}\right) \ .$$

Similarly, the maximum acceleration is achieved for $t = t_0$ or $t = t_1$ and has the value $\gamma_{\max} = 6\frac{|\xi_1 - \xi_0|}{(t_1 - t_0)^2}$. It satisfies $\gamma_{\max} \leq g$ if an only if $t_1 \geq t_0 + \sqrt{6\frac{|\xi_1 - \xi_0|}{g}}$. The current and voltage constraints are treated in a similar way, using the fact that since $\sup_t(i_1^*(t), i_2^*(t)) \leq \sup_t i_d^*(t)$, it suffices to set $\sup_t i_d^*(t) \leq I_{\max}$. An inferior bound for $t_1 - t_0$, function of $(I_{\max}, U_{\max}, v_{\max}, \gamma_{\max})$ is thus easily deduced.

4. TRACKING BY HIERARCHICAL CONTROL

Note that, though the system (3) can be transformed into the following linear controllable one:

$$y_1^{(3)} = v_1 \; , \quad \dot{y}_2 = v_2 \tag{13}$$

where v_1 and v_2 are the new input, we have preferred a hierarchical approach, using the fact that the open-loop response time of the mechanical subsystem (3a)-(3b) is much larger than the electromagnetic one. A natural control architecture, in this case, consists in dividing the system (3) into a high-level subsystem (the mechanical part) (3a)-(3b), controlled by the current i_d and a low-level subsystem (the electromagnetic part) (3c)-(3d) whose objective is to follow the current i_d, required by the high-level part, at a higher rate, while maintaining i_q close to 0.

Setting

$$\ddot{y}_1 - \ddot{y}_1^* = w_1 - w_1^* \; , \tag{14}$$

this subsystem is the image of the high-level part (3a)-(3b) and the subsystem made up with

$$\dot{w}_1 - \dot{w}_1^* = v_1 - v_1^* \; , \quad \dot{y}_2 - \dot{y}_2^* = v_2 - v_2^* \tag{15}$$

turns out to be the image of the low-level part.

It suffices then to set

$$w_1 - w_1^* = K_{1,0}(y_1 - y_1^*) + K_{1,1}(\dot{y}_1 - \dot{y}_1^*) \tag{16}$$

with $K_{1,0}$ and $K_{1,1}$ chosen to ensure the stability of the high-level (14) and to control, through the voltages, the error

$$\eta = w_1 - w_1^* - K_{1,0}(y_1 - y_1^*) - K_{1,1}(\dot{y}_1 - \dot{y}_1^*)$$

by the *high-gain* proportional controller

$$v_1 - v_1^* = -\frac{\kappa_1}{\varepsilon}\eta \tag{17}$$

where ε is an arbitrary small real positive number and κ_1 a positive real number.

Translating these formulas in the original coordinates, the high-level controller can be seen as a force control scheme and the low-level one as the high-gain current loop that realizes the desired force.

An elementary computation provides $\dot{\eta} = -\frac{\kappa_1}{\varepsilon}\eta - K_{1,0}(\dot{y}_1 - \dot{y}_1^*) - K_{1,1}(\ddot{y}_1 - \ddot{y}_1^*) = -(\frac{\kappa_1}{\varepsilon} + K_{1,1})\eta - K_{1,1}K_{1,0}(y_1 - y_1^*) - (K_{1,0} + K_{1,1}^2)(\dot{y}_1 - \dot{y}_1^*)$. Putting together the high and low levels, with the proportional controller $v_2 - v_2^* = -\frac{\kappa_2}{\varepsilon}(y_2 - y_2^*)$, again with κ_2 positive real so that $\frac{\kappa_2}{\varepsilon}$ is a high-gain, ensuring that i_q goes fastly to 0, and applying standard singular perturbation results (Kokotović *et al.*, 1986; Lévine and Rouchon, 1994), we get that the equilibrium point $\eta = 0$, $y_1 = y_1^*$ and $y_2 = y_2^*$ is unique, and that the whole system is stable at this point. The parameter ε is then tuned to suitably separate the respective time scales of the high and low levels, separation which indeed remains valid in the original coordinates, though the corresponding force and current loops are nonlinear.

Note that the main advantage, for real time computation, is that the only information shared by the high and low levels is the variable η. The controller architecture is displayed in figure 3.

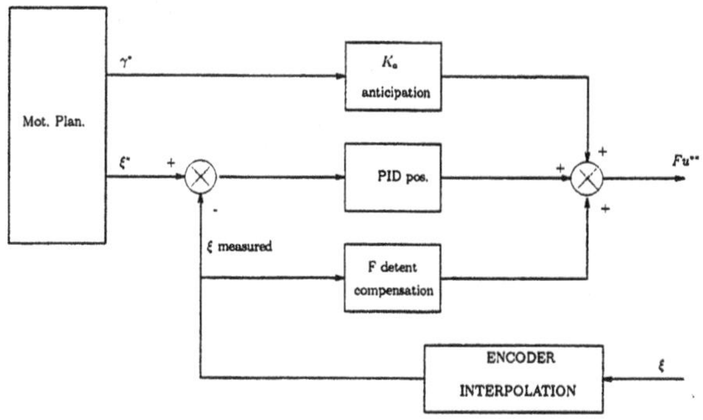

Fig. 3. High and low level controller architecture.

5. EXPERIMENTS

Real experiments showing two kinds of displacements are now presented for a mass of 6kg: a displacement of 25mm in approximately 0.2s, including the settling time (see figure figure 4, and figure 5 for the position error), and a displacement of 3μm in approximately 0.1s, including the settling time (see figure figure 6, and figure 7 for the position error). In both cases, the final positionning error is less than 1 count (4nm).

6. REFERENCES

Akmese, R. and J. F. Eastham (1992). Design of permanent magnet flat linear motor for standstill applications. *IEEE Trans. Magn.* **28**(5), 3042–3044.

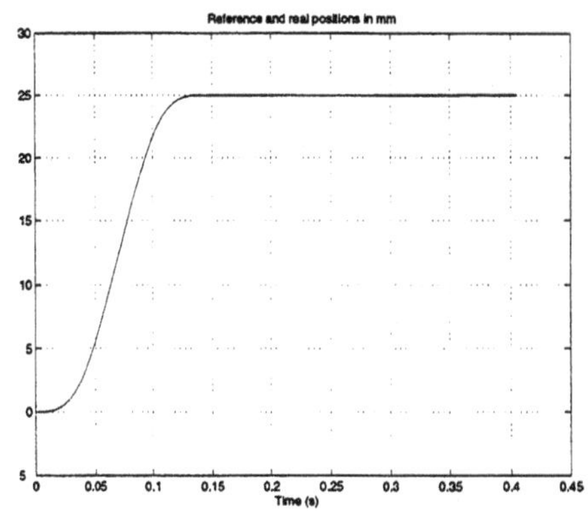

Fig. 4. Desired and real position during a 25mm displacement.

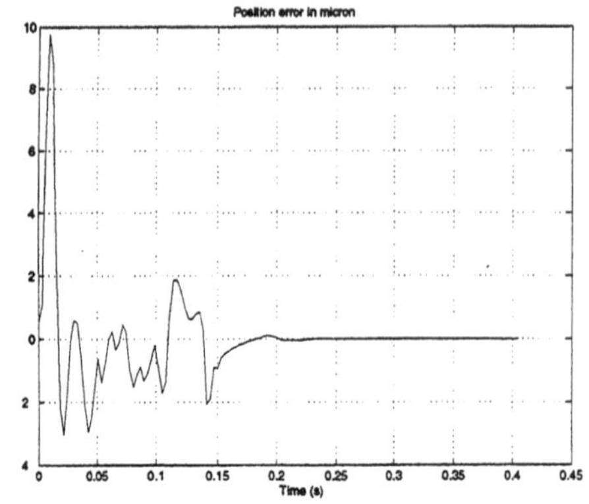

Fig. 5. Position error during a 25mm displacement.

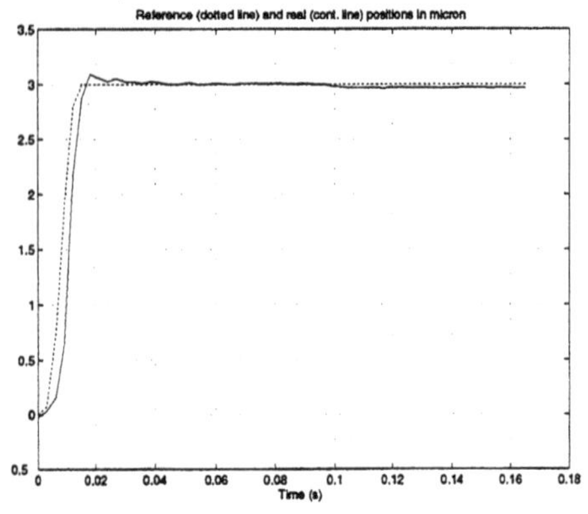

Fig. 6. Desired and real position during a 3μm displacement.

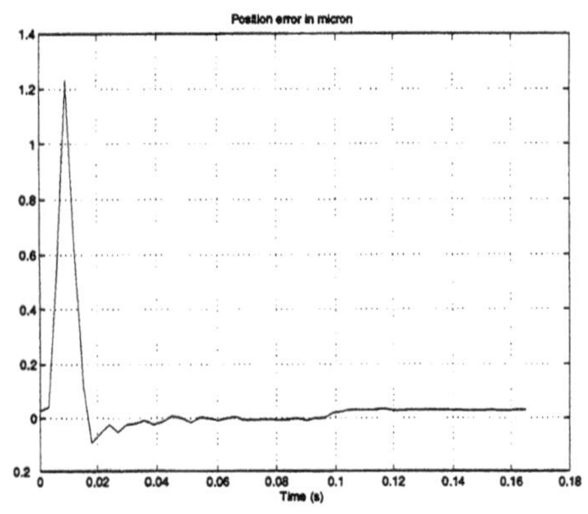

Fig. 7. Position error during a 3μm displacement.

Bartos, F. J. (1997). Linear motors and controls. Take a 'direct' route to top performance. *Control Engineering* pp. 90–100.

Bo, Li Chun and D. Pavelescu (1982). The friction-speed relation and its influence on the critical velocity of stick-slip motion. *Wear* **82**, 277–289.

Fliess, M., J. Lévine, P. Martin and P. Rouchon (1995). Flatness and defect of non-linear systems: introductory theory and examples. *Internat. J. Control* **61**, 1327–1361.

Fliess, M., J. Lévine, P. Martin and P. Rouchon (1999). A Lie-Bäcklund approach to equivalence and flatness of nonlinear systems. *IEEE Trans. Automatic Control.*

Futami, S. and A. Furutani (March 1989). Super precision positionning with an AC linear motor. In: *THK Reports, Technology Dept.* number 103–9056. pp. 26–31.

Gieras, J. F. (1994). *Linear Induction Drives.* Clarendon Press. Oxford.

Haessig, Jr., D. A. and B. Friedland (1991). On the modeling and simulation of friction. *J. Dynamic Systems, Measurement and Control. Trans. ASME* **113**, 354–362.

Jung, I.-S., S.-B. Yoon, J.-H. Shim and D.-S. Hyun (1997). Analysis of forces in a short primary type and short secondary type permanent magnet linear synchronous motor. In: *Proc. IEMDC'97.* pp. MC1.8.1–MC1.8.3.

Karnopp, D. (1985). Computer simulation of stick-slip friction in mechanical dynamic systems. *J. Dynamic Systems, Measurement and Control. Trans. ASME* **107**, 100–103.

Kokotović, P., H.K. Khalil and J. O'Reilly (1986). *Singular perturbation methods in control: analysis and design.* Academic Press. London.

LDIA (1995). *Proc. First International Symposium on Linear drives for Industry Applications.* The Institute of Electrical Engineers of Japan.

Lévine, J. and P. Rouchon (1994). An invariant manifold approach for robust control design. In: *Systems and Networks: Mathematical Theory and Applications (MTNS'93, Volume II, Invited and Contributed Papers)* (U. Helmke, R. Mennicken and J. Saurer, Eds.). Akademie Verlag. Berlin. pp. 309–312.

McLean, G. W. (1988). Review of recent progress in linear motors. *IEE Proceedings, Pt B* **135**(6), 380–416.

Mizumu, T. and H. Yamada (1992). Magnetic circuit analysis of a linear synchronous motor with permanent magnet. *IEEE Trans. Magn.* **28**(5), 3027–3029.

Poloujadoff, M. (1980). *Theory of linear induction machinery.* Clarendon Press. Oxford.

Xiong, G. Y. and S. A. Nasar (1989). Analysis of fields and forces in a permanent magnet linear synchronous machine based on the concept of magnetic charge. *IEEE Trans. Magn.* **25**(3), 2713–2719.

Yamamura, S. (1972). *Theory of linear induction motors.* Halstead.

Yoshimura, T., H. J. Kim, M. Watada, S. Torii and D. Ebihara (1995). Analysis of the reduction of the detent force in a permanent magnet linear synchronous motor. *IEEE Trans. Magn.* **31**(6), 3728–3730.

Zhu, Z. Q., Z. P. Xia, D. Howe and P. H. Mellor (1997). Reduction of cogging force in slotless linear permanent magnet motors. *IEE Proc. Electr. Power Appl.* **144**(4), 277–282.

MULTIVARIABLE NONLINEAR INTERNAL MODEL CONTROL APPLIED TO A BIOREACTOR

G. Roux
e-mail : roux@laas.fr

B. Dahhou
e-mail : dahhou@laas.fr

LAAS-CNRS, 7 Avenue du Colonel Roche, 31077 Toulouse Cedex, France.
Phone : 33 5 61 55 94 93 Fax : 33 5 61 55 94 40

Abstract : Nonlinear control has emerged as an area of extensive research activity during the last decade. An important class of problems concerns the study of disturbance inputs for analysis and controller synthesis purposes. This paper describes some aspects of the design of multivariable non-linear internal model control in the case of continuous flow-fermentation process with several types of modelling errors. The controller is based on the inverse of the process model and a non-linear filter. The approach used get rid of the assumption of full state feedback. Simulation results for a continuous stirred tank reactor illustrate the good behaviour of the control strategy. *Copyright © 1999 IFAC*

Keywords : Internal model control, nonlinear control, multivariable control, biotechnological process.

1. INTRODUCTION

It is well-known that the high nonlinearity of processes is one of the challenging control problems (Bastin and Dochain, 1990). The non-linear, strongly interacting nature of multivariable chemical processes necessitates the development of solid control methodologies that are capable of coping with both non-linearities and interactions. However, in the process in the process control field, the customary approach has been to neglect the non-linearities by approximating the non-linear model by a linear one and to apply linear theory to design linear controllers. From a control point of view, fermentation processes are characterised, by the uncertainties of their internal dynamics models, and randomness of the external disturbances. In the same time the control law must satisfy strong requirements on accuracy and quality of the output tracking. All these restrictions impose a complex control task. The controller should force a process to produce an output that tracks any desired output under action of unpredictable disturbances and in spite of its internal dynamics uncertainty. Therefore, the design of a control law dealing with these problems is a challenge; moreover, it could be used as well for a better understanding of microbial physiology and for the improvement of productivity. The main reason of using non-linear control techniques is that it takes into account the dynamics, strongly non-linear and non stationary behaviour of the living organisms. Since reasonably accurate non-linear models are available for a variety of process model, they are used as the basis for the controller design, which can be expected to yield significantly improved performance. In this paper the methodology of (Roux and Dahhou, 1996; Henson and Seborg, 1991) is extended to multi-input/multi-output systems. The approach used in this paper is a general extension of linear internal model control (Morari and Zafiriou, 1989) to open loop stable, non-linear systems with stable inverse. The non-linear process model is used directly in the design of the controller, and a non-linear filter (Henson and Seborg, 1991) is added to make the controller implementable and to improve robustness. Unlike most input-output linearization techniques, full-state feedback is not required and the approach uses only output-feedback. In fact the process model functions are used as an open-loop observer. A simulation evaluation is performed on a continuous stirred-tank fermentation process which exhibits non-linear and unstationary features. Several types of disturbance and modelling errors are considered in the simulation study (step disturbance on the yield and on other parameters, structural error in the growth rate). In the sequel, the paper is structured as follows. In section 2, the description of the internal model control strategy is presented in three step : model and plan formulation, control design, filter design. In section 3, the process model is described and the different control objectives are given. The efficiency of the proposed algorithms is compared and discussed via simulation results on a continuous stirred tank bioreactor. A general conclusion ends the paper.

2. DESCRIPTION OF THE INTERNAL MODEL CONTROL STRATEGY

Schematically, the non-linear internal model control structure can be represented as in Schema 1 (Economou and Morari, 1985). In this bloc diagram,

C represents a non-linear controller, F denotes a robustness filter, M designates the model and P symbolises the plant. The non-linear internal model controller Q is defined as the composition of the non-linear controller C and the robustness filter F. All these blocs, will be explained in the next sub-sections.

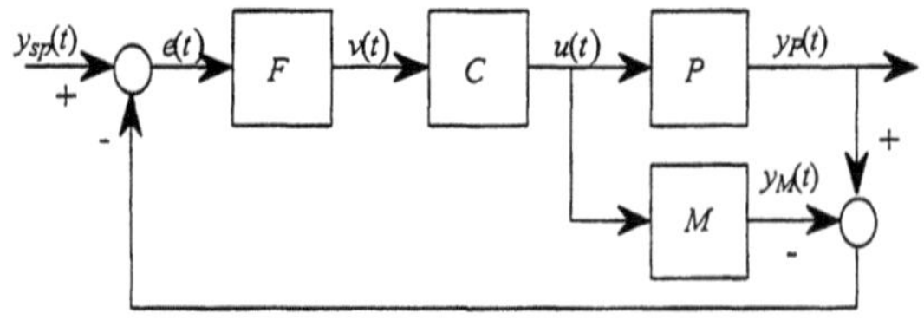

Scheme 1 : General structure of non-linear IMC.

2.1 Model and plant formulation

We consider the control of non-linear single-input single-output systems. Assume that the model M available for controller design has the form :

$$\begin{cases} \dot{x}_M = f_M(x_M) + \sum_{j=1}^{n_y} g_{M_j}(x_M)u_j \\ y_{M_i} = h_{M_i}(x_M) \qquad i = 1\cdots n_y \end{cases} . \qquad (1)$$

In this system, x_M is a state vector of dimension n_M, u is a manipulated input vector of dimension n_y, y_M is a controlled output vector of dimension n_y. $f_M(x_M)$ is a smooth vector field on $\Re^{n_M}$, $g_{M_1}(x_M),\cdots,g_{M_{n_y}}(x_M)$ are smooth vector fields on $\Re^{n_M}$ and $h_{M_1}(x_M),\cdots,h_{M_{n_y}}(x_M)$ are smooth scalar fields on $\Re^{n_M}$. The plant P is assumed to be of the form :

$$\begin{cases} \dot{x}_P = f_P(x_P) + \sum_{j=1}^{n_y} g_{P_j}(x_P)u_j \\ y_{P_i} = h_{P_i}(x_P) \qquad i = 1\cdots n_y \end{cases} . \qquad (2)$$

Here, x_P is a state vector of dimension n_P, u is a manipulated input vector of dimension n_y, y_P is a controlled output vector of dimension n_y. $f_P(x_P)$ is a smooth vector field on $\Re^{n_P}$, $g_{P_1}(x_P),\cdots,g_{P_{n_y}}(x_P)$ are smooth vector fields on $\Re^{n_P}$ and $h_{P_1}(x_P),\cdots,h_{P_{n_y}}(x_P)$ are smooth scalar fields on $\Re^{n_P}$.

2.2 Controller design

As in linear internal model control (Morari and Zafiriou, 1989), the first objective is to design the controller C such that the closed loop system is stable if a perfect model is available. In this part, we assume that the filter is the identity operator and the model is perfect ($M \equiv P$). The controller C is chosen to minimise the tracking vector error $(y_{sp}(t) - y_P(t))$, the problem can be formulated as follows (Henson and Seborg, 1991) :

$$\min_C \left\| \left(I_{n_y} - MC\right) y_{sp} \right\| . \qquad (3)$$

This criterion (3) is zero for any norm and any setpoint if the controller C is chosen to be the right inverse of the model ($C = M_r^{-1}$). The construction of this inverse model is based on the method of Hirschorn (1979). Assume that the non-linear system (1) is input/output linearisable with relative order r_i (Isidori, 1989; Kravaris and Soroush, 1990). Therefore, r_i is the smallest order of derivative of y_{M_i} that explicitly depends on the vector u. Then the first r_i time derivatives of the model output (1) are given by :

$$\begin{cases} \dfrac{d^k y_{M_i}}{dt^k} = L_{f_M}^k h_{M_i}(x_M) \qquad k = 1\cdots r_i - 1 \\ \dfrac{d^{r_i} y_{M_i}}{dt^{r_i}} = L_{f_M}^{r_i} h_{M_i}(x_M) + \sum_{j=1}^{n_y} L_{g_{M_j}} L_{f_M}^{r_i - 1} h_{M_i}(x_M) u_j \end{cases} . \qquad (4)$$

The Hirschorn inverse (Hirschon, 1979; Daoutidis and Kravaris, 1991) for systems with non singular characteristic matrix, is obtained by solving the last equation in system (4) for each input u_i and substituting the result in the system of equation (1) :

$$\begin{cases} \dot{x}_M = f_M(x_M) \\ \qquad + g_M(x_M) C(x_M)^{-1} \left(\begin{bmatrix} y_M^{(r_1)} \\ \vdots \\ y_M^{(r_{n_y})} \end{bmatrix} - \begin{bmatrix} L_f^{r_1} h(x_M) \\ \vdots \\ L_f^{r_{n_y}} h(x_M) \end{bmatrix} \right) \\ u = C(x_M)^{-1} \left(\begin{bmatrix} y_M^{(r_1)} \\ \vdots \\ y_M^{(r_{n_y})} \end{bmatrix} - \begin{bmatrix} L_f^{r_1} h(x_M) \\ \vdots \\ L_f^{r_{n_y}} h(x_M) \end{bmatrix} \right) \end{cases} . \qquad (5)$$

In the equation (5), the matrix $C(x_M)$ is called the characteristic matrix of the system (1) and it is defined :

$$C(x_M) = \begin{bmatrix} L_{g_1} L_f^{r_1 - 1} h_{M_1}(x_M) & \cdots & L_{g_{n_y}} L_f^{r_1 - 1} h_{M_1}(x_M) \\ \vdots & & \vdots \\ L_{g_1} L_f^{r_{n_y} - 1} h_{M_{n_y}}(x_M) & \cdots & L_{g_{n_y}} L_f^{r_{n_y} - 1} h_{M_{n_y}}(x_M) \end{bmatrix} \qquad (6)$$

The inverse model reconstruct the input $u_i(t)$ from the r_ith derivative of the output $y_{M_i}^{(r_i)}(t)$. To track the setpoint, the inverse model controller C can be chosen as in equation (5) where $y_{sp}^{(r)}(t)$ is the input vector :

$$u = C(x_M)^{-1} \left(\begin{bmatrix} y_{sp}^{(r_1)} \\ \vdots \\ y_{sp}^{(r_{n_y})} \end{bmatrix} - \begin{bmatrix} L_f^{r_1} h(x_M) \\ \vdots \\ L_f^{r_{n_y}} h(x_M) \end{bmatrix} \right) . \qquad (7)$$

The inverse model controller in equation (20) does not require state feedback but only an output feedback from the plant, since the model operates as an open-loop observer.

2.3 Filter design

To realise the controller implementation, the inverse model is augmented with a linear filter in the linear internal model control. The same approach can be used with the non-linear inverse model controller in equation (7) if a non-linear filter is used. The non-linear filter F is chosen of the form :

$$\begin{bmatrix} v^{(r_1)} \\ \vdots \\ v^{(r_{n_y})} \end{bmatrix} = \begin{bmatrix} -\alpha_{1r_1} L_{f_M}^{r_1-1} h_{M_i}(x_M) - & \cdots & -\alpha_{11} h_{M_i}(x_M) + \alpha_{11} e_1 \\ & \vdots & \\ -\alpha_{n_y r_{n_y}} L_{f_M}^{r_{n_y}-1} h_{M_i}(x_M) - & \cdots & -\alpha_{n_y 1} h_{M_i}(x_M) + \alpha_{n_y 1} e_{n_y} \end{bmatrix} \quad (8)$$

where $v^{(r_i)}$ are the output of the filter F, the $\alpha_{r_i i}$ are tuning filter parameters and the error signal e_i is defined as $\left(e_i = y_{sp_i} - y_{P_i} + y_{M_i}\right)$. The inverse model controller C is obtained by modifying the input to the controller in equation (7) as follows :

$$u = C(x_M)^{-1}\left(\begin{bmatrix} v^{(r_1)} \\ \vdots \\ v^{(r_{n_y})} \end{bmatrix} - \begin{bmatrix} L_f^{r_1} h(x_M) \\ \vdots \\ L_f^{r_{n_y}} h(x_M) \end{bmatrix}\right). \quad (9)$$

The classical structure of the non-linear internal model control (Scheme 1) remain still valid, it only takes to replace v by $v^{(r)}$. Finally the internal model controller $Q = CF$ is obtained by combining the two latest equations

$$u = \tilde{C}(x_M)^{-1}\left(\begin{bmatrix} \alpha_{11} e_1 \\ \vdots \\ \alpha_{n_y 1} e_{n_y} \end{bmatrix} - \begin{bmatrix} \sum_{k=1}^{r_1} \alpha_{1k} L_{f_M}^{k-1} h_{M_1}(x_M) \\ \vdots \\ \sum_{k=1}^{r_{n_y}} \alpha_{n_y k} L_{f_M}^{k-1} h_{M_{n_y}}(x_M) \end{bmatrix}\right) \quad (10)$$

where :

$$\tilde{C}(x_M) = \begin{bmatrix} \alpha_{1r_1} L_{g_1} L_f^{r_1-1} h_{M_1}(x_M) & \cdots & \alpha_{1r_1} L_{g_{n_y}} L_f^{r_1-1} h_{M_1}(x_M) \\ \vdots & & \vdots \\ \alpha_{n_y r_{n_y}} L_{g_1} L_f^{r_{n_y}-1} h_{M_{n_y}}(x_M) & \cdots & \alpha_{n_y r_{n_y}} L_{g_{n_y}} L_f^{r_{n_y}-1} h_{M_{n_y}}(x_M) \end{bmatrix} \quad (11)$$

3. APPLICATION TO A FERMENTATION PROCESS

Consider a continuous stirred tank reactor, evolution of the cells, substrate and metabolites inside the reactor, is governed by the following material balance set of non-linear equations (Maher, 1995) :

$$\begin{cases} \dot{c}(t) = \mu(s(t), p(t))c(t) - D(t)c(t) \\ \dot{s}(t) = -\dfrac{1}{Y_{c/s}}\mu(s(t), p(t))c(t) - D(t)s(t) + D(t)S_{in}(t) \\ \dot{p}(t) = \dfrac{1}{Y_{c/p}}\mu(s(t), p(t))c(t) - D(t)p(t) \end{cases} \quad (12)$$

where :

$$\mu(s(t), p(t)) = \left(\frac{\mu_m s(t)}{K_s + s(t)}\right)\left(1 - \frac{p(t)}{P_i}\right). \quad (13)$$

The variable $c(t)$, $s(t)$ and $p(t)$ represent the biomass, the substrate and the metabolite concentration respectively. The variable $D(t)$ is the dilution rate and $S_{in}(t)$ denotes the substrate concentration in the feed stream. The kinetics of the cell mass production is defined in terms of the specific growth rate (13) and the two yields coefficients $Y_{c/s}$ and $Y_{c/p}$. The kinetics contain five uncertain parameters : the maximum growth rate μ_m, the Michaelis-Menten constant K_s, the product inhibition constant P_i, the substrate/biomass conversion yield $Y_{c/s}$ and the product/biomass conversion yield $Y_{c/p}$. The specific growth rate $\mu(s(t), p(t))$, is a non-linear functions of state variables and parameters, the numerical values of the kinetics parameters are listed in the table 1.

Table 1 : Values of the kinetics parameters.

Figures	Step disturbance on the kinetics parameters
3-4	$t \le 060h \rightarrow \mu_m = 0.38\ 1/h$; $t > 060h \rightarrow \mu_m = 0.48\ 1/h$
3-4	$t \le 210h \rightarrow Y_{c/s} = 0.07\ g/g$; $t > 210h \rightarrow Y_{c/s} = 0.14\ g/g$
3-4	$t \le 360h \rightarrow K_s = 5\ g/l$; $t > 360h \rightarrow K_s = 12\ g/l$
3-4	$t \le 510h \rightarrow P_i = 100\ g/l$; $t > 510h \rightarrow P_i = 50\ g/l$
3-4	$t \le 660h \rightarrow Y_{c/p} = 0.16\ g/g$; $t > 660h \rightarrow Y_{c/p} = 0.16\ g/g$
1-2	$\mu_m = 0.38 1/h$; $K_s = 5 g/l$; $K_i = 22 g/l$; $P_i = 100 g/l$; $Y_{c/s} = 0.07 g/g$; $Y_{c/p} = 0.16 g/g$;

The state variables, the manipulated input and the controlled output are defined as $x_P = [c(t) \quad s(t) \quad p(t)]^T$, $u = [D(t) \quad S_{in}(t)]$ and $y_P = [c(t) \quad s(t)]$ respectively. The relative degree of each input was $r = 1$, the controllers equations were (Wang, *et al.*, 1993):

$$u_1 = -\frac{1}{y_{M_1}}\left(\alpha_{11}\left(y_{sp_1} - y_{P_1} + y_{M_1}\right) - \alpha_{11} y_{M_1} - \left(\frac{\mu_m x_{M_2}}{K_s + x_{M_2}}\left(1 - \frac{x_{M_3}}{P_i}\right)\right) y_{M_1}\right) \quad (14)$$

and

$$u_2 = -\frac{1}{y_{M_2}}\left(\alpha_{21}\left(y_{sp_2} - y_{P_2} + y_{M_2}\right) - \alpha_{21} y_{M_2} - \left(\frac{\mu_m x_{M_2}}{K_s + x_{M_2}}\left(1 - \frac{x_{M_3}}{P_i}\right)\right) y_{M_2}\right) \quad (15)$$

Table 2 : Different concentration setpoints.

Figures	Biomass concentration setpoint	Substrate concentration setpoint
1-2	$000h < t \le 150h \rightarrow c = 5g/l$	$000h < t \le 300h \rightarrow c = 6g/l$
1-2	$150h < t < 450h \rightarrow c = 4g/l$	$300h < t < 450h \rightarrow c = 7g/l$
3-4	$000h < t \le 150h \rightarrow c = 5g/l$	$000h < t \le 300h \rightarrow c = 6g/l$
3-4	$150h < t \le 600h \rightarrow c = 4g/l$	$300h < t \le 450h \rightarrow c = 7g/l$
3-4	$600h < t < 750h \rightarrow c = 5g/l$	$450h < t < 750h \rightarrow c = 6g/l$

Two series of simulation tests on a continuous stirred tank bioreactor have been conducted to evaluate the performance of the multivariable nonlinear IMC strategy. The table 2 gives the biomass and subtrate concentrations setpoint for the two series of simulation.

3.1 Modelling error

In this part (figure 1 and 2), the controller behaviour was tested face to a structural errors in the specific growth rate. The specific growth rate of the plant *P* change at different instant during the simulation (last three line in the table 3).

Table 3 : Expressions of the growth rate.

Figures	Structural error in the growth rate
3-4	$000h < t < 750h \rightarrow (\mu_m s(t)/(K_s + s(t)))(1 - p(t)/P_l)$
1-2	$000h < t \leq 060h \rightarrow (\mu_m s(t)/(K_s + s(t)))(1 - p(t)/P_l)$
1-2	$060h < t \leq 360h \rightarrow \left(\mu_m s(t) \big/ \left(K_s + s(t) + s(t)^2/K_i\right)\right)$
1-2	$360h < t < 450h \rightarrow (\mu_m s(t)/(K_s + s(t)))$

The results concerning the case without noise are shown in Fig. 1a-f and with noise in Fig. 2a-f. Here, the controlled variables are the biomass concentration (subplot (*a*)) and the substrate concentration (subplot (*b*)). The controller is able to achieve both desired setpoints quite satisfactory after a short transient of a few hours. A coupling effect appears on the substrate concentration at time t=150h when the other setpoint change. During the experiment no coupling effect is present on the biomass concentration because this output is less sensitive than the substrate concentration. The controller rejects the effect of the structural error on the biomass concentration slighty better than the substrate concentration (see subplot (a) and (b)). Subplot (*c*) show the control variables, these variables were very smooth even in the case with noise. The evolutions of the three plant and model concentration are reported on subplots (d), (*e*), (*f*). Despite the significant errors in the biomass, substrate and product concentration, the non-linear internal model control is able to provide excellent control action.

Fig. 1a : Biomass concentration evolution

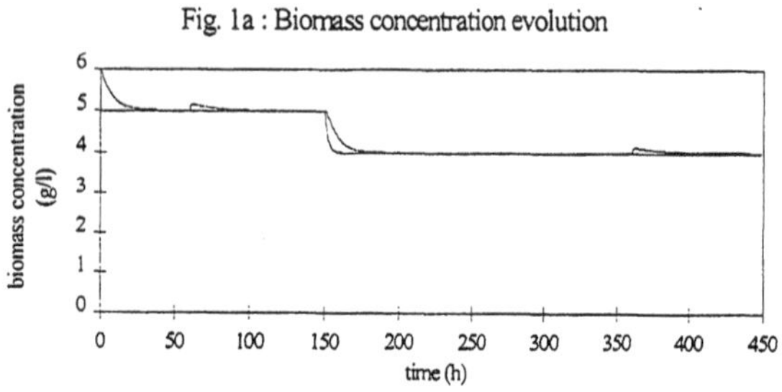

Fig. 1b : Substrate concentration evolution

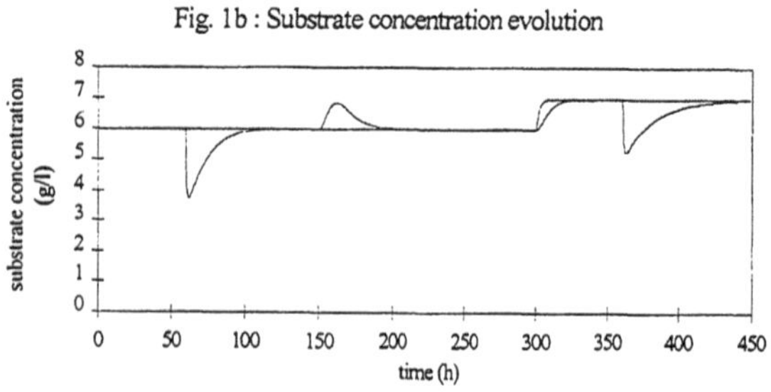

Fig. 1c : Evolution of the different input variables

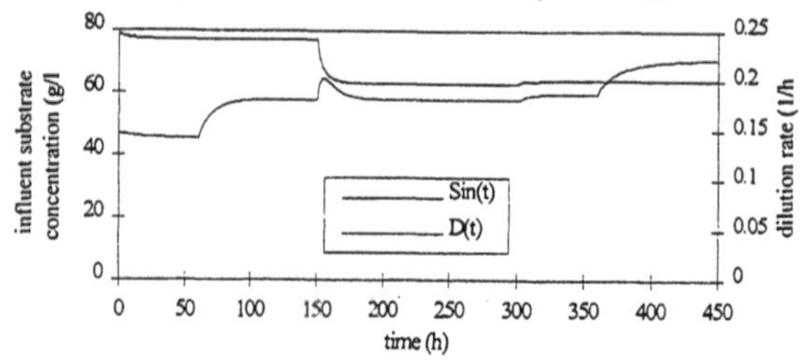

Fig. 1d : Biomass concentration evolution

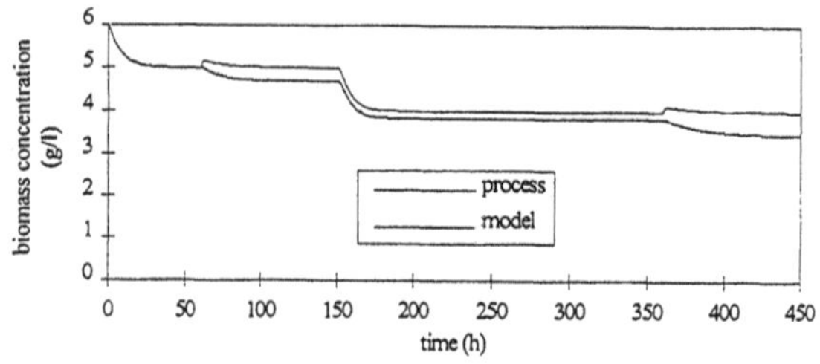

Fig. 1e : Substrate concentration evolution

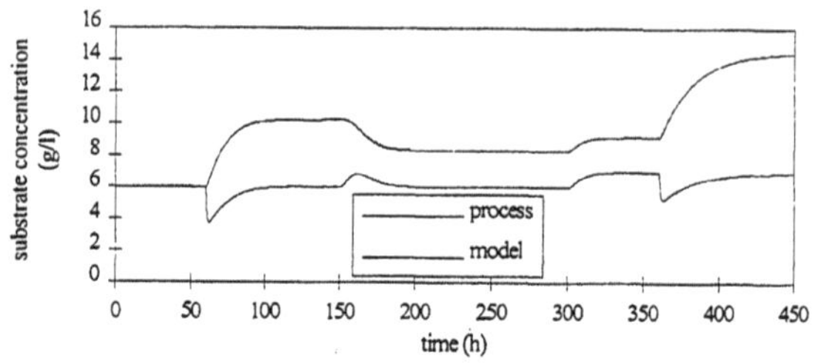

Fig. 1f : Product concentration evolution

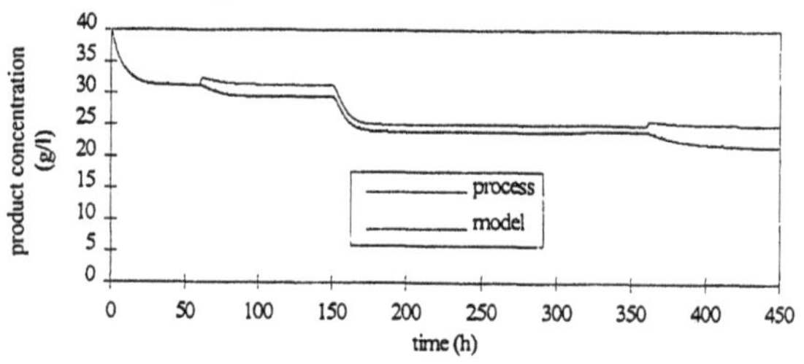

Fig. 2a : Biomass concentration evolution

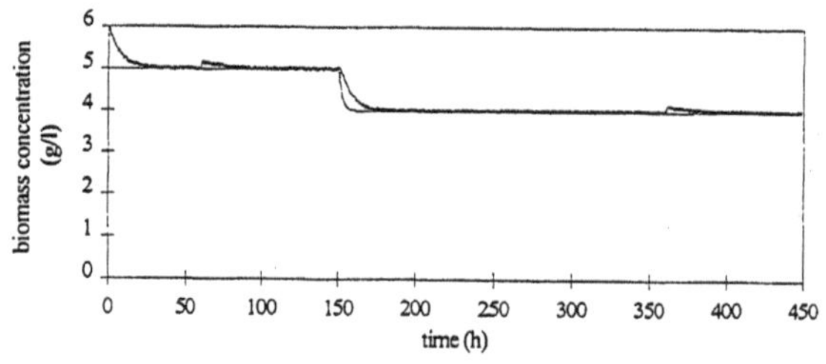

Fig. 2b : Substrate concentration evolution

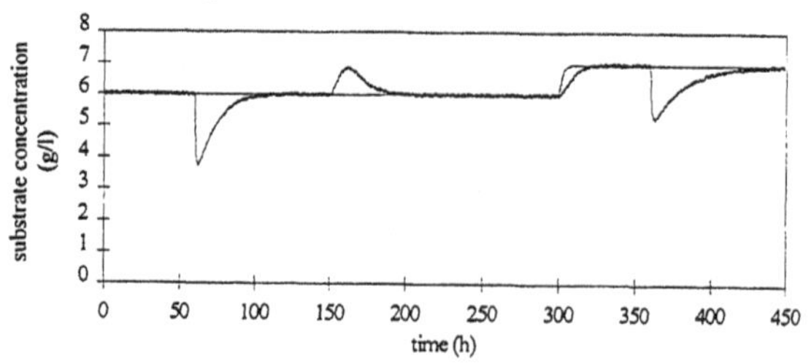

Fig. 2c : Evolution of the different input variables

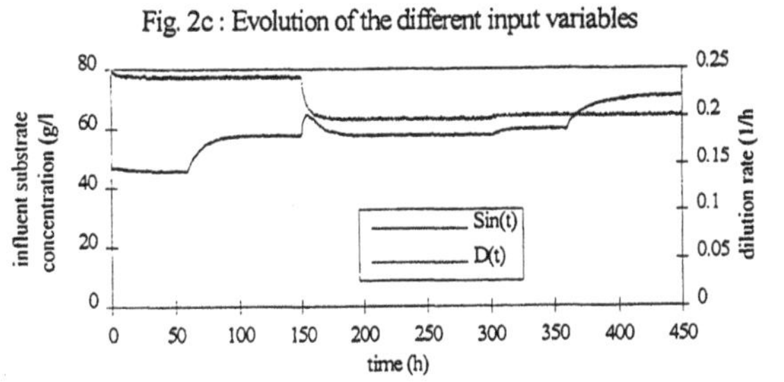

Fig. 2d : Biomass concentration evolution

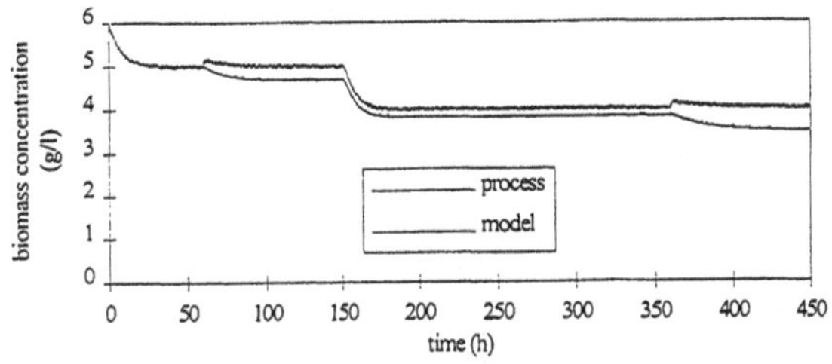

Fig. 2e : Substrate concentration evolution

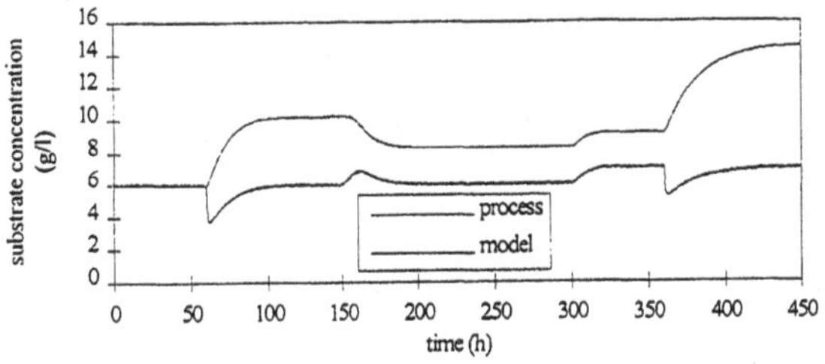

Fig. 2f : Product concentration evolution

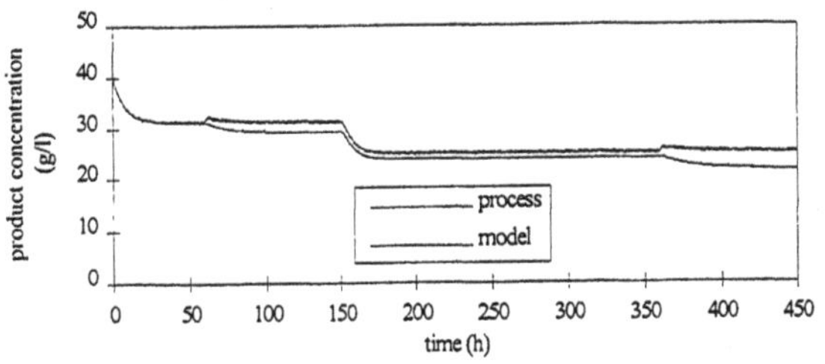

3.2 Disturbance on the kinetics parameters

In this case the specific growth rates $\mu(s(t),p(t))$ of the plant P and the model M are the same $(\mu_m s(t)/K_s + s(t))(1 - p(t)/P_i)$ (first line of the table 3). The controller behaviour was tested face to unmeasuerd step disturbance on the five kinetics parameters (see table 1). In the latter simulation Figs.3 and 4, we have shown the behaviour obtained with the multivariable non-linear internal model controller for unmeasured step disturbances. The regulated output variables are the biomass and substrate concentration represented on subplots (a) and (b) respectively. As in the first simulation, the biomass concentration is best regulated than the substrate concentration (subplot (b)). This controller eliminates the effect of disturbance, but it does not decouple both outputs. On subplot (b) the effect of the step disturbances, can be seen at different instant during the simulation run. Nevertheless, the effect of the step disturbance disappeared after a short transient of time. The two control variables evolution are shown in subplot (c) : the dilution rate $D(t)$ and the influent substrate concentration $S_{in}(t)$. In this experiment the input variables are slightly more disturbed than the first experiment in the structural error case. Subplot (d), (e) and (f) show the actual and the estimated value of the biomass, substrate and product concentration. The algorithm generates a correct control action in a given situation. Therefore, the algorithm fits itself efficiently to the system dynamics and to the bioreactor configuration changes. The simulation results demonstrate that the non-linear internal model control strategy is able to provide excellent performances for the required control specifications if significant modelling errors or step disturbances are present.

Fig. 3a : Biomass concentration evolution

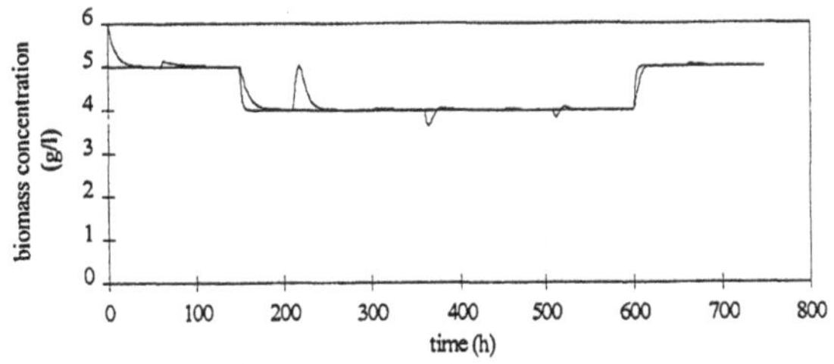

Fig. 3b : Substrate concentration evolution

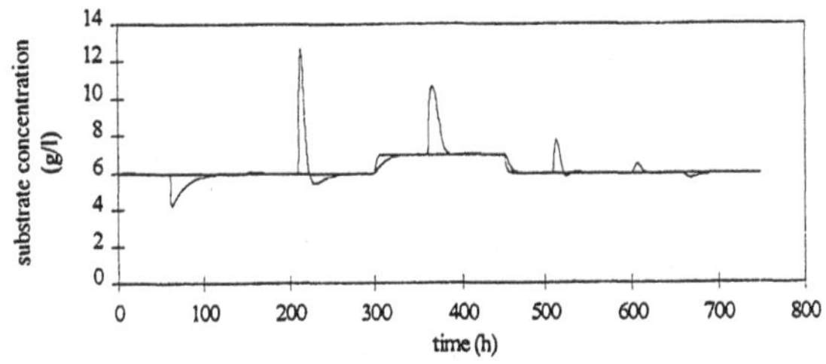

Fig. 3c : Evolution of the different input variables

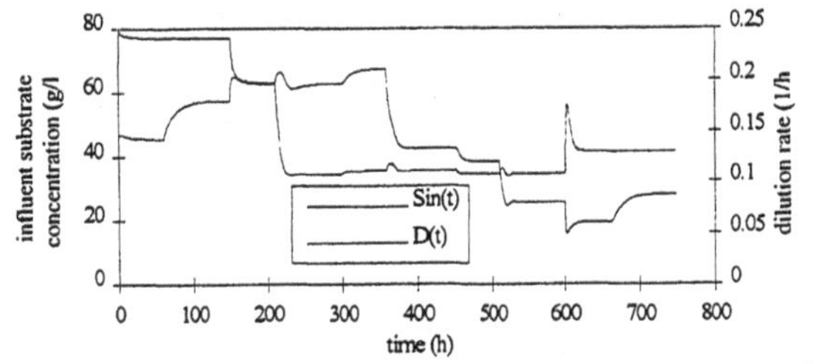

Fig. 3d : Biomass concentration evolution

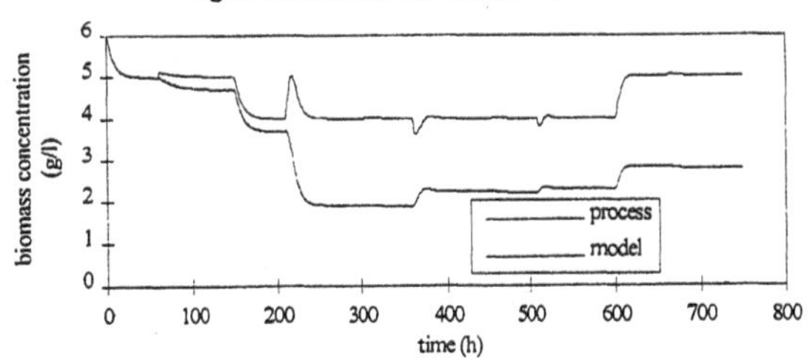

Fig. 3e : Substrate concentration evolution

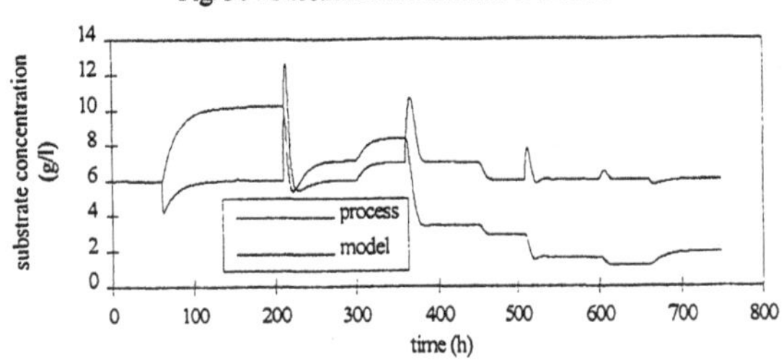

Fig. 3f : Product concentration evolution

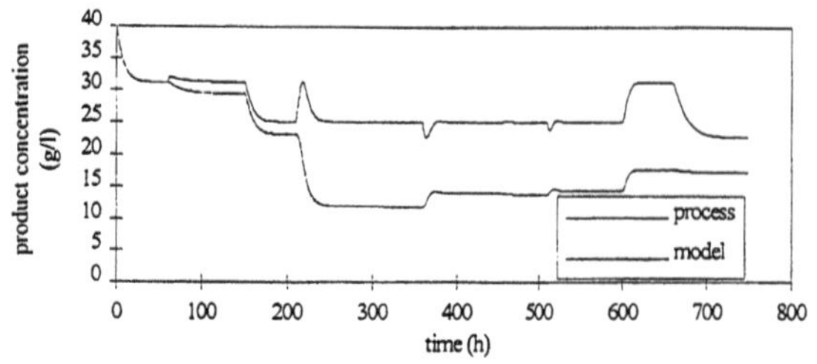

Fig. 4a : Biomass concentration evolution

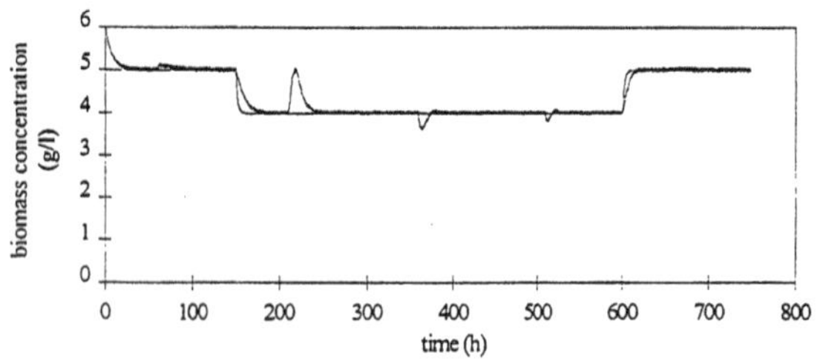

Fig. 4b : Substrate concentration evolution

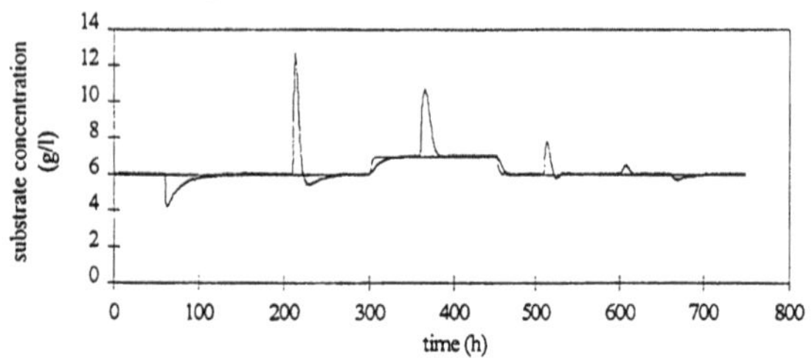

Fig. 4c : Evolution of the different input variables

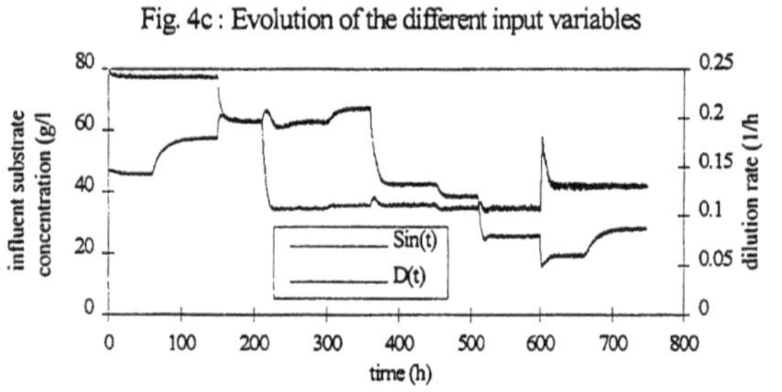

Fig. 4d : Biomass concentration evolution

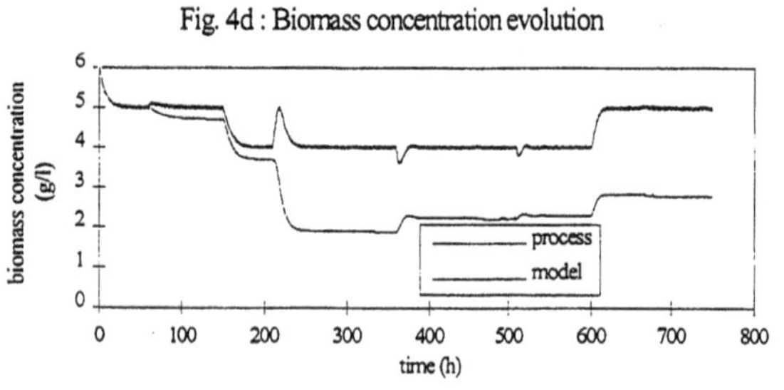

Fig. 4e : Substrate concentration evolution

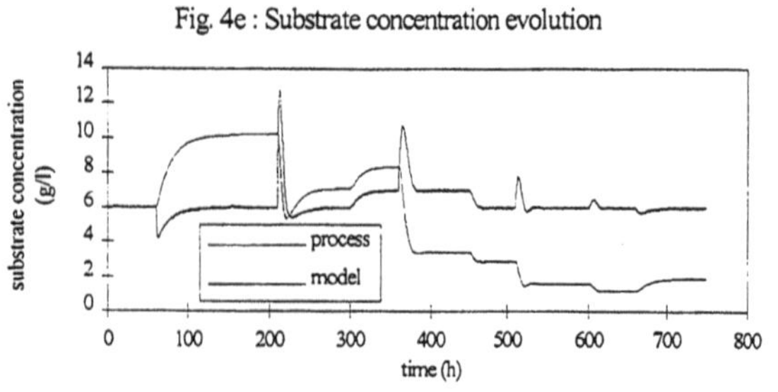

Fig. 4f : Product concentration evolution

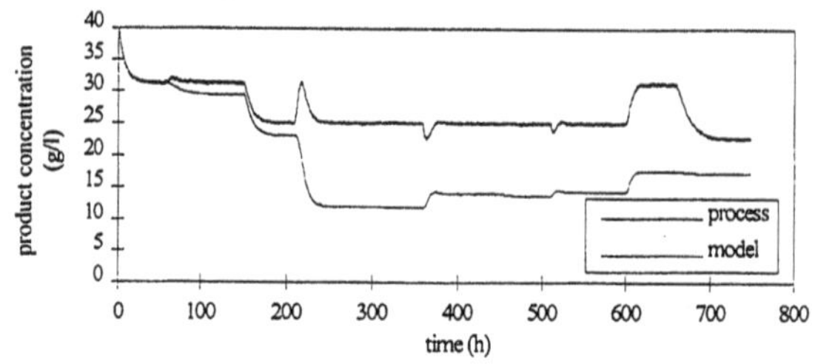

4. CONCLUSION

The aim of this paper was to investigate the performances of a multivariable non-linear internal model control strategy applied on a continuous stirred tank reactor. An important advantage of the used approach is that the assumption of full-state feedback is not required. Simulation results showed the good behaviour of this type of control law face to the disturbance rejection. Moreover the results proved that the algorithm is quite robust to structural modelling errors. After these encouraging results, the following step will be to confirm this internal model control strategy on an experiment context and see if the good performances are kept.

REFERENCES

Bastin, G. and D. Dochain (1990), *On-line estimation and adaptive control of bioreactors*, Elsevier Science Publisher B. V., Amsterdam.

Daoutidis, P. and C. Kravaris (1991), "Inversion and zero dynamics in nonlinear multivariable control", *AICHE Journal*, 37 (4), pp. 527-538.

Economou, C. G. and M. Morari (1985), "Newton control laws for non-linear controller design", *Proc. IEEE Conf. on Decision and Control*, pp 1361.

Henson, M. A. and D.E. Seborg (1991), "An internal model control strategy for nonlinear systems", *AICHE Journal*, 37 (7), pp. 1065-1081.

Hirschorn, R. M. (1979), "Invertibility of multivariable nonlinear control systems", *IEEE Trans. Auto. Contr.*, AC-24, pp. 855.

Isidori, A. (1989), *Nonlinear Control Systems : An Introduction*, 2nd ed., Springer-Verlag, New York.

Kravaris, C. and M. Soroush (1990), "Synthesis of multivariable nonlinear controller by input/output linearization", *AICHE Journal*, 36 (2), pp. 249-264.

Maher, M. (1995), "Modélisation et élaboration d'algorithmes d'estimation et de commande : application à un bioprocédé", Doctorate thesis, Université de Toulouse III, France.

Morari, M. and E. Zafiriou (1989), *Robust Process Control*, Prentice Hall, Englewood Cliffs, New Jersey.

Roux, G. and B. Dahhou (1996), "Commande par modèle interne non linéaire appliquée à un procédé de fermentation alcoolique", *Proc. SIMO'96*, pp 265-270.

Wang, Z. Q., S. Skogestad and Y. Zhao (1993), "Exact linearization control of continuous bioreactors : a comparison of various control structures", *Proc. IEEE Conf. on Control and Applications*, pp 107-112.

ON THE REGULATION OF A DOUBLE EFFECT EVAPORATOR: A TRAJECTORY PLANNING AND PASSIVITY APPROACH

Hebertt Sira-Ramírez[1] **and Gerardo Silva-Navarro**

Centro de Investigación y de Estudios Avanzados del IPN
Departamento de Ingeniería Eléctrica
Apartado Postal 14-740
C.P. 07000 México, D.F., México
Email: *hsira@mail.cinvestav.mx*, *gsilva@ctrl.cinvestav.mx*

Abstract: A dynamical feedback controller is proposed for the regulation of a simplified model of a double effect evaporator. The controller is based on feedback passivity considerations, complemented with a trajectory planning scheme in terms of the systems passive output. A drift vector field decomposition greately facilitates the direct application of the "energy shapping plus damping injection" feedback controller design methodology. The controller performance is enhanced with the aid of a trajectory planning scheme which effectively avoids singularity arcs in the state space, related to a loss of the relative degree of the passive output, while allowing for the satisfaction of physically meaningful state space and input space restrictions. The performance of the obtained controller design is evaluated by means of computer simulations. *Copyright © 1999 IFAC*

Keywords: Double Effect Evaporators, Trajectory Planning, Passivity.

1 INTRODUCTION

Passivity based control of nonlinear systems is rapidly gaining well deserved popularity due to the several advantages related to controller simplicity, robustness and the physically appealing features of the approach. General developments of passivity based control, into the realm of nonlinear affine systems, have been carried over relatively recently. The seminal contribution is that of Willems (1972) in the context of *dissipative systems*. The articles by Hill (1976) constitute also a general approach with emphasis on conditions for stability of feedback interconnected systems. A geometric approach to feedback equivalence of nonlinear passive systems was contributed by the work of Byrnes *et al* (1991). Recent contributions, within the context of adaptive systems, have been given by Serón *et al* (1995). A full perspective of the area is found in the book by Sepulchre *et al* (1997) and the book by Ortega *et al* (1998).

Section 2 presents some generalities regarding a canonical form for passivity based control presented in Sira-Ramírez (1998). Section 3 is devoted to derive a passivity based controller for a simplified double effect evaporator model. The performance of the designed controller is also illustrated by means of digital computer simulations. Section 4 contains the conclusions and suggestions for further research.

[1]This research was supported by the Centro de Investigación y de Estudios Avanzados (CINVESTAV) of the Instituto Politécnico Nacional, México, D.F. and by the Consejo Nacional de Ciencia y Tecnología of Venezuela (CONICIT) under Research Grant S1-95-000886

2 A CANONICAL FORM FOR PASSIVITY BASED CONTROL

2.1 Background Results

Consider the class of nonlinear single-input single-output systems described by

$$\begin{aligned} \dot{x}(t) &= f(x) + g(x)u, \ x \in \mathcal{X} \subset \mathbf{R}^n, \ u \in \mathcal{U} \subset \mathbf{R} \\ y &= h(x), \ y \in \mathcal{Y} \subset \mathbf{R} \end{aligned} \tag{1}$$

where $\mathcal{X}$ denotes the *operating region* of the system, constituted by a sufficiently large open set containing a continuum of equilibrium points, possibly parametrized by a constant control input value $u = U \in \mathcal{U}$, of the form $x = \overline{x}(U)$ and given by the solution of $f(\overline{x}) + g(\overline{x})U = 0$. In particular, for $u = 0$, we assume $f(\overline{x}) = 0$ implies $\overline{x} = 0$. However, motivated by a large class of real life systems, we are specifically interested in *nonzero* constant state equilibrium points $x = \overline{x}$, obtained by nonzero constant control inputs $u = U$.

We assume that a C^1 positive definite *storage function* $V : \mathbf{R}^n \to \mathbf{R}^+$ is given such that $V(0) = 0$.

By $\partial V/\partial x$ we denote the *column* vector field with components $\partial V/\partial x_i$ $i = 1, \ldots, n$. The transpose of this gradient field, $(\partial V/\partial x)^T$, is denoted by the *row* vector $\partial V/\partial x^T$. Let $L_g V(x)$ denote the directional derivative of the scalar function $V(x)$ with respect to the control input vector field $g(x)$ at the point x. We assume throughout the entire article that the following assumption holds valid:

$$L_g V(x) = \frac{\partial V}{\partial x^T} g(x) \neq 0 \quad \forall\, x \in \mathcal{X} \tag{2}$$

This last condition is usually known as the *transversality condition* and simply establishes that the vector field $g(x)$ is not orthogonal to the gradient of $V(x)$ at any point x in $\mathcal{X}$. In other words, the control vector field $g(x)$ is not tangential, at each x, to the storage function level sets, defined in the state space of the system as, $\{x \in \mathcal{X} \ : \ V(x) = \mathit{constant}\}$. This condition is quite familiar in sliding mode control of nonlinear systems (see Sira-Ramirez, 1988) and it amounts to having a storage function which is locally *relative degree* one in $\mathcal{X}$.

For each $x \in \mathcal{X}$, we define a *projection operator, along the span of the control vector field $g(x)$ onto the tangent space to the constant level sets of the energy function $V(x)$*, as the matrix $M(x)$ given by

$$M(x) = \left[I - \frac{1}{L_g V(x)} g(x) \frac{\partial V}{\partial x^T} \right] \tag{3}$$

The following are some properties of the matrix $M(x)$ which further justify the given name of "projection operator" (see Sira-Ramírez, 1998)

Proposition 1 *The matrix $M(x)$ enjoys the following properties:*

$$\begin{aligned} g(x) &\in \mathcal{K}er M(x) \\ \frac{\partial V}{\partial x} &\in \mathcal{K}er M^T(x) \\ M(x)\,[I - M(x)] &= 0. \end{aligned}$$

The following proposition depicts further properties of the projection matrix $M(x)$.

Proposition 2 *Let $f(x)$ be a smooth vector field, then the vector $M(x)f(x)$ can be written as*

$$M(x) f(x) = \tilde{\mathcal{J}}(x) \frac{\partial V}{\partial x}$$

where $\tilde{\mathcal{J}}(x)$ is a skew-symmetric matrix, i.e., $\tilde{\mathcal{J}}(x) + \tilde{\mathcal{J}}^T(x) = 0$. On the other hand the vector field $[I - M(x)]\, f(x)$ can be written as

$$[I - M(x)]\, f(x) = -\frac{1}{2} \tilde{\mathcal{J}}(x) \frac{\partial V}{\partial x} + \mathcal{S}(x) \frac{\partial V}{\partial x}$$

where $\mathcal{S}(x)$ is a symmetric matrix, i.e., $\mathcal{S}(x) = \mathcal{S}^T(x)$.

2.2 Vector field decompositions through projection operators

As a consequence of the above propositions and definitions we have the following decomposition of drift vector fields

Proposition 3 *A vector field $f(x)$ can be naturally decomposed in the following sum*

$$\begin{aligned} f(x) &= M(x) f(x) + (I - M(x)) f(x) \\ &= \mathcal{J}(x) \frac{\partial V}{\partial x} + \mathcal{S}(x) \frac{\partial V}{\partial x} \end{aligned} \tag{4}$$

where $\mathcal{J}(x)$ is a skew-symmetric matrix and $\mathcal{S}(x)$ is a symmetric matrix.

2.3 A canonical form for nonlinear systems

As a corollary to the above results, a nonlinear system of the form (1), with a positive definite storage function $V(x)$ which also satisfies the transversality condition $L_g V(x) \neq 0$, can always be rewritten as

$$\begin{aligned} \dot{x}(t) &= \mathcal{J}(x) \frac{\partial V}{\partial x} + \mathcal{S}_p(x) \frac{\partial V}{\partial x} \\ &\quad + \mathcal{S}_n(x) \frac{\partial V}{\partial x} + g(x) u \end{aligned} \tag{5}$$

with $\mathcal{J}(x)$ being skew-symmetric, $\mathcal{S}_p(x)$ being positive semi-definite and $\mathcal{S}_n(x)$ being negative semi-definite. However, if $\mathcal{S}_p(x)$ is positive definite, then $\mathcal{S}_n(x)$ is zero and conversely if $\mathcal{S}_n(x)$ is negative definite then $\mathcal{S}_p(x)$ is zero.

3 A PASSIVITY BASED FEEDBACK CONTROLLER DESIGN FOR THE DOUBLE EFFECT EVAPORATOR

Industrial evaporators are extensively used in the chemical and food industries to enhance final product concentrations from a given low concentration feed solution. Under standard assumptions, treated in Silva-Navarro and Alvarez-Gallegos (1995), a reasonable reduced order model for a double effect evaporator is given by,

$$\begin{aligned} \dot{x}_1 &= \delta_1 F_0 (C_0 - x_1) + \delta_2 x_1 u \\ \dot{x}_2 &= \delta_3 F_0 (x_1 - x_2) + (\delta_4 x_1 + \delta_5 x_2) u \\ y &= x_2 \end{aligned} \tag{6}$$

where x_1 and x_2 represent the weight fraction solution concentration in the first and the second constitutive parts of the evaporator, also called, for simplicity, the "first effect" and the "second effect" respectively. The state variables x_1, x_2 of the plant, naturally satisfy the following physically meaningful restrictions:

$$x \in \tilde{\mathcal{X}} = \{0 < C_0 \leq x_1 \leq x_2 < 1\} \tag{7}$$

These restrictions effectively bound the set of possible state trajectories to a triangular region in the two-dimensional state space (see Fig. 1).

The parameter C_0 is the constant weight fraction input concentration to the first effect, while F_0 is the solution feed flow rate expressed in Kg/min. The control input u is the steam feed flow rate, also expressed in Kg/min. The constants, $\delta_1, \ldots \delta_4$, are given by

$$\begin{aligned} \delta_1 &= \frac{1}{W_1},\ \delta_2 = \frac{k_1}{W_1},\ \delta_3 = \frac{1}{W_2} \\ \delta_4 &= -\frac{k_1}{W_2},\ \delta_5 = \frac{k_1(1+k_2)}{W_2} \end{aligned} \tag{8}$$

where W_1,W_2 are the hold-up masses in each one of the effects, respectively. These quantities are expressed in Kg and they satisfy $(0 < W_1 < W_2)$. The constant parameters k_1 and k_2 are positive constants denoting the ratio of the vapor flow rate produced with respect to the steam fed flow rate in both effects. They are such that the condition : $0 < k_2 < k_1 < 1$ holds valid. The control input variable u may be constrained by a set of the following form,

$$u \in \mathcal{U} = \{u \in R \ : \ U_{max} > u > 0\}$$

The family of equilibrium points, parametrized by a constant equilibrium value for u, set to be $\overline{u} = U$, is given by

$$\overline{x}_1 = \frac{F_0 C_0}{F_0 - k_1 U} \ ; \ \overline{x}_2 = \frac{F_0 C_0}{F_0 - k_1(1+k_2)U}$$

Eliminating the constant parameter U from the above relations, it readily follows that the family of physically meaningful equilibrium points, $(\overline{x}_1, \overline{x}_2)$, are all located on the first quadrant, in the corresponding branch of the following hyperbola,

$$x_2 = \frac{x_1 C_0}{(1+k_2)C_0 - k_2 x_1} \tag{9}$$

The manifold of physically valid equilibrium points is, necessarily, constrained to lay within the triangular region previously described by equation (7).

It is easy to verify that the output variable $y = x_2$ is a relative degree one output, as long as the singularity condition,

$$\delta_4 x_1 + \delta_5 x_2 = 0$$

is not valid. This singularity line is also given by,

$$x_2 = -\frac{\delta_4}{\delta_5} x_1 = \frac{1}{1+k_2} x_1 \tag{10}$$

The singularity line (10) is a straight line with positive slope crossing the origin and having two points in common with the manifold of equilibrium points, (9), one in the first quadrant and the second one in the third quadrant which is not physically meaningful. In the first quadrant the singularity line lies below the set of possible equilibria.

It has been shown in Silva-Navarro and Alvarez-Gallegos (1995) using Lyapunov stability theory, that the outupt variable x_2 is also a minimum phase output. Thus, the variable x_2 is a *passive output*. Stabilization of x_2 towards a desired constant or reference value can be achieved by means of system inversion, with a stable zero dynamics, as long as the resulting trajectory does not violate the relative degree one condition for this variable.

We next consider the storage function $V(x)$ given by,

$$V = \frac{1}{2}\left(x_1^2 + x_2^2\right)$$

The storage function directional derivative, $L_g V(x)$, along the control input vector field

$$g(x) = [\delta_2 x_1 \ \ (\delta_4 x_1 + \delta_5 x_2)]^T$$

is given by,

$$L_g V(x) = \delta_2 x_1^2 + \delta_4 x_1 x_2 + \delta_5 x_2^2$$

The transversality condition is valid everywhere in the region $\tilde{\mathcal{X}}$. and it fails to be valid on the set,

$$\{x \in R^2 : L_g V(x) = 0\} = \{x \in R^2 : \delta_2 x_1^2 + \delta_4 x_1 x_2 + \delta_5 x_2^2 = 0\} \quad (11)$$

This set actually represents the origin of coordinates $(0,0)$ in R^2 and, therefore, it is not a meaningful restriction to the proposed solution.

A control problem involving a transfer between two equilibria, or a maneuver involving a trajectory tracking task, must be defined in such a way that the transversality condition $L_g V(x) \neq 0$, is always satisfied by each and every point of the prescribed state trajectory in R^2.

The operating region, free of singularities, is constituted by the following set,

$$\begin{aligned} \mathcal{X} &= \{x \in R^2 : 0 < C_0 \leq x_1 \leq x_2 < 1\,; \\ & x_2 - \frac{1}{1+k_2} x_1 \neq 0\} \end{aligned} \quad (12)$$

This set is shown in Fig. 1, along with the manifold of equilibrium points and the line of singularity for the relative degree 1 condition of the passive output x_2. Note that on $\mathcal{X}$ the quantity $L_g V(x)$ is strictly positive.

The given system may be rewritten in the form,

$$\dot{x} = [\mathcal{J}(x) + \mathcal{S}(x)] \frac{\partial V}{\partial x} + g(x)u \quad (13)$$

where,

$$\begin{aligned} \mathcal{J}(x) &= \alpha(x) \begin{bmatrix} 0 & 1 \\ -1 & 0 \end{bmatrix} \\ \alpha(x) &= \\ & \frac{\delta_1 F_0 (C_0 - x_1)(\delta_4 x_1 + \delta_5 x_2) - \delta_2 \delta_3 F_0 x_1 (x_1 - x_2)}{2\,[\delta_2 x_1^2 + \delta_4 x_1 x_2 + \delta_5 x_2^2]} \end{aligned}$$

and

$$\mathcal{S}(x) = \beta(x) \begin{bmatrix} S_{11}(x) & S_{12}(x) \\ S_{12}(x) & S_{22}(x) \end{bmatrix} \quad (14)$$

with

$$\begin{aligned} S_{11}(x) &= 2\delta_1 \delta_2 F_0 (C_0 - x_1) x_1 \\ S_{12}(x) &= \delta_1 F_0 (C_0 - x_1)(\delta_4 x_1 + \delta_5 x_2) \\ & +\delta_2 \delta_3 F_0 x_1 (x_1 - x_2) \\ S_{22}(x) &= 2\delta_3 F_0 (x_1 - x_2)(\delta_4 x_1 + \delta_5 x_2) \\ \beta(x) &= \frac{1}{2\,[\delta_2 x_1^2 + \delta_4 x_1 x_2 + \delta_5 x_2^2]} \end{aligned} \quad (15)$$

The symmetric matrix $\mathcal{S}(x)$, with the state x constrained to the operating region $\mathcal{X}$, can be further decomposed into the sum of a positive semi-definite matrix $\mathcal{S}_p(x)$ and a negative semi-definite matrix $\mathcal{S}_n(x)$ The following input coordinate transformation makes invariant the positive term appearing in the contribution to $\dot{V}$. Its effect is that of neutralizing the corresponding vector field by making it tangent to the stored energy level sets. This procedure renders the system output passive, from the new input v towards the original output x_2, with respect to the proposed storage function $V(x)$.

Substituting the feedback expression

$$\begin{aligned} u &= \frac{1}{\delta_2 x_1^2 + \delta_4 x_1 x_2 + \delta_5 x_2^2} \times \\ & [x_2 v - F_0 x_1 (\delta_1 C_0 + \delta_3 x_2)] \end{aligned} \quad (16)$$

in equation (13) and after some straightforward algebraic manipulations and some simplifications, the partially closed loop system in PBCCF is seen to be expressed as,

$$\begin{aligned} \dot{x} &= m(x) \begin{bmatrix} 0 & 1 \\ -1 & 0 \end{bmatrix} \begin{bmatrix} x_1 \\ x_2 \end{bmatrix} \\ & + \begin{bmatrix} -\delta_1 F_0 & 0 \\ 0 & -\delta_3 F_0 \end{bmatrix} \begin{bmatrix} x_1 \\ x_2 \end{bmatrix} \\ & + \frac{1}{\delta_2 x_1^2 + \delta_4 x_1 x_2 + \delta_5 x_2^2} \begin{bmatrix} \delta_2 x_1 x_2 \\ (\delta_4 x_1 + \delta_5 x_2) x_2 \end{bmatrix} v \end{aligned} \quad (17)$$

$$\begin{aligned} y &= \frac{1}{\delta_2 x_1^2 + \delta_4 x_1 x_2 + \delta_5 x_2^2} \times \\ & [\delta_2 x_1 x_2 \quad (\delta_4 x_1 + \delta_5 x_2) x_2] \begin{bmatrix} x_1 \\ x_2 \end{bmatrix} = x_2 \end{aligned} \quad (18)$$

with

$$m(x) = \frac{\delta_1 F_0 C_0 (\delta_4 x_1 + \delta_5 x_2) - \delta_2 \delta_3 F_0 x_1^2}{\delta_2 x_1^2 + \delta_4 x_1 x_2 + \delta_5 x_2^2}$$

The "Energy Shapping plus Damping Injection" controller design methodology (see Ortega *et al*, 1998) yields the following dynamical feedback controller,

$$v = \frac{\delta_2 x_1^2 + \delta_4 x_1 x_2 + \delta_5 x_2^2}{x_2 (\delta_4 x_1 + \delta_5 x_2)} \times \quad (19)$$

$$\begin{aligned} & \Big\{ \dot{x}_{2d}^* - \frac{[\delta_2 \delta_3 F_0 x_1^2 - \delta_1 F_0 C_0 (\delta_4 x_1 + \delta_5 x_2)]}{\delta_2 x_1^2 + \delta_4 x_1 x_2 + \delta_5 x_2^2} \xi \\ & + \delta_3 F_0 x_{2d}^* - R_2 (x_2 - x_{2d}^*) \Big\} \end{aligned} \quad (20)$$

with ξ given by the solution of

$$\begin{aligned} \dot{\xi} &= \frac{[-\delta_2 \delta_3 F_0 x_1^2 + \delta_1 F_0 C_0 (\delta_4 x_1 + \delta_5 x_2)]}{\delta_2 x_1^2 + \delta_4 x_1 x_2 + \delta_5 x_2^2} x_{2d}^* \\ & -\delta_1 F_0 \xi + \frac{\delta_2 x_1 x_2}{\delta_2 x_1^2 + \delta_4 x_1 x_2 + \delta_5 x_2^2} v \\ & + R_1 (x_1 - \xi) \end{aligned} \quad (21)$$

The specification of a suitable trajectory for $x_{2d}^*(t)$ and its corresponding time derivative $\dot{x}_{2d}^*(t)$ is carried out so that the systems response evolves away from the singular line in R^2 where $L_g V(x) = 0$. The simulation example, presented next, illustrates this point in detail.

3.1 Simulation results

We let, the system parameters to be given by

$$\begin{aligned} F_0 &= 2.525 \text{ Kg/min} \\ C_0 &= 0.04 \text{ Kg.sugar/Kg.water} \\ W_1 &= 95.0 \text{ Kg.} \; ; \; W_2 = 105.0 \text{ Kg.} \\ k_1 &= 0.808458 \; ; \; k_2 = 0.338461 \end{aligned} \quad (22)$$

A controlled transfer from an initial equilibrium point, given by $\overline{x}_1(t_0) = 0.070$ Kg.sugar/Kg.water and $\overline{x}_2(t_0) = 0.093954$ Kg.sugar/Kg.water, towards a second equilibrium point given by $\overline{x}_1(t_0 + T) = 0.09549$ Kg.sugar/Kg.water and $\overline{x}_2(t_0+T) = 0.1800$ Kg.sugar/Kg.water was attempted by prescribing a desired reference trajectory, $x^*_{2d}(t)$, for the derived dynamical feedback controller. The initial and final equilibrium points are located well within the operating region. The transition time between the equilibrium points was set to be achieved in $T = 250$ minutes, with $t_0 = 300$ min and $t_0 + T = 550$ min.

The following open loop trajectory was prescribed for $x^*_{2d}(t)$,

$$x^*_{2d}(t) = \begin{cases} \overline{x}_2(t_0) \text{ for } t < t_0 \\ \overline{x}_2(t_0) + (\overline{x}_2(t_0+T) - \overline{x}_2(t_0)) \times \\ \frac{(t-t_0)^5}{(T)^5}\left[21 - 35\frac{t-t_0}{T} + 15\frac{(t-t_0)^2}{(T)^2}\right] \\ \quad \text{for } t_0 \le t \le t_0 + T \\ \overline{x}_2(t_0+T) \text{ for } t > T \end{cases}$$

The time derivative of the desired open loop trajectory is simply obtained as

$$\dot{x}^*_{2d}(t) = \begin{cases} 0 \text{ for } t < t_0 \\ (\overline{x}_2(T) - \overline{x}_2(t_0))\frac{(t-t_0)^4}{(T-t_0)^5} \times \\ \left[105 - 210\frac{t-t_0}{T-t_0} + 105\frac{(t-t_0)^2}{(T-t_0)^2}\right] \\ \quad \text{for } t_0 \le t \le t_0 + T \\ 0 \text{ for } t > t_0 + T \end{cases}$$

The planned trajectory guarantees a sufficiently smooth departure from the initial equilibrium point (the first four time derivatives being zero at such initial point) and a smooth arrival at the second equilibrium point (the first two time derivatives being also zero at the arrival time). The control input is maintained within reasonable bounds, which are physically acceptable (see Fig. 2). Indeed, typically, the maximum allowed value for the control input is $U_{max} = 3.4$ Kg/min. The simulations show that the controller maximum effort is well below this limiting input value. In the x_2-x_1 coordinates, the proposed open loop trajectory follows a curve joining the equilibrium points $(\overline{x}_1(t_0), \overline{x}_2(t_0)) = (0.070, 0.093954)$ and $(\overline{x}_1(t_0+T), \overline{x}_2(t_0+T)) = (0.095541, 0.18)$. The controlled trajectory, in fact, moves away from the singularity line and from the origin and it lies entirely inside the operating region $\mathcal{X}$ (see Fig. 3).

Other equilibrium transfers are also possible without crossing the singularity arc. However, using special trajectory planning techniques, transfers including crossings of the singularity arc are also possible.

4 CONCLUSIONS

In this article a passivity based control design option was complemented with a trajectory planning approach in the context of a double effect evaporator controller design example. The passivity based controller was based on the "energy shapping -plus- damping injection" methodology which thus far has been unnecessarily restricted to Euler-Lagrange type of systems. The advantages of the proposed approach exploit the natural phisically oriented features of passivity while at the same time allows for efficient circumvention of singularities and other forbidden regions of the state space thanks to the trajectory planning features of the approach.

REFERENCES

Byrnes, Ch., Isidori, A. and Willems, J. C. (1991). Passivity, feedback equivalence and the global stabilization of minimum phase nonlinear system. IEEE Trans. on Automatic Control **36**, 1228-1240.

Hill, D. and Moylan, P. (1976). The stability of nonlinear dissipative system. IEEE Trans. on Automatic Control **21**, 708-711.

Ortega, R., Loria, A., Nicklasson, P., and Sira-Ramírez, H. (1998). *Passivity Based Control of Euler-Lagrange Systems*, Springer-Verlag: London.

Sepulchre, R., Janković, M., and Kokotović, P. (1997) *Constructive Nonlinear Control*, Springer-Verlag, London.

Serón, M., Hill, D., and Fradkov, A. L. (1995). Nonlinear adaptive control of feedback passive systems. Automatica, **31**, 1053-1060.

Silva-Navarro, G. and Alvarez-Gallegos, J. (1995). Positiveness and Asymptotic Stability of a Double Effect Evaporator, *NOLCOS'95*, Lake Tahoe, California.

Sira-Ramírez, H (1988) Differential geometric methods in variable sturcture control. Int. Journal of Control **48**, 1359-1390.

Sira-Ramírez, H. (1998) A General Canonical Form for Passivity Based Control of Noninear Systems. Int. Journal of Control, Special Issue on Advances in Nonlinear Control (to appear).

Willems, J. C. (1972). Dissipative dynamical systems, part I and II. Archives of Rational Mechanics and Analysis, **45**, 321-393.

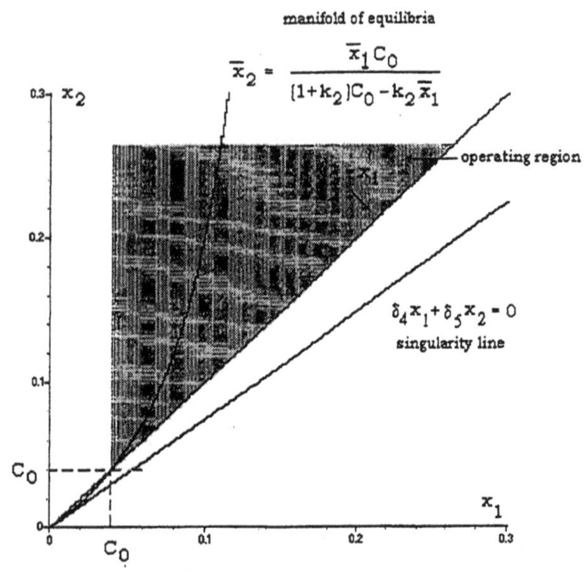

Figure 1: Singularity line, manifold of equilibria, and operating region in the state space.

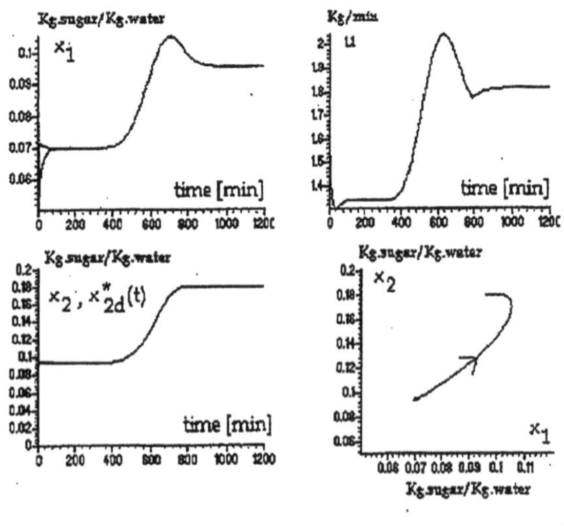

Figure 2: Simulations of passivity based control of the double effect evaporator system with output concentration trajectory planning.

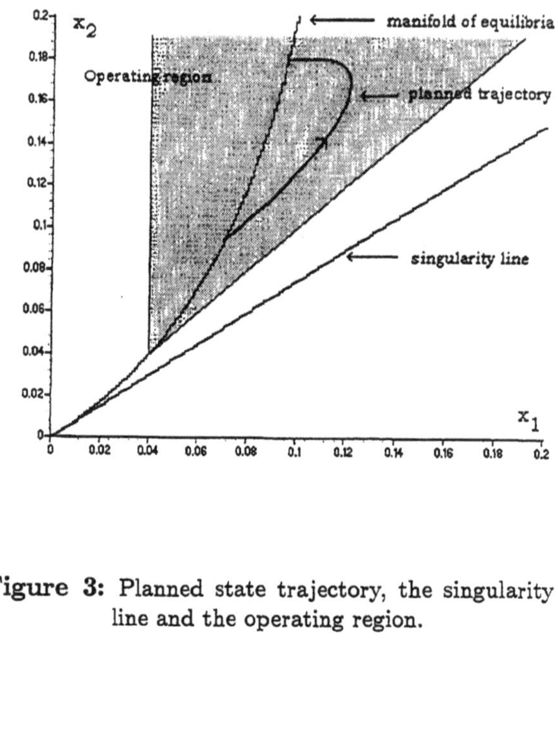

Figure 3: Planned state trajectory, the singularity line and the operating region.

STALL CONTROL FOR AXIAL COMPRESSORS *

Calin Belta[§], Guoxiang Gu[§,1], Andrew Sparks[†], and Siva Banda[†]

[§]Department of Electrical and Computer Engineering
Louisiana State University, Baton Rouge, LA 70803-5901, U.S.A.
[†]Flight Dynamics Directorate, Wright Laboratory
Wright-Patterson Air Force Base, Dayton, OH 45433-7531, U.S.A.

Abstract: Rotating stall is a primary constraint for the performance of flow compressors. This paper establishes a necessary and sufficient condition for a feedback controller to locally stabilize the critical equilibrium of the uniform flow at the inception of rotating stall. The explicit condition obtained in this paper provides an effective synthesis tool for the stall control. *Copyright © 1999 IFAC*

Keywords: Hopf Bifurcation, Rotating stall, local stability, stabilization.

1. INTRODUCTION

Axial flow compressors are subject to two distinct aerodynamic instabilities, rotating stall and surge, which can severely limit the compressor performances. Both these instabilities are disruptions of the normal operating condition which is designed for steady and axisymmetric flow. Rotating stall is a severely non-axisymmetric distribution of axial flow velocity, taking the form of a wave or "stall cell", that propagates steadily in the circumferential direction at a fraction of the rotor speed. Surge, on the other hand, is an axisymmetric oscillation of the mass flow along the axial length of the compressor. Both can result in catastrophic consequences.

Feedback control was proposed to improve compressor performance in [6], and has since received great attention in recent years. The existing results include linear control law [19], bifurcation stabilization [15, 20], and backstepping method [13, 4].

This paper focuses on rotating stall control based on the multi-mode Moore-Greitzer model [16], which in the limiting case converges to the full PDE (partial differential equation) model of Moore-Greitzer [18]. A feedback control law, consisting of both linear and quadratic feedback terms, will be investigated which is a generalization of the feedback control laws in [15, 20, 5] for the third order (or single-mode) Moore-Greitzer model. It will be shown that the critical operating point at the peak pressure rise of the performance characteristic curve is locally stabilizable with the proposed feedback control law. A necessary and sufficient condition will be derived for local stabilization of the underlying feedback compression system. This condition holds, and results in lumped feedback controllers for the full PDE Moore-Greitzer model in the limiting case. In a special case, the proposed feedback control law reduces to the linear feedback control law proposed in [5] which was also studied in [14] recently. In contrast to the global stabilization results in [4], our local stabilization conditions are explicit, and provide an effective synthesis tool for rotating stall control without distributed sensors. Hence our result compliments those in [4].

* This research was supported in part by AFOSR and ARO.
[1] The corresponding author with email: ggu@ee.lsu.edu, and fax: (225) 388-5200.

2. MULTI-MODE COMPRESSOR MODEL

A schematic axial compressor is shown in Figure 1.

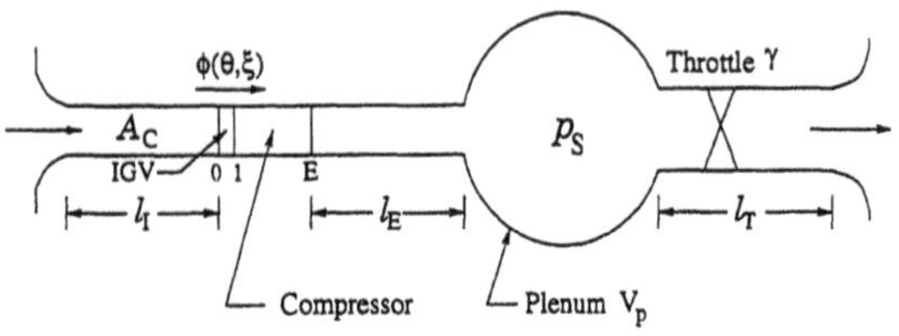

Figure 1 Schematic of compressor.

A post-stall model for compression systems was developed by Moore and Greitzer, and its full PDE form is described by [18]

$$\Psi + l_c\frac{d\Phi}{d\xi} = \psi_c(\phi) - m\left(\frac{\partial}{\partial\xi}\int_{-l_F}^{0}\varphi\, d\eta\right) - \left(\frac{1}{a}\frac{\partial\varphi}{\partial\xi} + \frac{1}{b}\frac{\partial\varphi}{\partial\theta}\right)\bigg|_{\eta=0}, \quad (1)$$

$$\dot{\Psi} = \frac{1}{4B^2 l_c}\left(\Phi - \gamma\sqrt{\Psi}\right), \quad \phi = \Phi + \varphi|_{\eta=0}. \quad (2)$$

All notations are consistent with [18]. The incompressible and irrotational assumption on the gas flow implies the existence of disturbance flow potential that satisfies Laplace's equation with zero boundary condition at $\eta = -l_F \le -l_c$, which in turn implies that the disturbance flow at station 0 has the form:

$$\varphi|_{\eta=0} = \sum_{n=1}^{N}\left(A_n\cos(n\theta) + B_n\sin(n\theta)\right), \quad (3)$$

where $N \to \infty$. Let $\{e^{j\theta_n}\}_{n=0}^{2N}$ be a set of $2N+1$ uniformly distributed samples on the unit circle. Denote

$$\phi_N^{\mathrm{T}} = \left[\begin{array}{ccc} \phi(\theta_1) & \phi(\theta_2) & \vdots & \phi(\theta_{2N+1}) \end{array}\right],$$

$$\frac{T}{t_n} = \begin{bmatrix} \frac{1}{\sqrt{2}} & \frac{1}{\sqrt{2}} & \cdots & \frac{1}{\sqrt{2}} \\ \cos\theta_1 & \cos\theta_2 & \cdots & \cos\theta_{2N+1} \\ \sin\theta_1 & \sin\theta_2 & \cdots & \sin\theta_{2N+1} \\ \vdots & \vdots & \ddots & \vdots \\ \cos N\theta_1 & \cos N\theta_2 & \cdots & \cos N\theta_{2N+1} \\ \sin N\theta_1 & \sin N\theta_2 & \cdots & \sin N\theta_{2N+1} \end{bmatrix},$$

$$E = \mathrm{diag}\left(l_c, m_1 I_2, m_2 I_2, \cdots, m_N I_2\right),$$

$$m_n = \frac{m\cosh(nl_F)}{n\sinh(nl_F)} + \frac{1}{a}, \quad I_2 = \begin{bmatrix} 1 & 0 \\ 0 & 1 \end{bmatrix},$$

$$F = \mathrm{diag}\left(0, F_1, \cdots, F_N\right), \quad F_n = \begin{bmatrix} 0 & -\frac{n}{b} \\ \frac{n}{b} & 0 \end{bmatrix},$$

$$M_1 = T^{\mathrm{T}}E^{-1}FT, \quad M_2 = T^{\mathrm{T}}E^{-1}T,$$

where

$$t_N = \sqrt{\frac{2}{2N+1}}, \quad T^{-1} = T^{\mathrm{T}}.$$

Then the full PDE model can be approximated by [16]

$$\dot{\phi}_N = M_1\phi_N + M_2\psi_c(\Delta\phi_N) - M_2\bar{e}_n, \quad (4)$$

$$\Delta\phi_N = \frac{\phi_N}{W} - \bar{e}_N \in \mathbf{R}^{2N+1}, \quad (5)$$

$$\dot{\Psi} = \frac{1}{4B^2 l_c}\left(\frac{\bar{e}_N^{\mathrm{T}}\phi_N}{2N+1} - \gamma\sqrt{\Psi}\right), \quad (6)$$

$$\bar{e}_N^{\mathrm{T}} = \left[\begin{array}{cccc} 1 & 1 & \cdots & 1 \end{array}\right]. \quad (7)$$

The above multi-mode model converge to the PDE model described in (1) and (2) uniformly as $N \to \infty$ under some rather mild assumption. It is worth to note that with $e_N^{\mathrm{T}} = \left[\begin{array}{cccc} 1 & 0 & \cdots & 0 \end{array}\right]$, there hold

$$T\bar{e}_N = e_N\sqrt{2N+1}, \quad T^{\mathrm{T}}e_N = \frac{\bar{e}_N}{\sqrt{2N+1}}.$$

The performance characteristic curve $\psi_c(\cdot)$ is assumed to be a cubic polynomial [18], and given by

$$\psi_c(\phi_N) = H\left[c_0\bar{e}_N + c_1\phi_N + c_3\phi_N^{\cdot 3}\right], \quad (8)$$

where $x^{\cdot 2} = x \cdot x$ denotes Hadmard product [9], and $x^{\cdot k}$ raising each element of x to the power of k. The uniform flow can be represented by $\phi_N = \phi_e\bar{e}_N$ with ϕ_e intensity of the flow rate.

Proposition 2.1 *The steady equilibria* $(\phi_N, \Psi) = (\phi_e\bar{e}_N, \Psi_e)$ *for the multi-mode compressor model in (4) and (6) are dependent on the throttle parameter* γ, *and governed by:*

$$\Psi_e = H\left[c_0 + c_1\left(\frac{\phi_e}{W} - 1\right) + c_3\left(\frac{\phi_e}{W} - 1\right)^3\right] = \frac{\phi_e^2}{\gamma^2}. \quad (9)$$

The first equality in (9) is obtained by setting $\dot{\phi}_N = 0$ in (4), while the second equality in (9) by setting $\dot{\Psi} = 0$ in (6). The proof is thus omitted. A local model around equilibria $(\phi_e\bar{e}_N, \Psi_e)$ has the form

$$\dot{x} = f(\gamma, x) = Lx + Q[x,x] + C[x,x,x] + \cdots, \quad (10)$$

where $x = \delta\phi_N \oplus \delta\Psi$ with $\delta\phi_N = \phi_N - \phi_e\bar{e}_N$, $\delta\Psi - \Psi_e$, and

$$L = S^{-1}\begin{bmatrix} \tau & \kappa e_N^{\mathrm{T}} \\ -\kappa e_N & E^{-1/2}(F + \alpha I)E^{-1/2} \end{bmatrix}S, \quad (11)$$

$$S = \begin{bmatrix} 0 & 2B\sqrt{(2N+1)l_c} \\ E^{1/2}T & 0 \end{bmatrix}, \quad (12)$$

$$Q[x,x] = \begin{bmatrix} \frac{3Hc_3}{W^2}\left(\frac{\phi_e}{W} - 1\right)M_2 & 0_{2N+1} \\ 0_{2N+1}^{\mathrm{T}} & -\frac{\tau}{4\Psi_e} \end{bmatrix}x^{\cdot 2}, \quad (13)$$

$$C[x,x,x] = \begin{bmatrix} \frac{Hc_3}{W^3}M_2 & 0_{2N+1} \\ 0_{2N+1}^{\mathrm{T}} & \frac{\tau}{8\Psi_e^2} \end{bmatrix}x^{\cdot 3}, \quad (14)$$

$$\kappa = \frac{1}{2Bl_c}, \quad \tau = -\frac{\gamma}{8B^2 l_c\sqrt{\Psi_e}}, \quad (15)$$

$$\alpha = H\psi_c' = H\frac{d\psi_c}{d\phi_e} = \frac{H}{W}\left[c_1 + 3c_3\left(\frac{\phi_e}{W} - 1\right)^2\right]. \quad (16)$$

While κ is independent of the throttle parameter, τ and α are functions of γ. By the form of e_N, the linear matrix can be written as:

$$L = S^{-1}DS, \qquad D = \mathrm{diag}\left(D_0, D_1, \cdots, D_N\right),$$

$$D_0 = \begin{bmatrix} \tau & \kappa \\ -\kappa & \alpha/l_c \end{bmatrix}, \qquad D_n = \begin{bmatrix} \alpha & -n/b \\ n/b & \alpha \end{bmatrix}m_n^{-1},$$

for $1 \le n \le N$. The eigenvalues of L can now be easily computed as

$$\lambda_{1,2} = 0.5\tau_c \pm 0.5j\sqrt{4(\kappa^2 + \tau\alpha/l_c) - \tau_c^2}, \quad (17)$$

$$\lambda_{2n+1,2n+2} = \left(\alpha \pm j\frac{n}{b}\right)m_n^{-1}, \quad \tau_c = \tau + \alpha/l_c \quad (18)$$

where $n = 1, 2, \cdots, N$. Hence eigenvalues are also functions of γ.

Assume S-shape for the performance characteristic curve $\psi_c(\cdot)$, and $c_1 + 3c_3 = 0$ as in [18]. A schematic compressor characteristic is shown in Figure 2. The designed operating point is uniquely determined by the intersection of the throttle line (dashed lines A-B or C-D for two different values of the throttle position) with the compressor performance curve $\psi_c(\cdot)$, according to (9). The maximum pressure rise takes place at $\alpha = 0$ (point A) because that is where the derivative of $\psi_c(\cdot)$ equals to zero. Since $\alpha < 0$ on the right side of the maximum pressure rise, the N pairs of eigenvalues of L matrix in (18) are stable. By $\tau < 0$ the pair of eigenvalues of (17) are also stable. Hence any point on right side of the peak of $\psi_c(\cdot)$ is a stable operating point, and the uniform flow is a stable equilibrium, as highlighted by the solid line. However if the throttle value γ decreases, then the flow rate intensity ϕ_e decreases, and the derivative of $\psi_c(\cdot)$ eventually changes its sign into positive. Thus left side of the peak point A on $\psi_c(\cdot)$ corresponds to unstable operating points, as indicated by dotted line. It is noted that the N pais of eigenvalues of L in (18) become imaginary precisely at $\alpha = 0$. The critical throttle value at which the N pairs of eigenvalues cross the imaginary axis simultaneously is given by, due to assumption $c_1 + 3c_3 = 0$,

$$\gamma_c = \left.\frac{\phi_e}{\sqrt{\Psi_e}}\right|_{\phi_e=\phi_c,\Psi_e=\Psi_c} = \left.\frac{\phi_c}{\sqrt{\psi_c(\phi_c)}}\right|_{\phi_c=2W} = \frac{2W}{\sqrt{H(c_0+c_1+c_3)}}, \tag{19}$$

and $\lambda_{2n+1,2n+2} = \pm j\omega_n$ with $\omega_n = \frac{n}{b} m_n^{-1}$ for $1 \le n \le N$. At the criticality $\gamma = \gamma_c$, Hopf bifurcations occur for the multi-mode model of the compression system in (4) and (6) that induce rotating stall. Furthermore if the underlying bifurcation is subcritical, or unstable, the A-C portion of the stall curve is unstable which is sketched for $N = 1$. Stall cells will be born at point A, and grow which will throttle the operating point from A to B quickly which is a stable operating point, although undesirable. There is a tremendous drop in both the pressure rise, and flow rate. Moreover increasing the throttle position and flow rate at point B does not increase the pressure rise. Rather the pressure rise becomes even lower before it comes to point C at which it jumps back to (due to again loss of stability) the performance curve $\psi_c(\cdot)$. The hysteresis loop A-B-C-D is the main cause for loss of compressor performance, and the potential damage to aeroengines.

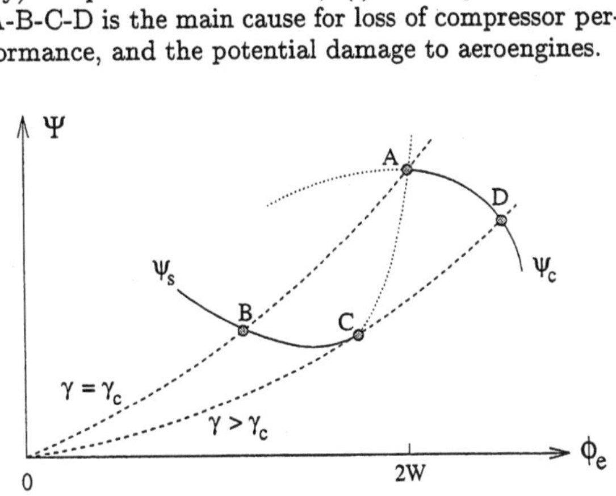

Figure 2 Compressor characteristic, and stall.

The above discussion indicates that stabilization of the critical operating point, and A-C portion of the stall curve is a key issue for rotating stall control which will eliminate the hysteresis loop, and allow smooth change of the operating point between axisymmetric flow and stalled flow for the compression system. In the next section, some of the existing results on Hopf bifurcations will be reviewed which will be used in Section 4 for the synthesis of a stabilizing feedback control law.

3. STABILITY FOR HOPF BIFURCATIONS

This section reviews briefly the projection and Lyapunov methods for local stability of Hopf bifurcations. For stability analysis of Hopf bifurcations, the system under consideration is the following nth order parametrized nonlinear system:

$$\dot{x} = f(\gamma, x), \quad f(\gamma, x_0) = 0 \;\; \forall\, \gamma \in \mathbf{R}, \quad x_0 \in \mathcal{S} \subset \mathbf{R}^n, \tag{20}$$

where $x \in \mathbf{R}^n$, γ is a real-valued parameter, and $\mathcal{S}$ a linear subspace of $\mathbf{R}^n$, to be clarified later. It is assumed that $f(\cdot,\cdot)$ is sufficiently smooth such that the equilibrium solution x_e, satisfying $f(\gamma, x_e) = 0$, is a smooth function of γ. The smoothness of $f(\cdot,\cdot)$ implies the existence of a Taylor series expansion near the origin of $\mathbf{R}^n$ in the form of

$$\dot{x} = f(\gamma, x) = L(\gamma)x + Q(\gamma)[x,x] + C(\gamma)[x,x,x] + \cdots \tag{21}$$

where $L(\gamma)x$, $Q(\gamma)[x,x]$, and $C(\gamma)[x,x,x]$ can each be expanded into

$$\begin{aligned} L(\gamma)x &= L_0x + \delta\gamma L_1x + (\delta\gamma)^2 L_2x + \cdots, \\ Q(\gamma)[x,x] &= Q_0[x,x] + \delta\gamma Q_1[x,x] + \cdots, \\ C(\gamma)[x,x,x] &= C_0[x,x,x] + \delta\gamma C_1[x,x,x] + \cdots, \end{aligned}$$

with $\delta\gamma = \gamma - \gamma_c$, and L_0, L_1, L_2 constant matrices of size $n \times n$. Suppose that $L(\gamma)$ possesses m $(< n/2)$ pairs of complex eigenvalues $\lambda_k(\gamma) = \alpha_k(\gamma) \pm j\beta_k(\gamma)$, dependent smoothly on γ. It is assumed that for $1 \le k \le m$,

$$\alpha_k(\gamma_c) = 0, \quad \beta(\gamma_c) = \omega_k \ne 0, \quad \alpha_k'(\gamma_c) = \frac{d\alpha_k}{d\gamma}(\gamma_c) \ne 0, \tag{22}$$

while all other eigenvalues of $L(\gamma)$ are stable at, and in a neighborhood of $\gamma = \gamma_c$. That is, m pairs of eigenvalues cross the imaginary axis *simultaneously*. Then each $\lambda_k(\gamma)$ is a critical eigenvalue, so is its conjugate. The center space, or the eigen-space for the m pairs of critical eigenvalues is denoted by $\mathcal{S}_c$, and is assumed to be orthogonally complement to $\mathcal{S}$ in the sense that $\mathcal{S} \oplus \mathcal{S}_c = \mathbf{R}^n$. Since projection of $x_0 \in \mathcal{S}$ to $\mathcal{S}_c$ is zero, the equilibria x_0 satisfying (20) will be called zero solution. The strict crossing assumption implies that the zero solution changes its stability as γ crosses γ_c. For instance $\alpha_k'(\gamma_c) > 0$ implies that the zero solution is locally stable for $\gamma < \gamma_c$, and becomes unstable for $\gamma > \gamma_c$. The crucial problem is the determination of local stability near the critical parameter γ_c at which Hopf bifurcations occur, and the periodic solutions are born.

3.1 The Projection Method

The center space $\mathcal{S}_c$, spanned by all the critical eigenvectors, is complete. If all the critical eigenvalues are distinct,

$$\mathcal{S}_c = \mathcal{S}_{c_1} \oplus \mathcal{S}_{c_2} \oplus \cdots \oplus \mathcal{S}_{c_m},$$

where $\mathcal{S}_{c_k}$ denotes the eigen-space spanned by kth pair of the critical eigenvectors. The projection method in [11] projects the nonlinear dynamics into the subspace of $\mathcal{S}_{c_k}$ so that stability of the projected dynamics can be analyzed. Associated with each pair of the critical eigenvalues $(\lambda_k, \bar{\lambda}_k)$, there can define a SCV (stability characteristic value) $\tilde{\lambda}_2^{(k)}$ whose sign determines local stability of the projected dynamics. See [11, 10, 1] for details. Let the Taylor series of $f(\gamma, x)$ be of the form in (21) where $L_0 = L(0)$. An algorithm to compute $\tilde{\lambda}_2^{(k)}$ is outlined next.

- Step 1: Compute left row eigenvector ℓ_k and right column eigenvector r_k of L_0 corresponding to the kth critical eigenvalue of $\lambda_k(0) = j\omega_k$. Normalize by setting $\ell_k r_k = 1$.
- Step 2: Solve column vectors μ_k and ν_k from the equations

$$\begin{aligned} -L_0\mu_k &= \frac{1}{2}Q_0[r_k, \bar{r}_k], & (23)\\ (2j\omega_k I - L_0)\nu_k &= \frac{1}{2}Q_0[r_k, r_k]. & (24) \end{aligned}$$

- Step 3: The coefficient $\tilde{\lambda}_2^{(k)}$ is given by

$$\begin{aligned} \tilde{\lambda}_2^{(k)} &= 2\mathrm{Re}\,\{2\ell_k Q_0[r_k, \mu_k] + \ell_k Q_0[\bar{r}_k, \nu_k] \\ &\quad + \frac{3}{4}\ell_k C_0[r_k, r_k, \bar{r}_k]\Big\}. \qquad (25) \end{aligned}$$

Theorem 3.1 *Suppose that L_0 has m pairs of nonzero critical eigenvalues on the imaginary axis with the rest on open left half plane. Then the projected dynamics along the kth pair of the eigenvectors is stable, if $\tilde{\lambda}_2 < 0$, and unstable if $\tilde{\lambda}_2 > 0$. For the case of $N = 1$, the associated Hopf bifurcation is supercritical or stable, if $\tilde{\lambda}_2 < 0$, and subcritical or unstable, if $\tilde{\lambda}_2 > 0$.*

It should be clear that if L_0 admits only a single pair of imaginary eigenvalues, then local stability of the bifurcated system is equivalent to $\tilde{\lambda}_2^{(k)} < 0$. However if L_0 admits more than one pair of imaginary eigenvalues, local stability of each projected dynamics does not guarantee local stability of the bifurcated systems. In this case, local stability of the bifurcated systems can be tackled by the Lyapunov method [7].

3.2 The Lyapunov Method

The Lyapunov method for multiple critical modes is developed by Fu [7]. It gives sufficient conditions on the existence of a Lyapunov function that guarantees local stability of the bifurcated system. The following notion is needed.

Definition 3.2 *Under the condition that L_0 admits $N > 1$ pairs of eigenvalues on the imaginary axis with the rest on open left half plane, the nonlinear system (21) is said to be locally center-symmetric in the sense of Lyapunov if* (i) *$\omega_i \neq q\omega_p$ for $q = 1, 2, 3$, whenever $m \geq 2$,* (ii) *$q\omega_i \neq \omega_p + \omega_k$, $\omega_i \neq q\omega_p + \omega_k$ for $q = 1, 2$, whenever $m \geq 3$, and* (iii) *$\omega_i \pm \omega_p \neq \omega_k + \omega_l$, whenever $m \geq 4$ where all indices are distinct.*

Theorem 3.3 *Suppose that L_0 admits $m > 1$ pairs of eigenvalues on the imaginary axis with the rest on open left half plane, and the nonlinear system (21) is locally center-symmetric in the sense of Lyapunov. Then there exists a Lyapunov function that guarantees local stability of (21) at the criticality, if*

$$\begin{aligned} \chi_{kk} &= 8\tilde{\lambda}_2^{(k)} < 0, \quad \text{for } k = 1, \cdots, m, & (26)\\ \chi_{kp} &= 16\mathrm{Re}\left\{\ell_k\left(Q_0[r_k, \mu_p] + Q_0[\bar{r}_p, \mu_{k_p}]\right.\right. & (27)\\ &\quad \left.\left. + Q_0[r_p, \nu_{k_p}] + \frac{3}{4}C_0[r_k, r_p, \bar{r}_p]\right)\right\} \leq 0, & \end{aligned}$$

for $k, p = 1, \cdots, m$ with $k \neq p$, where μ_{k_p}, ν_{k_p} satisfy

$$\begin{aligned} (j(\omega_k - \omega_p)I_n - L_0))\,\mu_{k_p} &= \frac{1}{2}Q_0[r_k, \bar{r}_p], & (28)\\ (j(\omega_k + \omega_p)I_n - L_0)\,\nu_{k_p} &= \frac{1}{2}Q_0[r_k, r_p]. & (29) \end{aligned}$$

Since $\chi_{kk} = 8\tilde{\lambda}_2^{(k)}$. local stability of Hopf bifurcation with multiple pairs of critical modes requires more than local stability of each projected dynamics. Roughly speaking, the nonpositivity of χ_{kp} for $k \neq p$ takes the coupling of different pairs of critical modes into account for local stability of the bifurcated system in (21).

4. LOCAL FEEDBACK STABILIZATION

Due to unpredictability, it is difficult to design axial flow compressors such that the critical operating condition corresponding to rotating stall is stable. Thus feedback control is proposed to suppress rotating stall [6]. This paper considers the use of bleedvalve as actuator for the compression system. It is noted that the throttle parameter is a composition of uncertainty γ_o, synthesized disturbance from inlet and combustion chamber, and the bleed valve position $\delta\gamma$ that can be employed as actuator. Assume that Ψ is measurable. Then the throttle parameter can be expressed as

$$\gamma = \gamma_o + \delta\gamma = \gamma_o - \frac{u}{\sqrt{\Psi}}, \qquad (30)$$

where u is the feedback control law to be designed. Our objective is local (asymptotic) stabilization around the critical equilibrium of the uniform flow $(\phi_c\bar{e}_N, \Psi_c)$ for the closed-loop compression system. The compression control system can now be written as

$$\dot{x} = f(\gamma, x) = Lx + Q[x, x] + C[x, x, x] + \cdots + \Gamma u, \quad (31)$$

where $\Gamma = \begin{bmatrix} 0 & \cdots & 0 & 1 \end{bmatrix}^{\mathrm{T}} \kappa^2 l_c$, and Lx, $Q[x, x]$, and $C[x, x, x]$ are linear, quadratic, and cubic terms respectively as given in (11)–(15) with γ replaced by γ_o. Applying similarity transform $z = Sx$ yields the linearized system

$$\begin{aligned} \dot{z} &= Dz + S\Gamma u, \quad S\Gamma = \begin{bmatrix} e_N \\ 0 \end{bmatrix} \kappa\sqrt{(2N+1)l_c}, \\ D &= SLS^{-1} = \begin{bmatrix} D_0 & 0 & \cdots & 0 \\ 0 & D_1 & \ddots & \vdots \\ \vdots & \ddots & \ddots & 0 \\ 0 & \cdots & 0 & D_N \end{bmatrix}. \end{aligned}$$

Thus the state variables corresponding to the subsystem of D_0 are deviations in pressure rise $\delta\Psi = \Psi - \Psi_e$, and in average flow rate $\delta\Phi = \Phi - \phi_e$ respectively. The two eigenvalues of D_0 as in (17) are associated with surge dynamics, and are linearly controllable. The state variables corresponding to subsystems $D_1 \sim D_N$ represent amplitudes of harmonics for disturbance flow rate in (3). Thus the N pairs of eigenvalues for $D_1 \sim D_N$ are associated with the rotating stall dynamics, and are linearly uncontrollable. While linear feedback law can be used for surge control, the uncontrollability for the critical modes of $D_1 \sim D_N$ implies that quadratic feedback law is more appropriate for rotating stall control [1]. The above consideration suggests the employment of the following feedback control law:

$$u = K_{\psi}\delta\Psi + \sum_{i=1}^{2N+1} K_{Q_i} x_i^2, \quad \delta\Psi = x_{2N+2}. \qquad (32)$$

The linear term needs be designed to take care of the surge dynamics, and the quadratic terms need be designed to stabilize the rotating stall dynamics locally. The linear feedback term involving the average disturbance flow is intentiously skipped because of the difficulty in measuring $x_1 \sim x_{2N+1}$ which requires distributed sensors as $N \to \infty$. For the same reason, more general quadratic form is avoided in the proposed feedback control law. Define discrete Fourier transform

$$\hat{K}_{Q_n} = \sum_{i=1}^{2N+1} K_{Q_i} e^{-jn\theta_i}, \quad 0 \le n \le 2N. \qquad (33)$$

It will be shown later that local stabilization requires that $\hat{K}_{Q_0} \neq 0$, rather than $K_{Q_i} \neq 0$ for all $i > 0$. Hence distributed sensors are unnecessary in the limiting case $N \to \infty$.

For the feedback control law as in (32), the closed-loop compression system can now be expanded into the following Taylor series:

$$\begin{aligned}
\dot{x} &= L^* x + Q^*[x,x] + C^*[x,x,x] + \cdots, \\
L^* &= S^{-1} D^* S, \quad D^* = \mathrm{diag}(D_0^*, D_1^*, \cdots, D_N^*), \\
D_0^* &= \begin{bmatrix} \tau + \kappa^2 l_c K_\Psi & \kappa \\ -\kappa & \frac{\alpha}{l_c} \end{bmatrix}, \quad D_n^* = D_n, \\
Q^*[x,x] &= \begin{bmatrix} \frac{3Hc_3}{W^2}\left(\frac{\phi_e}{W} - 1\right) M_2 & 0_{2N+1} \\ \kappa^2 l_c K_Q & -\frac{\tau}{4\Psi_e} \end{bmatrix} x^{\cdot 2}, \\
K_Q &= \begin{bmatrix} K_{Q_1} & \cdots & K_{Q_{2N+1}} \end{bmatrix},
\end{aligned}$$

and $C^*[x,x,x] = C[x,x,x]$. It is noted that D_0^* is associated with the surge dynamics, and D_n^* with the rotating stall dynamics for $n \ge 1$. The following result can be easily shown.

Lemma 4.1 *Under the feedback control law (32), the surge dynamics is locally stable at the criticality $\gamma_o = \gamma_c$, if and only if*

$$K_\Psi < -\left.\frac{\tau}{\kappa^2 l_c}\right|_{\gamma_o = \gamma_c} = \frac{\gamma_c}{2\sqrt{\Psi_c}}. \qquad (34)$$

The main result of this section is the following. Its proof is based on the projection method, and Lyapunov method in Section 3. Due to the space limit, the proof is omitted.

Theorem 4.2 *Assume that $c_3 < 0$, $\phi_c > W$, and the condition (34) is satisfied. Moreover suppose that the imaginary part of the eigenvalues in (18) are locally center-symmetric in the sense of Lyapunov. Then the rotating stall dynamics is locally asymptotically stable at the criticality $\gamma_o = \gamma_c$ for the feedback compression system described in (34) $\sim$ (34), if and only if*

$$\begin{aligned}
\hat{K}_{Q_0} &< \frac{1}{4W}\left(\frac{\phi_c}{W} - 1\right)^{-1} \qquad (35) \\
&\quad -\frac{3Hc_3}{W^2}\left(\frac{\phi_c}{W} - 1\right)\left(K_\Psi - \frac{\gamma_c}{2\sqrt{\Psi_c}}\right).
\end{aligned}$$

The significance of Theorem 4.2 lies in the fact that the stabilizability condition (35) requires only the sum of all K_{Q_i}'s to satisfy a bound determined by various compressor parameters and linear feedback gain K_Ψ. As a result, distributed sensing for the flow rate ϕ_N is unnecessary in the limiting case $N \to \infty$. Moreover precise information on the cubic performance characteristic curve, and on the critical operating condition is not required, provided that unpper and lower bounds on c_3, H, W, ϕ_c, and $\gamma_c/\sqrt{\Psi_c}$ are available.

Remark 4.3 In [4], backstepping method was used to derive a global stabilization condition on a more general state feedback control law than (32). It was shown that any given operating point (or set point) $\gamma_o \ge \gamma_c$ can be globally stabilized. It should be clear that if the set point is chosen as at the criticality, local stability will be ensured for $\gamma_o = \gamma_c$. However it requires distributed sensing as $N \to \infty$. Moreover the stabilizability condition in [4] is less explicit than the one in (35).

The stabilization condition (35) also indicates that the critical operating point can be locally stabilized by using linear feedback control law, i.e., by setting $K_Q = 0$ in (32). The next result follows easily from Theorem 4.2, and thus the proof is skipped.

Corollary 4.4 *Assume that $c_3 < 0$, $\phi_c > W$, and the condition (34) is satisfied. Moreover suppose that the imaginary part of the eigenvalues in (18) are locally center-symmetric in the sense of Lyapunov. Then the rotating stall dynamics is locally asymptotically stable at the criticality $\gamma_o = \gamma_c$ under feedback control $u = K_\Psi \delta\Psi$, if and only if*

$$\hat{K}_\Psi > \frac{\gamma_c}{2\sqrt{\Psi_c}} + \frac{W}{12Hc_3}\left(\frac{\phi_c}{W} - 1\right)^{-2}. \qquad (36)$$

By the fact that $c_3 < 0$, stabilizing gain K_Ψ exists that satisfies (34).

Remark 4.5 Feedback stabilization using only the measurement of pressure rise was considered in [8, 5] for the third order Moore-Greitzer model which is a crude approximation. Corollary 4.4 is surprising which indicates that linear control law is also effective in stabilization of the critical operating point for the multi-mode model, which in the limiting case converges to the PDE Moore-Greitzer model. It should be pointed that the same feedback control law was studied in [14] as well. Using equilibria dependent feedback control, [14] was able to obtain a global stabilization result for $\gamma_o > \gamma_c$[1]. In combination with the result in Corollary 4.4, global stabilization can be claimed for $\gamma_o \ge \gamma_c$.

[1]In [14], global stabilization condition was claimed for $\gamma_o \ge \gamma_c$ which is incorrect. In light of Corollary 4.4, condition (36) needs be enforced at $\gamma_o = \gamma_c$

References

[1] E.H. Abed and Jyun-Horng Fu, "Local feedback stabilization and bifurcation control, I. Hopf bifurcation," *Systems and Control Letters*, 7 (1986), 11-17.

[2] R.A. Adomaitis and E.H. Abed, "Bifurcation analysis of nonuniform flow patterns in axial-flow gas compressors," in *1st World Congress of Nonlinear Analysis*, Aug. 1992.

[3] O.O. Badmus, S. Chowdhury, and C.N. Nett, "Nonlinear control of surge in axial compression systems," *Automatica*, Vol. 32, 59-70, 1996.

[4] A. Banaszuk, H.A. Hauksson, and I. Mezic (1996), "A Backstepping controller for Moore-Greitzer PDE model describing stall and surge in compressors", preprint.

[5] X. Chen, G. Gu, P. Martin, and K. Zhou, "Rotating stall control via bifurcation stabilization," *Automatica*, vol 34, 437-443, April 1998.

[6] A.H. Epstein, J.E. Ffowcs Williams, and E.M. Greitzer, "Active suppression of aerodynamic instabilities in turbomachinery," *J. Propulsion*, vol. 5, 204-211, 1989.

[7] J.-H Fu, "Lyapunov functions and stability criteria for nonlinear systems with multiple critical modes", *Mathematics of Contr., Sig., and Syst.*, 7, 255-278, 1994.

[8] G. Gu, A. Sparks, and C. Belta (1997), "Stability analysis for rotating stall and surge in axial flow compressors", preprint.

[9] R.A. Horn and C.R. Johnson, *Topics in Matrix Analysis*, Cambridge University Press, 1991.

[10] L.N. Howard, "Nonlinear oscillations," in: F.C. Hoppensteadt, Ed., *Nonlinear Oscillations in Biology*, (Amer. Math. Soc., Providence, RI, 1979), pp. 1-68.

[11] G. Iooss and D.D. Joseph, *Elementary Stability and Bifurcation Theory*, Springer-Verlag, New York, 1980.

[12] Wei Kang, "Bifurcation control via state feedback for single input nonlinear systems: Part I and Part II", preprint, 1996.

[13] M. Krstic, J.M. Protz, J.D. Paduano, and P.V. Kokotovic, "Backstepping designs for jet engine stall and surge control," *Proc. of 34th IEEE Conf. Dec. and Contr.*, 3049-3055, 1995.

[14] A. Leonessa, V.-S. Chellaboina, and W.M. Haddad, "Globally stabilizing controllers for multimode axial flow compressors via equilibria-dependent Lyapunov functions," *Proc. 1997 Amer. Contr. Conf.*, 993-1002, 1997.

[15] D.-C. Liaw and E.H. Abed, "Active control of compressor stall inception: A bifurcation-theoretical approach," *Automatica*, vol. 32, 109-116, 1996.

[16] C.A. Mansoux, J.D. Setiawan, D.L. Gysling, and J.D. Paduano, "Distributed nonlinear modeling and stability analysis of axial compressor stall and surge," *Proc. 1994 Amer. Contr. Conf.*, 2305-2316, 1994.

[17] F.E. McCaughan, "Bifurcation analysis of axial flow compressor stability," *SIAM J. Applied Mathematics*, vol. 20, 1232-1253, 1990.

[18] F.K. Moore and E.M. Greitzer (1986), "A theory of post-stall transients in axial compressors: Part I - development of the equations," *ASME J. of Engr. for Gas Turbines and Power*, vol. 108, pp. 68-76.

[19] J.D. Paduano, A.H. Epstein, L. Valavani, J.P. Longley, E.M. Greitzer, and G.R. Guenette (1993), "Active control of rotating stall in a low-speed axial compressor," *J. Turbomachinery*, vol. 115, 48-56.

[20] H.O. Wang, R.A. Adomaitis and E.H. Abed, "Nonlinear analysis and control of rotating stall in axial flow compressors," in *American Control Conference*, 2317-2321, 1994.

NONLINEAR ADAPTIVE OBSERVER DESIGN BASED ON B-SPLINE NEURAL NETWORK

H. Y. Zhang†, C. W. Chan‡, K. C. Cheung‡, Hong Jin†

† Department of Automatic Control, Beijing University of Aeronautics and Astronautics
Beijing 100083, P. R. China, Email: buaa301@mimi.cnc.ac.cn

‡ Department of Mechanical Engineering, University of Hong Kong
Pokfulam road, Hong Kong, Email: mechan@hkucc.hku.hk

Abstract: A design approach of adaptive observer for a class of nonlinear systems with unknown time-varying nonlinearity is proposed. The nonlinear adaptive observer is constructed using a linear observer and a nonlinear compensation term. The latter term is determined recursively using a proposed "deconvolution procedure". The determined term is then approximated using a B-spline neural network with state estimate as its input. The convergence of state estimation error and weight error of the network is proved, and a weight updating rule is derived. Other weight updating rules, such as Instantaneous Gradient Descent method and Recursive Least-Squares method, are also discussed. Simulation results show that the proposed approach is effective. *Copyright © 1999 IFAC*

Keywords: Nonlinear time-varying systems, Adaptive observer, Neural networks

1. INTRODUCTION

The design of adaptive observer for linear time invariant systems has been well studied (Narendra , et al.,1989). In the case of nonlinear systems, Sastry and Isidori(1989) presented results on the use of parameter adaptive control for obtaining asymptotically exact cancellation for the class of nonlinear systems which can be feedback linearized. Marino et al. (1990; 1992) proposed to find a coordinate transformation so that the estimation error dynamics would be linearized in the new coordinates. Cho and Rajamani(1997) recently addressed the issue of state estimation in the presence of parameter uncertainty for Lipschitz nonlinear systems. They found the stability of observer is related to finding solutions to a linear matrix inequality. The convergence of parameter, however, is determined by a persistence -of-excitation type constraint.

Neural networks have been proven to be as powerful tools for analyzing static and dynamic highly nonlinear systems. Only few open literatures addressed the problem of state estimation using neural network for nonlinear systems. The problems of state estimation, identification and control of nonlinear systems using feedforward neural network are addressed in (Levin, 1996). However, the neural network based observer is trained using the state estimation error between the state of simulated system and state estimation given by network output. In (Rovithakis,1994),Unknown nonlinear state space model was approximated using dynamic neural networks. However, they assumed the input of network is state variable which is generally not available.

In (Zhang., et al.,1998), a nonlinear observer design method was proposed for system with unknown nonlinearity and known linear part. In this paper, the method is extended to design an adaptive nonlinear observer for systems with unknown time-varying nonlinearity. The stability of estimation error and the convergence of network weights updating are analyzed based on some results of (Cho,1997) and (Rovithakis,1994). The derived weight updating rule is based on observer output error which is available. Other weight updating rules based on network approximation error are also discussed.

The rest of the paper is organized as follows: section 2 is a brief description of design procedure; section 3 is the determination of nonlinear compensation term using 'deconvolution method'; section 4 is the analysis of stability and convergence of the designed adaptive observer and the weight updating rules; section 5 is the discussion of different weight updating rules; section 6 is simulation examples; section 7 is conclusion.

2. BRIEF DESCRIPTION OF DESIGN PROCEDURE

A class of systems with unknown time-varying nonlinearity is described by

$$\dot{x}(t) = Ax(t) + Bf(x,t) \qquad (1)$$
$$y(t) = Cx(t) \qquad (2)$$

where $A \in R^{n\times n}$, $B \in R^{n\times d}$, $d \le n$, $C \in R^{p\times n}$ are known, $f(x,t) = [f_1(x,t), \cdots, f_d(x,t)]'$ is unknown and slowly time-varying.

Assume $f_i(x,t)$ can be modeled exactly by B-spline neural network as

$$f_i(x,t) = a_i(x)w_i(t), \quad i = 1,2,\cdots,d \qquad (3)$$

where $a_i(x) = [a_{i1}(x), \cdots, a_{ip_i}(x)]$, $w_i' = [w_{i1}, \cdots, w_{ip_i}]$, $a_i(x)$ may be a multivariate spline function, w_i is ideal weight. Let

$$f(x,t) = a(x)w(t) \overset{\Delta}{=} S(t) \qquad (4)$$

where

$$\begin{cases} a(x) = diag(a_1(x), a_2(x), \cdots, a_d(x)) \\ w' = [w_1', \cdots, w_d'] \end{cases} \qquad (5)$$

A nonlinear adaptive observer can be formulated as

$$\dot{\hat{x}}(t) = A_0\hat{x}(t) + K\hat{y}(t) + BS(t) \qquad (6)$$
$$\hat{y}(t) = C\hat{x}(t) \qquad (7)$$

where $A_0 = A - KC$, K is gain matrix. $S(t)$ is a nonlinear compensation term which will be determined recursively using a proposed 'deconvolution method' described in section 3. The determined $S(t)$ is then approximated by a B-spline neural network, that is

$$\hat{S}(t) = a(\hat{x})\hat{w}(t) = \hat{f}(\hat{x},t) \qquad (8)$$

$\hat{f}(\hat{x},t)$ is used in the observer instead of $S(t)$. The updating rule of $\hat{w}(t)$ is derived in section 5 using Lyapunov stability theory.

3. DETERMINATION OF NONLINEAR COMPENSATION TERM USING "DECONVOLUTION PROCEDURE"

Since the deconvolution procedure for determing nonlinear compensation term is much simpler in discrete version of (6) which can be expressed as

$$\hat{x}(k+1) = \Phi\hat{x}(k) + K_d y(k) + \Gamma S(k) \qquad (9)$$
$$\hat{y}(k) = C\hat{x}(k) \qquad (10)$$

From (9) and (10), we have

$$\begin{aligned} \hat{y}(k) = {} & C\,\Phi^k\hat{x}(0) + \sum_{i=1}^{k} C\,\Phi^{i-1}K_d y(k-i) \\ & + \sum_{i=1}^{k} C\,\Phi^{i-1}\Gamma S\,(k-i) \end{aligned} \qquad (11)$$

$$\begin{aligned} \tilde{y}(k) &= y(k) - \hat{y}(k) \\ &= y^*(k) - \sum_{i=1}^{k} C\,\Phi^{i-1}\Gamma S(k-i) \end{aligned} \qquad (12)$$

$$y^*(k) = y(k) - C\,\Phi^k\hat{x}(0) - \sum_{i=1}^{k} C\,\Phi^{i-1}K_d y(k-i) \qquad (13)$$

$\hat{x}(0)$ can be assigned arbitrary, except for satisfying $C\hat{x}(0) = \hat{y}(0) = y(0)$, therefore $y^*(k)$ is known. Let $\tilde{y}(k) = 0$ and $C_i = C\,\Phi^{i-1}$, $k=1,2,\ldots,M+1$. From (12) one can obtain

$$\begin{cases} C_1\Gamma S(0) = y^*(1) \\ C_1\Gamma S(1) + C_2\Gamma S(0) = y^*(2) \\ \vdots \\ C_1\Gamma S(M) + \cdots + C_{M+1}\Gamma S(0) = y^*(M+1) \end{cases} \qquad (14)$$

when $rank(C) = n$, the solution of $\Gamma S(k)$ is

$$\begin{cases} \Gamma S(0) = C^+ y^*(1) \\ \Gamma S(k) = C^+\left[y^*(k+1) - \sum_{i=0}^{k-1} C_{k+1-i}\Gamma S(i)\right] \end{cases} \qquad (15)$$

where $C^+ = (C'C)^{-1}C'$, $k = 1, \cdots, M$

When $\Gamma S(k)$ can be expressed as

$$\Gamma S(k) = \begin{bmatrix} 0 \\ \bar{S}(k) \end{bmatrix} \qquad (16)$$

where $\bar{S}(k) \in R^d$. C can be partitioned accordingly as $C = [\tilde{C} \quad \bar{C}]$, where $\bar{C} \in R^{p \times d}$. When $rank(\bar{C}) = d$, one can obtain a solution for $\bar{S}(k)$ as

$$\begin{cases} \bar{S}(0) = \bar{C}^+ y^*(1) \\ \bar{S}(k) = \bar{C}^+ \left[y^*(k+1) - \sum_{i=0}^{k-1} C_{k+1-i} \Gamma S(i) \right] \end{cases} \quad (17)$$

where $\bar{C}^+ = (\bar{C}'\bar{C})^{-1}\bar{C}'$, $k = 1, \cdots, M$.

4. CONVERGENCE ANALYSIS

Define the state estimation error *e(t)* and weight estimation error $\tilde{w}(t)$ as

$$e(t) = x(t) - \hat{x}(t), \quad \tilde{w}(t) = w(t) - \hat{w}(t) \quad (18)$$

From (1), (2), (6) and (7), the error dynamic equation becomes

$$\dot{e}(t) = A_0 e(t) + B[a(x)w - a(\hat{x})\hat{w}] \quad (19)$$

where $\quad A_0 = A - KC \quad (20)$

Let $\quad \Delta a = a(x) - a(\hat{x}) \quad (21)$

(19) can be written as

$$\dot{e}(t) = A_0 e(t) + B\,\Delta a w + Ba(\hat{x})\tilde{w} \quad (22)$$

Assume that *a(x)* is Lipschits, and w is bound, i.e.,

$$\| \Delta a \| = \| a(x) - a(\hat{x}) \| \le \gamma_1 \| e \| \quad (23)$$

$$\| w \| \le \gamma_2 \quad (24)$$

Now a Lyapunov function is selected as

$$V(t) = e'(t) P e(t) + \rho \tilde{w}'(t) \tilde{w}(t) \quad (25)$$

P is a positive definite symmetric matrix, ρ is a positive number. Along the trajectory of (22),

$$\begin{aligned} \dot{V}(t) &= \dot{e}'Pe + e'P\dot{e} + \rho(\dot{\tilde{w}}'\tilde{w} + \tilde{w}'\dot{\tilde{w}}) \\ &= e'(A_0'P + PA_0)e + 2w'\Delta a'B'Pe \\ &\quad + 2\tilde{w}'a'(\hat{x})B'Pe + 2\rho\tilde{w}'\dot{\tilde{w}} \end{aligned} \quad (26)$$

Since A_0 is stable, one can find a positive definite symmetric matrix Q such that

$$A_0'P + PA_0 = -Q \quad (27)$$

Using (23), (24) and (27), it yields

$$\begin{aligned} \dot{V}(t) &\le (-\lambda_{q\min} + 2\gamma_1\gamma_2 \| B \| \lambda_{p\max}) \| e \|^2 \\ &\quad + 2\tilde{w}'a'(\hat{x})B'Pe + 2\rho\tilde{w}'\dot{\tilde{w}} \end{aligned} \quad (28)$$

where $\lambda_{q\min}$ is the smallest eigenvalue of Q, $\lambda_{p\max}$ is the largest eigenvalue of P.
If the following two conditions are satisfied

i. $\quad \lambda_{q\min} / \lambda_{p\max} > 2\gamma_1\gamma_2 \| B \| \quad (29)$

ii. $\quad 2\tilde{w}'a'(\hat{x})B'Pe + 2\rho\tilde{w}'\dot{\tilde{w}} = 0 \quad (30)$

then $\dot{V}(t) \le 0$

(1). Convergence of *e(t)*

Suppose

$$-\lambda_{q\min} + 2\gamma_1\gamma_2 \| B \| \lambda_{p\max} \le -\beta, \ (\beta > 0) \quad (31)$$

then

$$\dot{V}(t) \le -\beta e'(t) e(t) \quad (32)$$

$$V(t) \le V(0) - \beta \int_0^t e'e dt \quad (33)$$

Since $\dot{V}(t) \le 0$, $V(t) \in L_\infty$, (25) implies $e \in L_\infty$, $\tilde{w} \in L_\infty$. V(t) is a nonincreasing function of t and bounded from below(V(0) is finite), therefore $\lim_{t\to\infty} V(t) = V(\infty)$ exists. From (33), one can get

$$\int_0^\infty \| e \|^2 dt \le (V(0) - V(\infty))/\beta < \infty \quad (34)$$

which implies $e(t) \in L_2$.
From (22) it yields $\dot{e} \in L_\infty$, this is because $e \in L_\infty$, $\tilde{w} \in L_\infty$, B and w are bounded, $a(\hat{x})$ is a spline function which is also bounded, and $\| \Delta a \| \le \gamma_1 \| e \|$. Thus $e \in L_\infty \cap L_2$, $\dot{e} \in L_\infty$, using Barbalat's lemma (Narendra, 1989), it is concluded that $e \to 0$.

(2). Convergence of $\tilde{w}$

$$\int_0^\infty \dot{e} dt = \lim_{t\to\infty} e(t) - e(0) = -e(0)$$

which is finite. From (19), Using the Lipschitz continuity of *a(x)* and $a(\hat{x})$, $\dot{e}$ is uniformly continuous. Hence again by Barbalat's lemma $\dot{e} \to 0$. From (19), this implies that

$$B[a(x)w - a(\hat{x})\hat{w}] \to 0 \quad (35)$$

Because

$$B[a(x)w - a(\hat{x})\hat{w}] = B[(a(x)w - a(x)\hat{w}) + (a(x)\hat{w} - a(\hat{x})\hat{w})]$$

When $e \to 0$, $\hat{x} \to x$, the second term of the right hand side approaches zero, therefore one can get

$$Ba(x)(w - \hat{w}) \to 0 \quad (36)$$

From (36), it can be shown that $\hat{w} \to w$. Since $f(x,t)$ is a slowly time-varying function which implies *w(t)* is also a slowly time-varying function. Therefore, given a sufficient small time interval $[t_0, t_0 + \delta]$ within which $w(t) \approx$ constant. For arbitrary t_0. Let $t_i = t_0 + i\Delta t$, $i = 1,2,\cdots,M$, $M\Delta t = \delta$, From (4), one can get

$$\overline{BA}w = \overline{B}F \tag{37}$$

where

$$\overline{B} = diag(B,\cdots,B) \tag{38}$$

$$\overline{A} = \begin{bmatrix} a(x(t_0)) \\ a(x(t_1)) \\ \vdots \\ a(x(t_M)) \end{bmatrix}, \quad F = \begin{bmatrix} f(x(t_0),t_0) \\ f(x(t_1),t_1) \\ \vdots \\ f(x(t_M),t_M) \end{bmatrix} \tag{39}$$

If $\overline{BA} \in R^{(M+1)n\times(M+1)d}$ has full column rank, then w can be solved as

$$w = [(\overline{BA})'\overline{BA}]^{-1}(\overline{BA})'\overline{B}F \tag{40}$$

Since $\overline{B}F$ is not a zero vector, hence w has a nonzero unique solution. The nonsingularity of $(\overline{BA})'\overline{BA}$ is the condition of persistent excitation. Similarly, $\hat{w}$ has a non-zero unique solution. From (36), one can get $\overline{BA}(w-\hat{w}) \to 0$. Since $\overline{BA}$ has full column rank, and both w and $\hat{w}$ are nonzero generally, it is concluded that $\hat{w} \to w$.

5. WEIGHT UPDATING RULES

(1). Design of a weight updating rule

From (30), the weight updating rule for $\hat{w}$ is derived as

$$\dot{\hat{w}} = -a'(\hat{x})B'Pe/\rho \tag{41}$$

There are two problems in the implementation of (41):

i. e is not available;

ii. (41) is not suitable for computer implementation.

Following (Cho, 1997), problem i is solved by imposing a constraint

$$B'Pe = TCe = T\tilde{y} \tag{42}$$

where $\tilde{y} = y - \hat{y}$, and T is a coordinate transformation matrix. Now, (41) can be rewritten as

$$\dot{\hat{w}} = -a'(\hat{x})T\tilde{y}/\rho \tag{43}$$

where $\tilde{y}$ is available.

For solving problem ii of (40), notice that

$$\begin{aligned}\dot{\tilde{w}}(t) &= (\dot{w}(t) - \dot{\hat{w}}(t)) \\ &\approx [(w(t) - w(t-\Delta t)) - (\hat{w}(t) - \hat{w}(t-\Delta t))]/\Delta t\end{aligned} \tag{44}$$

The ideal weight is not available, however, it is assumed that the time-varying nonlinearity changes very slow, which implies $w(t) - w(t-\Delta t)$ is much smaller than $\hat{w}(t) - \hat{w}(t-\Delta t)$, and can thus be neglected.

Substituting (45) into (44) and imposing the above condition , an implementable weight updating rule is obtained as

$$\Delta\hat{w}(t) = \hat{w}(t) - \hat{w}(t-\Delta t) = a'(\hat{x})T\tilde{y}\Delta t/\rho \tag{45}$$

Now the design of such adaptive nonlinear observer is reduced to find matrixes P and T. Notice from (42),

$$B'P = TC \tag{46}$$

There are $d\times n$ equations and $n(n+1)/2$ unknowns in symmetric matrix P, when $d \le (n+1)/2$ there is always a solution for P, sometimes many solutions. The freedom left in finding P can be used to satisfy (29). T can be selected to ensure the positive definitness of P. Sometimes, T is an identity matrix. From (45), the final weight updating rule becomes

$$\Delta\hat{w}(t) = \eta a'(\hat{x})T\tilde{y} \tag{47}$$

where $\eta = \Delta t/\rho$ is adjusted to ensure the convergence.

(2). Comparison of different weight updating rules

It is interesting to compare the updating rule in (47) for training a network in observer with the Instantaneous Gradient Descent (IGD) rule for training solely a network (Brown, 1994). The latter one can be expressed as

$$\Delta\hat{w}_i(t) = \delta a_i'(x(t))\varepsilon_i(t)/\| a_i(x(t)) \|, \textit{for } i=1,2,\dots,d \tag{48}$$

where δ is the step size and ε_i is the ith component of error between the output and the desired output of network, x is the input of B-spline function. In the notation of this paper, from (4) and (8) one can get

$$\varepsilon(t) = (a(x)w - a(\hat{x})\hat{w}) = S(t) - \hat{S}(t) \tag{49}$$

One can see that (48) and (47) are similar. In (47), the error $\tilde{y}$ is the observer output error. Whilst in (48), the error is the network approximation error.

Since from (19), $\dot{e}(t) = A_0 e(t) + B\,\varepsilon(t)$. In S domain (Chen, 1984)

$$e(s) = (sI - A_0)^{-1}B\,\varepsilon(s) = \left[\sum_i \frac{1}{s-\lambda_i} q_i p_i\right] B\,\varepsilon(s) \tag{50}$$

where λ_i are the eigenvalues of A_0, assuming λ_i distinct, q_i and p_i are right and left eigenvector of A_0 associated with λ_i. From (50) it can be seen that *e(t)* is delayed from $\boldsymbol{\varepsilon}(t)$, therefore, updating weight according to *e(t)* (or $\tilde{y}(t)$) is delayed from using $\boldsymbol{\varepsilon}(t)$. Since $\boldsymbol{\varepsilon}(t)$ can be obtained from (49), where *S(t)* is determined from deconvolution procedure, and $\hat{S}(t)$ from network output, we may take this advantage to update weight using $\boldsymbol{\varepsilon}(t)$. Simulation examples in section 6 show that this will lead to faster convergence of state estimation error. Based on $\boldsymbol{\varepsilon}(t)$, the weight can be updated using recursive identification algorithms.

6. SIMULATION EXAMPLES

A time-varying nonlinear system is described as

$$\begin{bmatrix} \dot{x}_1(t) \\ \dot{x}_2(t) \end{bmatrix} = \begin{bmatrix} 0 & 1 \\ 0 & 0 \end{bmatrix} \begin{bmatrix} x_1(t) \\ x_2(t) \end{bmatrix} + \begin{bmatrix} 0 \\ 1 \end{bmatrix} f(x,t)$$

$$y(t) = x_1(t) + x_2(t)$$

The scalar time-varying nonlinear function $f(x,t)$ is unknown. The nonlinear observer has the form as

$$\dot{\hat{x}} = A\hat{x} + K(y - C\hat{x}) + B\hat{f}(\hat{x},t)$$

where $A = \begin{bmatrix} 0 & 1 \\ 0 & 0 \end{bmatrix}$, $K = \begin{bmatrix} 1 \\ 1 \end{bmatrix}$, $B = \begin{bmatrix} 0 \\ 1 \end{bmatrix}$, $C = [1 \;\; 1]$.

The corresponding discrete observer with sampling interval $\Delta t = 0.05$ has the following form

$$\hat{x}(k+1) = \boldsymbol{\Phi}\hat{x}(k) + \boldsymbol{K}_d y(k) + \boldsymbol{\Gamma}\hat{f}(\hat{x}(k),k)$$

where $\boldsymbol{\Phi} = \begin{bmatrix} 0.9512 & 0 \\ -0.0476 & 0.9512 \end{bmatrix}$, $\boldsymbol{K}_d = \begin{bmatrix} 0.0488 \\ 0.0476 \end{bmatrix}$, $\boldsymbol{\Gamma} = \begin{bmatrix} 0 \\ 0.0488 \end{bmatrix}$. $\hat{f}(\hat{x}(k),k)$ is determined using the above proposed approach. First, the "deconvolution procedure" described in section 5 is used to determine $\boldsymbol{\Gamma}S(k)$. $S(k)$ is a scalar function of k which has same time evolution as $f(\boldsymbol{x}(k),k)$. $S(k)$ then is approximated using B-spline neural network as

$$S(k) = \sum_{i=1}^{p} a_i(\hat{x}_1(k))\hat{w}_i(k) + \varepsilon(k)$$

$$= \hat{S}(k) + \varepsilon(k) = \hat{f}(\hat{x}(k),k) + \varepsilon(k)$$

Three weight updating rules are used:

i. weight updating rule 1, which is developed from Lyapunov stability theory and described in (47). In this example $\boldsymbol{T}$ is a scalar and can be included in η for adjustment.

ii. Weight updating rule 2 which is developed using Instantaneous Gradient Decent method for training sole B-spline network and can be described as (Brown, 1994).

$$\Delta w(k) = \eta a'(\hat{x})\varepsilon(k)/\| a(\hat{x}) \|$$

$$\varepsilon(k) = S(k) - \hat{S}(k) = S(k) - \sum a_i(\hat{x}_1(k))\hat{w}_i(k)$$

iii. Weight updating rule 3, which is the recursive least-squares algorithm with forgetting factor , can be described as

$$\Delta w(k) = \hat{w}(k) - \hat{w}(k-1) = \boldsymbol{L}(k)\varepsilon(k)$$

$$\varepsilon(k) = S(k) - \boldsymbol{\psi}'(k)\hat{w}(k-1)$$

$$\boldsymbol{\psi}'(k) = [a_1(k) \;\; \cdots \;\; a_p(k)],$$

$$\hat{w}'(k) = [\hat{w}_1(k) \;\; \cdots \;\; \hat{w}_p(k)]$$

$$\boldsymbol{L}(k) = \frac{\boldsymbol{P}(k-1)\,\boldsymbol{\psi}(k)}{\lambda(k) + \boldsymbol{\psi}'(k)\boldsymbol{P}(k-1)\,\boldsymbol{\psi}(k)}$$

$$\boldsymbol{P}(k) = \boldsymbol{P}(k-1) - \frac{\boldsymbol{P}(k-1)\,\boldsymbol{\psi}(k)\,\boldsymbol{\psi}'(k)\boldsymbol{P}(k-1)}{\lambda(k) + \boldsymbol{\psi}'(k)\boldsymbol{P}(k-1)\,\boldsymbol{\psi}(k)}$$

Example (1)

$$f(\boldsymbol{x}(k),k) = \sin x_1(k)\cdot 1.001^k, \quad k \le 1000$$

$$f(\boldsymbol{x}(k),k) = \sin x_1(k), \; k>1000$$

B-spline network is used to approximate the above time-varying nonlinear function. The network has 9 spline functions and even spaced knots. The above 3 rules are used to update the weights. All initial values $w_i(0) = 0.05, i = 1 \sim 9$.

(a). Using rule 1, the estimation results are shown in Fig.1 where $\eta = \delta\Delta t = 0.005$, $\delta = 0.005/0.05 = 0.1$.

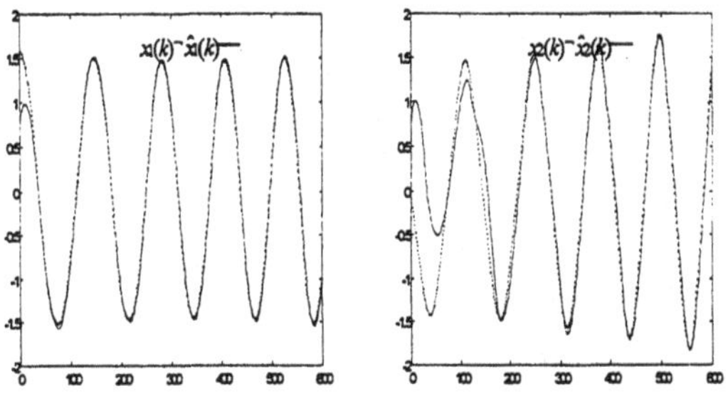

Fig.1: updating rule 1, $\delta = 0.1$

(b). Using rule 2, the results are shown in Fig.2 where $\eta = 1.5$, the estimate $\hat{x}_2$ is much better then that in Fig.1. This is because rule 2 adjusts weight according to $\varepsilon(k)$ not $\tilde{y}(k)$, and $\tilde{y}(k)$ is delayed from $\varepsilon(k)$ as explained in section 6. Fig.3 is for $\eta = 2.015$, the estimate $\hat{x}_2$ diverges. This verifies the theoretical result of Instantaneous Gradient Decent method that the stability region for δ is $0 < \delta < 2$.

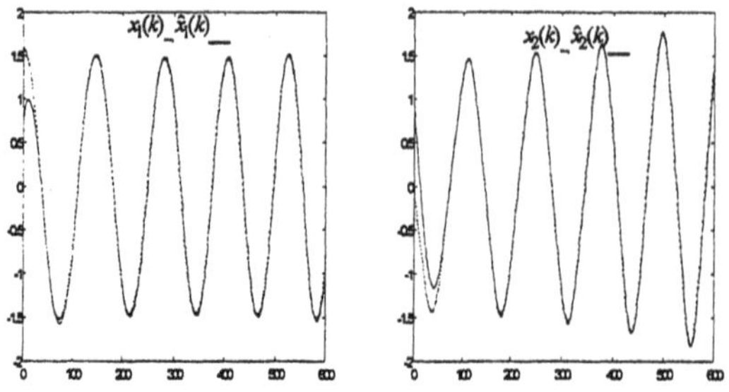

Fig.2: updating rule 2, $\eta = 1.5$

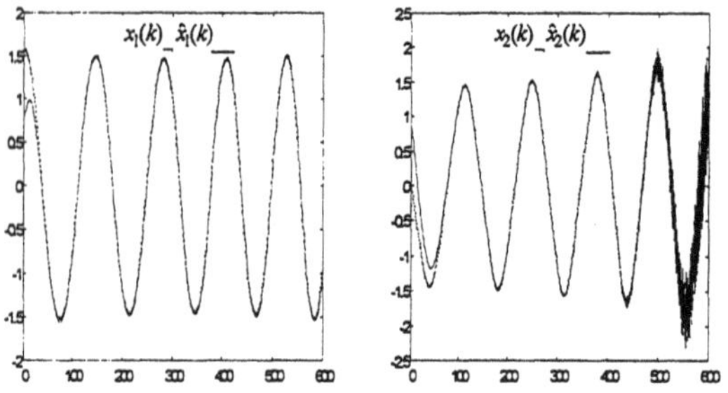

Fig.3: updating rule 2, $\eta = 2.015$

(c). Using rule 3, the results are shown in Fig.4 where $\lambda(k) = 0.98$, $\boldsymbol{P}(0) = 1000\boldsymbol{I}$. The estimation results are as good as in Fig.2.

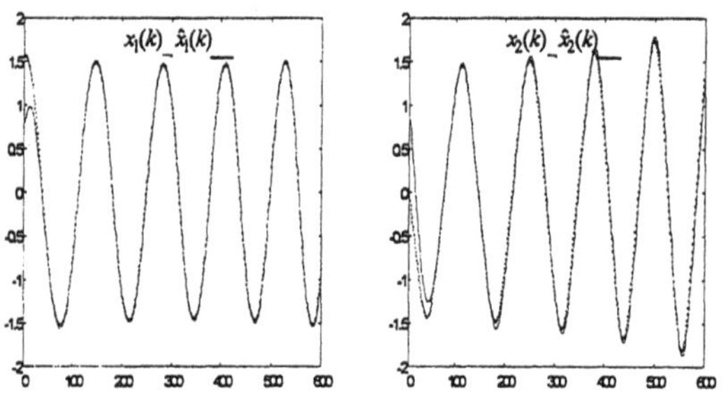

Fig.4: updating rule 3, $\lambda(k) = 0.98$

Example (2)

$$f(\boldsymbol{x}(k), k) = \sin x_1(k), \quad k < 250$$

$$f(\boldsymbol{x}(k), k) = 2\sin x_1(k), \quad k \geq 250$$

The estimation result is also good. Because the space limitation, it is not shown here.

7. CONCLUSION

A design method of nonlinear adaptive observer for unknown nonlinear time-varying system has been proposed. The nonlinear observer consists of a linear observer and a time-varying nonlinear compensation term. The linear observer can be designed using existing methods. The compensation term can be determined using a proposed "deconvolution procedure", The term then is approximated by the output of a B-spline neural network. The input of network can be chosen based on the knowledge of the nonlinearity or by trial and errors. Convergence of the state estimation error and weight error of the neural network based observer has been proved using Lyapunov stability theory. The weight updating rule derived from this theory is related to state estimation error which is not available. Therefore it was reformed to use observer output estimation error. This rule was compared with rules using network output error. The estimation accuracy of that using latter rules is much better than that using the former one.

The Instantaneous Gradient Decent (IGD) rule and Recursive Least Squares (RLS) rule with forgetting factor yield similar results. RLS algorithm is more complex than IGD, but it does not need to select a suitable learning rate, instead, needs to select a suitable forgetting factor. Simulation results have shown that the proposed design approach is effective.

REFERENCE

Brown M. and C. Harris(1994). *Neurofuzzy Adaptive Modelling and Control*, Prentice Hall.

Chen C. T.(1984). *Linear System Theory and Design*, Saunders College Publishing.

Cho Y. M. and R. Rajamani(1997). A systematic approach to adaptive observer for nonlinear systems, *IEEE Trans. Automat. Contr.*, **Vol. 42, No. 4, April** ,pp534-537.

Levin A. U. and K. S. Narendra(1996). Control of nonlinear dynamic systems using neural networks –part II: Observablility, identification, and control, *IEEE Trans. Neural Network*, **Vol. 7, No. 1, Jan,** pp30-42..

Marino R(1990). Adaptive Observer for Single-output Nonlinear Systems, *IEEE Trans. Automat. Contr.*, **Vol.35, No.9,** pp1054-1058.

Marino R. and P. Tomei(1992). Adaptive observers for a class of multi-output nonlinear systems, *Int. J. Adaptive Contr. Signal Processing,* **Vol.6,** pp35-365.

Narendra K. S. and A. Isidori(1989). *Stable Adaptive Systems*, Englewood Cliffs, NJ: Prentice-Hall.

Rovithakis G. A. and M. A. Christodoulou(1994). Adaptive control of unknown plants using dynamic neural network, *IEEE Trans. System, Man and Cybernetics,* **Vol. 24, No. 3,** pp400-411,.

Sastry S. and A. Isidori(1989). Adaptive Control of Linearizable Systems, *IEEE Trans. Automatic Control* **Vol.34, No.11,** pp1123-1131.

Zhang H. Y., C. W. Chan, K. C. Cheung and H. Jin (1998). Nonlinear observer design with unknown nonlinearity via B-spline network approach, *Proceedings of the American Control Conference,* Philadelphia, Pennsylvania, June, pp2339-2343.

E-2c-12-2
14th Triennial World Congress, Beijing, P.R. China

STATE OBSERVATION PROBLEM OF A CLASS OF SEMILINEAR HYPERBOLIC SYSTEMS

Liping Pan

Department of Mathematics, Fudan University, Shanghai, 200433, China
email: *lppan@ms.fudan.edu.cn*

K. L. Teo

School of Mathematics and Statistics, Curtin University of Technology, GPO Box U 1987, Perth 6845, Western Australia, Australia

and

Xu Zhang

Institute of Mathematics, Fudan University, Shanghai, 200433, China & Institute of Mathematics, Sichuan Union University, Chengdu, 610064, China

Abstract: In this paper, the state observation problem of a class of semilinear hyperbolic systems is considered. It is shown that the initial state of the system can be uniquely determined from its measurable boundary data, and then an effective algorithm is presented, which gives the approximate solutions of this problem by solving a series of finite dimensional nonlinear programming problems.

Keywords: State observers, algorithms, observability, nonlinear systems, nonlinear programming, wave equations, measured values.

1. INTRODUCTION

Consider the following semilinear hyperbolic system:

$$(1.1)\quad \begin{cases} y'' - a(x)y_{xx} \\ \quad = f(y, y', y_x) + b(t,x),\ (t,x) \in Q, \\ y(t,0) = y(t,l) = 0, \quad t \in (0,T), \end{cases}$$

This work is partially supported by the Natural Science Foundation of China, the Chinese Education Ministry Science Foundation, the Youth Science Foundation of Shanghai's University, and a Research Grant from the Australian Research Council.

$$(1.2)\quad y(0,x) = y_0(x),\ y'(0,x) = y_1(x),\ \ x \in I,$$

where $' = \frac{\partial}{\partial t}$, $'' = \frac{\partial^2}{\partial t^2}$, $Q \triangleq (0,T) \times I$, $T > 0$, $I \triangleq [0,l]$ with $l \in (0,\infty)$ being a given constant, $y_0(\cdot) \in H_0^1(I)$ and $y_1(\cdot) \in L^2(I)$. Fix a nonhomogeneous term $b(\cdot,\cdot)$ and a nonlinear function $f(\cdot,\cdot,\cdot) : \mathbf{R}^3 \to \mathbf{R}^1$. Denote the weak solution of system (1.1)–(1.2) correspond to initial value $(z_0(\cdot), z_1(\cdot)) \in X \triangleq H_0^1(I) \times L^2(I)$ by $y(\cdot,\cdot;z_0,z_1)$. The main purpose of this paper is to consider the following state observation problem:

(SOP): Let $T > 0$ be given. Determine the initial state (y_0, y_1) from the measurable boundary data

$$\left\{ \frac{\partial y}{\partial x}(t,0) \,\Big|\, t \in [0,T] \right\}. \tag{1.3}$$

The above problem is in essential an observability problem for system (1.1)–(1.2). As two of the basic notions, observability and controllability, for linear controlled system in finite dimension, were introduced by Kalman (1960). Thereafter, many authors have devoted to the generalization of these notions to more general system, for the case in infinite dimension and/or its nonlinear counterpart, for example, see (Bartosiewicz, 1995; Delfour and Mitter, 1972; Dolecki and Russell, 1977; Gauthier and Kupka, 1994; George, 1992; Hermann and Krener, 1977; Komornik, 1994; Li and Yong, 1995; Lions, 1988; Markus, 1965; Quinn and Carmichael, 1984–85; Russell, 1978; Sussmann, 1987; Zuazua, 1993) and the references cited therein.

It is well-known that controllability and observability are a pair of dual notions for *linear* controlled systems. For the linear case, the corresponding theory is well-developed for both abstract and concrete problems, see (Komornik, 1994; Li and Yong, 1995; Lions, 1988; Russell, 1978). However, there are no duality relationship between controllability and observability for *nonlinear* controlled systems. Consequently, one had to study the controllability and the observability problem separately for the nonlinear systems. For the controllability problem of semilinear controlled system, the readers are referred to (Markus, 1965; Zuazua, 1993) and the rich references cited therein. As to the observability problem for semilinear systems, the finite dimensional case were considered by some authors, see (Bartosiewicz, 1995; Gauthier and Kupka, 1994; Hermann and Krener, 1977) and the references cited therein. However, to our knowledge, there are few works, only two papers of George (1992) and Quinn and Carmichael (1984–85), concern such problems on infinite dimensional systems; furthermore these two papers only consider the case that $f(u,v,w) = g(u)$ for all $u,v,w \in \mathbf{R}^1$, where $g(\cdot) : \mathbf{R}^1 \to \mathbf{R}^1$ is continuously differentiable and Lipschitz continuous with small Lipschitz constants; These two papers do not give any algorithms.

The purpose of this paper is to show that the initial state of system (1.1)–(1.2) can be uniquely determined from its measurable boundary data (1.3), and then to present an effective algorithm which gives the approximate solutions of state observation Problem (SOP) by solving a series of finite dimensional nonlinear programming problems (the precise statement of the results and the proof details are given in Sections 2–4).

2. STATEMENT OF THE MAIN RESULT

Suppose that:

(H1) $a(\cdot) \in C^\infty(I)$ with $\min_{x\in I} a(x) > 0$;

(H2) $b(\cdot,\cdot) \in C([0,\infty), L^2(I))$;

(H3) $f(\cdot,\cdot,\cdot) : \mathbf{R}^3 \to \mathbf{R}^1$ is continuously differentiable with respect to all of its arguments and there is a constant $L > 0$ such that

$$|f(u_1,v_1,w_1) - f(u_2,v_2,w_2)| \leq L(|u_1-u_2| + |v_1-v_2| + |w_1-w_2|)$$

for any $(u_i,v_i,w_i) \in \mathbf{R}^3$, $i = 1,2$.

Using the well-known results and methods (for example, see (Lasiecka, Lions and Triggiani, 1986; Pazy, 1983)) and the so-called energy estimate method one can easily get the following theorem.

Theorem 2.1. *Let (H1)–(H3) hold and $T > 0$. Then:*

(i) For any initial value $(z_0(\cdot), z_1(\cdot)) \in X$, system (1.1)–(1.2) admits a unique weak solution $y(\cdot,\cdot;z_0,z_1) \in C([0,T]; H_0^1(I)) \cap C^1([0,T]; L^2(I))$, and there exists a constant $C = C(T,L) > 0$ such that

$$\begin{aligned} &\max_{0\le t\le T} \Big[|y(t,\cdot;\tilde z_0,\tilde z_1) - y(t,\cdot;\bar z_0,\bar z_1)|^2_{H_0^1(I)} \\ &\quad + |y'(t,\cdot;\tilde z_0,\tilde z_1) - y'(t,\cdot;\bar z_0,\bar z_1)|^2_{L^2(I)} \Big] \\ &\le C^2 \Big[|\tilde z_0(\cdot) - \bar z_0(\cdot)|^2_{H_0^1(I)} \\ &\quad + |\tilde z_1(\cdot) - \bar z_1(\cdot)|^2_{L^2(I)} \Big], \\ &\forall\, (\tilde z_0(\cdot),\tilde z_1(\cdot)), (\bar z_0(\cdot),\bar z_1(\cdot)) \in X; \end{aligned} \tag{2.1}$$

(ii) For any $(z_0(\cdot), z_1(\cdot)) \in X$,

$$\frac{\partial y}{\partial x}(\cdot,0;z_0,z_1),\ \frac{\partial y}{\partial x}(\cdot,l;z_0,z_1) \in L^2(0,T) \tag{2.2}$$

and there exists a constant $C = C(T,L) > 0$ such that

$$\begin{aligned} &\int_0^T \Big[\Big| \frac{\partial y}{\partial x}(t,0;\tilde z_0,\tilde z_1) \\ &\quad - \frac{\partial y}{\partial x}(t,0;\bar z_0,\bar z_1) \Big|^2 + \Big| \frac{\partial y}{\partial x}(t,l;\tilde z_0,\tilde z_1) \\ &\quad - \frac{\partial y}{\partial x}(t,l;\bar z_0,\bar z_1) \Big|^2 \Big] dt \\ &\le C^2 \Big[|\tilde z_0(\cdot) - \bar z_0(\cdot)|^2_{H_0^1(I)} \\ &\quad + |\tilde z_1(\cdot) - \bar z_1(\cdot)|^2_{L^2(I)} \Big], \\ &\forall\, (\tilde z_0(\cdot),\tilde z_1(\cdot)), (\bar z_0(\cdot),\bar z_1(\cdot)) \in X; \end{aligned} \tag{2.3}$$

(iii) If $b(\cdot,\cdot) \in C^1([0,+\infty) \times I)$, $z_0(\cdot) \in H^2(I) \cap H_0^1(I)$, $z_1(\cdot) \in H_0^1(I)$ and $f(u,v,w) = h(u)$ for all $u, v, w \in \mathbf{R}^1$, where $h(\cdot) : \mathbf{R}^1 \to \mathbf{R}^1$ is continuously differentiable and globally Lipschitz continuous, then

$$\frac{\partial y}{\partial x}(\cdot, 0; z_0, z_1),\ \frac{\partial y}{\partial x}(\cdot, l; z_0, z_1) \in H^1(0,T)(\subseteq C[0,T]). \tag{2.4}$$

According to (2.2), one can define a nonlinear map $\Gamma : X \to L^2(0,T)$ by

$$\Gamma(z_0, z_1)(\cdot) \triangleq \frac{\partial y}{\partial x}(\cdot, 0; z_0, z_1), \quad \forall\ (z_0(\cdot), z_1(\cdot)) \in X. \tag{2.5}$$

The following result is crucial in the sequel.

Theorem 2.2. *Let (H1)-(H3) hold and $T > l_0 \triangleq 2\int_0^l a^{-\frac{1}{2}}(x)dx$. Then there exists a constant $C = C(T, l, L)$ such that*

$$\begin{aligned} &\frac{1}{C}|\Gamma(\bar{z}_0, \bar{z}_1)(\cdot) - \Gamma(\bar{z}_0, \bar{z}_1)(\cdot)|_{L^2(0,T)} \\ &\leq |\bar{z}_0(\cdot) - \bar{z}_0(\cdot)|_{H_0^1(I)} \\ &\qquad + |\bar{z}_1(\cdot) - \bar{z}_1(\cdot)|_{L^2(I)} \\ &\leq C|\Gamma(\bar{z}_0, \bar{z}_1)(\cdot) - \Gamma(\bar{z}_0, \bar{z}_1)(\cdot)|_{L^2(0,T)}, \\ &\quad \forall\ (\bar{z}_0(\cdot), \bar{z}_1(\cdot)), (\bar{z}_0(\cdot), \bar{z}_1(\cdot)) \in X. \end{aligned} \tag{2.6}$$

Remark 2.3. One can find a counter example from (Komornik, 1994) which show the second inequality in (2.6) does not necessarily hold if $T < l_0$.

The proof of Theorem 2.2 will be given in Section 3.

By Theorem 2.2, one see that the map Γ defined by (2.5) is one-to-one. Thus Problem (SOP) admits an unique solution. Furthermore by the estimate (2.6), the solution of Problem (SOP) depend continuously on the measurable boundary data $\frac{\partial y}{\partial x}(\cdot, 0; y_0, y_1)$. Thanks to these facts, one can establish a numerical method solving Problem (SOP). For this purpose, define the following functional:

$$\begin{aligned} &J(z_0(\cdot), z_1(\cdot)) \\ &\triangleq \frac{1}{2}\int_0^T |\frac{\partial y}{\partial x}(t, 0; z_0(\cdot), z_1(\cdot)) \\ &\qquad - \frac{\partial y}{\partial x}(t, 0; y_0(\cdot), y_1(\cdot))|^2 dt, \\ &\quad \forall\ (z_0(\cdot), z_1(\cdot)) \in X. \end{aligned} \tag{2.7}$$

It is easy to see that Problem (SOP) can be reduced to the following minimization problem:

(MP): Find $(\hat{z}_0(\cdot), \hat{z}_1(\cdot)) \in X$ such that

$$\begin{aligned} &J(\hat{z}_0(\cdot), \hat{z}_1(\cdot)) = \hat{J} \\ &\triangleq \inf\{J(z_0(\cdot), z_1(\cdot)) \mid (z_0(\cdot), z_1(\cdot)) \in X\}. \end{aligned} \tag{2.8}$$

By (2.6), it is easy to see that the initial state $(y_0(\cdot), y_1(\cdot))$ to be determined in (SOP) is the unique solution of (MP).

Take a partition sequence $\{x_0^j, x_1^j, \cdots, x_{N_j}^j\}_{j=1}^{\infty}$ of I as follows:

$$\begin{cases} 0 = x_0^j < x_1^j < \cdots < x_{N_j}^j = l, \\ \Delta_j \triangleq \max_{1 \leq i \leq N_j}(x_i^j - x_{i-1}^j) \to 0 \\ \qquad (j \to \infty). \end{cases} \tag{2.9}$$

Let us consider the following nonlinear programming problems in finite dimension:

(MP)$_j$ Find $\{\hat{u}_i^j,\ \hat{v}_i^j,\ i = 1, 2, \cdots, N_j\} \subseteq \mathbf{R}^1$, such that

$$J_j(\hat{u}_1^j, \cdots, \hat{u}_{N_j}^j; \hat{v}_1^j, \cdots, \hat{v}_{N_j}^j) \leq \hat{J}_j + \varepsilon_j \tag{2.10}$$

where $\{\varepsilon_j\}$ is an arbitrary given sequence of positive numbers satisfying $\varepsilon_j \to 0$ $(j \to \infty)$,

$$\begin{aligned} &J_j(u_1^j, \cdots, u_{N_j}^j; v_1^j, \cdots, v_{N_j}^j) \\ &\triangleq \frac{1}{2}|\Gamma(z_0(\cdot; \{u_i^j\}_{i=1}^{N_j}), z_1(\cdot; \{v_i^j\}_{i=1}^{N_j})) \\ &\qquad - \Gamma(y_0(\cdot), y_1(\cdot))|_{L^2(0,T)}^2 \end{aligned} \tag{2.11}$$

with

$$\begin{cases} z_0(x; \{u_i^j\}_{i=1}^{N_j}) \\ = z(x; \{u_i^j\}_{i=1}^{N_j}) - \frac{x}{l}z(l; \{u_i^j\}_{i=1}^{N_j}), \\ z(x; \{u_i^j\}_{i=1}^{N_j}) \\ = \int_0^x \sum_{i=1}^{N_j} u_i^j \chi_{(x_{i-1}^j, x_i^j]}(s)ds, \\ z_1(x; \{v_i^j\}_{i=1}^{N_j}) = \sum_{i=1}^{N_j} v_i^j \chi_{(x_{i-1}^j, x_i^j]}(x), \\ \qquad 0 \leq x \leq l \end{cases} \tag{2.12}$$

and

$$\begin{aligned} \hat{J}_j = \inf\{&J_j(u_1^j, \cdots, u_{N_j}^j : v_1^j, \cdots, v_{N_j}^j) \mid \\ &u_i^j, v_i^j \in \mathbf{R}^1, i = 1, \cdots, N_j\}, \\ &j = 1,\ 2\ \cdots. \end{aligned} \tag{2.13}$$

The main result in this paper reads:

Theorem 2.4. *If the hypotheses of Theorem 2.2 hold, then*

$$\lim_{j \to \infty} \hat{J}_j = \hat{J} = J(y_0(\cdot), y_1(\cdot)) = 0 \tag{2.14}$$

and

$$(2.15)\quad \begin{cases} |z_0(\cdot;\{\hat{u}_i^j\}_{i=1}^{N_j}) - y_0(\cdot)|_{H_0^1(I)} \to 0, \\ |z_1(\cdot;\{\hat{v}_i^j\}_{i=1}^{N_j}) - y_1(\cdot)|_{L^2(I)} \to 0, \\ \qquad j \to \infty. \end{cases}$$

The proof of Theorem 2.4 will be given in Section 4.

3. PROOF OF THEOREM 2.2

In order to prove Theorem 2.2, let us present the following lemma, which is a simple generalization and improvement of Theorem 3.1 in (Kazemi and Klibanov, 1993) (so its proof is omitted).

Lemma 3.1. *Let $\Omega \subseteq \mathbf{R}^n$ be a bounded domain, and let $x_0 \in \mathbf{R}^n \setminus \bar{\Omega}$ and $T > 2R_1$ with $R_1 \triangleq \max_{x\in\bar{\Omega}} |x - x_0|$. Suppose that $u \in H^2(Q)$ satisfy*

$$(3.1)\quad \begin{cases} |u'' - \Delta u| \le A(|u| + |u'| + |\nabla u|), \\ \qquad (t,x) \in Q, \\ u(t,x) = 0, \qquad (t,x) \in \Sigma, \end{cases}$$

where A is a nonnegative constant, $Q = (0,T)\times\Omega$ and $\Sigma = (0,T)\times\partial\Omega$. Then there exists a constant $B = B(A, R_1, T, x_0)$ such that

$$(3.2)\quad |u|^2_{H^1(Q)} \le B\left|\frac{\partial u}{\partial \nu}\right|^2_{L^2(\Sigma_0)},$$

where $\nu = \nu(x)$ denotes the unit normal vector to $\partial\Omega$ at $x \in \partial\Omega$, directed towards the exterior of Ω; $\Sigma_0 = (0,T) \times \Gamma_0$ with

$$(3.3)\quad \Gamma_0 \triangleq \{x \in \partial\Omega \mid (x - x_0)\cdot\nu(x) > 0\}.$$

□

Corollary 3.2. *Suppose $u \in C([0,T]; H_0^1(I)) \cap C^1([0,T]; L^2(I))$ (with $I \triangleq (0,l)$) is a weak solution of the following equation*

$$(3.4)\quad \begin{cases} u'' - u_{xx} = q_1 u + q_2 u' + q_3 u_x, \\ u(t,0) = u(t,l) = 0, \\ u(0,x) = u_0(x),\ u'(0,x) = u_1(x), \\ \qquad (t,x) \in Q, \end{cases}$$

where $Q \triangleq (0,T) \times I$, $q_i \in L^\infty(Q)$, $i = 1,2,3$. Then for any $x_0 \in (l,\infty)$ and $T > 2x_0$, there is a constant $B = B(|(q_1,q_2,q_3)|_{L^\infty(Q)}, T, x_0)$ such that

$$(3.5)\quad |u_0|^2_{H_0^1(I)} + |u_1|^2_{L^2(I)} \le B\left|\frac{\partial u(\cdot,0)}{\partial x}\right|^2_{L^2(0,T)}.$$

Proof of Corollary 3.2. Using a special case of Lemma 3.1, in which one take $n = 1$, $\Omega = I$, $x_0 \in (l,\infty)$ (then $\Gamma_0 = \{0\}$), one conclude that there is a constant $B = B(A,T,x_0)$ with $A \triangleq |q_1|_{L^\infty(Q)} + |q_2|_{L^\infty(Q)} + |q_3|_{L^\infty(Q)}$ such that

$$(3.6)\quad |u|^2_{H^1(Q)} \le B\left|\frac{\partial u(\cdot,0)}{\partial x}\right|^2_{L^2(0,T)}, \quad \forall\, u \in H^2(Q) \text{ satisfying eq. (3.4).}$$

By the dense argument, one see that estimate (3.6) holds for any weak solution $u \in C([0,T]; H_0^1(I)) \cap C^1([0,T]; L^2(I))$ of eq. (3.4). On the other hand, using the usual energy estimate, one get

$$(3.7)\quad |u_0|^2_{H_0^1(I)} + |u_1|^2_{L^2(I)} \le B|u|^2_{H^1(Q)}.$$

Thus the desired estimate (3.5) follows from (3.6) and (3.7) immediately. The proof of Corollary 3.2 is completed. □

Proof of Theorem 2.2. Set

$$(3.8)\quad w(\cdot,\cdot) = y(\cdot,\cdot;\tilde{z}_0,\tilde{z}_1) - y(\cdot,\cdot;\bar{z}_0,\bar{z}_1).$$

It is easy to see that w is the unique weak solution of the following equation

$$(3.9)\quad \begin{cases} w'' - a(x)w_{xx} = q_1 w + q_2 w' + q_3 w_x, \\ w(t,0) = w(t,l) = 0, \\ w(0,x) = w_0(x),\ w'(0,x) = w_1(x), \\ \qquad (t,x) \in (0,T)\times I, \end{cases}$$

with

$$(3.10)\quad \begin{aligned} q_k(t,x) &\triangleq \int_0^1 f'_k\Big(y(t,x;\bar{z}_0,\bar{z}_1) + \theta w(t,x), \\ &\qquad y'(t,x;\bar{z}_0,\bar{z}_1) + \theta w'(t,x), \\ &\qquad y_x(t,x;\bar{z}_0,\bar{z}_1) + \theta w_x(t,x)\Big)d\theta, \\ &\qquad (t,x) \in Q, \quad k = 1,2,3 \end{aligned}$$

and

$$(3.11)\quad \begin{cases} w_0(x) \triangleq \tilde{z}_0(x) - \bar{z}_0(x), \\ w_1(x) \triangleq \tilde{z}_1(x) - \bar{z}_1(x), \\ \qquad x \in I. \end{cases}$$

Let us introduce the following coordinate transformation:

$$(3.12)\quad \hat{x} = r(x) \triangleq \int_0^x a^{-\frac{1}{2}}(s)ds.$$

Denote

$$(3.13)\quad u(t,\hat{x}) = w(t,x).$$

Then by equation (3.9), it is easy to see that $u \in C([0,T]; H_0^1(I_0)) \cap C^1([0,T]; L^2(I_0))$ (with $I_0 \triangleq$

$(0, l_0))$ satisfies

$$(3.14)\quad \begin{cases} u'' - u_{\hat{x}\hat{x}} = p_1 u + p_2 u' + p_3 u_{\hat{x}}, \\ u(t,0) = u(t,l_0) = 0, \\ u(0,\hat{x}) = u_0(\hat{x}),\ u'(0,\hat{x}) = u_1(\hat{x}). \\ (t,\hat{x}) \in (0,T) \times I_0 \end{cases}$$

with

$$(3.15)\quad \begin{cases} p_1(t,\hat{x}) \triangleq q_1(t,x), \\ p_2(t,\hat{x}) \triangleq q_2(t,x), \\ p_3(t,\hat{x}) \triangleq a(x) r_{xx}(x) \\ \qquad + a^{-\frac{1}{2}}(x) q_3(t,x), \\ \qquad (t,\hat{x}) \in (0,T) \times I_0 \end{cases}$$

and

$$(3.16)\quad u_0(\hat{x}) \triangleq w_0(x),\ u_1(\hat{x}) \triangleq w_1(x),\ \hat{x} \in I_0.$$

Now the second inequality in (2.6) follows from Corollary 3.2 immediately while the first inequality in (2.6) is an obvious consequence of (2.3) in Theorem 2.1. The proof of Theorem 2.2 is completed. □

4. PROOF OF THEOREM 2.4

Note that for any $u_i^j, v_i^j \in \mathbf{R}^1$, $i = 1, \cdots, N_j$, $j = 1, 2, \cdots$, one have

$$(4.1)\quad (z_0(\cdot; \{u_i^j\}_{i=1}^{N_j}), z_1(\cdot; \{u_i^j\}_{i=1}^{N_j})) \in X.$$

Thus

$$(4.2)\quad \hat{J}_j \geq \hat{J}, \quad j = 1, 2, \cdots,$$

which implies

$$(4.3)\quad \varliminf_{j\to\infty} \hat{J}_j \geq \hat{J}.$$

On the other hand, by means of the density of $C_0^\infty(I)$ in $H_0^1(I)$ and $C(I)$ in $L^2(I)$ one see that for any given $\delta > 0$, there exists a $y_{0,\delta}(\cdot)$ in $C_0^\infty(I)$ and an integer $K > 0$ such that

$$(4.4)\quad |y_{0,\delta}(\cdot) - y_0(\cdot)|_{H_0^1(I)} < \delta$$

and for all $j \geq K$ there exist $\tilde{u}_i^j$, $\tilde{v}_i^j \in \mathbf{R}^1$, $i = 1, \cdots, N_j$, $j = 1, 2, \cdots$ such that

$$(4.5)\quad \begin{cases} |\sum_{i=1}^{N_j} \tilde{u}_i^j \chi_{(x_{i-1}^j, x_i^j]}(\cdot) - \dot{y}_{0,\delta}(\cdot)|_{L^2(I)} < \delta, \\ |\sum_{i=1}^{N_j} \tilde{v}_i^j \chi_{(x_{i-1}^j, x_i^j]}(\cdot) - y_1(\cdot)|_{L^2(I)} < \delta. \end{cases}$$

Denote

$$(4.6)\quad \begin{cases} \tilde{z}^j(x) = \int_0^x \sum_{i=1}^{N_j} \tilde{u}_i^j \chi_{(x_{i-1}^j, x_i^j]}(s) ds, \\ \qquad \forall\, x \in I, \end{cases}$$

then

$$(4.7)\quad \begin{aligned} &|\tilde{z}^j(x) - y_{0,\delta}(x)| \\ &= |\int_0^x [\sum_{i=1}^{N_j} \tilde{u}_i^j \chi_{(x_{i-1}^j, x_i^j]}(s) - \dot{y}_{0,\delta}(s)] ds| \\ &\leq \sqrt{x} |\sum_{i=1}^{N_j} \tilde{u}_i^j \chi_{(x_{i-1}^j, x_i^j]}(\cdot) - \dot{y}_{0,\delta}(\cdot)|_{L^2(I)} \\ &\leq \sqrt{l}\delta, \qquad \forall\ x \in I, \end{aligned}$$

which implies that

$$(4.8)\quad \max_{0 \leq x \leq l} |\tilde{z}^j(x) - y_{0,\delta}(x)| \leq l^{1/2}\delta.$$

Thus

$$(4.9)\quad |\tilde{z}^j(l)| = |\tilde{z}^j(l) - y_{0,\delta}(l)| \leq l^{1/2}\delta.$$

Denote

$$(4.10)\quad \tilde{z}_0^j(x) = \tilde{z}^j(x) - \frac{x}{l}\tilde{z}^j(l), \ \forall\ x \in I.$$

Then $\tilde{z}_0^j(\cdot) \in H_0^1(I)$. So by the first inequality in (4.5) and (4.7)–(4.9), one get

$$(4.11)\quad \begin{aligned} &|\tilde{z}_0^j(\cdot) - y_{0,\delta}(\cdot)|_{H_0^1(I)}^2 \\ &= \int_0^l |\tilde{z}_0^j(x) - y_{0,\delta}(x)|^2 dx \\ &\quad + \int_0^l |\dot{\tilde{z}}_0^j(x) - \dot{y}_{0,\delta}(x)|^2 dx \\ &\leq l^2\delta^2 + \int_0^l |\sum_{i=1}^{N_j} \tilde{u}_i^j \chi_{(x_{i-1}^j, x_i^j]}(x) \\ &\quad - \dot{y}_{0,\delta}(x) - \frac{1}{l}\tilde{z}^j(l)|^2 dx \\ &\leq l^2\delta^2 + 2[\int_0^l |\sum_{i=1}^{N_j} \tilde{u}_i^j \chi_{(x_{i-1}^j, x_i^j]}(x) \\ &\quad - \dot{y}_{0,\delta}(x)|^2 dx + \frac{1}{l}|\tilde{z}^j(l)|^2] \\ &\leq l^2\delta^2 + 2(\delta^2 + \delta^2) \\ &= (l^2 + 4)\delta^2. \end{aligned}$$

Now by (4.11), the second inequality in (4.5) and

Theorem 2.2, one conclude that

$$
\begin{aligned}
&J_j(\breve{u}_1^j, \cdots, \breve{u}_{N_j}^j; \breve{v}_1^j, \cdots, \breve{v}_{N_j}^j) \\
&\triangleq \frac{1}{2}|\Gamma(\breve{z}_0^j(\cdot), \sum_{i=1}^{N_j} \breve{v}_i^j \chi_{(x_{i-1}^j, x_i^j]}(\cdot)) \\
&\quad - \Gamma(y_0(\cdot), y_1(\cdot))|_{L^2(0,T)}^2 \\
&\leq C^2[|\breve{z}_0^j(\cdot) - y_0(\cdot)|_{H_0^1(I)}^2 \\
&\quad + |\sum_{i=1}^{N_j} \breve{v}_i^j \chi_{(x_{i-1}^j, x_i^j]}(\cdot) - y_1(\cdot)|_{L^2(I)}^2] \\
&\leq C^2\{2[|\breve{z}_0^j(\cdot) - y_{0,\delta}(\cdot)|_{H_0^1(I)}^2 \\
&\quad + |y_{0,\delta}(\cdot) - y_0(\cdot)|_{H_0^1(I)}^2] \\
&\quad + |\sum_{i=1}^{N_j} \breve{v}_i^j \chi_{(x_{i-1}^j, x_i^j]}(\cdot) - y_1(\cdot)|_{L^2(I)}^2\} \\
&\leq C^2\{2[(l^2+4)\delta^2 + \delta^2] + \delta^2\} \\
&= C^2(2l^2+11)\delta^2,
\end{aligned}
\tag{4.12}
$$

which implies that

$$
\hat{J}_j \leq C^2(2l^2+11)\delta^2, \quad \forall\, j \geq K. \tag{4.13}
$$

Consequently

$$
\overline{\lim_{j\to\infty}}\, \hat{J}_j \leq 0 = J(y_0(\cdot), y_1(\cdot)) = \hat{J}. \tag{4.14}
$$

Combining (4.3) and (4.14), one obtain (2.14).

Finally, by Theorem 2.2, (2.10) and (2.14), one get

$$
\begin{aligned}
&\Big[|z_0(\cdot; \{\hat{u}_i^j\}_{i=1}^{N_j}) - y_0(\cdot)|_{H_0^1(I)} \\
&\quad + |z_1(\cdot; \{\hat{v}_i^j\}_{i=1}^{N_j}) - y_1(\cdot)|_{L^2(I)}\Big]^2 \\
&\leq C^2|\Gamma(z_0(\cdot; \{\hat{u}_i^j\}_{i=1}^{N_j}), z_1(\cdot; \{\hat{v}_i^j\}_{i=1}^{N_j})) \\
&\quad - \Gamma(y_0(\cdot), y_1(\cdot))|_{L^2(0,T)}^2 \\
&= 2C^2 J_j(\hat{u}_1^j, \cdots, \hat{u}_{N_j}^j; \hat{v}_1^j, \cdots, \hat{v}_{N_j}^j) \\
&\leq 2C^2(\hat{J}_j + \varepsilon_j) \to 2C^2\hat{J} = 0, \\
&\qquad j \to \infty,
\end{aligned}
\tag{4.15}
$$

namely (2.15). The proof is completed. □

REFERENCES

Bartosiewicz, L. (1995). Local observability of nonlinear systems, *Systems & Control Letters,* **25**, 295–298.

Delfour, M.C. and S.K. Mitter (1972). Controllability and observability for infinite dimensional systems, *SIAM J. Control & Optim.,* **10**, 329–333.

Dolecki, S. and D.L. Russell (1977). A general theory of observation and control, *SIAM J. Control & Optim.,* **15**, 185–220.

Gauthier, J.P. and I.A.K. Kupka (1994). Observability and observers for nonlinear systems, *SIAM J. Control & Optim.,* **32**, 975–994.

George, R.K. (1992). Observability of a class of nonlinear abstract systems, in: *Mathematical Theory of Control* (Joshi, M.C. and A.V. Balakrishnan. (Ed)), *Lecture Notes in Pure And Applied Mathematics,* **142**, 143–160.

Hermann, R. and A. Krener (1977). Nonlinear and observability, *IEEE Trans. Ac.,* **22**, 728–740.

Kalman, R.E. (1960). On the general theory of control systems, in: *Proc. 1st IFAC Congress,* 481–492, Butterworth, London.

Kazemi, M.A. and M.V. Klibanov (1993). Stability estimates for ill-posed Cauchy problems involving hyperbolic equations and inequalities, *Appl. Anal.,* **50**, 93–102.

Komornik, V. (1994). *Exact Controllability and Stabilization (the Multiplier Method),* RAM, John Wiley & Sons, Masson, Paris.

Lasiecka, I., J.L. Lions and R. Triggiani (1986). Nonhomogeneous boundary value problems for second-order hyperbolic operators, *J. Math. Pures Appl.,* **65**, 149–192.

Li, X. and J. Yong (1995). *Optimal Control Theory for Infinite Dimensional Systems,* Birkhäuser, Boston.

Lions, J.L. (1988). *Contrôlabilité exacte, stabilisation et perturbations de systémes distribués, Tomes 1 & 2,* Masson, RMA 8 & 9, Paris.

Markus, L. (1965). Controllability of semilinear control systems, *SIAM J. Control,* **3**, 78–90.

Pazy, A. (1983). *Semigroups of Linear Operators and Applications to Partial Differential Equations,* Springer–Verlag, Berlin.

Quinn, M.D. and N. Carmichael (1984-85). An approach to nonlinear control problems using fixed point methods, degree theory and pseudo- inverses, *Numer. Funct. Anal. & Optim.,* **7**, 197–230.

Russell, D.L. (1978). Controllability and stabilizability theory for linear partial differential equations: recent progress and open problems, *SIAM Rev.* **20**, 639–739.

Sussmann, H.J. (1987). A general theorem on local controllability, *SIAM J. Control & Optim.,* **25**, 158–194.

Zuazua, E. (1993). Exact controllability for semilinear wave equations in one space dimension, *Ann. Inst. Henri Poincaré Anal. non linéaire,* **10**, 109–129.

LINEAR MEASUREMENT FEEDBACK CONTROL OF NONLINEAR PLANTS

Salvador Padilla[(1)], Jesús Alvarez[(2)] and Eduardo Castellanos[(3)]
[(1)]Comercial Mexicana de Pinturas, Marcos Achar Lobatón 6, Tepexpan Edo. México,México. Email: spadilla@comex.com.mx.
[(2)]Universidad Autónoma Metropolitana-Iztapalapa, Depto. de Ingeniería de Procesos e Hidráulica, Apdo. 55534, 09340 México D.F. Email: jac@xanum.uam.mx
[(3)]Universidad de Guadalajara, Depto. Ingeniería Química, M. García Barragán 1451, Guadalajara, Jalisco 44430 México. Email: tobias@mati.net.mx

Abstract: For multi-input multi-output (MIMO) (possibly open-loop unstable) plants we consider the problem of designing a nonlinear feedforward-feedback controller driven by on-line measurements. The proposed controller stems from a nonlinear geometric control framework, guarantees closed-loop robust exponential stability, contains a feedforward-feedback compensator of dynamic interaction, tolerates modeling errors and has a tuning procedure that amounts to the tuning of independent SISO (single-input single-output) systems. The control of an azeotropic binary distillation column with tray hydraulics dynamics is considered as an application example. *Copyright © 1999 IFAC*

Keywords: nonlinear control systems, linearization, MIMO systems, modeling errors, signal filtering.

1. INTRODUCTION

In the last years several chemical processes have been studied using different nonlinear estimator based geometric control techniques (Kravaris and Chung, 1987; Soroush and Kravaris, 1993; Daoutidis and Kravaris, 1994). Recently Alvarez (1996), proposed an estimator-based nonlinear geometric control design with stability criteria and a construction-tuning procedure. However, in most industrial settings these nonlinear estimator-based control techniques may be difficult to implement and maintain. The aim of the present work is to find out whether a simpler, application-oriented, linear measurement-feedback (MF) control scheme can be designed. The present work is motivated by the *ad hoc* construction of a linear controller in a five-input five-output multivariable continuous copolymer reactor (Padilla and Alvarez, 1997), where the output decoupling with closed-loop (CL) stability property of a nonlinear SF controller was inherited to its linear output-feedback approximation.

In this work, the design of a linear MF controller for MIMO nonlinear plants is addressed. The point of departure is the geometric nonlinear SF-control problem. The candidate MF controller is obtained from a Taylor series expansion of the input-output SF linearizing map associated to the nonlinear SF controller, in conjunction with suitable filters to obtain smooth estimates of the measurements and of their time derivatives. The design features a RE(robust exponential)-stability criterion, and a systematic construction-tuning procedure. Since the design achieves almost linear non-interactive pole-assignable (LNPA) output dynamics, the tuning of gains amounts to the tuning with conventional techniques, and independent SISO systems. The two point control of an azeotropic binary distillation column with tray hydraulics dynamics is considered as an application example.

2. STATEMENT OF THE PROBLEM

Let us consider nonlinear MIMO plants

$$\dot{x} = f(x,u,d,b), \quad x \in X_w, u \in U_w \qquad (1)$$

$$y = h(x,b)$$

with n states (x), m control inputs (u), m outputs (y), n_d known time varying exogenous inputs d, and n_b model parameters b. The maps f and h are smooth about $(\bar{x},\bar{u})$, and the plant must operate at a (possibly open-loop unstable) nominal steady-state $\bar{x}$, with nominal inputs $\bar{u}$ and outputs $\bar{y}$. This is, $f(\bar{x},\bar{u},\bar{d},b) = 0$, $\bar{y} = h(\bar{x},b)$. The problem consists in designing a linear feedforward - feedback controller under the following requirements and considerations. The controller must: 1) be linear and driven by the measured signals y and d; 2) yield output regulation with internal CL stability; 3) compensate for dynamic interactions such that the CL output dynamics are approximately LNPA; 4) tolerate errors in the parameters of the model used in the design of the controller, and handle noisy measurements; and 5) a simple tuning procedure with appeal to the practitioner, meaning that it must be stated, carried out, and interpreted in terms of notions and tools from industrial-type conventional design techniques for linear SISO systems.

This work is restricted to the case of uncertainties in the parameters of the plant, in the understanding that in the future the study can be extended to a more general kind of robustness (say, including errors in the dimension of the state space model and in its functionalities). Specifically, our aim is to design a dynamic feedforward-feedback linear controller of the form

$$\dot{\chi} = A_c\chi + B_d(d-\bar{d}) + B_y(y-\bar{y}) \qquad (2)$$
$$u = \bar{u}_m + C_c\chi$$

where χ is the state of the controller, and $\bar{u}_m$ is the nominal control input according to the plant with the approximated parameter β. Given the approximated parameter β, the nominal input $\bar{d}$, and the nominal output $\bar{y}$; the model-based input $\bar{u}_m$ and its corresponding steady state $\bar{x}_m$ are the solution of the (n+m)-equation set

$$0 = f(\bar{x}_m, \bar{u}_m, \bar{d}, \beta),\ \bar{y} = h(\bar{x}_m, \beta)$$

3. CONSTRUCTION OF THE CONTROLLER

In this section a candidate linear output-feedback controller is built as follows. Assuming that the state of the plant is known, build a nominal SF nonlinear auxiliary controller in order to generate a perfect input-output interaction compensator with CL exponential (E) stability and LNPA output error dynamics. Then, generate a candidate linear measurement-driven controller (2) as a suitable approximation to the exact SF-controller, in conjunction with filters to obtain estimates of the measured signals and of their time-derivatives.

3.1 Nonlinear State Feedback Control

Let us recall the constructive solvability conditions of the nominal nonlinear geometric SF-control problem (Isidori, 1995) in a format (Alvarez, 1996) that highlights the properties of interest: state-stability and LNPA output dynamics, but introducing one modification: the presence of a time varying exogenous inputs (d) will be accounted for (Alvarez et al., 1994) in order to endow the controller with a feedforward path.

The next definition is a straightforward extension, to a plant with exogenous input d, of the solvability conditions given in Alvarez (1996). To state the definiton, let us introduce some notation and definitions. Let $L^i_f c$ denote the recursive directional derivative of the scalar field c(x,t) with respect to the vector field f(x,t),

$$L^{i+1}_f c = L_f(L^i_f c),\ L^0_f c = c,\ L_f c = L_f c + c_u\,\dot{u}, i \geq 1$$
$$L_f c = [\partial c/\partial x_1, \ldots, \partial c/\partial x_n]f,\ c_u = [\partial c/\partial u_1, \ldots, \partial c/\partial u_p]$$

and obtain the following expressions for the outputs $y_1,\ldots,y_m$ and their time-derivatives [$y_i^{(j)} = d^j y_i / dt^j$]:

$$\mathcal{y} = \phi_I(x,\mathcal{d},b)$$
$$v = \varphi(x,\mathcal{d},u,b)$$
$$\mathcal{y} = [y_1,\ldots, y_1^{(\kappa_1-1)};\ldots;\ y_m,\ldots, y_m^{(\kappa_m-1)}]'$$
$$v = [y_1^{(\kappa_1)},\ldots,\ y_m^{(\kappa_m)}]'$$

the maps ϕ_I and φ are given by:

$$\phi_I(x,\mathcal{d},b) = [h_1,.., L_f^{(\kappa_1-1)}h_1;..;h_m,.., L_f^{(\kappa_m-1)}h_m]' \qquad (3)$$
$$\varphi(x,\mathcal{d},u,b) = [L_f^{\kappa_1} h_1,\ldots,L_f^{\kappa_m} h_m]' \qquad (4)$$
$$\mathcal{d} = [d_1,\ldots, \mathcal{d}_1^{(\upsilon_1-1)};\ldots;d_m,\ldots, \mathcal{d}_m^{(\upsilon_m-1)}]'$$
$$v_d = [\mathcal{d}_1^{(\upsilon_1)},\ldots,\mathcal{d}_m^{(\upsilon_m)}]'$$

The map ϕ is x-invertible if there exists and inverse map $\alpha(x,\mathcal{y},\mathcal{d},b)$ such that, about $(\bar{x},\bar{\mathcal{y}},\bar{\mathcal{d}},b)$, $\phi^{-1}[x, \alpha(x,\mathcal{y},\mathcal{d},b), \mathcal{d}, b] = x$. Let φ be an Ru-invertible map, and $\phi_I(x,\mathcal{d},b) = \mathcal{y}$ determine a (n-κ) dimensional time-varying surface $\Sigma(t)$. The zero-dynamics are given by:

$$\dot{x} = f[\,x, \mu(x, 0, \mathcal{d}, b), d, b],\quad x \in \Sigma(t) \qquad (5)$$

$$\mu(x,v,\mathcal{d},b) = \varphi^{-1}(x,v,\mathcal{d},b)$$
$$\Sigma(t) = \{x \in X_w | \phi_I(x,\mathcal{d},b) = \phi_I(\bar{x},\bar{\mathcal{d}},b)\}.$$

Definition 1. The solvability property of the SF control problem is met if there are m strictly positive integers (output relative degrees), $\kappa_1,..,\kappa_m$ ($\kappa_1+..+\kappa_m = \kappa \leq n$, $\kappa_i > 0$) and a map $\phi_{II}(x,\mathcal{d},b) = [\phi_{\kappa+1},\ldots,\phi_m]'$ such that in some neighborhood of $(\bar{x},\bar{u},b)$:

i) $\partial\phi_I/\partial u = 0$.

ii) $\phi(x,\mathscr{d},b):=[\phi_I',\phi_{II}']'$ is x-invertible (ϕ_I in Eq. 3).

iii) $\varphi(x,\mathscr{d},b)$ (Eq. 4) is u-invertible.

iv) $\bar{x}$ is an E-stable steady-state of the zero-dynamics (5). ♦

If the plant meets the solvability property of definition 1, the PI (proportional-integral) controller

$$u = \mu(x,v,\mathscr{d},b) \tag{6a}$$

$$v = -K_c[\phi_I(x,\mathscr{d},b) - \phi_I(\bar{x},\bar{\mathscr{d}},b)]+z_m \tag{6b}$$

$$\dot{z}_m = -K_I\,[h(x,b)-h(\bar{x},b)],\ z_m(0) = 0 \tag{6c}$$

yields an E-stable CL plant with LNPA output error dynamics, provided the gain pair (K_c,K_I) is chosen in the block-diagonal parametrized form (Alvarez, 1996)

$$K_c(s_c)=bd[(s_c^{\kappa_1}k_{11}^c,\ ..,s_c k_{1\kappa_1}^c),..,(s_c^{\kappa_m}k_{m1}^c,..,s_c k_{m\kappa_m}^c)]$$

$$K_I(s_c) = diag(s_c^{\kappa_1+1}k_1^I,\ldots,\ s_c^{\kappa_m+1}k_n^I),\ s_c > 0 \tag{7}$$

and the entries of $[K_c(1),K_I(1)]$ are chosen so that the reference LNPA output error dynamics are stable:

$$\eta_i^{(\kappa_i)} + k_{i\kappa_i}^c\eta_i^{(\kappa_{i-1})}+\ldots+k_{i1}^c\eta_i^{(1)}+k_i^I\eta_i = 0$$
$$\eta_i = y_i - \bar{y}_i,\ 1\le i\le m \tag{8}$$

3.2 Linear MF-Controller

The map $\phi(x,\mathscr{d},b)$ is said to be Rx-invertible about $(\bar{x},\bar{\mathscr{d}},b)$ if it is x-invertible and ϕ and its inverse α are L (Lipschitz)-continuous about $(\bar{x},\bar{\mathscr{d}},b)$. Let us require the maps $\phi(x,\mathscr{d},b)$ and $\varphi(x,\mathscr{d},b)$ to be Rx and Ru-invertible, and express the IO [input(v)-output(y)]-linearizing controller (6a) as a Taylor series expansion about the model-based argument $(\bar{x}_m,0,\bar{\mathscr{d}},b)$:

$$u^* = \bar{u}_m + B_d(\mathscr{d}-\bar{\mathscr{d}}) + B_v v + B_x(x-\bar{x}) - B_b e_b + Q$$

where Q denotes quadratic and higher order terms, and $\bar{u}_m = \mu(\bar{x}_m,0,\bar{\mathscr{d}},b)$; $B_A = (\partial\mu/\partial A)(\bar{x}_m,0,\bar{\mathscr{d}},b)$

$A = x,\ v,\ \mathscr{d},\ e_b = \beta - b,\ e_d = \hat{\mathscr{d}} - \mathscr{d},\ e_v = \hat{v} - v$, and rewrite the exact controller u* as follows

$$u^* = u + B_d e_d + B_v e_v + B_x(x-\bar{x}_m) - B_b e_b + Q$$

$$u^\circ = \bar{u}_m + B_d(\mathscr{d}-\bar{\mathscr{d}}) + B_v\hat{v}$$

$$\hat{v} = -K_c(\hat{y}-\bar{y}) + z_m, \qquad \dot{z}_m = -K_I(y-\bar{y})$$

and consider its linear truncation u° as the candidate controller, in conjunction with the battery (9a,b) of m+p independent single-output filters, one for each measured signal:

$$\dot{\chi}_d = A_d\chi_d + K_d(d-\bar{d}), \qquad A_d = \Gamma_d - K_d\Delta_d \tag{9a}$$

$$\dot{\chi}_y = A_y\chi_y + K_y(y-\bar{y}), \qquad A_y = \Gamma_y - K_y\Delta_y \tag{9b}$$

$$\dot{\chi}_m = -K_I(y-\bar{y}) \tag{9c}$$

$$u = \bar{u}_m + B_d\chi_d + B_v(-K_c\chi_y + \chi_m) \tag{9d}$$

The matrix set (given in Appendix A) $\{\Gamma_d,\ \Gamma_y,\ \Delta_d,\ \Delta_y\}$ is determined by the sets $\{\kappa_1,\ldots,\kappa_m\}$ and $\{\nu_1,\ldots,\nu_p\}$ of Definition 1, and K_d, K_y are gain matrices in parametrized form (Alvarez, 1996):

$$K_d(s_d) = bd[(s_d k_{11}^d,..,s_d^{\nu_1}k_{\nu_1 1}^d),..,(s_d k_{1p}^d,\ldots,s_d^{\nu_p}k_{\nu_i p}^d)] \tag{10a}$$

$$K_y(s_y) = bd[(s_y k_{11}^y,..,s_y^{\kappa_1}k_{\kappa_1 1}^y),..,(s_y k_{1m}^y,\ldots,s_y^{\kappa_m}k_{\kappa_i m}^y)] \tag{10b}$$

with entries of $[K_d(1),\ K_y(1)]$ that set stable the reference LNPA output error dynamics $(1\le i\le m,\ p)$

$$\sigma_i^{(\nu_i)} + k_{1i}^u\sigma_i^{(\nu_i-1)}+..+k_{\nu_i i}^u\sigma_i = 0,\ \sigma_i = \hat{d}_i - d_i \tag{11a}$$

$$\upsilon_i^{(\kappa_i)} + k_{1i}^o\upsilon_i^{(\kappa_i-1)}+..+k_{\kappa_i i}^o\upsilon_i = 0,\ \upsilon_i = \hat{y}_i - y_i \tag{11b}$$

Summarizing, the candidate controller (9) has the linear form specified in (2), contains feedforward and feedback paths and has integral action to compensate for persistent model parameter errors.

4. ROBUST CLOSED-LOOP STABILITY

In the next theorem are given sufficient conditions to ensure the robust stability of the nonlinear plant (1) with the candidate linear controller (9). To state the theorem, let us introduce some definitions. The steady-state $\bar{x}$ of the CL nonlinear plant $\dot{x} = f_c[x,d(t),b]$ is RE (robustly exponentially)-stable if about $(\bar{x},\bar{\mathscr{d}},b)$ with $0 = f_c[\bar{x},\bar{d}(t),b]$, there are positive constants A_x, L_x, B_d and B_b such that the motions x(t) are bounded as follows

$$\|x(t)-\bar{x}\| \le A_x e^{-L_x(t-t_0)}\|x_0-\bar{x}\| + B_d\|d(t)-\bar{d}\|^s + B_b\|\beta - b\|$$

$[\|d(t)\| = \sup_{t\ge0}\|\alpha(t)\|]$. Let the plant meet, with Rx and Ru invertibility, the conditions (i)-(iii) of Definition 1, and introduce the restricted dynamics

$$\dot{x} = f[x,\bar{u},\bar{d},b], \qquad x\in\Sigma(t) \tag{12}$$

In virtue of the parametrizations (7,10) of the filter $(K_d,\ K_y)$ and controller $(K_c,\ K_I)$ gains, the reference gain set $[K_d,K_y,K_c,K_I](1)$ determine the $s_{d|y|c}$-independent amplitude-decay pair set $[(a_d,\lambda_d),\ (a_y,\lambda_y),\ (s_c,\lambda_c)]$ associated to the reference dynamics (8,11) (Alvarez, 1996). Let (A_v,L_{II}) be the amplitude-decay pair of the E-stability property of the restricted dynamics (12). Introduce the "instability" margins m_c, n_s and n_r

$$m_c = \|\bar{\varphi}_u\bar{\mu}_x\|,\ n_s = a_y(\|\bar{\varphi}_x\bar{\phi}_x^{-1}\| + \|\bar{\mu}_v K_c\|)$$

$$n_r = a_0\frac{a_r}{l_{II}}c_\phi\|\bar{\varphi}_x\bar{\phi}_x^{-1}\|$$
$$\|\bar{\mu}_v K_c\|^2\|(a_c\|\bar{\varphi}_u\|,a_{II}\|w_u\|)'\|$$

$c_\phi = \|\bar{\phi}_x\|\ \|\bar{\phi}_x^{-1}\|$, $w = \phi_{IIx}f + p_{IIx_d}(\Gamma_d x_d + \Pi v_d)$

$(a_r,l_{II}) = \lim\,(A_r,L_{II})$ as $\|x-\bar{x}\|$, $\|\beta-b\|$, $\|d(t)-\bar{d}\| \to 0$.

In the next theorem, the conditions for the RE-stability of the CL plant are given. The sketch of the proof of the theorem is given in Appendix B.

Theorem 1. Let the plant meet, with Rx and Ru-invertibility for ϕ and φ, respctively, the conditions (i)-(iii) of Definition 1; and let the (possibly open-loop unstable) nominal state $\bar{x}$ be a RE-stable steady-state of the restricted dynamics (12). Then, the linear controller (9) yields a CL plant with $\bar{x}$ as a RE-stable point and with quasi-LNPA output error dynamics if:

i) The parameter s_c of the controller (9c,d), is chosen so that the instability margin m_c is dominated,

$$s_c\lambda_c > m_c$$

ii) The parameter s_y of the y-filter (9b,10b) is tuned sufficiently large (fast) such that the self-destabilization (n_s) and the filter-control resonant (n_r) instability margins are dominated,

$$s_y\lambda_y > n_s+n_r$$

iii) The parameter s_d, of the d-filter (9a,10a), is tuned faster than the y-filter,

$$s_d\lambda_d > s_y\lambda_y - n_s. \qquad \blacklozenge$$

5. APPLICATION EXAMPLE

Consider a 15 tray binary distillation column, with one feed stream, total condensation and no lateral exit streams. The feed is a binary ethanol-water mixture, and the column operates at a pressure of 1 BAR. The nominal distillate composition is 84% mol (i.e., 97% of the highest attainable purity). The two point control problem consists in regulating simultaneously the top and bottom compositions, by manipulating two inputs (say vapor and reflux ratios). It is well known that the two-point control of azeotropic mixtures is a difficult problem because of the nonlinear input-output interaction of the column. The model is given by the following set of equations:

$$\dot{c}_B=[L_1(c_1-c_B)+V(c_B-\omega(c_B))]m_B^{-1},\ \dot{m}_1=L_2-L_1$$
$$\dot{c}_i=[L_{i+1}(c_{i+1}-c_i)+V(\omega(c_{i+1})-\omega(c_i))]m_i^{-1},\ \dot{m}_i=L_{i+1}-L_i$$
$$\dot{c}_4=[L_5(c_5-c_4)+F_A(c_A-c_4)+V(\omega(c_3)-\omega(c_4))]m_4^{-1}$$
$$\dot{m}_4=L_5-L_4+F_A$$
$$\dot{c}_{15}=[R/(R+1)(\omega(c_{15})-c_{15})+V(\omega(c_{14})-\omega(c_{15}))]m_{15}^{-1}$$
$$\dot{m}_{15}=RV/(R+1)-L_{15}$$

The states of the column are: c_B (mol fraction of the volatile component in the bottom), c_i (mol concentration in the i-th tray), and m_i (molar holdup in the i-th tray). The control imputs are V (vapor flowrate, or equivalently, heat load rate) and R (reflux ratio); ω is the liquid-vapor equilibrium expression. In compact notation, the column can be written in the form (1), where:

$$x=[c_B,\ c_1,\dots,c_{15},m_1,\dots,m_{15}]',\ y_1=T(c_1),$$
$$y_2=T(c_8),\ u_1=V,\ u_2=R,\ d_1=F_A,\ d_2=c_A$$

y_1 and y_2 are the temperatures of trays 1 and 8, respectively. F_A is the molar feedrate and c_A its composition. The modeling considerations, the model parameters, and the nominal operating conditions are given in Castellanos (1997). The conditions of Theorem 1 are met with the maps

$$\phi_I(x)=[y_1,y_2,L_f y_2,\dots,L_f^7 y_2]',\ (\kappa_1,\kappa_2,\nu_1,\nu_2)=(1,7,4,4)$$
$$\phi_{II}(x)=[c_B,c_2,\dots,c_7,m_1,\dots,m_{15}]'$$

The restricted-dynamics (12) are RE-stable, and their state is $x_R=[c_B,c_2,\dots,c_7,m_1,\dots,m_{15}]'$. With numerical simulations, the Rx and Ru-invertibilities of the maps ϕ and φ were verified. The reference gains of the d-filter (9a), of the y-filter (9b), and of the PI-controller (9c,d) were set according to the pole-placement given in Alvarez and López (1999), with damping factor ξ = 0.9 and the characteristic frequency ω = (1/9) hrs^{-1}. The acceleration/retardation parameters were (s_d,s_y,s_c) = (30,30,3), meaning that the filters were ten times faster than the controller, which in turn was three times faster than the restricted (uncontrolled) dynamics. In figure 1 is presented the open-loop response of the column to a +10% step change in the feedrate F_A, showing that the bottom (c_B) and distillate (c_D) compositions had significantly different static gains and settling times. In general, this step responses are not symmetric with respect to opposite changes in the input. In the same figure are shown the responses of the temperature-driven linear dynamic controller (9) showing that the controller is able to regulate both the bottom and distillate compositions; with settling times that are (as expected) about three times faster than the one of the open-loop column, and with an almost decoupled output dynamics.

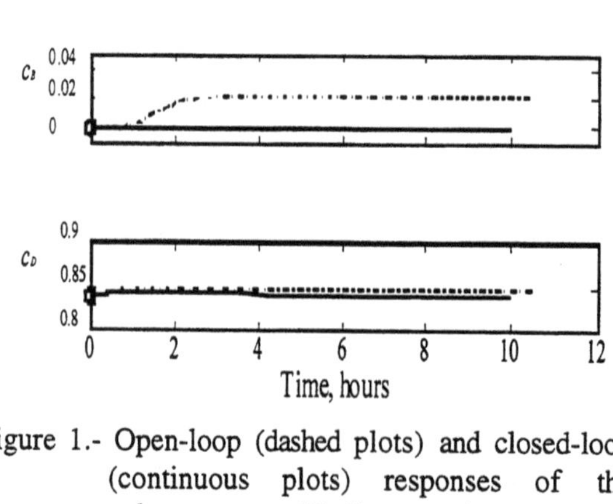

Figure 1.- Open-loop (dashed plots) and closed-loop (continuous plots) responses of the column to a +10 % step change in the feedrate.

6. CONCLUSIONS

A linear dynamic measurement-driven feedforward-feedback control design for MIMO nonlinear plants has been presented. The design has a RE-stability condition and a systematic construction-tuning procedure. The linear output-feedback controller is an approximation to a nonlinear SF-controller, inheriting the decoupling-with stability property of the nonlinear controller. The proposed controller handled successfully the strong interaction problem of a binary azeotropic distillation column.

REFERENCES

Alvarez, J., R. Suarez and A. Sanchez (1994). Semiglobal Nonlinear Control Based on Complete Input-Output Linearization and its Application to the Start-up of a Continuous Polymerization Reactor. *Chem. Eng. Sci.*, **49**(21), 3617.

Alvarez, J., (1996). Output-Feedback Control of Nonlinear Plants. *AIChEJ*, **42**(9), 2540.

Alvarez, J. (1997). A Robust State Estimator Design for Nonlinear Plants. *Proc. ACC*, **5**, 3058.

Alvarez, J. and T. Lopez (1999). Robust Dynamic State Estimation of Nonlinear Plants. To appear in *AIChEJ*.

Castellanos, E. (1997). *Nonlinear Geometric Control of Binary Distillation Columns by Measurement Feedback.* M. Sc. Thesis (in Spanish), Univ. de Guadalajara.

Daoutidis, P. and C. Kravaris. (1994). Dynamic Output Feedback Control of Minimum Phase Multivariable Nonlinear Processes. *Chem. Eng. Sci.* **49**(4), 433.

Isidori, A. (1995). *Nonlinear Control Systems.* Springer-Verlag, New York.

Khalil, H. (1996). *Nonlinear Systems*, Prentice Hall, New York.

Kravaris, C., and C. Chung, (1987). Nonlinear State-Feedback Synthesis By Global Input-Output Linearization. *AIChEJ.* **33,** 73.

Padilla, S., J. Alvarez (1997). Control of Continuous Copolymer Reactors. *AIChEJ.* **43**(2), 448.

Soroush, M. and C. Kravaris (1993). Multivariable Nonlinear Control of a Continuous Polymerization Reactor: An Experimental Study. *AIChEJ.* **39**(12), 1920.

APPENDIX A

bd M := a block-diagonal matrix M

$\Gamma_{d/y} = bd[\Gamma_1^{d/y},\ldots,\Gamma_{nd/m}^{d/y}]$, $\Pi_{d/y} = bd[\pi_1^{d/y},\ldots,\pi_{nd/m}^{d/y}]$

$\Delta_{d/y} = bd[\delta_1^{d/y},\ldots, \delta_{nd/m}^{d/y}], \delta_i^{d/y}{}_{(1x\nu_i)/(1x\kappa_i)} = [1,0,\cdot\cdot,0]$

$A_{d/y} = \Gamma_{d/y} - K_{d/y}\,\Delta_{d/y}$, $A_c^e = \Gamma_c^e - K_c^e \Delta_c^e$

$$\Gamma_i^{d/y}{}_{(\nu_i x\nu_i)/(\kappa_i x\kappa_i)} = \begin{bmatrix} 010\cdot\cdot0 \\ \vdots \quad\quad \vdots \\ 0\cdots01 \\ 0\cdots\cdots0 \end{bmatrix}, \ \pi_i^{d/y}{}_{(\nu_i x1)/(\nu_i x1)} = \begin{bmatrix} 0 \\ \vdots \\ 0 \\ 1 \end{bmatrix}$$

$$\Gamma_c^e = bd[\Gamma_1^e,\ldots, \Gamma_m^e], \qquad \Delta_c^e = bd[\delta_1^e,\ldots, \delta_m^e]$$

$$\Pi_c^e = bd[\pi_1^e,\ldots, \pi_m^e]; \qquad \Gamma_i^e = \Gamma_i^o{}_{(\kappa_i+1)x(\kappa_i+1)},$$

$$\delta_i^e = (\delta_i^o, 0), \qquad \pi_i^e = (\pi_i^{o\prime}, 0)'$$

$$K_c^e = bd[(s_c^{\kappa_1} k_{11}^c,..,s_c k_{1\kappa_1}^c, s_c^{\kappa_1+1} k_1^I)',..,$$
$$(s_c^{\kappa_m} k_{m1}^c,..,s_c k_{m\kappa_m}^c, s_c^{\kappa_m+1} k \ )']$$

$$T[z_I', z_m']' = [z_1^I, \ldots, z_{\kappa_1}^I, z_q^1;\ldots ;z_{\kappa_m-1}^I,\ldots,z_\kappa^I, z_m^q]';$$

$$z_I = [z_1^I,\ldots,z_{\kappa_1}^I ;\ldots;z_{\kappa_m-1}^I,\ldots,z_\kappa^I]', \ z_q = [z_q^1,\ldots,z_m^q]'$$

APPENDIX B

Here is sketched the proof of Theorem 1. Apply, to the CL plant (…,…) the error coordinate change

$$e_d = \hat{d} - d, \ e_y = \hat{y} - y, \ z_I = y - \bar{y}, \ z_{II} = \phi_{II}(x,d) - \phi_{II}(\bar{x},\bar{d})$$

$$z_m = K_I \int_{t_0}^{t} (y-\bar{y})d\tau; \ y = \phi_I(x,d) - \phi_I(\bar{x},\bar{d}), \ e_b = \beta\text{-}b$$

and obtain the filtering-regulation error dynamics

$$\dot{e}_d = A_d e_d - \Pi_d v_d(t), \ \dot{e}_y = A_o e_y - \Pi_o \theta_I$$

$$\dot{z}_e = A_o^e z_e + \Pi_o^e q_I, \ \dot{z}_{II} = w(z_{II}) + q_{II}, \ z_e = T[z_I', z_m']'$$

where A_d, A_o and A_o^e are stable matrices, and $\dot{z}_{II} = w(z_{II})$ is the RE-stable restricted dynamics (12). $\|v_d(t)\| \le \varepsilon_d$. T is a permutation of the identity matrix. θ_I, q_I and q_{II} are L-bounded disturbances,

$$\|\theta_I\| \le R_d\|z_d\| + R_e\|z_e\| + R_{II}\|z_{II}\| + R_{ed}\|e_d\| + R_y\|e_y\| + R_b\|e_r\|$$

$$\|q_I\| \le M_d\|z_d\| + M_e\|z_e\| + M_{II}\|z_{II}\| + M_{ed}\|e_d\| + M_y\|e_y\|$$
$$+ M_b\|e_r\|$$

$$\|q_{II}\| \le N_d\|z_d\| + N_e\|z_e\| + N_{ed}\|e_d\| + N_y\|e_y\| + N_b\|e_r\|$$

In a way that parallels the stability proofs in nonlinear estimation (Alvarez, 1999) and in nonlinear estimator-based controller (Alvarez, 1996), one obtains the inequalities of statements (i)-(ii) of Theorem 1. This conclusion follows from the application of (Khalil, 1992) Gronwall's Lemma, Lyapunov's converse Theorem, and the Comparison Principle, in order to mayorize the norms of the error motions with the scalar motions of a linear system. Then Hurwitz' stablity criterion is applied.

E-2c-12-4
14th Triennial World Congress, Beijing, P.R. China

OPTIMIZATION BASED NONLINEAR OBSERVERS REVISITED

Mazen Alamir *

* *Laboratoire d'Automatique de Grenoble, B.P. 46, 38402, Saint Martin d'Hères, France*
mazen.alamir@inpg.fr

Abstract: In this paper, we present some new theoretical results concerning Optimization Based Observers. Despite the title we use, our observer is not based on the search for a global minimum of a cost function. It is based on a descent-like approach without the use of the Hessian matrix. A nice feature is that the resulting observer takes a classical form in the sense that it is defined by a set of ordinary differential equations.
Beside the theoretical results, some implementation issues are discussed. In particular, we propose to use the post stabilization approach, well known when dealing with differential systems with invariant submanifolds. This enables to increase the sampling period while keeping a very good precision. An example is presented to illustrate the efficiency of the proposed method. *Copyright © 1999 IFAC*

Keywords: Nonlinear observer, post stabilization, detectability, gradient, output prediction error, invariant submanifold.

1 Introduction

Optimization Based Nonlinear Observers (see (Zimmer, 1994; Michalska and Mayne, 1995)) are those where the integrated output prediction error is used to define the dynamic of the observer's state. More precisely, in these approaches, a cost function is defined as the L_2-norm of the difference between the measured output and the output predicted based on the current estimated state. In (Michalska and Mayne, 1995), a nonlinear optimization problem has to be solved at each sampling time, this may present some serious drawbacks in the sense that one may not be able to control the time needed to achieve a step computation with a beforehand given accuracy. In (Zimmer, 1994), the Hessian matrix of the above mentioned cost function has to be computed in order to perform the Newton step, this may also be difficult to do in a real time context or impossible if the underlying functions are not twice differentiable.

In this paper, we are interested in optimization based nonlinear observers. These are observers whose dynamic is defined in order to dynamically decrease the L_2-norm of the **Output Prediction Error (OPE)**. The main attraction of such observers is that they are generically independent of any particular structure. Moreover, no coordinate transformations are needed to design the observer nor a special system dependent study.

Another consequence that follows from the use of the L_2 norm of the **OPE** is the possibility to relate the observation problem to that of **stabilizing invariant submanifolds** when integrating differential systems. This is because the 0-L_2 norm of the **OPE** can clearly defines an invariant submanifolds w.r.t joint system/observer set of differential equations. This makes possible the use of existing techniques to the nonlinear observer problem.

This paper is organized as follows : First, some

definitions and notations are introduced. These definitions concern the class of systems under consideration, the notion of **observability** and **detectability** and some technical notations used throughout the paper.
The main theoretical results are presented in section 3, namely the design of an observer based on the L_2 norm of the **OPE** without the use of the Hessian matrix (theorem 1).
The observer so obtained while theoretically irreproachable under the associated assumptions may present some numerical drawbacks in practice. A **regularized version** is therefore proposed that leads to an approximate observer (theorem 2).
The proposed observer, while simple to implement may correspond to a relatively high computational cost. Some implementation issues are discussed in section 4. Namely, **the post stabilization technique** is applied to the observation problem in order to derive a good "*precision/(computation cost)*" ratio. This is the aim of theorem 3. Finally, an illustrative example is proposed in section 5.
Due to the paper's size limitation, all proofs are dropped. Readers that are interested may consult (Alamir, 1998).

2 Definitions and Notations

2.1 System related definitions

We consider nonlinear systems given by :

$$\dot{x} = f(x) \quad ; \quad y = h(x) \qquad (1)$$

where $x \in \Re^n$ is the state, $y \in \Re^p$ is the measured output. The functions f and h are supposed to be **continuously differentiable.**
The following notations are used throughout the paper :

- $X(t;0;x_0)$: the solution at t of (1) with the initial condition $x(0) = x_0$,
- $Y(t;0;x_0) = h(X(t;0;x_0))$,
- For $\varphi : \Re^+ \to \Re^+$ we shall write $\varphi \in \mathcal{K}$ to designate the following properties :
  - · $\varphi(.)$ is continuous and strictly increasing,
  - · $\varphi(0) = 0$.
- For all piece-wise continuous function $g(.)$ defined over the time interval I, we shall use the following notations :

$$\| g(.) \|_{L_i}^{I} := \int_I \| g(\tau) \|^i \, d\tau \qquad (2)$$

$$\| g(.) \|_{\infty}^{I} := \sup_{t \in I} \| g(t) \| \qquad (3)$$

- For all matrix A, $\underline{\sigma}(A)$ stands for the smallest singular value of A.

2.2 Observability related definitions

Several definitions of observability for nonlinear systems have been considered (Hermann and Krener, 1977; Isidori, 1989; Fossard and Cyrot, 1995). The most commonly used is based on the notion of indistinguishable pair of states to derive a sufficient rank condition for local weak observability.
Such definitions, when expressing the observability rank constraint may introduce some restrictive regularity conditions that are far from being necessary. In many cases the higher derivatives involved in the expression of such structural observability conditions are used either in the design of the observer dynamic or in the design of the coordinate transformation that brings the system to some observable "*canonical form*".
In this work, we try to avoid as far as possible the use of such "*higher derivation*"- expressions.

For linear systems, the possibility of designing an observer depends on the detectability of the pair (A, C) such that the complete observability is not necessary. For nonlinear systems, we can do the same. Indeed, we can use the following definition :

definition 1. [Definition of detectable systems]
The system (1) is said to be detectable if there is a partition $x = \begin{pmatrix} x_I \\ x_{II} \end{pmatrix}$ of the state space such that for all $T > 0$, the following conditions hold :

(1) The L_2-norm of the output prediction error gives a good indication on the estimation error of x_I in the following sense,

$$\begin{aligned} &\exists \varphi_1 \in \mathcal{K} \quad \text{s.t} \quad \forall x^{(1)}, x^{(2)} \in \Re^n \\ &\| x_I^{(1)} - x_I^{(2)} \| \leq \\ &\varphi_1 \Big(\| Y(.;0;x^{(1)}) - Y(.;0;x^{(2)}) \|_{L_2}^{[0,T]} \Big) \end{aligned} \qquad (4)$$

(2) The substate x_{II} has no effect on the output, namely :

$$\forall \tau \quad \forall x \qquad \frac{\partial Y}{\partial x_{II}} \Big(\tau; 0; x \Big) . f_{II}(x) = 0 \quad (5)$$

(3) For any state evolutions $x^{(1)}(t)$ and $x^{(2)}(t)$ satisfying :

$$\| x_I^{(1)}(t) - x_I^{(2)}(t) \| \to 0$$

we have

$$\| X_{II}(t;0;x^{(1)}) - X_{II}(t;0;x^{(2)}) \| \to 0 \quad (6)$$

♠

Let us anticipate by emphasizing the fact that only the existence of the partition mentioned in

definition 1 is needed. The observer design does not use such partition in its formulation.
Recent definition of detectability has been proposed in (Sontag and Wang, 1997). A discussion concerning the connection between the two definitions can be found in (Alamir, 1998).

As for observability, we shall use the following definition

definition 2. [Definition of a completely observable system]

The system (1) is said to be completely observable if it is detectable in the sense of definition 1 with $x_I = x$.

2.3 Observer related definitions

Condition (4) suggests that an observer has to deal with the following key quantity :

$$J(z(t), x(t-T)) = \| Y(.;0;z(t)) - Y(.;0;x(t-T)) \|_{L_2}^{[0,T]} \quad (7)$$

where $z(t)$ is the current estimation of $x(t-T)$. Indeed, this quantity represents the output prediction error associated to the estimation $z(t)$ of $x(t-T)$. The observer design paradigm is therefore the definition of a dynamic evolution $\dot{z}(t)$ that :

(1) decreases $J(z(t), x(t-T))$ when it is > 0,
(2) makes $\Big\{(z(t), x(t-T)) \;\; | \;\; J(z(t), x(t-T)) = 0\Big\}$ invariant with respect to the combined dynamics of $x(t-T)$ and $z(t)$.

If these conditions hold then the detectability property guarantees the convergence of the estimation error. The current estimation of $x(t)$ is naturally given by :

$$\hat{x}(t) = X(T;0;z(t)) \quad (8)$$

We shall denote $G(z(t), x(t-T))$ the gradient of J with respect to $z(t)$:

$$G(z(t), x(t-T)) = \frac{\partial J}{\partial z}(z(t), x(t-T)) \quad (9)$$

which is therefore a continuous function.
In what follows, the state $x(t-T)$ is frequently omitted as an argument of J and G. More precisely, we shall sometimes write $J(z(t))$ and $G(z(t))$ instead of $J(z(t), x(t-T))$ and $G(z(t), x(t-T))$ respectively when no confusion is possible.

3 The Main theoretical results

In this section, the main theoretical results of this paper are presented. First, we give the assumptions under which these results can be obtained.

3.1 Assumptions

First, recall that throughout the paper, the system (1) is supposed to be detectable in the sense of definition 1. Beside this, we need some regularity condition on the Jacobian matrix $G(z(t), x(t-T))$ defined by (9) :

Assumption 1. [Uniform regularity of G]

There is a function $\sigma \in \mathcal{K}$ such that :

$$\underline{\sigma}\Big(G(z,x)\Big) \geq \sigma\Big(\| z_I - x_I \|\Big) \quad (10)$$

where z_I and x_I are the substates defined by the partition associated to the detectability assumption (see definition 1) ♠

In other words, the Jacobian matrix $G(z, x(t-T))$ is of rank 1 "*uniformly*" far from the diagonal $z_I(t) = x_I(t-T)$. This is a high level assumption that may be difficult to examine. It is worth noting that assumption 1 is the key feature that makes the whole thing work. Roughly speaking, it states that when $x(t)$ moves, for all $t \geq 0$, the cost functions $J(., x(t))$ has one global minimum. This is clearly a quite restrictive condition. However, such restrictive condition does always exist in all works that address the design of globally convergent nonlinear observers. (For instance, the high gain observer need some diffeomorphism and some triangular structure that must be globally defined).

Fortunately, assumption 1, while quite restrictive in the large can be locally related to some detectability assumption on the linearized system as it is shown by the following lemma :

lemma 1. [Local significance of assumption 1 in the case of C^2 systems]

Suppose that f and h of (1) are twice continuously differentiable.
Suppose also that for all $x \in \Re^n$, the linearized system :

$$\dot{\xi} = \frac{\partial f}{\partial x}\Big(X(t;0;x)\Big)\xi \quad ; \quad v = \frac{\partial h}{\partial x}\Big(X(t;0;x)\Big)\xi \quad (11)$$

is detectable,
then assumption 1 is locally satisfied. ♠

The second assumption is a **technical requirement** on the existence of some bounding function. namely, consider the function :

$$\Psi(\tau, z) := \frac{\partial Y}{\partial z}\Big(\tau; 0; z\Big) f(z) \qquad (12)$$

Since f and h are supposed to be continuously differentiable, we know that $\Psi(\tau, z)$ is continuous with respect to z and differentiable with respect to τ. The following assumption concerns the existence of a bounding function for $\| \Psi(\tau, x) - \Psi(\tau, z) \|$. Note that according to (5), we already know that

$$\| \Psi(\tau, x) - \Psi(\tau, z) \| = 0 \quad \text{whenever } x_I = z_I$$

So, let us consider the following assumption :

Assumption 2.

There are two continuous functions $U_1, U_2 : \Re^+ \to \Re^+$ with $U_2 \in \mathcal{K}$ such that :

$$\| \Psi(., x) - \Psi(., z) \|_\infty^{[0,T]} \leq U_1(\| z \|).U_2(\| z_I - x_I \|) \qquad (13)$$

♠

Note that assumption 2 generalizes the classical Lipschitz condition.

3.2 Main theoretical results

We have the following main result :

Theorem 1. [The observer design]

Suppose that :

(1) System (1) is detectable [resp. completely observable] in the sense of definition 1,
(2) Assumptions 1-2 are satisfied,

Let a positive real $\alpha > 0$ and two functions $\gamma(.)$ and $\Gamma(.)$ be such that for all $r > 0$,

$$\gamma(r) \geq U_1(r) + \alpha \quad ; \quad \Gamma(r) \geq U_2 \circ \varphi_1(r) \qquad (14)$$

where φ_1, U_1 and U_2 are those defined in assumption 1 and assumption 2 respectively.

then the dynamic system defined by :

$$\dot{z} = f(z) - 2\gamma(\| z \|) G^T \Big[GG^T\Big]^{-1} \Gamma(J)\sqrt{J} \qquad (15)$$

$$\hat{x} = X(T; 0; z) \qquad (16)$$

where $J(t) = J(z(t), x(t-T))$ and $G(t) = G(z(t), x(t-T))$,

is an asymptotic observer for (1). ♠

Note that the observer dynamic has a quite classical form being the summation of the system dynamic with an **innovation process**. The difference is that this later process is not based on the instantaneous output prediction error, but on the cost over a time interval in the past of this output prediction error.

Corollary 1. [A high gain observer as a particular case]

Suppose that :

(1) System (1) is detectable in the sense of definition 1,
(2) Assumption 1 is satisfied,
(3) There is a bounded domain $\mathcal{D}$ containing all possible evolutions of z and x,

Then there is sufficiently high gain γ such that the dynamic system defined by :

$$\dot{z} = f(z) - \gamma G^T \Big[GG^T\Big]^{-1} \sqrt{J} \qquad (17)$$

$$\hat{x} = X(T; 0; z(t)) \qquad (18)$$

where $J(t) = J(z(t), x(t-T))$ and $G(t) = G(z(t), x(t-T))$,

is an asymptotic observer for (1). ♠

The problem with the observer equations as proposed by theorem 1 is that they may lead to numerical instability when approaching the solution $J = 0$. Indeed, we know that at $J = 0$, $G = 0$ and therefore, the unavoidable truncation errors corresponding to computation sampling time may lead to situations where $J \approx 0$ (but $J \neq 0$) while $G = 0$. This justifies the following result

Theorem 2. [Approximate observer with regularized dynamic]

Choose an admissible bound $\eta > 0$ on the estimation error,
Suppose that :

(1) System (1) is completely observable in the sense of definition 1,
(2) Assumptions 1-2 are satisfied,

Let two positive reals $\alpha > 0$, $\beta > 0$ and two functions $\gamma(.)$ and $\Gamma(.)$ be such that for all $r > 0$,

$$\gamma(r) \geq (1 + \beta)U_1(r) + \alpha; \Gamma(r) \geq U_2 \circ \varphi_1(r) \qquad (19)$$

where φ_1, U_1 and U_2 are those defined in assumption 1 and assumption 2 respectively.

then, there exists sufficiently small $\epsilon > 0$ such that the dynamic system defined by :

$$\dot{z} = f(z) - 2\gamma(\| z \|) G^T [GG^T + \epsilon]^{-1} \Gamma(J)\sqrt{J} \qquad (20)$$

$$\hat{x} = X(t; 0; z) \quad (21)$$

where $J(t) = J(z(t), x(t-T))$ and $G(t) = G(z(t), x(t-T))$, is an approximate observer for (1) in the sense that :

(1) It reaches the set

$$E_\eta := \left\{ (\hat{x}, x) \in \Re^{2n} \quad | \quad \| \hat{x} - x \| < \eta \right\} \quad (22)$$

after a finite time

(2) It makes (22) invariant with respect to the system/observer dynamics

♠

4 Implementation issue : the post stabilization approach

In this section, we shall argue on the high gain observer (17) in order to simplify the presentation. The same ideas can be applied to the observer (15) of theorem 1 with straightforward technical modifications.

The on-line integration of the observer equations :

$$\dot{z} = f(z) - \gamma G^T(z) \left[GG^T(z) \right]^{-1} \sqrt{J(z)} \quad (23)$$

$$\hat{x} = X(T; 0; z(t)) \quad (24)$$

may be quite difficult because of the relative heavy cost associated to the gradient computation. Here, we propose to use the post stabilization technique in order to alleviate the computations while keeping a good final precision.

First, let us define some technical notations :

- $k > 0$ is a given sampling period,
- $t_1, t_2, \dots, t_n, \dots$ the sampling instants, namely $t_j = jk$,
- z_n is the state of the observer at instant $t_n = nk$. The estimated state is therefore given by :

$$\hat{x}(t_n + t) = X(t; 0; z_n) \quad (25)$$

- $z(t_{n+1}, t_n, z_n)$: Exact solution at instant $t = t_{n+1}$ of (23) with $z(t_n) = z_n$ as initial condition,
- $\hat{z}(t_{n+1}, t_n, z_n)$: solution at $t = t_{n+1}$ of the following differential system :

$$\dot{z}(t) = f(z(t)) - \gamma G^T(t_n) \left[G(t_n) G^T(t_n) \right]^{-1} \sqrt{J(t_n)}$$

$$t_n \le t < t_{n+1} \quad (26)$$

with initial condition $z(t_n) = z_n$. Note that in (26), $J(t_n)$ stands for $J(z_n, x(t_n - T))$ and $G(t_n)$ for $G(z_n, x(t_n - T))$

Note that (26) clearly represents an alleviated version of (23) which is more suitable for on-line computation. (The heavily computable innovation term is constant during a sampling period)

- $\tilde{z}^p_{n+1}$: The numerical solution at $t = t_{n+1}$ of (26) using an integration scheme of order p,
- The post-stabilized alleviated version of (23) is therefore given by the following equation in which z_{n+1} is the new state of the observer at $t = t_{n+1}$:

$$z_{n+1} = \tilde{z}^p_{n+1} - \tilde{G}^T(t_{n+1}) \left[\tilde{G}(t_{n+1}) \tilde{G}^T(t_{n+1}) \right]^{-1} \tilde{J}(t_{n+1}) \quad (27)$$

where $\tilde{J}(t_{n+1}) = J(\tilde{z}^p_{n+1}, x(t_{n+1} - T))$ and $\tilde{G}(t_{n+1}) = G(\tilde{z}^p_{n+1}, x(t_{n+1} - T))$.

Namely, the evolution of the observer state z_n is obtained as follows :

· First, integrate (26) to obtain $\tilde{z}^p_{n+1}$,
· Then, correct $\tilde{z}^p_{n+1}$ by the post stabilizing term :

$$-\tilde{G}^T(t_{n+1}) \left[\tilde{G}(t_{n+1}) \tilde{G}^T(t_{n+1}) \right]^{-1} \tilde{J}(t_{n+1})$$

This technique is used (see for example (Ascher, 1997)) when integrating differential systems over some **invariant submanifold** defined by algebraic equations. The post stabilization enables then to control the distance between the numerical solution and the invariant submanifold despite the truncation error associated to the numerical integration scheme.

The reason why post stabilization strategies are suitable for our problem yields in the fact that the submanifold :

$$\left\{ (z(t), x(t-T)) \in \Re^{2n} \quad | \quad J(z(t), x(t-T)) = 0 \right\}$$

is clearly an invariant submanifold for the differential system :

$$\dot{x}(t-T) = f(x(t-T))$$

$$\dot{z}(t) = f(z(t)) - \gamma G^T(z(t)) \left[GG^T(z(t)) \right]^{-1} \sqrt{J(z(t), x(t-T))}$$

this is a fundamental difference between an innovation process that is based on the integrated output prediction error J and those based on the instantaneous output prediction error $h(\hat{x}) - y$.

Here, the post stabilization step is used to compensate for two sources of error :

(1) The error introduced by using (26) instead of (23)
(2) The truncation error associated with any numerical scheme.

This last theorem underlines the advantage of the post stabilization step in the definition of the observer equations :

Theorem 3. [The advantage of post stabilization]

Suppose that the assumptions of corollary 1 hold. Suppose further that f and h are twice continuously differentiable, then there is a sufficiently high gain γ such that the post stabilization-based observer defined by (27), namely :

$$z_{n+1} = \tilde{z}^p_{n+1} - \tilde{G}^T(t_{n+1})\left[\tilde{G}(t_{n+1})\tilde{G}^T(t_{n+1})\right]^{-1}\tilde{J}(t_{n+1}) \quad (28)$$

$$\hat{x}(t_n) = X(T; 0; z_n) \quad (29)$$

is such that :

$$\lim_{n \to \infty} J(z_n) = O(k^4) \quad (30)$$

and this for all integration scheme order $p \geq 1$. ♠

5 An illustrative example

We consider the well known Van der Pol system given by :

$$\dot{x}_1 = x_2$$
$$\dot{x}_2 = -9x_1 + \mu(1 - x_1^2)x_2$$
$$y = x_1$$

where we take $\mu = 2$.

Figure 1.(a) presents the behavior of the observer when no post stabilization is performed while figure 1.(b) shows result under the same conditions with post stabilization. In these two figures, the sampling period $k = 0.005$ was taken. This seems too small to underline the effect of the post stabilization. Indeed, the observer works well even without the stabilization step.

In figures 2.(a) and 2.(b) however, a sampling period of $k = 0.05$ is adopted. Figure 2.(a) then shows that without post stabilization step, the observer does not converge. The advantage of post stabilization can therefore be noticed on the observer behavior shown on figure 2.(b).

Finally, figures 3 shows the same comparison with $k = 0.25$.

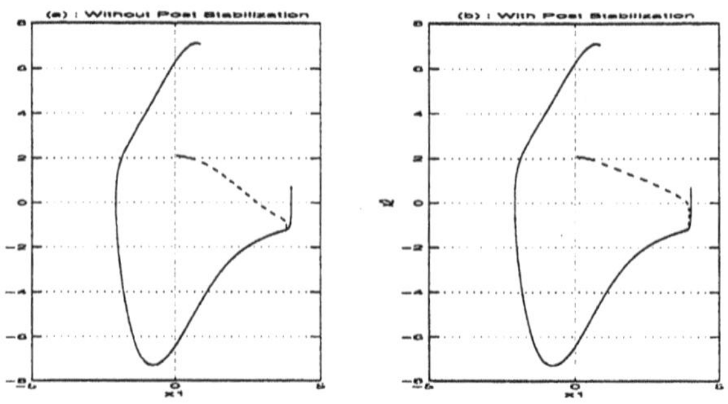

Fig. 1. Simulation with $k = 0.005$, $T = 0.02$ and $\gamma = 10$

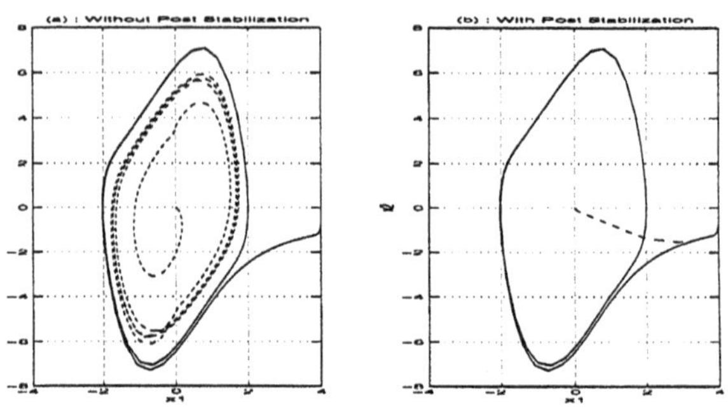

Fig. 2. Simulation with $k = 0.05$, $T = 0.1$ and $\gamma = 0.5$

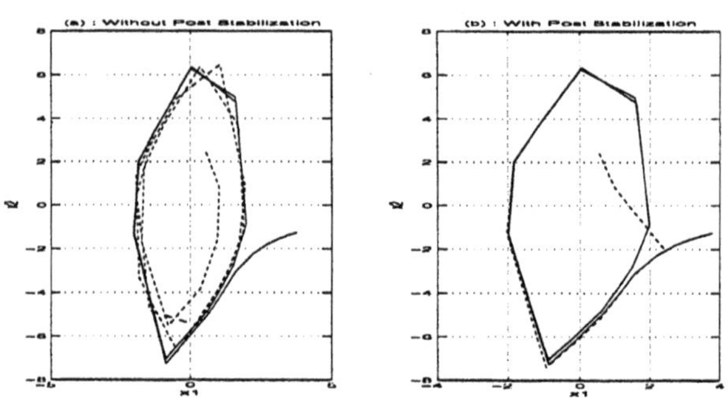

Fig. 3. Simulation with $k = 0.25$, $T = 0.25$ and $\gamma = 0.2$

6 REFERENCES

Alamir, M. (1998). Optimization based nonlinear observers revisited. Inernal report, no 98010. Laboratoire d'Automatique de Grenoble.

Ascher, U. M. (1997). Stabilization of invariants of discretized differential systems. *Numerical Algorithms* **14**, 1–24.

Fossard, A. and D. Normand Cyrot (1995). *Nonlinear Systems.* Chapman & Hall.

Hermann, R. and A. J. Krener (1977). Nonlinear controllability and observability. *IEEE trans. on Automatic Control* **22**, 728–740.

Isidori, A. (1989). *Nonlinear Control Systems.* Springer Verlag.

Michalska, H. and D. Q. Mayne (1995). Moving horizon observers and observer-based control. *IEEE Trans. on Automatic Control* **40**, 995–1006.

Sontag, E. D. and Y. Wang (1997). Output-to-state stability and detectability of nonlinear systems. *Systems & Control Letters* **29**, 279–290.

Zimmer, G. (1994). State observation by on-line minimization. *Int. Journal of Control* **60**, 595–606.

14th Triennial World Congress, Beijing, P.R. China

ADAPTIVE NONLINEAR OBSERVER/CONTROLLER DESIGN FOR UNCERTAIN NONLINEAR SYSTEMS [1]

Yaolong Tan and Ioannis Kanellakopoulos

UCLA Electrical Engineering
Los Angeles, CA 90095-1594, USA
E-mail: yltan@ee.ucla.edu, ioannis@ee.ucla.edu

Abstract. We present a procedure for global adaptive output-feedback stabilization and tracking of a class of nonlinear systems that can be transformed via a global diffeomorphism into systems that are linear up to output injection and linear with respect to the unknown parameters. Compared to the well-known class of *parametric output-feedback systems*, our extended class allows output-dependent nonlinearities to appear not only additively, but also multiplicatively. The procedure consists in the construction of a nonlinear coordinate transformation that allows us to design a nonlinear observer, which is then integrated with a backstepping controller and a parameter estimator to yield a globally stable closed-loop system.

Keywords. Nonlinear control systems, Observers, Adaptive control

1. INTRODUCTION

A great deal of research effort has been directed towards identifying practically interesting classes of nonlinear systems for which global results can be obtained in the presence of unknown parameters and in the absence of full-state measurement.

One representative example is the class of *parametric strict-feedback systems* (Kanellakopoulos *et al.*, 1991; Krstić *et al.*, 1995), which consists of nonlinear systems that can be transformed via a parameter-independent change of coordinates into a "triangular" form. The nonlinearities depend only on state variables that are "fed back" through the system. Under the full-state feedback assumption, one can achieve global stability and performance without any growth restrictions on the nonlinearities and without any information about the unknown constant parameters.

In the absence of full-state feedback, if the nonlinearities depend on the unmeasured variables, the current state of the art is that only *semiglobal* results (Teel and Praly, 1994; Khalil, 1996; Jankovic, 1997) can be obtained: given any compact region of initial conditions, a high-gain controller-observer pair with either a saturating observer or a saturating controller can be designed to guarantee stabilization.

However, when the nonlinearities depend only on the measured states, global results can be obtained. In particular, if only a single output is measured, then global stability and asymptotic tracking are achieved for the class of nonlinear systems which can be transformed via a global diffeomorphism into the *parametric output-feedback form* (Marino and Tomei, 1993; Krstić *et al.*, 1995). In the special case where there are no zero dynamics, namely when the relative degree is equal to the system order, parametric output-feedback systems can be expressed in the following form:

[1] This research was supported in part by NSF under Grant ECS-9502945.

$$\begin{aligned} \dot{x}_1 &= x_2 + \varphi_1(y) + \psi_1(y)^{\mathrm{T}}\theta \\ \dot{x}_2 &= x_3 + \varphi_2(y) + \psi_2(y)^{\mathrm{T}}\theta \\ &\vdots \\ \dot{x}_{n-1} &= x_n + \varphi_{n-1}(y) + \psi_{n-1}(y)^{\mathrm{T}}\theta \\ \dot{x}_n &= u + \varphi_n(y) + \psi_n(y)^{\mathrm{T}}\theta \\ y &= x_1, \end{aligned} \tag{1}$$

where only the output y is measured.

Currently, this is the largest class of nonlinear systems that can be globally stabilized using only output feedback. The case where the nonlinearities depend *linearly* on the unmeasured states was investigated recently in (Freeman and Kokotović, 1996); however, that procedure is based on the restrictive assumption that the unmeasured states are part of an already stabilized subsystem. This result represents an important first step towards solving more general output feedback problems, but only if the pre-stabilization assumption is removed.

In the case of parametric output-feedback systems, there are coordinate-free geometric conditions that are necessary and sufficient for the existence of a global diffeomorphism that transforms a given nonlinear system into the form (1) (Krener and Isidori, 1983; Marino and Tomei, 1993; Krstić *et al.*, 1995). With this equivalence, the nonlinear system can be treated as a linear system with *additive* nonlinearities of the measured output, for which a linear observer with the injection of output nonlinearities can be designed to estimate the unmeasured states. In this paper, we extend these results to the case of a general relative-degree-three nonlinear system in which the output-dependent nonlinearities appear not only additively, but also multiplicatively; as a result, this system does not satisfy the aforementioned geometric conditions, unless additional restrictions are imposed. Hence, our class strictly contains the class of parametric output-feedback systems.

The presentation is organized as follows: In Section 2, we introduce the extended class of nonlinear systems, followed by some examples that illustrate the differences between our class and that of parametric output-feedback systems. In Section 3, the nonlinear observer is designed via a coordinate transformation. The integration of the nonlinear observer into the controller design procedure is presented in Section 4. In Section 5, the closed-loop system stability and asymptotic performance are analyzed. Section 6 summarizes our design methodology and addresses future work.

2. A NEW CLASS OF NONLINEAR SYSTEMS

We consider the following relative-degree-three system whose nonlinearities are smooth functions that depend only on the output, some of which are multiplied by unknown parameters:

$$\begin{aligned} \dot{x}_1 &= \phi_1(y)x_2 + \varphi_1(y) + \psi_1^{\mathrm{T}}(y)\theta \\ \dot{x}_2 &= \phi_2(y)x_3 + \varphi_2(y) + \psi_2^{\mathrm{T}}(y)\theta \\ \dot{x}_3 &= \phi_3(y)\,u + \varphi_3(y) + \psi_3^{\mathrm{T}}(y)\theta \\ y &= x_1\,, \end{aligned} \tag{2}$$

where $x \in \mathbb{R}^3$ is the state, $u \in \mathbb{R}$ is the input, $y \in \mathbb{R}$ is the output, $\phi_i, \varphi_i,\ 1 \le i \le 3$ are smooth nonlinear functions,

$$\psi_i(y) = \begin{bmatrix} \psi_{i,1}(y) \\ \psi_{i,2}(y) \\ \vdots \\ \psi_{i,q}(y) \end{bmatrix} \in \mathbb{R}^q \ ,1 \le i \le 3\,, \tag{3}$$

are vectors of smooth functions, and $\theta = [a_1, \ldots, a_q]^{\mathrm{T}} \in \mathbb{R}^q$ is a vector of unknown constant parameters. The control objective is to track a given reference signal $y_r(t)$, while keeping all the signals in the closed-loop system globally bounded. Only the output y is assumed to be measured.

We have the following assumptions for the system (2):

Assumption 1. $\phi_i(y) \neq 0, 1 \le i \le 3$, for all $y \in \mathbb{R}$.

Assumption 2. The reference signal $y_r(t)$ is at least third-order differentiable and $y_r(t), \dot{y}_r(t), \ddot{y}_r(t), y_r^{(3)}(t)$ are computable, piecewise continuous and bounded.

The first assumption guarantees the local observability of the system (2); it is dictated by the presence of the nonzero smooth nonlinearities $\phi_i(y), 1 \le i \le 3$, which is what makes our class of systems more general than the parametric output-feedback form (1). These nonlinearities are not restricted by any growth conditions, but they are assumed to be always nonzero. The existence of a global parameter-independent diffeomorphism which transforms an nth-order nonlinear system to the system structure (1) requires that the following necessary condition be satisfied (Krstić *et al.*, 1995):

$$[\mathrm{ad}_f^i r,\ \mathrm{ad}_f^{i+1} r] = 0,\ 0 \le i \le n-2\,, \tag{4}$$

where r is the vector field defined by

$$L_r L_f^i h = \begin{cases} 0 & i \le n-2 \\ 1 & i = n-1 \end{cases}, \tag{5}$$

with f the drift vector of the nonlinear system, and $y = h(x)$. The third-order system (2) violates the condition (4), because $[\mathrm{ad}_f r,\ \mathrm{ad}_f^2 r] = 0$ only when both $\phi_1(y)$ and $\phi_2(y)$ are constant. This means that (2) can not be transformed into the parametric output-feedback form via a global diffeomorphism when $\phi_1(y), \phi_2(y)$ are arbitrary smooth nonlinear functions. On the other hand, it is straightforward to see that the class of nonlinear systems that can be transformed into (1) is a strict subset of the class of systems that can be transformed into (2) since we can always set $\phi_1(y) = \phi_2(y) = 1$ to obtain (1) with $n = 3$.

Let us now illustrate the differences between the system (2) and the parametric output-feedback form (1) using a few representative examples.

Example 1. Consider the second-order system

$$\begin{aligned} \dot{x}_1 &= x_2 \\ \dot{x}_2 &= u + \varphi(y)x_2 \\ y &= x_1 , \end{aligned} \tag{6}$$

where $\varphi(y)$ is an arbitrary smooth nonlinear function. This system satisfies the condition $[r, \mathrm{ad}_f r] = 0$. Thus, there exists a parameter-independent change of coordinates which transforms the system (6) to the parametric output-feedback form (1) with $\theta = 0$. One such transformation is

$$\xi_1 = x_1, \quad \xi_2 = x_2 - \int \varphi(y)dy , \tag{7}$$

which transforms the system (6) into

$$\begin{aligned} \dot{\xi}_1 &= \xi_2 + \int \varphi(y)dy \\ \dot{\xi}_2 &= u . \end{aligned} \tag{8}$$

In (Tan and Kanellakopoulos, 1998), we presented an explicit construction of the diffeomorphism that transforms any nth-order system whose nonlinearities depend linearly on the unmeasured output derivative, into the output-feedback form.

Example 2. Let us consider a more complicated nonlinearity such as x_2^2 in the $\dot{x}_2$-equation:

$$\begin{aligned} \dot{x}_1 &= x_2 \\ \dot{x}_2 &= u + \varphi(y)x_2^2 \\ y &= x_1 . \end{aligned} \tag{9}$$

This system does not satisfy the geometric conditions (4) for diffeomorphic equivalence to the parametric output-feedback form (1), since $[r, \mathrm{ad}_f r] = -2\varphi(y)\frac{\partial}{\partial x_2}$ generally is not zero as $\varphi(y) \neq 0$. Therefore, we can not transform it into (1). However we can still transform the above system into (2) via the change of coordinates:

$$\xi_1 = x_1 , \quad \xi_2 = e^{-\int \varphi(y)dy} x_2 . \tag{10}$$

The transformed system is

$$\begin{aligned} \dot{\xi}_1 &= e^{\int \varphi(y)dy} \xi_2 \\ \dot{\xi}_2 &= e^{-\int \varphi(y)dy} u . \end{aligned} \tag{11}$$

As we will see in the following section, it is possible to design a globally stabilizing observer-controller pair for the transformed system (11). To understand the significance of the second-order nonlinearity in (9), let us recall that it has been shown in (Mazenc *et al.*, 1994) that the system

$$\begin{aligned} \dot{x}_1 &= x_2 \\ \dot{x}_2 &= u + x_2^n \\ y &= x_1 , \end{aligned} \tag{12}$$

can not be globally asymptotically stabilized by output feedback for n strictly larger than 2. Therefore, the system (9) contains the highest-order nonlinearity of x_2 that can be globally stabilized by output feedback. It can be transformed into our class of systems (2) but not into the parametric output-feedback form (1).

Example 3. Consider the third-order system

$$\begin{aligned} \dot{x}_1 &= x_2 \\ \dot{x}_2 &= x_3 + \frac{2x_1}{1+x_1^2}x_2^2 \\ \dot{x}_3 &= \left(1+x_1^2\right)u + \frac{2x_1+4x_1^3+6x_1^5}{(1+x_1^2)(1+x_1^4)}x_2x_3 \\ y &= x_1 . \end{aligned} \tag{13}$$

This system does not satisfy the condition (4) because

$$[r, \mathrm{ad}_f r] = 0, \quad [\mathrm{ad}_f r, \mathrm{ad}_f^2 r] \neq 0 .$$

However, it can be transformed into our class of systems in (2) with the following change of coordinates:

$$\xi_1 = x_1, \; \xi_2 = \frac{x_2}{1+x_1^2}, \; \xi_3 = \frac{x_3}{(1+x_1^2)(1+x_1^4)} . \tag{14}$$

The transformed system under the new coordinates is

$$\begin{aligned} \dot{\xi}_1 &= (1+y^2)\xi_2 \\ \dot{\xi}_2 &= (1+y^4)\xi_3 \\ \dot{\xi}_3 &= \frac{1}{1+y^4}u . \end{aligned} \tag{15}$$

As can be seen from these examples, our class of systems strictly contains the class of parametric-output feedback systems. In this paper we present only the simplified case where no zero dynamics are present, but our results can easily be extended to the more general case.

3. COORDINATE TRANSFORMATION AND FILTER DESIGN

A novelty of our output feedback design is the coordinate transformation which generates dynamics that enable us to handle multiplicative output nonlinearities. To start our nonlinear observer design, we define the following coordinate transformation for the system state x_2:

$$\xi_2 = x_2 - \iota_2(y)\,, \tag{16}$$

where ξ_2 is a new coordinate variable and $\iota_2(y)$ is a design function of the output y yet to be determined. This transforms the $\dot{x}_2$-equation in (2) into

$$\begin{aligned}\dot{\xi}_2 &= \dot{x}_2 - \frac{d}{dt}\iota_2(y)\\ &= \phi_2(y)x_3 + \varphi_2(y) + \psi_2^{\mathrm{T}}(y)\theta - \frac{\partial\iota_2}{\partial y}\dot{y}\\ &= -\frac{\partial\iota_2}{\partial y}\phi_1(y)\xi_2 + \phi_2(y)x_3 + \varphi_2(y) - \frac{\partial\iota_2}{\partial y}\varphi_1(y)\\ &\quad -\frac{\partial\iota_2}{\partial y}\phi_1(y)\iota_2(y) + \left[\psi_2(y) - \frac{\partial\iota_2}{\partial y}\psi_1(y)\right]^{\mathrm{T}}\theta\,,\end{aligned} \tag{17}$$

where to obtain the last equality we used $x_2 = \xi_2 + \iota_2(y)$. In order to deal with the nonlinearity $\phi_2(y)x_3$, which would appear in our transformed system, we choose $\iota_2(y)$ to satisfy

$$-\frac{\partial\iota_2}{\partial y}\phi_1(y) = -k_2\phi_2(y)\,, \tag{18}$$

where k_2 is a design constant and will be chosen later. An explicit solution for the above differential equation is

$$\iota_2(y) = \int\frac{k_2\phi_2(y)}{\phi_1(y)}dy\,. \tag{19}$$

By assumption, $\phi_1(y)$ and $\phi_2(y)$ are smooth functions and are nonzero for all $y \in \mathbb{R}$. Therefore, the solution defined in (19) always exists. With the choice of ι_2 in (18), $\dot{\xi}_2$ becomes

$$\begin{aligned}\dot{\xi}_2 =& -k_2\phi_2(y)\xi_2 + \phi_2(y)x_3 + \varphi_2(y) - \frac{k_2\phi_2(y)}{\phi_1(y)}\varphi_1(y)\\ & -k_2\phi_2(y)\iota_2(y) + \left[\psi_2(y) - \frac{k_2\phi_2(y)}{\phi_1(y)}\psi_1(y)\right]^{\mathrm{T}}\theta\,.\end{aligned} \tag{20}$$

Similarly we define the coordinate transformation for the unmeasured state x_3:

$$\xi_3 = x_3 - \iota_3(y)\,, \tag{21}$$

where ι_3 is another design function of the output y to be determined. Its derivative is computed as

$$\begin{aligned}\dot{\xi}_3 =& -\frac{\partial\iota_3}{\partial y}\phi_1(y)\xi_2 + \phi_3(y)u + \varphi_3(y) - \frac{\partial\iota_3}{\partial y}\varphi_1(y)\\ & -\frac{\partial\iota_3}{\partial y}\phi_1(y)\iota_3(y) + \left[\psi_3(y) - \frac{\partial\iota_3}{\partial y}\psi_1(y)\right]^{\mathrm{T}}\theta\,.\end{aligned} \tag{22}$$

With the choice

$$\iota_3(y) = \int\frac{k_3\phi_2(y)}{\phi_1(y)}dy \tag{23}$$

with k_3 a design constant, (22) becomes

$$\begin{aligned}\dot{\xi}_3 =& -k_3\phi_2(y)\xi_2 + \phi_3(y)u + \varphi_3(y) - \frac{k_3\phi_2(y)}{\phi_1(y)}\varphi_1(y)\\ & -k_2\phi_2(y)\iota_3(y) + \left[\psi_3(y) - \frac{k_3\phi_2(y)}{\phi_1(y)}\psi_1(y)\right]^{\mathrm{T}}\theta\,.\end{aligned} \tag{24}$$

The coordinate transformation defined in (21) also transforms (20) into

$$\begin{aligned}\dot{\xi}_2 =& -k_2\phi_2(y)\xi_2 + \phi_2(y)\xi_3 + \varphi_2(y) + \phi_2(y)\iota_3(y)\\ & -\frac{k_2\phi_2(y)}{\phi_1(y)}\varphi_1(y) - k_2\phi_2(y)\iota_2(y)\\ & + \left[\psi_2(y) - \frac{k_2\phi_2(y)}{\phi_1(y)}\psi_1(y)\right]^{\mathrm{T}}\theta\,.\end{aligned} \tag{25}$$

Considering the system dynamics in (25) and (24) under the new coordinate variables ξ_2 and ξ_3, we now construct the following filters:

$$\dot{\nu}_2 = -k_2\phi_2\nu_2 + \phi_2\nu_3 + \varphi_2 + \phi_2\iota_3 - \frac{k_2\phi_2}{\phi_1}\varphi_1 - k_2\phi_2\iota_2 \tag{26}$$

$$\dot{\nu}_3 = -k_3\phi_2\nu_2 + \phi_3 u + \varphi_3 - \frac{k_3\phi_2}{\phi_1}\varphi_1 - k_2\phi_2\iota_3 \tag{27}$$

$$\dot{\Omega}_2^{\mathrm{T}} = -k_2\phi_2\Omega_2^{\mathrm{T}} + \phi_2\Omega_3^{\mathrm{T}} + \left[\psi_2 - \frac{k_2\phi_2}{\phi_1}\psi_1\right]^{\mathrm{T}} \tag{28}$$

$$\dot{\Omega}_3^{\mathrm{T}} = -k_3\phi_2\Omega_2^{\mathrm{T}} + \left[\psi_3 - \frac{k_3\phi_2}{\phi_1}\psi_1\right]^{\mathrm{T}}. \tag{29}$$

If θ were known, we would obtain the estimates for ξ_2 and ξ_3 by

$$\hat{\xi}_2 = \nu_2 + \Omega_2^{\mathrm{T}}\theta\,, \qquad \hat{\xi}_3 = \nu_3 + \Omega_3^{\mathrm{T}}\theta\,. \tag{30}$$

It is straightforward to verify that the state estimation errors $\tilde{\xi}_2 = \xi_2 - \hat{\xi}_2$ and $\tilde{\xi}_3 = \xi_3 - \hat{\xi}_3$ satisfy

$$\begin{aligned}\dot{\tilde{\xi}}_2 &= -k_2\phi_2(y)\tilde{\xi}_2 + \phi_2(y)\tilde{\xi}_3\,,\\ \dot{\tilde{\xi}}_3 &= -k_3\phi_2(y)\tilde{\xi}_2\,.\end{aligned} \tag{31}$$

Equivalently we can rewrite (31) as

$$\dot{\tilde{\xi}} = \phi_2(y)A_0\tilde{\xi}\,, \tag{32}$$

where

$$\tilde{\xi} = \begin{bmatrix} \tilde{\xi}_2 \\ \tilde{\xi}_3 \end{bmatrix}^{\mathrm{T}}, \qquad A_0 = \begin{bmatrix} -k_2 & 1 \\ -k_3 & 0 \end{bmatrix}. \tag{33}$$

Since $\phi_2(y)$ is nonzero for all $y \in \mathbb{R}$ by assumption, it is either positive or negative. Thus, we can choose the design constants k_2 and k_3 such that

$$PA_0 + A_0^{\mathrm{T}} P = -\mathrm{sgn}(\phi_2(y))I, \; P = P^{\mathrm{T}} > 0. \tag{34}$$

If we choose the observer Lyapunov function

$$V_{\mathrm{obs}} = \tilde{\xi}^{\mathrm{T}} P \tilde{\xi}, \tag{35}$$

then its derivative along the trajectories (32) can be computed as

$$\dot{V}_{\mathrm{obs}} = \phi_2(y)\tilde{\xi}^{\mathrm{T}}(A_0^{\mathrm{T}}P + PA_0)\tilde{\xi} \tag{36}$$

$$= -|\phi_2(y)|\tilde{\xi}^{\mathrm{T}}\tilde{\xi}. \tag{37}$$

From the non-positivity of the Lyapunov derivative, we know that the observer error $\tilde{\xi}$ asymptotically converges to zero. Since θ is not known, the state estimates (30) can not be implemented. However we still implement the filters (26)–(29) to provide filter dynamics for the controller design in the next section.

4. ADAPTIVE CONTROLLER DESIGN

The controller design proceeds along the lines of the output-feedback design methodology in (Krstić *et al.*, 1995). The nonlinearity $\phi_2(y)$ in the observer error dynamics is handled by adding nonlinear damping terms in the stabilizing functions. At each step, we design a stabilizing function for the virtual control variable to produce desired system behavior and update the tuning function for the parameter estimate. This procedure is repeated until the actual control u appears in the differential equation. While the last stabilizing function is used as the control and the parameter update law is chosen, the design is completed.

Following the notation in (Krstić *et al.*, 1995), the recursive expressions for the controller are as follows:

$$z_1 = y - y_r \tag{38}$$

$$z_i = \nu_i - \alpha_{i-1}, \; i = 2,3, \tag{39}$$

$$\alpha_1 = \frac{1}{\phi_1}\left[-c_1 z_1 - d_1 \frac{\phi_1^2}{|\phi_2|} z_1 - \varphi_1 - \phi_1 \iota_2 - \omega^{\mathrm{T}}\hat{\theta} + \dot{y}_r\right] \tag{40}$$

$$\alpha_2 = \frac{1}{\phi_2}\left[-c_2 z_2 - \phi_1 z_1 - \beta_2 - d_2\left(\frac{\partial \alpha_1}{\partial y}\right)^2 \frac{\phi_1^2}{|\phi_2|} z_2 + \frac{\partial \alpha_1}{\partial y}\omega^{\mathrm{T}}\hat{\theta} + \frac{\partial \alpha_1}{\partial \hat{\theta}}\Gamma\tau_2\right], \tag{41}$$

$$\beta_2 = -k_2\phi_2\nu_2 + \varphi_2 + \phi_2\iota_3 - \frac{k_2\phi_2}{\phi_1}\varphi_1 - k_2\phi_2\iota_2 + \frac{\partial \alpha_1}{\partial y}\left[\phi_1\nu_2 + \varphi_1 + \phi_1\iota_2\right] - \frac{\partial \alpha_1}{\partial y_r}\dot{y}_r - \frac{\partial \alpha_1}{\partial \dot{y}_r}\ddot{y}_r - \frac{\partial \alpha_1}{\partial \Omega_2}\dot{\Omega}_2 \tag{42}$$

$$\beta_3 = -k_3\phi_2\nu_2 + \varphi_3 - \frac{k_3\phi_2}{\phi_1}\varphi_1 - k_2\phi_2\iota_3 - \frac{\partial \alpha_2}{\partial y_r}\dot{y}_r - \frac{\partial \alpha_2}{\partial \dot{y}_r}\ddot{y}_r - \frac{\partial \alpha_2}{\partial \ddot{y}_r}y_r^{(3)} - \frac{\partial \alpha_2}{\partial \Omega_2}\dot{\Omega}_2 - \frac{\partial \alpha_2}{\partial \Omega_3}\dot{\Omega}_3 - \frac{\partial \alpha_2}{\partial \nu_2}\dot{\nu}_2 - \frac{\partial \alpha_2}{\partial y}\left[\phi_1\nu_2 + \varphi_1 + \phi_1\iota_2\right], \tag{43}$$

$$\omega = \psi_1 + \phi_1\Omega_2, \tag{44}$$

$$\tau_1 = \omega z_1 \tag{45}$$

$$\tau_i = \tau_{i-1} - \frac{\partial \alpha_{i-1}}{\partial y}\omega z_i, \; i = 2,3. \tag{46}$$

Adaptive control law:

$$u = \frac{1}{\phi_3}\left[-c_3 z_3 - \phi_2 z_2 - d_3\left(\frac{\partial \alpha_2}{\partial y}\right)^2 \frac{\phi_1^2}{|\phi_2|} z_3 - \beta_3 + \frac{\partial \alpha_2}{\partial y}\omega^{\mathrm{T}}\hat{\theta} + \frac{\partial \alpha_2}{\partial \hat{\theta}}\Gamma\tau_3 - z_2\frac{\partial \alpha_1}{\partial \hat{\theta}}\Gamma\frac{\partial \alpha_2}{\partial y}\omega\right]. \tag{47}$$

Parameter update law:

$$\dot{\hat{\theta}} = \Gamma\tau_3. \tag{48}$$

The candidate Lyapunov function is chosen as

$$V = \frac{1}{2}\sum_{i=1}^{3} z_i^2 + \sum_{i=1}^{3}\frac{1}{4d_i}\tilde{\xi}^{\mathrm{T}}P\tilde{\xi} + \frac{1}{2}\tilde{\theta}^{\mathrm{T}}\Gamma^{-1}\tilde{\theta}. \tag{49}$$

Its derivative is computed as

$$\begin{aligned}\dot{V} = &-\sum_{i=1}^{3} c_i z_i^2 - \sum_{i=1}^{3} d_i\left(\frac{\partial \alpha_{i-1}}{\partial y}\right)^2 \frac{\phi_1^2}{|\phi_2|} z_i^2 \\ &-\sum_{i=1}^{3} z_i\frac{\partial \alpha_{i-1}}{\partial y}\phi_1\tilde{\xi}_2 - \sum_{i=1}^{3}\frac{1}{4d_i}|\phi_2|\left(\tilde{\xi}_2^2 + \tilde{\xi}_3^2\right) \\ = &-\sum_{i=1}^{3} c_i z_i^2 - \sum_{i=1}^{3} d_i\left(\frac{\phi_1}{\sqrt{|\phi_2|}} z_i\frac{\partial \alpha_{i-1}}{\partial y} + \frac{\sqrt{|\phi_2|}}{2d_i}\tilde{\xi}_2\right)^2 \\ &-\sum_{i=1}^{3}\frac{|\phi_2|}{4d_i}\tilde{\xi}_3^2 \\ \le &-\sum_{i=1}^{3} c_i z_i^2,\end{aligned} \tag{50}$$

where we have introduced $\frac{\partial \alpha_0}{\partial y} \triangleq -1$ for notational convenience.

5. STABILITY ANALYSIS

From the choice of the Lyapunov function in (49) and the non-posititivity of the Lyapunov derivative $\dot{V}$ in (50), we know that $z, \tilde{\xi}$, and $\tilde{\theta}$ are globally uniformly bounded. Since θ is an unknown constant vector, we see that $\hat{\theta} = \theta - \tilde{\theta}$ is bounded. Since y_r is bounded by assumption and $z_1 = y - y_r$, the output y is also globally uniformly bounded. Since the output y is bounded, we conclude that Ω_2 and Ω_3 are globally uniformly bounded from (28) and (29). Since α_1 in (40) is a smooth function of $y, z_1, \hat{\theta}, \Omega_2, \dot{y}_r$ that are all bounded, we conclude that α_1 is bounded. This implies that ν_2 is globally uniformly bounded because $z_2 = \nu_2 - \alpha_1$ is bounded. From the choice of the stabilizing function α_2 in (41), we see that α_2 is also bounded, which implies the boundedness of ν_3 from $z_3 = \nu_3 - \alpha_2$ and the boundedness of z_3. From (30), we know that the virtual estimates $\hat{\xi}_2$, $\hat{\xi}_3$ are bounded. This implies that the variables ξ_2 and ξ_3 are bounded since $\tilde{\xi}$ is bounded and $\xi = \hat{\xi} + \tilde{\xi}$. From the coordinate transformations in (16) and (21), we conclude that the original states x_2 and x_3 are globally uniformly bounded. The boundedness of the control u is straightforward from (47). Thus, we have proven that all the internal signals are globally uniformly bounded. This implies that the derivative $\dot{z}$ is bounded. From the Lyapunov derivative (50), we see that $z \in \mathcal{L}_2$. Combining this with the fact that $z, \dot{z} \in \mathcal{L}_\infty$, we conclude that $z \to 0$ as $t \to +\infty$. In particular, this implies that the output y converges to the given reference signal y_r asymptotically.

6. CONCLUDING REMARKS

In this paper, we have designed an adaptive nonlinear controller for a third-order nonlinear system which can not be transformed into a parametric output-feedback form via a parameter-independent change of coordinates. Using a coordinate transformation, we "filter out" the nonlinearities from the system dynamics and design a virtual observer with a partial observer Lyapunov function. Although the derivative of this observer Lyapunov function depends on output nonlinearities, we compensate those nonlinearities in the controller design by adding nonlinear damping terms; Furthermore, our adaptive controller guarantees that the tracking error convergences to zero asymptotically. We expect that our result can be generalized to a class of nonlinear systems which can be transformed into a more general output-feedback form:

$$\begin{aligned}
\dot{x}_1 &= \phi_1(y)x_2 + \varphi_1(y) + \psi_1(y)^{\mathrm{T}}\theta \\
\dot{x}_2 &= \phi_2(y)x_3 + \varphi_2(y) + \psi_2(y)^{\mathrm{T}}\theta \\
&\vdots \\
\dot{x}_{n-1} &= \phi_{n-1}(y)x_n + \varphi_{n-1}(y) + \psi_{n-1}(y)^{\mathrm{T}}\theta \\
\dot{x}_n &= \phi_n(y)u + \varphi_n(y) + \psi_n(y)^{\mathrm{T}}\theta \\
y &= x_1.
\end{aligned} \tag{51}$$

This extension is a topic of further research, as is the coordinate-free characterization of this class of systems via geometric conditions that are necessary and sufficient for the existence of a global diffeomorphism that transforms a given uncertain nonlinear system into the form (51).

7. REFERENCES

Freeman, R. A. and P. V. Kokotović (1996). Tracking controllers for systems linear in the unmeasured states. *Automatica* **32**, 735–746.

Jankovic, M. (1997). Adaptive nonlinear output feedback tracking with with a partial high-gain observer and backstepping. *IEEE Transactions on Automatic Control* **42**, 106–113.

Kanellakopoulos I., P. V. Kokotović, and A. S. Morse (1991). Systematic design of adaptive controllers for feedback linearizable systems. *IEEE Transactions on Automatic Control* **36**, 1241–1253.

Khalil, H. K. (1996). Adaptive output feedback control of nonlinear systems represented by input-output models. *IEEE Transactions on Automatic Control* **41**, 177–188.

Krener, A. J. and A. Isidori (1983). Linearization by output injection and nonlinear observers. *Systems & Control Letters* **3**, 47–52.

Krstić, M., I. Kanellakopoulos, and P. V. Kokotović (1995), *Nonlinear and Adaptive Control Design*, John Wiley & Sons. New York, NY.

Marino, R. and P. Tomei (1993). Global adaptive output–feedback control of nonlinear systems, Part I: linear parameterization. *IEEE Transactions on Automatic Control* **38**, 17–32.

Mazenc, F., L. Praly, and W. P. Dayawansa (1994). Global stabilization by output feedback: examples and counterexamples. *Systems & Control Letters* **23**, 119–125.

Tan, Y. and I. Kanellakopoulos (1998). *Proceedings of 37th Conference on Control and Decision*, Tampa, Florida, 2503–2508.

Teel, A. R. and L. Praly (1994). Global stabilizability and observability imply semi-global stabilizability by output feedback. *Systems & Control Letters* **22**, 313–325.

14th Triennial World Congress, Beijing, P.R. China

E-2c-12-6

AN OBSERVER FOR M.I.M.O. NONLINEAR DELAY SYSTEMS

A. Germani[1,2] **C. Manes**[1,2] **P. Pepe**[1]

[1] *Dipartimento di Ingegneria Elettrica, Università dell'Aquila, Monteluco di Roio, 67040 L'Aquila, Italy, e-mail: manes@ing.univaq.it*

[2] *Istituto di Analisi dei Sistemi e Informatica del CNR, Viale Manzoni 30, 00185 Roma, Italy , e-mail: germani@iasi.rm.cnr.it*

Abstract: In this paper an asymptotic state observer for multi-input multi-output (M.I.M.O.) nonlinear delay systems is developed. On this matter many works have been published concerning the linear case, while for the nonlinear case to our knowledge the first result is the one in (Germani, *et al.*, 1998d), that considers single-input single-output nonlinear delay systems. A suitable modification of the standard tools of differential geometry is used in this paper for the observer construction and to work out conditions of convergence to zero of the observation error. The implementation of the proposed observation algorithm is very easy and computer simulations show its effectiveness. *Copyright © 1999 IFAC*

Keywords: nonlinear systems, delay analysis, state observers, delay systems.

1. INTRODUCTION

Many ecological systems, industrial processes, telerobotic systems, earth controlled satellite devices and other significant applications are modelled by delay differential equations. The main difficulty in dealing with delay systems is that the state should be properly considered in an infinite dimensional space (Banks and Kappel, 1979; Gibson, 1983; Bensoussan, *et al.*., 1992).

In (Germani, *et al.*, 1996) the authors showed that, through the introduction of a suitable mathematical formalism, this difficulty can be overcome for an interesting class of nonlinear delay systems, and solutions have been found for many classical problems in systems and control theory. The concept of *delay relative degree* has been introduced in (Germani, *et al.*, 1996) for single-input single-output nonlinear delay systems, and in (Germani, *et al.*, 1998b) for multi-input multi-output (M.I.M.O.) systems.

This work is supported by *CNR* (Italian National Research Council) and by *ASI* (Italian Aerospace Agency).

For systems that have delay relative degree a state feedback control law has been found that linearizes the input-output mapping and removes the effect of the delay. In (Germani, *et al.*, 1998d), a Luenberger-like observer has been developed for single-input single-output nonlinear delay systems.

In this paper an asymptotic observer is built up for M.I.M.O. nonlinear delay systems. Such observer is an *identity observer* (or a *Luenberger-like* observer) because it reproduces exactly the system equations when initialized by the true initial state in a suitable Hilbert space. The computation of the gain is very straightforward and the observer equations are easy to implement. Sufficient conditions are found for exponential convergence to zero of the observation error. All simulations made showed high effectiveness of the observation algorithm.

The paper is organized as follows: in section 2 some *extended* tools of differential geometry are reported in short, that allow the definition of a *delay vector relative degree* (Germani, *et al.*, 1996, 1998b); in section 3 the state observer is built up and simulation results are reported in section 4. Conclusions follow.

2. THE DELAY VECTOR RELATIVE DEGREE

The control system under investigation is described by the following equations

$$\dot{x}(t)=f(x(t),x(t-\Delta))+g(x(t),x(t-\Delta))u(t), \quad (1)$$
$$y(t)=h(x(t)), \qquad t\geq 0, \quad (2)$$

where $x(t) \in \mathbb{R}^n$, $u(t) \in \mathbb{R}^p$, $y(t) \in \mathbb{R}^q$, the terms f and h are C^∞ vector functions, the term $g(x(t),x(t-\Delta))$ can be written as

$$[g_1(x(t),x(t-\Delta)) \cdots g_p(x(t),x(t-\Delta))], \quad (3)$$

in which the terms g_j are C^∞ vector functions. The model description is completed by the knowledge of the function $x(\tau)$, $\tau \in [-\Delta, 0]$, which represents the initial state in the classical infinite dimensional description of delay systems. As usual, we suppose such function belongs to $W^{1,2}([-\Delta,0];\mathbb{R}^n)$, the space of absolutely continuous functions with derivative in L_2 (Banks and Kappel, 1979; Gibson, 1983; Bensoussan, *et al.*., 1992).

Let, for any integer i and for $j=1,2,\ldots,p$,

$$x_{i\Delta}(t)=x(t-i\Delta), \quad u_{j,i\Delta}(t)=u_j(t-i\Delta). \quad (4)$$

In the following we shall omit the time argument when it does not cause confusion.

In the papers (Germani, *et al.*, 1996, 1998c), in order to apply standard strategies of output control of nonlinear systems without delay to the case of S.I.S.O. nonlinear delay systems, the definition of a *relative degree* for such systems was mandatory. Here follows the definition of a *delay vector relative degree* for M.I.M.O. nonlinear delay systems, that holds in the case of systems with the same number of inputs and outputs, i.e. when $p=q$. It was given in (Germani, *et al.*, 1998b).

Definition 1. *Consider system* (1) (2) *in which* $p=m$ *and* $q=m$. *It is said to have delay vector relative degree*

$$r=[r_1 \quad r_2 \quad \ldots \quad r_m], \quad (5)$$

in an open set $\Omega_{\bar r} \in \mathbb{R}^{n(\bar r+1)}$, *with* $\bar r = \max_{l=1,\ldots,m}\{r_l\}$, *if, in* $\Omega_{\bar r}$, *the following conditions are verified*

$$L_G L_F^j H_i(X)=0 \qquad \begin{array}{l} j=0,1,\ldots,r_i-2,\\ i=1,2,\ldots,m\end{array}$$
$$\det\left(\begin{bmatrix} L_G L_F^{r_1-1}H_1(X)\\ \vdots \\ L_G L_F^{r_q-1}H_m(X)\end{bmatrix} S\right)\neq 0 \quad (6)$$

where

$$X=\begin{bmatrix} x\\ x_\Delta\\ \vdots\\ x_{\bar r\Delta}\end{bmatrix}, \qquad \overline{X}=\begin{bmatrix} x\\ x_\Delta\\ \vdots\\ x_{(\bar r-1)\Delta}\end{bmatrix}, \quad (7)$$

$$F(X)=\begin{bmatrix} f(x,x_\Delta)\\ f(x_\Delta,x_{2\Delta})\\ \vdots\\ f(x_{(\bar r-1)\Delta},x_{\bar r\Delta})\end{bmatrix}, \quad (8)$$

$$G(X)=[G_1(X) \quad \cdots \quad G_m(X)] \quad (9)$$
$$G_k(X)=\mathrm{diag}\left\{g_k(x,x_\Delta),\ldots,g_k(x_{(\bar r-1)\Delta},x_{\bar r\Delta})\right\}$$

$$H_k(X)=h_k(x) \qquad L_F^0 H_k(X)=H_k(X),$$
$$L_F^j H_k(X)=\left(\frac{d}{d\overline{X}}L_F^{j-1}H_k\right)F(X), \quad (10)$$
$$L_G L_F^j H_k(X)=\left(\frac{d}{d\overline{X}}L_F^j H_k\right)G(X), \quad j=0,1,2,\ldots$$

$$S=\mathrm{diag}\,[v_1 \quad \cdots \quad v_m], \quad (11)$$
$$v_k=[1 \quad 0 \quad \cdots \quad 0]^T \in \mathbb{R}^{\bar r}, \quad k=1,\cdots,m.$$

If $\Omega_{\bar r}=\mathbb{R}^{n(\bar r+1)}$, *then system* (1) (2) *is said to have global delay vector relative degree* r. ∎

This definition extends the well-known concept of *vector relative degree* for nonlinear systems without delay. In (Germani, *et al.*, 1998b) it is shown that the classical control problem of linearization and decoupling of input-output channels through static state-feedback can be solved for systems for which a delay vector relative degree can be defined.

3. THE STATE OBSERVER

As in the case of nonlinear systems without delay, existence of a delay vector relative degree is a strong condition on systems of the type (1) (2). However, for the solution of the state observation problem a weaker definition of relative degree is required. The definition of *observation delay vector relative degree* follows below. Note that in order to give such definition there is no need to have the same number of inputs and outputs.

Definition 2. *System* (1) (2) *is said to have observation delay vector relative degree* $r=[r_1 \cdots r_q]$ *in an open set* $\Omega_{\bar r} \in \mathbb{R}^{n(\bar r+1)}$, *with* $\bar r=\max_{l=1,\ldots,m}\{r_l\}$, *if the following conditions are verified for* $i=1,2,\ldots,q$,

$$\forall X \in \Omega_{\bar r},\ \ L_G L_F^j H_i(X)=0 \quad j=0,\ldots,r_i-2,$$
$$\exists X \in \Omega_r:\ \ L_G L_F^{r_i-1}H_i(X)\neq 0, \quad (12)$$

If $\Omega_{\bar r}=\mathbb{R}^{n(\bar r+1)}$, *then system* (1) (2) *is said to have global observation delay vector relative degree* r. ∎

As it will be illustrated next, the observer for nonlinear delay systems presented in this paper ensures state observation if some assumptions are satisfied. Among the assumptions, in the case in which the input function is not identically zero, the following is needed:

Hp_0) *system* (1), (2) *has global observation delay relative degree equal to* r, *with* $\sum_{k=1}^{n} r_k = n$ (*n is the dimension of vector x*).

Let the system (1) (2) have delay vector relative degree r in $\Omega_{\bar{r}}$. Let

$$z = \begin{bmatrix} y_1 \\ \vdots \\ y_1^{(r_1-1)} \\ y_2 \\ \vdots \\ y_2^{(r_2-1)} \\ \vdots \\ \vdots \\ y_q \\ \vdots \\ y_q^{(r_q-1)} \end{bmatrix} = \begin{bmatrix} H_1(\overline{X}) \\ \vdots \\ L_F^{r_1-1}H_1(\overline{X}) \\ H_2(\overline{X}) \\ \vdots \\ L_F^{r_2-1}H_2(\overline{X}) \\ \vdots \\ \vdots \\ H_q(\overline{X}) \\ \vdots \\ L_F^{r_q-1}H_q(\overline{X}) \end{bmatrix} = \phi(\overline{X}) \quad (13)$$

Using an extended notation the map ϕ can be rewritten as

$$z = \phi(x, x_\Delta, \cdots, x_{(\bar{r}-1)\Delta}). \quad (14)$$

If assumption Hp_0 holds, then $z \in \mathbb{R}^n$.

Let A_r, B_r and C_r be the block diagonal matrices

$$\begin{aligned} &A_r = \text{diag}_{j=1}^{q}\{A_j\}, \quad B_r = \text{diag}_{j=1}^{q}\{B_j\}, \\ &C_r = \text{diag}_{j=1}^{q}\{C_j\}. \end{aligned} \quad (15)$$

where the m triples (A_j, B_j, C_j) are Brunowsky matrices

$$A_j = \begin{bmatrix} 0 & 1 & \cdots & 0 \\ 0 & 0 & \cdots & 0 \\ \vdots & \vdots & \ddots & \vdots \\ 0 & 0 & \cdots & 1 \\ 0 & 0 & \cdots & 0 \end{bmatrix}, \qquad B_j = \begin{bmatrix} 0 \\ 0 \\ \vdots \\ 0 \\ 1 \end{bmatrix} \quad (16)$$

$$C_j = [0 \;\; 0 \;\; \cdots \;\; 1],$$
$$A_j \in \mathbb{R}^{r_j \times r_j}, \; B_j \in \mathbb{R}^{r_j}, \; C_j^T \in \mathbb{R}^{r_j}.$$

Remark 3. The pair A_r, C_r is observable and the pair A_r, B_r is controllable. ■

If assumption Hp_0 holds then $A_r \in \mathbb{R}^{n \times n}$, $B_r \in \mathbb{R}^{n \times q}$, $C_r \in \mathbb{R}^{q \times n}$.

It can be verified that, under assumption Hp_0, the vector $z \in \mathbb{R}^n$ satisfies the following equation:

$$\begin{aligned} \dot{z} &= A_r z + B_r \left(\begin{bmatrix} L_F^{r_1} H_1 \\ L_F^{r_2} H_2 \\ \vdots \\ L_F^{r_q} H_q \end{bmatrix} + \begin{bmatrix} L_G L_F^{r_1-1} H_1 \\ L_G L_F^{r_2-1} H_2 \\ \vdots \\ L_G L_F^{r_q-1} H_q \end{bmatrix} U \right), \\ y &= C_r z, \qquad t \geq (n-1)\Delta, \end{aligned} \quad (17)$$

where vector $U \in \mathbb{R}^{p\bar{r}}$ is defined as follows

$$U = \begin{bmatrix} U_1 \\ \vdots \\ U_p \end{bmatrix}, \quad \text{with} \quad U_j = \begin{bmatrix} u_j \\ u_{j,\Delta} \\ \vdots \\ u_{j,(\bar{r}-1)\Delta} \end{bmatrix}. \quad (18)$$

Note that eq. (17) represents a linear undelayed system with a nonlinear-delayed perturbation.

As stated before, the pair A_r, C_r is observable, and it is an easy matter to assign eigenvalues to the matrix $A_r - KC_r$. In particular, a suitable choice of matrix K preserves the block diagonal structure of matrix $A_r - KC_r$. The structure for K is

$$K = \text{diag}_{i=1}^{q}\{K_i\}, \quad \text{with } K_i \in \mathbb{R}^{r_i}, \quad (19)$$

and is such that the i-th block on the diagonal of matrix $A_r - KC_r$ has the following companion structure

$$A_i - K_i C_i = \begin{bmatrix} -k_{i,1} & 1 & \cdots & 0 \\ \vdots & \vdots & \ddots & \vdots \\ -k_{i,r_i-1} & 0 & \cdots & 1 \\ -k_{i,r_i} & 0 & \cdots & 0 \end{bmatrix}. \quad (20)$$

It follows that if a r_i-pla $\lambda_i = (\lambda_{i,1}, \ldots, \lambda_{i,r_i})$ of eigenvalues has to be assigned to the i-th block, the vector gain K_i must contains the coefficients of the monic polynomial that has λ_i as roots.

Definition 4. *A system of the form* (1), (2) *is said globally delay drift observable if the hypothesis Hp_0 is satisfied and there exists the inverse ϕ^{-1} of function ϕ defined in* (13) *with respect to x for all* $(z, x_\Delta, \ldots, x_{(\bar{r}-1)\Delta}) \in \mathbb{R}^{n\bar{r}}$, *that is*

$$x = \phi^{-1}(z, x_\Delta, \cdots, x_{(\bar{r}-1)\Delta}) \quad (21)$$

■

From the definition it follows that for globally delay drift observable systems the Jacobian of ϕ w.r.t. the first argument, denoted as $Q(\overline{X})$ and defined as

$$Q(\overline{X}) = \frac{\partial \phi(x, x_\Delta, \cdots, x_{(\bar{r}-1)\Delta})}{\partial x} \quad (22)$$

is nonsingular for all $(x, x_\Delta, \cdots, x_{(\bar{r}-1)\Delta}) \in \mathbb{R}^{n\bar{r}}$. Moreover, it is such that

$$\left. \frac{\partial \phi^{-1}(z, x_\Delta, \cdots, x_{(\bar{r}-1)\Delta})}{\partial z} \right|_{z=\phi(\overline{X})} = Q^{-1}(\overline{X}). \quad (23)$$

We will refer to matrix Q as the *delay drift-observability matrix* associated to system (1), (2).

The proposed observer for nonlinear delay systems that are globally delay drift observable is given by the following delay differential equation, that holds for $t \geq 0$,

$$\begin{aligned} \dot{\hat{x}}(t) &= f(\hat{x}(t), \hat{x}_\Delta(t)) + g(\hat{x}(t), \hat{x}_\Delta(t))u(t) + w(t), \\ w(t) &= Q^{-1}(\hat{\overline{X}}(t))K\left(y(t) - h(\hat{x}(t))\right) \\ &\quad - Q^{-1}(\hat{\overline{X}}(t)) \sum_{i=1}^{\bar{r}-1} \left(\frac{\partial \phi}{\partial x_{i\Delta}} \hat{\overline{X}}(t) \right) w_{i\Delta}(t), \end{aligned} \quad (24)$$

with initial conditions

$$\begin{aligned}\hat{x}(\tau) &= \xi(\tau),\\ w(\tau) &= \eta(\tau),\end{aligned} \quad \tau \in [-(\bar{r}-1)\Delta, 0], \tag{25}$$

$$\xi, \eta \in W^{1,2}([-(\bar{r}-1)\Delta, 0]; \mathbb{R}^n).$$

The gain matrix K is chosen such to assign the eigenvalues to the matrix $A_r - KC_r$. The part in $[-\Delta, 0]$ of the function ξ that initializes the observer represents the a priori knowledge on the system state. If the exact $x(t)$ is known for $t \in [-\Delta, 0]$, the observer initialization is made as follows

$$\begin{aligned}\hat{x}(\tau) &= \bar{\xi}(\tau),\\ w(\tau) &= 0,\end{aligned} \quad \tau \in [-(\bar{r}-1)\Delta, 0], \tag{26}$$

where $\bar{\xi}$ is any function in $W^{1,2}([-(\bar{r}-1)\Delta, 0]; \mathbb{R}^n)$ such that $\bar{\xi}(\tau) = x(\tau), \tau \in [-\Delta, 0]$. In this case the feedback terms in the summation in (24) are identically zero and for all t it is $\hat{x}(t) = x(t)$.

Note that if for a given $\bar{t}$ it is $\hat{x}(\tau) = x(\tau)$ for $\tau \in [\bar{t} - (\bar{r}-1)\Delta, \bar{t}]$, it follows that $\hat{x}(t) = x(t)$ for all $t > \bar{t}$ (it is easy to check that in this case $w(\tau) = 0$ and $y(\tau) = h(\hat{x}(\tau))$ for $\tau \in [\bar{t} - (\bar{r}-1)\Delta, \bar{t}]$), so that for $t \geq \bar{t}$ the feedback terms in (24) are identically zero). This means that the observer (24) is an *identity observer*, or a *Luenberger-like* observer, since it reproduces exactly the system state when initialized by the true initial state in the Hilbert space $W^{1,2}([\bar{t} - (\bar{r}-1)\Delta, \bar{t}], \mathbb{R}^n)$

The following lemma provides a nice representation for the observer (24).

Lemma 5. *For $t \geq (\bar{r}-1)\Delta$, the dynamics of the observer* (24) *can be written as*

$$\begin{aligned}\dot{\hat{z}} &= A_r\hat{z} + K(y - C_r\hat{z})\\ &+ B_r\left(\begin{bmatrix} L_F^{r_1}H_1(\hat{X})\\ \vdots\\ L_F^{r_q}H_q(\hat{X})\end{bmatrix} + \begin{bmatrix} L_G L_F^{r_1-1}H_1(\hat{X})\\ \vdots\\ L_G L_F^{r_q-1}H_q(\hat{X})\end{bmatrix} U\right),\\ \hat{X} &= (\hat{x}, \hat{x}_\Delta, \ldots, \hat{x}_{\bar{r}\Delta}),\\ \hat{x} &= \phi^{-1}(\hat{z}, \hat{x}_\Delta, \ldots, \hat{x}_{(\bar{r}-1)\Delta})\end{aligned} \tag{27}$$

where A_r, B_r, C_r is the Brunowsky triple defined in (15).

Proof. The equivalence of system equations (24) and (27) can be proved directly differentiating (21) w.r.t. time, and using (27) to obtain (24). The computations are not difficult, and are not reported here for brevity. ■

Equivalence of systems (24) and (27) must be intended in the sense that both equations give the same solution $\hat{x}(t), t \geq (\bar{r}-1)\Delta$, provided that both systems are initialized with the same initial condition $\hat{x}(\tau), \tau \in [-\Delta, (\bar{r}-1)\Delta]$.

For the observer (24) the following convergence theorem can be given.

Theorem 6. *Let the system* (1), (2) *be globally delay drift observable. Assume there exists u_M such that $\|u(t)\| \leq u_M$ $\forall t \geq 0$. Moreover assume the following Lipschitz hypotheses:*

H1) there exists a positive constant α_0 such that for $\|u(t)\| \leq u_M$

$$\begin{aligned}\sup_{i=1,\ldots,m} &\|L_F^{r_i}H_i(*) - L_F^{r_i}H_i(\hat{*}) +\\ &+ L_G L_F^{r_i-1}H_i(*) - L_G L_F^{r_i-1}H_i(\hat{*})U\| \leq\\ &\leq \alpha_0 \left\| \begin{matrix} z - \hat{z}\\ x_\Delta - \hat{x}_\Delta\\ \vdots\\ x_{\bar{r}\Delta} - \hat{x}_{\bar{r}\Delta}\end{matrix} \right\|,\end{aligned} \tag{28}$$

where the arguments $*$ and $\hat{*}$ stand for

$$\begin{aligned}* &= (\phi^{-1}(z, x_\Delta, \ldots, x_{(\bar{r}-1)\Delta}), x_\Delta, \ldots, x_{\bar{r}\Delta}),\\ \hat{*} &= (\phi^{-1}(\hat{z}, \hat{x}_\Delta, \ldots, \hat{x}_{(\bar{r}-1)\Delta}), \hat{x}_\Delta, \ldots, \hat{x}_{\bar{r}\Delta});\end{aligned}$$

H2) in the case $\bar{r} > 1$, there exist positive constants α and γ such that the function ϕ^{-1} satisfies

$$\begin{aligned}&\|\phi^{-1}(z, x_\Delta, \ldots, x_{(\bar{r}-1)\Delta})\\ &\quad - \phi^{-1}(\hat{z}, \hat{x}_\Delta, \ldots, \hat{x}_{(\bar{r}-1)\Delta})\|\\ &\leq \alpha\|z - \hat{z}\| + \gamma \left\| \begin{matrix} x_\Delta - \hat{x}_\Delta\\ \vdots\\ x_{(\bar{r}-1)\Delta} - \hat{x}_{(\bar{r}-1)\Delta}\end{matrix} \right\|,\end{aligned} \tag{29}$$

$$\text{with} \qquad \gamma < \frac{1}{\bar{r}-1}. \tag{30}$$

Then, there exists a gain vector K with the structure given in (19) *to be put in the observer* (24) *such that*

$$\lim_{t\to\infty} \|x(t) - \hat{x}(t)\| = 0. \tag{31}$$

Proof. The proof is quite long and needs some lemmas not reported in the paper and therefore is omitted. It exploits the same ideas used in the proof of convergence of the observer in (Ciccarella, *et al.*, 1993) for nonlinear systems without delay. Of course, the presence of delay complicates a lot the proof. The main idea is to write the dynamics of the observation error in the pseudo-coordinates z, given by the map $\phi(x, x_\Delta, \cdots, x_{(n-1)\Delta})$, exploiting equations (17) and (27) for the original system and for the observer, and to prove that $\|z(t) - \hat{z}(t)\| \to 0$. From this result, suitably exploiting the assumptions of the theorem, it can be proved that also $\|x(t) - \hat{x}(t)\| \to 0$. A simplified version of the proof, for the S.I.S.O. case, is reported in (Germani, *et al.*, 1998d). ■

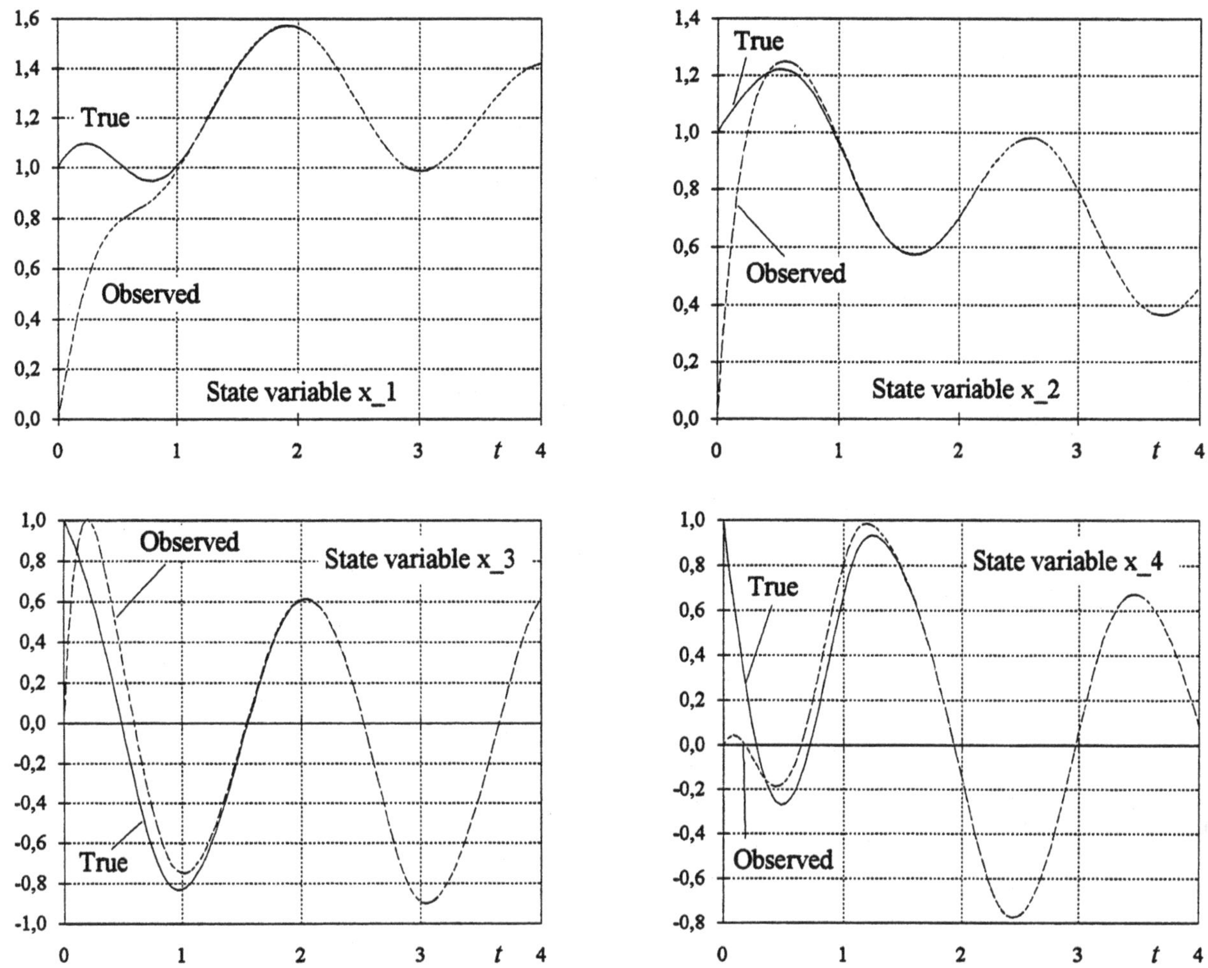

Figure 1. True (continuous lines) and observed (dashed lines) state variables.

4. SIMULATION RESULTS

As an example, let us consider the following system:

$$\begin{aligned}
\dot{x}_1(t) &= x_4(t) - 0.1x_2^2(t-\Delta),\\
\dot{x}_2(t) &= x_3(t) - 0.1x_4(t-\Delta)\exp\big(x_2(t-\Delta)\big),\\
\dot{x}_3(t) &= -x_3^3(t) - 0.1x_2^2(t-\Delta)\\
&\quad + (2+\cos(x_1(t)))u_1(t),\\
\dot{x}_4(t) &= -x_1(t)x_4(t-\Delta) - x_4^5(t)\\
&\quad + \big(1 + x_2^2(t) + \sin(x_2(t))\big)u_2(t),\\
y_1(t) &= x_1(t),\\
y_2(t) &= x_2(t).
\end{aligned} \tag{32}$$

with delay $\Delta = 0.1$ s.

Straightforward calculations show that the observer delay vector relative degree is $r = [2\ 2]$, and that the map ϕ defined in (13) is

$$z = \phi(x, x_\Delta) = \begin{bmatrix} x_1 \\ x_4 - 0.1x_{2,\Delta}^2 \\ x_2 \\ x_3 - 0.1x_{4,\Delta}\exp(x_{2,\Delta}) \end{bmatrix}. \tag{33}$$

This map is globally invertible for x, and its jacobian $Q(x, x_\Delta)$ is constant

$$Q = \frac{\partial\phi(x, x_\Delta)}{\partial x} = \begin{bmatrix} 1 & 0 & 0 & 0 \\ 0 & 0 & 0 & 1 \\ 0 & 1 & 0 & 0 \\ 0 & 0 & 1 & 0 \end{bmatrix}. \tag{34}$$

It is easy to check that the inverse of Q is its transpose Q^{T}. The gain matrix to use in the observer (24) has the block diagonal structure

$$K = \begin{bmatrix} k_1 & 0 \\ k_2 & 0 \\ 0 & k_3 \\ 0 & k_4 \end{bmatrix}. \tag{35}$$

If eigenvalues λ_1, λ_2, λ_3, λ_4 have to be assigned to matrix $A_r - KC_r$, the constants k_i in matrix K can be computed as follows

$$\begin{aligned} k_1 &= -\lambda_1 - \lambda_2, & k_2 &= \lambda_1\lambda_2, \\ k_3 &= -\lambda_3 - \lambda_4, & k_4 &= \lambda_3\lambda_4. \end{aligned} \tag{36}$$

In the simulations reported the input functions used are

$$u_1(t) = -\sin 3t, \qquad u_2(t) = -\cos 3t. \tag{37}$$

The eigenvalues assigned are $\lambda = (-1, -2, -3, -4)$. The initialization of the system (32) is

$$x(\tau) = [1\ 1\ 1\ 1]^{\mathrm{T}}, \quad \tau \in [-\Delta, 0], \qquad (38)$$

while the initialization of the observer (24) is made with

$$\hat{x}(\tau) = 0, \quad w(\tau) = 0, \quad \tau \in [-\Delta, 0]. \qquad (39)$$

Figure 1 reports the behavior of true and estimated state variables during the first 4 seconds. For all components, the true state starts from 1 (continuous line) and the observed state starts from 0 (dashed line). With the chosen eigenvalues the convergence is obtained after about one second.

5. CONCLUSIONS

It is well known that the analysis of nonlinear delay systems is heavily complicated by the infinite dimension of the system state and by the nonlinear structure of the differential equations. The contribution of this paper is the construction of a Luenberger-like observer for nonlinear delay multi-input multi-output systems, through a suitable modification of the classical tools of finite dimensional differential geometry. The observer algorithm is very easy to implement. Sufficient conditions are found that ensure convergence of the observed state to the real one. The proof is quite lengthy and is not reported in the paper. In all simulations made the observer proposed has always worked well, both with stable and unstable systems, also in cases in which the sufficient conditions of convergence were not satisfied. Also when a control loop was implemented, closed on the observed state, in all cases simulated the observer performed well.

REFERENCES

Banks, H. T. and F. Kappel, (1979). Spline Approximations for Functional Differential Equations, *J. Differential Equations*, No. 34, pp. 496–522.

Bensoussan, A., G. Da Prato, M. C. Delfour and S. K. Mitter, (1992). *Representation and Control of Infinite Dimensional Systems,* Birkhauser, Boston.

Ciccarella, G., M. Dalla Mora and A. Germani (1993). A Luenberger-like observer for nonlinear systems, *Int. Journal of Control*, **Vol. 57**, No. 3, pp. 537–556.

Dalla Mora, M., A. Germani and C. Manes (1997). A state observer for nonlinear dynamical systems, *Nonlinear Analysis: Theory, Methods and Applications,* **Vol. 30**, No. 7, pp. 4485-4496.

Fairman F. W. and A. Kumar, (1986). Delayless Observers for Systems with Delay, *IEEE Trans. on Automatic Control*, **Vol. AC-31**, No. 3, pp. 258–259.

Germani, A., L. Jetto, C. Manes and P. Pepe, (1994). The LQG Control Problem for a Class of Hereditary Systems: a Method for Computing its Approximate Solution, proc. of *33rd IEEE Conf. on Decision and Control (CDC94)*, Vol. 2, pp. 1362–1367, Orlando, Fl.

Germani, A., C. Manes and P. Pepe, (1996). Linearization of Input-Output Mapping for Nonlinear Delay Systems via Static State Feedback, proc. of *CESA IMACS Multiconf. con Comp. Eng. in Systems Appl.*, Vol. 1, pp. 599–602, Lille, France.

Germani, A., C. Manes and P. Pepe, (1998a). Local Asymptotic Stability for Nonlinear State Feedback Delay Systems, *6th IEEE Mediterranean Conf. on Control and Systems*, Alghero, Italy.

Germani, A., C. Manes and P. Pepe, (1998b). Linearization and Decoupling of Nonlinear Delay Systems, proc. of *1998 American Control Conference ACC*, Vol. 3, pp. 1948–1952, Philadelphia, PA.

Germani, A., C. Manes and P. Pepe (1998c). Tracking, model matching, disturbance decoupling for a class of nonlinear delay systems, proc. of *Large Scale Systems IFAC Conference*, LLS'98, Vol. 1, pp. 423–429, Patras, Greece.

Germani, A., C. Manes and P. Pepe, (1998d). A State Observer for Nonlinear Delay Systems, proc. of *37th Conf. on Decision and Control (CDC98)*, Tampa, Fl.

Gibson, J. S., (1983). Linear Quadratic Optimal Control of Hereditary Differential Systems: Infinite-Dimensional Riccati Equations and Numerical Approximations, *SIAM J. Control Optim.*, No. 31, pp. 95–139.

Olbrot, A. W., (1981). Observability and Observers for a Class of Linear Systems with Delays, *IEEE Trans. on Automatic Control*, **Vol. AC-26**, No. 2, pp. 513–517.

Pearson, A. E., and Y. A. Fiagbedzi, (1989). An Observer for Time Lag Systems, *IEEE Trans. on Automatic Control*, **Vol. 34**, No. 7, pp. 775–777.

Salamon, A. (1984). On Controllability and Observability of Time Delay Systems, *IEEE Trans. on Automatic Control*, **Vol. AC-29**, No. 5, pp. 432–439.

Pepe, P., (1996). Il Controllo LQG dei Sistemi con Ritardo, PhD thesis, (in italian), Department of Electrical Engineering, L'Aquila, Italy.

TRANSFER EQUIVALENCE AND REALIZATION OF NONLINEAR HIGHER ORDER I/O DIFFERENCE EQUATIONS USING MAPLE

Ü. Kotta*,[1] P. Liu** A.S.I. Zinober**

* *Institute of Cybernetics, Tallinn Technical University, 21 Akadeemia tee, Tallinn, 12618, Estonia; email: kotta@ioc.ee*
** *Applied Mathematics Section, School of Mathematics and Statistics, The University of Sheffield, Sheffield S10 2TN, UK*

Abstract: The paper presents a contribution to the development of symbolic computation tools for discrete-time nonlinear control systems. It presents certain modelling problems to show what can be achieved in nonlinear control with symbolic computation. A set of procedures is developed in Maple V that test if a given input-output (i/o) difference equation is realizable in the classical state space form, and for simple control systems also find the transformation. Other procedures test whether two (or more) input-output difference equations are equivalent, and find the minimal (irreducible) representation of the i/o model. The irreducible form is the basis for constructing the minimal, i.e. accessible and observable, realization. The application of the developed procedures is demonstrated on the example of the model of the grain-drying process, obtained via identification. *Copyright © 1999 IFAC.*

Keywords: nonlinear systems, discrete-time systems, input-output models, equivalence, state-space realization, Maple.

1. INTRODUCTION

The analysis and synthesis of nonlinear control systems involve a lot of symbolic computations not easily done by hand calculation. Because of this and the improvement of commercially available symbolic computation tools, the use of various computer algebra systems is becoming ever more popular in the control community. Various useful toolboxes and software packages have been developed using computer algebra systems such as Reduce (Amirat and Diop, 1996; Barbot, 1990), Macsyma (Akhrif and Blankenship, 1989; Birk and Zeitz, 1991), Maple (Van Essen and Jager, 1993; De Jager, 1995; Aranda-Bricaire, 1996; Post, 1995; Glumineau and Graciani, 1996), and Mathematica (Rothfuss and Zeitz, 1996; Besancon and Bornard, 1997). However, symbolic computation systems have been not practically used for the analysis and synthesis of nonlinear discrete-time systems (Aranda-Bricaire, 1996; Barbot, 1990; Kotta and Roomeldi, 1990; Post, 1995). Our purpose is to contribute to the development of such tools. Procedures in the computer algebra system Maple V have been developed which allow one to decide whether a given discrete-time nonlinear i/o system is irreducible, whether two (or more) nonlinear i/o systems are equivalent, and if the irreducible nonlinear i/o difference equation is realizable in the classical state space form. Our Maple procedures use some procedures (but in a modified form) implemented by Aranda-Bricaire (1996) and Post (1995) for the analysis of the accessibility and static state feedback linearizability of discrete-time nonlinear state equations.

[1] Partially supported by the Estonian Science Foundation Grant N 3137

The paper is organized as follows. In Section 2 we recall the basics of the (linear) algebraic theory for nonlinear discrete-time control systems. Moreover, the concepts of equivalence and realization theory are summarized briefly in Section 2. The material of this section is taken from (Aranda-Bricaire *et al.*, 1996; Kotta *et al.*, 1997; Kotta *et al.*, 1998). The Maple procedures are described in Section 3. Section 4 presents the model of a column-type grain drying process on which the procedures are demonstrated. Finally, some concluding remarks are given in Section 5.

2. PRELIMINARIES

Consider a discrete-time single-input single-output nonlinear system Σ described by a higher order input-output (i/o) difference equation relating the input u, the output y and a finite number of their time shifts

$$\varphi(y(t), \ldots, y(t+n), u(t), \ldots, u(t+s)) = 0 \quad (1)$$

where $u \in \mathbb{R}$, $y \in \mathcal{Y} \subset \mathbb{R}$, and φ is a real analytic function defined on $\mathcal{Y}^{n+1} \times \mathbb{R}^{s+1}$. This equation provides a natural generalization of the autoregressive moving-average representation which is very popular in linear systems theory.

In order to be able to use the mathematical tools from the linear algebraic framework, we assume that the following assumptions hold for system (1).

A1 Equation (1) can be solved uniquely (at least locally) for $y(t+n)$:

$$\begin{aligned} y(t+n) = \phi(y(t), \ldots, y(t+n-1), \\ u(t), \ldots, u(t+s)) \end{aligned} \quad (2)$$

Note that this can always be guaranteed about a regular point where $(y(t), \ldots, y(t+n), u(t), \ldots, u(t+s))$ is such that $\partial\varphi(\cdot)/\partial y(t+n) \neq 0$.

A2 $\varphi(0, ..., 0) = 0$. We make this assumption because the mathematical tools to be employed require that, instead of working with the equations themselves, we work with their differentials and therefore will not able to distinguish between the systems with $\varphi(\cdot) = 0$ and $\varphi(\cdot) + const = 0$. So both systems would be recognized as the same system (i.e. equivalent). This should be avoided.

A3 $\partial\varphi(\cdot)/\partial(y(t), u(t))$ is not equal to zero. In the case of shift-invariant systems this may be assumed without loss of generality since, if this is not the case, we can always apply the backward shift operator a sufficient number of times so that this condition is satisfied; and we still have the same system.

When A1 holds, we may associate with the system Σ an extended state-space system Σ_e with input $v(t) = u(t+s+1)$ and state $z(t) = [y(t), \ldots, y(t+n-1), u(t), ..., u(t+s)]^T$ defined as

$$z(t+1) = f_e(z(t), v(t)), \quad (3)$$

where $f_e(\cdot) = [z_2, \ldots, z_n, \phi(z), z_{n+2}, \ldots, z_{n+s+1}, v]^T$. The system (3) will play a key role in the definition of equivalence and in the solution of the realization problem.

Let $\mathcal{K}$ be the field of meromorphic functions in a finite number of the variables $\{z(0), v(t), t \geq 0\}$. Denote by $\mathcal{E}$ the formal vector space spanned by the differentials of the elements of $\mathcal{K}$; that is, $\mathcal{E} := \mathrm{span}_{\mathcal{K}}\{\mathrm{d}\varphi \mid \varphi \in \mathcal{K}\}$. The forward-shift operator $\delta : \mathcal{K} \to \mathcal{K}$ is defined by

$$\delta\varphi(z(0), v(j)) = \varphi(f_e(z(0), v(0)), v(j+1)).$$

The pair $(\mathcal{K}, \delta)$ is a difference field. $(\mathcal{K}, \delta)$ is not inversive in general. Nevertheless, it is always possible to imbed $\mathcal{K}$ into an inversive difference overfield $\mathcal{K}^*$, called the inversive closure of $\mathcal{K}$. In (Aranda-Bricaire *et al.*, 1996) an explicit construction of $(\mathcal{K}^*, \delta^*)$ is given which will be briefly recalled here.

Introduce a (nonunique) vector-valued function $p(0) = \varphi(z(0))$ such that $\mathrm{d}\varphi \in \mathrm{span}_{\mathcal{K}}\{\mathrm{d}z(0)\}$ and

$$\mathrm{span}_{\mathcal{K}}\{\mathrm{d}\bar{p}(0)\} \sim \frac{\mathrm{span}_{\mathcal{K}}\{\mathrm{d}z(0)\}}{\mathrm{span}_{\mathcal{K}}\{\mathrm{d}z(0)\} \cap \mathrm{span}_{\mathcal{K}}\{\mathrm{d}z(1)\}}$$

where $\mathrm{d}\bar{p}(0)$ denotes the coset associated to the element $\mathrm{d}p(0)$. Then there exist a vector-valued function $\psi(z(1), p(0))$ such that $z(0) = \psi(z(1), p(0))$ which defines the backward-shift operator

$$\delta^{-1}(z(0)) = \delta^{-1}\psi(z(1), p(0)) = \psi(z(0), p(-1))$$

and $\mathcal{K}^*$ is the field of meromorphic functions in a finite number of independent variables $\{z(0), v(t), v(-k), p(-k), t \geq 0, k \geq 1\}$. The operators δ and δ^{-1} induce the operators $\Delta : \mathcal{E} \to \mathcal{E}$ and $\Delta^{-1} : \mathcal{E} \to \mathcal{E}$, respectively, by

$$\sum_i a_i \mathrm{d}\varphi_i \mapsto \sum_i (\delta a_i)\mathrm{d}(\delta\varphi_i), \quad a_i, \varphi_i \in \mathcal{K}.$$

and

$$\sum_i a_i \mathrm{d}\varphi_i \mapsto \sum_i (\delta^{-1} a_i)\mathrm{d}(\delta^{-1}\varphi_i), \quad a_i, \varphi_i \in \mathcal{K}.$$

Hereinafter, we use the same symbol to denote the difference field $(\mathcal{K}, \delta)$ and its inversive closure.

A sequence of subspaces $\{\mathcal{H}_k\}$ of $\mathcal{E}$ is defined for $k \geq 1$ by

$$\begin{aligned} \mathcal{H}_1 &= \mathrm{span}_{\mathcal{K}}\{\mathrm{d}z(0)\} \\ &= \mathrm{span}_{\mathcal{K}}\{\mathrm{d}y(0), \ldots, \mathrm{d}y(n-1), \\ &\qquad \mathrm{d}u(0), \ldots, \mathrm{d}u(s)\}, \\ \mathcal{H}_{k+1} &= \mathrm{span}_{\mathcal{K}}\{\omega \in \mathcal{H}_k \mid \Delta\omega \in \mathcal{H}_k\}. \end{aligned} \quad (4)$$

The $\mathcal{H}_k$'s can alternatively be defined by

$$\mathcal{H}_{k+1} = \Delta^{-1}(\mathcal{H}_k \cap \Delta\mathcal{H}_k), \quad k \geq 1. \quad (5)$$

It is clear that the sequence (4) is decreasing. Denote by k^* the least integer such that

$$\mathcal{H}_1 \supset \cdots \supset \mathcal{H}_{k^*+1} = \mathcal{H}_{k^*+2} = \cdots =: \mathcal{H}_\infty \quad (6)$$

A $\mathcal{K}$-subspace of $\mathcal{E}$ spanned by a finite number of one-forms is called a Pfaffian system or a codistribution. Hereinafter we make use of some standard properties of Pfaffian systems (Abraham *et al.*, 1988). In particular, we make use of $\wedge$, the exterior or wedge product of differential forms, and d, the exterior derivative of a one-form.

Theorem (Frobenius). *Let $\mathcal{I}$ be a Pfaffian system generated by the one-forms $\{w_1, \ldots, w_s\}$. Suppose that for $k = 1, \ldots, s$ the condition*

$$\mathrm{d}w_k \wedge w_1 \wedge \ldots \wedge w_s = 0$$

is satisfied. Then there exists locally a system of coordinates $\{\xi_1, \ldots, \xi_s\}$ such that $\mathcal{I}$ is generated by $\{\mathrm{d}\xi_1, \ldots, \mathrm{d}\xi_s\}$. In this case the Pfaffian system is said to be completely integrable.

2.1 *Irreducible systems and transfer equivalence*

The sequence $\{u(t), y(t), t \geq 0\}$, is called a solution of (1), if, for any $t \geq 0$, $u(t), y(t)$ satisfy equation (1). Under assumption A1 the solution of (1) is uniquely determined for a given set of initial conditions $y(0), \ldots, y(n-1), u(0), \ldots, u(s)$ and a fixed input sequence. The converse does not hold. Higher order difference equations, with different functions φ and integers n and s, may describe the same i/o behaviour. This leads to the problem of equivalence of discrete-time nonlinear i/o systems: given an arbitrary recursive representation of Σ, the problem of finding a *minimal* recursive representation which is equivalent to Σ with the integers n and s as small as possible. The definition of transfer equivalence proposed in (Kotta *et al.*, 1998) is based upon the notion of the irreducible differential form of difference equations.

Definition (Irreducible system) (Kotta *et al.*, 1998) *The system Σ is said to be irreducible if the associated extended system Σ_e satisfies the accessibility condition $\mathcal{H}_\infty = 0$.*

If $\mathcal{H}_\infty = 0$, then Σ is already irreducible by definition. Of course, the system (3) does not necessarily satisfy the condition $\mathcal{H}_\infty = 0$. If the system is not irreducible, i.e. if $\mathcal{H}_\infty \neq 0$, then we may pick any nonzero exact form $\mathrm{d}\varphi_r \in \mathcal{H}_\infty$ which is called a reduced differential form of system (2). This can be done since $\mathcal{H}_\infty$ is integrable (Aranda-Bricaire *et al.*, 1996).

We may repeat the reduction procedure for the reduced form of system (2), i. e. for the system $\varphi_r(\cdot) = 0$, provided that this new system satisfies the assumptions A1–A3. At each step the order of the reduced form decreases and finally the procedure converges to an irreducible form.

Of course, even if the original system $\varphi(\cdot) = 0$ satisfies the assumptions A1–A3, this is not necessarily so for the reduced system $\varphi_r(\cdot) = 0$. In some cases, one can reduce an i/o equation and get a reduced (but not irreducible) form that does not satisfy the assumption $\varphi(0.\ldots, 0) = 0$.

If $\varphi_r(y(t), \cdots, y(t+n), u(t), \cdots, u(t+s)) = 0$ satisfying $\varphi_r(0, \cdots, 0) = 0$ is irreducible, then $\mathrm{d}\varphi_r$ is said to be an irreducible differential form.

Definition (Transfer equivalence) *(Kotta et al., 1998) Two systems Σ_1 and Σ_2, which are assumed to admit an irreducible form, are said to be transfer equivalent if they have the same irreducible differential form.*

2.2 *Realization*

The central problem of realization theory is one of constructing, for a given external description of the system (1), an internal description in the form of state equations

$$\begin{aligned} x(t+1) &= f(x(t), u(t)) \\ y(t) &= h(x(t)) \end{aligned} \quad (7)$$

It has been shown (Fliess, 1990) that a nonlinear i/o difference equation, in general, cannot be realized by the standard state space equations. One needs generalized state equations whose dynamics contain, in addition to the input, a finite number of its time shifts. Since the classical state representation is still dominant in the control literature, it is natural to consider the problem of whether there exists a classical state space realization of an input-output nonlinear difference equation.

Theorem (Kotta *et al.*, 1997) *The nonlinear system Σ described by the irreducible input-output difference equation (2) has a minimal (i.e. observable and accessible) state space realization iff for $1 \leq k \leq s+2$ the subspaces $\mathcal{H}_k$ defined by (4) are completely integrable.*

The sufficiency part of the proof of the theorem (see (Kotta *et al.*, 1997)) provides a constructive procedure (up to integration of one-forms) to obtain the state and output equations.

3. SYMBOLIC IMPLEMENTATION USING MAPLE V

In this section the description of the procedures, which have been implemented in Maple V for the study of irreducibility and realizability properties of nonlinear discrete-time i/o systems, are given. These procedures make extensive use of the algebraic framework presented in the previous section and rely partly on some modified procedures and theoretical results of (Post, 1995; Aranda-Bricaire, 1996). Moreover, they use the Maple V library LIESYMM. In order to make the computer algebra implementation more tractable, we use, as in (Post, 1995; Aranda-Bricaire, 1996), instead of the sequence of subspaces $\{\mathcal{H}_k\}$ a related sequence of decreasing subspaces $\{\mathcal{I}_k\}$ defined by

$$\begin{aligned} \mathcal{I}_1 &= \text{span}_{\mathcal{K}}\{dx(0)\} \\ \mathcal{I}_{k+1} &= \mathcal{I}_k \cap \Delta\mathcal{I}_k, k \geq 1 \end{aligned} \tag{8}$$

Lemma 1. $\mathcal{I}_k = \Delta^{k-1}\mathcal{H}_k$ *and* $\mathcal{I}_\infty = \mathcal{H}_\infty$.

Proof. Using repeatedly the definitions of subspaces $\mathcal{I}_k$ and $\mathcal{H}_k$ we get, $\mathcal{I}_1 = \mathcal{H}_1$, $\mathcal{I}_2 = \Delta\mathcal{H}_2$, and

$$\begin{aligned} \mathcal{I}_3 = \mathcal{I}_2 \cap \Delta\mathcal{I}_2 = \Delta\mathcal{H}_2 \cap \Delta^2\mathcal{H}_2 = \\ \Delta(\mathcal{H}_2 \cap \Delta\mathcal{H}_2) = \Delta(\Delta\mathcal{H}_3) = \Delta^2\mathcal{H}_3 \end{aligned} \tag{9}$$

Proceeding in this fashion, proves the first statement in the lemma. To prove the second statement, we note that $\Delta^r\mathcal{H}_{k^*+1} = \mathcal{H}_{k^*+1}$ for arbitrary $r > 0$, so

$$\mathcal{I}_\infty = \mathcal{I}_{k^*+1} = \Delta^{k^*}\mathcal{H}_{k^*+1} = \mathcal{H}_{k^*+1} = \mathcal{H}_\infty.$$

Lemma 2. $\mathcal{I}_k$ *is completely integrable iff* $\mathcal{H}_k$ *is completely integrable.*

So, alternatively, we may call the system Σ irreducible if the associated extended system Σ_e satisfies $\mathcal{J}_\infty = 0$, and we may give the alternative realizability conditions as follows:

Theorem *The nonlinear system* Σ *described by the irreducible input-output difference equation (2) has a minimal (i.e. observable and accessible) state space realization iff for* $1 \leq k \leq s+2$ *the subspaces* $\mathcal{I}_k$ *defined by (8) are completely integrable.*

The main difference between the sets of subspaces (4) and (8) is that (8) does not require the construction of the backward-shift operator, which in turn relies on the use of the implicit function theorem, and is therefore extremely difficult to implement in a computer algebra framework. The new subspaces require only standard linear algebraic techniques, and therefore can be straightforwardly implemented in Maple. The drawback of the sequence of new subspaces $\{\mathcal{I}_k\}$ is that these cannot be used to construct the coordinates of the state vector and therefore, for finding the state equations of the i/o difference equations.

The procedures that have been implemented in Maple V are described below.

IOExtended. This procedure finds the extended state-space system associated with the i/o difference equation.

ForwardShiftBasis(fe : *procedure*; B : *set*). This procedure computes the forward-shifted one-forms corresponding to a basis of a subspace in $\mathcal{E}$. Its input is fe, the state transition map of the extended system (3) and B, a basis of one-forms. Its output is *Bplus*, the set consisting of the forward-shifted one-forms of B.

BackwardShiftBasis(fe : *procedure*; B : *set*). This procedure computes the backward-shifted one-forms corresponding to a basis of a subspace in $\mathcal{E}$. Its input is fe, the state transition map of the extended system (3) and B, a basis of one-forms. Its output is *Bminus*, the set consisting of the backward-shifted one-forms of B.

IntersectionSpace($S1$: *set*; $S2$: *set*). This procedure computes a basis of the intersection of the two subspaces of $\mathcal{E}$, spanned by $S1$ and $S2$. Its inputs are two sets of one-forms $S1$ and $S2$ and its output is IS, a basis of the intersection of the spaces spanned by $S1$ and $S2$.

Integrability(P : *set*, *vars* : *list*). This procedure determines on the basis of the Frobenius theorem whether the Pfaffian system spanned by the set of one-forms $\mathcal{P}$ is completely integrable. Its inputs are P, a set of one-forms, and *vars*, a list of variables. Its output is a Boolean variable. Note that this procedure is implemented using the Maple V package LIESYMM.

IntegratePfaff. This procedure tries to integrate the completely integrable Pfaffian system.

ISpaces This procedure computes the sequence of subspaces $\mathcal{I}_k$. Since this sequence of subspaces has non-increasing dimensions, the computations stop when the dimensions do not increase. The first two spaces $\mathcal{I}_1$ and $\mathcal{I}_2$ are computed straightforwardly. The next spaces are computed using the procedures ForwardShiftBasis and IntersectionSpace.

HSpaces This procedure computes the sequence of subspaces $\mathcal{H}_k$. The first two spaces $\mathcal{H}_1$ and $\mathcal{H}_2$ are computed straightforwardly. The next spaces are computed with the procedures BackwardShiftBasis and IntersectionSpace.

Irreducibility. (fe : *procedure*) This procedure determines whether the nonlinear higher order i/o difference equation is irreducible or not. This is done by computing the sequence of subspaces $\mathcal{I}_k$. The system is irreducible if the space $\mathcal{I}_\infty$ contains only the null-vector; otherwise the system is not irreducible. The input is fe, the state transition map of the extended system and the output is a Boolean variable.

Reduction. This procedure finds the irreducible differential form, if possible, or advises that the system does not have an irreducible form.

ReducedForm. This procedure finds, on the basis of the reduced differential form, the reduced form (i.e. it integrates the differential form). Note that this can be done, at least in principle, since $\mathcal{I}_\infty$ is completely integrable.
Equivalence. This procedure determines whether two nonlinear i/o difference equations are transfer equivalent or not.
Realizability. (*fe* : *procedure*) This procedure determines whether the nonlinear higher order i/o difference equation can be realized in the classical state space form. This is done by computing the sequence of subspaces $\{\mathcal{I}_k\}$ for $k = 1, \ldots, s+2$ and checking if all the Pfaffian systems $\mathcal{I}_1, \ldots, \mathcal{I}_{s+2}$ are completely integrable or not. This procedure uses the procedure Integrability and is implemented using the Maple V package LIESYMM. The inputs are *fe*, the state transition map of the extended system. The output is a Boolean variable.
Realization. This procedure tries to find the state variables that transform the i/o system into the classical form, and transforms the system into the Kalmanian form.

4. EXAMPLE

Consider the model of a column-type grain drying process obtained via identification (Jaaksoo *et al.*, 1982):

$$\begin{aligned} y(t+3) = {} & 1.6389y(t+2) - 0.4397y(t+1) \\ & -0.1803y(t) - 0.0082y(t+2)u(t+2) \\ & -0.0042y(t+1)u(t+1) - 0.0074y(t)u(t) \\ & +0.0021u(t) + 0.0019u(t+2) \\ & -0.0041u(t+1) \end{aligned} \quad (10)$$

First we have to find the state-transition map *fe* of an extended state-space system that corresponds to (10)

$$\begin{aligned} fe := {} & [z2, z3, 1.6389z3 - 0.4397z2 - 0.1803z1 \\ & -0.0082z3z6 - 0.0042z2z5 - 0.0074z1z4 \\ & +0.0021z4 + 0.0019z6 - 0.0041z5, z5, z6, v] \end{aligned}$$

The Maple V session below corresponds to the analysis of the system (10). We first use the procedure Irreducibility to check whether the system (10) is irreducible or not, because the Realizability procedure works only on irreducible systems. We are interested in obtaining only the minimal realizations.

Irreducibility(fe)

The system is irreducible

The procedure Realizability checks whether the system is transformable into the classical state-space form or not.

Realizability(fe)

The system is realizable in the classical state space form

Once we know that the system is realizable in the classical state space form, we may try to find the state coordinates which place the system into the Kalmanian form. For this purpose one can use the procedure *Realization*. Below we will examine the process of realization in detail looking at the outputs of the subprocedures inside *Realization*.

The state coordinates, according to the results of (Kotta *et al.*, 1997), are defined by $\mathcal{H}_{s+2}$, i.e. in the case of our example, by $\mathcal{H}_4$. Note that the subspaces are computed recursively using the procedure *HSpaces(fe)*. We need only $\mathcal{H}_4$ which is as follows:

$$\begin{aligned} \mathcal{H}_4 = \text{span}_{\mathcal{K}}\{ & \mathrm{d}z_1(0), \mathrm{d}z_2(0) + 0.0082z_1(0)\mathrm{d}z_4(0) \\ & -0.0019\mathrm{d}z_4(0), \mathrm{d}z_3(0) + [(0.0082z_2(0) \\ & -0.0019)(0.001 + (0.0176 - 0.000067z_5(0)) \\ & z_1(0) + 0.000015z_5(0)]\mathrm{d}z_5(0)\}. \end{aligned}$$

In order to find the state coordinates we have to integrate the one-forms, which define the basis of $\mathcal{H}_4$. Note that these are integrable in principle, since according to the procedure Realizability the system can be transformed into the classical state space form. Nevertheless, the most complicated part of our procedures is to integrate the obtained one-forms. At present the procedure *IntegratePfaff* that has been written for this purpose, does not work with more complicated forms. Unfortunately, it is unable to integrate the one-forms in our $\mathcal{H}_4$. In order to complete our task, we have to simplify the basis of $\mathcal{H}_4$ interactively (manually) using linear transformation. We change the basis of $\mathcal{H}_4$ manually as shown below (where $\mathcal{H}_{4i}$ denotes the ith one-form in the span of the original $\mathcal{H}_4$, obtained via the computer):

$$\begin{aligned} \mathcal{H}_4 = \text{span}_{\mathcal{K}}\{ & \mathcal{H}_{41}, \mathcal{H}_{42} + 0.0082z_4(0)\mathcal{H}_{41}, \\ & \mathcal{H}_{43} + 0.0082z_5(0)\mathcal{H}_{42} + 0.0176z_4\mathcal{H}_{41}\}. \end{aligned}$$

Now *IntegratePfaff* is able to integrate the one-forms in $\mathcal{H}_4$ and we get the state coordinates in terms of the variables of the extended system. In terms of the input and output variables:

$$\begin{aligned} x_1(t) &= y(t) \\ x_2(t) &= y(t+1) + 0.0082y(t)u(t) - 0.0019u(t) \\ x_3(t) &= y(t+2) + 0.0082y(t+1)u(t+1) \\ & \quad - 0.0019u(t+1) + 0.0176y(t)u(t) \\ & \quad + 0.0010u(t) \end{aligned}$$

Finally, the classical state space form can be found:

$$x_1(t+1) = x_2(t) - 0.0082x_1(t)u(t) + 0.0019u(t)$$
$$x_2(t+1) = x_3(t) - 0.0010u(t) - 0.0176x_1(t)u(t)$$
$$x_3(t+1) = 0.0004u(t) - 0.1803x_1(t) - 0.4397x_2(t) + 1.6389x_3(t) - 0.0327x_1(t)u(t)$$

5. CONCLUSIONS

Maple V procedures have been developed that allow one to decide whether a given system can be realized in the classical state-space form. Our procedure, which finds the state variables that transforms the system into the Kalmanian form, is applicable only in case of relatively simple examples and not for the general case. The reason is that Maple is unable to solve the complicated partial differential equations which define the state variables. The possibility of widening the applicability of the realization procedure is currently under the study. The other problematic part is finding the backward shift operator which requires solving the system of nonlinear algebraic equations. These difficulties have already been reported (De Jager, 1995).

6. REFERENCES

Abraham, R., J.E. Marsden, and T. Ratiu. (1988). *Manifolds, Tensor Analysis, and Applications.* Springer-Verlag, New York.

Akhrif, O. and G.L. Blankenship. (1989). Symbolic computations in differential geometry applied to nonlinear control systems. In: *Symbolic Computation. Applications to Scientific Computing*, R. Grossman (Ed.), SIAM, Philadelphia, 53–96.

Amirat, Y.A. and S. Diop. (1996). Towards an algebraic system theory toolbox. *Proc. 13th IFAC World Congress*, San Francisco, v.F, 299–304.

Aranda–Bricaire, E. (1996). Computer algebra analysis of discrete–time nonlinear systems. *Proc. 13th IFAC World Congress*, San Francisco, v. F, 305–309.

Aranda-Bricaire, E., U. Kotta, and C.H. Moog. (1996). Linearization of discrete-time systems. *SIAM J. Control Optim.*, **34**, 1999–2023.

Barbot, J. P. (1990). A forward accessibility algorithm for nonlinear discrete-time systems. *Lecture Notes in Control and Inf. Sci*, **144**, 314–323.

Besancon, G. and G. Bornard. (1997) Symbolic computation in nonlinear control systems analysis: the example of bilinearization. *Mathematics and Computers in Simulation*, **44**, 11–27.

Birk, J. and M. Zeitz. (1991). Program for symbolic and rule-based analysis and design of nonlinear systems. In: *Algebraic Computing in Control*, Lecture Notes in Control and Inf. Sci., **165**, 115–128.

Van Essen, H. and B. de Jager. (1993). Analysis and design of nonlinear control systems with symbolic computation system MAPLE. *Proc. 2nd European Control Conference*, Groningen, 2081–2085.

Fliess, M. (1990). Automatique en temps discret et algebre aux differences. *Forum Mathematicum*, **2**, 213–232.

Glumineau, A. and L. Graciani. Smbolic nonlinear analysis and control package. *Proc. the 13th IFAC World Congress*, San Francisco, v. F, 295–298.

Jaaksoo, U.J., E.M. Talvis, and J.M. Ummer. (1982). Microprocessor-based grain dryer control. *Prep. 3rd IFAC/IFIP Symp. on Software for Computer Control*, Madrid, 469–473.

De Jager, B. (1994). Symbolic solution for a class of partial differential equations. *Proc. of the Rhine Workshop on Computer Algebra*, J. Calmet (Ed.), Karlsruhe, Germany, 84–88.

De Jager, B. (1995). Symbolics for control: MAPLE used in solving the exact linearization problem. In: *Computer Algebra in Industry 2: Problem Solving in Practice*, A.M. Cohen, M.van Gastel and S. Verduyn Lunel (Eds), John Wiley, Chichester, 291–311.

De Jager, B. (1995). The use of symbolic computation in nonlinear control: is it viable? *IEEE Trans. on Automatic Control*, **40**, 84–89.

Kotta, Ü., P. Liu and A. Zinober. (1997). State-space realization of input-output nonlinear difference equations. *Proc. ECC'97*, Brussels, paper N 851, 6pp.

Kotta, U., Liu, P. and A.S.I. Zinober. (1998). Transfer equivalence of nonlinear higher order difference equations. *Proc. of IFAC Conf. on Nonlinear Control Systems*, Enschede,

Kotta, Ü. and R. Roomeldi. (1990). An application of REDUCE for design of nonlinear control systems. *Proc. 4th Int. Conf. on Problems of Complex Automatization*, Kiev, 187–191.

Post, B.A. (1995). Computer algebra analysis of nonlinear discrete-time systems. MSc thesis, University of Twente.

Rothfuss, R. and M. Zeitz. (1996). A toolbox for symbolic nonlinear feedback design. *Proc. the 13th IFAC World Congress*, San Francisco, vol. F, 283–288.

Chaos in One-Dimensional Sampled-Data Systems with Saturating Controls

Xiao Fan Wang, Zhi Quan Wang

Department of Automatic Control, Nanjing University of Science & Technology
Nanjing, 210094 P. R. China
e-mail: wangxf@mail.njust.edu.cn

Abstract: A theoretical analysis of chaotic dynamics in a simple one-dimensional sampled-data control system with a saturating actuator is presented. It is shown that there is a sampling time interval in which chaos is generated. Analytical expression for the lower and the upper bound of such time interval are computed. The bounds are verified via an example. *Copyright © 1999 IFAC*

Keywords: Chaos, Discrete-time systems, Sampled-data systems, Saturation, Stability.

1. INTRODUCTION

Recently, eliminating or producing chaos in a controlled way has attracted a great deal of interests due to its potential applications in several technological fields (Chen and Dong, 1998, and references therein). On the other hand, it has been shown that chaos exists in many control systems. Indeed, in the past two decades, complex dynamical behaviors in feedback control systems have been intensively studied from control system perspectives (Chen, 1999, and references therein). It should be noted that, in spite of the complexity of chaotic motions, they commonly occur in systems which themselves are not complex and are even surprisingly simple.

A widely accepted mathematical definition of chaos was introduced by Li and Yorke (1975), where a surprisingly simple condition ensuring the existence of chaos in a one-dimensional discrete-time system was derived: "period three implies chaos". Later, Marotto (1978) extended their work to multi-dimensional systems by introducing the concept of snap-back repellers; that is, "snap-back repellers imply chaos in $\Re^n$". Shiraiwa and Kurata (1980) further generalized the theorem of Marotto to include not only repellers but also saddle points, and showed that their theorem is in a sense a generalization of Smale's results on transversal homoclinic points (Smale, 1968). Ushio and Hirai [1983, 1985] verified the existence of chaos in nonlinear sampled-data control systems by applying Marotto's theorem and Shiraiwa-Kurata's theorem. More recently, Chen and Aihara [1997] used Marotto's theorem to show that both transiently chaotic neural networks (TCNN) and discrete-time recurrent neural networks (DRNN) are chaotic in the sense of Li and Yorke.

Even though Ushio and Hirai(1983) showed that a class of nonlinear sampled-data control systems with sampling period T is chaotic if $T > T^*$ for sufficiently large T^*, they did not tell us how to obtain the critical sampling period T^*. Moreover, the state of a sampled-data control system can be unbounded if the sampling period T is too large. In other words, there exists another critical sampling period $T^{**}(> T^*)$ such that the state of the system is bounded only when $T < T^{**}$.

Saturation is one of the most commonly observed nonlinearities in the real world of control engineering.

In this paper, for a one-dimensional sampled-data control system with a saturating actuator, analytical expression for the two critical sampling periods T^* and T^{**} are obtained. By applying Marotto's theorem, the sampling signal is proved to be chaotic in the sense of Li and Yorke when the sampling period belongs to the interval (T^*, T^{**}). Numerical simulation is also provided as illustrating example.

2. PRELIMINARIES

Consider a one-dimensional discrete-time system of the form

$$x(k+1) = f(x(k), k = 0,1,..., \qquad (1)$$

where $x(k) \in \Re$ and $f: \Re \to \Re$ is continuous. Let f^t denote the composition of f with itself t times, then a point x is a p-periodic point if $f^p(x) = x$ but $f^l(x) \neq x$ for $1 \le l < p$. If $p=1$, that is, $f(x) = x$, x is called a fixed point. Let $f'(x)$ be the derivative of f at the point x. $B(x;r)$ be a closed interval in $\Re$ of radius r centered at the point x.

Li and Yorke (1975) have established analytically that the existence of a 3-periodic point is a sufficient condition for the existence of chaos. That is, there exist

(i) a positive integer n such that for each integer $p \ge n$, (1) has p-periodic points;
(ii) a scrambled set, i.e., an uncountable, invariant set S containing no periodic points such that:

$$\limsup_{k\to\infty} \left| f^k(x) - f^k(y) \right| > 0,$$

for all $x \in S$, $y \in S$, $y \neq x$ or y any periodic point of f;
(iii) an uncountable subset S_0 of S such that for every $x, y \in S_0$:

$$\liminf_{k\to\infty} \left| f^k(x) - f^k(y) \right| = 0.$$

Marotto (1978) extended Li-Yorke's work to arbitrary n-dimensional systems by defining snap-back repellers, that is, "snap-back repellers imply chaos in $\Re^n$". For the one-dimensional case, it can be shown that the existence of a snap-back repeller of f is equivalent to the existence of a 3-period point for the function f^n for some positive integer n. A fixed point x^* of the one-dimensional system (1) is said to be a snap-back repeller of (1) if

(i) There exists a number $r>0$ such that f is differentiable and $|f'(x)| > 1$ for all $x \in B(x^*;r)$.

(ii) There exists a point $x^0 \in B(x^*;r)$ with $x^0 \neq x^*$ such that, for some positive integer m, $f^m(x^0) = x^*$ and f^m is differentiable at x^0, $(f^m)'(x_0) \neq 0$.

3. CHAOS IN ONE-DIMENSIONAL SAMPLED-DATA SATURATING SYSTEMS

A one-dimensional sampled-data control system with a saturating actuator is described by

$$\dot{x}(t) = ax(t) + bu(t), \qquad (2)$$

where $x(t) \in \Re$, a, b are constants which satisfy $0 < a < b$. The saturated control input $u(t)$ is taken to be

$$\begin{aligned} u(t) &= -Sat(x(kT)) \\ &= -\tfrac{1}{2}\left[|x(kT) + \Lambda| - |x(kT) - \Lambda| \right] \qquad (3) \\ &= \begin{cases} -\Lambda, & x(kT) > \Lambda \\ -x(kT), & |x(kT)| \le \Lambda \\ \Lambda, & x(kT) < -\Lambda \end{cases} \end{aligned}$$

when $kT \le t \le (k+1)T$, $k = 0, 1, \cdots$, where T is a sampling period, $\Lambda > 0$ is a constant.

Let $x(k) = x(kT)$, $u(k) = u(kT)$. It is easy to shown that the sampling signal satisfies the following equation:

$$\begin{aligned} x(k+1) &= g(x(k)) \\ &= e^{aT}x(k) - \frac{b}{2a}\left(e^{aT} - 1\right)\left\{ |x(k) + \Lambda| - |x(k) - \Lambda| \right\} \end{aligned} \qquad (4)$$

Theorem 1. 0 is a snapback repeller of (4), if

$$T > T^* \equiv \frac{1}{a}\ln\left(\frac{b + \sqrt{ab}}{b - a} \right). \qquad (5)$$

Proof. By (4), we have

$$g'(y) = \left(1 - \tfrac{b}{a}\right)e^{aT} + \tfrac{b}{a}, \quad \text{for all } y \in B(0;\Lambda).$$

Condition (5) implies that

$$|g'(y)| > \sqrt{\tfrac{b}{a}} > 1, \quad \text{for all } y \in B(0;\Lambda). \quad (6)$$

Let

$$y_0 = \frac{(e^{aT}-1)b\Lambda}{e^{2aT}(a-b)+e^{aT}b},$$

it is easy to show that $-\Lambda < y_0 < 0$. Hence we have

$$\begin{aligned} y_1 &= g(y_0) = \left((1-\tfrac{b}{a})e^{aT}+\tfrac{b}{a}\right)y_0 \\ &= \tfrac{b}{a}(1-e^{-aT})\Lambda > \Lambda \end{aligned}$$

$$\begin{aligned} y_2 &= g^2(y_0) = g(y_1) \\ &= e^{aT}y_1 - \tfrac{b}{a}(e^{aT}-1)\Lambda = 0 \end{aligned},$$

and

$$(g^2)(y_0) = g(y_1)g(y_0) = e^{aT}\left((1-\tfrac{b}{a})e^{aT}+\tfrac{b}{a}\right) \neq 0$$

which implies that 0 is a snap-back repeller of (4).

Remark 1. Theorem 1 implies that, for $T > T^*$, the sampling signal $x(k) = x(kT)$ of (1), or equivalently, system (4) is chaotic if the sampling signal is bounded. It is surprising to note here that T^* is unrelated with Λ!

Theorem 2. Suppose that

$$\begin{aligned} &b \geq 2a, \Lambda \geq 1, \\ &T < T^{**} \equiv \frac{1}{a}\ln\frac{2b\Lambda}{b\Lambda-a}. \end{aligned} \quad (6)$$

If $|x(0)| < \tfrac{b}{a}\Lambda$, then

$$|x(k)| < \tfrac{b}{a}\Lambda, \quad k = 1, 2, \cdots.$$

Moreover, if $T^* < T^{**}$, then $x(k)$ is a chaotic signal for $T^* < T < T^{**}$.

Proof. We need only to prove that, if $|x(k)| < \tfrac{b}{a}\Lambda$, then $|x(k+1)| < \tfrac{b}{a}\Lambda$ for any *k*.

(i) If $|x(k)| \leq \Lambda$, then

$$x(k+1) = \left((1-\tfrac{b}{a})e^{aT}+\tfrac{b}{a}\right)x(k).$$

By (6), we have

$$\begin{aligned} \left|(1-\tfrac{b}{a})e^{aT}+\tfrac{b}{a}\right| &< \left(\frac{b}{a}-1\right)\left(\frac{2b\Lambda}{b\Lambda-a}\right)-\frac{b}{a} \\ &= \frac{2b}{a}\left(\frac{(b-a)\Lambda}{b\Lambda-a}\right)-\frac{b}{a} \leq \frac{b}{a} \end{aligned}$$

which implies that $|x(k+1)| < \tfrac{b}{a}\Lambda$.

(ii) If $\Lambda < x(k) < \tfrac{b}{a}\Lambda$, then

$$\begin{aligned} x(k+1) &= e^{aT}x(k) - \tfrac{b}{a}(e^{aT}-1)\Lambda \\ &< e^{aT}\left(\tfrac{b}{a}\Lambda\right) - \tfrac{b}{a}(e^{aT}-1)\Lambda < \tfrac{b}{a}\Lambda. \end{aligned}$$

On the other hand,

$$\begin{aligned} x(k+1) &> e^{aT}(1-\tfrac{b}{a})\Lambda + \tfrac{b}{a}\Lambda \\ &> \frac{b}{a}\Lambda - \frac{2b\Lambda}{a}\left(\frac{(b-a)\Lambda}{b\Lambda-a}\right) > -\frac{b}{a}\Lambda \end{aligned}$$

Therefore, $|x(k+1)| < \tfrac{b}{a}\Lambda$.

(iii) If $-\Lambda > x(k) > -\tfrac{b}{a}\Lambda$, the same result can be proved by a similar argument as in the proof of (ii).

4. EXAMPLE

Consider a special case of the system (1) by letting:

$$a = 0.2, b = 1, \ \Lambda = 1.$$

That is

$$\dot{x}(t) = 0.2x(t) + u(t)$$

$$u(t) = \begin{cases} -1, x(kT) > 1, \\ -x(kT), |x(kT)| \leq 1, \\ 1, x(kT) < -1, \end{cases}$$

when $kT \leq t < (k+1)T$, $k = 0, 1, \cdots$. For this example, we have

$$\begin{aligned} T^* &= \frac{1}{0.2}\ln\frac{1+\sqrt{0.2}}{1-0.2} \approx 2.9639, \\ T^{**} &= \frac{1}{0.2}\ln\frac{2}{1-0.2} \approx 4.5815. \end{aligned}$$

Theorem 1 and Theorem 2 reveal that chaos should occur for 2.9639<T<4.5815. Fig. 1 shows the sampling signal *x(kT)* (800≤k≤1000) and the corresponding Lyapunov exponent with increasing sampling period *T*. It is shown that chaos appears in two bands for T>2.03. However, the attractor widens into a single chaotic band for T>T*, implies that (5) is a sufficient condition for the existence of chaos.

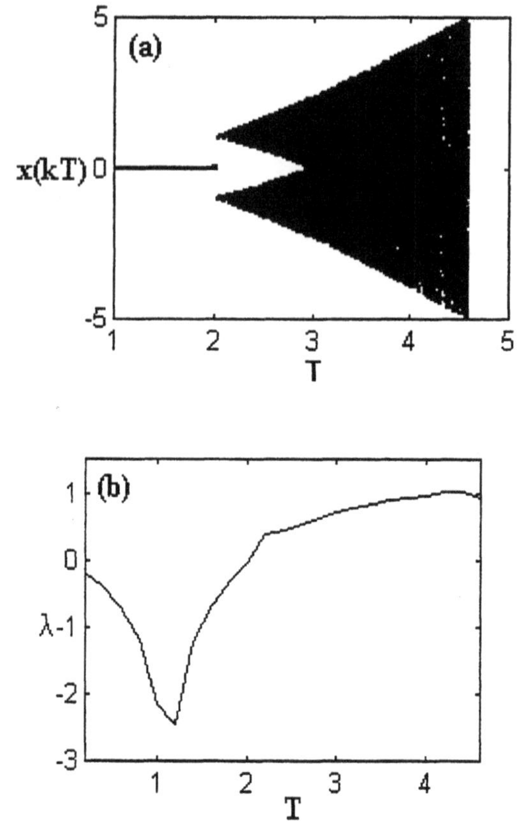

Fig.1. The sampling signal $x(kT)$ (Fig.1(a)) and the Lyapunov exponent (Fig.1(b)) with increasing sampling period T.

5. CONCLUSION

This paper deals with the theoretical analysis of one-dimensional sampled dynamics under control saturation. It is shown that there is a sampling time interval in which chaos is generated. In particular, analytical expression for the lower and the upper bound of such time interval are computed.

REFERENCES

Chen, G. (1999). Controlling Chaos and Bifurcations in Engineering Systems. CRC Press, Florida.

Chen, G. and X. Dong. (1998). *From Chaos to Order: Methodologies, Perspectives and Applications.* World Scientific, Singapore.

Chen, L. and K. Aihara. (1997). Chaos and asymptotical stability in discrete-time neural networks. *Physica D,* **104,** 286-325.

Li, T. Y. and J. A. Yorke (1975). Period three implies chaos. *Amer. Math. Monthly,* **82,** 985-992.

Marotto, F. R. (1978). Snap-back repellers imply chaos in R^n. *J. Math. Anal. Appl.*, **63,** 199-223.

Shiraiwa, K. and M. Kurata (1980). A generalization of a theorem of Marotto. *Proc. Japan Acad.*, **55,** 286-289.

Smale, S. (1968). Diffeomorphisms with many periodic points. In *Differential and Combinatorial Topology.* 63-80, Princeton University Press, Princeton, NJ.

Ushio, T. and K. Hirai (1983). Chaos in non-linear sampled-data control systems. *Int. J. Control,* 38, 1023-1033.

Ushio, T. and K. Hirai (1985). Chaotic behavior in piecewise-linear sampled-data control systems. *Int. J. Nonlinear Mechanics,* **20,** 493-506.

NONLINEAR H_∞ FILTERING OF SAMPLED-DATA SYSTEMS

Sing Kiong Nguang* Peng Shi**

* *The department of Electrical and Electronic Engineering*
The University of Auckland
Private Bag 92019, Auckland, New Zealand.
Email: sk.nguang@auckland.ac.nz
** *Center for Industrial and Applied Mathematics*
School of Mathematics, The University of South Australia
The Levels, SA 5095 Australia.
Email: matsp@phoenix.levels.unisa.edu.au

Abstract: This paper considers the problem of $\mathcal{H}_\infty$ filtering for nonlinear sampled-data systems. Sufficient conditions are obtained for the existence of filters that guarantee the $\mathcal{L}_2$ gain from an exogenous input to an estimation error is less than or equal to a prescribed value. These conditions are expressed in terms of both continuous and discrete Hamilton-Jacobi equations. *Copyright © 1999 IFAC*

Keywords: Sampled-data, Hamilton-Jacobi equations, and Nonlinear systems.

1. INTRODUCTION

Past years have seen a numerous industrial systems that are continuous time processes but controlled by a digital controller. In general, there are two classic approaches which have been adopted in engineering practice for the design of a discrete controller. The first approach is, first designing a continuous controller such that the closed-loop system has satisfactory performances, and then discretising the continuous controller into a discrete controller using one of the discretisation methods. This technique has the advantage that the synthesis is done in continuous time, but it may have a serious degradation in performances when the sampling is not fast enough. The second approach is, discretising the process first, then designing a discrete controller. The main benefit of the approach is that the synthesis procedure is again simplified. However, the discrete models obtained by discretisation fail to represent the intersample behaviour of the systems. The main drawback of these approaches is neither of them offers an adequate framework for analysis of the sampled-data system which requires the consideration of both sampled and intersample behaviours.

Recently, a number of different approaches have been proposed to incorporate intersample behaviour. These approaches include lifting technique (Bamieh *et al.*, 1991; Toivonen, 1992; Bamieh and Pearson, 1992; Yamamoto, 1991), descriptor system technique (Hara and Kabamba, 1990), and technique based on linear systems with jumps (Shi, 1996; Shi, 1998; Sun *et al.*, 1993; Shi, 1994; Sivashankar and Khargonekar, 1994). The lifting technique consists of transforming the original sampled-data system into an equivalent LTI discrete-time system with infinite-dimensional input-output signal space. Then $\mathcal{L}_2$ induced norm of the sampled-data system is shown to be less than one if and only if the $\mathcal{H}_\infty$ of this equivalent discrete system is less than one. In the descriptor system approach, on the other hand, the system is represented by a hybrid state space model, from which the descriptor system is formulated. The solution of the $\mathcal{H}_\infty$ sampled-data problem is then characterised by the solution of certain associated Hamilton-Jacobi equation. In contrast with these procedures, the theory of linear systems with jumps allows a direct characterisation of the problem in the similar terms to those of standard LTI $\mathcal{H}_\infty$ control problems, and leads to a pair of Riccati equations. Despite the procedural differences in all these approaches, the results obtained are mathematically equivalent.

At the same time, the problem of nonlinear $\mathcal{H}_\infty$ control problem has been studied by a number of researchers; see for instance, (Ball and Helton, 1989; Basar, 1991; van der Schaft, 1992; Isidori and Astolfi, 1992; Isidori, 1991). There are two commonly used approaches for providing solutions to nonlinear $\mathcal{H}_\infty$ control problems. One is based on the dissipativity theory and theory of differential games (see (Basar, 1991; Ball and Helton, 1989)). Another is based on the nonlinear version of classical Bounded Real Lemma as developed by Willems (Willems, 1972), and Hill and Moylan (Hill and Moylan, 1980); see, e.g., (van der Schaft, 1991; Isidori, 1991; Isidori and Astolfi, 1992). Both of these approaches convert the problem of nonlinear $\mathcal{H}_\infty$ control to the solvability of the so-called Hamilton-Jacobi equation (HJE). A nice feature of these results is that they are parallel to the linear $\mathcal{H}_\infty$ results. Further research along the line of the dissipativity theory and theory of differential games has been attempted (see, e.g., (Ball *et al.*, 1993; Isidori and Astolfi, 1992; Isidori, 1991)) where results on disturbance attenuation for nonlinear systems via state feedback and/or output feedback have been provided. In (Nguang and Fu, 1994; Berman and Shaked, 1996; Nguang and Fu, 1996), solutions to the nonlinear $\mathcal{H}_\infty$ filtering problem have been obtained. These continuous nonlinear $\mathcal{H}_\infty$ results have further been extended to discrete nonlinear systems (Lin and Byrnes, 1996; Lin and Byrnes, 1995). In (Suzuki *et al.*, 1995), sufficient conditions for the existence of a solution to the problem of $\mathcal{H}_\infty$ control nonlinear systems with sampled-data have been developed. Their sufficient conditions are expressed in terms of both Hamilton-Jacobi equations and Riccati differential equation with jump.

In the present work, the nonlinear $\mathcal{H}_\infty$ filtering theory is extended to a nonlinear $\mathcal{H}_\infty$ estimator for nonlinear sampled-data systems. This problem can be stated as follows: given a dynamic system with exogenous input and sampled output measurements at discrete instants of time, design a filter to estimate an unmeasured output such that the $\mathcal{L}_2$ norm of the mapping from the exogenous input to the estimation error is minimised or no larger than some prescribed level. The performance requirements are defined directly in terms of the continuous-time signals which means the intersample behaviour is incorporated in the design. Based on a nonlinear version of the classical bounded real lemma theorem for sampled-data systems, sufficient conditions for the existence of a filter that guarantees the $\mathcal{L}_2$ gain from an exogenous input to an estimation error is less than or equal to a prescribed value are derived. These conditions are expressed in terms of both continuous and discrete Hamilton-Jacobi equations. Since filtering results are closely related to output feedback results, following from (Suzuki *et al.*, 1995), it is easy to see that the filtering condition given in there is expressed in terms of Riccati differential equation with jump. However, sufficient conditions given here are expressed in terms of both continuous and discrete Hamilton-Jacobi equations, which are much more general than the filtering condition given in (Suzuki *et al.*, 1995). Another difference between the results of this paper and the result given in (Suzuki *et al.*, 1995) is the derivation of results. Here the results are derived based on classical bounded real lemma theorem for sampled-data systems, whereas in (Suzuki *et al.*, 1995) the results are derived based on dissipativity theory and theory of differential games.

The introduction will end with some remarks on the notations. The notations used in this paper are fairly standard. The superscript t denotes matrix transposition. $\mathcal{L}_2[0,T]$ stands for the space of square integrable vector functions over $[0,T]$, $\ell_2(0,T)$ is the space of square summable vector sequences over $(0,T)$. $f(\theta^-)$ and $f(\theta^+)$ denote for the left limit and right limit of a function $f(\theta)$, respectively. The gradient of a scalar-valued function $f(\xi)$ along the vector $\xi \in \mathbb{R}^n$ is a column vector vector denoted by $\nabla_\xi f(\xi)$, and the gradient of a column vector-valued function

$$\begin{bmatrix} \phi_1(\xi) \\ \phi_2(\xi) \\ \vdots \\ \phi_m(\xi) \end{bmatrix} \tag{1}$$

is a matrix defined by

$$\mathcal{D}_\xi \phi(\xi) = \begin{bmatrix} \frac{\partial \phi_1(\xi)}{\partial \xi_1} & \frac{\partial \phi_1(\xi)}{\partial \xi_2} & \cdots & \frac{\partial \phi_1(\xi)}{\partial \xi_n} \\ \frac{\partial \phi_2(\xi)}{\partial \xi_1} & \frac{\partial \phi_2(\xi)}{\partial \xi_2} & \cdots & \frac{\partial \phi_2(\xi)}{\partial \xi_n} \\ \vdots & \vdots & \ddots & \vdots \\ \frac{\partial \phi_m(\xi)}{\partial \xi_1} & \frac{\partial \phi_m(\xi)}{\partial \xi_2} & \cdots & \frac{\partial \phi_m(\xi)}{\partial \xi_n} \end{bmatrix}. \tag{2}$$

2. DEFINITION AND PRELIMINARY RESULTS

Consider the nonlinear time-varying system with finite discrete jumps:

$$\Sigma_1 : \begin{array}{rl} \dot{x}(t) & = A(x(t),t) + B(x(t),t)w(t), \\ & \quad t \in [0,T], t \neq ih;\ x(0) = 0 \\ x(ih) & = A_d(x(ih^-),ih) + \\ & \quad B_d(x(ih^-),ih)v(ih) \\ z(t) & = L(x(t),t), t \in [0,T], \neq ih \\ z_d(ih) & = L_d(x(ih^-),ih) \end{array} \quad (3)$$

where $x(t) \in \mathbb{R}^n$ is the state, $w(t) \in \mathbb{R}^p$ and $v(ih) \in \mathbb{R}^q$ are the continuous and discrete exogenous inputs which belong to $\mathcal{L}_2[0,T]$ and $\ell_2(0,T)$, respectively, $z(t) \in \mathbb{R}^r$ and $z_d(ih) \in \mathbb{R}^s$ are the continuous and discrete outputs, $0 < h \in \mathbb{R}$ is the sampling period, i is a positive integer, $A(x(t),t)$, $B(x(t),t)$, $A_d(x(ih^-),ih)$, $B_d(x(ih^-),ih)$ $L_d(x(ih^-),ih)$ and $L(x(t),t)$are known real time-varying bounded nonlinear matrix functions of appropriate dimensions with $A(x(t),t)$, $B(x(t),t)$ and $L(x(t),t)$ being piecewise continuous.

Definition 2.1. Suppose γ is a given positive real number. A system Σ_1 of the form (3) is said to have $\mathcal{L}_2$ gain less than or equal to γ if

$$\int_0^T z^t(t)z(t)\,dt + \sum_{i=1}^{k} z_d^t(ih)z_d(ih) \leq \gamma^2 \left[\int_0^T w^t(t)w(t)\,dt + \sum_{i=1}^{k} v^t(ih)v(ih) \right] \quad (4)$$

where k is the largest integer such that $k < T$.

Note that the left hand side of (4) can be viewed as a 'mixed L_2/l_2' norm for the output signals z and z_d, and the right hand side of (4) as a 'mixed L_2/l_2' norm of the input signals which comprise of w and v.

In the sequel, without loss of generality, we assume $\gamma = 1$.

Theorem 2.1. Consider the system Σ_1 of the form (3). Suppose there exist positive definite functions $\epsilon(x(t),t)$ and $Q(x(t),t)$ with $\epsilon(0,t) = 0$ and $Q(0,t) = 0$ which satisfy

C1)

$$HJE(x(t),t) \triangleq \dot{\epsilon}(x(t),t) + \nabla_x^t \epsilon(x(t),t)A(x(t),t) + \frac{1}{4}\nabla_x^t \epsilon(x(t),t)B(x(t),t)B^t(x(t),t)\nabla_x \epsilon(x(t),t) + L^t(x(t),t)L(x(t),t) + Q(x(t),t) = 0 \quad (5)$$

C2)

$$DHJE(x(ih^-),v^*(ih),ih) \triangleq \epsilon(A_d(x(ih^-),ih) + B_d(x(ih^-),ih)v^*(ih),ih^+) - \epsilon(x(ih^-),ih) - v^{t*}(ih)v^*(ih) + L_d^t(x(ih^-),ih)L_d(x(ih^-),ih) + Q(x(ih^-),ih) = 0 \quad (6)$$

where

$$\nabla_v \epsilon(A_d(x(ih^-),ih) + B_d(x(ih^-),ih)v^*(ih),ih^+) - 2v^*(ih) = 0 \quad (7)$$

and

$$\mathcal{D}_v \nabla_v \epsilon(A_d(x(ih^-),ih) + B_d(x(ih^-),ih)v^*(ih),ih^+) - 2I < 0 \quad (8)$$

Then the system Σ_1 is having $\mathcal{L}_2$-gain less than or equal to 1.

Proof of Theorem 2.1: For $\tau \in (ih, ih+h)$,

$$\int_{ih^+}^{\tau} \frac{d}{dt}\{\epsilon(x,t)\}\,dt = \epsilon(x(\tau),\tau) - \epsilon(x(ih^+),ih^+) \quad (9)$$

First let consider the left hand side of (9)

$$H(x(t),t) = \int_{ih^+}^{\tau} \frac{d}{dt}\{\epsilon(x,t)\}\,dt = \int_{ih^+}^{\tau} [\dot{\epsilon}(x(t),t) + \nabla_x^t \epsilon(x(t),t)\{A(x(t),t) + B(x(t),t)w(t)\}]\,dt \quad (10)$$

Next adding and subtracting $(w^t(t)w(t) - z^t(t)z(t))$ to and from (10), it follows that

$$H(x(t),t) = \int_{ih^+}^{\tau} [\dot{\epsilon}(x(t),t) + \nabla_x^t \epsilon(x(t),t)\{A(x(t),t) + B(x(t),t)w(t)\} + w^t(t)w(t) - z^t(t)z(t) - w^t(t)w(t) + z^t(t)z(t)]\,dt \quad (11)$$

Completing the square and knowing that $z(t) = L(x(t),t)$, one has

$$H(x(t),t) = \int_{ih^+}^{\tau} \Big[\dot{\epsilon}(x(t),t) + \nabla_x^t \epsilon(x(t),t) \times A(x(t),t) + L^t(x(t),t)L(x(t),t) \frac{\nabla_x^t \epsilon(x(t),t)B(x(t),t)B^t(x(t),t)\nabla_x \epsilon(x(t),t)}{4} - \left(\frac{1}{2}B^t(x(t),t)\nabla_x \epsilon(x(t),t)) - w(t)\right)^t \times \left(\frac{1}{2}B^t(x(t),t)\nabla_x \epsilon(x(t),t)) - w(t)\right) + w^t(t)w(t) - z^t(t)z(t)\Big]\,dt \quad (12)$$

Employing (5), we have

$$H(x(t),t) \leq \int_{ih^+}^{\tau} [w^t(t)w(t) - z^t(t)z(t) - Q(x(t),t)]\, dt \quad (13)$$

From (9)

$$\int_{ih^+}^{\tau} [w^t(t)w(t) - z^t(t)z(t) - Q(x(t),t)]\, dt \geq \epsilon(x(\tau),\tau) - \epsilon(x(ih^+),ih^+) \quad (14)$$

Now let us consider at the sampling instant

$$\epsilon(x(t),t)|_{ih^-}^{ih^+} = \epsilon(x(ih^+),ih^+) - \epsilon(x(ih^-),ih^-) = \epsilon(x(ih),ih^+) - \epsilon(x(ih^-),ih) \quad (15)$$

Next adding and subtracting the term $z_d^t(ih)z_d(ih) -v^t(ih)v(ih)$ to and from (15), we obtain

$$H_d(x(ih),ih) = \epsilon(x(ih),ih^+) -\epsilon(x(ih^-),ih) + z_d^t(ih)z_d(ih) - v^t(ih)v(ih) -z_d^t(ih)z_d(ih) + v^t(ih)v(ih) \quad (16)$$

Knowing that $z_d(ih) = L_d(x(ih^-),ih)$, we have

$$H_d(x(ih),ih) = \epsilon(A_d(x(ih^-),ih) + B_d(x(ih^-),ih)v(ih),ih^+) - \epsilon(x(ih^-),ih) + L_d^t(x(ih^-),ih)L_d(x(ih^-),ih) - v^t(ih)v(ih) -z_d^t(ih)z_d(ih) + v^t(ih)v(ih) \quad (17)$$

Using (6)and the maximiser $v^*(ih)$ given in (7) and (8), we get

$$H_d(x(ih),ih) \leq v^t(ih)v(ih) - z_d^t(ih)z_d(ih) -Q(x(ih^-),ih) \quad (18)$$

By combining (13) and (18) over all possible ih on $[0,T]$, one has

$$\epsilon(x(T),T) - \epsilon(0,0) \leq \int_0^T [w^t(t)w(t) - z^t(t)z(t)]\, dt + \sum_{i=1}^k v^t(ih)v(ih) - \sum_{i=1}^k z_d^t(ih)z_d(ih) - \int_0^T Q(x(t),t)\, dt - \sum_{i=1}^k Q(x(ih^-),ih) \quad (19)$$

Knowing that $\epsilon(0,t) = 0$ and $Q(x(t),t) > 0,\ \forall x(t) \neq 0$, we obtain

$$\sum_{i=1}^k z_d^t(ih)z_d(ih) + \int_0^T z^t(t)z(t)\, dt \leq \int_0^T w^t(t)w(t)\, dt + \sum_{i=1}^k v^t(ih)v(ih) - \epsilon(x(T),T) \leq \int_0^T w^t(t)w(t)\, dt + \sum_{i=1}^k v^t(ih)v(ih) \quad (20)$$

This shows that the system Σ_1 has the $\mathcal{L}_2$-gain less than or equal to 1. To prove the stability, we rearrange (19) to

$$\epsilon(x(T),T) - \epsilon(0,0) + \int_0^T Q(x(t),t)\, dt + \sum_{i=1}^k Q(x(ih^-),ih) \leq \int_0^T [w^t(t)w(t) - z^t(t)z(t)]\, dt + \sum_{i=1}^k v^t(ih)v(ih) - \sum_{i=1}^k z_d^t(ih)z_d(ih) \quad (21)$$

Since $\epsilon(0,0) = 0$, we obtain

$$\int_0^T Q(x(t),t)\, dt + \sum_{i=1}^k Q(x(ih^-),ih) \leq \int_0^T w^t(t)w(t)\, dt + \sum_{i=1}^k v^t(ih)v(ih) \quad (22)$$

This means that

$$\lim_{T\mapsto\infty} \left[\int_0^T Q(x(t),t)\, dt + \sum_{i=1}^k Q(x(ih^-),ih)\right] \leq \lim_{T\mapsto\infty} \left[\int_0^T w^t(t)w(t)\, dt + \sum_{i=1}^k v^t(ih)v(ih)\right] < \infty \quad (23)$$

Since $Q(x(t),t)$ is a positive definite function, we conclude that the system Σ_1 is bounded input and bounded state stable. ∇∇∇

3. NONLINEAR $\mathcal{H}_\infty$ FILTERING

Consider the following class of nonlinear sampled-data systems

$$\Sigma_2 : \begin{array}{l} \left.\begin{array}{rcl} \dot{x}(t) &=& A(x(t)) + B(x(t))w(t) \\ z(t) &=& L(x(t)) \end{array}\right\} \\ \forall\, t \in [0\ T],\ t \neq ih,\ x(0) = 0 \\ z_d(ih^-) = L_d(x(ih^-),ih) \\ y(ih) = C(x(ih^-)) + D(x(ih^-))v(ih) \end{array} \quad (24)$$

where $x(t) \in \mathbb{R}^n$ is the state, $w(t) \in \mathbb{R}^p$ and $v(ih) \in \mathbb{R}^q$ are the continuous and discrete inputs which belong to $\mathcal{L}_2[0,T]$ and $\ell_2(0,T)$, respectively, $z(t) \in \mathbb{R}^r$ and $z_d(ih) \in \mathbb{R}^s$ are the continuous and discrete regulated outputs, $0 < h \in \mathbb{R}$ is the sampling period, i is a positive integer, $A(x(t),t)$, $B(x(t),t)$, $C(x(ih^-),ih)$, $D(x(ih^-),ih)$, $L_d(x(ih^-),ih)$ and $L(x(t),t)$ are known real time-varying bounded nonlinear matrix functions of appropriate dimensions with $A(x(t),t)$, $B(x(t),t)$ and $L(x(t),t)$ being piecewise continuous.

Assumption 3.1.

$$R(x(ih^-),ih) = D(x(ih^-),ih)D^t(x(ih^-),ih) > 0 \tag{25}$$

Problem Formulation: Given a scalar $\gamma > 0$, design a nonlinear filter of the form

$$\mathcal{K}: \begin{array}{ll} \dot{\xi}(t) &= a(\xi(t)) \;\; t \in [0,\, T], \;\; t \neq ih \\ \xi(ih) &= \xi(ih^-) + b(\xi(ih^-),ih)\times \\ & \quad [y(ih) - C(\xi(ih^-))] \\ \hat{z}(t) &= l(\xi(t)) \;\forall\, t \in [0,T]\; t \neq ih \\ \hat{z}_d(ih) &= L_d(\xi(ih^-),ih) \end{array} \tag{26}$$

such that

$$\int_0^T (z(t) - \hat{z}(t))^t(z(t) - \hat{z}(t))\, dt + \sum_{i=1}^{k} (z_d(ih) - \hat{z}_d(ih))^t(z_d(ih) - \hat{z}_d(ih)) \leq \int_0^T w^t(t)w(t)\, dt + \sum_{i=1}^{k} v^t(ih)v(ih) \tag{27}$$

holds.

Theorem 3.1. Consider the system Σ_2 satisfying Assumption 3.1. Then, there exists a nonlinear filter $\mathcal{K}$ such that the closed-loop system of Σ_2 with $\mathcal{K}$ satisfies (27) if there exist $\epsilon(\tilde{x},t) > 0$, $Q(\tilde{x},t) > 0$ and $b(\xi(ih^-),ih)$ such that the following conditions hold

C1)

$$\begin{aligned} HJ &\triangleq \dot{\epsilon}(\tilde{x}(t),t) + \nabla^t_{\tilde{x}}\epsilon(\tilde{x}(t),t)\tilde{A}(\tilde{x}(t),t) \\ &+\frac{1}{4}\nabla^t_{\tilde{x}}\epsilon(\tilde{x}(t),t)\tilde{B}(\tilde{x}(t),t)\tilde{B}^t(\tilde{x}(t),t)\nabla_{\tilde{x}}\epsilon(\tilde{x}(t),t) \\ &+ \tilde{L}^t(\tilde{x}(t),t)\tilde{L}(\tilde{x}(t),t) + Q(\tilde{x}(t),t) = 0 \end{aligned} \tag{28}$$

for all $\tilde{x}(t) \in \mathbb{R}^{2n}$, where $\tilde{x}(t) = \begin{bmatrix} x(t) \\ \xi(t) \end{bmatrix}$, $\tilde{A}(\tilde{x}(t),t) = \begin{bmatrix} A(x(t),t) \\ A(\xi(t),t)) \end{bmatrix}$, $\tilde{L}(\tilde{x}(t),t) = L(x(t),t) - L(\xi(t),t)$ and $\tilde{B}^t(\tilde{x}(t),t) = [B^t(x(t),t) \;\; 0]$.

C2)

$$\begin{aligned} &\epsilon(\tilde{A}_d(\tilde{x}(ih^-),ih) + \tilde{B}_d(\tilde{x}(ih^-),ih)v^*(ih),ih^+) \\ &\quad -\epsilon(\tilde{x}(ih^-),ih) - v^{t*}(ih)v^*(ih) + \\ &\tilde{L}^t_d(\tilde{x}(ih^-),ih)\tilde{L}_d(\tilde{x}(ih^-),ih) + Q(x(ih^-),ih) \\ &= 0 \end{aligned} \tag{29}$$

where

$$\nabla_v\epsilon(\tilde{A}_d(\tilde{x}(ih^-),ih) + \tilde{B}_d(\tilde{x}(ih^-),ih)v(ih),ih^+) - 2v^*(ih) = 0, \tag{30}$$

$$\mathcal{D}_v\nabla_v\epsilon(\tilde{A}_d(\tilde{x}(ih^-),ih) + \tilde{B}_d(\tilde{x}(ih^-),ih) \times v(ih),ih^+) - 2I < 0, \tag{31}$$

$$\tilde{A}_d(\tilde{x}(ih^-),ih) = \begin{bmatrix} x(ih^-) \\ \xi(ih^-) + b(\xi(ih^-),ih)\times \\ [C(x(ih^-),ih) - \\ C(\xi(ih^-),ih)] \end{bmatrix},$$

$\tilde{L}_d(\tilde{x}(ih^-),ih) = L_d(x(ih^-),ih) - L_d(\xi(ih^-),ih)$ and

$$\tilde{B}^t_d(\tilde{x}(ih^-),ih) = [0 \; b(\xi(ih^-),ih)D(x(ih^-),ih)]$$

If this is the case, then a suitable filter of the form $\mathcal{K}$ is given by

$$\begin{array}{ll} \dot{\xi}(t) &= A(\xi(t),t) \;\; t \in [0,\, T], \\ & \quad t \neq ih, \; \xi(0) = 0 \\ \xi(ih) &= \xi(ih^-) + b(\xi(ih^-),ih)\times \\ & \quad [y(ih) - C(\xi(ih^-),ih)] \\ \hat{z}(t) &= L(\xi,t) \;\; t \in [0,\, T], \;\; t \neq ih \\ \hat{z}_d(ih) &= L_d(\xi(ih^-),ih) \end{array} \tag{32}$$

The filter given above is a nonlinear counterpart of the filter given in (Sun *et al.*, 1993; Shi, 1994).

Proof of Theorem 3.1: Rewrite the augmented system Σ_2 with (32) as follows:

$$\Sigma_{as}: \begin{array}{l} \left.\begin{array}{l} \dot{x}(t) = A(x(t),t) + B(x(t),t)w(t) \\ \dot{\xi}(t) = A(\xi(t),t) \\ z(t) = L(x(t),t) \\ \hat{z}(t) = L(\xi(t),t) \end{array}\right\} \\ \forall\, t \in [0\; T],\; t \neq ih,\; x(0) = 0 \\ \xi(ih) = \xi(ih^-) + b(\xi(ih^-),ih)\times \\ \big[C(x(ih^-),ih) + D(x(ih^-),ih)v(ih) \\ -C(\xi(ih^-),ih)\big] \\ z_d(ih^-) = L_d(x(ih^-),ih) \\ \hat{z}_d(ih) = L_d(\xi(ih^-),ih) \end{array} \tag{33}$$

Defining $\tilde{x} = \begin{bmatrix} x(t) \\ \xi(t) \end{bmatrix}$, the system Σ_{as} can be recast into the following form:

$$\dot{\tilde{x}}(t) = \tilde{A}(\tilde{x}(t),t) + \tilde{B}(\tilde{x}(t),t)w(t), \quad t \neq ih; \;\; \tilde{x}(0) = 0$$

$$\tilde{x}(ih) = \tilde{A}_d(\tilde{x}(ih^-), ih) + \tilde{B}_d(\tilde{x}(ih^-), ih)v(ih),$$
$$\forall\, ih \in (0, T)$$
$$\tilde{z}(t) = z(t) - \hat{z}(t) = \tilde{L}(\tilde{x}(t), t)$$
$$\tilde{z}_d(ih) = z_d(ih) - \hat{z}_d(ih) = \tilde{L}_d(\tilde{x}(ih^-), ih)$$

where $\tilde{A}(\tilde{x}(t), t)$, $\tilde{B}(\tilde{x}(t), t)$, $\tilde{A}_d(\tilde{x}(ih^-), ih)$, $\tilde{B}_d(\tilde{x}(ih^-), ih)$, $\tilde{L}(\tilde{x}(t), t)$, and $\tilde{L}_d(\tilde{x}(ih^-), ih)$ are defined in Theorem 3.1. Applying Theorem 2.1, we obtain Theorem 3.1. $\nabla\nabla\nabla$.

4. CONCLUSION

In this paper the problem of nonlinear $\mathcal{H}_\infty$ filtering for sampled-data systems has been considered. Sufficient conditions are obtained for the existence of filters that guarantee the $\mathcal{L}_2$ gain from an exogenous input to an estimation error is less than or equal to a prescribed value. The underlying problem can be solved if certain continuous and discrete Hamilton-Jacobi equations can be solved. This result can be viewed as an extension of (Sun *et al.*, 1993) and (Shi, 1994) which treat linear systems.

5. REFERENCES

Ball, J. A. and J. W. Helton (1989). H_∞ control for nonlinear plants: Connection with differential games. In: *Proc. 28th IEEE Conf. Decision Contr.*. pp. 956–962.

Ball, J. A., J. W. Helton and M. L. Walker (1993). H_∞ control for nonlinear systems with output feedback. *IEEE Trans. Automat. Contr.* **38**, pp. 546–559.

Bamieh, B. and J. B. Pearson (1992). A general framework for linear periodic systems with application to $\mathcal{H}_\infty$ sampled-data control. *IEEE Trans. Automat. Contr.* **37**, 418–435.

Bamieh, B., J. Pearson, B. Francis and A. Tannenbaum (1991). A lifting technique for linear periodic systems with applications to sampled-data control. *Systems & Control Letters* **17**, 78–88.

Basar, T. (1991). Optimum performance levels for minimax filters, predictors and smoothers. *Systems & Control Letters* **16**, pp. 309–317.

Berman, N. and U. Shaked (1996). $\mathcal{H}_\infty$ nonlinear filtering. *Int. J. Robust and Nonlinear Contr.* in press.

Hara, S. and P. T. Kabamba (1990). Worst case analysis and design of sampled-data control systems. In: *Proc. of 29th IEEE Conf. Decision Contr., Honolulu, HI.* pp. 202–203.

Hill, D. J. and P. J. Moylan (1980). Dissipative dynamical systems: Basic input-output and state properties. *J. Franklin Inst.* **309**, pp. 327–357.

Isidori, A. (1991). Feedback control of nonlinear systems. In: *Proc. First European Contr. Conf.* pp. 1001–1012.

Isidori, A. and A. Astolfi (1992). Disturbance attenuation and H_∞- control via measurement feedback in nonlinear systems. *IEEE Trans. Automat. Contr.* **37**, pp. 1283–1293.

Lin, W. and C. I. Byrnes (1995). Discrete-time nonlinear $\mathcal{H}_\infty$ control with measurement feedback. *Automatica* **31**, 419–434.

Lin, W. and C. I. Byrnes (1996). $\mathcal{H}_\infty$ control of discrete time nonlinear systems. *IEEE Trans. Automat. Contr.* **41**, 494–510.

Nguang, S. K. and M. Fu (1994). H_∞ filtering for known and uncertain nonlinear systems. In: *Proc. of IFAC, Symposium Robust Control Design.* pp. 347–352.

Nguang, S. K. and M. Fu (1996). Robust nonlinear $\mathcal{H}_\infty$ filtering. *Automatica* **32**, 1195–1199.

Shi, P. (1994). Issues in robust filtering and control of sampled-data systems. Technical report. The University of Newcastle. Ph.D Thesis.

Shi, P. (1996). Robust filtering for uncertain systems with sampled measurements. *Int. J. Systems Science* **27**(12), 1403–1415.

Shi, P. (1998). Filtering on sampled-data systems with parametric uncertainty. *IEEE Trans. Automat. Control.* to appear.

Sivashankar, N. and P. Khargonekar (1994). Characterization of the $\mathcal{L}_2$-induced norm for linear systems with jumps with application to sampled-data systems. *SIAM J. Control and Optimisation* **32**, 1128–1150.

Sun, W., K. M. Nagpal and P. P. Khargonekar (1993). $\mathcal{H}_\infty$ control and filtering for sampled-data systems. *IEEE Trans. Automat. Contr.* **38**, 1162–1175.

Suzuki, S., A. Isidori and T. J. Tarn (1995). $\langle_\infty$ control of nonlinear systems with sampled measurements. *J. Math. Systems, Estimation, and Control* **5**, 1–12.

Toivonen, H. T. (1992). Sampled-data control of continuous-time system with an $\mathcal{H}_\infty$ optimality criterion. *Automatica* **28**, 45–54.

van der Schaft, A. J. (1991). A state-space approach to nonlinear H_∞ control. *Systems & Control Letters* **16**, pp. 1–8.

van der Schaft, A. J. (1992). $\mathcal{L}_2$-gain analysis of nonlinear systems and nonlinear state feedback H_∞ control.. *IEEE Trans. Automat. Contr.* **37**, pp. 770–784.

Willems, J. C. (1972). Dissipative dynamical systems Part I: Genernal theory.. *Arch. Rational Mech. Anal.* **45**, pp. 321–351.

Yamamoto, Y. (1991). On the state space and frequency domain characterization of $\mathcal{H}_\infty$ norm of sampled-data systems. *Systems & Control Letters* **21**, 163–172.

Nonlinear Analysis of Sampled-data Control Systems

[a] Luonan Chen [b]Kazuyuki Aihara
[a] Osaka Sangyo University,3-1-1 Nakagaito, Daito, Osaka 574-8530,Japan.
Fax:81-798-70-8189;E-mail:chen@elec.osaka-sandai.ac.jp
[b]The University of Tokyo,7-3-1 Hongo, Bunkyo-Ku, Tokyo 113-8656,Japan

Abstract: This paper treats a digital-controlled Control system with both continuous-time and discrete-time variables as a model of differential- difference-algebraic equations (DDA), presents a fundamental analysing method of such a nonlinear DDA for local sampling, asymptotical stability, singular perturbations and bifurcations, and further shows that there exist five types of generic bifurcations at the equilibria in contrast to two types in continuous-time dynamical systems and three types in discrete-time dynamical systems. Finally the theoretical results are applied to digital control of power systems. Numerical simulations have verified that our results are useful. *Copyright © 1999 IFAC*

Keywords: asymptotic stability, singular perturbations, sampled-data control, nonlinear analysis, nonlinear theory, power system, bifurcation

1 Introduction

Many physical systems with digital devices require mathematical descriptions not only with ordinary differential-algebraic equations, but also with difference equations. As an example, for the power system with digital controllers, the dynamics of the generators as well as their continuous-time controllers and the load dynamics together define the ordinary differential equations while algebraic equalities are defined by the power balance equations of the transmission network[3]. On the other hand, the difference equations describe the dynamics of the digital devices. Therefore, the digital control systems of power grids are typical models mixed with both continuous and discrete time sequences which can mathematically be formulated as differential- difference-algebraic equations (DDA). In this paper, we define the constrained sampled-data system as a model of DDA. So far, both continuous-time and discrete-time nonlinear systems have attracted considerable attention and a variety of techniques have been developed[3, 7, 5]. This paper aims at analysing the asymptotical stability and bifurcations of the DDA [5, 6] and further applying the theoretical results to digital control of power systems. We first define a DDA model and two types of equilibria, namely quasi-equilibria and strict equilibria due to its mixing nature of continuous and discrete time, and summarize the recent results for locally sampling, asymptotical stability and singular perturbations of DDA[5, 6]. Then this paper shows that there exist five types of generic bifurcations at the strict equilibria in contrast to two types at an equilibrium in continuous-time dynamical systems and three types at a fixed point in discrete-time dynamical systems, and provides their bifurcation conditions for the DDA. Furthermore, this paper indicates that most of locally nondegenerate bifurcations for a DDA at strict equilibria are related with the local bifurcations of the sampling instants under some generic assumptions. Finally, the theoretical results are applied to the evaluations and designs of digital control for power systems.

2 Definition of DDA

The following notation is used throughout the paper. Let

$$\dot{x}(t) = f(x(t), x_n, y_n, z(t), z_n, \alpha) \quad (1)$$

$$y_{n+1} = g(x_n, y_n, z_n, \alpha) \quad (2)$$

$$h(x(t), x_n, y_n, z(t), z_n, \alpha) = 0 \quad (3)$$

$$(n\tau \leq t < (n+1)\tau; n = 0, 1, 2, ...)$$

be the **differential-difference-algebraic equations** (DDA) under study, where vectors x, x_n, y_n, z and z_n belong to the Euclidean spaces $\mathcal{R}^{n_x}, \mathcal{R}^{n_x}, \mathcal{R}^{n_y}$, $\mathcal{R}^{n_z}$ and $\mathcal{R}^{n_z}$, respectively. τ is a sampling interval (non-negative real number), and $\alpha \in \mathcal{R}$ is a parameter. Define $x_n = x(n\tau), z_n = z(n\tau)$ and y_n is the value at the instant $n\tau$, where x_n, y_n and z_n are constantly hold during $n\tau \leq t < (n+1)\tau$ for eqns.(1) and (3). Then eqns.(1)-(3) are an ordinary differential - difference - algebraic equation model (DDA). Let $f_x = \frac{\partial f(x,x_n,y_n,z,z_n)}{\partial x}$ and $|f_x|$ is the determinant of f_x. $||\cdot||$ is the usual Euclidean norm of a vector or the induced matrix norm of a matrix. E is an identity matrix. e^A is the exponential function of a matrix A.

In this paper, a *non-equality* is an equation with the form of $u \neq 0$ while an *inequality* is the form of $u < 0$ or $u > 0$. On the other hand, an *equality* is an equation with the form of $u = 0$.

$x(t)$ is continuous due to eqn.(1) while $z(t)$ is generally discontinuous at $t = n\tau, n = 0, 1, 2, ...,$ namely $\lim_{t \to n\tau - 0} x(t) = x(n\tau) = x_n = \lim_{t \to n\tau + 0} x(t)$ and $\lim_{t \to n\tau - 0} z(t) \neq z(n\tau) = z_n = \lim_{t \to n\tau + 0} z(t)$, as far as $|h_z| \neq 0$ and $|h_z + h_{z_n}| \neq 0$. In this paper, sampling instants mean $t = n\tau$ while inter-sampling instants imply $n\tau < t < (n+1)\tau$, where $n = 0, 1, 2, ...$

Definition 2. 1 *Vectors* $\bar{p} = (\bar{x}, \bar{x}, \bar{y}_n, \bar{z}, \bar{z})$ *are called a strict equilibrium of DDA eqns.(1)-(3) if* $f(\bar{x}, \bar{x}_n, \bar{y}_n, \bar{z}, \bar{z}_n, \alpha) = 0$, $h(\bar{x}, \bar{x}_n, \bar{y}_n, \bar{z}, \bar{z}_n, \alpha) = 0$ *and* $\bar{y}_n = g(\bar{x}_n, \bar{y}_n, \bar{z}_n, \alpha)$.

Next, without confusion, t is occasionally dropped from the related variables, e.g., $x(t)$ or $z(t)$ is simply expressed by x or z.

3 Local Sampling and Stability

Let $\hat{p} = (\hat{x}_n, \hat{x}_n, \hat{y}_n, \hat{z}_n, \hat{z}_n)$ satisfy eqn.(3) and $|h_z + h_{z_n}|_{\hat{p}} \neq 0$. Then, according to the implicit function theorem there exists an unique C^∞ mapping in open neighborhoods of $\hat{p}$ such that at sampling instants

$$z_n = q_n(x_n, y_n). \tag{4}$$

Furthermore, let $\hat{p}$ satisfy eqn.(3) and $|h_z|_{\hat{p}} \neq 0$. Then by using the implicit function theorem again, there exists an unique C^∞ mapping in open neighborhoods of $\hat{p}$ such that at the intersampling instants

$$z(t) = q(x(t), x_n, y_n) \quad n\tau \leq t < (n+1)\tau \tag{5}$$

where $q_n(x_n, y_n) = q(x(n\tau), x_n, y_n)$ due to $z_n = \lim_{t \to n\tau+0} z(t)$.
Thus the reduced system of eqns.(1) and (3) at the neighbour of $\hat{p}$ becomes

$$\dot{x} = f_R(x, x_n, y_n) \equiv f(x, x_n, y_n, q(x, x_n, y_n), q_n(x_n, y_n))$$

and further the reduced system of eqns.(2) and (3) at the neighbour of $\hat{p}$ becomes

$$y_{n+1} = g_R(x_n, y_n) \equiv g(x_n, y_n, q_n(x_n, y_n)) \tag{6}$$

for $n = 0, 1, ...$
Let $\frac{\partial \hat{f}_R}{\partial x}$, $\frac{\partial \hat{f}_R}{\partial x_n}$ and $\frac{\partial \hat{f}_R}{\partial y_n}$ are partial derivatives of $f_R(x, x_n, y_n)$ at $(\hat{x}, \hat{x}_n, \hat{y}_n)$ with respect to (x, x_n, y_n), respectively. Then we obtain a locally sampling theorem [5, 6].

Theorem 3. 1 *Assume that $\hat{p} = (\hat{x}, \hat{x}, \hat{y}, \hat{z}, \hat{z})$ is an arbitrary point which satisfies eqn.(3). Assume that the partial derivatives of f unto appropriate order exist, and $|h_z| \neq 0, |h_z + h_{z_n}| \neq 0$ at any solution of eqns.(1) and (3) starting from $\hat{p}$. Then the discrete-time system for eqns.(1) and (3) can be expanded in a neighbourhood of $\hat{p}$ for (x_n, y_n),*

$$x_{n+1} = \phi_\tau + D_{xy}\phi_\tau\, \Delta_{xy} + \frac{1}{2} D^2_{xy}\phi_\tau\, \Delta_{xy} \cdot \Delta_{xy} + \dots \quad (n = 0, 1, ...)$$

where $\phi_\tau = \phi_\tau(\hat{x}, \hat{y}) = \hat{x} + \int_0^\tau \hat{f}_R dt$ with $\hat{f}_R = f_R(\phi_t, \hat{x}, \hat{y})$ and $\phi_0 = \hat{x}$.

$$\Delta_{xy} = \begin{bmatrix} x_n - \hat{x} \\ y_n - \hat{y} \end{bmatrix}, D_{xy}\phi_\tau = [\frac{\partial \phi_\tau}{\partial x_n}, \quad \frac{\partial \phi_\tau}{\partial y_n}],$$

$$D^2_{xy}\phi_\tau = \begin{bmatrix} \frac{\partial^2 \phi_\tau}{\partial x_n^2} & \frac{\partial^2 \phi_\tau}{\partial x_n \partial y_n} \\ \frac{\partial^2 \phi_\tau}{\partial y_n \partial x_n} & \frac{\partial^2 \phi_\tau}{\partial y_n^2} \end{bmatrix}.$$

$\frac{\partial \phi_\tau}{\partial x_n} = e^{\int_0^\tau \frac{\partial \hat{f}_R}{\partial x} dt} + \int_0^\tau e^{\int_t^\tau \frac{\partial \hat{f}_R}{\partial x} dt} \frac{\partial \hat{f}_R}{\partial x_n} dt$, $\frac{\partial \phi_\tau}{\partial y_n} = \int_0^\tau e^{\int_t^\tau \frac{\partial \hat{f}_R}{\partial x} dt} \frac{\partial \hat{f}_R}{\partial y_n} dt$, $\frac{\partial^2 \phi_\tau}{\partial x_n^2} = \int_0^\tau e^{\int_t^\tau \frac{\partial \hat{f}_R}{\partial x} dt} [\frac{\partial^2 \hat{f}_R}{\partial x^2} (\frac{\partial \phi_t}{\partial x_n})^2 + \frac{\partial^2 \hat{f}_R}{\partial x_n \partial x} \frac{\partial \phi_t}{\partial x_n} + \frac{\partial^2 \hat{f}_R}{\partial x \partial x_n} \frac{\partial \phi_t}{\partial x_n} + \frac{\partial^2 \hat{f}_R}{\partial x_n^2}] dt$, $\frac{\partial^2 \phi_\tau}{\partial x_n \partial y_n} = \int_0^\tau e^{\int_t^\tau \frac{\partial \hat{f}_R}{\partial x} dt} (\frac{\partial^2 \hat{f}_R}{\partial x^2} \frac{\partial \phi_t}{\partial y_n} \frac{\partial \phi_t}{\partial x_n} + \frac{\partial^2 \hat{f}_R}{\partial x \partial y_n} \frac{\partial \phi_t}{\partial x_n} + \frac{\partial^2 \hat{f}_R}{\partial x \partial x_n} \frac{\partial \phi_t}{\partial y_n} + \frac{\partial^2 \hat{f}_R}{\partial x_n \partial y_n}) dt$, $\frac{\partial^2 \phi_\tau}{\partial y_n \partial x_n} = \int_0^\tau e^{\int_t^\tau \frac{\partial \hat{f}_R}{\partial x} dt} (\frac{\partial^2 \hat{f}_R}{\partial x^2} \frac{\partial \phi_t}{\partial x_n} \frac{\partial \phi_t}{\partial y_n} + \frac{\partial^2 \hat{f}_R}{\partial x \partial x_n} \frac{\partial \phi_t}{\partial y_n} + \frac{\partial^2 \hat{f}_R}{\partial y_n \partial x} \frac{\partial \phi_t}{\partial x_n} + \frac{\partial^2 \hat{f}_R}{\partial x_n \partial y_n}) dt$, $\frac{\partial^2 \phi_\tau}{\partial y_n^2} = \int_0^\tau e^{\int_t^\tau \frac{\partial \hat{f}_R}{\partial x} dt} (\frac{\partial^2 \hat{f}_R}{\partial x^2} (\frac{\partial \phi_t}{\partial y_n})^2 + \frac{\partial^2 \hat{f}_R}{\partial x \partial y_n} \frac{\partial \phi_t}{\partial y_n} + \frac{\partial^2 \hat{f}_R}{\partial y_n \partial x} \frac{\partial \phi_t}{\partial y_n} + \frac{\partial^2 \hat{f}_R}{\partial y_n^2}) dt.$

Theorem 3. 1 ensures that the discretized system can locally approximate the original continuous system at any order at sampling instants. At a strict equilibrium $\bar{p}$, if h_z and $h_z + h_{z_n}$ are nonsingular

$$A = f_x - f_z h_z^{-1} h_x \tag{7}$$

$$B = f_{x_n} - f_z h_z^{-1} [h_{x_n} - h_{z_n} (h_z + h_{z_n})^{-1} h_{x_n}] - f_{z_n} (h_z + h_{z_n})^{-1} (h_x + h_{x_n}) \tag{8}$$

$$C = f_{y_n} - f_z h_z^{-1} [h_{y_n} - h_{z_n} (h_z + h_{z_n})^{-1} h_{y_n}] - f_{z_n} (h_z + h_{z_n})^{-1} h_{y_n} \tag{9}$$

Assume $|A| \neq 0$ and $|h_z| \neq 0, |h_z + h_{z_n}| \neq 0$ at $\bar{p}$. Define a Jacobian

$$J = \begin{bmatrix} e^{\tau A} + (e^{\tau A} - E) A^{-1} B & (e^{\tau A} - E) A^{-1} C \\ GX & GY \end{bmatrix}$$

where $GX = g_{x_n} - g_{z_n} (h_z + h_{z_n})^{-1} (h_x + h_{x_n})$ and $GY = g_{y_n} - g_{z_n} (h_z + h_{z_n})^{-1} h_{y_n}$. Then we have the following theorem for the stability of a strict equilibrium $\bar{p}$ for all of instants.

Theorem 3. 2 *Assume that $\bar{p} = (\bar{x}, \bar{x}, \bar{y}, \bar{z}, \bar{z})$ is a strict equilibrium of DDA eqns.(1)-(3), and $|h_z|_{\bar{p}} \neq 0$, $|h_z + h_{z_n}|_{\bar{p}} \neq 0$, $|A| \neq 0$.*

1. *If all the eigenvalues of J at $\bar{p}$ have moduli less than 1, then $\bar{p}$ is an asymptotically stable strict-equilibrium of DDA eqns.(1)-(3).*
2. *If any one of the eigenvalues of J at $\bar{p}$ has modulus more than 1, then $\bar{p}$ is an unstable strict-equilibrium of DDA eqns.(1)-(3).*

Actually, we can prove that Theorem 3. 2 still holds even if $|A| = 0$.

4 Singular Perturbation

For the DDA eqns.(1)-(3), we define an **associated singularly perturbed system** as

$$\dot{x}(t) = f(x(t), x_n, y_n, z(t), z_n) \tag{10}$$

$$y_{n+1} = g(x_n, y_n, z_n) \tag{11}$$

$$\epsilon \dot{z}(t) = h(x(t), x_n, y_n, z(t), z_n) \tag{12}$$

$$(n\tau \leq t < (n+1)\tau; n = 0, 1, ...)$$

where ϵ is a sufficiently small positive number. That is, the algebraic eqn.(3) corresponds to a limit of the fast dynamics eqn.(12). Obviously, eqn.(12) will approach the algebraic manifold when $\epsilon \to 0$. It is evident that eqns.(10)-(12) has the same equilibrium as eqns.(1)-(3) as far as $h = 0$.

Theorem 4. 1 *Assume that $\bar{p} = (\bar{x}, \bar{x}, \bar{y}, \bar{z}, \bar{z})$ are a strict equilibrium of the DDA eqns.(1)-(3), and all real parts of eigenvalues for $h_z|_{\bar{p}}$ are negative (or positive). Furthermore, assume that no eigenvalue of J at $\bar{p}$ has the modulus 1, and $|h_z + h_{z_n}|_{\bar{p}} \neq 0, |A|_{\bar{p}} \neq 0$.*
Then there exists $\bar{\epsilon} > 0$ such that for all positive $\epsilon < \bar{\epsilon}$, stability of eqns.(10)-(12) at $\bar{p}$ is identical to that of eqns.(1)-(3).

5 Local Bifurcations

We will show that there are five types of bifurcations for a strict equilibrium under certain generic conditions, i.e., singularity-1 induced bifurcations at the intersampling instants, and fold, flip, Neimark-Sacker and singularity-2 induced bifurcations at the sampling instants, in codimension one parametric space.

5.1 Sampling instants

For the sampling instants, eqns.(1)-(3) at the neighbourhood of a strict equilibrium $\bar{p} = (\bar{x}, \bar{x}, \bar{y}, \bar{z}, \bar{z}, \bar{\alpha})$ can be expressed as

$$x_{n+1} = \phi_\tau + D_{xy\alpha}\phi_\tau\, \Delta_{xy\alpha} + \tfrac{1}{2} D^2_{xy\alpha}\phi_\tau\, \Delta_{xy\alpha} \cdot \Delta_{xy\alpha} + \dots \tag{13}$$

$$y_{n+1} = g(x_n, y_n, q_n(x_n, y_n, \alpha), \alpha) \quad (n = 0, 1, ...), \tag{14}$$

where $\phi_\tau = \bar{x} + \int_0^\tau f_R(\phi_t, \bar{x}, \bar{y}, \bar{\alpha})dt = \bar{x} + \tau f_R(\bar{x}, \bar{x}, \bar{y}, \bar{\alpha})$ and $\Delta_{xy\alpha} = \begin{bmatrix} x_n - \bar{x} \\ y_n - \bar{y} \\ \alpha - \bar{\alpha} \end{bmatrix}$, when $|h_z|_{\bar{p}} \neq 0$ and $|h_z + h_{z_n}|_{\bar{p}} \neq 0$.

Since a strict equilibrium of the DDA eqns.(1)-(3) is a fixed point of discrete time system eqns.(13)-(14), the bifurcations of eqns.(1)-(3) at the sampling instants for the strict equilibria are basically the same as local bifurcations of the difference-algebraic model eqns.(13)-(14). Next we examine the fixed point bifurcations of eqns.(13)-(14).

We first summarize the fold, flip and Neimark-Sacker bifurcations [1], and then present the existence of singularity-2 induced bifurcation as well as their conditions.

Theorem 5. 1 *Suppose that $\bar{p} = (\bar{x}, \bar{x}, \bar{y}, \bar{z}, \bar{z}, \bar{\alpha})$ is a strict equilibrium of eqns.(1)-(3) and $|h_z|_{\bar{p}} \neq 0$, $|h_z + h_{z_n}|_{\bar{p}} \neq 0$ and $|A| \neq 0$. Assume J at $\bar{p}$ has a geometrically simple 1 eigenvalue with right eigenvector U and left eigenvector $U^\dagger$ and there is no other eigenvalue with modulus 1. Then a fold bifurcation [1, 4, 6] occurs for (x_n, y_n, z_n) of eqns.(1)-(3) if at $\bar{p}$*

$$U^\dagger \begin{bmatrix} D^2_{xy}\phi_\tau \\ D^2_{xy}g_R \end{bmatrix} (U \cdot U) \neq 0 \text{ and } U^\dagger \begin{bmatrix} \frac{\partial \phi_\tau}{\partial \alpha} \\ \frac{\partial g_R}{\partial \alpha} \end{bmatrix} \neq 0,$$

where $D^2_{xy}g = \begin{bmatrix} \frac{\partial^2 g_n}{\partial x_n^2} & \frac{\partial^2 g_n}{\partial x_n \partial y_n} \\ \frac{\partial^2 g_n}{\partial y_n \partial x_n} & \frac{\partial^2 g_n}{\partial y_n^2} \end{bmatrix}$. $\frac{\partial g_R}{\partial \alpha} = g_\alpha - g_{z_n}(h_z + h_{z_n})^{-1} h_\alpha$ *and* $\frac{\partial \phi_\tau}{\partial \alpha} = \int_0^\tau e^{\int_t^\tau \frac{\partial f_R}{\partial x} dt} \frac{\partial f_R}{\partial \alpha} dt.$

Theorem 5. 2 *Suppose that $\bar{p} = (\bar{x}, \bar{x}, \bar{y}, \bar{z}, \bar{z}, \bar{\alpha})$ is a strict equilibrium of eqns.(1)-(3) and $|h_z|_{\bar{p}} \neq 0$, $|h_z + h_{z_n}|_{\bar{p}} \neq 0$ and $|A| \neq 0$. Assume J at $\bar{p}$ has a geometrically simple -1 eigenvalue with right eigenvector U and left eigenvector $U^\dagger$ and there is no other eigenvalue with moduli 1. Then a flip bifurcation [1, 4, 6] occurs for (x_n, y_n, z_n) of eqns.(1)-(3) if at $\bar{p}$*

$$[\frac{U^\dagger}{2} \begin{bmatrix} D^2_{xy}\phi_\tau \\ D^2_{xy}g_R \end{bmatrix} (U \cdot U)]^2 + \frac{U^\dagger}{6} \begin{bmatrix} D^3_{xy}\phi_\tau \\ D^3_{xy}g_R \end{bmatrix} (U \cdot U \cdot U) \neq 0,$$

$$2U^\dagger \frac{\partial D_{xy}\phi_\tau}{\partial \alpha} U + U^\dagger \begin{bmatrix} D_{xy}\phi_\tau \\ D_{xy}g_R \end{bmatrix} U^\dagger \begin{bmatrix} \frac{\partial \phi_\tau}{\partial \alpha} \\ \frac{\partial g_R}{\partial \alpha} \end{bmatrix} (U \cdot U) \neq 0.$$

Theorem 5. 3 *Suppose that $\bar{p} = (\bar{x}, \bar{x}, \bar{y}, \bar{z}, \bar{z}, \bar{\alpha})$ is a strict equilibrium of eqns.(1)-(3) and $|h_z|_{\bar{p}} \neq 0$, $|h_z + h_{z_n}|_{\bar{p}} \neq 0$ and $|A| \neq 0$. Assume that J at $\bar{p}$ has an algebraically simple pair of eigenvalues in the form of $\lambda = r(\alpha - \bar{\alpha})e^{\pm j\phi(\alpha - \bar{\alpha})}$ where $r(\bar{\alpha}) = 1$ for all sufficiently small $|\alpha - \bar{\alpha}|$, and there is no other eigenvalue with modulus 1.*
If $\frac{dr(\bar{\alpha})}{d\alpha} \neq 0$ and $e^{jk\phi(\bar{\alpha})} \neq 1$ for $k = 1, 2, 3, 4$, then there exists the Neimark-Sacker bifurcation which generates an unique closed invariant curve from the strict equilibrium, for (x_n, y_n, z_n) of eqns.(1)-(3).

Theorems 5. 1 , 5. 2 and 5. 3 are derived from eqns.(13)-(14) and the definitions of Fold, Flip and Neimark-Sacker bifurcations.

A singular surface is $\{(x_n, x_n, y_n, z_n, z_n, \alpha) \| h_z(x_n, x_n, y_n, z_n, z_n, \alpha)| = 0)\}$. At a strict equilibrium $\bar{p}$, from eqns.(1)-(3), we have

$$f(\bar{x}, \bar{x}, \bar{y}, \bar{z}, \bar{z}, \bar{\alpha}) = 0 \quad (15)$$

$$g(\bar{x}, \bar{y}, \bar{z}, \bar{\alpha}) - \bar{y} = 0 \quad (16)$$

$$h(\bar{x}, \bar{x}, \bar{y}, \bar{z}, \bar{z}, \bar{\alpha}) = 0. \quad (17)$$

Suppose that the full Jacobian matrix

$$\begin{pmatrix} f_x + f_{x_n} & f_{y_n} & f_z + f_{z_n} \\ g_{x_n} & g_{y_n} - E & g_{z_n} \\ h_x + h_{x_n} & h_{y_n} & h_z + h_{z_n} \end{pmatrix}_{\bar{p}} \quad (18)$$

for $(\bar{x}, \bar{y}, \bar{z})$ of eqns.(15)-(17) is nonsingular. Then there exists an unique *equilibrium locus* $(x(\alpha), y(\alpha), z(\alpha), \alpha) \in \mathcal{R}^{n_x+n_y+n_z+1}$ locally near $\alpha = \bar{\alpha}$, according to implicit function theorem and eqns.(15)-(17).

Before presenting the singularity-2 induced bifurcation, we define a coordinate condition for $h = 0$ at point p.

Assumption 5. 1 (Coordinate Condition) *Assume that there exist a column β_z in matrix $\begin{pmatrix} h_z \\ h_z + h_{z_n} \end{pmatrix}$ and a column β_x in matrix $\begin{pmatrix} h_x \\ h_x + h_{x_n} \end{pmatrix}$ such that after β_x replaces β_z into $\begin{pmatrix} h_z \\ h_z + h_{z_n} \end{pmatrix}$, $h_{z'}$ and $h_{z'} + h_{z'_n}$ are nonsingular at a point p for the alternated matrix $\begin{pmatrix} h_{z'} \\ h_{z'} + h_{z'_n} \end{pmatrix}$, where $\begin{pmatrix} h_{x'} \\ h_{x'} + h_{x'_n} \end{pmatrix}$ is constructed by $n_z - 1$ columns of $\begin{pmatrix} h_z \\ h_z + h_{z_n} \end{pmatrix}$ except the column β_z, and by the column β_x.*

Then we describe the singularity-2 induced bifurcation at sampling instants.

Theorem 5. 4 *Suppose that $\bar{p} = (\bar{x}, \bar{x}, \bar{y}, \bar{z}, \bar{z}, \bar{\alpha})$ is a strict equilibrium of eqns.(1)-(3) with one-dimensional parameter $\alpha \in \mathcal{R}$ and the Coordinate Condition (Assumption 5. 1) holds at $\bar{p}$. Assume that $|h_z|_{\bar{p}}$ has an algebraically simple zero eigenvalue and $b_z = -trace\ [\ f_z adj(h_z)h_x\ ]$ is nonzero, $||b_z|| > ||b_{z_n}||$ where $b_{z_n} = -trace\ [\ f_{z_n} adj(h_z)h_x\ ]$, and furthermore the following conditions are satisfied,*

$$\begin{vmatrix} f_x + f_{x_n} & f_{y_n} & f_z + f_{z_n} \\ g_{x_n} & g_{y_n} - E & g_{z_n} \\ h_x + h_{x_n} & h_{y_n} & h_z + h_{z_n} \end{vmatrix} \neq 0, \text{and}$$

$$\begin{vmatrix} f_x + f_{x_n} & f_{y_n} & f_z + f_{z_n} & f_\alpha \\ g_{x_n} & g_{y_n} - E & g_{z_n} & g_\alpha \\ h_x + h_{x_n} & h_{y_n} & h_z + h_{z_n} & h_\alpha \\ \delta_x + \delta_{x_n} & \delta_{y_n} & \delta_z + \delta_{z_n} & \delta_\alpha \end{vmatrix} \neq 0.$$

Then there exists a smooth curve of equilibria in $\mathcal{R}^{n_x+n_y+n_z+1}$ which passes through $\bar{p}$ and is transversal to the singular surface at $\bar{p}$. When α increases through $\bar{\alpha}$, one eigenvalue of the system for (x_n, y_n, z_n) of eqns.(1)-(3) (i.e., an eigenvalue of J evaluated along the equilibrium locus) moves from less than modulus 1 to more than modulus 1 if $b_z/c > 0$ (respectively, from more than modulus 1 to less than modulus 1 if $b_z/c < 0$) by diverging through $\pm\infty$. The other $(n_x + n_y - 1)$ eigenvalues remain bounded and change in the order of $\mathcal{O}(\alpha - \bar{\alpha})$, where $c = \delta_\alpha - (\delta_x + \delta_{x_n}, \delta_{y_n}, \delta_z + \delta_{z_n}) \begin{pmatrix} f_x + f_{x_n} & f_{y_n} & f_z + f_{z_n} \\ g_{x_n} & g_{y_n} - E & g_{z_n} \\ h_x + h_{x_n} & h_{y_n} & h_z + h_{z_n} \end{pmatrix}^{-1} \begin{pmatrix} f_\alpha \\ g_\alpha \\ h_\alpha \end{pmatrix}$ evaluated at $\bar{p}$.

We call the bifurcations depicted in Theorem 5. 4 as *singularity-2 induced bifurcations*, in distinction from the singularity-1 induced bifurcations described in Theorem 5. 5 . The singularity-2 induced bifurcations occur at sampling instants for

(x_n, y_n, z_n). Before proving Theorem 5.4, we present the following lemma[5, 6].

Lemma 5.1 *Assume the same conditions as Theorem 5.4, then there exists a local coordinates (x', z') for (x, z) with (x'_n, z'_n) for x_n, z_n such that the matrices A, B, C for eqns.(7)-(9) are diffeomorphically transformed into the form A', B', C' respectively, along the equilibrium locus at the neighbourhood of $\bar{p}$*

$$A' = \begin{pmatrix} \frac{b_z}{c(\alpha-\bar{\alpha})} + \mathcal{O}(\alpha-\bar{\alpha}) & 0 \\ 0 & C_x + \mathcal{O}(\alpha-\bar{\alpha}) \end{pmatrix},$$

$$B' = \begin{pmatrix} \frac{b_{z_n}}{c(\alpha-\bar{\alpha})} + \mathcal{O}(\alpha-\bar{\alpha}) & 0 \\ C_{z_n} + \mathcal{O}(\alpha-\bar{\alpha}) & C_{x_n} + \mathcal{O}(\alpha-\bar{\alpha}) \end{pmatrix},$$

$$C' = \begin{pmatrix} 0 & 0 \\ C_{y_{1,n}} + \mathcal{O}(\alpha-\bar{\alpha}) & C_{y'_n} + \mathcal{O}(\alpha-\bar{\alpha}) \end{pmatrix}$$

where $x' = x'(x, x_n, y_n, z, z_n, \alpha) \in \mathcal{R}^{n_x}, z' = z'(x, x_n, y_n, z, z_n, \alpha) \in \mathcal{R}^{n_z}$ are C^∞ mappings in the open neighbourhoods of $\bar{p}$ and C_x is a nonsingular matrix. $y'_n = \{y_{2,n}, ..., y_{n_y,n}\}$. $C_x, C_{z_n}, C_{x_n}, C_{y_{1,n}}$ and $C_{y'_n}$ are $(n_x-1)\times(n_x-1), (n_x-1)\times 1, (n_x-1)\times(n_x-1), (n_x-1)\times 1$ and $(n_x-1)\times(n_y-1)$ matrices respectively, which are all independent of $(\alpha-\bar{\alpha})$.

Proof of Theorem 5.4. We first show that the equilibrium locus is transversal to the singular surface. Since the matrix of eqn.(18) is nonsingular, there exists an unique *equilibrium locus* $(x(\alpha), y(\alpha), z(\alpha), \alpha) \in \mathcal{R}^{n_x+n_y+n_z+1}$ locally near $\alpha = \bar{\alpha}$. Then along the equilibrium locus at $\bar{\alpha}$,

$$\begin{aligned} c = & \tfrac{d}{d\alpha}\delta(x(\alpha), x(\alpha), y(\alpha), z(\alpha), z(\alpha), \alpha) \\ = & \delta_\alpha + (\delta_x + \delta_{x_n}, \delta_{y_n}, \delta_z + \delta_{z_n})(\tfrac{dx}{d\alpha}, \tfrac{dy}{d\alpha}, \tfrac{dz}{d\alpha})^T \quad (19) \\ & \neq 0 \quad (20) \end{aligned}$$

where $(\frac{dx}{d\alpha}, \frac{dy}{d\alpha}, \frac{dz}{d\alpha})^T = -\begin{pmatrix} f_x + f_{x_n} & f_{y_n} & f_z + f_{z_n} \\ g_{x_n} & g_{y_n} - E & g_{z_n} \\ h_x + h_{x_n} & h_{y_n} & h_z + h_{z_n} \end{pmatrix}^{-1} \begin{pmatrix} f_\alpha \\ g_\alpha \\ h_\alpha \end{pmatrix}$ derived from eqns.(15)-(17) and eqn.(18) is substituted into eqn.(19). In eqn.(20), Schur's Theorem is also used from the assumptions. Therefore, eqn.(20) means that the equilibrium locus is transversal to the singular surface $|h_z| = 0$.

Without loss of generality, next we take the local coordinates (x', z') with (x'_n, z'_n) of Lemma 5.1 at $\bar{p}$. Then at the new local coordinates, $e^{\tau A}$ can be transformed into

$$\begin{pmatrix} e^{\frac{\tau b_z}{c(\alpha-\bar{\alpha})}} + \mathcal{O}(\alpha-\bar{\alpha}) & 0 \\ 0 & e^{\tau C_x} + \mathcal{O}(\alpha-\bar{\alpha}) \end{pmatrix}$$

according to Lemma 5.1, and furthermore $(e^{\tau A} - E)A^{-1}B$ and $(e^{\tau A} - E)A^{-1}C$ have the forms

$$\begin{pmatrix} \frac{b_{z_n}}{b_z}(e^{\frac{\tau b_z}{c(\alpha-\bar{\alpha})}} - 1) + \mathcal{O}(\alpha-\bar{\alpha}) & 0 \\ C_0 C_x^{-1} C_{z_n} + \mathcal{O}(\alpha-\bar{\alpha}) & C_0 C_x^{-1} C_{x_n} + \mathcal{O}(\alpha-\bar{\alpha}) \end{pmatrix}$$

and

$$\begin{pmatrix} 0 & 0 \\ C_0 C_x^{-1} C_{y_{1,n}} + \mathcal{O}(\alpha-\bar{\alpha}) & C_0 C_x^{-1} C_{y'_n} + \mathcal{O}(\alpha-\bar{\alpha}) \end{pmatrix}$$

respectively, due to Lemma 5.1, where $C_0 = e^{\tau C_x} - E$. Hence at the transformed coordinates $(x', x'_n, y_n, z', z'_n)$, the Jacobian matrix J for (x'_n, y_n) has the form J',

$$J' = \begin{pmatrix} \frac{b_z + b_{z_n}}{b_z} e^{\frac{\tau b_z}{c(\alpha-\bar{\alpha})}} - \frac{b_{z_n}}{b_z} + \mathcal{O}(\alpha-\bar{\alpha}) & 0 \\ \star & C_1 + \mathcal{O}(\alpha-\bar{\alpha}) \end{pmatrix} \quad (21)$$

where

$$C_1 = \left(\begin{array}{c|cc} C_0 C_x^{-1} C_{x_n} & C_0 C_x^{-1} C_{y_{1,n}} & C_0 C_x^{-1} C_{y'_n} \\ \frac{\partial \bar{g}_R}{\partial v'_n} & & \frac{\partial \bar{g}_R}{\partial y_n} \end{array} \right),$$

and $v'_n = \{x'_{2,n}, ..., x'_{n_x,n}\}$, $\frac{\partial \bar{g}_R}{\partial v'_n} = g_{v'_n} - g_{z'_n}(h_{z'} + h_{z'_n})^{-1}(h_{x'} + h_{v'_n})$, $\frac{\partial \bar{g}_R}{\partial y_n} = g_{y_n} - g_{z'_n}(h_{z'} + h_{z'_n})^{-1} h_{y_n}$. $\star$ is an appropriate matrix with each element having the form of *constant* $+ \mathcal{O}(\alpha-\bar{\alpha})$.

Therefore, when α increases through $\bar{\alpha}$, one eigenvalue of the system (i.e., an eigenvalue of J evaluated along the equilibrium locus) moves from less than modulus 1 to more than modulus 1 if $b_z/c > 0$ (respectively, from more than modulus 1 to less than modulus 1 if $b_z/c < 0$) by diverging through $\pm\infty$. The other $(n_x + n_y - 1)$ eigenvalues remain bounded, change in the order of $\mathcal{O}(\alpha-\bar{\alpha})$ and stay close to those of C_1 due to $b_z \neq 0, c \neq 0$ and $||b_z|| > ||b_{z_n}||$. □

Assuming the same conditions as Theorem 5.4 but with $||b_z|| \leq ||b_{z_n}||$, it is evident that the singularity-2 induced bifurcations still occur but are degenerate according the proof of Theorem 5.4. According to Theorem 5.4 and the Jacobian J' of eqn.(21), we have the following remark.

Remark 5.1 *Assume the same conditions of Theorem 5.4. When α increases through $\bar{\alpha}$, one eigenvalue of J evaluated along the equilibrium locus moves*

- *from less than modulus 1 to more than 1 by diverging through ∞ if $\frac{b_z + b_{z_n}}{b_z} > 0$ & $\frac{b_z}{c} > 0$;*
- *from less than modulus 1 to less than -1 by diverging through $-\infty$ if $\frac{b_z + b_{z_n}}{b_z} < 0$ & $\frac{b_z}{c} > 0$;*
- *from more than 1 (or ∞) to less than modulus 1 if $\frac{b_z + b_{z_n}}{b_z} > 0$ and $\frac{b_z}{c} < 0$;*
- *from less than -1 (or $-\infty$) to less than modulus 1 if $\frac{b_z + b_{z_n}}{b_z} < 0$ and $\frac{b_z}{c} < 0$;*

while the other $(n_x + n_y - 1)$ eigenvalues remain bounded and change in the order of $\mathcal{O}(\alpha-\bar{\alpha})$.

It can be shown from Theorem 5.4 that the dynamics can be locally decomposed into n^s-dimension invariant stable manifold, a n^u-dimension invariant unstable manifold and one dimension invariant singular manifold near $\bar{p}$ where n^s and n^u be the numbers of stable and unstable bounded eigenvalues of J[5, 6].

Example 5.1

$$\dot{x} = \alpha^2 - z + y_n - \alpha \quad (22)$$

$$y_{n+1} = 0.5 y_n + x_n - \alpha^4 + 0.5\alpha \quad (23)$$

$$x - z^2 = 0 \quad (24)$$

$$(n\tau \leq t < (n+1)\tau; n = 0, 1, 2, ...)$$

where α is a bifurcation parameter, and x, x_n, y_n, z, z_n and $\alpha \in \mathcal{R}^1$.

Obviously $\bar{x} = \bar{x}_n = \alpha^4$, $\bar{y}_n = \alpha$ and $\bar{z} = \bar{z}_n = \alpha^2$ are a strict equilibrium of eqns.(22)-(24) for any $\alpha \in \mathcal{R}^1$. Let $\bar{\alpha} = 0$ and $\bar{p} = (0, 0, 0, 0, 0, 0)$. $(x(\alpha), y(\alpha), z(\alpha), \alpha) = (\alpha^4, \alpha, \alpha^2, \alpha)$ is an unique equilibrium locus in $\mathcal{R}^4$ locally near $\alpha = 0$.

It is easy to check that eqns.(22)-(24) at $\bar{p}$ satisfies the conditions of Theorem 5.4 which means that the singularity-2 induced bifurcation occurs at $\bar{p}$ when α changes.

5.2 Intersampling instants

In this section, we will show that all of the bifurcations except a "degenerate" singularity induced bifurcation for a strict equilibrium at the intersampling instants are related to those at the sampling instants. In other words, there is no independently local bifurcations at the intersampling instants, except the singularity induced bifurcations which include both independent and dependent bifurcations on those of the sampling instants.

The stability of the DDA is determined by J which is the Jacobian of the equivalent discrete system for eqns.(1)-(3) according to Theorem 3. 2 . Therefore, there is no bifurcation at the intersampling instants as far as $|h_z| \neq 0$ at $\bar{p}$.

Although the stability of eqns.(1)-(3) is reined by J, the trajectory of $(x(t), z(t))$ at the intersampling instants follows eqns.(1) and (3) which is a differential-algebraic model.

Theorem 5. 5 *Suppose that $\bar{p} = (\bar{x}, \bar{x}, \bar{y}, \bar{z}, \bar{z}, \bar{\alpha})$ is a strict equilibrium of eqns.(1)-(3) with one-dimensional parameter $\alpha \in \mathcal{R}$. Assume that $|h_z|_{\bar{p}}$ has an algebraically simple zero eigenvalue and $b_z = -trace\ [\ f_z adj(h_z) h_x\ ]$ is nonzero, there is no purely imaginary eigenvalues for Jacobian matrix A and furthermore the following conditions are satisfied,*

$$\begin{vmatrix} f_x & f_z \\ h_x & h_z \end{vmatrix} \neq 0, \quad \text{and} \quad \begin{vmatrix} f_x & f_z & f_\alpha \\ h_x & h_z & h_\alpha \\ \delta_x & \delta_z & \delta_\alpha \end{vmatrix} \neq 0.$$

Then there exists a smooth curve of equilibria (of eqns.(1) and (3)) in $\mathcal{R}^{n_x+n_z+1}$ which passes through $\bar{p}$ and is transversal to the singular surface at $\bar{p}$. When α increases through $\bar{\alpha}$, one eigenvalue of the system for (x, z) of eqns.(1)-(3) (i.e., an eigenvalue of A evaluated along the equilibrium locus) moves from C^- to C^+ if $b_z/c > 0$ (respectively, from C^+ to C^- if $b_z/c < 0$) along the real axis by diverging through ∞. The other $(n_x - 1)$ eigenvalues remain bounded and change in the order of $\mathcal{O}(\alpha - \bar{\alpha})$, where $c = \delta_\alpha - (\delta_x, \delta_z) \begin{pmatrix} f_x & f_z \\ h_x & h_z \end{pmatrix}^{-1} \begin{pmatrix} f_\alpha \\ h_\alpha \end{pmatrix}$ evaluated at $\bar{p}$.

We call the bifurcations depicted in Theorem 5. 5 as *singularity-1 induced bifurcations* which occur at intersampling instants.

Remark 5. 2 *All of the local bifurcations except singularity-induced bifurcations at the intersampling instants are related with those of the sampling instants.*

5.3 Genericity of local bifurcations

This section will show that the DDA, eqns. (1)-(3) has five generic bifurcations described in Theorems 5. 1-5. 5 under certain generic assumptions.

We first give several definitions for structurally stability and bifurcation set [1].

Definition 5. 1 *A diffeomorphism map or vector field $F(x) : \mathcal{X} \to \mathcal{X} \in \mathcal{R}^m$ is called structurally stable in $\mathcal{X}$ if there is an $\epsilon > 0$ such that whenever $\sup_{x \in \mathcal{X}}\{||F(x) - G(x)|| + ||F_x(x) - G_x(x)|| < \epsilon\}$ for $G : \mathcal{X} \to \mathcal{X}$, F is topologically equivalent (conjugate) to G, in other words, there exists homeomorphism $H : \mathcal{X} \to \mathcal{X}$ such that $H \circ F = G \circ H$.*

Definition 5. 2 *The bifurcation set is the set of points for which the dynamics is structurally unstable under variation of parameters.*

Let $EP = \{x_n, y_n, z_n, \alpha\}$ and $BP = \{x_n, y_n, z_n, \alpha\}$ denote the set of strict equilibria and the bifurcation set of the strict equilibria for eqns.(1)-(3), respectively. Therefore $EP = \{(x_n, y_n, z_n, \alpha) : f(p) = 0, y_n = g(p), h(p) = 0$ where $p = (x_n, x_n, y_n, z_n, z_n, \alpha)\}$ which is clearly an open subset of $\mathcal{R}^{n_x+n_y+n_z+1}$. Define the Hurwitz matrix H_{m-1} as

$$\begin{pmatrix} a_1 & a_3 & a_5 & \cdots & a_{2m-3} \\ a_0 & a_2 & a_4 & \cdots & a_{2m-4} \\ 0 & a_1 & a_3 & \cdots & a_{2m-5} \\ 0 & a_2 & a_4 & \cdots & a_{2m-6} \\ \vdots & & & & \vdots \\ 0 & 0 & 0 & \cdots & a_{m-1} \end{pmatrix}, \qquad (25)$$

and $H_{m-1}(\lambda E - \mathcal{A})$ means that the elements a_i correspond to the efficients of polynomial $|\lambda E - \mathcal{A}| = a_0\lambda^m + a_1\lambda^{m-1} + \ldots + a_m$ for $m \times m$ matrix $\mathcal{A}$, while $H_{m-1}(\frac{1+\lambda}{1-\lambda}E - \mathcal{A})$ means that the elements a_i correspond to the efficients of $|\frac{1+\lambda}{1-\lambda}E - \mathcal{A}| = (a_0\lambda^m + a_1\lambda^{m-1} + \ldots + a_m)/(1-\lambda)^{m-1}$ for $m \times m$ matrix $\mathcal{A}$. $|H_{m-1}(\lambda E - \mathcal{A})| = 0$ implies that $\mathcal{A}$ has purely imaginary eigenvalues while $|H_{m-1}(\frac{1+\lambda}{1-\lambda}E - \mathcal{A})| = 0$ means that $\mathcal{A}$ has modulus 1 eigenvalues.

Then we directly have the following lemma.

Lemma 5. 2 *The bifurcation points for strict equilibria of eqns.(1)-(3) are composed by four sets*

$$BP = BP \cap (S_{SI} \cup S_{FD} \cup S_{FP} \cup S_{NS}) \qquad (26)$$

where

$$S_{SI} = \{(x_n, y_n, z_n, \alpha) \in EP : |h_z| = 0\} \qquad (27)$$
$$S_{FD} = \{(x_n, y_n, z_n, \alpha) \in EP : |J - E| = 0\} \qquad (28)$$
$$S_{FP} = \{(x_n, y_n, z_n, \alpha) \in EP : |J + E| = 0\} \qquad (29)$$
$$S_{NS} = \{(x_n, y_n, z_n, \alpha) \in EP : |H_{n_x+n_y-1}(\tfrac{1+\lambda}{1-\lambda}E - J)| = 0\}. \qquad (30)$$

Notice that the system does not generally undergo bifurcations at the set $\{(x_n, y_n, z_n, \alpha) \in EP : |h_z + h_{z_n}| = 0\}$, because one of the bifurcations in Theorems 5. 1 - 5. 4 will occur before $|h_z + h_{z_n}|$ approaches zero.

It is evident that each set in Lemma 5. 2 is restricted by only one equality on EP. Let $S_{si1}, S_{fd}, S_{fp}, S_{ns}$ and S_{si2} be the sets of points satisfying the conditions of 5. 5 ,5. 1 ,5. 2 ,5. 3 and 5. 4 , respectively. Then, we suppose the following generic assumptions.

Assumption 5. 2 *$S_{si1}, S_{fd}, S_{fp}, S_{ns}$ and S_{si2} are open and dense in S_{SI}, S_{FD}, S_{FP}, S_{NS} and S_{SI}, respectively.*

Theorem 5. 6 *The generic bifurcations of eqns.(1)-(3) are constituted by the bifurcations of Theorems 5. 5 ,5. 1 ,5. 2 ,5. 3 and 5. 4 under Assumption 5. 2 .*

Proof of Theorem 5. 6 . Since only additional independent non-equalities are restricted on $S_{si1}, S_{fd}, S_{fp}, S_{ns}$ from $S_{SI}, S_{FD}, S_{FP}, S_{NS}$, due to Assumption 5. 2 we get

$$\overline{S_{fd}} = S_{FD}, \overline{S_{fp}} = S_{FP}, \overline{S_{ns}} = S_{NS}, \overline{S_{si1}} = S_{SI} \qquad (31)$$

where $\overline{S}$ means the closure of a set S. Moreover compared with S_{si1}, S_{si2} is also constrained by an inequality $||b_z|| > ||b_{z_n}||$ in addition to several non-equalities from S_{SI}. Hence $\overline{S_{si2}} \subset \overline{S_{si1}}$. Considering that S_{si2} is also a dense and open subset of S_{SI} according to Assumption 5. 2 , we express S_{SI} by

$$\overline{S_{si1}} \cup \overline{S_{si2}} = S_{SI}.$$

Therefore according to Lemma 5. 2 , we have

$$BP = BP \cap (\overline{S_{si1}} \cup \overline{S_{si2}} \cup \overline{S_{fd}} \cup \overline{S_{fp}} \cup \overline{S_{ns}}) \qquad (32)$$

which proves this theorem. □

Singularity-1 induced bifurcation can generally be expected to occur when the system undergoes the singularity-2 induced bifurcation due to $\overline{S_{si2}} \subset \overline{S_{si1}}$, although they are restricted by different inequalities and non-equality, i.e. S_{si2} is not a subset of S_{si1}. However, most of the independently locally bifurcations at the intersampling instants are composed of the degenerate singularity-1 induced bifurcations as shown in the proof of Theorem 5. 5 , according to the genericity from Theorem 5. 6 .

Remark 5. 3 *Most of locally nondegenerate bifurcations for the DDA at strict equilibria are related with the local bifurcations of the sampling instants under Assumption 5. 2 .*

6 Application to Power Systems

We show how a power system with digital PSS can be treated as a DDA in this section. For a power system, the dynamics of the generators and analog controllers(e.g. AVR) is defined as eqn.(1) of the DDA, while eqn.(3) is composed of the power balance equations. Then eqn.(2) corresponds to the discrete-time controllers (or digital PSS). In this paper, the parameters of the digital PSS are first designed by imitating the conventional analog PSS, and then are adjusted and evaluated by using Theorems 3. 2 -5. 4 .

Next in order to obtain the digital PSS by imitating the designing of analog PSS, we derive eqn.(2) from an analog PSS with a hold ciruit and a sampler. Generally, this process is not necessary if the digital PSS or eqn.(2) is given.

Generally, the digital PSS consists of a sampler, a discrete-time controller and a hold circuit. In this paper, we adopt three-order digital PSS with a zero-order hold circuit which is obtained by discretizing the analog PSS. i.e. "Control device". The state $y(t) \in \mathcal{R}^3$ and input $u_n \in \mathcal{R}$ of the PSS follow

$$\begin{aligned} \dot{y}(t) &= D_1 y(t) + D_2 u_n \\ u_n &= u(x_n, z_n) = u(x(n\tau), z(n\tau)) \\ & (n\tau \le t < (n+1)\tau; n = 0, 1, ...) \qquad (33) \end{aligned}$$

and the output $v_n \in \mathcal{R}$ of the digital PSS can be expressed as

$$v(t) = \left\{ \begin{array}{cc} v_{max} & \text{if } v(t) \ge v_{max} \\ \multicolumn{2}{c}{-\frac{T_3T_5}{T_2T_4} y_1(t) + \frac{T_5}{T_4} y_2(t) + y_3(t) + \frac{KT_3T_5}{T_1T_2T_4} u_n} \\ v_{min} & \text{if } v(t) \le v_{min} \end{array} \right\}$$

$$v_n = v(n\tau)$$

$$(n\tau \le t < (n+1)\tau; n = 0, 1, ...), \qquad (34)$$

where, $y(t) = (y_1(t), y_2(t), y_3(t))^T$. $u \in \mathcal{R}$ is the real power output of the generator which is also a given function of (x, z). $v \in \mathcal{R}$ (or $v_n \in \mathcal{R}$) is the PSS output, and v_{max}, v_{min} are upper and lower bounds for the internal output of the PSS. u_n and v_n are the values of $u(t)$ and $v(t)$ at instant $t = n\tau$ respectively, which are constantly hold during $n\tau \le t < (n+1)\tau$. Define D_1 and D_2 as follows

$$D_1 = \begin{bmatrix} -\frac{1}{T_1} & 0 & 0 \\ -\frac{T_2-T_3}{T_2^2} & -\frac{1}{T_2} & 0 \\ -\frac{T_3}{T_2}\frac{T_4-T_5}{T_4^2} & \frac{T_4-T_5}{T_4} & -\frac{1}{T_4} \end{bmatrix}, D_2 = \begin{bmatrix} \frac{1}{T_1^2} \\ \frac{K}{T_1}\frac{T_2-T_3}{T_2^2} \\ \frac{KT_3}{T_1T_2}\frac{T_4-T_5}{T_4^2} \end{bmatrix}.$$

Since eqn.(33) is a linear differential equation, it can be analytically solved. Therefore, we have following difference equation by integrating eqn.(34) from $n\tau$ to $(n+1)\tau$,

$$y_{n+1} = e^{D_1\tau} y_n + D_1^{-1}[e^{D_1\tau} - E] D_2 u(x_n, z_n) \quad (35)$$

which has the same form with eqn.(2). That is, eqn.(35) is the discrete-time controller whose design imitates that of the continuous-time controller, i.e. a three-order analog PSS. Note that the real power output $u(x_n, z_n)$ of the generator is a given function of (x_n, z_n).

The **PSS output** v_n is obviously a function of y_n and $u(x_n, z_n)$ according to eqn.(34), i.e.,

$$v_n = \eta_n(x_n, y_n, z_n) \equiv \eta(y_n, u(x_n, z_n)) \qquad (36)$$

where v_n is not a linear function because of the saturation v_{min}, v_{max} of $v(t)$ shown in eqn.(34).

On the other hand, the dynamics of generators and analog controllers corresponding to eqn.(1), the power balance equations corresponding to eqn.(3) can be written as follows

$$\dot{x}(t) = f(x(t), \eta_n(x_n, y_n, z_n), z(t)) \qquad (37)$$

$$h(x(t), \eta_n(x_n, y_n, z_n), z(t)) = 0. \qquad (38)$$

Notice that x_n, y_n and z_n can be viewed as constants during $n\tau \le t < (n+1)\tau$ for eqns.(37)-(38) because v_n is constantly hold during that period. Eqns.(37)-(38) are generally nonlinear.

Therefore, combining eqns.(37)(35)(38), we can use the Theorems 3. 2 -5. 4 to analyse the power systems with the digital PSS numerically in a nonlinear way. Several numerical simulations have verified that our results are useful[5, 6].

7 Conclusion

We first described locally sampling, asymptotical stability and singular perturbation for DDA. And then, this paper showed that there are five types of generic bifurcations at strict equilibria which cause the system unstable, and derived their bifurcation conditions. Finally, the theoretical results were applied to the evaluations of digital control for power systems.

References

[1] J.Guckenheimer, P.Holmes.(1983). Nonlinear Oscillations, Dynamical Systems, and Bifurcations of Vector Fields. *Springer-Verlag,* New York.

[2] V.Venkatasubramanian, H.Schattler, J.Zaborszky. (1992). A stability theory for large differential algebraic systems. *Report SSM 9201-Part I,* Department of Systems Science and Mathematics, Washington University, St.Louis,MO.

[3] V.Venkatasubramanian, H.Schattler, J.Zaborszky. (1995). Local Bifurcations and Feasibility Regions in Differential-Algebraic Systems. *IEEE Trans. on AC*, **40**, No.12, pp.1992-2013.

[4] L.Chen, K.Aihara. (1997). Chaos and asymptotical stability in discrete-time neural networks. *Physica D*, **104**, pp.286-325.

[5] L.Chen, K.Aihara. (1997). Local Stability in Hybrid Dynamical Systems. *1997 International Symposium on Nonlinear Theory and its Applications (NOLTA'97)*, pp.73-76.

[6] L.Chen, K.Aihara. (1998). Stability and Bifurcation Analysis of Hybrid Dynamical Systems. *Technical Reports*, **METR 98-02**, The University of Tokyo.

[7] R.L.Grossman, A.Nerode, A.P.Ravn, H.Rishel (edited). (1993). Hybrid Systems. *Springer-Verlag*, New York.

14th Triennial World Congress, Beijing, P.R. China

E-2c-13-5

GLOBAL STABILIZATION OF A CLASS OF DISCRETE-TIME NONLINEAR STOCHASTIC SYSTEMS

Ruisheng Li and Mehrdad Saif

School of Engineering Science, Simon Fraser University, Burnaby, British Columbia, Canada V5A 1S6. e-mail: saif@cs.sfu.ca

Abstract: For a class of nonlinear stochastic systems with parametric-strict-feedback form, global stabilization of closed-loop system is established by designing an adaptive controller based on the weighted least-squares algorithm and by using a backstepping method where nonlinearity vectors appearing in the last state equation are permitted to have the growth rate faster than linear. *Copyright ©1999 IFAC*

Keywords: Nonlinear systems, Random noise, least-squares algorithm, adaptive control, global stability.

1. INTRODUCTION

In the past three decades a lot of theoretical and applied progress has been made in the direction of adaptive control for linear systems. However, when output of a true system changes too fast, an adaptive controller designed on a hypothetical linear model could hardly be used to control the system since global stabilization of input and output signal might not be achieved. Hence, to satisfy with complicated practical application it is necessary to design adaptive nonlinear controller and to develop related theoretical results. Further many scientists in the field of control have focused their attention on the adaptive control for nonlinear systems in recent decade.

Many significant results have been obtained in the direction of discrete-time adaptive nonlinear systems (Kanellakopoulos 1994), (Zhang and Lang 1989), (Lin and Byrnes 1995), (Yeh and Kokotovic 1995*a*), (Yeh and Kokotovic 1995*b*). By use of multilayer neural network method, local stabilities of adaptive discrete-time nonlinear systems were established (Chen and Khail 1992), (Chen and Khail 1995), (Ruisheng 1996). The restriction of linear growth rate on the nonlinearities was used by (Yeh and Kokotovic 1995*a*), (Yeh and Kokotovic 1995*b*), (Byrnes C.I. and Ghosh 1993), (Byrnes and Lin 1994) and (Wei and Guo 1996) for system stabilization. By designing an adaptive controller based on the least mean algorithm, the global adaptive stabilization was achieved by (Yeh and Kokotovic 1995*a*). The global stabilization of discrete-time nonlinear where the nonlinearities were permitted to have growth rate faster than linear was obtained via some adaptive algorithm based on the modified least squares (Kanellakopoulos 1994) and the standard least squares (LS) (Guo and Wei 1996). All results as above mentioned were achieved in the framework of deterministic systems, that is, free of system noise.

The first paper studing adaptive nonlinear stochastic systems with nonlinear function having a growth rate faster than linear appears in the recent work (Guo 1997), where it is successfully found that in the scalar unknown parameter case if the growth rate of a nonlinear function $f(x)$ is

$f(x) = O(||x||^b)$ with $b \geq 0$, then the discrete-time adaptive nonlinear stochastic systems are globally stable whenever $b < 4$, and is unstable in general if $b \geq 4$. It is also found an interesting phenomenon (Guo 1997) that the linear case does not have: for some LS-based certainty equivalence adaptive controls, even if the LS parameter estimates are strongly consistent, the closed-loop system may still be unstable. The work (Guo 1997) provided the fundamental analytical methods and results including a crucial technical lemma, towards a more comprehensive investigation of discrete-time adaptive nonlinear stochastic control systems.

In this paper the global stabilization of adaptive discrete-time nonlinear stochastic systems with parametric-strict-feedback form is established, inspired by the idea provided by (Guo 1997). The paper is organized as follows. The design procedure is presented in Section 2. The main results are stated in Section 3. The proofs are given in Section 4. The conclusions are provided in Section 5.

2. ADAPTIVE CONTROL

Consider the nonlinear stochastic systems:

$$\begin{aligned} x_i(k+1) &= x_{i+1}(k) + w_i(k+1) \\ &\quad +\theta^{\tau}\alpha_i(x_1(k), \cdots x_i(k)) \\ i &= 1, 2, \cdots, n-1 \\ x_n(k+1) &= \alpha_0(x_1(k), \cdots, x_n(k)) + u_k \\ &\quad + \theta^{\tau}\alpha_n(x_1(k), \cdots x_n(k)) + w_n(k+1) \\ y_k &= x_1(k) \end{aligned} \tag{1}$$

where $\{u_k\}$ and $\{y_k\}$ are respectively the system input and output sequences with $y_k = u_k \equiv 0$ for any $k < 0$, $\{x_i(k), i = 1, 2, \cdots, n\}$ is known state, $\{w_i(k), i = 1, 2, \cdots, n\}$ is state noise, with $w_i(k) \equiv 0$ and $x_i(k) \equiv 0$ for any $k < 0$ and each $i = 1, 2, \cdots, n$, $\theta \in R^d$ is a deterministic unknown nonlinear vector to be estimated, $\alpha_0(.)$ and $\alpha_i(.)$ $(1 \leq i \leq n)$ are known nonlinear function and vectors, respectively.

The estimates for a $d-$dimensional unknown parameter vector θ in the system (1) are recursively generated from a weighted least-squares (WLS) algorithm used by (Bercu 1995) and (Guo 1996):

$$\widehat{\theta}_{k+1} = \widehat{\theta}_k + P_k\varphi_k S_k^{-1}(X_{k+1} - \widehat{X}_{k+1}) \tag{2}$$

$$P_{k+1} = P_k - P_k\varphi_k S_k^{-1}\varphi_k^{\tau} P_k \tag{3}$$

$$S_k = \alpha_k^{-1} I + \varphi_k^{\tau} P_k \varphi_k \tag{4}$$

$$\alpha_k = \frac{1}{log^{1+\delta} r_k}, \quad \delta > 0 \tag{5}$$

$$r_k = r_0 + \sum_{i=0}^{k} tr(\varphi_i^{\tau}\varphi_i) \tag{6}$$

$$\begin{aligned} \widehat{X}_{k+1} &\triangleq A_n X_k + \varphi_k^{\tau}\widehat{\theta}_k + b_n u_k \\ &\quad +b_n\alpha_0(x_1(k), x_2(k), \cdots, x_n(k)) \end{aligned} \tag{7}$$

with $r_0 \triangleq max\{e, tr(P_0^{-1})\}$ where I is a $n \times n$ unit matrix, the positive integer n is a state dimension, the initial values $\widehat{\theta}_0$ and $P_0 > 0$ are arbitrarily chosen and φ_k is a known $d \times n$ matrix with

$$\varphi_k^{\tau} \triangleq \begin{bmatrix} \alpha_1^{\tau}(x_1(k)) \\ \alpha_2^{\tau}(x_1(k), x_2(k)) \\ \vdots \\ \alpha_n^{\tau}(x_1(k), x_2(k), \cdots, x_n(k)) \end{bmatrix} \tag{8}$$

$$A_n \triangleq \begin{bmatrix} 0 & 1 & 0 & \cdots & 0 \\ & 0 & & \ddots & \vdots \\ \vdots & \vdots & \ddots & & 0 \\ & & & \ddots & 1 \\ 0 & 0 & \cdots & & 0 \end{bmatrix}, \quad b_n \triangleq= \begin{bmatrix} 0 \\ \vdots \\ 0 \\ 1 \end{bmatrix}$$

$$X_k \triangleq [x_1(k), x_2(k), \cdots, x_n(k)]^{\tau}$$

(1) is rewritten as

$$\begin{aligned} X_{k+1} &= A_n X_k + \varphi_k^{\tau}\theta + W_{k+1} + b_n u_k \\ &\quad +b_n\alpha_0(x_1(k), x_2(k), \cdots, x_n(k)) \end{aligned} \tag{9}$$

where

$$W_k \triangleq [w_1(k), w_2(k), \cdots, w_n(k)]^{\tau} \tag{10}$$

It follows from (7) and (9) that the one-step-ahead prediction error $X_{k+1} - \widehat{X}_{k+1}$ is adapted to

$$X_{k+1} - \widehat{X}_{k+1} = \varphi_k^{\tau}\tilde{\theta}_k + w_{k+1} \tag{11}$$

where $\tilde{\theta}_k$ is an estimation error defined by

$$\tilde{\theta}_k = \theta - \widehat{\theta}_k$$

For designing an adaptive controller, new state vectors are introduced:

$$\begin{aligned} z_1(k) &\triangleq x_1(k) \\ z_j(k) &\triangleq x_j(k) + \widehat{\theta}_k^{\tau}\alpha_{j-1}(k) + \sum_{i=1}^{j-2}\widehat{\theta}_k^{\tau}\overline{\alpha}_{ij-1}(k) \\ j &= 2, 3, \cdots, n \end{aligned} \tag{12}$$

where and in the squel write

$$\alpha_j(k) \triangleq \alpha_j(x_1(k), \cdots, x_j(k))$$

for $j = 0, 1, \cdots, n$. $\widehat{\theta}_k$ is determined by (2) – (8) and

$$\begin{aligned}\overline{\alpha}_{11} &\triangleq \alpha_1(z_1(k)) \\ \overline{\alpha}_{ij}(k) &\triangleq \alpha_i(z_{j-i+1}(k),\ z_{j-i+2}(k) \\ &\quad -\widehat{\theta}_k^{\tau}\overline{\alpha}_{1j-i+1}(k), \\ &\quad z_{j-i+3}(k) - \widehat{\theta}_k^{\tau}\overline{\alpha}_{1j-i+2}(k) \\ &\quad -\widehat{\theta}_k^{\tau}\overline{\alpha}_{2j-i+2}(k),\ \cdots, \\ &\quad z_{j-1}(k) - \sum_{s=1}^{i-2}\widehat{\theta}_k^{\tau}\overline{\alpha}_{sj-2}(k), \\ &\quad z_j(k) - \sum_{s=1}^{i-1}\widehat{\theta}_k^{\tau}\overline{\alpha}_{sj-1}(k)), \\ &\forall\ 1 \le i \le j-1, \quad j = 2, 3, \cdots, n \end{aligned} \tag{13}$$

Further define an adaptive controller

$$u_k \triangleq y_{k+n}^* - \alpha_0(k) - \widehat{\theta}_k^{\tau}\alpha_n(k) - \sum_{i=1}^{n-1}\widehat{\theta}_k^{\tau}\overline{\alpha}_{in}(k) \tag{14}$$

By the same argument as that given by (Yeh and Kokotovic 1995*a*) the state equation is obtained from (1) and (12) – (14):

$$\begin{aligned} z_1(k+1) &= z_2(k) + \tilde{\theta}_k^{\tau}\alpha_1(k) + w_1(k+1) \\ z_i(k+1) &= z_{i+1}(k) + \tilde{\theta}_k^{\tau}\alpha_i(k) + \beta_i(k+1) \\ &\quad + w_i(k+1) \\ i &= 2, 3, \cdots, n-1 \\ z_n(k+1) &= y_{k+n}^* + \tilde{\theta}_k^{\tau}\alpha_n(k) + \beta_n(k+1) \\ &\quad + w_n(k+1) \end{aligned} \tag{15}$$

where

$$\beta_j(k+1) \triangleq \sum_{i=1}^{j-1}\widehat{\theta}_{k+1}^{\tau}\overline{\alpha}_{ij-1}(k+1) - \sum_{i=1}^{j-1}\widehat{\theta}_k^{\tau}\overline{\alpha}_{ij}(k), \quad j = 2, 3, \cdots, n \tag{16}$$

For simplicity (15) is rewritten as a state-space form:

$$\begin{aligned} Z_{k+1} &= A_n Z_k + \varphi_k^{\tau}\tilde{\theta}_k + b_n y_{k+n}^* \\ &\quad + \gamma_{k+1} + W_{k+1} \end{aligned} \tag{17}$$

where A_n, b_n, φ_k, $\tilde{\theta}_k$ and W_k are denoted as before,

$$Z_k \triangleq [z_1(k),\ z_2(k),\ \cdots, z_n(k)]^{\tau}$$

and

$$\gamma_k \triangleq [0,\ \beta_2(k),\ \beta_3(k),\ \cdots, \beta_n(k)]^{\tau} \tag{18}$$

Set

$$\widehat{Z}_{k+1} \triangleq A_n Z_k + b_n y_{k+1}^* + \gamma_{k+1} \tag{19}$$

Thus it is easy to see from (17) and (19) that the one-step-ahead prediction error $Z_{k+1} - \widehat{Z}_{k+1}$ satifies (11).

It is not difficult to get a closed-loop system from (8), (17) and (18) and $y_k = x_1(k) = z_1(k)$ that

$$\begin{aligned} y_{k+1} - y_{k+1}^* &= \sum_{i=1}^{n}\alpha_i(k-i+1)^{\tau}\tilde{\theta}_{k-i+1} \\ &\quad + \sum_{i=1}^{n-1}\beta_{i+1}(k-i+1) + \overline{w}_{k+1} \end{aligned} \tag{20}$$

where

$$\overline{w}_k \triangleq \sum_{i=1}^{n} w_i(k-i+1) \tag{21}$$

3. MAIN RESULTS

The following two assumptions are needed in this paper.

Assumption A. $\{W_t,\ F_t\}$ is a martingale difference sequence, i.e., $W_t \in F_t$ and $E(W_{t+1}|F_t) = 0$ for any t, and

$$\sup_{t\ge 0} E(||W_{t+1}||^{\beta}|F_t) < \infty \quad a.s. \tag{22}$$

with $\beta > 2$ and

$$\sum_{i=1}^{t} ||W_i||^{2b} = O(t) \quad a.s. \tag{23}$$

Assumption B. There exist $b \in [1, 4)$, $K_i > 0$ and $M_i > 0$ such that

$$\begin{aligned} &||\alpha_i(X_i) - \alpha_i(\overline{X}_i)||^2 \le K_i + M_i||X_i - \overline{X}_i||^2 \\ &\forall\ X_i = (x_1,\ x_2,\ \cdots, x_i) \in R^i \\ &\forall\ \overline{X}_i = (\overline{x}_1,\ \overline{x}_2,\ \cdots, \overline{x}_i) \in R^i \\ &i = 1,\ 2,\ \cdots, n-1 \end{aligned} \tag{24}$$

$$\begin{aligned} &||\alpha_j(x_1, x_2, \cdots, x_n)||^2 \le K_j + M_j\left(\sum_{i=1}^{n}|x_i|^{2b}\right) \\ &\forall\ (x_1,\ x_2,\ \cdots, x_n) \in R^n,\ j = 0,\ n \end{aligned} \tag{25}$$

The imposed Lipschiz condition on $\alpha_i(.)$ $(i = 1, 2, \cdots, n-1)$ is the same as that given by (Yeh and Kokotovic 1995*a*) . It is necessary to deal with $\beta_j(k-j+2)$ $(j = 2, \cdots, n)$ in (20). The nonlinearity vectors $\alpha_0(.)$ and $\alpha_n(.)$ here are permitted to have the growth rate faster than linear: $||\alpha_n(X)|| = O(||X||^b)$ and $||\alpha_0(X)|| = O(||X||^b)$ with $b \in [1, 4)$. However, they are restricted to the linear growth rate corresponding with $b = 1$ (Yeh and Kokotovic 1995*a*). It is obvious that the following two-order system satifies Assumption B and not satisfies the condition used by (Yeh and Kokotovic 1995*a*).

$$\begin{aligned} x_1(k+1) &= x_2(k) + \theta x_1(k)\, sin(x_1(k)) \\ &\quad + w_1(k+1) \\ x_2(k+1) &= 4 + 2[|x_1(k)|^3 + |x_2(k)|^3] \\ &\quad + \theta\{10 + 5[|x_1(k)|^3 + |x_2(k)|^{\frac{7}{2}}]\} \\ &\quad + u_k + w_2(k+1) \\ y_k &= x_1(k) \end{aligned}$$

The main results are given in the following.

Theorem 1. Applying the adative control algorithm (2) – (8) and (12) – (14) to the system (1) with $d = 1$ if the assumptions A and B are satisfied and $\liminf_{t\to\infty} \frac{r_t}{t} \neq 0$, where r_t is defined by (6), then
(i). the adaptive tracking is sub-optimal in the sense

$$\sum_{k=1}^{t} [y_k - y_k^* - Q_k]^2 = O[(logt)^{2+\delta}] \quad (26)$$

with $\delta > 0$ where

$$Q_k \triangleq \sum_{i=2}^{n} \beta_i(k-i+1) + w^*(k-1) \quad (27)$$

$$w^*(k) \triangleq \sum_{i=2}^{n} w_i(k-i+2)$$

where $\beta_i(.)$ is given by (16), and further, there exists a constant $\xi_0 > 0$ such that

$$\limsup_{t\to\infty} \frac{1}{t} \sum_{k=1}^{t} |y_k - y_k^*|^2 \le \xi_0 \quad (28)$$

(ii). the closed-loop system is globally stable in the sense

$$\limsup_{t\to\infty} \frac{1}{t} \sum_{i=1}^{t} (||y_i||^2 + ||u_i||^2) < \infty \quad (29)$$

4. ANALYSIS

For Theorem 1 five lemmas are needed to be established.

Lemma 1. Under Assumption A the algorithm (2) – (8) has the properties:
(i). $\tilde{\theta}_t^T P_t^{-1} \tilde{\theta}_t = O(1)$ *a.s.* $\forall\, t$
(ii). $\sum_{t=0}^{\infty} ||S_t^{-\frac{1}{2}} \varphi_t^T \tilde{\theta}_t||^2 < \infty$ *a.s.*
(iii). $\hat{\theta}_t \longrightarrow \overline{\theta}$ *a.s.* as $t \to \infty$
where

$$\overline{\theta} \triangleq \hat{\theta}_0 + \sum_{t=0}^{\infty} P_t \varphi_t S_t^{-1} (X_{t+1} - \hat{X}_{t+1}) \quad (30)$$

The simialr proof was given by Guo (1996).

Lemma 2. Assumption B guarantees that

$$\begin{aligned} |\beta_j(k)|^{2b} &= O(1) + o[|z_1(k)|^{2b} + |z_j(k-1)|^{2b}] \\ &\quad + o\left\{\sum_{i=2}^{j-1} [|z_i(k)|^{2b} + |z_i(k-1)|^{2b}]\right\} \\ &\quad + O\left\{\sum_{i=1}^{j-1} [|\tilde{\theta}_{k-1}^T \alpha_i(k-1)|^{2b}\right\} \\ &\quad + O\left\{\sum_{i=1}^{j-1} |w_i(k)|^{2b}\right\} \\ &\quad \forall\, j = 2, 3, \cdots, n, \ \forall\, b \in [1, 4) \end{aligned}$$

where $\{\beta_j(k),\ 2 \le j \le n\}$ is given by (16).

Lemma 3. Assumption B implies that

$$\begin{aligned} ||\varphi_k||^2 &= O(1) + o\left\{\sum_{j=1}^{n-1} \sum_{i=0}^{n-1} |z_j(k-i)|^{2b}\right\} \\ &+ o\left\{\sum_{i=1}^{n} |z_n(k-i)|^{2b}\right\} \\ &+ o\left\{\sum_{j=1}^{n-1} \sum_{i=0}^{n} |z_j(k-i)|^2 + \sum_{i=1}^{n} |z_n(k-i)|^2\right\} \\ &+ O\left\{\sum_{i=1}^{n} \sum_{j=1}^{i} |\tilde{\theta}_{k-j}^T \alpha_i(k-j)|^{2b}\right\} \\ &+ O\left\{\sum_{i=1}^{n} \sum_{j=1}^{i} |\tilde{\theta}_{k-j}^T \alpha_i(k-j)|^2\right\} \\ &+ O\left\{\sum_{i=1}^{n} \sum_{j=0}^{i-1} [|w_i(k-j)|^{2b} + |w_i(k-j)|^2]\right\} \end{aligned}$$

The proof is given in Appendix C.

Lemma 4. Under the assumptions A and B

$$r_t = O(t) + O\left[\sum_{k=0}^{t-1}(||\varphi_k^\tau\tilde{\theta}_k||^{2b} + ||\varphi_k^\tau\tilde{\theta}_k||^2)\right]$$

Lemma 5. The algorithm (2) – (8) with $d = 1$ yields that

$$tr(S_k) = O\left[(logr_k)^{2(1+\delta)}\frac{r_k}{r_{k-1}}\right] \quad \forall\, k \geq 1$$

The conclusion follows from (3) and (4), omitted here.

The proof of Theorem 1. By Lemmas 1 (ii), 4 and 5

$$r_t = O(t) + O\left(\sum_{k=0}^{t-1}||S_k^{-\frac{1}{2}}\varphi_k^\tau\tilde{\theta}_k||^2\frac{r_k^{b+\varepsilon}}{r_{k-1}^b}\right) +O\left(\sum_{k=0}^{t-1}||S_k^{-\frac{1}{2}}\varphi_k^\tau\tilde{\theta}_k||^2\frac{r_k^{1+\varepsilon}}{r_{k-1}}\right) \tag{31}$$

for any $\varepsilon > 0$. Note that from $b \geq 1$

$$r_k^\varepsilon\left(\frac{r_k^b}{r_{k-1}^b} + \frac{r_k}{r_{k-1}}\right) \leq 2\frac{r_k^{b+\varepsilon}}{r_{k-1}^b}$$

Therefore (31) leads to

$$r_t = O(t) + O\left(\sum_{k=0}^{t-1}||S_k^{-\frac{1}{2}}\varphi_k^\tau\tilde{\theta}_k||^2\frac{r_k^{b+\varepsilon}}{r_{k-1}^b}\right)$$

for any $\varepsilon > 0$. Hence, by use of Lemma 4.1 given by Guo (1997), Lemma 1 (ii) here, the condition $\limsup_{t\to\infty}\frac{r_t}{t} \neq 0$ and take that $\varepsilon \in (0, \sqrt{b}(2-\sqrt{b}))$ an assertion is yielded:

$$r_t = O(t), \ \sup_t \frac{r_{t+1}}{r_t} < \infty \tag{32}$$

Further, from Lemma 1 (ii), (2), (8) and (20)

$$\sum_{k=0}^{t-1}|y_{k+1} \ -y_{k+1}^* - Q_{k+1}|^2 = O\left(\sum_{k=0}^{t-1}||\varphi_k^\tau\tilde{\theta}_k||^2\right) = O\left[(logr_{t-1})^{2(1+\delta)}\right] \tag{33}$$

for any $\delta > 0$ Furthermore, by (32), (26) holds true. It follows from Lemma 2 with $j = n$, (8), (20), and (23) that

$$\sum_{k=1}^{t}|y_k - y_k^*|^2 = O(t) + O\left[\sum_{k=1}^{t}||\varphi_k^\tau\tilde{\theta}_k||^2\right] +o\left[\sum_{k=1}^{t}|y_k|^2\right] + O\left[\sum_{k=1}^{t}\sum_{j=2}^{n-1}|\beta_j(k-j+1)|^2\right] +o\left\{\sum_{k=1}^{t}\left[\sum_{j=2}^{n-1}|z_j(k-n+1)|^2 + |z_n(k-n)|^2\right]\right\} \tag{34}$$

Via (15)

$$|z_n(k-n)|^2 = O[|z_{n-1}(k-n+1)|^2 +||\varphi_{k-n}^\tau\tilde{\theta}_{k-n}||^2 + |\beta_{n-1}(k-n+1)|^2 +|w_{n-1}(k-n+1)|^2] \tag{35}$$

By substituting (35) into (34) and using Lemma 2 and (15) repeatly, it is derived that

$$\sum_{k=1}^{t}|y_k - y_k^*|^2 = O(t) + O\left(\sum_{k=1}^{t}||\varphi_k^\tau\tilde{\theta}_k||^2\right) +o\left(\sum_{k=1}^{t}|y_k - y_k^*|^2\right)$$

Also by (32) and (33)

$$\sum_{k=1}^{t}|y_k - y_k^*|^2 = O(t) + O\left(\sum_{k=1}^{t}||\varphi_k^\tau\tilde{\theta}_k||^2\right) = O(t)$$

Therefore (28) is valid. By use of Lemma 1 (iii) and 3, Assumption B, (8), (13), (14) and (23) that

$$\sum_{k=1}^{t}|u_k|^2 = O(t) + o\left[\sum_{k=1}^{t}\sum_{i=1}^{n}|z_n(k-i)|^{2b}\right] +o\left[\sum_{k=1}^{t}\sum_{j=1}^{n-1}\sum_{i=0}^{n-1}|z_j(k-i)|^{2b}\right] +o\left[\sum_{k=1}^{t}\sum_{i=1}^{n}|z_n(k-i)|^2\right] +o\left[\sum_{k=1}^{t}\sum_{j=1}^{n-1}\sum_{i=0}^{n-1}|z_j(k-i)|^2\right] +O\left[\sum_{k=1}^{t}\sum_{j=1}^{n}||\varphi_{k-j}^\tau\tilde{\theta}_{k-j}||^{2b}\right] +O\left[\sum_{k=1}^{t}\sum_{j=1}^{n}||\varphi_{k-j}^\tau\tilde{\theta}_{k-j}||^2\right] \tag{36}$$

By the argument simialr to (28) it is not difficult to assert from (36) and (31) – (33) that

$$\sum_{k=1}^{t}|u_k|^2 = O(t) + O\left(\sum_{k=1}^{t}||\varphi_k^T\tilde{\theta}_k||^{2b}\right)$$
$$+O\left(\sum_{k=1}^{t}||\varphi_k^T\tilde{\theta}_k||^2\right)$$
$$= O(t) + O(r_t) = O(t) \quad a.s.$$

Finally, (29) follows from this and (28) immediately. The proof is completed.

5. CONCLUSIONS

The global stabilization of a class of discrete-time-time nonlinear stochastic systems has been achieved by using the WLS-based adaptive controller where the nonlinearity vectors $\alpha_0(.)$ and $\alpha_n(.)$ are permitted to have the growth rate faster than linear. An open problem is whether or not the global stabilization of the closed-loop system is still obtained in the case of when $d > 1$, that is, the number of unknown parameters is larger than one. This is needed to investigate further in the future.

6. APPENDIX C

The proof of Lemma 3.
From (13) and Assumption B

$$\sum_{i=1}^{n-2}|\overline{\alpha}_{in-1}(k)|^{2b} = O(1) + O\left[\sum_{i=1}^{n-1}|z_i(k)|^{2b}\right] \quad (37)$$

Consequently, form (12) and (37)

$$|x_n(k)|^{2b} = O(1) + O\left[\sum_{i=1}^{n}|z_i(k)|^{2b}\right]$$
$$+O\left[\sum_{i=1}^{n}|x_i(k)|^{2b}\right]$$

Hence, by Assumption B and (8)

$$||\varphi_k||^2 = O(1) + O\left[\sum_{i=1}^{n}|z_i(k)|^{2b}\right]$$
$$+O\left\{\sum_{i=1}^{n-1}[|x_i(k)|^{2b} + |x_i(k)|^2]\right\} \quad (38)$$

Finally, the conclusion follows from Lemma 2 and (38). The proof in detail is omitted here.

7. REFERENCES

Bercu, B. (1995). Weighted estimation and tracking for armax models. *SIAM J. Control Optimiza.* **33**, 89–106.

Byrnes, C.I. and W. Lin (1994). Losslessness, feedback equivalence and the global stabilization of discrete-time nonlinear systems. *IEEE Trans. on Auto. Control* **39**, 83–97.

Byrnes C.I., W. Lin and B.K. Ghosh (1993). Stabilization of discrete-time nonlinear systems by smooth state feedback. *Systems and Control Letters* **21**, 255–263.

Chen, F.C. and H.K. Khail (1992). Adaptive control of systems using neural networks. *Int. J. Control* **55**, 1299–1317.

Chen, F.C. and H.K. Khail (1995). Adaptive control of a class of nonlinear discrete-time systems using neural networks. *IEEE Trans. on Auto. Control.*

Guo, L. (1996). Self-convergence of weighted least-squares with application to stochastic adaptive control. *IEEE Trans. on Auto. Control* **41**, 79–89.

Guo, L. (1997). On critial stability of discrete-time adaptive nonlinear control. *IEEE Trans. on Auto. Control* **42**, 1488–1499.

Guo, L. and C. Wei (1996). Ls-based discrete-time adaptive nonlinear control: Feasibility and limitations. *Science in China, (Series E)* **39**, 255–269.

Kanellakopoulos, I. (1994). A discrete-time adaptive nonlinear system. *IEEE Trans. on Auto. Control.*

Lin, W. and C.I. Byrnes (1995). Passivity and absolute stabilization of a class of disrete-time nonlinear systems. *Automatica* **31**, 263–268.

Ruisheng, L. (1996). Adaptive control of systems using neural network and least-squares algorithm. *Proc. Chinese Control Conference, QingDao, P.R. CHINA* **1**, 513–518.

Wei, C. and L. Guo (1996). Adaptive control of a class of nonlinear stochastic systems. *Proc. Chinese Control Conference, QingDao, P.R. CHINA* **1**, 508–512.

Yeh, P.C. and P.V. Kokotovic (1995*a*). Adaptive control of a class of nonlinear discrete-time systems. *Int. J. Control* **62**, 303–324.

Yeh, P.C. and P.V. Kokotovic (1995*b*). Adaptive output-feedback design for a class of nonlinear discrete-time systems. *IEEE Trans. on Auto. Control* **40**, 1663–1667.

Zhang, J. and S. Lang (1989). Adaptive control of a class of multivariable nonlinear systems and convergence analysis. *IEEE Trans. on Auto. Control* **34**, 787–791.

14th Triennial World Congress, Beijing, P.R. China

E-2c-13-6

INPUT-TO-STATE STABILITY FOR DISCRETE-TIME NONLINEAR SYSTEMS

Zhong-Ping Jiang * **Eduardo Sontag** **,[1] **Yuan Wang** ***,[2]

* *Department of Electrical Engineering, Polytechnic University, Six Metrotech Center, Brooklyn, NY 11201, USA*
** *Department of Mathematics, Rutgers University, New Brunswick, NJ 08903, USA*
*** *Department of Mathematical Sciences Florida Atlantic University, USA Boca Raton, FL 33431. Email:* ywang@math.fau.edu

Abstract: In this paper the input-to-state stability (ISS) property is studied for discrete-time nonlinear systems. We show that many ISS results for continuous-time nonlinear systems in earlier papers (Sontag, 1989; Sontag, 1990; Sontag and Wang, 1996; Jiang *et al.*, 1994; Coron *et al.*, 1995) can be extended to the discrete-time case. More precisely, we provide a Lyapunov-like sufficient condition for ISS, and we show the equivalence between the ISS property and various other properties. Utilizing the notion of ISS, we present a small gain theorem for nonlinear discrete time systems. ISS stabilizability is discussed and connections with the continuous-time case are made. As in the continuous time case, where the notion ISS found wide applications, we expect that this notion will provide a useful tool in areas related to stability for nonlinear discrete time systems as well. *Copyright © 1999 IFAC*

Keywords: discrete-time nonlinear systems, input-to-state stability, Lyapunov methods.

1. INTRODUCTION

The notion of input-to-state stability (for short, ISS) for nonlinear control systems was proposed by one of the authors (Sontag, 1989; Sontag, 1990) and has been used in the stability analysis and control synthesis of nonlinear systems by several authors – see, e.g., (Sontag, 1989; Sontag, 1990; Sontag and Wang, 1996; Tsinias, 1993; Kazakos and Tsinias, 1994; Praly and Jiang, 1993; Jiang *et al.*, 1994; Coron *et al.*, 1995; Krstić *et al.*, 1995). Basically, ISS gives a quantitative bound of the state trajectories in terms of the magnitude of the control input and their initial conditions.

The purpose of this paper is to study the ISS property for discrete-time nonlinear systems. We show that many ISS results for continuous-time nonlinear systems in earlier papers (Sontag, 1989; Sontag, 1990; Sontag and Wang, 1996; Jiang *et al.*, 1994; Coron *et al.*, 1995) can be extended to the discrete-time case. Among these extensions, we prove that various equivalent characterizations of the ISS condition proposed in (Sontag and Wang, 1996) also hold for discrete-time nonlinear systems. It is also shown that ISS small gain theorems in (Jiang *et al.*, 1994; Coron *et al.*, 1995) are extendable to the discrete-time setting. However, new phenomena arise in the extension from continuous-time to discrete-time. For continuous-

[1] Supported in part by US Air Force Grant F49620-98-1-0242
[2] Corresponding author. Supported in part by NSF Grant DMS-9457826

time affine systems, continuous stabilization implies ISS stabilization by means of state-feedback change $u = K(x) + v$. For discrete-time affine or nonaffine systems, a more complex feedback transformation of the form $u = K_1(x) + K_2(x)v$ is in general required. The construction of the feedback terms K_1 and K_2 turns out to be nontrivial; nevertheless, we show that they can be explicitly obtained, for a class of feedback linearizable systems.

Section 2 starts with basic definitions and states some preliminary results for ISS systems. Section 3 gives some equivalent definitions of ISS, and closes with the statement of the main theorem. Section 4 proposes a nonlinear small gain theorem for discrete-time interconnected systems. Section 5 proves the important fact that continuous stabilization implies ISS stabilization. An illustration is given via a class of feedback linearizable systems.

2. BASIC DEFINITIONS AND RESULTS

We consider general nonlinear systems:

$$x(k+1) = f(x(k), u(k)), \qquad (1)$$

where states $x(k)$ are in $\mathbb{R}^n$, and control values $u(k)$ in $\mathbb{R}^m$, for some n and m, for each time instant $k \in \mathbb{J}_+$, the set of all nonnegative integers. We assume that $f : \mathbb{R}^n \times \mathbb{R}^m \to \mathbb{R}^n$ is continuous and satisfies $f(0,0) = 0$.

Controls or *inputs* are functions $u : \mathbb{J}_+ \to \mathbb{R}^m$. The set of all such functions with the supremum norm $\|u\| = \sup\{|u(k)| : k \in \mathbb{J}_+\} < \infty$ is denoted by l^m_∞, where $|\cdot|$ denotes the usual Euclidean norm. For a given system, we often consider the same system but with controls restricted to take values in some subset $\Omega \subset \mathbb{R}^m$; we use $\mathcal{M}_\Omega$ for the set of all such controls. We use Id to denote the identity function.

For each $\xi \in \mathbb{R}^n$ and each input u, we denote by $x(\cdot, \xi, u)$ the trajectory of system (1) with initial state $x(0) = \xi$ and the input u. Clearly such a trajectory is defined uniquely on $\mathbb{J}_+$, and for each u and each $k \in \mathbb{J}_+$, $x(k, \xi, u)$ depends on ξ continuously.

Recall that a function $\gamma : \mathbb{R}_{\geq 0} \longrightarrow \mathbb{R}_{\geq 0}$ is a $\mathcal{K}$-*function* if it is continuous, strictly increasing and $\gamma(0) = 0$; it is a $\mathcal{K}_\infty$-*function* if it is a $\mathcal{K}$-function and also $\gamma(s) \to \infty$ as $s \to \infty$; and it is a *positive definite* function if $\gamma(s) > 0$ for all $s > 0$, and $\gamma(0) = 0$. A function $\beta : \mathbb{R}_{\geq 0} \times \mathbb{R}_{\geq 0} \longrightarrow \mathbb{R}_{\geq 0}$ is a $\mathcal{KL}$-*function* if, for each fixed $t \geq 0$, the function $\beta(\cdot, t)$ is a $\mathcal{K}$-function, and for each fixed $s \geq 0$, the function $\beta(s, \cdot)$ is decreasing and $\beta(s, t) \to 0$ as $t \to \infty$.

Definition 2.1. System (1) is (*globally*) *input-to-state stable* (ISS) if there exist a $\mathcal{KL}$-function $\beta : \mathbb{R}_{\geq 0} \times \mathbb{R}_{\geq 0} \longrightarrow \mathbb{R}$ and a $\mathcal{K}$-function γ such that, for each input $u \in l^m_\infty$ and each $\xi \in \mathbb{R}^n$, it holds that

$$|x(k, \xi, u)| \leq \beta(|\xi|, k) + \gamma(\|u\|) \qquad (2)$$

for each $k \in \mathbb{J}_+$. □

Note that, by causality, the same definition would result if one would replace (2) by

$$|x(k, \xi, u)| \leq \beta(|\xi|, k) + \gamma(\|u_{[k-1]}\|) \qquad (3)$$

where $k \geq 1$ and, for each $l \geq 0$, $u_{[l]}$ denotes the truncation of u at l; i.e., $u_{[l]}(j) = u(j)$ if $j \leq l$, and $u_{[l]}(j) = 0$ if $j > l$.

Definition 2.2. A continuous function V on $\mathbb{R}^n$ is called an ISS-*Lyapunov function* for system (1) if

$$\alpha_1(|\xi|) \leq V(\xi) \leq \alpha_2(|\xi|) \qquad (4)$$

holds for some $\alpha_1, \alpha_2 \in \mathcal{K}_\infty$, and

$$V(f(\xi, \mu)) - V(\xi) \leq -\alpha_3(|\xi|) + \sigma(|\mu|), \qquad (5)$$

for some $\alpha_3 \in \mathcal{K}_\infty$, $\sigma \in \mathcal{K}$.

A smooth ISS-Lyapunov function is one which is smooth. □

Observe that if V is an ISS-Lyapunov function for (1), then V is a Lyapunov function for the 0-input system $x(k+1) = f(x(k), 0)$.

Proposition 2.3. If system (1) admits an ISS-Lyapunov function, then it is ISS.

The converse of this result, as well as a converse Lyapunov theorem regarding uniform global asymptotic stability for systems with disturbances taking values in compacts, are also true, and will be proved in a forthcoming paper.

Sketch of the proof of Proposition p-liss-iss. Assume that system (1) admits an ISS-Lyapnov function V. Let α_i $(i = 1, 2, 3)$ and σ be as in (4) and (5). First observe that (5) can be rewritten as

$$V(f(\xi, \mu)) - V(\xi) \leq -\alpha_4(V(\xi)) + \sigma(|\mu|),$$

for all ξ and μ, where $\alpha_4 = \alpha_3 \circ \alpha_2^{-1}$. To prove the lemma, one needs the following:

Lemma 2.4. For any $\mathcal{K}_\infty$-function α, there is a $\mathcal{K}_\infty$-function $\widehat{\alpha}$ satisfying the following:

- $\widehat{\alpha}(s) \leq \alpha(s)$ for all $s \geq 0$; and

• $\mathrm{Id} - \widehat{\alpha} \in \mathcal{K}$. □

We now return to the proof of Lemma 2.3. For the $\mathcal{K}_\infty$-function α_4, let $\widehat{\alpha}_4$ be a function picked as in Lemma 2.4. We then have

$$V(f(\xi, \mu)) - V(\xi) \leq -\widehat{\alpha}_4(V(\xi)) + \sigma(|\mu|). \quad (6)$$

Fix a point $\xi \in \mathbb{R}^n$ and pick an input u. Let $x(k)$ denote the corresponding trajectory $x(k, \xi, u)$ of (1). Let c be any number in $(0,1)$, and consider the set defined by $D = \{\xi : V(\xi) \leq b\}$, where $b = \widehat{\alpha}_4^{-1}\left(\frac{\sigma(\|u\|)}{c}\right)$.

Claim: If there is some $k_0 \in \mathbb{J}_+$ such that $x(k_0) \in D$, then $x(k) \in D$ for all $k \geq k_0$.

Proof. Assume that $x(k_0) \in D$. Then $V(x(k_0)) \leq b$, that is, $c\,\widehat{\alpha}_4(V(x(k_0))) \leq \sigma(\|u\|)$. By (6),

$$\begin{aligned} V(x(k_0+1)) \leq& -(1-c)\widehat{\alpha}_4(V(x(k_0))) \\ &+ \widetilde{\alpha}_4(V(x(k_0))) + \sigma(\|u\|) \end{aligned}$$

where $\widetilde{\alpha}_4 = \mathrm{Id} - c\,\widehat{\alpha}_4$. Observe that $\widetilde{\alpha}_4 \in \mathcal{K}$ because $\mathrm{Id} - \widehat{\alpha}_4 \in \mathcal{K}$. Combining this fact with the assumption that $V(x(k_0)) \leq b$, one can show that $V(x(k_0+1)) \leq -(1-c)\widehat{\alpha}_4(V(x(k_0)) + b \leq b$. Using induction, one can show that $V(x(k_0+j)) \leq b$ for all $j \in \mathbb{J}_+$, that is, $V(x(k)) \in D$ for all $k \geq k_0$. □

We now let $j_0 = \min\{k \in \mathbb{J}_+ : x(k) \in D\} \leq \infty$. Then it follows from the above conclusion that $V(x(k)) \leq \widehat{\gamma}(\|u\|)$ for all $k \geq j_0$, where $\widehat{\gamma}(r) = \widehat{\alpha}_4^{-1}(\sigma(r)/c)$. For $k < j_0$, it holds that $c\,\widehat{\alpha}_4(V(x_k)) > \sigma(\|u\|)$, and hence, $V(x(k+1)) - V(x(k)) \leq -(1-c)\widehat{\alpha}_4(V(x(k)))$. We let $\widehat{\beta}(s,k)$ be the solution of the scalar difference equation

$$z(k+1) = z(k) - (1-c)\widehat{\alpha}_4(z(k))$$

with initial condition $z(0) = s$. Observe that, for any $s > 0$, the sequence $z(t)$ decreases to zero (it never crosses zero, because $\mathrm{Id} - \widehat{\alpha}_4 \geq 0$); thus, $\widehat{\beta}$ is a $\mathcal{KL}$-function. It follows, by induction on k, that $V(x(k)) \leq \widehat{\beta}(V(x(0)), k)$ for all $0 \leq k \leq j_0 + 1$. Thus, $V(x(k)) \leq \max\{\widehat{\beta}(V(\xi), k),\ \widehat{\gamma}(\|u\|)\}$. From this one gets (2) with $\beta(s,r) = \alpha_1^{-1}(\widehat{\beta}(\alpha_2(|\xi|), r))$ and $\gamma(s) = \alpha_1^{-1}\widehat{\gamma}(s)$. □

Remark 2.5. From the proof of Lemma 2.3, one sees that if the ISS-Lyapunov function satisfies the inequality $V(x(k+1)) - V(x(k)) \leq -\alpha(V(x(k))) + \sigma(\|u\|)$, with the property that $\mathrm{Id} - \alpha \in \mathcal{K}$, then for any $0 < c < 1$, the function $(\alpha_1^{-1} \circ \alpha^{-1})(\sigma(s)/c)$ can be taken as a gain function of the system. That is, for any $0 < c < 1$, there is a $\mathcal{KL}$-function β such that $|x(k, \xi, u)| \leq \beta(|\xi|, k) + (\alpha_1^{-1} \circ \alpha^{-1})(\sigma(\|u\|)/c)K$ for all ξ and all u. □

3. EQUIVALENT NOTIONS OF ISS

In this section we show that, as in the continuous time case, there are various notions that are equivalent to ISS.

3.1 *Asymptotic Gains*

Consider system (1). We say that the system has *$\mathcal{K}$-asymptotic gain* if there exists some $\gamma \in \mathcal{K}$ such that

$$\overline{\lim_{k\to\infty}} |x(k, \xi, u)| \leq \overline{\lim_{k\to\infty}} \gamma(|u(k)|) \quad (7)$$

for all $\xi \in \mathbb{R}^n$.

We say that system (1) is *uniformly bounded-input bounded-state* (UBIBS) stable if bounded initial states and controls produce uniformly bounded trajectories, i.e., there exist two $\mathcal{K}$-functions σ_1, σ_2 such that

$$\sup_k |x(k, \xi, u)| \leq \max\{\sigma_1(|\xi|),\ \sigma_2(\|u\|)\} \quad (8)$$

Again, by the causality property of the system, the above is equivalent to $\sigma_1(s) \geq s$ and

$$|x(k, \xi, u)| \leq \max_{0 \leq j \leq k-1} \{\sigma_1(|\xi|),\ \sigma_2(|u(j)|)\} \quad (9)$$

It is not hard to see that if a system is ISS, then it is UBIBS and it admits a $\mathcal{K}$-asymptotic gain. The converse is also true. It will be proved after we introduce another related notion in the next section.

3.2 *Robust Stability Margins*

As in the case of continuous time systems, there turns out to be also an interesting connection between the ISS and the robust stability.

By a (possibly time-varying) *feedback law* for a system (1) we will mean any function $w : \mathbb{J}_+ \times \mathbb{R}^n \longrightarrow \mathbb{R}^n$ such that for each fixed k, $w(k, \cdot)$ is a continuous function.

Let ρ be any $\mathcal{K}_\infty$ function. For any $d \in \mathcal{M}_\Omega$ with Ω as the closed unit ball in $\mathbb{R}^m$, we view the system

$$x(k+1) = f(x(k), d(k)\rho(|x(k)|)) \quad (10)$$

as a system $x(k+1) := g(x(k), d(k))$ with time-varying parameters $d \in \mathcal{M}_\Omega$. We use $x_\rho(k, \xi, d)$ to denote the solution to (10) corresponding to the initial state ξ and the disturbance signal d.

We will say that the system (1) is *robustly stable* if there exists a $\mathcal{K}_\infty$ function ρ (called a *stability margin*) such that system (10) is globally asymptotically stable uniformly in d. To be more precise, this means that there exists $\beta \in \mathcal{KL}$ such that the following holds for all trajectories of (10):

$$|x_\rho(k,\xi,d)| \le \beta(|\xi|,k)$$

for all $k \ge 0$, all $d \in \mathcal{M}_\Omega$.

Note that for a nonlinear GAS system, in general only small perturbations can be tolerated while preserving stability. The requirement $\rho \in \mathcal{K}_\infty$ is thus nontrivial.

Following the same steps as in Section V of (Sontag and Wang, 1996), one may show the next result:

Lemma 3.1. System (1) is ISS if and only if it is robustly stable. □

(As a matter of fact, the proof turns out to be far simpler, because a key technical point, that asymptotic stability of systems with disturbances is equivalent to uniform asymptotic stability, is much easier in discrete-time.)

The following lemma provides a key link in our main results. Its proof is analogous to the proof of the corresponding continuous time result in (Sontag and Wang, 1996, Section V)).

Lemma 3.2. If system (1) is UBIBS and if it admits $\mathcal{K}$-asymptotic gain, then it is robustly stable. □

To close this section, we summarize our results in this section:

Theorem 1. Consider system (1). The following are equivalent:

(1) It is ISS.
(2) It is UBIBS and it admits $\mathcal{K}$-asymptotic gain.
(3) It is robustly stable.

Proof. We have: [1 ⇒ 2] (clear by Definitions), [2 ⇒ 3] (see Lemma 3.2), [3 ⇒ 1] (see Lemma 3.1). □

4. NONLINEAR SMALL GAIN THEOREMS

In this section, we will discuss the ISS property for interconnected systems of the following type:

$$\begin{aligned} x_1(k+1) &= f_1(x_1(k), x_2(k), u_1(k)), \\ x_2(k+1) &= f_2(x_1(k), x_2(k), u_2(k)), \end{aligned} \tag{11}$$

where for $i = 1,2$ and for each $k \in \mathbb{J}_+$, $x_i(k) \in \mathbb{R}^{n_i}$, $u_i(k) \in \mathbb{R}^{m_i}$, and $f_i : \mathbb{R}^{n_1} \times \mathbb{R}^{n_2} \times \mathbb{R}^{m_i} \to \mathbb{R}^{n_i}$ is continuous.

It turns out that, in the continuous time case, ISS small gain theorems are very powerful in treating stability problems for such interconnected systems. The first such results were obtained in (Jiang *et al.*, 1994). Later the proofs of the main results were considerably simplified, using asymptotic gains, in (Coron *et al.*, 1995); in conjunction with the results in (Sontag and Wang, 1996) this gives ISS results. In (Jiang *et al.*, 1996), a small gain theorem was presented in terms of Lyapunov functions. By following the continuous-time approach used in (Coron *et al.*, 1995, Sec. 4), one can get the following small theorem. For the interconnected system (11), we assume that both of the subsystems are ISS. To be more precise, we assume that, for the following systems,

$$x_1(k+1) = f_1(x_1(k), v_1(k), u_1(k)), \tag{12}$$
$$x_2(k+1) = f_2(x_2(k), v_2(k), u_2(k)), \tag{13}$$

the trajectories satisfy

$$|x_1(k)| \le \max\left\{\beta_1(|\xi_1|,k),\ \gamma_1^x(\|v_1\|),\ \gamma_1^u(\|u_1\|)\right\}$$

$$|x_2(k)| \le \max\left\{\beta_2(|\xi_2|,k),\ \gamma_2^x(\|v_2\|),\ \gamma_1^u(\|u_2\|)\right\}$$

Theorem 2. Assume that systems (12) and (13) are ISS, with $\gamma_1^x(\gamma_2^x(s)) < s$ (or, $\gamma_2^x(\gamma_1^x(s)) < s$) for all $s > 0$. Then, the interconnected system (11) is ISS with (u_1, u_2) as input. □

The above small gain theorem can also be stated in terms of ISS-Lyapunov functions. Assume that both subsystems in (11) admit ISS-Lyapunov functions. Let V_1 (respectively, V_2) be an ISS-Lyapunov function of (12) (respectively, (13)). That is, there exist class $\mathcal{K}_\infty$-functions α_{ij}, σ_i, ρ_i^x and ρ_i^u $(1 \le i,j \le 2)$ such that $\alpha_{i1}(|\xi|) \le V_i(\xi) \le \alpha_{i2}(|\xi|)$, and

$$\begin{aligned} V_1(f_1(\xi_1,\xi_2,\mu_1)) - V_1(\xi_1) &\le -\sigma_1(V_1(\xi_1)) \\ &\quad + \rho_1^x(V_2(\xi_2)) + \rho_1^u(|\mu_1|), \qquad (14) \\ V_2(f_2(\xi_2,\xi_1,\mu_2)) - V_2(\xi_2) &\le -\sigma_2(V_2(\xi_2)) \\ &\quad + \rho_2^x(V_1(\xi_1)) + \rho_2^u(|\mu_2|). \qquad (15) \end{aligned}$$

In view of Lemma 2.4, we may assume that $\mathrm{Id} - \sigma_i \in \mathcal{K}$ for $i = 1,2$.

Theorem 3. Assume that systems (12) and (13) admit ISS-Lyapunov functions V_1 and V_2 respectively that satisfy (14)-(15), with $\mathrm{Id} - \sigma_i \in \mathcal{K}$ for $i = 1,2$. If there exists a $\mathcal{K}_\infty$-function ρ such that $\sigma_1^{-1} \circ (\mathrm{Id} + \rho) \circ \rho_1^x \circ \sigma_2^{-1} \circ (\mathrm{Id} + \rho) \circ \rho_2^x < \mathrm{Id}$, then the interconnected system (11) is ISS with (u_1, u_2) as input.

5. ISS-STABILIZABILITY

Consider system (1). We say that the system is *continuously stabilizable* if there is a continuous function $w : \mathbb{R}^n \to \mathbb{R}^m$ with $w(0) = 0$ such that under the feedback $u = w(x)$, the closed-loop system

$$x(k+1) = f(x(k), w(x(k)))$$

is globally asymptotically stable (GAS), i.e., there exists some $\mathcal{KL}$-function $\beta(\cdot,\cdot)$ such that the following holds for all trajectories of the system:

$$|x(k,\xi)| \leq \beta(|\xi|, k)$$

for all $k \geq 0$.

We say that system (1) is *continuously* ISS *stabilizable* if there exist a continuous map $w : \mathbb{R}^n \to \mathbb{R}^m$ with $w(0) = 0$ and an $n \times n$ matrix Γ of continuous functions, invertible for each x, such that under the control law $u = w(x) + \Gamma(x)v$ the closed-loop system

$$x(k+1) = f(x(k), w(x(k)) + \Gamma(x(k))v(k)) \qquad (16)$$

is ISS (with v as new input).

Clearly, if a system is ISS-stabilizable, then it is stabilizable. By following the approach used in the continuous case (see (Sontag, 1990)), one can prove the following:

Theorem 4. System (1) is continuously stabilizable if and only if it is ISS-stabilizable. □

Remark 5.1. In contrast to the continuous case, the construction of an ISSstabilizing control law is far from explicit even for affine systems. Nevertheless, it is shown in the next section that an alternative and simple design of ISS controllers is possible for a class of feedback linearizable systems. □

5.1 *Feedback Linearizable Systems*

The main purpose of this section is to illustrate our main results in the preceding section via feedback linearizable systems. As done in (Sontag, 1989) for continuous-time, we derive explicit formulas for the special case of such systems. Both continuous- and discrete-time linearizable systems have been fairly studied in the recent control literature, see, e.g., (Isidori, 1995; Nijmeijer and van der Schaft, 1990; Sontag, 1998) and references therein. It is worthing noting that a large body of research papers on nonlinear adaptive control have been based on globally feedback linearizable systems, see the recent texts (Krstić *et al.*, 1995; Marino and Tomei, 1995) and the references cited therein.

We restrict ourselves to systems (1) which are *affine* in u, that is,

$$x(k+1) = a(x(k)) + b(x(k))u \qquad (17)$$

A system (17) is said to be globally feedback linearizable if there exist a state-coordinate change, that is, an isomorphism $\phi : \mathbb{R}^n \to \mathbb{R}^n$ which vanishes at zero,

$$z = \phi(x) \qquad (18)$$

and a regular state-feedback transformation

$$u = \vartheta_1(x) + \vartheta_2(x)v \qquad (19)$$

with $\vartheta_2(x)$ a globally invertible matrix, such that system (17) is brought to a linear controllable system

$$z(k+1) = Az(k) + Bv \qquad (20)$$

Take any constant matrix K_0 so that $A + BK_0$ is asymptotically stable. The linear feedback $v(k) = K_0 z(k)$ globally asymptotically stabilizes the linear system (20) and therefore its original nonlinear system (17). In addition, the feedback $v(k) = K_0 z(k) + w(k)$ makes the closed-loop system

$$z(k+1) = (A + BK_0)z(k) + Bw(k)$$

ISS (with respect to w). This can be seen through the following:

$$|z(k)| \leq \lambda_{\max}(A + BK_0)^k |z(0)| + \frac{|B|}{1 - \lambda_{\max}(A + BK_0)} \|w\| \qquad (21)$$

where $\lambda_{\max}(A + BK_0)$ denotes the spectral radius of $A + BK_0$.

Since ϕ is a global isomorphism and vanishes at zero, there exist two class $\mathcal{K}_\infty$-functions $\widehat{\phi}_1$ and $\widehat{\phi}_2$ such that

$$\widehat{\phi}_1(|x|) \leq |\phi(x)| \leq \widehat{\phi}_2(|x|) \qquad (22)$$

This together with (21) and (18) implies that

$$|x(k)| \leq \widehat{\phi}_1^{-1}\left(2\lambda_{\max}(A + BK_0)^k \widehat{\phi}_2(|x(0)|)\right) + \widehat{\phi}_1^{-1}\left(\frac{2|B|}{1 - \lambda_{\max}(A + BK_0)} \|w\|\right) \qquad (23)$$

Therefore, system (17) is made ISS with a feedback law $u = \vartheta_1(x(k)) + \vartheta_2(x(k))K_0\phi(x(k)) + \vartheta_2(x(k))w(k)$.

By analogy with the case of continuous-time systems (Sontag, 1989), we would ask whether or not it is possible to render system (17) ISS using a nonlinear feedback of the form

$$u = K(x) + w \qquad (24)$$

where $K : \mathbb{R}^n \to \mathbb{R}^m$ is continuous with $K(0) = 0$.

First of all, we point out that, similar as in the continuous time case, a linear feedback does not necessarily produce a system which is ISS with respect to additive disturbances occuring at the level of real control input u. More precisely, when taking K as the linearizing feedback, i.e. $K(x(k)) = \vartheta_1(x(k)) + \vartheta_2(x(k))K_0\phi(x(k))$, the corresponding closed-loop system $x(k+1) = a(x(k)) + b(x(k))(K(x(k)) + w)$ may fail to be ISS with respect to the signal w.

As an illustration of this important point, we consider a discrete-time scalar system

$$x(k+1) = x(k)^3 + (|x(k)| + 1)u \qquad (25)$$

which is globally rendered linear by a nonlinear feedback of the form

$$K(x) = -\frac{x^3}{|x|+1} + \frac{1}{|x|+1}v. \qquad (26)$$

A further choice of $v(k) = \frac{1}{2}x(k)$ gives a feedback law that globally asymptotically stabilizes system (25). However, this linear feedback controller does not make the closed-loop system ISS with respect to additive disturbances w at the true control input u:

$$x(k+1) = \frac{1}{2}x(k) + (|x(k)| + 1)w. \qquad (27)$$

Indeed, letting $w = 1$ in (27) yields that the solutions $x(k)$ with $x(0) = 0$ of (27) diverge to $+\infty$ as $k \to +\infty$.

To compare with the continuous time case, a natural question to ask is if there is *any* feedback that make the closed-loop system ISS? Unfortunately, in contrast to the continuous time situation, there does generally not exist a feedback of the form (24) that produces an ISS system. As a counterexample, we still consider system (25).

The system (25) in closed-loop with (24) is as the following

$$x(k+1) = G(x(k)) + (|x(k)| + 1)w(k)$$

with $G(x(k)) = x(k)^3 + (|x(k)| + 1)K(x(k))$. Now consider the signal $w(k)$ given by:

$$w(k) = \begin{cases} 1 \text{ if } G(x(k)) \geq 0, \\ -1 \text{ if } G(x(k)) < 0. \end{cases}$$

It can be seen that with such a choice of w, $|x(k+1)| \geq |x(k)| + 1$, and hence, $x(k+1) \to \infty$.

The above argument shows that the system fails to be ISS no matter what G is. Observe that this is still true even if G is dead-beat stable, i.e., G is such a function that every trajectory of $x(k+1) = G(x(k))$ can reach the origin in finite time (e.g., $G(x) \equiv 0$).

6. REFERENCES

Coron, J.-M., L. Praly and A. Teel (1995). Feedback stabilization of nonlinear systems: sufficient conditions and Lyapunov and input-output techniques. In: *Trends in Control* (A. Isidori, Ed.). Springer-Verlag. London.

Isidori, A. (1995). *Nonlinear Control Systems: An Introduction.* 3rd ed., Springer-Verlag. New York.

Jiang, Z.-P., A. Teel and L. Praly (1994). Small-gain theorem for ISS systems and applications. *Mathematics of Control, Signals, and Systems* **7**, 95–120.

Jiang, Z.-P., I. M. Y. Mareels and Y. Wang (1996). A Lyapunov formulation of the nonlinear small gain theorem for interconnected ISS systems. *Automatica* **32**, 1211–1215.

Kazakos, D. and J. Tsinias (1994). The input to state stability conditions and global stabilization of discrete-time systems. *IEEE Trans. Automat. Control* **39**, 2111–2113.

Krstić, M., I. Kanellakopoulos and P. V. Kokotović (1995). *Nonlinear and Adaptive Control Design.* John Wiley & Sons. New York.

Marino, R. and P. Tomei (1995). *Nonlinear Control Design: Geometric, Adaptive and Robust.* Prentice Hall Europe. London.

Nijmeijer, H. and A. van der Schaft (1990). *Nonlinear Dynamical Control Systems.* Springer-Verlag. New York.

Praly, L. and Z. P. Jiang (1993). Stabilization by output feedback for systems with ISS inverse dynamics. *Systems Control Letters* **21**, 19–33.

Sontag, E. D. (1989). Smooth stabilization implies coprime factorization. *IEEE Trans. Automat. Control* **34**, 435–443.

Sontag, E. D. (1990). Further facts about input to state stabilization. *IEEE Trans. Automat. Control* **35**, 473–476.

Sontag, E. D. (1998). *Mathematical Control Theory, 2nd ed..* Springer. New York.

Sontag, E. D. and Y. Wang (1996). New characterizations of the input to state stability property. *IEEE Trans. Automat. Control* **41**, 1283–1294.

Tsinias, J. (1993). Versions of Sontag's "input to state stability condition" and the global stabilization problem. *SIAM J. Control and Optimization* **31**, 928–941.

ON THE STABILIZATION OF HOMOGENEOUS SYSTEMS IN THE PLANE

Ph. Adda and H. Zenati

INRIA-Lorraine (CONGE project) & University of Metz
ISGMP, Bât A, Ile du Saulcy 57045 - METZ (FRANCE)
e-mail: {adda, zenati}@saulcy.loria.fr

Abstract: In this paper we consider the stabilization problem of two-dimensional homogeneous systems. Some sufficient conditions for the asymptotic stabilization by homogeneous feedbacks are derived. A new approach inspired by the well-known Kalman's method (pole assignment) is developed to construct explicitly the stabilizing feedback laws. Local stabilization result is obtained for a class of nonlinear systems.

Keywords: nonlinear systems, pole assignment, feedback stabilization

1. INTRODUCTION

We consider a single input homogeneous system,

$$\dot{x} = f(x) + u\, g(x) \tag{1}$$

where $x \in \mathbb{R}^2$, $u \in \mathbb{R}$, f and g are homogeneous (rational) vector fields of degree, respectively, $2k_1 + 1$ and k_2 with $k_1 \geq 1$. The problem is to find a feedback function $u(x)$, which is homogeneous, smooth on $\mathbb{R}^2 - \{0\}$ and which asymptotically stabilizes the control system (1).

In the literature, global stabilization results are often obtained by the Lyapunov machinery, see for instance (Kawski, 1990; Jurdjevic and Quin, 1978; Rosier, 1992; Sontag, 1989; Tsinias, 1990). Another stabilization result (Chabour *et al.*, 1993) is from the bilinearization point of view. Whereas, since the systems studied here may satisfy $g(0) = 0$, we are not in the scope of (Boothby and Marino, 1989; Dayawansa *et al.*, 1990). Our approach to the stabilization problem of (1) is inspired by the well-known Kalman's linear pole assignment (LPA) technique and is not a Lyapunov like method. Then our results are completing the preceding works. The idea of variable pole assignment (VPA) was first introduced by (Clavier, 1995) for planar bilinear systems and recently discussed in (Adda and Zenati, 1998) for nonlinear systems in the plane. In this paper, we apply VPA technique to our system (1) and we prove that under some conditions the homogeneous constructed feedback laws globally asymptotically stabilize the system at the origin.

The paper is organized as follows. Section 2 considers preliminaries where VPA method is developed. Section 3 contains our main result. In section 4 a homogeneous approximation for nonlinear systems in the plane is discussed, a local result of stabilization is given. Section 5 contains some concluding remarks.

2. PRELIMINARIES

In this section we develop the VPA method applied to (1) in the plane and recall a known stability result related to planar homogeneous systems.

First of all, let us briefly recall the LPA theory. Given a controllable linear system $\dot{x} = A\,x + ub, x \in \mathbb{R}^n, u \in \mathbb{R}$. For all monic real polynomial p of degree n, there is some line matrix F such that the characteristic polynomial of the matrix $A + bF$ is the desired p. Then if the spectrum of $A + bF$ is chosen to have negative real part, the linear feedback $u = Fx$ stabilizes the system.

So, let us consider the homogeneous system (1). It is known that by Euler's formula

$$f(x) = \frac{1}{2k_1+1}\,(Df)_x\, x$$

where $(Df)_x$ is the jacobian matrix of f, $(2k_1+1$ is the homogeniety degree of f). Then (1) can be written in the following form

$$\dot{x} = A(x)\, x + u\, g(x) \qquad (2)$$

where $A(x) = \dfrac{1}{2k_1+1}\,(Df)_x$ and its coefficients are homogeneous of degree $2k_1$.

The main idea here is to apply the pole assignment for the homogeneous system (2) in the plane. It means that at each point $x \in \mathbb{R}^2 - \{0\}$, we apply the above pole assignment technique to the pair $(A(x), g(x))$. More precisely, for each $x \neq 0$ we construct, exactly in the same way than for linear systems, a functional matrix $F(x)$ such that the matrix $A(x)+g(x)F(x)$ has a desired preassigned spectrum. The method is as follows:

Assume that the following rank condition is satisfied

$$\text{(RC)} \quad rank(g(x), A(x)\; g(x)) = 2\;, \quad \forall x \neq 0.$$

For each $x \in \mathbb{R}^2 - \{0\}$, define the set of desired spectrum $S_p = \{\lambda_1(x), \lambda_2(x)\}$, where $\lambda_1(x)$ and $\lambda_2(x)$ are homogeneous functions of the same degree $2k_1$, and let $P_A(X) = X^2 + \alpha_1(x)X + \alpha_2(x)$ and $P_S(X) = X^2 + \beta_1(x)X + \beta_2(x)$ the characteristic polynomials respectively associed to the matrix $A(x)$ and to the spectrum S_p. Put $F'(x) = (\alpha_2(x) - \beta_2(x), \alpha_1(x) - \beta_1(x))$ a line matrix and let the functional invertible matrix $P(x) = (A(x)g(x) + \alpha_1(x)g(x), g(x))$. Then, we construct the functional line-matrix $F(x) = F'(x)P(x)^{-1}$ and we obtain the closed-loop matrix $M(x) = A(x) + g(x)F(x)$ which has the preassigned spectrum S_p. Finally, putting

$$u(x) = F(x)\; x$$

we get two different expressions of the closed-loop system:

$$\dot{x} = A(x)x + u(x)g(x),$$

$$\dot{x} = (A(x) + g(x)F(x))x = M(x)x$$

because $u(x) = F(x)x \in \mathbb{R}$.

We can easily verify that, by construction, the functional matrix $M(x)$ is homogeneous of degree $2k_1$ (its coefficients are homogeneous of degree $2k_1$) and, hence the right-hand side of the closed-loop system $\dot{x} = M(x)x$ is homogeneous of degree $2k_1 + 1$.

It should be noted that an analogous construction holds also in $\mathbb{R}^n$, $n > 2$.

Working on $\mathbb{R}^2$, we shall make a constant use of the following result.

Theorem 1 (Hahn, 1967) *Consider the planar system*

$$\begin{cases} \dot{x}_1 = f_1(x_1, x_2) \\ \dot{x}_2 = f_2(x_1, x_2) \end{cases} \qquad (3)$$

where $f = (f_1, f_2)^T$ is Lipschitz and homogeneous polynomial vector field of degree $k \geq 1$. The system (3) is asymptotically stable if and only if one of the following is satisfied:

(i) The restriction of the system to each of its one dimensional invariant homogeneous rays is asymptotically stable;

(ii) The system does not have any one dimentional invariant subspaces and

$$\int_0^{2\pi} \frac{\cos\theta f_1(\cos\theta, \sin\theta) + \sin\theta f_2(\cos\theta, \sin\theta)}{\cos\theta f_2(\cos\theta, \sin\theta) - \sin\theta f_1(\cos\theta, \sin\theta)}\, d\theta < 0$$

where (ρ, θ) denote the polar coordinates $x_1 = \rho\cos\theta, x_2 = \rho\sin\theta$.

3. GLOBAL STABILIZATION

In this section we apply, under RC condition, VPA to (2) with negative definite spectrum $(\lambda_1(x), \lambda_2(x))$ for $M(x)$. We prove that if there exists an invariant homogeneous ray, then the choice $\lambda_1(x)$ and $\lambda_2(x)$ negative definite ensures that the restriction of the closed-loop system to this invariant ray is asymptotically stable. Then, according to (i) of theorem 1, proving global asymptotic stability of the closed-loop system amounts to prove the existence of (at least) one invariant ray.

So, let us consider the two-dimensional homogeneous system (2) with

$$A(x) = \begin{pmatrix} a(x) & c(x) \\ b(x) & d(x) \end{pmatrix}\;,\; g(x) = \begin{pmatrix} g_1(x) \\ g_2(x) \end{pmatrix}$$

where the coefficients of $A(x)$ are homogeneous rational functions of the same degree $2k_1$ and $g_i, i = 1, 2$, are also homogeneous rational functions of degree k_2; $x = (x_1, x_2) \in \mathbb{R}^2$.

From now on, we assume that the following assumptions hold

Assumption (a): (RC) $rank(g(x), A(x)g(x)) = 2, \forall x \in \mathbb{R}^2 - \{0\}$, i.e.

$$det(g(x), A(x)g(x)) \neq 0, \forall x \neq 0.$$

(where *det* denotes determinant of matrix).

Assumption (b): the desired spectrum is $S_p = (\alpha||x||^{2k_1},\; \beta||x||^{2k_1})$, where α and β are arbitrary negative real constants. ($||.||$ denotes the euclidian norm).

Then, the constructed matrix $F(x)$ given in section 2 takes the following form

$$F(x) = \frac{1}{det(g(x), A(x)g(x))}(F_1(x), F_2(x))$$

where

$$F_1(x) = -b(a+d-\alpha||x||^{2k_1} + \beta||x||^{2k_1})g_1 + (a^2+bc-a(\alpha||x||^{2k_1}+\beta||x||^{2k_1}) - \alpha\beta||x||^{2k_1})g_2,$$

$$F_2(x) = -(d^2+bc-d(\alpha||x||^{2k_1}+\beta||x||^{2k_1}) + \alpha\beta||x||^{4k_1})g_1 + c(a+d-\alpha||x||^{2k_1} - \beta||x||^{2k_1})g_2$$

Easy computation shows that the components of $F(x)$ are homogeneous of degree $2k_1 - k_2$ and analytic on $\mathbb{R}^2 - \{0\}$. Then, setting

$$u(x) = F(x)x$$

on obtains the following homogeneous closed-loop system of degree $2k_1 + 1$

$$\dot{x} = (A(x) + g(x)F(x))\, x = M(x)x \qquad (4)$$

where $M(x)$ has, by construction, the preassigned spectrum $(\alpha||x||^{2k_1},\ \beta||x||^{2k_1})$.

Remark 1: When $2k_1 + 1 \leq k_2$, the constructed homogeneous feedback may fail to be continuous at the origin. But, this does not cause problems, since the right-hand side of the closed-loop system (4) is homogeneous of degree $2k_1 + 1 \geq 1$ and, hence, at least continuous at the origin.

Theorem 2 *Consider the homogeneous system (2) with $A(x)$ and g as defined above. Under the assumptions (a) and (b), if the closed-loop system (4) admits at least one invariant homogeneous ray, then it has $x = 0$ as a globally asymptotically stable equilibruim point.*

Proof: Let $\mathbb{R}^+ x_0$ be an invariant homogeneous ray of (4). It means that the vector field $M(x)\, x$ is radial at x_0, i.e. $\exists\ \rho_{x_0} \in \mathbb{R}$ such that $M(x_0)x_0 = \rho_{x_0}\, x_0$, from where x_0 is an eigenvector of the matrix $M(x_0)$ which has the spectrum $(\alpha||x_0||^{2k_1}, \beta||x_0||^{2k_1})$ with $\alpha < 0$, $\beta < 0$. So, $\rho_{x_0} = \alpha||x_0||^{2k_1}$ or $\beta||x_0||^{2k_1}$. it follows that the restriction of (4) to the ray $\mathbb{R}^+ x_0$ is

$$\dot{x} = r||x||^{2k_1}\, x \qquad (5)$$

where $r = \alpha$ (or β). Now, it is obvious that (5) has the same trajectories than $\dot{x} = r\, x$ which is asymptotically stable (because $\alpha < 0, \beta < 0$). So, the theorem follows from (i) of theorem 1.

We will now discuss about the existence of invariant homogeneous rays for (4). Two cases will be distinguished: the case when the degree of homogeneity of g, k_2, is even and the case when k_2 is odd. In both cases, the problem amounts to prove the existence of (at least) one invariant ray. Therefore, we have the following theorem.

Theorem 3 *Given the homogeneous system (2) with $A(x)$ and g as defined above. Under the assumptions (a) and (b), we have*

Case 1: the degree of homogeneity k_2 is even. For all $(\alpha < 0, \beta < 0)$, the closed-loop system (4) has at least one invariant homogeneous ray.

Case 2: k_2 is odd. For a suitable choice of (α, β), system (4) has at least one invariant homogeneous ray if the following extra assumption holds:

assumption (c): $det\,(g(x), x)$ has no constant sign on $\mathbb{R}^2 - \{0\}$.

Proof: Case 1: The property in theorem 3 is satisfied if the equation

$$D(x) = det\,(M(x)x, x) = 0 \qquad (6)$$

has at least one real root.

When g_1 and g_2 are rational fractions, i.e.

$$g_1(x) = \frac{p_1(x)}{q_1(x)},\ g_2(x) = \frac{p_2(x)}{q_2(x)}$$

with $q_1(x)q_2(x) \neq 0, \forall x \in \mathbb{R}^2 - \{0\}$, $D(x)$ takes the form

$$D(x) = \frac{P_\alpha(x)\ P_\beta(x)}{q_1^2 q_2^2\ det(g(x), A(x)g(x))}$$

which is well defined on $\mathbb{R}^2 - \{0\}$ (because of the (RC) condition) and where

$$P_\alpha(x) = (b(x)x_1 + (d(x) - \alpha||x||^{2k_1})x_2)p_1q_2 + ((\alpha||x||^{2k_1} - a(x))x_1 - c(x)x_2)p_2q_1,$$

$$P_\beta(x) = (b(x)x_1 + (d(x) - \beta||x||^{2k_1})x_2)p_1q_2 + ((\beta||x||^{2k_1} - a(x))x_1 - c(x)x_2)p_2q_1$$

Now, since g_1 and g_2 have the same even degree k_2, it is easy to see that $p_1\ q_2$ and $p_2\ q_1$ are homogeneous polynomials with the same even degree k_2'. Therefore, the numerator of $D(x)$ is a product of two homogeneous polynomials of odd degree $k_2' + 2k_1 + 1$ and, so, the equation (6) has at least two distinct real roots if $\alpha \neq \beta$. Thus, the system (4) has at least two invariant homogeneous rays.

Case 2: A straightforword computation yields

$$D(x) = \frac{Q_\alpha(x) \times Q_\beta(x)}{det(g(x), A(x)g(x))} \qquad (7)$$

where

$$Q_\alpha(x) = (b(x)x_1 + (d(x) - \alpha||x||^{2k_1})x_2)g_1(x) + ((\alpha||x||^{2k_1} - a(x))x_1 - c(x)x_2)g_2(x),$$

$$Q_\beta(x) = (b(x)x_1 + (d(x) - \beta||x||^{2k_1})x_2)g_1(x) + ((\beta||x||^{2k_1} - a(x))x_1 - c(x)x_2)g_2(x)$$

Observe that Q_α and Q_β are rational homogeneous functions of even degree $2k_1 + k_2 + 1$ which can be written as follows:

$$Q_\alpha(x) = \langle(\tilde{A}(x) - \alpha B\ ||x||^{2k_1})\ x, g(x)\rangle,$$
$$Q_\beta(x) = \langle(\tilde{A}(x) - \beta B||x||^{2k_1})\ x, g(x)\rangle$$

where

$$\tilde{A}(x) = \begin{pmatrix} b(x) & d(x) \\ -a(x) & -c(x) \end{pmatrix},\ B = \begin{pmatrix} 0 & 1 \\ -1 & 0 \end{pmatrix}.$$

Now, by homogeneity, the signs of Q_α and Q_β do not change along any ray issuing from the origin. So, it is sufficient to study their signs along the unit circle $S^1 = \{x \in \mathbb{R}^2 /\ ||x|| = 1\}$. Namely, we have to show that $Q_\alpha(x)$ (or $Q_\beta(x)$) vanishes at some point of S^1. Indeed, observe that

$$\forall\ x \in S^1,\ \lim_{\alpha \to -\infty} \frac{(\tilde{A}(x) - \alpha B||x||^{2k_1})\ x}{-\alpha||x||^{2k_1}} = Bx$$

This in turn implies that for all $x \in S^1$

$$\lim_{\alpha \to -\infty} \frac{Q_\alpha(x)}{-\alpha||x||^{2k_1}} = < Bx, g(x) > = det\,(g(x), x)$$

Then, it follows that

$i)$ $\exists\ y \in S^1$, $\exists\ \alpha_1 < 0$ such that

$$\alpha < \alpha_1 \Longrightarrow \frac{Q_\alpha(y)}{-\alpha\ ||y||^{2k_1}} > 0$$

$ii)$ $\exists\ z \in S^1$, $\exists\ \alpha_2 < 0$ such that

$$\alpha < \alpha_2 \Longrightarrow \frac{Q_\alpha(z)}{-\alpha\ ||z||^{2k_1}} < 0.$$

Putting $\alpha_0 = \min(\alpha_1, \alpha_2)$, then for all $\alpha < \alpha_0$ we get:

$$Q_\alpha(y) > 0 \text{ and } Q_\alpha(z) < 0$$

Finally, by continuity one has

$$\forall \alpha < \alpha_0, \exists x_0 \in S^1 :\ exitQ_\alpha(x_0) = 0$$

Obviously, we prove in the same way that

$$\exists \beta_0,\ \forall \beta < \beta_0,\ \exists\ y_0 \in S^1 :\ Q_\beta(y_0) = 0$$

Thus, the system (4) admits two invariant rays for a suitable choice of (α, β) and the theorem follows.

Remark 2: The closed-loop system (4) is homogeneous of degree $2k_1 + 1$ but not necessary polynomial as it has been assumed in theorem 1. However, this does not cause problems, since (4) has the same trajectories and the same invariant homogeneous rays that the system

$$\dot{x} = |det(g(x), A(x)g(x))|\ M(x)x$$

which is polynomial and so theorem 1 applies.

Remark 3: When $k_1 = 0$ the spectrum of $M(x)$, S_p, becomes constant ($S_p = (\alpha, \beta)$) and the closed-loop system (4) turns out to be homogeneous of degree one. However, contrary to the linear systems, the fact that α and β are chosen strictly negative does not imply the asymptotic stability of (4). (see the counter-example in (Clavier, 1995)).

On the other hand, as (4) is homogeneous of degree one, asymptotic stability implies exponential stability see (Hahn, 1967).

4. LOCAL STABILIZATION BY HOMOGENEOUS APPROXIMATION

Consider a single input, nonlinear system defined on a neighborhood U of the origin

$$\dot{x} = f_1(x) + u\ g_1(x) \tag{8}$$

where f_1 and g_1 are sufficiently regular (say C^∞) vector fields such that $D^i f_1(0) = (D^i f_1)_0 = 0$, $\forall i \in \{0, 1, ..., 2k_1\}, k_1 \geq 0$ and $D^j g_1(0) = (D^j g_1)_0 = 0$, $\forall j \in \{0, 1, ..., k_2 - 1\}, k_2 \geq 1$. Using the Taylor expansion of $f_1(x)$ and $g_1(x)$ near $x = 0$, there exist f_2 and g_2 such that

$$f_1(x) = (D^{2k_1+1} f_1)_0\ x^{2k_1+1} + f_2(x)$$

and

$$g_1(x) = (D^{k_2} g_1)_0\ x^{k_2} + g_2(x)$$

where

$$||f_2(x)|| \leq M_1\ ||x||^{2k_1+1} \text{ and } ||g_2(x)|| \leq M_2\ ||x||^{k_2}$$

($M_1 > 0, M_2 > 0$), for every x in a neighborhood $U' \subseteq U$ of the origin.

Now, with system (8) we associate the "homogeneous" approximation

$$\dot{x} = f(x) + u\ g(x) \tag{9}$$

where $f(x) = (D^{2k_1+1} f_1)_0\ x^{2k_1+1}$ and $g(x) = (D^{k_2} g_1)_0\ x^{k_2}$. Therefore, we prove the following theorem.

Theorem 4 *If the homogeneous approximation (9) is stabilizable by means of a positively homogeneous feedback of degree $2k_1 - k_2 + 1$, then the system (8) is locally stabilizable by the same feedback.*

Proof: Let $u = u(x)$ be a positively homogeneous stabilizing feedback of degree $2k_1 - k_2 + 1$ ($u(\lambda x) = \lambda^{2k_1-k_2+1}u(x)$, for all x and any positive number λ) for system (9). Set

$$F(x) = f(x) + u(x)g(x)$$

and

$$G(x) = f_2(x) + u(x)g_2(x).$$

Since u is C^1 on $U' - \{0\}$, F and G are locally Lipschitz. Moreover, we have

$$|u(x)| \leq M_0 \, ||x||^{2k_1-k_2+1} \qquad \forall x \in U'.$$

So, F is positively homogeneous function of degree $2k_1 + 1$ and G satisfies

$$||G(x)|| \leq M \, ||x||^{2k_1+1} \qquad \forall x \in U'.$$

where $M = M_1 + M_0 M_2$. It follows from Massera's theorem (Massera, 1956) that the solution $x = 0$ of the closed-loop system

$$\dot{x} = F(x) + u(x) \, G(x)$$

is asymptoticaly stable. Thus, (8) is locally asymptotically stabilizable.

5. CONCLUSION

Explicit construction of stabilizing feedbacks for a class of two-dimensional homogeneous systems is given. The application field of this result is of course limited by the low dimension and the restrictive rank condition (RC). Nevertheless, this work may seem of some interest. Indeed, it turns out that classical methods that include the construction of a Lyapunov function are difficult to apply to systems that we consider. So, our approach which provides feedbacks that are analytic except at zero, allows to remove this difficulty.

6. REFERENCES

Adda, P. and H. Zenati (1998). Variable pole assignment for a class of nonlinear systems. In: *2nd Conference on Computational Engineering in Systems Applications* (Pierre Borne, Ed.). Vol. 1. pp. 167–170. CESA98. Hammamat, Tunisia.

Boothby, W. and R. Marino (1989). Feedback stabilization of planar nonlinear systems. *Systems and Control Lett.* **12**, 87–92.

Chabour, R., G. Sallet and J. C. Vivalda (1993). Stabilization of nonlinear two dimensional systems: a bilinear approch. *Math. Control Signals Systems* **6**, 224–246.

Clavier, M. (1995). *Stabilisation des Systèmes Bilinéaires et des Systèmes mécaniques*. Thesis. Metz, France.

Dayawansa, W. P., C. F. Martin and G. Knowles (1990). Asymptotic stabilization of a class of smooth two-dimentional systems. *SIAM J. Control and Optimization* **28**, 1321–1349.

Hahn, W. (1967). *Stability of motion*. Springer Verlag. New York.

Jurdjevic, V. and J. P. Quin (1978). Controllability and stability. *J. Diff. Equations* **28**, 381–389.

Kawski, M. (1990). Homogeneous stabilizing feedback. *Control-Theory and Advanced Technology* **6**, 497–516.

Massera, J. L. (1956). Contribution to stability theory. *Anals of Mathematics* **64**, 1182–206.

Rosier, L. (1992). Homogeneous lyapunov function for homogeneous continuous vector field. *Systems Control Lett.* **19**, 467–473.

Sontag, E. D. (1989). A universal construction of arstein's theorem on nonlinear stabilization. *Systems Control Lett.* **10**, 263–284.

Tsinias, J. (1990). Remarks on feedback stabilization of homogeneous systems. *Control-Theory and Advanced Technology* **6**, 533–541.

14th Triennial World Congress, Beijing, P.R. China

ON CONSTRAINED EXPONENTIAL STABILIZABILITY OF MARGINAL AND/OR UNSTABLE LINEAR SYSTEMS

Jinhoon Choi

Department of Chemical Engineering, Sogang University
1 Sinsoo-Dong, Mapo-Koo, Seoul 121-742 Korea
Tel: +82-2-705-8917, Fax: +82-2-711-0439, E-mail: jchoi@ccs.sogang.ac.kr

Abstract: It was established that the constrained asymptotic stabilization of marginal and/or unstable linear systems on the constrained asymptotically stabilizable set is possible by nonlinear control and the constrained exponential stabilization on any compact subset of the constrained asymptotically stabilizable set by linear control. However, we show in this paper that the constrained exponential stabilization of such systems on the constrained asymptotically stabilizable set is impossible by any control law. *Copyright ©1999 IFAC*

Keywords: Constraints, Stabilization, Exponentially Stable, Discrete-Time Systems, Marginal Linear Systems, Unstable Linear Systems

1. INTRODUCTION

It is shown in (Sontag, 1984; Yang, 1993a) that an input constrained linear discrete time system can be globally asymptotically stabilizable iff all its poles are located inside or on the unit circle. In (Yang, 1993a), a nonlinear globally stabilizing control law for such a system is constructed. Moreover, in (Choi and Lee, in preparation), the mixed constrained infinite horizon linear quadratic optimal control is shown to asymptotically stabilize constrained marginal and/or unstable systems on the constrained asymptotically stabilizable set. However, Yang (Yang, 1993b) showed a negative result that the constrained global stabilization of a marginal system is in general impossible through linear feedback. Nevertheless, Lin and Saberi (Lin and Saberi, 1995; Lin, 1994) showed that there exists a linear control that exponentially stabilizes such a system on any bounded subset of the state space. Moreover, it is shown in (Choi, in print) that any unstable systems can be exponentially stabilized by a linear periodic variable structure controller on any compact subset of the constrained asymptotically stabilizable set.

In this paper, we show that a constrained linear system with its poles inside or on the unit circle cannot be globally exponentially stabilized and a constrained unstable linear system cannot be exponentially stabilized on the constrained asymptotically stabilizable set.

2. MAIN RESULTS

Consider an input constrained linear system:

$$x(k+1) = Ax(k) + B\sigma(u(k)), \qquad x(0) = x_0, \quad (1)$$

$$y(k) = Cx(k),$$

where $x(k) \in \mathbf{R}^n$, $u(k) \in \mathbf{R}^m$, $y(k) \in \mathbf{R}^l$, and

$$\sigma_i(u_k) := \begin{cases} -u_i^{sat} & \text{if } u_{ki} < -u_i^{sat} \\ u_i & \text{if } -u_i^{sat} \leq u_{ki} \leq u_i^{max} \\ u_i^{sat} & \text{if } u_{ki} > u_i^{sat} \end{cases} .$$

Throughout the paper, the stabilizability of (A, B) is assumed that is necessary for the constrained stabilizability.

Let

$$U := \{u \in \mathbf{R}^m : -u^{sat} \leq u \leq u^{sat}\}.$$

Clearly, U is a rectangle in $\mathbf{R}^m$ that is compact, convex and symmetric with respect to the origin.

We now define the constrained N step null controllable set as:

$$C_N := \{x_0 \in \mathbf{R}^n \,|\exists u(k) \in U, x(N) = 0\}$$

and the constrained asymptotically null controllable set as:

$$C_\infty := \left\{ x_0 \in \mathbf{R}^n \,\middle|\, \exists u(k) \in U, \lim_{k\to\infty} x(k) = 0 \right\}.$$

The detailed proofs of the results can be found in (Choi, submitted). Hence, we only illustrate the main idea of the results with examples here.

2.1 Marginal Plants

Theorem 2.1: The constrained global exponential stabilization of a marginal linear system is not possible.

To demonstrate the main idea of this theorem, consider a simple first order dynamic system

$$x(k+1) = x(k) + u(k)$$

subject to

$$-1 \le u(k) \le 1.$$

For exponential stability of the closed loop system, we need

$$|x(k)| \le a|x(0)|\rho^k$$

where $a \ge 1$ and $\rho \in (0,1)$. Consider $x(0) \gg 1$. Then regardless of control law, the smallest possible $x(k)$ is $x(0) - k$. Now since $\rho < 1$, there exists k^* such that $1 - a\rho^{k^*} > 0$. Suppose $x(0) > \frac{k^*}{1-a\rho^{k^*}}$. Then

$$x(0) - k^* > a x(0)\rho^{k^*}.$$

Hence the constrained global exponential stabilization of the system is impossible.

2.2 Exponentially Unstable Plants

Theorem 2.2: The constrained exponential stabilization of an exponentially unstable linear system on C_∞ is not possible.

To demonstrate the main idea of this theorem, consider an exponentially unstable linear system

$$x(k+1) = 2x(k) + u(k)$$

subject to

$$-1 \le u(k) \le 1.$$

Then

$$C_\infty = \{-1 < x < 1\}.$$

Again for exponential stability of the closed loop system, we need

$$|x(k)| \le a|x(0)|\rho^k$$

where $a \ge 1$ and $\rho \in (0,1)$. Consider $x(0) \in (0,1)$ and $x(0) \approx 1$. Then irrespective of control law, the smallest possible $x(k)$ is $2^k x(0) - \sum_{i=1}^{k-1} 2^i$. Now since $\rho < 1$, there exists k^* such that $a\rho^{k^*} < \frac{1}{2}$. Then $a|x(0)|\rho^{k^*} < \frac{1}{2}$. Suppose

$$x(0) > \frac{\frac{1}{2} + \sum_{i=1}^{k^*-1} 2^i}{2^{k^*}}.$$

Notice that

$$\frac{\frac{1}{2} + \sum_{i=1}^{k^*-1} 2^i}{2^{k^*}} \in (0,1)$$

since $2^{k^*} - \sum_{i=1}^{k^*-1} 2^i = 1$. Then

$$x(k^*) = 2^{k^*} x(0) - \sum_{i=1}^{k^*-1} 2^i > \frac{1}{2} a|x(0)|\rho^{k^*}.$$

Hence the constrained exponential stabilization of the system on C_∞ is impossible.

REFERENCES

Choi, J. (in print). On the Stabilization of Linear Discrete Systems Subject to Input Saturation. *Systems & Control Letters.*

Choi, J. (submitted). Constrained Exponential Stabilization of Marginal And/or Unstable Linear Systems. *International Journal of Control.*

Choi, J. and K. S. Lee (in preparation). Mixed Constrained Infinite Horizon Linear Quadratic Optimal Control.

Lin, Z. (1994). *Global and semi-global control problems for linear systems subject to input saturation and minimum-phase input-output linearizable systems.* Ph. D. Dissertation, Washington State University.

Lin, Z. and A. Saberi (1995). Semi-global Exponential Stabilization of Linear Discrete-Time Systems Subject to Input Saturation via Linear Feedbacks. *Systems & Control Letters*, **24**, 125-132.

Sontag, E. D. (1984). An Algebraic Approach to Bounded Controllability of Linear Systems. *International Journal of Control*, **39**, 181-188.

Yang, Y. (1993a). *Global stabilization of linear systems with bounded feedback.* Ph. D. Dissertation, the State University of New Jersey.

Yang, Y. (1993b). *Unpublished work.*

14th Triennial World Congress, Beijing, P.R. China

STABILIZATION AND CONTROL OF NONLINEAR SYSTEMS IN STRICT FEEDBACK FORM

Krishna K. Busawon, Mehrdad Saif [1]

School of Engineering Science
Simon Fraser University
8888 University Dr.
Vancouver, British Columbia V5A 1S6
CANADA

Abstract: In this paper, we consider the problem of global feedback stabilization of a class of single input nonlinear systems. The system under consideration is composed of a linear, and a triangular structured nonlinear parts. Furthermore, the linear part is in the form of the well known Brunovsky structure. In this article a control design methodology for such systems based on linear design techniques using some special transformations is proposed. Consequently, the proposed design is simple, systematic, and easy to implement. In addition, the paper addresses robustness properties of the proposed control law and possible extensions to the multi-input case. Finally, control problem for a flexible joint robot is used as an application example to show the design and performance of the proposed controller. *Copyright ©1999 IFAC*

Keywords: Nonlinear systems, stabilization, strict feedback systems.

1. INTRODUCTION

We consider the problem of global feedback stabilization of single input nonlinear systems which are of the following form :

$$\begin{cases} \dot{x}_1 = x_2 + f_1(x_1) \\ \dot{x}_2 = x_3 + f_2(x_1, x_2) \\ \cdots \\ \dot{x}_i = x_{i+1} + f_i(x_1, \ldots, x_i) \\ \cdots \\ \dot{x}_{n-1} = x_n + f_{n-1}(x_1, \ldots, x_{n_1}) \\ \dot{x}_n = \varphi(x) + \psi(x)u \end{cases} \tag{1}$$

where $x \in R^n$ is the state vector, $u \in R$ is the input, and nonlinearities f_i $(i = 1, \ldots, n-1)$ are continuously differentiable.

System (1) can be written in a condensed form as:

$$\dot{x} = Ax + f(x) + B[\varphi(x) + \psi(x)u] \tag{2}$$

where

$$A = \begin{bmatrix} 0 & 1 & & 0 \\ \vdots & \ddots & \ddots & \\ \vdots & & \ddots & 1 \\ 0 & \ldots & \ldots & 0 \end{bmatrix}, \quad B = \begin{bmatrix} 0 \\ \vdots \\ 0 \\ 1 \end{bmatrix}$$

and $f(x) = \begin{bmatrix} f_1(x) & \cdots & f_{n-1}(x) & 0 \end{bmatrix}^T$. We assume that $f(0) = 0$ and $\varphi(0) = 0$. It is important to note that when $f_1 = f_2 = \ldots = f_{n-1} = 0$, then the design of a feedback law, which globally stabilize the system at 0, is trivial provided that $\psi(x) \neq 0$, for all $x \in R^n$. Indeed, such feedback can be designed as:

$$u(x) = -\frac{1}{\psi(x)}[\varphi(x) + Kx] \tag{3}$$

[1] Corresponding Author
saif@cs.sfu.ca, Fax:(604) 291-4951, Tel: (604) 291-3119

where K is chosen such that $A - BK$ is Hurwitz. This control is the well known feedback linearizing control law. However, when f_i's are not all zero, a systematic design procedure for the construction of the stabilizing feedback law is not obvious at all. The paper addresses this precise problem.

A slightly more general class of systems than the one described by (1) was considered in [11] where it is shown that under some general assumptions, there exist a linear feedback law which globally exponentially stabilizes the system at its equilibrium point. However, the construction of the linear feedback law is not direct and a procedure for the construction of such a linear feedback law is not given.

The class of systems described by (1) are also referred to as nonlinear system in the "strict feedback form" (see e.g., [8,5]). In fact, in [5], it is shown that all strict feedback system are input to state stabilizable. A stabilizing feedback is also proposed. However, once again what is lacking is a systematic and direct design procedure for construction of the control law. Another way to deal with the stabilization of system (1) is to use the integrator backstepping. However, this design requires the existence of a suitable Lyapunov function which is not always easy to derive (see e.g.,[8]).

In this paper, we give a simple algorithm for constructing a global stabilizing feedback for system (1) under very mild assumptions. The construction uses linear design techniques involving transformations based on "controllability matrices". Consequently, the design is simple and easy to implement. We show that when all the f_i's are zero then the control law reduces to the one given by equation (3).

2. PRELIMINARY RESULTS

In this section, we give some preliminary results which are induced by the particular structure of system (1). These results are crucial for the controller design for system (1) and will be used in the subsequent sections.

Consider system (1) and define the following matrices successively:

i)

$$U(x) = \left[B,\ (A + F(x))\,B,\ \ldots, (A + F(x))^{n-1}B\right]$$

where $F(x) = \dfrac{\partial f}{\partial x}(x)$. It is easy to check that $U(x)$ is of full rank for all $x \in R^n$.

ii)

$$V_0 = \left[B,\ AB,\ \ldots,\ \ A^{n-1}B\right]$$

iii)

$$M_0(x) = V_0 U^{-1}(x)$$

It also can be verified that $M_0(x)$ is lower triangular and its diagonal elements are all equal to 1; i.e. $\det M_0(x) = 1$.

iv)

$$Q(x) = M_0(x)[A + F(x)]M_0^{-1}(x) - A$$

It can also be checked that the entries of the matrix $Q(x)$ are all zero except the first column. More precisely $Q(x)$ is of the following form:

$$Q(x) = \begin{bmatrix} q_1(x) & 0 \ldots 0 \\ q_2(x) & 0 \ldots 0 \\ \vdots & \vdots \quad \vdots \\ q_{n-1}(x) & 0 \ldots 0 \\ 0 & 0 \ldots 0 \end{bmatrix} \tag{4}$$

v)

$$R(x) = V_0 Q(x)$$

It is easy to see that

$$R(x) = \begin{bmatrix} 0 & 0 \ldots 0 \\ q_{n-1}(x) & 0 \ldots 0 \\ \vdots & \vdots \quad \vdots \\ q_2(x) & 0 \ldots 0 \\ q_1(x) & 0 \ldots 0 \end{bmatrix}.$$

$R(x)$ can also be written as $R(x) = \Gamma(x)C$ where $C = (1, 0 \ldots, 0)$ and $\Gamma(x) = \begin{bmatrix} 0 \\ q_{n-1}(x) \\ \vdots \\ q_1(x) \end{bmatrix}$.

vi)

$$\bar{A}(x) = A + BL(x)$$

where $L(x) = \Gamma^T(x)$. Clearly $\bar{A}(x)$ is given by:

$$\bar{A}(x) = \begin{bmatrix} 0 & 1 & 0 & \ldots & 0 \\ 0 & 0 & 1 & \ddots & \vdots \\ \vdots & \vdots & & \ddots\ \ddots & 0 \\ \vdots & \vdots & & \ddots & 1 \\ 0 & q_{n-1}(x) & \ldots & \ldots & q_1(x) \end{bmatrix}$$

vii)

$$V(x) = \left[B,\ \bar{A}(x)B,\ \ldots,\ \ \bar{A}^{n-1}(x)B\right]$$

viii)

$$M(x) = V(x)U^{-1}(x)$$

Here again, it can be verified that $M(x)$ is lower triangular and its diagonal elements are all equal to 1; i.e. $\det M(x) = 1$.

Based on the above results, a number of observations can be made at this point:

Remark 1.1) Matrix $M(x)$ has the following special properties:

$$M[A + F(x)]M^{-1} = A + BL(x); M(x)B = B$$

Remark 1.2) Let

$$\Delta_\alpha = diag\left[\alpha^n, \alpha^{n-1}, \ldots, \alpha\right] \quad (5)$$

with $\alpha > 0$ is a parameter. A straightforward computation shows that $\Delta_\alpha B = \alpha B$ and $\Delta_\alpha A \Delta_\alpha^{-1} = \alpha A$.

Remark 1.3) For any lower triangular bounded matrix $W(x) = (w_{i,j}(x))$; i.e. $\|W(x)\| \leq \delta$ with $\delta > 0$, the matrix $S(x) = \|\Delta_\alpha W(x) \Delta_\alpha^{-1}\|$ is also upper bounded by some δ' which is independent of α provided that $\alpha \geq 1$. Indeed, it is easy to see that the elements of $S(x)$ are given by $s_{i,j}(x) = \dfrac{w_{ij}(x)}{\alpha^{(i-j)}}$ for $j \leq i$ and $s_{i,j}(x) = 0$ otherwise. Therefore, if $\alpha \geq 1$, it is obvious that $S(x)$ is upper bounded by a constant which is independent of α.

3. MAIN RESULT

Consider the nonlinear system in (1) and assume that the functions f_i and ψ satisfy the following assumptions:

A1) $\psi(x) \neq 0$, for all $x \in R^n$.
A2) For all $i = 1, \ldots, n$ and all $x \in R^n$, the functions $f_i(x)$ and their respective time derivative are bounded.

Based on the above, the main result on the controller design and proof of stability can be stated:

Theorem 1. *Assume that system (2) satisfies A1) and A2), and consider the following control law*

$$u(x) = -\frac{1}{\psi(x)}[L(x)M(x)x + K\Delta_\alpha M(x)x + \varphi(x)] \quad (6)$$

where K is a row vector chosen such that the matrix $A - BK$ is Hurwitz, $M(x)$ and $L(x)$ are defined as in the previous section, and Δ_α is defined as in (5). In this case, there exists an $\alpha_0 > 0$ such that for all $\alpha > \alpha_0$ the origin $x = 0$ of the system in (2) subject to the above control law is globally asymptotically stable.

Proof:
Rewrite system (1) as :

$$\begin{aligned}\dot{x} &= Ax + \frac{\partial f}{\partial x}(x)x + B[\varphi(x) + \psi(x)u(x)] + r(x)\\ &= [A + F(x)]x + B[\varphi(x) + \psi(x)u(x)] + r(x)\end{aligned}$$

where $r(x) = f(x) - F(x)x$. Hereafter, the dependence of all the variables on x will be suppressed for the sake of simplicity and they will be mentioned only when necessary. Let $\bar{x} = \Delta_\alpha M(x)x$. Then using the Remark 1.1, we get:

$$\begin{aligned}\dot{\bar{x}} &= \Delta_\alpha M(x)\dot{x} + \Delta_\alpha \dot{M}(x)x\\ &= \Delta_\alpha M[A + F(x)]M^{-1}\Delta_\alpha^{-1}\bar{x} + \Delta_\alpha MB[\varphi(x) +\\ &\quad \psi(x)u(x)] + \Delta_\alpha \dot{M}M^{-1}\Delta_\alpha^{-1}\bar{x} + \Delta_\alpha Mr(x)\\ &= \Delta_\alpha[A + BL(x)]\Delta_\alpha^{-1}\bar{x} + \Delta_\alpha B[\varphi(x) +\\ &\quad \psi(x)u(x)] + \Delta_\alpha \dot{M}M^{-1}\Delta_\alpha^{-1}\bar{x} + \Delta_\alpha Mr(x)\end{aligned}$$

Using Remark 1.2:

$$\begin{aligned}\dot{\bar{x}} &= \alpha A\bar{x} + \alpha BL(x)\Delta_\alpha^{-1}\bar{x} + \alpha B[\varphi(x) + \psi(x)u(x)]\\ &\quad + \Delta_\alpha \dot{M}(x)M^{-1}\Delta_\alpha^{-1}\bar{x} + \Delta_\alpha M(x)r(x)\end{aligned}$$

Now, since $u(x) = -\dfrac{1}{\psi(x)}[L(x)\Delta_\alpha^{-1}\bar{x} + K\bar{x} + \varphi(x)]$, we get:

$$\dot{\bar{x}} = \alpha(A - BK)\bar{x} + \Delta_\alpha \dot{M}(x)M^{-1}\Delta_\alpha^{-1}\bar{x} + \Delta_\alpha M(x)r(x)$$

Since $A - BK$ is Hurwitz, there exists a symmetric positive definite matrix P such that:

$$(A - BK)^T P + P(A - BK) = -I_n$$

where I_n is the $n \times n$ identity matrix. Choosing a Lyapunov function as $V(\bar{x}) = \bar{x}^T P \bar{x}$, then:

$$\begin{aligned}\dot{V}(\bar{x}) &= 2\bar{x}^T P \dot{\bar{x}}\\ &= -\alpha\|\bar{x}\|^2 + 2\bar{x}^T P \Delta_\alpha \dot{M} M^{-1} \Delta_\alpha^{-1}\bar{x}\\ &\quad + 2\bar{x}^T P \Delta_\alpha M r(x)\\ &\leq -\alpha\|\bar{x}\|^2 + 2\|P\bar{x}\|\|\Delta_\alpha \dot{M} M^{-1}\Delta_\alpha^{-1}\|\|\bar{x}\|\\ &\quad + 2\|P\bar{x}\|\|\Delta_\alpha M r(x)\|\end{aligned}$$

Since f is continuously differentiable, $r(x) = [F(\xi) - F(x)]x = \Psi(x, \xi)x$ for some ξ lying in the segment with end points 0 and x. Since $\Psi(x, \xi)$ is lower triangular, it can easily be shown, using A2) and Remark 1.3), that :

$$\sup_{\alpha \geq 1} \|\Delta_\alpha M(x)\Psi(x, \xi)M^{-1}(x)\Delta_\alpha^{-1}\| \leq k_r$$

for some positive constant k_r which depend on the upper bound of F and which is independent of α.

Similarly, using A2) and Remark 1.3) it can be shown that :

$$\sup_{\alpha\geq 1} \|\Delta_\alpha \dot{M} M^{-1}\Delta_\alpha^{-1}\| \leq k_m$$

for some positive constant k_m which is independent of α. Hence,

$$\begin{aligned}\dot{V}(\bar{x}) \leq & -\alpha\|\bar{x}\|^2 + 2k_m\|P\bar{x}\|\|\bar{x}\| \\ & +2\|P\bar{x}\|\|\Delta_\alpha M\Psi M^{-1}\Delta_\alpha^{-1}\|\|\bar{x}\| \\ \leq & -\alpha\|\bar{x}\|^2 + 2k_m\|P\bar{x}\|\|\bar{x}\| + 2k_r\|P\bar{x}\|\|\bar{x}\|\end{aligned}$$

By using the fact that $\lambda_i\|\bar{x}\|^2 \leq V(\bar{x}) \leq \lambda_s\|\bar{x}\|^2$ and that $\|P\bar{x}\| \leq \sqrt{\lambda_s}\sqrt{V(\bar{x})}$ where λ_i and λ_s are the smallest and the largest eigenvalues of P respectively, we get:

$$\dot{V}(\bar{x}) \leq -\frac{\alpha}{\lambda_s}V + 2c\sigma V$$

where $c = (k_m + k_r)$ and $\sigma = \sqrt{\frac{\lambda_s}{\lambda_i}}$. Therefore, choosing $\alpha > \alpha_0 = 2c\sigma\lambda_s$ results in $\dot{V}(\bar{x}) < 0$. □

4. DISCUSSION AND EXTENSIONS

The following remarks either shed additional light on the above proposed controller, or discuss various extensions of the above idea to other closely related forms of the nonlinear system described in (1).

1.0– It is important to note that the basic idea of the design was to find a transformation $\bar{x} = M(x)x$ with $M(x)[A+F(x)]M^{-1}(x) = A+BL(x)$ and $M(x)B = B$ to transform the nonlinear system into:

$$\begin{aligned}\dot{\bar{x}} = & A + BL(x)\bar{x} + B[\varphi(x) + \psi(x)u(x)] + \\ & \dot{M}M^{-1}\bar{x} + Mr(x)\end{aligned}$$

The matrix Δ_α was then introduced in order to deal with the remaining nonlinearities $\dot{M}M^{-1}\bar{x}$ and $Mr(x)$ through a high gain approach. In addition, Δ_α brings some flexibility into the controller gain and enabling to shift the eigenvalues of the close-loop system as desired.

2.0– It is interesting to note that when $f(x) = 0$, the matrix $M(x)$ is equal to the identity matrix and $L(x) = 0$. Consequently, in this case, α can be chosen to be equal to 1 and the control law (6) reduces to the feedback linearizing controller given by (3).

3.0– The following control law:

$$u(x) = -\frac{1}{\psi(x)}[K\Delta_\alpha M_0(x)x + \varphi(x)] \qquad (7)$$

where $M_0(x)$ is defined as in iii) Section 2, is also a global stabilizing law for system (1). However, this control law is of "higher" gain than the control law (6). To see this let $\bar{x} = \Delta_\alpha M_0(x)x$. Then, proceeding as in the proof of Theorem 1, we obtain

$$\begin{aligned}\dot{\bar{x}} = & \Delta_\alpha[A + Q(x)]\Delta_\alpha^{-1}\bar{x} + \Delta_\alpha B[\varphi(x) + \psi(x)u(x)] + \\ & \Delta_\alpha\dot{M}_0M_0^{-1}\Delta_\alpha^{-1}\bar{x} + \Delta_\alpha M_0 r(x)\end{aligned}$$

where $Q(x)$ is as in (4). Setting $V(\bar{x}) = \bar{x}^T P\bar{x}$, we get

$$\begin{aligned}\dot{V}(\bar{x}) \leq & -\alpha\|\bar{x}\|^2 + 2c'\|P\bar{x}\|\|\bar{x}\| + \\ & 2\|P\bar{x}\|\|\Delta_\alpha Q(x)\Delta_\alpha^{-1}\|\|\bar{x}\|\end{aligned}$$

for some positive constant c' which is independent of α, since $M_0(x)$ has similar boundedness properties as $M(x)$. Owing to the particular form of $Q(x)$ and using Assumption A2), it can be shown that:

$$\sup_{\alpha\geq 1}\|\Delta_\alpha Q(x)\Delta_\alpha^{-1}\| \leq k_q$$

for some positive constant k_q which is independent of α. Hence,

$$\dot{V}(\bar{x}) \leq -\frac{\alpha}{\lambda_s}V + 2(c + k_q)\sigma V$$

Therefore, by choosing $\alpha > 2\lambda(c' + k_q)\sigma$ we obtain $\dot{V}(\bar{x}) < 0$.

4.0– The control laws in (6) and (7) assume that all the state variables are available for feedback. However, this is not always true in practice. In [2], an observer design methodology is proposed for a general class of nonlinear systems and that observer can be used here for the purpose of estimating the unmeasurable state variables. Furthermore, by using similar techniques as used in [1] a separation principle can be proved for this class of systems provided the control law is bounded.

5.0–The above design methodology can be extended in a blockwise way to a class of multi-inputs systems, especially when the dimension of the state is a multiple of the number of input variables.

6.0– Finally, it is important to remark that system (1) covers a large number of nonlinear systems. Indeed, consider the following class of strict feedback systems (see e.g., [5]) :

$$\begin{cases}\dot{x}_1 & = b_1(x_1)x_2 \quad + a_1(x_1) \\ \dot{x}_2 & = b_2(x_1, x_2)x_3 \quad + a_2(x_1, x_2) \\ & \cdots \\ \dot{x}_i & = b_i(x_1, \ldots, x_i)x_{i+1} + a_i(x_1, \ldots, x_i) \\ & \cdots \\ \dot{x}_{n-1} & = b_{n-1}(x_1, \ldots, x_{n_1})x_n \quad + a_{n-1}(x_1, \ldots, x_{n_1}) \\ \dot{x}_n & = b_n(x)u + a_n(x)\end{cases} \qquad (8)$$

where $0 < \mu \leq |b_i(x_1, \ldots, x_i)| \leq \eta < +\infty$ and μ, η are positive constants. It can be easily verified that system (8) can be brought into the form (1) by the following transformation

$$z = \Lambda(x)x$$

where

$$\Lambda(x) = diag[1,\ b_1(x_1),\ b_1(x_1)b_2(x_1,x_2), \ldots, \prod_{i=1}^{n-1} b_i(x_1, \ldots, x_i)]$$

provided that the functions a_i are differentiable (see [3]).

5. APPLICATION TO A SINGLE-LINK ROBOT WITH FLEXIBLE JOINT

In this section, we shall apply the previous result to a single-link robot with flexible joint. The dynamics of such a system rotating in a vertical plane is given (see [9], [10]) in state space representation with $q_1 = x_1$, $\dot{q}_1 = x_2$, $q_2 = x_3$, $\dot{q}_2 = x_4$, as :

$$\begin{cases} \dot{x}_1 = x_2 \\ \dot{x}_2 = -\dfrac{mgl}{J_1}\sin x_1 - \dfrac{k}{J_1}(x_1 - x_3) \\ \dot{x}_3 = x_4 \\ \dot{x}_4 = \dfrac{k}{J_2}(x_1 - x_3) + \dfrac{1}{J_2}u \end{cases} \tag{9}$$

where q_1 and q_2 are the link displacement and the rotor displacement, respectively. The link inertia J_1, the motor rotor inertia J_2, the elastic constant k, the link mass m, the gravity constant and the center of mass l are all positive. It is easy to see that the above system is of the form (9). Therefore to transform the system into the form (1), we make the following change of variables : $z = \Lambda x$ where $\Lambda = diag[1, 1, \frac{k}{J_1}, \frac{k}{J_1}]$ (see Remark (*vii*), Section 4). Note that in this particular case Λ is constant. We then obtain :

$$\begin{cases} \dot{z}_1 = z_2 \\ \dot{z}_2 = z_3 - \dfrac{mgl}{J_1}\sin z_1 - \dfrac{k}{J_1}z_1 \\ \dot{z}_3 = z_4 \\ \dot{z}_4 = \dfrac{k}{J_2}(\dfrac{k}{J_1}z_1 - z_3) + \dfrac{k}{J_1 J_2}u \end{cases} \tag{10}$$

The above system is now of the form given in (1). The equilibrium point of system (13) is given by: $z_{1,0} = z_1^0$; $z_{2,0} = 0$; $z_{3,0} = \frac{mgl}{J_1} sin(z_1^0) + \frac{k}{J_1} z_1^0$; $z_{4,0} = 0$ and $u_0 = -kz_1^0 + J_1 z_{3,0}$.

Our objective here is to design a controller so as to stabilize the system at a particular given equilibrium point. For this we make the following further change of variables : $\xi_1 = z_1 - z_1^0$; $\xi_2 = z_2 - z_{2,0} = z_2$; $\xi_3 = z_3 - z_{3,0}$; $\xi_4 = z_4 - z_{4,0} = z_4$ and $v = u - u_0$.

We then obtain :

$$\begin{cases} \dot{\xi}_1 = \xi_2 \\ \dot{\xi}_2 = \xi_3 + \dfrac{mgl}{J_1}[\sin(z_1^0) - \sin(\xi_1 + z_1^0)] - \dfrac{k}{J_1}\xi_1 \\ \dot{\xi}_3 = \xi_4 \\ \dot{\xi}_4 = \dfrac{k}{J_2}\left[\dfrac{k}{J_1}\xi_1 - \xi_3\right] + \dfrac{k}{J_1 J_2}v \end{cases} \tag{11}$$

System (14) is also of the form (1) with $\varphi(\xi) = \frac{k}{J_2}\left[\frac{k}{J_1}\xi_1 - \xi_3\right]$, $\psi(\xi) = \frac{k}{J_1 J_2}$ and

$$f(\xi) = \begin{bmatrix} 0 \\ \dfrac{mgl}{J_1}[\sin(z_1^0) - \sin(\xi_1 + z_1^0)] - \dfrac{k}{J_1}\xi_1 \\ 0 \\ 0 \end{bmatrix}.$$

Therefore, a controller of the form (6) can be constructed so as to achieve $\lim_{t \to \infty} \|\xi\| = 0$. Following the procedure given in Section 2, it can be shown that:

$$L(\xi) = [0,\ 0,\ \lambda,\ 0];$$

where

$$\lambda = \frac{-mgl\cos(\xi_1 + z_1^0) - k}{J_1}$$

$$M(\xi) = \begin{bmatrix} 1 & 0 & 0 & 0 \\ 0 & 1 & 0 & 0 \\ \lambda & 0 & 1 & 0 \\ 0 & \lambda & 0 & 1 \end{bmatrix}$$

and

$$u(\xi) = u_0 - \frac{J_1 J_2}{k} \times \left(L(\xi)M(\xi)\xi + K\Delta_\alpha M(\xi)\xi + \frac{k}{J_2}\left[\frac{k}{J_1}\xi_1 - \xi_3\right]\right)$$

with

$$L(\xi)M(\xi)\xi = \left(\frac{mgl\cos(\xi_1 + z_1^0) + k}{J_1}\right)^2 \xi_1 - \left(\frac{mgl\cos(\xi_1 + z_1^0) + k}{J_1}\right)\xi_3$$

$$K\Delta_\alpha M(\xi, \bar{y})\xi = \left(k_1\alpha^4 - \frac{k_3\alpha^2(mgl\cos(\xi_1 + z_1^0) + k)}{J_1}\right)\xi_1 + \left(k_2\alpha^3 - \frac{k_4\alpha(mgl\cos(\xi_1 + z_1^0) + k)}{J_1}\right)\xi_2 + k_3\alpha^2\xi_3 + k_4\alpha\xi_4$$

5.1 *Simulation results*

The following numerical values was used for the purpose of the simulation study: $z_1^0 = 45\%$, $J_2 = 0.2kgm^2$, $k = 10N$, $m = 1kg$, $g = 10m/s^2$, $J_1 = 1kgm^2$, $l = 1m$, $\alpha = 2$, $k_1 = 1$, $k_2 = 4$, $k_3 = 6$, $k_4 = 4$. The result of the simulation is depicted in Figure 1 which shows the rapid convergence of the closed-loop system to the equilibrium position $x_{1,0} = z_1^0 = 0.79$; $x_{2,0} = x_{2,0} = 0$; $x_{3,0} = \frac{J_1}{k} z_{3,0} = 1.497$; $x_{4,0} = \frac{J_1}{k} z_{4,0} = 0$ and hence the effectiveness of the controller.

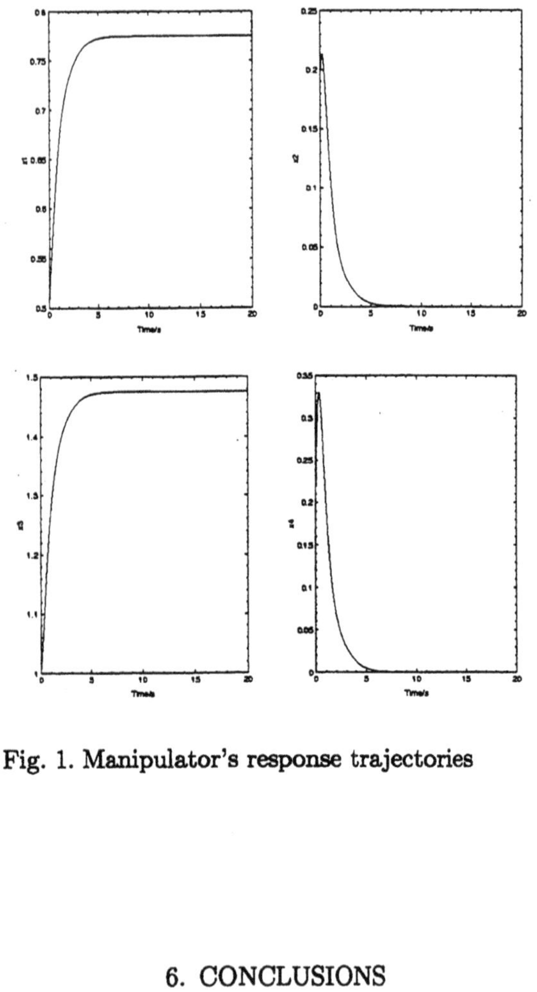

Fig. 1. Manipulator's response trajectories

6. CONCLUSIONS

In this paper, we proposed a control design methodology for a class of single-input nonlinear systems. The design is simple and easy to implement since it is based on linear design techniques. We have also shown that the proposed controller is robust with respect to certain types of uncertainties and disturbances. Finally, we have applied the results to a single link flexible joint mechanism in order to illustrate the design methodology. We have not treated the problem of plant parameter variations and model inaccuracies. This is a topic of continuing research.

ACKNOWLEDGEMENTS

This research was sponsored in part by Natural Sciences and Engineering Research Council (NSERC) of Canada through its *Research* and *Strategic* grants program

7. REFERENCES

[1] Busawon, K., A. Elassoudi and H. Hammouri, "Dynamic output feedback stabilization of a class of nonlinear systems", *Proc. of the 32nd IEEE CDC*, San Antonio, TX, 1993.

[2] Busawon, K. and M. Saif, "An observer design methodology for nonlinear systems", to appear in the *American Control Conference*, 1998.

[3] Busawon, K., M. Farza and H. Hammouri, "An simple observer for a class of nonlinear systems", *Applied Maths. Letters*, Vol. 11, n0 3, 27-31, 1998.

[4] Chen, C.T., "Introduction to linear system theory", *HRW series in Electrical Engineering, Electronics, and Systems*, 1970.

[5] Freeman R. A. and P.V. Kokotovic, "Global robustness of nonlinear systems to state measurement disturbances", *Proc. of the 32nd IEEE CDC*, San Antonio, TX, 1993.

[6] Isidori A. , "Nonlinear control systems", *Springer Verlag*, London, 1995.

[7] Jurdjevic, V, and J.P. Quinn, "Controllability and stability", *J. Diff. Equations*, Vol. 28, pp. 381-389, 1978.

[8] Krstić, M., I. Kanallakopoulos, P. Kokotović, "Nonlinear and adaptive control design", *Wiley-Interscience Publication*, New York, 1995.

[9] Lewis, F. L., Abdallah, C. T. and Dawson, D. M. "Control of robot manipulators", *Macmillan*, New York, 1993.

[10] Marino, R. and Tomei, P., "Nonlinear control design", *Prentice Hall*, U.K., 1995.

[11] Tinias, J. , "A theorem on global stabilization of nonlinear systems by linear feedback", *Systems and Control Letters*, Vol. 17, pp. 357-362, 1991.

STABILIZATION OF NONLINEAR SYSTEMS USING THE ASSOCIATED ANGULAR SYSTEM

S.P.Banks* and M.Salamci**

**Department of Automatic Control and Systems Engineering,*
University of Sheffield, Mappin Street,
Sheffield S1 3JD, UK
e-mail: s.banks@sheffield.ac.uk
***Middle East Technical University, Mechanical Engineering Department,*
Inonu Bulvari, Ankara, Turkey.

Abstract: The stabilizability of general nonlinear systems is studied by considering the associated angular system, which leads to a simple stabilizing controller in many cases. Bilinear systems are studied in detail and general nonlinear systems are reduced to infinite dimensional bilinear form by using Lie theory.

Keywords:Stabilization, Nonlinear Systems, Angular System.

1.INTRODUCTION

The control of nonlinear systems has been approached by a variety of techniques, including linearization (Banks, 1988), sliding mode control (Luk'yanov and Utkin, 1981;Utkin, 1992) and optimal state-space and H^∞ control (Knobloch,*et al*, 1993). Morever, several authors have studied the frequency-domain theory of nonlinear systems with a view to establishing stabilizing controllers (Banks,1985;Banks,1997). In this paper we shall use the angular form of a differential equation to reduce the problem to the control of the radial component of the state. This representation has been used by a number of authors to study the Lyapunov spectrum of a nonlinear system (see, for example, (Colonius and Kliemann,1996) and the references cited there). In this paper we shall consider systems of the form

$$\dot{x} = A(x)x + B(x)u \qquad (1)$$

and write $z = x/\|x\|$ and $r = \|x\|$ which implies that we can obtain two equations of the form

$$\dot{z} = \overline{A}z + u\overline{B}z$$

$$\dot{r} = \lambda_A r + u\lambda_B r$$

from which we can obtain a control

$$u = -\frac{(\lambda_A + \alpha)}{\lambda_B} \qquad (2)$$

for any $\alpha > 0$ where $\lambda_B \neq 0$, which will drive $r = \|x\|$ to zero. If the set $\Sigma = \{x : \lambda_B(x) = 0\}$ is not just the origin then the control will drive the system to Σ while reducing r.

In section 2 we shall consider first the special case of bilinear systems and in section 3 the more general systems (1). In both cases we shall use the control (1) (or a simple modification) to drive the system to Σ. The results depend strongly on the choice of α as we shall see. Finally in section 4, we consider general nonlinear systems of the form

$$\dot{x} = f(x,u)$$

and use Lie series techniques (Banks, 1992) to reduce the problem to an infinite-dimensional bilinear system.

2. THE ANGULAR FORM AND CONTROL OF BILINEAR SYSTEMS

In this section we shall consider the bilinear system

$$\dot{x} = A(x)x + uBx \qquad (2)$$

The first result is well-known, but we shall prove it here for the convenience of the reader and so that we can generalize it later.

Lemma 2.1 *The equation (2.1) can be written in the form*

$$\begin{aligned} \dot{z} &= \overline{A}(z)z + u\overline{B}(z)z \\ \dot{r} &= \lambda_A(z)r + u\lambda_B(z)r \end{aligned} \tag{3}$$

where

$$z = x/\|x\|, r = \|x\|,$$
$$\lambda_A(z) = z^T Az,\ \lambda_B(z) = z^T Bz$$
$$\overline{A}(z) = A - \lambda_A(z)I\ ,\ \overline{B}(z) = B - \lambda_B(z)I.$$

Proof We have

$$\dot{z} = \frac{\dot{x}}{\|x\|} - \frac{x\frac{d}{dt}\|x\|}{\|x\|^2}$$

and since

$$\begin{aligned} \frac{1}{2}\frac{d}{dt}(\|x\|^2) &= \frac{1}{2}\frac{d}{dt}(x^T x) = x^T\dot{x} \\ &= x^T(Ax + uBx) \end{aligned}$$

it follows that

$$\frac{d}{dt}(\|x\|) = \frac{x^T Ax}{\|x\|} + u\frac{x^T Bx}{\|x\|}$$

and so

$$\dot{z} = Az + uBz - (z^T Az + uz^T Bz)z$$

and the result follows.

Corollary 2.2 *The system is stable if we choose u to satisfy*

$$\int_0^t (\lambda_A(z) + u\lambda_B(z))dt \to -\infty$$

as $t \to \infty$.

Proof We simply integrate the scalar equation for r in (4) to give

$$r(t) = e^{\int_0^t (\lambda_A(z)+u\lambda_B(z))dt} r_0 .$$

The above results suggest the use of the control

$$u = -\frac{(\lambda_A(z) + \alpha)}{\lambda_B(z)} \tag{5}$$

for any $\alpha > 0$ provided $\lambda_B(z) \neq 0$. In fact, we have

Theorem 2.3 *If $\lambda_B(z) \neq 0$ for $z \neq 0$, then the control (5) will drive the system (3) to the origin. The system with this control is*

$$\dot{x} = A(x)x - \frac{(\lambda_A(x) + \alpha\|x\|^2)}{\lambda_B(x)} B(x). \tag{6}$$

Proof We note that

$$\begin{aligned} \dot{r} &= \lambda_A(z)r - \frac{(\lambda_A(z) + \alpha)}{\lambda_B(z)}\lambda_B(z)r \\ &= -\alpha r \end{aligned}$$

and so

$$r(t) = e^{-\alpha t} r_0 .$$

Remark 2.4 In the case where $\lambda_B(z) = 0$ on some set Σ with nonzero elements, we can use the control

$$u = -\frac{(\lambda_A(z) + \alpha)}{(\lambda_B(z) + \varepsilon\lambda_B(z))/|\lambda_B(z)|}, z \notin \Sigma,$$

for some $\varepsilon > 0$. Then we have

$$\begin{aligned} \dot{r} &= \lambda_A(z)r - \\ &\quad \frac{(\lambda_A(z) + \alpha)}{(\lambda_B(z) + \varepsilon\lambda_B(z))/|\lambda_B(z)|}\lambda_B(z)r \\ &= -\frac{(\alpha\lambda_B(z) - \varepsilon\sigma(z))}{\lambda_B(z) + \varepsilon\sigma(z)} r \end{aligned}$$

where

$$\sigma(z) = \frac{\lambda_B(z)}{|\lambda_B(z)|} = \mathrm{sgn}(\lambda_B(z))$$

and r will decay for all z for which

$$\alpha\,|\lambda_B(z)| > \varepsilon . \tag{7}$$

Example 2.5 Consider the system

$$\begin{aligned} \begin{pmatrix} \dot{x}_1 \\ \dot{x}_2 \end{pmatrix} &= \begin{pmatrix} 0 & 1 \\ 1 & 1 \end{pmatrix}\begin{pmatrix} x_1 \\ x_2 \end{pmatrix} + u\begin{pmatrix} 0 & 0 \\ 0 & 1 \end{pmatrix}\begin{pmatrix} x_1 \\ x_2 \end{pmatrix} \\ &= Ax + uBx \end{aligned} \tag{8}$$

Then,

$$\lambda_A(z) = 2z_1 z_2 + z_2^2$$
$$\lambda_B(z) = z_2^2$$

and the controlled system becomes

$$\begin{pmatrix} \dot{x}_1 \\ \dot{x}_2 \end{pmatrix} = \begin{pmatrix} 0 & 1 \\ 1 & 1 \end{pmatrix}\begin{pmatrix} x_1 \\ x_2 \end{pmatrix} - \begin{pmatrix} 0 \\ (2x_1 x_2 + x_2^2 + \alpha(x_1^2 + x_2^2))/x_2 \end{pmatrix} \tag{9}$$

This system is not globally stabilizable, but we can drive the velocity $x_2 = \dot{x}_1$ by using this control. By choosing α to depend on the initial state, we get effective control reducing the state to a neighbourhood of zero for many initial states. Results for $\alpha = 0.84$ and $\alpha = 1$ are shown in figures 1,2. Of course, if we have a small control effect in the x_1 direction, then λ_B is nonsingular and we get global control as in figures 3,4. These correspond to the systems

$$\begin{pmatrix} \dot{x}_1 \\ \dot{x}_2 \end{pmatrix} = \begin{pmatrix} 0 & 1 \\ 1 & 1 \end{pmatrix}\begin{pmatrix} x_1 \\ x_2 \end{pmatrix} + u\begin{pmatrix} 0.1 & 0 \\ 0 & 1 \end{pmatrix}\begin{pmatrix} x_1 \\ x_2 \end{pmatrix} \tag{10}$$

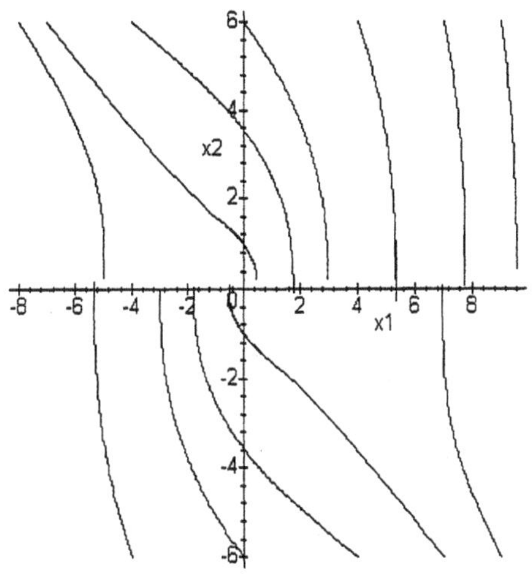

1 Fig. 1. Controlled Dynamics for System (8) with α=0.84

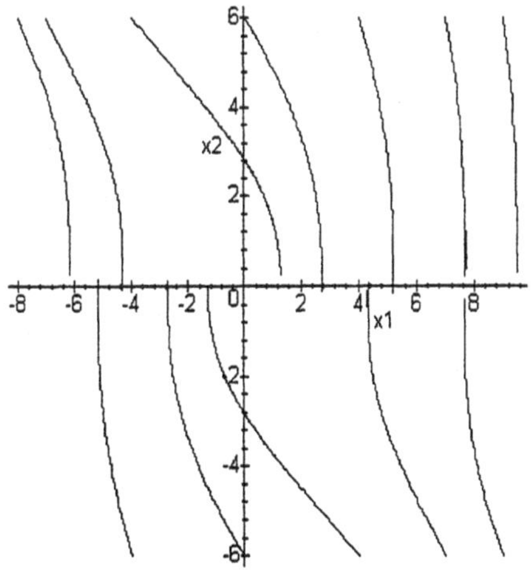

Fig. 2 Controlled Dynamics for System (8) with α=1

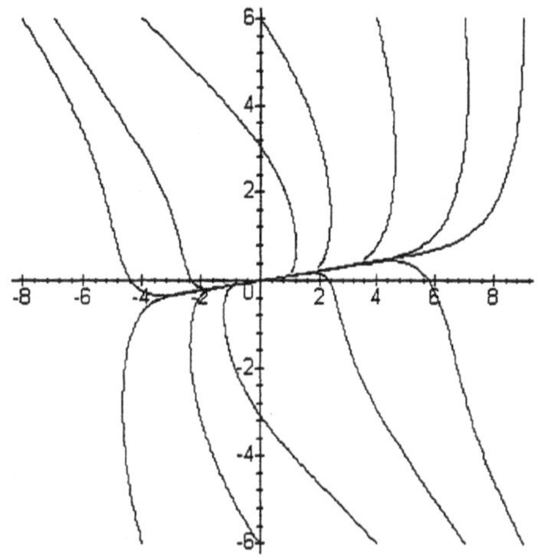

Fig. 3. Dynamics for the Controlled System (10), α=1

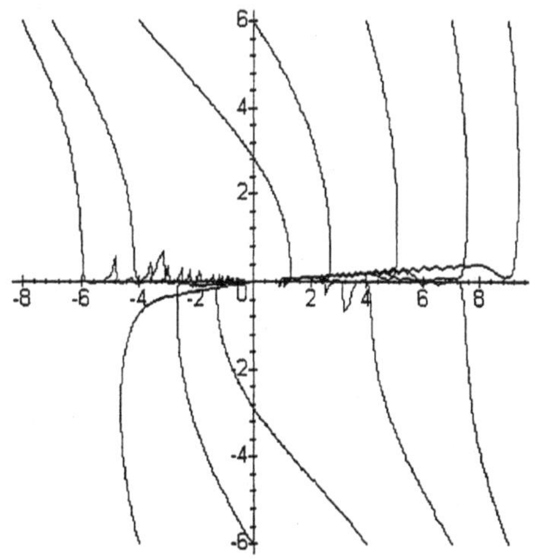

Fig. 4. Dynamics for the Controlled System (11), α=1

and

$$\begin{pmatrix}\dot{x}_1\\ \dot{x}_2\end{pmatrix}=\begin{pmatrix}0 & 1\\ 1 & 1\end{pmatrix}\begin{pmatrix}x_1\\ x_2\end{pmatrix}+u\begin{pmatrix}0.01 & 0\\ 0 & 1\end{pmatrix}\begin{pmatrix}x_1\\ x_2\end{pmatrix} \quad (11)$$

with the respective controls

$$u=-(2x_1x_2+x_2^2+\alpha(x_1^2+x_2^2))/(0.1x_1^2+x_2^2)$$

$$u=-(2x_1x_2+x_2^2+\alpha(x_1^2+x_2^2))/(0.01x_1^2+x_2^2)$$

In figure 4, the crossing trajectories are due to inaccuracies in the numerical computation of the solution and is not present in the actual system.

PSEUDO-LINEAR SYSTEMS

Now consider systems of the form

$$\dot{x}=A(x)x+uB(x)x \quad (12)$$

Lemma 2.1 can be generalized to

Lemma 3.1 *The equation (12) can be written in the form*

$$\dot{z}=\overline{A}(z)z+u\overline{B}(z)z$$

$$\dot{r}=\lambda_A(z)r+u\lambda_B(z)r$$

where

$$z=x/\|x\|,\ r=\|x\|$$

$$\lambda_A(z) = z^T A(rz)z \ , \ \lambda_B(z) = z^T B(rz)z$$
$$\overline{A}(z) = A(rz) - \lambda_A(z)I,$$
$$\overline{B}(z) = B(rz) - \lambda_B(z)I \quad .$$

Generalizing theorem 2.3, we have

Theorem 3.2 *If* $\lambda_B(z) \neq 0$ *for* $z \neq 0$, *then the control*

$$u = -\frac{(\lambda_A(z) + \alpha)}{\lambda_B(z)}$$

will drive the system (12) to the origin and the controlled system is

$$\dot{x} = Ax - \frac{(\lambda'_A(x) + \alpha\|x\|^2)}{\lambda'_B(x)} B(x)x$$

where

$$\lambda'_A(x) = x^T A(x)x \ , \ \lambda'_B(x) = x^T B(x)x \, .$$

Example 3.3 Consider the system

$$\begin{pmatrix} \dot{x}_1 \\ \dot{x}_2 \\ \dot{x}_3 \end{pmatrix} = \begin{pmatrix} 0 & 1 & 0 \\ 0 & 0 & 1 \\ x_2 & 0 & 0 \end{pmatrix} \begin{pmatrix} x_1 \\ x_2 \\ x_3 \end{pmatrix} + u \begin{pmatrix} 0 & 0 & 0 \\ 0 & 0 & 0 \\ 0 & 0 & x_1^2 + x_2^2 \end{pmatrix} \begin{pmatrix} x_1 \\ x_2 \\ x_3 \end{pmatrix}$$

The controlled system is

$$\begin{pmatrix} \dot{x}_1 \\ \dot{x}_2 \\ \dot{x}_3 \end{pmatrix} = \begin{pmatrix} x_2 \\ x_3 \\ c(x) \end{pmatrix} \qquad (13)$$

where

$$c(x) = x_1 x_2 - (x_1 x_2 + x_2 x_3 + x_1 x_2 x_3 + \alpha(x_1^2 + x_2^2 + x_3^2)) / x_3 \, .$$

Again, by a judicious choice of α depending on the initial state, we can obtain regulation of x_3 and even for x_1, x_2 in certain regions (fig. 5).

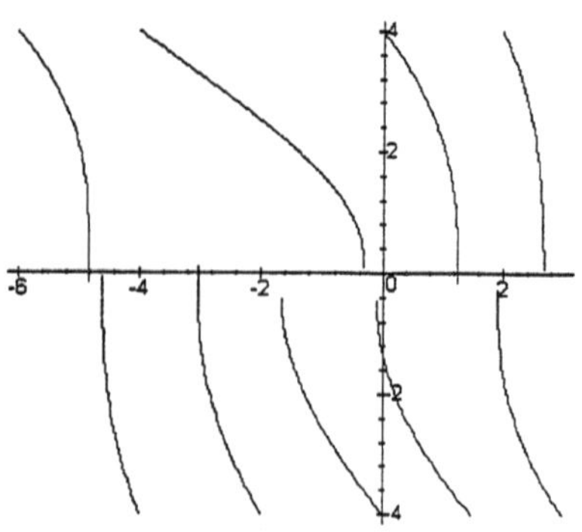

Fig. 5. Controlled Trajectories for (13) in the x_1, x_2 plane.

As before, if we have a component of control in the x_1, x_2 directions, then the control is globally stabilizing; e.g. the system

$$\begin{pmatrix} \dot{x}_1 \\ \dot{x}_2 \\ \dot{x}_3 \end{pmatrix} = \begin{pmatrix} 0 & 1 & 0 \\ 0 & 0 & 1 \\ x_2 & 0 & 0 \end{pmatrix} \begin{pmatrix} x_1 \\ x_2 \\ x_3 \end{pmatrix} + u \begin{pmatrix} 0.1 & 0 & 0 \\ 0 & 0.1 & 0 \\ 0 & 0 & x_1^2 + x_2^2 \end{pmatrix} \begin{pmatrix} x_1 \\ x_2 \\ x_3 \end{pmatrix}$$

with control

$$u = -\frac{(x_1 x_2 + x_2 x_3 + x_1 x_2 x_3 + \alpha(x_1^2 + x_2^2 + x_3^2))}{((x_1^2 + x_2^2)x_3^2 + 0.1x_1^2 + 0.1x_2^2)} \qquad (14)$$

for any $\alpha > 0$ (fig. 6).

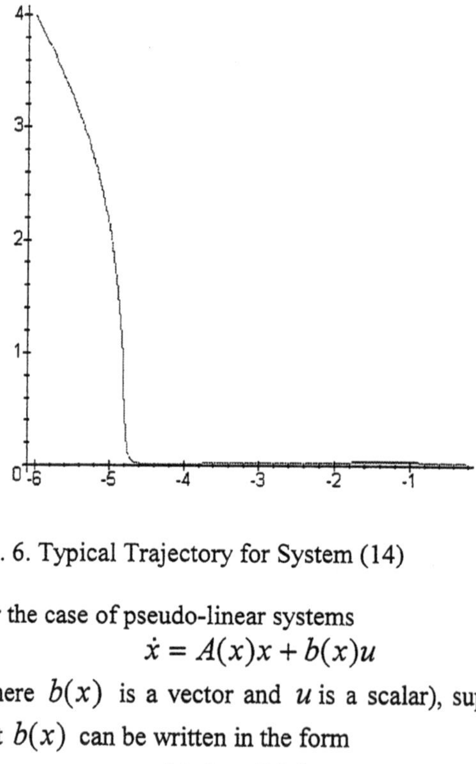

Fig. 6. Typical Trajectory for System (14)

For the case of pseudo-linear systems

$$\dot{x} = A(x)x + b(x)u \qquad (15)$$

(where $b(x)$ is a vector and u is a scalar), suppose that $b(x)$ can be written in the form

$$b(x) = B(x)x.$$

Then from theorem 3.2, we have

Corollary 3.4 *If* $x^T b(x) \neq 0$ *for* $x \neq 0$, *then the control*

$$u = -\frac{(\lambda_A(z) + \alpha)}{z^T b(rz)}$$

will drive the system (15) to the origin.

Example 3.5 Consider the system

$$\begin{pmatrix} \dot{x}_1 \\ \dot{x}_2 \end{pmatrix} = \begin{pmatrix} 0 & 1 \\ \sin x_2 & x_1 \end{pmatrix} \begin{pmatrix} x_1 \\ x_2 \end{pmatrix} + \begin{pmatrix} 0 \\ x_1^2 + 3x_2^4 \end{pmatrix} u$$
$$= \begin{pmatrix} 0 & 1 \\ \sin x_2 & x_1 \end{pmatrix} \begin{pmatrix} x_1 \\ x_2 \end{pmatrix} + u \begin{pmatrix} 0 & 0 \\ x_1 & 3x_2 \end{pmatrix} \begin{pmatrix} x_1 \\ x_2 \end{pmatrix}.$$

Then we have the control

$$u = -\frac{(x_1x_2 + x_1x_2\sin x_2 + x_1x_2^2 + \alpha(\|x\|^2))}{x_2(x_1^2 + 3x_2^4)} \tag{16}$$

Some of trajectories of the controlled system are shown in fig. 7.

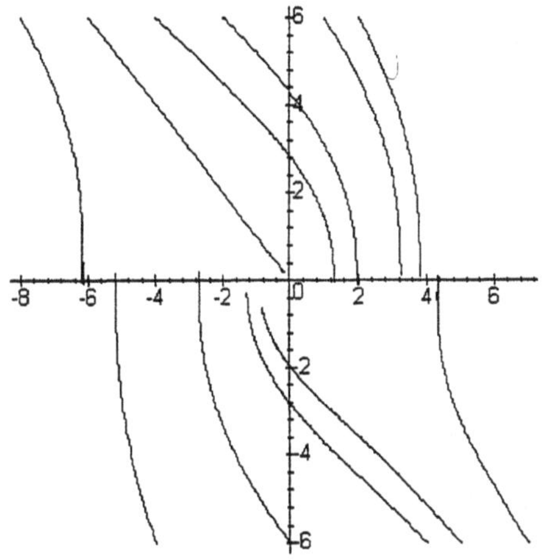

Fig. 7. Controlled Trajectories for System (16)

GENERAL NONLINEAR SYSTEMS

In this section we shall consider the system

$$\dot{x} = f(x,u) \tag{17}$$

for scalar control u and some analytic function f. We can write the system in the form

$$\dot{x} = f(x,u)$$
$$\dot{u} = v$$

i.e.

$$\dot{y} = F(y) + \begin{pmatrix} 0 \\ 1 \end{pmatrix} v. \tag{18}$$

Hence there is no loss of generality in considering the linear-analytic system

$$\dot{x} = f(x) + ug(x). \tag{19}$$

As is well-known, this system can be written in the form of an infinite-dimensional bilinear system in the following way (see Banks, 1992):

Put $\phi_1(x) = x$ and

$$\phi_i(x) = \frac{\partial \phi_{i/2}(x)}{\partial x} \cdot f(x) \stackrel{\Delta}{=} L_f \phi_{i/2}(x)$$

if i even

$$\phi_i(x) = \frac{\partial \phi_{(i-1)/2}(x)}{\partial x} \cdot g(x) \stackrel{\Delta}{=} L_f \phi_{(i-1)/2}(x)$$

if i odd. Then,

$$\dot{\phi}_i = \frac{\partial \phi_i}{\partial x}\dot{x}$$
$$= \frac{\partial \phi_i}{\partial x} f + u \frac{\partial \phi_i}{\partial x} g$$
$$= \phi_{2i}(x) + u\phi_{2i+1}(x)$$

and so, if $\Phi = (\phi_1, \phi_2, \cdots)^T$, we have

$$\dot{\Phi} = A\Phi + uB\Phi$$

where

$$A = (a_{ij}),\ B = (b_{ij})$$

with

$$a_{ij} = \delta_{2i,j} \quad , \quad b_{ij} = \delta_{2i+1,j}$$

i.e.

$$A = \begin{pmatrix} 0 & 1 & 0 & \cdots & \cdots & \cdots & \cdots & \cdots \\ 0 & 0 & 0 & 1 & 0 & \cdots & \cdots & \cdots \\ 0 & 0 & 0 & 0 & 0 & 1 & 0 & \cdots \\ \cdots & \cdots & \cdots & \cdots & \cdots & \cdots & \cdots & \cdots \end{pmatrix},$$

$$B = \begin{pmatrix} 0 & 0 & 1 & 0 & \cdots & \cdots & \cdots & \cdots & \cdots \\ 0 & 0 & 0 & 0 & 1 & 0 & \cdots & \cdots & \cdots \\ 0 & 0 & 0 & 0 & 0 & 0 & 1 & 0 & \cdots \\ \cdots & \cdots & \cdots & \cdots & \cdots & \cdots & \cdots & \cdots & \cdots \end{pmatrix}$$

In a similar way to (Banks, 1992) we can define a map $N_{f,g} : R^n \to \widetilde{S}$, where $\widetilde{S}$ is the set of all (unrestricted) sequences, by

$$N_{f,g}(x) = (L_f^0 x, L_f^1 x, L_g^1 x, L_f^2 x, L_g L_f x, L_f L_g x, L_g^2 x, \cdots)^T$$

where $L_f^i = L_f L_f \cdots L_f$ (i times) and the inner product and norm on $\widetilde{S}$ by

$$\langle \sigma, \tau \rangle_S = \sum_{i=0}^{\infty} \frac{\langle \sigma_i, \tau_i \rangle}{(i!)^2}$$
$$\|\sigma\|_S = (\langle \sigma, \sigma \rangle_S)^{1/2}$$

where $\langle \cdot,\cdot \rangle$ is the usual inner product on R^n and

$$\sigma = (\sigma_0, \sigma_1, \sigma_2, \cdots)^T,$$
$$\tau = (\tau_0, \tau_1, \tau_2, \cdots)^T \quad .$$

Then it can be shown, as in (Banks, 1992), that $S = \{\sigma \in \widetilde{S} : \|\sigma\|_S < \infty\}$ is a Banach space.

As before, put $\psi = \phi / \|\phi\|_S$, $\rho = \|\phi\|_S$ and we then obtain the system

$$\dot{\psi} = \overline{A}(\psi)\psi + u\overline{B}(\psi)\psi$$
$$\dot{\rho} = \lambda_A(\psi)\rho + u_B(\psi)\rho \tag{20}$$

where

$$\lambda_A(\psi) = \langle \psi, A\psi \rangle_S ,\ \lambda_B(\psi) = \langle \psi, B\psi \rangle_S$$
$$\overline{A}(\psi) = A - \lambda_A(\psi)I ,\ \overline{B}(\psi) = A - \lambda_B(\psi)I .$$

The control is then

$$u = -\frac{(\lambda_A(\psi) + \alpha)}{\lambda_B(\psi)}$$

$$= -\frac{(\lambda_A(\phi) + \alpha\|\phi\|^2)}{\lambda_B(\phi)}\quad .$$

Typical trajectories for this control applied to system (16) with only two terms in the Lie series included are shown in fig. 8. These should be compared with those in fig. 7.

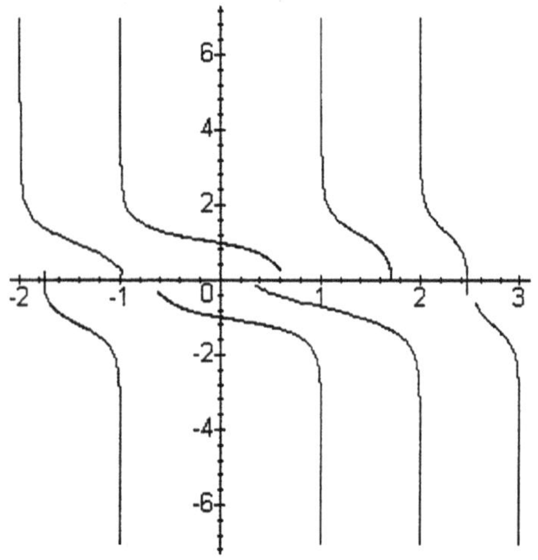

Fig. 8. Lie Series Control for System (16)

The controller described here will be improved with the order of the Lie series. However, a small number of terms will usually be satisfactory, especially in the region where the Lie series converges rapidly.

CONCLUSIONS

In this paper we have studied the control of nonlinear systems via the angular system associated with the given system. The controls are seen to be effective for the controllable part of the system and provide global stabilizing controls when the system is globally stabilizable. The Lie series approach to general nonlinear systems has been shown to be particularly effective. The main drawback with the latter approach is that the computation time is significantly greater than with the other methods.

REFERENCES

Banks, S.P. (1985). On Nonlinear Systems and Algebraic Geometry, *Int. J. Control,* **42,** 333-352.

Banks, S.P. (1988). *Mathematical Theories of Nonlinear Systems,*:Prentice-Hall, London.

Banks, S.P.(1992). Nonlinear Systems, the Lie Series and the Left-Shift Operator: Applications to Nonlinear Optimal Control,*IMA J. Math. Control and Inf.*, **9**, 23-34.

Banks, S.P.(1997). The Schwartz Kernel Theorem and the Frequency-Domain Theory of Nonlinear Systems, *Arch. Cont. Sci.*, **6**(XLII),29-45.

Colonius, F.and Kliemann, W.(1996). The Lyapunov Spectrum of Families of Time-Varying Matrices, *Trans. Amer. Math. Soc.*, **348**, 4389-4407.

Knobloch, H.W., Isidori, A. and Flockerzi, D. (1993) *Topics in Control Theory,* DMV Seminar Band 22,Birkhauser, Berlin.

Luk'yanov, A.G.and Utkin, V.I.. (1981) Methods of Reducing Equations for Dynamic Systems to a Regular Form, Automation and Remote Control, **42**, 413-420.

Utkin, V.I..(1992) *Sliding Modes in Control and Optimization*,Springer-Verlag, Berlin.

14th Triennial World Congress, Beijing, P.R. China

E-2c-14-5

Exponential Stabilization of Uncertain Nonlinear Systems in Chained Form

Y. A. Jiang D. J. Clements T. Hesketh
Email: T.Hesketh@unsw.edu.au

School of Electrical Engineering, University of New South Wales,
Sydney 2052, Australia

Abstract: This paper considers systems in chained form with structured uncertainty. We show that such systems can be robustly and exponentially stabilized.

Keywords: Nonholonomic system, robust control, exponential stability, back-stepping.

1. INTRODUCTION

The control of nonholonomic systems in chained form has been the subject of much research in recent years (Kolmanovsky and McClamroch, 1995; M'Closkey and Murray, 1997). Examples of such systems include a tractor with trailers, a tricycle-type mobile robot and a rigid spacecraft with two torque actuators (Murray and Sastry, 1993; Kolmanovsky and McClamroch, 1995; Samson, 1995; M'Closkey and Murray, 1997). Standard tools of linear and nonlinear control cannot be directly used for the feedback stabilization of chained systems since the controllability of such systems is control input dependent. It is shown in (Brockett, 1983) that it is not possible to stabilize such systems by smooth static-state feedback. Moreover, exponential stability cannot be obtained with any Lipschitz feedback control law (Kolmanovsky and McClamroch, 1995; Sordalen and Egeland, 1995).

A number of approaches to feedback stabilization of chained systems have been proposed. Discontinuous time-invariant control laws are proposed in (Lafferriere and Tsakiris, 1993; Astolfi and Schaufelberger, 1996; Bloch and Drakunov, 1996). Asymptotic stabilization is achieved with sliding mode control in (Bloch and Drakunov, 1996). Piecewise continuous stabilization is used in (Lafferriere and Tsakiris, 1993; Sordalen and Egeland, 1995) to obtain a variant of exponential stability, and controllers, discontinuous at the origin, are used in (Astolfi and Schaufelberger, 1996) to obtain exponential convergence.

Another class of methods uses smooth time-periodic stabilizing control laws (Coron, 1992; Pomet, 1992; Jiang, 1996), however, which can only achieve asymptotic Al stability. For exponential convergence rates, non-smooth control law is necessary (Samson, 1995; M'Closkey and Murray, 1997). Hybrid feedback control laws are also available (Canudas de Wit, Berghuis and Nijmeijer, 1994; Sordalen and Egeland, 1995; Kolmanovsky and McClamroch, 1996). In these methods, a controller switches at discrete time-instants between various continuous-time controllers.

The literature is almost devoid of papers studying robustness in the control of chained systems. The one major exception is (Jiang and Pomet, 1996), which designs an asymptotically stable (not exponentially stable) controller that adapts to a constant, but unknown, parameter. In this paper, we design an exponentially stable controller for a chained system with lower triangular, Lipschitz uncertainty using a combination of back-stepping and scaling.

2. THE SYSTEM

A system in chained form with modeling uncertainties may be written as

$$\begin{aligned} \dot{x}_0 &= u_0 \\ \dot{x} &= u_0 N x + \Delta(x,t)x + bu \end{aligned} \tag{1}$$

where x_0 and x are states and u_0 and u are control inputs, $\Delta(x,t)x$ represents modeling uncertainties, and

$$x = \begin{bmatrix} x_1 \\ \vdots \\ x_n \end{bmatrix} \quad N = \begin{bmatrix} 0 & 1 & & \\ & \ddots & \ddots & \\ & & \ddots & 1 \\ & & & 0 \end{bmatrix} \quad b = \begin{bmatrix} 0 \\ \vdots \\ 0 \\ 1 \end{bmatrix}.$$

Assumption 1 We assume that $\Delta(x,t)$ is a lower triangular matrix and is upper-bounded

$$\|\Delta(x,t)\|_2 \le \gamma$$

with $\gamma > 0$ known.

The objective is to choose control laws for u_0 and u so that the states x_0 and x of the system converge exponentially to the origin in the presence of the modeling uncertainties. Note that the x-subsystem is scaled by the control input u_0 so that this subsystem is uncontrollable when $u_0 = 0$. This is the major difficulty for control design for systems in chained form.

In this paper, a scaling function is used to eliminate the effect of u_0 on the x-subsystem so that exponential stability can be obtained, and a state transformation method is used for control law development and stability analysis. Note that only one chain is considered in this paper. The extension to multi-input chained form systems is straightforward.

3. A SCALED STATE TRANSFORMATION

We introduce a scaled state transformation which together with a feedback control law can transform the system (1) into a special form so that stability analysis can be easily carried out.

First, a closed-loop system matrix in a skew-symmetric form is constructed. This is a special case of backstepping (Kanellakopoulos, *et al.*, 1991; Krstić, *et al.*, 1995).

Lemma 1 *For two sets of positive constants α_i, $i = 1, \cdots, n$ and β_i, $i = 1, \cdots, n-1$, there exist a nonsingular lower triangular matrix T_ξ and a row vector K_ξ such that*

$$A_\xi = T_\xi(N - bK_\xi)T_\xi = \begin{bmatrix} -\alpha_1 & \beta_1 & 0 & \cdots & 0 \\ -\beta_1 & -\alpha_2 & \beta_2 & \ddots & \vdots \\ 0 & -\beta_2 & \ddots & \ddots & 0 \\ \vdots & \ddots & \ddots & \ddots & \beta_{n-1} \\ 0 & \cdots & 0 & -\beta_{n-1} & -\alpha_n \end{bmatrix}. \tag{2}$$

Moreover, T_ξ^{-1} and K_ξ are defined by

$$T_\xi^{-1} = \begin{bmatrix} e_1^T \\ e_1^T A_\xi \\ \vdots \\ e_1^T A_\xi^{n-1} \end{bmatrix}, \quad e_1 = \begin{bmatrix} 1 \\ 0 \\ \vdots \\ 0 \end{bmatrix} \tag{3}$$

and

$$K_\xi = e_1^T A_\xi^n T_\xi. \tag{4}$$

Proof: Since $\beta_i > 0$, it is clear from the structure of A_ξ that T_ξ^{-1} is a lower triangular non-singular matrix. Next, we note from the structure of N and the definition of T_ξ^{-1} that

$$T_\xi^{-1}A_\xi = \begin{bmatrix} e_1^T A_\xi \\ \vdots \\ e_1^T A_\xi^{n-1} \\ e_1^T A_\xi^n \end{bmatrix} \quad \text{and} \quad NT_\xi^{-1} = \begin{bmatrix} e_1^T A_\xi \\ \vdots \\ e_1^T A_\xi^{n-1} \\ 0 \end{bmatrix}.$$

It follows that

$$T_\xi^{-1}A_\xi - NT_\xi^{-1} = be_1^T A_\xi^n = bK_\xi T_\xi^{-1}$$

which implies

$$A_\xi = T_\xi(N - bK_\xi)T_\xi^{-1}.$$

□□□

Since A_ξ is skew-symmetric, we have

$$A_\xi + A_\xi^T = -2\begin{bmatrix} \alpha_1 & & \\ & \ddots & \\ & & \alpha_n \end{bmatrix} < 0.$$

If $u_0 > 0$ and is constant, a stabilizing control law can be given as $u = -K_\xi x$ with sufficiently large α_i. However, since $u_0(t)$ needs to converge to zero, some scaling is required to obtain exponential stability of $x(t)$. For a scaling function $\lambda(t) \ne 0$, we introduce

$$D = \begin{bmatrix} 1 & & & \\ & \lambda^{-1} & & \\ & & \ddots & \\ & & & \lambda^{1-n} \end{bmatrix}. \tag{5}$$

Lemma 2 *Consider the system (1) with the control law*

$$u = -u_0\lambda^n K_\xi Dx. \tag{6}$$

Then, with the state transformation

$$\xi = T_\xi Dx, \tag{7}$$

the closed-loop system becomes

$$\begin{aligned} \dot{x}_0 &= u_0 \\ \dot{\xi} &= u_0\lambda A_\xi\xi - \tfrac{\dot{\lambda}}{\lambda}T_\xi M T_\xi^{-1}\xi + \Delta_\xi\xi \end{aligned} \tag{8}$$

where

$$M = \operatorname{diag}\{0\ 1\ \cdots\ n-1\}, \qquad \Delta_\xi = T_\xi D\Delta D^{-1}T_\xi^{-1}.$$

Proof: First, consider the state transformation $z = Dx$. It follows from (1) that

$$\begin{aligned} \dot{z} &= D\dot{x} + \dot{D}x \\ &= u_0 DND^{-1}z + \lambda^{1-n}bu + D\Delta D^{-1}z + \dot{D}D^{-1}z. \end{aligned}$$

From the structure of D, we have

$$DND^{-1} = \lambda N \quad \text{and} \quad \dot{D}D^{-1} = -\frac{\dot{\lambda}}{\lambda}MD,$$

and so

$$\dot{z} = u_0\lambda Nz + \lambda^{1-n}bu - \frac{\dot{\lambda}}{\lambda}Mz + D\Delta D^{-1}z.$$

Now define a second transformation

$$\xi = T_\xi z = T_\xi Dx.$$

It follows from (4) and Lemma 1 that

$$\begin{aligned} \dot{\xi} &= T_\xi \dot{z} \\ &= u_0\lambda T_\xi NT_\xi^{-1} - u_0\lambda bK_\xi T_\xi^{-1} - \frac{\dot{\lambda}}{\lambda}T_\xi MT_\xi^{-1}\xi + \Delta_\xi\xi \\ &= u_0\lambda A_\xi\xi - \frac{\dot{\lambda}}{\lambda}T_\xi MT_\xi^{-1}\xi + \Delta_\xi\xi. \end{aligned}$$

□□□

From (8), it is clearly necessary to choose λ so that $u_0\lambda > 0$. The transformed modeling uncertainty Δ_ξ is now a function of λ. However, it can be shown that an upper-bound (which is independent of λ) on the induced 2-norm of Δ_ξ is available.

Lemma 3 *Consider the scaling matrix D defined in (5). For any lower triangular matrix Δ and any $|\lambda| \geq 1$, we have*

$$\|D\Delta D^{-1}\|_2 \leq \sqrt{n}\|\Delta\|_2. \tag{9}$$

Proof: See the Appendix.

□□□

4. CONTROLLER DESIGN

We first assume that $x_0(0) \neq 0$ and choose the control law as

$$u_0 = -k_0x_0 \qquad u = u_0\lambda^n K_\xi Dz(t) \tag{10}$$

with scaling

$$\lambda = \frac{k_1 + |u_0|}{u_0}.$$

Note that as u_0 (or equivalently x_0) tends to zero, λ tends to infinity. We will show that the components x_i of the state x also tend to zero in such a way that the control signal u is bounded and also tends to zero.

Theorem 1 *Consider the system (1) with $x_0(0) \neq 0$ and control law (10). Suppose that the design parameters k_0 and k_1 are chosen according to*

$$2k_1a - k_0b - \sqrt{n}c - k_0(n-1) > 0 \tag{11}$$

where

$$a = \min_i\{\alpha_i\}, \quad b = \|T_\xi MT_\xi^{-1} - M\|_2, \quad c = \|T_\xi\|_2\|T_\xi^{-1}\|_2\gamma. \tag{12}$$

Then the states x_0 and x, and the control signals u_0 and u, all converge to zero exponentially.

Proof: For simplicity, the proof assumes $x_0(0) < 0$, but holds equally well for $x_0(0) > 0$. So, we immediately have $u_0 = -k_0x_0 > 0$ which implies that the scaling function λ is smooth and has the form

$$\lambda = 1 + \frac{k_1}{u_0}.$$

Since

$$\dot{u}_0 = -k_0\dot{x}_0 = -k_0u_0,$$

we also have

$$\frac{\dot{\lambda}}{\lambda} = \frac{k_0k_1}{k_1 + u_0} \leq k_0.$$

Note that since $u_0 > 0$, D is well defined and thus it follows from Lemma 2 and Lemma 3 that with the control law (10), the system (1) can be written as

$$\dot{x}_0 = -k_0x_0$$

and

$$\dot{\xi} = (k_1 + u_0)A_\xi\xi - \frac{k_0k_1}{k_1 + u_0}T_\xi MT_\xi^{-1}\xi + \Delta_\xi\xi. \tag{13}$$

The first equation implies that x_0 decays to zero exponentially, i.e.,

$$x_0(t) = x_0(0)e^{-k_0t}.$$

For the ξ-subsystem, consider a Lyapunov function

$$V = \frac{1}{2}\xi^\top\xi.$$

Then, it follows from Lemma 3 and (12) that

$$\begin{aligned} \dot{V} &= -2(k_1 + u_0)\sum_{i=1}^n \alpha_i\xi_i^2 - \frac{k_0k_1}{k_1 + u_0}\xi^\top T_\xi MT_\xi^{-1}\xi + \xi^\top\Delta_\xi\xi \\ &\leq -(2k_1a - k_0b - \sqrt{n}c)\xi^\top\xi \\ &= -\rho\xi^\top\xi \\ &= -2\rho V \end{aligned}$$

where

$$\rho = 2k_1a - k_0b - \sqrt{n}c > k_0(n-1) > 0$$

by assumption. This implies that

$$V(t) \leq V(0)e^{-2\rho t}$$

and thus

$$\|\xi(t)\|_2 \leq \|\xi(0)\|_2 e^{-\rho t}.$$

We now need to obtain a similar exponentially decaying estimate for the original coordinates x. First, note from (7) that

$$x = D^{-1}T_\xi^{-1}\xi.$$

Let t_i^T be the ith row of the matrix T_ξ^{-1}. Then, it follows from the structure of D that

$$x_i = \lambda^{i-1}t_i^\mathsf{T}\xi.$$

Since

$$\lambda(t) = 1 + \frac{k_1}{u_0} = 1 + \frac{k_1}{u_0(0)}e^{k_0 t} < \lambda(0)e^{k_0 t}$$

we have

$$\begin{array}{rcl} |x_i(t)| & \leq & \lambda(0)^{i-1}e^{k_0(i-1)t}|t_i^\mathsf{T}\xi(0)|e^{-\rho t} \\ & \leq & B_i(|x_0(0)|)\|x(0)\|_2 e^{-\hat{\alpha}_i t} \end{array} \quad (14)$$

where

$$\begin{array}{rcl} B_i(|x_0(0)|) & = & \|t_i\|\|T_\xi\|\left|1+\frac{k_1}{|k_0 x_0(0)|}\right|^{i-1} \\ \hat{\alpha}_i & = & \rho - k_0(i-1) \\ & = & 2k_1 a - k_0 b - \sqrt{n}c - k_0(i-1). \end{array} \quad (15)$$

Observe that each bounding function B_i is monotonic decreasing but converges to infinity as $|x_0(0)|$ approaches zero. Also note from (11) that

$$\rho = \hat{\alpha}_1 > \cdots > \hat{\alpha}_n > 0.$$

We conclude that $x(t)$ is exponentially stable in any region for which $|x_0(0)|$ is bounded away from zero in the presence of the uncertainty Δx.

Finally, we show that the control signal u is bounded and converges to zero exponentially for $x_0(0) \neq 0$. From (6) and (10), we have that

$$u = u_0\lambda^n K_\xi Dx = u_0\sum_{i=1}^{n} K_i\lambda^{n-i}x_i$$

where K_i is the ith element of the row vector K_ξ. It easily follows that

$$|u(t)| \leq M|u_0(0)|u_e^{-\alpha_n t}$$

where

$$M = \|x(0)\|\sum_{i=1}^{n}|K_i|\|t_i\|_2.$$

This completes the proof. □□□

Remark 1 The diagonal structure of D and M and the lower triangular structure of T_ξ imply that the diagonal part of $T_\xi MT_\xi^{-1}$ is M which then implies that $T_\xi MT_\xi^{-1} - M$ is strictly lower triangular and

$$\|T_\xi MT_\xi^{-1} - M\|_2 < \|T_\xi MT_\xi^{-1}\|_2.$$

Remark 2 The stability condition (11) is conservative. It can be seen from (13) that, for a given system, a smaller k_1 than the one given by (11) is possible.

Remark 3 From (15), we have

$$\hat{\alpha}_i \geq \hat{\alpha}_n + k_0(n-i).$$

It follows from (14) that the states $x_i(t)$ converge faster than $x_0(t)$.

Remark 4 If the scaling matrix D is chosen as

$$D = \begin{bmatrix} \lambda^{1-n} & & & \\ & \lambda^{2-n} & & \\ & & \ddots & \\ & & & 1 \end{bmatrix}, \quad (16)$$

then with the control law

$$u = -\frac{u_0}{\lambda}K_\xi Dx \quad (17)$$

and the state transformation (7), the closed-loop system can be written as

$$\begin{array}{rcl} \dot{x}_0 & = & u_0 \\ \dot{\xi} & = & \frac{u_0}{\lambda}A_\xi\xi - \frac{\dot{\lambda}}{\lambda}T_\xi MT_\xi^{-1}\xi + \Delta_\xi\xi \end{array} \quad (18)$$

Now, in the control law (10), if the scaling function is chosen as

$$\lambda = -x_0,$$

then, the ξ-subsystem can be written as

$$\dot{\xi} = k_0 A_\xi\xi + k_o T_\xi MT_\xi^{-1}\xi + \Delta_\xi\xi. \quad (19)$$

When $\Delta_\xi = 0$, we have the control law given in (Astolfi and Schaufelberger, 1996).

Note that the theorem only proves exponential stability for initial condition values of $x_0(0)$ bounded away from zero. If $x_0(0)$ is small, then it is possible to drive x_0 away from zero initially so that the theorem then applies. Since the modeling uncertainty Δx is Lipschitz, the dynamics of the system (1) in a sufficiently small interval $0 \leq t < t_o$ can be effectively ignored. A starting control signal can be used to obtain a suitable $x_0(t_o)$.

Remark 5 A possible control law, suitable for any initial value of $x_0(0)$, can be chosen as

$$\begin{aligned} u_0(t) &= -k_0 x_0 + c_0 \text{sign}(x_0) g(t) H(t) \\ u(t) &= u_0 \lambda^n(u_0) K_\xi D(\lambda) x(t)(1 - H(t)) \\ \lambda(u_0) &= \text{sign}(u_0) + \tfrac{k_1}{u_0} \end{aligned} \quad (20)$$

where

$$H(t) = \begin{cases} 1, & t < t_0 \\ 0, & t \geq t_0 \end{cases} \quad (21)$$

is the unit pulse of length t_0. The function $g(t)$ is any non-negative, smooth function with $g(0) = 0$.

In the control law, t_0 is a sufficiently small start-up time, c_0 is chosen to obtain a suitable initial value of $x_0(t_0)$ for the trade-off between the control efforts of u_0 and u, and the smooth function $g(t)$ is used to reduce the impact of the impulse function $H(t)$ on the system.

5. EXAMPLE

Consider the 3rd-order chained system

$$\begin{aligned} \dot{x}_0 &= u_0 \\ \dot{x}_1 &= \delta(t) x_1 + u_0 x_2 \\ \dot{x}_2 &= u_0 x_3 \\ \dot{x}_3 &= u \end{aligned} \quad (22)$$

where $|\delta| \leq 1$ is some bounded perturbation. For the design of A_ξ we let $\alpha_{1,2,3} = \beta_{1,2} = 1$. We then calculate T_ξ and T_ξ^{-1} to be

$$T_\xi = \begin{bmatrix} 1 & 0 & 0 \\ 1 & 1 & 0 \\ 2 & 2 & 1 \end{bmatrix}, \quad T_\xi^{-1} = \begin{bmatrix} 1 & 0 & 0 \\ -1 & 1 & 0 \\ 0 & -2 & 1 \end{bmatrix}$$

and we calculate K_ξ to be

$$K_\xi = [\ 3 \quad 5 \quad 3\ ].$$

To choose k_0 and k_1, we first calculate

$$T_\xi M T_\xi^{-1} = \begin{bmatrix} 0 & 0 & 0 \\ -1 & 1 & 0 \\ -2 & -2 & 2 \end{bmatrix},$$

and

$$\Delta_\xi = T_\xi D \Delta D^{-1} T_\xi^{-1} = \delta(t) \begin{bmatrix} 1 & 0 & 0 \\ 1 & 0 & 0 \\ 2 & 0 & 0 \end{bmatrix}.$$

We note that $\|T_\xi M T_\xi^{-1} - M\|_2 \leq 3$. The control signal u_0 is given by (20) with

$$k_0 = 0.3, \quad c_0 = 60, \quad t_0 = 0.05, \quad g(t) = \sin(10\pi t),$$

and the control signal u is given by (21) with $k_1 = 2.5$ in view of Remark 2. The initial values for simulation are given as $x_0(0) = 0$ and $x(0) = [\ 1 \quad 0.1 \quad -0.1]\ ]^{\mathsf{T}}$.

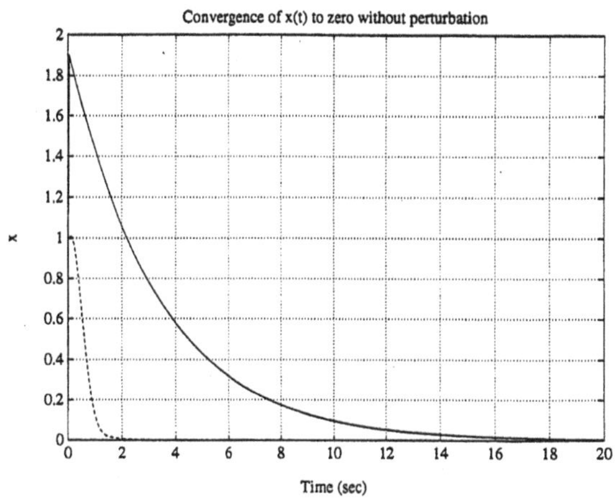

Figure 1: Solid: $x_0(t)$, Dashed: $x_1(0)$ with $\delta = 0$.

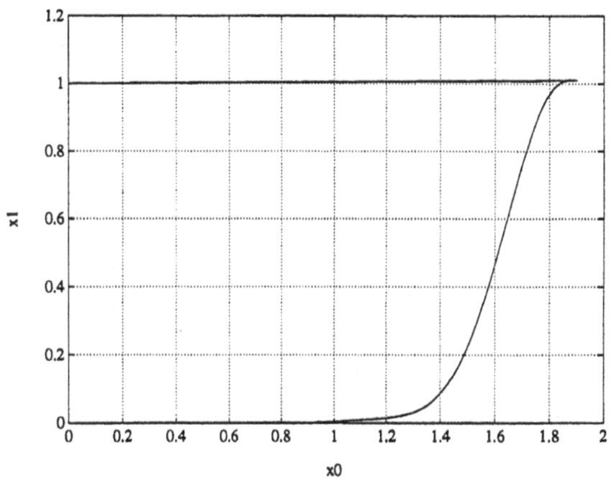

Figure 2: Trajectory of $x_1(t)$ vs. $x_0(t)$ with $\delta = 0$.

With this control law, the simulation results for $\delta = 0$ are shown in Figures 1-3. Figure 1 shows that both $x_0(t)$ and $x_1(t)$ converge to zero. Note, however, that $x_0(0)$ has a slower rate of convergence than $x_1(t)$. This coincides with the relation given in Remark 3. The trajectory of x_1 versus x_0 is shown in Figure 2 which may be interpreted as a parallel parking trajectory (Kolmanovsky and McClamroch, 1995).

Now, to show the robustness, the same control law is used for the case $\delta(t) = 1$ and the simulation results are shown in Figures 3.

APPENDIX

Proof of Lemma 3: Let $\Delta = [\delta_{i,j}]$ and let $\hat{\Delta} = D \Delta D^{-1} = [\hat{\delta}_{i,j}]$. It follows from the structure of D that

$$\hat{\delta}_{i,j} = \lambda^{j-i} \delta_{i,j}.$$

Since Δ is lower triangular, we have $\delta_{i,j} = 0$ for $j > i$,

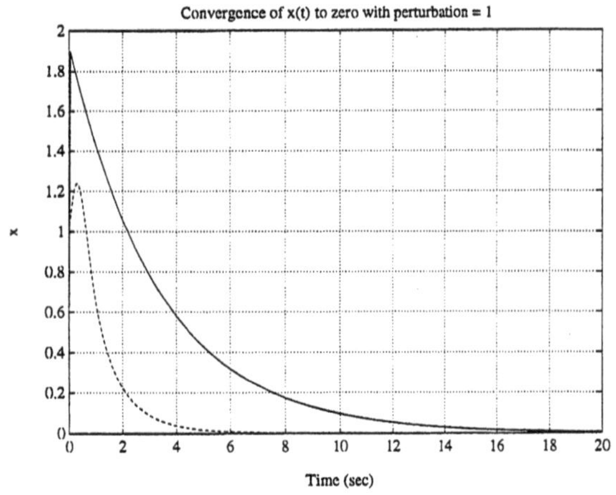

Figure 3: Solid: $x_0(t)$, Dashed: $x_1(0)$ with $\delta = 1$.

and since $\lambda \geq 1$, it follows that

$$|\hat{\delta}_{i,j}| \leq |\delta_{i,j}| \quad \text{and} \quad \mathrm{tr}(\hat{\Delta}\hat{\Delta}^{\mathsf{T}}) \leq \mathrm{tr}(\Delta\Delta^{\mathsf{T}}).$$

Next, by definition, the induced 2-norm of the matrix Δ is the largest singular value of the matrix:

$$\|\Delta\|_2^2 = \sigma_{\max}^2(\Delta) = \lambda_{\max}(\Delta\Delta^{\mathsf{T}}).$$

The following sequence of relationships then follows:

$$\begin{aligned}\lambda_{\max}(\hat{\Delta}\hat{\Delta}^{\mathsf{T}}) &\leq \sum_{i=1}^{n} \lambda_i(\hat{\Delta}\hat{\Delta}^{\mathsf{T}}) = \mathrm{tr}(\hat{\Delta}\hat{\Delta}^{\mathsf{T}}) \\ &\leq \mathrm{tr}(\Delta\Delta^{\mathsf{T}}) = \sum_{i=1}^{n} \lambda_i(\Delta\Delta^{\mathsf{T}}) \leq n\|\Delta\|_2^2.\end{aligned}$$

□□□

REFERENCES

Astolfi, A. and W. Schaufelberger, "State and output feedback stabilization of multiple chained systems with discontinuous control", *Proc. 35th IEEE Conf. Decision Control*, Kobe, Japan, pp. 1443–1447, 1996.

Bloch, A. and S. Drakunov, "Stabilization and tracking in the nonholonomic integrator via sliding modes", *Systems & Control Letters*, Vol. 29, pp. 91–99, 1996.

Brockett, R. W., "Asymptotic stability and feedback stabilization", in: R. W. Brockett, R. S. Millman and H. J. Sussmann, eds., *Differential Geometric Control Theory*, pp. 181–191, 1983.

Coron, J-M., "Global asymptotic stabilization for controllable nonlinear systems without drift", *Math. Control Systems Signals*, Vol. 5, pp. 295–312, 1992.

Canudas de Wit, C., H. Berghuis and H. Nijmeijer, "Practical stabilization of nonlinear systems in chained form" *Proc. 33rd IEEE Conf. Decision Control*, pp. 3475–3480, Orlando, 1994.

Jiang, Z. P., "Iterative design of time-varying stabilizers for multi-input systems in chained form", *Systems & Control Letters*, Vol. 28, pp. 255–262, 1996.

Jiang, Z. P., and J-B. Pomet, "Global stabilization of parametric chained-form systems by time-varying dynamic feedback", *Int. J. Adaptive Contr. Signal Processing*, Vol. 10, pp. 47–59, 1996.

Kanellakopoulos, I., P. V. Kokotovic, and A. S. Morse, "Systematic design of adaptive controllers for feedback linearizable systems", *IEEE Trans. Automat. Control*, Vol. 36, pp. 1241–1253, 1991.

Kolmanovsky, I. and N. H. McClamroch, "Developments in nonholonomic control problems", *IEEE Control Systems*, Vol. 15, No. 6, pp. 20–36, 1995.

Kolmanovsky, I. and N. H. McClamroch, "Hybrid feedback laws for a class of cascade nonlinear control systems", *IEEE Trans. Automat. Control*, Vol. 41, pp. 1271–1282, 1996.

Krstić, M., I. Kanellakopoulos, and P. V. Kokotović, *Nonlinear and Adaptive Control Design*, New York, NY: John Wiley & Sons, 1995.

Lafferriere, G. A. and D. P. Tsakiris, "Remarks on control Lyapunov functions for discontinuous stabilizing feedback", *Proc. 32nd IEEE Conf. Decision Control*, pp. 2398–2403, 1993.

M'Closkey, R. and M. Murray, "Exponential stabilization of drift-less nonlinear control systems using homogeneous feedback", *IEEE Trans. Automat. Control*, Vol. 42, pp. 614–628, 1997.

Murray, R. M. and S. Sastry, "Nonholonomic motion planning: steering using sinusoids", *IEEE Trans. Automat. Control*, Vol. 38, pp. 700–716, 1993.

Pomet, J-P., "Explicit design of time-varying stabilizing control law for a class of controllable system without drift", *Systems & Control Letters*, Vol. 18, pp. 147–158, 1992.

Samson, C., "Control of chained systems application to path following and time-varying point-stabilization of mobile robots", *IEEE Trans. Automat. Control*, Vol. 40, pp. 64–77, 1995.

Sordalen, O. J. and O. Egeland, "Exponential stabilization of nonholonomic chained systems", *IEEE Trans. Automat. Control*, Vol. 40, pp. 35–49, 1995.

USING SMALL FEEDBACK TO STABILIZE A WIDER CLASS OF FEEDFORWARD SYSTEMS [1]

Chunjiang Qian and **Wei Lin**

Department of Electrical Engineering and Computer Science
Case Western Reserve University, Cleveland, OH 44106, U.S.A.
Phone:+1-216-368-4493 Fax:+1-216-368-3123
E-mail: linwei@nonlinear.cwru.edu

Abstract. This paper presents a further improvement of the previous work (Mazenc, 1997; Lin and Qian, 1998). By a delicate analysis and an appropriate separation of the vector field, we show how a novel synthesis approach that combines the forwarding design (Mazenc and Praly, 1996; Mazenc, 1997) and the passivity-based small control technique (Lin, 1995 and 1996) can be used to solve the problem of global stabilization, for a significant class of feedforward systems that cannot be adequately controlled using existing methods. *Copyright © 1999 IFAC*

Keywords. Feedforward system, smooth state feedback, small controls, global asymptotic stabilization, Lyapunov-based recursive design.

1. INTRODUCTION

Since the initial work (Teel, 1992), the problem of global stabilization of feedforward systems has received considerable attention from the control systems society. In the past few years, a number of synthesis approaches have been developed towards the systematic design of globally stabilizing feedback controllers for various feedforward systems. Among them, the nested saturation design (Teel, 1992; R. Sepulchre, 1997; Lin and Li, 1998) and Lyapunov-based forwarding design (Mazenc and Praly, 1996; Mazenc, 1997; Jankovic et al., 1996; Lin and Qian, 1998) are two most important and powerful methods. The significance of these design methodologies lies in that they provide not only a systematic procedure for the design of a globally stabilizing feedback control law but also an elegant way to take advantage of genuinely nonlinear effects of the system in the feedback design.

In the recent paper (Lin and Qian, 1998), we studied an upper-triangular system of the form

$$
\begin{aligned}
\dot{x}_n &= f_{n-1}(y_{n-1}) + \phi_n(y_n; u) \\
\dot{x}_{n-1} &= f_{n-2}(y_{n-2}) + \phi_{n-1}(y_n; u) \\
&\vdots \\
\dot{x}_2 &= f_1(y_1) + \phi_2(y_n; u) \\
\dot{x}_1 &= u + \phi_1(y_n; u)
\end{aligned}
\tag{1}
$$

where the control input $u \in R$, $y_i = (x_i, \cdots, x_1)^T$, the functions f_i and ϕ_i, $i = 1, \cdots, n$, are smooth. It was showed in (Lin and Qian, 1998) that (1) is globally asymptotically stabilizable at $(x_1, \cdots, x_n) = 0$ by the small smooth state feedback control law

$$u = -\alpha_1(y_1)f_1(y_1) - \cdots - \alpha_n(y_n)f_n(y_n), \tag{2}$$

if the following conditions are fulfilled:

(H1) $f_i(x_i, \cdots, x_1) = x_i^{p_i} g_i(y_i), \quad with \quad g_i(y_i) > 0;$

(H2) $|\phi_i(y_n; u)| \leq (1+|x_i|)(F_{i-1}^T(y_{i-1})y_{i-1}+u^2)\gamma_i(\left\| \begin{matrix} y_{i-1} \\ u \end{matrix} \right\|)$ $i = 2, \cdots, n,$ $|\phi_1(y_n; u)| \leq u^2\gamma_1(\left\| \begin{matrix} x_1 \\ u \end{matrix} \right\|)$, where $F_i^T(y_i) = (f_i(y_i), \cdots, f_1(y_1));$

(H3) $\{p_i\}$ is a non-decreasing sequence of odd positive integers.

This result generalized the work of Mazenc (1997) on global stabilization of feedforward systems having a *nonlinear* chain of integrators in several aspects. Among these aspects are the relaxation of the restriction on the *power* chain integrators and the development of a set of *weaker* growth conditions for a larger class of feedforward systems to be globally stabilizable by smooth state feedback. Using the stabilization theorem established in (Lin and Qian, 1998), it is possible to prove,

[1] This research is supported in part by NSF under Grant DMS-9634395, and by the Wareen E. Rupp Endowed Professorship.

for instance, that the upper-triangular system

$$\dot{x}_3 = x_2^3 + x_1 x_2^2 + (x_1^2 + u^2)\ln(1 + x_3^2)$$
$$\dot{x}_2 = x_1$$
$$\dot{x}_1 = u \tag{3}$$

does not satisfy the conditions in (Mazenc, 1997) but the hypotheses (H1)–(H3), and (3) is therefore stabilizable in the large by small controls.

In this paper we continue to pursue a line of work started in (Mazenc and Praly, 1996; Mazenc, 1997; Lin and Qian, 1998), and further refine the results presented in (Lin and Qian, 1998). This is accomplished by significantly relaxing the sufficient conditions in (Lin and Qian, 1998) so that a wider class of upper-triangular nonlinear systems can still be globally stabilizable via small state feedback. A bottom-up, recursive design procedure is presented for the construction of a small state feedback law. The key ideas behind our design are to fully exploit the physical structure of the upper-triangular system, and to take advantage of the composite Lyapunov function with *nonlinear weights* proposed in (Mazenc and Praly, 1996; Mazenc, 1997). The proposed approach allows us to make an appropriate separation and treatment of the nonlinear vector fields, thus being able to develop a novel small feedback design strategy and establish stronger global stabilization theorems for rather general upper-triangular non-affine systems. Unlike our previous work (Lin and Qian, 1998) where some restrictive conditions are imposed on the nonlinear functions $f_i(y_i)$ (e.g. (H1)), we show here that they are not necessary for achieving global stabilization of the system (1) and (H1) can be weakened. Indeed, (H1) is replaced by some mild conditions in this paper. As we shall see, the relaxation of the restrictions on the nonlinear functions $f_i(y_i)$ significantly enlarges the class of feedforward systems that can be stabilized by small state feedback. Moreover, the new structural conditions proposed in this work (see (A1) or (B1)–(B2)) can be easily verified.

The paper is organized as follows: Section 2 introduces a set of assumptions and some useful technical lemmas to be used in the sequel. Section 3 contains the main result of this paper and its proof. Further extensions are discussed in Section 4. Section 5 gives three illustrative examples. Concluding remarks are drawn in Section 6.

2. ASSUMPTIONS AND TECHNICAL LEMMAS

To solve the problem of global stabilization of (1), we introduce a set of sufficient conditions which are essential for achieving a successful design. The conditions below are a refinement of those given in (Mazenc, 1997) and (Lin and Qian, 1998).

(A1) $D^k f_i(0) = 0, k = 1, \cdots, p-1, \ \frac{\partial^p f_i}{\partial x_i^p}(0) \neq 0, \forall i = 1, \cdots, n-1,$

(A2) $|\phi_i(y_n; u)| \leq (1 + |x_i|)(\|y_{i-1}\|^{p+1} + u^2)\gamma_i\left(\left\| \begin{matrix} y_{i-1} \\ u \end{matrix} \right\|\right),$

$$|\phi_1(y_n; u)| \leq u^2 \gamma_1\left(\left\| \begin{matrix} x_1 \\ u \end{matrix} \right\|\right), \ \forall i = 2, \cdots, n,$$

(A3) p is an odd positive integer.

Under assumption A1, $f_i(y_i)$ can be decomposed as

$$f_i(y_i) = \hat{f}_i(y_i) + \bar{f}_i(y_i) \tag{4}$$

where $\hat{f}_i(y_i)$ is a $p-th$ order polynomial of y_i with the coefficient of x_i^p being $\frac{\partial^p f_i}{\partial x_i^p}(0) \neq 0$, i.e.

$$\hat{f}_i(y_i) = \sum_{r_{i,i}+\cdots+r_{i,1}=p} \left. \frac{\partial^p f_i(y_i)}{\partial x_i^{r_{i,i}} \cdots \partial x_1^{r_{i,1}}} \right|_{y_i=0} x_i^{r_{i,i}} \cdots x_1^{r_{i,1}},$$

$$i = 1, \cdots, n-1, \qquad \hat{f}_n(y_n) := x_n^p,$$

and $\bar{f}_i(y_i)$ represents the higher order terms ($\geq p+1$) of $f_i(y_i)$. Without loss of generality, we suppose

$$\frac{\partial^p f_i}{\partial x_i^p}(0) = 1 \qquad i = 1, \cdots, n-1.$$

Hence, $\hat{f}_1(x_1) = x_1^p$. It is important to observe that $\hat{f}_i(y_i)$ has the following nice property which plays a crucial role in the proof of the main result of the paper.

Lemma 2.1 Assume that A1 and A3 hold. Then for $i = 2, \cdots, n$, there exist positive constants ρ_i, $\lambda < 1$ and $\mu < 1$, such that

$$\theta|\hat{f}_i(y_i)|\|y_{i-1}\| \leq \left(\theta^\lambda \|y_{i-1}\|^{p+1} + \theta^{\mu+1} x_i^{p+1}\right)\rho_i, \tag{5}$$

where $\theta(\cdot) < 1$ is any *nonnegative function or constant,* and $\hat{f}_n(y_n) := x_n^p$.

Proof By the Young's inequality, we have

$$\begin{aligned}
\theta|\hat{f}_i(y_i)|\|y_{i-1}\| &\leq \theta\Big(\sum_{r_{i,i}+\cdots+r_{i,1}=p} |a_{r_{i,i}\cdots r_{i,1}} x_i^{r_{i,i}} \cdots x_1^{r_{i,1}}|\Big)\|y_{i-1}\| \\
&\leq \theta(b_0|x_i|^p + \sum_{j=1}^{i-1} b_j|x_j|^p)\|y_{i-1}\| \\
&\leq c_2\theta^\lambda\|y_{i-1}\|^{p+1} + \theta^{\mu+1}x_i^{p+1} + c_1\theta\|y_{i-1}\|^{p+1} \\
&\leq \rho_i\left(\theta^\lambda\|y_{i-1}\|^{p+1} + \theta^{\mu+1}x_i^{p+1}\right),
\end{aligned}$$

where $a_{r_{i,i}\cdots r_{i,1}} = \frac{\partial^p f_i}{\partial x_i^{r_{i,i}} \cdots \partial x_1^{r_{i,1}}}(0)$, c_1, c_2, b_j are positive constants, and we can choose $\lambda = \frac{1}{2}$, $\mu = \frac{1}{2p}$.

The following Lemma and its proof can be found in (Lin and Qian, 1998).

Lemma 2.2 Let $g(x) < 0$ and $R(x)$ be smooth scalar functions, and $\varepsilon > 0$ be a real constant. For any $\beta > 0$ and smooth function $\delta(x) \in (0, \beta)$ $\forall x$, there exists a smooth function $\alpha(x)$, $0 < \alpha(x) < \delta(x)$, such that

$$g(x)\alpha(x) + R(x)\alpha^{1+\varepsilon}(x) \leq \frac{\alpha(x)g(x)}{2}.$$

3. MAIN RESULTS

The main result of this paper is summarized in the following statement, which improves the recent work (Mazenc, 1997; Lin and Qian, 1998) by relaxing the restriction on the nonlinear functions $f_i(y_i)$ in (1).

Theorem 3.1 Under Assumptions 1–2–3, feedforward system (1) is globally asymptotically stabilized by the arbitrarily small feedback controller

$$u = -\alpha_1(y_1)\hat{f}_1(y_1) - \cdots - \alpha_n(y_n)\hat{f}_n(y_n), \quad (6)$$

where $\alpha_i(\cdot)$ is a strictly positive and smooth function, bounded by $\frac{1}{n}$

Proof The proof is based on a step-by-step, bottom-up, inductive argument which simultaneously constructs a composite Lyapunov function with *nonlinear weights* as well as an *small* state feedback control law. According to the decomposition of f_i (see (4)), define

$$\bar{\phi}_i(y_n, u) = \phi_i(y_n, u) + \bar{f}_{i-1}(y_{i-1}).$$

Initial Step: To begin with, consider the bottom equation of (1), i.e. x_1–subsystem. Let

$$V_1(x_1) = \frac{x_1^2}{2}, \qquad u = v_1 + u_2.$$

The time derivative of V_1 along the x_1–subsystem is

$$\begin{aligned}\dot{V}_1(x_1) &= x_1(u + \phi_1(y_1, u)) \\ &\le x_1 v_1 + x_1 u_2 + v_1^2 \bar{\gamma}_1(|x_1|) + u_2^2 \bar{\gamma}_1(|x_1|) \quad (7)\end{aligned}$$

where $\bar{\gamma}_1 \ge \{|x_1|(1+|x_1|)2\gamma_1(|u|) \mid |u| \le 1\}$. As we shall see in what follows, our construction guarantees that the feedback control law $u(\cdot)$ is always bounded by 1. Hence there is a non-negative function $\bar{\gamma}_1(\cdot)$ such that (7) holds. Choose $0 < \beta < \frac{1}{n}$ and

$$v_1 = -\frac{\beta}{(1+x_1^{2p})(1+\bar{\gamma}_1^2(x_1))}\hat{f}_1(x_1) := -\alpha_1(x_1)\hat{f}_1(x_1).$$

Then it is easy to show that

$$\dot{V}_1(x_1) \le \frac{\beta - 2}{2}\alpha_1(x_1)x_1^{p+1} + L_1(y_1)u_2 + N_1(y_1)u_2^2 \quad (8)$$

where $L_1(y_1) = x_1$, $N_1(y_1) = \bar{\gamma}_1(|x_1|)$. Obviously, $L_1(0) = 0$. This property will be used in the next step design.

Inductive Step: We assume that for the last i-equations of (1), there are a *small* controller of the form

$$\begin{aligned} u &= v_1 + v_2 + \cdots + v_{i-1} + u_i \\ &= -\alpha_1(y_1)\hat{f}_1(y_1) - \cdots - \alpha_i(y_i)\hat{f}_i(y_i) + u_{i+1} \\ u_i &= -\alpha_i(y_i)\hat{f}_i(y_i) + u_{i+1}, \ 0 < |v_i|, \ \alpha_i(y_i) < \beta, \quad (9)\end{aligned}$$

and a change of coordinates

$$z_1 = x_1, \ z_j = \alpha_{j-1}(0)x_j + z_{j-1}, \ j = 2, \cdots, i, \quad (10)$$

such that after the transformation (10) and the state feedback (9), the subsystem can be represented as

$$\frac{d}{dt}\begin{bmatrix} z_i \\ \vdots \\ z_2 \\ z_1 \end{bmatrix} = \begin{bmatrix} -1 & \cdots\cdots & 0 & 0 \\ \vdots & \cdots\cdots & \vdots & \vdots \\ -1 & \cdots\cdots & -1 & 0 \\ -1 & \cdots\cdots & -1 & -1 \end{bmatrix}\begin{bmatrix} \alpha_i(y_i)\hat{f}_i(y_i) \\ \vdots \\ \alpha_2(y_2)\hat{f}_2(y_2) \\ \alpha_1(y_1)\hat{f}_1(y_1) \end{bmatrix} + [1 \cdots 1\ 1]^T u_{i+1} + \Psi_i(y_n, u_{i+1}) \quad (11)$$

where

$$\Psi_i(y_n, u_{i+1}) = \begin{bmatrix} \psi_i(y_n, -\alpha_i(y_i)\hat{f}_i(y_i) + u_{i+1}) \\ \vdots \\ \psi_2(y_n, -\sum_{j=2}^{i}\alpha_j(y_j)\hat{f}_j(y_j) + u_{i+1}) \\ \psi_1(y_n, -\sum_{j=1}^{i}\alpha_j(y_j)\hat{f}_j(y_j) + u_{i+1}) \end{bmatrix}$$

with $\psi_1(y_n, u) = \phi_1(y_n, u)$. Suppose

$$\psi_i(y_n, u_i) \le (1 + |z_i|)(\|y_{i-1}\|^{p+1} + u_i^2)\bar{\gamma}_i\left(\left\| \begin{matrix} y_{i-1} \\ u_i \end{matrix} \right\|\right) \quad (12)$$

and assume that there is a proper and positive definite Lyapunov function $V_i(z_1, \cdots, z_i)$ such that the time derivative of V_i along the trajectories of (11) satisfies

$$\begin{aligned}\frac{d}{dt}V_i(x_1, z_2, \cdots, z_i) \le &-\beta_1(y_1)x_1^{p+1} - \cdots - \beta_i(y_i)x_i^{p+1} \\ &+ L_i(y_i)u_{i+1} + N_i(y_i)u_{i+1}^2\end{aligned}$$

where $\beta_k(y_k) > 0, k = 1, \cdots, i$, $N_i(\cdot) \ge 0$ and $L_i(\cdot)$ are all smooth functions, with $L_i(0) = 0$.

We now augment the system (11) by adding the $(i+1)$–th equation of (1) into (11). Using the change of coordinate

$$z_{i+1} = \alpha_i(0)x_{i+1} + z_i,$$

we obtain

$$\dot{z}_{i+1} = u_{i+1} + \psi_{i+1}(y_n, u_{i+1}), \quad (13)$$

where $\psi_{i+1}(y_n, u_{i+1}) := \psi_i(y_n, -\alpha_i(y_i)\hat{f}_i(y_i) + u_{i+1}) + \hat{f}_i(y_i)(\alpha_i(0) - \alpha_i(y_i)) + \alpha_i(0)\bar{\phi}_{i+1}(y_n; u)$. By Lemma 1,

$$|\hat{f}_i(y_i)(\alpha_i(0) - \alpha_i(y_i))| \le \|y_i\|^{p+1}\gamma_{i+1,1}(\|y_i\|) \quad (14)$$

By construction, it is clear from (9) that $\exists \bar{\rho}(y_i) > 0$ s.t. $u^2 \le \|y_i\|^{p+1}\bar{\rho}(y_i) + 2u_{i+1}^2$. Thus by A2 and the higher order property of the function $\bar{f}_i(y_i)$,

$$|\bar{\phi}_{i+1}(y_n; u)| \le (1 + |z_{i+1}|)(\|y_i\|^{p+1} + u_{i+1}^2)\gamma_{i+1,2}\left(\left\| \begin{matrix} y_{i+1} \\ u_{i+1} \end{matrix} \right\|\right). \quad (15)$$

Using (14)-(15), the inductive condition (12) and $u_i = v_i + u_{i+1}$, one can show that

$$|\psi_{i+1}(y_n; u_{i+1})| \le (1+|z_{i+1}|)(\|y_i\|^{p+1} + u_{i+1}^2)\bar{\gamma}_{i+1}\left(\left\| \begin{matrix} y_i \\ u_{i+1} \end{matrix} \right\|\right)$$

Moreover,

$$\frac{d}{dt}\ln(1+z_{i+1}^2) \le \frac{2z_{i+1}}{1+z_{i+1}^2}u_{i+1} + (\|y_i\|^{p+1} + u_{i+1}^2)\bar{r}_{i+1}(\|y_i\|)$$

where $\bar{r}_{i+1}(\|y_i\|) > \gamma_{i+1}\left(\left\| \begin{matrix} y_i \\ u_{i+1} \end{matrix} \right\|\right)$ whenever $|u_{i+1}| \leq 1$.

Define the positive definite function

$$W_i(y_i) = \beta_1(y_1)x_1^{p+1} + \cdots + \beta_i(y_i)x_i^{p+1}.$$

Clearly, there exist two smooth functions $\bar{\sigma}_{i+1}(\|y_i\|) > 0$ and $\bar{\gamma}_{i+1}(\|y_i\|) > 0$ satisfying

$$\bar{\sigma}_{i+1}(\|y_i\|)\|y_i\|^{p+1} \leq W_i(y_i) \leq \bar{\gamma}_{i+1}(\|y_i\|)\|y_i\|^{p+1} \quad (16)$$

for instance, $\bar{\sigma}_{i+1}(\|y_i\|) = \prod_{k=1}^{i} \frac{\beta_k(y_k)}{1+\beta_k^2(y_k)}$ and $\bar{\gamma}_{i+1}(\|y_i\|) = \sum_{k=1}^{i} \beta_k(y_k)$ do the job.

The inequality (16), together with the property of $V_i(\cdot)$ being positive definite and proper, ensures the existence of a smooth function $T_i(V_i(y_i)) > 0$ satisfying

$$T_i(V_i(y_i)) > 2\frac{\|y_i\|^{p+1}\bar{r}_{i+1}(\|y_i\|)}{W_i(y_i)}. \quad (17)$$

Construct a $\mathcal{K}_\infty$ function $K_i(s) = \int_0^s T_i(\tau)d\tau$ and consider the positive definite and proper Lyapunov function

$$V_{i+1}(y_{i+1}) = K_i(V_i(y_i)) + \ln(1 + z_{i+1}^2).$$

Then, the time derivative of V_{i+1} along the augmented system (13)–(11) is

$$\begin{aligned}\dot{V}_{i+1}(y_{i+1}) \leq & -\frac{1}{2}T_i(V_i(y_i))W_i(y_i) + M_{i+1}(y_{i+1})v_{i+1} \\ & + N_{i+1}(y_{i+1})v_{i+1}^2 + \frac{2\alpha_i(0)x_{i+1}}{1+z_{i+1}^2}v_{i+1} \\ & + L_{i+1}(y_{i+1})u_{i+2} + N_{i+1}(y_{i+1})u_{i+2}^2 \quad (18)\end{aligned}$$

where $u_{i+1} = v_{i+1} + u_{i+2}$, $N_{i+1}(y_{i+1}) = 2T_i(V_i(y_i))N_i(y_i) + 2\bar{r}_{i+1}(\|y_i\|) \geq 0$ and

$$M_{i+1}(y_{i+1}) = T_i(V_i(y_i))L_i(y_i) + \frac{2z_i}{1+z_{i+1}^2}$$

$$L_{i+1}(y_{i+1}) = T_i(V_i(y_i))L_i(y_i) + \frac{2z_{i+1}}{1+z_{i+1}^2} \quad L_{i+1}(0) = 0.$$

Since $M_{i+1}(y_{i+1})|_{y_i=0} = 0$, we have $|M_{i+1}(y_{i+1})| \leq \tilde{M}_{i+1}(y_{i+1})\|y_i\|$ for some smooth function $\tilde{M}_{i+1}(y_{i+1}) \geq 0$. Let

$$v_{i+1} = -\alpha_{i+1}(y_{i+1})\hat{f}_{i+1}(y_{i+1}),$$

where $\alpha_{i+1}(\cdot)$ is a smooth function satisfying $0 < \alpha_{i+1}(\cdot) < \beta < \frac{1}{n}$, to be determined below.

From Lemma 1 and (17), it can be proved that

$$\begin{aligned}& |M_{i+1}(y_{i+1})v_{i+1}| \\ & \leq \tilde{M}_{i+1}(y_{i+1})\rho_{i+1}(\|y_{i+1}\|)\left(\alpha_{i+1}^{\lambda}(y_{i+1})\|y_i\|^{p+1}\right. \\ & \quad \left. + \alpha_{i+1}^{1+\mu}(y_{i+1})x_{i+1}^{p+1}\right) \\ & \leq A_{i+1}(y_{i+1})T_i(V_i(y_i))W_i(y_i)\alpha_{i+1}^{\lambda}(y_{i+1}) \\ & \quad + B_{i+1}(y_{i+1})\alpha_{i+1}^{1+\mu}(y_{i+1})x_{i+1}^{p+1} \quad (19)\end{aligned}$$

where $A_{i+1}(y_{i+1}) = \frac{\tilde{M}_{i+1}(y_{i+1})\rho_{i+1}(\|y_{i+1}\|)}{2\bar{r}_{i+1}(\|y_i\|)}$ and $B_{i+1}(y_{i+1}) = \tilde{M}_{i+1}(y_{i+1})\rho_{i+1}(\|y_{i+1}\|)$. In addition, there is a C^∞ function $C_{i+1}(y_{i+1}) = \frac{(i+1)N_{i+1}(y_{i+1})\sum_{k=1}^{i} b_{i+1,k}^{2p}x_k^{p-1}}{\bar{2}r_{i+1}(y_{i+1})}$ such that

$$\begin{aligned}N_{i+1}(y_{i+1})v_{i+1}^2 \leq & (i+1)N_{i+1}(y_{i+1})\alpha_{i+1}^2(y_{i+1})x_{i+1}^{2p} \\ & + C_{i+1}(y_{i+1})\alpha_{i+1}^2(y_{i+1})T_i(V_i(y_i))W_i(y_i). \quad (20)\end{aligned}$$

Using the Young's inequality

$$|x_{i+1}(\hat{f}_{i+1}(y_{i+1}) - x_{i+1}^p)| \leq \frac{x_{i+1}^{p+1}}{2} + \sum_{k=1}^{i} a_{i+1,k}x_{i+1}x_k^p,$$

we have

$$\begin{aligned}\frac{2\alpha_i(0)x_{i+1}}{1+z_{i+1}^2}v_{i+1} & \leq \frac{2\alpha_i(0)\alpha_{i+1}(y_{i+1})}{1+z_{i+1}^2}\left[\frac{-x_{i+1}^{p+1}}{2} + c\sum_{k=1}^{i} x_k^{p+1}\right] \\ & \leq \frac{\alpha_i(0)}{1+z_{i+1}^2}\alpha_{i+1}(y_{i+1})(-x_{i+1}^{p+1}) \\ & \quad + D_{i+1}(y_{i+1})T_i(V_i(y_i))W_i(y_i), \quad (21)\end{aligned}$$

where $D_{i+1}(y_{i+1}) := \frac{c}{2\bar{r}_{i+1}(y_{i+1})}$. It follows from (19)-(21) that

$$\begin{aligned}\dot{V}_{i+1}(y_{i+1}) \leq & T_i(V_i(y_i))W_i(y_i)[-\frac{1}{2} + A_{i+1}(\cdot)\alpha_{i+1}^{\lambda}(\cdot) \\ & + C_{i+1}(\cdot)\alpha_{i+1}^2(\cdot) + D_{i+1}(\cdot)\alpha_{i+1}(\cdot)] \\ & + x_{i+1}^{p+1}\left[(i+1)N_{i+1}(y_{i+1})\alpha_{i+1}^2(y_{i+1})x_{i+1}^{p-1}\right. \\ & \left. - \frac{\alpha_i(0)\alpha_{i+1}(y_{i+1})}{1+z_{i+1}^2} + B_{i+1}(y_{i+1})\alpha_{i+1}^{1+\mu}(y_{i+1})\right] \\ & + L_{i+1}(y_{i+1})u_{i+2} + N_{i+1}(y_{i+1})u_{i+2}^2 \\ \leq & T_i(V_i(y_i))W_i(y_i)\left(-\frac{1}{2}\bar{A}_{i+1}(y_{i+1})\alpha^{\lambda}(\cdot)\right) \\ & + x_{i+1}^{p+1}\left[g(\cdot)\alpha(\cdot) + \alpha^{1+\varepsilon}(\cdot)R(\cdot)\right] \\ & + L_{i+1}(y_{i+1})u_{i+2} + N_{i+1}(y_{i+1})u_{i+2}^2 \quad (22)\end{aligned}$$

where $\varepsilon = min\{1, \mu\} = \mu$, $\bar{A}_{i+1}(y_{i+1}) = A_{i+1}(y_{i+1}) + D_{i+1}(y_{i+1}) + C_{i+1}(y_{i+1})$ and

$$\alpha(\cdot) = \alpha_{i+1}(y_{i+1}), \quad g(\cdot) = -\frac{\alpha_{i+1}(0)}{1+z_{i+1}^2},$$

$$R(\cdot) = B_{i+1}(y_{i+1}) + (i+1)N_{i+1}(y_{i+1})x_{i+1}^{p-1}.$$

The inequality (22) is obtained by using the facts that $0 < \alpha_{i+1}(y_{i+1}) < 1$, $0 < \lambda < 1$ and $0 < \mu < 1$. According to Lemma 2, there exists a smooth function $\alpha_{i+1}(y_{i+1})$ such that

$$\begin{aligned} & 0 < \alpha_{i+1}(y_{i+1}) \leq \alpha_{i+1}^{\lambda}(y_{i+1}) < \delta(x), \\ & g(\cdot)\alpha(\cdot) + \alpha^{1+\varepsilon}(\cdot)R(\cdot) \leq \frac{\alpha(\cdot)g(\cdot)}{2}, \\ & \delta(x) = \frac{1}{1+\hat{f}_{i+1}^2}\frac{\beta/2}{(1+\bar{A}_{i+1}^2(y_{i+1}))}.\end{aligned}$$

With this in mind, we deduce from (22) that

$$\frac{d}{dt}V_{i+1}(y_i) \leq L_{i+1}(y_{i+1})u_{i+2} + N_{i+1}(y_{i+1})u_{i+2}^2$$

$$-\frac{1}{4}T_i(V_i(y_i))W_i(y_i) - \frac{\alpha_i(0)}{2(1+z_{i+1}^2)}\alpha_{i+1}(y_{i+1})x_{i+1}^{p+1}$$

Define

$$W_{i+1}(y_{i+1}) := -\frac{T_i(V_i(y_i))W_i(y_i)}{4} - \frac{\alpha_i(0)\alpha_{i+1}(y_{i+1})x_{i+1}^{p+1}}{1+z_{i+1}^2}$$
$$= -\bar{\beta}_1(y_1)x_1^{p+1} - \cdots - \bar{\beta}_{i+1}(y_{i+1})x_{i+1}^{p+1}$$

where $\bar{\beta}_k(y_k) > 0, k = 1, \cdots, i+1$, are smooth functions. Thus we arrive at

$$\dot{V}_{i+1}(y_{i+1}) \le -\bar{\beta}_1(y_1)x_1^{p+1} - \cdots - \bar{\beta}_{i+1}(y_{i+1})x_{i+1}^{p+1}$$
$$+L_{i+1}(y_{i+1})u_{i+2} + N_{i+1}(y_{i+1})u_{i+2}^2.$$

Using repeatedly the argument in the inductive step, it is not difficult to prove that at Step n, there is a smooth state feedback controller of the form

$$u = v_1 + v_2 + \cdots + v_n$$
$$= -\alpha_1(y_1)\hat{f}_1(y_1) - \cdots - \alpha_n(y_n)\hat{f}_n(y_n), \quad (23)$$

such that the equilibrium $(z_1, z_2, \cdots, z_n) = 0$ of the closed-loop system is globally asymptotically stable. This, in turn, implies GAS of the equilibrium $(x_1, x_2, \cdots, x_n) = 0$ of (1) (because the transformation (10) is a global diffeomorphism). Finally, by construction it is trivial to verify that $|u| \le n\beta < 1$.

Remark 3.2 From Theorem 3.1, it is straightforward to deduce the following result: consider a system (1) in which $f_i(y_i) = \sum_{j=1}^{i} a_{i,j}x_j^p$, $a_{i,i} \neq 0$ and Assumptions A2 and A3 hold. Then there is a controller of the form (6) such that the closed-loop system is GAS at the origin. As a consequence, the stabilization result (Theorem 2.1, Mazenc, 1997) follows immediately.

Remark 3.3 When $p = 1$, (A3) holds automatically and (A1) implies $\frac{\partial f_i}{\partial x_i}(0) \neq 0$, and therefore the linearized system of (1) is stabilizable. In this case, under the condition (A2), Theorem 3.1 reduces to the known result by Mazenc and Praly (1996).

4. EXTENSIONS

So far we have established a global stabilization result for (1) under (A1)-(A3). In this section, we further relax the structural conditions by allowing different orders p_i of the partial derivatives of $f_i(\cdot)$'s. In the case when the sequence $\{p_i\}$ is non-decreasing, one can take full advantage of the nonlinearity so that a GAS result can be built for more general upper-triangular systems.

To be precise, we suppose that $\forall i = 1, \cdots, n-1$, there exist odd integers $\{p_i\}$ satisfying the following:

(B1) $\frac{\partial^{p_i} f_i}{\partial x_i^{p_i}}(0) \neq 0$ *and* $\quad p_{n-1} \ge \cdots \ge p_i \ge \cdots \ge p_1.$

Remark 4.1 By the Taylor expansion theorem, the smooth function $f_i(y_i)$ can be decomposed as

$$f_i(y_i) = \bar{f}_i(y_i) + h_i(y_i).$$

where $|h_i(y_i)| \le \|y_i\|^{p_i+1}\gamma(y_i)$ is a higher order term, while $\bar{f}_i$ is a polynomial with a order $\le p_i$, having the form

$$\bar{f}_i(y_i) = \sum_{r_{i,i}+\cdots+r_{i,1} \le p_i} \left.\frac{\partial^{r_{i,i}+\cdots+r_{i,1}} f_i(y_i)}{\partial x_i^{r_{i,i}} \cdots \partial x_1^{r_{i,1}}}\right|_{y_i=0} x_i^{r_{i,i}} \cdots x_1^{r_{i,1}}$$

Based on the representation above, we introduce the following structural conditions.

(B2) If $\left.\frac{\partial^{r_{i,i}+\cdots+r_{i,1}} f_i(y_i)}{\partial x_i^{r_{i,i}} \cdots \partial x_1^{r_{i,1}}}\right|_{y_i=0} \neq 0$, the corresponding orders $r_{i,i}, \cdots, r_{i,1}$ satisfy either

(B2.1) $\sum_{j=1}^{i} \frac{r_{i,j}}{p_j+1} \ge 1$; or

(B2.2) $\sum_{j=1}^{i} \frac{r_{i,j}}{p_j+1} + \frac{1}{p_i+1} \ge 1$, if $\sum_{j=1}^{i} \frac{r_{i,j}}{p_j+1} < 1$;

(B3) $|\phi_i(y_n;u)| \le (1+|x_i|)(x_i^{p_i+1}+\cdots+x_i^{p_1+1}+u^2)\gamma_i\left(\left\|\begin{matrix} y_{i-1} \\ u \end{matrix}\right\|\right)$

$|\phi_1(y_n;u)| \le u^2\gamma_1\left(\left\|\begin{matrix} x_1 \\ u \end{matrix}\right\|\right)$ $\forall i = 2, \cdots, n.$

Theorem 4.2 Under Assumptions B1-B3, the upper-triangular system (1) is globally asymptotically stabilized by the arbitrarily small feedback control law

$$u = -\alpha_1(y_1)\hat{f}_1(y_1) - \cdots - \alpha_n(y_n)\hat{f}_n(y_n),$$

where $\alpha_i(\cdot)$ is a smooth function satisfying $0 < \alpha_i(\cdot) < \beta < \frac{1}{n}$, and $\hat{f}_i(y_i)$ is the summation of all the terms which satisfy Assumption (B2.2), and $\hat{f}_n(y_n) = x_n^{p_{n-1}}$.

Sketched Proof To prove this theorem, we decompose $\bar{f}_i(x_i, \cdots, x_1)$ as follows:

$$\bar{f}_i(x_i, \cdots, x_1) = \hat{f}_i(x_i, \cdots, x_1) + \tilde{f}_i(x_i, \cdots, x_1)$$

where $\hat{f}_i$ is the summation of all the terms which satisfy (B2.2), $\hat{f}_n(y_n) = x_n^{p_{n-1}}$, and $\tilde{f}_i$ is the summation of all the terms which satisfy (B2.1). With the help of this decomposition and Remark 4.1, all the *"higher order terms"* of $f_i(y_i)$, i.e. $\tilde{f}_i$ and h_i are singled out, and can therefore be included in ϕ_i. In this way, the upper-triangular system (1) can be treated as a lower order polynomial system perturbed by a higher-order vector field. A detailed proof of Theorem 4.2 will be reported in a companion paper.

5. ILLUSTRATIVE EXAMPLES

We now present three examples to demonstrate some of significant features of our stabilization theorems.

Example 5.1

$$\dot{x}_3 = x_2^3 + x_1^2 x_2 + x_1 x_2^2 + \frac{x_3 \sin x_3}{1 + x_2^2} u^2$$
$$\dot{x}_2 = x_1^3$$
$$\dot{x}_1 = u \qquad (24)$$

This system satisfies neither the conditions of Theorem 2.1 in (Mazenc, 1997) nor Theorem 1 in (Lin and Qian, 1998). Due to the existence of the terms $x_1^2 x_2$ and $x_1 x_2^2$, it can be verified that the system is not globally diffeomorphic to a chain of *power* integrators plus a higher order vector field. However, (24) is globally asymptotically stabilizable by the small feedback law (6). Indeed, system (24) satisfies A1-A3 of Theorem 3.1.

The next example illustrates a powerful application of Theorem 4.2 to a upper-triangular system with different p_i's.

Example 5.2

$$\dot{x}_3 = x_2^5 + x_1^3 x_2 + x_3(u - \sin u)$$
$$\dot{x}_2 = x_1^3$$
$$\dot{x}_1 = u \qquad (25)$$

It is not difficult to prove that there is no a change of coordinates that transforms (25) to the solvable feedforward forms in (Mazenc, 1997; Lin and Qian, 1998). The term $x_1^3 x_2$ cannot be included in the higher order term $\phi_3(y_2, u)$. In fact, suppose there exists a smooth function $\rho(x_1, x_2) > 0$ such that

$$|x_1^3 x_2| \le (x_1^4 + x_2^6)\rho(x_1, x_2).$$

This would cause a contradiction, simply by setting $x_1 = x_2^{\frac{5}{4}}$ and $x_1 \to 0$. In conclusion, (25) cannot be handled by the previous methods (Mazenc, 1997; Lin and Qian, 1998). On the other hand, $x_1^3 x_2$ satisfies (B2.2), i.e.

$$\frac{3}{3+1} + \frac{1}{5+1} < 1, \quad but \quad \frac{3}{3+1} + \frac{1}{5+1} + \frac{1}{5+1} > 1.$$

So GAS of (25) is guaranteed by Theorem 4.2.

Example 5.3

$$\dot{x}_3 = x_1^3 x_2$$
$$\dot{x}_2 = x_1^3$$
$$\dot{x}_1 = u \qquad (26)$$

This example shows that without Assumption (A1), it may not be possible to achieve asymptotic stabilization of the system (1), even locally. In fact, (26) cannot be stabilizable by any smooth state feedback, due to the well known Brockett's necessary condition (Brockett, 1983).

6. CONCLUSIONS

In this paper we have explicitly constructed a *small* state feedback control law for a rather general class of upper-triangular nonlinear systems. This was accomplished by fully exploiting the physical structure of the systems, as well as by effectively coupling the forwarding design (Mazenc and Praly, 1996; Mazenc, 1997) and the passivity-based small feedback technique (Lin, 1995 and 1996). The results presented here complemented the previous work (Mazenc, 1997; Lin and Qian, 1998) and enlarged the class of feedforward systems that can be GAS by small controls.

7. REFERENCES

Barbu, C., R. Sepulchre, W. Lin and P. V. Kokotović (1997). Global asymptotic stabilization of the ball–and–beam system, Proc. of the *36th IEEE CDC*, San Diego, 2351-2355.

Brockett, R.W. (1983). Asymptotic stability and feedback stabilization, in Differential Geometric Control Theory, R.W. Brockett, R.S. Millman and H.J. Sussmann, Eds., Birkäuser, Basel-Boston, 181–191.

Lin, W. (1995). Bounded smooth state feedback and a global separation principle for non-affine nonlinear systems, *Syst. Contr. Lett.*, Vol. 26, 41-53.

Lin, W. (1996). Global asymptotic stabilization of general nonlinear systems with stable free dynamics via passivity and bounded feedback, *Automatica*, Vol. 32, 915-924.

Lin, W. and X. Li. (1998). Global stabilization of nonlinear systems with marginally unstable free dynamics by small controls, *Proc. of 1998 ACC, Philadelphia*, 303-307. Also, in press in *Int. J. Contr.* (1999).

Lin, W. and C. Qian. (1998), New Results on Synthesis of An Enlarged Class of Feedforward Systems Via Small Feedback, *Proc. of the 37th IEEE CDC, Tampa, FL.*

Mazenc, F. and L. Praly. (1996). Adding integrations, saturated controls, and stabilization for feedforward systems, *IEEE Trans. Automat. Contr.*, Vol. 41, 1559-1578.

Mazenc, F. (1997). Stabilization of feedforward systems approximated by a nonlinear chain of integrators, *Syst. Contr. Lett*, Vol. 32, 223-229.

Jankovic, M., R. Sepulchre, and P. Kokotović. (1996). Constructive Lyapunov stabilization of nonlinear cascade systems, *IEEE Trans. on Autom. Control*, 1723-1735.

Sepulchre, R. (1997). Slow-peaking and low-gain design for global stabilization of nonlinear systems, Proc. of the 36th IEEE CDC, 3491–3496.

Teel, A.R.(1992), Feedback stabilization: nonlinear solution to inherently nonlinear problems, Ph.D dissertation, University of California, Berkeley.

Teel, A.R.(1996), A nonlinear small gain theorem for the analysis of control systems with saturation, *IEEE Trans. Automat. Contr.*, Vol. 41, 1256-1270.

E-2c-15-1
14th Triennial World Congress, Beijing, P.R. China

NONLINEAR DISTURBANCE ATTENUATION WITH GLOBAL STABILITY VIA OUTPUT FEEDBACK

Zhong-Ping Jiang

Department of Electrical Engineering, Polytechnic University, Six Metrotech Center, Brooklyn, NY 11201, U.S.A.

Abstract: This paper deals with the problem of disturbance attenuation with stability for a class of minimum-phase nonlinear uncertain systems by output feedback. In contrast to the state-feedback case, a *nonlinear*, not necessarily linear L_2, gain is employed to characterize the disturbance rejection level. A constructive output feedback design is described to achieve global asymptotic stability while the effect of the disturbance on the output is diminished to an arbitrary degree of accuracy. Sufficient conditions are additionally provided to yield a nonlinear output feedback solving the standard L_2-gain disturbance attenuation problem with internal stability.

Keywords: Nonlinear control; Output feedback; Partial-state observer; Disturbance attenuation; Uncertain systems.

1. INTRODUCTION

Since Willems addressed the problem of almost disturbance decoupling in his seminal paper (Willems 1982), considerable attention has been paid to the construction of a feedback law that solves this problem and, at the same time, attains internal stability for both linear and nonlinear systems – see, for instance, (Saberi 1987, Saberi and Sannuti 1988, Marino *et al.* 1994, Isidori 1996, Isidori 1995, van der Schaft 1996, Battilotti 1996, Jiang and Jiang 1997). When full state information is available, it is now well understood that the L_2-gain almost disturbance decoupling with internal stability is solvable for a large class of certain or uncertain nonlinear systems. However, when only the output feedback is allowed, fewer constructive results exist in the literature.

In this paper, the problem of global output feedback disturbance attenuation with internal stability is addressed for a class of time-varying and nonlinear uncertain systems. As shown in (Saberi and Sannuti 1988, Marino and Tomei 1995, Battilotti 1996), this problem is much more involved than its state-feedback counterpart even for systems whose nonlinearities depend only on the output. In earlier work (Saberi 1987, Saberi and Sannuti 1988), a singular perturbation approach was proposed to yield a linear output feedback compensator which achieves global exponential stability and rejects the disturbance to any specified degree of accuracy (in the L_p-gain sense). Unfortunately, the system vector fields were imposed to satisfy a global Lipschitz condition and a matching condition was required. In (Marino and Tomei 1995, Battilotti 1996), genuinely nonlinear systems were considered though the system nonlinearities were only dependent on the output. The output feedback solution to the disturbance attenuation problem with stability in (Marino and Tomei 1995) is based on some filtered transformation and backstepping technique while (Battilotti 1996) requires the knowledge of solutions to two uncoupled Hamilton-Jacobi inequalities together with a coupling condition. Here, we will employ the innovative concept of ISS (input-to-state stability) and its Lyapunov characterization proposed by Sontag and his co-workers (Sontag

1995) to relax the linear dependence of the system vector field on the unmeasured states. The price to be paid for this generalization is that, instead of the usual L_2-gain almost disturbance decoupling property as in (Saberi and Sannuti 1988, Battilotti 1996, van der Schaft 1996), we obtain a *nonlinear-gain* disturbance attenuation property with global stability by output feedback. Nonetheless, it is of interest to point out that, for a subclass of the systems considered here, (Marino and Tomei 1995) established a polynomial-gain disturbance attenuation property with global stability via output feedback. Although the standard L_2-gain disturbance attenuation problem is not solvable for general nonlinear systems using output feedback (see (Ball *et al.* 1993)), additional conditions are given in Section 4 of the present paper to guarantee the possibility of constructing a nonlinear (resp. linear) output feedback compensator that solves the standard problem of L_2-gain disturbance attenuation with internal global asymptotic stability (resp. global exponential stability).

2. SYSTEM DESCRIPTION

In this paper, a class of time-varying and nonlinear uncertain systems is considered, which can be put into a normal form as follows:

$$\begin{aligned} \dot{z} &= q_0(t,z,y) + q_1(t,z,y)w \\ \dot{x}_1 &= x_2 + f_1(t,z,y) + g_1(t,z,y)w \\ &\vdots \\ \dot{x}_i &= x_{i+1} + f_i(t,z,y) + g_i(t,z,y)w \\ &\vdots \\ \dot{x}_n &= u + f_n(t,z,y) + g_n(t,z,y)w \\ y &= x_1 \end{aligned} \tag{1}$$

where $(z,x) \in \mathbb{R}^{n_0+n}$, $u \in \mathbb{R}$ and $y \in \mathbb{R}$ represent the system state, control input and to-be-controlled output respectively, $w \in \mathbb{R}^m$ is the disturbance input of the system (1). Given this system structure (1), we assume that the output y is measurable and that the functions q_0, q_1, f_i and g_i $(1 \le i \le n)$ are uncertain but Lipschitz continuous.

It is observed that time-invariant uncertain nonlinear systems in a form close to (1) have appeared in a number of earlier research work. As we shall see below, this paper extends those classes of systems to the time-varying case and does not restrict the nonlinearities to be dependent on the output y-only. As mentioned above, the objective of this paper is to design an output feedback controller in such a way that the closed-loop system is internally globally asymptotically stable while the disturbance w is rejected to any specified degree of accuracy. More precisely, the following nonlinear disturbance attenuation problem will be addressed in the paper.

Disturbance Attenuation Problem with Stability: There is a class K-function γ such that, for any positive real number ϵ, an output feedback control law of the form

$$\dot{\chi} = \nu(\chi,y), \quad u = \mu(\chi,y) \tag{2}$$

achieves the following properties:

(*i*) When $w=0$, the closed-loop system (1)-(2) is globally uniformly asymptotically stable (GUAS) at the origin.

(*ii*) When $w \in L^m_{\infty e}$, the closed-loop system solution issued from the origin satisfies the following inequality $(t \ge t_0 \ge 0)$

$$\int_{t_0}^{t} |y(\tau)|^2 d\tau \le \epsilon \int_{t_0}^{t} \gamma(|w(\tau)|) d\tau \tag{3}$$

Notice that, when (3) holds with $\gamma(r) = kr^2$ for all $r \in \mathbb{R}_+$ and some $k>0$, we say that the L_2-gain disturbance attenuation problem with stability is solvable for system (1) using output feedback. In this case, letting $y_{[t_0,t]}$ and $w_{[t_0,t]}$ denote the truncation of y and w over the interval $[t_0,t]$, respectively, the inequality (3) can be rewritten as $\|y_{[t_0,t]}\|_2 \le \sqrt{\epsilon k}\|w_{[t_0,t]}\|_2$. The smallest value of $\sqrt{\epsilon k}$ such that (3) holds is referred to as the L_2-gain of the closed-loop system.

In order to find such a nonlinear output feedback law (2), the following assumptions will be made about the class of systems (1).

(H1) There exists an ISS (input-to-state stable) Lyapunov function V_0 for the z-system in (1). Namely, there are class K_∞-functions $\underline{\alpha}$, $\overline{\alpha}$, α_0, γ_0 and γ_1 satisfying the following inequalities

$$\underline{\alpha}(|z|) \le V_0(t,z) \le \overline{\alpha}(|z|) \tag{4}$$

$$\dot{V}_0 \le -\alpha_0(|z|) + \gamma_0(|y|) + \gamma_1(|w|) \tag{5}$$

for all $t \in \mathbb{R}_+$, $z \in \mathbb{R}^{n_0}$, $y \in \mathbb{R}$ and $w \in \mathbb{R}^m$.

(H2) For every $1 \le i \le n$, there exist a positive constant ψ_{i0}, three smooth nonnegative functions φ_{i1}, ψ_{i1} and ψ_{i2}, and a continuous nonnegative function φ_{i2}, all vanishing at zero, such that

$$|f_i(t,z,y)| \le \varphi_{i1}(|y|) + \varphi_{i2}(|z|) \tag{6}$$

$$|g_i(t,z,y)| \le \psi_{i1}(y) + \psi_{i2}(z) + \psi_{i0} \tag{7}$$

for all $t \in \mathbb{R}_+$, $z \in \mathbb{R}^{n_0}$ and $y \in \mathbb{R}$.

Borrowing the terminology from (Sontag 1995), Assumption (H1) implies that the z-system with inputs y and w is input-to-state stable (ISS). In particular, the z-system with zero inputs (i.e. $y = 0$ and $w = 0$) is GUAS at $z = 0$. In other words, the nominal model of uncertain system (1) is minimum-phase. The terms of ISS and ISS Lyapunov functions used here are variants of those originally proposed by (Sontag 1995) for *time-invariant* nonlinear controlled systems, also see (Jiang and Mareels 1997, Jiang *et al.* 1994). The interested reader is referred to (Sontag 1995) for various characterizations and applications of this useful ISS property. For instance, any stable linear system is ISS. Examples of control applications by means of the ISS property are found in the recent literature on adaptive control, robust nonlinear control by state and output feedbacks and control of interconnected systems using small gain theorems. Assumption (H2) requires that the uncertain system nonlinearities are bounded by some known functions which are nonlinear in the unmeasured state component z and the measured output y. (H2) significantly relaxes the severe assumptions imposed on the uncertainty considered in earlier work (Saberi and Sannuti 1988, Marino and Tomei 1995, Battilotti 1996).

In the sequel, an observer-based output feedback design will be proposed, that solves the above disturbance attenuation problem. Furthermore, additional sufficient conditions are given to yield a solution to the usual L_2-gain disturbance attenuation problem.

3. ROBUST CONTROL DESIGN

The main purpose of this section is to develop a robust output feedback law design procedure to achieve the control task stated in Section 2. Firstly, an observer-like dynamical system is introduced. Then, based on the new augmented system, it is shown how to recursively construct a dynamic output feedback law of the form (2) and establish a differential dissipation inequality. Finally, an appropriate choice of the design parameters and functions gives a solution of the aforementioned disturbance attenuation problem.

3.1 *Observer design*

In light of the system structure (1), the following reduced-order observer is introduced to rebuild the unmeasured partial-states $(x_2, \ldots, x_n)$.

$$\begin{aligned} \dot{\xi}_i &= \xi_{i+1} + \ell_{i+1}y - \ell_i(\xi_1 + \ell_1 y) \\ & \qquad 1 \le i \le n-2 \\ \dot{\xi}_{n-1} &= u - \ell_{n-1}(\xi_1 + \ell_1 y) \end{aligned} \tag{8}$$

Letting

$$\begin{aligned} \zeta &= (x_2 - \xi_1 - \ell_1 y, \ \ldots, \ x_n - \xi_{n-1} - \ell_{n-1}y) \\ f^* &= (f_2 - \ell_1 f_1, \ \ldots, \ f_n - \ell_{n-1} f_1)^T \\ g^* &= (g_2 - \ell_1 g_1, \ \ldots, \ g_n - \ell_{n-1} g_1)^T \end{aligned}$$

it holds :

$$\dot{\zeta} = A\zeta + f^*(t,z,y) + g^*(t,z,y)w \tag{9}$$

where A is a stable matrix due to the suitable choice of the real numbers ℓ_i's. Let $P = P^T > 0$ be the solution to the equation

$$PA + A^T P = -2I \tag{10}$$

Using (H2) and completing the squares, the time derivative along the solutions of (9) of the quadratic function $V_\zeta = \zeta^T P \zeta$ satisfies:

$$\begin{aligned} \dot{V}_\zeta \le & -\zeta^T\zeta + \phi_1(y) + \phi_2(z) \\ & + \phi_0 w^2 + 6|P|^2 w^4 \end{aligned} \tag{11}$$

where ϕ_1 and ϕ_2 are two continuous nonnegative function and ϕ_0 is a positive constant.

The disturbance attenuation problem is solved for the original system (1) if it is solvable for the following new augmented system from w to y using partial-state $(y, \xi_1, \cdots, \xi_{n-1})$ feedback:

$$\begin{aligned} \dot{z} &= q_0(t,z,y) + q_1(t,z,y)w \\ \dot{\zeta} &= A\zeta + f^*(t,z,y) + g^*(t,z,y)w \\ \dot{y} &= \xi_1 + \ell_1 y + \zeta_1 + f_1 + g_1 w \\ \dot{\xi}_i &= \xi_{i+1} + \ell_{i+1}y - \ell_i(\xi_1 + \ell_1 y), \ 1 \le i \le n-2 \\ \dot{\xi}_{n-1} &= u - \ell_{n-1}(\xi_1 + \ell_1 y) \end{aligned} \tag{12}$$

The latter is the objective of the next subsections. We first establish a differential dissipation inequality for the (ζ, y, ξ)-subsystem of the interconnected system (12) in subsection 3.2. In subsection 3.3, techniques are borrowed from (Sontag and Teel 1995) in modifying supply functions for a single dissipative nonlinear system in order to form a storage function for the interconnected nonlinear system (12).

3.2 *Recursive design*

We begin with the (ζ, y)-subsystem of (12):

$$\begin{aligned} \dot{\zeta} &= A\zeta + f^*(t,z,y) + g^*(t,z,y)w \\ \dot{y} &= \xi_1 + \ell_1 y + \zeta_1 + f_1 + g_1 w \end{aligned} \tag{13}$$

with ξ_1 regarded as the virtual control input.

With the help of (11), differentiating $V_1 = V_\zeta(\zeta) + \frac{1}{2}y^2$ with respect to the solutions of (12) gives

$$\dot{V}_1 \leq -\frac{1}{2}\zeta^T\zeta + \phi_1(y) + \phi_2(z) + \phi_0 w^2 + 6|P|^2 w^4 + y\left(\xi_1 + \ell_1 y + \frac{1}{2}y + f_1 + g_1 w\right) \quad (14)$$

By completing the squares, with (H2), there exist some functions $\widehat{\phi}_1(y)$ and $\widehat{\varphi}_1(y)$ such that

$$\dot{V}_1 \leq -\frac{1}{2}\zeta^T\zeta - \epsilon^{-1}y^2 - y^2\nu_1(y^2) + y\widetilde{\xi}_1 + \phi_2(z) + \varphi_{12}^2(z) + \psi_{12}^4(z) + (1 + \phi_0 + \psi_{10}^2)w^2 + 6|P|^2 w^4 \quad (15)$$

with $\widetilde{\xi}_1 = \xi_1 - \vartheta_1(y)$ and

$$\vartheta_1(y) = -\epsilon^{-1}y - y\nu_1(y^2) - \ell_1 y - \frac{1}{2}y - y\widehat{\phi}_1(y) - y\widehat{\varphi}_1(y) \quad (16)$$

where $\epsilon > 0$ and ν_1 is a nondecreasing smooth function with $\nu_1(0) > n-1$ to be determined later.

By application of integrator backstepping, we can recursively construct a series of virtual control functions ϑ_{i+1} $(1 \leq i \leq n-1)$ and obtain a change of coordinates $\widetilde{\xi}_{i+1} = \xi_{i+1} - \vartheta_{i+1}(y, \xi_1, \ldots, \xi_i)$ where we let $\xi_n = u - \ell_{n-1}y$ and $\xi_0 = \widetilde{\xi}_0 = y$. The following result can be shown by induction.

Lemma 1. Along the solutions of $(\zeta, y, \ldots, \xi_{i-1})$-subsystem, the time derivative of the positive definite and proper storage function

$$V_i = V_\zeta(\zeta) + \frac{1}{2}y^2 + \sum_{j=1}^{i-1}\frac{1}{2}\widetilde{\xi}_j^2 \quad (17)$$

satisfies

$$\dot{V}_i \leq -\frac{1}{2^i}\zeta^T\zeta - \epsilon^{-1}y^2 - y^2(\nu_1(y^2) - i + 1) - \sum_{j=1}^{i-1}(c_j - i + j + 1)\widetilde{\xi}_j^2 + \widetilde{\xi}_{i-1}\widetilde{\xi}_i + \phi_2(z) + i\varphi_{12}(|z|)^2 + i\psi_{12}(z)^4 + (i + \phi_0 + i\psi_{10}^2)w^2 + 6|P|^2 w^4 \quad (18)$$

with $c_j > n - j - 1$.

Thanks to Lemma 1, setting $u = \vartheta_n + \ell_{n-1}y$ (i.e. $\widetilde{\xi}_n = 0$) yields:

$$\dot{V}_n \leq -\frac{1}{2^n}\zeta^T\zeta - \epsilon^{-1}y^2 - y^2(\nu_1(y^2) - n + 1) - \sum_{i=1}^{n-1}(c_i - n + i + 1)\widetilde{\xi}_i^2 + (n + \phi_0 + n\psi_{10}^2)w^2 + 6|P|^2 w^4 + \phi_2(z) + n\varphi_{12}(|z|)^2 + n\psi_{12}(z)^4 \quad (19)$$

It remains to choose the design function ν_1 appropriately so that a storage function can be generated for the whole interconnected nonlinear system (12). This will be achieved by means of (H1) and techniques in (Sontag and Teel 1995) and under additional sufficient conditions.

3.3 *Changing supply functions*

We will make use of some techniques recently proposed in (Sontag and Teel 1995) in changing supply functions for input-to-state stable systems and construct a storage function for the complete system (12) on the basis of differential dissipation inequalities (19) and (5).

To this end, let us define

$$U_0(t, z) = \int_0^{V_0(t,z)} \rho(\tau)d\tau \quad (20)$$

where $\rho : \mathbb{R}_+ \to \mathbb{R}_+$ is a smooth nondecreasing function such that $\rho(\tau) > 0$ for all $\tau \geq 0$. Clearly, U_0 is a positive definite and proper function.

Under the conditions of Lemma 2 below, the design functions ρ and ν_1 can be selected appropriately so that the following differential dissipation inequality holds with $V_c = V_n + U_0$

$$\dot{V}_c \leq -\alpha(|z|) - \frac{1}{2^n}\zeta^T\zeta - \epsilon^{-1}y^2 - \sum_{i=1}^{n-1}(c_i - n + i + 1)\widetilde{\xi}_i^2 + \gamma(|w|) \quad (21)$$

where α and γ are some class K-functions.

Lemma 2. If the functions γ_0, α_0 and φ_{i2} in (H1) and (H2) satisfy the following local property

$$\left.\begin{array}{l} \limsup_{s\to 0+} \dfrac{\gamma_0(s)}{s^2} < +\infty \\ \limsup_{s\to 0+} \dfrac{\varphi_{i2}(s)^2}{\alpha_0(s)} < +\infty \end{array}\right\} \quad \forall 1 \leq i \leq n \quad (22)$$

then the inequality (21) holds with

$$\alpha(r) = \frac{1}{4}\rho \circ \underline{\alpha}(r)\alpha_0(r)\,, \quad (23)$$

$$\gamma(r) = \rho \circ \overline{\alpha} \circ \alpha_0^{-1} \circ 4\gamma_1(r)\gamma_1(r) + (n + \phi_0 + n\psi_{10}^2)r^2 + 6|P|^2 r^4 \quad (24)$$

4. MAIN RESULTS

We can now formulate our main theorem of this paper.

Theorem 1. Under the Assumptions (H1) and (H2) and the conditions of Lemma 2, the Disturbance Attenuation Problem with Stability is solvable for the uncertain nonlinear system (1) via a dynamic output feedback of form (2).

Proof. Thanks to Lemma 2, we have established (21). Whenever $w = 0$, it is seen that $\dot{V}_c$ is bounded by a negative definite function. Since $V_c = V_n + U_0$ is a positive definite and proper function of the variable $(z, \zeta, y, \tilde{\xi})$, or (z, x, ξ), it follows from LaSalle-Yoshizawa Theorem that the origin is a GUAS equilibrium point of the resulting closed-loop system. The second property (3) follows directly from (21). □

As seen above and also illustrated in (Marino and Tomei 1995), it is generally impossible to establish an L_2-gain disturbance attenuation property with global stability for systems (1) via measurement feedback. The interested reader is referred to (Ball *et al.* 1993) for some necessary conditions for the existence of solutions to the H_∞ control of general nonlinear systems. Hereafter, some extra conditions are given on the system (1) under which the L_2-gain disturbance attenuation problem with internal stability is solvable. We first consider the case where the zero-dynamics z-equation of (1) is not affected by the disturbance input w. Then, we consider the case where the zero-dynamics is affected by w but is globally exponentially stable and uncertain nonlinearities are dominated by some linear functions. The first case is described by the following hypothesis.

(H3) the zero-dynamics z is not affected by w, i.e. $(q_1, \gamma_1) \equiv (0, 0)$. The functions ψ_{i1} and ψ_{i2} $(1 \leq i \leq n)$ in (H2) are zero.

Under (H3), we obtain, instead of (11),

$$\dot{V}_\zeta \leq -\zeta^T\zeta + \phi_1(y) + \phi_2(z) + \phi_0 w^2 \quad (25)$$

As a consequence, in this case, the control design procedure presented in Section 3 gives, instead of (21),

$$\dot{V}_c \leq -\alpha(|z|) - \frac{1}{2^n}\zeta^T\zeta - \epsilon^{-1}y^2 - \sum_{i=1}^{n-1}(c_i - n + i + 1)\tilde{\xi}_i^2 + (n + \phi_0 + n\psi_{10}^2)w^2 \quad (26)$$

Therefore, an L_2-gain disturbance attenuation property with stability is guaranteed. The following result has been proved.

Corollary 1. Under the Assumption (H3) and the conditions of Theorem 1, the L_2-gain Disturbance Attenuation Problem with Stability is solvable for the uncertain nonlinear system (1) by a dynamic output feedback (2).

The next hypothesis describes the second situation where the L_2-gain almost disturbance decoupling problem has a solution via *linear* output feedback.

(H4) Assumption (H1) holds with $\underline{\alpha}(|z|) = \underline{\alpha}^*|z|^2$, $\overline{\alpha}(|z|) = \overline{\alpha}^*|z|^2$, $\alpha_0(|z|) = \alpha_0^*|z|^2$, $\gamma_0(|z|) = \gamma_0^*|z|^2$ and $\gamma_1(|z|) = \gamma_1^*|z|^2$ for some constants $\overline{\alpha}^* \geq \underline{\alpha}^* > 0$, $\alpha_0^* > 0$, $\gamma_0^* \geq 0$ and $\gamma_1^* \geq 0$. Assumption (H2) holds with $\psi_{i1} = \psi_{i2} = 0$, $\varphi_{i1}(|x|) = k_{i1}|x|$ and $\varphi_{iz}(|z|) = k_{i2}|z|$ for some $k_{i1} \geq 0$ and $k_{i2} \geq 0$.

The following corollary can be established, which recovers part of the results in (Saberi 1987, Saberi and Sannuti 1988).

Corollary 2. Under the Assumption (H4) and the conditions of Theorem 1, the L_2-gain Disturbance Attenuation Problem with Stability is solvable for the uncertain nonlinear system (1) by a *linear* dynamic output feedback (2). Furthermore, the closed-loop system is globally exponentially stable at the origin if $w = 0$.

5. AN ILLUSTRATIVE EXAMPLE

As an illustration of the above design method, consider a simple nonlinear system of the form

$$\begin{aligned} \dot{z} &= -z^3 + y^2 \\ \dot{x}_1 &= x_2 + z^p + w_1 \\ \dot{x}_2 &= u + w_2 \end{aligned} \quad (27)$$

with $p \geq 2$. Denote $y := x_1$.

Clearly, this system belongs to the class of systems (1) with $w = (w_1, w_2)$. It can be pointed out that none of the known approaches in (Battilotti 1996, Marino and Tomei 1995, Saberi and Sannuti 1988, van der Schaft 1996) are applicable to (27) to provide an output feedback solution to the L_2-gain disturbance attenuation problem with global stability.

It is directly verified that Assumptions (H1) and (H2) hold with

$$\begin{aligned} &V_0(z) = z^2,\ \underline{\alpha}(|z|) = \overline{\alpha}(|z|) = z^2,\ \alpha_0(|z|) = z^4 \\ &\gamma_0(|y|) = \frac{3}{4^{2/3}}|y|^{\frac{8}{3}},\ \gamma_1 \equiv 0, \varphi_{12}(|z|) = |z|^p \\ &\varphi_{i1} = \varphi_{22} = \psi_{i1} = \psi_{i2} \equiv 0,\ \psi_{i0} = 1,\ i = 1, 2 \end{aligned}$$

It is easily seen that (22) and (H3) are satisfied. Thus, all the conditions of Corollary 1 are checked. So, the output feedback L_2-gain Disturbance Attenuation Problem with Stability is solvable for the system (27).

To derive such an output feedback law, the control design procedure developed in Section 3 is invoked. Firstly, a reduced-order partial-state observer is designed as follows

$$\dot{\xi}_1 = u - \ell_1(\xi_1 + \ell_1 y), \quad \ell_1 > 0 \quad (28)$$

Letting $\zeta = x_2 - \xi_1 - \ell_1 y$, we have

$$\dot{\zeta} = -\ell_1\zeta - \ell_1 z^p + w_2 - \ell_1 w_1 \quad (29)$$

With respect to the solutions of (29), the time derivative of $V_\zeta = \zeta^2/\ell_1$ satisfies

$$\dot{V}_\zeta \leq -\zeta^2 + 2z^{2p} + 4\max\left\{1, \frac{4}{\ell_1^2}\right\} w^2$$

When applying the recursive control design procedure in Section 3 to the new augmented system

$$\begin{aligned}\dot{z} &= -z^3 + y^2 \\ \dot{\zeta} &= -\ell_1\zeta - \ell_1 z^p + w_2 - \ell_1 w_1 \\ \dot{y} &= \xi_1 + \ell_1 y + \zeta + z^p + w_1 \\ \dot{\xi}_1 &= u - \ell_1(\xi_1 + \ell_1 y),\end{aligned}$$

a desired dynamic output feedback is obtained, that solves the L_2-gain property with global stability:

$$\begin{aligned}u = &-\tilde{\xi}_1 - y + \ell_1(\xi_1 + \ell_1 y) - (1 + \ell_1 + \epsilon^{-1}) \times \\ &\left(1 + \frac{\partial(y\nu_1(y^2))}{\partial y}\right)\left[\xi_1 + \ell_1 y \right. \\ &\left. + \frac{3}{2}\tilde{\xi}_1(1 + \ell_1 + \epsilon^{-1})\left(1 + \frac{\partial(y\nu_1(y^2))}{\partial y}\right)\right]\end{aligned}$$

where $\epsilon > 0$ is arbitrary, $\tilde{\xi}_1$ and the smooth nondecreasing function ν_1 are given by

$$\begin{aligned}\tilde{\xi}_1 &= \xi_1 + (1 + \ell_1 + \epsilon^{-1})y + y\nu_1(y^2), \\ \nu_1(r) &\geq \frac{3 \times 4^{\frac{1}{3}}}{4} r^{\frac{1}{3}} \rho\left(\sqrt{\frac{3 \times 4^{\frac{1}{3}}}{2}} r^{\frac{2}{3}}\right)\end{aligned}$$

with $\rho(r) \geq 16\ell_1^{p-2} r^{p-2}$ for all $r \geq 0$.

6. CONCLUDING REMARKS

In this paper, the output feedback Disturbance Attenuation Problem with Stability has been addressed for a class of minimum-phase nonlinear uncertain systems. The main contribution is to provide a constructive solution to the problem for uncertain plants under more relaxed assumptions than earlier work, i.e. that the system nonlinearities can be genuinely nonlinear functions of unmeasured states and the measured output. Extra conditions were given to guarantee the solvability of the standard L_2-gain Disturbance Attenuation Problem with Stability.

7. REFERENCES

Ball, J., J. Helton and M. Walker (1993). H^∞ control for nonlinear systems with output feedback. *IEEE TAC* **38**, 546–559.

Battilotti, S. (1996). Global output regulation and disturbance attenuation with global stability via measurement feedback for a class of nonlinear systems. *IEEE TAC* **41**, 315-327.

Isidori, A. (1995). *Nonlinear Control Systems.* 3rd Ed., Springer-Verlag. London.

Isidori, A. (1996). A note on almost disturbance decoupling for nonlinear minimum-phase systems. *Systems Control Letters* **27**, 191–194.

Jiang, D. C. and Z. P. Jiang (1997). H_∞ almost disturbance decoupling with stability for uncertain nonlinear systems. *Proc. 4th European Contr. Conf.*

Jiang, Z. P., A. Teel and L. Praly (1994). Small-gain theorem for ISS systems and applications. *Maths Control Sig. Syst.* **7**, 95–120.

Jiang, Z. P. and I. Mareels (1997). A small gain control method for nonlinear cascaded systems with dynamic uncertainties. *IEEE Trans. Automat. Contr.* **42**, 292–308.

Marino, R. and P. Tomei (1995). Nonlinear output feedback almost disturbance decoupling. *Proc. 34th Conf. Dec. Control* pp. New Orleans , 1–6.

Marino, R., W. Respondek, A. J. van der Schaft and P. Tomei (1994). Nonlinear $\mathcal{H}_\infty$ almost disturbance decoupling. *Systems and Control Letters* **23**, 159–168.

Saberi, A. (1987). Output-feedback control with almost-disturbance-decoupling property – a singular perturbation approach. *Int. J. Control* **45**, 1705–1722.

Saberi, A. and P. Sannuti (1988). Global stabilization with almost disturbance decoupling of a class of uncertain non-linear systems. *Int. J. Control* **47**, 717–727.

Sontag, E. D. (1995). On the input-to-state stability property. *European J. Control* **1**, 24–36.

Sontag, E. D. and A. Teel (1995). Changing supply functions in input/state stable systems. *IEEE TAC* **40**, 1476–1478.

van der Schaft, A. J. (1996). *L_2-Gain and Passivity Techniques in Nonlinear Control.* Springer-Verlag. London.

Willems, J. C. (1982). Almost invariant subspaces: an approach to high gain feedback design - part ii: almost conditionally invariant subspaces. *IEEE Trans. Automat. Contr.* **27**, 1071–1085.

SOLUTION OF THE PROBLEM OF ABSOLUTE STABILITY OF NONLINEAR CONTROLLED SYSTEM

F.A. Ovsepian

Institute of Control Science, Russian Academy of Sciences,
Profsoyuznaya str. 65, Moscow 117806, Russia
E-mail: ovsepian@ipu.rssi.ru

Abstract: It is proved that the properties of a nonlinear system are at the interface of two theories - deterministic and stochastic. Hence, one has to consider a steady random process at the output of the system when a similar random process forms an input of the system. In this case one has a determining factor which is the symmetry. It is the symmetry that makes a dynamic system as a frequency amplifier which allows determing the angle of absolute stability. *Copyright © 1999 IFAC*

Keywords: absolute stability, nonlinear system, random functions, physical model, autocorrelation function.

1. INTRODUCTION

It is obvious that from the stability of a nonlinear system for the arbitrary initial conditions follows the stability of a linear system. It is in this form that a problem was first stated by Lourier (1944), for the solution of which the Lyapunov's function was suggested as "quadratic form plus an integral of nonlinearity".

This problem has long transformed to that of absolute stability. It has been discussed in hundreds of articles and dozens of monographs but has not been solved yet.
The difficulty of solving the problem of an absolute stability is due to not only the complexity of the problem itself but also due to the unclear nature of the motion stability. It seems obvious that the properties of a nonlinear system should depend on its linear part. Until recently however, these properties have not been identified.

For the first time the idea of these properties being at the boundary of two theories - deterministic and stochastic was put forward by Ovsepian (1996) and was based on the identity of the properties of the solution of a linear deterministic system and the correlation functions of the stationary random processes. But that theorem, however, was not constructive. This presentation pursues the same approach using the example of the absolute stability problem statement. When a stationary random process takes place in a linear system the absolute stability angle turns out to be less than the angle of an asymptotic stability of a deterministic linear system and this narrowing of the angle is sufficient for the stability of a nonlinear system as a whole.

To determine the value of the angle, in this case, a stationary real differentiable random process is fed to the input of a linear system while the output of this system is adjusted to have a stationary random process. The symmetry will "force" such a linear dynamic system to function as an amplifier, i.e. when $\mathrm{Im}W(i\omega) \equiv 0$. In this mode the gain coefficient sign is determined, i.e. the sign of $\mathrm{Re}W(i\omega)$.
The criterion is stated using the language of frequency parameters widely applied to the control

theory. A stability in this presentation is understood in the asymptotic sense.

2. PROBLEM STATEMENT

The *absolute stability problem* consists in determining the conditions when the equilibrium position $\mathbf{u} = 0$ of the system

$$\mathbf{du} / dt = \mathbf{Au} + \mathbf{b}\varphi(y), \quad y = (\mathbf{c},\mathbf{u}), \quad \varphi(0) = 0 \qquad (1)$$

is asymptotically stable on the whole for any uni-valued continuous nonlinear function $\varphi(y)$, which meets the inequality

$$0 \leq y\varphi(y) \leq k_m y^2 \qquad (2)$$

where k_m - is a finite positive value.

In (1) matrix $\mathbf{A}$ is Hurwitzian, while $\mathbf{b}$ and $\mathbf{c}$ are specified constant vectors, the variable y is a result of scalar product of $\mathbf{c}$ and $\mathbf{u}$. It is assumed that for (1) the usual conditions of unique existence of solutions for all $t \geq 0$ for any initial conditions.

Excluding in (1) all variables but for y and $\varphi(y)$, one can obtain

$$N(p)y + Q(p)\varphi(y) = 0 \qquad (3)$$

where N(p) and Q(p) are the polynomials of p of the n-th and q-th degree, respectively ($n > q$), and p is a differentiation operator d/dt.
One will discuss here only systems where n and q are even numbers. The frequency characteristic of system (3) from the input $\varphi(y)$ to the output $-y$ can denote as $W(i\omega)$.

Assuming in (3) that $\varphi(y) = ky$, where k belongs to (2), one can obtain the asymptotically stable linear systems

$$N(p)y + kQ(p)y = 0. \qquad (4)$$

The problem is to obtain k_m, when the system (3) is stable for any $\varphi(y)$ from (2) and for the arbitrary initial conditions. The value of k_m is found by the Nyquist criterion or in accordance with the criterion determined below.

3. AUXILIARY PROPOSITIONS

Statement. A stationary random process

$$\gamma(t) = \xi\cos\omega t + \delta\sin\omega t$$

is real, if ξ and δ are the real random values with

$$E\xi = E\delta = 0, \; E\xi^2 = E\delta^2 = b \text{ and } E\xi\delta = 0.$$

The proof is in the calculation of the autocorrelation function $K_\gamma(\tau)$ of the process $\gamma(t)$ which equals

$$K_\gamma(\tau) = b\cos\omega\tau.$$

Corollary. If in the harmonic function

$$y(t) = b_1\cos\omega t + b_2\sin\omega t$$

the parameters b_1 and b_2 are changed in accordance with the table

1) $b_1 = +b, \; b_2 = -b,$ 3) $b_1 = -b, \; b_2 = -b,$
2) $b_1 = +b, \; b_2 = +b,$ 4) $b_1 = -b, \; b_2 = +b,$

the realization of the harmonic function y(t) one can be treated as the process $\gamma(t)$ realization.

Proof. The role of random values ξ and δ in the function y(t) is fulfilled by the parameters b_1 and b_2 respectively. Since y(t) should have characteristics of the process $\gamma(t)$, there should be an equality

$$Eb_1 = Eb_2 = 0, Eb_1^2 = Eb_2^2 = b, Eb_2 b_1 = 0.$$

The table given for the variation of the parameters b_1 and b_2 of the function y(t) meets the above requirements.
This proves Corollary.
Definition 1. The linear non-homogeneous system

$$N(p)z = Q(p)x(t)$$

will term a α-system, if :

a) x(t) is an arbitrary random stationary process;

b) a cross correlation function $K_{zx}(\tau)$ of the processes x(t) and z(t) is even, i.e.

$$K_{zx}(\tau) = K_{zx}(-\tau);$$

c) $K_{zx}(0) < 0$.

Definition 2. The inequality

$$-\mathrm{Re}W(i\omega) + 1/k_m > 0,$$

which holds at frequencies of the solution of the equation $\mathrm{Im}W(i\omega) = 0$, excluding the point $\omega = 0$, will term a modified Nyquist criterion.

Definition 3. The function $P(\omega)$ one will call sign - constant, if it assumes values of only one sign in the domain of the argument ω variation.

4. THEOREM

Theorem. The nonlinear system (3) is absolutely stable in the angle $[0, k_m]$ iff any linear non-homogeneous system in $[0, k_m]$ is a α-system.

The Proof of the Theorem.
Necessity. To prove the necessity it is sufficient to consider the solutions of the linear systems (4) in the angle
of a stability. Since the nonlinear system (3) is absolutely stable, then the linear system (4) in the same angle is also stable. Having chosen the initial conditions one has a solution y(t) of this system and after substituting it in (4) one obtains the identity

$$N(p)y(t) + kQ(p)y(t) \equiv 0. \quad (5)$$

The solution of the linear homogeneous system y(t) is at the same time a solution of the linear non-homogeneous system

$$N(p)y_1 + kQ(p)y(t) = 0 \quad (6)$$

where Q(p)y(t) is the concrete ime function.
A solution y(t) of this non-homogeneous system can be presented as a sum of two functions

$$y(t) = y_h(t) + y_p(t), \quad (7)$$

where $y_h(t)$ is a solution of a stable system $N(p)y_1 = 0$ for the same initial conditions which have been used for obtaining y(t), and $y_p(t)$ is a particular solution is of the same non-homogeneous system under the effect of the time function Q(p)y(t) holding for the zero initial conditions.

After the substitution of (7) to (6) one can obtain the identity

$$N(p)y_p(t) + kQ(p)y(t) \equiv 0 \quad (8)$$

because $N(p)y_h(t) \equiv 0$ by construction.
According to (5)

$$kQ(p)y(t) \equiv - N(p)y(t)$$

that is why one has the other identity

$$N(p)y_p(t) - N(p)y(t) \equiv 0. \quad (9)$$

The initial conditions in (4) are arbitrary. Since one considers a set of solutions $Y=\{y_i(t)\}$ of the system (4) and (6) obtained for the arbitrary randomly selected initial conditions, and the respective sets Y(t) and $Y_p(t)$. These sets are treated as random functions which have an arbitrary distribution of probabilities (arbitrary probability density). So hereinafter the functions y(t) and $y_p(t)$ can interprete as the random functions.

In sthe ystem (4) $y(t) \to 0$ for $t \to +\infty$, consequently the even continuation of the solution of this system to the negative time semi-axis should also possess a similar property, i.e. $y(-t) \to 0$ for $t \to -\infty$. Hence, one can build a linear non-homogeneous system

$$N_1(p)y_2 + kQ_1(p)y(-t) = 0, \quad (10)$$

where the polynomials $N_1(p)$ and $Q_1(p)$ differ from the polynomials N(p) and Q(p) in the sign before odd derivatives. For the system (9) one obtains the identity

$$N_1(p)y_p(-t) + kQ_1(p)y(-t) \equiv 0. \quad (11)$$

The solutions of (6) and (10) are the function $y_p(t)$ which is differentiable on the time interval $(-\infty, +\infty)$, while in the point $t = 0$ the function and its derivatives up to the $(n - 1)$th order are equal to zero.
The identity (11) spreads the respective sets Y(t) and $Y_p(t)$ to $t \le 0$.

Since for the solution y(t) the function $-y(t)$ is also solution of the same equation (4), it is possible put

$$E\,Y(t) = E\,Y_p(t) = 0.$$

Multiplying (8) by y(s) and, having calculated the mathematical expectation, one obtains

$$N(p)E[y_p(t)y(s)] + kQ(p)E[y(t)y(s)] \equiv 0 \quad (12)$$

which links the autocorrelation function E[y(t)y(s)] of the function y(t) with the cross correlation function
$E[y_p(t)y(s)]$ of the functions $y_p(t)$ and y(t).
Multiplying (9) by y(s) and, having calculated the mathematical expectation, can obtains other identity between the same correlation functions

$$N(p)E[y_p(t)y(s)] - N(p)E[y(t)y(s)] \equiv 0. \quad (13)$$

The identities (12) and (13) are used for the calculation of the angle of the absolute stability.

Supply to the input ky(t) of the system (6) a random stationary arbitrary process x(t):

$$N(p)\,z = Q(p)\,x(t) \quad (14)$$

So $N(p)y_1 = 0$ is stable and $n > q$ at the output of this system forms the similar random process z(t).
After the substitution z(t) to (14), can obtains the identity

$$N(p)z(t) \equiv Q(p)x(t). \qquad (15)$$

Multiplying (15) by y(s) and, having calculated the mathematical expectation, can obtains

$$N(p)E[z(t)x(s)] = Q(p)E[x(t)x(s)]. \qquad (16)$$

It is not difficult to prove that processes z(t) and x(t) not only stationary, as well as stationary bound.
This signifies that the autocorrelation function

$$K_x(\tau) = E[x(t)x(s)]$$

of the process x(t) and the cross correlation function

$$K_{zx}(\tau) = E[z(t)x(s)]$$

of the processes x(t) and z(t) depends only on $\tau = t - s$.
Having substituted these correlation functions to (16), one has

$$N(p)\, K_{zx}(\tau) \equiv Q(p)\, K_x(\tau). \qquad (17)$$

Comparing the identities (12) and (17), one concludes

$$E[y_p(t)y(s)] = K_{zx}(\tau),$$
$$-k\, E[y(t)y(s)] = K_x(\tau),$$

what accounting (13) produce

$$-k\, K_{zx}(\tau) = K_x(\tau), \qquad (18)$$

i.e. the cross correlation function $K_{zx}(\tau)$ is the even function.
Systems (6) and (14) are described by identic differential equations, but their inputs and outputs are different. To distinguish these systems the frequency characteristic of system (14) denote as $W_t(i\omega)$.
According to the Nyquist criterion the angle of the stability of the linear deterministic system (6) is defined in coordinate plane $W(i\omega)$ on frequency $\omega = \omega_{cr}$, when $\text{Im}W(i\omega_{cr}) = 0$ so that

$$\text{Re}W(i\omega_{cr}) + 1/k_m > 0. \qquad (19)$$

Since the angle of the stability is defined on the negative real semi-axis of plane $W(i\omega)$, so

$$\text{signRe}W(i\omega_{cr}) < 0. \qquad (20)$$

Now it is necessary to define the sign of $\text{Re}W_t(i\omega)$, for what an equation for the correlation functions (16) rewrite to the other form, multiplying (15) upon $Q(p)x(s)$:

$$[N(p)z(t)][Q(p)x(s)]\}=E\{[Q(p)x(t)][Q(p)x(s)]\}. \qquad (21)$$

Substituting the correlation functions to (21) by their spectral decomposition, computing the expectation and, assuming $\tau = 0$, one obtains the identity

$$N(i\omega)Q(-i\omega)S_{zx}(\omega) \equiv |Q(i\omega)|^2 S_x(\omega),$$

where $S_x(\omega)$ and $S_{zx}(\omega)$ are, respectively, the spectral density of the process x(t) and the cross spectral density of the processes x(t) and z(t). Dividing the identity obtained by $|N(i\omega)|^2$ and identifying the imaginary component one has two identities:

$$S_{zx}(\omega)\text{Re}W_t(i\omega) \equiv |W_t(i\omega)|^2 S_x(\omega), \qquad (22)$$

$$\text{Im}W_t(i\omega) \equiv 0. \qquad (23)$$

The frequency characteristic $W(i\omega)$ is a particular case of the transmission Laplace function W(p) for $p = i\omega$. When the parameter $p \to 0$, i.e. $\omega \to 0$, $t \to \infty$ due to the limit properties of the Laplace transformations. The identities (22) and (23) are obtained for $s - t = \tau = 0$, i.e. for an arbitrary finite t. Hence, in the modified Nyquist criterion which is going to be used below for obtaining the stability angle the frequency $\omega \neq 0$.

The system (17) operates in the mode of amplifier, i.e. the processes x(t) and z(t) in (14) do not change their own signs one respect to another. Since $K_{zx}(\tau)$ is even, it follows that a cross spectral density $S_{zx}(\omega)$ is a sign-constant function. From (23) it immediately follows that $\text{Re}W_t(i\omega)$ is also a sign-constant function. Hence, it is required to determine the sign of one of these functions at an arbitrary frequency ω. The sign of $\text{Re}W(i\omega_{cr})$ is known according (20), consequently is known the sign of $\text{Re}W_t(i\omega)$, i.e. $\text{Re}W_t(i\omega) < 0$ therefore $S_{zx}(\omega) < 0$. That is why

$$K_{zx}(0) < 0. \qquad (24)$$

according to the well-known Khinchine theorem.

The condition (24) is the nessesary condition for stability of the system (14).

The double iterated differentiation changes the sign of a stationary random process. This readily follows by definition of $\gamma(t)$ and implies the following assertion.

1. When $n - q = 2, 6, 10, \ldots$ one has

$$K_{zx}(0) < 0.$$

According to (24). the system (14) is a α-system. Therefore the stability angle of a of the system (14) is defined by the ordinary Nyquist criterion.

2. When $n - q = 4, 8, 12, \ldots$ one has

$$K_{z(-x)}(0) < 0.$$

According to (24) the system (14) has a form

$$N(p)\, z = [-Q(p)][-x(t)]. \qquad (25)$$

and the system (25) is α-system. Therefore the angle of the stability of the system (14) is defined two times:

a) first once is defined by the ordinary Nyquist criterion because the linear deterministic system is stable;

b) second once is defined by the modified Nyquist criterion according to (25).

Minimum of two angles defines the angle, in which system (25) stables.
The necessity is proved.

The proof of the sufficiency is not cited here.

5. RESULTS APPLIED TO THEORETICAL PHYSICS AND ASTROPHYSIC

For the further study supply on input of system (6) real stationary random process $x(t, \omega) = \gamma(t)$ from statement. Define the steady oscillation at the output

$$z(t, \omega) = |W_t(i\omega)|\, [\xi \cos(\omega t - \alpha) + \delta \sin(\omega t - \alpha)],$$

where $z(t, \omega) = -y_1(t, \omega)$ and $\alpha = \arg W_t(i\omega)$.
According to (23) one has

$$z_1(t, \omega) = |W_t(i\omega)|\, (\xi \cos\omega t + \delta \sin\omega t) = |W_t(i\omega)|\, x(t, \omega)$$

Variation of $\mathrm{Re}W_t(i\omega)$ in the $W(i\omega)$ plane occurs along the real axis. Since $k_m \neq \infty$, one can transfer the axis $\mathrm{Im}W(i\omega)$ in this plane to the point ω_{cr}. Magnifying the value of k_m to k_{cr} so that

$$[0, k_m] = [0, k_{cr}),$$

the system is taken to the boundary of the stability and from the inequality (19) one has the equality

$$\mathrm{Re}W_t(i\omega_{cr}) + 1/k_{cr} = 0, \qquad (26)$$

It follows from the equality (26) that in a new coordinate system the axis of coordinates is set on the boundary of the stability linear system where a real component of the frequency characteristic in the linear system equals zero. The frequency characteristic of the system considered cannot be shifted more to the right on the new ordinate axis since in that part of the plane the sign of $\mathrm{Re}W_t(i\omega)$ is positive. Consequently, at the point

$$\omega = \omega_{cr}$$

one can obtain

$$z(t, \omega_{cr}) = 0, \quad x(t, \omega_{cr}) = 0.$$

According to the corrolary of statement one has

$$x(t, \omega_{cr}) = b\,(\cos\omega_{cr} t + \sin\omega_{cr} t) = 0$$

or

$$x(t, \omega_{cr}) = b\,(\cos\omega_{cr} t - \sin\omega_{cr} t) = 0.$$

It means that not only the processes are the zero on the frequency $\omega = \omega_{cr}$ but are the zero and all their derivatives. It is the prove of two next corollary.

Corollary 1. The coordinate system at the boundary of stability is the priority-based coordinate system.

Corollary 2. Electromagnetic waves cannot overcome the motion stability boundary.

The coordinate system at the boundary of stability of the Universe motion is exactly the priority-based coordinate system which makes real Newton's absolute space. In the light of the above of interest are the excerpts from the book by two well-known physicists.

Lanczos (1965) wrote: "... Now is physics this question concerning a "frame of reference" arises specifically in connection of measurements ... Let us suppose we have to measure the position of a planet in astronomy... We need certain fixed points in the universe. But where are these fixed points? No matter what we may choose as "fixed points" we cannot help operating certain entities which have no absolute significance. There are no natural landmarks in space and thus we are forced to set up some artifical landmarks. But then we have to be awase of the fact this landmarks were set up by ourselves. There are not have absolute qualities within the frame of the physical universe."

Born (1962) wrote: " ... A concept refers to a physical reality only when there is something ascertainable by measurement corresponding to it in the world of the phenomena ... Every concept that does not satisfy it has gradually been eliminated from the structure of physics. We see at once that in this sense a "fixed spot" in Newton's absolute space has no (physical) reality... In this way the absolute space of Newton loses a considerable part of its weird existense. A space in which there is not place that can be marked by any physical means whatsoever is it any rate a very subtle and abstract idea, and not simply a box into which material things are crammed. "

According to the Corollary 1 a natural landmark or a "fixed spot" in Newton's absolute space is the coordinate system at the boundary of the stability.

According to the Corollary 2 the velocity of an electromagnetic wave is constant.

In the theoretical physics the constant velocity of an electromagnetic wave in the Universe is just postulated as a result of experimental observations. The author proves that the constant velocity of the electromagnetic wave is a specific condition of a stable system. The author believes that the result obtained by him leads to the following conclusion: we live in a stationary Universe where there can be no such processes as "expansion", "narrowing", etc.

REFERENCES

Born, M. (1962) Albert Einstein and the Cosmic World Order, 69-70. Dover publications, NY .

Lanczos, C. (1965) Einstein's Theory of Relativity, 20-21. Interscience publishers,.NY..

Lourier, A.I. (1944). On the theory of stability of controled systems. *Applied mathematics and mechanics*, **8** , 17 - 44. Nauka, Moscow.

Ovsepian, F.A. (1996). Necessary and sufficient conditions for absolute stability of nonlinear controlled systems. Abstract. *IY International Workshop "Stability and oscillations of nonlinear control systems"*, 9 - 11. Institute of Control Sciences, Moscow.

14th Triennial World Congress, Beijing, P.R. China

LMI REPRESENTATION OF THE SHIFTED POPOV CRITERION FOR STABILITY OF NONLINEAR FEEDBACK SYSTEMS

Yuji Kakutani†, Tomomichi Hagiwara‡ and Mituhiko Araki††

Department of Electrical Engineering, Kyoto University,
Yoshida, Sakyo-ku, Kyoto 606-8501, Japan
† *Email:* kakutani@kuee.kyoto-u.ac.jp
‡ *Email:* hagiwara@kuee.kyoto-u.ac.jp
†† *Email:* araki@kuee.kyoto-u.ac.jp
Phone: +81-75-753-5334, Fax: +81-75-753-5332

Abstract. This paper is concerned with a stability criterion of nonlinear feedback systems, called the shifted Popov criterion. It includes the Popov and circle criteria as special cases, and was originally derived as a frequency-domain condition to check stability in a graphical manner. Recently, it was shown that this stability criterion can be reduced to a convex optimization problem in the frequency domain. Motivated by this fact, this paper shows that the shifted Popov criterion, both in the Lyapunov stability context and the input-output stability context, can be transformed into an LMI representation when the linear part of the nonlinear feedback system has a state-space realization. *Copyright ©1999 IFAC*

Key Words. Nonlinear systems, Stability criteria, Lyapunov stability, Input-output stability, Linear Matrix Inequality

1 INTRODUCTION

As for the stability analysis of nonlinear feedback systems, two types of stability conditions are well known: the Popov criterion and the circle criterion. The Popov criterion can be applied only to such a system whose nonlinear part is memoryless and time-invariant. On the other hand, the circle criterion can deal with a more general system whose nonlinear part has memory (dynamics) and/or time-varying characteristics. The condition the circle criterion requires, however, is much stronger than that required by the Popov criterion.

Now, suppose that the nonlinear characteristic of the system is almost memoryless and time-invariant, and with only slight dynamics and/or time-variations. Although the stability condition for such a system is, intuitively speaking, expected to be close to the one obtained by applying the Popov criterion, we can actually apply only the circle criterion, so that we are led to a much stronger stability condition.

In order to bridge such a gap between the two criteria, a more general stability criterion was derived, both in the Lyapunov stability context [Furutani, Saeki and Araki(1992), Furutani, Saeki and Araki (1994)] and the input-output stability context [Hagiwara, Furutani and Araki (1997), Hagiwara, Miyake, Furutani and Araki (1997)]. It involves two free parameters, includes the above two famous criteria as special cases, and enables us to examine stability in the frequency domain in a graphical manner. Since we plot the shifted Popov locus (i.e., the Popov locus with each point shifted to the left by the amount determined by the two parameters), this stability criterion is called the shifted Popov criterion. In Hagiwara, Miyake, Furutani and Araki (1997), it was further shown that the problem of checking the existence of the two parameters that fulfill the required graphical property of the resulting shifted Popov locus, can be reduced to a convex optimization problem in the frequency domain. This convex optimization approach is quite important since we are released from the trial-and-error search for the appropriate values of the two parameters. However, since it is impossible for a computer to deal with all the infinitely many frequencies, this is, strictly speaking, an approximate way of checking the stability condition, from the numerical point of view.

Now, in this paper, we show that the shifted Popov criterion, both in the Lyapunov stability context and the input-output stability context, can be transformed into an LMI representation in the state space when the linear part of the nonlinear feedback system is finite-dimensional. With this LMI approach, we are free from the numerical difficulties mentioned above. Furthermore, it would be useful for extending the stability criterion to take account of the parametric uncertainties in the linear part of the nonlinear feedback system.

The notation used in this paper is standard. L_2 denotes the space of (Lebesgue) square integrable functions over the nonnegative time interval, that is,

$$L_2 = \left\{ z(t) \mid \|z\| = \left\{ \int_0^\infty \|z(t)\|^2 dt \right\}^{1/2} < \infty \right\} . \quad (1)$$

L_{2e} denotes its extended space, that is,

$$L_{2e} = \left\{ z(t) \mid \|z_T\| = \left\{ \int_0^T \|z(t)\|^2 dt \right\}^{1/2} < \infty \quad ({}^\forall T \geq 0) \right\} \quad (2)$$

where

$$z_T(t) := \begin{cases} z(t) & (0 \leq t < T) , \\ 0 & (t \geq T) . \end{cases} \quad (3)$$

2 REVIEW OF THE SHIFTED POPOV CRITERION

In this section, we review the stability criterion of nonlinear feedback systems known as the shifted Popov criterion, both in the Lyapunov stability context Furutani, Saeki and Araki(1992), Furutani, Saeki and Araki (1994) and the input-output stability context Hagiwara, Furutani and Araki (1997), Hagiwara, Miyake, Furutani and Araki (1997).

2.1 *Nonlinear Feedback System*

Let us consider the scalar nonlinear feedback system Σ shown in Fig. 1. Here, G denotes the linear part of the system, and we assume that it is time-invariant, finite-dimensional, and has the following state-space realization:

$$\frac{dx}{dt} = Ax + Bu , \quad (4)$$

$$y = Cx . \quad (5)$$

We assume that A is stable. The transfer function of G is given by

$$\widehat{G}(s) = \left[\begin{array}{c|c} A & B \\ \hline C & 0 \end{array} \right] = C(sI - A)^{-1}B . \quad (6)$$

The nonlinear part of the system, denoted by N, is assumed to be modeled as the parallel connection of a memoryless time-invariant nonlinearity N_0 and another type of nonlinearity N_1 as in Fig. 2. Namely, $f = Ne$ satisfies

$$f = f_0 + f_1 \quad (7)$$

$$f_0 = N_0 e , \ f_1 = N_1 e . \quad (8)$$

The idea behind this assumption is that the nonlinear part N can be approximated by N_0 to within an 'error' of N_1; it is often the case that the nonlinearity encountered in practice is 'almost memoryless and time-invariant' (so that it can be approximated by N_0) but with 'slight dynamics and/or time-variations' as modeled by N_1 (See e.g. Furutani, Saeki and Araki(1992), Hagiwara, Furutani and Araki (1997), Hagiwara, Miyake, Furutani and Araki (1997)).

Here, we assume that N_0 belongs to sector $[a, b]$ $(a \leq b < \infty)$ in the sense defined below.

Definition 1 The memoryless time-invariant nonlinearity N_0 is said to belong to sector $[a, b]$ if there exists a nonlinear characteristic function $\nu_0(\cdot)$ such that

$$f_0(t) = \nu_0(e(t)) \quad (9)$$

and

$$a \leq \frac{\nu_0(e)}{e} \leq b \quad (e \neq 0) , \quad \nu_0(0) = 0 \quad (10)$$

Without loss of generality, we assume $a = 0$ in the following.

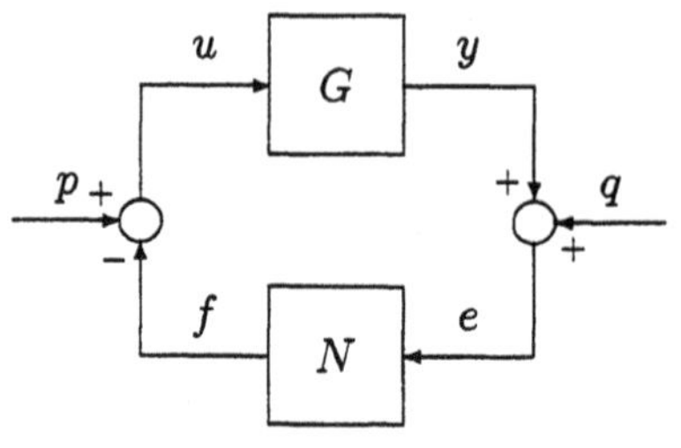

Fig. 1. Nonlinear feedback system Σ.

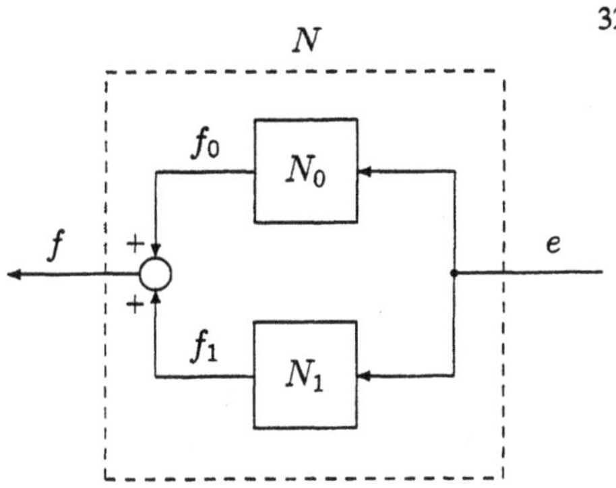

Fig. 2. Structure of the nonlinear part N.

As for the nonlinearity N_1, we make the following assumption. In the Lyapunov stability context, we assume that N_1 is memoryless and time-varying, and the input e and output f_1 of N_1 satisfy

$$\|f_1(t)\| \le \varepsilon \|e(t)\| \quad (^\forall t \ge 0) . \tag{11}$$

Likewise, in the input-output stability context, we assume that N_1 is memoryless and time-varying, and belongs sector $[-\varepsilon, \varepsilon]$ in the sense defined below.

Definition 2 The dynamical and time-varying nonlinearity N_1 is said to belong to sector $[-\varepsilon, \varepsilon]$ if

$$\|f_{1T}\| \le \varepsilon \|e_T\| \quad (^\forall T \in [0, \infty)\ ) . \tag{12}$$

The parameter ε (> 0) above could be interpreted as a measure of the dynamics and/or time-variations involved in the nonlinear part N, which cannot be modeled by N_0. Also note that the assumption (12) is slightly more general than the assumption (11) in the sense that the latter implies the former.

2.2 *Shifted Popov Criterion in the Frequency Domain*

The shifted Popov criterion in the Lyapunov stability context reads as follows [Furutani, Saeki and Araki(1992), Furutani, Saeki and Araki (1994)].

Theorem 1 (Shifted Popov Criterion for the Lyapunov Stability) Suppose that (A, B) is controllable and (C, A) is observable. Then, the nonlinear feedback system Σ is stable in the sense of Lyapunov if there exist a real number θ and a positive number ρ satisfying the following two conditions:

(a)

$$\begin{aligned} 1 &+ b \cdot \mathrm{Re}\{(1 + j\omega\theta)\widehat{G}(j\omega)\} \\ &- \frac{b\varepsilon}{2\rho}|(1 + j\omega\theta)\widehat{G}(j\omega)|^2 \\ &- \varepsilon\left(\varepsilon + \frac{b\rho}{2}\right)|\widehat{G}(j\omega)|^2 > 0 \\ &\qquad (^\forall \omega \in [0, \infty]) \end{aligned} \tag{13}$$

(b) $(C(I + \theta A), A)$ is observable when the matrix A has repeated eigenvalues.

Furthermore, under the same constraint(b) on θ, Σ is asymptotically stable in the large if there exists a positive number δ such that (13) remains true under the replacement of $j\omega$ with $j\omega - \delta$.

Remark 1 In stead of the inequality (13), the original form of the shifted Popov criterion, given in Furutani, Saeki and Araki(1992), Furutani, Saeki and Araki (1994), requires the existence of a real number θ and positive numbers κ, μ such that

$$\kappa\mu \ge \varepsilon^2 \tag{14}$$

$$\begin{aligned} 1 &+ b \cdot \mathrm{Re}\{(1 + j\omega\theta)\widehat{G}(j\omega)\} \\ &- \frac{b\kappa}{2}|(1 + j\omega\theta)\widehat{G}(j\omega)|^2 \\ &- \varepsilon\left(\varepsilon + \frac{b\mu}{2}\right)|\widehat{G}(j\omega)|^2 \ge 0 \\ &\qquad (^\forall \omega \in [0, \infty)\ ) \end{aligned} \tag{15}$$

$$1 - \frac{b}{2}\{\kappa\theta^2(CB)^2 - 2\theta CB\} > 0 . \tag{16}$$

The condition of Theorem 1 is actually a slightly strengthened version of this original condition. The gap between these conditions, however, is quite small; the condition of Theorem 1 is obtained if the nonstrict inequality in (15) is replaced by a strict one. To see this, observe the following two points:

1. The lefthand side of (16) is nothing but the limit of the lefthand side of (15) for $\omega \to \infty$.
2. Without loss of generality, the search for κ and μ may be carried out over the 'boundary' of (14) (i.e., $\kappa\mu = \varepsilon^2$) [Furutani, Saeki and Araki (1994)]. In this case, they can be parameterized by

$$\kappa = \frac{\varepsilon}{\rho}, \quad \mu = \varepsilon\rho \tag{17}$$

where ρ is a positive parameter†. Substituting the above into the lefthand side of (15) yields the lefthand side of (13).

Next, the shifted Popov criterion in the input-output stability context reads as follows [Hagiwara, Furutani and Araki (1997), Hagiwara, Miyake, Furutani and Araki (1997)].

Theorem 2 (Shifted Popov Criterion for the Input-Output Stability) The nonlinear feedback system Σ is pseudo L_2-stable if there exist a real number θ and a positive number ρ such that (13) holds.

†In Furutani, Saeki and Araki (1994), $\psi \left(= \rho^{-1}\right)$ was used instead of ρ. In this paper, we use ρ in order to exploit a convexity property.

Here, pseudo L_2-stability (the same stability notion as in the Popov criterion) is defined as follows.

Definition 3 The nonlinear feedback system Σ is said to be pseudo L_2-stable if there exists a positive number γ such that

$$\left\| \begin{bmatrix} f \\ y \end{bmatrix} \right\| \leq \gamma \left\| \begin{bmatrix} p \\ q \\ \dot{q} \end{bmatrix} \right\| \quad \text{(whenever } p,\ q,\ \dot{q} \in L_2) \ , \tag{18}$$

where $\dot{q}$ denotes the derivative of q.

3 MAIN RESULTS—LMI REPRESENTATION OF THE SHIFTED POPOV CRITERION

In this section, we show that the shifted Popov criterion, reviewed in the preceding section, and represented as a frequency-domain condition, can be transformed into an LMI representation. One possible way to do this might be to modify the arguments of Furutani, Saeki and Araki(1992), Furutani, Saeki and Araki (1994) with a Lyapunov function, but we make use of simpler arguments in Hagiwara, Furutani and Araki (1997), Hagiwara, Miyake, Furutani and Araki (1997). This approach is also convenient since the condition dealt with in this paper is a slightly stronger one than that in Furutani, Saeki and Araki(1992), Furutani, Saeki and Araki (1994) (see Remark 1).

According to the arguments of Hagiwara, Furutani and Araki (1997),Hagiwara, Miyake, Furutani and Araki (1997), the condition (13) is true if and only if the system

$$G_s = \begin{bmatrix} \varepsilon^{1/2} G(1 + \frac{b}{2} M_\theta G)^{-1} \\ \left(\frac{b}{2\rho}\right)^{1/2} M_\theta G(1 + \frac{b}{2} M_\theta G)^{-1} \end{bmatrix} \tag{19}$$

is stable and

$$\|\widehat{G}_s\|_\infty < \left(\varepsilon + \frac{b\rho}{2}\right)^{-1/2} \tag{20}$$

where M_θ denotes the multiplier with transfer function

$$\widehat{M}_\theta(s) = 1 + \theta s \tag{21}$$

and $\|\widehat{G}_s\|_\infty$ denotes the H_∞-norm of the transfer function $\widehat{G}_s(s)$ associated with G_s.

Here, $\widehat{G}_s(s)$ has a state-space realization

$$\widehat{G}_s(s) = \left[\begin{array}{c|c} A_s & B_s \\ \hline C_s & D_s \end{array}\right] \tag{22}$$

where

$$A_s = A - \frac{b}{2 + b\theta CB} BC(I + \theta A) \ ,$$
$$B_s = \frac{2}{2 + b\theta CB} B \ ,$$
$$C_s = \begin{bmatrix} \varepsilon^{1/2} C \\ \left(\frac{b}{2\rho}\right)^{1/2} \frac{2}{2 + b\theta CB} C(I + \theta A) \end{bmatrix} ,$$
$$D_s = \begin{bmatrix} 0 \\ \left(\frac{b}{2\rho}\right)^{1/2} \frac{2\theta CB}{2 + b\theta CB} \end{bmatrix} . \tag{23}$$

Therefore, applying the bounded real lemma [Boyd, Ghaoui, Feron and Balakrishnan (1994)], the condition (13) is true if and only if

$$R_s = \left(\varepsilon + \frac{b\rho}{2}\right)^{-1} I - D_s D_s^T > 0 \tag{24}$$

and there exists $X > 0$ such that

$$A_s X + X A_s^T + (X C_s^T + B_s D_s^T) R_s^{-1} \times (X C_s^T + B_s D_s^T)^T + B_s B_s^T < 0 \, . \tag{25}$$

Alternatively, we can rewrite (25) as

$$P A_s + A_s^T P + (C_s^T + P B_s D_s^T) R_s^{-1} \times (C_s^T + P B_s D_s^T)^T + P B_s B_s^T P < 0 \tag{26}$$

where $P = X^{-1} > 0$. By substituting (23) into (26) and rearranging the result, we obtain

$$PA + A^T P + \varepsilon\left(\varepsilon + \frac{b\rho}{2}\right) C^T C + \frac{\left(\frac{2}{2 + b\theta CB}\right)^2}{1 - \frac{b}{2\rho}\left(\varepsilon + \frac{b\rho}{2}\right)\left(\frac{2\theta CB}{2 + b\theta CB}\right)^2} \times \begin{bmatrix} PB & (I + \theta A^T) C^T \end{bmatrix} \times \begin{bmatrix} 1 & \frac{b\varepsilon}{2\rho}\theta CB - \frac{b}{2} \\ \frac{b\varepsilon}{2\rho}\theta CB - \frac{b}{2} & \frac{b}{2\rho}\left(\varepsilon + \frac{b\rho}{2}\right) \end{bmatrix} \times \begin{bmatrix} B^T P \\ C(I + \theta A) \end{bmatrix} < 0 \, . \tag{27}$$

Further rearranging the above, we are led to the following inequality.

$$PA + A^T P + \varepsilon\left(\varepsilon + \frac{b\rho}{2}\right) C^T C - \begin{bmatrix} PB - \frac{b}{2}(I + \theta A^T) C^T & \left(\frac{b\varepsilon}{2}\right)^{1/2} (I + \theta A^T) C^T \end{bmatrix} \times \begin{bmatrix} -(1 + b\theta CB) & \left(\frac{b\varepsilon}{2}\right)^{1/2} \theta CB \\ \left(\frac{b\varepsilon}{2}\right)^{1/2} \theta CB & -\rho \end{bmatrix}^{-1}$$

$$\times \begin{bmatrix} B^TP - \frac{b}{2}C(I+\theta A) \\ \left(\frac{b\varepsilon}{2}\right)^{1/2} C(I+\theta A) \end{bmatrix} < 0 \quad (28)$$

Then, by using the Schur complement argument, we finally obtain the following LMI representation of the condition (26):

$$\begin{bmatrix} PA + A^TP + \varepsilon\left(\varepsilon + \frac{b\rho}{2}\right)C^TC & PB - \frac{b}{2}(I+\theta A^T)C^T & \left(\frac{b\varepsilon}{2}\right)^{1/2}(I+\theta A^T)C^T \\ B^TP - \frac{b}{2}C(I+\theta A) & -(1+b\theta CB) & \left(\frac{b\varepsilon}{2}\right)^{1/2}\theta CB \\ \left(\frac{b\varepsilon}{2}\right)^{1/2} C(I+\theta A) & \left(\frac{b\varepsilon}{2}\right)^{1/2}\theta CB & -\rho \end{bmatrix} < 0 \quad (29)$$

Here, the above LMI condition implies, in particular, that

$$\begin{bmatrix} -(1+b\theta CB) & \left(\frac{b\varepsilon}{2}\right)^{1/2}\theta CB \\ \left(\frac{b\varepsilon}{2}\right)^{1/2}\theta CB & -\rho \end{bmatrix} < 0 \quad (30)$$

and it is easy to verify that the condition (24) is equivalent to this inequality. Hence, we can conclude that the condition (29) is the LMI representation of the frequency-domain condition (13).

Thus, from Theorem 2, we obtain the following theorem.

Theorem 3 (Shifted Popov Criterion for the Input-Output Stability) The nonlinear feedback system Σ is pseudo L_2-stable if there exist a real number θ, a positive number ρ, and a positive-definite matrix P such that the LMI condition (29) holds.

Next, let us consider the condition of Theorem 1. In this case, it is important to fulfill the additional constraint on the parameter θ regarding the observablity of $(C(I+\theta A), A)$ when A has repeated eigenvalues. But, since $(C(I+\theta A), A)$ is observable if and only if $1+\lambda\theta \neq 0$ for each eigenvalue λ of A, this observablity constraint can always be satisfied by slightly modifying the value of θ. Also, for each fixed $\rho > 0$ and $P > 0$, the set of θ satisfying the strict inequality (29) is open. Therefore, we are led to the conclusion that we may actually drop this additional constraint on θ.

Furthermore, in a similar fashion (i.e., from the strictness of the inequality (29)), we can show the following: if (29) is satisfied, then it is always possible to find a positive δ such that the condition (13) with $j\omega$ replaced by $j\omega - \delta$ is also true.

Hence, from Theorem 1, we obtain the following theorem.

Theorem 4 (Shifted Popov Criterion for the Lyapunov Stability) Suppose that (A, B) is controllable and (C, A) is observable. Then, the nonlinear feedback system Σ is asymptotically stable in the large in the sense of Lyapunov if there exist a real number θ, a positive number ρ, and a positive-definite matrix P such that the LMI condition (29) holds.

Remark 2 As shown in the previous studies [Furutani, Saeki and Araki(1992), Furutani, Saeki and Araki (1994), Hagiwara, Furutani and Araki (1997), Hagiwara, Miyake, Furutani and Araki (1997)], the shifted Popov criterion includes the Popov and circle criteria as special cases. This can be verified also from (29): if $\varepsilon = 0$, then (29) reduces to the LMI representation of the Popov criterion [Boyd, Ghaoui, Feron and Balakrishnan (1994)], and if $b = 0$, then (29) reduces to that of the circle criterion [Boyd, Ghaoui, Feron and Balakrishnan (1994)].

4 EXAMPLE

Let us consider the nonlinear feedback system of Fig. 1, where we assume that the linear part G is given by

$$A = \begin{bmatrix} 0 & 1 & 0 \\ 0 & 0 & 1 \\ -2 & -5 & -4 \end{bmatrix}, \quad B = \begin{bmatrix} 0 \\ 0 \\ 1 \end{bmatrix},$$
$$C = \begin{bmatrix} 5 & 0 & 0 \end{bmatrix}. \quad (31)$$

Here, (A, B) is controllable and (C, A) is observable. Suppose that the parameters associated with the nonlinear part N are $b = 3$ and $\varepsilon = 0.25$.

Now we examine the stability of the above system, by rewriting the LMI (29) to a convex optimization problem

$$\min_{P>0,\rho>0,\theta} \lambda \ \text{ subject to } \ S(P,\theta,\rho) \leq \lambda I \ , \quad (32)$$

where $S(P,\theta,\rho)$ denotes the lefthand side of (29). Obviously the LMI (29) is satisfied if and only if

$$\lambda_{\min} < 0 \quad (33)$$

where $\lambda_{\min}$ is the minimum value of λ.

In our example, we obtain

$$\lambda_{\min} = -0.0182748 < 0 \quad (34)$$

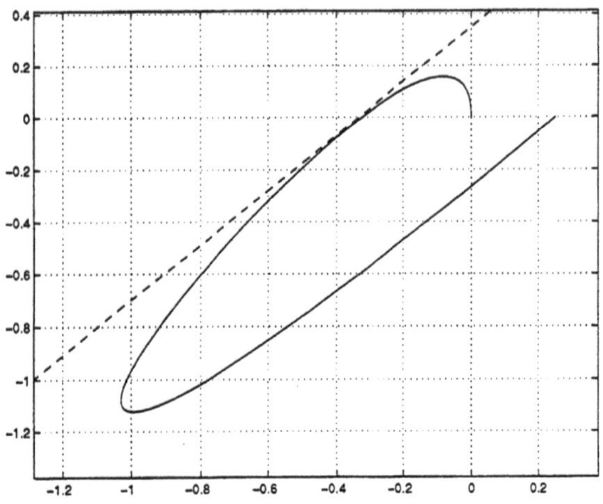

Fig. 3. The shifted Popov locus for $(\theta, \rho) = (0.953808, 2.26932)$

at

$$\theta = 0.953808 \ , \ \rho = 2.26932 \ , \tag{35}$$

$$P = \begin{bmatrix} 60.1551 & 36.3104 & 10.0223 \\ 36.3104 & 38.7872 & 7.67110 \\ 10.0223 & 7.67110 & 4.05211 \end{bmatrix} . \tag{36}$$

Namely, we can conclude that the nonlinear feedback system considered here satisfies the LMI (29), so that it is stable both in the Lyapunov stability sense and the input-output stability sense.

For reference, the shifted Popov locus with the parameters (35) is shown in Fig. 3 (where the solid line is the shifted Popov locus and the dotted line is the Popov line). Since the shifted Popov locus lies to the right of the Popov line, we can also conclude stability of this system in a graphical manner (see Furutani, Saeki and Araki(1992), Furutani, Saeki and Araki (1994) for details).

5 CONCLUSION

In this paper, we studied a stability condition of nonlinear feedback systems, called the shifted Popov criterion, and showed that it can be transformed into an LMI representation in the state space when the linear part of the nonlinear control system is finite-dimensional.

With this LMI approach, we can get around the difficulty of the convex optimization approach in the frequency domain: we no longer have to deal with all the infinitely many frequencies. None the less, the results of this paper will never diminish the significance of the results of Hagiwara, Furutani and Araki (1997), Hagiwara, Miyake, Furutani and Araki (1997) since the linear part that the LMI approach can deal with is only a finite-dimensional time-invariant one, while the frequency-domain convex optimization approach can deal with an infinite-dimensional time-invariant one such as the system with a pure delay.

6 REFERENCES

Boyd, S., L. E. Ghaoui, E. Feron and V. Balakrishnan (1994). *Linear Matrix Inequarity in System and Control Theory*, Society for Industrial and Applied Mathematics.

Furutani, E., M. Saeki and M. Araki (1992). Shifted Popov Criterion and Stability Analysis of Fuzzy Control Systems. *Proc. 31st Conference on Decision and Control*, Tucson, Arizona, pp. 2790–2795.

Furutani, E., M. Saeki and M. Araki (1994). Shifted Popov Criterion (in Japanese). *Trans. Institute of Systems, Control and Information Engineers*, 7-7, pp. 247–254.

Hagiwara, T., E. Furutani and M. Araki (1997). Stability Condition of a Class of Nonlinear Feedback Systems: Reduction to a Convex Problem. *European Control Conference '97*, FR-A C4, Brussels, Belgium.

Hagiwara, T., Y. Miyake, E. Furutani and M. Araki (1997). Input-Output Stability of a Class of Nonlinear Feedback Systems: A Convexity Approach. *26th SICE Symposium on Control Theory*, pp. 169–174.

ON CONSTRUCTION OF NON-SMOOTH LYAPUNOV FUNCTIONS FOR NON-SMOOTH SYSTEMS

Q. Wu* and **N. Sepehri**

Department of Mechanical & Industrial Engineering
University of Manitoba
Winnipeg, Manitoba
Canada R3T 5V6
*cwu@cc.umanitoba.ca
(Fax) (204) 275-7507

Abstract: Lyapunov's stability theory has been extended to the stability analysis of non-smooth systems in which non-smooth Lyapunov functions are often constructed. It is believed that the main difficulty in Lyapunov stability analysis of non-smooth systems is the determination of the generalized derivative on a discontinuity surface, which involves the estimation of the intersection of a number of convex sets. Although in this work, it still requires above determination, the form of the sets is simplified, and the number of the sets involved in the generalized derivative is significantly reduced as compared with previous work by Shevitz and Paden (1994). The current extension does not require much advanced mathematical machinery and therefore, makes the stability analysis of non-smooth systems practically easier. Two examples, including a system with stick-slip friction and a robotic system having interaction with the environment, are used to demonstrate the applicability of the method proposed in this paper.

Keywords: Lyapunov stability, Lyapunov function, non-smooth systems, Filippov's solution theory, Clarke's generalized gradient.

1. INTRODUCTION

Non-smooth dynamic systems, described by differential equations with discontinuous terms, appear naturally and frequently in the control field and require stability analysis. Typical examples include systems with stick-slip friction, contact tasks, variable structure and optimal control systems. Two issues involved in the analysis of non-smooth systems are firstly; classical solution theories to ordinary differential equations are no longer valid. Secondly, Lyapunov's stability theory has been developed for smooth systems.

Filippov's solution theory (1960, 1979) is one of the earliest and conceptually straightforward approaches developed for analysis of non-smooth systems. The theory has been used by many researchers (Slotine and Sastry, 1983; Paden and Sastry, 1987; DeCarlo *et al.* 1988; Southward *et al.* 1991; Corless, 1993; Shevitz and Paden, 1994; Wu *et al*, 1998a and b). Consequently, extensions of Lyapunov's second method to non-smooth dynamic systems, based on Filippov's solution theory, have been studied (Hahn, 1963; Paden and Sastry, 1987; Shevitz and Paden, 1994 and Wu *et al.* 1998a). The most recent and systematic extensions of Lyapunov's second method to non-smooth dynamic systems were developed by Shevitz and Paden (1994) and Wu *et al* (1998a). In Shevitz and Paden's extension, piecewise smooth Lyapunov functions were utilized. The important step in constructing a non-smooth Lyapunov function is to examine the sign of its derivative with respect to time on the discontinuity surface. This is done by estimating the sign of each element of the generalized derivative, which is the intersection of a large number of convex sets. Such estimation requires much heavier mathematical machinery and can become unmanageable for systems with more than two state variables or when the discontinuity surface is the intersection of several discontinuity surfaces. The difficulty in determining the generalized derivative has restricted the application of the above extended Lyapunov's stability theory. Wu *et al.* (1998a) extended Lyapunov's stability theory to non-smooth systems in that smooth Lyapunov functions were used and the conditions for the construction of smooth Lyapunov functions for non-smooth systems have been established. The significance of the above work is that it simplifies the construction of Lyapunov functions and therefore, makes the stability analysis of non-smooth systems practically easier. Such an extension has been used for stability control of a base-excited inverted pendulum (Wu *et al*, 1998b) and modeling human locomotion (Wu *et al*, 1998c). However, as pointed out by the authors (1998a), for some non-smooth systems, non-smooth Lyapunov functions seem to be the only choices.

The work presented in this paper follows the track of Shevitz and Paden's extension (1994) in that non-smooth Lyapunov functions are used for non-smooth systems. The objective of this work is to simplify the estimation of the generalized derivative on the discontinuity surface. Thus, more non-smooth systems can be studied under Lyapunov's stability theory. First, a brief review of relevant mathematical concepts is given. Next, it is shown how the estimation of the generalized derivative can be simplified. Examples from mathematical and engineering systems are presented to demonstrate the feasibility of the proposed method for the construction of non-smooth Lyapunov functions.

2. MATHEMATICAL REVIEW

The extension of Lyapunov's stability theory presented here is based on Filippov's solution theory and Clarke's generalized gradient concept. In this section, we review the relevant mathematical framework. Since Filippov's solution concept has been reviewed by Wu *et al.* (1998a). Clarke's generalized gradient and the definition of the generalized derivative of a function on a discontinuity surface will be reviewed here.

Clarke's Generalized Gradient: For a locally Lipschitz function V: $R^n \times R \to R$ define the generalized gradient of V at (x, t) by

$$\partial V(x,t) = \overline{co}\{\lim \nabla V(x,t)|\ (x_i, t_i) \to (x,t), (x_i, t_i \notin \Omega_V\} \quad (1)$$

where Ω_V is the set of measure zero where the gradient of V is not defined. $\overline{co}$ denotes a convex set. For example, Clarke's generalized gradient for function $V(x) = |x|$ is

$$\partial V(x) = \begin{cases} -1 & x < 0 \\ 1 & x > 0 \\ \overline{co}\{-1,\ 1\} & x = 0 \end{cases}$$

where $\overline{co}\{-1,\ 1\}$ is equivalent to $[-1,\ 1]$, i.e., any real number between -1 and 1 belongs to set $\overline{co}\{-1,\ 1\}$.

Derivative of function V with respect to time on the discontinuity surface (Shevitz and Paden, 1994): Let x(.) be a Filippov's solution to $\dot{x} = f(x,t)$ on an interval containing t and V: $R^n \times R \to R$ be a Lipschitz and in addition, regular function. Then V(x(t),t) is absolutely continuous, $(d/dt)V(x(t),t)$ exists almost everywhere and

$$\frac{d}{dt} V(x(t),t) \in^{a.e.} \dot{\tilde{V}}(x,t) \quad (2)$$

where $\dot{\tilde{V}}$ is the generalized derivative defined as

$$\dot{\tilde{V}}(x,t) := \bigcap_{\xi \in \partial V(x(t),t)} \xi^T K[f](x(t),t) \quad (3)$$

In the above relation, ξ will range all the elements of the convex set $\partial V(x(t),t)$.

3. STABILITY THEOREM

One of the key-steps using the extended Lyapunov's second method by Shevitz and Paden (1994) is to prove that every element of set $\dot{\tilde{V}}(x,t)$, defined in (3) is negative or zero. Since $\dot{V}(x(t),t) \in \dot{\tilde{V}}(x,t)$, it is guaranteed that $\dot{V}(x(t),t) \le 0$. However, from authors' experience, it is very difficult to prove that each element belonging to set $\dot{\tilde{V}}(x,t)$ is negative or zero. This is because the form of $\dot{\tilde{V}}(x,t)$ can be very complicated as the system involves many state-space variables or when the discontinuity surface is the intersection of several surfaces. Furthermore, $\dot{\tilde{V}}(x,t)$ is generally the intersection of infinite convex sets. The determination of such an intersection of convex sets requires much heavier mathematical machinery.

In this section, Lyapunov's second method is extended to non-smooth systems with the objective of simplifying the determination of the generalized derivative of a Lyapunov function on the discontinuity surface. We first consider the first kind of discontinuity, i.e., as the solution trajectory approaches the discontinuity surface, there exists finite limits of ∇V, the gradient of a Lyapunov function V, which do not coincide. Based on the definition of Clarke's generalized gradient, the above assumptions can be denoted as:

$$\partial V(x,t) := \overline{co}\{\nabla V^1(x,t), \ldots, \nabla V^n(x,t)\} \quad (4)$$

Relation (4) shows that $\nabla V^i (i = 1,2,\ldots,n)$ is the base vector of the convex set ∂V. Note that if ∇V^i and ∇V^j $(1 \le i, j, \le n)$ coincide, the base vectors of the convex set, $\partial V(x,t)$, need to be determined first. Lyapunov's second method can be extended below:

Modified Lyapunov's second method: Let $\dot{x} = f(x,t)$ be locally bounded and $0 \in K[f](0,t)$ in a region $Q \supset \{x \in R^n \mid \|x\| < r\} \times \{t \mid t_0 \le t < \infty\}$. Also, let $V : R^n \times R \rightarrow R$ be a regular function satisfying that (*i*) V is continuous, positive and definite and, (*ii*) $\dot{V}(x,t)$ is piecewise continuous and each element of convex set $\overline{\dot{V}}$, defined below, is negative or zero, then x(t) is a stable solution in the sense of Lyapunov.

In the above stability theory, convex set $\overline{\dot{V}}$ is defined as:

$$\overline{\dot{V}}(x,t) := \bigcap_{i=1}^{n} \nabla V^i K[f](x(t),t) \qquad (5)$$

where n is the number of the base vectors of Clarke's generalized gradient.

The proof of the extension by Shevitz and Paden (1994) is valid for the proposed modified Lyapunov's second method since the convex set $\tilde{\dot{V}}(x,t)$ is a sub-set of the set $\overline{\dot{V}}(x,t)$. From relation (2), we have

$$\frac{d}{dt} V(x(t),t) \in^{a.e.} \tilde{\dot{V}}(x,t) \subset \overline{\dot{V}}$$

The above extension of Lyapunov's second method simplifies the determination of the derivative of a Lyapunov function on the discontinuity surface, which is believed one of the main difficulties in using the previous extension of Lyapunov's stability theory (Shevitz and Paden, 1994). This can be shown on two aspects: firstly, the convex set $\overline{\dot{V}}$ has a simple form since only the base vectors of convex set $\partial V(x,t)$ are involved; secondly, set $\overline{\dot{V}}$ is the intersection of n convex sets, rather than a large number of (probably infinite) sets as defined in $\tilde{\dot{V}}(x,t)$ (see (3)). Here, n is the number of the base vectors of Clarke's generalized gradient $\partial V(x,t)$. Such a simplification makes the stability analysis using non-smooth Lyapunov functions feasible, and therefore, more non-smooth systems can be under study in the frame of Lyapunov's stability theory.

It is important to point out that the presented extension is a sufficient condition for the stability proof, that is, if every element of set $\overline{\dot{V}}$, defined in (5), is negative or zero, every element of set $\tilde{\dot{V}}$, defined in (3), is negative or zero since set $\tilde{\dot{V}}$ is a sub-set of set $\overline{\dot{V}}$ and, therefore, $\dot{V} \le 0$. However, if there is a positive element in set $\overline{\dot{V}}$, it is still possible that all the elements of set $\tilde{\dot{V}}$ are negative or zero.

4. EXAMPLES

Two examples, a mechanical system with stick-slip friction and a system having a contact interaction with the environment, are presented to demonstrate the feasibility of the proposed extended stability theorem.

<u>Example 1</u> Design of Bounce-less Controller for Contact Tasks (Payandeh, 1996)

Let us consider a one-dimensional compliant mechanism with the actuation force shown in Figure 1. The dynamic model of the object in the contact and non-contact regions can be written in term of the state-space model assuming $x_1 = x$:

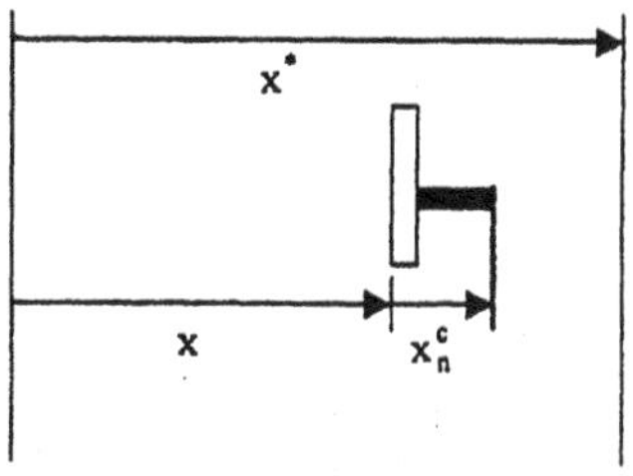

Fig.1 A one-dimensional compliant mechanism.

$$\dot{x}_2 = -\frac{r}{m} + \frac{u(t)}{m} \qquad (6)$$

where $r = k_c(x_n^c - x^* + x_1)$ for $x^* - x_1 < x_n^c$ is the reaction force due to the interaction with the rigid environment and k_c is the compliant. r=0 in the non-contact region. Figure 1 shows all the position variables involved. A controller u(t), proposed by Payandeh (1996) for both contact and non-contact phases of the mechanisms is shown below:

$$u(t) = \begin{cases} u^1(t) = -k_p(x_1 - x^d) - k_d x_2 & \text{for } r = 0 \\ u^2(t) = -k_{pf} k_c (x - x^d) - k_{df} x_2 + r^d & \text{for } r > 0 \end{cases} \qquad (7)$$

where x^d and $\dot{x}^d$ are desired trajectories of the end-point and r^d is the desired contact force to be exerted on the environment.

The above dynamic system is a non-smooth system. The discontinuity arises from the controller. The discontinuity surface can be represented as $S := \{(x_1, x_2), x_1 - x^* + x_n^c = 0\}$. The existence and uniqueness of the solution have been proven, but not detailed here (see Wu 1998d). In this work, we focus on the stability analysis. Assuming $e_1 = x_1 - x^d$, the state-space model can be written as:

$$\dot{e}_1 = e_2$$
$$\dot{e}_2 = -\frac{k_p}{m} e_1 - \frac{k_d}{m} e_2 \qquad (8)$$

and

$$\dot{e}_1 = e_2$$
$$\dot{e}_2 = -\frac{k_{pf}}{m}(k_c + 1)e_1 - \frac{k_{df}}{m}k_c e_2 \tag{9}$$

Equations (8) and (9) describe the object in both non-contact region and contact region, respectively. The discontinuity surface is defined as:

$$S := \{(e_1, e_2); e_1 + x^d + x_n^c - x^* = 0\}$$

When $\bar{e} \in \Omega^+$, the object is in contact with the environment and when $\bar{e} \in \Omega^-$, it is not in contact with the environment. Regions Ω^- and Ω^+ are defined as follows:

$$\Omega^+ := \{(e_1, e_2); e_1 > x^* - x^d - x_n^c\}$$
$$\Omega^- := \{(e_1, e_2); e_1 < x^* - x^d - x_n^c\}$$

A Lyapunov function is now constructed in each region as:

$$V = \begin{cases} V_1 = \frac{1}{2}e_2^2 + \frac{k_p}{2m}e_1^2 + \frac{(k_c+1)k_{pf} - k_p}{2m}(x^* - x_n^c - x^d)^2 & e \in \Omega^- \\ V_2 = \frac{1}{2}e_2^2 + \frac{k_c+1}{2m}k_{pf}e_1^2 & e \in \Omega^+ \\ V_3 = \frac{1}{2}e_2^2 + \frac{k_c+1}{2m}k_{pf}e_1^2 & e \in S \end{cases} \tag{10}$$

Function V is positive through the entire region under the condition that $(k_{pf} + 1)k_c - k_p > 0$. Function V is definite since V=0 (i.e., $V_2 = 0$) if and only if $e_1 = 0$ and $e_2 = 0$. We next estimate $\dot{V}$ on the discontinuity surface S where $e_1 = x^* - x_n^c - x^d$.

$$\partial V(e \in S) = \overline{co}\left[\begin{matrix} \left\{\begin{matrix} \frac{k_p}{m}(x^* - x_n^c - x^d) \\ e_2 \end{matrix}\right\}, \\ \left\{\begin{matrix} \frac{k_c+1}{m}k_{pf}(x^* - x_n^c - x^d) \\ e_2 \end{matrix}\right\} \end{matrix}\right] \tag{11}$$

Accordingly, Filippov's differential inclusion is

$$K[f] = \overline{co}\left[\begin{matrix} \left\{\begin{matrix} e_2 \\ -\frac{k_p}{m}x_T - \frac{k_d}{m}e_2 \end{matrix}\right\}, \\ \left\{\begin{matrix} e_2 \\ -\frac{k_c+1}{m}k_{pf}x_T - \frac{k_{df}}{m}k_c e_2 \end{matrix}\right\} \end{matrix}\right] \tag{12}$$

where $x_T = (x^* - x_n^c - x^d)$. For the above system, $\dot{\bar{V}} = \bigcap_{i=1}^{2} \nabla V^i K[f]$. It is now required to determine set $\nabla V^1 K[f]$ and $\nabla V^2 K[f]$, respectively.

$$\nabla V^1 K[f] = \overline{co}\left[-\frac{k_d}{m}e_2^2, \left(\frac{k_p}{m} - \frac{k_c+1}{m}k_{pf}\right)x_T e_2 - \frac{k_{df}}{m}k_c e_2^2\right]$$

Similarly, we have

$$\nabla V^2 K[f] = \overline{co}\{-\left(\frac{k_p}{m} - \frac{k_c+1}{m}k_{pf}\right)x_T e_2 - \frac{k_d}{m}e_2^2, -\frac{k_{df}}{m}k_c e_2^2\}$$

where x_T is defined in equation (12).

If we assume the relation $k_d = k_{df}k_c$, we have that $\dot{\bar{V}} = \nabla V^1 K[f] \cap \nabla V^2 K[f] = \overline{co}\left[-\frac{k_d}{m}e_2^2\right]$, that is, set $\dot{\bar{V}}$ contains only one element as $-\frac{k_d}{m}e_2^2$. Thus, $\dot{V}(e \in S) = -\frac{k_d}{m}e_2^2$, which is negative. Therefore, the system shown in equations (8) and (9) is stable.

<u>Example 2</u>: *Robust Nonlinear Stick-Slip Friction Compensation (Southward et al., 1991)*

The following state-space model represents a one-degree-of-freedom mechanical system subjected to stick-slip friction (F_{fric}). A nonlinear compensation force for stick-slip friction, F_c, was developed to supplement a PD control law to keep the mechanical system having a unique equilibrium point. Assuming that x_1 is the displacement of the mass (m) and x_2 is the velocity. We have

$$\dot{x}_1 = x_2$$
$$\dot{x}_2 = \frac{1}{m}(F - F_{fric}) \tag{13}$$

where the friction model is defined as F_{fric}

$$F_{fric} = F_{slip}\,\mathrm{sgn}(x_2) + F_{stick}(1 - \mathrm{sgn}(x_2)) \tag{14}$$

$$\mathrm{sgn}(x_2) = \begin{cases} 1 \\ 0 \\ -1 \end{cases}$$

In (14), F_{slip} is a positive constant and F_{stick} is defined as follows:

$$F_{stick}(F) = \begin{cases} F_S^+ & F \geq F_S^+ > 0 \\ F & F_S^- < F < F_S^+ \\ F_S^- & F \leq F_S^- < 0 \end{cases} \tag{15}$$

where F_S^+ and F_S^- are the positive and negative limits on the static friction forces. F, the control law, proposed by Southward *et al.* (1991), is designed to keep the system having a unique equilibrium point at $x_1 = 0$ and $x_2 = 0$ as shown below:

$$F = -k_p x_1 - k_d x_2 - k_p \begin{cases} 0 & x_1 > \tilde{x}_H \\ (\tilde{x}_H - x_1) & 0 < x_1 \leq \tilde{x}_H \\ 0 & x_1 = 0 \\ (\tilde{x}_L - x) & \tilde{x}_L \leq x_1 < 0 \\ 0 & x_1 < \tilde{x}_L \end{cases} \tag{16}$$

where $\tilde{x}_L = -\frac{F_S^+}{k_p}$ and $\tilde{x}_H = -\frac{F_S^-}{k_p}$.

The state-space model shown in equations (13) together with equations (14) and (16) is a non-smooth dynamic system. The discontinuity is caused by (*i*) friction which is discontinuous with respect to the velocity x_2 and, (*ii*) control law which is discontinuous at $x_1 = 0$. Therefore, the discontinuity surfaces can be defined as

$$S_1 := \{x, x_1 = 0, x_2 \neq 0\}$$
$$S_2 := \{x, x_2 = 0, x_1 \neq 0\}$$
$$S_3 := \{x, x_1 = 0 \text{ and } x_2 = 0\}$$

The discontinuity surface S_3 is the intersection of surfaces S_1 and S_2. In this paper, Filippov's solution concept is used. The existence and uniqueness of Filippov's solution have been proven, but not detailed here (see Wu, 1998d).

In their work, Southward *et al.* (1991) constructed a non-smooth Lyapunov function to prove the stability. The derivative of their Lyapunov function on the discontinuity surface was replaced with Dini derivative since such a derivative is not well defined based on conventional solution theories. In this work, a special case is studied for the discontinuity surface of S_1. Other cases of S_2 and S_3 can be investigated in a similar manner. A Lyapunov function shown below is used here.

$$V(x_1, x_2) = \frac{1}{2}k_p x_1^2 + \frac{1}{2}k_d x_2^2 + \begin{cases} \frac{1}{2}k_p \tilde{x}_H^2 & x > \tilde{x}_H \\ k_p(\tilde{x}_H x_1 - \frac{1}{2}x_1^2) & 0 \le x_1 \le \tilde{x}_H \\ k_p(\tilde{x}_L x_1 - \frac{1}{2}x_1^2) & \tilde{x}_L \le x_1 \le 0 \\ \frac{1}{2}k_p \tilde{x}_L^2 & x < \tilde{x}_L \end{cases} \tag{17}$$

However, its derivative with respect to time is replaced with Clark's generalized derivative. Following the framework described in Section 3, we need to prove that each element of the following convex set $\dot{\tilde{V}}$ is negative. For this problem, Clark's generalize gradient ∇V and Filippov's inclusion, on the discontinuity surface S_1, have the following forms:

$$\nabla V := \overline{co}\left[\begin{Bmatrix} k_p \tilde{x}_L \\ m x_2 \end{Bmatrix}, \begin{Bmatrix} k_p \tilde{x}_H \\ m x_2 \end{Bmatrix}\right] \tag{18}$$

$$K[f] := \overline{co}\left[\begin{Bmatrix} x_2 \\ \frac{1}{m}(-k_d x_2 - k_p \tilde{x}_L - F_{slip}\,\mathrm{sgn}(x_2)) \end{Bmatrix} \begin{Bmatrix} x_2 \\ \frac{1}{m}(-k_d x_2 - k_p \tilde{x}_H - F_{slip}\,\mathrm{sgn}(x_2)) \end{Bmatrix}\right] \tag{19}$$

We only need to prove that all the elements of the convex set $K[\dot{V}] := \xi_1 K[f] \cap \xi_2 K[f]$ are negative or zero. Here we have that $\xi_1 = \{k_p \tilde{x}_L \;\; m x_2\}^T$ and $\xi_2 = \{k_p \tilde{x}_H \;\; m x_2\}^T$. The proof is as follows:

$$\xi_1 K[f] = K[\xi_1 \cdot f]$$
$$= K\begin{bmatrix} \{k_p x_L, m x_2\}\begin{Bmatrix} x_2 \\ \frac{1}{m}(-k_d x_2 - k_p \tilde{x}_L - F_{slip}\,\mathrm{sgn}(x_2)) \end{Bmatrix} \\ \{k_p \tilde{x}_L \;\; m x_2\}\begin{Bmatrix} x_2 \\ \frac{1}{m}(-k_d x_2 - k_p \tilde{x}_H - F_{slip}\,\mathrm{sgn}(x_2)) \end{Bmatrix} \end{bmatrix}$$
$$= K[-k_d x_2^2 - F_{slip}|x_2|, -k_d x_2^2 - F_{slip}|x_2| - k_p x_2(\tilde{x}_H - \tilde{x}_L)]$$

Similarly,

$$\xi_2 K[f] = K[-k_d x_2^2 - F_{slip}|x_2| - k_p x_2(\tilde{x}_L - \tilde{x}_H), -k_d x_2^2 - F_{slip}|x_2|]$$

We now prove that all the elements of set $K[\dot{V}] := \xi_1 K[f] \cap \xi_2 K[f]$ are either negative or zero. Note $-k_d x_2^2 - F_{slip}|x_2| < 0$. If $-k_p x_2(\tilde{x}_L - \tilde{x}_H) \le 0$, $-k_p x_2(\tilde{x}_H - \tilde{x}_L) \ge 0$. Therefore, the intersection of sets $\xi_1 K[f] \cap \xi_2 K[f]$ is

$$\dot{\tilde{V}} := \xi_1 K[f] \cap \xi_2 K[f] = \overline{co}[-k_d x_2^2 - F_{slip}|x_2|] \tag{20}$$

Since $\dot{V} \in \dot{\tilde{V}}$, we have that $\dot{V} = -k_d x_2^2 - F_{slip}|x_2| < 0$ on the discontinuity surface.

It is also possible to use a Lyapunov function based on Shevitz and Paden's extension. However, it is more complicated to analyze the signs of the elements of the intersection of the convex sets.

5. CONCLUSIONS

In this paper, Lyapunov's second method was extended to non-smooth systems based on Filippov's solution theory and Clarke's generalized gradient. The extension presented here has the advantages, over the previous work by Sheviz and Paden (1994), in that it simplifies the procedure of the construction of non-smooth Lyapunov functions and, it requires less advanced mathematical machinery to analyze the properties of the generalized derivative of a Lyapunov function on the discontinuity surface, which is believed to be very difficult. Therefore, the proposed extension of Lyapunov's second method makes the stability analysis of non-smooth systems practically easier.

Two examples were chosen to demonstrate the feasibility of the extension of Lyapunov's theory proposed in this paper. It is shown that, at least for the classes of non-smooth dynamic systems presented here, our extension is effective and easier to use. This advantage allows more non-smooth systems to be under study using Lyapunov's stability theory. The extended Lyapunov's second method, proposed in this paper, should find applications in analysis of many other non-smooth systems, such as variable structure control systems, optimum control systems, robotic systems having impact with the environment, and mechanical systems with friction.

It is important to point out that the presented extension is a sufficient condition for the stability proof, that is, if set $\dot{\tilde{V}}$, defined in (5), only contains negative or zero elements, the derivative of a Lyapunov function on the discontinuity surface is negative or zero. However, if set $\dot{\tilde{V}}$ contains positive elements, it is still possible that the derivative of a Lyapunov function is negative or zero. For this case, the elements of set $\dot{\tilde{V}}$ have to be determined.

6. REFERENCES

Clarke, F. H. (1983). *Optimization and Nonsmooth Analysis*. John Wiley & Sons, New York.

Corless, M. (1993) Control of uncertain nonlinear systems, *ASME Journal of Dynamic Systems, Measurement, and Control*, **115**, 362-372.

DeCarlo, R. A., S. H. Zak and P. Matthews (1988) Variable structure control of nonlinear multivariable systems: a tutorial, *Proceedings of the IEEE*, **76**, 212-232.

Filippov, A. F. (1960) Differential equations with discontinuous right-hand side, *Math. Sbornik* **51**, 99-128. (English Translation: *American Mathematical Society Translations* **42**, 191-231, 1964).

Filippov, A. F. (1979) Differential equations with second members discontinuous on intersecting surfaces, *Differentsial'nye Uravneniya*, **15**, 1814-1832 (English Translation: *Differential Equations*, **15**, 1292-1299, 1980).

Hahn, W. (1963) *Theory and Application of Lyapunov's Direct Method*. Prentice-Hall, Englewood Cliffs.

Paden, B. E. and S. S. Sastry (1987) A calculus for computing Filippov's differential inclusion with application to the variable structure control of robot manipulators, *IEEE Transactions on Circuits and Systems* **34**, 73-82.

Payandeh,S. (1996) A method for controlling robotic contact tasks, *Robotica*,**14**, 281-288.

Shevitz, D and B. Paden (1994) Lyapunov stability theory of nonsmooth systems, *IEEE Transactions on Automatic Control* **39**, 1910-1914.

Slotine, J. J. and S. S. Sastry (1983) Tracking control of nonlinear systems using sliding surfaces, with application to robot manipulators, *International Journal of Control* **38**, 465-492.

Southward, S. C., C. J. Radcliffe and C. R., MacCluer (1991) Robust nonlinear stick-slip friction compensation, *ASME Journal of Dynamic Systems, Measurement, and Control* **113**, 639-645.

Wu, Q., S. Onyshko, N. Sepehri and A. B. Thornton-Trump (1998a) On construction smooth Lyapunov functions for non-smooth systems, *International Journal of Control* **69**, 443-457.

Wu, Q., A.B. Thornton-Trump and N. Sepehri (1998b) Lyapunov stability control of inverted pendulums with general base point motion, *International Journal of Non-Linear Mechanics*, **33**, 801-818.

Wu, Q., N. Sepehri, A.B. Thornton-Trump and M. Alexander (1998c) Stability and control of human trunk movement during walking, *Computer Methods in Biomechanics and Biomedical Engineering*, **1**, 247-259.

Wu, Q. (1998d) Lyapunov's stability analysis of non-smooth systems, Report 98-1, Department of Mechanical & Industrial Engineering, University of Manitoba, Winnipeg, Manitoba, Canada.

MATRIX-THEORETIC CONDITIONS FOR THE REALIZABILITY OF SLIDING MANIFOLDS [1]

L. Hsu * E. Kaszkurewicz * A. Bhaya *

* *Dept. of Electrical Engineering, COPPE, Federal University of Rio de Janeiro, P.O. Box 68504, RJ 21945-970, BRAZIL*
E-mail: {LIU, EUGENIUS, AMIT}@COEP.UFRJ.BR

Abstract: Systems of the type $\dot{s} = K\,\text{sgn}(s)$, $s \in \mathbb{R}^m$, where $\text{sgn}(\cdot)$ is the vector sign function, determine the realizability of sliding modes in control systems with high frequency gain K. A new condition on K is established for the global stability of this system using a nonsmooth Liapunov function. This condition is used to establish necessary and sufficient stability conditions for second order systems. *Copyright © 1999 IFAC*

Keywords: Stability, sliding modes, variable structure systems, matrix diagonal stability, nonsmooth Liapunov functions.

1. INTRODUCTION

Consider a nonlinear control system affine in the control given by

$$\dot{x} = f(x,t) + B(x,t)u \qquad (1)$$

where, $x \in \mathbb{R}^n$, $f : \mathbb{R}^n \times \mathbb{R} \to \mathbb{R}^n$ and $B : \mathbb{R}^n \times \mathbb{R} \to \mathbb{R}^{n\times m}$ are continuous in their arguments, and the control u is discontinuous on the smooth discontinuity surfaces $s_i(x) = 0$, $i = 1, ..., m$. Let us define the vector $s = [s_1, ..., s_m]^T$. Then, $s(x) = 0$ defines the intersection manifold of all discontinuity surfaces.

This paper establishes conditions under which a globally stable sliding mode motion along the manifold $s = 0$ is produced, i.e., according to definition given in Utkin (1992, p. 49), the origin of the s-space is globally asymptotically stable in the usual sense of Liapunov.

The equations of motion in the s-space can be easily derived giving

$$\dot{s} = K(x,t)(u - u_{eq}) \qquad (2)$$

where $K(x,t) = G(x,t)B(x,t)$, with $G = \frac{\partial s(x)}{\partial x}$; $u_{eq} = -[G(x)B(x,t)]^{-1}G(x)f(x,t)$ is the equivalent control corresponding to a sliding mode on $s = 0$, assuming that GB is nonsingular. The focus will henceforth be on linear time-invariant plants with linear discontinuity surfaces, i.e.,

$$\dot{x} = Ax + r(t) + Bu \quad ; \quad s = Gu \qquad (3)$$

where, A, B, and G are now constant matrices. The transfer function from u to s is $H(s) = G(sI - A)^{-1}B$. The high frequency gain is thus $K = GB$. The equation corresponding to (2) in the linear case is

$$\dot{s} = K(u - u_{eq}); \quad K = GB, \qquad (4)$$

where $u_{eq} = -[GB]^{-1}G(Ax + r(t))$. Without loss of generality, one may consider the *homogeneous* case where $u_{eq} = 0$. For more details about the nonhomogeneous case see Utkin (1992). It is also assumed that the discontinuous control is given by $u = -\text{sgn}(s)$. In fact, one should consider the case of different amplitudes for each element of u, i.e., $u_i = -F_i\,\text{sgn}(s_i)$. However, with F_d denoting the diagonal matrix $\text{diag}(-F_1, \cdots, -F_m)$, the matrix

[1] This research was partially supported by the Brazilian Government agencies CAPES and CNPq, as well as MCT under the PRONEX program. Corresponding author: L. Hsu.

K can be redefined as KF_d. Thus, the problem is reduced to determining conditions for the global asymptotic stability of $s = 0$ for the system

$$\dot{s} = K\,\mathrm{sgn}(s) \tag{5}$$

The objective of this paper is to extend the following results in Utkin (1992) : (i) Persidskii type results, based on integral-of-nonlinearity nonsmooth Liapunov functions; (ii) Dominance principle based results, which depend on the observation that dominance of the negative diagonal element in a particular row allows a sliding mode that zeros the variable associated to that row.

Preliminaries and definitions: The class of **diagonally stable** matrices $\mathcal{D}$ is defined as the subset of the set of Hurwitz matrices for which the Liapunov equation admits a positive diagonal solution, $P = \mathrm{diag}\,(p_1, \ldots, p_n)$, $p_i > 0, \forall i$. When it is desired to refer to a positive diagonal solution P, the notation $A \in \mathcal{D}(P)$ is used. It means that there exists a diagonal matrix $P > 0$ such that $A^T P + PA$ is negative definite. The **class** $\mathcal{S}$ of functions is defined as follows: $\mathcal{S} := \{f(\cdot) \,|\, f : \mathbb{R} \to \mathbb{R};\; f(\xi)\cdot\xi > 0;\; f(0) = 0;$ and $\int_0^x f(\tau)d\tau \to \infty$ as $|x| \to \infty\}$.. A dynamical system $\dot{x} = f(x)$, $x \in \mathbb{R}^n$, $f : \mathbb{R}^n \to \mathbb{R}^n$ is said to be of **Persidskii type** if it is of the form:

$$\dot{x}_i = \sum_{j=1}^{n} a_{ij} f_j(x_j(t)), \quad f_j \in \mathcal{S}, \forall j, \tag{6}$$

for $i = 1, \ldots, n$. It is also written compactly as follows: $\dot{x} = AF(x)$, where $A = (a_{ij})$ and $F = (f_1, \cdots, f_n)^T$ is a *diagonal* or *separable nonlinearity* such that $F \in \mathcal{S}^n$. Persidskii (1969) derived the following sufficient condition for global asymptotic stability of the zero solution of (6).

Theorem 1.1. The zero solution of the nonlinear system (6) is globally asymptotically stable *for all* $F(\cdot) \in \mathcal{S}^n$ if $A \in \mathcal{D}(P)$. Equivalently, for each member of the class of nonlinear systems $\{\dot{x} = AF(x),\;\; x(0) = x_0 : F \in \mathcal{S}^n\}$, the positive definite function

$$V(x) = 1/2 \sum_{i=1}^{n} p_i \int_0^{x_i} f_i(\tau)d\tau \tag{7}$$

is a diagonal type Liapunov function establishing global asymptotic stability of the zero solution of the corresponding system.

2. PERSIDSKII TYPE RESULTS

In this section, theorem 1.1, valid for continuous systems, is extended to systems with discontinuous right hand side of the type (5). Solutions of such systems will be understood as defined by Filippov (1964) – such a solution, say $x(t)$, is an absolutely continuous function of t, thus almost everywhere differentiable, i.e., $\dot{x}(t)$ is almost everywhere defined. It is also important to emphasize that a Filippov solution is identical to the usual definition of solution of differential equations, except on a sliding mode on some discontinuity surface. The following definition is convenient.

Definition 2.1. *(Extended Equivalent Control) (Hsu and Costa, 1996)* Consider a Filippov solution $x(t)$ of system (1), defined for t in a certain interval T. Then, the extended equivalent control $u_{eq}(t)$ is an integrable function, defined almost everywhere in T, given by:

$$u_{eq}(t) = -[GB(x(t),t)]^{-1}[Gf(x(t),t) - \frac{d}{dt}s(x(t),t)]. \tag{8}$$

The following stability theorem can now be demonstrated.

Theorem 2.2. Consider the discontinuous system (5). Then $s(t) = 0$ is a globally asymptotically stable solution, if K is a *diagonally stable* matrix, i.e., there exists a positive diagonal matrix $P = \mathrm{diag}(p_1, \cdots, p_m)$ and a positive definite Q such that

$$K^T P + PK = -Q \tag{9}$$

Moreover, in this case the origin is reached in finite time and the function

$$V(s) = 2\sum_{i=1}^{m} p_i \int_0^{s_i} \mathrm{sgn}(z)dz = 2\sum_{i=1}^{m} p_i |s_i| \tag{10}$$

has time-derivative almost everywhere given by $\dot{V} = -u_{eq}^T Q u_{eq}$ on the solutions of (5).

Proof. (*Outline only*) Consider the Persidskii type Liapunov function (7). The time-derivative of $V(s(t))$ along a Filippov solution $s(t)$ of (5) exists almost everywhere. If we had a continuously differentiable function $V(s)$, then the derivative would be given by

$$\frac{dV}{dt} = \left(\frac{\partial V}{\partial s}\right)^T (t)\,\dot{s} = \left(\frac{\partial V}{\partial s}\right)^T (t)\,Ku_{eq}(t) \tag{11}$$

In order to calculate $\dot{V}$ along a solution $s(t)$, one can set

$$\left(\frac{\partial V}{\partial s}\right)(t) = Pu_{eq}(t) \tag{12}$$

Rewrite $\dot{V} (:= \frac{dV}{dt})$ as

$$\dot{V}(t) = 2\sum_{i=1}^{m}\left(\frac{\partial V}{\partial s_i}\right)(t)\dot{s}_i(t) \qquad (13)$$

If $s_i \neq 0,\ i = 1, \cdots, m$ then

$$\left(\frac{\partial V(s)}{\partial s_i}\right)(t) = p_i \mathrm{sgn}(s_i(t)) = p_i u_{ieq}(t) \quad (14)$$

Now, if for some open interval of t, there exists i such that $s_i(t) \equiv 0$, then $\dot{s}_i(t) \equiv 0$ in that interval. Thus one trivially has

$$\left(\frac{\partial V(s)}{\partial s_i}\right)(t)\dot{s}_i(t) \equiv p_i u_{ieq}(t)\dot{s}_i(t) \equiv 0$$

in the given interval. Since all other possible situations involve some $s_i = 0$ only in sets of measure zero, it follows that the equality (12) holds. Then, from (13), (5) and (9) one gets

$$\dot{V} = 2u_{eq}^T PKu_{eq} = -u_{eq}^T Q u_{eq} \qquad (15)$$

Moreover, the vector $u_{eq}(t)$ is null only when $s(t) = 0$. In fact, $\|\mathrm{sgn}(s)\|_2 \geq 1,\ \forall s \neq 0$. Hence, $(dV/dt) \leq -\lambda_{min}^{1/2}(Q),\ \forall s \neq 0$. Hence $V(s(t))$ must converge to zero in finite time and the time needed to reach the sliding mode $s = 0$ is less than $V_0/(\lambda_{min}^{1/2}(Q))$, V_0 being the initial value of V. □

Discussion: This theorem generalizes a stability condition obtained by Utkin (1992) who utilized the nonsmooth Liapunov function $V(s) = |s_1| + |s_2| + \cdots + |s_m|$ which is a particular case of (10) (setting $d_i = 1$, for all i) to obtain the following sufficient stability condition $K + K^T < 0$. Note that the Liapunov theory of nonsmooth systems developed in Shevitz and Paden (1994), that uses the machinery of generalized gradients, allows a simple proof of the above theorem.

2.1 *The Second Order Case*

In the following, a complete characterization of the behavior of system (5) in the second order case ($m = 2$) is obtained. The case of either $k_{12} = 0$ and/or $k_{21} = 0$ is quite trivial. When both off-diagonal elements are nonzero, the key idea is to use the scaling invariance property of (5) to reduce the system to a normal form with the change of variables $s \to z = Ds$ where $D = \mathrm{diag}(1/k_{12}, 1/|k_{21}|)$ the following class of systems is obtained

$$\dot{z} = \begin{bmatrix} * & 1 \\ \pm 1 & * \end{bmatrix} \mathrm{sgn}(z)$$

The system is then said to be in normal form. The main result is that stability of $s = 0$ holds if and only if the normalized system matrix is Hurwitz. Since the normalized (scaled) K can be expressed in terms of the elements of the original K, the final result is as follows. Proofs are in the appendix.

Theorem 2.3. Consider the discontinuous system (5) in the second order case, i.e. $K = \begin{pmatrix} k_{11} & k_{12} \\ k_{21} & k_{22} \end{pmatrix}$. Then $s(t) = 0$ is a globally asymptotically stable solution *if and only if* one of the following conditions holds:

(1) $k_{12}k_{21} = 0$, $k_{11} < 0$ and $k_{22} < 0$.
(2) $k_{12}k_{21} \neq 0$ and

$$k_{11}/|k_{12}| + k_{22}/|k_{21}| < 0, \qquad (16)$$

as well as

$$\det(K) = k_{11}k_{22} - k_{12}k_{21} > 0. \qquad (17)$$

The statement of this theorem in terms of the normalized or scaled system matrix is also of interest, since it leads to a connection with D-stable matrices.

Alternative statement of Theorem 2.3 *For $K \in \mathbb{R}^{2\times 2}$, the zero solution of $\dot{s} = K sgn(s)$ is globally asymptotically stable, if and only if the diagonally scaled matrix DK is Hurwitz, where $D = diag((1/|k_{12}|), (1/|k_{21}|))$, if the off-diagonal entries are both nonzero, and $D = I$, otherwise.*

Recalling the definition that a matrix A is **D-stable** if PA is Hurwitz for all positive diagonal matrices P (Hershkowitz, 1992), the following corollary can be stated.

Corollary 2.4. If $K \in \mathbb{R}^{2\times 2}$ is D-stable, then the zero solution of (5) is globally asymptotically stable.

Note also that this corollary requires the matrix K to be Hurwitz, while the theorem does not: it is enough to have a positive diagonal scaling that results in a Hurwitz matrix. For example, the matrix $K_2 = \begin{pmatrix} -0.5 & 1 \\ -5 & 1 \end{pmatrix}$ is not Hurwitz and hence not D-stable, yet when diagonally scaled by the reciprocals of the moduli of its off-diagonal elements, it becomes Hurwitz: thus, theorem 2.3 ensures stability of the zero solution of (5), although corollary 2.4 does not.

3. DOMINANCE PRINCIPLE BASED RESULTS

This section considers a method to design a stabilizing discontinuous control u proposed by Utkin (1992). In this method, the components u_i of the input vector u are designed so that sliding modes are produced on each surface $s_i = 0$, successively according to some order determined by what is referred to as the "control hierarchy".

The dominance principle is actually a restatement, in the context of the system (5), of the fundamental principle of sliding mode theory that a sliding mode occurs on a surface if the vector field on either side of it points towards it. It is stated as follows:
Dominance principle: *If the matrix K in (5) is ith-row negative diagonally dominant, then there exists a sliding mode on the surface $s_i = 0$.*

The truth of this principle can be seen by looking at the ith component of the differential equation, namely,

$$\dot{s_i} = k_{i1}\text{sgn}(*) + \cdots + \cdots + k_{im}\text{sgn}(*),$$

where the asterisks $*$ denote any combination of the variables s_i, or, indeed, any other exogenous variables. Since k_{ii} is negative and, in modulus, dominates the sum of the moduli of the k_{ij}s, clearly $\dot{s}_i$ is always negative, i.e. the vector field on both sides of the surface $s_i = 0$ points towards it. The following definition is needed.

Definition 3.1. A matrix A in $\mathbb{R}^{n\times n}$ is called ***i*th row negative diagonally dominant** if $a_{ii} < 0$ and $|a_{ii}| > \sum_j |a_{ij}|$. If A is *ith* row negative diagonally dominant for all i, then it is called **negative diagonally dominant.**

Theorem 3.2. Consider the discontinuous system (5). The zero solution $s(t) = 0$ of this system is globally asymptotically stable if, after an appropriate permutation matrix P is applied to the rows and columns of matrix K, the permuted matrix PKP^T can be transformed, by elementary row operations, to an upper triangular negative diagonally dominant matrix U. In this case, stable sliding modes of system (5) will occur successively starting from $s_{i_1} = 0$, according to the ordering $\{s_{i_1}, \cdots, s_{i_m}\}$ determined by the permutation P.

Proof. Consider that a reordering of the variables induced by P has been applied so that matrix can is transformed to $H = PKP^T$. Denote $\text{sgn}(s_i)$ by v_i. Assuming that K_P can be reduced to the matrix U by elementary row operations only (i.e. no further permutations) means that the first row of H is negative diagonally dominant. Since sliding will occurs on the surface $s_1 = 0$, v_1 can be replaced by its equivalent control $(v_1)_{eq} = -h_{11}^{-1}(h_{12}v_2 + \cdots + h_{1m}v_m)$ in all subsequent rows. This substitution in the reordered system (5) zeros all elements of the first column of H , below the first diagonal element and the second row is coincident with the second row of U, which is negative diagonally dominant, by assumption. Hence $s_2 = 0$ is a sliding mode that will be eventually attained after the sliding mode $s_1 = 0$ is attained. This reasoning can be pursued leading to the conclusion that in the system

$$\dot{s} = U\,\text{sgn}(s) \tag{18}$$

the rows represent the equations for each variable s_i after the sliding modes s_j, $,j = 1, \cdots, i-1$ have been reached. Since all rows of U are assumed negative diagonally dominantall sliding modes $s_i = 0$ will be attained (in finite time) according to the hierarchy of control $s_1 \to s_2 \cdots \to s_m$, after the reordering induced by the permutation matrix P is performed, which proves the theorem. □

Recall the following facts (see Strang (1980)): (i) the diagonal elements of the upper triangular matrix that results from Gaussian elimination are called pivots and are all nonzero if and only if the original matrix is nonsingular, (ii) the kth pivot is given by the ratio of the determinants of the principal leading minors of orders k and $(k-1)$, and (iii) by Sylvester's law of inertia, for a symmetric matrix, the signs of the pivots match the signs of the eigenvalues. Given these facts, the new insights into Theorem 3.2 obtained from the Gaussian elimination interpretation are as follows.

Proposition 3.3. (i)A necessary condition for a matrix K to be reducible by permutations and elementary row operations to a negative diagonally dominant matrix is that it be nonsingular, since all pivots are nonzero.
(ii) Since all pivots are actually negative, a more refined necessary condition is that all leading principal minors of odd order must have negative determinants and all leading principal minors of even order must have positive determinants.
(iii) If K is symmetric, the signs of the pivots (all negative) coincide with the signs of the eigenvalues, thus all eigenvalues of K must be negative. In other words, a symmetric matrix K satisfies the conditions of Theorem 3.2, only if it is negative definite.

We can generalize Theorem 3.2 by requiring only Block-Triangularizability, which is less drastic than requiring pointwise negative diagonal dominance. To do this, we simply need to redefine the concept of block-row diagonal dominance. We explain this concept on a two block example, for simplicity. Assume that K is transformable, by reordering and by block-row operations, to the form

$$K \to \begin{bmatrix} T_{11} & T_{12} \\ 0 & T_{22} \end{bmatrix} \tag{19}$$

The vector is correspondingly reordered and partitioned as $s^T = ((s^1)^T, (s^2)^T)$, where $s^i \in \mathbb{R}^{m_i}$, $i = 1, 2$. Now consider the subsystem

$$\dot{s}^1 = T_{11}\,\mathrm{sgn}(s^1) \quad (20)$$

Then we say that T_{11} dominates T_{12} if the above subsystem, as well as the system

$$\dot{s}^1 = T_{11}\,\mathrm{sgn}(s^1) + T_{12}\,\mathrm{sgn}(s^2) \quad (21)$$

are stable at the manifold $s^1 = 0$, irrespective of s^2.

Two kinds of dominance can be outlined, which correspond to the use of a Liapunov function of type sum-of-moduli-of-elements (10) or of quadratic type. In the first case, due to the obvious association, we speak of L_1 dominance, while in the second we speak of L_2 dominance. Let us now establish explicit conditions for each of the two kinds of diagonal dominance just mentioned.

L_1-diagonal dominance
Assume that system (20) is in the class of systems defined in Theorem 2.2. Evaluating the derivative of V (10) on the trajectories of (21), we find

$$\begin{aligned}\dot{V} = &-\mathrm{sgn}(s^1)^T Q \mathrm{sgn}(s^1) + \\ &2\mathrm{sgn}(x^1)^T P T_{12} \mathrm{sgn}(s^2)\end{aligned} \quad (22)$$

$$\begin{aligned}\dot{V} \leq &\|\mathrm{sgn}(s^1)\|_2 \left(-\lambda_{min}(Q)\|\mathrm{sgn}(s^1)\|_2 + \right. \\ &\left. 2\|PT_{12}\|_2\|\mathrm{sgn}(s^2)\|_2\right)\end{aligned} \quad (23)$$

Since we are interested in the motion outside the manifold $s^1 = 0$, then $\|\mathrm{sgn}(s^1)\|_2 \geq 1$. Moreover $\|\mathrm{sgn}(s^2)\|_2 \leq \sqrt{m_2}$ and thus (23) leads to

$$\dot{V} \leq \|\mathrm{sgn}(s^1)\|_2(-\lambda_{min}(Q) + 2\|PT_{12}\|_2\sqrt{m_2}) \quad (24)$$

Thus, L_1 negative diagonal dominance can be expressed as

$$-\lambda_{min}(Q) + 2\|PT_{12}\|_2\sqrt{m_2} < 0 \quad (25)$$

L_2-diagonal dominance
Now, instead of the nonsmooth Liapunov function, we consider a quadratic one $V = (s^1)^T W s^1$ and let $L = WT_{11}$, $M = WT_{12}$. Then, calculating its derivative on the trajectories of (21), we obtain

$$\begin{aligned}\dot{V} = &-\sum_{k=1}^{m} |s_k^1|(l_{kk} + \sum_{i=1,\ i\neq k}^{m_1} l_{ki}\mathrm{sgn}(s_k^1 s_i^1) + \\ &\sum_{j=1}^{m_2} m_{kj}\mathrm{sgn}(s_j^2 s_k^1))\end{aligned} \quad (26)$$

It is clear from the above expression that stability of $s^1 = 0$ will be preserved, in spite of the coupling term $T_{12}\mathrm{sgn}(s^2)$, if the negative diagonal dominance extends to the additional coupling terms, i.e., the L_2 diagonal dominance is given by

$$l_{kk} < 0;\ |l_{kk}| > \sum_{i=1,\ i\neq k}^{m_1} |l_{ki}| + \sum_{j=1}^{m_2} |m_{kj}|; \quad (27)$$

for $m = 1, \cdots, m_1$. We can now state the generalization of Theorem 3.2.

Theorem 3.4. Consider the discontinuous system (5). Then $s(t) = 0$ is a globally asymptotically stable solution if K can be transformed to an upper block-triangular block diagonal dominant form T, by means of reordering and block-row operations similar to those for the scalar-row case of Theorem 3.2. This property is equivalent to the existence of permutation transformation $\mathcal{R}$ and a lower block-triangular matrix with identity diagonal blocks $\mathcal{L}$, such that

$$T = \mathcal{L}\mathcal{R}^T K \mathcal{R} = \begin{bmatrix} T_{11} & T_{12} & \cdots & T_{1m} \\ 0 & T_{22} & \cdots & T_{2m} \\ \cdot & \cdot & \cdot & \cdot \\ & & & T_{kk} \end{bmatrix} \quad (28)$$

where T_{ii}, $i = 1, \cdots, k$ is a Hurwitz matrix for the ith block and dominates i-th block-row. In this case, stable sliding modes of system (5) will occur successively, starting from the manifold $s^1 = 0$, according to the order determined by the block partitioning and the permutations.

Proof. The proof follows the same lines as that for Theorem 3.2. Successive sliding is guaranteed by the property of block-diagonal dominance introduced above. □

4. CONCLUDING REMARKS

In conclusion, this paper put various results due to Utkin in a matrix-theoretic framework – specifically using Persidskii type Liapunov functions generalized to the nonsmooth case and the concept of diagonal stability. This led to some generalizations and a characterization of the class of stable matrices in the 2×2 case, as well as a unified perspective using matrix diagonal stability (well researched in the matrix theory literature (Hershkowitz, 1992)) and negative diagonal dominance. Connections of matrix diagonal stability with the dominance principle as well as with quadratic stability results (not discussed in this paper) are being investigated and will be reported on elsewhere.

APPENDIX: PROOF OF THEOREM 2.3

The stability results for each possible case are presented below. Non-negative entries of K are denoted by "+", while strictly negative entries are denoted by "−". Proofs are only briefly outlined due to space limitations.

1. $k_{21} = -1$, $k_{11} < 0$, $k_{22} < 0$: Apply Theorem 2.2 with the nonsmooth Liapunov function $V = |s_1| +$

$|s_2|$ to conclude that the origin ($s = 0$) is always stable. Note that the normalized K is Hurwitz.

2. $k_{21} = -1$, $k_{ii} \geq 0$, $k_{jj} < 0$, $i,j \in \{1,2\}$: Consider only the first subcase $k_{22} < 0$ since the second one is entirely similar. Since $k_{11} \geq 0$, $s_1 = 0$ cannot be a sliding surface. Using the same Liapunov function as in case 1 and applying the same technique as in the proof of Theorem 2.2 one has:

$$\dot{V}/2 = k_{11}u_{1eq}^2 + k_{22}u_{2eq}^2 = k_{11} + k_{22}u_{2eq}$$

The following subcases arise: (i)trace(K) < 0, $|k_{22}| < 1$: then $s_2 = 0$ is not a sliding surface either. Consequently, $u_{2eq}^2 = 1$. Then $\dot{V}/2 = \text{trace}(K) < 0$ and thus asymptotic stability holds; (ii) If trace(K) < 0, $|k_{22}| > 1$, then $s_2 = 0$ is a sliding surface where $u_{2eq}^2 = a^{-2}u_1^2 = a^{-2}$; $\dot{V}/2 = k_{11} + k_{22}^{-1} = \det(K)k_{22}^{-1}$. Thus, asymptotic stability is guaranteed if $\det(K) > 0$. Further, if $\det(K) < 0$ the origin $s = 0$ is *unstable*; (iii) trace(K) $>$ 0, $|k_{22}| < 1$: no sliding regime occurs and thus $\dot{V}/2 = \text{trace}(K) > 0$ implying instability; (iv) trace(K) > 0, $|k_{22}| > 1$: then $s_2 = 0$ is a sliding surface where $u_{2eq}^2 = a^{-2}u_1^2 = a^{-2}$; $\dot{V}/2 = k_{11} + k_{22}^{-1} = \det(K)k_{22}^{-1}$. Now, if trace($K$) > 0 then $k_{11} > -k_{22} > 1$ and thus $\det(K) < 0$ and $\dot{V} > 0$ which implies instability. Thus, here stability holds only if trace(K) < 0. Again, in this case, K is Hurwitz.

3. $k_{21} = -1$, $k_{11} \geq 0$, $k_{22} \geq 0$: since here sliding cannot occur, use Theorem 2.2 to get $\dot{V}/2 = \text{trace}(K)$ and thus the system is either weakly stable, if $k_{11} = k_{22} = 0$ or unstable, otherwise. Note that K is not Hurwitz here.

4. $k_{21} = 1$, $k_{11} \geq 0$, $k_{22} \geq 0$: there exist diverging solutions in the first and third quadrants. The origin is always unstable. Again, matrix K is not Hurwitz.

5. $k_{21} = 1$, $k_{11} < 0$, $k_{22} < 0$: the following subcases arise: (i) $|k_{11}| < 1$, $|k_{22}| < 1$: we can show instability using Chetaevs Theorem with the function $V = s_1 s_2$ and noting that $|s_i| \leq s_i sgn(s_i)$, $(i = 1,2)$, which implies $\dot{V} > 0$; (ii) $|k_{11}| < 1$, $|k_{22}| > 1$: here, matrix K is diagonally stable if $\det(K) > 0$ (Kaszkurewicz and Hsu, 1984). Thus, from Theorem 2.2, with $V = |s_1| + c|s_2|$ one has $\dot{V} = u_{eq}^T Q u_{eq}$, where

$$Q = \begin{bmatrix} k_{11} & \dfrac{(1+c)}{2} \\ \dfrac{(1+c)}{2} & ck_{22} \end{bmatrix}$$

which is positive definite if $c(1-c^{-1})^2 < 4\det(K)$. Obviously there exists c sufficiently close to 1 satisfying such an equality. Hence, stability is concluded for $\det(K) > 0$. On the other hand, if $\det(K) < 0$, one can easily verify that the sliding mode $s_2 = 0$ has an unstable motion given by $\dot{s}_1 = \frac{\det(K)}{k_{22}}\text{sgn}(s_1)$. Thus, in this case, asymptotic stability holds if and only if $\det(K) > 0$; (iii)$|k_{11}| > 1$, $|k_{22}| > 1$: This is the trivial case of diagonal dominance. Stability holds, K is Hurwitz and both surfaces $s_1 = 0$, $s_2 = 0$ are sliding surfaces.

All remaining cases correspond to instability. In all such cases K is not Hurwitz since $\det(K) < 0$.

6. $k_{21} = 1$, $k_{11} < 0$, $k_{22} > 0$: The following subcases arise: (ii)$|k_{11}| < 1$, $|k_{22}| < 1$; $|k_{11}| < 1$, $|k_{22}| > 1$; $|k_{11}| > 1$, $|k_{22}| > 1$: in these subcases there exist diverging motions in the 1rst and 3rd quadrants; $|k_{11}| > 1$, $|k_{22}| < 1$: here $s_1 = 0$ is an unstable sliding mode.

From the above results, which cover all possible cases of second order systems (5), we conclude that the stability of $s = 0$ holds if and only if the normalized system matrix is Hurwitz. The normalized (scaled) K matrix, with scaling factors $d_1 = 1/k_{12}$ and $d_2 = 1/|k_{21}|$, can be expressed in terms of the elements of the original K, so that the final result is as stated above.

5. REFERENCES

Filippov, A. F. (1964). Differential equations with discontinuous right-hand side. *American Mathematical Society Translations* **42**(2), 199–231.

Hershkowitz, D. (1992). Recent directions in matrix stability. *Lin. Algebra Appl.* **171**, 161–186.

Hsu, L. and R. R. Costa (1996). Adaptive control with sliding modes: Theory and applications. In: *Minicourse Proceedings of the XI Congresso Brasileiro de Automática.* São Paulo, Brazil. pp. 39–60.

Kaszkurewicz, E. and L. Hsu (1984). On two classes of matrices with positive diagonal solutions to the Lyapunov equation. *Lin. Algebra Appl.* **59**, 19–27.

Persidskii, S.K. (1969). Problem of absolute stability. *Automation and Remote Control* **12**, 1889–1895.

Shevitz, D. and B. Paden (1994). Lyapunov stability theory of nonsmooth systems. *IEEE Trans. Automat. Control* **39**(9), 1910–1914.

Strang, G. (1980). *Linear Algebra and its Applications,* 2nd. ed.. Academic Press. New York.

Utkin, V. I. (1992). *Sliding Modes In Control And Optimization.* Springer-Verlag. Berlin.

DYNAMIC OUTPUT FEEDBACK STABILIZATION OF PARTIALLY LINEAR COMPOSITE SYSTEMS

Pengnian Chen* Huashu Qin+ Quan M. Zhu++

* *Division of Mathematics, China Institute of Metrology, Hangzhou 310034, P.R. C., E-mail: pnchen@public1.hz.zj.cn*

\+ *Institute of Systems Science, Chinese Academy of Sciences, Beijing 100080, P.R. C., E-mail:qin@iss 03. iss. ac. cn*

\++ *School of Engineering, Aston University, Aston Triagle, Birmingham, B47 ET, UK, E-mail: q.m.zhu@aston.ac.uk*

Abstract: The problem of stabilization by dynamic output feedback is investigated for partially linear composite systems. The main result is that if the linear subsystem has no unstable zero points and the nonlinear subsystem is homogeneously asymptotically stable, then the system is stabilizable by dynamic output feedback.. A corollary of the result is that a minimum phase nonlinear system with the homogeneously asymptotically stable zero dynamics is stabilizable by dynamic output feedback. The controller is linear and can be constructed explicitly. *Copyright © 1999 IFAC*

Keywords: Stabilization, nonlinear systems, dynamic output feedback, minimum-phase systems, closed-loop systems

1. INTRODUCTION

In the paper, we study the problem of dynamic output feedback stabilization of a nonlinear system of the form

$$\begin{cases} \dot{z} = f(z,\xi) \\ \dot{\xi} = A\xi + B(G(z,\xi)u + F(z,\xi)) \\ y = C\xi \end{cases} \qquad (1)$$

where $z \in R^n$, $\xi \in R^r$, $u \in R^m$ and $y \in R^p$; $A \in R^{r\times r}$, $B \in R^{r\times m}$ and $C \in R^{p\times r}$; $f{:}U \to R^n$, $G{:}U \to R^r$, and $F{:}U \to R^r$ are smooth functions, $f(0,0) = 0$, $\det(G(0,0)) \neq 0$, $F(0,0) = 0$, $F(0,\xi) = O(\|\xi\|^2)$ and U is aneighborhood of $(z,\xi) = (0,0)$.

In the last decade, the problem of stabilization of nonlinear systems by dynamic output feedback has been extensively studied. Esfandiari and Khalil (1992) investigated dynamic output feedback stabilization of full linearizable systems. Chen et al. (1997) and Tornambé (1992) showed. that systems which have stabilizability by state feedback and local complete observability are stabilizable by dynamic output feedback. Khalil and Esfandiari (1993) and Teel and Praly (1994,1995) proved thatsystems having global stabilizabilty by state feedback and uniform complete observability are semi- globally stabilizable by dynamic output feedback.. A key assumption made explicitly or implicitly in most of the previous researches is that the system has the local complete observability. But many important systems, for example, minimum phase nonlinear systems and partially linearizable systems etc., do not have the local complete observability. Byrnes and Isidori (1989) gave a example of a planar nonlinear system which is not uniformly completely observable, but is stabilizable by dynamic output feedback. But, for general nonlinear systems without uniform complete observability, the problem of stabilization by dynamic output feedback has not been deeply studied up to now.

System (1) is an important class of nonlinear systems without uniform complete observability. Many authors have studied stabilization of system (1) by state feedback , see (Byrnes and Isidori, 1988,1991; Saberi, et al.,1990). Although system (1) is locally stabilizable by state feedback if the "nonlinear zero dynamics" $\dot{z} = f(z,0)$ is asymptotically stable, the extension of such results to dynamic output feedback is not immediate.

Using the concept of homogeneous asymptotic stability and the tool of center manifold theory, we investigate the problem of stabilization of system (1) by dynamic output feedback. Some sufficient conditions for stabilizability are given. In particular, it was shown that if the zero dynamics of a minimum phase nonlinear system is homogeneously asymptotically stable, then it is stabilizable by dynamic output feedback.

The paper is organized as follows. Section 2 is the preliminaries on detectability of dynamically extended systems and on stability of nonlinear systems. The main result is presented in Section 3. Section 4 gives two examples to illustrate the applications of the stabilizability results. Some concluding remarks are given in Section 5.

2. PRELIMINARIES

In this section, some lemmas on detectability of dynamically extended systems and on stability of nonlinear systems will be presented.

Throughout the paper it is assumed that $\Sigma(C,A,B)$ is complete, Rank $B = m$, Rank $C = p$, and $m \le p$.

2.1 Detectability

Let C^- denote the open left half complex plane. Let $\sigma^*(A,B;\operatorname{Ker} C)$ denote the set of zero points of $\Sigma(C,A,B)$. $\sigma^*(A,B;\operatorname{Ker} C)$ is just the list of zeros of nontrivial numerator polynomials in Smith-McMillan form of the transfer matrix of $\Sigma(C,A,B)$, see (Wonham, 1979). When $p \ge m$, $\sigma^*(A,B;\operatorname{Ker} C)$ is the set of common zeros of C_{r+p}^{r+m} polynomials and, in general, is empty.

Consider the dynamically extended system

$$\begin{cases} \dot{x} = Ax + B\lambda_1 \\ \dot{\lambda}_1 = \lambda_2 \\ \cdots \\ \dot{\lambda}_{l-1} = \lambda_l \\ \dot{\lambda}_l = \sum_{i=1}^{l} \alpha_i \lambda_i + u \\ y = Cx \end{cases} \tag{2}$$

where $\lambda_i \in R^m, \alpha_i \in R, i = 1,2,\cdots,l; u$ is the input and y is the output.

Lemma 2.1 If $\sigma^*(A,B;\operatorname{Ker} C) \subset C^-$, then system (2) is controllable and detectable.
The proof of Lemma 2.1 is easy and here omitted

Lemma 2.2 Let $m = p$. If $\Sigma(C,A,B)$ has a relative degree $\{r_1, r_2, \cdots, r_m\}$ with $\sum_{i=1}^{l} r_i = r$, then system (2) is controllable and detectable.

Proof This is an immediate consequence of Lemma 2.1 and the fact that under the condition of Lemma 2.2, $\sigma^*(A,B;\operatorname{Ker} C) = \varnothing$.

2.2. Stability

Definition 2.1. Consider the differential equation

$$\dot{x} = F(x), \qquad F(0) = 0,$$

where $x \in R^n, F: U \to R^n$ is smooth function and U a neighborhood of $x = 0$. The system is said to be N-th order homogeneously asymptotically (HA-) stable if for any $\varphi \in C^1(U,R^n)$, $\varphi(x) = O(\|x\|^N)$, the system

$$\dot{x} = F(x) + \varphi(x)$$

is asymptotically stable. Moreover, the system is said to be homogeneously asymptotically (HA-) stable if there is a positive integer N such that it is N-th order HA-stable.

A system with HA-stability can remain its asymptotic stability under perturbation if the order of the perturbed term is high enough. In other words, HA-stability means some robustness with respect to high degree perturbed terms. Therefore ,this concept is natural for practical systems since the performance of an asymptotically stable system, which is not HA-stable, may be very bad. Throughout the paper, only the local stability at zero is considered ,the statement "at zero" will be omitted.

Consider now the following system

$$\begin{cases} \dot{x} = F(x,y) \\ \dot{y} = By + G(x,y) \end{cases} \tag{3}$$

where $x \in R^n, y \in R^m$, and $B \in R^{m\times m}$; F and G are smooth functions, $F(0,0) = 0$, and $G(x,y) = \mathrm{O}(\|(x,y)\|^2)$.

Lemma 2.3. Supposed the following conditions hold:

(i) B is Hurwitz, i.e. $\mathrm{Re}\,\sigma(B) < 0$;

(ii) The system $\dot{x} = F(x,0)$ is N-th order HA-stable;

(iii) $G(x,0) = \mathrm{O}(\|x\|^{N+1})$.

Then system (3) is asymptotically stable.

The proof of Lemma 23 is too long and will appear in other paper.

3. STABILIZATION

In this section, we will study the problem of dynamic output feedback stabilization of system (1). For convenience, system (1)is changed by coordinate transformation into the following from

$$\begin{cases} \dot{z} = f(z,\xi) \\ \dot{\xi}_1 = A_{11}\xi_1 + A_{12}\xi_2 \\ \dot{\xi}_2 = A_{21}\xi_1 + A_{22}\xi_2 + G(z,\xi)u + F(z,\xi) \\ y = C\xi \end{cases} \tag{4}$$

where $\xi_1 \in R^{r-m}$, $\xi_2 \in R^m$, $\xi = (\xi_1^T, \xi_2^T)^T$, $G(0,0) = I_m$, $F(0,z) = \mathrm{O}(\|\xi\|^2)$,

$$A = \begin{bmatrix} A_{11} & A_{12} \\ A_{21} & A_{22} \end{bmatrix}, \quad B = \begin{bmatrix} 0 \\ I_m \end{bmatrix}.$$

The "nonlinear zero dynamics" of system (1) (or (4)) is

$$\dot{z} = (z,0). \tag{5}$$

Theorem 3.1 Assume that (i) $\sigma^*(A,B;\mathrm{Ker}\,C) \subset C^-$, and (ii) system (5) is homogeneously asymptotically stable. Then system (4) is stabilizable by dynamic output feedback.

Proof Let system (4) be N-th order homogeneously asymptotically stable. Construct a preliminary controller of the following form

$$\begin{cases} \dot{\lambda}_1 = \lambda_2 \\ \cdots \\ \dot{\lambda}_{l-1} = \lambda_l \\ \dot{\lambda}_l = \hat{u} \\ u = \lambda_1 \end{cases} \tag{6}$$

where $\lambda_i \in R^m$, $i = 1,2,\cdots,l$; $\hat{u}$ is the new input; $u = \lambda_1$ is the output; l is a positive integer to be determined later.

Let

$$F_0(z,\xi) = A_{12}\xi_1 + A_{22}\xi_2 + F(z,\xi).$$

Then the composite system consisting of (4) and (6) can be written as

$$\begin{cases} \dot{z} = f(z,\xi) \\ \dot{\xi}_1 = A_{11}\xi_1 + A_{12}\xi_2 \\ \dot{\xi} = F_0(z,\xi) + G(z,\xi)\lambda_1 \\ \dot{\lambda}_1 = \lambda_2 \\ \cdots \\ \dot{\lambda}_{l-1} = \lambda_l \\ \dot{\lambda}_l = \hat{u} \\ y = C\xi. \end{cases} \tag{7}$$

Let

$$\eta_1 = F_0(z,\xi) + G(z,\xi)\lambda_1. \tag{8}$$

Since $G(z,\xi)$ is locally invertible, λ_1 can be solved as

$$\lambda_1 = G^{-1}(z,\xi)(\eta_1 - F_0(z,\xi)).$$

It is easy to show that can be written as

$$\dot{\eta}_1 = F_1(z,\xi,\eta_1) + G(z,\xi)\lambda_2 \tag{9}$$

where $F_1(z,\xi,\eta_1)$ is a certain C^∞ function.

Inductively, it can be defined that

$$\eta_i = F_{i-1}(z,\xi,\eta_1,\cdots\eta_{i-1}) + G(z,\xi)\lambda_i, \quad i = 2,\cdots,l, \tag{10}$$

$$\dot{\eta}_i = F_i(z,\xi,\eta_1,\cdots,\eta_i) + G(z,\xi)\lambda_{i+1}, \quad i = 2,\cdots,l, \tag{11}$$

where $\lambda_{l+1} = \hat{u}$.

Denote by $Z(N)$ a basis of the set S of polynomials, with

$$S = \{p(z) \mid p(0), \deg(p(z)) \le N\}$$

and $q = \dim(S)$. Then there exist constant $m \times q$ matrices C_i, $i = 1,2,\cdots,l$, such that F_i can be expressed as

$$F_i(z,\xi,\eta_1,\cdots,\eta_i)=C_iZ(N)+$$
$$\varphi_i(z,\xi,\eta_1,\cdots,\eta_i),\quad i=1,2,\cdots,l,\quad (12)$$

where $\varphi_i(z,\xi,\eta_1,\cdots,\eta_i)$ has the property that

$$\varphi_i(z,0,0,\cdots,0)=O(\|z\|^{N+1}),\quad i=1,2,\cdots,l.$$

Since C_i has fixed dimension, there exist a positive integer l and a set of constant numbers $\alpha_i, i=1,2,\cdots,l-1$, such that

$$C_l=\sum_{i=1}^{l-1}C_i.\quad (13)$$

Fix l as the smallest positive number, which makes (13) true. Then, by (8)--(13),

$$\begin{aligned}C_lZ(N)&=\sum_{i=1}^{l-1}\alpha_iC_iZ(N)\\&=\sum_{i=1}^{l-1}\alpha_i(F_i(z,\xi,\eta_1,\cdots,\eta_i)\\&\quad-\varphi_i(z,\xi,\eta_1,\cdots,\eta_i))\quad (14)\\&=\sum_{i=1}^{l-1}\alpha_i\,(\eta_{i+1}-\varphi_i(z,\xi,\eta_1,\cdots,\eta_i)\\&\quad-G(z,\xi)\,\lambda_{i+1}).\end{aligned}$$

It follows from (8)--(11) and (14) that

$$\begin{aligned}\dot\eta_l&=C_lZ(N)+\varphi_l(z,\xi,\eta_1,\cdots,\eta_l)+G(z,\xi)\hat u\\&=\sum_{i=1}^{l-1}\alpha_i\,(\eta_{i+1}-\varphi_i(z,\xi,\eta_1,\cdots,\eta_i)\\&\quad-G(z,\xi)\,\lambda_{i+1})+G(z,\xi)\hat u\\&\quad+\varphi_l(z,\xi,\eta_1,\cdots,\eta_l)\quad (15)\\&=\sum_{i=1}^{l-1}\alpha_i\,(\eta_{i+1}-\varphi_i(z,\xi,\eta_1,\cdots,\eta_i))\\&\quad+\varphi_l(z,\xi,\eta_1,\cdots,\eta_l)\\&\quad+G(z,\xi)\,(\hat u-\sum_{i=1}^{l-1}\alpha_i\lambda_{i+1})\\&=\hat F(z,\xi,\eta_1,\cdots,\eta_l)+G(z,\xi)v\end{aligned}$$

where

$$\begin{aligned}&\hat F(z,\xi,\eta_1,\cdots,\eta_l)\\&=\sum_{i=1}^{l-1}\alpha_i\,(\eta_{i+1}-\varphi_i(z,\xi,\eta_1,\cdots,\eta_i))\\&\quad+\varphi_l(z,\xi,\eta_1,\cdots,\eta_l),\\&v=\hat u-\sum_{i=1}^{l-1}\alpha_i\lambda_{i+1}.\quad (16)\end{aligned}$$

Clearly, by (12) and (16),

$$\begin{aligned}\hat F(z,0,0,\cdots,0)&=-\sum_{i=1}^{l-1}\alpha_i\varphi_i(z,0,0,\cdots,0)\\&\quad+\varphi_l(z,0,0,\cdots,0)\\&=O(\|z\|^{N+1}).\quad (17)\end{aligned}$$

By (8), (15) and (16), system (7) can be written as

$$\begin{cases}\dot z=f(z,\xi)\\\dot\xi_1=A_{11}\xi_1+A_{12}\xi_2\\\dot\xi_2=\eta_1\\\dot\eta_1=\eta_2\\\cdots\\\dot\eta_{l-1}=\eta_l\\\dot\eta_l=\hat F(z,\xi,\eta_1,\cdots,\eta_l)+G(z,\xi)v\\y=C\xi.\end{cases}\quad (18)$$

The system consisting of the linear part of system (3.1), system (3,4) and feedback law (16) is

$$\begin{cases}\dot\xi_1=A_{11}\xi_1+A_{12}\xi_2\\\dot\xi_2=A_{21}\xi_1+A_{22}\xi_2+G(0,0)\lambda_1\\\dot\lambda_1=\lambda_2\\\cdots\\\dot\lambda_{l-1}=\lambda_l\\\dot\lambda_l=\sum_{i=1}^{l-1}\alpha_i\lambda_{i+1}+v\\y=C\xi.\end{cases}\quad (19)$$

According to Lemma 2.1, system (19) is controllable and detectable. Then there exists a linear dynamic compensator (Wonham, 1997)

$$\begin{cases}\dot w=Sw+Ry\\v=Ky+Qw\end{cases}\quad (20)$$

such that the closed loop system formed by (18) and (19) is asymptotically stable.

In terms of (17), $\hat F$ can written as the form

$$\hat F(z,\xi,\eta_1,\cdots,\eta_l)=H_1\xi+H_2\eta$$
$$+\varphi(z,\xi,\eta)+O(\|z\|^{N+1}),$$

where H_1 and H_2 are constant matrices, $\eta=(\eta_1^T,\eta_2^T,\cdots,\eta_l^T)^T$, $\varphi(z,0,0)=0$, $\varphi(0,\xi,\eta)=O(\|(\xi,\eta)\|^2)$.

Then the closed loop system formed by (18) and (20) is

$$\begin{cases} \dot{z} = f(z,\xi) \\ \dot{\xi}_1 = A_{11}\xi_1 + A_{12}\xi_2 \\ \dot{\xi}_2 = \eta_1 \\ \dot{\eta}_1 = \eta_2 \\ \cdots \\ \dot{\eta}_{l-1} = \eta_l \\ \dot{\eta}_l = H_1\xi + H_2\eta + G(z,\xi)(KC\xi + Qw) \\ +\bar{\varphi}(z,\xi,\eta) \\ \dot{w} = Sw + RC\xi, \end{cases} \tag{21}$$

where $\bar{\varphi}$ has the properties that $\bar{\varphi}(z,0.0.) = O(\|z\|^{N+1})$, and $\bar{\varphi}(0,\xi,\eta) = O(\|(\xi,\eta)\|^2)$.

From (8)--(16),it is easy to see that the " linear part " of (21) is

$$\begin{cases} \dot{\xi}_1 = A_{11}\xi_1 + A_{12}\xi_2 \\ \dot{\xi}_2 = \eta_1 \\ \dot{\eta}_1 = \eta_2 \\ \cdots \\ \dot{\eta}_{l-1} = \eta_l \\ \dot{\eta}_l = H_1\xi + H_2\eta + KC\xi + Qw \\ \dot{w} = Sw + RC\xi, \end{cases} \tag{22}$$

which is obtained from (19) and (20) by invertible linear transformation. Therefore system (22) is asymptotically stable. By Lemma 2.3, system (21) is asymptotically stable. This completes the proof.

Consider now the system

$$\begin{cases} \dot{x} = f(x) + g(x)u \\ y = h(x) \end{cases} \tag{23}$$

where $x \in R^n, u, y \in R^m$; f, g and h are all C^∞ functions, $f(0) = 0$, $h(0) = 0$.

Theorem 3.2 If system (23) has a relative degree $\{r_1, r_2, \cdots, r_m\}$ and its zero dynamics is homogeneously asymptotically stable, then it is stabilizable by dynamic output feedback.

Proof Since system (23) have the relative degree $\{r_1, r_2, \cdots, r_m\}$, it can be transformed into the form (Byrnes and Isidori,1991)

$$\begin{cases} \dot{z} = f_0(z,\xi) \\ \dot{\xi} = A\xi + B(G(z,\xi)u + F(z,\xi)) \\ y = C\xi, \end{cases} \tag{24}$$

where A, B and C are constant matrices; $G(0,0)$ is invertible, and $F(0,\xi) = O(\|z\|^2)$.

The zero dynamics of system (24) is

$$\dot{z} = f(z,0),$$

which is homogeneously asymptotically stable. By Lemma 2.2, $\sigma^*(A,B;\mathrm{Ker}\,C) = \varnothing$, since $\Sigma(C,A,B)$ has the relative degree $\{r_1, r_2, \cdots, r_m\}$. Therefore Theorem 3.2 follows directly from Theorem 3.1.

Example 3.1 Consider the system

$$\begin{cases} \dot{z} = -z^5 + \xi_1^2 + \xi_2^2 \\ \dot{\xi}_1 = \xi_2 \\ \dot{\xi}_2 = \xi_1 + 2\xi_2 + z\sin z + z^4 + u \\ y = \xi_1 + \xi_2 \end{cases} \tag{25}$$

with $A = \begin{bmatrix} 0 & 1 \\ 1 & 2 \end{bmatrix}, B = \begin{bmatrix} 0 \\ 1 \end{bmatrix}, C = \begin{bmatrix} 1 & 1 \end{bmatrix}$.

First of all, note that the "nonlinear zero dynamics" of system (4.1)

$$\dot{z} = -z^5$$

is 5-th order homogeneously asymptotically stable and $\sigma^*(A,B;\mathrm{Ker}\,C) = \{\lambda \mid 1 + \lambda\} = \{-1\} \subset C^-$.

Then by Theorem 3.1, system (25) is stabilizable by dynamic output feedback. The dynamic compensator of system (25) can be constructed in the light of the method given in the proof of Theorem 3.1. Let

$$F_0(z,\xi_1,\xi_2) = \xi_1 + 2\xi_2 + z\sin z + z^4,$$

$$\eta_1 = F_0(z,\xi_1,\xi_2) + \lambda_1.$$

Then en $\dot{\eta}_1$ can be expressed as

$$\begin{aligned} \dot{\eta}_1 &= \xi_2 + 2\eta_1 \\ &\quad + (\sin z + z\cos z + 4z^3)(-z^5 + \xi_1^2 + \xi_2^2) + \lambda_2 \\ &= F_1(z,\xi_1,\xi_2,\eta_1) + \lambda_2 \end{aligned}$$

where

$$F_1(z,\xi_1,\xi_2,\eta_1) = \xi_2 + 2\eta_1 + (\sin z + z\cos z + 4z^3)(-z^5 + \xi_1^2 + \xi_2^2).$$

Since

$$F_1(z,0,0,0) = -z^5(\sin z + z\cos z + 4z^3) = O(\|z\|^6),$$

it is sufficient to take $l = 1$. Then a dynamic compensator of (25) can be constructed based on the linear system

System (26) is controllable and observable. Finally, a dynamic output feedback control law proposed is

$$\begin{cases} u = \lambda_1 \\ \dot{\lambda}_1 = Fw \\ \dot{w} = (\bar{A} - K\bar{C})w + Ky + \bar{B}Fw \end{cases}$$

where

$$\bar{A} = \begin{bmatrix} 0 & 1 & 0 \\ 1 & 2 & 1 \\ 0 & 1 & 2 \end{bmatrix}, \quad \bar{B} = \begin{bmatrix} 0 \\ 0 \\ 1 \end{bmatrix},$$

$$\bar{C} = \begin{bmatrix} 1 & 1 & 0 \end{bmatrix}, \quad K = \begin{bmatrix} 6.6 \\ 0.4 \\ 0.92 \end{bmatrix},$$

$$F = \begin{bmatrix} -6 & -15 & -7 \end{bmatrix}.$$

4. CONCLUSIONS

In this paper, we present a constructive method of stabilizing a partially linear composite system. The closed loop system is HA-stable. How to simplify the controller or even get the one with minimum size is a topic for further study.

ACKNOWLEDGMENTS

This paper is supported by the National Natural Science Foundation of China and the National Key Project of China.

REFERENCES

Byrnes, C.I. and A. Isidori (1988). Local stabilization of minimum phase nonlinear systems. *Syst. Contr. Lett.*, **11**, 9-17.

Byrnes, C.I. and A. Isidori (1989). New results and examples in nonlinear feedback stabilization. Syst. *Contr. Lett.*, **12**, 437- 442.

Byrnes, C.I. and A. Isidori (1991). Asymptotic stabilization of minimum phase nonlinear systems. *IEEE Trans. Automat. Contr.*, **36**, 1122-1137.

Carr, J. (1981). *Applications of Centre Manifold Theory*. Springer-Verlag, New York.

Chen, P., Z. Han and Z. Zhang (1997). Dynamic output feedback stabilization of affine nonlinear systems. *Acta Automatica Sinica*, **23** (1997), 338-344.

Esfandiari, F. and H.K. Khalil (1992). Output feedback stabilization of full linearizable Systems. *Int. J. Contr.*, **56**, 1007-1037.

Hahn, W. (1967). *Stability of Motion*. Springer-Verlag, New York.

Khalil, H.K. and F. Esfandiari (1993). Semi-global stabilization of a class of nonlinear systems using output feedback. *IEEE Trans. Automat. Contr.*, **38**, 1412 --1415.

Saberi, A., P.K. Kokotovic and H.J. Sussmann (1990). Global stabilization of partially linear composite systems. *SIAM J. Contr. Optimiz.*, **28**, 1491-1503.

Teel, A. and L. Praly (1994). Global stabilizability and observability imply semi-global stabilizability by output feedback. *Syst. Contr. Lett.*, **22**, 313-325.

Teel, A. and L. Praly (1995). Tools for semi-global stabilization by partial state and output feedback. *SIAM J. Contr. Optimiz.*, **33**, 1443-1488.

Tornambé, A. (1992). Output feedback stabilization of a class of non-minimum phase nonlinear systems, *Syst. Contr. Lett.*, **19**, 193-204.

Wonham, W.M. (1979). *Linear Multivariable Control: A Geometric Approach*. Springer-Verlag, New York.

14th Triennial World Congress, Beijing, P.R. China

ADAPTIVE CONTROL OF A TIME-VARYING PARABOLIC SYSTEM: AVERAGING ANALYSIS

Keum-Shik Hong*, Victor Solo, and Joseph Bentsman‡**

** School of Mechanical Engineering, Pusan National University, 30 Changjeon-dong, Kumjeong-ku, Pusan, 609-735, Korea. Tel. +82(51)510-2454, Fax. (51)514-0685, E-mail: kshong@hyowon.pusan.ac.kr*
***Department of Statistics, Macquarie University, Sydney, NSW 2109, Australia, Fax. +61(2)850-7669, E-mail: v.solo@zen.efs.mq.edu.au*
‡Department of Mechanical and Industrial Enginering, University of Illinois at Urbana-Champaign, 1206 W. Green St. Urbana, IL, 61801, USA. Tel. (217)244-1076, Fax. (217)244-6534, E-mail: jbentsma@uiuc.edu

Abstract: Averaging theorems are developed for a coupled dynamic system which consists of ordinary differential equations and a parabolic partial differential equation, and applied in the convergence analysis of parameter estimate errors to zero in the model reference adaptive control of a nonautonomous parabolic partial differential equation with slowly varying parameters.

Keywords: Adaptive Control, Averaging, Parabolic Partial Differential Equation, Slow Varying System

1. INTRODUCTION

In recent control literature the adaptive control/identification of distributed parameter systems are getting more attention (Balas, 1983; Baumeister and Scondo, 1987; Kobayashi, 1988; Banks and Kunisch, 1989; Miyasato, 1990; Hong and Bentsman, 1992; Bentsman et al., 1992; Hong and Bentsman, 1994a,b; Demetriou and Rosen, 1994; Hong, 1997; Baumeister et al., 1997). There have been increasing efforts for the last several years in explicit incorporation of time-varying parameters into adaptive control analysis. Also averaging methods have been emerged as a powerful tool for the analysis of adaptive algorithms. The aim of this paper is to bring these two streams together with the hope that averaging methods can yield extra insights on the adaptive control process in the presence of time-varying parameters. The control law and adaptive estimates design follows that in (Hong and Bentsman, 1994a) which dealt with an autonomous parabolic plant, but the presence of time-varying parameters considered here calls for new averaging theorems. It is shown that a similar design methodology can be extended to the systems with time-varying coefficients.

The model reference adaptive control of a time-varying parabolic system is firstly introduced with the Lyapunov stability analysis. Next the convergence of parameter errors to zero is investigated through averaging. Even both the treatment of time-varying parameters and averaging for parabolic systems are new, the main focus of this paper is to show the convergence of parameter errors to zero through averaging. For detailed construction of adaptive controller or related issues refer to (Astrom and Wittenmark, 1995; Narendra and Annaswamy, 1989; Sastry and Bodson, 1989). In this paper, B, C and L denote generic constants.

2. PROBLEM SETUP

Consider a nonautonomous parabolic partial differential equation as

$$\xi_t = a(\varepsilon t)\xi_{xx} + b(\varepsilon t)\xi + u,\ 0 \le x \le 1,\ t \ge 0, \quad (1)$$
$$\xi(0,t) = g_a(t),\ \xi(1,t) = g_b(t),$$
$$\xi(x,0) = 0,$$

where $u = u(x,t)$ is the control input, $\xi = \xi(x,t)$ is the observed distributed state, $\xi_t = d\xi/dt$, and $\xi_{xx} = \partial^2\xi/\partial x^2$. The initial state is set to zero for simplicity since it only matters in transient dynamics and leads to no difference in the averaging analysis. The coefficients a(·), b(·) are bounded slowly varying parameters in the system. Specifically in heat transfer they are referred to as a heat conductivity and a radiation factor, respectively. Assume $0 < \varepsilon << 1$ so that the variation is slow, and $da(\varepsilon t)/dt = \theta(\varepsilon)$, $db(\varepsilon t)/dt = \theta(\varepsilon)$. However the amplitude variations of the parameters are large, i.e. $a(\varepsilon t)=O(1)$, $b(\varepsilon t)=O(1)$. Hence, $a(\varepsilon t) = 2-\sin(\varepsilon t)$, $b(\varepsilon t) = -2+\cos(\varepsilon t)$ would be plausible examples.

Along with system (1) consider a reference model with the same boundary and initial conditions as

$$\xi_{mt} = a_m\xi_{mxx} + b_m\xi_m + r, \; 0 \le x \le 1, \; t > 0 \quad (2)$$

where the subscript m stands for model, and $r = r(x,t)$ is the reference signal. Note that $g_a(t)$ and $g_b(t)$ could also be thought as the boundary reference signals. $a_m > 0$, $b_m < 0$ are constant coefficients.

Now adopting the procedure in (Hong and Bentsman, 1994a) consider a control law of the form

$$u = (a_m - \hat{a})\xi_{xx} + (b_m - \hat{b})\xi + r \quad (3)$$

where $\hat{a}$ and $\hat{b}$ are adaptive estimates to be specified. Substituting (3) into (1) yields the closed loop plant equation as

$$\xi_t = (a_m - \tilde{a})\xi_{xx} + (b_m - \tilde{b})\xi + r, \quad (4)$$

where $\tilde{a} = \hat{a} - a, \tilde{b} = \hat{b} - b$ are the parameter estimation errors. Note that if $\tilde{a},\tilde{b} = 0$, then (4) is exactly the same as (2). Introducing the state error $e = \xi - \xi_m$, the following state error equation is obtained

$$e_t = (a_m - \tilde{a})e_{xx} + (b_m - \tilde{b})e - (\tilde{a}\xi_{mxx} + \tilde{b}\xi_m),$$
$$e(0,t) = e(1,t) = 0, e(x,0) = 0. \quad (5)$$

Now, consider a functional $V: L_2(0,1)\times R^2 \to R^+$ such that

$$V(t) = \tfrac{1}{2} < e,e > + \tfrac{1}{2\varepsilon}(\tilde{a}^2 + \tilde{b}^2) \quad (6)$$

where $<\cdot,\cdot>$ is the inner product in space $L_2(0,1)$ defined as $<h,g> = \int_0^1 h(x,t)g(x,t)dx$, and with the induced norm $\|\cdot\|$. Differentiating (6) with respect to t along (5) yields

$$\begin{aligned} \dot{V}(t) &= < e,\dot{e} > + \tfrac{1}{\varepsilon}(\tilde{a}\dot{\tilde{a}} + \tilde{b}\dot{\tilde{b}}) \\ &= a_m < e_x,e_{xx} > + b_m < e,e > \\ &\quad + \tilde{a}\left[- < e,e_{xx} > - < e,\xi_{mxx} > + \frac{1}{\varepsilon}\dot{\tilde{a}}\right] \\ &\quad + \tilde{b}\left[- < e,e > - < e,\xi_m > + \frac{1}{\varepsilon}\dot{\tilde{b}}\right] \end{aligned} \quad (7)$$

Therefore choosing

$$\dot{\tilde{a}} = \varepsilon(< e,e_{xx} > + < e,\xi_{mxx} >) \quad (8a)$$

$$\dot{\tilde{b}} = \varepsilon(< e,e > + < e,\xi_m >) \quad (8b)$$

(7) becomes

$$\begin{aligned} \dot{V}(t) &= -a_m < e_x,e_x > + b_m < e,e > \\ &\le (-a_m\pi^2 + b_m)\|e\|^2 \\ &\le 0. \end{aligned} \quad (9)$$

Hence (6) is a Lyapunov function for the whole error systems (5) and (8a,b) in the state space $H = L_2(0,1)\times R^2$. The existence and uniqueness of the solutions for systems (5) and (8a,b) are addressed in Appendix A.

Now the above development is summarized in the following lemma.

Lemma: Consider equations (5) and (8a,b), then $\|e\| \to 0$ as $t \to \infty$.

Proof: Note first that $(e,\tilde{a},\tilde{b}) = (0,0,0)$ is an equilibrium point for the system. Since the functional (6) with (9) is a Lyapunov function in $H=L_2(0,1)xR^2$, the uniform stability of (0,0,0) in H is established. LaSalle's Invariance principle may not be applicable in this case since system (5) is time-varying due to the reference signal. The conclusions of the lemma can be shown in two different ways. One is applying Barbalat's Lemma (Narendra and Annaswamy, 1989), and the other is utilizing the uniqueness and semigroup properties of the solution (Hong, 1997).

Now we turn to the analysis of parameter error convergence through averaging. Note that the differential equations for the controller parameters in (3) are written as

$$\begin{aligned} \dot{\hat{a}} &= \varepsilon[< e,e_{xx} > + < e,\xi_{mxx} >] + \varepsilon\dot{a}(\varepsilon t), \\ \dot{\hat{b}} &= \varepsilon[< e,e > + < e,\xi_m >] + \varepsilon\dot{b}(\varepsilon t). \end{aligned} \quad (10)$$

Define $\theta = [\hat{a} \; \hat{b}]^T$, then (10) can be written as follows

$$\dot{\theta} = \varepsilon f(t,\theta,e,\varepsilon t). \quad (11)$$

Note that $f(\cdot)$ is a functional of e rather than a function, and the explicit appearance of time t as an argument in f comes from the exogenous signal ξ. The last argument εt denotes the existence of slow-varying parameters.

In (Bentsman et al., 1992) the averaged system corresponding to (10) has been computed explicitly (without linearization). Associated with (5) a frozen state $e_{\tau,\tilde{\theta}}(\sigma)$ is defined through

$$e_{t|\tau,\tilde{\theta}} = (a_m - \tilde{a})e_{xx|\tau,\tilde{\theta}} + (b_m - \tilde{b})e_{\tau,\tilde{\theta}} - [\xi_{mxx} \; \xi_m]\tilde{\theta} \quad (12)$$

where parameters $\tilde{a},\tilde{b}$ are assumed to be frozen, and

$$e_{t|\tau,\tilde\theta} = \partial e_{\tau,\tilde\theta}(x,t)/\partial t,$$

$$e_{xx|\tau,\tilde\theta} = \partial e^2_{\tau,\tilde\theta}(x,t)/\partial x^2.$$

3. AVERAGING ANALYSIS

Consider a coupled system as

$$\dot\theta^\varepsilon(t) = \varepsilon f(t,\theta^\varepsilon,e^\varepsilon,\varepsilon t), \quad (13)$$

$$e_t^\varepsilon = \left(a_m - \tilde a^\varepsilon\right)e_{xx}^\varepsilon + \left(b_m - \tilde b^\varepsilon\right)e^\varepsilon - \left[\tilde a^\varepsilon \xi_{mxx} + \tilde b^\varepsilon \xi_m\right]. \quad (14)$$

(13) and (14) correspond to (11) and (5), respectively. The superscript ε is affixed to denote the variables in fast time t prior to time scaling. An averaged system associated with (13) is introduced as

$$\dot\theta_{av}(t) = \varepsilon f_{av}\left(\theta_{av},\varepsilon t\right), \quad (15)$$

where the averaged function is defined by

$$f_{av}(\theta_{av},\varepsilon t) = \lim_{T\to\infty}\frac{1}{T}\int_t^{t+T} f\left(\sigma,\theta_{av},e^\varepsilon(\sigma),\varepsilon t\right)d\sigma \quad (16)$$

if the limit exists uniformly in t. It is assumed that $|f(\sigma,0,0,0)| \le B$, and $|f_{av}(0,0)| \le B$. Wth a new time scale (slow time) $\tau = \varepsilon t$, (15) can be rewritten as

$$\dot\theta_a(\tau) = f_{av}\left(\theta_a(\tau),\tau\right). \quad (17)$$

To investigate the closeness of the two solutions of primary system (13) and averaged system (15) for sufficiently small ε, (13) is integrated in t and is rescaled to the slow time $\tau = \varepsilon t$ as follows

$$\theta_\varepsilon(\tau) = \theta^\varepsilon(\tau\varepsilon^{-1}) = \theta^\varepsilon(0) + \varepsilon\int_o^{\tau\varepsilon^{-1}} f_\sigma d\sigma \quad (18)$$

where $f_\sigma = f\left(\sigma,\theta^\varepsilon(\sigma),e^\varepsilon(\sigma),\varepsilon\sigma\right)$. The subscript ε denotes the rescaled variable in the slow time. Similarly for averaged system (15), the following is obtained

$$\begin{aligned}\theta_a(\tau) &= \theta_{av}(\tau\varepsilon^{-1})\\ &= \theta_a(0) + \varepsilon\int_o^{\tau\varepsilon^{-1}} f_{av}(\theta_a(\varepsilon\sigma),\varepsilon\sigma)d\sigma\end{aligned} \quad (19)$$

Now subtract (19) from (18), then we have

$$\Delta_\varepsilon(\tau) = \theta_\varepsilon(\tau) - \theta_a(\tau) = \Delta_\varepsilon(0) + I_\varepsilon(\tau), \quad (20)$$

where

$$\Delta_\varepsilon(0) = \theta^\varepsilon(0) - \theta_a(0), \quad (21)$$

$$I_\varepsilon(\tau) = \varepsilon\int_o^{\tau\varepsilon^{-1}} [f_\sigma - f_{av}(\theta_a(\varepsilon\sigma),\varepsilon\sigma)]d\sigma. \quad (22)$$

By introducing a frozen state as an intermediate state between f_σ and f_{av} the evaluation of (22) can be carried out in two stages as follows

$$I_\varepsilon(\tau) = I_{\varepsilon A}(\tau) + I_{\varepsilon B}(\tau), \quad (23)$$

where

$$I_{\varepsilon A}(\tau) = \varepsilon\int_o^{\tau\varepsilon^{-1}} [f_\sigma - f(\sigma,\theta_a(\varepsilon\sigma),e_{\varepsilon\sigma,\theta_a(\varepsilon\sigma)}(\sigma),\varepsilon\sigma)]d\sigma, \quad (24)$$

$$I_{\varepsilon B}(\tau) = \varepsilon\int_o^{\tau\varepsilon^{-1}} [f(\sigma,\theta_a(\varepsilon\sigma),e_{\varepsilon\sigma,\theta_a(\varepsilon\sigma)}(\sigma),\varepsilon\sigma) - f_{av}(\theta_a(\varepsilon\sigma),\varepsilon\sigma)]d\sigma \quad (25)$$

Note that the first integral $I_{\varepsilon A}(\tau)$ deals with the approximation of the fast system f_σ by the frozen system $f(\cdot)$, whereas the second integral $I_{\varepsilon B}(\tau)$ deals with the approximation of the frozen system $f(\cdot)$ by the averaged system $f_{av}(\cdot)$.

Now, specific bounds for both integrals (24), (25) are obtained. However for proper implication a bound for the second integral $I_{\varepsilon B}(\tau)$ is firstly obtained. Detailed derivations are gathered in Appendices. Appendix B shows that

$$|I_{\varepsilon B}(\tau)| \le C_T(\varepsilon), \quad 0 \le \tau \le T, \quad (26)$$

where $C_T(\varepsilon) \to 0$ as $\varepsilon \to 0$. In obtaining (26) a general Lipschitz condition on f has been assumed

$$\begin{aligned}|f(\sigma,\theta,e,\tau) - f(\sigma,\theta',e',\tau')| &\le L_1|\theta-\theta'| + L_1|e-e'|\\ &+ L_1|\tau-\tau'|,\ L_1 = const > 0.\end{aligned} \quad (27)$$

The satisfaction of (27) for our case is obvious from (A.5) with the uniform stability obtained. Now for the first integral $I_{\varepsilon A}(\tau)$, a bound as following is obtained

$$\begin{aligned}|I_{\varepsilon A}(\tau)| &\le \varepsilon\int_o^{\tau\varepsilon^{-1}}\\ &\left|\begin{aligned}&f(\sigma,\theta^\varepsilon(\sigma),e^\varepsilon(\sigma),\varepsilon\sigma)\\ &\quad - f(\sigma,\theta_a(\varepsilon\sigma),e_{\varepsilon\sigma,\theta_a(\varepsilon\sigma)}(\sigma),\varepsilon\sigma)\end{aligned}\right|d\sigma\\ &\le \varepsilon L_1\int_o^{\tau\varepsilon^{-1}}\\ &\quad\left(|\Delta_\varepsilon(\varepsilon\sigma)| + \left\|e^\varepsilon(\sigma) - e_{\varepsilon\sigma,\theta_a(\varepsilon\sigma)}(\sigma)\right\|\right)d\sigma.\end{aligned} \quad (28)$$

It is shown in Appendix D that (27) again ensures

$$\int_o^{\tau\varepsilon^{-1}}\left\|e^\varepsilon(\sigma) - e_{\varepsilon\sigma,\theta_a(\varepsilon\sigma)}(\sigma)\right\|d\sigma \le B. \quad (29)$$

Therefore, finally combining (20), (26) and (28)-(29) together yields

$$|\Delta_\varepsilon(\tau)| \le |\Delta_\varepsilon(0)| + L_1\int_o^\tau |\Delta_\varepsilon(s)|ds + C_T(\varepsilon) + \varepsilon L_1 B. \quad (30)$$

The Bellman-Gronwall inequality then gives

$$|\Delta_\varepsilon(\tau)| \le e^{L_1\tau}\left(\Delta_\varepsilon(0) + C_T(\varepsilon) + \varepsilon L_1 B\right), 0 \le \tau \le T \quad (31)$$

which again implies

$$\sup_{0\le\tau\le T}|\Delta_\varepsilon(\tau)| = \sup_{0\le\tau\le T}|\theta_\varepsilon(\tau) - \theta_a(\tau)| \le B_T(\varepsilon), \quad (32)$$

and $B_T(\varepsilon) \to 0$ as $\varepsilon \to 0$.

Finally all above development is summarized as follows.

Theorem 1: Consider (13)-(14) and (15) with appropriate regularity conditions. Then for fixed T and sufficiently small ε

$$\sup_{0\le t\le T/\varepsilon}|\theta^\varepsilon(t) - \theta_{av}(t)| \le B_T(\varepsilon)$$

where $B_T(\varepsilon) \to 0$ as $\varepsilon \to 0$.

Theorem 1 does not yet connect the stability properties between (1)-(2) and (3). However with further assumptions on f_{av} the following theorem can be stated.

Theorem 2: Consider (13)-(14) and (15) with appropriate regularity conditions. Further assume

that the averaged system (15) is exponentially stable, then the trivial solutions $\theta^{\varepsilon}(t)=0, e^{\varepsilon}(t)=0$ are exponentially stable for sufficiently small ε.
Proof: The proof follows exactly that of Theorem 4.3 in (Hong and Bentsman, 1994a).

4. CONCLUSIONS

In this paper averaging theorems are developed for a coupled ordinary and partial differential equations and applied to the asymptotic convergence analysis of parameter estimates errors to zero in the model reference adaptive control of a time-varying parabolic partial differential equation.

ACKNOWLEDGEMENTS

This work was supported by the Korea Science and Engineering Foundation under KOSEF Grant 971-0920-137-2.

REFERENCES

Astrom, K. and B. Wittenmark (1995). *Adaptive Control.* Addison-Wesley, Reading, MA.

Balas, M. J. (1983). Some critical issues in stable finite dimensional adaptive control of linear distributed parameter system. *Proc. 4th Yale Conf. Adaptive Contr.*

Banks, H. T. and K. Kunisch (1989). *Estimation Techniques for Distributed Parameter Systems,* Birkhauser, Boston.

Baumeister, J. and W. Scondo (1987). Asymptotic embedding methods for parameter estimation. in *Proc. 26th IEEE Conf. Dec. Contr.*, Los Angeles, CA, pp.170-174.

Baumeister, J., W. Scondo, M. A. Demetriou, and I. G. Rogen (1997). On-line parameter estimation for infinite-dimensional dynamical systems. *SIAM J. Control Optim.*, Vol. **35**, No. 2, pp.678-713.

Demetriou, M. A. and I. G. Rosen (1994). On the persistence of excitation in the adaptive estimation of distributed parameter systems. *IEEE Trans. Automat. Contr.*, Vol. **39**, No.5, pp.1117-1123.

Bentsman, J., V. Solo, and K. Hong (1992). Adaptive control of a parabolic system with time varying parameters: Averaging analysis. *Proceedings of the 31st IEEE CDC*, pp. 710-711.

Henry, D. (1981). *Geometric Theory of Semilinear Parabolic Equations.* Lecture Notes in Mathematics, 840, Springer-Verlag.

Hong, K. S. and J. Bentsman (1992). Nonlinear control of diffusion processes with uncertain parameters using MRAC approach. *Proceedings American Control Conference*, Chicago, IL, pp.1343-1347.

Hong, K.S. and J. Bentsman (1994a). Direct adaptive control of parabolic systems: Algorithm synthesis, convergence and stability analysis. *IEEE Trans. Automat. Contr.*, Vol. **39**, No.10, pp.2018-2033.

Hong, K. S. and J. Bentsman (1994b). Application of averaging method for integro-differential equations to model reference adaptive control of parabolic systems. *Automatica*, Vol. **30**, No. 9, pp.1415-1419.

Hong, K. S., J. Wu, and K. I. Lee (1994). New conditions for the exponential stability of evolution equations. *IEEE Trans. Automat. Contr.*, Vol. **39**, No. 7, pp. 1432-1436.

Hong, K. S. (1997). Asymptotic behavior analysis of a coupled time-varying system: Application to adaptive systems. *IEEE Trans. Automat. Contr.*, Vol. **42**, No. 12, pp.1693-1697.

Kobayashi, T. (1988). Finite dimensional adaptive control for infinite dimensional systems. *Int. J. Contr.*, Vol. **48**, No.1, pp.289-302.

Middleton, R. H. and G. C. Goodwin (1988). Adaptive control of time varying linear plants. *IEEE Trans. Auto. Contr.*, Vol. **33**, pp.150-155.

Miyasato, Y. (1990). Model reference adaptive control for distributed parameter systems of parabolic type by finite dimensional controller. *Proc. 29th IEEE CDC*, Honolulu, pp.1459-1464.

Narendra, K. S. and A. M. Annaswamy (1989). *Stable Adaptive Systems*, Prentice Hall.

Pazy, A. (1983). *Semigroups of Linear Operators and Applications to Partial Differential Equations*, Springer-Verlag, New York.

Sastry, S. and M. Bodson (1989). *Adaptive Control: Stability, Convergence and Robustness*, Prentice Hall.

Solo, V. (1996). Deterministic adaptive control with slowly varying parameters: An averaging analysis. *Int. J. Control*; Vol. **64**, No.1, pp.99-125.

Walker, J. A. (1980). *Dynamical Systems and Evolution Equations*, New York, Plenum Press.

APPENDIX A. EXISTENCE AND UNIQUENESS

Rewrite the nonlinear error equations (5) and (2.8a,b) in the following form

$$\dot{z}=Az+F(t,z),\ z(0)=z_0 \tag{A.1}$$

where $z=(e,\tilde{a},\tilde{b})^T$ and

$$A=\begin{bmatrix}(a_m-\tilde{a})\dfrac{\partial^2}{\partial x^2}+(b_m-\tilde{b}) & 0 & 0\\ 0 & -1 & 0\\ 0 & 0 & -1\end{bmatrix}$$

$$=\begin{bmatrix}A_0 & 0 & 0\\ 0 & -1 & 0\\ 0 & 0 & -1\end{bmatrix}$$

$$F(t,z)=\begin{bmatrix}-\tilde{a}\xi_{mxx}-\tilde{b}\xi_m\\ \tilde{a}+\varepsilon<e,e_{xx}>+\varepsilon<e,\xi_{mxx}>\\ \tilde{b}+\varepsilon<e,e>+\varepsilon<e,\xi_m>\end{bmatrix}$$

Define a state space as $H\equiv L_2(0,1)\times R^2$, and

$$D(A)=\left\{\begin{matrix}(e,\tilde{a},\tilde{b})\in H: e\in H^2(0,1)\cap H_0^1(0,1)\\ \text{with } e(0)=0=e(1), \text{ and } \tilde{a},\tilde{b}\in R\end{matrix}\right\}. \tag{A.2}$$

Note that the boundary conditions of (5) has been incorporated in the space $H^2(0,1)\cap H_0^1(0,1)$, which is the domain of the differential operator A_0. D(A) is dense, and A is a closed operator. See (Walker, 1980, p.68, Example 2.5). For $z\in D(A)$

$$\begin{aligned}<z,Az>_H &= \int_0^1 e(x)\left[(a_m-\tilde{a})\frac{\partial^2 e(x)}{\partial x^2}+(b_m-\tilde{b})e(x)\right]dx\\ &\quad -\tilde{a}^2-\tilde{b}^2\\ &\le\left[-\pi^2(a_m-\tilde{a})+(b_m-\tilde{b})\right]\int_0^1 e^2(x)dx-\tilde{a}^2-\tilde{b}^2\\ &\le -C_1<z,z>_H\end{aligned} \tag{A.3}$$

where $C_1=\min\left\{\pi^2(\tilde{a}-a_m)+(b_m-\tilde{b}),1\right\}$, and $a_m-\tilde{a}\ge\Delta a>0$ is assumed. Now by the linearity of A, we see that $\omega I-A$ is monotone (accretive) for every $\omega\le C_1$. Hence $A: D(A)\subset H\to H$ is the infinitesimal generator of a linear dynamical system $\{S(t)\}_{t\ge0}=\left\{(E(t),\tilde{A}(t),\tilde{B}(t))\right\}_{t\ge0}$ on H See (Walker, 1980, Theorem 3.2, p.92). Note that the first component E(t) is generated by A_0 and the specific form of semigroup E(t) can be found in (Hong and Bentsman, 1994b). Note also that $E(t)e_0$ is the strong solution of the evolution equation $\dot{e}(t)=A_0e(t)$ for every $e_0\in D(A_0)$.

Now set $z=(e,\tilde{a},\tilde{b})$ and $z'=(e',\tilde{a}',\tilde{b}')$. Then

$$\begin{aligned}|F(t,z)-F(t,z')>|_H^2 &= \left\|-\tilde{a}\xi_{mxx}-\tilde{b}\xi_m+\tilde{a}'\xi_{mxx}+\tilde{b}'\xi_m\right\|^2\\ &+\left|\tilde{a}-\tilde{a}'+\varepsilon\begin{pmatrix}<e,e_{xx}>+<e,\xi_{mxx}>-<e',e_{xx}>\\ +<e',e_{xx}>-<e',e'_{xx}>-<e',\xi_{mxx}>\end{pmatrix}\right|^2\\ &+\left|\tilde{b}-\tilde{b}'+\varepsilon\begin{pmatrix}<e,e>+<e,\xi_m>\\ -<e',e>+<e',e>-<e',e'>-<e',\xi>\end{pmatrix}\right|^2\\ &\le\|\xi_{mxx}\|^2|\tilde{a}-\tilde{a}'|^2+|\xi_m|^2\left|\tilde{b}-\tilde{b}'\right|^2+|\tilde{a}-\tilde{a}'|^2\\ &+\varepsilon\left(|e_{xx}|^2+|\xi_{mxx}|^2+|e'_{xx}|^2\right)|e-e'|^2\\ &+\left|\tilde{b}-\tilde{b}'\right|^2+\varepsilon\left(|e|^2+|\xi_m|^2+\|e'\|^2\right)|e-e'\|^2.\end{aligned}$$

Hence

$$|F(t,z)-F(t,z')|_H\le C_2\|z-z'\|_H \tag{A.4}$$

where C_2 is a constant in the bounded set considered. Therefore $F: H\to H$ is locally Lipshchitz continuous in H. Therefore a unique solution exists. Finally the solution of (5) can be written in the following variation of constant formula (Henry, 1981; Pazy, 1983)

$$\begin{aligned}e(t) &= E(t,0)e(0)\\ &+\int_0^t E(t,\tau)(-\tilde{a}(\tau)\xi_{mxx}(\tau)-\tilde{b}(\tau)\xi_m(\tau))\,d\tau\end{aligned} \tag{A.5}$$

where E(t,s) is the evolution operator associated with A_0 in the space $L_2(0,1)$.

APPENDIX B. BOUND (16)

Following the work of (Solo, 1996), the integral (15) is divided into small pieces as follows. Introduce a sequence of increasing integers $N_\varepsilon\to\infty$ as $\varepsilon\to0$ such that $\{n_\varepsilon\}$ is an increasing sequence:

$$n_\varepsilon=\frac{1}{\varepsilon N_\varepsilon}=\frac{\delta_\varepsilon}{\varepsilon}\to\infty \text{ as } \varepsilon\to0.$$

Then, (15) can be written as

$$I_{\varepsilon B}(s)=\delta_\varepsilon\sum_{\ell=1}^{N_\varepsilon}\varphi_\varepsilon\left(s(\ell-1)\delta_\varepsilon\right), \tag{B.1}$$

where

$$\begin{aligned}\varphi_\varepsilon(\tau) &= n_\varepsilon^{-1}\int_{\tau\varepsilon^{-1}}^{\tau\varepsilon^{-1}+n_\varepsilon s}\\ &[f(\sigma,\theta_a(\varepsilon\sigma),e_{\varepsilon\sigma,\theta_a(\varepsilon\sigma)}(\sigma),\varepsilon\sigma)\\ &-f_{av}(\theta_a(\varepsilon\sigma),\varepsilon\sigma)]d\sigma\end{aligned}$$

Also in the sequel it is shown that

$$\varphi_\varepsilon(\tau)\to0, \tag{B.2}$$

as $\varepsilon\to0$ uniformly in $0\le\tau\le T$. Therefore for $0\le s\le T$ the following is obtained

$$|I_{\varepsilon B}(s)|\le C_T(\varepsilon)\equiv T\sup_{0\le\tau\le T}|\varphi_\varepsilon(\tau)|\to0, \tag{B.3}$$

as $\varepsilon\to0$. To prove (B.2) $\varphi_\varepsilon(\tau)$ is considered into three pieces as follows

$$\varphi_\varepsilon(\tau)=\varphi_{\varepsilon a}(\tau)+\varphi_{\varepsilon b}(\tau)+\varphi_{\varepsilon c}(\tau),$$

where

$$\begin{aligned}\varphi_{\varepsilon a}(\tau) &= n_\varepsilon^{-1}\int_{\tau\varepsilon^{-1}}^{\tau\varepsilon^{-1}+n_\varepsilon s}\\ &[f(\sigma,\theta_a(\varepsilon\sigma),e_{\varepsilon\sigma,\theta_a(\varepsilon\sigma)}(\sigma),\varepsilon\sigma)\\ &-f(\sigma,\theta_a(\tau),e_{\tau,\theta_a(\tau)}(\sigma),\tau)]d\sigma,\end{aligned} \tag{B.4a}$$

$$\begin{aligned}\varphi_{\varepsilon b}(\tau) &= n_\varepsilon^{-1}\int_{\tau\varepsilon^{-1}}^{\tau\varepsilon^{-1}+n_\varepsilon s}\\ &\left[f(\sigma,\theta_a(\tau),e_{\tau,\theta_a(\tau)}(\sigma),\tau)-f_{av}(\theta_a(\tau),\tau)\right]d\sigma,\end{aligned} \tag{B.4b}$$

$$\begin{aligned}\varphi_{\varepsilon c}(\tau) &= n_\varepsilon^{-1}\int_{\tau\varepsilon^{-1}}^{\tau\varepsilon^{-1}+n_\varepsilon s}[f_{av}(\theta_a(\tau),\tau)\\ &-f_{av}(\theta_a(\varepsilon\sigma),\varepsilon\sigma)]d\sigma.\end{aligned} \tag{B.4c}$$

From the Lipschitz condition (17), the first term (B.4a) is bounded as

$$\begin{aligned}|\varphi_{\varepsilon a}(\tau)| &\le L_1\sup_{\tau\varepsilon^{-1}\le\sigma\le\tau\varepsilon^{-1}+n_\varepsilon s}|\theta_a(\varepsilon\sigma)-\theta_a(\tau)|\\ &+L_1\sup_{\tau\varepsilon^{-1}\le\sigma\le\tau\varepsilon^{-1}+n_\varepsilon s}\left|e_{\varepsilon\sigma,\theta_a(\varepsilon\sigma)}(\sigma)-e_{\tau,\theta_a(\tau)}(\sigma)\right|\\ &+L_1\delta_\varepsilon.\end{aligned} \tag{B.5}$$

To bound the first term in (B.5) we use the following

$$\theta_a(\varepsilon\sigma)-\theta_a(\tau)=\int_\tau^{\varepsilon\sigma}f_{av}(\theta_a(\tau'),\tau')d\tau'.$$

Since $f_{av}(\cdot)$ also obeys the Lipschitz condition, it follows that

$$|\theta_a(\varepsilon\sigma)-\theta_a(\tau)|\le L_1(h+|\varepsilon\sigma-\tau|)|\varepsilon\sigma-\tau|,$$

which again implies that

$$\sup_{\tau\varepsilon^{-1}\le\sigma\le\tau\varepsilon^{-1}+n_\varepsilon s}|\theta_a(\varepsilon\sigma)-\theta_a(\tau)|\le\delta_\varepsilon sL_1(h+\delta_\varepsilon s)$$
$$\le\delta_\varepsilon TL_1(h+\delta_\varepsilon T),$$

for $0\le\tau\le T$. For the second term in (B.5) the following bounds are obtained in Appendix C.

$$\left|\partial e_{\tau,\theta}(\sigma)/\partial\tau\right|\le B, \text{ and} \qquad \text{(B.6)}$$
$$\left|\partial e_{\tau,\theta}(\sigma)/\partial\theta\right|\le B$$

Then the second term in (B.5) is bounded by

$$L_1 B \sup_{\tau\varepsilon^{-1}\le\sigma\le\tau\varepsilon^{-1}+n_\varepsilon s}\left(|\varepsilon\sigma-\tau|+|\theta_a(\varepsilon\sigma)-\theta_a(\tau)|\right)$$
$$\le\delta_\varepsilon C$$

by arguments just given. Finally, we have

$$|\varphi_{\varepsilon a}(\tau)|\le\delta_\varepsilon B,$$

for $0\le\tau\le T$. Therefore, (B.2) holds for $\varphi_{\varepsilon a}(\tau)$. Through a similar analysis it can be shown that (B.2) holds for $\varphi_{\varepsilon c}(\tau)$ and $\varphi_{\varepsilon b}(t)$, respectively. Q.E.D.

APPENDIX C. BOUNDS ON ERROR TERMS

Differentiating (12) with respect to θ gives

$$e^{\theta}_{t|\tau,\theta}=\left(a_m-\tilde{a}\right)e^{\theta}_{xx|\tau,\theta}+\left(b_m-\tilde{b}\right)e^{\theta}_{\tau,\theta}$$
$$-\begin{bmatrix}e_{xx|\tau,\theta}\\ e_{\tau,\theta}\end{bmatrix}-\begin{bmatrix}\xi_{mxx}\\ \xi_m\end{bmatrix}$$

where $e^{\theta}_{\tau,\theta}=\partial e_{\tau,\theta}(\sigma)/\partial\theta$, etc. and the other terms with subscripts are defined similarly. A similar equation holds for $e^{\tau}_{\tau,\theta}$. The results now follow from the parabolic boundedness lemma of Appendix E.

APPENDIX D. FAST STATE DYNAMICS

In this appendix a bound for the fast state e is obtained. Rewrite (5) as

$$e_t=\left(a_m-\tilde{a}\right)e_{xx}+\left(b_m-\tilde{b}\right)e-h\left(t,\tilde{\theta}\right),$$

where $h\left(t,\tilde{\theta}\right)=\left[\xi_{mxx}\ \xi_m\right]\tilde{\theta}$.

Denoting $\hat{e}(t)=e_{\varepsilon t,\theta_a(\varepsilon)}(t)$., the following approximate system is introduced using (12)

$$d\hat{e}/dt=\partial\hat{e}/\partial t+\left[\frac{\partial\hat{e}}{\partial\tilde{\theta}}\right]^T\frac{\partial\tilde{\theta}}{\partial t}$$
$$=\left(a_m-\tilde{a}\right)\hat{e}_{xx}+\left(b_m-\tilde{b}\right)\hat{e}-h\left(t,\tilde{\theta}\right)+\varepsilon f^T\left[\frac{\partial\hat{e}}{\partial\tilde{\theta}}\right]$$

where $\hat{e}_{xx}=e_{xx|\varepsilon t,\theta_a(\varepsilon)}(t)$. Thus the error dynamics between e(t) and $\hat{e}(t)$ is obtained as

$$\tilde{e}_t=e_t-\hat{e}_t$$
$$=\left(a_m-\tilde{a}\right)\tilde{e}_{xx}+\left(b_m-\tilde{b}\right)\tilde{e}-\varepsilon f_t^T\partial\hat{e}/\partial\tilde{\theta}.$$

Now apply the parabolic boundedness lemma of Appendix E to deduce

$$\|\tilde{e}\|\le\varepsilon B^2$$

provided $|f|\le B$, $\left\|\partial\hat{e}/\partial\tilde{\theta}\right\|\le B$. The first inequality follows from (27). The second one follows from Appendix C, since $|\theta_a|\le h$. Q.E.D

APPENDIX E. PARABOLIC BOUNDEDNESS LEMMA

Lemma E1: Consider the following parabolic time varying system:

$$v_t=\left[a_m-\alpha(t)\right]v_{xx}+\left[b_m-\beta(t)\right]v+h(x,t),$$
$$v(0,t)=v(1,t)=0, v(x,0)=0. \qquad \text{(E.1)}$$

Assume that $a_m-\alpha(t)\ge\Delta_a$ = constant, $b_m-\beta(t)\ge 0$, and $\|h\|\le B$ for all t where B is a constant. Then

$$|v|\le B\Delta_a^{-1}\left(\Sigma_1^\infty\lambda_n^{-4}\right)^{\frac{1}{2}}.$$

Proof. Parabolic system (E.1) has a solution

$$v(x,t)=\sum_{n=1}^{\infty}\varphi_n(x)v_n(t), \qquad \text{(E.2)}$$

where $v(t)$ satisfies

$$\dot{v}_n(t)=-\bar{k}_n(t)\,v_n(t)+<\varphi_n,h>, \qquad \text{(E.3)}$$

and $\bar{k}_n(t)=\left(a_m-\alpha(t)\right)\lambda_n^2+\left(b_m-\beta(t)\right)$. Also the solution to (E.3) is given by

$$v_n(t)=\int_0^t e^{-\int_s^t\bar{k}_n(u)du}<h,\varphi_n>ds.$$

Since $\bar{k}_n(t)\ge\Delta_a\lambda_n^2$ and $\|h\|\le B$, we have

$$|v_n|\le\int_0^t e^{-\Delta_a\lambda_n^2(t-s)}ds\ B\le B\Delta_a^{-1}/\lambda_n^2.$$

Which finally implies

$$|v(x,t)|=\left(\Sigma\,v_n^2(t)\right)^{\frac{1}{2}}\le B\Delta_a^{-1}\left(\Sigma_1^\infty\lambda_n^{-4}\right)^{\frac{1}{2}}$$

as required.

GAIN SCHEDULLED CONTROLLERS FOR DYNAMIC SYSTEMS WITH SECTOR NONLINEARITIES

Eduardo F. Costa **Vilma A. Oliveira**

Departamento de Engenharia Elétrica
Universidade de São Paulo
Caixa Postal 359, 13560-970, São Carlos, SP, BRASIL
vilmao@sel.eesc.sc.usp.br

Abstract: A gain scheduling approach to the control of dynamic systems with sector nonlinearities is given. The gain scheduled controller proposed uses the system state to switch linear robust controllers in subsets of the system domain of attraction around the operating point. The procedure guarantees the stability of the feedback closed-loop system and reduces the amount of conservatism that results when a large region of attraction around the operating point is considered. In addiction, the problem of guaranteed transition between operating points by using a pre-specified path in the state space for the system operating points is considered. *Copyright © 1999 IFAC*

Keywords: non-linear systems, structured uncertainties, gain scheduled controllers, Lyapunov functions, attraction domains

1. INTRODUCTION

Control design via the linearization approach has the limitation that the controller is guaranteed to work only in a neighbourhood of the equilibrium point (Khalil, 1996). In this paper, an uncertain linear system is used to describe a non-linear system in a region around the equilibrium point, extending the validity of the linearization approach. The uncertain linear system belongs to a class of linear systems with structured uncertainties (Petersen and Hollot, 1986 and Costa and Oliveira 1998). A Lyapunov function for the uncertain linear system is used to estimate subsets of the domain of attraction of the non-linear system in sectors, which are used in two gain scheduling design procedures. In the first the scheduling is on the system state. The controllers are associated with smaller regions around the equilibrium and are switched as they enter their region. The objective of this is to reduce the conservatism of the solution with a stabilising robust controller. The second procedure addresses the problem of scheduling on a pre-specified state function trajectory. The construction of the trajectory is inspired in the work of Shamma and Athans (1990).

2. NON-LINEAR SYSTEMS IN SECTORS

Consider the non-linear system

$$\dot{x} = f(x,u) \tag{1}$$

where $f: R^n \times R^p \to R^n$ is a C^1 function with $f(0,0)=0$, $x \in R^n$ the state and $u \in R^p$ the control. The associated linearized system around the equilibrium point is given by $\dot{x} := f^{\ell}(x,u) = Ax + Bu$ with A and B the Jacobian matrices of f whose elements in the ith row and jth column are $\partial f_i / \partial x_j$ and $\partial f_i / \partial u_j$ evaluated at the equilibrium point, respectively. The description of f in a sector is given by (Vidyasagar, 1993)

$$(1-\varphi_j) f_j^{\ell}(x,u) \le f_j(x,u) \le (1+\varphi_j) f_j^{\ell}(x,u), j = 1,\dots,n \tag{2}$$

where φ_j are scalars and f_j^{ℓ} the jth row of f_j^{ℓ} =[$a_{j\,1}$ ... $a_{j\,n}$] x + [$b_{j\,1}$ ... $b_{j\,p}$] u with $a_{i,j}$ and $b_{i,j}$ the elements of A and B, respectively. Define region $R\varphi$ as:

$$R\varphi = \{[x\,u] \in R^{n+p} : (1-\varphi_j)f_j^{\ell}(x,u) \le f_j(x,u) \le (1+\varphi_j)f_j^{\ell}(x,u), j=1,\ldots,n\} \quad (3)$$

A description of the non-linear system (1) in region $R\varphi$ may be obtained by rewriting (2) as

$$f_j(x,u) = (1+\varphi_j\beta_j(x,u))f_j^{\ell}(x,u) \quad (4)$$

with $\beta_j(x,u)$ scalar functions in [-1,1]. If in (4) scalars β_j are replaced by scalar functions of measurable Lebesgue type, denoted $\alpha_j(t)$, using (1), one can write

$$\dot{x}_j(t) = [a_{j1}\ldots a_{jn}]x + [b_{j1}\ldots b_{jp}]u + \varphi_j\alpha_j(t)[a_{j1}\ldots a_{jn}]x + \varphi_j\alpha_j(t)[b_{j1}\ldots b_{jp}]u \quad (5)$$

or, in matrix form:

$$\dot{x}(t) = (A+\Delta A)x + (B+\Delta B)u \quad (6)$$

with $\Delta A(t)$ and $\Delta B(t)$ such that its elements are $\varphi_j\, a_{j\,k}\, \alpha_j(t)$ and $\varphi_j\, b_{j\,m}\, \alpha_j(t)$, $k, j=1, \ldots, n$, $m=1, \ldots, p$, respectively. In the specialised literature system (6) is referred to as an uncertain linear system with structured uncertainties. The linear description of f given is referred to as linear differential inclusion (LDI) (Boyd *et al.*, 1994).

Finally, a special case of (1) is considered:

$$\dot{x} = f(x) + Bu \quad (7)$$

with $f{:}R^n \to R^n$ in C^1 and $B \in R^{p\times p}$ a constant matrix. Similarly, a region $R\varphi'$ may be defined:

$$R\varphi' = \{x \in R^n : (1-\varphi_j)f_j^{\ell}(x) \le f_j(x) \le (1+\varphi_j)f_j^{\ell}(x)\} \quad (8)$$

which yields the uncertain linear system:

$$\dot{x} = (A+\Delta A)x + Bu \quad (9)$$

with ΔA having elements $\varphi_j\, a_{j\,k}\, \alpha_j(t)$ for $k, j=1, \ldots, n$.

3. ATTRACTION DOMAINS

In Section 2 a description of a class of non-linear systems in sectors in terms of an uncertain linear system was given. In this section Lyapunov functions, invariant sets and attraction domains for system (1) using its linear description are given.

Theorem 1: Any solution of non-linear system (1) in $R\varphi$ is a solution of uncertain linear system (6).
Proof. Let $s(t,x_0,u) \in R^n$ and $s_u(t,x_0,u) \in R^n$ be solutions of (1) and (6) respectively, and let $[x_0\ u] \in R\varphi$. By construction we can make $\alpha_j(t)=\beta_j(x,u)$, $\forall\,[x\ u] \in R\varphi$. Using (4) one may write

$$f_j(x,u) = (1+\varphi_j\beta_j(x,u))f_{lj}(x,u) = (1+\varphi_j\alpha_j(t))f_{lj}(x,u)$$

Thus $f(x,u) = (A+\Delta A)x + (B+\Delta B)u$, $\forall [x\ u] \in R\varphi$ and $s_u(t,x_0,u) = s(t,x_0,u)$. □

Theorem 2: A Lyapunov function $V(x)$ for uncertain linear system (6) and a control law u, with $-\dot{V}(x)$ a positive definite function (pdf) is a Lyapunov function for state feedback non-linear system (1) in $R\varphi$ with $-\dot{V}(x)$ a locally positive definite function (lpdf). In addiction, if $V(x)$ is decrescent, state feedback non-linear system (1) in $R\varphi$ is asymptotically stable at the origin.

Proof: a) Suppose $\dot{V}(x)<0$, $\forall [xu] \in R^n$, $\forall \Delta A(t)$, $\forall \Delta B(t)$, that is,

$$(\partial V/\partial x).[(A_l+\Delta A(t))x + (B+\Delta B(t))u(x)] < 0.$$

From Theorem 1 $f(x,u)=(A+\Delta A))x + (B+\Delta B))u$ for $[x\ u] \in R\varphi$ and $\Delta A(t)$, $\Delta B(t)$ admissibles. Then $(\partial V/\partial x)$ $f(x,u)<0$, $[x\ u] \in R\varphi$, that is $-\dot{V}(x)$ is lpdf. b) Suppose $V(x)$ is decrescent. Since $-\dot{V}$ is lpdf, the equilibrium 0 is asymptotically stable (Vidyasagar, 1993). □

Theorem 3: Consider systems (1) and (6), a control law $u(t)$ and a set of states and controls denoted by Z. If $[s_u(t,x_0,u)\ u(t)] \in Z \subset R\varphi$, $\forall t>0$, $[x_0\ u(0)] \in Z$, then $[s(t,x_0,u)\ u(t)] \in Z \subset R\varphi$, $\forall t>0$. In addiction, if (6) is asymptotically stable, then $s(t, x_0,u) \to 0$, $t \to \infty$.

Proof. a) Suppose that for some particular t $[s(t,x_0,u)\ u(t)] \notin Z$, $[x_0\ u(0)] \in Z$. Since $[s(0,x_0,u(0))\ u(0)] \in Z$, at some instant t_1 $[s(t_1,x_0,u)\ u(t_1)] \in Z$, $[s(t_1+\delta,x_0,u)\ u(t_1+\delta)] \notin Z$, $\delta>0$ arbitrarily small. At instant t_1, from Theorem 1 $f(x,u) = (A+\Delta A(t_1))x + (B+\Delta B(t_1))u$ for $[x\ u] \in R\varphi$ and some $\Delta A(t_1)$ and $\Delta B(t_1)$ admissibles which yields $[s_u(t_1+\delta,x_0,u)\ u(t_1+\delta)] \notin Z$, which is a contradiction. b) For any $[s(t,x_0,u)\ u(t)] \in Z$ we have $s_u(t,x_0,u) = s(t,x_0,u)$ for some $\Delta A(t)$ and $\Delta B(t)$ admissibles. Then, since $s_u(t,x_0,u) \to 0$, $t \to \infty$ for $\Delta A(t_l)$ and $\Delta B(t_1)$ admissibles $s(t, x_0,u) \to 0$, $t \to \infty$. □

Corollary 3: Consider systems (1) and (6) and a control law $u(x)$. If $Z \subset R\varphi'$ is an invariant set of (6), then Z is an invariant set of (1). Further, if (6) is asymptotically stable, then Z is a subset of the attraction domain of (1).

Remark 1: A level set L_v defined as the connected component of $M=\{x \in R^n: V(x) \le c\}$ with $V(x)$ a Lyapunov function for system (6) with control law $u(x)$ is a subset of the attraction domain and a invariant set of system (6) (Vidyasagar, 1993). Therefore, having $V(x)$, to obtain invariant sets and estimates of attraction domains for system (6) contained in $R\varphi'$, one can simply search for a convenient value of c such that $L_v \subset R\varphi'$.

Remark 2: For system (7) the results follow straightforward from Theorems 1, 2 and Corollary 3.

4. GAIN SCHEDULING IN THE REGULATOR PROBLEM WITH STATE FEEDBACK

To reduce the amount of conservatism that results when a large region of attraction around the equilibrium point is considered in this section a procedure to switch linear robust controllers in subsets of the system domain of attraction is proposed. An uncertain linear system is used to describe the non-linear system (1) in sectors. Invariant subsets of the attraction domain of (1) are obtained using the results of the previous section. The procedure guarantees the asymptotic stability of the non-linear closed-loop system. The key idea is to design controllers to work in small regions Z_i, i=2,...,T around the equilibrium point, with T the total number of controllers.

With the scheduling variable being the system state a global control law may be given by

$$u_G(x)=\begin{cases} u_1, x\in Z_1-\bigcup_{k=2}^{T} Z_k \\ u_i, x\in Z_i-\bigcup_{k=i+1}^{T} Z_k, i=2,...,T-1 \\ u_T, x\in Z_T \end{cases} \quad (10)$$

where Z_i is the region of controller i and u_i is the associated control. In the switching scheme (10) there is a hierarchy among the controllers. The less peripheral one, with higher subscript i, has priority over the remaining.

As stated in Remark 1, it is possible to obtain invariant sets and subsets of the attraction domain of (1) of the form $L=\{x\in R^n : x^T Px\le c_{max}\}$ where P and c_{max} are a symmetric definite positive matrix and a positive scalar, respectively. To see if x belongs to a given set L one may simply verify if $x'Px\le c_{max}$ and the global control law (10) may be rewritten as

$$u_G(x)=\begin{cases} K_i x, \; x^T P_i x\le c_{\max i}, \\ \qquad xP_k x > c_{\max k}, \quad k=i+1,...,T, i=1,...,T-1 \\ K_T x, \quad x^T P_T x\le c_{\max T} \end{cases} \quad (11)$$

An advantage of (11) is that it is simpler to verify $x^T Px\le c_{max}$ than $x\in L$. The remaining of this section is devoted to the analysis of the closed-loop system with state feedback law given by (11). To not be overwhelming by notation we shall consider only two controllers are considered, that is, $T=2$ in (10).

Lemma 1: If Z_1 and Z_2 are subsets of the attraction domain and invariant sets of (1), respectively, with control law u_1 and u_2, then $Z_1\cup Z_2$ is a subset of the attraction domain and an invariant set of (1) with global control law (10) for $T=2$.

Proof: (a) Consider $x_o\in Z_2$. According to (12) the control is u_2. Then $s(t,x_0)\in Z_2$, $\forall\, t$, and $s(t,x_0)\to 0$, $t\to\infty$. (b) Consider now $x_o\in Z_1$-Z_2. By (12) as long as $s(t,x_0)\in Z_1$- Z_2 the control is u_1 and $s(t,x_0)$ does not leaves Z_1 and $s(t,x_0)\to 0$ as $t\to\infty$. In case $s(t,x_0)$ enters Z_2 for some $t_1>t_0$ we have situation (a) again. Then $s(t,x_0)\in Z_1\cup Z_2$, $\forall\, t$, e $s(t,x_0)\to 0$, $t\to\infty$. □

5. GAIN SCHEDULING ON A PRE-SPECIFIED STATE TRAJECTORY

The transition between operating conditions of state non-linear processes is an interesting control problem. In this section, an approach to operating condition transition based on a gain scheduling procedure is presented. The main issue dealt with here is the guarantee of transition.

5.1 Problem Formulation and Design Procedure

Consider the non-linear system (1) where the output is the state. It is assumed that there is a unique pair of functions of one or more state variables in C^1 called here x_{EQ}, u_{EQ} such that $f(x_{EQ}, u_{EQ})$=0. This means that there is a family of equilibrium points that are possible operating points. Let $x=x_I$, $u=u_I$ and $x=x_F$ and $u=u_F$ define operating points of system (1). It is assumed that there exists a path parameterised by one or more state variables linking x_I to x_F which is used as the scheduling variables. The linearized system about the family of operating points is assumed to be controllable. The state transition from x_I to x_F is thus established by using a pre-specified path in the state space for the system operating points as follows.

Transition algorithm

Step 1: Find the set $X_{EQ}=\{x_{EQ}\in R^n : f(x_{EQ}, u_{EQ})=0\}$.
Step 2: Check if x_I and x_F are connected in X_{EQ}.
Step 3: Specify a parameterized path $x_{EQ}(\zeta)$, $u_{EQ}(\zeta)$, $0\le\zeta\le 1$ connecting x_I to x_F such that $x_{EQ}(0)=x_I$, $u_{EQ}(0)=u_I$, $x_{EQ}(1)=x_F$, $u_{EQ}(1)=u_F$.
Step 4: Do i = 1, $x_{eq\,1}=x_F$, $u_{eq\,1}=u_F$ and $\xi_0=1$.
Step 5: Define $x_p=x-x_{eqi}$ and $u_p=u-u_{eqi}$.
Step 6: Linearize the system at x_p = 0, u_p = 0 and obtain $\dot{x}_p=Ax_p+Bu_p$.
Step 7: Choose values for φ and use (5) to obtain

$$\dot{x}_u=(A_i+\Delta A_i)x_u+(B+\Delta B)u_p \quad (12)$$

Step 8: Obtain $u_p(x_p)=-R^{-1}B^T P^{-1}x_p$ with P given by a quadratic stabilisation solution and R a control weighting matrix. $V(x_p)=x_p{}^T P^{-1}x_p$ is a Lyapunov function candidate for the feedback linearized system at x_p=0, u_p=0.
Step 9: Find c_{max} such that $L_v(c_{max})=\{x_p\in R^n : V(x_p)\le c_{max}\}$ is contained in $R\varphi'$ (8). Using $L_v(c_{max})$ obtain $D_i(x_{eq\,i})=\{x\in R^n: (x-x_{eqi})\in L_v(c_{max})\}$.

Remark 3: It can be easily shown that $L_v(c_{max})$ is connected. $L_v(c_{max})\subset R\varphi'$ is an invariant subset of attraction of the feedback linearized system at x_p=0,

u_p=0. In addiction, since $u_p(x_p)$ guarantees asymptotic stability of (12), we have that $L_v(c_{max})$ is an invariant subset of the attraction domain of the feedback non linear system (1).

Step 10: If $x_{eq\,i} = x_I$ the control law is given by (13) else if x_I is inside $D_i(x_{eqi})$ do $i = i+1$, $x_{eq\,i} = x_I$, $u_{eqi} = u_I$ and return to step 5.

Step 11: Obtain $\zeta_{min} = min(\zeta)$ subject to $x_{EQ}(\zeta) \in D_i(x_{eqi})$ and $0 \leq \zeta \leq 1$. Do $\xi_i = \zeta_{min} + \delta$ for δ $0<\delta<\xi_{i-1}-\zeta_{min}$. Do $i = i+1$, $x_{eq\,i} = x_{EQ}(\xi_i)$ and return to step 5. *Hint.* Make $\delta = \kappa.(\xi_{i-1} - \zeta_{min})$, $0<\kappa<1$.

Remark 4: The parameter δ guarantees that $x_{eq\,i}$ enters into D_{i-1}, otherwise the state trajectory could never move from D_i to D_{i-1}.

$$u_G(x) = \begin{cases} u_T = -R_T^{-1} B^T W^{-1}{}_T (x - x_{eqT}) + u_{eqT}, x \in R^n - \bigcup_{i=1}^{T-1} D_i \\ u_j = -R_j^{-1} B^T W^{-1}{}_j (x - x_{eqj}) + u_{eqj}, x \in R^n - \bigcup_{i=1}^{j-1} D_i, j = 2,\ldots,T-1 \\ u_1 = -R_1^{-1} B^T W^{-1}{}_1 (x - x_{eq1}) + u_{eq1}, x \in D_1 \end{cases} \qquad (13)$$

with T the total of controllers.

5.2. Guaranteed Transition

In this section the problem of guaranteed transition between operating points is studied.

Lemma 2: Consider a set S and an $x_{eq} \in R^n$ in S. If a trajectory $s(t) \rightarrow x_{eq}$ then $s(t_1) \in$ S for some $t_1 \in R$.

Proof: Suppose $s(t) \notin$ S, $\forall\, t \in R$. For δ sufficiently small such that $B_\delta(x_{eq}) \subseteq$ S it follows that $s(t) \notin B_\delta(x_{eq})$, $\forall\, t$, where B_δ is the ball

$$B_\delta(x_{eq}) = \{x \in R^n : \|x - x_{eq}\| < \delta\}$$

Since $x_{eq} \in S$ a δ always exists, then $\|s(t) - x_{eq}\| \geq \delta$, $\forall\, t \in R$, which contradicts the hypothesis $s(t) \rightarrow x_{eq}$.□

Lemma 3: Consider any $x_{eq} \in R^n$ and a set S. If $s(t) \notin$ S, $\forall\, t \in R$ and if x_{eq} is in S then $s(t)$ does not tend to x_{eq}.
Proof: The proof is immediate from Lemma 2.

Theorem 4: If $x(t_0) \in D_T(x_I)$ then the family of controllers $u_G(x)$ (13) implemented as in the transition algorithm guarantees system (1) transition from equilibrium x_I to equilibrium x_F.

Proof. Suppose $x(t_0) \in D_T(x_{eqT})$ (remind that $x_{eqT} = x_I$). It is proved that for some $t_1 > t_0$ $s(t_1, x(t_0)) \in D_{(j)}(x_{eq(j)})$ for some j:$1 \leq j \leq$T-1. Suppose that $s(t_1, x(t_0)) \notin D_j(x_{eq(j)})$, $\forall\, j$:$1 \leq j \leq$T-1, $\forall\, t > t_0$. By (13) only the controller u_T is on and thus $s(t_1, x(t_0)) \rightarrow x_{eqT}$. By construction x_{eqT} is in $D_{(T-1)}(x_{eq(T-1)})$ and by Lemma 3 the fact that $s(t_1, x(t_0)) \notin D_j(x_{eq(j)})$, $\forall\, j$:$1 \leq j \leq$T-1, $\forall\, t > t_0$ implies that $s(t_1, x(t_0))$ does not tend to x_{eqT}, which is an absurd. In the same manner, it can be shown that $x(t_1) := s(t_1, x(t_0)) \in D_{(T-1)}(x_{eq(T-1)})$ leads to $s(t_2, x(t_1))$ $\in D_{(j)}(x_{eq\,(j)})$ for some j:$1 \leq j \leq$T-2 and some instant $t_2 > t_1$. Repeating this reasoning it can be shown that for some instant $t_{(T-1)}$, $x(t_{(T-1)}) := s(t_{(T-1)}, x(t_{(T-2)}))$ $\in D_{(2)}(x_{eq(2)})$ and thus for some $t_T > t_{(T-1)}$, $s(t_{(T)}, x(t_{(T-1)})) \in D_1(x_{eq1})$ and consequently $s(t_{(T)}, x(t_{(T-1)})) \rightarrow x_{eq1}$. Given that $x_{eqT} = x_I$ and $x_{eq1} = x_F$ the transition from x_I to x_F is guaranteed. □

6. A NUMERICAL EXAMPLE

In this section results for the control of an magnetic suspension system are presented. Results of a feedback linearizing controller is also presented. In the literature, different approaches to the magnetic suspension control problem are found (Trumper *et al.*, 1997; Kim and Kim 1994; Bittar and Sales, 1997). An attraction type electromagnetic steel ball suspension system is considered here. The system equations can be given by

$$\begin{aligned} \dot{x}_1 &= x_2 \\ \dot{x}_2 &= g - \frac{L_{b0}}{2\,a\,m} \frac{(x_3 + x_{3c})^2}{(1 + (x_1 + x_{1c})/a)^2} \\ \dot{x}_3 &= -(R_b / L_b) x_3 + (1/L_b) u \end{aligned} \qquad (14)$$

with x_1 the ball vertical position, x_2 the ball vertical velocity, x_3 the coil current, u the coil applied voltage, g the acceleration due to gravity, m the ball mass, R the coil resistance, L the coil inductance, L_{b0} and a constants and x_{1c} and x_{3c} constants related with a change of variables. Note that (14) is in the form $\dot{x} = f(x) + Bu$, where f:$R^3 \rightarrow R^3$ is a C^1 function, $f(0)=0$ and $B \in R^{3\times1}$. Table 1 yields numerical values of the physical parameters used.

Table 1. Suspension system parameters.

$g[m/s^2]$	$m[kg]$	$R[\Omega]$	$L[mH]$	$L_{b0}[H]$	$a[m]$
9.81	.0226	19.9	.470	.0245	0.00607

6.1 State Regulation

Consider a stabilising quadratic control with cost function $J(x,u)$ and weighting matrices R_0 and Q_0.

The controller shall have the subscript "1" and the feedback control law is of the form $u_1(x) = -K_1x$. The system sector description is made by choosing the values of φ_1, φ_2 and φ_3 in (4) denoted as φ_{11}, φ_{21} and φ_{31} to refer to controller "1". Since f_1 and f_3 are linear choose $\varphi_{11}=0$ and $\varphi_{31}=0$. To study the effect of different values for φ_{21} obtain the representation for $x_{1c}=0.01\ m$ and $x_{3c}=0.874\ A$.

$$\dot{x}_l = \begin{bmatrix} 0 & 1 & 0 \\ 1219 & 0 & -22.4 \\ 0 & 0 & -42.3 \end{bmatrix} x_l +$$

$$\begin{bmatrix} 0 & 0 & 0 \\ 1219\varphi_{21}\alpha_2(t) & 0 & -22.4\varphi_{21}\alpha_2(t) \\ 0 & 0 & 0 \end{bmatrix} x_l + \begin{bmatrix} 0 \\ 0 \\ 2.13 \end{bmatrix} u_l$$

Adopt $\varphi_{21} = 0.8$ and choose $Q_1 = Q_0 = [\ 10^6\ 0\ 0\ ;\ 0\ 1\ 0\ ;\ 0\ 0\ 1\ ]$, $R_1 = R_0 = 0.001$. The control is found as $u_1 \cong [\ 3.098\times10^5\ \ 6.425\times10^3 - 4.803\times10^2\ ]\ x$ (Costa and Oliveira, 1998) for

$$P_1 \cong \begin{bmatrix} 1{,}728\times10^5 & 3{,}258\times10^3 & -1{,}456\times10^2 \\ 3{,}258\times10^3 & 5{,}518\times10^1 & -3{,}020\times10^0 \\ -1{,}456\times10^2 & -3{,}020\times10^0 & 2{,}257\times10^{-1} \end{bmatrix}$$

with $L_{V1}(.)$ calculated at $c_{max1}\cong0.075$. Consider now the gain scheduled controller design with abrupt switching and adopt the global controller

$$u_G(x) = \begin{cases} u_1, x \in R^n - Z_2 \cup Z_3 \\ u_2, x \in Z_2 - Z_3 \\ u_3, x \in Z_3 \end{cases}$$

where $u_i = -R_i^{-1}B_i^T P_i x$, i=1,...,3. Choosing Q_i=Q_0, R_i=Q_0, i = 2.3 and φ_{22}=0.5 and φ_{23}=0.2 it yields:

$$P_2 = \begin{bmatrix} 3.930\times10^4 & 5.257\times10^2 & -3.737\times10^1 \\ 5.257\times10^2 & 8.851\times10^0 & -6.956\times10^{-1} \\ -3.737\times10^1 & -6.956\times10^{-1} & 9.376\times10^{-2} \end{bmatrix}$$

$K_2 \cong$ [-7,951$\times10^4$ 1.480$\times10^3$ 1.995$\times10^2$]

$$P_3 \cong \begin{bmatrix} 2.461\times10^4 & 2.568\times10^2 & -2.227\times10^1 \\ 2.568\times10^2 & 3.704\times10^0 & -3.766\times10^{-1} \\ -2.227\times10^1 & -3.766\times10^{-1} & 5.989\times10^{-2} \end{bmatrix}$$

$K_3 \cong$ [-4.738$\times10^4$ -8.012$\times10^2$ 1.274$\times10^2$];

$Z_2 = L_{v2}(c_{max2}) = \{x \in R^n\ :\ x'P_2x \leq c_{max2}\}$ and $Z_3 = L_{v3}(c_{max3}) = \{x \in R^n\ : x'P_3x \leq c_{max3}\}$, with $c_{max2}\cong0.0180$ and $c_{max3}\cong0.00175$ and $J(x,u_G) \cong 5.24\times10^{-2}$. For x_0=[0.0025 –0.15 –0.4]^T the system responses are shown in Figures 1 and 2.

Finally, a feedback linearizing control law u_{FL} is given. The simulation response results are shown in Table 2 and Figure 4 show the cost values for u_1, u_G and u_{FL}.

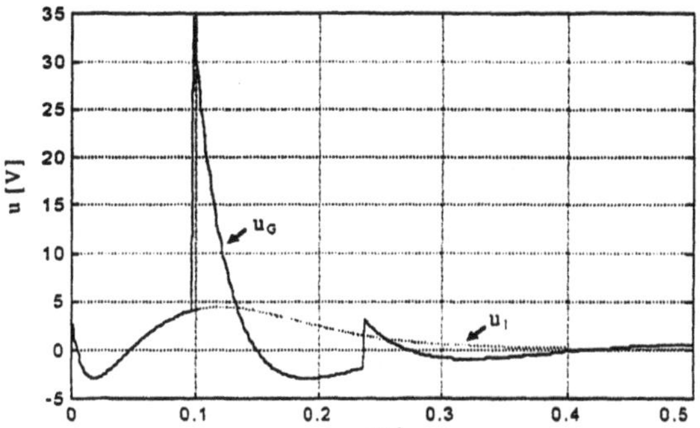

Fig. 1. u_1 and u_G: x_0=[0.0025 –0.15 –0.4]^T.

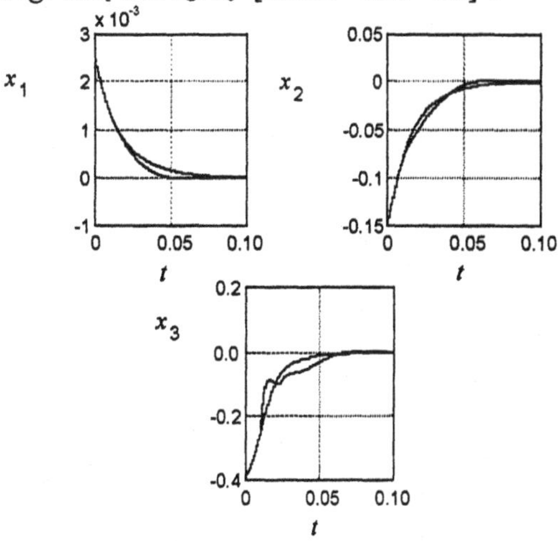

Fig. 2. Responses for u_1 and u_G: x_0=[0.0025 –0.15 –0.4]^T.

Table 2. Costs for u_1, u_G and u_{FL}.

x_0	$J(x_0,u_1)$	$J(x_0,u_G)$	Reduction (%)	$J(x_0,u_{FL})$
[2.5$\times10^{-3}$– .15 - .4]^T	.054	.052	2.6	4.9$\times10^{-2}$
[1.3$\times10^{-3}$ –.083 –.023]^T	.16	.015	10.1	1.5$\times10^{-2}$
[6.9$\times10^{-4}$ –.052 –.093]^T	.0069	.003	60.0	3.0$\times10^{-3}$
[2.9$\times10^{-4}$ – 029 –.068]^T	.0029	4.6$\times10^{-4}$	84.0	4.9$\times10^{-4}$
[8.3$\times10^{-5}$ –.014 –.056]^T	8.5$\times10^{-4}$	5.4$\times10^{-5}$	93.6	5.8$\times10^{-5}$

6.2 Guaranteed Transition

By making $f(x_{EQ}, u_{EQ})$= 0 it can be found:

$$x_{EQ2}(\zeta) = 0$$

$$x_{EQ3}(\zeta) = \sqrt{\frac{2amg}{L_{bo}}}(1 + x_{EQ1}(\zeta)/a)$$

$$u_{EQ}(\zeta) = R_b\sqrt{\frac{2amg}{L_{bo}}}(1 + x_{EQ1}(\zeta)/a)$$

Therefore the scheduling variable can be chosen as $x_{EQ1}(\zeta)$ and one can make $x_{EQ1}(\zeta)=\zeta(x_{F1}-x_{I1})+x_{I1}$ and $x_{I1}=x_{EQ1}(\zeta)$. Adopt x_{I1}=0.005 and x_{F1}=0.010, φ_{21}=0.8; Q_0=[10^6 0 0 ; 0 1 0 ; 0 0 1]; R_0=0.001 and find the equilibrium points x_I=[0.005 0 0.60380], x_F=[0.010 0 0.8765], u_I=12.02 and u_F=17.44. To obtain $\zeta_{min}=min(\zeta)$ subject to $x_{EQ}(\zeta)\in D_1(x_{eq1})$ for $0\leq\zeta\leq1$, use the bisection method. Start with ζ_A=0 and $\zeta_B=\xi_{i-1}=\xi_0$ obtain ζ_T=0,5(ζ_A+ζ_B). If $x_{EQ}(\zeta_T)\in D_1(x_{eq1})$ make $\zeta_A=\zeta_T$, otherwise make $\zeta_B=\zeta_T$. This procedure is repeated until $\zeta_A\cong\zeta_B$.

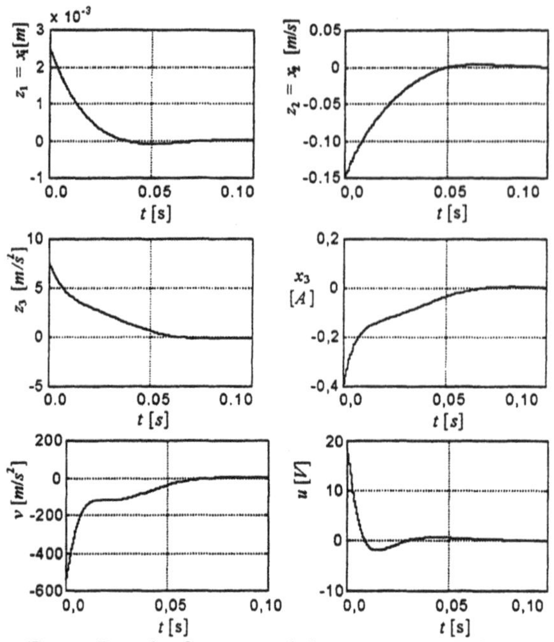

Fig. 4. Results for u_{FL} and the same x_0 as before.

Figure 5 shows responses for different values of κ. The parameter κ is related with the equilibrium family used. Table 3 yields the corresponding number of controllers implemented.

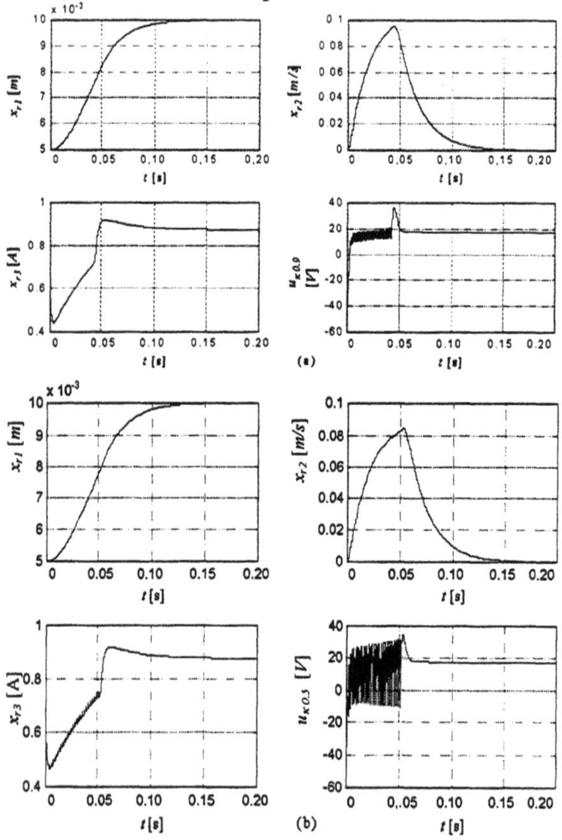

Fig. 5. Responses for φ=0.8; $x_r=[x_{r1}\ x_{r2}\ x_{r3}]$ is not referenced to the origin (a) κ=0.9, (b) κ=0.5.

Table 3 - T versus κ for φ=0.8, Q_0 and R_0 as before.

κ	0.10	0.20	0.50	0.80	0.90
T	19	21	32	78	169

The weighting role of the matrices R_0 used in the robust controller design can be used to reduce the control chattering but at the expense of increasing the system response time.

7. CONCLUSIONS

This paper address the problem of implementing a robust linear controller for an enlarged region of attraction around equilibrium points. The non-linear system is described as a linear system with parametric uncertainties. The attractiveness of the linear description is that the parametric uncertainty of the non-linear system around the equilibrium can be modelled. The use of a gain scheduling procedure in such an application is justified as long as trajectories closer to the equilibrium point have a lower associated control cost. In addiction, a gain scheduling procedure for a guaranteed transition from pre-specified operating conditions is given. In this case the scheduling is on a state function trajectory.

ACKNOWLEDGMENT

The authors would like to gratefully acknowledge the support of the Fundação de Amparo à pesquisa do Estado de São Paulo.

REFERENCES

Bittar, A. and Sales, R.M. (1997). H_2 and H_∞ control for MagLev vehicles. *IEEE Control Systems*, **16**, 18-25.

Costa, E.F. and Oliveira, V.A. (1998). Using linear matrix inequalities for guaranteed cost control with arbitrary rank uncertainty matrices. In *Proc. American Contr. Conf.*, pp. 99-103.

Boyd, S., El Gharow, L. Feron, E. and Balakridrman, V. (1994) *Linear Matrix Inequalities and Control Theory*, SIAM.

Khalil, H.K. (1996). *Nonlinear Systems.* Prentice-Hall, Upper Saddle River, NJ.

Kim, C.Y. and Kim, K.H. (1994). Gain scheduled control of magnetic bearings technology. In *Proc. American Contr. Conf.*, pp.3127-3131.

Petersen, I.R. and Hollot, C.V. (1986). A Riccati equation approach to the stabilization of uncertain linear systems. *Automatica*, **22**, 397-411.

Shamma, J.S. and Athans, M. (1990). Analysis of gain scheduled control for non-linear plants. *IEEE Trans. Automat. Contr.*, **35**, 898-907.

Trumper, D.L., Olson, S.M. and Subrahmanyan, K. (1997). Linearizing control of magnetic suspension systems. *IEEE Trans. Contr. Systems Tech.*, **5**, 427-438.

Vidyasagar, M. (1993). *Nonlinear systems analysis.* Prentice Hall International Editions, Englewood Cliffs, NJ.

DESIGN AND ANALYSIS OF A NEW ADAPTIVE ROBUST CONTROL SCHEME FOR A CLASS OF NONLINEAR UNCERTAIN SYSTEMS

Jian-Xin Xu * Qing-Wei Jia * Tong-Heng Lee *

* *Department of Electrical Engineering*
National University of Singapore
10 Kent Ridge Crescent
Singapore 119260

Abstract: This paper presents a new adaptive robust control scheme which is the extension of (Xu *et al.*, 1997) in the sense that more general classes of nonlinear uncertain dynamical systems are under consideration. To reduce the robust control gain and widen the application scope of adaptive techniques, the system uncertainties are classified into two different categories: the structured and non-structured uncertainties. The structured uncertainty is estimated with adaptation and compensated. Meanwhile, the adaptive robust method is applied to deal with the non-structured uncertainty by estimating unknown parameters in the upper bounding function. The μ-modification scheme is used to cease parameter adaptation in accordance with the adaptive robust control law. The new control scheme guarantees the uniform boundedness of the system and at the same time, the tracking error enters an arbitrarily designated zone in a finite time. The new control scheme also improves the result of (Xu *et al.*, 1997) in that the unknown input distribution matrix of the system can be state-dependent instead of being a constant matrix.

Keywords: Adaptive robust control, nonlinear system, uncertain system

1. INTRODUCTION

Numerous adaptive robust control algorithms for systems containing uncertainties have been developed (Xu *et al.*, 1997; Liao *et al.*, 1990; Brogliato and Neto, 1995; Ioannou and Sun, 1996; Taylor *et al.*, 1989; Sastry and A.Isidori, 1989; Narendra and Annaswamy, 1989). In (Liao *et al.*, 1990) variable structure control with an adaptive law is developed for an uncertain input-output linearizable nonlinear system, where linearity-in-parameter condition for uncertainties is assumed. The unknown gain of the upper bounding function is estimated and updated by adaptation law so that the sliding condition can be met and the error state reaches the sliding surface and stays on it. To deal with a class of nonlinear systems with partially known uncertainties, in (Brogliato and Neto, 1995) an adaptive law using a dead-zone and a hysteresis function is proposed to guarantee both the uniform boundedness of all the closed-loop signals and uniform ultimate boundedness of the system states. In both control schemes, it is assumed that the system uncertainties are bounded by a bounding function which is a product of a set of known functions and unknown positive constants. The objective of adaptation is to estimate these unknown constants.

In (Xu *et al.*, 1997), a new adaptive robust control scheme is developed for a class of nonlinear uncertain systems with both parameter uncertainties and exogenous disturbances. Including the categories of system uncertainties in (Liao *et al.*, 1990) and (Brogliato and Neto, 1995) as its subsets, it is assumed that the disturbances is bounded by a known upper bounding function. Furthermore, the input distribution matrix is assumed to be constant but unknown.

In this paper we proposed a continuous adaptive robust control scheme which is the extension of (Xu *et al.*, 1997) in the sense that more general classes of nonlinear uncertain dynamical systems are under consideration. The unknown input distribution matrix of the system input can be state-dependent here instead of being a constant matrix in (Xu *et al.*, 1997). To reduce the robust control gain and widen the application scope of adaptive techniques, the system uncertainties are supposed to be composed of two different categories: the first can be separated and expressed as the product of known function of states and a set of unknown constants, and the other category is not separable but with partially known bounding functions. It is further assumed that the bounding function is convex to the set of unknown parameters, i.e., the bounding function is no longer linear in parameters. The first category of uncertainties is dealt with by means of the well used adaptive control method. Meanwhile an adaptive robust method is applied to deal with the second category of uncertainties, where the unknown parameters in the upper bounding function are estimated with adaptation.

This paper is organized as follows. Section 2 describes the class of nonlinear uncertain systems to be controlled. Section 3 gives the design procedure of the adaptive robust control and the stability analysis. Simulation-based studies on the proposed method are given in Section 4.

2. PROBLEM FORMULATION

Consider a class of uncertain dynamical system described by

$$\dot{\mathbf{x}} = \mathbf{f}(\mathbf{x}, t) + B(\mathbf{p})\{[I + E(\mathbf{x}, \mathbf{p}, t)]\mathbf{u}(t) + \mathbf{g}(\mathbf{x}, \mathbf{p}, t) + \Delta\mathbf{g}(\mathbf{x}, \mathbf{p}, \omega, t)\} \tag{1}$$

where $\mathbf{x} = [x_1, x_2, \cdots, x_n]^\top \in \mathcal{R}^n$ is the measurable state vector of the system; $\mathbf{u} = [u_1, u_2, \cdots, u_n]^\top \in \mathcal{R}^n$ is the control input of the system; $\mathbf{p} \in \mathcal{P}$ is an unknown system parameter vector and $\mathcal{P}$ is the set of admissible system parameters; $\mathbf{f}(\mathbf{x}, t) \in \mathcal{R}^n$ is a nonlinear function vector; $B(\mathbf{p}) \in \mathcal{R}^{n\times n}$ is the input distribution matrix; $\mathbf{g}(\mathbf{x}, \mathbf{p}, t)$ and $\Delta\mathbf{g}(\mathbf{x}, \mathbf{p}, \omega, t)$ are nonlinear uncertain function vectors of the state $\mathbf{x}$, unknown parameter $\mathbf{p}$, time t as well as a set of random variables ω. Here we make the following assumptions:

Assumption A1: $\mathbf{f}(\mathbf{x}, t) = [f_1(\mathbf{x}, t), f_2(\mathbf{x}, t), \cdots, f_n(\mathbf{x}, t)]^\top \in \mathcal{R}^n$ is a known nonlinear function vector. The matrix $B(\mathbf{p})$ is unknown but $B(\mathbf{p}) > 0 \ \ \forall \mathbf{p} \in \mathcal{P}$.

Assumption A2: For $E \in \mathcal{R}^{n\times n}$,

$$\forall t \in [0, \infty) \quad \forall \mathbf{x} \in \mathcal{D} \quad \forall \mathbf{p} \in \mathcal{P}$$
$$r_{max} \geq \lambda(\frac{1}{2}E + \frac{1}{2}E^\top) \geq r_{min} > -1 \tag{2}$$

where $\lambda(\cdot)$ indicates the eigenvalues of "$\cdot$".

Assumption A3: The structured uncertainty $\mathbf{g}(\mathbf{x}, \mathbf{p}, t) = [g_1(\mathbf{x}, \mathbf{p}, t), \cdots, g_n(\mathbf{x}, \mathbf{p}, t)]^\top \in \mathcal{R}^n$ is a nonlinear function vector which can be expressed as

$$\begin{aligned} \mathbf{g}(\mathbf{x}, \mathbf{p}, t) &= \Theta(\mathbf{p})\xi(\mathbf{x}, t) \\ \Theta &= diag(\theta_1^\top, \cdots, \theta_n^\top) \\ \xi &= [\xi_1, \xi_2, \cdots, \xi_n]^\top \end{aligned} \tag{3}$$

where $\theta_i^\top = [\theta_{i1}, \theta_{i2}, \cdots, \theta_{ip_i}]$ is an unknown parameter vector and $\xi_i^\top = [\xi_{i1}, \xi_{i2}, \cdots, \xi_{ip_i}]$ is a known function vector. The non-structured uncertainty $\Delta\mathbf{g}(\mathbf{x}, \mathbf{p}, \omega, t) = [\Delta g_1(\mathbf{x}, \mathbf{p}, \omega, t), \Delta g_2(\mathbf{x}, \mathbf{p}, \omega, t), \cdots, \Delta g_n(\mathbf{x}, \mathbf{p}, \omega, t)]^\top$ is bounded such that

$$\forall t \in \mathcal{R}^+ \quad \forall \mathbf{x} \in \mathcal{D} \quad \forall p \in \mathcal{P}$$
$$||\Delta\mathbf{g}(\mathbf{x}, \mathbf{p}, \omega, t)|| \leq \rho_d(\mathbf{x}, \mathbf{q}, t)$$

where $||\cdot||$ represents the Euclidean norm; $\mathcal{D}$ is a compact subset of $\mathcal{R}^n$ in which the solution of (1) uniquely exists with respect to the given desired state trajectory $\mathbf{x}_d(t)$. $\rho_d(\mathbf{x}, \mathbf{q}, t)$ is an upper bounding function with unknown parameter vector $\mathbf{q} \in \mathcal{P}$. Here $\rho_d(\mathbf{x}, \mathbf{q}, t)$ is differentiable and convex to $\mathbf{q}$, that is

$$\rho_d(\mathbf{x}, \mathbf{q}_2, t) - \rho_d(\mathbf{x}, \mathbf{q}_1, t) \leq (\mathbf{q}_2 - \mathbf{q}_1)^\top \left.\frac{\partial \rho_d}{\partial \mathbf{q}}\right|_{\mathbf{q}_1} \tag{4}$$

The control objective is to find a suitable control input $\mathbf{u}$ for state tracking with the desired state trajectory $\mathbf{x}_d \in \mathcal{R}^n$, where $\mathbf{x}_d$ is at least continuously differentiable.

Remark 1: In assumption A1 we assume that $\mathbf{f}(\mathbf{x}, t)$ is a known function vector. If it is unknown, but able to be decomposed into three parts: known, structured uncertain and non-structured uncertain vectors as below

$$\mathbf{f}(\mathbf{x}, \mathbf{p}, t) = \mathbf{f}'(\mathbf{x}, t) + \mathbf{g}'(\mathbf{x}, \mathbf{p}, t) + \Delta\mathbf{g}'(\mathbf{x}, \mathbf{p}, \omega, t) \tag{5}$$

which are analogous to $\mathbf{f}(\mathbf{x}, t)$, $\mathbf{g}(\mathbf{x}, \mathbf{p}, t)$ and $\Delta\mathbf{g}(\mathbf{x}, \mathbf{p}, \omega, t)$ respectively in (1) and satisfying definitions in *Assumption A1* and *A3*, then the plant can be rearranged into

$$\dot{\mathbf{x}} = \mathbf{f}'(\mathbf{x}, t) + B(\mathbf{p})\{(I + E)\mathbf{u} + \mathbf{g}''(\mathbf{x}, \mathbf{p}, t) + \Delta\mathbf{g}''(\mathbf{x}, \mathbf{p}, \omega, t)\} \tag{6}$$

which is exactly in the form of (1) with

$$\mathbf{g}''(\mathbf{x},\mathbf{p},t) \triangleq [\Theta, B^{-1}\Theta'][\xi^\top, \xi'^\top]$$
$$||\Delta\mathbf{g}''(\mathbf{x},\mathbf{p},\omega,t)|| = ||\Delta\mathbf{g} + B^{-1}\Delta\mathbf{g}'|| \le ||\Delta\mathbf{g}|| + ||B^{-1}\Delta\mathbf{g}'|| \le \rho_d(\mathbf{x},\mathbf{q},t) + q_0\rho_d'(\mathbf{x},\mathbf{q}',t)$$

where $\mathbf{g}' = \Theta'\xi'$, $||\Delta\mathbf{g}'|| \le \rho_d'(\mathbf{x},\mathbf{q}',t)$ which is convex to $\mathbf{q}'$, and $q_0 = \lambda_{max}(B^{-1})$ is an unknown constant.

Remark 2: $\Delta\mathbf{g}$ is the only system uncertainty taken into account in (Liao *et al.*, 1990) and (Brogliato and Neto, 1995), which are particular cases of (1). Besides, in (Liao *et al.*, 1990) and (Brogliato and Neto, 1995), ρ_d is a linear combination of unknown parameters and known functions, namely $\rho_d = \sum q_i\rho_i(\mathbf{x},t)$, which corresponds to the case of "=" in (4).

Remark 3: (Xu *et al.*, 1997) is also a particular case of (1) in the following three aspects:

1. The input distribution matrix $B(\mathbf{p})$ considered in (Xu *et al.*, 1997) is assumed to be constant, while in this paper we deal with a more general class of input distribution matrix expressed as $B(\mathbf{p})[I + E(\mathbf{x},\mathbf{p},t)]$. Obviously $B(\mathbf{p})$ is a subset of $B(\mathbf{p})[I+E(\mathbf{x},\mathbf{p},t)]$.
2. The bounding function $\rho_d(\mathbf{x},t)$ in (Xu *et al.*, 1997) is a particular case of $\rho_d(\mathbf{x},\mathbf{q},t)$ in which $\rho_d(\mathbf{x},\mathbf{q},t) = \sum q_i\rho_i(\mathbf{x},t)$ with $q_i = 1$, i.e., the bounding function is fully known.
3. The function $\mathbf{f}(\mathbf{x},\mathbf{p},t)$ in (5) can have the uncertain term $\Delta\mathbf{g}'(\mathbf{x},\mathbf{p},\omega,t)$, whereas (Xu *et al.*, 1997) cannot, as the upper bounding function of $B^{-1}\Delta\mathbf{g}'$ still contains unknown parameters q_0 and unknown parameter vector $\mathbf{q}'$.

3. ADAPTIVE ROBUST CONTROL WITH μ − *MODIFICATION*

The adaptive robust technique is used in this section to develop a controller which guarantees the global boundedness of the system. The design procedures are presented in detail as follows.

A. *The Adaptive Robust Control Law*

Define the measured state tracking error vector as

$$\mathbf{e} = \mathbf{x} - \mathbf{x}_d \tag{7}$$

and the parameter error matrices as

$$\tilde{\mathbf{q}} = \mathbf{q} - \hat{\mathbf{q}} \tag{8}$$
$$\tilde{\Theta} = \Theta - \hat{\Theta} \tag{9}$$
$$\tilde{\Phi} = \Phi - \hat{\Phi} \tag{10}$$

where $\hat{\mathbf{q}}, \hat{\Theta}, \hat{\Phi}$ are the estimates of $\mathbf{q}, \Theta, \Phi$ respectively; and

$$\Phi \triangleq B^{-1}. \tag{11}$$

The control law $\mathbf{u}$ and the corresponding adaptive law are chosen to be

$$\mathbf{u} = \mathbf{u}_c + \mathbf{u}_v \tag{12}$$
$$\mathbf{u}_c = -K\mathbf{e} - \hat{\Theta}\xi - \hat{\Phi}[\mathbf{f} - \dot{\mathbf{x}}_d] - \mathbf{v}_d$$
$$\mathbf{u}_v = -\frac{r_{max}^2||\mathbf{u}_c||^2\mathbf{e}}{(1+r_{min})(r_{max}||\mathbf{e}^\top\mathbf{u}_c|| + \varepsilon_1)}$$
$$\mathbf{v}_d = \frac{\hat{\rho}_d^2}{\hat{\rho}_d||\mathbf{e}|| + \varepsilon_2}\mathbf{e}$$
$$\dot{\hat{\Theta}} = \Gamma_1(\mathbf{e}\xi^\top - \mu_1\hat{\Theta})$$
$$\dot{\hat{\Phi}} = \Gamma_2[\mathbf{e}(\mathbf{f} - \dot{\mathbf{x}}_d)^\top - \mu_2\hat{\Phi}]$$
$$\dot{\hat{\mathbf{q}}} = \Gamma_3(||\mathbf{e}|| \left.\frac{\partial\rho_d}{\partial\mathbf{q}}\right|_{\hat{\mathbf{q}}} - \mu_3\hat{\mathbf{q}}) \tag{13}$$

where ε_1 and ε_2 are positive constants; $K \in \mathcal{R}^{n\times n}$ is a gain matrix; $\hat{\rho}_d \triangleq \rho_d(\mathbf{x},\hat{\mathbf{q}},t)$; Γ_i, $i = 1,2,3$ are positive definite matrices chosen to be

$$\Gamma_i = diag(\gamma_{i1}, \gamma_{i2}, \cdots, \gamma_{in}). \tag{14}$$

μ_i, $i = 1,2,3$, which constitute the μ-modification scheme, are defined as

$$\mu_1 = \begin{cases} k_i(\varepsilon_0 - ||\mathbf{e}||) & \mathbf{e} \in E_0 \\ 0 & elsewhere \end{cases} \tag{15}$$

where k_1, k_2 and k_3 are positive constants.

$$E_0 \triangleq \{\mathbf{e} : ||\mathbf{e}|| < \varepsilon_0\} \tag{16}$$

where ε_0 is a positive constant specifying the desired tracking error bound.

B. *Convergence Analysis*

For the above adaptive robust controller, we have the following theorem.

Theorem : By choosing the control gain matrix such that $\lambda_{min}(K) \ge \frac{\varepsilon + c_1}{\varepsilon_0^2}$ with $c_1 > 0$, the proposed adaptive robust control law (12)-(15) ensures that the system trajectory enters the set E_0 in a finite time. Moreover, the tracking errors as well as the parameter estimation errors are bounded by the set

$$D = \{\mathbf{e}, \tilde{\Theta}, \tilde{\Phi}, \tilde{\mathbf{q}} : \mathbf{e}^\top\mathbf{e} + trace\{\tilde{\Theta}^\top\tilde{\Theta}\} + trace\{\tilde{\Phi}^\top\tilde{\Phi}\} + \tilde{\mathbf{q}}^\top\tilde{\mathbf{q}} < k'[\frac{1}{2}k_1\varepsilon_0\, trace\{\Theta^\top\Theta\} + \frac{1}{2}k_2\varepsilon_0\, trace\{\Phi^\top\Phi\} + \frac{1}{2}k_3\varepsilon_0\, \mathbf{q}^\top\mathbf{q} + \varepsilon]\} \tag{17}$$

where k' is defined to be

$$k' = \frac{1}{k'' \; min\{\lambda_{min}(B^{-1}), \lambda_{min}(\Gamma_i^{-1})\}}$$

$$k'' = \frac{2min\{\lambda_{min}(K), k_1\delta, k_2\delta, k_3\delta\}}{max\{\lambda_{max}(B^{-1}), \lambda_{max}(\Gamma_i^{-1})\}}, \quad i = 1, 2, 3.$$

$\lambda_{max}(A)$ and $\lambda_{min}(A)$ indicates the maximum and minimum eigenvalues of the matrix A respectively, and ε and δ are positive values to be defined later. □□

Proof:

The following positive definite function V is selected

$$V = \frac{1}{2}\mathbf{e}^\top B^{-1}\mathbf{e} + \frac{1}{2}trace\{\tilde{\Theta}^\top \Gamma_1^{-1}\tilde{\Theta}\} + \frac{1}{2}trace\{\tilde{\Phi}^\top \Gamma_2^{-1}\tilde{\Phi}\} + \frac{1}{2}\tilde{\mathbf{q}}^\top \Gamma_3^{-1}\tilde{\mathbf{q}} \,. \quad (18)$$

Taking the derivative of V along the trajectory of the dynamic system (1) with the control law (12)-(13), we have

$$\begin{aligned}
\dot{V} = & \mathbf{e}^\top B^{-1}\dot{\mathbf{e}} - trace\{\tilde{\Theta}^\top \Gamma_1^{-1}\dot{\hat{\Theta}}\} \\
& -trace\{\tilde{\Phi}^\top \Gamma_2^{-1}\dot{\hat{\Phi}}\} - \tilde{\mathbf{q}}^\top \Gamma_3^{-1}\dot{\hat{\mathbf{q}}} \\
= & \mathbf{e}^\top[\Theta\xi + (I+E)(\mathbf{u}_c + \mathbf{u}_v) + \Delta\mathbf{g} + \Phi(\mathbf{f}(\mathbf{x}, t) \\
& -\dot{\mathbf{x}}_d)] - trace\{\tilde{\Theta}^\top(\mathbf{e}\xi^\top - \mu_1\hat{\Theta})\} \\
& -trace\{\tilde{\Phi}^\top(\mathbf{e}(\mathbf{f} - \dot{\mathbf{x}}_d)^\top - \mu_2\hat{\Phi})\} - \tilde{\mathbf{q}}^\top \Gamma_3^{-1}\dot{\hat{\mathbf{q}}} \\
= & \mathbf{e}^\top[\tilde{\Theta}\xi + \tilde{\Phi}(\mathbf{f} - \dot{\mathbf{x}}_d) + (I+E)\mathbf{u}_v + E\mathbf{u}_c \\
& -K\mathbf{e} - \mathbf{v}_d + \Delta\mathbf{g}] - \mathbf{e}^\top\tilde{\Theta}\xi - \mathbf{e}^\top\tilde{\Phi}(\mathbf{f} - \dot{\mathbf{x}}_d) \\
& +\mu_1 trace\{\tilde{\Theta}\hat{\Theta}^\top\} + \mu_2 trace\{\tilde{\Phi}\hat{\Phi}^\top\} - \tilde{\mathbf{q}}^\top \Gamma_3^{-1}\dot{\hat{\mathbf{q}}} \\
\leq & -\mathbf{e}^\top K\mathbf{e} - \mathbf{e}^\top\mathbf{v}_d + ||\Delta\mathbf{g}|| \cdot ||\mathbf{e}|| \\
& -\mathbf{e}^\top(I+E)\frac{r_{max}^2||\mathbf{u}_c||^2\mathbf{e}}{(1+r_{min})(r_{max}||\mathbf{e}^\top\mathbf{u}_c|| + \varepsilon_1)} \\
& +r_{max}||\mathbf{e}^\top\mathbf{u}_c|| + \mu_1 trace\{\tilde{\Theta}^\top\hat{\Theta}\} \\
& +\mu_2 trace\{\tilde{\Phi}^\top\hat{\Phi}\} - \tilde{\mathbf{q}}^\top \Gamma_3^{-1}\dot{\hat{\mathbf{q}}} \qquad (19)
\end{aligned}$$

In above derivation the following property of *trace* is used:

$$trace\{Q^\top \mathbf{v}\mathbf{w}^\top\} = \mathbf{v}^\top Q\mathbf{w} \qquad (20)$$

where $\mathbf{v} \in \mathcal{R}^{n\times 1}$, $\mathbf{w} \in \mathcal{R}^{l\times 1}$, and $Q \in \mathcal{R}^{n\times l}$.

Using the fact that

$$\mathbf{e}^\top(I+E)\mathbf{e} = \mathbf{e}^\top(I + \frac{E + E^\top}{2})\mathbf{e}$$

$$\frac{I + \frac{1}{2}E + \frac{1}{2}E^\top}{1 + r_{min}} \geq 1,$$

it follows that

$$\begin{aligned}
\dot{V} \leq & -\mathbf{e}^\top K\mathbf{e} - \frac{\hat{\rho}_d^2||\mathbf{e}||^2}{\hat{\rho}_d||\mathbf{e}|| + \varepsilon_2} + \rho_d||\mathbf{e}|| \\
& -||\mathbf{e}||(\mathbf{q} - \hat{\mathbf{q}})^\top \left.\frac{\partial\rho_d}{\partial\mathbf{q}}\right|_{\hat{\mathbf{q}}} + \mu_1 trace\{\tilde{\Theta}^\top\hat{\Theta}\} \\
& +\mu_2 trace\{\tilde{\Phi}^\top\hat{\Phi}\} + \mu_3\tilde{\mathbf{q}}^\top\hat{\mathbf{q}} + \varepsilon_1 \\
\leq & -\mathbf{e}^\top K\mathbf{e} - \frac{\hat{\rho}_d^2||\mathbf{e}||^2}{\hat{\rho}_d||\mathbf{e}|| + \varepsilon_2} + \rho_d||\mathbf{e}|| \\
& +\mu_1 trace\{\tilde{\Theta}^\top\hat{\Theta}\} + \mu_2 trace\{\tilde{\Phi}^\top\hat{\Phi}\} \\
& -||\mathbf{e}||(\rho_d - \hat{\rho}_d) + \mu_3\tilde{\mathbf{q}}^\top\hat{\mathbf{q}} + \varepsilon_1 \\
\leq & -\mathbf{e}^\top K\mathbf{e} + \mu_1 trace\{\tilde{\Theta}^\top(\Theta - \tilde{\Theta})\} + \varepsilon_1 + \varepsilon_2 \\
& +\mu_2 trace\{\tilde{\Phi}^\top(\Phi - \tilde{\Phi})\} + \mu_3\tilde{\mathbf{q}}^\top(\mathbf{q} - \tilde{\mathbf{q}}) \\
= & -\mathbf{e}^\top K\mathbf{e} - \mu_1 trace\{\tilde{\Theta}^\top\tilde{\Theta}\} + \mu_1 trace\{\tilde{\Theta}^\top\Theta\} \\
& -\mu_2 trace\{\tilde{\Phi}^\top\tilde{\Phi}\} + \mu_2 trace\{\tilde{\Phi}^\top\Phi\} \\
& -\mu_3\tilde{\mathbf{q}}^\top\tilde{\mathbf{q}} + \mu_3\tilde{\mathbf{q}}^\top\mathbf{q} + \varepsilon \\
\leq & -\mathbf{e}^\top K\mathbf{e} - \frac{1}{2}\mu_1 trace\{\tilde{\Theta}^\top\tilde{\Theta}\} - \frac{1}{2}\mu_3\tilde{\mathbf{q}}^\top\tilde{\mathbf{q}} \\
& -\frac{1}{2}\mu_2 trace\{\tilde{\Phi}^\top\tilde{\Phi}\} + \frac{1}{2}\mu_1 trace\{\Theta^\top\Theta\} \\
& +\frac{1}{2}\mu_2 trace\{\Phi^\top\Phi\} + \frac{1}{2}\mu_3\mathbf{q}^\top\mathbf{q} + \varepsilon. \qquad (21)
\end{aligned}$$

where $\varepsilon = \varepsilon_1 + \varepsilon_2$.

By choosing K such that

$$\lambda_{min}(K) \geq \frac{\varepsilon + c_1}{\varepsilon_0^2}, \qquad (22)$$

where c_1 is an arbitrary positive constant, then from (15) we have

$$\dot{V} \leq -\mathbf{e}^\top K\mathbf{e} + \varepsilon \leq -c_1, \;\; \forall\, \mathbf{e} \in \mathcal{R}^n - E_0. (23)$$

Note that, in terms of the adaptive robust control law (12)-(15), $\dot{V}$ is a continuous function. We can show that there exists a constant $0 < \varepsilon_0' < \varepsilon_0$ such that (See Appendix A)

$$\forall\mathbf{e} \in \mathcal{R}^n - E_0' \qquad \dot{V} < 0 \qquad (24)$$

where $E_0' \triangleq \{\mathbf{e} : \;\; ||\mathbf{e}|| < \varepsilon_0'\}$ is a subset of E_0. Noting the relation $E_0' \subset E_0$, (24) implies that the system will enter the set E_0 in a finite time.

When $\mathbf{e} \in E_0'$, it is obvious that

$$k_i(\varepsilon_0 - \varepsilon_0') \leq \mu_i \leq k_i\varepsilon_0, \qquad i = 1, 2, 3 \;. (25)$$

Define $\delta = \varepsilon_0 - \varepsilon_0' > 0$, then from (15), (21) and (25) we obtain

$$\begin{aligned}
\dot{V} \leq & -\mathbf{e}^\top K\mathbf{e} - \frac{1}{2}k_1\delta\; trace\{\tilde{\Theta}^\top\tilde{\Theta}\} - \frac{1}{2}k_3\delta\;\tilde{\mathbf{q}}^\top\tilde{\mathbf{q}} \\
& -\frac{1}{2}k_2\delta\; trace\{\tilde{\Phi}^\top\tilde{\Phi}\} + \frac{1}{2}k_1\varepsilon_0\; trace\{\Theta^\top\Theta\} \\
& +\frac{1}{2}k_2\varepsilon_0\; trace\{\Phi^\top\Phi\} + \frac{1}{2}k_3\varepsilon_0\;\mathbf{q}^\top\mathbf{q} + \varepsilon \\
\leq & -k''V + \frac{1}{2}k_1\varepsilon_0\; trace\{\Theta^\top\Theta\} \\
& +\frac{1}{2}k_2\varepsilon_0\; trace\{\Phi^\top\Phi\} + \frac{1}{2}k_3\varepsilon_0\;\mathbf{q}^\top\mathbf{q} + \varepsilon \;(26)
\end{aligned}$$

By solving (26) we can establish that

$$V(t) \leq e^{-k''t}V(t=0) + \frac{1}{k''}[\frac{1}{2}k_1\varepsilon_0\ trace\{\Theta^\top\Theta\} + \frac{1}{2}k_2\varepsilon_0\ trace\{\Phi^\top\Phi\} + \frac{1}{2}k_3\varepsilon_0\ \mathbf{q}^\top\mathbf{q} + \varepsilon]$$

which implies that $\mathbf{e}$, $\tilde{\Theta}$, $\tilde{\Phi}$ and $\mathbf{q}$ converge exponentially to the residual set (17). □□

4. SIMULATION STUDY

Consider an uncertain system given as

$$\dot{\mathbf{x}} = \mathbf{f}(\mathbf{x}, t) + B[(I + E(\mathbf{x}, \mathbf{p}, t)\mathbf{u} + \mathbf{g}(\mathbf{x}, t) + \triangle\mathbf{g}(\mathbf{x}, t)] \quad (27)$$

where

$$\mathbf{f}(\mathbf{x}, t) = \begin{bmatrix} x_1x_2 \\ sin(t)x_2^2 \end{bmatrix}, \quad B = \begin{bmatrix} b_{11} & b_{12} \\ b_{21} & b_{22} \end{bmatrix},$$

$$\mathbf{g}(\mathbf{x}, t) = \begin{bmatrix} \theta_1\xi_1 \\ \theta_2\xi_2 \end{bmatrix} = \begin{bmatrix} \theta_1x_1^2 \\ \theta_2x_2sinx_1 \end{bmatrix},$$

$$\triangle\mathbf{g}(\mathbf{x}, t) = \begin{bmatrix} 0.5sin(x_1) \\ 0.8sin(5t)x_2 \end{bmatrix},$$

$$E(\mathbf{x}, \mathbf{p}, t) = \begin{bmatrix} asin(t) & 0 \\ \frac{ct}{1+t^2} & bcos(x_1) \end{bmatrix}.$$

The real values of the plant chosen for simulation are as below

$$\theta_1 = 1.0, \qquad \theta_2 = 1.0,$$

$$b_{11} = b_{22} = 1.0, \qquad b_{12} = b_{21} = 0.5\ ,$$

$$a = 0.4, \qquad b = -0.15, \qquad c = 0.6\ .$$

The control and adapting parameters are chosen to be

$$K = diag[80, 80], \qquad k_1 = k_2 = k_3 = 10.0,$$

$$\varepsilon_1 = \varepsilon_2 = 0.5, \qquad \Gamma_1 = \Gamma_2 = \Gamma_3 = diag(1, 1)$$

$$\rho_d = \sqrt{q_1 + q_2x_2^2}\ .$$

Note that the actual upper bound of $\triangle\mathbf{g}$ is

$$||\triangle\mathbf{g}|| = \sqrt{0.25sin^2(x_1) + 0.64sin^2(5t)x_2^2} \leq \sqrt{0.25 + 0.64x_2^2}$$

which is a subset of the given ρ_d.

The given trajectory is

$$\mathbf{x}_d = [5sint, 5cost]^\top \quad (28)$$

and the desired tracking accuracy is $\varepsilon_0 = 0.1$ for both $e_1 = x_1 - x_{1d}$ and $e_2 = x_2 - x_{2d}$. The initial states are $x_0 = [-2.0, 3.0]^\top$, and the initial values of the parameter estimates $\hat{\Phi}$, $\hat{\Psi}$ and $\hat{\mathbf{q}}$ are zeros.

The simulation results are given in Fig. 1-2. It can be seen that the control scheme has achieved a good tracking performance in the sense that the system enters the set E_0 after a short transient period, and the control signal is quite smooth.

Remark 4: The control input is highly dependent on the choice of the design parameters ε_0, ε_1 and ε_2. The greater the parameters ε_1 and ε_2, the smoother the input. Notice that if ε_1 and ε_2 are set to zero, the control scheme is discontinuous. From (22), we can also see that ε_0 has great influence on the control gain K, and a smaller ε_0 results in a larger control gain. Fig. 3-4 give the simulation results when the design parameters are $\varepsilon_1 = \varepsilon_2 = 0.05$ and $\varepsilon_0 = 0.08$. It is shown that the system responses and control signals are quite similar to, but not as smooth as the previous case with larger magnitudes.

5. CONCLUSION

In this paper, an adaptive robust control scheme is developed for a class of uncertain systems with both unknown parameters and system disturbances. The uncertainties are assumed to be composed of two categories: the structured category and the non-structured category with partially known bounding functions. The structured uncertainty is estimated with adaptive method. Meanwhile, the adaptive robust method is applied to deal with the non-structured uncertainty, where the unknown parameters in the upper bounding function are estimated with adaptation. It is shown that the control scheme developed here can guarantee the uniform boundedness of the system and assure that the tracking error enter the arbitrarily designated zone in a finite time. The effectiveness of the control scheme is verified by both theoretical analysis and simulation studies.

6. REFERENCES

Brogliato, B. and A. Trofino Neto (1995). Practical stabilization of a class of nonlinear systems with partially known uncertainties. *Automatica* **31**, 145–150.

Ioannou, P.A. and Jing Sun (1996). *Robust Adaptive Control.* Prentice-Hall, Inc.

Liao, Teh-Lu, Li-Chen Fu and Chen-Fa Hsu (1990). Adaptive robust tracking of nonlinear systems and with an application to a robotic manipulator. *Systems & Control Letters* **15**, 339–348.

Narendra, K.S. and A.M. Annaswamy (1989). *Stable Adaptive Systems.* Prentice-Hall, Englewood Cliffs, NJ.

Sastry, S.S. and A.Isidori (1989). Adaptive control of linearizable systems. *IEEE trans. on automatic control* **34**, 1123–1131.

Taylor, D.G., P.V. Kokotovic, R.Marino and I.Kanellakopoulos (1989). Adaptive regulation of nonlinear systems with unmodeled dynamics. *IEEE trans. on automatic control* **34**, 405–412.

Xu, Jian-Xin, Tong-Heng Lee and Qing-Wei Jia (1997). An adaptive robust control scheme for a class of nonlinear uncertain systems. *International Journal of Systems Science* **28**, 429–434.

APPENDIX A: The solution of ε_0'

From (21), it can be seen that $\dot{V} < 0$ if

$$\mathbf{e}^\top K\mathbf{e} > \frac{1}{2}\mu_1 trace\{\Theta^\top\Theta\} + \frac{1}{2}\mu_2 trace\{\Phi^\top\Phi\} + \frac{1}{2}\mu_3\mathbf{q}^\top\mathbf{q} + \varepsilon \tag{29}$$

then ε_0' can be easily determined by solving the following equation

$$\frac{\varepsilon + c_1}{\varepsilon_0^2}||\mathbf{e}||^2 = \frac{1}{2}\mu_1 trace\{\Theta^\top\Theta\} + \frac{1}{2}\mu_2 trace\{\Phi^\top\Phi\} + \frac{1}{2}\mu_3\mathbf{q}^\top\mathbf{q} + \varepsilon$$

Substituting μ_1, μ_2 and μ_3 in terms of (15) and letting $||\mathbf{e}|| = \varepsilon_0'$ yields

$$\frac{(\varepsilon + c_1){\varepsilon_0'}^2}{\varepsilon_0^2} = \frac{1}{2}k_1(\varepsilon_0 - \varepsilon_0')trace\{\Theta^\top\Theta\} + \frac{1}{2}k_2(\varepsilon_0 - \varepsilon_0')trace\{\Phi^\top\Phi\} + \frac{1}{2}k_3(\varepsilon_0 - \varepsilon_0')\mathbf{q}^\top\mathbf{q} + \varepsilon\,. \tag{30}$$

Denote

$$a = \frac{\varepsilon + c_1}{\varepsilon_0^2}$$
$$b = \frac{1}{2}(k_1 trace\{\Phi^\top\Phi\} + k_2 trace\{\Psi^\top\Psi\} + k_3\mathbf{q}^\top\mathbf{q})$$
$$c = \frac{1}{2}k_1\varepsilon_0 trace\{\Phi^\top\Phi\} + \frac{1}{2}k_2\varepsilon_0 trace\{\Psi^\top\Psi\} + \frac{1}{2}k_3\varepsilon_0\mathbf{q}^\top\mathbf{q} + \varepsilon\,, \tag{31}$$

then equation (30) can be transformed to the following

$$a{\varepsilon_0'}^2 + b\varepsilon_0' - c = 0.$$

The solutions of above equation are

$$\varepsilon_0' = \frac{-b \pm \sqrt{b^2 + 4ac}}{2a} \tag{32}$$

It is obvious that the solutions $\in \mathcal{R}$. Notice that ε_0' is positive, hence the desired solution is

$$\varepsilon_0' = \frac{-b + \sqrt{b^2 + 4ac}}{2a} > 0. \tag{33}$$

□□

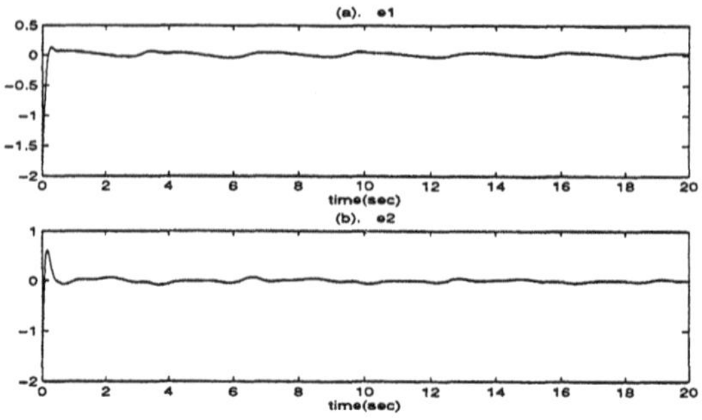

Fig. 1. Tracking Error

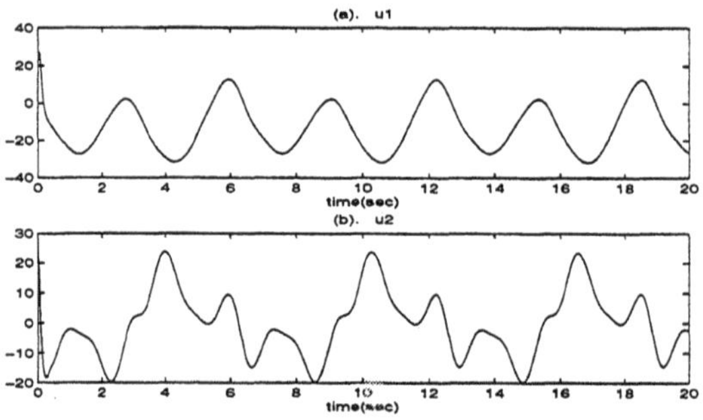

Fig. 2. Control Input

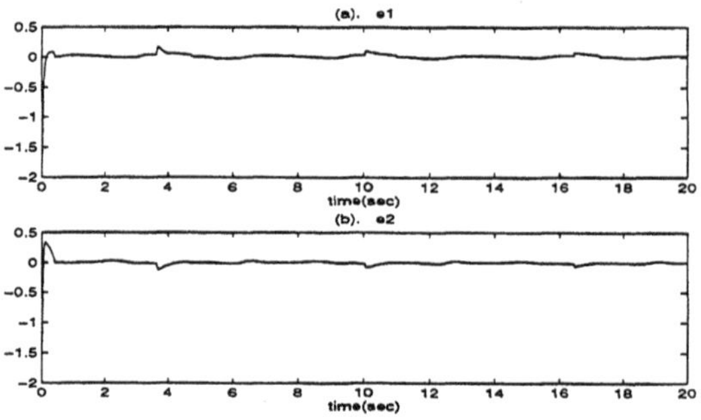

Fig. 3. Tracking Error

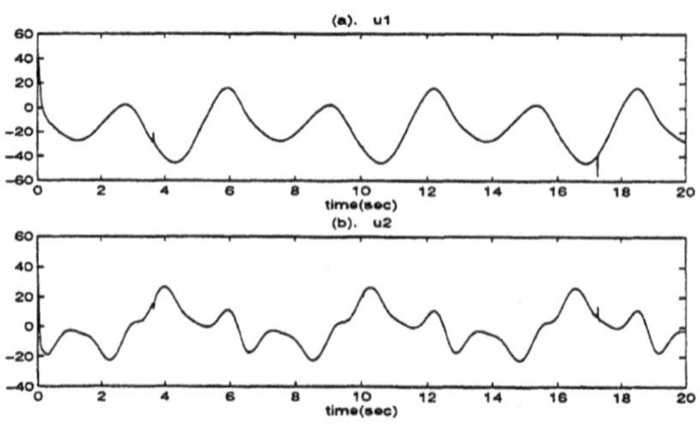

Fig. 4. Control Input

MULTIVARIABLE FUZZY ADAPTIVE CONTROL APPROACH OF NONLINEAR SYSTEMS[1]

Zhang Huaguang * ** Cai Lilong** Zeungnam Bien***

*Department of Automatic Control,Northeast University
Shenyang,110006 ,P.R.China.
E-mail: hgzhang@mail.neu.edu.cn.
**Department of Mechanical Engineering,HKUST,Clear Water Bay,Hong Kong
***Department of Electrical Engineering,KAIST,Taejon,305-701,Korea

Abstract:This paper presents a new fuzzy basis function approach for adaptive control of Multi-input and Multi-output (MIMO) system. In this method the nonlinear plants is first linearised and the fuzzy basis function vector is introduced to adaptively learn the upper bound of the system uncertainty vector,while its output is used as the parameters of the compensator in the sense that both the robustness and the asymptotic error convergence can be obtained for the closed loop nonlinear control system.

Keywords: Stability analysis, Fuzzy hybrid systems, Nonlinear systems, Adaptive algorithms, Multivariable control.

1. INTRODUCTION

Fuzzy controller has been utilized in many fields where the controlled systems are complex or model-free.The successful application of fuzzy controller in those field owing to the fact that it employs the expert control rules of conditional linguistic statements on the relationship of system variables and has the merit of emulating the behaviour of a skilled operator.

However,some important theoretical and practical issues have not been fully dealt with. For example,error convergence,stability and robustness problems are yet to be further studied for fuzzy logic control schemes.The problem of designing fuzzy robust controllers for uncertain nonlinear systems have been recognized as an important one and some interested results have been reported in the literature (Lee,1998; Su,1994; Spooner,1996; Lu,1994). Especially,integrating fuzzy base function and robust control theory into fuzzy control design to acquire stability and consistent performance is vigorous area of fuzzy control.

In (Zhihong,et al.,1997), a fuzzy basis function is used to approximate an unknown system parameter vector. The weights of the fuzzy basis function network are adaptively adjusted. But the proof of stability is not given out. On the other hand,a stable adaptive control approach is given by Spooner (1996) in which fuzzy forms and neural networks are used,its control is composed of a bounding-control term,a sliding-mode-control term,and a certainty-control term. Morever the emphasis in both of the technique

[1] This work was supported by the national natural science foundation of China and the special doctoral branch foundation of C.E.C

is placed on the control of single-input single-output (SISO) plants.

In this paper, a fuzzy basis function vector based adaptive control is proposed for control of MIMO square nonlinear systems.

Notations.

$|A|$ denotes the modulus matrix of matrix A,i.e.,matrix with modulus elements of A.

$\|A\| = \max_{1\le i\le m}(\sum_{j=1}^{n}|a_{ij}|)$, is the norm of matrix A, and $A = \{a_{ij}\}_{m\times n}$.

$|A|<|B|$ means $|a_{ij}|<|b_{ij}|$, $\forall a_{ij} \in A$ and $\forall b_{ij} \in B$, where $A=[a_{ij}],B=[b_{ij}]$.

$|A|_m$ is the matrix,all of whose elements are equal to $a_{\max}$,where $a_{\max} = \max_{i,j}|a_{ij}|$ and $A = [a_{ij}]$.

sign(A) is the sign matrix,which meets the condition of Asign(A)=|A|,where A is a vector.

tr(A) is the trace of the matrix A.

2. PROBLEM STATEMENTS

Consider the following MIMO square nonlinear system given by

$$\dot{X} = f(X) + \sum_{j=1}^{m} g_j(X)u_j(t)$$
$$y_1(t) = h_1(X)$$
$$\cdots \tag{1}$$
$$y_m(t) = h_m(X)$$

where $X \in R^n$ is the plant state vector, $U = [u_1, u_2 \cdots, u_m] \in R^m$ is the control input vector, and $Y = [y_1, y_2 \cdots, y_m] \in R^m$ is the output vector. $f(.), g_i(.) : R^n \to R^n$, i=1,2,...,m are smooth functions vector, $h_i(.) : R^n \to R$, i=1,2,...,m are smooth scalar functions. For convenience, the above equation will be rewritten in a condensed form:

$$\dot{X} = f(X) + G(X)U \tag{2}$$
$$Y = h(X) \tag{3}$$

where

$$U = col(u_1, u_2, \ldots u_m)$$
$$Y = col(y_1, y_2, \ldots y_m)$$
$$G(X) = [g_1(X), g_2(X), \ldots g_m(X)]$$
$$h(X) = col(h_1(X), h_2(X), \ldots h_m(X))$$

To proceed conveniently,some necessary notations and terminologies are introduced and recalled in the following.

The Lie derivative of a scalar function ϕ along a vector $f=[f_1,\ldots,f_n]^T$ is defined as:

$$L_f\phi(X) = \frac{\partial\phi}{\partial X} f(X) = [\frac{\partial\phi}{\partial x_1}, \ldots, \frac{\partial\phi}{\partial x_n}] \begin{bmatrix} f_1(X) \\ \cdots \\ f_n(X) \end{bmatrix}$$
$$= \sum_{i=1}^{n} \frac{\partial\phi}{\partial x_i} f_i(X) \tag{4}$$

where $X=[x_1,\ldots,x_n]^T$. The derivative of ϕ taken first along f and then along a vector g is

$$L_g L_f \phi(X) = \frac{\partial(L_f\phi)}{\partial X} g(X) \tag{5}$$

If ϕ is being differentiated j times along f, the notation $L_f^j\phi$ is used with $L_f^0\phi(X) = \phi(X)$.

Definition 1 (Isidori,1989): A MIMO nonlinear system described by the following equations

$$\dot{X} = f(X) + \sum_{j=1}^{m} g_j(X)u_j(t)$$
$$y_i = h_i(X)\ i = 1, \ldots, m \tag{6}$$

has a (vector) relative degree $[r_1,\ldots r_m]$ at X_0 if
(i) $L_{g_j} L_f^k h_i(X) = 0$ for all $1 \le i,j \le m, k<r_i-1$ and for all X in a neighbourhood of X_0.
(ii) the coupling matrix

$$J(X,t) = \begin{bmatrix} L_{g1} L_f^{r_1-1} h_1(X) \cdots L_{gm} L_f^{r_1-1} h_1(X) \\ L_{g1} L_f^{r_2-1} h_2(X) \cdots L_{gm} L_f^{r_2-1} h_2(X) \\ \cdots \quad \cdots \\ L_{g1} L_f^{r_m-1} h_m(X) \cdots L_{gm} L_f^{r_m-1} h_m(X) \end{bmatrix} \tag{7}$$

is nonsingular at $X=X_0$.

By differentiating the output y_i with respect to time t in Eqn.(6), $\dot{y}_i$ can be gotten as

$$\dot{y}_i = L_f h_i(X) + \sum_{j=1}^{m} L_{g_j} h_i(X) u_j(t) \tag{8}$$

where, $L_f h_i(X)$ and $L_{g_j} h_i(X)$ are the Lie derivatives of $h_i(X)$ with respect to f(X) and $g_j(X)$, respectively.

If $L_{g_j} h_i(X) = 0$, then $\dot{y}_i = L_f h_i(X)$ and

$$y_i^{(r_i)} = L_f^{r_i} h_i(X) + \sum_{j=1}^{m} L_{g_j} L_f^{r_i-1} h_i(X) u_j(t) \quad (9)$$

here $L_{g_j} L_f^{l} h_i(X)=0$, $l<r_i-1$ and $L_{g_j} L_f^{r_i-1} h_i(X) \neq 0$

In this way, the plant's input-output equation may be rewritten as:

$$\underbrace{\begin{bmatrix} y_1^{(r_1)} \\ \vdots \\ y_m^{(r_m)} \end{bmatrix}}_{Y(t)} = \underbrace{\begin{bmatrix} L_f^{r_1} h_1 \\ \vdots \\ L_f^{r_m} h_m \end{bmatrix}}_{B(X,t)} + \underbrace{\begin{bmatrix} L_{g_1} L_f^{r_1-1} h_1(X) \cdots L_{g_m} L_f^{r_1-1} h_1(X) \\ \vdots \quad \vdots \\ L_{g_1} L_f^{r_m-1} h_m(X) \cdots L_{g_m} L_f^{r_m-1} h_m(X) \end{bmatrix}}_{J(X,t)} \underbrace{\begin{bmatrix} u_1(t) \\ \vdots \\ u_m(t) \end{bmatrix}}_{U(t)} \quad (10)$$

An ideal static-state feedback linearizing control law can be obtained by

$$U^* = J^{-1}(-B+V) \quad (11)$$

Note that,for convenience, the references to X and t are dropping in Eqn.(11).

The term V in Eqn.(11) is to be designed below.In order for U^* to be defined,some assumptions about the plant have to be met. In particular, the following assumptions is needed:

Assumption 1. The matrix J as defined above is nonsinqular,i.e.,J^{-1} exists and has bounded norm for all $X \in S_x$, $t \geq 0$,where $S_x \in R^n$ is some compact set of allowing state trajectories. This is equivalent to assuming that

$$\sigma_p(J) \geq \sigma_{\min} > 0 \quad (12)$$

$$\|J\|_2 = \sigma_1(J) \leq \sigma_{\max} < \infty \quad (13)$$

where $\sigma_p(J)$ and $\sigma_1(J)$ are, respectively, the smallest and largest singular values of J.

Assumption 2. The plant has a general vector relative degree $[r_1,\ldots,r_p]^T$, and its zero dynamics are exponentially attractive.

The assumption 2 is needed to guarantee state boundedness under state feedback linearization.Also please refer to (Sastry,1989) for a review on the concept of zero dynamics of square MIMO systems.

Following transformation can be obtained from Eqn.(10)

$$E(X)Y=F(X)+U \quad (14)$$

where $F(X)=J^{-1}(X) \in R^{m\times m}$ (15)

$F(X)=J^{-1}(X)B(X) \in R^m$ (16)

and $E(X)^{-1}$ and F(X) are assumed to be bounded.

In practical situation,E(X) and F(X) may not be exactly known,i.e.,

$$E(X) = E_0(X) + \Delta E(X) \quad (17)$$

$$F(X) = F_0(X) + \Delta F(X) \quad (18)$$

where $E_0(X)$ (nonsinqular) and $F_0(X)$ are known parts, and $\Delta E(X)$ and $\Delta F(X)$ are unknown parts.

Based on above analysis,Eqn.(14) can be written as the following form:

$$E_0(X)Y = F_0(X) + U + \rho(t) \quad (19)$$

where $\rho(t) = \Delta F(X) - \Delta E(X)Y$ (20)

is defined as the system uncertainty vector.

The following system without uncertainty is defined as a nominal system:

$$E_0(X)Y = F_0(X) + U \quad (21)$$

For the nominal system in Eqn.(21), take

$$U = E_0(X)V - F_0(X) \quad (22)$$

where $V=[v_1,v_2,\ldots,v_m]^T$ is chosen to provide stable tracking.

Consider the system in Eqn.(19) with uncertain dynamics,the control input of the closed loop system can be set in the following form:

$$U=U_0+U_1 \quad (23)$$

where U_0 is given in Eqn.(22) to control the nominal system(21) stably and U_1 is used to deal with the effects of the uncertainty.

Lemma. If the control input U is designed in the sense that the modulus vector of the control signal is upper bounded by a positive function vector $U_{\max}(X)$,then the modulus vector of the system uncertainty vector $\rho(t)$ in (20) is upper bounded by a positive function vector $\bar{\rho}(X)$.

In this paper, the following fuzzy basis function network will be used to learn $\bar{\rho}(X) = [\bar{\rho}_1(X), \bar{\rho}_2(X), \ldots, \bar{\rho}_m(X)]^T$,the upper bound of the system uncertainty vector:

$$\hat{\bar{\rho}}_i(X,\theta_i) = \hat{\theta}_i^T \phi(X) \quad (i=1,2,\ldots,m) \quad (24)$$

where $\hat{\theta}_i \in R^M$ is the weight vector of the fuzzy basis expansion vector $\phi(X)$.

3. COMPENSATOR DESIGN

Let $\varepsilon = Y - Y_d$, Y_d is the desired output vector of Y, then $E = [\varepsilon, \dot{\varepsilon}, \ldots, \varepsilon^{n-1}]^T$ can be written as

$$\dot{E} = \Lambda E + \Psi E_0(X)^{-1} \rho(t) + \Psi E_0(X)^{-1} U_1 \quad (25)$$

where

$$\Lambda = \begin{bmatrix} O & I & O & \dots O \\ O & O & I & \dots O \\ \dots \\ -A_0 - A_1 - A_2 \dots - A_{n-2} \end{bmatrix}$$

$$\Psi = [O \;\; O \;\; O \dots \;\; I]^T$$

$$E \in R^{mn\times 1}, \Lambda \in R^{mn\times mn}, \Psi \in R^{mn\times m}. \quad (26)$$

Theorem. Consider the error dynamics in Eqn.(25), *If* (a) the compensator U_1 is designed as follows

$$U_1 = E_0(X)(C\Psi)^{-1}[-C\Lambda E - sign(S^T)\|C\Psi E_0(X)^{-1}\|\hat{\theta}^T\Phi(X) \quad (27)$$

where the vector $S = CE = [s_1,...,s_m]^T$, $S \in R^m$, $C \in R^{m\times mn}$;

(b) the matrix C should be chosen such that the polynomial s_i is Hurwitz about ε_i (i=1,2,...,m);

(c) the matrix $\hat{\theta}$ in Eqn.(27) is updated by the following adaptive mechanism:

$$\dot{\hat{\theta}} = \eta\phi(X)|E^T C^T|\|C\Psi E_0^{-1}(X)\| \quad (28)$$

with $\eta > 0$ and arbitrary positive initial values vector $\hat{\theta}$,

then the output tracking errors vector ε asymptotically converges to zero vector.

Proof. Considering following Lyapunov function

$$\aleph = \frac{1}{2}S^T S + \frac{1}{2}\eta^{-1}tr[\tilde{\theta}^T\tilde{\theta}] \quad (29)$$

where $\tilde{\theta} = \theta^* - \hat{\theta}$ (30)

$$\dot{\tilde{\theta}} = -\dot{\hat{\theta}} \quad (31)$$

Then $\dot{\aleph} = S^T\dot{S} + \eta^{-1}tr[\tilde{\theta}^T\dot{\tilde{\theta}}]$ (32)

The first term in Eqn.(32) is

$$\begin{aligned} S^T\dot{S} &= E^T C^T C\dot{E} \\ &= E^T C^T C[\Lambda E + \Psi J_0(X,t)\rho(t) + \Psi J_0(X,t)U_1] \\ &= E^T C^T C\Lambda E + E^T C^T C\Psi J_0(X,t)\rho(t) \\ &\quad + E^T C^T C\Psi J_0(X,t)U_1 \\ &= E^T C^T C\Lambda E + E^T C^T C\Psi J_0(X,t)\rho(t) \\ &\quad + E^T C^T C\Psi J_0(X,t)E_0(X)(C\Psi)^{-1}[-C\Lambda E \\ &\quad - sign(S^T)\|C\Psi J_0(X,t)\|\hat{\theta}^T\Phi(X)] \\ &= E^T C^T C\Psi J_0(X,t)\rho(t) \\ &\quad - E^T C^T sign(S^T)\|C\Psi J_0(X,t)\|\hat{\theta}^T\Phi(X) \\ &= E^T C^T C\Psi J_0(X,t)\rho(t) \\ &\quad - |E^T C^T|\|C\Psi J_0(X,t)\|\hat{\theta}^T\Phi(X) \end{aligned} \quad (33)$$

The second term in Eqn.(32) is

$$\begin{aligned} \eta^{-1}tr[\tilde{\theta}^T\dot{\tilde{\theta}}] &= -\eta^{-1}tr[\tilde{\theta}^T\dot{\hat{\theta}}] = -\eta^{-1}tr\{[\theta^* - \hat{\theta}]^T\dot{\hat{\theta}}\} \\ &= -\eta^{-1}tr[\theta^{*T}\dot{\hat{\theta}}] + \eta^{-1}tr[\hat{\theta}^T\dot{\hat{\theta}}] \\ &= -\eta^{-1}tr[\theta^{*T}\eta\phi(X)|E^T C^T|\|C\Psi J_0(X,t)\|] \\ &\quad + \eta^{-1}tr[\hat{\theta}^T\eta\phi(X)|E^T C^T|\|C\Psi J_0(X,t)\|] \\ &= -tr[\theta^{*T}\phi(X)|E^T C^T|\|C\Psi J_0(X,t)\|] \\ &\quad + tr[\hat{\theta}^T\phi(X)|E^T C^T|\|C\Psi J_0(X,t)\|] \end{aligned} \quad (34)$$

Therefore we have:

$$\begin{aligned} \dot{\aleph} &= E^T C^T C\Psi J_0(X,t)\rho(t) \\ &\quad - tr[\theta^{*T}\phi(X)|E^T C^T|\|C\Psi J_0(X,t)\|] \\ &= E^T C^T C\Psi J_0(X,t)\rho(t) \\ &\quad - |E^T C^T|\|C\Psi J_0(X,t)\|\theta^{*T}\phi(X) \\ &\le |E^T C^T|\|C\Psi J_0(X,t)\|(|\rho(t)| - \theta^{*T}\phi(X)) \\ &< 0 \end{aligned} \quad (35)$$

4. NUMERICAL SIMULATION

A simulation model is given by

$$\dot{X} = f(X) + G(X)U \quad (2)$$

$$Y = h(X) \quad (3)$$

where

$$f(X) = \begin{bmatrix} x_4 - x_1 \\ x_2 + x_1x_3 \\ x_1^2 - x_2 - x_3 \\ x_3 \end{bmatrix}, G(X) = \begin{bmatrix} 0 & 0 \\ 1 + 2x_3 & 1 \\ 2x_3 & 1 \\ 0 & 0 \end{bmatrix}$$

$$h(X) = \begin{bmatrix} x_2 - x_3 \\ x_4 \end{bmatrix}, U = \begin{bmatrix} u_1 \\ u_2 \end{bmatrix}.$$

The relative degrees of the model are $r_1 = 1, r_2 = 2$.

If taking a transform that $\xi_1 = x_2 - x_3, \xi_2 = x_4$, $\xi_3 = x_3, \psi = x_1$, Eqn(10) can be written as

$$\underset{Y(t)}{\begin{bmatrix} \dot{y}_1 \\ \ddot{y}_2 \end{bmatrix}} = \underset{B(X,t)}{\begin{bmatrix} -x_1^2 + 2x_2 + x_3 + x_1x_3 \\ x_1^2 - x_2 - x_3 \end{bmatrix}} + \underset{J(X,t)\,U(t)}{\begin{bmatrix} 1 & 0 \\ 2x_3 & 0 \end{bmatrix}\begin{bmatrix} u_1 \\ u_2 \end{bmatrix}} \quad (36)$$

and $\dot{\psi} = \xi_2 - \psi$ (37)

Thus, the zero dynamics of the given system is

$$\dot{\psi} = -\psi \quad (38)$$

Obviously, the error dynamics of the given system is exponentially attractive in equilibrium point.

From Eqn.(25),(26), $E(X)$ and $F(X)$ can be obtained by

$$E(X) = J^{-1}(X) = \begin{bmatrix} 1 & 0 \\ -2x_3 & 1 \end{bmatrix}$$

$$F(X) = E(X)B(X,t) = \begin{bmatrix} -x_1^2 + 2x_2 + x_3 + x_1x_3 \\ 2x_1^2x_3 - 4x_2x_3 - 2x_3^2 - 2x_1x_3^2 + x_1^2 - x_2 - x_3 \end{bmatrix}$$

If taking

$$E_0(X) = J_0^{-1}(X) = \begin{bmatrix} 1 & x_3 \\ 0 & 1 \end{bmatrix}$$

$$F_0(X) = \begin{bmatrix} -x_1^2 + 2x_2 + x_3 \\ 2x_1^2x_3 - 4x_2x_3 - 2x_3^2 - 2x_1x_3^2 - x_2 - x_3 \end{bmatrix}$$

then

$$\Delta E = \begin{bmatrix} 0 & -x_3 \\ -2x_3 & 0 \end{bmatrix}$$

$$\Delta F = \begin{bmatrix} x_1 x_3 \\ x_1^2 \end{bmatrix}$$

By taking $E = [\varepsilon_1, \varepsilon_2, \dot{\varepsilon}_2]$, the error state equation of the closed-loop system is

$$\dot{E} = \Lambda E + \Psi J_0(X,t)\rho(t) + \Psi J_0(X,t)U_1 \quad (39)$$

where $\Lambda = \begin{bmatrix} -a_0^1 & 0 & 0 \\ 0 & 0 & 1 \\ 0 & -a_0^2 & a_1^2 \end{bmatrix}$, $\Psi = \begin{bmatrix} 1 & 0 \\ 0 & 0 \\ 0 & 1 \end{bmatrix}$

and the parameters a_i^j can be chosen by Eqn.(22).

Table 1 gives all relative parameters in the simulation study, which can be chosen subjectively according to above method.

Table 1 Relative parameters and initial conditions

Parameters	Values	in Eqn.
a_0^1	2	(22)
a_0^2	6	
a_1^2	5	
C	$\begin{bmatrix} 0.5 & 2 & 2.5 \\ 0.25 & 1.25 & 2 \end{bmatrix}$	(51)
σ	0.6	(44)
M	81	(43)
n	4	(43)
η	1.5	(52)
X(0)	{0}	(2)
$\theta(0)$	{0.001}	(52)
$y_1(0)$	0.1	(60)
$y_2(0)$	-0.1	
y_{d1}	0.0	(35)
y_{d2}	0.0	

Fig.1 shows the membership functions of variable x_1, x_2, x_3, x_4, which consist of *small, medium, large.* therefore, the number of the fuzzy rules should be 3^4=81.

Fig.2 and Fig.3 give the response curves of the output $y_1(t), y_2(t)$ respectively. And it can be observed that the nonlinear control system will reach stable state after 15 seconds. The greater η is, the faster the response is. But too greater η maybe cause a small vibrating motion in the meantime.

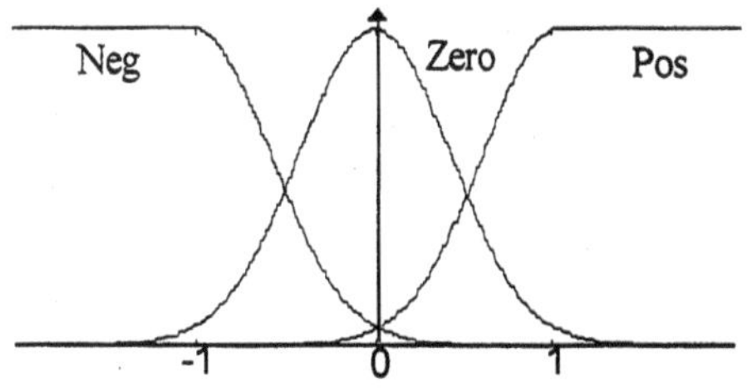

Fig. 1 The membership functions of variable x_1, x_2, x_3, x_4

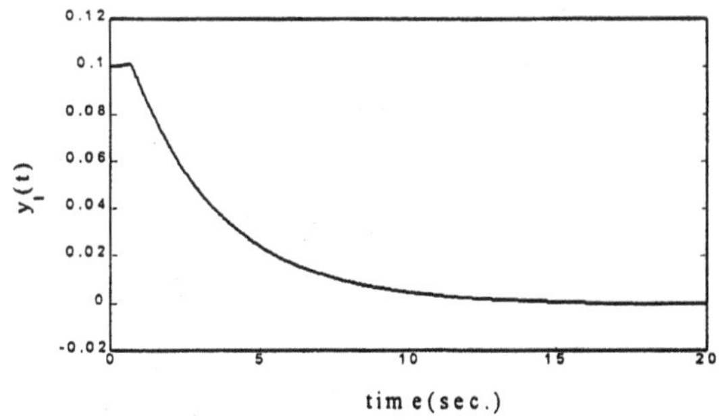

Fig. 2 The response curve of the output $y_1(t)$

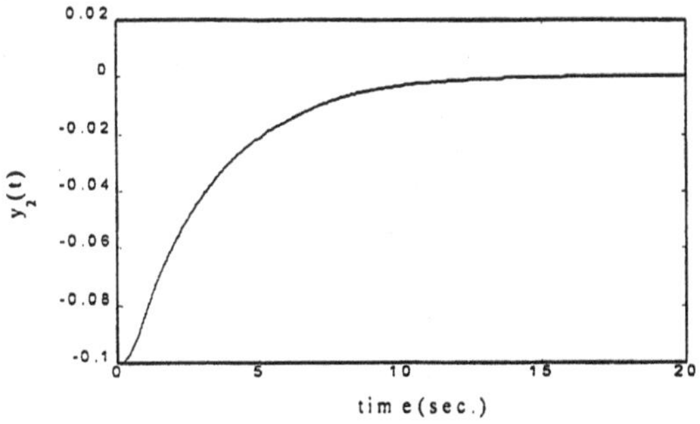

Fig. 3 The response curve of the output $y_2(t)$

5. CONCLUSION

An adaptive controller using fuzzy basis function vector is proposed in this paper.Our analysis and simulation demonstrate that the fuzzy basis function expansion vector can be used to learn the MIMO system uncertainty bounds in Lyapunov sense, and a fuzzy function vector-based adaptive controller can be designed to eliminate the effects of dynamical uncertainties and guarantees that the output tracking errors asymptotically converges to zero and it has a better robustness.

REFERENCES

Changguo, D. (1993). *Linear algebra in control systems*, Southeastern university publisher, Nanking.

Isidori, A. (1989). *Nonlinear control systems: An introduction* ,Springer,Berlin.

Lee, T. S., *et al.* (1998). Fuzzy modelling and uncertain-based control for nonlinear systems.*Proc. 1998 ACC,* Philodelphia, pp:2088-2092.

Lu, Y. S. and J. S. Chen, (1994). A self-organizing fuzzy sliding-mode controller design for a class of nonlinear serve systems.*IEEE Trans Industrial Electronics*, **Vol. 41**, pp:492-502.

Sastry, S. S.and A. Isidor, (1989). Adaptive control of linearizable systems. *IEEE Trans Automatic Control,* **Vol.34**, pp:1123-1131.

Spooner, J. T. and K. M. Passino, (1996). Stable adaptive using fuzzy systems and neural networks.*IEEE Trans Fuzzy Systems*,**Vol.4**, pp:339-359.

Su, C. Y. and Y. Stepanenko, (1994). Adaptive controller of a class of nonlinear systems with fuzzy logic. *IEEE Trans. Fuzzy Systems*, **Vol.2**, pp.285-294.

Zhihong, M., X. H. Yu,. and Q. P. Ha,.(1997). Adaptive control using fuzzy basis function expansion for SISO linearizable nonlinear systems. *Proc. 2nd ASCC*, Seoul , pp:695-698.

14th Triennial World Congress, Beijing, P.R. China

ROBUST ADAPTIVE OUTPUT-FEEDBACK CONTROL OF A CLASS OF UNCERTAIN NONLINEAR SYSTEMS

Chen Weitian[+,1] **Chu Xuedao**[+] **Shi Songjiao**[‡] **Liu Xiaohua**[+‡]

[+] *Institute of Automation, Qufu Normal University, Shandong Province, P.R.China 273165*

[‡] *Dept. of Automation, Shanghai Jiaotong University, Shanghai, P.R.China 200030*

[+‡] *Department of Mathematics, Yantai Teachers College, Shandong Province, P.R.China 264025*

Abstract: With an attempt to simplify the controller design and to reduce the control effort, a new robust output-feedback controller is designed to solve the tracking problem of a class of nonlinear uncertain systems by using observer plus backstepping. In the controller design, a simple observer is constructed, by which only a coarse estimate of the unknown states is provided, and a positive function used to generate a new normalization signal and the positive nonlinear functions used in nonlinear damping terms are all left to be chosen freely. Thanks to these techniques, the new controller can be designed with a simpler structure and reduced control effort under weaker assumptions on both the uncertainties and the reference signal. The global boundness of all closed-loop signals can be guaranteed, and the output tracking error can be made as small as possible. *Copyright* © 1999 *IFAC*

Key words: nonlinear system, adaptive control, robust control, output feedback, tracking

1. INTRODUCTION

Recently, much attention has been paid to the nonlinear adaptive output-feedback control problems. Detailed discussions in such a direction can be found in both Marino and Tomei (1995) and Krstic, et al. (1995). Most of the early results were obtained under linear parameterization assumption. To remove the linear parameterization assumption, an important control scheme which allows some nonlinearly parametrized uncertainties to present was proposed in Marino and Tomei (1993b). However, the control scheme requires the unknown constant parameter vector to lie in a known compact set and can only be applied

[1] This work was supported by Natural Science Foundation of Shandong Province (Q97G02159). The E-mail of the first author is: cwt@qfnu.edu.cn.

applied to solve the set point regulation problems. Although the partial-state feedback control problem formulated and solved in Jiang and Praly (1996) is more general than the output-feedback control problem, the existence of an exp-ISS Lyapunov function which can not be ascertained a priori for an unknown system is required. This implies that the existence of c, γ and ψ_1 defined in their paper can not be ascertained a priori either. Even if the existence of c, γ and ψ_1 could be ascertained a priori, it would be still hard for them to be known. If c, γ can not be known a priori, the normalization signal introduced in their paper can not be defined.

Till now, the problems of how to simplify the controller design and how to reduce control effort have not been enough studied yet. The only attempt to simplify the controller design was made in Kanellakopoulos et al., 1993, and the only attempt to reduce control effort was made in Freeman and Kokotovic, 1993, in which non adaptive state-feedback robust controller was designed.

Inspired by Kanellakopoulos et al. (1993) and Freeman and Kokotovic (1993), the main purpose of this paper is to design a simple robust adaptive output-feedback controller with reduced control effort while guaranteeing the global boundness of the closed-loop system signals.

Consider a nonlinear system of the form

$$\dot{x}_i = x_{i+1} + \phi_{0i}(y) + \sum_{j=1}^{p} \theta_j \phi_{ji}(y) + \psi_i(x,u,t), 1 \le i \le \rho - 1 \tag{1a}$$

$$\dot{x}_i = x_{i+1} + \phi_{0i}(y) + \sum_{j=1}^{p} \theta_j \phi_{ji}(y) + b_{n-i}\sigma(y)u + \psi_i(x,u,t), \rho \le i \le n-1 \tag{1b}$$

$$\dot{x}_n = \phi_{0n}(y) + \sum_{j=1}^{p} \theta_j \phi_{jn}(y) + b_0\sigma(y)u + \psi_n(x,u,t) \tag{1c}$$

$$y = x_1 \tag{1d}$$

where y, u, $x = (x_1, \dots, x_n)^T$, and $\theta = (\theta_1, \dots, \theta_p)^T \in R^p$ denote the output, input, state vector and the vector of unknown constant parameters respectively. $\psi_i(x,u,t), 1 \le i \le n$ represent the system uncertainties.

Assuming that only the output y is measured, the control objective of our paper is to design a robust adaptive output-feedback controller for the system (1) to guarantee the global boundness of all closed-loop signals, and to make the tracking error $|y - y_r|$ as small as possible, where y_r is a given reference signal.

The following assumptions will be made:

Assumption A1 The sign of $b_{n-\rho}$ is known.

Assumption A2 $B(s) = b_{n-\rho}s^{n-\rho} + \dots + b_1 s + b_0$ is known to be Hurwitz.

Assumption A3 $\sigma(y)$ is known and $\sigma(y) \ne 0$, $\forall y \in R$ and $\sigma^{(n-\rho)}(y)$ exists.

Assumption A4 $\phi_{ji}, 0 \le i \le n, 1 \le j \le p$ are all known functions, and ϕ_{ji} is $n-i$ times differentiable.

Assumption A5 Each uncertainty $\psi_i(x,u,t)$ is a real-valued function with continuous partial derivatives of order $n-i$, and there exists a known continuous function $\rho_i(y)$ such that $\frac{|\psi_i(x,u,t)|}{\rho_i(y)}$ is bounded.

Assumption A6 y_r and its first $n-\rho+1$ derivatives are known and bounded, and $y^{(n-\rho+1)}{}_r$ is continuous.

Remark 1 Assumption A1 ~ A4 are also required in Kanellakopoulos et al., 1993. In Marino and Tomei, 1993a, y_r is required to be smooth. Though only the first ρ derivatives of y_r are assumed to be known and bounded in Kanellakopoulous et al., 1993, it is only enough for their controller design but not enough to prove their stability results. In fact, to design our controller, we only require y_r and $\dot{y}_r$ to be known and bounded. The reason why assumption A6 is used is that one needs it to prove the stability results. The assumption A6 is in fact the weakest assumption on y_r.

2. FILTER DESIGN

Rewrite (1) in the form

$$\dot{x} = A_0 x + k_0 y + \phi_0(y) + \sum_{j=1}^{p} \theta_j \phi_j(y)$$
$$+ b\sigma(y)u + \psi(x,u,t) \qquad (2)$$
$$y = C^T x$$

where $A_0 = (-k_0 \quad e_1 \quad \cdots \quad e_{n-1})$ is chosen to be Hurwitz with e_i the i th coordinate vector in R^n, and $k_0 = (k_{01} \quad \cdots \quad k_{0n})^T$, $C^T = (1 \quad 0 \quad \cdots \quad 0)$, $\phi_j = [\phi_{j1}, \cdots, \phi_{jn}]^T$ and $\psi = [\psi_1, \cdots, \psi_n]^T$.

The following filters will be first introduced

$$\begin{aligned} \dot{\xi}_j &= A_0 \xi_j + \phi_j(y), 1 \le j \le p \\ \dot{v}_j &= A_0 v_j + e_{n-j}\sigma(y)u, 0 \le j \le n-\rho \end{aligned} \qquad (3)$$

Remark 2 Unlike in Kanellakopoulos, et al. (1993), ξ_0 is not introduced in the above filters in order to reduce the dynamic order of the resulting controller and to simplify the controller design.

Using (3), a coarse estimate $\hat{x}$ of the state can be given as

$$\hat{x} = \sum_{j=1}^{p} \theta_j \xi_j + \sum_{j=0}^{n-\rho} b_j v_j \qquad (4)$$

and let $\varepsilon = x - \hat{x}$, it is easy to get

$$\dot{\varepsilon} = A_0 \varepsilon + \psi + k_0 y + \phi_0(y) \qquad (5)$$

The following results can be derived.

Lemma 2.1 If the filters (3) is applied, and if (4) is used to estimate the unmeasured states, then one obtains

$$\|\varepsilon(t)\| \le N_2 + N_1 e^{-g_1 t} \int_0^t e^{g_1 \tau} \|\psi + k_0 y + \phi_0(y)\| d\tau \qquad \cdots \cdots \quad (6)$$

Particularly

$$|\varepsilon_1| \le N_2 + N_1 e^{-g_1 t} \int_0^t e^{g_1 \tau} \|\psi + k_0 y + \phi_0(y)\| d\tau \qquad \cdots \cdots \quad (7)$$

where $0 < g_1 < \min_{1 \le i \le n} |\text{Re}\, \lambda_i(A_0)|$.

To deal with the state estimation error and to simplify the controller design, a normalization signal $m(t)$ is introduced as follows

$$\dot{m}(t) = -g_1 m(t) + \beta(y) \qquad (8)$$

where $m(0) = \delta > 0$, $\beta(y)$ is a positive smooth design function such that $\frac{\|\psi + k_0 y + \phi_0(y)\|}{\beta}$ and $\frac{\|\phi_j(y)\|}{\beta}, 1 \le j \le p$ are bounded and $\beta(y) \ge g_1 \delta > 0$.

Lemma 2.2 If the normalization signal $m(t)$ is defined as in (8), then $\frac{|\varepsilon_2(t)|}{m(t)}$ and $\frac{\|\xi_j(t)\|}{m(t)}, 1 \le j \le p$ are bounded.

3. ROBUST ADAPTIVE OUTPUT-FEEDBACK CONTROLLER DESIGN

The design objective is to make a Lyapunov function satisfy the following inequality:

$$\dot{V} \le -2\lambda V + M \qquad (9)$$

where λ and M are both positive constants.

By using similar design procedure given in Kanellakopoulos, et al. (1993), the robust adaptive control law can be designed as

$$u = \frac{1}{\sigma(y)}\left(-c_\rho z_\rho + \frac{\partial \alpha_{\rho-1}}{\partial y} \omega^T \hat{\theta} - k_\rho z_\rho \beta_\rho\right) \qquad (10)$$

with the parameter update laws given as

$$\dot{\hat{\theta}} = \Gamma \tau_\rho \tag{11}$$

$$\dot{\hat{q}} = \gamma \mathrm{sgn}(b_{n-\rho}) z_1 \pi_1 - l_q \hat{q} \tag{12}$$

where the functions π_1, τ_ρ, $\alpha_{\rho-1}$ are defined by the recursive expressions

$$z_1 = y - y_r,\; z_i = v_{n-\rho,i} - \alpha_{i-1}$$

$$\pi_1 = c_1 z_1 + k_1 z_1 \beta_1(y,m) + \overline{\omega}^T \hat{\theta}$$

$$\theta \overset{\Delta}{=} [\theta_1, \cdots, \theta_p, b_0, \cdots, b_{n-\rho}]^T, \quad q = \tfrac{1}{b_{n-\rho}}$$

$$\overline{\omega} \overset{\Delta}{=} [\phi_{11} + \xi_{12}, \cdots, \phi_{p1} + \xi_{p2}, v_{0,2}, \cdots, v_{n-\rho-1,2}, 0]^T$$

$$\omega \overset{\Delta}{=} [\phi_{11} + \xi_{12}, \cdots, \phi_{p1} + \xi_{p2}, v_{02}, \cdots, v_{n-\rho-12}, v_{n-\rho 2}]^T$$

$$\alpha_1(y, y_r, m, v_{0,2}, \cdots, v_{n-\rho-1,2}, \hat{q}, \hat{\theta}) = -\hat{q}\pi_1$$

$$\tau_1 = z_1 \overline{\omega} - l_\theta \hat{\theta}, \quad \tau_i = \tau_{i-1} - \frac{\partial \alpha_{i-1}}{\partial y} \omega z_i$$

$$\alpha_i(y, y_r, v_{0,1}, \cdots, v_{n-\rho,1}, \cdots, v_{0,i}, \cdots, v_{n-\rho,i}, v_{0,i+1}, \cdots, v_{n-\rho-1,i+1}) = -c_i z_i - k_i z_i \beta_i + \frac{\partial \alpha_{i-1}}{\partial y} \omega^T \hat{\theta}$$

where $i = 2, \cdots, \rho$, and $c_i, k_i, 1 \le i \le \rho$, Γ, γ, l_θ and l_q are positive design constants. $\beta_1(y,m)$, $\beta_2(y, m, v_{0,1}, \cdots, v_{n-\rho,1}, v_{0,3}, \cdots, v_{n-\rho-1,3}, \hat{q}, \hat{\theta})$ ... $\beta_\rho(y, m, v_{0,1}, \cdots, v_{n-\rho,1}, \cdots, v_{0,\rho+1}, \cdots, v_{n-\rho,\rho+1}, \hat{q}, \hat{\theta})$ are positive smooth functions which are chosen such that $\frac{|\varepsilon_2 + \psi_1 + \phi_{01} - \dot{y}_r|^2}{\beta_1(y,m)}$,

$$\frac{\left| -\frac{\partial \alpha_1}{\partial y}(\varepsilon_2 + \psi_1 + \phi_{01}) - \frac{\partial \alpha_1}{\partial \hat{\theta}} \Gamma \tau_2 + \pi_2 + b_{n-\rho} z_1 \right|^2}{\beta_2}$$

$$\frac{\left| -\frac{\partial \alpha_{i-1}}{\partial y}(\varepsilon_2 + \psi_1 + \phi_{01}) - \frac{\partial \alpha_{i-1}}{\partial \hat{\theta}} \Gamma \tau_i + \sum_{j=2}^{i-1} z_j \frac{\partial \alpha_{i-1}}{\partial \hat{\theta}} \Gamma \frac{\partial \alpha_{i-1}}{\partial y} \omega + \pi_i + z_{i-1} \right|^2}{\beta_i}$$

$, 3 \le i \le \rho$ are all bounded, and $\pi_2, \cdots, \pi_\rho$ are defined as follows

$$\pi_i = -\frac{\partial \alpha_{i-1}}{\partial y_r} \dot{y}_r - \sum_{j=1}^{p} \frac{\partial \alpha_{i-1}}{\partial \xi_j}(A_0 \xi_j + \phi_j(y)) - k_{0i} v_{n-\rho 1} - \sum_{l=1}^{i-1} \frac{\partial \alpha_{i-1}}{\partial v_{n-\rho l}}(v_{n-\rho l+1} - k_{0l} v_{j1}) - \frac{\partial \alpha_{i-1}}{\partial m} \dot{m} - \sum_{j=0}^{n-\rho-1} \sum_{l=1}^{i} \frac{\partial \alpha_{i-1}}{\partial v_{jl}}(v_{jl+1} - k_{0l} v_{j1}) - \frac{\partial \alpha_{i-1}}{\partial \hat{q}} \dot{\hat{q}}$$

Remark 3 To guarantee the global boundness of the closed-loop system, σ-modification is introduced in the update law of the unknown parameters, which is well known in linear robust adaptive control, see (Narendra and Annaswamy, 1989).

4. THE CHOICE OF THE POSITIVE SMOOTH FUNCTIONS

The choice of $\beta, \beta_1, \cdots, \beta_\rho$ will be given in following steps.

Step 0: Choose $\beta(y)$ as any positive smooth function defined in the following set:

$$S_0 = \{\beta \big| \beta' / \beta \quad \text{is bounded and} \quad \beta \ge g_1 \delta\}$$

where $\beta' = \|k_0 y + \phi_0\| + \sum_{j=1}^{p} \|\phi_j\| + \sum_{j=1}^{n} \rho_j$.

Step 1: Choose $\beta_1(y,m)$ as any positive smooth function defined in the following set:

$$S_1 = \{\beta_1(y,m) \big| (\beta^2 + m^2) / \beta_1 \quad \text{is bounded}\}$$

Step 2: Define

$$f_2 = -\frac{\partial\alpha_1}{\partial y}(\varepsilon_2 + \psi_1 + \phi_{01})$$
$$-\frac{\partial\alpha_1}{\partial\hat{\theta}}\Gamma\tau_2 + \pi_2 + b_{n-\rho}z_1$$

Choose s_2 as

$$s_2 = [f_{2,y} f_{2,m} f_{2,\hat{q}} f_{2,\hat{\theta}} \cdot \prod_{i=0,\cdots,n-\rho,j=1,2} f_{i,j}^{v} \cdot \prod_{i=0,\cdots,n-\rho-1} f_{i,3}^{v}]^2 + 1$$

where $f_{2,y}(y)$ is chosen such that $\frac{|f_2|}{f_{2,y}}$ is bounded when other variables are fixed, other terms in (23) are defined in the same fashion. Now choose β_2 as any positive smooth function defined in the following set:

$$S_2 = \{\beta_2(y,m) \Big| {}^{s_2}\!/_{\beta_2} \quad \textit{is bounded}\}$$

Step $3 \le h \le \rho$ Define

$$f_h = -\frac{\partial\alpha_{h-1}}{\partial y}(\varepsilon_2 + \psi_1 + \phi_{01}) - \frac{\partial\alpha_{h-1}}{\partial\hat{\theta}}\Gamma\tau_h$$
$$+\sum_{j=2}^{h-1} z_j \frac{\partial\alpha_{h-1}}{\partial\hat{\theta}}\Gamma\frac{\partial\alpha_{h-1}}{\partial y}\omega + \pi_h + z_{h-1}$$

Choose β_h as any positive smooth function in the set:

$$S_h = \{\beta_h(y,m) \Big| {}^{s_h}\!/_{\beta_h} \quad \textit{is bounded}\}$$

where S_h and the term $f_{h,y}$ are defined in the same fashion as in step 2.

Remark 4 The freedom of choosing $\beta, \beta_1, \cdots, \beta_\rho$ is utilized to simplify the controller design in the following three aspects: 1) By choosing them properly, it is no need to compute the known terms $\phi_{01} - \dot{y}_r$, $-\frac{\partial\alpha_1}{\partial y} - \frac{\partial\alpha_1}{\partial\theta}\Gamma\tau_2 + \pi_2$ and $\phi_{0,1} - \frac{\partial\alpha_{i-1}}{\partial\hat{\theta}}\Gamma\tau_i + \sum_{j=2}^{i-1} z_j \frac{\partial\alpha_{i-1}}{\partial\hat{\theta}}\Gamma\frac{\partial\alpha_{i-1}}{\partial y}\omega + \pi_i + z_i, 3 \le i \le \rho$, and these terms can be easily dealt with by viewing them as uncertainties. This reduces greatly the computation. 2) β is required to make

$$\frac{\|\psi + k_0 y + \phi_0(y)\|}{\beta} \quad \text{and} \quad \frac{\|\phi_j(y)\|}{\beta}, 1 \le j \le p$$

bounded so that $m(t)$ can be used to avoid the controller to depend on $\xi_j, 1 \le j \le p$. This also simplifies greatly the design procedure since the controller can be designed with much less variables. 3) Since β and β_1 are chosen not to depend on y_r and its derivatives and $\beta_2, \cdots, \beta_\rho$ are chosen not to depend on $\dot{y}_r, \cdots y^{(\rho)}{}_r$, the controller can be designed not to depend on $\dot{y}_r, \cdots y^{(\rho)}{}_r$. This will also greatly simplify of the controller design since it may be much more involved if $\dot{y}_r, \cdots y^{(\rho)}{}_r$ are allowed to present in the controller.

Remark 5 Since there infinite candidate choices of $\beta, \beta_1, \cdots, \beta_\rho$ and different choices may result in different controllers with different control effort, there must exist one choice which achieves the best tracking performance with the least control effort. This choice may be difficult to be found. However, one can indeed use this freedom to select a satisfactory choice of $\beta, \beta_1, \cdots, \beta_\rho$ to design a simpler controller with reduced control effort. If one only wants to find the best one among a finite groups of positive smooth functions, one may simply choose them by test or simulations.

5. ROBUSTNESS

Theorem 5.1. Under Assumptions A1 ~ A6, and suppose that the robust adaptive controller given by (10) ~ (12) is applied to system (1). If we choose β and $\beta_i, 1 \le i \le \rho$ are positive smooth functions chosen as in Section 4.

Then, all closed-loop signals are globally bounded, and there exist two positive constants λ and M such that the output tracking error satisfies

$$|y-y_r| \le \sqrt{2V_p(0)e^{-\lambda t}} + \sqrt{\tfrac{M}{\lambda}}$$

where $V_p = \sum_{i=1}^{p} \frac{1}{2} z_i^2 + \frac{1}{2}(\theta-\hat{\theta})^T \Gamma^{-1}(\theta-\hat{\theta})$

$+\frac{|b_{n-p}|}{2\gamma}(q-\hat{q})^2$ and $\lambda = \min\left\{c_1,\cdots,c_p,\frac{l_q}{2},\frac{l_\theta}{2}\right\}$

and $\dfrac{|\varepsilon_2+\psi_1+\phi_{0,1}-\dot{y}_r|^2}{4k_1\beta_1} + \sum_{i=2}^{p}\dfrac{|f_i|^2}{4k_i\beta_i} +$

$$\frac{|b_{n-p}|}{2\gamma} l_q q^2 + \frac{l_\theta}{2}\theta^T\Gamma^{-1}\theta \le M$$

Remark 6. From the definition of λ and M, it is obvious that if $c_1,\cdots,c_p,l_\theta,l_q,\gamma,\Gamma$ and $\beta,\beta_1,\cdots,\beta_p$ are chosen properly, $\dfrac{M}{\lambda}$ can be made arbitrarily small which implies that the output tracking error can be made arbitrarily small.

6. SIMULATIONS

In the simulations, the following system is considered:

$$\dot{x}_1 = x_2 + \theta x_1^2 + \psi_1(x_1,u,t), \dot{x}_2 = u; y = x_1$$

where $\psi_1(x,u,t) = \eta_1 x_1^3 \cos(x_2 u) + \eta_2 \sin \omega t$, η_1, η_2 and ω unknown parameters.

Two groups of choices of $\beta(y)$, $\beta_1(y,m)$, $\beta_2(y,m,v_{0,1},v_{0,2},\hat{\theta})$ are given to show different choices may result in quite different results for the same system. In the simulations, one can see that similar tracking performances are achieved for both groups of choices while the control effort needed by one group is much less than that needed by the other group. This implies that, by choosing β,β_1,β_2 properly, one can indeed design a simpler controller to achieve the same performance with less control effort. Due to lack of space, the simulation results are all omitted.

7. CONCLUSION

By introducing a new normalization signal $m(t)$ and leaving a group of positive smooth functions $\beta,\beta_1,\cdots,\beta_p$ to be chosen properly, a simpler robust adaptive output-feedback controller is designed for a class of nonlinear uncertain systems. The global boundness of all closed-loop signals can be guaranteed and the output tracking error can be made arbitrarily small. The freedom of choosing $\beta,\beta_1,\cdots,\beta_p$ enables us to design a simpler controller to achieve the same or better performance with reduced control effort. To find the best one among the infinite choices of $\beta,\beta_1,\cdots,\beta_p$ seems to be more interesting and more difficult.

REFERENCES

Freeman, R. A. and P.V. Kokotovic (1993). Design of ‘softer’ robust nonlinear control laws. Automatica, **29**, 1425-1437.

Jiang Z. P. and L. Praly (1996). A self-tuning robust nonlinear controller. 1996 IFAC World Congress, San Francisco, USA,169-174.

Kanellakopoulous I., M. Krstic and P.V. Kokotovic (1993). k-Adaptive control of output-feedback nonlinear systems. Proc. 32th CDC, San Antonio, Texas. pp.1061 ~ 1066.

Krstic, M., I. Kanellakopoulos and P.V. Kokotovic (1995). Nonlinear and Adaptive Control Design. New York: John Wiley & Sons.

Marino R., and P. Tomei (1995). Nonlinear Control Design: Gemetric, Adaptive and Robust. London: Prentice-Hall.

Marino R. and P. Tomei P. (1993b). Global adaptive output-feedback control of nonlinear systems, Part I : linear parameterization. IEEE Trans. Automat. Contr., 38(1), pp. 17 ~ 32.

Marino R. and P. Tomei P. (1993b). Global adaptive output-feedback control of nonlinear systems, Part II : nonlinear parameterization. IEEE Trans. Automat. Contr. , 38(1), pp. 33 ~ 48.

Narendra, K. S. and A. M. Annaswamy (1989). *Stable Adaptive Systems*, Englewood Cliffs, NJ: Prentice-Hall.

14th Triennial World Congress, Beijing, P.R. China

E-2c-16-6

ADAPTIVE OUTPUT FEEDBACK TRACKING FOR A CLASS OF NONLINEAR SYSTEMS WITH TIME-VARYING PARAMETERS

Riccardo Marino and Patrizio Tomei

Dipartimento di Ing. Elettronica, Università di Roma 'Tor Vergata', Via di Tor Vergata, 00133 Roma, Italy, e-mail: tomei@eln.uniroma2.it

Abstract: We address the tracking problem for minimum phase single-input, single-output nonlinear systems with uncertain differentiable bounded time-varying parameters (whose time derivatives are not restricted to be small or to have known bounds) whose zero dynamics are linear time-invariant with unknown constant parameters. We design a global output feedback adaptive control so that for any initial condition: i) all closed loop signals are bounded, ii) the effects of parameter uncertainties on the tracking error are arbitrarily attenuated, iii) when all parameters are constant and belong to a given compact set, asymptotic tracking is achieved with arbitarily good transient performance.

Keywords: Nonlinear control, Time-varying systems, Output feedback, Adaptive control.

1. INTRODUCTION

Given a smooth bounded reference signal $y_r(t)$ to be tracked by the output $y(t)$, we address the problem of designing an output feedback adaptive control algorithm for nonlinear systems with bounded uncertain time-varying parameters and linear time-invariant zero-dynamics with unknown constant parameters so that, for any initial condition: (i) all closed loop signals are bounded; (ii) the effects of parameter uncertainties on the tracking error are arbitrarily attenuated in both L_2 and L_∞ sense; (iii) when all uncertain parameters are constant and belong to a given compact set, asymptotic tracking is achieved with arbitrarily good transient performance in terms of both L_2 and L_∞ tracking error norms.

The problem posed is well rooted in adaptive control design whose main motivation is to cope with uncertain time-varying parameters in linear or nonlinear plants. For minimum phase systems with constant parameters, known and constant relative degree, known sign of the high frequency gain and known upper bound on the plant order, output feedback adaptive controls guaranteeing boundedness of all signals and asymptotic tracking were given in Feuer and Morse (1978), Morse (1980) and Narendra *et al.* (1980), for linear systems, and in Marino and Tomei (1993), Kanellakopoulos *et al.* (1991), for nonlinear ones with output dependent nonlinearities; more recently, adaptive algorithms achieving also arbitrarily improved transient performance are given in Krstic *et al.* (1993), Datta and Ioannou (1994), for linear systems and in Krstic *et al.* (1995), Krstic and Kokotovic (1996) for nonlinear ones. Adaptive tracking controls for linear time-varying systems with uncertain slowly varying parameters are given in Tsakalis and Ioannou (1987), Middleton and Goodwin (1988), without guaranteed transient behavior and more recently in Zang and Ioannou (1996) where performance bounds, which can not be arbitrarily improved, are explicitly obtained. Robust adaptive state feedback controls for nonlinear systems with time-varying parameters or disturbances have been designed in Yao and Tomizuka (1997), Marino and Tomei (1998).

In this paper we address the posed problem for a class of observable, minimum phase, nonlinear systems with output dependent nonlinearities multiplying time-varying parameters, which is a generalization of the one introduced in Marino and Tomei (1993) since some parameters are allowed to be time-varying. For this class of systems, under the assumption of known zero dynamics, an adaptive output feedback tracking control was designed in Marino and Tomei (1997) without restricting the parameters to be slowly time-varying. In this paper we allow the zero dynamics to contain constant unknown parameters and we solve the output feedback adaptive tracking problem.

We consider the following class of single input, single output uncertain nonlinear systems with time-varying parameters

$$\begin{aligned} \dot{x} &= A_c x + b\sigma(y)u + \psi_0(y) + \theta(t)\psi_1(y) \\ y &= C_c x \end{aligned} \tag{1}$$

with (A_c, b, C_c) in observer canonical form

$$A_c = \begin{bmatrix} 0 & 1 & \cdots & 0 \\ \vdots & \vdots & \ddots & \vdots \\ 0 & 0 & \cdots & 1 \\ 0 & 0 & \cdots & 0 \end{bmatrix}, \quad b = \begin{bmatrix} b_1 \\ \vdots \\ b_{n-1} \\ b_n \end{bmatrix}$$
$$C_c = \begin{bmatrix} 1 & 0 & 0 & \cdots & 0 \end{bmatrix}$$

and $b = [b_1, \ldots, b_n]^T$ a Hurwitz vector (the associated polynomial $b_1 s^{n-1} + b_2 s^{n-2} + \cdots + b_n$ has all its zeros with negative real part) of unknown constant parameters . If the relative degree of system (1) is ρ, then $b_i = 0$, $1 \le i \le \rho - 1$, and $b_\rho \neq 0$. We assume that the relative degree ρ and the sign of b_ρ are known (without loss of generality b_ρ is assumed to be positive). In (1) $x \in R^n$ is the state, $u \in R$ is the control, $y \in R$ is the output variable to be controlled and the only variable to be measured; σ is a smooth function, ψ_0 and ψ_1 are smooth vector fields in R^n, $\theta(t)$ is a time-varying differentiable bounded uncertain parameter ($\theta \in R^p$, $p = 1$ without loss of generality) with $\theta = 0$ being the nominal value; $\dot{\theta}(t)$ is assumed to be bounded but is not restricted to be small or to have known bounds.

2. PRELIMINARY RESULTS

Lemma 1. Consider system (1). Given $\rho - 1$ arbitrary positive reals λ_i, $1 \le i \le \rho - 1$, the filtered transformation

$$\begin{aligned} z_1 &= x_1 \\ z_j &= x_j - \sum_{i=2}^{\rho} \bar{b}_j[i]\mu_{i-1} \end{aligned} \tag{2}$$

depending on the filters ($\mu \in R^{\rho-1}$)

$$\dot{\mu} = \begin{bmatrix} -\lambda_1 & 1 & \ldots & 0 \\ 0 & -\lambda_2 & \ldots & 0 \\ \vdots & \vdots & \ldots & \vdots \\ 0 & 0 & \ldots & 1 \\ 0 & 0 & \ldots & -\lambda_{\rho-1} \end{bmatrix} \mu + \begin{bmatrix} 0 \\ 0 \\ \vdots \\ 0 \\ u \end{bmatrix} \tag{3}$$

with

$$\begin{aligned} \bar{b}[\rho] &= b \\ \bar{b}[i-1] &= A_c \bar{b}[i] + \lambda_{i-1}\bar{b}[i], \qquad \rho \ge i \ge 2 \end{aligned}$$

transforms system (1) into

$$\begin{aligned} \dot{z} &= A_c z + \bar{b}[1]\mu_1 + \psi_0(y) + \theta(t)\psi_1(y) \\ &\triangleq A_c z + \frac{1}{\nu} b' \mu_1 + \psi_0(y) + \theta(t)\psi_1(y) \\ y &= C_c z \end{aligned} \tag{4}$$

in which b' is a Hurwitz vector with $b'_1 = 1$ and $\frac{1}{\nu} = b_\rho$.

Lemma 2. Consider system (4). The filtered transformation ($\bar{\mu}[i] \in R^i$)

$$\zeta = z - \sum_{i=2}^{n} \sum_{j=2}^{i} d[j]\bar{\mu}_{j-1}[i-1]$$
$$\dot{\bar{\mu}}[i] = \begin{bmatrix} -\bar{\lambda}_1 & 1 & \ldots & 0 \\ 0 & -\bar{\lambda}_2 & \ldots & 0 \\ \vdots & \vdots & \ldots & \vdots \\ 0 & 0 & \ldots & 1 \\ 0 & 0 & \ldots & -\bar{\lambda}_i \end{bmatrix} \bar{\mu}[i] + \begin{bmatrix} 0 \\ 0 \\ \vdots \\ 0 \\ 1 \end{bmatrix} \mu_1$$
$$y[i] = \begin{bmatrix} 1 & 0 & \ldots & 0 \end{bmatrix} \bar{\mu}[i], \quad 1 \le i \le n-1 \tag{5}$$

with $d[n] = \begin{bmatrix} 0 & \ldots & 0 & 1 \end{bmatrix}^T$, $d[j-1] = A_c d[j] + \bar{\lambda}_{j-1} d[j]$, $n \ge j \ge 2$, the constants δ_i implicitly defined by $b' = \sum_{i=1}^{n} \delta_i d[i]$ and $\bar{\lambda}_i$, $1 \le i \le n-1$, arbitrary positive reals, transforms (5) into

$$\begin{aligned} \dot{\zeta} &= A_c \zeta + d\left(\frac{1}{\nu}\mu_1 + \sum_{i=2}^{n} \delta_i y[i-1]\right) \\ &\quad + \psi_0(y) + \theta(t)\psi_1(y) \\ y &= C_c \zeta \end{aligned} \tag{6}$$

in which $d = d[1] = [1, d_2, \ldots, d_n]^T$ is a Hurwitz vector.

Proof. See Marino and Tomei (1995), p. 310. By construction we have

$$\left[d[1] \ldots d[n] \right] = \begin{bmatrix} 1 & 0 & \ldots & 0 \\ d_2[1] & 1 & \ldots & 0 \\ \vdots & \vdots & \ddots & \vdots \\ d_n[1] & d_n[2] & \ldots & 1 \end{bmatrix} . \quad (7)$$

Lemma 3. Consider system (6). The filtered transformation ($\xi[i] \in R^{n-1}$, $i = 0, 1$)

$$\begin{aligned} \bar{\zeta}_1 &= \zeta_1 \\ \bar{\zeta}_j &= \zeta_j - \xi_{j-1}[0] - \theta(t)\xi_{j-1}[1], \quad 2 \le j \le n \\ \dot{\xi}[i] &= \begin{bmatrix} -d_2 & 1 \ldots 0 \\ -d_3 & 0 \ldots 0 \\ \vdots & \vdots \; \vdots \; \vdots \\ -d_{n-1} & 0 \ldots 1 \\ -d_n & 0 \ldots 0 \end{bmatrix} \xi[i] \\ &+ \begin{bmatrix} -d_2 & 1\,0 \ldots 0\,0 \\ -d_3 & 0\,1 \ldots 0\,0 \\ \vdots & \vdots\;\vdots\;\vdots\;\vdots\;\vdots \\ -d_{n-1} & 0\,0 \ldots 1\,0 \\ -d_n & 0\,0 \ldots 0\,1 \end{bmatrix} \psi_i(y_r) \end{aligned} \quad (8)$$

transforms system (6) into

$$\begin{aligned} \dot{\bar{\zeta}} &= A_c\bar{\zeta} + d\Bigg(\frac{1}{\nu}\mu_1 + \sum_{i=2}^{n} \delta_i y[i-1] + \psi_{01}(y_r) \\ &+\xi_1[0] + \theta(t)(\psi_{11}(y_r) + \xi_1[1])\Bigg) \\ &+\theta(t)[\psi_1(y) - \psi_1(y_r)] - \begin{bmatrix} 0 \\ \xi[1] \end{bmatrix} \dot{\theta}(t) \\ y &= C_c\bar{\zeta} \; . \end{aligned} \quad (9)$$

Proof. See Marino and Tomei (1995), p. 312, with obvious modifications. Note that $\xi[i](t)$ are bounded since d is a Hurwitz vector.

3. MAIN RESULT

Perform the change of coordinates

$$\begin{aligned} y &= \bar{\zeta}_1 \\ \eta_j &= \bar{\zeta}_{j+1} - d_{j+1}\bar{\zeta}_1, \; 1 \le j \le n-1 \end{aligned} \quad (10)$$

so that, in new coordinates, we have

$$\begin{aligned} \dot{y} &= \eta_1 + d_2 y + \frac{1}{\nu}\mu_1 + \sum_{i=2}^{n} \delta_i y[i-1] + \psi_{01}(y) \\ &+\xi_1[0] + \theta(t)(\psi_{11}(y) + \xi_1[1]) \\ \dot{\eta} &= \Gamma\eta + \beta y + (\bar{\psi}_0(y) - \bar{\psi}_0(y_r)) + \theta(t)(\bar{\psi}_1(y) \\ &-\bar{\psi}_1(y_r)) - \xi[1]\dot{\theta}(t) \end{aligned} \quad (11)$$

with

$$\Gamma = \begin{bmatrix} -d_2 & 1\;0 \cdots 0 \\ -d_3 & 0\;1 \cdots 0 \\ \vdots & \vdots\;\vdots\;\ddots\;\vdots \\ -d_{n-1} & 0\;0 \cdots 1 \\ -d_n & 0\;0 \cdots 0 \end{bmatrix}, \quad \beta = \begin{bmatrix} d_3 - (d_2)^2 \\ d_4 - d_3 d_2 \\ \vdots \\ d_n - d_{n-1}d_2 \\ -d_n d_2 \end{bmatrix}$$

$$\bar{\psi}_i = \begin{bmatrix} \psi_{i2} - d_2\psi_{i1} \\ \vdots \\ \psi_{in} - d_n\psi_{i1} \end{bmatrix}, \quad i = 0, 1 \; .$$

The following lemma will be useful to prove the main result. For simplicity, we consider systems with relative degree $\rho = 2$.

Lemma 4. Consider system (11) with relative degree $\rho = 2$. Let $y_r(t) \in Y \subset R$ be a smooth bounded reference signal with bounded time derivatives $\dot{y}_r(t)$, $\ddot{y}_r(t)$ and let

$$\dot{\eta}_r = \Gamma\eta_r + \beta y_r, \qquad \eta_r \in R^{n-1} \quad (12)$$

be the corresponding reference dynamics. Given the adaptation dynamics

$$\begin{aligned} \dot{\hat{\theta}}_e &= f_{\hat{\theta}_e}(y, \xi, \mu_1, \eta_r, y_r, \dot{y}_r, \bar{M}_{n-1}, \hat{\theta}_e, \hat{\nu}) \\ \dot{\hat{\nu}} &= f_{\hat{\nu}}(y, y_r, \hat{\theta}_e, \bar{M}_1) \end{aligned} \quad (13)$$

with $\theta_e = [\theta, \delta_2, \ldots, \delta_n, 1/\nu, k_0]^T$, $\hat{\theta}_e$ an estimate of θ_e, $\bar{M}_i = [\bar{\mu}_1[1], \bar{\mu}_1[2], \bar{\mu}_2[2], \ldots, \bar{\mu}_i[1], \ldots, \bar{\mu}_i[i]]^T$, $\xi = [\xi^T[0], \xi^T[1]]^T$, $f_{\hat{\theta}_e}$, $f_{\hat{\nu}}$ Lipschitz continuous functions, and given two smooth functions

$$\begin{aligned} &v_1(y, \xi, \bar{M}_1, y_r, \eta_r, \hat{\theta}_e) \\ &v_2(y, \xi, \mu_1, \bar{M}_2, y_r, \dot{y}_r, \eta_r, \hat{\theta}_e, \hat{\nu}) \end{aligned} \quad (14)$$

there exist a global filtered transformation

$$\begin{aligned} \tilde{y} &= y - y_r \\ \tilde{\mu}_1 &= \mu_1 - \mu_1^*(y, \xi, \bar{M}_1, \eta_r, y_r, \hat{\theta}_e, \hat{\nu}, v_1) - \dot{y}_r \\ \tilde{\eta} &= \eta - \eta_r \end{aligned} \quad (15)$$

and a dynamic output feedback control

$$u = u^*(y, \xi, \mu_1, \bar{M}_2, \eta_r, y_r, \dot{y}_r, \hat{\theta}_e, \hat{\nu}, v_2) + \ddot{y}_r \quad (16)$$

with μ_1^* and u^* suitable smooth functions, transforming system (11) into

$$\begin{aligned} \dot{\tilde{y}} &= -c_1\tilde{y} - k_0\bar{\psi}_{1a}^2(y, y_r)\tilde{y} - k_1\bar{\psi}_{0a}^2(y, y_r)\tilde{y} + \frac{1}{\nu}\tilde{\mu}_1 \\ &+\tilde{\eta}_1 + \tilde{\theta}_e^T\gamma_1(y, y_r, \bar{M}_1, \xi) \\ &-\frac{\tilde{\nu}}{\nu}\phi(y, y_r, \bar{M}_1, \hat{\theta}_e, v_1) \\ &-\frac{1}{4\nu}\phi^2(y, y_r, \bar{M}_1, \hat{\theta}_e, v_1)\tilde{y} + v_1 \\ \dot{\tilde{\mu}}_1 &= -c_2\tilde{\mu}_1 - \frac{1}{\mu}\tilde{y} + \tilde{\theta}_e^T\gamma_2(y, y_r, \eta_r, \bar{M}_2, \xi, \hat{\theta}_e, \hat{\nu}) \\ &+\tilde{\eta}_1\omega_1(y, y_r, \eta_r, \bar{M}_1, \xi, \hat{\theta}_e, \hat{\nu}) + v_2 \end{aligned}$$

$$\dot{\tilde{\eta}} = \Gamma\tilde{\eta} + \beta\tilde{y} + \bar{\psi}_{0a}(y, y_r)\tilde{y} + \theta\bar{\psi}_{1a}(y, y_r)\tilde{y} - \dot{\theta}\xi[1] \quad (17)$$

in which $\tilde{\theta}_e = \theta_e - \hat{\theta}_e$, $\tilde{\nu} = \nu - \hat{\nu}$, k_0, c_1 and c_2 are arbitrary positive reals, α_i, α_i', γ_i and ω_i are smooth functions, and $\bar{\psi}_{ia}(y, y_r) = (\bar{\psi}_i(y) - \bar{\psi}_i(y_r))/(y - y_r)$, $i = 0, 1$.

Theorem 5. Consider system (1). For any smooth bounded trajectory $y_r(t) \in Y \subset R$ with bounded time derivatives $y_r^{(1)}, \ldots, y_r^{(\rho)}$, a dynamic output feedback control, parameterized by the positive scalars k, r_{Ω_1}, r_{Ω_2}, exists such that for any initial condition of the closed loop system, provided that $\|\hat{\theta}_e(t_0)\| \le r_{\Omega_1}$, $\|\hat{\nu}(t_0)\| \le r_{\Omega_2}$:

a) all signals are bounded $\forall t \ge t_0$;
b) the following two inequalities hold $\forall t \ge t_0$, $\forall t_0 \ge 0$:

$$|\tilde{y}(t)| \le \beta_1(\|[\tilde{y}, \tilde{\mu}_1, \ldots, \tilde{\mu}_{\rho-1}, \tilde{\eta}]\|, t - t_0) + \frac{a_1}{\sqrt{k}} \sup_{\tau\in[t_0,t]} \|\tilde{\theta}_e(\tau), \tilde{\nu}(\tau)]\| + \frac{a_2}{\sqrt{k}} \sup_{\tau\in[t_0,t]} \|\dot{\theta}(\tau)\| \quad (18)$$

$$\int_{t_0}^{t} \tilde{y}^2(\tau)d\tau \le \frac{1}{k}\Big\{\beta_2(\|[\tilde{y}, \tilde{\mu}_1, \ldots, \tilde{\mu}_{\rho-1}, \tilde{\eta}]\|) + a_3 \int_{t_0}^{t} \|[\tilde{\theta}_e(\tau), \tilde{\nu}(\tau)]\|^2 d\tau + a_4 \int_{t_0}^{t} \|\dot{\theta}(\tau)\|^2 d\tau\Big\} \quad (19)$$

with a_i positive constants, β_1 a class KL function and β_2 a class K function;

c) if $\|\hat{\theta}_e(t)\| \le r_{\Omega_1}$, $\forall t \ge t_1 \ge t_0$, $\|\hat{\nu}\| \le r_{\Omega_2}$ and $\dot{\theta} \in L_1$, then

$$\lim_{t\to\infty} \tilde{y}(t) = 0$$
$$\lim_{t\to\infty} \tilde{\mu}_i(t) = 0, \quad 1 \le i \le \rho - 1$$
$$\lim_{t\to\infty} \|\tilde{\eta}(t)\| = 0.$$

Proof. For simplicity the proof is given for $\rho = 2$. The extension to systems with any relative degree is straightforward. System (1) may be transformed into (17), by means of Lemmas 1, 2, 3 and 4. Define

$$v_1 = -\frac{1}{4}\tilde{y}\gamma_1^T\gamma_1 - k\tilde{y}$$
$$v_2 = -\frac{1}{4}\tilde{\mu}_1\gamma_2^T\gamma_2 - \omega_1^2\tilde{\mu}_1 - \frac{1}{4}\tilde{\mu}_1\|\alpha_1\|^2\|\gamma_1\|^2 - \tilde{\mu}_1[1 + \|\alpha_1\|^2\|\gamma_2\|^2]^{1/2} - \frac{1}{4}\tilde{\mu}_1(\alpha_1'\phi)^2$$
$$f_{\hat{\theta}_e} = \mathrm{Proj}(\gamma_1\tilde{y} + \gamma_2\tilde{\mu}_1, \hat{\theta}_e)$$
$$f_{\hat{\nu}} = \mathrm{Proj}(-\phi\tilde{y}, \hat{\nu}) \quad (20)$$

where $\mathrm{Proj}(w, \hat{\theta})$ is the smooth projection algorithm introduced in Pomet and Praly (1992) and defined in our case by:

$$\mathrm{Proj}(w, \hat{\theta}) = w, \text{ if } p(\hat{\theta}) \le 0$$
$$\mathrm{Proj}(w, \hat{\theta}) = w, \text{ if } p(\hat{\theta}) \ge 0 \text{ and } \langle \mathrm{grad}\, p(\hat{\theta}), w\rangle \le 0$$
$$\mathrm{Proj}(w, \hat{\theta}) = \left[I - \frac{p(\hat{\theta})\mathrm{grad}\, p(\hat{\theta})^T \mathrm{grad}\, p(\hat{\theta})}{\|\mathrm{grad}\, p(\hat{\theta})\|^2}\right] w, \quad \text{if } p(\hat{\theta}) > 0 \text{ and } \langle \mathrm{grad}\, p(\hat{\theta}), w\rangle > 0$$

with $p(\theta) = \frac{\|\theta\|^2 - r_\Omega^2}{\epsilon^2 + 2\epsilon r_\Omega}$ and $\epsilon > 0$, so that if $\|\hat{\theta}(t_0)\| \le r_\Omega$ then:

i) $\|\hat{\theta}(t)\| \le r_\Omega + \epsilon$, $\forall t \ge t_0$,
ii) $\mathrm{Proj}(w, \hat{\theta})$ is Lipschitz continuous,
iii) $\|\mathrm{Proj}(w, \hat{\theta})\| \le \|w\|$,
iv) if $\|\theta\| \le r_\Omega$ then $\tilde{\theta}^T\mathrm{Proj}(w, \hat{\theta}) \ge \tilde{\theta}^T w$.

The boundedness of $\hat{\theta}_e$ and $\hat{\nu}$ is implied by property (i). The closed loop system becomes

$$\dot{\tilde{y}} = -c_1\tilde{y} - k_0\bar{\psi}_{1a}^2\tilde{y} - k_1\bar{\psi}_{0a}^2\tilde{y} + \frac{1}{\nu}\tilde{\mu}_1 + \tilde{\eta}_1 + \tilde{\theta}_e^T\gamma_1 - \frac{1}{4}\gamma_1^T\gamma_1\tilde{y} - \frac{\tilde{\nu}}{\nu}\phi - \frac{1}{4\nu}\phi^2\tilde{y} - k\tilde{y}$$
$$\dot{\tilde{\mu}}_1 = -c_2\tilde{\mu}_1 - \frac{1}{\nu}\tilde{y} + \tilde{\theta}_e^T\gamma_2 - \frac{1}{4}\tilde{\mu}_1\gamma_2^T\gamma_2 + \tilde{\eta}_1\omega_1 - \omega_1^2\tilde{\mu}_1 + \alpha_1^T\mathrm{Proj}(\gamma_1\tilde{y} + \gamma_2\tilde{\mu}_1, \hat{\theta}_e) + \alpha_1'\mathrm{Proj}(-\phi\tilde{y}, \hat{\nu}) - \frac{1}{4}\tilde{\mu}_1\|\alpha_1\|^2\|\gamma_1\|^2 - \tilde{\mu}_1[1 + \|\alpha_1\|^2\|\gamma_2\|^2]^{1/2} - \frac{1}{4}\tilde{\mu}_1(\alpha_1'\phi)^2$$
$$\dot{\tilde{\eta}} = \Gamma\tilde{\eta} + \beta\tilde{y} + \bar{\psi}_{0a}\tilde{y} + \theta(t)\bar{\psi}_{1a}\tilde{y} - \dot{\theta}\xi[1]\,. \quad (21)$$

Consider the function

$$W = \frac{1}{2}\tilde{y}^2 + \frac{1}{2}\tilde{\mu}_1^2 + \tilde{\eta}^T P\tilde{\eta} \quad (22)$$

with P solution of $\Gamma^T P + P\Gamma = -2I$. The time derivative of (22) is given by

$$\dot{W} = -c_1\tilde{y} + \tilde{y}\tilde{\eta}_1 + \frac{1}{\nu}\tilde{y}\tilde{\mu}_1 - k_0\tilde{y}^2\bar{\psi}_{1a}^2 - k_1\tilde{y}^2\bar{\psi}_{0a}^2 + \tilde{\theta}_e\gamma_1\tilde{y} - \frac{1}{4}\gamma_1^T\gamma_1\tilde{y}^2 - \frac{\tilde{\nu}}{\nu}\phi\tilde{y} - \frac{1}{4\nu}\tilde{y}^2\phi^2 - k\tilde{y}^2 - c_2\tilde{\mu}_1^2 - \frac{1}{\nu}\tilde{y}\tilde{\mu}_1 + \tilde{\mu}_1\tilde{\theta}_e^T\gamma_2 - \frac{1}{4}\tilde{\mu}_1^2\gamma_2^T\gamma_2 + \tilde{\mu}_1\tilde{\eta}_1\omega_1 - \omega_1^2\tilde{\mu}_1^2 + \tilde{\mu}_1\alpha_1^T\mathrm{Proj}(\gamma_1\tilde{y} + \gamma_2\tilde{\mu}_1, \hat{\theta}_e) + \tilde{\mu}_1\alpha_1'\mathrm{Proj}(-\phi\tilde{y}, \hat{\nu}) - \frac{1}{4}\tilde{\mu}_1^2\|\alpha_1\|^2\|\gamma_1\|^2 - \tilde{\mu}_1^2[1 + \|\alpha_1\|^2\|\gamma_2\|^2]^{1/2} - \frac{1}{4}\tilde{\mu}_1^2(\alpha_1'\phi)^2$$

$$
\begin{aligned}
&-2\tilde{\eta}^T\tilde{\eta} + 2\tilde{y}\tilde{\eta}^T P\beta + 2\tilde{y}\tilde{\eta}^T P\bar{\psi}_{0a} \\
&+2\theta\tilde{y}\tilde{\eta}^T P\bar{\psi}_{1a} - 2\tilde{\eta}^T P\xi[1]\dot{\theta} \\
= &-\begin{bmatrix} |\tilde{y}| \\ \|\tilde{\eta}\| \end{bmatrix}^T \begin{bmatrix} c_1 & -\frac{1}{2} - \|P\beta\| \\ -\frac{1}{2} - \|P\beta\| & \frac{1}{4} \end{bmatrix} \begin{bmatrix} |\tilde{y}| \\ \|\tilde{\eta}\| \end{bmatrix} \\
&-\begin{bmatrix} |\tilde{y}| \\ \|\tilde{\eta}\| \end{bmatrix}^T \begin{bmatrix} k_1\bar{\psi}_{0a}^2 & -\|P\bar{\psi}_{0a}\| \\ \|P\bar{\psi}_{0a}\| & \frac{1}{4} \end{bmatrix} \begin{bmatrix} |\tilde{y}| \\ \|\tilde{\eta}\| \end{bmatrix} \\
&-\begin{bmatrix} |\tilde{y}| \\ \|\tilde{\eta}\| \end{bmatrix}^T \begin{bmatrix} k_0\bar{\psi}_{1a}^2 & -\|P\bar{\psi}_{1a}\||\theta| \\ \|P\bar{\psi}_{1a}\||\theta| & \frac{1}{4} \end{bmatrix} \begin{bmatrix} |\tilde{y}| \\ \|\tilde{\eta}\| \end{bmatrix} \\
&-\left(\frac{\gamma_1}{2}\tilde{y} - \tilde{\theta}_e\right)^T \left(\frac{\gamma_1}{2}\tilde{y} - \tilde{\theta}_e\right) + \|\tilde{\theta}_e\|^2 \\
&-\frac{1}{\nu}\left(\frac{\phi}{2} - \tilde{\nu}\right)^2 + \frac{\tilde{\nu}^2}{\nu} - k\tilde{y}^2 - c_2\tilde{\mu}_1^2 \\
&-\left(\frac{\gamma_2}{2}\tilde{\mu}_1 - \tilde{\theta}_e\right)^T \left(\frac{\gamma_1}{2}\tilde{\mu}_1 - \tilde{\theta}_e\right) + \|\tilde{\theta}_e\|^2 \\
&-\left(\frac{\tilde{\eta}}{2} + 2P\xi[1]\dot{\theta}\right)^T \left(\frac{\tilde{\eta}}{2} + 2P\xi[1]\dot{\theta}\right) \\
&+4\dot{\theta}^2\|P\xi[1]\|^2 \\
&-\left(\omega_1\tilde{\mu}_1 - \frac{\tilde{\eta}_1}{2}\right)^2 - \frac{3}{4}\|\tilde{\eta}\|^2 \\
&+\tilde{\mu}_1\alpha_1^T \mathrm{Proj}(\gamma_1\tilde{y} + \gamma_2\tilde{\mu}_1, \hat{\theta}_e) \\
&+\tilde{\mu}_1\alpha_1' \mathrm{Proj}(-\phi\tilde{y}, \hat{\nu}) \\
&-\frac{1}{4}\tilde{\mu}_1^2\|\alpha_1\|^2\|\gamma_1\|^2 - \tilde{\mu}_1^2[1 + \|\alpha_1\|^2\|\gamma_2\|^2]^{1/2} \\
&-\frac{1}{4}\tilde{\mu}_1^2(\alpha_1'\phi)^2 \,. \qquad (23)
\end{aligned}
$$

Choosing

$$
\begin{aligned}
&c_1 - 4\left(\frac{1}{2} + \|P\beta\|\right)^2 \triangleq \bar{c}_1 > 0 \\
&k_1 \geq 4\|P\|^2 \\
&k_0 \geq 4\|P\|^2 \left(\sup_{t\in[t_0,\infty)} |\theta(\tau)|\right)^2 \qquad (24)
\end{aligned}
$$

and recalling property (iii) of the operator Proj, we obtain

$$
\dot{W} \leq -(\bar{c}_1 - 2)\tilde{y}^2 - k\tilde{y}^2 - \frac{3}{4}\|\tilde{\eta}\|^2 - c_2\tilde{\mu}_1^2 + 2\|\tilde{\theta}_e\|^2 + \frac{\tilde{\nu}^2}{\nu} + M\dot{\theta}^2 \qquad (25)
$$

with $M = 4(\sup_{t\in[0,\infty]} \|P\xi[1]\|)^2$. By choosing $\bar{c}_1 > 2$, from (22) and (25) we obtain that $\tilde{y}$, $\tilde{\eta}$ and $\tilde{\mu}_1$ are bounded, since $\tilde{\theta}_e$, $\tilde{\nu}$ and $\dot{\theta}$ are bounded. Since y_r is bounded, from (12) it follows that also η_r is bounded and, consequently, y and η are bounded. From (10) it follows that $\bar{\zeta}$ is bounded and, by virtue of (8), also ζ is bounded. On the other hand, by making the change of coordinates

$$
y = z_1
$$

$$
q_j = z_{j+1} - b'_{j+1}z_1, \qquad 1 \leq j \leq n-1 \qquad (26)
$$

from (4) we obtain

$$
\begin{aligned}
\dot{y} &= q_1 + b'_2 y + \frac{1}{\nu}\mu_1 + \psi_{01}(y) + \theta\psi_{11}(y) \\
\dot{q} &= \bar{\Gamma}q + \bar{\beta}y + \gamma_0(y) + \gamma_1(y)\theta, \quad q \in R^{n-1} \qquad (27)
\end{aligned}
$$

with $\bar{\Gamma}$ a Hurwitz matrix since b' is a Hurwitz vector with $b'_1 = 1$. Since y is bounded, (27) implies that q is also bounded which, in turn, implies by (26) that z is bounded. Recalling (5), the vector

$$
\sum_{i=2}^{n} \delta_i \sum_{j=2}^{i} d[j]\bar{\mu}_{j-1}[i-1] \qquad (28)
$$

is bounded. The second component of (28), recalling (7), is given by

$$
\sum_{i=2}^{n} \delta_i\bar{\mu}_1[i-1](t) = \sum_{i=2}^{n} \delta_i y[i-1](t) \qquad (29)
$$

which, therefore, is bounded. Now, recalling the first in (11) and the first in (5), we can write

$$
\frac{d}{dt}\left(y - \frac{\bar{\mu}_1[1]}{\nu}\right) = -\bar{\lambda}_1\left(y - \frac{\bar{\mu}_1[1]}{\nu}\right) + w(t) \qquad (30)
$$

with $w(t)$ a bounded signal. As a consequence, $\bar{\mu}_1[1]$ is bounded. Proceeding the same way, we can prove the boundedness of all $\bar{\mu}[i]$ which imply the boundedness of μ_1^* and, consequently, of μ_1. Therefore $u*$ and u are also bounded, so that statement a) is proved. Integrating (25), we obtain

$$
\int_{t_0}^{t} \tilde{y}^2(\tau)d\tau \leq \frac{1}{k}\left[W(t_0) + \int_{t_0}^{t}\left(2\|\tilde{\theta}_e(\tau)\|^2 + \frac{\tilde{\nu}^2(\tau)}{\nu}\right)d\tau + M\int_{t_0}^{t}\dot{\theta}^2(\tau)d\tau\right] \qquad (31)
$$

which is in the form (19). Consider the function $W_1 = \frac{1}{2}\tilde{y}^2$ whose time derivative in view of (21) and (25) satisfies the inequality

$$
\begin{aligned}
\dot{W}_1 &= \dot{W}_1 - 2\tilde{\eta}^T\tilde{\eta} - \frac{4\nu^2}{c_1}\tilde{\mu}_1^2 + 2\tilde{\eta}^T\tilde{\eta} + \frac{4\nu^2}{c_1}\tilde{\mu}_1^2 \\
&\leq -2kW_1 + c_W W + \|\tilde{\theta}_e\|^2 + \frac{\tilde{\nu}^2}{\nu} \qquad (32)
\end{aligned}
$$

with

$$
c_W = \max\left\{\frac{8\nu^2}{c_1}, \frac{2}{\lambda_m(P)}\right\}
$$

where $\lambda_m(P)$ is the smallest eigenvalue of P. On the other hand, (25) implies

$$\dot{W} \leq -k_W W + 2\|\tilde{\theta}_e\|^2 + \frac{\tilde{\nu}^2}{\nu} + M\dot{\theta}^2 \quad (33)$$

with a suitable value for k_W. The inequalities (32) and (33) imply (18). To show property c), consider the function

$$V = W + \frac{1}{2}\tilde{\theta}_e^T\tilde{\theta}_e + \frac{1}{2}\tilde{\nu}^2 \quad (34)$$

and recall that $\|\theta_e(t)\| \leq r_{\Omega_1}$, $\forall t \geq t_1 \geq t_0$, $\|\nu\| \leq r_{\Omega_2}$. The time derivative of V, in view of (20), (21), (23) and property (iv) of Proj is such that

$$\dot{V} \leq -\bar{c}_1\tilde{y}^2 - k\tilde{y}^2 - \frac{3}{4}\|\tilde{\eta}\|^2 - c_2\tilde{\mu}_1^2 + \tilde{\theta}\dot{\theta}$$
$$-2\tilde{\eta}^T P\xi[1]\dot{\theta} \ . \quad (35)$$

Since $\tilde{\theta}$, $\tilde{\eta}$ and $\xi[1]$ are bounded and $\dot{\theta} \in L_1$, integrating (35) we obtain

$$\lim_{t\to\infty} \int_{t_0}^{t} (k\tilde{y}^2 + \bar{c}_1\tilde{y}^2 + c_2\tilde{\mu}_1^2 + \frac{3}{4}\|\tilde{\eta}\|^2)d\tau)d\tau < \infty$$

which, since $\dot{\tilde{y}}$, $\dot{\tilde{\mu}}_1$, $\dot{\tilde{\eta}}$ are bounded by (21), implies by Barbalat Lemma (Narendra and Annaswamy, 1989; Marino and Tomei, 1995) property c).

ACKNOWLEDGEMENT

This work has been supported by MURST and by ASI.

REFERENCES

Datta, A. and P. A. Ioannou (1994). Performance analysis and improvement in Model Reference Adaptive Control. *IEEE Trans. Automatic Control*, 39:2370–2387.

Feuer, A. and A. S. Morse (1978). Adaptive control of single-input, single-output linear systems. *IEEE Trans. Automatic Control*, 23:557–569.

Kanellakopoulos, I., P. V. Kokotovic, and A. S. Morse (1991). Adaptive output feedback control of a class of nonlinear systems. In *IEEE 30th Conf. on Decision and Control*, pages 1082–1087, Brighton.

Krstic, M., I. Kanellakopoulos, and P. V. Kokotovic (1995). *Nonlinear and Adaptive Control Design*. J. Wiley, New York.

Krstic, M. and P. V. Kokotovic (1996). Adaptive nonlinear output feedback schemes with Marino-Tomei controller. *IEEE Trans. Automatic Control*, 41:274–280.

Krstic, M., P.V. Kokotovic, and I. Kanellakopoulos (1993). Transient-performance improvement with a new class of adaptive controllers. *Systems and Control Letters*, 21:451–461.

Marino, R. and P. Tomei (1993). Global adaptive output-feedback control of nonlinear systems. Part I: linear parameterization. *IEEE Trans. Automatic Control*, 38:17–32.

Marino, R. and P. Tomei (1995). *Nonlinear Control Design - Geometric, Adaptive and Robust.* Prentice Hall, London.

Marino, R. and P. Tomei (1997). Adaptive output feedback tracking for nonlinear systems with time-varying parameters. In *IEEE 36th Conf. on Decision and Control*, San Diego, CA.

Marino, R. and P. Tomei (1998). Robust adaptive state-feedback tracking for nonlinear systems. *IEEE Trans. Automatic Control*, 43:84–89.

Middleton, R. H. and G. C. Goodwin (1988). Adaptive control of time-varying linear systems. *IEEE Trans. Automatic Control*, 33:150–155.

Morse, A. S. (1980). Global stability of parameter adaptive control systems. *IEEE Trans. Automatic Control*, 25:433–439.

Narendra, K. S. and A. M. Annaswamy (1989). *Stable Adaptive Systems*. Prentice Hall, Englewood Cliffs.

Narendra, K. S., Y. H. Lin, and L. S. Valavani (1980). Stable adaptive controller design - Part II: proof of stability. *IEEE Trans. Automatic Control*, 25:440–448.

Pomet, J. B. and L. Praly (1992). Adaptive nonlinear regulation: estimation from the Lyapunov equation. *IEEE Trans. Automatic Control*, 37:729–740.

Tsakalis, K. S. and P. A. Ioannou (1987). Adaptive control of linear time-varying plants. *Automatica*, 23:459–468.

Yao, B. and M. Tomizuka (1997). Adaptive robust control of SISO nonlinear systems in a semi-strict feedback form. *Automatica*, 33:893–900.

Zang, Y. and P. A. Ioannou (1996). Adaptive control of linear time-varying systems. In *IEEE 35th Conf. on Decision and Control*, pages 837–842, Kobe, Japan.

CHARACTERIZATION OF LYAPUNOV FUNCTIONS FOR SMOOTH NONLINEAR SYSTEMS USING LMIS

Tor A. Johansen *

* *Department of Engineering Cybernetics, Norwegian University of Science and Technology, N-7034 Trondheim, Norway. Email: Tor.Arne.Johansen@itk.ntnu.no*

Abstract: A generic parameterization of Lyapunov function candidates is suggested for a quite general class of non-autonomous smooth nonlinear systems. It is shown that conditions for the existence of a Lyapunov function that guarantees uniform exponential stability can be formulated as a set of linear matrix inequalities. Hence, computation of the Lyapunov function involves the solution of a convex optimization problem for which reasonably efficient algorithms exist. The tradeoff between accuracy and computational complexity as well as possible conservativeness of the procedure is given particular attention.

Keywords: Nonlinear systems; stability; linear matrix inequalities.

1. INTRODUCTION

This work describes a procedure for computing a Lyapunov functions (if one exists) for the equilibrium point $x = x_0$ for the class of non-autonomous non-linear systems that can be written in the form

$$\dot{x} = A(x, \theta)(x - x_0) \tag{1}$$

where $x \in R^n$ is the state vector, $\theta \in R^d$ is a possibly time-varying parameter vector, and the $n \times n$ matrix-valued function $A : R^n \times R^d \to R^{n \times n}$ is smooth.

There are significant amounts of research on quadratic stability (i.e. conditions for the existence of a quadratic Lyapunov function) of closely related classes of systems, e.g. (Kamenetskii and Pyatnitskii 1987, Becker and Packard 1994, Boyd *et al.* 1994, Gahinet *et al.* 1995). Indeed, restriction to the class of quadratic Lyapunov function candidates may lead to significant conservativeness. In particular, linear parameter varying (LPV) systems $\dot{x} = A(\theta)x$ (where θ is time-varying but known in real time) have recently attracted great interest, and in addition to the quadratic stability results mentioned above, there are stability conditions available that utilize quadratic parameter-dependent Lyapunov-functions of the form $V(x) = x^T P(\theta) x$ and information on the rate of variation of the parameter θ to reduce the conservativeness (Gahinet *et al.* 1996, Wu *et al.* 1996, Watanabe *et al.* 1996). However, in many practical cases $A(\cdot)$ depends on the states rather than external parameters. To embed nonlinear systems into the LPV framework, the states that $A(\cdot)$ depend on are typically being viewed as time-varying parameters and the nonlinear dynamics characterized by a uniform bound on $\dot{\theta}$ (Gahinet *et al.* 1996, Wu *et al.* 1996, Watanabe *et al.* 1996). This may obviously lead to conservative results when utilizing these procedures. The purpose of the present work is to utilize the explicit knowledge of the dependence of $A(\cdot)$ on x in our formulation to reduce the conservativeness by introducing a parameterized class of non-quadratic Lyapunov functions. The procedure presented in this paper reduces the problem of checking the existence of a suitable Lyapunov function of the form $V(x) = (x - x_0)^T P(x)(x - x_0)$ to solving a state-dependent matrix inequality (SDMI) that can be approximated by a finite set of linear matrix inequalities (LMIs). Hence, reasonably efficient convex optimization algorithms (Boyd *et al.* 1994, Gahinet *et al.* 1995) can be applied. This characterization of the Lyapunov function is related to the SDMI characterizations of H_∞ optimal controllers studied in (Lu and Doyle 1995). However, the present approach to approximating the SDMI characterization by a finite number of LMIs differs considerably from the computational procedures suggested in the cited work.

Introducing a parameterized Lyapunov-function is certainly not a new idea, e.g. (Brayton and

Tong 1980, Michel *et al.* 1984, Borisov and Diligenskii 1986, Molchanov and Pyatnitskii 1986, Kamenetskii and Pyatnitskii 1987, Molchanov 1987, Barabanov 1988, Barabanov 1989, Ohta *et al.* 1993, Zelentsovsky 1994, Gahinet *et al.* 1996, Wu *et al.* 1996, Johansson and Rantzer 1998, Petterson and Lennartson 1997, Watanabe *et al.* 1996). The contribution of the present procedure is the utilization of a generic parameterization that does not introduce significant conservativeness and allows the problem to be reduced to a convex optimization problem involving LMI constraints. This leads to a tight and efficient computational procedure for systems of sufficiently low order.

2. LMI CONDITIONS FOR UNIFORM EXPONENTIAL STABILITY

Uniform exponential stability of the system (1) is considered. In other words, let $X^0 \subset R^n$ be a compact and convex region of the state space such that $x_0 \in X^0$, and $\Theta \subset R^d$ be a compact region of the parameter space. A Lyapunov function that guarantees that $x(0) \in X \subset X^0$ implies $x(t) \to x_0$ as $t \to \infty$ at an exponential rate for any parameter trajectory that satisfies $\theta(t) \in \Theta$ for all t is sought. Knowledge of θ is not assumed to be available.

Consider a Lyapunov function candidate $V : X^0 \to R$ of the form

$$V(x) = (x - x_0)^T P(x)(x - x_0) \tag{2}$$

where the symmetric matrix valued function $P : X^0 \to R^{n \times n}$ is defined by the following linear parameterization

$$P(x) = \sum_{i=1}^{N} P_i \rho_i(x) \tag{3}$$

where $P_i = P_i^T > 0$ for all $i = 1, 2, ..., N$, and $\rho_i : X^0 \to R$ are smooth basis-functions. The attention is restricted to positive semi-definite basis-functions that form a partition-of-unity:

$$\sum_{i=1}^{N} \rho_i(x) = 1, \text{ for all } x \in X^0 \tag{4}$$

These conditions ensure that V is indeed a Lyapunov function candidate since it satisfies $V(x_0) = 0$, $V(x) > 0$ for $x \in X^0 - \{x_0\}$, and it is smooth.

The time-derivative of V along the trajectory of the system for a given parameter trajectory is

$$\frac{d}{dt}V(x) = (x - x_0)^T \left(A^T(x,\theta)P(x) + P(x)A(x,\theta) + \frac{d}{dt}P(x)\right)(x - x_0) \tag{5}$$

A condition for uniform exponential stability is now that there exists a function P such that the matrix-valued function $L : X^0 \times \Theta \to R^{n \times n}$

$$\begin{aligned} L(x,\theta) &= A^T(x,\theta)P(x) + P(x)A(x,\theta) + \frac{d}{dt}P(x) \\ &= \sum_{i=1}^{N} \Big(A^T(x,\theta)P_i\rho_i(x) + P_iA(x,\theta)\rho_i(x) \\ &\quad + P_i\left(\left(\frac{d\rho_i(x)}{dx}\right)^T A(x,\theta)x\right)\Big) \end{aligned}$$

satisfies

$$L(x,\theta) < -\gamma P(x) \tag{6}$$

for some constant $\gamma > 0$ and all $x \in X^0$ and $\theta \in \Theta$. Eq. (6) defines an infinite number of matrix inequalities. Below, a finite discretization of the state and parameter spaces is introduced in order to reduce this condition to a finite number of LMIs.

The following theorem gives conditions on P_1, P_2, ..., P_N that ensures V is a Lyapunov function, and forms the basis for the computational procedure that will be described below.

Theorem 1. Let X^0 be a compact and convex set. Suppose $P_i = P_i^T > 0$ for all $i = 1, 2, ..., N$ and there exists a scalar $\gamma > 0$ such that $L(x,\theta) < -\gamma P(x)$ for all $x \in X^0$ and $\theta \in \Theta$. Then for all parameter trajectories $\theta(t) \in \Theta$ and initial conditions $x(0) \in X$, the equilibrium point $x_0 \in X^0$ is uniformly exponentially stable, i.e.

$$||x(t) - x_0||_2 \le \sqrt{\frac{c_2}{c_1}}||x(0) - x_0||_2 e^{-\gamma t/2} \tag{7}$$

where $c_1 = \min_i \underline{\sigma}(P_i)$, $c_2 = \max_i \overline{\sigma}(P_i)$. The region of attraction X is estimated by

$$X = \left\{x \in X^0 \mid V(x) \le \inf_{\xi \in \partial X^0} V(\xi)\right\} \tag{8}$$

Proof: Since $L(x,\theta) < 0$ on X^0 is is clear that $x(0) \in X$ guarantees $x(t) \in X^0$ for all $t \ge 0$. Now

$$\begin{aligned} \alpha_1(||x - x_0||_2) &= c_1||x - x_0||_2^2 \le V(x) \\ &\le c_2||x - x_0||_2^2 = \alpha_2(||x - x_0||_2) \end{aligned}$$

and

$$\frac{d}{dt}V(x) < -\gamma V(x) = -\alpha_3(||x - x_0||_2) \tag{9}$$

It is clear that α_1, α_2 and α_3 are class $\mathcal{K}$ functions and the result follows from Theorem 4.1 and Corollary 4.2 of (Khalil 1992).

□

3. COMPUTATIONAL PROCEDURE

The SDMI $L(x,\theta) < -\gamma P(x)$ on $X^0 \times \Theta$ defines an infinite number of LMIs in the finite number of parameters $P_1, P_2, ..., P_N$. A finite number of LMIs results from discretization of the compact sets X^0 and Θ by defining finite design sets X_d^0 and Θ_d at which the following constraints on the matrices $P_1, P_2, ..., P_N$

$$L(x,\theta) < -\alpha P(x), \text{ for all } (x,\theta) \in X_d^0 \times \Theta_d \quad (10)$$

are imposed for some $\alpha > 0$. This is a finite number of LMI constraints in a finite number of variables, which are computationally feasible.

However, because of the finite discretization, the existence of positive definite symmetric matrices $P_1, ..., P_N$ that satisfies (10) is not sufficient to guarantee that V is a Lyapunov function. It must be checked that there exists a $\gamma > 0$ such that $L(x,\theta) < -\gamma P(x)$ holds for all $x \in X^0$ and $\theta \in \Theta$. Then it can be argued that $x(0) \in X$ guarantees uniform exponential stability of x_0. In practise, this condition must be checked in a sufficiently dense set of checking points $X_c^0 \times \Theta_c \subset X^0 \times \Theta$. If it does not hold, the density of the set of design points X_d^0 and the parameterization of P should be made richer, and the above procedure iterated.

A computational procedure for searching for a Lyapunov function is:

- Input data: The system function A, a compact and convex set X^0 that contains an interior point x_0, and a compact set Θ.
- Step 1: Select a parameterization of P.
- Step 2: Select finite design sets $X_d^0 \subset X^0$ and $\Theta_d \subset \Theta$.
- Step 3: Solve the convex optimization problem of maximizing α subject to the constraints

$$P_i = P_i^T > 0 \text{ and } L(x,\theta) < -\alpha P(x)$$
$$\text{for all } x \in X_d^0 \text{ and } \theta \in \Theta_d$$

- Step 4: Generate "sufficiently dense" but finite checking sets $X_c \subset X^0$ and $\Theta_c \subset \Theta$.
- Step 5: If $L(x,\theta) < -\gamma P(x)$ does not hold on $X_c \times \Theta_c$ for some $\gamma > 0$, go to either Step 1 or Step 2.
- Output data: If the procedure converges, a Lyapunov function has been found, and the the region of attraction X can be estimated according to (8).

The above procedure is not an algorithm, in the sense that several of the steps requires further specification. In order to keep the procedure computationally efficient, these steps should rely on engineering insight. "Very dense" uniform grids of design and checking points should be avoided, in particular in high-dimensional state and parameter spaces since they will make the problem computationally intractable. Important information about the required density of design and checking points in the state and parameter spaces as well as the parameterization of P can be determined by analyzing the nonlinearity of A over different regions in the state and parameter spaces. The idea is that if A is a highly nonlinear function in some regions of the state or parameter spaces, it makes sense to allow P to be highly non-quadratic in these regions, and to increase the density of design and checking points in these regions. Define the checking set granularity function $\varepsilon : X^0 \times \Theta \to R$

$$\varepsilon(x,\theta) = \inf_{(\xi,\zeta)\in X_c^0\times\Theta_c} ||(x,\theta) - (\xi,\zeta)||_2 \quad (11)$$

The above considerations can be formalized theoretically by assuming A, ρ_i and $d\rho_i/dx$ to be locally Lipschitz functions in the sense that for every $(x_1,\theta_1),(x_2,\theta_2)$ in an open ball with radius $\varepsilon(x,\theta)$ around a point (x,θ) there exist bounded functions $L_A : X^0 \times \Theta \to R$, $L_\rho : X^0 \to R$ and $L_{\rho'} : X^0 \to R$ that satisfy

$$||A(x_1,\theta_1) - A(x_2,\theta_2)||_2 \leq L_A(x,\theta)||(x_1,\theta_1) - (x_2,\theta_2)||_2$$
$$|\rho_i(x_1) - \rho_i(x_2)| \leq L_\rho(x)||x_1 - x_2||_2$$
$$\left|\left|\frac{d\rho_i}{dx}(x_1) - \frac{d\rho_i}{dx}(x_2)\right|\right|_2 \leq L_{\rho'}(x)||x_1 - x_2||_2$$

Furthermore, define

$$K_1(x,\theta) = \sup_{(\xi,\zeta)\in B_{\varepsilon(x,\theta)}((x,\theta))} ||A(\xi,\zeta)||_2$$
$$K_2(x,\theta) = \sup_{(\xi,\zeta)\in B_{\varepsilon(x,\theta)}((x,\theta))} ||A(\xi,\zeta)\xi||_2$$

Theorem 2. Suppose X^0 and Θ are compact sets, and there exists an $\alpha > 0$ such that for all $(\xi,\zeta) \in X_c^0 \times \Theta_c$

$$L(\xi,\zeta) < -\alpha P(\xi) \quad (12)$$

Assume A, ρ_i and $d\rho_i/dx$ are locally Lipschitz and assume there exists a $\alpha > \gamma > 0$ such that for all $x \in X^0$ and $\theta \in \Theta$ the checking grid granularity $\varepsilon(x,\theta)$ is so fine that

$$\varepsilon(x,\theta) \leq \frac{\alpha - \gamma}{n(x,\theta)} \quad (13)$$

where

$$n(x,\theta) = \kappa(P(x))(2L_A(x,\theta) + (2K_1(x,\theta) + \alpha)L_\rho(x) + K_2(x,\theta)L_{\rho'}(x)) > 0$$

and $\kappa(P) = \overline{\sigma}(P)/\underline{\sigma}(P)$ is the condition number of P. Then for all $x \in X^0$ and $\theta \in \Theta$

$$L(x,\theta) < -\gamma P(x) \quad (14)$$

Proof: Let $(x,\theta) \in X^0 \times \Theta$ be arbitrary. It follows from (12) and (11) that there exists a $(\xi,\zeta) \in (X_c^0 \times \Theta_c) \cap B_{\varepsilon(x,\theta)}((x,\theta))$ such that

$$L(x,\theta) < -\alpha P(x) + Q(x,\xi,\theta,\zeta) \quad (15)$$

where

$$\begin{aligned} Q(x,\xi,\theta,\zeta) &= (A(x,\theta) - A(\xi,\zeta))^T P(\xi) \\ &+ P(x)(A(x,\theta) - A(\xi,\zeta)) \\ &+ A^T(x,\theta)(P(x) - P(\xi)) \\ &+ (P(x) - P(\xi))A(\xi,\zeta) \\ &+ \frac{d}{dt}(P(x) - P(\xi)) + \alpha(P(x) - P(\xi)) \end{aligned}$$

It is straightforward to see that $\alpha > \gamma > 0$ and

$$||Q(x,\xi,z,\zeta)||_2 \leq \frac{\alpha-\gamma}{\kappa(P(x))} ||P(x)||_2 \quad (16)$$

implies

$$-\alpha P(x) + Q(x,\xi,\theta,\zeta) \leq -\gamma P(x) \quad (17)$$

and the theorem follows from (15) by verifying that the stated local Lipschitz conditions and constraint on the granularity $\varepsilon(x,\theta)$ is sufficient for (16) to hold.

□

This theorem relates the required local checking set granularity to local nonlinearity measures (local Lipschitz constants) of the system function and the Lyapunov function. Introducing more restrictive semi-global Lipschitz conditions a simplified version of this results presents a uniform bound $\overline{\varepsilon}$ on the required granularity of a regular checking set grid:

Corollary 1

Suppose X^0 and Θ are compact sets, and there exists an $\alpha > 0$ such that for all $(\xi,\zeta) \in X_c^0 \times \Theta_c$

$$L(\xi,\zeta) < -\alpha P(\xi) \quad (18)$$

Furthermore, suppose there exists constants $\overline{L}_A = \sup_{x,\theta} L_A(x,\theta)$, $\overline{L}_\rho = \sup_x L_\rho(x)$, $\overline{L}_{\rho'} = \sup_x L_{\rho'}(x)$, $\overline{K}_1 = \sup_{x,\theta} K_1(x,\theta)$, $\overline{K}_2 = \sup_{x,\theta} K_2(x,\theta)$, $\overline{\kappa} = \sup_x \kappa(P(x))$ and $\overline{\varepsilon} = \sup_{x,\theta} \varepsilon(x,\theta)$ such that

$$\overline{\varepsilon} \leq \frac{\alpha-\gamma}{\overline{\kappa}\left(2\overline{L}_A + (2\overline{K}_1 + \alpha)\overline{L}_\rho + \overline{K}_2\overline{L}_{\rho'}\right)} \quad (19)$$

Then for all $x \in X^0$ and $\theta \in \Theta$

$$L(x,\theta) < -\gamma P(x) \quad (20)$$

□

The parameterization of P is of great practical and theoretical importance in order to find a tight Lyapunov function with a computationally efficient procedure. In order to argue that this parameterization does not introduce significant conservativeness, it is required that the basis-functions can be selected from a set of functions that guarantees that any smooth Lyapunov function candidate and its gradient can be approximated to arbitrary accuracy on X^0. By selecting the basis-functions from a basis that is complete in a Sobolev norm this can be satisfied:

Theorem 3. Suppose X^0 is a compact set and $x_0 \in X^0$. Let $W : X^0 \to R$ be an arbitrary smooth Lyapunov function candidate, i.e. $W(x_0) = 0$, $W(x) > 0$ for any $x \in X^0 - \{x_0\}$. Suppose the set of functions $\mathcal{F}(X^0)$ defines a complete basis for the set of smooth functions in the Sobolev p-norm

$$||f||_{1,p} = ||f||_p + \left|\left|\frac{df}{dx}\right|\right|_p$$

where $p \geq 1$ is arbitrary. Then for any $\delta > 0$ there exist basis-functions $\rho_1, \rho_2, ..., \rho_N \in \mathcal{F}(X^0)$ and matrices $P_1, P_2, ..., P_N$ such that

$$V(x) = \sum_{i=1}^{N} (x-x_0)^T P_i (x-x_0) \rho_i(x) \quad (21)$$

satisfies

$$||W-V||_p \leq \delta, \quad \left|\left|\frac{dW}{dx} - \frac{dV}{dx}\right|\right|_p \leq \delta$$

□

4. NUMERICAL EXAMPLE

Consider the autonomous non-linear system with equilibrium $x_0 = 0$ defined by the matrix-valued function

$$A(x) = \begin{pmatrix} -3 & 1 \\ \dfrac{2x_1}{0.3 + (x_2+0.4)(x_2-0.6)} & -2 \end{pmatrix}$$

We apply normalized local basis-functions of the form [1]

$$r_i(x) = \exp\left(-0.5(x-x_i)^T \Sigma_i^{-1} (x-x_i)\right)$$
$$\rho_i(x) = \frac{r_i(x)}{\sum_{j=1}^{N} r_j(x)}$$

and define the subset of the state-space $X^0 = [-1,1] \times [-1,1] \subset R^2$. Consider the following two parameterizations of P:

[1] Notice that this class of basis-functions is closely related to a class for which Sobolev norm approximation properties are known (Rovatti 1996).

$\mathcal{P}_1$: Only one basis-function, which leads to a quadratic Lyapunov function on X^0.

$\mathcal{P}_2$: 4 basis-functions uniformly covering the set X^0.

Next, we select a total of 441 design points regularly distributed over X^0. The LMI problem with constraints (11) does not have a feasible solution with parameterization $\mathcal{P}_1$, while it leads to a Lyapunov function with parameterization $\mathcal{P}_2$, cf. Figure 1, under the assumptions that the checking set is "sufficiently dense" for this conclusion to be made. In Figure 2 $\log_{10}(\varepsilon(x))$ is illustrated from which the uniform bound $\bar{\varepsilon} = 0.4466 \cdot 10^{-3}$ can be found for the choice $\gamma = \alpha/2$. The contributions to the expression (13) for $\varepsilon(x)$ are illustrated in Figure 3. Application of a regular checking grid thus requires a discretization interval $\Delta x_1 = \Delta x_2 = 0.632 \cdot 10^{-3}$ which lead to the order of 10^7 checking point. It is verified for this example that the computed function is indeed a Lyapunov function using this grid size. However, from Figure 2 it is also clear that the checking point density need not be so high in all regions of the state space.

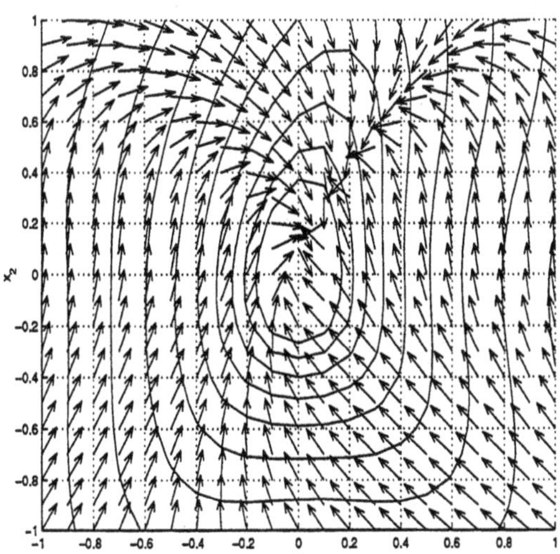

Fig. 1. The vectors (normalized to the same length) illustrates the flow of the dynamic system, while the contour lines illustrate the Lyapunov-function.

From Figure 1 it is clear that the region of attraction is $X = X^0$ (since the flow vector never points out of X^0), although this is not possible to derive from knowledge of the Lyapunov function only. Hence, some conservativeness with respect to the estimates region of attraction must be expected.

5. DISCUSSION AND CONCLUSIONS

A numerical procedure for computation of Lyapunov functions using an LMI solver has been proposed. It is argued that the procedure can be

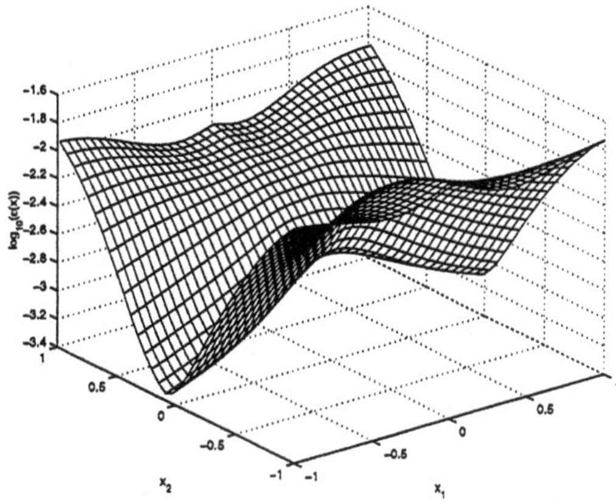

Fig. 2. Checking grid granularity $\varepsilon(x)$.

applied to a quite general class of smooth nonlinear systems.

The applicability of the procedure is somewhat limited by the fact that the number of LMIs will grow exponentially with both the dimension of the state-space and the required accuracy if one applies regular grids of design and checking points. In order to handle high-dimensional systems, the procedure relies on engineering insight in order to avoid regular grids. It is shown that the design and checking points should be concentrated in regions of the state-space where the nonlinearities are most pronounced. A lower bound on the required granularity of the checking point set is characterized in terms of regularity properties of the system function and basis functions in the Lyapunov function parameterization.

For the purpose of checking stability, conservativeness of the approach is due to finite parameterization of the set of Lyapunov function candidates, and finite discretization of the state-space. Also, any prior knowledge on the rate of change of the parameters θ is not required, but combining the present results with the results in (Gahinet *et al.* 1996, Wu *et al.* 1996) should be feasible in order to incorporate this. It would reduce the conservativeness of the method when θ are time-varying parameters known in real time.

The estimated region of attraction may be conservative. The reason for this is that we seek the Lyapunov-function that gives the least conservative estimate of the exponential convergence-rate $\gamma/2$, rather than the Lyapunov-function with the largest region of attraction.

6. REFERENCES

Barabanov, N. E. (1988). The Lyapunov indicator of discrete inclusions - III. *Automation and Remote Control* **49**, 558–565.

Barabanov, N. E. (1989). Method for the computation of the Lyapunov exponent of a differential inclusion. *Automation and Remote Control* **50**, 475–479.

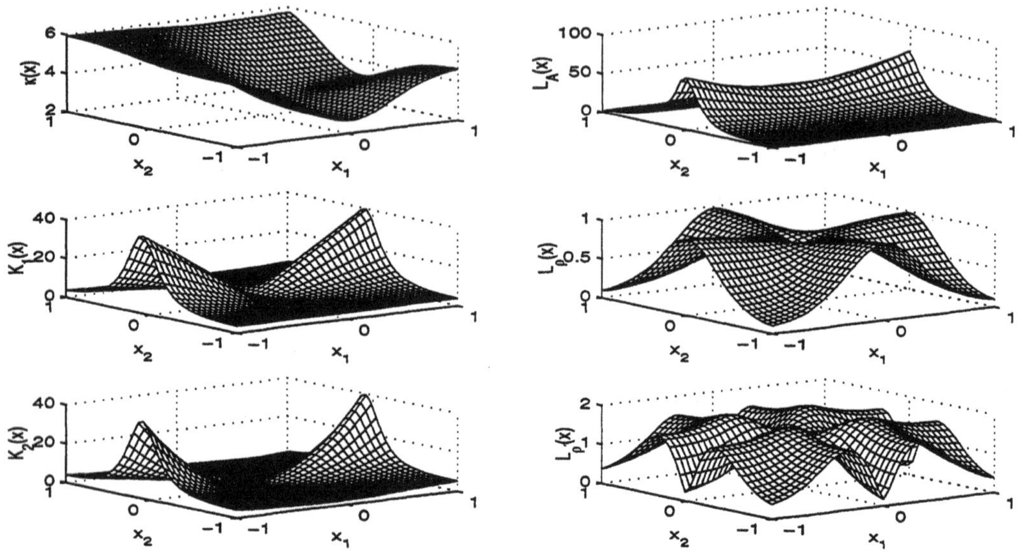

Fig. 3. Contributions to the checking grid granularity.

Becker, G. and A. Packard (1994). Robust performance of linear parametrically varying systems using parametrically-dependent linear feedback. *Systems and Control Letters* **23**, 205–215.

Borisov, V. G. and S. N. Diligenskii (1986). Numerical method of stability analysis of nonlinear systems. *Automation and Remote Control* **47**, 1373–1380.

Boyd, S., L. El Ghaoui, E. Feron and V. Balahrishnan (1994). *Linear Matrix Inequalities in System and Control Theory.* SIAM, Philadelphia.

Brayton, R. K. and C. H. Tong (1980). Constructive stability and asymptotic stability of dynamical systems. *IEEE Trans. Circuits and Systems* **27**, 1121–1130.

Gahinet, P., A. Nemirovski, A. J. Laub and M. Chilali (1995). *LMI Control Toolbox - For Use with MATLAB.* The MathWorks, Inc.

Gahinet, P., P. Apkarian and M. Chilali (1996). Affine parameter-dependent Lyapunov function and real parametric uncertainty. *IEEE Trans. Automatic Control* **41**, 436–442.

Johansson, M. and A. Rantzer (1998). Computation of piecewise quadratic Lyapunov functions for hybrid systems. *IEEE Trans. Automatic Control* **43**, 555–559.

Kamenetskii, V. A. and E. S. Pyatnitskii (1987). Gradient method of constructing Lyapunov functions in problems of absolute stability. *Automation and Remote Control* **48**, 1–9.

Khalil, H. K. (1992). *Nonlinear Systems.* Macmillan, New York.

Lu, W.-M. and J. C. Doyle (1995). H_∞ control of nonlinear systems: A convex characterization. *IEEE Trans. Automatic Control* **40**, 1668–1675.

Michel, A. N., B. H. Nam and V. Vittal (1984). Computer generated Lyapunov functions for interconnected systems: Improved results with application to power systems. *IEEE Trans. Circuits and Systems.*

Molchanov, A. P. (1987). Lyapunov functions for nonlinear discrete-time control systems. *Automation and Remote Control* **48**, 728–736.

Molchanov, A. P. and E. S. Pyatnitskii (1986). Lyapunov functions that specify necessary and sufficient conditions of absolute stability of nonlinear nonstationary control systems. parts I - III. *Automation and Remote Control* **47**, 344–354, 443–451, 620–630.

Ohta, Y., H. Imanishi, L. Gong and H. Haneda (1993). Computer generated Lyapunov functions for a class of nonlinear systems. *IEEE Trans. Circuits and Systems* **40**, 343–354.

Petterson, S. and B. Lennartson (1997). Exponential stability analysis of nonlinear systems using lmis. In: *Proc. IEEE Conf. Decision and Control, San Diego.*

Rovatti, R. (1996). Takagi-Sugeno models as approximators in Sobolev norms - the SISO case. In: *Proc. IEEE Conf. Fuzzy Systems, New Orleans.* pp. 1060–1066.

Watanabe, R., K. Uchida and M. Fujita (1996). A new LMI approach to analysis of linear systems with scheduling parameter - Reduction to finite number of LMI conditions. In: *Proc. 35th Conference on Decision and Control, Kobe.* pp. 1663–1665.

Wu, F., X. H. Yang, A. Packard and G. Becker (1996). Induced L_2-norm control for LPV systems with bounded parameter variation rates. *Int. J. Nonlinear and Robust Control* **6**, 983–998.

Zelentsovsky, A. L. (1994). Nonquadratic Lyapunov function for robust stability analysis of linear uncertain systems. *IEEE Trans. Automatic Control* **39**, 135–138.

V-DICHOTOMY OF PASSIVE NON-AFFINE NONLINEAR SYSTEMS

A.S. Shiriaev *,**

* *Department of Engineering Cybernetics, Norwegian University of Science and Technology, N-7034 Trondheim, NORWAY. email: Anton.Shiriaev@itk.ntnu.no (Corresponding address)*
** *Institute of Problems of Mechanical Engineering, RAS, St.Petersburg, Russia.*

Abstract: The method of the invariant set stabilization for non-affine nonlinear passive system is proposed. New important notion, weak V-detectability of nonlinear system, is introduced. It is shown that this notion plays the fundamental role in the problem of the invariant set stabilization of nonlinear system. Conditions implying the weak V-detectability of passive nonlinear system is obtained. *Copyright © 1999 IFAC*

Keywords: nonlinear systems, invariant set stabilizability, zero-state detectability, V-detectability

1. INTRODUCTION

The stability of a prescribed nonlinear system invariant set, which is possible not an equilibrium or a compact set, is equivalent under mild assumptions to the existence of the Lyapunov function, see (Krasovskii, 1963; Zubov, 1964; Rumyantsev and Oziraner, 1987; Lin *et al.*, 1996). Due to such an equivalence it is natural to consider in general the stabilization of a given invariant set for a nonlinear control system in terms of a control-Lyapunov function, which finally leads to the Lyapunov function of the closed-loop system with respect to the desired attractor.

This scheme originally contains a obstacle preventing from its direct implementation in many examples: the strict control-Lyapunov function corresponding to a prescribed invariant set of a given nonlinear control system is very often unknown.

From the other hand the detailed consideration of the system may result in the construction of some control-Lyapunov function candidate V such that, first, the zero-level set of V is larger than the desired attractor; second, the derivative of V along the solutions of the system cannot be made strictly negative definite using any available control values; third, V may depend only on part variables of the state vector and may be non-proper; and so on. The simplest example of such a function V comes from the knowledge of one or several conserved quantities of the investigated system, see (Fradkov, 1996; Åström and Furuta, 1996; Lozano and Fantoni, 1998). Another interesting examples are due to the partial stability theory, see (Rumyantsev and Oziraner, 1987; Vorotnikov, 1993).

This produces the natural questions: *When is it sufficient to use such a function V to define a regulator and to establish stability of the desired attractor (goal set)?*

For some cases the answer is positive and known. For example, Byrnes *et al.* (1991) proved new sufficient conditions implying the stabilizability of the zero-state equilibrium of the passive affine nonlinear system, which is based on the zero-state detectability property. Moreover, Byrnes *et al.* (1991) suggested the way to verify this property for a given passive affine nonlinear system. See the extension of this test in (Shiriaev, 1998*b*).

Another important case, the stabilization of the prescribed invariant set for passive nonlinear system, was covered by (Shiriaev and Fradkov, 1998*b*; Shiriaev and Fradkov, 1998*a*; Shiriaev, 1998*a*; Shiriaev, 1998*c*). As it is shown in (Shiriaev, 1998*c*) the stabilization of the prescribed invariant set Γ for the passive nonlinear system with a storage function V,

$$\Gamma \supset V_0 = \{x \in R^n : V(x) = 0\},$$

can be achieved due to a V-detectability of the nonlinear system. The main disadvantage of (Shiriaev, 1998*c*) was the necessity to provide the existence of a non-empty ω-limit set for any trajectory of the closed loop system with the regulators formed by the speed-gradient methodology, see (Fradkov, 1990).

This paper is addressed to the problem: the nonlinear control system is described by the equation

$$\dot{x} = f(x, u), \tag{1}$$

where $x \in X = R^n$ is a state vector, $u \in U = R^m$ is a control input, f is a smooth vector field. Given a nonnegative C^r-smooth function $V : R^n \to R^1_{\geq 0}$, suppose that

$$\frac{\partial V(x)}{\partial x} f(x, 0) \leq 0, \quad \forall x.$$

When does a regulator exist providing that along any bounded trajectory

$$\lim_{t \to T^*} V(x(t)) = 0, \tag{2}$$

where T^* is a finite time or $+\infty$. In this paper the limit relation (2) is called V-dichotomy of the nonlinear system (1). It is worth to mention that it serves the natural extension of dichotomy notion, see (Gelig *et al.*, 1978).Thus in contrast to (Shiriaev, 1998*c*) in this paper it does not assumed that all trajectories of the closed-loop system are well defined on $[0, +\infty)$ and have a non-empty ω-limit sets. Such a situation may occur if, for example, some part of the state variables tends to infinity as $t \to +\infty$.

It will be shown that essentially the existence of a stabilizing in a sense (2) regulator, formed by the speed-gradient methodology, is equivalent to the property of some auxiliary affine in control nonlinear system to be *weak V-detectable*, see definition below. Moreover, the sufficient conditions, providing this property for the passive affine nonlinear system, are established.

The paper is organized as follows. In section 2 the problem of invariant set stabilization for affine in the control nonlinear system is considered. For the sake of completeness the notions of V-dichotomy, V-detectability and weak V-detectability of nonlinear system are introduced. The sufficient conditions implying these properties are proposed. Section 3 contains the solution of the stabilization problem (1), (2) for non-affine in the control systems. The conclusions are given in section 4.

2. INVARIANT SET STABILIZATION OF PASSIVE AFFINE IN CONTROL NONLINEAR SYSTEMS

In this section the invariant set stabilization problem for affine in the control nonlinear system

$$\dot{x} = f_a(x) + g_a(x)u, \tag{3}$$
$$y = h_a(x), \tag{4}$$

is considered. Here x, u are the vectors with the same dimensions as in (1), y is a vector of a measurable outputs, $y \in R^m$; f_a, g_a, h_a are the smooth vector fields. In section 3 it will be shown that in some cases the criteria of stabilizability of invariant set of (3), (4) can be extended for general nonlinear systems of the form (1).

Definition 1. (Willems (1972)). The system (3), (4) is *passive* if there exists a C^0 nonnegative storage function $V : X \to R$, $V(0) = 0$, such that for all piece-wise continuous control vector functions $u(s)$ along the solution $x(t) = x(t, x_0, u(s))$ of (3), defined on the maximal interval $[0, t_*)$, for all $t \in [0, t_*)$ the inequality

$$V(x(t)) - V(x_0) \leq \int_0^t y(s)^T u(s) ds \tag{5}$$

holds. □

It is well known, see (Hill and Moylan, 1976), that in the case of a C^1 smooth function V the notion of passivity of the system (3), (4) can be expressed in infinitesimal form

$$L_{f_a} V(x) \leq 0, \quad \forall x \in R^n \tag{6}$$
$$L_{g_a} V(x) = h_a(x)^T, \quad \forall x \in R^n. \tag{7}$$

Definition 2. (Shiriaev (1998*c*)). Let $V(x)$ be a nonnegative continuous function. The system (3), (4) is said to be *locally V-detectable* if there exists a positive constant $c > 0$ such that for all $x_0 \in \mathcal{U} = \{x \in R^n : V(x) < c\}$

$$h_a(x(t, x_0, 0)) = 0 \text{ for all } t \geq 0 \Rightarrow \lim_{t \to +\infty} V(x(t, x_0, 0)) = 0, \tag{8}$$

where $x(t, x_0, 0)$ is a solution of (3) with $u(t) \equiv 0$. If $\mathcal{U} = R^n$, the system (3), (4) is *V-detectable.* □

Definition 3. a). Let $V(x)$ be a nonnegative continuous function. The system (3) to be *V-dichotomous* if along any bounded solution $x(t) = x(t, x_0, 0)$ of (3)

$$\lim_{t \to +\infty} V(x(t, x_0, 0)) = 0; \tag{9}$$

b). The system (3), (4) is said to be *weakly V-detectable* if along any bounded solution $x(t) = x(t, x_0, 0)$ of (3) consistent with the constraint $h_a(x(t)) = 0$ for all $t \geq 0$, the limit relation (9) is valid. □

As it was pointed out in (Shiriaev, 1998*c*) the notion of V-detectability is the natural extension of the zero-state detectability notion which was introduced in (Byrnes *et al.*, 1991). Indeed, if the function V is positive definite then these notions are equivalent.

The next Theorem reflects one of the main properties of the passive affine nonlinear control systems and, in particular, incorporates Theorem 3.2 (Byrnes *et al.*, 1991), Theorem 2.3 (Shiriaev, 1998*c*). Recall that a function V is said to be *proper* if the set $V_c = \{x \in R^n : V(x) \leq c\}$ is compact for all $c \geq 0$.

Theorem 4. Suppose the system (3), (4) is passive with a continuous storage function V. Let

$$u = -\phi(y), \tag{10}$$

where $\phi : Y \to U$ be C^1-smooth function such that $y^T\phi(y) > 0$ for each nonzero y, $\phi(0) = 0$. Then

1.) If the set $V_0 = \{x \in R^n : V(x) = 0\}$ is compact and the system (3), (4) is locally V-detectable then V_0 is a locally asymptotically stable set of the closed loop system (3), (4), (10);

2.) If the function V is proper and the system (3), (4) is V-detectable then V_0 is a globally asymptotically stable set of the closed loop system (3), (4), (10);

3). If the system (3), (4) is weakly V-detectable then the closed loop system (3), (4), (10) is V-dichotomous. □

Remark 5. The feedback control law of the form (10) is well known in the Western and Russian literature as a passification control or speed gradient method, see (Byrnes *et al.*, 1991; Fradkov, 1990) and others. So theorem 4 just provides the new conditions of its applicability. □

The condition of (weak) V-detectability involved in Theorem 4 is nonconstructive and requires additional investigation. In the next statements three different sufficient conditions, which imply (weak) V-detectability of (3), (4), are suggested. Given a C^r, $r \geq 1$, smooth function $V(x)$, introduce the set

$$S = \{x \in R^n : L_{f_a}^m L_\tau V(x) = 0, \quad \text{for all } \tau \in D,\ 0 \leq m \leq r-1\}, \tag{11}$$

where D is a distribution which is defined by

$$D = \text{span}\{ad_{f_a}^k g_i, 0 \leq k \leq n-1, 1 \leq i \leq m\} \tag{12}$$

where g_i, $1 \leq i \leq m$ are the vector components of the smooth vector field $g_a(x)$.

Proposition 6. Suppose that the system (3), (4) is passive with a C^r, $r \geq 1$, smooth storage function V. If the set

$$S \setminus \{x \in R^n : V(x) = 0\} \tag{13}$$

does not contain any whole bounded trajectory of the unforced system (3) then the system (3), (4) is weakly V-detectable. If, in addition, V is proper then (3), (4) is V-detectable. □

To formulate another criteria of (weak) V-detectability let us assume that the C^r-smooth scalar function $V(x)$, $r \geq 1$, admits the factorization

$$V(x) = \frac{1}{2}|w(x)|^2, \tag{14}$$

where $w : R^n \to R^l$, $l \leq n$, is a C^r-smooth vector function. Introduce the distribution $\hat{S}(x)$ as follows

$$\hat{S}(x) = \text{span}\{L_{f_a}^k L_{g_a} w(x),\ k = 0, \ldots, r-1\}. \tag{15}$$

It is easy to see that the dimension of the distribution $\hat{S}(x)$ cannot be greater than l for any point $x \in R^n$. Denote the set

$$\Lambda_1 = \{x \in R^n : (L_{f_a} w(x))^T L_{f_a}^k L_{g_a} w(x) = 0, \quad k = 0, 1, \ldots, r-1\},$$

and consider its subset P_1 where the distribution $\hat{S}(x)$ has the maximal dimension, i. e.

$$P_1 = \{x \in \Lambda_1 : \dim \hat{S}(x) = l\}.$$

Let the sets X_1, Ω_1 be defined as follows

$$X_1 = R^n \setminus (P_1 \cup \{x \in R^n : V(x) = 0\}),$$

$$\Omega_1 = \{x_0 \in X_1 : \text{the whole bounded trajectory } x = x(t, x_0) \text{ of (3) with } u = 0 \text{ lies in } X_1\}.$$

Proposition 7. Suppose the system (3), (4) is passive with a C^r, $r \geq 1$, smooth storage function V, which can be presented in the form (14). If the set $\Omega_1$is empty,

$$\Omega_1 = \emptyset,$$

then (3), (4) is weakly V-detectable. If, in addition, V is proper then (3), (4) is V-detectable. □

Similar to the sets involved in Proposition 7 introduce the sets

$$\Lambda_2 = \{x \in R^n : w(x)^T L_{f_a}^k L_{g_a} w(x) = 0, \\ k = 0, 1, \dots, r-1\},$$

$$P_2 = \{x \in \Lambda_2 : \dim \hat{S}(x) = l\},$$

$$X_2 = R^n \setminus (P_2 \cup \{x \in R^n : V(x) = 0\}),$$

$$\Omega_2 = \{x_0 \in X_2 : \text{the whole bounded trajectory } x = x(t, x_0) \text{ of (3) with } u = 0 \text{ lies in } X_2\}.$$

Proposition 8. Suppose the system (3), (4) is passive with a C^r, $r \geq 1$, smooth storage function V which can be presented in the form (14). If the set $\Omega_2$is empty,

$$\Omega_2 = \emptyset,$$

then (3), (4) is weakly V-detectable. If, in addition, V is proper then (3), (4) is V-detectable. □

Remark 9. It is worth to point out that by definition the sets S, Λ_1, Λ_2 satisfy the inclusions

$$S \subset \Lambda_1, \quad S \subset \Lambda_2$$

This is due to the fact that Λ_1, Λ_2 are only defined by some part of the equations involved in the definition of the set S. □

Remark 10. The factorization of the function $V(x)$, see (14), can be performed in several different ways. In particularly, even the dimension of the vector space R^l is not originally defined. □

Remark 11. It is interesting also to mention that under some assumptions Proposition 6 may be equivalent to Proposition 7 and Proposition 8. The advantage of such an equivalence is a computational aspect: the numbers of the equations involved in the definitions of the sets S and Λ_1, Λ_2 are strictly different. One of such sufficient conditions ensuring the mentioned equivalence are

$$1).\ V(x) \in C^\infty (R^n, R^1),$$
$$2).\ L_{g_a(x)} V(x) = 0 \Rightarrow L_{f_a(x)} w(x) = 0,$$

where the function $w(x)$ comes from the factorization (14). See for details (Shiriaev, 1998*a*). □

3. V-DICHOTOMY OF WEAKLY PASSIVE NON-AFFINE NONLINEAR SYSTEMS

The notion of passivity for non-affine in the control nonlinear system

$$\dot{x} = f(x, u), \tag{16}$$
$$y = h(x, u). \tag{17}$$

with a C^0-smooth storage function V is the same as in the case of affine in the control system, see definition 1. But the infinitesimal variant of passivity with a C^1-smooth storage function differs from the relations (6), (7).

The next Lemma contains an almost obvious necessary condition for passivity of (16), (17). Given a C^1-smooth function $V(x)$, denote

$$f_a(x) = f(x, 0), \quad g_a(x) = \frac{\partial f(x, 0)}{\partial u},$$
$$\mathcal{S} = \{x \in R^n : L_{f_a} V(x) = 0\}.$$

Lemma 12. (Shiriaev and Fradkov (1998*b*)). Let the system (16), (17) be passive with a C^1-smooth storage function V. Then the following relations

$$L_{f_a} V(x) \leq 0, \quad \forall x \in R^n, \tag{18}$$
$$L_{g_a} V(x) = h(x, 0)^T, \quad \forall x \in \mathcal{S} \tag{19}$$

hold. □

By the assumption the full state vector x of the system (16) is measurable. Thus one can try to define the output (17) such that the system (16), (17) becomes passive with a given storage function V. But again in contrast to the affine nonlinear systems for non-affine ones the defining of such a output maybe nontrivial problem.

Fortunately, due to the smoothness of vector field $f(x, u)$ in (16) it is possible to provide a stabilizability conditions (in a sense (1), (2)) for the non-affine nonlinear control system (16), (17), which satisfy only relaxed passivity assumption. By this reason introduce

Definition 13. The system (16), (17) is said to be *weakly passive* with nonnegative C^1-smooth storage function V if the the relations (18), (19) hold.

Remark 14. One can easily introduce the analogs for the notions of (local) V-detectability, weak V-detectability, V-dichotomy for non-affine nonlinear control system (16), (17), which in general repeat Definitions 2–3. Although these notions are of interest, they, except the V-dichotomy, will not be used in the further analysis, and are omitted. □

Assume that the right side of (16) has at least third continuous partial derivative with respect to u, i. e. the system (16) admits the representation in the form

$$\dot{x} = f_a(x) + g_a(x)u + \sum_{j=1}^{m} u_j(R_j(x,u)u), \quad (20)$$

where u_j are the components of the vector u; the functions R_j, $j = 1, \ldots, m$, are smooth in x, C^1-smooth in u and, in particular, for any x the functions R_j, $j = 1, \ldots, m$, are bounded in u if $|u| \leq \varepsilon$ with some $\varepsilon > 0$. Without loss of generality it will be assumed that $\varepsilon = 1$.

Due to the last assumption there exists a scalar smooth function $\rho(x)$ such that $\forall\, x$ and for any $j = 1, \ldots, m$ the following estimation

$$\max_{|u|\leq 1} |R_j(x,u)| \leq \rho(x) \quad (21)$$

holds. Obviously, such a function $\rho(x)$ is not uniquely defined.

The analysis of V-dichotomy of (16), (17) essentially relies on the properties of the affine part of (20) so we will also consider the auxiliary affine in control nonlinear system

$$\dot{x} = f_a(x) + g_a(x)u, \quad (22)$$

$$y = h_a(x) = h(x,0). \quad (23)$$

The next Theorem contains the main result of the paper.

Theorem 15. Suppose the system (16), (17) is weakly passive with a C^1-smooth storage function V and (16) can be rewritten in the form (20). Let $\Phi : Y \to U$ be a C^1-smooth function such that $\Phi(0) = 0$ and

$$y^T\Phi(y) \geq \psi(y)|\Phi(y)|^2 > 0, \quad \forall y \neq 0, \quad ||\Phi(y)|| \leq 1, \quad (24)$$

where $\psi(y)$ is scalar smooth function, $\psi(y) \geq \psi_0 > 0$, $\forall y$. Suppose $\alpha(x)$ is a scalar C^1-smooth function such that $0 < \alpha(x) \leq \hat{\alpha}(x)$, where

$$\hat{\alpha}(x) = \frac{2 \cdot \beta}{m\left(1 + [||\partial V(x)/\partial x|| \rho(x)]^2\right)}, \quad (25)$$

and β is a constant, $\min\{\psi_0, m\} > \beta > 0$. Take the regulator

$$u = -\alpha(x)\Phi([L_{g_a}V(x)]^T). \quad (26)$$

Then

1.) If the set $V_0 = \{x \in R^n : V(x) = 0\}$ is compact and the system (22), (23) is locally V-detectable then V_0 is a locally asymptotically stable set of the closed loop system (16), (17), (26);

2.) If the function V is proper and the system (22), (23) is V-detectable then V_0 is a globally asymptotically stable set of the closed loop system (16), (17), (26).

3). If the system (22), (23) is weakly V-detectable then the closed loop system (16), (17), (26) is V-dichotomous. □

Remark 16. One of the possible choice of the functions $\Phi(y) = [\Phi_1(y), \ldots, \Phi_m(y)]^T$, $\psi(y)$ and the constant ψ_0 involved in (24) is the following

$$\Phi_j(y) = \frac{1}{\pi}\arctan(y_j), \quad j = 1, \ldots, m,$$

$$\psi(y) = \psi_0 = 1,$$

where y_j, $j = 1, \ldots, m$, are the components of the vector y.

4. CONCLUSIONS

In this paper the global (local) problem of the invariant set stabilization is considered. The main idea of our approach is to extend the passification scheme for an equilibrium point stabilization introduced by Byrnes *et al.* (1991), to the problem of invariant set stabilization. The basis of such an extension consists of the natural generalization of the zero-state detectability notion to the weak V-detectability of the nonlinear system. Moreover, the sufficient conditions of the last property are derived.

It is shown that if the non-affine nonlinear system is *weakly* passive with some storage function V; if the desired attractor (goal invariant set) Γ contains the set $V_0 = \{x \in R^n : V(x) = 0\}$ and if some auxiliary affine nonlinear system is weak V-detectable, then any bounded solution of the closed-loop system with the feedback control formed by the modified speed-gradient algorithm, will tend to the desired attractor Γ.

This will be due to the fact that along any bounded solution of the closed-loop system the function the value of $V(x)$ will tend to zero. This property of the system is introduced as V-dichotomy of the nonlinear control system. It serves the natural extension for the dichotomy notion, see (Gelig *et al.*, 1978).

The suggested ideas explicitly provide several interesting results, for example, for the partial stabilization theory and for the stabilization of the non-compact invariant manifolds for the nonlinear systems with several known conserved quantities. The standard examples of such a nonlinear systems is a system with a conserved energy (the pendulum, the double pendulum, the spherical pendulum). For these examples the proposed results provide the stablization of the prescribed energy level (Shiriaev *et al.*, 1998; Shiriaev *et al.*, 1999).

ACKNOWLEDGEMENTS

The author would like to thank Prof. A.L. Fradkov and Prof. O. Egeland for helpful discussions.

This work was supported by research grant of the Norwegian research council, by RFBR (grant 96-0101151) and by the Russian Federation Programme "Integration" (project 2.1-589).

5. REFERENCES

Byrnes, C.I., A. Isidori and J.C. Willems (1991). Passivity, feedback equivalence, and the global stabilization of minimum phase nonlinear systems. *IEEE Transactions on Automatic Control* **36**, 1228–1240.

Fradkov, A.L. (1990). *Adaptive control of complex systems.* Moscow: Nauka. (in Russian).

Fradkov, A.L. (1996). Swinging control of nonlinear oscillations. *International Journal of Control* **64**(6), 1189–1202.

Gelig, A.Kh., G.A. Leonov and V.A. Yakubovich (1978). *Stability of nonlinear systems with nonunique equilibrium states.* Nauka. Moscow. (in Russian).

Hill, D. and P. Moylan (1976). The stability of nonlinear dissipative systems. *IEEE Transactions on Automatic Control* **21**, 708–711.

Krasovskii, N.N. (1963). *Stability of Motion.* Stanford University Press. Stanford.

Lin, Y., E.D. Sontag and Y. Wang (1996). A smooth converse Lyapunov theorem for robust stability. *SIAM Journal of Control & Optimization* **34**, 124–160.

Lozano, R. and I. Fantoni (1998). Passivity based control of the inverted pendulum. In: *Proceedings of the 4th IFAC Symposium on Nonlinear Control Systems Design.* Vol. 1. IFAC. Enschede, The Netherlands. pp. 145–150.

Åström, K.J. and K. Furuta (1996). Swinging up a pendulum by energy control. In: *Proceedings of the 13th IFAC World Congress.* Vol. E. San Francisco. pp. 37–42.

Rumyantsev, V.V. and A.S. Oziraner (1987). *Partial Stability and Stabilization.* IL. Moscow.

Shiriaev, A.S. (1998*a*). Control of oscillations in affine nonlinear systems. In: *Proceedings of the IFAC Conference on System Structure and Control.* Vol. 3. Nantes, France. pp. 789–794.

Shiriaev, A.S. (1998*b*). New tests of zero-state detectability. In: *Proceedings of the 4th IFAC Symposium on Nonlinear Control Systems Design.* Vol. 2. Enschede, The Netherlands. pp. 533–538.

Shiriaev, A.S. (1998*c*). The notion of *V*-detectability and stabilization of invariant sets of nonlinear systems. In: *Proceedings of the 37th CDC.* Tampa. pp. 2509–2514.

Shiriaev, A.S. and A.L. Fradkov (1998*a*). Stabilization of invariant manifolds for nonaffine nonlinear systems. In: *Proceedings of the 4th IFAC Symposium on Nonlinear Control Systems Design.* Vol. 1. Enschede, The Netherlands. pp. 215–220.

Shiriaev, A.S. and A.L. Fradkov (1998*b*). Stabilization of invariant set of nonlinear systems. In: *Nonlinear dynamical systems.* St. Petersburg State University. St. Petersburg, Russia. (in Russian).

Shiriaev, A.S., H. Ludvigsen and O. Egeland (1999). Swinging up of the spherical pendulum. In: *Proceedings of the 14th IFAC World Congress.*

Shiriaev, A.S., O. Egeland and H. Ludvigsen (1998). Global stabilization of unstable equilibrium point of pendulum. In: *Proceedings of 37th CDC.* IEEE. Tampa. pp. 4584–4585.

Vorotnikov, V.I. (1993). Stability and stabilization of motion: research approaches, results, distinctive characteristics. *Automation and Remote Control* (3), 3–62.

Willems, J.C. (1972). Dissipative dynamic systems. Part 1. General Theory. *Arch. Rat. Mech. Anal.* **45**(5), 321–351.

Zubov, V.I. (1964). *Methods of A.M. Lyapunov and their applications.* P. Noordhoff. Groningen.

E-2c-17-3
14th Triennial World Congress, Beijing, P.R. China

NECESSARY AND SUFFICIENT CONDITIONS FOR PASSIVITY OF THE LUGRE FRICTION MODEL [1]

Nikita Barabanov

Electrotechnical University 197376, St. Petersburg
nikita@freya.etu.ru *RUSSIA*

Romeo Ortega [2]

Lab. des Signaux et Systèmes CNRS-SUPELEC Gif-sur-Yvette
91192 rortega@lss.supelec.fr *FRANCE*

Abstract: Friction is a nonlinear phenomenon difficult to describe analytically. To capture its effect in mechanical systems a bristle–based dynamical model, known as the LuGre model, was recently proposed in the literature. It is difficult to assess whether this (or any other) mathematical model constitutes a *bona fide* friction model. It should however reflect the dissipative nature of friction, which mathematically translates into the requirement of defining a passive operator from velocity to friction force. In this paper we give *necessary and sufficient* conditions for this property to hold for the LuGre model. The conditions are expressed in terms of a simple algebraic inequality involving the parameters of the model. If this inequality does not hold we construct an input signal that generates a periodic orbit along which the passivity inequality is violated. *Copyright © 1999 IFAC*

1. INTRODUCTION AND PROBLEM FORMULATION

It is well known that one of the major limitations to achieve good performance in mechanical systems is the presence of friction, which is a nonlinear phenomenon difficult to describe analytically. Different models have been proposed to capture this phenomenon, starting with the classical descriptions via static maps between velocity and friction force. However, it appears that classical models are unable to capture some of the behaviour experimentally observed in systems subject to strong friction effects. It is argued in [Armstrong et al, 1994] that *dynamic* models are necessary to describe the friction phenomena accurately. For a review of some of the existing dynamic friction models see [Olsson et al, 1998; Armstrong et al, 1994] and references cited therein.

Inspired by the works of [Haessig,Friedland, 1991] and [Bliman,Sorine, 1995], a new dynamical model of the friction force —based on a bristle deflection interpretation— is proposed in [Canudas et al, 1995] as follows [3]

$$y = \sigma_0 x + \sigma_1 \dot{x} + \alpha_2 u \qquad (1)$$

$$\dot{x} = -\sigma_0 a(u) x + u \qquad (2)$$

[1] This work was supported in part by the INTAS cooperation agreement 94-965.
[2] Author to whom all correspondence should be addressed.

[3] The notation used here differs from the one adopted in [Canudas et al, 1995] as follows: $F \to y$, $v \to u$, $z \to x$. Our choice is motivated by the standard notation of a dynamical system with input u, output y and state x.

$$a(u) \triangleq \frac{|u|}{\alpha_0 + \alpha_1 \, e^{-(\frac{u}{v_0})^2}} \tag{3}$$

where y is the friction force, x is the average deflection of the bristles, u is the relative velocity between the surfaces, α_i, $i = 0,1,2$, σ_i, $i = 0,1$ and v_0 are some positive coefficients.

It is difficult to assess whether the equations above (or for that matter any other mathematical model) constitute a *bona fide* friction model. Particularly because of the distressing fact that it was not known whether the model above satisfies the fundamental property of defining a passive operator $u \mapsto y$. This property captures the dissipation nature of friction and should be reflected in any sensible model of it.[4] This problem has been extensively studied in [Olsson, 1996] and [Gafvert, 1996]. In [Olsson, 1996] it is shown that passivity is ensured for a *modified* model where the coefficient σ_1 is made velocity–dependent. See also [Olsson et al, 1998]. A numerical example for which the passivity condition is violated is presented in [Gafvert, 1996]. In the latter reference the following analysis is carried out. To simulate the scenario of a simple (unitary) mass with friction, the friction model above is placed in closed–loop with an integrator, e. g. $\dot{u} = -y$. For the resulting second order dynamical system it is then shown that there exists a positive definite function whose derivative is strictly positive in some non–empty subset of the state space provided some condition on the friction parameters is satisfied. Unfortunately this fact does not allows us to make any claim about passivity of the operator, –nor of stability of the system for that matter–, whose rigourous analytical proof remained an open issue. It is interesting, however, to point out that the condition of [Gafvert, 1996] exactly coincides with the necessary and sufficient condition for passivity that we derive below.

In this paper we provide a complete answer to this question. Speciffically, we give *necessary and sufficient* conditions for the passivity property to hold. The conditions are expressed in terms of a simple algebraic inequality involving the parameters of the model. If this inequality does not hold we construct an input signal that generates a periodic orbit along which the passivity inequality is violated.

2. MAIN RESULT

We shall solve the problem for the more general functions $a(u)$. Namely, we assume only that the function $a(u)$ in (2) is even and it satisfies following conditions:

$$\gamma_1|u| \leq a(u) \leq \gamma_2|u| \quad \text{for all } u, \tag{4}$$

and

$$\gamma_1 = \inf \frac{a(u)}{|u|} \in (0,\infty); \qquad \gamma_2 = \varlimsup_{u\to\infty} \frac{a(u)}{|u|} \in (0,\infty). \tag{5}$$

In the case (3) we have

$$\gamma_1 = \frac{1}{\alpha_0 + \alpha_1}, \quad \gamma_2 = \frac{1}{\alpha_0}.$$

Proposition The dynamical system (1),(2),(4) defines a passive operator $\Sigma : \mathcal{L}_{2e} \to \mathcal{L}_{2e} : u \mapsto y$, that is, along the solutions of (1),(2),(4) with zero initial conditions we have

$$I(0,T) = \int_0^T uy dt \geq 0 \tag{6}$$

for all $u \in \mathcal{L}_{2e}$ and all $T \geq 0$, *if and only if*

$$\gamma_2 \leq \gamma_1 (1 + \frac{\alpha_2}{\sigma_1}). \tag{7}$$

Proof
(Sufficiency)
It is clear from (3), (4) that for all solutions $x(\cdot)$ with zero initial data and all $t \geq 0$

$$|x(t)| \leq \frac{1}{\sigma_0 \gamma_1} \tag{8}$$

Now, we will evaluate (6) splitting it into two terms $I = I_1 + I_2$ with

$$I_1(0,T) \triangleq \int_0^T u(\sigma_1 \dot{x} + \alpha_2 u) dt$$

and

$$I_2(0,T) \triangleq \sigma_0 \int_0^T x u dt$$

Replacing u from (2) in I_2 we get

$$\begin{aligned} I_2(0,T) &= \sigma_0 \int_0^T x[\dot{x} + \sigma_0 a(u) x] dt \\ &= \frac{\sigma_0}{2} x^2(T) + \sigma_0^2 \int_0^T x^2 a(u) dt \\ &\geq 0 \end{aligned}$$

For the other term we replace $\dot{x}$ from (2) and use the bounds (8)

[4] It is worth pointing out that the passivity established in [Canudas et al, 1995] pertains to the map $u \mapsto x$, which is not the one of physical interest.

$$I_1(0,T) = \int_0^T u[-\sigma_1\sigma_0 a(u)x + (\sigma_1+\alpha_2)u]dt$$
$$\geq \int_0^t u^2(-\sigma_1\sigma_0 \cdot \gamma_2 \cdot \frac{1}{\gamma_1\sigma_0} + \sigma_1 + \alpha_2)dt$$
$$= \int_0^T u^2 \frac{\alpha_2\gamma_1 - \sigma_1(\gamma_2-\gamma_1)}{\gamma_1}dt$$
$$\geq 0$$

where we have used (7) to establish the last inequality. This completes the proof of sufficiency.

(Before proceeding with the necessity proof we bring to the readers attention the fact that, if (7) is satisfied with strict inequality, then the operator is actually *input strictly passive.* This follows immediately from the second before last inequality above).

(Necessity)
To prove necessity we will construct a periodic solution along which (6) is violated.

It is easy to see that for the system (2) any value $x \in (1/\gamma_2\sigma_0, 1/\gamma_1\sigma_0)$ is reachable from zero in a finite time (use the control $u \equiv u_*$ such that $\sigma_0 a(u_*)x < u_*$).

Hence, it is sufficient to show that there exists a number $\bar{T}$ and a $\bar{T}$-periodic functions u, x satisfying (2) such that

$$\int_0^{\bar{T}} uydt < 0.$$

We will create the function u such that

$$u \equiv const = u_0 \;\; if \;\; t \in [0, T_0)$$
$$u \equiv const = u_1 \;\; if \;\; t \in [T_0, T_0+T_1)$$
$$u(t+T_0+T_1) = u(t) \;\; \text{for all } t,$$

with $\bar{T} = T_0 + T_1$ and the system (2) has a periodic solution. The values T_0 and T_1 will be chosen below.

To this end consider a number

$$x_0 \in (\frac{1}{\sigma_0\gamma_2}(1+\frac{\alpha_2}{\sigma_1}), \frac{1}{\sigma_0\gamma_1}). \qquad (9)$$

Such a number exists due to the inequality (7). Now we take a number u_0 such that

$$\frac{\sigma_0 a(u_0)}{u_0}x_0 > 1 + \frac{\alpha_2}{\sigma_1}. \qquad (10)$$

Note that there exists arbitrary big such values u_0 due to conditions (5) and inequality (9). We choose

$$T_0 = \frac{1}{\sigma_0 a(u_0)} ln[\frac{\sigma_0\sigma_1 a(u_0)}{\alpha_2 u_0}(x_0 - \frac{u_0}{\sigma_0 a(u_0)})].$$

The term in square brackets is greater than 1 due to the inequality (10). So, $T_0 > 0$. It is easy to check that the solution x of the system (2) with $u \equiv u_0$ and initial condition $x(0) = x_0$ at point T_0 is equal to

$$x(T_0) = \frac{u_0}{\sigma_0 a(u_0)}(1+\frac{\alpha_2}{\sigma_1}).$$

Define $x_1 = x(T_0)$. Routine calculations show that

$$I(0,T_0) =$$
$$= \frac{u_0^2\alpha_2}{\sigma_0 a(u_0)}[ln(1+\frac{\sigma_0\sigma_1 a(u_0)}{\alpha_2 u_0}(x_0-x_1)) -$$
$$\frac{\sigma_0\sigma_1 a(u_0)}{\alpha_2 u_0}(x_0-x_1)] +$$
$$+ \frac{u_0^2}{\sigma_0 a(u_0)^2} ln(1+\frac{\sigma_0\sigma_1 a(u_0)}{\alpha_2 u_0}(x_0-x_1)) +$$
$$+ \frac{u_0}{a(u_0)}(x_0-x_1).$$

Note that

$$I(0,T_0)/u_0 \to const < 0 \;\; \text{as} \;\; u_0 \to \infty.$$

and

$$x_1 \to (1+\alpha_2/\sigma_1)/\sigma_0\gamma_2) \;\; \text{as} \;\; u_0 \to \infty.$$

It is important that the integral $I(0,T_0)$ tends to ∞ as $u_0 \to \infty$ while the value x_1 remains bounded.

Now we choose the value u_1 such that

$$x_0 < \frac{u_1}{\sigma_0 a(u_1)}.$$

Such value u_1 exists due to the inequality (9). Consider t he solution x of the equation (2) with $u \equiv u_1$ and initial conditions $x(T_0) = x_1$. We choose the positive number T_1 such that $x(T_0 + T_1) = x_0$. Simple calculations show that

$$T_1 = \frac{1}{\sigma_0 a(u_1)} ln[1 + \frac{x_0 - x_1}{\frac{u_1}{\sigma_0 a(u_1)} - x_0}].$$

The functional I increases on the interval $[T_0, T_0 + T_1)$ by the value

$$\int_{T_0}^{T_0+T_1} uydt =$$

$$= \frac{u_1^2}{\alpha_0 a(u_1)}(\alpha_2 + \frac{1}{a(u_1)} ln[1+\frac{x_0-x_1}{\frac{u_1}{\sigma_0 a(u_1)} - x_0}] +$$

$$+(\sigma_1 - \frac{1}{a(u_1)})(x_0 - x_1).$$

This value remains bounded if $u_0 \to \infty$ (i.e. $x_1 \to (1 + \alpha_2/\sigma_1)/\sigma_0\gamma_2$).

Thus, for sufficiently big u_0 and the control u pointed out above

$$I = \int_0^{T_0+T_1} uydt < 0.$$

while $x(0) = x_0 = x(T_0 + T_1)$. Therefore we construct the $T_0 + T_1$-periodic solution for which

$$\int_0^{T_0+T_1} uydt < 0.$$

Hence, the system (1), (2), with the function $a(u)$ satisfying conditions (4),(5) is not passive. □

Remark
A preliminary version of this paper was reported in [Barabanov,Ortega, 1998]. The sufficiency part of this proof appears also in [Ortega et al, 1998], where the work cited above is quoted.

Remark
If we impose some additional restrictions on the function u such as it's uniform boundedness or boundedness of it's derivative than the inequality (7) becomes not necessary (but certainly it reamains sufficient) for the dissipativity of the system.

3. CONCLUSIONS

We have provided in this paper a conclusive answer to the question of passivity of the generalyzed LuGre friction model (1),(2),(4),(5). Namely we have shown that the model defines a passive operator *if and only if* the coefficients satisfy the inequality (7). We have furthermore shown that, if the inequality is strict, the operator is actually input strictly passive.

It is interesting to try to check the passivity condition (7) in a real application. To this end, let us consider the experimental setup (consisting of a DC motor connected to a gearbox with significant friction) installed in the Laboratoire d'Automatique de Grenoble, see [Olsson et al, 1998] for a detailed description. The identified parameters of the friction model (1)–(3) are

$$[\sigma_0,\ \sigma_1,\ \alpha_2,\ \alpha_0,\ \alpha_1,\ v_0]^\top =$$
$$= [260, 0.6, 0.0176, 0.28, 0.05, 0.01]^\top.$$

Unfortunately, the key passivity inequality (7) is not satisfied for these data. Since the diference is of an order of magnitude (i.e. $0.0176 < 0.107$) the violation of the inequality can hardly be attributed to (unavoidable but small) identification errors. In spite of this discrepancy the model seems to provide a good fit for the experimental data, see, e.g., [Olsson et al, 1998].

In conclusion we hope the analysis presented here provides some guidelines for the refinement of the basic LuGre model. To some extent, this has already been done in [Olsson, 1996] and [Olsson et al, 1998] where a passive *modified* model with a velocity–dependent coefficient σ_1 is proposed.

4. REFERENCES

[1] Armstrong, B., P. Dupont and C. Canudas, A survey of analysis tools and compensation methods for the control of machines with friction, *Automatica*, Vol. 30, No. 7, July 1994, pp. 1083-1138.

[2] Barabanov, N. and R. Ortega, Necessary and Sufficient Conditions for Passivity of the LuGre Friction Model, *Supelec Techn. Report*, France, March 1998.

[3] Bliman, P. and M. Sorine, Easy–to–use realistic dry friction models for automatic control, *Proc. European Cont. Conf.*, Rome, Italy, pp. 3788-3794, 1995.

[4] Haessig, D. and B. Friedland, On the modelling and simulation of friction, *J Dyn Syst Meas. Control Trans. ASME*, 113(3): 354–362, Sept. 1991.

[5] Canudas, C., H. Olsson, K. Astrom and P. Lischinsky, A new model for control of systems with friction, *IEEE Trans. Automat. Contr.*, Vol. 40, No. 3, March 1995.

[6] Gafvert, Comparison of two friction models, *Master Thesis*, Lund Inst. of Techn., ISRN LUTFD2/ TFRT–5561–SE, 1996.

[7] Olsson, H., Control systems with friction, *Ph. D. Thesis*, Lund Inst. of Techn., ISRN LUTFD2/ TFRT–1045–SE, 1996.

[8] Olsson, H., K. J. Astrom, C. Canudas, M. Gafvert and P. Lischinky, Friction models and friction compensation, *European J. of Control*, (to appear), 1998.

[9] R. Ortega, A. Loria, P. J. Nicklasson and H. Sira–Ramirez, *Passivity–based control of Euler–Lagrange systems*, Springer-Verlag, Berlin, Communications and Control Engineering, Sept. 1998.

E-2c-17-4
14th Triennial World Congress, Beijing, P.R. China

FURTHER RESULTS ON A COMPARISON BETWEEN VARIOUS NONLINEAR CONTROLLER DESIGNS

M. Torres * R. Ortega *

* *Laboratoire des signaux et systèmes, C.N.R.S. – Supélec, 91192 – Gif-sur-Yvette CEDEX, FRANCE*
{torres,ortega}@lss.supelec.fr

Abstract: The objective of this paper is to compare in a simple example three controller design techniques for stabilization of nonlinear systems which have emerged in the last few years: feedback linearization, integrator backstepping and passivity-based control. Application of these, seemingly unrelated, methodologies will typically lead to the definition of different control schemes. For some specific examples these differences blur and some interesting connections and similarities between the controller design techniques emerge. The careful study of such cases will improve our understanding of their common ground fostering cross–fertilization. This paper continue the work done in (Torres and Ortega, 1998) using a more complete model.

Keywords: Nonlinear systems, Magnetic bearings.

1. INTRODUCTION

Among the various controller design techniques for stabilization of nonlinear systems that have emerged in the last few years we can distinguish three which are probably the most widely applicable and systematic: feedback linearization control (FLC), integrator backstepping control (IBC) and passivity–based control (PBC). The three techniques have evolved from different considerations and exploit different properties of the system.

It is clear then that the three techniques will typically lead to the definition of different control schemes. For some specific examples these differences blur and some interesting connections and similarities between the controller design techniques emerge. It is our belief that the detailed study of such cases will improve our understanding of their common ground fostering cross–fertilization. In this paper we compare these three controller design techniques in an example which is simple enough to reveal the connections between the controllers, but yet requires the application of nonlinear control to achieve a satisfactory performance.

Magnetic bearing systems have been studied from the perspective of flatness and feedback linearization in (Levine *et al.*, 1996) and with a backstepping approach in (Queiroz and Dawson, 1996). An output feedback adaptive IBC for the levitated ball studied here, albeit with a simpler model, has been reported in (Krstic *et al.*, 1995). In (Ortega *et al.*, 1998) several PBCs are developed for this system. Here, we adapt for our simplified example the schemes of (Levine *et al.*, 1996) for FLC, of (Queiroz and Dawson, 1996) for IBC, and of (Ortega *et al.*, 1998) for PBC.

In this paper the flux linkage model used in (Torres and Ortega, 1998) is completed by adding a leakage inductance term. The remaining of the paper is organized as follows. In section 2 we present the model, the control problem and the controller structure that we adopt for the three techniques. In sections 3, 4 and 5 we derive the FLC, PBC and IBC, respectively. Their comparison is presented in section 6. Simulations are given in section 7, and we wrap up the paper with concluding remarks and open problems in section 8.

2. MODEL, CONTROL PROBLEM AND CONTROLLER STRUCTURE

Consider the system consisting of an iron ball in a vertical magnetic field created by a single electromagnet. Denote by y the vertical position of the ball measured with respect to the nominal position with the y-axis oriented upwards. If we assume linearity of the magnetic circuit the flux linkage is given by $\lambda(y,i) = L(y)i$, with $L(y)$ the inductance. A good approximation for the latter, in the domain $-\infty < y \le c_2 - \epsilon$ where $c_2 > 0$ is the nominal air gap, is given by

$$L(y) = \frac{c_1}{c_2 - y} + L_0 = \frac{c}{c_2 - y}$$

where ϵ is some small positive number, c_1 is some positive constant that depends on the number of coil turns, air permeability and the cross–sectional area of the electromagnet, L_0 is the leakage inductance and $c \equiv c_1 + L_0(c_2 - y)$. Then the behaviour of the system can be decribed by (see (Torres and Ortega, 1998) for a detailed derivation)

$$\Sigma : \begin{cases} \dot{\lambda} = -R\dfrac{c_2 - y}{c}\lambda + u \\ F = \dfrac{1}{2}\dfrac{c_1 \lambda^2}{c^2} \\ m\ddot{y} = F - mg \end{cases} \tag{1}$$

where λ is the flux and u is the voltage input. The *control objective* is to track a bounded reference signal $y_*(t)$ (with known and bounded first, second and third order derivatives) with all signals bounded. We will assume that the *full state* is available for measurement.

Notice that the system Σ can be decomposed as a feedback interconnection of an electrical (Σ_e) and a mechanical (Σ_m) subsystem as depicted in Fig. 1. Since the latter is just a double integrator it is reasonable to concentrate our attention on the problem of controlling force F, in the understanding that if force is suitably regulated then position y will be easy to control with classical linear techniques, e.g., PID control. An additional motivation for this approach stems from the fact that typically there is a natural time–scale decomposition between the electrical and mechanical dynamics. For these reasons the so–called nested–loop control is the prevailing structure in practical applications of electromechanical systems which will be adopted in the PBC and IBC designed below.

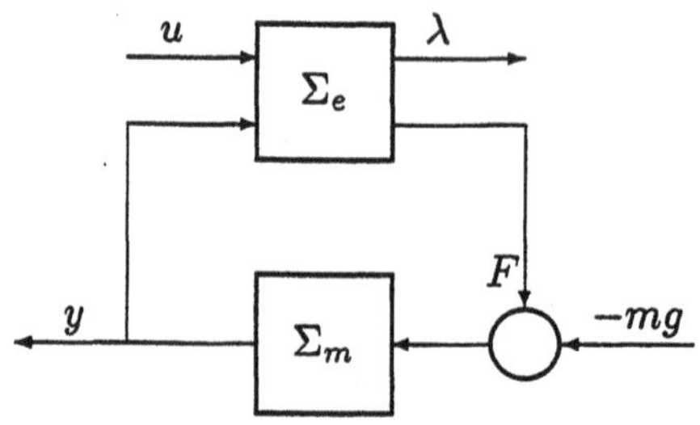

Fig. 1. Feedback decomposition of the system

The nested–loop control configuration is shown in Fig. 2, where C_{il}, C_{ol} are the inner and outer–loop controllers, respectively. The inner–loop controller is designed to achieve asymptotic tracking of the desired force F_d. The outer–loop controller takes care of the position tracking, and essentially generates a signal F_d such that *if* $F \to F_d$ then position tracking, i.e., $y \to y_*$, is ensured.

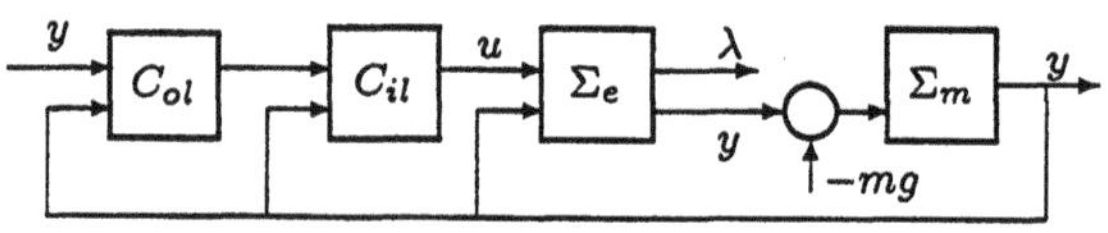

Fig. 2. Nested–loop control configuration.

3. FEEDBACK–LINEARIZATION CONTROL

We start our study with FLC. To this end, we find convenient to introduce a change of coordinates and we write the system dynamics in terms of force, position and velocity as

$$\begin{aligned} \dot{F} &= \frac{2F}{c}[L_0\dot{y} - R(c_2 - y)] + \frac{\sqrt{2c_1F}}{c}u \\ m\ddot{y} &= F - mg \end{aligned} \tag{2}$$

Differentiating once more $\ddot{y}$ we get

$$my^{(3)} = \frac{2F}{c}[L_0\dot{y} - R(c_2 - y)] + \frac{\sqrt{2c_1F}}{c}u$$

It is clear from the equations above that, taking as output y and provided $F > 0$, the system has relative degree 3 and no zero dynamics. Thus, an FLC can be immediately defined as $u = u_{\rm FL}$ with

$$u_{\rm FL} = \frac{cm\,v_{\rm FL}}{\sqrt{2c_1F}} + \sqrt{\frac{2F}{c_1}}[R(c_2 - y) - L_0\dot{y}] \tag{3}$$

which yields

$$y^{(3)} = v_{\rm FL}(y, \dot{y}, F)$$

A suitable choice for the outer loop signal is then

$$v_{\rm FL}(y, \dot{y}, F) = y_*^{(3)} - k_2\ddot{\tilde{y}} - k_1\dot{\tilde{y}} - k_0\tilde{y} \tag{4}$$

where the tracking error is defined as $\tilde{y} \equiv y - y_*$ and the aceleration must be obtained from system dynamics (1). The FLC (3), (4) is a static state feedback law which ensures in closed-loop the third order linear dynamics

$$\tilde{y}^{(3)} - k_2\ddot{\tilde{y}} - k_1\dot{\tilde{y}} - k_0\tilde{y} = 0 \qquad (5)$$

where the coefficients $k_i > 0,\ i = 1,2,3$ are chosen to ensure that the caracteristic polynomial of (5) is Hurwitz.

Remark 1 The controller (3) has a singularity in $F = 0$. As we will see later PBC and IBC are also only locally defined. This problem is related with the poor controllability properties of levitation systems when just one electromagnet is used.

4. PASSIVITY-BASED CONTROL

In PBC we will ensure stability of the closed-loop system by rendering passive (actually strictly passive to enforce *asymptotic* stability) a suitably defined map. Depending on the particular passivity property that we want to impose to the closed-loop there are several ways to carry out the design.

4.1 *Controller design*

The procedure to design the cascaded PBC consists of the following steps: 1) Choice of u to ensure flux tracking, i.e., $\lambda \to \lambda_d$; 2) Choice of λ_d so that $\lambda \to \lambda_d \Rightarrow F \to F_d$; 3) Proof that $F \to F_d \Rightarrow y \to y_*$ with all signals bounded.

1. Flux tracking

In systems described by Euler-Lagrange equations the passivity property is always satisfied with storage function the total energy. We propose then the (scaled) magnetic energy in the inductance as storage function candidate

$$H = \frac{1}{2}\lambda^2$$

whose derivative along the trajectories of (1) satisfies

$$\begin{aligned} \dot{H} &= -R\frac{c_2 - y}{c}\lambda^2 + u\lambda \\ &\leq -\alpha\lambda^2 + \lambda u \end{aligned}$$

with $\alpha \equiv \frac{R\epsilon}{c} > 0$. Integrating the last inequality from 0 to t and using the fact that $H \geq 0$ yields

$$\int_0^t u(s)\lambda(s)ds \geq \alpha\int_0^t \lambda^2(s)ds + \beta$$

where we have set $\beta \equiv -H(0)$. This proves (local) *output strict passivity* of the map $u \mapsto \lambda$.

The passivity property above suggests to try to assign to the closed-loop the *desired storage function*

$$H_d = \frac{1}{2}\tilde{\lambda}^2 \qquad (6)$$

where $\tilde{\lambda} \equiv \lambda - \lambda_d$, with λ_d the desired flux that induces on the mechanical subsytem the force required to regulate y, this signal is defined in the next step. With some abuse of notation this step is commonly refered as *energy shaping.*

It is easy to see that if we set $u = u_{\text{PB}}$ with

$$u_{\text{PB}} = \dot{\lambda}_d + \frac{c_2 - y}{c}\lambda_d + v \qquad (7)$$

then we get the error dynamics

$$\dot{\tilde{\lambda}} = -\frac{c_2 - y}{c}\tilde{\lambda} + v$$

Proceeding as above we can establish the bound

$$\dot{H}_d \leq -\alpha\tilde{\lambda}^2 + \tilde{\lambda}v \qquad (8)$$

which proves that the map $v \mapsto \tilde{\lambda}$ is output strictly passive as desired. Of course, if we set $v \equiv 0$ we have that $\tilde{\lambda} \to 0$, exponentially fast. Given that the rate of convergence α may be small to enhance performance it is convenient to add a so-called *damping injection* term into the electrical subsytem, setting for instance

$$v = -R_{\text{DI}}\tilde{\lambda}$$

with $R_{\text{DI}} > 0$. It is clear that this additional term will speed up the electrical transient. However, as will be shown in the simulations of section 7, increasing R_{DI} does not necessarily leads to transient performance improvement. This stems from the fact that the overall closed-loop system under PBC is still nonlinear and we are confronted with unpredictable *peaking* effects.

2. From flux tracking to force tracking

The force of magnetical origin F can be written in terms of the flux error $\tilde{\lambda}$ and the desired flux λ_d as

$$F = \frac{1}{2}\frac{c_1}{c^2}[\lambda_d^2 + \tilde{\lambda}(\tilde{\lambda} + 2\lambda_d)] \qquad (9)$$

From the analysis above we have that $\tilde{\lambda} \to 0$, hence it is reasonable to choose λ_d as the solution of

$$F_d = \frac{1}{2}\frac{c_1}{c^2}[\lambda_d^2] \qquad (10)$$

where $F_d > 0$ is some desired force, whose derivative is assumed to be known. We can now write the control (7) in terms of F_d and $\dot{F}_d$ as

$$u_{PB} = \frac{c\dot{F}_d}{\sqrt{2c_1F_d}} + \sqrt{\frac{2F_d}{c_1}}[R(c_2-y) - L_0\dot{y}] \quad (11)$$
$$+v \quad (12)$$

If F_d is bounded we can conclude from (9) and (10) that the force tracking controller (11) ensures $F \to F_d$. The problem is that in position tracking problems F_d will not be *a priori* bounded, since it is generated by $\mathcal{C}_{ol}$. This leads to the third step of our design.

3. From force tracking to position tracking

In this final step we have to design $\mathcal{C}_{ol}$ and prove convergence of the position error to zero. We start by defining F_d which, we set as

$$F_d = m[\ddot{y}_* - k_2\dot{\tilde{y}} - k_1\tilde{y} - k_0\int_0^t \tilde{y}(s)ds] \quad (13)$$

with $k_0, k_1, k_2 > 0$ chosen as in section 3. Notice that $\dot{F}_d$ is used in (11). From the definition of F_d above, this implies that $\ddot{y}$ (or equivalently F) is required for implementation of the PBC (11), (13). Remark also that we have introduced in (13) an integral term. This takes care of the steady-state error due to the constant disturbance mg, but it could be removed and simply replaced by g. Furthermore it simplifies the comparison of the various schemes because in this way the closed-loop mechanical dynamics is of the same order. For the stability analysis refers to section 4 of (Torres and Ortega, 1998).

5. INTEGRATOR BACKSTEPPING CONTROL

As pointed out in section 8.3 of (Krstic *et al.*, 1995) the system (1) is *not in any of the special forms* required for application of integrator backstepping. However, as shown in (Queiroz and Dawson, 1996) (see also (Krstic *et al.*, 1995)) this does not preclude a backstepping-like design. Since almost any design can be interpreted as backstepping-like we point out from the outset that we will take-off from the approach used in (Queiroz and Dawson, 1996) for a magnetic bearing system as applied to our simple levitated ball. Motivated by our final objective of comparison of techniques we will also present a variation of the basic scheme of (Queiroz and Dawson, 1996) that incorporates features of the PBC.

In IBC we start from the mechanical equation of (1) and, *assuming* F is the control, define an F_d that stabilizes this subsytem. For the sake of comparison we choose as in (13). Adding and substracting $\dot{F}_d$ to $my^{(3)} = \dot{F}$ and substituting (9) and (10) yields the error equation

$$\dot{x} = Ax + B\frac{c_1}{2mc^2}\tilde{\lambda}(\tilde{\lambda} + 2\lambda_d) \quad (14)$$

with $x \equiv [\tilde{y},\ \dot{\tilde{y}},\ z_2]^T$, $\dot{z}_2 = \tilde{y}$, $B \in \mathcal{R}^3$, and the characteristic polynomial of $A \in \mathcal{R}^{3\times3}$ equal to that of (5). We start constructing our Lyapunov function with the quadratic function

$$V_1 = \frac{1}{2}x^TPx \quad (15)$$

whose derivative yields

$$\dot{V}_1 = -\frac{1}{2}x^TQx + \frac{1}{2m}\frac{c_1}{c^2}x^TPB\tilde{\lambda}(\lambda + \lambda_d)$$

Let us now look at the dynamic equation of $\tilde{\lambda}$, that is

$$\dot{\tilde{\lambda}} = -R\frac{c_2 - y}{c^2}\lambda + u - \dot{\lambda}_d$$

At this stage we decide a control u that stabilizes $\tilde{\lambda}$ plus a term to compensate for the cross term in $\dot{V}_1$ above. We have here the choice of adopting an FLC–like approach to cancel the term $-\frac{R}{c}(c_2-y)\lambda$ and add a damping term $-R_{DI}\tilde{\lambda}$, or follow the PBC route, use the existing damping, and set $u = u_{IB}$ with

$$u_{IB} = u_{PB} \quad (16)$$

and u_{PB} as in (7). We can complete our Lyapunov function with H_d (6) as

$$V = V_1 + \frac{1}{\beta}H_d$$

where we have added a tuning parameter $\beta > 0$. Differentiating V and using (8) we get

$$\dot{V} \le -\frac{1}{2}x^TQx - \frac{\alpha}{\beta}\tilde{\lambda}^2 + \tilde{\lambda}[\frac{1}{\beta}v + \frac{c_1}{2mc^2}x^TPB(\lambda + \lambda_d)]$$

The IBC design is completed setting

$$v = -\frac{\beta c_1}{2mc^2}x^TPB(\lambda + \lambda_d) \quad (17)$$

which removes the cross terms and yields the desired strict Lyapunov function

$$\dot{V} \le -\frac{1}{2}x^TQx - \frac{\alpha}{\beta}\tilde{\lambda}^2$$

6. COMPARISON OF THE SCHEMES

We will now compare FLC (3), (4), PBC (11), (13) (without damping injection, i.e., with $v = 0$), and IBC (11), (16), (17). To this end, we notice that, (13), is related to (4) as $\dot{F}_d = mv_{\rm FL}(y, \dot{y}, F)$ Hence, the three control inputs can be written as

$$\begin{aligned} u_{\rm FL} &= \frac{c\, m v_{\rm FL}}{\sqrt{2c_1 F}} + \sqrt{\frac{2F}{c_1}}[R(c_2 - y) - \dot{y}L_0] \\ u_{\rm PB} &= \frac{c\, m v_{\rm FL}}{\sqrt{2c_1 F_d}} + \sqrt{\frac{2F_d}{c_1}}[R(c_2 - y) - \dot{y}L_0] \\ u_{\rm IB} &= u_{\rm PB} - \frac{\beta}{mc}\sqrt{\frac{c_1}{2}}x^T PB(\sqrt{F} + \sqrt{F_d}) \end{aligned}$$

The following remarks are in order

- FLC and PBC differ only on the utilisation of the forces in some terms: desired forces F_d for PBC instead of the actual forces F for FLC.
- One additional difference comes from the generation of F_d, which involves the inclusion of an additional state. Hence, PBC and IBC are *dynamic* state feeback controllers while FLC is static.
- The IBC equals the PBC plus the signal that achieves the cancellation of the cross terms in $\dot{V}$. The two schemes approach as $\beta \to 0$.
- In contrast with FLC, the closed–loop equations for PBC and IBC are nonlinear. In PBC we get a *cascade* structure where the first subsystem is exponentially stable and some growth conditions on the input to the second subsystem allows us to complete the stability proof. In IBC we get a closed–loop system of the form

$$\begin{bmatrix} \dot{\tilde{\lambda}} \\ \dot{x} \end{bmatrix} = K \begin{bmatrix} \tilde{\lambda} \\ x \end{bmatrix}$$

$$\begin{aligned} K &= \begin{bmatrix} -\frac{R}{c}(c_2 - y) & -\frac{\beta}{mc}\sqrt{\frac{c_1}{2}}(\sqrt{F} + \sqrt{F_d})B^T P \\ \frac{1}{mc}\sqrt{\frac{c_1}{2}}(\sqrt{F} + \sqrt{F_d})B & A \end{bmatrix} \\ &\equiv (A_{\rm DI} + A_{\rm SK}) \end{aligned}$$

  where the block–diagonal matrix $A_{\rm DI}$ is "stable" and $A_{\rm SK}$ satisfies the skew–symmetry property

$$\begin{bmatrix} \frac{1}{\beta} & 0 \\ 0 & P \end{bmatrix} A_{\rm SK} = -A_{\rm SK}^T \begin{bmatrix} \frac{1}{\beta} & 0 \\ 0 & P \end{bmatrix}$$

  This pattern is present in all IBC designs.
- Given that their closed–loop dynamics are nonlinear, predicting the effect of the tuning parameters for PBC and IBC is less obvious than for FLC.
- While there is an *output feedback* version of PBC, the presence of the extra term in IBC hampers this possibility. However, if we want to add damping to the PBC we need the full state.
- As explained in the previous section an FLC–like version of IBC may be obtained as

$$\begin{aligned} u_{\rm IB} = &\frac{c\, m v_{\rm FL}}{\sqrt{2c_1 F_d}} + \sqrt{\frac{2F}{c_1}}[R(c_2 - y) - \dot{y}L_0] \\ &- \frac{\beta}{mc}\sqrt{\frac{c_1}{2}}x^T PB(\sqrt{F} + \sqrt{F_d}) \\ &- R_{\rm DI}(\sqrt{F} - \sqrt{F_d}) \end{aligned}$$

Notice that it combines terms from PBC and FLC plus the damping injection and the additional "decoupling" term.

7. SIMULATION RESULTS

In this section we present some simulation results of the levitated ball model (1) in closed–loop with FLC, PBC and IBC. To facilitate the comparison, and unless otherwise stated, we have in all cases fixed $D(p) = (p + 10)^3$. System parameters were choosen as follows: $c = 10$, $m = 1$, $R = 1$ and $g = 10$. We tried a series of small step references in position. As expected, when we applied larger steps we cross through controller singularities (i.e. negative real or desired force). We have to point out that for the feedback linerization controller the "admisibles" references were larger than in the other cases. This is explained by the fact the this controller not depend on F_d but on F. Away from singularities the control signals do not differ significantly, therefore we show in all cases only the position and current responses.

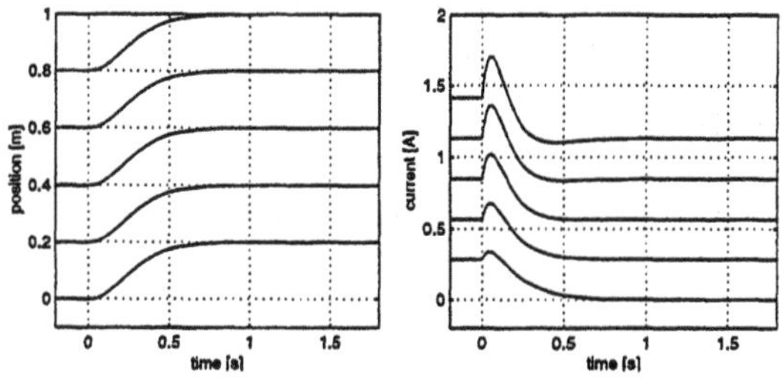

Fig. 3. Feedback-linearization controller

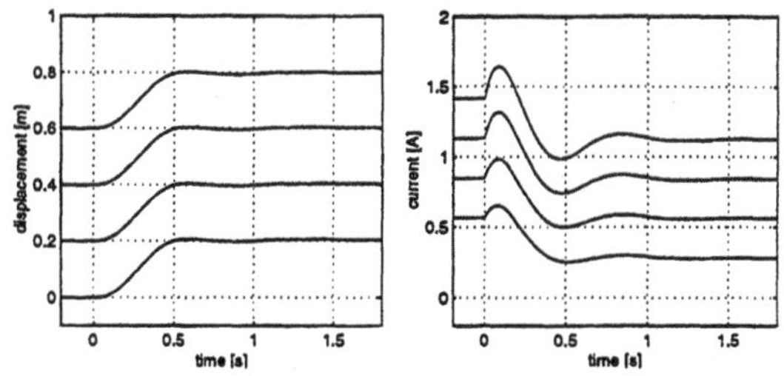

Fig. 4. Passivity Based Controller

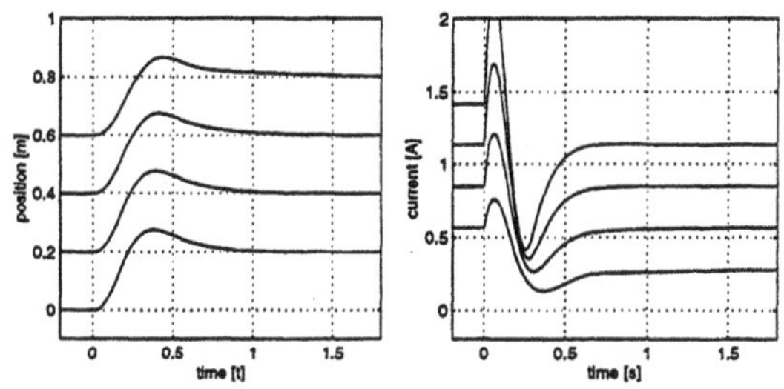

Fig. 5. PBC plus damping injection

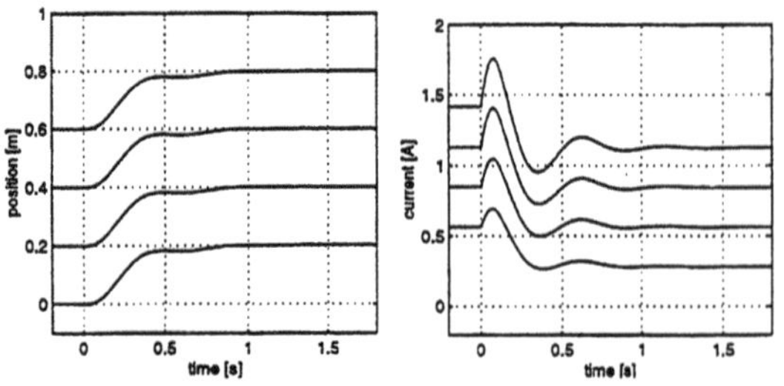

Fig. 6. Integrator backstepping-based controller

The response of FLC (3), (4), which is shown for the sake of reference in Fig. 3, is of course the same for all reference levels. It is interesting to note that we observed very little stability degradation when the actual parameters R and c were replaced by estimates within practically meaningful ranges of $\pm 50\%$. Of course, some steady state error was observed. Comparable robustness properties were observed for the other schemes.

PBC was tried in different variations. Fig. 4 shows the the response of (11), (13), while for Fig. 5 we included also the damping injection. For the latter we have observed that, for large values of R_{DI}, the response overshoots, due to the peaking phenomenon. Finally in Fig. 6 we present the responses of the IBC (11), (16), (17) for a value of $\beta = 100$.

In summary we can say that in must of the cases, and as long as we kept away from the singularity region, the observed responses were consistent with the theoretical predictions. Also, despite the significant (analytical) differences of the three controllers that we pointed out above we could not observe big discrepancies in the simulation responses. We believe this stems from the fact that the system is rather benign (it is passive with respect to λ, and y is a flat output!) particularly for the choice of parameters given here.

8. CONCLUSIONS AND FURTHER RESEARCH

The following concluding remarks concerning general aspects of FLC, PBC and IBC are in order

- While in FLC we aim at a linear system, in PBC we are satisfied with assigning a certain passivity property to a suitably defined map, –consequently enforcing a certain storage function to the closed–loop dynamics–. In IBC we aim at a somehow intermediate objective (more ambitious than passivation but less demanding than linearization) of assigning a *strict* Lyapunov function to the closed–loop.
- Both, in IBC and PBC, the main question is "Which storage–Lyapunov function you (can) want to assign?" In PBC, at least of Euler–Lagrange systems, the answer is provided by the systems energy function (either of the full system or just a part of it). The power of IBC is that, for a certain class of cascaded systems, the Lyapunov function can be *recursively* constructed. The example above shows that IBC can be profitably combined with PBC to answer this question.
- It is often argued and reasonable to expect, though hard to prove rigourously, that avoiding cancellation of nonlinearities enhances the robustness of the scheme in the face of parameter uncertainties. Our simulation evidence in the present example, however, showed that all schemes are highly insensitive to these uncertainties.

The three schemes will in general exhibit different transients and posses different robustness properties, a challenging research problem is to establish some common framework to compare their robustness and performance properties. One particularly interesting question is to assess the degrees of freedom provided to the designer to enhance the systems response.

The skew–symmetry property of IBC is related, though not so clear how, with the so–called workless–forces of Euler–Lagrange systems. In PBC we essentially disregard these forces and concentrate on the damping injection to increase the convergence rate. In this class of systems these coupling terms are related with the transformation of energy from one form to the other, e.g., from electrical to magnetical in electrical circuits or from potential to kinetic in mechanical systems.

9. REFERENCES

Krstic, M., I. Kanellakopoulos and P. Kokotovic (1995). *Nonlinear and Adaptive Control Design.* John Willey. New York.

Levine, J., J. Lottin and J. Ponsart (1996). A nonlinear approach to the control of magnetic bearings. *IEEE Trans. on Cont. Syst. Techn.* 4, 524–544.

Ortega, R., A. Loria, H. Sira and P. Nicklasson (1998). *Nonlinear Dynamical Control Systems.* Springer Verlag. Berlin.

Queiroz, M. and D. Dawson (1996). Nonlinear control of magnetic bearing: A backstepping approach. *IEEE Tran. on Contr. Syst. Tech.* 4, 545–552.

Torres, M. and R. Ortega (1998). Feedback linearization, integrator backstepping and passivity based controller design: A comparison example. In: *Perspective in Control: Theory and Application* (D. Normand-Cyrot, Ed.). pp. 97 – 115. Springer Verlag. London.

14th Triennial World Congress, Beijing, P.R. China

E-2c-17-5

AN ENERGY-BASED DERIVATION OF LYAPUNOV FUNCTIONS FOR FORCED SYSTEMS WITH APPLICATION TO STABILIZING CONTROL.

B.M.J. Maschke [(1)], **R.Ortega** [(2)], **A.J. van der Schaft** [(3)], **G.Escobar** [(2)]

[(1)] *Laboratoire d'Automatisme Industriel, Conservatoire National des Arts et Métiers Paris, FRANCE. e-mail: maschke@paris.ensam.fr*

[(2)] *Laboratoire des Signaux et Systèmes, Supelec, Gif-sur-Yvette, FRANCE*

[(3)] *Department of Mathematical Sciences, University of Twente Enschede, the NETHERLANDS*

Abstract. In this paper we show how the Hamiltonian formulation of systems may be used in order to derive Lyapunov functions for forced systems with constant input. In a first part the structure of Hamiltonian systems with dissipation is recalled and it is shown that, under certain integrability assumptions on the input vector fields, a candidate Lyapunov function for the forced system may be derived. In a second part, an interpretation of this candidate Lyapunov function is given in terms of the total energy of the forced system, composed of the sum of its internal energy and the energy function associated with the source element. In the third part, such a candidate Lyapunov function is used to derive a stabilizing controller for the boost converter. *Copyright © 1999 IFAC*

Keywords: Hamiltonian systems, Lyapunov stability, feeedback stabilization, switching network.

1. INTRODUCTION

In this paper we shall address the problem of generating Lyapunov candidate functions for forced systems in connection with an interpretation in terms of energy. Therefore we shall consider the generalized Hamiltonian formulation of the dynamics of physical systems which has the advantage of representing explicitly, in its definition, the internal energy of the system, as well as an interconnection which is reflected in the generalized Poisson bracket on the space of state variables (Maschke and van der Schaft, 1992) (van der Schaft and Maschke, 1995) (Maschke et al. 1995). For passive physical systems, the stability property is ensured by the boudedness of the energy function which may often be used directly as a Lyapunov function. However if the system is forced, the internal energy of the system is not, in general, a Lyapunov function. The first aim of this paper is to give condition under which a candidate Lyapunov function may be derived from the internal energy. Then using the concept of the network interconnection, we shall show that it is possible to give an interpretation of this function in terms of some energy functions. The second aim is to use this energy function as a Lyapunov function in order to derive stabilizing controllers for forced Hamiltonian systems with dissipation. This construction is illustrated by the control of a boost-converter.

2. LYAPUNOV FUNCTION CANDIDATES

The dynamics of (finite-dimensional) energy conserving physical systems may be formulated as a *port controlled Hamiltonian system* as follows (Maschke and van der Schaft, 1992) (van der Schaft and Maschke, 1995):

$$\left.\begin{aligned} \dot{x} &= J(x)\,\frac{\partial H_0(x)}{\partial x} + \sum_{i=1}^{m} g_i(x)\,u_i \\ y_i &= g_i^t(x)\,\frac{\partial H_0(x)}{\partial x} \qquad i = 1,\dots,m \end{aligned}\right\} \tag{1}$$

where the state space $\mathbb{R}^n$ is equipped with a generalized Poisson bracket (Libermann and Marle, 1987) defined by the skew-symmetric structure matrix $J(x)$, the drift dynamics is defined by a generalized Hamiltonian vectorfield (van der Schaft and Maschke, 1995) generated by the smooth Hamiltonian function $H_0(x)$ with respect to the structure matrix $J(x)$ and $g_i(x)$ are the input vector fields.

This structure stems from the network modeling of physical systems where the Hamiltonian function represents the total energy stored in the system and the structure matrix, defining the generalized Poisson bracket, corresponds to the interconnection structure of the network model representing interdomain couplings (Maschke et al. 1992). Such Poisson brackets may be systematically derived in the case of LC networks (Maschke et al. 1995) or multibody systems (Maschke 1996), but appear also more generally for instance in thermal interactions (Maschke 1998). An important departure with respect to standard Hamiltonian systems (Libermann and Marle, 1987) is, on the one hand, that the structure matrix may be degenerated (Maschke et al. 1992) and, on the other hand, that it does not necessarily verifies the Jacobi identities (van der Schaft and Maschke 1994) (Maschke 1998). In the case that it verifies the Jacobi identities the kernel of the structure matrix $J(x)$ indicates the existence of invariants, called Casimir functions (Maschke and van der Schaft, 1992) (Libermann and Marle, 1987), which may be used for stability analysis (Marsden 1992) and will play a key role in the construction proposed hereafter.

The input vectorfields define the interaction of the system with its environment in terms of two adjoint maps (associated with the inputs u and outputs y, corresponding actually to the port variables of the network model) and corresponds also to an interconnection structure relating the energy storing elements to the ports of the system in the network model (Maschke and van der Schaft, 1992) (van der Schaft and Maschke, 1995).

Energy dissipation may be taken into account by terminating some ports with dissipative elements yielding the *port controlled system with dissipation* (Dalsmo and van der Schaft 1999):

$$\left.\begin{aligned} \dot{x} &= (J(x)-S(x))\,\frac{\partial H_0(x)}{\partial x} + g(x)\,u \\ y &= g^t(x)\,\frac{\partial H_0(x)}{\partial x} \end{aligned}\right\} \quad (2)$$

where $S(x)$ is a non-negative symmetric matrix corresponding to the energy dissipation and $g(x)$ is the input matrix with column vectors $g_i(x)$, $i = 1..,m$. At the ports of the system the following *power balance* holds:

$$\frac{dH_0}{dt} = \sum_{i=1}^{m} y_i\,u_i - \frac{\partial H_0}{\partial x}^t(x)\;S(x)\;\frac{\partial H_0(x)}{\partial x} \quad (3).$$

It may be noted that the matrix $J(x)-S(x)$ defines a bracket (sum of a skew-symmetric and a symmetric bracket) generalizing the metriplectic bracket defined in (Morrison 1986).

It may furthermore be noted that, if the Hamiltonian function $H_0(x)$ is bounded from below, the power balance (3) implies that the port controlled Hamiltonian system with dissipation is *passive* (Willems, 1972).

Moreover, if the Hamiltonian function $H_0(x)$ admits a strict minimum at x^*, then from (3) and the symmetry of $S(x)$ it follows that it may be used as Lyapunov function for the stability analysis of the uncontrolled system ($u = 0$) at x^* (Khalil, 1996).

Consider now a *forced port controlled Hamiltonian system with dissipation* (2) driven by a constant (non-zero) input vector: $\bar{u}$. A *forced equilibrium* $\bar{x}$ is the solution of the *static model*:

$$\left(J(\bar{x})-S(\bar{x})\right)\,\frac{\partial H_0}{\partial x}(\bar{x}) + g(\bar{x})\;\bar{u} = 0 \quad (4).$$

In the sequel we shall moreover make the following assumption.

Assumption H1:

the structure matrix $\left(J(x)-S(x)\right)$ *is regular for any* x.

According to this assumption, any input $\bar{u}$ defines uniquely the gradient $v = \frac{\partial H_0}{\partial x}(\bar{x})$:

$$v = K(\bar{x})\;\bar{u} \quad (5)$$

where: $K(x) = -\left(J(x)-S(x)\right)^{-1} g(x)$ (6).

It is clear that in general the Hamiltonian function $H_0(x)$ cannot serve anymore as Lyapunov function as it does not admit in general the forced equilibrium $\bar{x}$ as minimum.

However assume that the input vectorfields $g_i(x)$ are generated by interaction potentials $C_i(x)$.

Assumption H2: *the input vectorfield are generated by interaction Hamiltonian functions* $-C_i(x)$ *with respect to the structure matrix* $\left(J(x)-S(x)\right)$ *of the system (2):*

$$g_i(x) = (J(x)-S(x))\,\frac{\partial C_i(x)}{\partial x} \quad (7).$$

Then the system (2) driven by the constant input vector $\bar{u}$ may be written as an autonomous Hamiltonian system with dissipation:

$$\dot{x} = (J(x)-S(x))\,\frac{\partial H_{\bar{u}}(x)}{\partial x} \quad (8)$$

generated by the Hamiltonian function, that we will call *forced Hamiltonian function*:

$$H_{\bar{u}}(x) = H_0(x) - \sum_{i=1}^{m} \bar{u}_i\;C_i(x) \quad (9).$$

In the case when there is no dissipation (i.e. $S(x) = 0$), this assumption means that the input vectorfields are Hamiltonian, generated by the interaction Hamiltonian functions $C_i(x)$ with respect to the generalized Poisson bracket defined by the structure matrix $J(x)$. Then the system (8) corresponds to the definition of controlled Hamiltonian system (Brockett, 1977, van der Schaft, 1989). However there exist important classes of (energy conserving) physical systems for which the input vectorfields are not Hamiltonian (Maschke and van der Schaft, 1992) (Maschke 1998).

Now a forced equilibrium of (2) correspond to an equilibrium of the system (8) and due to the assumption H1, any forced equilibrium associated with the input $\bar{u}$, corresponds to an extremum of $H_{\bar{u}}(x)$. The system (8) also verifies a power balance equation (3) (setting $u_i = 0$):

$$\frac{dH_{\bar{u}}}{dt} = -\frac{\partial H_{\bar{u}}}{\partial x}^t(x)\;S(x)\;\frac{\partial H_{\bar{u}}(x)}{\partial x} \quad (10).$$

Hence the Hamiltonian function $H_{\bar{u}}(x)$, defined in (9), may be used as Lyapunov function for the forced system if in addition the forced equilibrium is a minimum of $H_{\bar{u}}(x)$.

3. ENERGY-BASED INTERPRETATION

In this section, stemming from a network approach, we shall give an interpretation of the candidate Lyapunov function proposed in (9) in terms of some energy. This interpretation will be exposed as a two step procedure consisting of embedding the forced system in a higher dimensional system and then restricting this system using some invariant functions (exposed in more details in (Maschke et al., 1998)).

In the first step we shall embed the system in higher dimensional Hamiltonian system by considering the environment delivering the constant input $\bar{u}$ as an infinite energy reservoir which may be represented by the elementary port controlled Hamiltonian system, called *source system*:

$$\left.\begin{aligned} \dot{\varsigma} &= u_s \\ y_s &= dH_s(\varsigma) \end{aligned}\right\} \qquad (11)$$

where $\varsigma \in \mathbb{R}^m$ is the state vector of the source, the Poisson bracket is the zero-Poisson bracket (indicating that the drift dynamics is zero) and the Hamiltonian (corresponding to the unbounded energy of the reservoir) is:

$$H_s(\varsigma) = -\bar{u}\,\varsigma \qquad (12).$$

The forced Hamiltonian system with dissipation (2) with constant input $\bar{u}$ may then be represented as the system composed of the two systems (2) and (11) *interconnected* through the power conserving interconnection (Dalsmo and van der Schaft, 1999):

$$\left.\begin{aligned} u_s &= y \\ y &= -y_s \end{aligned}\right\} \qquad (13).$$

The composed system is again a Hamiltonian system with dissipation with Hamiltonian function being the total energy, i.e. the sum of the internal energy of the forced system and the energy of the source system:

$$H_a(x,\varsigma) = H_0(x) - \bar{u}\,\varsigma \qquad (14)$$

and defined with respect to the structure matrix $(J'-S')$ where:

$$J'(x,\varsigma) = \begin{pmatrix} J(x) & 0 \\ 0 & 0 \end{pmatrix}, \text{ and : } S'(x,\varsigma) = \begin{pmatrix} S(x) & 0 \\ 0 & 0 \end{pmatrix} \qquad (15).$$

However some very simple example (e.g. a parallel RLC circuit) shows that in general the total energy (14) does not remain bounded (Maschke et al. 1998).

Therefore one may consider a more general *embedding system* by modifying the bracket of the source system and simultaneously the interconnection, what comes down to consider, instead of (15), the following structure matrices:

$$\left.\begin{aligned} J_a(x,\varsigma) &= \begin{pmatrix} J(x) & J(x)K(x) \\ -\big(J(x)K(x)\big)^t & J_s(x,\varsigma) \end{pmatrix} \\ S_a(x,\varsigma) &= \begin{pmatrix} S(x) & S(x)K(x) \\ \big(S(x)K(x)\big)^t & S_s(x,\varsigma) \end{pmatrix} \end{aligned}\right\} \qquad (16).$$

With these structure matrices, the forced system is indeed embedded which may be seen by considering the first part of the state equations: (17)

$$\dot{x} = (J(x)-S(x))\,\frac{\partial H_0(x)}{\partial x} + (J(x)-S(x))\;K(x)\,\frac{\partial H_s(\varsigma)}{\partial \varsigma}$$

And using the definition (6) and (12) this becomes the forced system:

$$\dot{x} = (J(x)-S(x))\,\frac{\partial H_0(x)}{\partial x} + g(x)\,\bar{u} \qquad (18).$$

The second step consists in restricting the embedding system to the forced system by using some invariant or Casimir functions.

Therefore the two structure matrices $J_s(x)$ and $S_s(x)$ are chosen in such a way that the structure matrices of the embedding system admit a kernel corresponding to the static model (4), i.e. that the kernel is spanned by:

$$P(x) = \left\{ \begin{bmatrix} -K(x)\,u \\ u \end{bmatrix},\; u \in \mathbb{R}^m \right\} \qquad (19).$$

The unique solution is then:

$$\left.\begin{aligned} J_s(x,\varsigma) &= K^t(x)\;J(x)\;K(x) \\ S_s(x,\varsigma) &= K^t(x)\;S(x)\;K(x) \end{aligned}\right\} \qquad (20).$$

However the embedding system may be restricted if and only if there exists some invariant functions associated with the kernel $P(x)$. This depends on some integrability conditions which, in the case when the bracket is skew-symmetric (i.e. $S_a = 0$) are called the Jacobi identities (Maschke and van der Schaft, 1992) (Libermann and Marle, 1987). A necessary and sufficient condition may be stated as follows (Maschke et al., 1998).

Proposition: *Consider the column vectors $K_i(x)$ of the matrix $K(x)$ defined in (6). The conditions:*

$$\frac{\partial K_{ji}}{\partial x_k} = \frac{\partial K_{ki}}{\partial x_j}, \quad j,k = 1,\, \dots,\, n \qquad (21).$$

are necessary and sufficient conditions for the (local) existence of the following invariant functions of the embedding Hamiltonian system:

$$\varsigma_i - C_i(x) \qquad (22)$$

where $C_i(x)$ are smooth functions such that:

$$K_i(x) = \frac{\partial C_i(x)}{\partial x} \qquad (23).$$

On every invariant submanifolds defined by:

$$\varsigma_i - C_i(x) = k_i \in \mathbb{R} \quad i = 1,\, \dots,\, n \qquad (24)$$

the embedding Hamiltonian system with dissipation may be reduced to the Hamiltonian with dissipation, defined with respect to the bracket $(J(x)-S(x))$ and generated by the Hamiltonian function:

$$\begin{aligned} H_{\bar{u}}(x) &= H_a(x, C(x) + k) \\ &= H_0(x) - \sum_{i=1}^{m} \bar{u}_i\,\big(C_i(x) + k_i\big) \end{aligned} \qquad (25).$$

Actually this dynamics is exactly the forced system (18) and the existence of functions $C_i(x)$ satisfying (23) amounts to say that they are the interaction Hamiltonian functions generating the input vector fields $g_i(x)$ (with respect to the bracket $(J(x)-S(x))$) according to the assumption H2.

Actually this result gives an interpretation of the Lyapunov candidate function $H_{\bar{u}}(x)$ obtained in section 2 as the total energy function consisting in the sum of the internal energy function $H_0(x)$ of the forced system and the energy $H_s(\varsigma)$ of the source system. However this interpretation is only valid if the source system is endowed with nonzero structure matrices defined in (20) and interconnected to the system (2) by the interconnection associated with the antidiagonal terms of the structure matrix (16). A detailed example may be found in(Maschke et al., 1998).

4. APPLICATION TO THE CONTROL OF A BOOST-CONVERTER

In this section we shall show on the example of a boost converter how the condition for the existence of a Lyapunov function proposed here above may be achieved by state feedback control. As a consequence such a feedback control provide a stabilizing controller.
Indeed consider the boost converter circuit in figure 1. It consists of a constant voltage source, an inductor, a capacitor and a resistor (representing the output load) interconnected by a switching topological interconnection structure depending on the position of the switch.

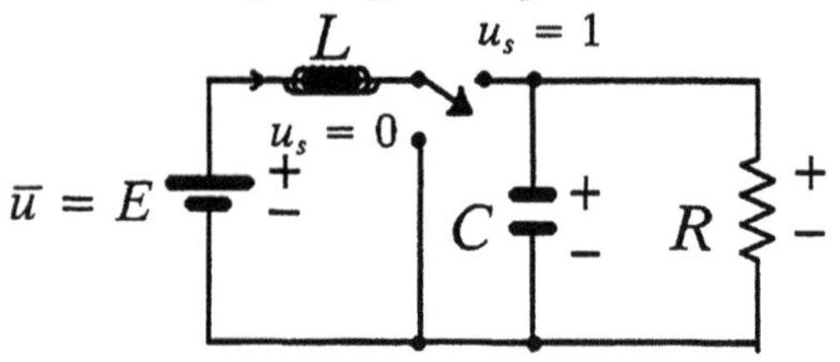

- Fig. 1 - Boost converter

The system's dynamical model may be formulated as a Hamiltonian system with dissipation of the form (2). In (Escobar et al. 1998) further details on the modeling of switched systems in the Hamiltonian formalism may be found. The state space is the space of the energy variables $x = (\varphi, q)^t \in \mathbb{R}^2$ where q denotes the charge of the capacitor and φ denotes the total magnetic flux of the inductor. The Hamiltonian function is the total electromagnetic energy $H_0(\varphi, q) = H_{mg}(\varphi) + H_{el}(q)$ where $H_{el}(q)$ denotes the electrical energy stored in the capacitor and $H_{mg}(\varphi)$ denotes the magnetic energy stored in the inductor. Although the procedure is identical for nonlinear constitutive relations of the inductor and capacitor, for the sake of simplicity of the final discussion, in the sequel we shall assume that the elements are linear:

$$H_{el}(q) = \frac{1}{2}\frac{q^2}{C} \text{ and: } H_{mg}(\varphi) = \frac{1}{2}\frac{\varphi^2}{L} \tag{26}$$

The differential of the Hamiltonian function is then the vector of the coenergy variables composed of the capacitor's voltage v_C and the inductor's current i_L:

$$dH_0(q, \varphi) = \begin{pmatrix} i_L \\ v_C \end{pmatrix} = \begin{bmatrix} \frac{\varphi}{L} \\ \frac{q}{C} \end{bmatrix} \tag{27}$$

The skew symmetric structure matrix is defined by the topological interconnection structure relating the inductor and the capacitor (Maschke et al. 1995) and is:

$$J(u_s) = \begin{pmatrix} 0 & -u_s \\ u_s & 0 \end{pmatrix} \tag{28}$$

where u_s depends on the position of the switch and takes values in $\{0, 1\}$. When $u_s = 1$, there is a loop containing the capacitor and the inductor and the structure matrix is the symplectic matrix and when $u_s = 0$ this loop does no more exist and the structure matrix is null.

Note that contrary to the general model (2), the skew-symmetric structure matrix does not depend on the state (i.e. the capacitor's charge and inductor's flux) which is a general property of electrical circuits (Maschke et al. 1995), but on an external variable being the position of the switch which modifies the topological interconnection network of the circuit.
Note moreover that for the two positions of the switch, the capacitor remains in a maximal tree of the port connection graph of the circuit and the inductor remain in its complementary cotree, hence there the continuity constraints at the switching instants are verified.
The (non-negative) symmetric structure matrix is defined by the interconnection structure relating the energy storing elements with the resistor and the constitutive relation of the resistor which leads to:

$$S = \begin{pmatrix} 0 & 0 \\ 0 & \frac{1}{R} \end{pmatrix} \tag{29}$$

And finally the input vector, defined by the interconnection topology relating the energy storing elements and the source, is:

$$g = \begin{pmatrix} 1 \\ 0 \end{pmatrix} \tag{30}$$

And the system is forced by a constant input $\bar{u} = E$, i.e. by applying a constant voltage source to the circuit.
For the control problem we shall consider that the the switch is controlled by Pulse Width Modulation signal what yields to consider that the control variable u_s in the skew-symmetric structure matrix $J(u_s)$ represents the slew rate of the Pulse Width Modulation (instead of the state of the switch) and takes values in the interval $[0, 1]$.
The control objective is to find a state feedback $u_s = u_s(x)$, such that the capacitor's voltage v_C is regulated to a desired voltage V_d. For this purpose we shall construct a feedback such that the closed-loop system satisfies firstly the assumption H2, ensuring the existence of the Hamiltonian function proposed in (9), and secondly that this function is a Lyapunov for a desired equilibrium ensuring the desired capacitor's voltage. The construction proceeds in two steps:
i) synthesize a family of feedback laws such that condition (21) is satisfied,
ii) select parameters such that the equilibrium is a minimum of the Lyapunov function.
Before starting the construction of the feedback control, let us define the forced equilibrium associated with a desired capacitor's voltage. Note first that the desired voltage V_d imposes, through the constitutive relation of the capacitor, the equilibrium charge: $\bar{q} = CV_d$. Then the static model (4) may be solved in the equilibrium flux $\bar{\varphi}$ and the equilibrium slew rate $\bar{u}_s$, yielding *the equilibrium associated with the desired voltage* V_d:

$$\left.\begin{aligned} \bar{x} &= \begin{pmatrix} \bar{\varphi} \\ \bar{q} \end{pmatrix} = \begin{bmatrix} \frac{L\,V_d^2}{R\,E} \\ C\,V_d \end{bmatrix} \\ \bar{u}_s &= \frac{E}{V_d} \end{aligned}\right\} \tag{31}$$

In the first step let us compute feedbacks such that condition (21) is satisfied:

$$K(x) = -\left(J(u_s(x))-S\right)^{-1} g = \begin{bmatrix} \dfrac{1}{R\, u^2(x)} \\ \dfrac{1}{u(x)} \end{bmatrix} = \begin{pmatrix} k_1(x) \\ k_2(x) \end{pmatrix} \tag{32}$$

Noting that $k_1(x) = \frac{1}{R}\, k_2(x)$, the conditions (21) on $K(x)$ imply that $k_2(x)$ should verify the following partial differential equation:

$$\frac{\partial k_2(x)}{\partial \phi} - \frac{2}{R}\, k_2(x) \frac{\partial k_2(x)}{\partial q} = 0 \tag{33}$$

which have the solutions:

$$k_2(x) = \frac{-\frac{R}{2}\, c_1\, q + c_3}{c_1\, \phi + c_2} \tag{34}$$

where c_1, c_2, c_3 are reals.

Using (32), the control is easily obtained as:

$$u(x) = \frac{1}{k_2(x)} = \frac{c_1\, \phi + c_2}{-\frac{R}{2}\, c_1\, q + c_3} \tag{35}$$

Note that the three parameters of the controller are related by the condition that, at the forced equilibrium, $u(\bar{x}) = \bar{u}$, where $\bar{u}$ and $\bar{x}$ are given by (31), what leads to the linear constraint:

$$c_2 = -\left(\frac{ERC}{2} + \frac{L\, V_d^2}{RE}\right) c_1 + \frac{E}{V_d}\, c_3 \tag{36}$$

Using the expression of the control (35), it may be shown that the vector $K(x)$ is found to be the coordinate vector of an exact 1-form, that means is the differential $K(x) = dC(x)$ of the function:

$$C(x) = \frac{-\frac{R}{2}\, c_1\, q^2 + c_3\, q - \frac{c_3^2}{R c_1}}{c_1\, \phi + c_2} + c_4 \tag{37}$$

From this expression, the Hamiltonian function of the forced system with control (35) becomes, according to (9):

$$H_{\bar{u}}(x) = \frac{1}{2}\frac{\varphi^2}{L} + \frac{1}{2}\frac{q^2}{C} + \frac{E}{R\, c_1}\, \frac{\left(\frac{R}{2}\, c_1\, q - c_3\right)^2}{c_1\, \phi + c_2} - E c_4 \tag{38}$$

where the constant c_4 may be chosen for instance in such a way that $H_{\bar{u}}(x)$ is equal to zero at the equilibrium.

The second step consists in choosing parameters of the controller in such a way that the forced equilibrium defined in (31) is indeed a local minimum for the forced Hamiltonian function (38). Therefore we shall consider the Hessian of the forced Hamiltonian function (38):

$$\frac{\partial^2 H_{\bar{u}}(x)}{\partial x^2} = \begin{bmatrix} \left(\dfrac{1}{L} + \dfrac{2E\, c_1}{R}\, \dfrac{\left(\frac{R}{2}\, c_1\, q - c_3\right)^2}{\left(c_1\, \phi + c_2\right)^3}\right) & \left(-E\, c_1\, \dfrac{\left(\frac{R}{2}\, c_1\, q - c_3\right)}{\left(c_1\, \phi + c_2\right)^2}\right) \\ \left(-E\, c_1\, \dfrac{\left(\frac{R}{2}\, c_1\, q - c_3\right)}{\left(c_1\, \phi + c_2\right)^2}\right) & \left(\dfrac{1}{C} + \dfrac{E\, c_1}{2\left(c_1\, \phi + c_2\right)}\right) \end{bmatrix} \tag{39}$$

After some tedious but straightforward computations, the Hessian of the forced Hamiltonian function may be shown to be positive definite if the parameters c_1, c_2 belong to one of the two domains (represented in figure 2):

$$\left.\begin{aligned} & c_1 < 0\,,\ c_2 > -\left(\frac{ERC}{2} + 3\,\bar{\phi}\right) c_1 \\ & c_1 > 0\,,\ c_2 > -\,\bar{\phi}\, c_1 \end{aligned}\right\} \text{domain 1} \tag{40}$$

$$\left.\begin{aligned} & c_1 < 0\,,\ c_2 < -\,\bar{\phi}\, c_1 \\ & c_1 > 0\,,\ c_2 < -\left(\frac{ERC}{2} + 3\,\bar{\phi}\right) c_1 \end{aligned}\right\} \text{domain 2} \tag{41}$$

In summary, we have constructed a family of control laws which ensure that the Hamiltonian function in (38) is locally a Lyapunov function for the closed-loop system at the equilibrium (31) associated with the desired output voltage V_d, hence it stabilizes the closed-loop system at the desired equilibrium.

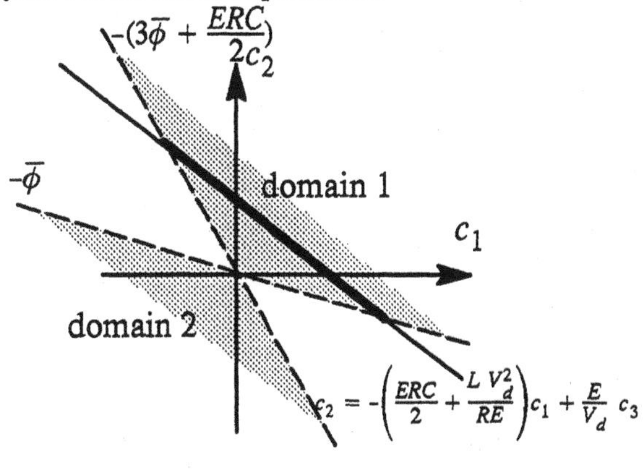

Figure 2 : admissible parameters of the controller in (c_1, c_2) plane.

This family of control laws defined in (35) is parametrized by three reals c_1, c_2, c_3 related by the linear relation (36) and the inequality constraints (40) or (41). The family of admissible parameters may be represented in the parameters c_1, c_2 (see figure 2) by a family of segments (parametrized by c_3) being the intersection of the line defined by (36) with the domains 1 and 2.

In figure 3, some simulation results are shown for the boost converter in closed-loop with the controller (35). The following parameters have been considered for the simulation: physical parameters: $L = 20$ mH, $C = 20$ µF, $R = 30\ \Omega$, $E = 15$ V, control parameters: $V_d = 37.7$ V, $c_1 = 5$, $c_3 = 5$ and initial conditions: $i_L(0) = 0$ Amp, $v_C(0) = 15$ V.

5. CONCLUSION

In this paper we have first shown that, under certain integrability conditions on the input vector fields, the stability of a forced Hamiltonian system with dissipation may be discussed with the help of a candidate Lyapunov function directly derived from the Hamiltonian function of the system. Then we have given an interpretation of this Hamiltonian function in terms of the total energy function composed of the sum of the internal energy of the system and the energy associated with the source. This interpretation is actually only possible by embedding the forced system in a higher dimensional system when the infinite energy reservoir representing the source, is connected in a particular way to the forced system. Finally we have given a constructive procedure leading to a stabilizing controller of a boost converter, where the total energy of closed-loop system is indeed a Lyapunov function for the forced system.

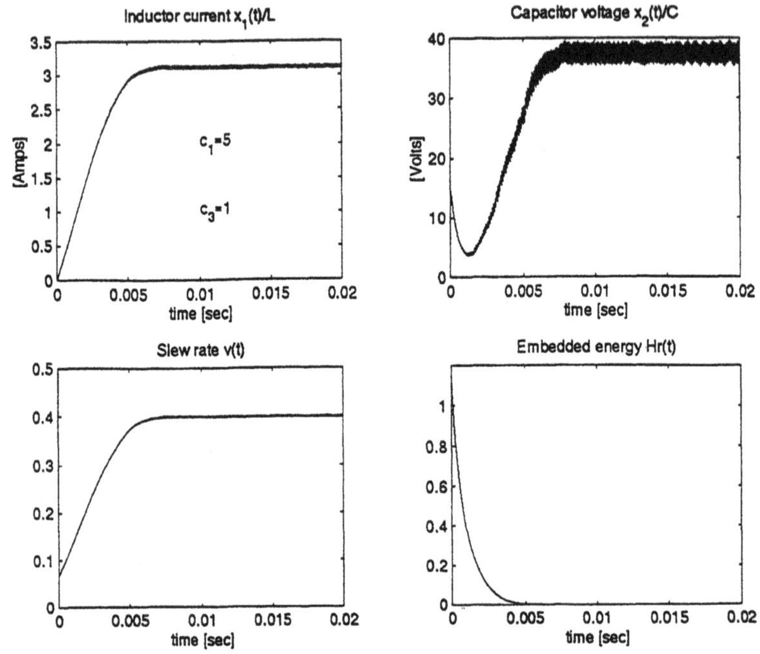

Figure 3 : Simulation of the closed-loop boost converter under PWM control with $c_1 = 5$, $c_3 = 5$.

6. REFERENCES

Brockett R.W., "Control theory and analytical mechanics", , in *Geometric Control Theory*, C.Martin and R.Herman eds., pp.1–46, Math.Sci.Press, Brookline, 1977

Dalsmo M. and A.J.van der Schaft (1999), "On the representations and integrability of mathematical structures in energy conserving physical systems", *SIAM J. on Optimization and Control*, **Vol.37**, n°1

Escobar G., A.J. van der Schaft and R.Ortega (1998), "A Hamiltonian viewpoint in the modeling of switching power converters", Int. Symp. on the Mathematical Theory of Networks and Systems, Padova (to appear in Automatica: special issue on hybrid systems)

Khalil H.K., *Nonlinear Systems*, MacMillan, New-York, 2nd edition, 1996

Libermann P. and C.M.Marle (1987), *Symplectic geometry and Analytical Mechanics*, Reidel, Dordrecht

Marsden J.E. (1992), *Lecture Notes on Mechanics*, London Math. Soc. Lecture Notes Series, 174, Cambridge Un. Press, Great Britain

Maschke B.M. (1996), "Elements on the Modelling of Multibody sytems" in *Modelling and Control of Mechanisms and Robots*, C.Melchiorri and A.Tornambè editors, World Scientific Publishing Ltd.

Maschke B.M. (1998), "Interconnection and structure in physical systems' dynamics", *Proc. 4rth IFAC Symp. on Nonlinear Control Systems Design*, NOLCOS'98, Enschede, the Netherlands, July 1-3, pp. 291-296

Maschke B.M.J., R.Ortega and A.J. van der Schaft (1998), "Energy-based Lyapunov functions for forced Hamiltonian systems with dissipation", submitted to the IEEE Trans. on Automatic Control

Maschke B.M. and A.J. van der Schaft (1992), "Port controlled Hamiltonian systems: modeling origins and system theoretic properties", *Proc. 2nd IFAC Symp. on Nonlinear Control Systems Design*, NOLCOS'92, Bordeaux, pp.282–288.

Maschke B.M., A.J.van der Schaft and P.C.Breedveld (1992), "An Intrinsic Hamiltonian Formulation of Network Dynamics: Nonstandard Poisson Structures and Gyrators", *Journal of the Franklin Institute*, **Vol. 329**, n. 5, pp. 923–966

Maschke B.M., A.J. van der Schaft and P.C.Breedveld (1995), "An intrinsic Hamiltonian formulation of the dynamics of LC-circuits", *Trans. IEEE on Circuits and Systems, I: Fundamental Theory and Applications*, Vol.42, n°2, pp.73–82, February.

Morrison P.J. (1986), "A paradigm for joined Hamiltonian and dissipative systems", *Physica D*, **Vol. 18**, pp.410–419

van der Schaft A.J. (1989), "System theory and mechanics", in *Three Decades of Mathematical System Theory*, H.Nijmeier and J.M.Schumacher eds., Lect. Notes Contr. Inf. Sci., Vol. 135, pp. 426–452, Springer, Berlin.

van der Schaft A.J. and B.M.Maschke (1995), "The Hamiltonian formulation of energy conserving physical systems with ports", *Archiv für Elektronik und Übertragungstechnik*, **Vol.49**, n°5/6, pp.362-371.

van der Schaft A.J. and B.M.Maschke (1994), "On the Hamiltonian formulation of non-holonomic mechanical systems", *Reports on Mathematical Physics*, **Vol.34**, n°2, pp.225-233

Willems J.C. (1972), "Dissipative Dynamical Systems, Part 1: General Theory", *Arch. Rat. Mech. Anal.*, Vol.45, pp.321-351.

14th Triennial World Congress, Beijing, P.R. China

NONLINEAR GENERALISED PREDICTIVE CONTROL AND OPTIMAL DYNAMICAL INVERSION CONTROL

Wen-Hua Chen Donald J. Ballance Peter J. Gawthrop

Centre for Systems & Control, and Department of Mechanical Engineering, University of Glasgow, Glasgow G12 8QQ, UK,
{WChen, D.Ballance and P.Gawthrop}@mech.gla.ac.uk

Abstract: This paper addresses nonlinear Generalised Predictive Control (GPC) for nonlinear mechanical systems and its relationship with dynamical inversion control. First continuous-time generalised predictive control for nonlinear systems is developed and the stability of the closed-loop systems is shown. It is pointed out that the dynamical inversion control is a special case of continuous-time nonlinear GPC. This result also provides an optimal design method for dynamical inversion control in term of the generalised predictive control performance index. The nonlinear GPC developed is successfully applied to the control of a two-link robotic manipulator. *Copyright © 1999 IFAC*

Keywords: Nonlinear systems, generalised predictive control, computed torque control, robotics, stability, mechanical systems

1. INTRODUCTION

Generalised Predictive Control (GPC) is one of a wide set of methods referred to as Model Based Control. It has received a great deal of attention and is widely applied in engineering (for example see (Clarke, 1994) and the references therein). Most work in Generalised Predictive Control is based on a discrete-time approach. The continuous-time GPC was developed by Gawthrop and Demircioglu (Demircioglu and Gawthrop, 1991; Demircioglu and Gawthrop, 1992) using a transfer function approach. Although Ordys and Clarke (Ordys and Clarke, 1993) suggests that there is no significant different between the transfer function approach and the state space approach, Morari (Morari, 1994) points that the state space approach has both conceptual and numerical advantages. Moreover as discussed by Orday and Clarke (Ordys and Clarke, 1993) in the context of nonlinear systems the state space approach is essential. Recently a state space approach to the continuous-time Generalised Predictive control has been developed by Gawthrop *et al.* (1998).

Model based predictive control for the tracking problem of nonlinear continuous-time systems has been proposed by several authors Lu (1994; 1995; 1997), Singh (Singh *et al.*, 1995; Yim and Singh, 1997) and Soroush and Kravaris (1992; 1993). In all of these a generalised minimum variance like cost (Clarke and Gawthrop, 1975; Clarke and Gawthrop, 1979), which is referred to as Predictive Generalised Minimum Variance (PGMV) (Gawthrop *et al.*, 1998), is employed. That is, the objective is to optimise the performance index at one point in the future. In this paper the generalised predictive control performance cost is adopted. The average performance over a prescribed time interval is optimised in generalised predictive control. The advantage of the generalised predictive control cost function has been extensively discussed (Clarke *et al.*, 1987*a*; Clarke *et al.*, 1987*b*; Demircioglu, 1989). Moreover when the minimum and maximum predictive times are chosen as the same, the generalised predictive control performance cost is reduced to that of generalised predictive minimum variance. In addition to adopting a different performance index, the approach in this paper is quite different from them.

The stability of GPC is one of the main issues in recent generalised predictive control research. Despite its advantages and success, GPC is deficient in that it does not offer a general stability result. A number of stable generalised predictive control algorithms have been proposed (Mosca and Zhang, 1992; Demircioglu and Clarke, 1992; Rossiter and Kouvaritakis, 1993; Kouvaritakis and Rossiter, 1993; Rawlings and Muske, 1993; Scotaert and Rawlings, 1994). In most of those the stability of GPC is achieved by adding terminal constraints in the design procedure. That is the stability of the closed-loop system is obtained at the price of the suboptimal design.

Dynamical inversion control provides a systematic alternative to gain scheduling for nonlinear systems (Slotine and Li, 1991; Singh and Rugh, 1972). The procedure is to rearrange the nonlinear equations to compute control inputs to cancel the nonlinear terms and replace them with desired linear terms. It has been widely applied in robotics and flight control (Lane and Stengel, 1988; Snell *et al.*, 1992; Meyer *et al.*, 1984; Bugajski and Enns, 1992). It is also known as computed torque control in robotics (Luca and Siciliano, 1993; Yim and Singh, 1997). The relationship between the dynamical inversion control and the generalised predictive control will be shown in this paper.

This paper is organised as follows. A brief description of dynamical inversion control is given in Section 2. In Section 3 the problem of the continuous-time generalised predictive control for nonlinear systems is formulated and the nonlinear GPC controller is designed. The relationship between nonlinear GPC and dynamical inversion control is highlighted. Stability of the closed-loop system under nonlinear GPC is shown in Section 4, and an application to the control of a two-link robotic manipulator is shown in Section 5. Finally a brief conclusion is given in Section 6.

2. DYNAMICAL INVERSION CONTROL

The general form of a mechanical system is given by the following second order matrix equation

$$J\Big(\theta(t),\dot{\theta}(t)\Big)\ddot{\theta}(t)+G\Big(\theta(t),\dot{\theta}(t)\Big) = B\Big(\theta(t),\dot{\theta}(t)\Big)U(t) \quad (1)$$

where $\theta \in R^n$ denotes the displacement of the mechanical systems and n is the degree of the freedom of the mechanical system. $U(t) \in R^n$ denotes the vector of the generalised torque and forces. $J \in R^{n\times n}$ denotes the inertia matrix. $G(\theta(t),\dot{\theta}(t))$ consists of viscous friction, Coriolis and centrifugal terms and the gravitational term etc. In general, the matrix $J(\theta(t),\dot{\theta}(t))$ is positive definite for all allowable $\theta(t)$ and $\dot{\theta}(t)$. It is supposed that the control $U(t)$ can be fully recovered from Equation (1). As discussed in Luca and Siciliano (1993), this assumption is satisfied by many physical systems. For simplicity, only the case where the number of the input is equal to the degree of the freedom of the mechanical systems is considered in this paper. However the conclusion also holds for the systems where the number of the output is less than that of the freedom of the mechanical systems and the zero dynamics is stable.

The controlled output is the displacement of the mechanical system, i.e.,

$$y = \theta(t) \quad (2)$$

Suppose the desired output, i.e., the command signal, is given by $\theta_d(t)$. The basic idea in the dynamic inversion control is to calculate the input $U(t)$ which is required to drive the system (1) to produce the desired response. That is

$$U(t) = B(\theta_d,\dot{\theta}_d)^{-1}\Big(J(\theta_d,\dot{\theta}_d)\ddot{\theta}_d + G(\theta_d,\dot{\theta}_d)\Big) \quad (3)$$

However due to the uncertainties and the disturbance etc., this ideal control law (3) is open loop control and hence not good in engineering application. When the states of the system (1) are available, the control law (3) is modified as

$$U(t) = B(\theta,\dot{\theta})^{-1}\Big(J(\theta,\dot{\theta})v + G(\theta,\dot{\theta}\Big) \quad (4)$$

where $\ddot{\theta}_d$ in (3) is replaced by a new input v, which is, in general, determined by

$$v(t) = \ddot{\theta}_d + K_v(\dot{\theta}_d - \dot{\theta}) + K_p(\theta_d - \theta) \quad (5)$$

The control law (4) and (5) is also known as computed torque control in robotics. The matrices K_v and K_p are called as the velocity feedback gain matrix and displacement feedback gain matrix respectively. In general, there is no systematic method for choosing these parameters and they are designed by trial and error according to the designer experience and system specifications.

3. GENERALISED PREDICTIVE CONTROL FOR NONLINEAR SYSTEMS

Consider the nonlinear system in (1) with the following performance index

$$J = \frac{1}{2}\int_{T_1}^{T_2} (y(t+\tau) - \theta_d(t+\tau))^T$$
$$(y(t+\tau) - \theta_d(t+\tau))d\tau \qquad (6)$$

where T_1 and T_2 are the minimum and maximum predictive times respectively. Following Gawthrop *et al.* (1998) the control weighting term is not included in the performance index (6) because it is believed that it is not a very useful form of control weighting. As will be shown later, the effect of control weighting can be accomplished by T_2 in Generalised Predictive Control(GPC). Another way to limit the control effect is to use P polynomials (Gawthrop *et al.*, 1998).

Differentiating the output $y(t)$ twice with respect to time, together with substitution of the system (1) gives

$$\dot{y}(t) = \dot{\theta}(t)$$

$$\ddot{y}(t) = \ddot{\theta} = J(\theta,\dot{\theta})^{-1}(-G(\theta,\dot{\theta}) + B(\theta,\dot{\theta})U(t))$$

Combining the above equation with the output equation (2) yields

$$\bar{Y} = \begin{bmatrix} y \\ \dot{y} \\ \ddot{y} \end{bmatrix} = M(\theta,\dot{\theta}) + \begin{bmatrix} 0 \\ 0 \\ N(\theta,\dot{(\theta)}) \end{bmatrix} U \qquad (7)$$

where

$$N = J(\theta,\dot{\theta})^{-1}B(\theta,\dot{\theta}) \qquad (8)$$

and

$$M = \begin{bmatrix} \theta \\ \dot{\theta} \\ -J(\theta,\dot{\theta})^{-1}G(\theta,\dot{\theta}) \end{bmatrix} \qquad (9)$$

A Taylor expansion of the output $y(t+\tau)$ and the command signal $\theta_d(t+\tau)$ at time t up to second order yields

$$y(t+\tau) \doteq \mathcal{T}(\tau)\bar{Y}(t) \qquad (10)$$

and

$$\theta_d(t+\tau) \doteq \mathcal{T}(\tau)\bar{W}(t) \qquad (11)$$

where

$$\bar{W}(t) = \begin{bmatrix} \theta_d(t) \\ \dot{\theta}_d(t) \\ \ddot{\theta}_d(t) \end{bmatrix} \qquad (12)$$

and the $n \times n$ matrix $\bar{\tau}$ and the $n \times 3n$ matrix $\mathcal{T}(\tau)$ are given by

$$\bar{\tau} = \mathrm{diag}\{\tau, \ldots, \tau\} \qquad (13)$$

and

$$\mathcal{T}(\tau) = \left[I \;\; \bar{\tau} \;\; \bar{\tau}^2/2! \right] \qquad (14)$$

respectively.

Hence the performance index (6) can be written as

$$J = \frac{1}{2}\int_{T_1}^{T_2} (\bar{Y}(t) - \bar{W}(t))^T \mathcal{T}(\tau)^T \mathcal{T}(\tau)$$
$$(\bar{Y}(t) - \bar{W}(t))d\tau$$
$$= \frac{1}{2}(\bar{Y}(t) - \bar{W}(t))^T \bar{\mathcal{T}}(T_1,T_2)$$
$$(\bar{Y}(t) - \bar{W}(t)) \qquad (15)$$

where

$$\bar{\mathcal{T}}(T_1,T_2) = \int_{T_1}^{T_2} \mathcal{T}(\tau)^T \mathcal{T}(\tau)d\tau \qquad (16)$$

Let

$$\bar{\mathcal{T}} = \begin{bmatrix} \bar{\mathcal{T}}_{11} & \bar{\mathcal{T}}_{12} & \bar{\mathcal{T}}_{13} \\ \bar{\mathcal{T}}_{21} & \bar{\mathcal{T}}_{22} & \bar{\mathcal{T}}_{23} \\ \bar{\mathcal{T}}_{31} & \bar{\mathcal{T}}_{32} & \bar{\mathcal{T}}_{33} \end{bmatrix} \qquad (17)$$

It follows from (14) that

$$\bar{\mathcal{T}}_{ij}(T_1,T_2) = \frac{\bar{T}_2^{i+j-1} - \bar{T}_1^{i+j-1}}{(i-1)!(j-1)!(i+j-1)} \qquad (18)$$

where $\bar{T}_2$ and $\bar{T}_1$ are obtained by replacing τ with T_2 and T_1 in (13) respectively. Substitution of Equation (7) into the cost (15) and minimisation of the cost (15) with respect to the control, $U(t)$, gives

$$U^{\star}(t) = -\left(\begin{bmatrix} 0 \\ 0 \\ N(\theta,\dot{\theta}) \end{bmatrix}^T \bar{\mathcal{T}}(T_1,T_2) \begin{bmatrix} 0 \\ 0 \\ N(\theta,\dot{\theta}) \end{bmatrix} \right)^{-1}$$
$$\begin{bmatrix} 0 \\ 0 \\ N \end{bmatrix}^T \bar{\mathcal{T}}(T_1,T_2)(M(\theta,\dot{\theta}) - \bar{W})) \qquad (19)$$

It can be shown that the matrix $\bar{\mathcal{T}}_{33}$ is positive definite and hence invertible. Also noting $B(\theta,\dot{\theta})$ for all allowable θ and $\dot{\theta}$ is invertible and using (9) and (8), the optimal control (19) can be written as

$$U^{\star}(t) = -B(\theta,\dot{\theta})^{-1}J(\theta,\dot{\theta})\bar{\mathcal{T}}_{33}^{-1}\left[\bar{\mathcal{T}}_{31} \;\; \bar{\mathcal{T}}_{32} \;\; \bar{\mathcal{T}}_{33} \right]$$
$$\begin{bmatrix} \theta - \theta_d \\ \dot{\theta} - \dot{\theta}_d \\ -J(\theta,\dot{\theta})^{-1}G(\theta,\dot{\theta}) - \ddot{\theta}_d \end{bmatrix} \qquad (20)$$

Let

$$K_1 = \bar{\mathcal{T}}_{33}^{-1}\bar{\mathcal{T}}_{31} \qquad (21)$$

and

$$K_2 = \bar{T}_{33}^{-1}\bar{T}_{32} \tag{22}$$

The control further becomes

$$U^*(t) = -B(\theta,\dot{\theta})^{-1}J(\theta,\dot{\theta})\left\{K_1\left(\theta-\theta_d\right)+ + K_2\left(\dot{\theta}-\dot{\theta}_d\right) - J(\theta,\dot{\theta})^{-1}G(\theta,\dot{\theta}) - \ddot{\theta}_d\right\}$$
$$= B(\theta,\dot{\theta})^{-1}\left\{G(\theta,\dot{\theta} + J(\theta,\dot{\theta})\left(K_1(\theta_d-\theta) + +K_2(\dot{\theta}_d-\dot{\theta}) + \ddot{\theta}_d\right)\right)\right\} \tag{23}$$

Obviously, generalised predictive control (23) is the same as the dynamical inversion control (4) in Section 2 except that the gain matrices are optimised in term of the generalised predictive control performance index (6). Hence the dynamical inversion control is a special case of GPC. That is, for the system (1), only the Taylor expansion up to second order is employed to predict the future output $y(t+\tau)$ for $T_1 \le \tau \le T_2$.

It should be noticed that the parameter matrices K_1 and K_2 are constant and only depend on the predictive times T_1 and T_2.

4. STABILITY OF CLOSED-LOOP SYSTEMS

The stability of the closed loop system is one of the main problems in GPC. This section will show that the closed-loop system under GPC (23) is asymptotically stable. For simplicity, choose T_1=0 and $T_2 = T$, but all results below can be extended to the case where $T_1 \neq 0$.

The closed-loop system of the plant (1) and the control (23) can be written as

$$J(\theta,\dot{\theta})\left(\ddot{\theta} - \ddot{\theta}_d + K_1(\dot{\theta}_d - \dot{\theta}) + K_2(\theta_d-\theta)\right) = 0 \tag{24}$$

Let

$$e(t) = \theta_d(t) - y(t) \tag{25}$$

It follows from $J(\theta(t),\dot{\theta}(t))$ being invertible for all allowable $\theta(t)$ and $\dot{\theta}(t)$ that

$$\ddot{e}(t) + K_1\dot{e}(t) + K_2 e(t) = 0 \tag{26}$$

Combining (21),(22) and (18) yields

$$K_2 = \frac{5}{2}\bar{T}^{-1} \tag{27}$$

and

$$K_1 = \frac{10}{3}\bar{T}^{-2} \tag{28}$$

Since $\bar{T}$ is a diagonal matrix, it implies

$$K_1(i,i) = \frac{5}{2T}, i = 1,\ldots,n; \qquad K_1(i,j) = 0, \quad i,j = 1,\ldots,n, i \neq j \tag{29}$$

and

$$K_2(i,j) = \frac{10}{3T^2}, i = 1,\ldots,n; \qquad K_2(i,j) = 0, \quad i,j = 1,\ldots,n, i \neq j \tag{30}$$

Thus the closed-loop error equations (26) are decoupled and the stability of the closed-loop system is determined by the equations

$$\ddot{e}_i(t) + 2.5/T\dot{e}_i(t) + 10/(3T^2)e_i(t) = 0, \quad i = 1,2,\ldots,n \tag{31}$$

The stability of the closed-loop system can be determined by letting $\theta_d = 0$ in (31) since the stability of the linear system (31) is not effected by the command signal. The poles of the above equation are $(-1.25 \pm 1.3307j)/T$. Hence $y(t)$ and $\dot{y}(t)$, which can be chosen as the states of the system (1), approach zero. This is a sufficient condition for the closed-loop system under the control (23) to be asymptotically stable.

It is interesting to notice that the poles of the equation linearly depend on the predictive time T. With a small T, the controlled output $y(t)$ will track the command signal $\theta_d(t)$ rapidly, which in general needs a large control effort. Hence as shown in the example in Section 5, the control effort can be adjusted by selecting the predictive time T.

5. GENERALISED PREDICTIVE CONTROL OF A TWO-LINK ROBOT

In this section the GPC controller developed in this paper is applied in control of a robotic manipulator. This manipulator consists of two links. Each link is directly driven by a D.C. motor independently. The displacement and the velocity of each link are measured by potentiometers and tachometers respectively. The dynamics of the two-link arms including the first order dynamics of the D. C. motors is given by (Roberts, 1993; Gawthrop and Smith, 1996). It can be put into the format (1).

where

This example has two degrees of the freedom, i.e., $n = 2$. First T_1 and T_2 in the performance index (6) are chosen as $T_1 = 0$ and $T_2 = 0.5$. Following, the resultant GPC control law is given by (23) with

$$K_1 = \mathrm{diag}\{5,5\}, \quad K_2 = \mathrm{diag}\{40/3, 40/3\} \tag{32}$$

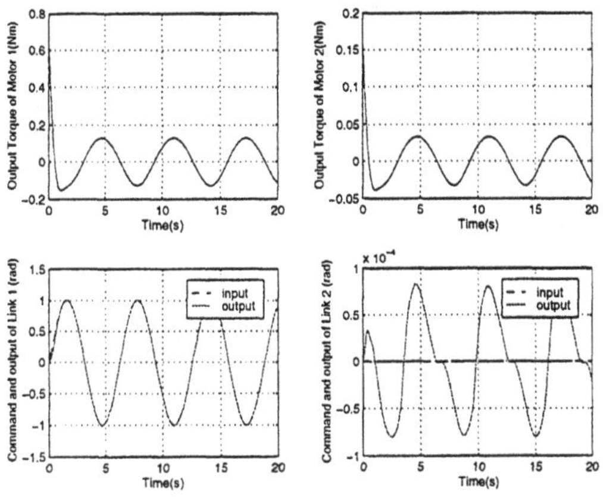

Fig. 1. The tracking performance with the predictive time T_2=0.5s

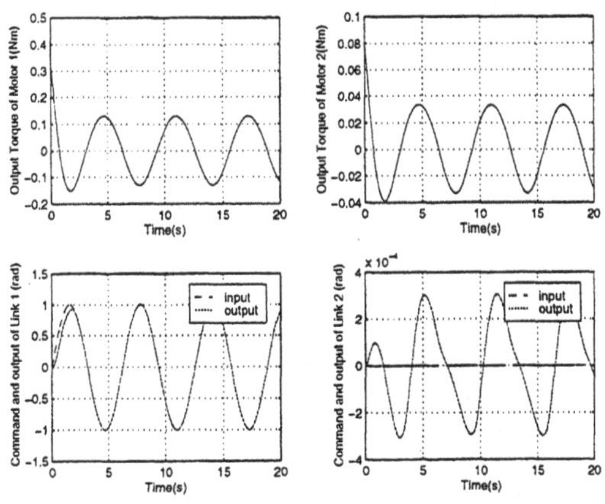

Fig. 2. The tracking performance with the predictive time T_2=1s

When the command signals on the first link and the second link are a sinusoidal signal and zero respectively, i.e.,

$$w_1(t) = \sin(t)(rad), \qquad w_2 = 0(rad) \quad (33)$$

Thus the first link is to track the given sinusoidal signal and the second link is to keep the relative angle to zero. The tracking performance of the robot under the control (23) with (32) are shown in Fig. 1. The dot-dashed line and the solid line are the reference signal and the response respectively. The first link tracks the command signal almost immediately and the second link keeps its angle error within a very small range (10^{-4} rad).

To show the effect of the predictive time, choose $T_1 = 0$ and $T_2 = 1$. The control law is the same as (23) but with the parameter matrices

$$K_1 = \text{diag}\{2.5, 2.5\}, \quad K_2 = \text{diag}\{10/3, 10/3\}$$

The output and the control are shown in Fig. 2. It is shown that the control is much smaller than that in the control law (32). This is achieved at a cost of slightly poorer performance.

6. CONCLUSION

A nonlinear continuous-time generalised predictive control has been derived in a state space formulation. It is shown that the generalised predictive control can ensure the closed-loop system is asymptotically stable. The analytical expression for the poles of the closed-loop systems are given and it is shown that they only depend on the predictive times. It is pointed out that the dynamical inversion control can be considered to be a special kind of Generalised Predictive Control. It is well known that the dynamical inversion control has a close relation with the feedback linearisation theory. Hence this result, from another angle, supports the view point in Gawthrop *et al.* (1998) that Generalised Predictive Control has a close relation with the feedback linearisation theory in nonlinear systems. Application of the nonlinear GPC in the two-link robotic manipulator shows that the developed method exhibits good tracking performance. Further work is required to develop the nonlinear generalised predictive control for general nonlinear systems particularly systems with unstable zero dynamics.

ACKNOWLEDGEMENTS

This work has been supported by the UK Engineering and Physical Sciences Research Council under the grant No. GR/L 62665.

7. REFERENCES

Bugajski, D. J. and D. F. Enns (1992). Nonlinear control law with application to high angle-of-attack flight. *Journal of Guidance, Control and Dynamics* **15**(3), 761–767.

Clarke, D. W. (1994). *Advances in Model-based Predictive Control.* Oxford University Press.

Clarke, D. W. and P. J. Gawthrop (1975). Self-tuning controller. *Proc. IEE Part D* **122**(9), 929–934.

Clarke, D. W. and P. J. Gawthrop (1979). Self-tuning Control. *Proc. IEE Part D* **126**(6), 633–640.

Clarke, D. W., C. Mohtadi and P. S. Tuffs (1987*a*). Generalised Predictive Control—Part I. The Basic Algorithm. *Automatica* **23**(2), 137–148.

Clarke, D. W., C. Mohtadi and P. S. Tuffs (1987*b*). Generalised Predictive Control—Part II. Extensions and Interpretations. *Automatica* **23**(2), 149–160.

Demircioglu, H. (1989). Continuous-time Self-tuning Algorithms. Ph.D. Thesis. Glasgow University, Faculty of Engineering. Supervised by P. J. Gawthrop.

Demircioglu, H. and D. W. Clarke (1992). CGPC with guarenteed stability properties. *Proc. IEE* **139**(4), 371–380.

Demircioglu, H. and P. J. Gawthrop (1991). Continuous-time generalised predictive control. *Automatica* **27**(1), 55–74.

Demircioglu, H. and P. J. Gawthrop (1992). Multivariable continuous-time generalised predictive control. *Automatica* **28**(4), 697–713.

Gawthrop, P. J. and L. P. S. Smith (1996). *Metamodelling: Bond Graphs and Dynamic Systems*. Prentice Hall. Hemel Hempstead, Herts, England.

Gawthrop, P. J., H. Demircioglu and I. Siller-Alcala (1998). Multivariable continuous-time generalised predictive control: A state-space approach to linear and nonlinear systems. *Proc. IEE Pt. D: Control Theory and Applications* **145**(3), 241–250.

Kouvaritakis, B. and J.A Rossiter (1993). Multivariable stable generalised predictive control. *IEE Proceedings-D* **140**(5), 364–372.

Lane, S. H. and R. F. Stengel (1988). Flight control design using non-linear inverse dynamics. *Automatica* **24**(4), 471–483.

Lu, Ping (1994). Nonlinear predictive controllers for continuous systems. *Journal of Guidance, Control and Dynamics* **17**(3), 553–560.

Lu, Ping (1995). Optimal predictive control for continuous nonlinear systems. *Int. J. of Control* **62**(3), 633–649.

Lu, Ping (1997). Tracking control of nonlinear systems with bounded controls and control rates. *Automatica* **33**(6), 1199–1202.

Luca, A. De and B. Siciliano (1993). Inversion based nonlinear control of robot arms with flexible links. *Journal of Guidance, Control and Dynamics* **16**(6), 1169–1176.

Meyer, G., R. Su and R. Hunt (1984). Application of nonlinear transformation to autimatic flight control. *Automatica* **20**(1), 103–107.

Morari, M. (1994). Model predictive control: Multivariable control technique of choice in the 1990s?. In: *Advances in Model-based Predictive Control*. pp. 22–37. Oxford University Press.

Mosca, E. and J. Zhang (1992). Stable redesign of predictive control. *Automatica* **28**(6), 1229–1233.

Ordys, A. W. and D. W. Clarke (1993). A state-space description for GPC controllers. *International Journal of Systems Science* **29**(9), 1727–1744.

Rawlings, J. B. and K. R. Muske (1993). Nonlinear predictive control of feedback linearisable systems and flight control system design. *IEEE Trans. on Automatic Control* **38**(10), 1512–1516.

Roberts, D. W. (1993). Bond Graph Model Based Control of Robotic Manipulators. Ph.D. Thesis. Glasgow University, Faculty of Engineering. Supervised by P. J. Gawthrop.

Rossiter, J.A. and B. Kouvaritakis (1993). Constrained stable predictive control. *IEE Proceedings-D* **140**(4), 243–254.

Scotaert, P. O. M. and J. B. Rawlings (1994). Infinite horizon linear quadratic control with constrains. In: *Proceedings of IFAC World Congress*. Vol. M. pp. 109–114.

Singh, S. N. and W. J. Rugh (1972). Decoupling in a class of nonlinear systems by state variable feedback. *Journal of Dynamic System, Measurement and Control* **94**(4), 323–329.

Singh, S. N., M. Steinberg and R. D. DiGirolamo (1995). Nonlinear predictive control of feedback linerisable systems and flight control system design. *Journal of Guidance, Control and Dynamics* **18**(6), 1023–1028.

Slotine, J. E. and W. Li (1991). *Applied Nonlinear Control*. Prentice-Hall. Englewood Cliffs.

Snell, S. A., D. F. Enns and W. L. Garrard (1992). Nonlinear inversion flight control for a supermaneuver aircraft. *Journal of Guidance, Control and Dynamics* **15**(4), 976–984.

Soroush, M. and C. Kravaris (1992). A continuous-time formulation of nonlinear model predictive control. In: *Proceedings of the 1992 American Control Conference*. Chicago, U.S.A.. pp. 393–399.

Soroush, M. and C. Kravaris (1993). Model predictive control of multivariable nonlinear processes in continuous-time. In: *Proceedings of the 1993 American Control Conference*. San Francisco, U.S.A.. pp. 1562–1567.

Yim, W. and S. N. Singh (1997). Nonlinear inverse and predictive end point trajectory control of flexible macro-micro manipulators. *Journal of Dynamic Systems, Measurement and Control* **119**(3), 365–604.

E-2c-18-2
14th Triennial World Congress, Beijing, P.R. China

MULTIPLE MAGNITUDE SCALE DLC SCHEME FOR HIGH ORDER SYSTEMS

Jian-Xin Xu * **Yanbin Song and Ji Qian**

* *National University of Singapore*
Department of Electrical Engineering
10 Kent Ridge Crescent Singapore 119260
Email: *elexujx@nus.edu.sg*

Abstract: In this paper, we propose two multi-scale Direct Learning Control (DLC) schemes for time-varying high order systems, one for linear systems and the other for nonlinear systems. DLC schemes developed hitherto only consider trajectories with a single-scale relation in that any two trajectories are spatially proportional to each other with a single scale for all coordinates (dimensions). In comparison, the new schemes are capable of handling trajectories with multiple scales, *i.e.*, each spatial coordinate of a trajectory is proportional to that of another trajectory associated with a distinct scale. Simulation results confirm the effectiveness of these methods.

Keywords: Direct Learning Control, Multiple Magnitude Scale, Linear and Nonlinear system

1. INTRODUCTION

Learning control methods have been developed for quite a long period and numerous algorithms have been proposed, e.g., (Arimoto *et al.*, 1984), (Xu, 1997). However, there still exist a number of problems which hinder the implementation of learning control techniques. The major obstacle is perhaps the requirement that the trajectory to be learned must be strictly repeatable over the operations. If any change occurs due to the variation of control objectives or task specifications, no matter how small it might be, the control system has to start the learning process from the very beginning and the previously learned control input profiles can no longer be used. Besides, from practical point of view, a control environment may even not allow any repeated operations at all.

Generally speaking, there are two kinds of non-repeatable problems encountered in learning control: non-repeatability of a motion task and non-repeatability of a process. The *non-repeatable motion task* could be shown through the following example: an XY-table draws two circles with the same period but different radii. The *non-repeatability of a process* could be due to the nature of system such as welding different parts in a manufacturing line. Without loss of generality, we refer to these two kinds of problems as non-repeatable control problems which result in extra difficulty when a learning control scheme is to be applied.

From practical point of view, non-repeatable learning control is more important and indispensable. In order to deal with non-repeatable learning control problems, we need to explore the inherent relations of different motions trajectory patterns. The resulting learning control scheme might be both plant-dependent and trajectory-dependent. On the other hand, since learning control task is essentially to drive the system tracking the given trajectories, the inherent spatial and speed relationships among distinct motion trajectories actually provide useful information. Moreover, in spite of the variations in the trajectory patterns, the underlying dynamic properties of the controlled system remain the same. Therefore it is possible for us to deal with non-repeatable learning control problems.

In practice, a control system may have plenty of prior control knowledge obtained through all the past control actions although they may correspond to different tasks. In order to effectively utilize all the prior control knowledge so as to remove the iterative learning process, a number of *direct learning control* schemes have been developed (Xu, 1997). *Direct Learning Control* is defined as the direct generation of the desired control profile from existing control inputs without any repeated learning. The ultimate goal of DLC is to fully utilize all the pre-stored control profiles and eliminate the time consuming iteration process thoroughly, even though these control profiles may correspond to different motion patterns and be obtained using different control methods.

In this paper we will propose a new DLC scheme for a class of non-repeatable learning control problems: all the trajectories are different in magnitude with multiple scales. In other words, a trajectory can no longer be simply expressed by another trajectory multiplying with a scalar. Consider again the XY-table as the example, the existing single scale DLC schemes can only learn and handle patterns with similar shapes but different sizes, such as circles with different radii. Multi-scale DLC schemes is able to learn and handle patterns with different shapes, such as from circles to ellipses or vice versa. That is, different scales are assigned to X and Y coordinates respectively. Obviously the multi-scale task is more general and challenging and worthwhile for us to explore new learning control methods.

This paper is organized in the following manner. Section 2 provides a multi magnitude scale DLC scheme for a class of linear time-varying high order system. Section 3 formulates a multi magnitude scale DLC for certain classes of nonlinear systems. The schemes are then applied to a two-link manipulator to verify their effectiveness in Section 4 as illustrative examples.

2. MULTI MAGNITUDE SCALE DLC SCHEME FOR LINEAR TIME VARYING SYSTEMS

2.1 *Problem statement*

Consider a linear time varying high order system given by the following equations

$$\begin{aligned}\dot{\mathbf{x}}_i &= \mathbf{x}_{i+1} \quad i = 1, \ldots, n-1\\ \dot{\mathbf{x}}_n &= A_1(t)\mathbf{x}_1 + \ldots + A_n(t)\mathbf{x}_n + B(t)\mathbf{u} \quad (1)\end{aligned}$$

where $\mathbf{x}_1 = [\, x_1 \ \ldots \ x_n \,]^T$ is the system states vector which is measurable. x_i is the ith coordinate of the system and $\mathbf{x}_i = [\, x_1^{(i-1)} \ \ldots \ x_n^{(i-1)} \,]^T$, $i = 2, \ldots, n$ is the $(i-1)$th differential of system states. $A_i(t), B(t) \in R^{n\times n}, i = 1, \ldots, n$ are unknown time-varying matrices. $B(t)$ is full rank for $\forall t \in [0, T]$ where $[0, T]$ is the tracking period.

The control objective is to find the control input for a new tracking control trajectory $\mathbf{x}_{1,d}$ over the given time period $t \in [0, T]$, where

$$\mathbf{x}_{1,d}(t) = [\, x_{1,d}(t) \ \ldots \ x_{n,d}(t) \,]^T$$

represents the desired system state trajectory.

[Assumption 4.1]
There are at least n known tracking control input profiles $\mathbf{u}_i(t)$ available with respect to n trajectories related to the desired one through the following multi magnitude scale relations

$$\mathbf{x}_{1,i} = K_i \mathbf{x}_{1,d}, \quad i = 1, \ldots, n$$

where $\mathbf{x}_{1,i} = [\, x_{1,i} \ \ldots \ x_{n,i} \,]^T$ is the system state vector for the ith trajectory, and K_i is a constant diagonal matrix taking the form

$$K_i = diag\,[\, k_{1,i} \ \ldots \ k_{n,i} \,].$$

[Assumption 4.2]
The n learned trajectories are chosen in such a way that

$$R_1 = \begin{bmatrix} k_{1,1} & \cdots & k_{n,1} \\ \vdots & \ddots & \vdots \\ k_{1,n} & \cdots & k_{n,n} \end{bmatrix},$$

is full rank.

Remark 4.1
Assumption 4.2 implies that the n trajectories must be "linearly independent". In other words, if the only solution to $\sum_{i=1}^{n} c_i K_i = 0$ is $c_i = 0, \forall i = 1, \ldots, n$, then the augmented matrix $\mathbf{K}$ is nonsingular. This will be further explained in Lemma 4.1.

To facilitate the derivation of DLC in subsequent section, two lemmas are given below.

Lemma 4.1 *For the multiplication of a matrix $A \in R^{n\times n}$ with a diagonal matrix $K_i \in R^{n\times n}$, the following relation holds*

$$AK_i = \sum_{l=1}^{n} K_i^l A^l \qquad (2)$$

where

$$K_i^l \triangleq diag\,[\, k_{1,i}^l \ \ldots \ k_{m,i}^l \ \ldots \ k_{n,i}^l \,]$$

$$k^l_{m,i} \triangleq k_{[l+m-1],i} \quad m = 1, \ldots, n$$

$$A^l \triangleq \begin{bmatrix} a^l_{1,1} & \cdots & a^l_{1,n} \\ \vdots & \ddots & \vdots \\ a^l_{n,1} & \cdots & a^l_{n,n} \end{bmatrix}$$

$$a^l_{m,j} \triangleq \begin{cases} 0, & others \\ a_{m,j}, \; j = [l+m-1] \end{cases}.$$

The notations $a_{m,j}, a^l_{m,j}$ denote the (m,j)th element of matrices A and A_l respectively.

The operator $[l]$ is defined as

$$[l] \triangleq \begin{cases} l & l \le n \\ l-n & l > n \end{cases}.$$

Lemma 4.2 *The singularity of the augmented matrix*

$$\mathbf{K} = \begin{bmatrix} K^1_1 & \cdots & K^n_1 \\ \vdots & \ddots & \vdots \\ K^1_n & \cdots & K^n_n \end{bmatrix}. \quad (3)$$

$\in R^{n^2 \times n^2}$ is equivalent to the singularity of the following matrix

$$R_1 = \begin{bmatrix} k_{1,1} & \cdots & k_{n,1} \\ \vdots & \ddots & \vdots \\ k_{1,n} & \cdots & k_{n,n} \end{bmatrix}.$$

where $k_{i,j}$ is the ith diagonal element of the multiscale matrix $K_j \in R^{n \times n}$ as defined in Lemma 4.1.

2.2 *Derivation of multi magnitude scale DLC scheme*

The following theorem relates the multi-scale DLC scheme.

Theorem 4.1 *For the plant given by Eq.(1), the desired control input $\mathbf{u}_d(t)$ with respect to a new trajectory $\mathbf{x}_{1,d}(t)$, $t \in [0,T]$ can be directly obtained using past control inputs according to the following relations*

$$\mathbf{u}_d(t) = [\,I \ \ldots \ I\,]\mathbf{K}^{-1} \begin{bmatrix} \mathbf{u}_1(t) \\ \vdots \\ \mathbf{u}_n(t) \end{bmatrix}$$

where $\mathbf{u}_i(t)$ is the ith known control input profile.

Proof:
Because $B(t)$ is of full rank, Eq.(1) can be written as follows,

$$\mathbf{u}(t) = B(t)^{-1}\dot{\mathbf{x}}_n - B(t)^{-1}A_1(t)\mathbf{x}_1 - \ldots \quad (4)$$
$$-B(t)^{-1}A_n(t)\mathbf{x}_n. \quad (5)$$

Since all trajectories are correlated as $\mathbf{x}_{1,i} = K_i\mathbf{x}_{1,d}$, $i = 1, \ldots, n$, it is straightforward to obtain the following equations

$$\mathbf{x}_{1,i} = K_i\mathbf{x}_{1,d}$$
$$\vdots$$
$$\mathbf{x}_{n,i} = K_i\mathbf{x}_{n,d}$$
$$\dot{\mathbf{x}}_{n,i} = K_i\dot{\mathbf{x}}_{n,d}.$$

By substituting the above equation into Eq.(4), the following relation can be obtained,

$$\begin{aligned} \mathbf{u}_i(t) &= B(t)^{-1}\dot{\mathbf{x}}_{n,i} - B(t)^{-1}A_1(t)\mathbf{x}_{1,i} - \ldots \\ &\quad -B(t)^{-1}A_n(t)\mathbf{x}_{n,i} \\ &= B(t)^{-1}K_i\dot{\mathbf{x}}_{n,d} - B(t)^{-1}A_1(t)K_i\mathbf{x}_{1,d} - \ldots \\ &\quad -B(t)^{-1}A_n(t)K_i\mathbf{x}_{n,d} \\ &= D_1(t)K_i\dot{\mathbf{x}}_{n,d} + D_2(t)K_i\mathbf{x}_{1,d} + \ldots \\ &\quad +D_{n+1}(t)K_i\mathbf{x}_{n,d} \end{aligned}$$

where

$$\begin{aligned} D_1(t) &= B(t)^{-1} \\ D_2(t) &= -B(t)^{-1}A_1(t) \\ &\cdots \\ D_{n+1} &= -B(t)^{-1}A_n(t). \end{aligned}$$

According to Lemma 4.1, the following relation exists,

$$\mathbf{u}_i(t) = \left(\sum_{j=1}^n K^j_i D^j_1(t)\right)\dot{\mathbf{x}}_{n,d} + \left(\sum_{j=1}^n K^j_i D^j_2(t)\right)\mathbf{x}_{1,d} + \ldots + \left(\sum_{j=1}^n K^j_i D^j_{n+1}(t)\right)\mathbf{x}_{n,d}$$

where

$$D^l_m(t) = \begin{bmatrix} (d^l_{1,1})_m & \cdots & (d^l_{1,n})_m \\ \vdots & \ddots & \vdots \\ (d^l_{n,1})_m & \cdots & (d^l_{n,n})_m \end{bmatrix}$$

$$(d^l_{i,j})_m = \begin{cases} 0, & others \\ (d_{ij})_m, \; j = [l+i-1] \end{cases},$$

$(d_{i,j})_m$ is the (i,j)th element of matrix D_m, $m = 1, \ldots, n$.

By rearranging the above equation, we have

$$\mathbf{u}_i(t) = \sum_{j=1}^n K^j_i \mathbf{z}_j(t) \quad (6)$$

$$\mathbf{z}_j(t) = D^j_1(t)\dot{\mathbf{x}}_{n,d} + D^j_2(t)\mathbf{x}_{1,d} + \ldots + D^j_{n+1}(t)\mathbf{x}_{n,d}.$$

It should be noted that although $D_i^j(t)$ is time-varying, it remains identical at any given time instant for different trajectories, so does the desired trajectory and its derivatives $\mathbf{x}_{i,d}(t)$. Therefore, the unknown terms $\mathbf{z}_j(t)$ can be calculated in a point-wise manner with the known coefficient matrix K_i^j and control inputs $\mathbf{u}_i(t)$.

From Assumption 4.1 we know that there are n previously stored control profiles. Eq. (6) can be written in an augmented form

$$\begin{bmatrix} \mathbf{u}_1(t) \\ \vdots \\ \mathbf{u}_n(t) \end{bmatrix} = \mathbf{K} \begin{bmatrix} \mathbf{z}_1(t) \\ \vdots \\ \mathbf{z}_n(t) \end{bmatrix} \tag{7}$$

where $\mathbf{K}$ is defined in Eq.(3).

If n pre-stored trajectories meet Assumption 4.2, according to Lemma 4.2 the terms $\mathbf{z}_j(t)$ can be obtained as

$$\begin{bmatrix} \mathbf{z}_1(t) \\ \vdots \\ \mathbf{z}_n(t) \end{bmatrix} = \mathbf{K}^{-1} \begin{bmatrix} \mathbf{u}_1(t) \\ \vdots \\ \mathbf{u}_n(t) \end{bmatrix}. \tag{8}$$

On the other hand, from Eq.(4) it can be concluded that the desired control input is

$$\mathbf{u}_d(t) = B(t)^{-1}\dot{\mathbf{x}}_{n,d} - B(t)^{-1}A_1(t)\mathbf{x}_{1,d} - \dots$$
$$-B(t)^{-1}A_n(t)\mathbf{x}_{n,d}.$$

Note that because $B(t)$ and $A_i(t)$ are unknown, the control input is not available from the above equation.

Utilizing the Definition of $\mathbf{z}_j(t)$ in Eq.(6), we have

$$\begin{aligned}
\mathbf{u}_d(t) &= D_1(t)K_d\dot{\mathbf{x}}_{n,d} + D_2(t)K_d\mathbf{x}_{1,d} + \dots \\
&\quad + D_{n+1}(t)K_d\mathbf{x}_{n,d} \\
&= \left(\sum_{j=1}^{n} K_d^j D_1^j(t)\right)\dot{\mathbf{x}}_{n,d} + \left(\sum_{j=1}^{n} K_d^j D_2^j(t)\right)\mathbf{x}_{1,d} \\
&\quad + \dots + \left(\sum_{j=1}^{n} K_d^j D_{n+1}^j(t)\right)\mathbf{x}_{n,d} \\
&= \sum_{j=1}^{n} K_d^j \mathbf{z}_j(t)
\end{aligned}$$

where K_d is an identity matrix and $K_d^j, j = 1, \dots, n$ are also identity matrices according to Lemma 4.1.

Therefore the desired control input is directly available as below

$$\mathbf{u}_d(t) = \sum_{j=1}^{n} \mathbf{z}_j(t)$$

$$= [\,I \,\dots\, I\,]\mathbf{K}^{-1} \begin{bmatrix} \mathbf{u}_1(t) \\ \vdots \\ \mathbf{u}_n(t) \end{bmatrix}. \tag{9}$$

□□

Remark 4.2
It is obvious that at least n sets of distinct previously achieved control profiles are required by this scheme. If two multi-scale matrices K_i and K_j are quite close to each other and make the augmented matrix $\mathbf{K}$ near singular, extra control input profiles should be added to improve the rank condition of $\mathbf{K}$. The multi-scale DLC schemes remains almost the same, and the terms $\mathbf{z}_j(t)$ can be obtained in the sense of *Least Squares*,

$$\begin{bmatrix} \mathbf{z}_1(t) \\ \vdots \\ \mathbf{z}_n(t) \end{bmatrix} = (\mathbf{K}_e^T\mathbf{K}_e)^{-1}\mathbf{K}_e^T \begin{bmatrix} \mathbf{u}_1(t) \\ \vdots \\ \mathbf{u}_n(t) \\ \mathbf{u}_{n+1}(t) \end{bmatrix}$$

$$\mathbf{K}_e = \begin{bmatrix} K_1^1 & \dots & K_1^n \\ \vdots & \ddots & \vdots \\ K_n^1 & \dots & K_n^n \\ K_{n+1}^1 & \dots & K_{n+1}^n \end{bmatrix}$$

where $\mathbf{u}_{n+1}(t)$ is the extra control input and K_{n+1}^i, $i = 1, \dots, n$ are generated from the extra multi-scale matrix K_{n+1} according to the rule in Lemma 4.1. This approach is applicable when more than one extra previously obtained control profiles are needed.

Remark 4.3
In application the DLC scheme can be simplified by grouping and sorting Eq.(7) such that only the inverse of a reduced matrix

$$R_1 = \begin{bmatrix} k_{1,1} & \dots & k_{n,1} \\ \vdots & \ddots & \vdots \\ k_{1,n} & \dots & k_{n,n} \end{bmatrix}$$

need to be calculated and the procedure is similar to that of Lemma 4.2's proof.

3. DLC SCHEME FOR NONLINEAR SYSTEMS

In this section, we will propose a multi magnitude scale DLC scheme for nonlinear system. Suppose a nonlinear system is described by the following equations

$$\mathbf{x}^{(p)} = R(t)\xi(\mathbf{x}, t) + B(t)\mathbf{u} \tag{10}$$

$$R = \begin{bmatrix} \sum_{j=1}^{n} a_{11j}(t)\eta_j & \dots & \sum_{j=1}^{n} a_{1lj}(t)\eta_j \\ \vdots & \ddots & \vdots \\ \sum_{j=1}^{n} a_{m1j}(t)\eta_j & \dots & \sum_{j=1}^{n} a_{mlj}(t)\eta_j \end{bmatrix}$$

$$\eta_j = \eta_j(\mathbf{x}, t) \qquad (11)$$

where $\mathbf{x} \in R^m$ is measurable state vector and $\mathbf{x}^{(p)}$ is its highest differential order; $a_{kij}(t)$ are unknown parameters of t only; $\eta_j(\mathbf{x}, t)$ are known nonlinear functions; $B(t)$ are $m \times m$ unknown full rank square matrix; $\boldsymbol{\xi}(\mathbf{x}, t) \in R^l$ are partially know homogeneous functions defined below.

Definition 4.1
The elements of $\boldsymbol{\xi}(\mathbf{x})$ are homogeneous function vectors of $\mathbf{x}$ with known orders. Usually its elements take the following form

$$\xi_i(\mathbf{x}) = \chi_1^{i_1}\chi_2^{i_2}\dots\chi_i^{i_i}\dots\chi_m^{i_m}, \qquad (12)$$

where χ_i is the product of system states x_i and its higher order derivatives.

Remark 4.4
The homogeneous functions have the property that during the same time period $t \in [0, T]$, for two related states $\mathbf{x}_1 = K\mathbf{x}_2$, $K = diag[k_1, k_2, \dots, k_m]$, there exists the following equation

$$\xi_i(\mathbf{x}_1) = \kappa_i\xi_i(\mathbf{x}_2)$$
$$\kappa_i = k_1^{i_1}k_2^{i_2}\dots k_m^{i_m}$$

where k_i, $i = 1, \dots, m$ is the corresponding element of the diagonal matrix of relationship matrix K.

[Assumption 4.3]
There are sufficient pre-stored trajectories $\mathbf{x}_i(t) \in [0, T]$. The corresponding desired control input signals $\mathbf{u}_i(t)$ have been achieved *a priori* through learning or other control methods. All the pre-stored trajectories including the desired trajectory $\mathbf{x}_d$ are inherently related with each other through a set of known constants multiple scale matrices K_i such that $\mathbf{x}_i(t) = K_i\mathbf{x}_d(t)$ where $t \in [0, T], K_i \neq I$.

The following theorem relates the multi scale DLC scheme for nonlinear systems.

Theorem 4.2 *For the nonlinear system described in Eq.(10), the desired control input $\mathbf{u}_d(t)$ with respect to a new trajectory $\mathbf{x}_d(t)$, $t \in [0, T]$ can be directly obtained using past control inputs directly.*

Proof:
Since matrix $B(t)$ is invertible, Eq. (10) can be rewritten as

$$\mathbf{u} = B^{-1}\mathbf{x}^{(p)} - B^{-1}R(t)\boldsymbol{\xi}(\mathbf{x}, t). \qquad (13)$$

It can be noted that $\mathbf{x}^{(p)}$ is a homogeneous function vector with homogeneous order of 1 according to Definition 4.1. Because for each element x_i, we have

$$x_i = \chi_1^0\chi_2^0\dots\chi_i^1\dots\chi_m^0$$

where $\chi_i = x_i^{(p)}$. Therefore $\mathbf{x}^{(p)}$ can be combined into $\boldsymbol{\xi}(\mathbf{x}), t)$. Denoting the number of the homogeneous functions after the combination as q $(q \leq m + l)$, then Eq.(13) can be further simplified as

$$\mathbf{u}(t) = \begin{bmatrix} \sum_{j=1}^{n} a'_{11j}(t)\eta_j & \dots & \sum_{j=1}^{n} a'_{1qj}(t)\eta_j \\ \vdots & \ddots & \vdots \\ \sum_{j=1}^{n} a'_{m1j}(t)\eta_j & \dots & \sum_{j=1}^{n} a'_{mqj}(t)\eta_j \end{bmatrix} \times \begin{bmatrix} \xi'_1(\mathbf{x}) \\ \vdots \\ \xi'_q(\mathbf{x}) \end{bmatrix} \qquad \eta_j = \eta_j(\mathbf{x}, t).$$

To facilitate discussion, we will process the above relations line by line. Consider the first element of the desired control input vector u_1, the following equation exists,

$$\begin{aligned} u_{1,d} &= \xi'_1(\mathbf{x}_d)\sum_{j=1}^{n} a'_{11j}(t)\eta_j(\mathbf{x}_d, t) + \dots \\ &+ \xi'_q(\mathbf{x}_d)\sum_{j=1}^{n} a'_{1qj}(t)\eta_j(\mathbf{x}_d, t) \\ &= \eta_1(\mathbf{x}_d, t)\sum_{j=1}^{q} a'_{1j1}(t)\xi'_j(\mathbf{x}_d) + \dots \\ &+ \eta_n(\mathbf{x}_d, t)\sum_{j=1}^{q} a'_{1jq}(t)\xi'_j(\mathbf{x}_d). \end{aligned}$$

Obviously the $u_{1,d}$ is not available now because of the unknown $a'_{1ij}(t)$ and $\xi_j(\mathbf{x})$.

On the other hand, the corresponding control input for the ith trajectory is

$$\begin{aligned} u_{1,i} &= \xi'_1(\mathbf{x}_i)\sum_{j=1}^{n} a'_{11j}(t)\eta_j(\mathbf{x}_i, t) + \dots \\ &+ \xi'_q(\mathbf{x}_i)\sum_{j=1}^{n} a'_{1qj}(t)\eta_j(\mathbf{x}_i, t) \\ &= \kappa_{1,i}\xi'_1(\mathbf{x}_d)\sum_{j=1}^{n} a'_{11j}(t)\eta_j(\mathbf{x}_i, t) + \dots \\ &+ \kappa_{q,i}\xi'_q(\mathbf{x}_d)\sum_{j=1}^{n} a'_{1qj}(t)\eta_j(\mathbf{x}_i, t) \end{aligned}$$

$$= \kappa_{1,i}\eta_1(\mathbf{x}_i,t)a'_{111}(t)\xi'_1(\mathbf{x}_d) + \dots$$
$$+\kappa_{1,i}\eta_n(\mathbf{x}_i,t)a'_{11n}(t)\xi'_1(\mathbf{x}_d) + \dots +$$
$$\kappa_{q,i}\eta_1(\mathbf{x}_i,t)a'_{1q1}(t)\xi'_q(\mathbf{x}_d) +$$
$$\dots + \kappa_{q,i}\eta_n(\mathbf{x}_i,t)a'_{1qn}(t)\xi'_q(\mathbf{x}_d).$$

Since we have assumed there are enough previous trajectories with known control inputs available, by using $N = q{\times}n$ such trajectories , the unknown parts $a'_{1ij}(t)\xi'_j(\mathbf{x})$ can be calculated as follows

$$\mathbf{v}_1 = \mathbf{K}_1^{-1}\mathbf{u}_1 \qquad (14)$$

where

$$\mathbf{u}_1(t) \triangleq [u_{1,1}(t), \dots, u_{1,N}(t)]$$

$$\mathbf{K}_1(t) \triangleq \begin{bmatrix} k_{1,1}\eta_1(\mathbf{x}_1,t) & \dots & k_{1,N}\eta_1(\mathbf{x}_N,t) \\ \vdots & & \vdots \\ k_{1,1}\eta_n(\mathbf{x}_1,t) & \dots & k_{1,N}\eta_n(\mathbf{x}_N,t) \\ \vdots & & \vdots \\ k_{q,1}\eta_1(\mathbf{x}_1,t) & \dots & k_{q,N}\eta_1(\mathbf{x}_N,t) \\ \vdots & & \vdots \\ k_{q,1}\eta_n(\mathbf{x}_1,t) & \dots & k_{q,N}\eta_n(\mathbf{x}_N,t) \end{bmatrix}^T$$

$$\mathbf{v}_1(t) \triangleq \Big[a'_{111}(t)\xi'_1(\mathbf{x}_d), \dots, a'_{11n}(t)\xi'_q(\mathbf{x}_d), \dots,$$
$$a'_{1q1}(t)\xi'_1(\mathbf{x}_d), \dots, a'_{1qn}(t)\xi'_q(\mathbf{x}_d)\Big]$$

Therefore, the first element of the desired control input vector can be calculated as

$$u_{1,d} = [\eta_1(t), \dots, \eta_n(t), \dots, \eta_1(t), \dots, \eta_n(t)] \times \mathbf{v}_1.$$

For the remaining $m-1$ elements in the desired control input vector similar steps can be take to calculate them. The resultant control input can thus be obtained as

$$\mathbf{u}_d = [u_{1,d}, \dots, u_{n,d}]^T.$$

□□

Remark 4.5
If the non homogeneous functions $\eta_j(\mathbf{x}_j,t)$ do not exist, *i.e.* they are all ones, then the system states can be not measurable.

4. ILLUSTRATIVE EXAMPLES

In this section, the multi magnitude scale DLC schemes are applied to the control of a two-link manipulator respectively for the illustrative purpose.

In the simulation we choose the sampling period to be $1ms$ and eight previous trajectories with different magnitude matrices relative to the desired trajectory specified as follows

$$K_1 = \begin{bmatrix} 1 & 0 \\ 0 & 2 \end{bmatrix} \; K_2 = \begin{bmatrix} 1.5 & 0 \\ 0 & 2.5 \end{bmatrix} \; K_3 = \begin{bmatrix} 2 & 0 \\ 0 & 2.5 \end{bmatrix}$$
$$K_4 = \begin{bmatrix} 2.5 & 0 \\ 0 & 3 \end{bmatrix} \; K_5 = \begin{bmatrix} 3 & 0 \\ 0 & 3.5 \end{bmatrix} \; K_6 = \begin{bmatrix} 3.5 & 0 \\ 0 & 4 \end{bmatrix}$$
$$K_7 = \begin{bmatrix} 4 & 0 \\ 0 & 4.5 \end{bmatrix} \; K_8 = \begin{bmatrix} 4.5 & 0 \\ 0 & 5 \end{bmatrix}$$

Multiple magnitude scale DLC scheme for nonlinear system (Eq.14) is applied here and the simulation results are shown in Fig. 1. From Fig. 1 it is obvious that the DLC learned control signals are exactly the same as the desired ones.

5. CONCLUSION

The multi magnitude scale Direct Learning Control scheme is developed for trajectory tracking problems. The new DLC control scheme is capable of learning from pre-stored control inputs and generating the desired control profile for a new trajectory which may differ from the previous ones along all the coordinates. In other words, learning has been generalized and applied to trajectories with rather different spatial patterns. Simulations has been carried out with satisfying results.

6. REFERENCES

Arimoto, S., S. Kawamura and F. Miyazaki (1984). Bettering operation of robots by learning. *Journal of Robotic Systems* **1**, 123–140.

Xu, J.-X. (1997). Direct learning of control efforts for trajectories with different magnitude scales. *Automatica* **33**(12), 2191–2195.

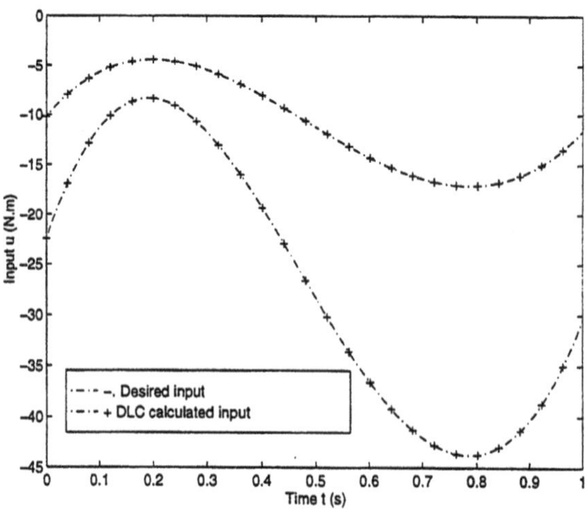

Fig. 1. DLC obtained Control Input

DUAL-SCALE DIRECT LEARNING CONTROL OF TRAJECTORY TRACKING FOR A CLASS OF NONLINEAR UNCERTAIN SYSTEMS

Jian-Xin Xu[1] **and Tao Zhu**

National University of Singapore, Department of Electrical Engineering
10 Kent Ridge Crescent, Singapore 119260
[1]*E-mail. elexujx@ nus.edu.sg*

Abstract: In this paper, a new learning control scheme: *dual-scale direct learning control* for trajectory tracking problems is presented. Direct learning control (DLC) is defined as the direct generation of the desired control input profile from existing control input profiles without any repetitive learning. For a class of nonlinear uncertain dynamic systems, the proposed dual-scale direct learning control scheme is able to generate the desired control input profile for a new trajectory which differs from all the previously controlled and stored trajectories in *both time and magnitude scales*. *Copyright © 1999 IFAC*

Keywords: Direct Learning Control, Dual-Scale, Non-uniform Trajectory Tracking

1. INTRODUCTION

Although learning control methods have been developed for quite a long period (Frueh J.A and M.Q.Phan, 1997; J.X Xu, 1997a; S.Arimoto, *et al.*, 1984; S.Saab,1994; T.Y.Kuc, *et al.*, 1992; Z.Geng, *et al.*, 1990), there still exist a number of problems which hinder further applications of learning control techniques. One of the major obstacles is the requirement that the trajectory to be learnt must be strictly identical over the repeated operations. If any change occurs due to the variation of control objectives or task specifications, no matter how small it might be, the learning process has to resume from very beginning and the previously learnt control knowledge can no longer be used to update the control signal. Compared with the learning capability of human beings, conventional learning control methods (iterative learning control, repetitive control) is rather primitive because it can only tackle exactly the same control problem (identical motion trajectories), whereas human beings can effectively utilize all the learnt knowledge to solve different but somehow correlated problems.

In practice, a control system may have plenty of prior control knowledge acquired through all the past control actions which correspond to different but highly correlated motion tasks. Consider the following example where a XY-table draws a number of circles in specified time periods. There are three different kinds of operation specifications: i) draw all circles with the same radius but different periods; ii) draw all the circles with the same period but different radii; or iii) draw all circles which differ one from another in both radii and periods. A number of control input profiles can be obtained after execution of the circle drawing using any existing control approaches. Obviously, those control signals are inherently correlated because 1) they are generated for the same dynamics; and 2) each motion pattern is related (or proportional) to another in spatial distribution and/or time scale. Now the control problem is, for a new trajectory which is different from all the previous ones in both magnitude scale and time scale, to find the desired control input signals so as to drive the system to follow the given trajectory in general, and make use of all the pre-obtained control knowledge in particular. This kind of control problem can be defined as *a non-repeatable learning control problem.* Obviously, what the conventional learning control can solve is a subset: repeatable learning control problems. The main focus of the non-repeatable learning control is how to effectively incorporate pre-stored control information into the generation of the desired control signals for different trajectories, and in the sequel remove the iterative learning control process which is time consuming and limited by the repeatability condition.

From the practical point of view, non-repeatable learning control is more important and indispensable, and some relevant research work has been reported in this area (J.X Xu, 1997b; J.X Xu 1998; S.Kawamura and N.Fukao, 1995). The difficulty of realizing non-repeatable learning control lies in the fact that it is impossible to exhaust all the inherent relations among different motion trajectories, hence impossible to find a universal learning control scheme which can solve all the non-repeatable learning control problems. On the contrary, for a repeatable pattern, it is straightforward to fix the relationship between any two consecutive operations as they are exactly the same. In order to deal with non-repeatable learning control problems, the inherent relations of different trajectory patterns need to be explored. And the resulting learning control schemes might be both plant-dependent and trajectory-dependent.

In this paper, a particular class of non-repeatable learning control problem: dual-scale direct learning control for trajectories which are proportional to each other *in both time scales and magnitude scales* with known scaling factors is discussed. This work is the extension of the previous work in which direct learning control schemes have been applied to time-scale problems (J.X Xu, 1998) and magnitude-scale problems (J.X Xu, 1997) respectively. The dual-scale DLC can thus be defined as the direct generation of desired control profiles from different past control profiles without repeated learning control processes. It is interesting to note that, under the general frame of non-repeatable learning, the dual-scale direct learning and conventional learning (iterative learning, repetitive control) are actually functioning in a complementary manner. The main features of the conventional learning are: 1) little prior control knowledge is required; 2) only effective for single trajectory; and 3) repeated learning process is needed. In comparison, the main features of the dual-scale DLC are: 1) rather accurate and sufficient prior control knowledge is required; 2) able to learn from different trajectories; and 3) not necessary to repeat learning process.

For the existing learning control, the desired control effort is obtained in a trial and error fashion, i.e. it is learnt from the previous control efforts and the resultant errors. While in dual-scale direct learning control, the desired control effort for tracking control of a given trajectory can be directly obtained from a set of pre-stored different pairs of control efforts/trajectories. An obvious advantage of the dual-scale direct learning control method is, control profiles obtained by means of any other existing control methods can be utilized equally to produce new control profiles. It will be shown that with proper choices of past output trajectories and corresponding control profiles, the desired learning control profile can be generated directly.

Generally speaking, learning can be classified into two categories: pattern based learning and repetition based learning. A typical example of the pattern based learning is neural network with properly selected basis functions. A typical example of the repetition based learning is iterative learning control (ILC). DLC is analogous to the first category – pattern based learning but also different in the sense that it is a partial-model based learning method. One of the main objectives of DLC is to fully use system model knowledge whenever available. In practice most systems to be controlled are partially known. As far as it is possible to partition the unknowns from the rest, and the unknowns are invariant over iteration and target trajectories, DLC schemes can be applied to achieve exact learning.

It is worth to point out that DLC schemes are different from other control methods which use system identification or model approximation. When a system to be controlled is totally unknown, system identification based control is perhaps the only effective approach. Such a scheme was presented by Frueh J.A and M.Q.Phan (1997). It employs the Least Squares method and uses the input-output data from repetitive trails to identify impulse responses of the unknown system so that the desired control input can be generated. A set of basis functions is further employed to reduce the number of input-output data set required for system identification. In comparison, DLC schemes apply to the system which is partially known in the sense that the unknown parts can be separated from the rest of the system as a set of basis functions invariant over iterations and target trajectories. DLC then learns those unknown basis functions. The main features of DLC are the applicability to certain classes of highly nonlinear systems, and the accurate generation of the desired control profile.

This paper is organized in the following manner. Section 2 gives formulation of the nonlinear dynamic system to be discussed and introduces important concepts and relations with regard to different trajectories. In Section 3, a detailed dual-scale direct learning method is derived and analyzed. In Section 4 the dual-scale direct DLC scheme is applied to a nonlinear dynamic system as the illustrative example.

2. PROBLEM STATEMENT

Consider a class of nonlinear uncertain dynamic systems described by the following equations

$$\frac{d\mathbf{x}(t)}{dt}=\sum_{j=1}^{l}A_j\xi_{j-r}(\mathbf{x}(t))+B\mathbf{u}(t) \quad (1)$$

$$\mathbf{y}(t)=C\mathbf{x}(t) \quad (2)$$

where $\mathbf{u}(t)\in R^n$ is the input vector and $\mathbf{y}(t)\in R^n$ is the output vector. $\mathbf{x}(t)\in R^n$ is a vector of the system state variables. $A_j\in R^{n\times n_j}$, $B\in R^{n\times n}$ and $C\in R^{n\times n}$ are unknown constant matrices. $\xi_{j-r}(\mathbf{x}(t))\in R^{n_j}$ is a partially known function vector to be defined later. Numbers l, $r\in Z_+$ where $Z_+=\{1,2,3,\cdots\}$ is a finite set of integers. For notational convenience, in the paper $Z(N)$ is used to denote the set $\{1,\ldots,N\}$ for any integer N.

Remark 1: Note that $\mathbf{y}(t)$ is required to track desired trajectory $\mathbf{y}_d(t)\in R^n$. This consequently requires $\mathbf{u}(t)\in R^n$. As shown by H.T. Banks, *et al.* (1975), the system is completely function-space controllable if and only if rank$(B)=n$.

Remark 2: In (1) and (2) the assumption that the state vector $\mathbf{x}(t)$ is not measurable and the matrix C is unknown is made, hence the usual parameter identification and state estimation methods are not applicable. Nevertheless the direct learning control scheme can still be applied using the only measurable output quantity $\mathbf{y}(t)$.

This nonlinear system is assumed to satisfy the following assumptions.

[*Assumption 1*]
Matrix CB is of full rank.

[*Assumption 2*]
$\forall j\in Z(l)$ and $r\in Z_+$, each element of $\xi_{j-r}(\mathbf{x}(t))$ is a homogenous function with a power of j-r of $\mathbf{x}(t)$, i.e. for a time varying magnitude scaling factor $\kappa(t)\neq 0$,

$$\xi_{j-r}(\kappa(t)\mathbf{x}(t))=\kappa^{j-r}(t)\xi_{j-r}(\mathbf{x}(t)).$$

Remark 3: In terms of *Assumption 2*, what is needed to know about the vector $\xi_{j-r}(\mathbf{x}(t))$ is, for a magnitude scale $\kappa(t)\neq 0$,
$\xi_{j-r}(\kappa(t)\mathbf{x}(t))=\kappa^{j-r}(t)\xi_{j-r}(\mathbf{x}(t))$, $\forall j\in Z(l)$ and $\forall r\in Z_+$. For example, consider a two dimensional state vector $\mathbf{x}=[x_1, x_2]^T$, some possible vectors $\xi_{j-r}(\mathbf{x}(t))$ would be

$$\xi_{-1}(\mathbf{x}(t))=\begin{bmatrix}(x_1+x_2)^{-1}\\ x_1^{-1}\\ x_2^{-1}\end{bmatrix} \text{ and } \xi_2(\mathbf{x}(t))=\begin{bmatrix}x_1^{\,2}\\ x_1x_2\end{bmatrix}, \text{ etc.}$$

The only available information here is that the order of $\xi_{-1}(\mathbf{x}(t))$ is -1 and that of $\xi_2(\mathbf{x}(t))$ is 2. Note that the elements of ξ_{-1} and ξ_2 may take complete different forms, for instance,

$$\xi_2(\mathbf{x}(t))=\begin{bmatrix}x_2^{\,2}\\ x_1^2\end{bmatrix}.$$

Before introducing the dual-scale DLC scheme, it is necessary to define the inherent relations among trajectories with different magnitude and time scales.

<u>*Definition 1*</u>
Trajectory $\mathbf{y}_i(t_i)$, $t_i\in[0,T_i]$ is said to be *proportional* to another trajectory $\mathbf{y}(t)$, $t\in[0,T]$ both in *magnitude* and in *time* scales if and only if

$$\kappa_i^{-1}(t_i)\mathbf{y}_i(t_i)=\mathbf{y}(t),$$

where $\kappa_i(t_i)\neq 0$ is the time varying magnitude scaling factor and $\rho_i(t)=t_i$ is the time scaling factor which is continuously differentiable and satisfying $\rho_i(0)=0$ and $\rho_i(T)=T_i$.

Remark 4: The definition *proportional both in magnitude and time scales* describes quite general inherent relations among different trajectories, and includes the following two as the special cases.

<u>*Case 1*</u> $\kappa_i(t_i)=1$
$\mathbf{y}_i(t_i)=\mathbf{y}_i(\rho_i(t))=\mathbf{y}(t)$, which means that the pre-stored trajectories and the new one have the same spatial path with different time scales. This problem has been addressed in J.X Xu (1998).

<u>*Case 2.*</u> $\rho_i(t)=t$
$\kappa_i^{-1}(t_i)\mathbf{y}_i(t_i)=\mathbf{y}(t)$, $\kappa_i(t_i)$ is non-zero, which means that the pre-stored trajectories and the new one are proportional in magnitude while the time scales are the same. This problem has been partially addressed in J.X Xu (1997b), where κ_i is a constant.

An illustration example for the general cases is shown in Fig.1, where $2^{-1}\mathbf{y}_i(t_i)=\mathbf{y}(t)$ and the time mapping is linear $t_i=\rho_i(t)=\frac{1}{2}t$, hence $t_i\in[0,1]$ and $t\in[0,2]$.

The following assumption further specifies the pre-stored trajectories.

[*Assumption 3*]
There are $N>l+2$ pre-stored trajectories $\mathbf{y}_i(t_i)$, $t_i\in[0,T_i]$. For any pre-stored trajectories $\mathbf{y}_i$ and $\mathbf{y}_j$ ($i\neq j$), their mapping functions satisfy $\forall t\in[0,T]$, i, $j=1,\ldots, N$.

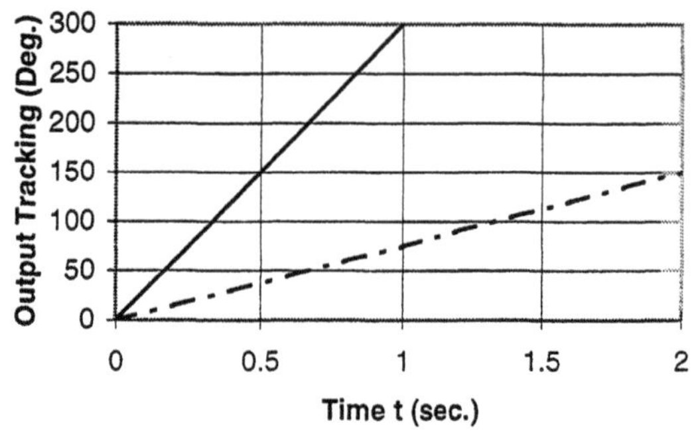

Fig. 1. Trajectories of $\mathbf{y}_i(t_i)$ and $\mathbf{y}(t)$: (1) Solid-line — $\mathbf{y}_i(t_i)$; (2) dash-dot line— $\mathbf{y}(t)$.

$$\frac{d\rho_i(t)}{dt} \neq 0, \qquad \frac{d\rho_j(t)}{dt} \neq 0.$$

In addition, for any $t \in [0, T]$, *there exist at least two trajectories* $\mathbf{y}_i(t_i)$ *and* $\mathbf{y}_j(t_j)$ *such that* $\left|\frac{d\rho_i(t)}{dt} - \frac{d\rho_j(t)}{dt}\right| > \delta$, *where* $\delta>0$ *is an arbitrary constant.*

The control objective summarized as the following.

Control Objective
The ultimate control objective is, for a new trajectory $\mathbf{y}_d(t_d)$, $t_d \in [0, T_d]$, *which is related to other pre-stored trajectories through the relation*
$\mathbf{y}_d(t_d) = \kappa_i^{-1}(t_i)\mathbf{y}_i(t_i) = \kappa_i^{-1}(\rho_i(t_d))\mathbf{y}_i(\rho_i(t_d))$,
$\kappa_i(t_i) \neq 1$, *generating the desired control signal profile* $\mathbf{u}_d(t_d)$ *directly from the pre-stored* $\mathbf{u}_i(t_i)$ $i \in Z(N)$, *instead of using conventional learning control techniques which are time consuming.*

3. DUAL-SCALE DLC SCHEME

In this section the dual-scale DLC scheme which is summarized by the following theorem is presented.

Theorem
For the nonlinear uncertain system (1) (2) associated with the new trajectory $\mathbf{y}_d(t_d)$, $t_d \in [0, T_d]$, the corresponding control signal profile $\mathbf{u}_d(t_d)$ can be directly calculated from the pre-stored $\mathbf{u}_i(t_i)$ (i=1, 2, ⋯, l+2) as follows

$$\mathbf{u}_d(t_d) = [I \cdots I \ 0]W^{-1}\bar{\mathbf{u}}(t_d) \tag{3}$$

where the regressor matrix W is defined to be

$$W = \begin{bmatrix} \kappa_1(\rho_1(t_d))p_1^{-1}I & \kappa_1^{l-r}(\rho_1(t_d))I & \cdots & \kappa_1^{l-r}(\rho_1(t_d))I & \kappa_1^{2}(\rho_1(t_d))p_1^{-1}\frac{d\kappa_1^{-1}(\rho_1(t_d))}{dt_d}I \\ \cdot & \cdot & & \cdot & \cdot \\ \cdot & \cdot & & \cdot & \cdot \\ \kappa_{l+2}(\rho_{l+2}(t_d))p_{l+2}^{-1}I & \kappa_{l+2}^{l-r}(\rho_{l+2}(t_d))I & \cdots & \kappa_{l+2}^{l-r}(\rho_{l+2}(t_d))I & \kappa_{l+2}^{2}(\rho_{l+2}(t_d))p_{l+2}^{-1}\frac{d\kappa_{l+2}^{-1}(\rho_{l+2}(t_d))}{dt_d}I \end{bmatrix},$$

$\bar{\mathbf{u}}(t_d) = [\mathbf{u}_1^T(\rho_1(t_d)) \ldots \mathbf{u}_{l+2}^T(\rho_{l+2}(t_d))]^T$,

$p_i \overset{\Delta}{=} \frac{d\rho_i(t_d)}{dt_d}$, i=1, 2, ⋯, l+2 and $I \in R^{n\times n}$ is an identity matrix.

Proof:
Consider the output equation (2). Differentiating $\mathbf{y}(t)$ with respect to t yields

$$\frac{d\mathbf{y}(t)}{dt} = C\frac{d\mathbf{x}(t)}{dt}$$

Substituting the dynamics (1) into the above equation for $\frac{d\mathbf{x}(t)}{dt}$, $\frac{d\mathbf{y}(t)}{dt} = \sum_{j=1}^{l} CA_j\xi_{j-r}(\mathbf{x}(t)) + CB\mathbf{u}(t)$. (4)

Since CB is full rank, control input can be expressed as

$$\mathbf{u}(t) = (CB)^{-1}\frac{d\mathbf{y}(t)}{dt} - (CB)^{-1}\left(\sum_{j=1}^{l} CA_j\xi_{j-r}(\mathbf{x}(t))\right). \tag{5}$$

The desired control input signal with respect to the new trajectory $\mathbf{y}_d(t_d)$, $t_d \in [0, T_d]$ can thus be expressed as

$$\mathbf{u}_d(t_d) = (CB)^{-1}\frac{d\mathbf{y}_d(t_d)}{dt_d} - (CB)^{-1}\left(\sum_{j=1}^{l} CA_j\xi_{j-r}(\mathbf{x}_d(t_d))\right). \tag{6}$$

Note that $\mathbf{u}_d(t_d)$ are not available directly in terms of the above formula due to the existence of system uncertainties in A_j, B, C and $\xi_{j-r}(\mathbf{x}_d(t_d))$.

Now choose pre-stored trajectories $\mathbf{y}_i(t_i)$ $t_i \in [0, T_i]$, $i \in Z(N)$ whose control profiles have been obtained *a priori*, that is

$$\mathbf{u}_i(t_i) = (CB)^{-1}\frac{d\mathbf{y}_i(t_i)}{dt_i} - (CB)^{-1}C\left(\sum_{j=1}^{l} A_j\xi_{j-r}(\mathbf{x}_i(t_i))\right). \tag{7}$$

Noting the relations $t_i = \rho_i(t_d)$, differentiating $\mathbf{y}_d(t_d)$ with respect to t_d to obtain

$$\frac{d\mathbf{y}_d(t)}{dt_d} = \frac{d\kappa_i^{-1}(t_i)\mathbf{y}_i(t_i)}{dt_i} \times \frac{d\rho_i(t_d)}{dt_d}$$

$$= \left(\mathbf{y}_i(t_i)\frac{d\kappa_i^{-1}(t_i)}{dt_i} + \kappa_i^{-1}(t_i)\frac{d\mathbf{y}_i(t_i)}{dt_i}\right) \times p_i \tag{8}$$

Note that, under the assumption that all the pre-stored trajectories $\mathbf{y}_i(t_i)$ and the new trajectories $\mathbf{y}_d(t_d)$ are *proportional both in magnitude and time scales* associated with known magnitude scales $\kappa_i(t_i)$ and time mapping $\rho_i(t_d)$, the following result is obtained:

$$\mathbf{y}_i(t_i)=\mathbf{y}_i(\rho_i(t_d))=\kappa_i(\rho_i(t_d))\mathbf{y}_d(t_d).$$

Similarly from $\mathbf{x}(t)=C^{-1}\mathbf{y}(t)$,

$$\mathbf{x}_i(t_i)=\mathbf{x}_i(\rho_i(t_d))=\kappa_i(\rho_i(t_d))\mathbf{x}_d(t_d).$$

From *Assumption* 2

$$\xi_{j-r}(\mathbf{x}_i(t_i))=\xi_{j-r}(\mathbf{x}_i(\rho_i(t_d))=\kappa_i^{j-r}(\rho_i(t_d))\xi_{j-r}(\mathbf{x}_d(t_d))$$

Substituting above relations into (7) leads to

$$\mathbf{u}_i(t_i)=(CB)^{-1}\left[\kappa_i(t_i)\left(\frac{d\mathbf{y}_d(t_d)}{dt_d}\times p_i^{-1}-\mathbf{y}_i(t_i)\frac{d\kappa_i^{-1}(t_i)}{dt_i}\right)\right]+\sum_{j=1}^{l}\kappa_i^{j-r}\mathbf{d}_j(\mathbf{x}_d(t_d))$$

$$=\kappa_i(\rho_i(t_d))p_i^{-1}\times(CB)^{-1}\frac{d\mathbf{y}_d(t_d)}{dt_d}+\kappa_i^2(\rho_i(t_d))\frac{d\kappa_i^{-1}(\rho_i(t_d))}{dt_d}p_i^{-1}\times(CB)^{-1}(-\mathbf{y}_d(t_d))+\sum_{j=1}^{l}\kappa_i^{j-r}\mathbf{d}_j(\mathbf{x}_d(t_d)) \quad (9)$$

where $\mathbf{d}_j(\mathbf{x}_d(t_d))=-(CB)^{-1}CA_j\xi_{j-r}(\mathbf{x}_d(t_d))$ and $j\in Z(N)$. Now it can be clearly seen from (9) that, vectors $(CB)^{-1}\left[\frac{d\mathbf{y}_d(t_d)}{dt_d}\right]$, $(CB)^{-1}(-\mathbf{y}_d(t_d))$ and $\mathbf{d}_j(\mathbf{x}_d(t_d))$, though unknown, are functions of the new trajectory and are invariant with respect to the pre-stored trajectories. Consequently all of them can be treated as a set of unknown basis function vectors and be calculated in point-wise manner according to (9). This is because all the coefficients $\kappa_i(\rho_i(t_d))p_i^{-1}$, $\kappa_i^{j-r}(\rho_i(t_d))$ and $\kappa_i^2(\rho_i(t_d))\frac{d\kappa_i^{-1}(\rho_i(t_d))}{dt_d}p_i^{-1}$ as well as $\mathbf{u}_i(t_i)=\mathbf{u}_i(\rho_i(t_d))$ are known.

On the other hand, from (6) it is easy to find

$$\mathbf{u}_d(t_d)=(CB)^{-1}\frac{d\mathbf{y}_d(t_d)}{dt_d}+\sum_{j=1}^{l}\mathbf{d}_j(\mathbf{x}_d(t_d)). \quad (10)$$

Further define the following augmented vector

$$\overline{\mathbf{v}}(\mathbf{x}_d(t_d))=[(CB)^{-1}\frac{d\mathbf{y}_d(t_d)}{dt_d}\quad \mathbf{d}_1(\mathbf{x}_d(t_d))\ \ldots\mathbf{d}_l(\mathbf{x}_d(t_d))\quad (CB)^{-1}(-\mathbf{y}_d(t_d))]^T$$

and use the pre-defined regressor matrix W and the augmented vector $\overline{\mathbf{u}}(t_d)^T$, then equation (9) can be written in a matrix form for l+2 distinct trajectories

$$W\cdot\overline{\mathbf{v}}(\mathbf{x}_d(t_d))=\overline{\mathbf{u}}(t_d). \quad (11)$$

Finally from (10) the desired control profile $\mathbf{u}_d(t_d)$ can be solved directly as below

$$\mathbf{u}_d(t_d)=[I\cdots I\ 0]\overline{\mathbf{v}}(\mathbf{x}_d(t_d))=[I\cdots I\ 0]W^{-1}\overline{\mathbf{u}}(t_d),$$

$\forall\, t_d\in[0, T_d]$, (12)

where $[I\cdots I\ 0]\in R^{p\times n(l+2)}$.

Remark 5: In equation (9) there are l+2 unknown vectors, which are $(CB)^{-1}\left[\frac{d\mathbf{y}_d(t_d)}{dt_d}\right]$, $(CB)^{-1}(-\mathbf{y}_d(t_d))$ and $\mathbf{d}_j(\mathbf{x}_d(t_d))$, i=1, 2, ..., l. Hence at least l+2 pre-stored trajectories are needed to solve these l+2 unknown vectors.

In practice, a simple way to implement the dual-scale direct learning control is to utilize all the pre-stored trajectories. Constructing the augmented vector $\overline{\mathbf{u}}_N(t_d)=[\mathbf{u}_1\cdots\cdots\mathbf{u}_N]^T\in R^{nN}$ and the augmented matrix

$$W_N=\begin{bmatrix}\kappa_1(\rho_1(t_d))p_1^{-1}I & \kappa_1^{1-r}(\rho_1(t_d))I & \cdots & \kappa_1^{l-r}(\rho_1(t_d))I & \kappa_1^2(\rho_1(t_d))p_1^{-1}\frac{d\kappa_1^{-1}(\rho_1(t_d))}{dt_d}I\\ \cdot & \cdot & & \cdot & \cdot\\ \cdot & \cdot & & \cdot & \cdot\\ \kappa_N(\rho_N(t_d))p_N^{-1}I & \kappa_N^{1-r}(\rho_N(t_d))I & \cdots & \kappa_N^{l-r}(\rho_N(t_d))I & \kappa_N^2(\rho_N(t_d))p_N^{-1}\frac{d\kappa_N^{-1}(\rho_N(t_d))}{dt_d}I\end{bmatrix}$$

where $W_N\in R^{nN\times n(l+2)}$, $\overline{\mathbf{u}}_N(t)=W_N\overline{\mathbf{v}}(\mathbf{x}_d(t_d))$. In terms of *Assumption* 3, the augmented matrix W_N has full column rank. Accordingly, the vector $\overline{\mathbf{v}}(\mathbf{x}_d(t_d))$ can be calculated as $\overline{\mathbf{v}}(\mathbf{x}_d(t_d))=(W_N^TW_N)^{-1}W_N^T\overline{\mathbf{u}}_N(t_d)$, $\forall t_d\in[0,T_d]$. Finally, the desired control input is $\mathbf{u}_d(t_d)=[I\cdots I\ 0]\overline{\mathbf{v}}(\mathbf{x}_d(t_d))$ according to the formula (12).

Remark 6: When there exists modeling approximation error, it can be equivalently expressed as a residue vector ε on the right hand side of equation (11), i.e.:

$$W\cdot\overline{\mathbf{v}}(\mathbf{x}_d(t_d))=\overline{\mathbf{u}}(t_d)+\varepsilon \quad (13)$$

where ε represents the modeling approximation error. If the signal to noise ration (SNR) is relatively high (ε is smaller as compared to $\overline{\mathbf{u}}(t_d)$) which is well assumed for most modeling and identification problems, the proposed dual-scale direct learning control scheme can still work effectively.

4. ILLUSTRATIVE EXAMPLE

Consider the nonlinear uncertain dynamic system described by

$$\frac{d\mathbf{x}(t)}{dt} = A_1\xi_1(\mathbf{x}(t)) + A_2\xi_2(\mathbf{x}(t)) + B\mathbf{u}(t)$$

$$\mathbf{y}(t) = C \cdot \mathbf{x}(t)$$

where

$$\mathbf{x}(t) = \begin{bmatrix} x_1 \\ x_2 \end{bmatrix}, \quad \xi_1(\mathbf{x}(t)) = \begin{bmatrix} x_1 \\ x_2 \end{bmatrix}, \quad \xi_2(\mathbf{x}(t)) = \begin{bmatrix} x_1 x_2 \\ x_2^2 \end{bmatrix},$$

$$A_1 = \begin{bmatrix} 0.5 & 0 \\ -1 & 1 \end{bmatrix}, A_2 = \begin{bmatrix} 1 & 0 \\ 0 & 1 \end{bmatrix}, B = \begin{bmatrix} 3 & 1 \\ 0 & 2.4 \end{bmatrix}, C = \begin{bmatrix} 1 & 2 \\ 2 & 3 \end{bmatrix}.$$

The described trajectory is specified as below:

$$\mathbf{x}_d(t_d) = \begin{bmatrix} 1 - \cos(\omega t_d) \\ \sin^2(\omega t_d) \end{bmatrix}, \quad t_d \in [0, 2], \ \omega = \pi \ .$$

In the simulation, it is assumed that control knowledge concerning the following four trajectories has been achieved *a priori.*

1). $x_1(t_1) = \begin{bmatrix} 1 - \cos(3\omega t_1 / 2) \\ \sin^2(3\omega t_1 / 2) \end{bmatrix},$

$$t_1 = \rho_1(t_d) = \frac{2}{3} t_d; \quad t_1 \in [0, \frac{4}{3}];$$

2). $x_2(t_2) = 0.25\cos(t_2)\begin{bmatrix} 1 - \cos(\omega t_2^2 / 4) \\ \sin^2(\omega t_2^2 / 4) \end{bmatrix},$

$$t_2 = \rho_2(t_d) = 2\sqrt{t_d}; \quad t_2 \in [0, 2\sqrt{2}];$$

3). $x_3(t_3) = (6 + t_3^2)\begin{bmatrix} 1 - \cos(\omega t_3^3) \\ \sin^2(\omega t_3^3) \end{bmatrix},$

$$t_3 = \rho_3(t_d) = \sqrt[3]{t_d}; \quad t_3 \in [0, \sqrt[3]{2}];$$

4). $x_4(t_4) = 0.8\begin{bmatrix} 1 - \cos(\omega t_4) \\ \sin^2(\omega t_4) \end{bmatrix},$

$$t_4 = \rho_4(t_d) = t_d; \quad t_4 \in [0, 2];$$

In this example the magnitude scales are $\kappa_1=1$, $\kappa_2=0.25\cos(t_2)$, $\kappa_3=6+t_3^2$, and $\kappa_4=0.8$ respectively. The vector $\overline{\mathbf{v}}(\mathbf{x}_d(t_d))$ consists of four basis function vectors $(CB)^{-1}\frac{d\mathbf{y}_d(t_d)}{dt_d}$, $\mathbf{d}_1 = -(CB)^{-1}CA_1\xi_1(\mathbf{x}_d(t_d))$, $\mathbf{d}_2 = -(CB)^{-1}CA_2\xi_2(\mathbf{x}_d(t_d))$ and $(CB)^{-1}(-\mathbf{y}_d(t_d))$. In order to solve $\mathbf{u}_d(t_d)$, a minimum of four pre-stored trajectories are required. Fig. 2 shows the simulation results of input vector $\mathbf{u}_d(t_d)$ with direct learning. The resultant input vector is compared with the desired input vector. Satisfactory tracking performance can be observed from the above figure, which confirms the effectiveness of the proposed *dual-scale direct learning control* scheme.

5. CONCLUSION

A dual-scale direct learning control scheme for a class of nonlinear uncertain dynamic systems is developed. The proposed control scheme is able to learn from pre-stored control profiles with different magnitude scales and time scales and to generate the desired control profile directly without any repeated learning process. Theoretical analysis and simulation results confirm the validity of the proposed direct learning control schemes.

6. REFERENCE

Frueh J.A. and M.Q.Phan (1997). System Identification and Inverse Control Using Input-Output Data from Repetitive Trials, *Proceedings of The 2nd Asia Control Conference*, **II**, pp.257-254.

H.T. Banks, M.Q.Jacobs, and C.E.Langenhop (1975). *SIAM J.Control Optimization* **13**, pp.611.

J. Kurek and M. Zaremba (1993). Iterative learning control synthesis based on 2-D system theory, *IEEE Trans. on Automat. Contr.*, **38**(1), pp.121-125.

J.X. Xu (1997a). Analysis of iterative learning control for a class of nonlinear discrete-time systems, *To appear on Automatica*.

J.X. Xu (1997b). Direct learning of control efforts for trajectories with different magnitude scales, *Automatica*, **33**(12), pp.2191-2195.

J.X. Xu (1998). Direct learning of control efforts for trajectories with different time scales, *To appear on IEEE Trans. on Automat. Contr.*

S. Arimoto, S. Kawamura and F. Miyazaki, (1984). Better operation of robots by learning. *Journal of Robotic Systems*, **1**, pp.123-140.

S. Kawamura and N. Fukao (1995). A time-scale interpolation for input torque patterns obtained through learning control on constrained robot motions, *Proc. of 1995 IEEE Int. Conf. on Robotics and Automation*, pp.2156-2161.

S. Saab (1994). On the P-type learning control, *IEEE Trans. Automat. Contr.*, **39**(11), pp.2298-2302.

T.Y. Kuc, J.S. Lee, and K. Nam (1992). An iterative learning control theory for a class of nonlinear dynamic systems, *Automatica* **28**(6), pp.1215-1221.

Z. Geng, R. Carroll, and J. Xie (1990). Two-dimensional model and algorithm analysis for a class of iterative learning control systems. *Int. J. Contr.*, **52**, pp.833-862.

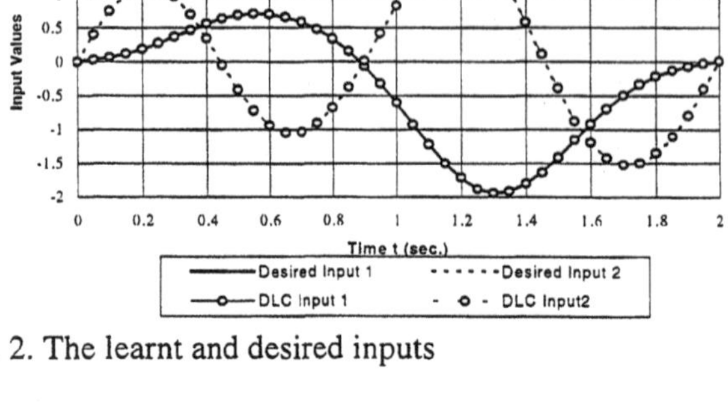

Fig. 2. The learnt and desired inputs

14th Triennial World Congress, Beijing, P.R. China

APPROXIMATE OUTPUT REGULATION OF NONLINEAR SYSTEMS

Zhengtao Ding * **Tai C. Yang** **

* *Department of Mechanical Engineering, Ngee Polytechnic,*
535 Clementi Road, Singapore 588489
email: dzt@np.edu.sg fax: +65 4677384
** *School of Engineering, University of Sussex, Falmer,*
Brighton, BN1 9QT, UK

Abstract: This paper deals with output regulation of nonlinear systems driven by exogenous signals whose dynamics are not exactly known. Output regulation with high order approximation is formulated and the solvability is characterized by the existence of certain invariant manifolds. Under the proposed control, the closed-loop system exhibits ultimate boundedness properties with respect to high order time-derivatives of the exogenous signals. Besides the high order approximation, another advantage of the proposed control design is that the measurement of derivatives of the exogenous signals is not essential for control design. *Copyright © 1999 IFAC*

Keywords: Output regulation; Nonlinear control systems; Multivariable feedback control; Model approximation; Disturbance rejection

1. INTRODUCTION

A classical problem in control theory is the design of a feedback control to impose a prescribed response subject to every external command in a prescribed family. This includes, for instance, the problem of designing a feedback control law to track a prescribed reference signal, and the problem of rejecting undesired disturbances. Both cases can be unified by the design of feedback control law to impose the asymptotic decay of the output tracking error (Isidori, 1995), and both reference signals and disturbances are referred as exogenous signals.

Output regulation of linear systems are posed and solved in (Smith and Davison, 1972; Francis, 1977). Only recently, solutions to output regulation of nonlinear systems become available. When the exogenous signals are generated by a known exosystem, necessary and sufficient conditions for the existence of a local full information solution are that the linearized system is stabilizable and that there exists a certain invariant manifold (Isidori and Byrnes, 1990). The later one is a completely nonlinear treatment, even though the stability solution is local in general. Output regulation of error feedback is solved recently (Isidori, 1995) with the application of system immersion technique (Fliess, 1982).

The assumption that the exosystem is exactly known is very restrictive in many situations. Also the robustness properties with respect to the variation in the exosystem need to be addressed. Efforts to relax the assumption have been reported in (Huang and Rugh, 1990; Sureshbabu and Rugh, 1995). The output regulation subject to a slowly-varying exogenous system is considered in (Huang and Rugh, 1990). Their solution is based on the measurement of instantaneous value of the exogenous signal and an underlying integrator-exosystem, and the existence of of an equilibrium manifold is shown to be sufficient for the existence of a control law. With the measurement of the first order time-derivative of the exogenous signal, the closed system with the proposed control of (Sureshbabu and Rugh, 1995) can track ramp signals or reject ramp disturbances. In both papers, the robustness properties are established with the application of the stability results shown in (Kelemen, 1986).

In this paper, a solution is proposed for output regulation with unknown exosystems to achieve tracking of high order polynomials of time or rejecting polynomial disturbances. First, a generalization to the results shown in (Sureshbabu and Rugh, 1995) is presented with the measurement of high order time derivatives of the exogenous signal. Even though D control mode is commonly used in industrial processes, measurement of time derivatives may not be feasible. To overcome this potential limitation, the second solution is proposed for high order tracking without measurement of derivatives. The related robustness properties are addressed.

2. PROBLEM FORMULATION

Consider a nonlinear system

$$\begin{aligned} \dot{x}(t) &= f(x(t), u(t), w(t)) \\ y(t) &= h(x(t), u(t), w(t)) \end{aligned} \tag{1}$$

where $x(t) \in R^n$ is the state variable; $u \in R^m$, the input; $y \in R^p$, the output; $t \in R^+$, the time; $w(t) \in R^q$, pointwise in time, is a C^v exogenous signal. Assume that $f : R^n \times R^m \times R^p \to R^n$ and $h : R^n \times R^m \times R^p \to R^m$ are smooth (C^∞) functions, satisfying $f(0,0,0) = 0$ and $h(0,0,0) = 0$ respectively.

Approximate Output Regulation Problem I (AORP-I): Given a system (1), show that there exist compact sets $\mathcal{W}_i \subset R^q$, $i = 0, 1, \ldots, v-1$, $\mathcal{O} \subset R^n$, all with the origin as an interior point, and a C^r ($r \geq 2$) map $k : R^n \times \mathcal{W}_0 \times \mathcal{W}_1 \times \ldots \mathcal{W}_{v-1} \to R^m$, with $k(0, 0, \ldots, 0) = 0$, such that under the control law

$$u(t) = k(x(t), w(t), \dot{w}(t), \ldots, w^{(v-1)}(t)) \tag{2}$$

the closed-loop system satisfies the following requirements:

i) If the exogenous signal is identically zero, then $x = 0$ is a locally exponentially stable equilibrium.
ii) Given any $\epsilon > 0$, there exist $T, \delta > 0$ such that for any initial state $x(0) \in \mathcal{O}$ and any exogenous signal satisfying $w^{(i)} \in \mathcal{W}_i$, $i = 0, 1, \ldots, v-1$, pointwise in time for all $t \geq 0$, if $\sup_{t\geq 0} \|w^{(v)}\| < \delta$, then there exists a unique bounded solution $x(t)$ defined for all $t \geq 0$ and the output satisfies $\|y(t)\| \leq \epsilon$ for all $t > T$.

Remark 1: The notations $w^{(0)} = w$ and $w^{(1)} = \dot{w}$ are used for convenience. When $v = 1$ the problem is equivalent to the Approximate Regulation Problem I in (Sureshbabu and Rugh, 1995) and the problem with $v = 2$ is equivalent to the Approximate Regulation Problem II in (Sureshbabu and Rugh, 1995), i.e., regulation with derivative information.

Even though the measurements of derivatives of signals are used in some occasions in control application, for example, the D-mode in PID control, in many situations, such measurement, especially for high order derivatives, may not be feasible or accurate. To circumvent this potential problem, dynamic feedback becomes necessary, based on the measurements of instantaneous values of the exogenous signal. Considering this, the second problem on approximate output regulation is formulated.

Approximate Output Regulation Problem II (AORP-II): Given a system (1), show that there exist compact sets $\mathcal{W}_i \subset R^q$, $i = 0, 1, \ldots, v-1$, $\mathcal{O} \subset R^n$, all with the origin as an interior point, and a C^r ($r \geq 3$) map $k : R^n \times \mathcal{W} \to R^m$, with $k(0,0) = 0$, with $\mathcal{W} = \mathcal{W}_0 \times \mathcal{W}_1 \times \ldots \mathcal{W}_{v-1}$, such that under the control law

$$u(t) = k(x(t), \Omega(t)) \tag{3}$$

$$\dot{\Omega} = g(\Omega, w) \tag{4}$$

the closed-loop system satisfies the following requirements:

i) If the exogenous signal is identically zero, then $x = 0$ is a locally exponentially stable equilibrium.
ii) Given any $\epsilon > 0$, there exist $T, \delta, \delta_w > 0$ such that for any initial state $x(0) \in \mathcal{O}$ and any exogenous signal satisfying $w^{(i)} \in \mathcal{W}_i$, $i = 0, 1, \ldots, v-1$, pointwise in time for all $t \geq 0$, if $\sup_{t\geq 0} \|w^{(v)}(t)\| < \delta$ and $\|[w^T(0), \ldots, w^{(v-1)}(0)]^T - \hat{W}(0)\| \leq \delta_w$, with $\hat{W}(0) \in R^{nq}$ being a known constant, then there exists a unique bounded solution $x(t)$ defined for all $t \geq 0$ and the output satisfies $\|y(t)\| \leq \epsilon$ for all $t > T$.

In the sequel, the following notations are used to denote system and control input linearization coefficients:

$$\begin{aligned} A &= \frac{\partial f}{\partial x}(0,0,0) \quad & B &= \frac{\partial f}{\partial u}(0,0,0) \\ E &= \frac{\partial f}{\partial w}(0,0,0) \quad & C &= \frac{\partial h}{\partial x}(0,0,0) \\ D &= \frac{\partial h}{\partial u}(0,0,0) \quad & F &= \frac{\partial h}{\partial w}(0,0,0) \\ K &= \frac{\partial k}{\partial x}(0,\ldots,0) \quad & L_i &= \frac{\partial k}{\partial w^{(i)}}(0,\ldots,0) \end{aligned} \tag{5}$$

3. APPROXIMATE OUTPUT REGULATION WITH DERIVATIVE INFORMATION

In this section, the necessary and sufficient conditions will be examined for a solution to AORP-I using measurements of high order derivatives of the exogenous signal.

Theorem 1: If the linearization of the system (1) is stabilizable at the origin, then AORP-I is solvable if and only if there exist compact sets $\mathcal{P}_i$, $i = 0, 1, \ldots, v-1$, all with the origin as an interior point, and a C^r maps $\pi : \mathcal{P}_0 \times \mathcal{P}_1 \times \ldots \mathcal{P}_{v-1} \to R^n$ and $\alpha : \mathcal{P}_0 \times \mathcal{P}_1 \times \ldots \mathcal{P}_{v-1} \to R^m$, with $\pi(0, 0, \ldots, 0) = 0$ and $\alpha(0, 0, \ldots, 0) = 0$, such that

$$\frac{\partial \pi(w, \omega_1, \ldots, \omega_{v-1})}{\partial w}\omega_1 + \sum_{i=1}^{v-2} \frac{\partial \pi(w, \omega_1, \ldots, \omega_{v-1})}{\partial \omega_i}\omega_{i+1} = f(\pi(w, \omega_1, \ldots, \omega_{v-1}), \alpha(w, \omega_1, \ldots, \omega_{v-1}), w) \quad (6)$$

$$0 = h(\pi(w, 0, \ldots, 0), \alpha(w, 0, \ldots, 0), w) \quad (7)$$

for $w \in \mathcal{P}_0$ and $\omega_i \in \mathcal{P}_i$, $i = 1, \ldots, v-1$.

Proof: For necessity, suppose the control (2) solves AORP-I. The closed-loop system (1) and (2) is exponentially stable at the origin. The solution is applicable if the exogenous signal satisfies $w^{(v)}(t) = 0$, $\forall t \geq 0$, which implies that the exogenous signal is generated by

$$\dot{W} = SW \quad (8)$$

where

$$W = \begin{bmatrix} w \\ \dot{w} \\ \vdots \\ w^{(v-1)} \end{bmatrix}, \quad S = \begin{bmatrix} 0 & I_q & & \\ & \ddots & \ddots & \\ & & 0 & I_q \\ & & & 0 \end{bmatrix}$$

The center manifold theorem in (Khalil, 1992) can be applied to an augmented closed-loop system

$$\begin{aligned} \dot{x}(t) &= f(x(t), u(x(t), W), w(t)) \\ \dot{W} &= SW \end{aligned} \quad (9)$$

The above augmented system can be rewritten, using (5), as

$$\begin{bmatrix} \dot{x}(t) \\ \dot{W} \end{bmatrix} = \begin{bmatrix} A + BK & BL_0 + E, \ BL_1 \ \ldots \ BL_{v-1} \\ 0 & S \end{bmatrix} \times \begin{bmatrix} x \\ W \end{bmatrix} + \begin{bmatrix} \tilde{f}(x, W) \\ 0 \end{bmatrix} \quad (10)$$

where the C^r map

$$\tilde{f}(x, W) = f(x(t), u(x(t), W), w(t)) - (A + BK)x - [BL_0 + E, \ BL_1 \ \ldots \ BL_{v-1}]W$$

vanishes at the origin along with its first-order partial derivatives. Since the eigenvalues of S are zeros, and different form those of $(A + BK)$ which are all with negative real parts by assumption, there exists an $M \in R^{n \times (qv)}$ (Horn and Johnson, 1990) satisfying

$$[A + BK]M - MS = [BL_0 + E, \ BL_1 \ \ldots \ BL_{v-1}] \quad (11)$$

Hence, a change of coordinate

$$z = x + MW$$

transforms the augmented system into

$$\begin{bmatrix} \dot{z}(t) \\ \dot{W} \end{bmatrix} = \begin{bmatrix} A + BK & 0 \\ 0 & S \end{bmatrix} \begin{bmatrix} z \\ W \end{bmatrix} + \begin{bmatrix} g(z, W) \\ 0 \end{bmatrix} \quad (12)$$

where

$$g(z, W) = \tilde{f}(z - MW, W)$$

Therefore, from the center manifold theorem, there exists an open neighborhood of the origin, $\mathcal{W}$, and a C^r map $z^a : \mathcal{W} \to R^n$ vanishing at the origin together with its partial derivatives, such that

$$\frac{\partial z^a(W)}{\partial W} SW = [A + BK]z^a(W) + g(z^a(W), W) \quad (13)$$

Let

$$\pi(w, \dot{w}, \ldots, w^{(v-1)}) = z^a(W) - MW$$

$$\alpha(w, \dot{w}, \ldots, w^{(v-1)}) = k(\pi(w, \dot{w}, \ldots, w^{(v-1)}), W) \quad (14)$$

and the condition (6) is obtained.

To show (7) is necessary, consider the case when $w(t) = w_c$ is a small, constant signal and $w(t) \in \mathcal{P} \subset \mathcal{W}_0$. From the initial state

$$w(0) = w_c, \quad x(0) = \pi(w_c, 0, \ldots, 0)$$

the solution of the closed-loop system is a constant given by

$$x(t) = \pi(w_c, 0, \ldots, 0) \quad \forall t \geq 0$$

which results in a constant output, $\forall t \geq 0$,

$$y(t) = h(\pi(w_c, 0, \ldots, 0), \alpha(w, 0, \ldots, 0), 0)$$

Therefore the necessity of (7) can be concluded from the property of AORP-I.

Sufficiency. Since $\{A, B\}$ is stabilizable, there exists a constant matrix K such that all the eigenvalues of $(A + BK)$ have negative real parts. Consider the control defined by the C^r map k,

$$k(x, w, \dot{w}, \ldots, w^{(v-1)}) = \alpha(w, \dot{w}, \ldots, w^{(v-1)}) + K[x - \pi(w, \dot{w}, \ldots, w^{(v-1)})] \quad (15)$$

Let

$$A_c(w, \dot{w}, \ldots, w^{(v-1)}) = \frac{\partial f}{\partial x}(\pi, \alpha, w) + \frac{\partial f}{\partial u}(\pi, \alpha, w)K \quad (16)$$

which represents the first order approximation of the closed-loop system on the manifold. It follows that $A_c(w, \dot{w}, \ldots, w^{(v-1)})$ is C^r on $\mathcal{P}_0 \times \mathcal{P}_1 \times \ldots \mathcal{P}_{v-1}$ and $A_c(0, 0, \ldots, 0)$ is Hurwitz. Applying results of small perturbation (Rugh, 1993), it can be obtained that

$$\|\Phi_c(t, \sigma)\| \leq \gamma e^{-2\eta(t-\sigma)}, \quad t \geq \sigma \geq 0 \quad (17)$$

where $\Phi_c(t, \sigma)$ is the transition matrix for

$$A_c(w, \dot{w}, \ldots, w^{(v-1)})$$

and γ, η are positive reals. Therefore the closed-loop system is exponentially stable on the manifold.

A stability result given in (Kelemen, 1986) can be used to analyze the behaviour of the closed-loop system in a small neighborhood of the manifold. A specialized version of the result in (Kelemen, 1986) is presented by Lemma A.2 in (Sureshbabu and Rugh, 1995) which is directly applicable with the following identifications:

$$p \leftrightarrow (w, \dot{w}, \ldots, w^{(v-1)}),$$

$$\mathcal{N} \leftrightarrow \mathcal{P}_0 \times \mathcal{P}_1 \times \ldots \mathcal{P}_{v-1},$$

$$z(p) \leftrightarrow \pi(w, \dot{w}, \ldots, w^{(v-1)}),$$

$$f_c(x, p) \leftrightarrow f(x, k(x, w, \dot{w}, \ldots, w^{(v-1)}), w),$$

$$r(p) \leftrightarrow \frac{\partial \pi(w, \dot{w}, \ldots, w^{(v-1)})}{\partial W} SW,$$

$$\tilde{p}(t) \leftrightarrow (w, \dot{w}, \ldots, w^{(v-1)}).$$

Applying Lemma A.2 in (Sureshbabu and Rugh, 1995) to the closed-loop system, it can be shown that the closed-loop system has a unique solution $x(t)$ defined for all $t \geq 0$, satisfying

$$\|x(t) - \pi(w(t), \dot{w}(t), \ldots, w^{(v-1)}(t))\| \leq \gamma e^{-\eta t}\|x(0) - \pi(w(0), \dot{w}(0), \ldots, w^{(v-1)}(0))\| + \gamma \int_0^t e^{-\eta(t-s)} \| \frac{d}{ds}\pi(w(s), \dot{w}(s), \ldots, w^{(v-1)}(s)) - \frac{\partial \pi}{\partial W} SW \| ds \quad (18)$$

if there exist positive real constants θ_1 and θ_2 such that, $\forall t \geq 0$,

$$\|x(0) - \pi(w(0), \dot{w}(0), \ldots, w^{(v-1)}(0))\| \leq \theta_1$$

$$\|\frac{d}{ds}\pi(w(t), \dot{w}(t), \ldots, w^{(v-1)}(t)) - \frac{\partial \pi}{\partial W} SW(t)\| \leq \theta_2$$

In fact,

$$\frac{d}{ds}\pi(w(t), \dot{w}(t), \ldots, w^{(v-1)}(t)) - \frac{\partial \pi}{\partial W} SW(t) = \frac{\partial \pi}{\partial w^{(v-1)}} w^{(v)}(t),$$

from (18), it can be obtained that

$$\|x(t) - \pi(w(t), \dot{w}(t), \ldots, w^{(v-1)}(t))\| \leq \gamma e^{-\eta t}\|x(0) - \pi(w(0), \dot{w}(0), \ldots, w^{(v-1)}(0))\| + \gamma\beta \int_0^t e^{-\eta(t-s)}\|w^{(v)}(s)\| ds \quad (19)$$

where

$$\beta = \max_{w^{(i)} \in \mathcal{P}_i, i=0,\ldots,v-1} \|\frac{\partial \pi}{\partial w^{(v-1)}}\|$$

If, $\forall t \geq 0$,

$$\|x(0) - \pi(w(0), \dot{w}(0), \ldots, w^{(v-1)}(0))\| \leq \theta_1$$

$$\|w^{(v)}(t)\| \leq \theta_3$$

then an argument similar to that in the proof of Theorem 2.1 in (Sureshbabu and Rugh, 1995) can be applied here to complete the proof of sufficiency.

4. APPROXIMATE OUTPUT REGULATION WITH HIGH ORDER APPROXIMATION

Since the measurements of high order derivatives of the exogenous signal may not be feasible in many situations, AORP-II is formulated to address the approximate output regulation only with the measurement of the exogenous signal itself. Assume that the signal is smooth enough, then it can be thought as the output of the following system

$$\dot{W} = SW + B_w w^{(v)}$$
$$w = C_w W \quad (20)$$

where

$$C_w = [I_q, 0, \ldots, 0] \quad B_w = [0, \ldots, I^q]^T$$

Since $\{C_w, S\}$ is observable, an observer can be designed as

$$\frac{d\hat{W}}{dt} = S\hat{W} + B_w w^{(v)} + K_w(w - C_w\hat{W}) \quad (21)$$

where K_w is chosen such that $(S - K_w C_w)$ is Hurwitz. Let $\tilde{W} = W - \hat{W}$ and it can be shown that

$$\frac{d\tilde{W}}{dt} = (S - K_w C_w)\tilde{W}.$$

A problem with (21) is that the measurement of $w^{(v)}$ is needed. In the spirit of approximate output regulation, if $\|w^{(v)}\|$ is small, the observer (21) can be approximated by

$$\frac{d\hat{W}}{dt} = S\hat{W} + K_w(w - C_w\hat{W}) \qquad (22)$$

In this case, it yields that

$$\frac{d\tilde{W}}{dt} = (S - K_w C_w)\tilde{W} + B_w w^{(v)} \qquad (23)$$

which means that estimation error can be small if $\|w^{(v)}\|$ is.

The result regard to the use of $\hat{W}$ in the control design for approximate output regulation is presented in the following theorem.

Theorem 2: If the linearization of the system (1) is stabilizable at the origin, then AORP-II is solvable if and only if the conditions (6) and (7) specified in Theorem 1 are satisfied.

Proof: Necessity. It follows the proof of necessity in Theorem 1.

Sufficiency. Use the estimate $\hat{W}$ to replace W in the control law (15). The closed-loop system is then described by

$$\begin{aligned} \dot{x}(t) &= f(x(t), u(x(t), \hat{W}), w(t)) \\ \frac{d\hat{W}}{dt} &= S\hat{W} + K_w(w - C_w\hat{W}) \\ \dot{W} &= SW \end{aligned} \qquad (24)$$

Consider another system

$$\begin{aligned} \frac{d\bar{x}(t)}{dt} &= f(\bar{x}(t), u(\bar{x}(t), W), w(t)), \qquad \bar{x}(0) = x(0) \\ \dot{W} &= SW \end{aligned} \qquad (25)$$

which describes the system response if W is used in the control. The existence of a unique solution of $\bar{x}(t)$ of (25) is guaranteed by Theorem 1 in the last section. The proof is completed if it can be shown that the difference between the state variables $\lim_{t\to\infty} \|x(t) - \bar{x}(t)\|$ can be set to be arbitrarily small.

Let

$$g(x, \hat{W}) = f(x, u(x, \hat{W}), w).$$

It implies

$$g(x, W) = f(x, u(x, W), w)$$

$$\frac{\partial g}{\partial x} = \frac{\partial f}{\partial x} + \frac{\partial f}{\partial u}K$$

From the system dynamics and using Taylor's theorem, it follows that

$$\begin{aligned} \frac{d}{dt}(x - \bar{x}) &= g(x, \hat{W}) - g(\bar{x}, W) \\ &= g(x, \hat{W}) - g(\bar{x}, \hat{W}) \\ &\quad + g(\bar{x}, \hat{W}) - g(\bar{x}, W) \\ &= \frac{\partial g}{\partial x}(\bar{x}, W)[x - \bar{x}] \\ &\quad + R_1(x, \bar{x}, W)[x - \bar{x}] \\ &\quad + R_2(\bar{x}, \hat{W}, W)\tilde{W} \end{aligned} \qquad (26)$$

where

$$R_1(x, \bar{x}, W) =$$

$$\int_0^1 \left[\frac{\partial g}{\partial x}(\bar{x} + \theta(x - \bar{x}), W) - \frac{\partial g}{\partial x}(\bar{x}, W)\right] d\theta \quad (27)$$

$$R_2(\bar{x}, \hat{W}, W) = \int_0^1 \frac{\partial g}{\partial W}(\bar{x}, \hat{W} + \theta(W - \hat{W}))d\theta$$

Using the Taylor's theorem again yields

$$\begin{aligned} \frac{\partial g}{\partial x}(\bar{x}, W) &= \frac{\partial g}{\partial x}(\pi(W), W) \\ &\quad + R_3(\bar{x}, \pi(W), W)[\bar{x} - \pi(W)] \end{aligned} \quad (28)$$

where

$$\begin{aligned} R_3(\bar{x}, \pi(W), W) = \int_0^1 &\frac{\partial}{\partial x}\frac{\partial g}{\partial x}(\pi(W) \\ &+ \theta(\bar{x} - \pi(W)), W)d\theta \end{aligned}$$

Noticing that

$$\frac{\partial g}{\partial x}(\pi(W), W) = A_c(W) \qquad (29)$$

and combining the above results, (26) can be rearranged as

$$\begin{aligned} \frac{d}{dt}(x - \bar{x}) &= A_c[x - \bar{x}] + R_1[x - \bar{x}] + R_2\tilde{W} \\ &\quad + R_3[\bar{x} - \pi(W)][x - \bar{x}] \end{aligned} \qquad (30)$$

With $x(0) = \bar{x}(0)$, the above equation results in

$$\begin{aligned} x(t) - \bar{x}(t) = \int_0^t &\Phi_c(t, s)\{R_1[x - \bar{x}] + R_2\tilde{W} \\ &+ R_3[\bar{x} - \pi(W)][x - \bar{x}]\}ds \end{aligned} \qquad (31)$$

The bound of $\|\bar{x} - \pi(W)\|$ has been discussed in Theorem 1, depending on the initial values and $\|w^{(v)}(t)\|$. From (23), it follows that

$$\|\tilde{W}\| \le \lambda_w e^{-\eta_w t}\|\tilde{W}(0)\| + \lambda_w \int_0^t e^{-\eta_w (t-s)}\|w^{(v)}(s)\|ds \quad (32)$$

where λ_w and η_w are positive real constants depending on $\{(S - K_w C_w), B_w\}$. Using (31) and following an analysis similar to the proofs of Lemma A.2 (Sureshbabu and Rugh, 1995) and Theorem in (Kelemen, 1986), it can be concluded that the solution $x(t)$ exists for all $t \geq 0$ and the bound of $\lim_{t\to 0} \|x_t - \bar{x}(t)\|$ can be set arbitrarily small if $\|w^{(v)}(s)\|$ and $\|\tilde{W}(0)\|$ are small enough.

Remark 2: Although Theorem 2 shows that the conditions specified in Theorem 1 are also the necessary and sufficient conditions for the solvability of AORP-II, the two problems are formulated under different assumptions. Additional conditions have been specified in the formulation of AORP-II.

Remark 3: The result shown in the proof of the sufficiency of Theorem 2 looks like an application of separation principle which works for linear systems. In general, it is not applicable to output regulation of nonlinear systems with error (output) feedback. When the exosystem is exactly known, output regulation via error feedback is solved in (Isidori, 1995) using system immersion. For the problems discussed in this paper, the state variable is available, and therefore the origin can always be stabilized by incorporating a term Kx in the control input. The estimation is only needed for derivatives of the exogenous signal.

5. CONCLUSION

Approximate output regulation of nonlinear systems has been investigated with the approximation to the exosystem by high order polynomials. Two problems are formulated and solved. The first problem (*AORP-I*) concerns with output regulation using sate feedback and measurement of the exogenous signal and its high order derivatives, while the second one (*AORP-II*) does not need the measurement of the derivatives. The proposed control strategy of *AORP-I* follows the format of the standard regulation problem, i.e., a feedforward term is added to the state feedback control to guarantee approximate tracking or rejection of the exogenous signal. The solution to *AORP-II* relies on the solution to the first problem and further approximation of high order derivatives by filtered signals generated from an observer.

With the proposed control strategies, output tracking or rejection of polynomials can be achieved, which includes regulation of constant output, tracking of ramp signals etc. Since an exogenous signal can be approximated by a polynomial within a given interval, approximate regulation to other exogenous signals can also be achieved. The stability results of the proposed algorithms are local, similar to the standard results of output regulation of nonlinear systems. The region of contraction may be even smaller than the regions given by standard regulation results (Isidori and Byrnes, 1990; Huang and Rugh, 1990), because more approximations are involved in the proposed algorithms. Further investigation is needed for the conditions for the existence of the extended regulation equations used in the two proposed problems, and for approximate solutions to those extended regulation equations, in case the accurate solutions are difficult to obtain.

REFERENCES

Fliess, M. (1982). Finite dimentional observation spaces for nonlinear sysems. In: *Feedback Control of Linear and Nonlinear Systems*(D. Hinrichsen and A. Isidori (Ed)), pp. 73–77. Springer Verlag, Berlin.

Francis, B. A. (1977). The linear multivariable regulator problem. *SIAM J. Contr. Optimiz.*, **15**, 486–505.

Horn, R. A. and C. A. Johnson (1990). *Matrix Analysis*. Cambridge Univ. Press, Cambridge, England.

Huang, J. and W. J. Rugh (1990). On a nonlinear multivariable servomechanism problem. *Automatica*, **26**(6), 963–972.

Isidori, A. (1995). *Nonlinear Control Systems*, 3 edn. Springer-Verlag, Berlin.

Isidori, A. and C. I. Byrnes (1990). Output regulation of nonlinear systems. *IEEE Trans. Automa. Contr.*, **35**(2), 131–140.

Kelemen, M. (1986). A stability property. *IEEE Trans. Automa. Contr.*, **31**(8), 766–768.

Khalil, H. K. (1992). *Nonlinear Systems*. Macmillan, New York.

Rugh, W. J. (1993). *Linear Systems Theory*. Prentice-Hall, Englewood Cliffs, NJ.

Smith, H. W. and E. J. Davison (1972). Design of industrial regulators: Integral feedback and feedforward control. *Proc. IEE*, **119**, 1210–1216.

Sureshbabu, N. and W. J. Rugh, (1995). Output regulation with derivative information. *IEEE Trans. Automa. Contr.*, **40**(10), 1755–1766.

14th Triennial World Congress, Beijing, P.R. China

E-2c-18-5

VARIABLE STRUCTURE CONTROL OF CONTINUOUS-TIME SYSTEM WITH SLIDING SECTOR

Katsuhisa Furuta* Yaodong Pan**

* *Department of Mechanical and Environmental Informatics*
Tokyo Institute of Technology
2-12-1 Oh-okayama, Meguro-ku, Tokyo 152, Japan
furuta@mei.titech.ac.jp
** *Department of Computers and Systems Engineering*
Tokyo Denki University, Hiki-gun, Saitama 350-0394, Japan
pan@k.dendai.ac.jp

Abstract: For any control systems, it is shown that in the state space there exists a subset around a hyperplane, inside which a norm of the state decreases without any control action. Therefore it is possible to use this natural feature of control systems to design a VS controller. In this paper, such subset is named the sliding sector. An algorithm to design the sliding sector using Riccati equation is given for controllable systems. A variable structure (VS) controller with the sliding sector for continuous-time system is proposed. With the VS control law designed in the paper, the state will move to the inside of the sliding sector and the norm of the state, i.e. some Lyapunov function decreases in all of the state space. Therefore the VS controller with the sliding sector quadratically stabilizes the system. A rotational pendulum system is used as a simulation model. The simulation results show that the proposed VS controller is chattering-free and with good control performance.

Keywords: Variable structure control, Sliding sector, Sliding mode, Quadratic stability, Lazy Control

1. INTRODUCTION

The continuous-time variable structure (VS) controller is known robust to parameter uncertainty and external disturbance because of the sliding motion on a predefined hyperplane (Utkin, 1964) (DeCarlo *et al.*, 1988) (Zinober, 1990) (Utkin, 1992) (Hung *et al.*, 1993). The control law has employed a nonlinear feedback varying the system structure by switching. This switching frequency is theoretically infinite. Therefore, there have been inherent chattering phenomena in practically realized VSC systems due to the finite switching frequency or switching delay. One approach proposed to remove the chattering is to introduce a boundary layer around the hyperplane and use a continuous control law inside the boundary layer(Slotine and Li, 1991)(Kachroo and Tomizuka, 1996). Another approach to due with the chattering is to replace the sliding mode by a sliding sector (Furuta, 1990)(Furuta and Pan, 1995)(Pan and Furuta, 1997)(Hara *et al.*, 1998).

As an effective method to design a robust controller for uncertainty systems, Lyapunov function has also been used, e.g., see (Barmish *et al.*, 1983), (Barmish, 1983) and (Corless, 1994). They have designed the controller so that the linear plant is quadratically stabilized for the Lyapunov Function with positive definite matrix P in the kernel of a $B^T P$ (B is the input matrix of the controlled system). Their controllers have the resemblance to the sliding mode controller since the closed loop system becomes variable structure system.

In this paper, the VS controller with sliding sector is designed such that a norm of the state, which is considered as a Lyapunov function candidate keeps decreasing in all of the state space with specified negativity of its derivative. Inside the sliding sector the Lyapunov function decreases for zero input with specified velocity and in the outside, the VS control law is used to move the state to the inside of the sliding sector. An algorithm to design the sliding sector by using Algebraic Riccati equation is given.

The organization of the paper is as follows. Section 2 gives the definition of the sliding sector and shows its existence. Section 3 designs the sliding sector using Riccati equation. Section 4 proposes a VS controller based on the sliding sector. Section 5 gives the simulation results on a rotational pendulum system.

2. DEFINITION OF *PR*-SLIDING SECTOR

For the clarity of the presentation, a linear time-invariant continuous-time single input system will be considered.

$$\dot{x}(t) = Ax(t) + Bu(t) \tag{1}$$

where $x(t) \in R^n$ and $u(t) \in R^1$ are state and input vectors, respectively. A and B are constant matrices of appropriate dimensions, pair (A, B) is controllable.

Definition 1. The *P-Norm* $||\cdot||_P$ of the system state is defined as

$$||x||_P = (x^T Px)^{\frac{1}{2}}, \quad x \in R^n \tag{2}$$

where $P \in R^{n\times n}$ is a positive definite symmetric matrix.

The square of the P-norm is denoted as

$$L = ||x||_P^2 = x^T Px > 0, \quad \forall x \in R^n, x \neq 0 \tag{3}$$

which will be considered as a Lyapunov function candidate. If the autonomous system of (1) is quadratically stable, then there exist a positive definite symmetric matrix P and a positive semi-definite symmetric matrix $R = C^T C$ such that

$$\dot{L} = x^T(A^T P + PA)x \leq -x^T Rx, \quad \forall x \in R^n$$

where $P \in R^{n\times n}$, $R \in R^{n\times n}$, and $C \in R^{l\times n}$, $l \geq 1$, (C, A) is observable pair. But for an unstable one, this inequality does not hold. It is possible to decompose all state space into two parts such that one satisfies $\dot{L} > -x^T Rx$ for some element $x \in R^n$, and the other one satisfies $\dot{L} \leq -x^T Rx$ for some other element $x \in R^n$. The latter ones form a special subset in which the P-norm $||x||_P$ decreases. Accordingly this special subset is defined as a PR-sliding sector.

Definition 2. The *PR-Sliding Sector* is a subset of R^n defined as

$$\mathcal{S} = \{x|\ x^T(A^T P + PA)x \leq -x^T Rx, x \in R^n\ \} \tag{4}$$

where $P \in R^{n\times n}$ is a positive definite symmetric matrix, $R \in R^{n\times n}$ is a positive semi-definite symmetric matrix, and $R = C^T C$, $C \in R^{l\times n}$, $l \geq 1$, (C, A) is observable pair.

Inside the PR-sliding sector, the P-norm $||x||_P$ of the plant (1) without any control action decreases with the rate of

$$\dot{L} \leq -x^T Rx \leq 0, \quad \forall x \in \mathcal{S}, x \neq 0. \tag{5}$$

Theorem 1. For the plant (1), the PR-sliding sector defined in (4) exists for any positive definite symmetric matrix P and any positive semi-definite symmetric matrix R described in Definition 2, and can be rewritten as

$$\mathcal{S} = \{x|\ s^2(x) \leq \delta^2(x)\ \} \tag{6}$$

where

$$s^2(x) = x^T P_1 x \geq 0 \tag{7}$$
$$\delta^2(x) = x^T P_2 x \geq 0, \tag{8}$$

P_1 and P_2 are $n \times n$ positive semi-definite symmetric matrices.

Proof: Denote

$$\Omega = A^T P + PA + R. \tag{9}$$

Then the PR-sliding sector defined in (4) is determined by

$$x^T \Omega x \leq 0.$$

For the matrix Ω, there exists a real orthogonal matrix $U \in R^{n\times n}$ such that

$$U^T \Omega U = diag(r_1, r_2, ..., r_n)$$

holds where $r_i (i = 1, 2, ...n)$ are the characteristic roots of Ω, which are all real because Ω is a symmetric matrix.

Assume

$$\bar{P}_1 = diag(\frac{|r_1| + r_1}{2}, \frac{|r_2| + r_2}{2}, ..., \frac{|r_n| + r_n}{2})$$

$$\bar{P}_2 = diag(\frac{|r_1| - r_1}{2}, \frac{|r_2| - r_2}{2}, \ldots, \frac{|r_n| - r_n}{2})$$

i.e, $\bar{P}_1$ and $\bar{P}_2$ are composed of the positive eigenvalues and the negative eigenvalues of Ω, respectively. Then the following holds.

$$U^T \Omega U = \bar{P}_1 - \bar{P}_2$$

where $\bar{P}_i \geq 0$ $(i = 1, 2)$. Denoting $P_i = U\bar{P}_i U^T$ $(n = 1, 2)$ gives

$$\Omega = P_1 - P_2, \quad P_i \geq 0, \quad (i = 1, 2).$$

The PR-sliding sector defined in (4) thus can be rewritten as (6) where $s^2(t)$ and $\delta^2(t)$ are determined by (7) and (8), respectively. Which implies that such PR-sliding sector exists for any controllable plant. ■

Corollary 2. Let $n_i = rank(P_i)$ $(i = 1, 2)$ and $n_3 = n - n_1 - n_2$. Then n_1 and n_2 are the numbers of the positive eigenvalues and the negative eigenvalues of Ω, respectively, n_3 is the number of the eigenvalues of Ω in the origin.

Example 3. Consider a third order plant represented by

$$\dot{x} = \begin{bmatrix} -1 & 0 & 0 \\ 0 & -2 & 0 \\ 0 & 0 & -4 \end{bmatrix} x + \begin{bmatrix} 1 \\ 1 \\ 1 \end{bmatrix} u. \qquad (10)$$

If R is chosen to be the 3×3 identity matrix I_3, i.e. $R = I_3$, P is chosen to be σI_3 $(\sigma > 0)$, σ is chosen to be a small positive constant $(\sigma < \frac{1}{8})$ in the beginning and then increased little by little until $\sigma > \frac{1}{2}$, the ranks of matrices P_1 and P_2 will change from 3, 2, 1, to 0 and from 0, 1, 2, to 3, respectively. Correspondingly, the PR-sliding sector is the origin for $\sigma < \frac{1}{8}$, a line for $\sigma = \frac{1}{8}$, a cone for $\frac{1}{4} < \sigma < \frac{1}{8}$, a subset between two planes for $\sigma = \frac{1}{4}$, a complementary set of a cone in the state space for $\frac{1}{2} < \sigma < \frac{1}{4}$, and the state space for $\sigma \geq \frac{1}{2}$, respectively, as shown in Table 1.

3. DESIGN OF PR-SLIDING SECTOR

When $n_1 = 1$ and $n_2 \geq 1$, the PR-sliding sector has a form shown in Figure 1 and Figure 2. A hyperplane $s(x) = 0$ is included in the PR-sliding sector. In the PR-sliding sector guaranteeing $|s(x) \leq |\delta(x)|$ including the hyperplane of $s(x) = 0$ with the dimension of $n - 1$, the P-norm decreases without any control action. such PR-sliding sector is called the simplified PR-sliding sector which can be written as

$$\mathcal{S} = \{x : \ |s(x)| \leq |\delta(x)| \ \} \qquad (11)$$

where,

$$s(x) = Sx, S = \begin{bmatrix} c_1 & c_2 & \ldots & c_n \end{bmatrix}, S^T S = P_1 \quad (12)$$

$$\delta(x) = \sqrt{x^T P_2 x}. \qquad (13)$$

The following Riccati equation is one of useful tools to design the PR-sliding sector.

$$PA + A^T P - PBB^T P = -Q. \qquad (14)$$

where $Q \in R^{n \times n}$ is a positive definite symmetric matrix.

As pair (A, B) in plant (1) is assumed to be controllable, the symmetric positive definite solution P of the Riccati equation (14) exists. If the solution P of the Riccati equation (14) is chosen to define the P-norm as in Definition 1, then for the zero input, (14) gives

$$\begin{aligned} \dot{L}(t) &= x^T(t)(A^T P + PA)x(t) \\ &= x^T(t)PBB^T Px(t) - x^T(t)Qx(t), \quad (15) \end{aligned}$$

As the direct result of the above equation, a design algorithm of the simplified PR-sliding sector is given by using the Riccati equation as shown in the following theory.

Theorem 4. For any controllable plant (1), if the solution P of the Riccati equation (14) with a symmetric positive definite matrix $Q \in R^{n \times n}$ is used to define the P-norm and the matrix R defined in Definition 2 is chosen so that $\Delta = Q - R \geq 0$, then the PR-sliding sector defined in (6) can be rewritten as

$$\mathcal{S} = \{x | \ |s(x)| \leq |\delta(x)| \ \} \qquad (16)$$

where

$$s(x) = Sx(t) \quad , \quad S = B^T P$$

$$\delta(x) = \sqrt{x^T(t) \Delta x(t)}$$

4. VS CONTROLLER WITH PR-SLIDING SECTOR

Inside the PR-sliding sector defined in Section 2 and designed in Section 3, the P-norm decreases without any control action, which is one of natural features of any control system. It is reasonable to design a VS controller with the PR-sliding sector so that the system state moves into the PR-sliding sector with some VS control law and then inside it the control input is zero. Such VS controller makes the best use of the natural feature. The control action is used only to move the system state into the inside of the PR-sliding sector. The VS controller with the PR-sliding sector is called a *lazy controller* because in most of time no control action is needed.

Table 1. PR-Sliding Sector

σ	$(0,\frac{1}{8})$	$\frac{1}{8}$	$(\frac{1}{8},\frac{1}{4})$	$\frac{1}{4}$	$(\frac{1}{4},\frac{1}{2})$	$\frac{1}{2}$	$(\frac{1}{2},+\infty)$
n_1	3	2	2	1	1	0	0
n_2	0	0	1	1	2	2	3
Sliding Sector	the origin	a straight line	a cone	a subset between two planes	complement of a cone	state space	state space
Figure			3	2	1		

To design a chattering free VS controller, an inner sector $\mathcal{S}_i$ of the PR-sliding sector is defined as a subset of $\mathcal{S}$, which is determined by

$$\mathcal{S}_i = \{x||s(x)| \le \alpha|\delta(x)|\} \quad (17)$$

where α is a real positive constant satisfying $0 < \alpha < 1$, $s(x)$ and $\delta(x)$ are the parameters of the PR-sliding sector (16). The VS controller in the paper thus is designed so that the VS control input outside the PR-sliding sector is active to move the system state into the inner sector, and then the VS control input becomes zero as soon as the inner sector is reached and will not be active until the system state goes out of the PR-sliding sector which is shown in Figure (4) for a second order system with the transfer function being $\frac{1}{s^2}$.

Theorem 5. Corresponding to the PR-sliding sector $\mathcal{S}$ (16) and the inner sector $\mathcal{S}_i$ (17) with $\Delta = rQ$ and $R = (1-r)Q$ for some positive constant r $(0 < r < 1)$, the VS control law

$$u(t) = -\sigma(x(t),\delta(x))(SB)^{-1}(SAx + ks(x)) \quad (18)$$

ensures the movement of the state from the outside of the PR-sliding sector to the inside of the inner sector and the decreasing of the P-norm at any state if the positive constant k satisfies

$$k > \max\{\frac{(SB)^{-1}}{2}, \frac{k_0}{\alpha}\} \quad (19)$$

where positive constant k_0 satisfies the following quadratic inequality.

$$2k_0 rQ + S^TSA + A^TS^TS > 0.$$

$\sigma(s(x),\delta(x))$ is a hysteresis dead-zone function on $s(x)$ and $\delta(x)$ and defined as

$$\sigma(x) = \begin{cases} 0 & x \in \mathcal{S}_i \\ \text{unchange} & x \in \mathcal{S} - \mathcal{S}_i \\ 1 & x \bar{\in} \mathcal{S}. \end{cases}$$

Therefore the control law (18) results in a quadratic stable VS controller, where parameters Q, P, R, Δ, $s(x)$, $\delta(x)$, and α are defined in Theorem 4 and Equation (17), SB is equal to B^TPB which is nonsingular according to the theorem.

Proof: Consider the square of the P-norm as a Lyapunov function, i.e.:

$$L(t) = x^T(t)Px(t) > 0, \quad \forall x \in R^n, x \ne 0. \quad (20)$$

It is assumed that the initial state is outside the PR-sliding sector. In this case the VS control law is determined by

$$u(t) = -(SB)^{-1}(SAx + ks(x)). \quad (21)$$

which will be active until the system state is moved into the inner sector. With the above control, the derivate of the linear function $s(x)$ satisfies that

$$\dot{s}(x) = S\dot{x}(t) = SAx(t) + SBu(t) = -ks(x)$$

Thus it follows that

$$\frac{d}{dt}s^2(x) = 2s(x)\dot{s}(x) = -2ks^2(x) < 0$$

which means that the absolute value of the linear function $s(x)$ will decrease so that the system state moves toward the inside of the inner sector and will move inside it in a finite time if the decreasing rate of $\delta(x)$ is slower than the one of $|s(x)|$ for some large enough k.

While the system state is being moved from the outside of the PR-sliding sector to the inside of the inner sector by the VS control law (18), the P-norm is decreasing because according to the definition of the PR-sliding sector (16), it follows some calculating that

$$\begin{aligned}\dot{L}(t) &= (Ax(t) + Bu(t))^TPx(t) \\ &\quad + x^T(t)P(Ax(t) + Bu(t)) \\ &< -x^T(t)Rx(t) \\ &< 0\end{aligned}$$

where $SB = B^TPB > 0$.

After being moved into the inner sector and until moving out of the PR-sliding sector, the control input $u(t) = 0$. In this case, $|s(x)| \le |\delta(x)|$ and the derivate of the Lyapunov function $L(t)$ is

$$\begin{aligned}\dot{L}(t) &= s^2(x) - \delta^2(x) - x^T(t)Rx(t) \\ &\le -x^T(t)Rx(t) \le 0.\end{aligned}$$

As long as the system state moves out of the PR-sliding sector, the control input (21) will let it back to the inside of the inner sector while the P-norm keeps decreasing.

Thus the system state will be moved from the outside of the PR-sliding sector to the inside of the inner sector if the system state is outside the PR-sliding sector and the Lyapunov function (20) keeps decreasing in all of the state space with the VS control law (18), i.e.

$$\dot{L}(t) \leq -x^T(t)Rx(t), \quad \forall x \in R^n$$

which means that the VS control law (18) quadratically stabilize the plant (1). ∎

5. EXAMPLES

A rotational inverted pendulum system (Hara *et al.*, 1998) is taken as the simulation model. An additional dynamics is included in the inverted pendulum system as the uncertainty. For detail about the inverted pendulum system, see (Hara *et al.*, 1998).

Based on the state equation of pendulum system without additional dynamics, a VS controller with PR-sliding sector is designed. The positive definite symmetric matrix Q is chosen as

$$Q = diag(1, 1, 1, 1)$$

then the solution P of the Riccati equation (14) and matrix R are determined by

$$P = \begin{bmatrix} 1.32132368 & -2.87041542 & 0.372473930 & -0.284276152 \\ -2.87041542 & 33.0669334 & -3.44522372 & 2.89077316 \\ 0.372473930 & -3.44522372 & 0.433959219 & -0.335507650 \\ -0.284276152 & 2.89077316 & -0.335507650 & 0.290947897 \end{bmatrix}$$

$$R = diag(0.5, 0.5, 0.5, 0.5)$$

The parameters S and Δ of the PR-sliding sector thus are shown as follows with $r = 0.5$:

$$S = \begin{bmatrix} -1.0 & 20.6041295 & -1.35212449 & 2.41640558 \end{bmatrix}$$
$$\Delta = diag(0.5, 0.5, 0.5, 0.5)$$

In the simulation, the sampling interval T is chosen to be 0.01 Second. The VS control law (18) is used with $\alpha = 0.2$ and $K = 100$. Figure 5 shows the simulation results with the VS controller and the LQ optimal controller, where the optimal feedback control law is $K = -B^T P$ and P is the same as the one used in the VS controller.

6. CONCLUSION

The paper proposed a VS controller with sliding sector for continuous-time systems. The sliding sector is defined such that inside it, a given norm decreases. The VS control law designed in the paper ensures the system state to move from outside to the inside of the sliding sector and the given norm of the system state or a Lyapunov functions keep decreasing in all of the state space. Thus the proposed VS controller quadratically stabilizes the system without chattering. The simulation results show the effectiveness of the proposed VS controller.

7. REFERENCES

Barmish, B.R. (1983). Stabilization of uncertain systems vis linear control. *IEEE Trans. on Automatic Control* **28**, 848–850.

Barmish, B.R., M. Corless and G. Leitmann (1983). A new class of stabilizing controllers for uncertain dynamical systems. *SIAM J. Control and Optimization* **21**, 246–255.

Corless, M. (1994). Robust stability analysis and controller design with quadratic lyapunov functions. *Lecture Notes in Control and Information Sciences: Variable Structure and Lyapunov Control* **193**, 181–203.

DeCarlo, R.A., S.H. Zak and G.P. Matthews (1988). Variable structure control of nonlinear multivariable systems: A tutorial. *Proc. of the IEEE* **76**, 212–232.

Furuta, K. (1990). Sliding mode control of a discrete system. *System & Control Letters* **14**, 145–152.

Furuta, K. and Y. Pan (1995). A new approach to design a sliding sector for vss controller. In: *Proc. of the American Control Conference.* Seattle. pp. 1304–1308.

Hara, M., K. Furuta, Y. Pan and T. Hoshino (1998). Evaluation of discrete-time vsc on inverted pendulum apparatus with additional dynamics. *International Journal of Applied Mathematics and Computer Science* **7**, 101–123.

Hung, J.Y., W. Gao and J.C. Hung (1993). Variable structure control: A survey. *IEEE Trans. on Industrial Electronics* **40**, 2–22.

Kachroo, P. and M. Tomizuka (1996). Chattering reduction and error convergence in the sliding-mode control of a class of nonlinear systems. *IEEE Transactions on Automatic Control* **AC-41**, 1063–1068.

Pan, Y. and K. Furuta (1997). Discrete-time vss controller design. *International Journal of Robust and Nonlinear Control* **7**, 373–386.

Slotine, J.-J.E. and W. Li (1991). *Applied Nonlinear Control.* Prentice Hall. Englewood Cliffs, NJ.

Utkin, U.I. (1992). *Sliding Modes in Control and Optimization.* Springer-Verlag.

Utkin, V.I. (1964). Variable structure system with sliding mode:a survey. *IEEE Transactions on Automatic Control* **AC-22**, 212–222.

Zinober, A.S.I. (1990). *Deterministic Control of Uncertain Systems.* Peter Peregrinus Ltd.. London, United Kingdom.

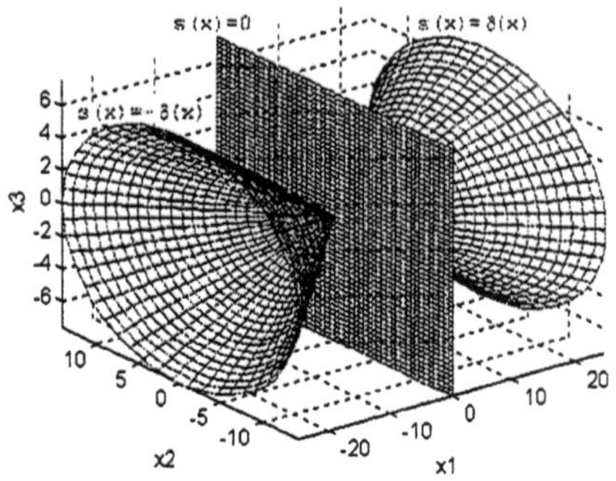

Fig. 1. PR-Sliding Sector : Complementary Set of a Cone

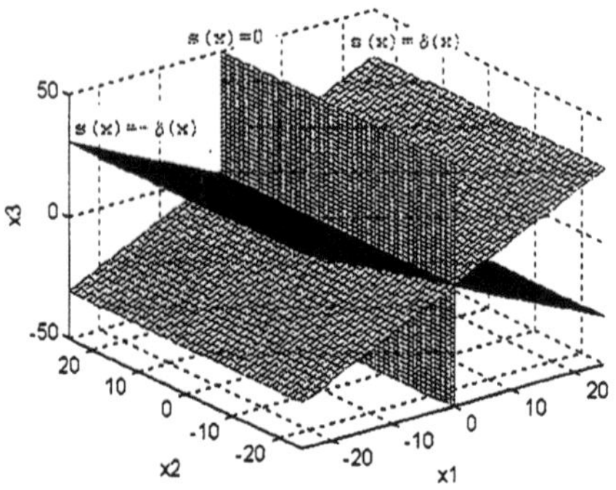

Fig. 2. PR-Sliding Sector including $s(x) = 0$ with Two Boundary Planes

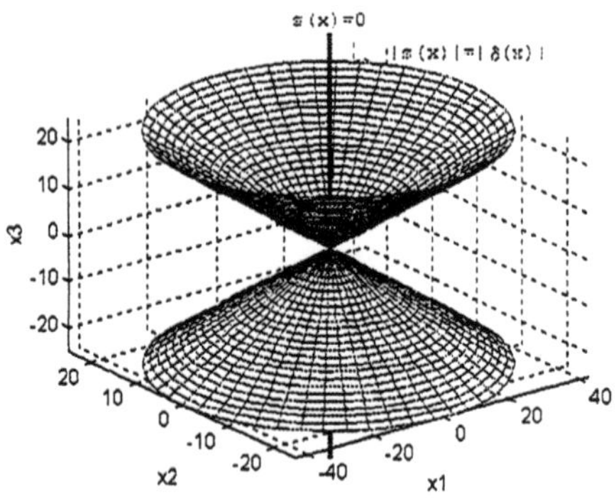

Fig. 3. PR-Sliding Sector : a Cone

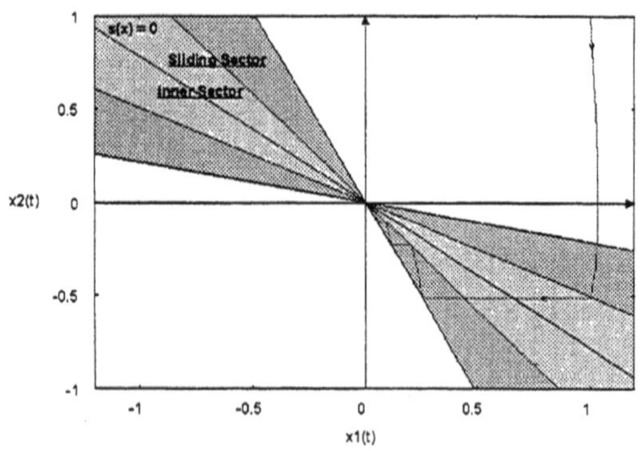

Fig. 4. PR-Sliding Sector and Its Inner Sector

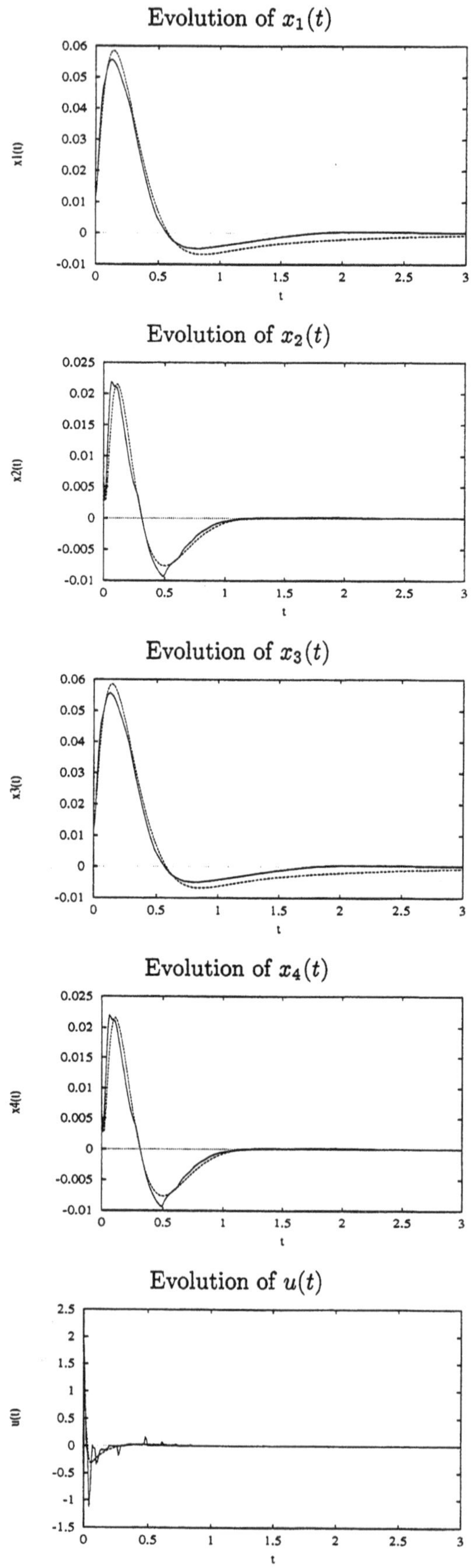

Fig. 5. Simulation result with VS Controller (Solid Line) and LQ Optimal Controller (Dash Line)

E-2c-18-6
14th Triennial World Congress, Beijing, P.R. China

COMPENSATION AND CONTROL OF NONLINEAR SYSTEMS IN FREQUENCY DOMAIN

Zhikui ZHAO, Chongzhao HAN, Jianhua MU, and **Fanglin DENG**

Faculty 301, Xi'an Institute of High Thechnology
Xi'an, Shaanxi 710049, P.R.China
Email: zkzhao_xian@163.net

Abstract: This paper studies the control problem of single-input and single-output nonlinear dynamic systems whose input-output relation is described by the Wolterra series, via the frequency-domain approach. Firstly, based on the generalized frequency response functions(GFRF) of Volterra functional series of nonlinear systems, develops a kind of synthesis approach of compensator of inverse nonlinear systems in frequency domain, and presents general structures and mathematical descriptions of the compensator. Then, discusses differential equations realization of GFRF of the compensator. Finally, the effectiveness of the method is verified in simulations. *Copyright©1999IFAC*

Keywords: Nonlinear systems; frequency-response methods; compensation; inverse system; control

1. INTRODUCTION

In recent years, with the developing of nonlinear system research, application of Volterra series theory has played an ever-increasing role in nonlinear system analysis, identification and prediction (Nam and Powers,1994). This is due, on the one hand, to the fact that Volterra series has a firm mathematical foundation (Boyd, et al, 1984; Schethzen, 1980; Sandberg, 1983), and on the other hand, a gently nonlinear time-invariant system can be described with reasonable accuracy by a truncated version of the Volterra series, which considerably reduces the complexity of the problem. In practice, many important nonlinear effects in engineering and science such as mechanical, biological, power conversion, optical, communication systems, and so on can be approximated by a Volterra series of second or third order (Chua and Liao, 1989; Nam and Powers,1990; Huang, 1997).

Volterra systems are nonlinear systems whose input-output relation is described by a Volterra series. There has been extensive research on the Volterra systems focus on such issues as controllability, observability, realizablity, stability, convergence, and so on (Chua and Ng, 1979; Li and Feng, 1991). However, little attention has been paid to the problem of designing control laws to control the behavior of Volterra systems. It is the problem to discuss in this paper.

The idea in this paper is, firstly, trying to obtain the inverse system of a nonlinear plant in frequency domain, and then, making use of the inverse system to compensate nonlinear effects of the plant so that the plant with the compensator is linear or approximating to linear in wide range, finally, using the achievements of linear systems theory to design the system compensated. The rest of this paper is organized as follows. Section 2 is mathematical description of the problem. Section 3 establishes the synthesis approach in domain-frequency based on GFRF model of plants. And section 4 discusses the differential equations realization of GFRF of the compensator. The simulation results and conclusion are in section 5 and section 6, respectively.

2. MATHEMATICAL DISCRIPTIONS OF PROBLEM

In practice and theory, the input-output relation of most time invariant nonlinear control systems can be expressed by Volterra series as follows

$$y(t) = \sum_{n=1}^{\infty} y_n(t),$$

where

$$y_n(t) = \int_{-\infty}^{\infty}\cdots\int_{-\infty}^{\infty} h_n(\tau_1,\cdots,\tau_n)\prod_{i=1}^{n} u(t-\tau_i)d\tau_i, \quad n \in N. \tag{1}$$

Where $u(t)$ and $y(t)$ are the input and the output of system, respectively, N is the nature set and the function $\{h_n(\tau_1,\cdots,\tau_n), n \in N\}$, is the so-called generalized pulse response function, or Volterra kernel associated with the system. The n-dimensional Fourier transforms of function (1) is

$$\hat{y}(\omega) = \sum_{n=1}^{\infty} \hat{y}_n(\omega),$$

where

$$\hat{y}_n(\omega) = (2\pi)^{-(n-1)}\int_{-\infty}^{\infty}\cdots\int_{-\infty}^{\infty} \hat{h}_n(\omega-\omega_2-\cdots-\omega_n,\omega_2,\cdots,\omega_n)\cdot\hat{u}(\omega-\omega_2-\cdots-\omega_n)\hat{u}(\omega_2)\cdots\hat{u}(\omega_n)d\omega_2,\cdots d\omega_n, \quad n \in N. \tag{2}$$

where $\hat{y}, \hat{y}_n$ and $\hat{u}$ are the Fourier transforms of y, y_n and u, respective, while $\{\hat{h}_n, n \in N\}$, which is called generalized frequency response function (GFRF) or Volterra kernel of frequency-domain, is the n-dimensional Fourier transforms of $\{h_n, n \in N\}$, i.e.

$$\hat{h}_n(\omega_1,\cdots,\omega_n) = \int_{-\infty}^{\infty}\cdots\int_{-\infty}^{\infty} h_n(\tau_1,\cdots,\tau_n) e^{-j(\omega_1\tau_1+\cdots+\omega_n\tau_n)}d\tau_1\cdots d\tau, \quad n \in N \tag{3}$$

Assuming the GFRFs of plant P and compensator G can be defined as followings function (4) and function (5), respectively,

$$P = \{\hat{p}_n(\omega_1,\cdots,\omega_n)\}, \quad n \in N \tag{4}$$

$$G = \{\hat{g}_n(\omega_1,\cdots,\omega_n)\}, \quad n \in N \tag{5}$$

then, the general structure of nonlinear system with compensators (to be called H) is as shown in Fig.1, its GFRF model can be written as

$$H = G \circ P = \{\hat{h}_n(\omega_1,\cdots,\omega_n)\}, \quad n \in N \tag{6}$$

where

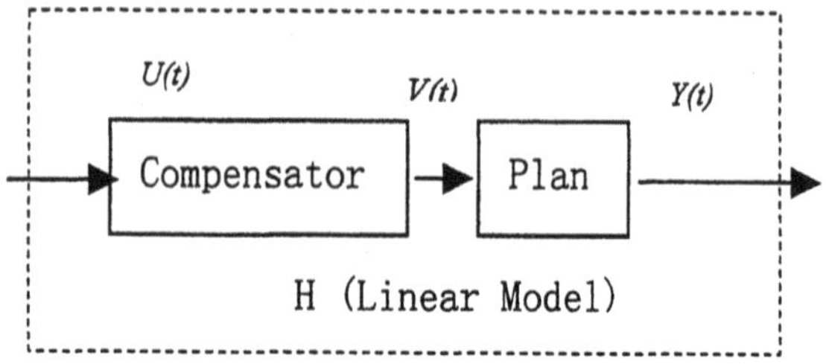

Fig.1 General structure of a nonlinear system with a compensator

$$\hat{h}_1(\omega_1) = \hat{p}_1(\omega_1)\hat{g}_1(\omega_1),$$

$$\hat{h}_2(\omega_1,\omega_2) = \hat{p}_1(\omega_1+\omega_2)\hat{g}_2(\omega_1,\omega_2) + \hat{p}_2(\omega_1,\omega_2)\hat{g}_1(\omega_1)\hat{g}_1(\omega_2),$$

$$\cdots\cdots$$

$$\hat{h}_n(\omega_1,\omega_2,\cdots,\omega_n) = \hat{p}_1(\omega_1+\omega_2+\cdots+\omega_n)\hat{g}_n(\omega_1,\omega_2,\cdots,\omega_n)+\cdots+\hat{p}_n(\omega_1,\omega_2,\cdots,\omega_n)\hat{g}_1(\omega_1)\hat{g}_1(\omega_2)\cdots\hat{g}_1(\omega_n)$$

$$\cdots\cdots \tag{7}$$

If we can get a compensator G on the basis of the plant P such that

$$\hat{h}_n(\omega_1,\cdots,\omega_n) = 0, \ \forall n \geq 2, \tag{8}$$

are satisfied, that is to say, only are linear terms left in the GFRF of model H, then the G is called as completely compensator. If the function (8) are approximately satisfied, then the G is also called approximately compensator.

If completely or approximately compensator of a nonlinear system are existed and can be obtained, then the synthesis problem of the nonlinear system will change into that of the linear system for which there have been many satisfied synthesis approaches. On the existence of compensators of nonlinear systems, in fact, is existence problem of inverse systems of nonlinear systems, which has been discussesed in (Li and Feng, 1991).

To sum up, there will be following synthesis steps for this method:

(1) Calculating GFRF model of compensators.

(2) Getting the differential realization of GFRF model of compensators.

(3) Design control laws for compensated system.

3. SYNTHESIS OF COMPENSATORS

According to section 2, it seems like possible that any nonlinear system can completely be compensated, however, it is not true. There are at least two reasons. (1). The GFRF of describing plant is a differential series with infinitude terms, it seldom happen that the series can be brought into closed. (2). Although GFRFs of compensators can be gotten, it is very difficult to describe them using differential functions. But as Taylor series can approximate a system by a truncated series in practice, Volterra series can also be treated as Taylor series under some conditions.

3.1 GFRF Model of Compensators

The synthesis approach in this paper is based on following assumptions and definitions.

Definition A1. The norms of GFRF of general nonlinear system are defined as

$$\|\hat{h}_1\|_\infty = \sup_{\omega_1 \in R} |\hat{h}_1(\omega_1)|,$$
$$\|\hat{h}_2\|_\infty = \sup_{\omega_1,\omega_2 \in R} |\hat{h}_2(\omega_1,\omega_2)|,$$
$$\cdots\cdots$$
$$\|\hat{h}_n\|_\infty = \sup_{\omega_1,\cdots,\omega_n \in R} |\hat{h}_n(\omega_1,\cdots,\omega_n)|,$$
$$\cdots\cdots \tag{9}$$

Assumption A1. The norms of GFRF of a plant satisfy following expression of (10), i.e. it is damply with increasing $n!$.

$$\|\hat{p}_1\|_\infty > \|\hat{p}_2\|_\infty \cdot 2! > \|\hat{p}_3\|_\infty \cdot 3! > \cdots > \|\hat{p}_n\|_\infty \cdot n! > \cdots. \tag{10}$$

In terms of literatures (Han and Cao, 1996, 1997), a system is BIBO stable if first order kernel of its GFRF (i.e. frequency response function of linear part) is of Hurwitz and the others order kernels are damply with increasing $n!$.

Assumption A2. If GFRF of a plant satisfies with Assumption A1 and its compensator G is existent, then the GFRF of the compensator G also satisfies expression of (10)

If GFRFs of a plant and its compensator satisfy Assumptions A1 and A2 respectively, then both of them can be approximated by truncated series. So we might as well only consider first three order kernels of GFRF of them for decreasing complexity of the calculation. This kind of approximation is proved to be reasonable for many important nonlinear effects in engineering and science. Under those above assumptions, GFRF model of the compensator G can calculated by the following theorem.

Theorem A.1 Assuming a plant can be described by the first three orders kernel of its GFRF, and $\hat{p}_1(\omega)$, $\hat{p}_2(\omega_1,\omega_2)$ and $\hat{p}_3(\omega_1,\omega_2,\omega_3)$ are first one, two and three terms of its GFRF respectively. $\hat{h}_1(\omega)$ is the frequency response function of the plant compensated by compensator G. Then first three terms of GFRF of the compensator G can be obtained by the following equations.

$$\hat{g}_1(\omega) = \frac{\hat{h}_1(\omega)}{\hat{p}_1(\omega)}, \tag{11}$$

$$\hat{g}_2(\omega_1,\omega_2) = -\frac{\hat{g}_1(\omega_1)\cdot\hat{g}_1(\omega_2)}{\hat{p}_1(\omega_1+\omega_2)}\cdot\hat{p}_2(\omega_1,\omega_2), \tag{12}$$

$$\begin{aligned}\hat{g}_3(\omega_1,\omega_2,\omega_3) = -\hat{p}_1^{-1}(\omega_1+\omega_2+\omega_3)\\ [\hat{p}_2(\omega_1+\omega_2,\omega_3)\hat{g}_2(\omega_1,\omega_2)\cdot\hat{g}_1(\omega_3)\\ +\hat{p}_2(\omega_1,\omega_2+\omega_3)\hat{g}_1(\omega_1)\hat{g}_2(\omega_2,\omega_3)\\ +\hat{p}_3(\omega_1,\omega_2,\omega_3)\hat{g}_1(\omega_1)\hat{g}_1(\omega_2)\hat{g}_1(\omega_3)]\end{aligned}. \tag{13}$$

Where $\hat{g}_1(\omega)$, $\hat{g}_2(\omega_1,\omega_2)$ and $\hat{g}_3(\omega_1\omega_2\omega_3)$ are first one, two and three terms of GFRF of the G, respectively.

Proof. Assuming the output of a plant P can be expressed as

$$\hat{y}(\omega) = \sum_{n=1}^{3}\hat{y}_n(\omega), \tag{14}$$

where

$$\hat{y}_1(\omega) = \hat{p}_1(\omega)\hat{v}(\omega), \tag{15}$$

$$\hat{y}_2(\omega) = (2\pi)^{-1}\int_{-\infty}^{\infty}\hat{p}_2(\omega-\omega_2,\omega_2) \bullet \hat{u}(\omega-\omega_2)\hat{u}(\omega_2)d\omega_2 \tag{16}$$

and

$$\begin{aligned}\hat{y}_3(\omega) = (2\pi)^{-2}\int_{-\infty}^{\infty}\int_{-\infty}^{\infty}\hat{p}_3(\omega-\omega_2-\omega_3,\omega_2,\omega_3)\\ \cdot\hat{v}(\omega-\omega_2-\omega_3)\hat{v}(\omega_2)\hat{v}(\omega_3)d\omega_2 d\omega_3,\end{aligned} \tag{17}$$

respectively, and at the same time assuming the characteristics of a compensator with three orders GFRF, as shown in Fig.2, can be described as

$$\hat{v}(\omega) = \sum_{n=1}^{3}\hat{v}_n(\omega), \tag{18}$$

where

$$\hat{v}_1(\omega) = \hat{g}_1(\omega)\hat{u}(\omega), \tag{19}$$

$$\hat{v}_2(\omega) = (2\pi)^{-1}\int_{-\infty}^{\infty}\hat{g}_2(\omega-\omega_2,\omega_2)\hat{u}(\omega-\omega_2)\hat{u}(\omega_2)d\omega_2, \tag{20}$$

and

$$\begin{aligned}\hat{v}_3(\omega) = (2\pi)^{-2}\int_{-\infty}^{\infty}\int_{-\infty}^{\infty}\hat{g}_3(\omega-\omega_2-\omega_3,\omega_2,\omega_3)\\ \cdot\hat{u}(\omega-\omega_2-\omega_3)\hat{u}(\omega_2)\hat{u}(\omega_3)d\omega_2 d\omega_3,\end{aligned} \tag{21}$$

respectively. Defining H as the transmission function from u to y, as shown in Fig.1, it is the cascade connection of the P and G. Based on the GFRFs of the G and P are known and the GFRF of the H can be

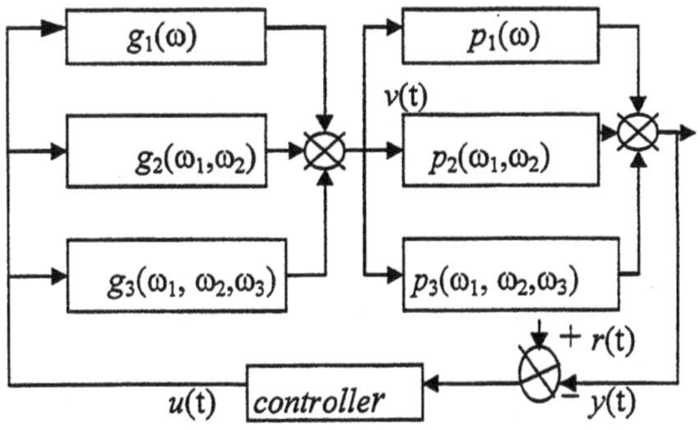

Fig.2 Structure of System with Compensator

approximated by a truncated series, and can be described by following expressions.

$$\hat{y}(\omega)=\sum_{n=1}^{3}\hat{y}_n(\omega), \quad (22)$$

where

$$\hat{y}_1(\omega)=\hat{h}_1(\omega)\hat{u}(\omega), \quad (23)$$

$$\hat{y}_2(\omega)=(2\pi)^{-1}\int_{-\infty}^{\infty}\hat{h}_2(\omega-\omega_2,\omega_2)\bullet\hat{u}(\omega-\omega_2)\hat{u}(\omega_2)d\omega_2, \quad (24)$$

and

$$\hat{y}_3(\omega)=(2\pi)^{-2}\int_{-\infty}^{\infty}\int_{-\infty}^{\infty}\hat{h}_3(\omega-\omega_2-\omega_3,\omega_2,\omega_3)\cdot\hat{u}(\omega-\omega_2-\omega_3)\hat{u}(\omega_2)\hat{u}(\omega_3)d\omega_2 d\omega_3. \quad (25)$$

respectively.

First order GFRF $\hat{g}_1(\omega)$ of the compensator. In terms of expressions of (15), (19) and (23), we can obtain

$$\hat{y}_1(\omega)=\hat{h}_1(\omega)\hat{u}(\omega)=\hat{p}_1(\omega)\cdot\hat{v}_1(\omega)=\hat{p}_1(\omega)\cdot\hat{g}_1(\omega)\cdot\hat{u}(\omega),$$

i.e.,

$$\hat{g}_1(\omega)=\frac{\hat{h}_1(\omega)}{\hat{p}_1(\omega)}.$$

.*Second order GFRF $\hat{g}_2(\omega_1,\omega_2)$ of the compensator.* According to expressions of (16), (19) and (20), following expression can be established

$$\hat{y}_2(\omega)=(2\pi)^{-1}\int_{-\infty}^{\infty}\hat{p}_2(\omega-\omega_2,\omega_2)\bullet\hat{v}_1(\omega-\omega_2)\hat{v}_1(\omega_2)d\omega_2+\hat{p}_1(\omega)\hat{v}_2(\omega)$$
$$=(2\pi)^{-1}\int_{-\infty}^{\infty}[\hat{p}_2(\omega-\omega_2,\omega_2)\bullet\hat{g}_1(\omega-\omega_2)\hat{g}_1(\omega_2)+\hat{p}_1(\omega)\hat{g}_2(\omega-\omega_2,\omega_2)]\bullet\hat{u}(\omega-\omega_2)\hat{u}(\omega_2)d\omega_2. \quad (27)$$

Substituting (24) into (27), and letting $\omega=\omega_1+\omega_2$ and $\hat{h}_2(\omega_1,\omega_2)$=0, we can obtain

$$\hat{g}_2(\omega_1,\omega_2)=-\frac{\hat{g}_1(\omega_1)\cdot\hat{g}_1(\omega_2)}{\hat{p}_1(\omega_1+\omega_2)}\cdot\hat{p}_2(\omega_1,\omega_2) \quad (28).$$

Third order GFRF $\hat{g}_3(\omega_1,\omega_2,\omega_3)$ of the conpensator. On the basis of expressions of (5.3.6) ~ (5.3.7), following expressions can be gotten

$$\hat{y}_3(\omega)=(2\pi)^{-2}\int_{-\infty}^{\infty}\int_{-\infty}^{\infty}\hat{p}_3(\omega-\omega_2-\omega_3,\omega_2,\omega_3)\cdot\hat{v}_1(\omega-\omega_2-\omega_3)\hat{v}_1(\omega_2)\hat{v}_1(\omega_3)d\omega_2 d\omega_3$$
$$+(2\pi)^{-1}\int_{-\infty}^{\infty}[\hat{p}_2(\omega-\omega_2,\omega_2)\ \hat{v}_1(\omega-\omega_2)\hat{v}_2(\omega_2)d\omega_2+\hat{p}_2(\omega-\omega_3,\omega_3)\ \hat{v}_2(\omega-\omega_3)\hat{v}_1(\omega_3)d\omega_3]+\hat{p}_1(\omega)\cdot\hat{v}_3(\omega),$$

$$\hat{y}_3(\omega)=(2\pi)^{-2}\int_{-\infty}^{\infty}\int_{-\infty}^{\infty}[\hat{p}_3(\omega-\omega_2-\omega_3,\omega_2,\omega_3)\cdot\hat{g}_1(\omega-\omega_2-\omega_3)\cdot\hat{g}_1(\omega_2)\hat{g}_1(\omega_3)$$
$$+\hat{p}_2(\omega-\omega_3,\omega_3)\hat{g}_2(\omega-\omega_2-\omega_3,\omega_2)\cdot\hat{g}_1(\omega_3)$$
$$+\hat{p}_2(\omega-\omega_2-\omega_3,\omega_2+\omega_3)\hat{g}_1(\omega-\omega_2-\omega_3)\cdot\hat{g}_2(\omega_2,\omega_3)$$
$$+\hat{p}_1(\omega)\cdot\hat{g}_3(\omega-\omega_2-\omega_3,\omega_2,\omega_3)]\cdot\hat{u}(\omega-\omega_2-\omega_3)\hat{u}(\omega_2)\hat{u}(\omega_3)d\omega_2 d\omega_3 \quad (29)$$

Substituting expression of (25) into expression of (34), and letting $\omega=\omega_1+\omega_2+\omega_3$ and $\hat{h}_3(\omega_1,\omega_2,\omega_3)$=0, we can obtain

$$\hat{g}_3(\omega_1,\omega_2,\omega_3)=-\hat{p}_1^{-1}(\omega_1+\omega_2+\omega_3)[\hat{p}_2(\omega_1+\omega_2,\omega_3)\hat{g}_2(\omega_1,\omega_2)\cdot\hat{g}_1(\omega_3)+\hat{p}_2(\omega_1,\omega_2+\omega_3)\hat{g}_1(\omega_1)\hat{g}_2(\omega_2,\omega_3)+\hat{p}_3(\omega_1,\omega_2,\omega_3)\hat{g}_1(\omega_1)\hat{g}_1(\omega_2)\hat{g}_1(\omega_3)] \quad (35).$$

On the basis of the theorem A.1, approximate compensator G of a nonlinear plant P can be obtained and its characteristics can be described by finite terms of its GFRF.

3.2 Differential Equations Realization of GFRF Model

In regard to the differential equations realization of GFRF, in fact, is a problem how to obtain the parameters of differential equations realization of describing GFRF model based on GFRF model is known. Generally, up to now, there are not good methods to solve this problem since the numbers of kernel of GFRF are infinite. However, under the assumptions in section 3, we need only discuss the differential equations realization of finite terms of GFRF.

In terms of literatures (Han, 1992), polynomial nonlinear systems, which includes only composed of input nonlinear, can be represented as

$$\sum_{p_1=0}^{M}a_{1,p_1}D^{p_1}y(t)+\sum_{n=1}^{N}\{\sum_{p_1=0}^{M}\cdots\sum_{p_n=0}^{M}[c_{n,p_1,\ldots,p_n}\prod_{i=1}^{n}D^{p_i}u(t)]\}=0, \quad (31)$$

where M is the maximum order and N is the maximum power of the differential equations. If N< ∞, then the GFRF model of system is definitely, which can be represented as

$$\hat{h}_n(\omega_1,\cdots,\omega_n)=-A_1^{-1}(j(\omega_1+\cdots+\omega_n))C_n(j\omega_1,\cdots,j\omega_n),\ n<N \quad (32)$$

where,

$$A_1(j\omega)=\sum_{p_1=0}^{M}a_{1,p_1}(j\omega)^{p_1},$$

$$C_n(j\omega_1,\cdots,j\omega_n)=\sum_{p_1=0}^{M}\cdots\sum_{p_n=0}^{M}c_{n,p_1,\cdots,p_n}\prod_{i=1}^{n}(j\omega_i)^{p_i}. \quad (33)$$

Therefor, for the sake of calculation as simple as possible, the polynomial nonlinear systems of containing input nonlinear are used as the differential equations realization of the compensator G. In this situation, its coefficients can be estimated by Least Mean Squares (LMS).

Parameters estimation for linear terms. When n=1, expression of (32) can be written as

$$-\sum_{p_1=0}^{M} a_{1,p_1}(j\omega)^{p_1} \bullet \hat{h}_1(\omega) = \sum_{p_1=0}^{M} c_{1,p_1}(j\omega)^{p_1}. \quad (34)$$

To certify the uniqueness of parameters estimated, assuming $a_{1,0}=1$ and moving it to right side of equation (34), i.e.

$$-\hat{h}_1(\omega) = \sum_{p_1=1}^{M} a_{1,p_1}(j\omega)^{p_1} \bullet \hat{h}_1(\omega) + \sum_{p_1=0}^{M} c_{1,p_1}(j\omega)^{p_1}. \quad (35)$$

The parameters of Equation (35) can be estimated by LMS.

Parameters estimation for two order nonlinear term. When n=2, expression of (32) can be written as

$$\sum_{p_1=0}^{M} a_{1,p_1}(j(\omega_1+\omega_2))^{p_1} \bullet \hat{h}_2(\omega_1,\omega_2) = \sum_{p_1=0}^{M}\sum_{p_2=0}^{M} c_{2,p_1,p_2}(j\omega_1)^{p_1}(j\omega_2)^{p_2}, \quad (36)$$

The parameters of equation (36) can also be estimated by LMS. Since $a_{1,p1}$ has been gotten in the step (1), so only $c_{2,p1,p2}$ is estimated in this step. Based on the Kronecker products, equation (36) can be written as following compressed form

$$\sum_{p_1=0}^{M} a_{1,p_1}(j(\omega_1+\omega_2))^{p_1} \bullet \hat{h}_2(\omega_1,\omega_2) = [v(\omega_1)\otimes v(\omega_2)]\cdot\theta_2. \quad (37)$$

Where $v(\omega)$ and θ_2 with the vectors of 1×(M+1) dimension and $(M+1)^2$×1 dimension separately, are as following expressions,

$$v(\omega) = [1, j\omega, (j\omega)^2, \cdots\cdots, (jw)^M],$$

and

$$\theta_2 = [c_{2,0,0}, c_{2,0,1}, \cdots\cdots, c_{2,0,M}, c_{2,1,0}, \cdots\cdots c_{2,M,M}]^T.$$

Parameter estimation of n order nonlinear term. Similar to the step (2), the following equation (36) to satisfy with LMS can be obtained

$$\sum_{p_1=0}^{M} a_{1,p_1}(j(\omega_1+\omega_2+\cdots\omega_n))^{p_1} \bullet \hat{h}_n(\omega_1,\cdots,\omega_n) = \bigotimes_{i=1}^{n} V(\omega_i)\cdot\theta_n. \quad (38)$$

Where θ_n with vector of $(M+1)^n$×1 dimension is the same form as the θ_2 i.e.

$$\theta_n = [c_{2,0,\cdots0},\ c_{2,0,\cdots0,1},\ \cdots c_{2,0,\cdots0,M},\ c_{2,0,\cdots1,0},\ \cdots\cdots c_{2,M,M,\cdots,M,M}]^T$$

It is clearly shown that the numbers of parameter of nonlinear term are increasing exponentially with the increase of the M and N based on expressions of (34)~(38). Thus it is necessary to restrict the value of the M and N in specified compensator design

4. SIMULATION EXAMPLES

Example. Consider the following pure input nonlinear system

$$\dot{y} + 5y - 5u + u^2 = 0 \quad (39)$$

Fig.3 shows simulation results of the system with input nonlinear under the sinusoidal input. Where the real lines, the dashed lines and the dashdotted lines correspond to the linear part of the nonlinear system, the compensated nonlinear system and the uncompensated nonlinear system, respectively. According to Fig.3, the properties of the system compensated are more approximately to its linear part and the compensated results weaken in some extent with nonlinear effect getting stronger and the increase of the input scope but it is obviously that the compensated nonlinear is more approximately to linear. Fig.4 is the comparisons of control results between the system compensated and uncompensated using PID as controllers, where the dashdotted lines, the real lines and the dashed lines correspond to the desired track cure, the compensated system and uncompensated that, respectively. According to Fig. 4 (a), the control results of the system compensated is obviously better than uncompensated that under the

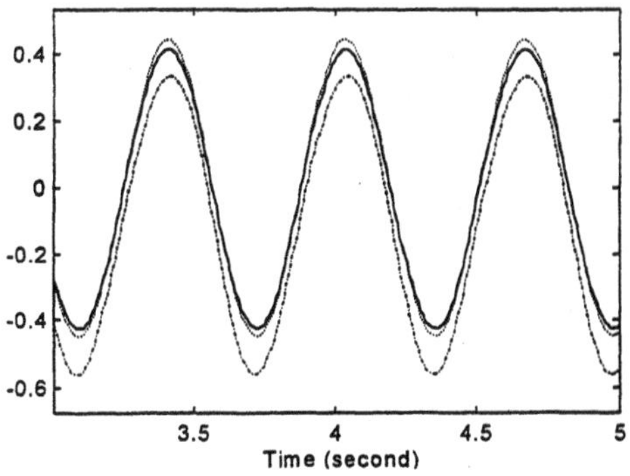

(a) u(t)=1*sin(10t)

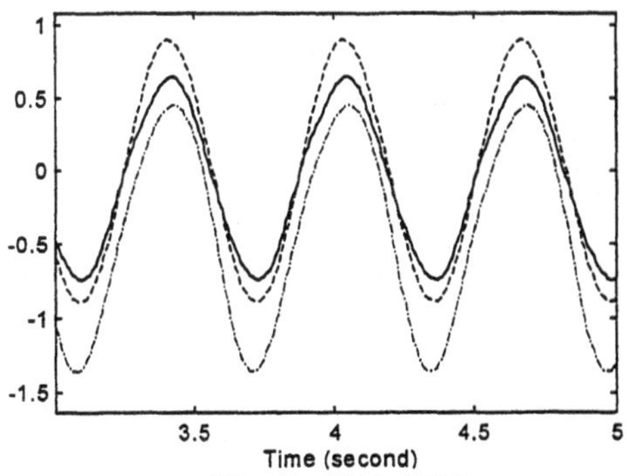

(b) u(t)=2*sin(10t)

Fig.3 Compensated results for the pure input nonlinear system

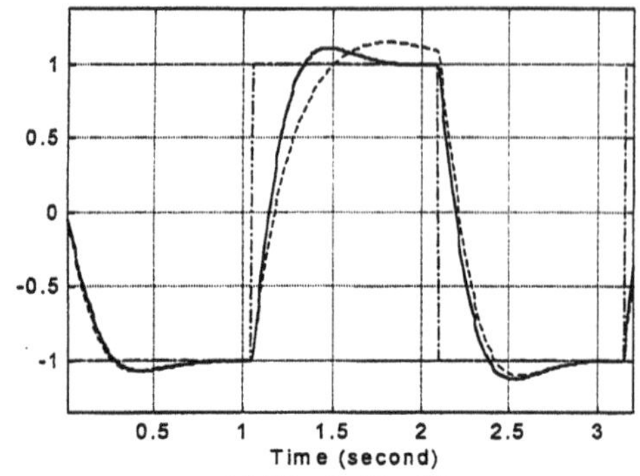

(a) $K_P=1, K_I=10$

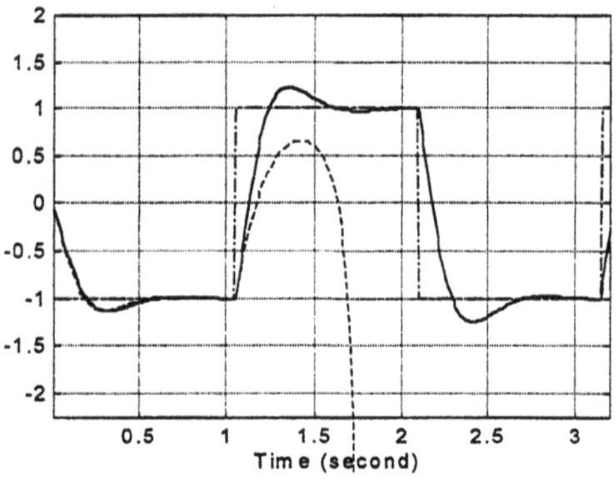

(b) $K_P=1, K_I=15$

Fig.4 Comparisons of control results between the pure input nonlinear system compensated and uncompensated using PID control

same control law is taken. Fig.4 (b) shows that the system compensated can be controlled well when the uncompensated system is unstable for the same controllers. It is illustrated that the compensator expends stable range of nonlinear systems.

5. CONCLUSION

This paper studies the control problem of single-input and single-output nonlinear dynamic systems, whose input—output relation is described by the Wolterra series, via the frequency-domain approach. Firstly, based on the generalized frequency response functions (GFRF) of Volterra functional series of nonlinear systems, develops a kind of synthesis approach of compensator of inverse nonlinear systems in frequency domain, and presents general structures and mathematical descriptions of the compensator. Then, under some assumptions, discusses the calculating of first three Wolterra kernel and differential equations realization of GFRF of the compensator. Finally, the effectiveness of the method is verified in simulations.

REFERENCES

S. W. Nam and E. G. Powers(1994). Application of Higher Order Spectral analysis to Cubically Nonlinear System Identification. *IEEE Trans., Signal Processing*, **Vol. 42,** pp. 1746-1765.

S. Boyd, L. O. Chua, and C. A. Desoer(1984). Analytical foundation of Volterra series, *IMAJ. Math. Contr.,Inform.*, **Vol. 1**, pp. 243-282.

M. Schethzen(1994). *The Volterra and Wiener theories of Nonlinear Systems.* Wiley, New York.

I. W. Sandberg(1983). The mathematical fundations of associated expansions for mildly nonlinear systems, *IEEE trans. Circuits Syst.*, **Vol. CAS-30**, pp. 441-455.

L. O. Chua and Y Liao(1989), Measuring Volterra Kernels (II), *Int.J.Circuit Theory, Applicat.*, **Vol. 17**, pp. 151-190.

S. W. Nam and E. G. Powers(1990), On the linearization of Volterra nonlinear systems using third-order inverses in the digital frequency-domain, in *Proc. ISCAS'90* (New Orleans, LA), May 1-3, pp. 869-891.

Jie Huang(1997), Asymptotic tracking in uncertain Volterra systems, *Syst. & Contr. Lett.* **Vol. 31**, pp. 215-223.

L. Chua, C. Ng(1979). Frequency-domain analysis of nonlinear systems: general theory.Formulation of functions, *IEE J. Electron. Circuits Systems*, **Vol. 3**, pp. 165-185.

C. W. Li and Y. K. Feng(1991), *The inverse system control approaches of multivariable nonlinear systems*, Tsinghua University Press, Beijing,.

Han Chongzhao. and Cao Jianfu(1996), Stability of nonlinear control systems on generalized frequency response functions, *Control Theory and Its Applications*, **Vol. 13**, No. 5.

Han Chongzhao. and Cao Jianfu(1997)., Stability of nonlinear closed-loop control systems on generalized frequency response functions, *Control Theory and Its Applications*, **Vol. 14**, No. 1.

Han C..(1992) A General formula of Generalized Frequency Response Functions for Nonlinear Differential Equation. *Research Report of Xi'an Jiaotong University.* pp. 1-12.

Zhao Zhikui(1997), Study on Control Algorithms for a Class of Nonlinear Dynamic Systems and Applications to Three-Axis Motion Simulator, *Ph.D. dissertation*, Xi'an Jiaotong University, Xian. pp. 68-78.

Copyright © 1999 IFAC
14th Triennial World Congress, Beijing, P.R. China

A SYMMETRY GROUP AND CONTROLLABILITY OF NONLINEAR HOMOGENEOUS SYSTEMS[1]

Jun Zhao[2], **Yuzhong Liu** and **Xiaoli Zhang**

School of Information Science and Engineering,
Northeastern University,
Shenyang, Liaoning, 110006, P.R.China.
E-mail: zhao@maggie.csl.uiuc.edu

Abstract: A 1-dimensional *Lie* group and its left action on convex cones are chosen, with respect to which nonlinear homogeneous control systems possess symmetries. Controllability of this class of systems is equivalently simplified, via this symmetry *Lie* group and its action, into two lower dimensional problems. *Copyright ©1999 IFAC*

Keywords: Symmetries, *Lie* group, Controllability, Nonlinear control, Homogeneous systems.

1. INTRODUCTION

Symmetries of nonlinear control systems play an important role in simplifying many control problem, and thus have been widely studied in the past decade. Grizzle and Marcus (1985) introduced the concepts of symmetries into nonlinear control systems and studied the decomposition problem. Van der Schaft (1984, 1981) derived the relationship between symmetries and conservation laws for Hamiltonian systems. Optimal control problems were considered by Grizzle and Marcus (1984). Sussman (1985), Crouch (1986), Zhao and Zhang (1992) gave results on controllability via symmetries. Invertibil-

[1]This work was supported in part by NSFC under Grant 69774005, the University Doctoral Foundation of Chinese State Education Commission, National Lab. of Industrial Control Technology at Zhejiang University under Grant K97K07, Liaoning NSF, and Complex System Engineering Laboratory of Institute of Automation, the Chinese Academy of Science.

[2]Present address of the first author:
Coordinated Science Laboratory
University of Illinois at Urbana-Champaign
1308 W.Main street, Urbana, IL, 61801, USA

ity of symmetric systems is also studied by Zhao and Zhang (1995). More recently, Gross and Chen (1996) studied applications of symmetries to a kind of large guest-host network. Applications to the control of underwater vehicle were reported by Leonard (1997).

On the other hand, controllability of nonlinear systems is a very difficult problem. Unlike linear systems, there have been no general methods to verify controllability for nonlinear systems by now. However, many efforts have been made to study controllability of nonlinear systems of special forms. For example, Hermes (1996), Hermes and Kawski (1987) derived results for nonlinear homogeneous systems. Zhao and Zhang (1992) gave a result of transforming controllability of nonlinear systems possessing symmetries into two lower dimensional problems which are much easier to verify. However the major difficulty in applying this result is the lack of knowledge of symmetry *Lie* group and its action. Generally speaking, one has no way to know whether a given nonlinear system has symmetries or not and what the symmetry *Lie* group and the left action are if the system indeed has symmetries. In other word, the construction of symmetry *Lie* group and a left action is extremely difficult.

This paper focuses on controllability of homogeneous systems on a convex cone in R^n. It is shown that for such systems, a 1-dimensional symmetry *Lie* group and a left action are always available. Thus, the n-dimensional controllability problem is transformed into an (n–1)-dimensional and a 1-dimensional ones.

2. PRELIMINARIES

This section gives a brief review of a few basic concepts and results on symmetries which will be used later.

Consider a general nonlinear control system

$$\dot{x} = f(x, u), \tag{1}$$

where $x \in M$, an n-dimensional smooth manifold, $u \in U \subset R^m$.

Assume G is a k-dimensional connected *Lie* group with a left action on M

$$\Phi : G \times M \longrightarrow M, \quad (g, x) \longrightarrow \Phi(g, x) = \Phi_g x. \tag{2}$$

Throughout this paper Φ is assumed to be free and proper. Therefore M/G and $Gx = \{\Phi_g x | g \in G\}$, for any fixed $x \in M$, are $(n-k)$-dimensional and $k-$dimensional manifolds. Let $P : M \longrightarrow M/G, \quad x \longrightarrow \tilde{x}$ be the natural projection. In addition, we suppose P admits a global cross section σ.

Definition 1 (see Grizzle and Marcus, 1985) System(1) is said to possess state-space symmetry(G, Φ) if

$$(\Phi_g)_* f(x, u) = f(\Phi_g x, u) \tag{3}$$

holds for all $x \in M, u \in U$ and $g \in G$.

Definition 2 (see Grizzle and Marcus, 1985, Zhao and Zhang, 1992) If system (1) has symmetries, then its quotient system is defined on the quotient manifold M/G by

$$\dot{y} = \tilde{f}(y, u) \stackrel{\text{def}}{=} P_* f(m, u), \tag{4}$$

where $y \in M/G, \; m \in M, \; P(m) = y$.

The following theorem (Zhao and Zhang, 1992) gives a result to verify controllability of nonlinear systems by means of symmetries.

Theorem 1 Suppose system (1) has symmetries. Then it is locally (or globally) controllable at x_0 if and only if the following two conditions hold.

(a) There exists a neighborhood V of x_0, such that $V \bigcap Gx_0 \subset R_{(1)}(x_0)$, where $R_{(1)}(x_0)$ is the reachable set of system (1) at x_0.

(b) The quotient system (4) is locally (or globally) controllable at $P(x_0)$

3. CONTROLLABILITY OF HOMOGENEOUS SYSTEMS

Now, consider system (1) in which $x \in M$, a connected open convex cone in R^n, $f(x,u)$ satisfies that

$$f(\lambda x, u) = \lambda f(x,u), \quad \forall \lambda \in R^+, \; x \in M. \quad (5)$$

In this case, system (1) is said to be homogeneous.

Choose $G = (0, \infty)$, which is obviously a connected *Lie* group with respect to the multiplication of real numbers. Let $\Phi : G \times M \longrightarrow M$, $(g,x) \longrightarrow \Phi_g x = gx$, where gx means the vector x multiplied by positive real number $g \in G$

It is easy to verify that Φ is indeed a left action of G on M. Moreover if the origin is removed, Φ is free and proper, and P: $M \longrightarrow M/G$ admits a global section $\sigma : M/G \longrightarrow M$, $y \longrightarrow x/\|x\|$, where $x \in P^{-1}(y)$. One can easily know that system (1) possesses symmetries for these specified G and Φ.

Now one can study controllability of system (1) by applying Theorem 1.

Since f is homogeneous with respect to x, $f(0,u) = 0$ for all $u \in U$ if $0 \in M$. Then system (1) does not satisfy the controllability rank condition (Isidory, 1995) at $x = 0$. Therefore the origin is removed.

Theorem 2 Suppose system (1) is homogeneous. Then it is locally (or globally) controllable at $x_0 \neq 0$, if and only if the following two conditions hold.

(a) There exists a positive constant $\delta > 0$, such that

$$\lambda x_0 \in R_{(1)}(x_0) \quad (6)$$

for all $\lambda \in (1-\delta, 1+\delta)$.

(b) Quotient system (4) is locally (or globally) controllable at $P(x_0)$.

Proof The Φ -orbit through x_0, in this case, is just $Gx_0 = \lambda x_0, \lambda \in (0,\infty)$. The conclusion then follows directly from Theorem 1.

Remark Though a necessary and sufficient condition for controllability has been obtained in Theorem 2, it is still required to derive an expression of quotient systems. This is not very easy in general. Fortunately, quotient systems can be easily constructed in the case of $n = 2$. Without loss of generality, one can suppose $x_0 = (a,b)^T$ with $a > 0$ and $b > 0$. As far as local controllability is concerned, M can be taken as $M = \{x | x_1 > 0, x_2 > 0\}$. Thus the quotient manifold can be expressed as

$$M/G = (0, \frac{\pi}{2}) \quad (7)$$

by the diffeomorphism

$$\phi : \; y \longrightarrow \mathrm{tg}^{-1}(\frac{x_2}{x_1}), \quad (8)$$

where $x = (x_1, x_2)^T \in P^{-1}(y)$. The quotient system then becomes

$$\dot{\theta} = (-\sin\theta \quad \cos\theta) f(\cos\theta \quad \sin\theta, u). \quad (9)$$

4. EXAMPLE

To illustrate how to apply Theorem 2 to verify controllability, let us consider an example.

$$\begin{aligned} \dot{x}_1 &= x_1 \, u \, \mathrm{tg}^{-1}\frac{x_2}{x_1}, \\ \dot{x}_2 &= x_2 \, u^5 \, \mathrm{tg}^{-1}\frac{x_2}{x_1}, \end{aligned} \quad (10)$$

where $x \in M = \{x | x_1 > 0, x_2 > 0\}$. This system is obviously a homogeneous system. By straightforward calculation one gets the quotient system as follows.

$$\begin{aligned} \dot{\theta} &= (-\sin\theta \quad \cos\theta) \begin{bmatrix} u \, \theta \cos\theta \\ u^5 \, \theta \sin\theta \end{bmatrix} \\ &= (u^5 - u) \, \theta \sin\theta \cos\theta \end{aligned} \quad (11)$$

which is obviously locally controllable for all $\theta \in (0, \frac{\pi}{2})$.

To check condition (a) one can choose $u = 1$ and $u = -1$ respectively. The corresponding interval curves are

$$x_1(t) = a \exp(t \text{ tg}^{-1}\frac{b}{a}), \quad x_2(t) = b \exp(t \text{ tg}^{-1}\frac{b}{a}) \tag{12}$$

and

$$x_1(t) = a \exp(-t \text{ tg}^{-1}\frac{b}{a}), \quad x_2(t) = b \exp(-t \text{ tg}^{-1}\frac{b}{a}). \tag{13}$$

From (12) and (13) one can see condition (a) holds and thus system (10) is locally controllable on M.

REFERENCES

Crouch, P.E. and C.I. Byrnes (1986). Symmetries and local controllability. In: *Algebraic and Geometric Methods in onlinear Control Theory* (M.Fliess, Ed.). pp. 55-75. M D. Reidel Publishing Com., Holland.

Grizzle, J.W. and S.I. Marcus (1985). The structure of nonlinear control systems possessing symmetries. *IEEE Trans.Automat. Control,* **30,** 248-258.

Grizzle, J.W. and S.I. Marcus (1984). Optimal control of systems possessing symmetries. *IEEE Trans.Automat. Control,* **29,** 1037-1040.

Gross, J.L. and J.Chen (1996). Algebraic specification of interconnection network relationship by permutation voltage graph mappings. *Math. Systems Theory,* **29,** 451-470.

Hermes, H. (1996). Large-time local controllability via homogeneous approximation. *SIAM Control & optimization,* **34,** 1291-1299.

Hermes, H and M.Kawski (1987). Local controllability of a single-input affine systems. *Nonlinear Analysis & Applications, Lecture Notes in Pure and Applied Maths,* (Lakshmikantham, Ed.). pp.235-248. Marcel Dekkar, Newyork.

Isidori, A. (1995). *Nonlinear Control Systems,* 3rd ed. Springer-Verlag.

Leonard, N.E. (1997). Stabilization of underwater vehicle dynamics with symmetry-breaking potential. *Systems & Control Letts.,* **32,** 35-42.

Sussmann,H.J. (1985). A general theorem on symmetries and local controllability. *Proc. 24th IEEE Conf.Decision and Control,* pp. 27-32.

Van der Schaft,A.J. (1984). Conservation laws and symmetries for Hamiltonian systems with inputs. *Proc.23rd IEEE Conf. on Decision and Control,* pp. 1583-1586.

Van der Schaft,A.J. (1981). Symmetries and conservation laws for Hamiltonian systems with inputs and outputs: A generalization of Noether's theorem. *Systems & Control Letts.,* **1,** 108-115.

Zhao, J. and S.Zhang (1992). On the controllability of nonlinear systems with symmetry. *Systems & Control Letts.,***18,** 445-448.

Zhao, J. and S.Zhang (1995). A simplified condition for the invertibility of symmetric systems. *Chinese Journal of Automation,* **7,** 172-175.

NONLINEAR SYSTEM DECOMPOSITION IN DIFFERENTIAL VECTOR SPACE [1]

Yufan Zheng* Daniel W.C. Ho Cishen Zhang*****

* *Institute of Systems Science, East China Normal University, Shanghai 200062, P.R. China*
** *Department of Mathematics City University of Hong Kong, Hong Kong.*
*** *Department of Electrical and Eclectronic Engineering, The University of Melbourne, Australia*

Abstract: This paper studies controllability and observability of nonlinear systems in differential vector space. It generalizes the coordinate invariant state space decomposition of linear systems to nonlinear systems. As a result, the dual-state space of nonlinear systems can be decomposed into controllable-observable, controllable-unobservable, uncontrollable-observable and uncontrollable-unobservable subspaces in the differential vector space.

Keywords: Controllability, Differential vector space, Dual space, Nonlinear systems, Observability.

1. INTRODUCTION

Controllability and observability are fundamental properties of linear and nonlinear systems and have been extensively studied for about four decades. For linear systems, it is well known that the system controllable and observable states can be decomposed by linearly transforming the system equation. There have been a number of procedures for the linear state transform in literature, see (Kailath, 1980) and references cited therein. A common property of these results is that the decomposition is for the state variables and the linear transform is dependent on the system state coordinates.

A state space decomposition for linear systems using a geometric approach is presented in (Wonham, 1979), where the system state space is decomposed into four subspaces, i.e. controllable-observable, controllable-unobservable, uncontrollable-observable and uncontrollable-unobservable subspaces. Such a decomposition provides a state space framework for presentation of the system controllability and observability properties, and is different from the results on linear state transforms in (Kailath, 1980) in that the decomposition is for the state space but not state variables and is independent of the system state coordinates.

Controllability (strong accessibility) and observability of nonlinear systems are also extensively studied, and a number of results on transform and decomposition of controllable and observable state variables are presented using the differential geometric approach, for example (Isidori, 1995),(Nijmeijer, *et al.*, 1989). These results are dependent on the system state coordinates and are extensions of the linear state transforms to nonlinear systems.

Recently, a differential vector space approach based on the theory of differential algebra is developed to nonlinear systems (Benedetto, *et*

[1] This work was supported by Natural Science Foundation of China, Doctoral Education Foundation of China, the Australia Research Council, Hong Kong RGC.

al., 1989),(Fliess, 1990),(Fliess, *et al.*, 1989). It is applied to the problems of controllability, observability and realizations of nonlinear systems and a number of results have been presented (Aranda-Bricaire, *et al.*, 1995),(Fliess, 1990),(Zheng,*et al.*, 1993),(Zheng,*et al.*, 1995).

The original result of this paper is to use the differential vector space approach to decompose the state space of nonlinear systems into the controllable- observable, controllable- unobservable, uncontrollable- observable and uncontrollable-unobservable subspaces. The decomposition is carried out in the dual-state space of the differential vector space and is invariant under the system state coordinates. Such a result extends the state space decomposition for linear systems in (Wonham, 1979) to provide a generalized framework for presentation of controllability and observability of nonlinear systems, and is, therefore, different form the results of coordinate dependent state transforms of linear and nonlinear systems.

2. DIFFERENTIAL VECTOR SPACE OF NONLINEAR SYSTEMS

In this paper, $\frac{\partial}{\partial a}$ denotes the partial derivative with respect to a variable a. $\frac{\partial}{\partial x} = \left(\frac{\partial}{\partial x_1} \cdots \frac{\partial}{\partial x_n}\right)$ denotes the partial derivative with respect to an n-dimensional vector x. L_v denotes the Lie derivative along a vector field v. $[\cdot,\cdot]$ denotes the Lie bracket. $\mathcal{L}^*$ denotes the dual space of a vector space $\mathcal{L}$. $< v^*, w >$ denotes the inner product between the vectors $v^* \in \mathcal{L}^*$ and $w \in \mathcal{L}$. $\mathcal{L}^{\perp}$ denotes the orthogonal space (annihilator) of a linear vector space $\mathcal{L}$. $\mathcal{L}_a \cong \mathcal{L}_b$ denotes that two vector spaces $\mathcal{L}_a$ and $\mathcal{L}_b$ are isomorphic. $\underline{n}$ denotes the closed integer set $[1 \cdots n]$.

Consider an nth order nonlinear system written as

$$\begin{aligned}\dot{x} &= f(x) + g(x)u = f(x) + (g_1(t) \ \cdots \ g_m(t))u \\ y &= h(x) = (h_1(x) \ \cdots \ h_p(x))^{\tau}\end{aligned} \tag{1}$$

where $x \in \mathcal{R}^n$, $u \in \mathcal{R}^m$, and $y \in \mathcal{R}^p$, are the system state, control input and output, respectively, and $f(\cdot)$ and $g_i(x)$, for $i \in \underline{m}$, are meromorphic vector fields in $\mathcal{R}^n$, and $h(x)$ is a meromorphic mapping from $\mathcal{R}^n$ to $\mathcal{R}^p$. In this paper, we assume that the nonlinear system (1) satisfies a regularity condition in the sense that the system is not singular over an open dense set of the system states. For example, if the dimension of a distribution spanned by $\{g_1(x) \cdots g_m(x)\}$ is m at x_o then the dimension of the distribution is m for all $x \in X$ where X is the neighborhood of the x_o. Under such a regularity condition, the results in this paper on the differential vector space of nonlinear systems are generic but may not cover certain special cases that occur at singular points of the system.

For the nonlinear system (1), let $\mathcal{K}$ be the field of meromorphic functions of x_i and $u_j^{(k)}$, for $i \in \underline{n}$, $j \in \underline{m}$ and $k \geq 0$ and $l \in \underline{r}$, see (Benedetto, *et al.*, 1989). Then for a function $\phi(x, u, u^{(1)}, \cdots, u^{(k)}) \in \mathcal{K}$, we have

$$\begin{aligned}\dot{\phi} &= \frac{\partial\phi}{\partial x}\dot{x} + \frac{\partial\phi}{\partial u}\dot{u} + \cdots + \frac{\partial\phi}{\partial u^{(k)}}u^{(k+1)} \\ &= \frac{\partial\phi}{\partial x}(f(x) + g(x)u) + \frac{\partial\phi}{\partial u}\dot{u} + \cdots + \frac{\partial\phi}{\partial u^{(k)}}u^{(k+1)}\end{aligned}$$

Thus $\dot{\phi} \in \mathcal{K}$ and $\mathcal{K}$ is a differential field defined by the nonlinear system (1).

Over the differential field $\mathcal{K}$, the system (1) defines a linear vector space $\mathcal{D}$ with infinitely large dimension written as

$$\mathcal{D} = span_{\mathcal{K}}\left\{\frac{\partial}{\partial x_i}, \frac{\partial}{\partial u_j^{(k)}};\ i \in \underline{n},\ j \in \underline{m},\ k \geq 0\right\}$$

Over the differential field $\mathcal{K}$, the dual space of $\mathcal{D}$ is

$$\mathcal{D}^* = span_{\mathcal{K}}\{dx_i, du_j^{(k)};\ i \in \underline{n},\ j \in \underline{m},\ k \geq 0\} \tag{2}$$

It is apparent that the dual space $\mathcal{D}^*$ defined by the nonlinear system (1) satisfies the properties of differential vector space over $\mathcal{K}$. Thus we define

The dual space $\mathcal{D}^*$ in (2) is defined as the differential vector space of the nonlinear system (1).

Remarks:

(i) The definition of the dual space over the differential field $\mathcal{K}$ of meromorphic functions follows from the standard mathematical formulation (Wagner, 1979),(Benedetto, *et al.*, 1989),(Fliess, 1990).

(ii) In the differential vector space $\mathcal{D}^*$, a vector $\omega \in \mathcal{D}^*$ is called a differential one-form. A vector $\omega \in \mathcal{D}^*$ is called an exact one-form if there exists a $\varphi \in \mathcal{K}$ such that $\omega = d\varphi$. It follows that the exact one-form ω is an integrable vector.

Over the differential field $\mathcal{K}$, the state space, input space and output space of the nonlinear system (1), denoted by $\mathcal{X}$, $\mathcal{U}$ and $\mathcal{Y}$, respectively, are defined as follows.

$$\begin{aligned}\mathcal{X} &= span_{\mathcal{K}}\left\{\frac{\partial}{\partial x_i};\ i \in \underline{n}\right\} \\ \mathcal{U} &= span_{\mathcal{K}}\left\{\frac{\partial}{\partial u_i^{(k)}};\ i \in \underline{m},\ k \geq 0\right\} \\ \mathcal{Y} &= span_{\mathcal{K}}\left\{\frac{\partial}{\partial y_i^{(k)}};\ i \in \underline{p},\ k \geq 0\right\}\end{aligned} \tag{3}$$

The dual-state space, dual-input space and dual-output space of the nonlinear system (1) over $\mathcal{K}$, denoted by $\mathcal{X}^*$, $\mathcal{U}^*$ and $\mathcal{Y}^*$, respectively, are further defined as

$$\mathcal{X}^* = span_{\mathcal{K}}\{dx_i;\ i \in \underline{n}\} \tag{4}$$

$$\mathcal{U}^* = span_{\mathcal{K}}\{du_i^{(k)};\ i \in \underline{m},\ k \geq 0\}$$

$$\mathcal{Y}^* = span_{\mathcal{K}}\{dy_i^{(k)};\ i \in \underline{p},\ k \geq 0\}$$

It follows that the dual-state and dual-input spaces are subspaces of the differential vector space and satisfy

$$\mathcal{X}^* \oplus \mathcal{U}^* = \mathcal{D}^*$$

The dual-output space is generated by the dual-state and dual-input spaces and satisfy

$$\mathcal{Y}^* \subseteq \mathcal{X}^* \oplus \mathcal{U}^* = \mathcal{D}^*$$

Remarks:

(i) The above defined dual-state, dual-input and dual-output spaces naturally extend the notion of dual space for linear systems (Wonham, 1979) over the real field $\mathcal{R}$ to the differential field $\mathcal{K}$. When the system is linear time invariant, the differential field $\mathcal{K}$ becomes $\mathcal{R}$ and the dual spaces of the system are consistent with the definitions and results in (Wonham, 1979).

(ii) Over an open set $X \in \mathcal{R}^n$, the state variables $\{x_1, x_2 \cdots x_n\}$ set up a local coordinate system and form a differential manifold. Thus the state space $\mathcal{X}$ represents the tangent bundle over the differential manifold X. Since the dual-state space $\mathcal{X}^*$ is orthogonal to $\mathcal{X}$, it represents the co-tangent bundle over the differential manifold X. Further, a subspace of $\mathcal{X}$ forms a distribution and a subspace of $\mathcal{X}^*$ forms a co-distribution over X.

For the decomposition of nonlinear systems in the differential vector space $\mathcal{D}^*$, we require certain integrability properties of subspaces in $\mathcal{D}^*$. These properties follow directly from the well known *Frobenis Theorem.*

In the dual-state space $\mathcal{X}^*$, an integrable subspace represents a set of regular submanifolds and can be spanned by exact one-forms. The integrals of these exact one-forms yield functions of the state x that can form a coordinate system over these submanifolds. In the sequel of this paper, we will decompose the dual-state spaces into integrable subspaces and use these to present controllability and observability properties of the system in the differential vector space.

3. CONTROLLABILITY IN THE DIFFERENTIAL VECTOR SPACE

Consider the nonlinear control system (1). For a function of the system states $\phi \in \mathcal{K}$ satisfying $d\phi \in \mathcal{X}^*$, it is possible that $d\phi^{(k)} \notin \mathcal{X}^*$ since $\phi^{(k)}$ may become a function of the control u. With regard to such an observation, we define

For a set of functions $\phi \in \mathcal{K}$, the relative degree of each ϕ is defined as $deg\ \phi = k$; if $d\phi^{(k-1)} \in \mathcal{X}^*$ and $d\phi^{(k)} \notin \mathcal{X}^*$, $k \geq 1$. Note that a function $\phi \in \mathcal{K}$ is autonomous if $deg\ \phi = \infty$. Since an autonomous function is not subject to the control u, it can be used to define the system controllability that is stated as follows.

The nonlinear system (1) is controllable (strongly accessible) if there exists no function $\phi \in \mathcal{K}$ with $\phi \neq$ constant such that $deg\ \phi = \infty$.

Let $\bar{C}$ be the set of all the autonomous functions of the system (1) in $\mathcal{K}$, namely,

$$\bar{C} = \{\phi \in \mathcal{K};\ deg\ \phi = \infty\}$$

We define the uncontrollable dual-state space of the system (1), denoted by $\mathcal{X}_{\bar{c}}^*$ as

$$\mathcal{X}_{\bar{c}}^* = span_{\mathcal{K}}\{d\phi;\ \phi \in \bar{C}\}$$

Apparently $\mathcal{X}_{\bar{c}}^* \subseteq \mathcal{X}^*$.

The controllable dual-state space of the system (1), denoted by $\mathcal{X}_c^*$, is defined as a subspace of $\mathcal{X}^*$ that satisfies

$$\mathcal{X}_c^* \cong \mathcal{X}^* / \mathcal{X}_{\bar{c}}^*, \qquad \mathcal{X}_{\bar{c}}^* \oplus \mathcal{X}_c^* = \mathcal{X}^* \tag{5}$$

For nonlinear systems, the factor space $\mathcal{X}^*/\mathcal{X}_{\bar{c}}^*$ in the above definition precisely describes the system controllability. Whereas the definition also allows the controllable subspace $\mathcal{X}_c^*$ to be non-unique in the dual-state space $\mathcal{X}^*$. We can state the system controllability in terms of the dual-state spaces $\mathcal{X}_{\bar{c}}^*$ and $\mathcal{X}_c^*$ as follows.

Proposition 3.1: *The following statements are equivalent:*

(i) The nonlinear system (1) is controllable;

(ii) $\mathcal{X}_{\bar{c}}^* = 0$;

(iii) $\mathcal{X}_c^* = \mathcal{X}^*$.

□

We now study the integrability of the dual-state spaces $\mathcal{X}_{\bar{c}}^*$ and $\mathcal{X}_c^*$. With the notion of relative degree, we define subspaces $\mathcal{X}_k^*$ of the dual-state space $\mathcal{X}^*$, for $k \geq 1$, as

$$\mathcal{X}_k^* = span_{\mathcal{K}}\{d\phi;\ deg\ \phi \geq k,\ k \geq 1\}$$

The definitions of the uncontrollable dual-state space and subspaces $\mathcal{X}_k^*$ lead to the following results.

Proposition 3.2: *(Zheng,et al., 1995)*

(i)

$$\mathcal{X}_\infty^* \subseteq \cdots \subseteq \mathcal{X}_{k+1}^* \subseteq \mathcal{X}_k^* \subseteq \cdots \subseteq \mathcal{X}_1^* = \mathcal{X}^* \quad (6)$$

(ii) The sequence of the subspaces $\mathcal{X}_k^*$ converges in the sense that there exists an integer $q \leq n$ such that $\mathcal{X}_i^* = \mathcal{X}_j^*$, for all $i, j, > q$.

(iii) $\mathcal{X}_{\bar{c}}^* = \mathcal{X}_{q+1}^* = \mathcal{X}_\infty^*$.

□

Proposition 3.2 (iii) addresses the problem of integrability of $\mathcal{X}_{\bar{c}}^*$ to that of the subspace $\mathcal{X}_i^*$ for some $i > q$. Using this we present a constructive result on integrability of $\mathcal{X}_{\bar{c}}^*$ and $\mathcal{X}_c^*$ in the following theorem.

Theorem 3.1: *The uncontrollable dual-state space $\mathcal{X}_{\bar{c}}^*$ and the controllable dual-state space $\mathcal{X}_c^*$ of the nonlinear system (1) are integrable and can be constructed.*

Proof: Since each subspace $\mathcal{X}_i^*$ is integrable by definition, the integrability of $\mathcal{X}_{\bar{c}}^*$ follows from Proposition 3.2 (iii).

For construction of $\mathcal{X}_{\bar{c}}^*$, we recursively use the Lie algebra operation to define the distributions $\{\mathcal{G}_k;\ k \geq 0\}$, as

$$\begin{aligned} \mathcal{G}_0 = 0, \ \ \mathcal{G}_1 = span_{\mathcal{K}}\{g_i;\ i \in \underline{m}\}, \\ \mathcal{G}_{k+1} = \bar{\mathcal{G}}_k + span_{\mathcal{K}}\{[f, l];\ l \in \bar{\mathcal{G}}_k\} \end{aligned} \quad (7)$$

where $\bar{\mathcal{G}}_k$ denote the involutive closure of the distribution $\mathcal{G}_k$. It follows from the definition of the subspaces $\mathcal{X}_k^*$ and the well known results on the distributions $\mathcal{G}_k$ (Isidori, 1995) that

$$\mathcal{X}_k^* = \bar{\mathcal{G}}_{k-1}^\perp; \qquad k \geq 1$$

Hence, $\mathcal{X}_k^*$ can be constructed using the the orthogonal space $\bar{\mathcal{G}}_{k-1}^\perp$ of $\bar{\mathcal{G}}_{k-1}$, for each $k \geq 1$. This immediately provides the uncontrollable dual-state space

$$\mathcal{X}_{\bar{c}}^* = \mathcal{G}_q^\perp$$

where $\mathcal{G}_q$ satisfies $\mathcal{G}_q = \bar{\mathcal{G}}_q$.

Since the dual-state space $\mathcal{X}^*$ and the uncontrollable dual-state space $\mathcal{X}_{\bar{c}}^*$ are integrable, it follows from Proposition 2.1 (ii) and (5) that the controllable dual-state space $\mathcal{X}_c^*$ is integrable and can be constructed. □

In this paper, we will call any integrable subspace $\mathcal{X}_{cs}^*$ of the controllable dual-state space $\mathcal{X}_c^*$ a controllable subspace of $\mathcal{X}^*$. This is in the sense that for each vector $d\phi \in \mathcal{X}_{cs}^*$ the relative degree of ϕ is finite.

4. OBSERVABILITY IN THE DIFFERENTIAL VECTOR SPACE

For observability of the nonlinear system (1) in the differential vector space, we make the following definitions.

The system (1) is observable if for any two initial states $x_a(0) \neq x_b(0)$ there exists a control $u(t)$, $t > 0$ such that the system output satisfies

$$y(t, x_a(0), u(t)) \neq y(t, x_b(0), u(t)), \quad \text{for } t \geq 0$$

A function of the system state $\varphi \in \mathcal{K}$ satisfying $d\varphi \in \mathcal{X}^*$ is an observable function of the system (1) if, for $\varphi(x_a(0)) \neq \varphi(x_b(0))$ with the initial states $x_a(0)$ and $x_b(0)$, there exists a control $u(t)$ such that the system output satisfies

$$y(t, x_a(0), u(t)) \neq y(t, x_b(0), u(t))$$

An observable function $\varphi \in \mathcal{K}$ can be used to form a new coordinate function over $\mathcal{R}^n$. Thus the observability of φ implies that along the coordination of φ any two different initial states can be distinguished from the measurement of the output subject to a suitably chosen control input u. In (Zheng,*et al.*, 1993), it is shown that the above defined system observability is equivalent to that in (Isidori, 1995) in terms of differential geometry.

Let O be the set of all the observable functions of the system (1) in $\mathcal{K}$, namely,

$$O = \{\varphi;\ \varphi \in \mathcal{K},\ \varphi \text{ is observable}\}$$

We define the observable dual-state space of the system (1), denoted by $\mathcal{X}_o^*$ as

$$\mathcal{X}_o^* = span_{\mathcal{K}}\{d\varphi;\ \varphi \in O\}$$

The unobservable dual-state space of the system (1), denoted by $\mathcal{X}_{\bar{o}}^*$, is defined as a subspace of $\mathcal{X}^*$ that satisfies

$$\mathcal{X}_{\bar{o}}^* \cong \mathcal{X}^*/\mathcal{X}_o^*, \qquad \mathcal{X}_{\bar{o}}^* \oplus \mathcal{X}_o^* = \mathcal{X}^* \quad (8)$$

where the factor-space $\mathcal{X}^*/\mathcal{X}_o^*$ provides a precise description of the system unobservability in the dual-state space $\mathcal{X}^*$, and the definition allows the unobservable subspace $\mathcal{X}_{\bar{o}}^*$ to be non-unique. Using the subspaces $\mathcal{X}_{\bar{o}}^*$ and $\mathcal{X}_o^*$, we can present the system observability as follows.

Proposition 4.1: *((Zheng,et al., 1993)) The following statements are equivalent:*

(i) The nonlinear system (1) is observable;

(ii) $\mathcal{X}^* \subseteq \mathcal{Y}^* + \mathcal{U}^*$;

(iii) $\mathcal{X}_o^* = \mathcal{X}^*$;

(iv) $\mathcal{X}_{\bar{o}}^* = 0$.

□

We now present, constructively, integrability of the observable and unobservable dual-state spaces in the following theorem.

Theorem 4.1: *The observable dual-state space $\mathcal{X}_o^*$ is integrable and can be constructed as follows*

$$\mathcal{X}_o^* = \mathcal{X}^* \cap (\mathcal{Y}^* + \mathcal{U}^*) \tag{9}$$

The unobservable dual-state space $\mathcal{X}_{\bar{o}}^*$ of the nonlinear system (1) is integrable and can be constructed, too.

Proof: In (Zheng, *et al.*, 1993) it is shown that the subspace $\mathcal{X}^* \cap (\mathcal{Y}^* + \mathcal{U}^*)$ is integrable and is indirectly proved that $\mathcal{X}^* \cap (\mathcal{Y}^* + \mathcal{U}^*) = span_{\mathcal{K}}\{d\varphi;\ \varphi \in O\}$.

Since the dual-state space $\mathcal{X}^*$ and the observable dual-state space $\mathcal{X}_o^*$ are integrable and can be constructed, the unobservable dual-state space $\mathcal{X}_{\bar{o}}^*$ is integrable and can be constructed following from Proposition 2.1 (ii) and (8). □

For the integrable dual-state space $\mathcal{X}_{\bar{o}}^*$, we will call any integrable subspace $\mathcal{X}_{\bar{o}s}^* \subset \mathcal{X}_{\bar{o}}^*$ an unobservable subspace of $\mathcal{X}^*$. This is in the sense that for any vector $d\varphi \in \mathcal{X}_{\bar{o}s}^*$ there exists two initial states $x_a(0)$ and $x_b(0)$ such that $\varphi(x_a(0)) \neq \varphi(x_b(0))$ and $y(t, x_a(0), u(t)) = y(t, x_b(0), u(t))$ for any input $u(t)$.

5. STATE SPACE DECOMPOSITION OF NONLINEAR SYSTEMS

This section presents nonlinear system decomposition in the differential vector space. Such a decomposition extends the linear system state space decomposition in (Wonham, 1979) to nonlinear systems and is invariant under the system state coordinates. As a result, the dual-state space $\mathcal{X}^*$ is decomposed into four integrable subspaces, i.e. controllable-observable subspace $\mathcal{X}_{co}^*$, controllable-unobservable subspace $\mathcal{X}_{c\bar{o}}^*$, uncontrol- lable-observable subspace $\mathcal{X}_{\bar{c}o}^*$ and uncontrollable- unobservable subspace $\mathcal{X}_{\bar{c}\bar{o}}^*$. These subspaces satisfy

$$\begin{aligned} \mathcal{X}_{co}^* &= \mathcal{X}_c^* \cap \mathcal{X}_o^* \\ \mathcal{X}_{c\bar{o}}^* &= \mathcal{X}_c^* \cap \mathcal{X}_{\bar{o}}^* \\ \mathcal{X}_{\bar{c}o}^* &= \mathcal{X}_{\bar{c}}^* \cap \mathcal{X}_o^* \\ \mathcal{X}_{\bar{c}\bar{o}}^* &= \mathcal{X}_{\bar{c}}^* \cap \mathcal{X}_{\bar{o}}^* \end{aligned} \tag{10}$$

and

$$\mathcal{X}^* = \mathcal{X}_{co}^* \oplus \mathcal{X}_{c\bar{o}}^* \oplus \mathcal{X}_{\bar{c}o}^* \oplus \mathcal{X}_{\bar{c}\bar{o}}^* \tag{11}$$

To initialize the decomposition, we present a technical result as follows.

Lemma 5.1: *$\mathcal{X}_{\bar{c}}^* \cap \mathcal{X}_o^*$ is an integrable subspace of $\mathcal{X}^*$ and can be constructed.*

Proof: It is shown in Proposition 3.2 and Theorem 3.1 that $\mathcal{X}_{\bar{c}}^* = \mathcal{G}_q^{\perp}$ where $\mathcal{G}_q$ is involutive (integrable). It is also shown in Theorem 4.1 that $\mathcal{X}_o^*$ is invariant under each vector in $\{f, g_1, \cdots, g_m\}$. Thus $\mathcal{X}_o^*$ and, consequently, $(\mathcal{X}_o^*)^{\perp}$ are invariant under each vector in $\mathcal{G}_q$. It follows that $(\mathcal{G}_q + (\mathcal{X}_o^*)^{\perp})$ is integrable and can be constructed. Hence $\mathcal{X}_{\bar{c}}^* \cap \mathcal{X}_o^* = (\mathcal{G}_q + (\mathcal{X}_o^*)^{\perp})^{\perp}$ is integrable and can be constructed. □

We now use the uniquely defined uncontrollable dual-state space $\mathcal{X}_{\bar{c}}^*$ and observable dual-state space $\mathcal{X}_o^*$ to define subspaces $\mathcal{X}_{co}^*$, $\mathcal{X}_{c\bar{o}}^*$, $\mathcal{X}_{\bar{c}o}^*$ and $\mathcal{X}_{\bar{c}\bar{o}}^*$. These subspaces are defined to satisfy

$$\mathcal{X}^* = \mathcal{X}_{c\bar{o}}^* \oplus (\mathcal{X}_o^* + \mathcal{X}_{\bar{c}}^*) \tag{12}$$

$$\mathcal{X}_o^* = \mathcal{X}_{co}^* \oplus (\mathcal{X}_o^* \cap \mathcal{X}_{\bar{c}}^*) \tag{13}$$

$$\mathcal{X}_{\bar{c}}^* = \mathcal{X}_{\bar{c}\bar{o}}^* \oplus (\mathcal{X}_o^* \cap \mathcal{X}_{\bar{c}}^*) \tag{14}$$

$$\mathcal{X}_{\bar{c}o}^* = \mathcal{X}_{\bar{c}}^* \cap \mathcal{X}_o^* \tag{15}$$

The following lemma presents integrability of these subspaces.

Lemma 5.2: *The subspaces $\mathcal{X}_{co}^*$, $\mathcal{X}_{c\bar{o}}^*$, $\mathcal{X}_{\bar{c}o}^*$ and $\mathcal{X}_{\bar{c}\bar{o}}^*$ as defined in (12) - (15) are all integrable and can be constructed, and satisfy equation (10).*

Proof: Since $\mathcal{X}^*$, $\mathcal{X}_{\bar{c}}^*$, $\mathcal{X}_o^*$, $\mathcal{X}_{\bar{c}}^* + \mathcal{X}_o^*$ and $\mathcal{X}_{\bar{c}}^* \cap \mathcal{X}_o^*$ are integrable and can be constructed, the result follows from the definitions (12) - (15) and Proposition 2.1 (ii).

Let

$$\begin{aligned} \mathcal{X}_c^* &= \mathcal{X}_{co}^* \oplus \mathcal{X}_{c\bar{o}}^* \\ \mathcal{X}_{\bar{o}}^* &= \mathcal{X}_{\bar{c}o}^* \oplus \mathcal{X}_{\bar{c}\bar{o}}^* \end{aligned} \tag{16}$$

Then $\mathcal{X}_c^*$ in (16) satisfies the definition of the controllable dual-state space and is a representation of the non-unique controllable dual-state space. Similarly, $\mathcal{X}_{\bar{o}}^*$ is a representation of the non-unique unobservable dual-state space. $\mathcal{X}_c^*$ and $\mathcal{X}_{\bar{o}}^*$ in (16) together with the definitions (12) - (15) yield (10). □

With the result of Lemma 5.2, we define the integrable subspaces $\mathcal{X}_{co}^*$, $\mathcal{X}_{\bar{c}o}^*$, $\mathcal{X}_{c\bar{o}}^*$ and $\mathcal{X}_{\bar{c}\bar{o}}^*$ as *controllable-observable, uncontrollable-observable, controllable-unobservable* and *uncontrollable-unobservable* subspace of the dual-state space $\mathcal{X}^*$, respectively.

The main result of this paper on state space decomposition of nonlinear systems is to decompose the dual-state space $\mathcal{X}^*$ into the four integrable subspaces $\mathcal{X}^*_{co}$, $\mathcal{X}^*_{\bar{c}o}$, $\mathcal{X}^*_{c\bar{o}}$ and $\mathcal{X}^*_{\bar{c}\bar{o}}$. This result is presented in the following theorem.

Theorem 5.1: *The dual-state space $\mathcal{X}^*$ of the nonlinear system (1) can be decomposed into the direct sum of the four integrable subspaces $\mathcal{X}^*_{co}$, $\mathcal{X}^*_{\bar{c}o}$, $\mathcal{X}^*_{\bar{c}o}$ and $\mathcal{X}^*_{\bar{c}\bar{o}}$ such that (10) and (11) are satisfied.*

Proof: The result of (10) has been established in Lemma 5.2. It follows from (14) and (15) that

$$\mathcal{X}^*_{\bar{c}} = \mathcal{X}^*_{\bar{c}o} \oplus \mathcal{X}^*_{\bar{c}\bar{o}}$$

Hence the result of (11) follows from

$$\mathcal{X}^* = \mathcal{X}^*_c \oplus \mathcal{X}^*_{\bar{c}}, \quad \mathcal{X}^*_c = \mathcal{X}^*_{co} \oplus \mathcal{X}^*_{c\bar{o}}$$

□

Remark: The solution to the decomposition (11) of the dual-state space $\mathcal{X}^*$ may not be unique as the same as that of linear systems.

We point out that the decomposition of the dual-state space $\mathcal{X}^*$ in (11) is invariant under state coordinate transform and is, therefore, a coordinate invariant result. To clarify this, we suppose that $z = T(x)$ is a regular state coordinate transform and $\mathcal{Z}^*$ is the dual-state space of the system under the transform. Then, there exists an open set $U \in \mathcal{R}^n$ with a coordinate chart $(U; x_1, \cdots, x_n)$ and the coordinate transformation T uniquely defines a homomorphism from the chart $(U; x_1, \cdots, x_n)$ to $V = T(U)$ with the chart $(V; z_1, \cdots, z_n)$. Therefore, $\mathcal{X}^*$ and $\mathcal{Z}^*$ present, respectively, the co-tangent bundles over the two charts $(U; x_1, \cdots, x_n)$ and $(V; z_1, \cdots, z_n)$.

Let T^* be the co-tangent map induced by the map T(Wagner, 1979). This yields four subspaces of $\mathcal{Z}^*$ that can be written as

$$\begin{aligned} T^*(\mathcal{Z}^*_{co}) = \mathcal{X}^*_{co}, \quad T^*(\mathcal{Z}^*_{c\bar{o}}) = \mathcal{X}^*_{c\bar{o}}, \\ T^*(\mathcal{Z}^*_{\bar{c}o}) = \mathcal{X}^*_{\bar{c}o}, \quad T^*(\mathcal{Z}^*_{\bar{c}\bar{o}}) = \mathcal{X}^*_{\bar{c}\bar{o}} \end{aligned} \tag{17}$$

It can be denoted by

$$\begin{aligned} \mathcal{Z}^*_{co} \cong \mathcal{X}^*_{co}, \quad \mathcal{Z}^*_{c\bar{o}} \cong \mathcal{X}^*_{c\bar{o}}, \\ \mathcal{Z}^*_{\bar{c}o} \cong \mathcal{X}^*_{\bar{c}o}, \quad \mathcal{Z}^*_{\bar{c}\bar{o}}) \cong \mathcal{X}^*_{\bar{c}\bar{o}} \end{aligned} \tag{18}$$

Under the state coordinate transform $z = T(x)$, the nonlinear system (1) can be written in the form

$$\dot{z} = f_z(z) + g_z(z)u$$

$$y = h_z(z)$$

We can accordingly decompose the dual-state space $\mathcal{Z}^*$ of the system to obtain

$$\mathcal{Z}^* = \mathcal{Z}^*_{co} \oplus \mathcal{Z}^*_{c\bar{o}} \oplus \mathcal{Z}^*_{\bar{c}o} \oplus \mathcal{Z}^*_{\bar{c}\bar{o}} \tag{19}$$

Hence we can state the following coordinate invariant property of the state space decomposition.

Theorem 5.2: *The state space decomposition of the nonlinear system (1) is coordinate invariant under the regular state transform $z = T(x)$ in the sense of (11), (17) and (19).* □

6. CONCLUSION

The main result of this paper is the decomposition of nonlinear systems in the differential vector space. The state space decomposition is coordinate invariant that extends the coordinate invariant result for linear systems to nonlinear systems and is different from the coordinate dependent state transformations using differential geometry. Hence the result of this paper provides new insights into the structure and properties of the nonlinear system state space and a useful framework for analysis and representation of controllability and observability

7. REFERENCES

Aranda-Bricaire, E., C.H. Moog and J-B Pomet (1995) A linear algebraic framework for dynamic feedback linearization, *IEEE Trans. Automatic Control*, AC-40, pp. 127-132.

Di Benedetto, M.D., J.W. Grizzle and C.H. Moog(1989) Rank invariant of nonlinear systems, *SIAM. J. Control and Optimization*, Vol. 27, pp. 658-672

Fliess, M.(1990) Generalized controller canonical forms for linear and nonlinear dynamics, *IEEE Trans. Automatic Control*, Vol. 28, pp. 994-1001.

Fliess, M. and S.T. Glad, (1993), An algebraic approach to linear and nonlinear control, *Essays on Control: Perspectives in the Theory and its Applications*, Birkhäuser, Boston

Isidori, A., (1995) *Nonlinear control systems: an introduction, 3nd Edition, Springer-Verlag, Berlin*

Kailath, T.(1980) *Linear systems, Englewood Cliffs, Prentice-Hall, N.J.*

Nijmeijer, H. and A.J. van der Schaft (1990) *Nonlinear Dynamical Control Systems, Springer-Verlag*

Wonham, M.W. (1979) *Linear Multivariable Control:A Geometric Approach, Springer-Verlag*

Wagner, F.W. (1979) Fundamentals of differential manifolds and lie groups, *Scott, Foresman, Hill*

Zheng, Y.F. and L. Cao (1993), Reduced inverse for controlled systems, *Mathematics for Control, Signals and Systems,* Vol. 6, pp. 363-379

Zheng, Y.F. and L. Cao(1995) Complete invariant of nonlinear control systems *Proc. IFAC NOLCOS-95*

14th Triennial World Congress, Beijing, P.R. China

LINEARIZATION BY PROLONGATIONS: NEW BOUNDS FOR THREE INPUT SYSTEMS

Jaume Franch and Enric Fossas

Dpt. de Mat. Apl. i T. - Universitat Politècnica de Catalunya
Jordi Girona 1-3, 08034 BARCELONA, SPAIN
email:jfranch@mat.upc.es

Abstract: This communication deals with the problem of linearization by prolongations. A necessary and sufficient condition for a three input system to be linearizable by prolongations is derived. This condition concerns the maximum number of integrators needed to linearize the system. The communication improves the bounds that have appeared in the literature and shows a natural way to continue for systems with m inputs, with $m > 3$.

Keywords: Nonlinear control, multiinput systems, differential geometric methods, feedback linearization, mathematical systems theory,

1. INTRODUCTION

Since 1980, the problem of feedback linearization for continuous nonlinear control systems has been considered by different authors in several frameworks. Different routes to linearization have been traced; namely, linearization by static feedback; linearization by prolongations; linearization by dynamic feedback, and finally flatness. The problem of linearization by prolongations was introduced in (Charlet *et al.*, 1989), (Charlet *et al.*, 1991), were some sufficient conditions and necessary conditions were given for some cases as, for example, the single input systems. These works have been made in the framework of differential geometry.

More recently, another approach was given in (Sluis and Tilbury, 1996), where differential forms and Pfaffian systems were the main tool. This framework permits not only the study of the problem of linearization by prolongations, but also the study of general dynamic feedback linearizations. See, for instance, (van Niewstadt *et al.*, 1994). This problem, also related with flatness, has been atacked in other settings as differential algebra (Fliess *et al.*, 1995),(Fossas *et al.*, 1998),(Martin, 1992). Other frameworks were considered in (Aranda-Bricaire *et al.*, 1995),(Pomet, 1997). Another way to solve the problem of dynamic feedback linearization could be traced using differential geometry of infinite dimension, that was introduced in (Fliess *et al.*, 1997),(Fliess *et al.*, 1996).

The results found in this paper can be generalized to a system with m inputs.

This communication is organized as follows: section 2 and 3 are devoted to some background. In section 4 the main result is stated and proven. Some explanations about the bounds are found in section 5. And, finally, the conclusions and suggestions for future work.

2. BACKGROUND

Definition: Consider Σ the m-input system on R^n

$$\dot{x} = f(x) + \sum_{i=1}^{m} g_i(x)u_i$$

[1] Partially supported by MEC under grant CICYT-1997-0969-CO3-01

A prolongation of Σ is given by

$$\begin{array}{lcl} \dot{x} & = & F(x) + \sum_{i=1}^{m} g_i(x) u_i^0 \\ \dot{u}_i^0 & = & u_i^1 \\ \vdots & & \\ \dot{u}_i^{k_i-2} & = & u_i^{k_i-1} \\ \dot{u}_i^{k_i-1} & = & v_i \end{array} \qquad i = 1 \dots m$$

where u_i^j, which corresponds to $u_i^{(j)}$, are new state variables $i = 1 \dots m$, $j = 0 \dots k_i - 1$, and the new inputs are v_i.

Definition: Let Σ be the m-input system on R^n

$$\dot{x} = f(x) + \sum_{i=1}^{m} g_i(x) u_i$$

Σ is said to be linearizable by prolongations if there exists a prolongation of the original system which is static feedback linearizable.

In fact, this is a particular case of dynamic feedback linearization. Let us recall, from (Hunt *et al.*, 1983), the conditions for a system to be static feedback linearizable :

Proposition: Let

$$\dot{x} = f(x) + \sum_{i=1}^{m} g_i(x) u_i$$

be a nonlinear system with m inputs. This system is static feedback linearizable if, and only if, the following distributions have constant rank and are involutive:

$$D_0 = < g_1, \dots, g_m >$$

$$D_i = < D_{i-1}, ad_f g_1, \dots, ad_f g_m >$$

$\forall i = 1, \dots, n-1$; and D_{n-1} is full rank.

Let now introduce a technical lemma which will be useful in order to lighten the proof of the main result:

Lemma: Let

$$\frac{\partial}{\partial y_i}, \; i \in I$$

a family of coordinate vector fields in R^N;

$$D_1 = < \alpha_1, \dots, \alpha_j ; \{\frac{\partial}{\partial y_k}\}_{k=n+1}^{n+l} >$$

$$D_2 = < \{\frac{\partial}{\partial y_h}\}_{k=n+l+1}^{n+s} >$$

where

$$\alpha_1, \dots, \alpha_j \in < \frac{\partial}{\partial y_1}, \dots, \frac{\partial}{\partial y_n} >$$

Then $D_1 \oplus D_2$ is involutive if, and only if, D_1 is involutive.
Proof: It is straightforward.

3. PROLONGATIONS OF M INPUTS ARE NOT NECESSARY

The following lemma, which proof can be found in (Sluis and Tilbury, 1996), states that, in order to find a linearization by prolongations of a m input system, it suffices to add to the original system derivatives of $m-1$ inputs. Therefore, in a 3 input system, derivatives of just two inputs will be considered.

Lemma: If the system

$$\dot{x} = f(x) + \sum_{i=1}^{m} g_i(x) u_i$$

is linearizable by prolongations of u_1 n_1-times, u_2 n_2-times, ..., u_m n_m-times (with $n_1 \geq n_2 \geq \dots \geq n_m$), then the system is linearizable by prolongations of u_1 $(n_1 - n_m)$-times, u_2 $(n_2 - n_m)$-times, ..., u_{m-1} $(n_{m-1} - n_m)$-times and zero derivatives of u_m.

4. MAIN RESULT

The following theorem, which states a necessary and sufficient condition for the linearizability by prolongations of a three input system, also provides an algorithm to decide whether or not a system is linearizable by prolongations.

Theorem: The system Σ defined by

$$\dot{x} = f(x) + \sum_{i=1}^{3} g_i(x) u_i$$

is linearizable by prolongations if, and only if, at least one of the following systems $\Sigma_{k_1}^{k_2}$ is static feedback linearizable. $\Sigma_{k_1}^{k_2}$ is given by

$\forall i = 1, \dots, n$
$\dot{y}_i = f_i(y_1, \dots, y_n) + g_{h_1 i}(y_1, \dots, y_n) y_{n+1} + g_{h_2 i}(y_1, \dots, y_n) y_{n+k_1+1} + g_{h_3 i}(y_1, \dots, y_n) w_3$

$$\begin{array}{lll} \dot{y}_{n+j} & = y_{n+j+1}; & j = 1, \dots, k_1 - 1 \\ \dot{y}_{n+k_1} & = w_1 & \\ \dot{y}_{n+k_1+l} & = y_{n+k_1+l+1}; & l = 1, \dots, k_2 - 1 \\ \dot{y}_{n+k_1+k_2} & = w_2 & \end{array}$$

$k_1 \in \{1, \dots, 2r+1\}$, $k_2 \in \{1, \dots, 2n-2+r\}$. The new state variables y_s and inputs w_j are related to the old ones by

$$\begin{aligned}
y_i &= x_i & i=1,\dots,n\\
y_{n+j} &= u_{h_1}^{(j-1)} & j=1,\dots,k_1\\
y_{n+k_1+l} &= u_{h_2}^{(l-1)} & l=1,\dots,k_2\\
w_1 &= u_{h_1}^{(k_1)}\\
w_2 &= u_{h_2}^{(k_2)}\\
w_3 &= u_{h_3}
\end{aligned}$$

being $\{h_1,h_2,h_3\} = \{1,2,3\}$. And, finally, r is defined in the following way: let consider the prolongation Σ_n^n. Compute the Lie brackets

$$\eta_i = ad_{\overline{f}}^i g_{h_3}$$

where $\overline{f}$ is the drift associated to the prolonged system, and define

$$r = \min\{i \in \{0, n-1\} \mid \eta_{i+1} \in <\eta_0,\dots,\eta_i>\}$$

Proof: There is no loss of generality to assume $h_i = i$, $i = 1,2,3$. The proof is based on the lemma of section 3 and on the justification of the following items:

(1) The static feedback linearizability conditions for the system Σ_{2r+1}^{2n-2+r} and $\Sigma_{r_1}^{r_2}$ (with $r_1 \geq 2r+1$ and $r_2 \geq 2n-2+r$) are the same.
(2) The static feedback linearizability conditions for the system $\Sigma_{k_1}^{2n-2+r}$ and $\Sigma_{k_1}^{r_2}$ (with $k_1 < 2r+1$ and $r_2 > 2n-2+r$) are the same.
(3) The static feedback linearizability conditions for the system $\Sigma_{2r+1}^{k_2}$ and $\Sigma_{r_1}^{k_2}$ (with $k_2 < 2n-2+r$ and $r_1 > 2r+1$) are the same.

So, this being proven, in order to check if a system is linearizable by prolongations, a finite algorithm can be applied: it is only necessary to check if any of the systems $\Sigma_{k_1}^{k_2}$ (with $k_1 \leq 2r+1$ and $k_2 \leq 2n-2+r$) is static feedback linearizable. Obviously, this fact have to be checked for any permutation of the inputs.

[1] First of all, let us study the static feedback linearizability conditions for the system Σ_{2r+1}^{2n-2+r}. Let

$$\begin{aligned}
\overline{f} &= (f + g_1 y_{n+1} + g_2 y_{n+2r+2})\frac{\partial}{\partial x} + \\
&+ \sum_{j=1}^{2r} y_{n+j+1}\frac{\partial}{\partial y_{n+j}} + \\
&+ \sum_{l=1}^{2n-2+r} y_{n+2r+1+l+1}\frac{\partial}{\partial y_{n+2r+1+l}}
\end{aligned}$$

be the drift of the system Σ_{2r+1}^{2n-2+r} (x must be understood as $(y_1,\dots,y_n)$). And let

$$\overline{g_1} = \frac{\partial}{\partial y_{n+2r+1}} \quad \overline{g_2} = \frac{\partial}{\partial y_{3n+3r-1}} \quad \overline{g_3} = g_3\frac{\partial}{\partial x}$$

be the vector fields associated to the new inputs. As in proposition 1, Σ_{2r+1}^{2n-2+r} is static feedback linearizable if, and only if, the following distributions are involutive and constant rank:

$$D_0^* = < \frac{\partial}{\partial y_{n+2r+1}}, \frac{\partial}{\partial y_{3n+3r-1}}, g_3\frac{\partial}{\partial x} >$$

$$D_{i+1}^* = < D_i^{2r+1,2n-2+r}, [\overline{f}, D_i^{2r+1,2n-2+r}] >$$

for all i such that $dim\ D_i^{2r+1,2n-2+r} < 3n+3r-1$, where $* = (2r+1, 2n-2+r)$.

Let us examine these distributions in some detail. It is a straightforward computation to check:

$$\begin{aligned}
ad_{\overline{f}}^i\left(\frac{\partial}{\partial y_{n+2r+1}}\right) &= \frac{\partial}{\partial y_{n+2r+1-i}} \quad \forall i \leq 2r\\
ad_{\overline{f}}^{2r+1}\left(\frac{\partial}{\partial y_{n+2r+1}}\right) &= g_1\frac{\partial}{\partial x}\\
ad_{\overline{f}}^i\left(\frac{\partial}{\partial y_{3n+3r-1}}\right) &= \frac{\partial}{\partial y_{3n+3r-1-i}}\\
&\quad \forall i \leq 2n-3+r\\
ad_{\overline{f}}^{2n-2+r}\left(\frac{\partial}{\partial y_{3n+3r-1}}\right) &= g_2\frac{\partial}{\partial x}
\end{aligned}$$

On the other hand:

- Denote $\eta_k = ad_{\overline{f}}^k g_3 \quad \forall k \leq r$. Note that $\eta_k \in S = < \frac{\partial}{\partial x} >$.
- Note also that η_k depends upon the variables $x, y_{n+1},\dots,y_{n+k}, y_{n+2r+2},\dots,y_{n+2r+1+k}$.

Let us enumerate the conditions to be checked in order to be involutive the corresponding distribution. From now on, when the involutivity of D_j^h will be studied, we will assume the involutivity of D_l^h, $l \leq j-1$.

1. Distributions $D_k^{2r+1,2n-2+r}$, $k \leq r$.
   Conditions: $0 \leq i \leq k$

$$[\eta_i, \eta_k] \in < \eta_0,\dots,\eta_k >$$

2. Distributions $D_{r+k}^{2r+1,2n-2+r} =$

$$= < \eta_0,\dots,\eta_r, \frac{\partial}{\partial y_{n+2r+1}},\dots,\frac{\partial}{\partial y_{n+r+1-k}}, \frac{\partial}{\partial y_{3n+3r-1}},\dots,\frac{\partial}{\partial y_{3n+2r-1-k}} > \quad 1 \leq k \leq r$$

   Conditions $\forall 1 \leq j \leq k$:

$$\left[\frac{\partial}{\partial y_{n+r+1-k}}, \eta_{r+1-j}\right] \in < \eta_0,\dots,\eta_r >$$

3. Distribution $D_{2r+1}^{2r+1,2n-2+r} =$

$$< \eta_0,\dots,\eta_r, \frac{\partial}{\partial y_{n+2r+1}},\dots,\frac{\partial}{\partial y_{n+1}}, g_1, \frac{\partial}{\partial y_{3n+3r-1}},\dots,\frac{\partial}{\partial y_{3n+r-2}} >$$

   Conditions $\forall 0 \leq k \leq r$:

$$[\eta_k, g_1] \in < \eta_0,\dots,\eta_r >$$

In order to derive the condition, remark that η_r depends upon the variables

$$x, y_{n+1}, \dots, y_{n+r}, y_{n+2r+2}, \dots, y_{n+3r+1}$$

and it can be assumed $r \leq n-2$, so $3n+r-2 > n+3r+1$ (if $r = n-1$, then $< \eta_0, \dots, \eta_r >= S$).

Note also that, thanks to the assumption of the involutivity of the preceding distributions, $D^{2r+1,2n-2+r}_{2r+1+j} =$

$$< D^{2r+1,2n-2+r}_{2r+j}, \frac{\partial}{\partial y_{3n+r-2-j}}, ad^j_f g_1 =$$

$$< D^{2r+1,2n-2+r}_{2r+j}, \frac{\partial}{\partial y_{3n+r-2-j}}, ad^j_{\tilde{f}} g_1 =$$

where

$$\tilde{f} = (f + g_2 y_{n+2r+2}) + \sum_{l=1}^{2n-2+r} y_{n+2r+1+l+1} \frac{\partial}{\partial y_{n+2r+1+l}}$$

So, let define $\delta_j = ad^j_{\tilde{f}} g_1$. This vector field depends upon the variables $x, y_{n+2r+2}, \dots, y_{n+2r+1+j}$ and belongs to S. Define also

$$h = \min\{j \in \{0, n-r-2\} | \\ \delta_{j+1} \in < \eta_0, \dots, \eta_r, \delta_0, \dots, \delta_j >\}$$

4. Distributions
$D^{2r+1,2n-2+r}_{2r+1+j} =< D^{2r+1,2n-2+r}_{2r}, \delta_0, \dots, \delta_j,$

$$\frac{\partial}{\partial y_{3n+r-2}}, \dots, \frac{\partial}{\partial y_{3n+r-2-j}} >$$

Conditions

$$\left.\begin{matrix}[\delta_j, \eta_k] \\ [\delta_j, \delta_i]\end{matrix}\right\} \in < \eta_0, \dots, \eta_r, \delta_0, \dots, \delta_j >$$

$\forall 0 \leq k \leq r \quad \forall 0 \leq i \leq j-1$.
Note that $3n + r - 2 - j > n + 3r + 1$ and $3n + r - 2 - j > n + 2r + 1 + j$, which are the maximum subindex of the variables appearing in η_r and δ_j respectively.

Now, the static feedback linearizability conditions for $\Sigma^{r_2}_{r_1}$ (with $r_1 \geq 2r+1$ and $r_2 \geq 2n-2+r$) will be studied. More precisely, the systems Σ^{2n-2+r}_{2r+2}, Σ^{2n-1+r}_{2r+1} and Σ^{2n-1+r}_{2r+2} will be detailed (that is to say, systems where an extra derivative of u_1 and/or u_2 have been added).

(1) System Σ^{2n-1+r}_{2r+1}.
The next equalities are satisfied:

$$D^{2r+1,2n-1+r}_j \oplus < \frac{\partial}{\partial y_{3n+3r-1-j}} >= D^{2r+1,2n-2+r}_j \oplus < \frac{\partial}{\partial y_{3n+3r}} >$$
$\forall j \leq 2r+1+h$

$$D^{2r+1,2n-1+r}_{j+1} = D^{2r+1,2n-2+r}_j \oplus < \frac{\partial}{\partial y_{3n+3r}} >$$
$\forall j \geq 2r+1+h$

Thanks to the technical lemma from section 2, the conditions for both systems are the same.

(2) System Σ^{2n-2+r}_{2r+2}.
Denoting

$$y_{3n+3r} = u_1^{(2r+1)} \quad \dot{y}_{3n+3r} = w_1$$

which appears in the equation

$$\dot{y}_{n+2r+1} = y_{3n+3r}$$

the following equalities hold:

$$D^{2r+2,2n-2+r}_j \oplus < \frac{\partial}{\partial y_{n+2r+1-j}} >=$$

$$= D^{2r+1,2n-2+r}_j \oplus < \frac{\partial}{\partial y_{3n+3r}} > \quad \forall j \leq r$$

$D^{2r+2,2n-2+r}_{j+1} =$

$$= D^{2r+1,2n-2+r}_j \oplus < \frac{\partial}{\partial y_{3n+3r}}, \frac{\partial}{\partial y_{3n+3r-2-j}} >$$

$\forall r \leq j \leq 2r+1+h$
$D^{2r+2,2n-2+r}_j =$

$$= D^{2r+1,2n-2+r}_j \oplus < \frac{\partial}{\partial y_{3n+3r}} >$$

$\forall j \geq 2r+2+h$

Therefore, taking into account the technical lemma, both systems fulfill the same conditions.

(3) Finally, the relationship between the system Σ^{2n-2+r}_{2r+2} and the system system Σ^{2n-1+r}_{2r+2} is the same than the one between system Σ^{2n-2+r}_{2r+1} and system Σ^{2n-1+r}_{2r+1}.

In other words, it has been proven that the static feedback linearizability conditions for the system Σ^{2n-2+r}_{2r+1} and the system where one more derivative of u_1 and/or u_2 have been added, are the same. The same proof is also valid for addition of more derivatives of the inputs.

[2] The systems $\Sigma^{2n-2+r}_{k_1}$ and $\Sigma^{r_2}_{k_1}$ (with $k_1 < 2r+1$ and $r_2 > 2n-2+r$) will be compared. In fact, in order to clarify the proof, it will be studied the systems $\Sigma^{2n-2+r}_{k_1}$ and $\Sigma^{2n-1+r}_{k_1}$. Extra additions of u_2 does not affect the proof. Using the same notations as before, let define:

$$h_1 = min\{i \in \{0, r\} \mid \eta_{i+1} \in < \eta_0, \dots, \eta_i >\}$$

The reason of this definition is that h_1 could be smaller than r. There are two possibilities, namely $h_1 < k_1$ and $h_1 \geq k_1$.

The distributions for the case $h_1 < k_1$ are:

- $\forall j \leq h_1$ $D_j^{k_1,2n-2+r} =$

$$=< \eta_0, \dots, \eta_j, \frac{\partial}{\partial y_{n+k_1}}, \dots, \frac{\partial}{\partial y_{n+k_1-j}}, \frac{\partial}{\partial y_{3n+k_1+r-2}}, \dots, \frac{\partial}{\partial y_{3n+k_1+r-2-j}} >$$

$\forall j \leq h_1$; $D_j^{k_1,2n-1+r} =$

$$=< \eta_0, \dots, \eta_j, \frac{\partial}{\partial y_{n+k_1}}, \dots, \frac{\partial}{\partial y_{n+k_1-j}}, \frac{\partial}{\partial y_{3n+k_1+r-1}}, \dots, \frac{\partial}{\partial y_{3n+k_1+r-1-j}}, >$$

They satisfy the same involutivity conditions because η_j does not depend on the variables $y_{3n+k_1+r-j-2}$.

- For the distributions $\forall h_1 < j < k_1$, $D_j^{k_1,2n-2+r} =$

$$=< \frac{\partial}{\partial y_{n+k_1}}, \dots, \frac{\partial}{\partial y_{n+k_1-j}}, \frac{\partial}{\partial y_{3n+k_1+r-2}}, \dots, \frac{\partial}{\partial y_{3n+k_1+r-2-j}}, \eta_0, \dots, \eta_{h_1} >$$

the equality

$$D_j^{k_1,2n-2+r} \oplus < \frac{\partial}{\partial y_{3n+k_1+r-1}} >=$$

$$= D_j^{k_1,2n-1+r} \oplus < \frac{\partial}{\partial y_{3n+k_1+r-2-j}} >$$

holds. Therefore, because of the technical lemma, they provide the same conditions.

- Let be $h_2 = \min\{i | \delta_{i+1} \in < \eta_0, \dots, \eta_{h_1}, \delta_0, \dots, \delta_i >\}$ and $\forall i \leq h_2$, $D_{k_1+i}^{k_1,2n-2+r} =$

$$=< \frac{\partial}{\partial y_{n+k_1}}, \dots, \frac{\partial}{\partial y_{n+1}}, \frac{\partial}{\partial y_{3n+k_1+r-2}}, \dots, \frac{\partial}{\partial y_{3n+r-2-i}}, \eta_0, \dots, \eta_{h_1}, \delta_0, \dots, \delta_i >$$

the following distributions to be studied. Note that $h_2 \leq n - h_1 - 3$, because when $h_2 = n - h_1 - 2$, then

$$< \eta_0, \dots, \eta_{h_1}, \delta_0, \dots, \delta_{n-h_1-2} >= S$$

and $D_{k_1+n-h_1-2}^{k_1,2n-2+r}$ is trivially involutive, as well as all the distributions after it.

The keypoint for these distributions is again the equality

$$D_{k_1+i}^{k_1,2n-2+r} \oplus < \frac{\partial}{\partial y_{3n+k_1+r-1}} >=$$

$$= D_{k_1+i}^{k_1,2n-1+r} \oplus < \frac{\partial}{\partial y_{3n+r-2-i}} >$$

But now, to assure the decoupling property, it has to be proven that $3n + r - 2 - i > n + k_1 + i$, which is the maximum subindex of the variables of which δ_i depends. It is enough to check this inequality when $i = n - h_1 - 3$, which is the highest possible value. Then, it must be proven

$$2n + r + 1 + h_1 > 2n + k_1 - h_1 - 3$$

or, equivalently, $r - k_1 + 4 > -2h_1$. This is obvious when $k_1 \leq r$. If $k_1 > r$, note that $h_1 = r$. In this case, the inequality to be proven becomes $r - k_1 + 4 > -2r$, also trivial.

- Finally, for all $i \geq k_1 + h_2$, $D_{i+1}^{k_1,2n-1+r} =$

$$= D_i^{k_1,2n-2+r} \oplus \frac{\partial}{\partial y_{3n+k_1+r-1}}$$

So, they fulfill the same conditions.

Now, for the case $h_1 \geq k_1$,

- $\forall j \leq n + k_1 - h_1 - 3$

$$D_j^{k_1,2n-2+r} \oplus \frac{\partial}{\partial y_{3n+k_1+r-1}} =$$

$$D_j^{k_1,2n-1+r} \oplus \frac{\partial}{\partial y_{3n+k_1+r-2-j}}$$

The following inequalities must to be fulfilled in order to satisfy the decoupling property (just studying the case $j = n + k_1 - h_1 - 3$):

$$2n + r + h_1 + 1 > n + h_1$$
$$2n + r + h_1 + 1 > n + k_1 + n - h_1 - 3$$

or, equivalently, the trivial inequalities

$$n + r + 1 > 0$$
$$r + 2h_1 + 4 > k_1$$

- And $\forall j \geq n + k_1 - h_1 - 3$, $D_{j+1}^{k_1,2n-2+r} =$

$$= D_j^{k_1,2n-2+r} \oplus \frac{\partial}{\partial y_{3n+k_1+r-1}}$$

Therefore, both systems satisfy the same static feedback linearizability conditions.

[3] The third case to compare is the static feedback linearizability conditions for the systems $\Sigma_{2r+1}^{k_2}$ and $\Sigma_{r_1}^{k_2}$ (with $k_2 < 2n - 2 + r$ and $r_1 > 2r + 1$). In fact it will be compared the systems $\Sigma_{2r+1}^{k_2}$ and $\Sigma_{2r+2}^{k_2+1}$ (which still satisfies the restriction $k_2 + 1 \leq 2n - 2 + r$). Denoting

$$y_{n+2r+1+k_2+1} = u_1^{(n+2r+1)}$$
$$y_{n+2r+1+k_2+2} = u_2^{(k_2)}$$

the following equalities hold

$$D_j^{2r+2,k_2+1} \oplus$$
$$\oplus < \frac{\partial}{\partial y_{n+2r+1-j}}, \frac{\partial}{\partial y_{n+2r+1+k_2-j}} >=$$

$$= D_j^{2r+1,k_2} \oplus$$
$$\oplus < \frac{\partial}{\partial y_{n+2r+1+k_2+1}}, \frac{\partial}{\partial y_{n+2r+1+k_2+2}} >$$

for all $j \leq r$. And for $j \geq r$

$$D_{j+1}^{2r+2,k_2+1} = D_j^{2r+1,k_2} \oplus$$
$$\oplus < \frac{\partial}{\partial y_{n+2r+1+k_2+1}}, \frac{\partial}{\partial y_{n+2r+1+k_2+2}} >$$

Moreover, we are under the hypothesis of the technical lemma. So, the conditions are the same for both systems. This ends the proof.

5. WHERE DO THE BOUNDS $2R+1$ AND $2N-2+R$ COME FROM?

If the bound $2r+1$ is relaxed to $2r$, then the distribution $D_r^{2r,2n-2+r}$ of the system Σ_{2r}^{2n-2+r} is

$$D_r^{2r,2n-2+r} =< \eta_0, \ldots, \eta_r,$$
$$\frac{\partial}{\partial y_{n+2r}}, \ldots, \frac{\partial}{\partial y_{n+r}},$$
$$\frac{\partial}{\partial y_{3n+3r-2}}, \ldots, \frac{\partial}{\partial y_{3n+2r-2}} >$$

Therefore, it appears a new involutivity condition, different from the ones appeared in $D_r^{2r+1,2n-2+r}$:

$$[\eta_r, \frac{\partial}{\partial y_{n+r}}] \in < \eta_0, \ldots, \eta_r >$$

So, it is not possible to relax that bound.

In the same manner, the reason of the bound $2n-2+r$ is to avoid extra involutivity conditions among η_j $(j \leq r)$, δ_i $(i \leq h)$ and the other elements of D_{2r+1+i}. In spite of that, it is not clear that this last bound can not be improved.

6. CONCLUSION AND FUTURE WORK

Using elementary differential geometry, it has been given a proof of the fact that there exists an upper bound on the number of integrators needed to linearize a three input nonlinear control system. Moreover, the bounds of this paper improves other bounds that have appeared in the literature (Sluis and Tilbury, 1996).

This procedure can also be applied for systems with more than 3 inputs. This will be the object of a more detailed paper. Another interesting work to do is to design an efficient algorithm to check the involutivity conditions of all the distributions. Although it seems there are a lot of conditions to satisfy, a detailed study can reduce the number of conditions to a reasonable one.

7. REFERENCES

Aranda-Bricaire, E., C. H. Moog and J.B. Pomet (1995). A linear algebraic framework for dynamic feedback linearization. *IEEE Transactions on Automatic Control* **40-1**, 127–132.

Charlet, B., J. Lévine and R. Marino (1989). On dynamic feedback linearization. *Syst. Contr. Lett.* **13**, 143–151.

Charlet, B., J. Lévine and R. Marino (1991). Sufficient conditions for dynamic state feedback linearization. *SIAM J. Contr. Opt.* **29**, 38–57.

Fliess, M., J. Lévine, Ph. Martin and P. Rouchon (1995). Flatness and defect of nonlinear systems: Introductory theory and examples. *Int. J. Contr.* **61**, 1327–1361.

Fliess, M., J. Lévine, Ph. Martin and P. Rouchon (1996). A lie-bäcklund approach to dynamic feedback equivalence and flatness. In: *Robust control via variable structure and Lyapunov techniques* (L. Glielmo F. Garofalo, Ed.). first ed.. Vol. LNCIS-217. pp. 245–268. Springer-Verlag. London.

Fliess, M., J. Lévine, Ph. Martin, F. Ollivier and P. Rouchon (1997). On nonlinear controllability, infinite jets and prolongations. In: *Proceedings of the ECC97*. Vol. CD-Rom. ECC97. Brussels.

Fossas, E., J. Franch and A. Palau (1998). Flatness, tangent systems and flat outputs. In: *Proceedings of the ACC98*. Vol. CD-Rom. American automatic control council. Philadelphia.

Hunt, L. R., R. Su and G. Meyer (1983). Design for multi-input nonlinear systems. In: *Differential geometric control theory* (H. J. Sussmann R. W. Brockett, R. S. Millmann, Ed.). first ed.. Vol. PM-27. pp. 268–298. Birkhäuser. Boston.

Martin, Ph. (1992). *Contribution à létude des systèmes différentiallement plats*. Thèse de doctorat. École Nationale Supérieure des Mines de Paris.

Pomet, J. B. (1997). On dynamic feedback linearization of four-dimensional affine control systems with two inputs. *ESAIM: Control, optimisation and calculus of variations* **2**, 151–230.

Sluis, W. M. and D. M. Tilbury (1996). A bound on the number of integrators needed to linearize a control system. *System and control letters* **29**, 43–50.

van Niewstadt, M., M. Rathinam and R. M. Murray (1994). Differential flatness and absolute equivalence. In: *Proceedings of the CDC94*. pp. 326–332. IEEE CSS. San Francisco.

14th Triennial World Congress, Beijing, P.R. China

ON OSCILLATIONS IN COUPLED DYNAMICAL SYSTEMS

Alexander Yu. Pogromsky *,[1] **Torkel Glad** **
Henk Nijmeijer ***,[2]

* *Department of Electrical Engineering, Technical University of Eindhoven, The Netherlands, sasha@isy.liu.se*
** *Department of Electrical Engineering, Linköping University, Sweden, torkel@isy.liu.se*
*** *Faculty of Mathematical Sciences, University of Twente, The Netherlands, H.Nijmeijer@math.utwente.nl*

Abstract: The paper deals with the problem of destabilization of diffusively coupled identical systems. It is shown that the globally asymptotically stable systems being diffusively coupled may exhibit an oscillatory behavior. It is shown that if the diffusive medium consists of hyperbolically nonminimum phase systems and the diffusive factors exceed some threshold value the origin of the overall system undergoes a Poincaré-Andronov-Hopf bifurcation resulting in oscillatory behavior. *Copyright © 1999 IFAC*

Keywords: Nonlinear systems, nonminimum phase, oscillation.

1. INTRODUCTION

Oscillations play an important role in nature. Every life form exhibits an oscillatory behavior at every level of biological organization with periods ranging from milliseconds (neurons) to seconds (cardiac sells), minutes (oscillatory enzymes), hours, days, weeks and even years (epidemiological processes and predator-prey interactions in ecology). Of course biological systems are very complex and difficult to model, nevertheless several attempts were made to explain the mechanism of how persistent oscillations occur in complex interconnected or distributed biological systems.

A recognized starting point in related research activities is the paper by A. Turing [1952] who studied a possible mechanism for morphogenesis – the creation of forms and patterns in the development of an embryo. A crucial role in the study is played by the diffusive medium which can be described in two ways – continuous, involving partial differential equations, or discrete, involving diffusion of enzymes (morphogenes in the terminology of Turing) between separate cells described by ordinary differential equations.

The diffusion which is very important for cooperative behavior of living cells was usually considered as a smoothening or trivializing process. However it turns out that it can result in non-trivial oscillatory behavior in different systems. Some phenomena occurring in distributed systems written in form of PDE are described in the paper by We-Ming Ni [1998] and references therein. In this paper we are concerned with oscillatory phenomena occurring in systems consisting of diffusively coupled subsystems described by ordinary differential equations. A motivation of our study

[1] On leave from the Institute for Problems of Mechanical Engineering, 61, Bolshoi, VO, St.Petersburg, 199178, Russia, e-mail: sasha@ccs.ipme.ru. The work was performed while the author was with Department of Electrical Engineering of Linkoping University, Sweden.
[2] Also at Faculty of Mechanical Engineering, Technical University of Eindhoven, The Netherlands

is the paper by Smale [1976] who proposed an example of two 4th order diffusively coupled differential systems. Each system describes a mathematical cell and by itself is inert or dead in the sense that it is globally asymptotically stable. In interaction, however "the cellular system pulses (or expressed perhaps over dramatically, becomes alive!) in the sense that the concentrations of the enzymes in each cell will oscillate infinitely". In his paper Smale posed the problem to find conditions under which globally asymptotically stable systems being diffusively coupled will oscillate. A solution to this problem based on a frequency domain formalism was proposed in [Tomberg & Yakubovich, 1989] where diffusive interaction of two Lur'e systems with scalar nonlinearity was studied. Our approach is based on the related concepts of passivity and minimum phaseness and the main contribution of the paper is that we will show that the Turing instability occurs as a result of a Poincaré-Andronov-Hopf bifurcation. We will show that this scenario occurs in diffusively coupled nonminimum phase systems. At the same time in [Pogromsky, 1998] it has been shown that diffusive coupling of minimum phase systems results in other interesting oscillatory phenomenon - synchronization. Therefore the results of this paper and [Pogromsky, 1998] allows one to understand better oscillatory behavior occurring in diffusively coupled systems.

The paper is organized as follows. First we will give some relevant definitions and recall some useful results. The problem statement is presented in Section 3. Section 4 contains some results regarding boundedness and ultimate boundedness of solutions of interconnected systems. In Section 5 we will present conditions resulting in the Turing instability and, finally,Section 6 contains the conclusions.

2. BASIC DEFINITIONS

The Euclidean norm in $\mathbb{R}^n$ is denoted simply as $|\cdot|$, $|x|^2 = x^\top x$, where $^\top$ stands for the transpose operation.

Let x_e be an equilibrium of the system $\dot{x} = f(x)$, $f : \mathbb{R}^n \to \mathbb{R}^n$. If the Jacobian $\partial f(x_e)/\partial x$ has no eigenvalues on the imaginary axis then the point x_e is called a *noncritical point.*

Definition 1. A scalar function $x : \mathbb{R}^1 \to \mathbb{R}^1$ is called oscillatory in the sense of Yakubovich (or Y-oscillation) for $t \to +\infty$ if $x(t)$ is bounded on $\mathbb{R}^+$ and

$$\overline{\lim_{t\to+\infty}} x(t) \geq \beta, \qquad \underline{\lim_{t\to+\infty}} x(t) \leq \alpha$$

for some $\beta > \alpha$.

Consider a general autonomous continuous-time system:

$$\dot{x} = F(x), \tag{1}$$

where $x(t) \in \mathbb{R}^n$ is an n-dimensional state vector and the vector field F satisfies conditions guaranteeing existence and uniqueness of the solutions of (1) for all $t \geq 0$ and all initial conditions $x(0) = x_0 \in \mathbb{R}^n$, and h is the output map.

Consider the following nonlinear system

$$\dot{x} = F(x), \quad x(t) \in \mathbb{R}^n, \tag{2}$$

$$y = h(x), \quad y(t) \in \mathbb{R}^l. \tag{3}$$

where as before F satisfies assumptions guaranteeing existence of a unique solution on the infinite time interval.

Definition 2. (Yakubovich, 1973). The system (2), (3) is called *Y-oscillatory with respect to the scalar output y* if it is Lagrange stable and for almost all initial conditions

$$\underline{\lim_{t\to+\infty}} y(t) < \overline{\lim_{t\to+\infty}} y(t).$$

Recalling Definition 1 of Y-oscillation given for an arbitrary scalar function (not necessary being a solution of some differential equation) we call the system with vector output y Y-oscillatory if it is Y-oscillatory with respect to at least one of the components of the vector y. If the system is Y-oscillatory with respect to the output $y \equiv x$ then it is referred to as just Y-oscillatory.

The following result can be proved similarly to the proof of Theorem 1.1, $3°, 5°$ [Tomberg & Yakubovich, 1989].

Theorem 1. Assume that

A1. The equation $F(x) = 0$ has only isolated solutions $\bar{x}_j, j = 1, 2....$

A2. The system (2) is ultimately bounded.

A3. $\bar{x}_j$ are noncritical points and each matrix $\frac{\partial F}{\partial x}(\bar{x}_j)$ has at least one eigenvalue with positive real part.

Then the system (1) is Y-oscillatory.

3. PROBLEM STATEMENT

Let $\mathbb{P}^n$ be the closed nonnegative orthant in $\mathbb{R}^n$: $\mathbb{P}^n = \{x \in \mathbb{R}^n,\ x = \mathrm{col}(x^{(1)}, \ldots, x^{(n)}),\ x^{(j)} \geq 0, j = 1, \ldots, n\}$. Let $\mathcal{R}$ be a smooth mapping $\mathcal{R} : \mathbb{P}^n \to \mathbb{P}^n$, such that the system

$$\dot{x} = \mathcal{R}(x), \quad x(t) \in \mathbb{P}^n \tag{4}$$

is globally (in $\mathbb{P}^n$) asymptotically stable, that is there is a point $x_e \in \mathbb{P}^n$, such that $\mathcal{R}(x_e) = 0$, x_e is Lyapunov stable and all solutions starting in $\mathbb{P}^n$ tend to x_e. The problem studied by Smale [1976] is to find (if possible) a positive definite diagonal $n \times n$ matrix D such that the system

$$\begin{cases} \dot{x}_1 = \mathcal{R}(x_1) + D(x_2 - x_1) \\ \dot{x}_2 = \mathcal{R}(x_2) + D(x_1 - x_2) \end{cases} \quad (5)$$

has a nontrivial periodic solution with a trajectory attracting almost all other solutions of (5). The state vector for each system consists of the concentrations of the chemicals which are nonnegative. The dynamics (4) describes a chemical reaction between some chemicals ("morphogenes") inside one cell. The system (5) describes a possible diffusive interaction between two cells and is referred to as the Turing equation. The paper by Smale posed a "sharp problem, namely to 'axiomatize' the properties necessary to bring about oscillation via diffusion. In the 2-cell case, just what properties does the pair $(\mathcal{R}, D)$ need to possess to make the Turing interacting system oscillate? In the many cell case, how does the topology contribute?"

It was shown by Smale that to design a system (5) it is sufficient to design a system with the same properties but assuming that $\mathcal{R} : \mathbb{R}^n \to \mathbb{R}^n$. In this case we can assume that the unique equilibrium point of (4) is the origin.

Definition 3. Given the systems

$$\begin{cases} \dot{x}_j = f(x_j) + Bu_j \\ y_j = Cx_j \end{cases} \quad (6)$$

where $j = 1, \ldots, k$, $x_j(t) \in \mathbb{R}^n$ is the state, $u_j(t) \in \mathbb{R}^m$ is the input, $y_j(t) \in \mathbb{R}^m$ is the output, $f(0) = 0$ and B, C are constant matrices of corresponding dimension. We will say that the systems (6) are *diffusively coupled* if the matrix CB is similar to a positive definite matrix and the systems (6) are closed by the following feedback

$$u_j = -\gamma_{j1}(y_j - y_1) - \ldots - \gamma_{jk}(y_j - y_k) \quad (7)$$

where $\gamma_{ij} = \gamma_{ji} \geq 0$ are constants such that $\sum_{j=1}^{k} \gamma_{ji} > 0$ for all $i = 1, \ldots, k$.

4. BOUNDEDNESS OF SOLUTIONS OF INTERCONNECTED SYSTEMS

Consider the nonlinear time-invariant affine in the control system:

$$\begin{cases} \dot{x} = f(x) + g(x)u \\ y = h(x) \end{cases} \quad (8)$$

where $x(t) \in \mathbb{R}^n$ is the state, $u(t) \in \mathbb{R}^m$ is the input which is assumed to be a continuous and (essentially) bounded function of time: $u \in \mathcal{C}^0 \cap \mathcal{L}_\infty$; $y(t) \in \mathbb{R}^l$ is the output; $f : \mathbb{R}^n \to \mathbb{R}^n$ and the columns of the matrix $g : \mathbb{R}^n \to \mathbb{R}^{n \times m}$ are smooth vector fields, $f(0) = 0$; and $h : \mathbb{R}^n \to \mathbb{R}^l$ is a smooth mapping, $h(0) = 0$.

If the system (8) possesses a zero dynamics system with noncritically stable zero solution it is referred to as *noncritically minimum phase*. If the origin is an hyperbolically unstable solution of the zero dynamics system for (8) then the system (8) is referred to as *hyperbolically nonminimum phase*. More about the concepts of zero dynamics, minimumphaseness and related normal forms can be found in [Isidori, 1995; Byrnes & Isidori, 1991; Byrnes *et. al.*, 1991]

Associated with the system (8) consider a real-valued function w defined on $\mathbb{R}^n \times \mathbb{R}^m$ called the *supply rate*. We will assume that this function is well defined on any compact subset of the set $\mathbb{R}^n \times \mathbb{R}^m$.

Definition 4. The system (8) with supply rate w is said to be $\mathcal{C}^r$-dissipative in the sense of Willems (or $\mathcal{W}$-dissipative) if there exists a $\mathcal{C}^r$-smooth, $(r \geq 0)$ nonnegative function $V : \mathbb{R}^n \to \mathbb{R}^+$, called the *storage function*, such that the following *dissipation inequality* holds:

$$V(x(t)) - V(x(0)) \leq \int_0^t w(x(s), u(s))ds \quad (9)$$

for all $u \in \mathcal{C}^0 \cap \mathcal{L}_\infty$, $x(0) \in \mathbb{R}^n$, $0 \leq t < T_{u,x_0}$, where T_{u,x_0} is the upper time limit for which the solution corresponding to the input u and initial conditions $x(0) = x_0$ exists.

Definition 5. (Pogromsky, 1998). The system (8) is called $\mathcal{C}^r$*-semipassive* if it is $\mathcal{C}^r$-dissipative with supply rate $w(x, u) = y^\top u - H(x)$ where the scalar function H is nonnegative outside some ball:

$$\exists \rho > 0 \ \forall |x| \geq \rho \Longrightarrow H(x) \geq \varrho(|x|)$$

for some continuous nonnegative function ϱ defined for $|x| \geq \rho$. If the function H is positive outside some ball:

$$\exists \rho > 0 \ \forall |x| \geq \rho \Longrightarrow H(x) \geq \varrho(|x|)$$

for some continuous positive function ϱ then the system (8) is referred to as *strictly semipassive*.

Consider the following k systems:

$$\begin{cases} \dot{x}_j = f_j(x_j) + g_j(x_j)u_j \\ y_j = h_j(x_j) \end{cases} \quad (10)$$

where $j = 1, \ldots, k$, $x_j(t) \in \mathbb{R}^{n_j}$ are the states, $u_j(t) \in \mathbb{R}^m$ are the inputs, $y_j(t) \in \mathbb{R}^m$ are the outputs and the functions f_j, g_j, h_j are smooth enough to ensure existence at least on some time interval of all solutions considered below.

Compose the symmetric matrix Γ as follows

$$\Gamma = \begin{pmatrix} \sum_{i=1}^{k} \gamma_{1i} & -\gamma_{12} & -\gamma_{13} & \cdots & -\gamma_{1k} \\ -\gamma_{21} & \sum_{i=1}^{k} \gamma_{2i} & -\gamma_{23} & \cdots & -\gamma_{2k} \\ \vdots & \vdots & \vdots & \ddots & \vdots \\ -\gamma_{k1} & -\gamma_{k2} & -\gamma_{k3} & \cdots & \sum_{i=1}^{k} \gamma_{ki} \end{pmatrix} \quad (11)$$

where $\gamma_{ij} = \gamma_{ji} \geq 0$. The matrix Γ is symmetric and therefore all its eigenvalues are real numbers. Moreover applying Gerschgorin's theorem about localization of eigenvalues (see, e.g. [Stewart & Sun, 1990]) one can see that all eigenvalues of Γ are nonnegative, that is the matrix Γ is positive semidefinite.

The following result gives conditions under which the solutions of the interconnected systems (10) are bounded.

Lemma 2. Consider the systems (10) closed by the following feedback

$$u_j = -\gamma_{j1}(y'_j - y'_1) - \ldots - \gamma_{jk}(y'_j - y'_k) \quad (12)$$

where $y'_j = h'_j(x_j)$ and $h'_j : \mathbb{R}^{n_j} \to \mathbb{R}^m$ is a smooth mapping such that $h'_j(0) = 0$. Suppose that the systems (10) are C^r-semipassive ($r \geq 0$) with respect to input u_j and output $y_j = h_j(x_j)$ with radially unbounded storage functions $V_j : \mathbb{R}^{n_j} \to \mathbb{R}^+$ and the function

$$H(x) = \sum_{j=1}^{k} H_j(x_j) + y^\top (I_m \otimes \Gamma) y'$$

where $x = \text{col}(x_1, \ldots, x_k)$, $y = \text{col}(y_1, \ldots, y_k)$, $y' = \text{col}(y'_1, \ldots, y'_k)$ and the functions $H_j, j = 1, \ldots, k$ are as in the definition of semipassivity, satisfies

$$\exists \rho > 0 \; \forall |x| \geq \rho \Longrightarrow H(x) \geq \varrho(|x|) \quad (13)$$

for some continuous function ϱ defined for $|x| \geq \rho$. Then if the function ϱ is nonnegative then all solutions of the closed loop system (10), (12) exist on the infinite time interval and are bounded, that is the system (10), (12) is Lagrange stable, moreover if the function ϱ is positive then the closed loop system (10), (12) is ultimately bounded.

5. DIFFUSION DRIVEN OSCILLATIONS

5.1 *Infinite interval of instability*

The purpose of this section is to give an explicit construction of diffusively coupled globally asymptotically stable systems that become oscillatory being interconnected. We assume that the topology of the interconnection is described by the matrix Γ defined in Eq. (11) with entries γ_{ij} as in Definition 3. In this subsection we consider the case when the zero dynamics for each system is nontrivial. This condition will be relaxed in the next subsection.

Let A be an $n \times n$ matrix, $n \geq 3$

$$A = \begin{pmatrix} A_{11} & A_{12} \\ A_{21} & A_{22} \end{pmatrix}$$

where A_{11} is an $(n-m) \times (n-m)$ matrix, $1 \leq m \leq n-2$ and the other matrices are of appropriate dimensions. Let B and C be full rank $n \times m$ and $m \times n$ matrices such that the product CB is a positive definite matrix.

Theorem 3. Assume that

A1 The matrix A is Hurwitz.

A2 The matrix A_{11} has an even nonzero number of eigenvalues with positive real parts.

A3 The matrix $(CB)^{-1}T$ where $T = (A_{22} - A_{21}A_{11}^{-1}A_{12})$ has no positive real eigenvalues.

Then there exists a function $f : \mathbb{R}^n \to \mathbb{R}^n$, $f(z,y) = \text{col}(q(z,y), a(z,y))$, where $z \in \mathbb{R}^{n-m}, y \in \mathbb{R}^m$, $q : \mathbb{R}^{n-m} \times \mathbb{R}^m \to \mathbb{R}^{n-m}$, $a : \mathbb{R}^{n-m} \times \mathbb{R}^m \to \mathbb{R}^m$, such that

$$A_{11} = \frac{\partial q}{\partial z}(0,0), \quad A_{12} = \frac{\partial q}{\partial y}(0,0),$$

$$A_{21} = \frac{\partial a}{\partial z}(0,0), \quad A_{11} = \frac{\partial a}{\partial y}(0,0)$$

and

1) The system

$$\begin{cases} \dot{z} = q(z,y) \\ \dot{y} = a(z,y) \end{cases} \quad (14)$$

has a unique globally asymptotically stable equilibrium (the origin).

2) For all $\gamma_{ij} \geq 0$ the system consisting of k diffusively coupled systems

$$\begin{cases} \dot{z}_j = q(z_j, y_j) \\ \dot{y}_j = a(z_j, y_j) + CBu_j \\ u_j = -\gamma_{j1}(y_j - y_1) - \ldots - \gamma_{jk}(y_j - y_k) \end{cases} \quad (15)$$

is ultimately bounded, has a unique equilibrium and there exists a positive number $\bar{\gamma}$ such that for all $\gamma > \bar{\gamma}$, where γ stands for the maximal eigenvalue of the matrix Γ, the closed loop system is Y-oscillatory.

Now let us show that the condition $n \geq 3$ is also necessary.

Proposition 4. Consider a locally Lipschitz continuous function $f : \mathbb{R}^n \to \mathbb{R}^n$, $f(x) = -f(-x)$ such that

i. The system $\dot{x} = f(x)$ has a globally noncritically stable zero solution.
ii. The closed loop system

$$\begin{cases} \dot{x}_1 = f(x_1) + B(x_2 - x_1) \\ \dot{x}_2 = f(x_2) + B(x_1 - x_2) \end{cases} \qquad (16)$$

where B is similar to a positive semidefinite $n \times n$ matrix, is Y-oscillatory and has a unique zero equilibrium.

Then $n \geq 3$.

5.2 *Finite interval of instability*

As we have seen the analysis of oscillatory diffusively coupled systems consists of two problems: analysis of boundedness of solutions of interconnected systems and local stability analysis of existing equilibria. A solution of the first problem can be obtained with the help of results of Section 4. To solve the second problem in the previous theorem we made a crucial assumption about instability of the zero dynamics (assumption A2). So, one may wonder how restrictive this assumption is in the analysis of diffusively coupled systems. At first glance this assumption is quite restrictive. Indeed, in the example proposed by Smale the matrix CB has rank n and therefore the system (15) has trivial zero dynamics in this case. It motivates our subsequent study. First we will show that Smale's understanding of diffusive coupling is equivalent to our definition. Our definition includes Smale's problem as a special case, therefore to show equivalence between them we need to prove that a solution to the problem posed in this paper implies existence of solution to the Smale problem. This fact will be established by the next theorem. Second we will relax the conditions imposed on the zero dynamics system.

Theorem 5. Assume that there exists an $n \times n$ matrix D such that D is singular and is similar to a positive semidefinite matrix and the zero equilibrium of the system $\dot{x} = (A - D)x$ is hyperbolically unstable. Then there exists a positive definite matrix B such that $\dot{x} = (A - B)x$ is hyperbolically unstable and the number of positive eigenvalues of $A - B$ coincides with the number of positive eigenvalues of $A - D$.

Now we can present a result guaranteeing existence of oscillatory behavior in the diffusively coupled systems in case when the largest eigenvalue $\gamma > 0$ of the matrix Γ is not greater than some limit value γ_0.

Theorem 6. Suppose that $n \times m$ and $m \times n$, $m \leq n$ matrices B and C are such that CB is similar to a positive definite matrix. Assume that

A1. The $n \times n$ matrix A is Hurwitz.

A2. There exists $0 < \gamma_1 \leq \gamma_0$ such that for all $\gamma \in [\gamma_1, \gamma_0]$ the matrix $A - \gamma BC$ has at least two eigenvalues with positive real parts and the matrix Γ has at least one eigenvalue in the region $[\gamma_1, \gamma_0]$.

A3. The matrix $A^{-1}BC$ has no real eigenvalues in the interval $[\gamma_0^{-1}, \infty)$.

Then there exist a function $f : \mathbb{R}^n \to \mathbb{R}^n$ such that

$$A = \frac{\partial f}{\partial x}(0)$$

and

1) The system $\dot{x} = f(x)$ has a unique globally asymptotically stable equilibrium (origin).

2) The system consisting of k diffusively coupled systems

$$\begin{cases} x_j = f(x_j) + Bu_j \\ y_j = Cx_j \\ u_j = -\gamma_{j1}(y_j - y_1) - \ldots - \gamma_{jk}(y_j - y_k) \end{cases}$$

is ultimately bounded, has a unique equilibrium and Y-oscillatory.

As one can see conditions imposed in Theorems 3, 6 ensure that the trivial solution of the coupled systems undergoes a Poincaré-Andronov-Hopf bifurcation for some critical value of γ. In the proof the cubic type nonlinearity we chose guarantees that this bifurcation is a global bifurcation. So one may wonder whether the cubic type nonlinearity is crucial to guarantee oscillations. Definetely not. As soon as the local behavior of the nonlinearity is described by the conditions from the theorems its global behavior should ensure 1) ultimate boundedness of solutions of coupled systems and 2) uniqueness of the equilibrium. A solution to the first problem can be derived from the results of section 4.

When parameters change a parametrized system with a unique equilibrium can gain an equilibrium in two ways. It can bifurcate from the existing one and it can appear far from the origin. The nonlinearity we chose provides only the first way which significantly simplifies the proofs and allows us to concentrate on the local behavior of the coupled systems. So in general the condition we presented are local and not sufficient, although they give a good understanding of the mechanism to bring about oscillations via diffusion. Now let us briefly scketch a way to solve problem 2 for general systems. A region in the state space where the graph of nonlinearity lies can be described for example by a set of linear constraints. Therefore for the *class* of nonlinearities satisfying these constraints the system has a unique equilibrium if and only if the set of linear strict and nonstrict inequalities has no solutions. This problem can be easily solved by linear programming.

It is worth mentioning that the methodology presented for the analysis of diffusively coupled systems can be extended to the case of multiple equilibria. However, in this case the complexity to check all conditions increases with the number of equilibria. Nevertheless in such systems it is possible to observe a quite interesting phenomenon when the coupled systems exhibit chaotic oscillations [Kocarev & Janić, 1995].

6. CONCLUSION

In the paper we presented conditions guaranteeing existence of oscillatory behavior in diffusively coupled systems while each system alone is globally asymptotically stable. A motivation for our study was the problem posed by Smale [1976]. He wrote: "There is a paradoxical aspect to the example. One has two dead (mathematically dead) cells interacting by a diffusion process which has a tendency in itself to equalize the concentrations. Yet in interaction, a state continues to pulse indefinitely".

From the proof of Theorem 3 one can make some important observations. First it can been shown that if all $\gamma_{ij}, i \neq j$ are equal to each other then the gain which ensures destabilization decreases with the growth of the number of interacting systems. Therefore the number of interacting systems can be considered as a bifurcation parameter. In other words, it is possible to create an oscillatory behavior in the "colony of dead cells" if we add into the colony an additional cell which diffusively interacts to the other cells. Moreover, in the diffusive medium consisting of diffusively coupled systems it is possible to observe locally generated spatial phenomena. Indeed, we have shown that the interconnected systems will exhibit an oscillatory behavior if the largest eigenvalue of the matrix Γ exceeds some threshold bifurcation value. Therefore the local changes in one parameter γ_{ij} which describes connection between ith and jth cells can result in propagation of oscillatory behavior through the whole diffusive medium. With this in mind it is interesting to notice that the real living cells have an ability to control diffusion between them. This control involves such processes as *facilitated diffusion, active transport* and transport through *coupled channel* (see [Raven & Johnson, 1992]).

REFERENCES

Byrnes, C.I. and A.Isidori [1991] "Asymptotic stabilization of minimum- phase systems," *IEEE Trans. on Automatic Control,* vol. **36**(10), pp. 1122-1137.

Byrnes, C.I., A. Isidori and J.C. Willems [1991] "Passivity, feedback equivalence and the global stabilization of minimum phase nonlinear systems", *IEEE Trans. Aut. Contr.*, vol. **36**(11), pp. 1228–1240.

Coddington, E.A. and N. Levinson [1955] *The theory of ordinary differential equations.* (New York: McGraw Hill).

Isidori, A. [1995] *Nonlinear control systems.* 3nd edition. (Springer-Verlag).

Kocarev, L.M. and P.A. Janić [1995] "On Turing instability in two diffusively coupled systems" *IEEE Trans. Circ. Sys., I,* vol. **42**(10), pp. 779–784.

Levinson, N. [1944] "Transformation theory of nonlinear differential equations of the second order", *Ann. Math.*, vol. **45**(4), pp. 723–737.

Pogromsky, A. Yu. [1998] "Passivity based design of synchronizing systems" *Int. J. Bifurcation and Chaos*, vol. **8**(2), pp. 295–319.

Raven, P.H. and G.B. Johnson [1992] *Biology*, Third edition, (Mosby Year Book–St. Louis).

Smale, S. [1976] "A mathematical model of two cells via Turing's equation," pp. 354–367, in Marsden, J.E. and M. McCracken (eds.) [1976] *The Hopf bifurcation and its applications*, Springer-Verlag, New York.

Stewart, G.W. and Ji-Guang Sun [1990] *Matrix Perturbation Theory,* Academic Press.

Tomberg, E.A. and V.A. Yakubovich [1989] "Self-oscillatory conditions in nonlinear systems", *Siberian Math. Journal,* v. **30**(4), pp. 180–194, in Russian.

Turing, A. [1952] "The chemical basis of morphogenesis", *Phylos. Trans. Royal Soc.* (B), vol. **237**, pp. 37–72.

We Ming-Ni [1998] "Diffusion, cross-diffusion, and their spike-layer steady states" *Notices of the American Mathematical Society,* vol. **45**(1), pp. 9–18.

Yakubovich [1973] "Frequency-domain conditions of self-oscillations in nonlinear systems with one stationary nonlinearity," *Siberian Math. Journal,* v. **14**(5), pp. 1100–1129, in Russian.

14th Triennial World Congress, Beijing, P.R. China

EMPIRICAL MODEL REDUCTION OF CONTROLLED NONLINEAR SYSTEMS

Sanjay Lall, Jerrold E. Marsden, Sonja Glavaški

Control and Dynamical Systems
California Institute of Technology
Pasadena, CA 91125, USA.
sanj@cds.caltech.edu

Abstract: In this paper we introduce a new method of model reduction for nonlinear systems with inputs and outputs. The method requires only standard matrix computations, and when applied to linear systems results in the usual balanced truncation. For nonlinear systems, the method makes used of the Karhunen-Loève decomposition of the state-space, and is an extension of the method of empirical eigenfunctions used in fluid dynamics. We show that the new method is equivalent to balanced-truncation in the linear case, and perform an example reduction for a nonlinear mechanical system. *Copyright © 1999 IFAC*

Full length paper: http://www.cds.caltech.edu/~sanj/home/publications/ifac99_empmr

Keywords: Karhunen-Loève decomposition, balanced truncation, model reduction, principal component analysis, total least-squares, Hankel norm.

1. Introduction

One of the difficulties in designing controllers for complex physical systems is in the problem of *modeling*. Very often, we have mathematical models which are highly nonlinear partial differential equations, for which control design techniques are little understood.

An example is given by formation flight; accurate models are known for the fluid dynamics, and for the aircraft motion, which together can be accurately simulated. However, on paper, the equations of motion consist of partial and ordinary differential equations coupled via their boundary conditions, a formulation which offers little to the control designer. It is therefore a problem of considerable interest to construct explicit low-order models.

The key difference between modeling for control, and modeling for analysis of dynamical system behavior is that in the latter case the system behavior is *autonomous*, with no external inputs driving the system. However, in the former case of controlled systems, we are concerned with preserving the relationship between the system behavior and the system inputs and outputs, or actuators and sensors. *Linear* input-output systems enjoy several model reduction techniques, including modal truncation, balanced truncation, and optimal Hankel, H_∞ and H_2 norm methods.

One method which has been used extensively for dynamical systems analysis is the Karhunen-Loève decomposition, in particular in the construction of models for complex fluid flows [6]. In this paper, we extend this procedure to show how it may be used for construction of explicit, low order nonlinear models for complex *controlled* nonlinear systems. The low order model is constructed in such a way as to preserve important features of the input-output behavior, so that it may be used for control design.

The central idea of applying the Karhunen-Loève decomposition for model reduction is to search for a low dimensional affine subspace of the state space, in which the dynamics of interest of the original system are contained. In general a subspace containing the interesting state trajectories may not have low (or finite) dimension, and the decomposition provides a way of finding the best approximating low-dimensional subspace. Once such a subspace is found, a Galerkin projection can be applied to project the dynamics onto it, so that the high-dimensional system is approximated by a small number of nonlinear ordinary differential equations.

Once an appropriate measure is assigned to the space of trajectories of the system, the problem of finding the subspace can be formulated as a 2-norm optimization problem. This has the advantage of requiring only linear matrix computations, despite its usefulness in application to nonlinear systems. The optimization need not be performed analytically; instead data from experiment or simulation can be used directly.

The model reduction method we describe in this paper combines the features of the Karhunen-Loève decomposition and balanced-truncation. This provides an immediate computable procedure for model reduction of nonlinear input-output systems, requiring only linear matrix computations. For autonomous systems, it corresponds to finding the subspace which most nearly contains the dynamics of the system. When applied to linear input-output systems, it results in the usual balanced truncation of the system, with the accompanying a-priori error bounds in the H_∞ and Hankel norms.

We conclude the paper with an example to illustrate the application of the method to a nonlinear mechanical system.

2. Model reduction for autonomous systems

In this section we give an outline of existing techniques using the Karhunen-Loève decomposition for model reduction of nonlinear autonomous uncontrolled systems of the form

$$\dot{x}(t) = f(x(t)) \qquad (1)$$

where $x(t) \in \mathbb{R}^n$. This procedure was used for the analysis of turbulence by Lumley [9].

The method makes essential use of *empirical data*, taken either from experiments or from numerical simulation, consisting of sampled measurements $\{x^{(1)}, \ldots, x^{(N)}\}$ of $x(t)$. The next step is to perform a principal component analysis of this data, to find how well it may be approximated by projection onto a k-dimensional subspace of the original n-dimensional state space.

The Karhunen-Loève decomposition. The Karhunen-Loève decomposition provides a method for finding this best approximating subspace. We give here a brief outline of the method, which goes back to Karl Pearson [11]; for details see [6].

We can characterize the subspace $S \subset \mathbb{R}^n$ by the projection operator Q mapping $\mathbb{R}^n$ onto S. We would like to find Q to minimize

$$H(Q) = \sum_{i=1}^{N} \|x^{(i)} - Qx^{(i)}\|_2^2$$

the total squared perpendicular distance of the points from the k-plane. The following result is standard.

Theorem 1. (Total Least Squares) *Let R be the correlation matrix of the data, defined by*

$$R := \sum_{i=1}^{N} x^{(i)} x^{(i)*}$$

and let $\lambda_1 \geq \lambda_2 \geq \cdots \lambda_n$ be the ordered eigenvalues of R. Then

$$\min_Q H(Q) = \sum_{i=n-k+1}^{n} \lambda_i$$

where the minimum is over all rank k projections Q.

In general, R may not have rank n, if the given data lies within a strict subspace of $\mathbb{R}^n$. Let $s = \operatorname{rank} R$, and let $\phi_1, \phi_2, \ldots, \phi_s$ be orthonormal eigenvectors of R, corresponding to the nonzero λ_i. Each $x^{(i)}$ can be written as

$$x^{(i)} = \sum_{j=1}^{s} a_{ij}\phi_j$$

where $a_{ij} = \langle x^{(i)}, \phi_j \rangle$, and $\langle \phi_i, \phi_j \rangle = \delta_{ij}$. The optimal k-dimensional subspace approximant is given by

$$\hat{x}^{(i)} = \sum_{j=1}^{k} a_{ij}\phi_j$$

Denote by P the $k \times n$ matrix whose rows are $\phi_1, \ldots, \phi_k$, so that $PP^* = I$. The projected approximant to x is given by $P^*Px \in S$, and $y = Px$ is a representation in terms of the new coordinates ϕ_i on S. This subspace approximant is then optimal, in the sense that the total energy (2-norm) in the subspace is given by

$$\sum_{i=1}^{N} \|Px^{(i)}\|_2^2 = \sum_{j=1}^{k} \lambda_j$$

and this is the maximum achievable by any k-plane.

Model reduction. The above procedure finds the optimal subspace; in order to allow affine variation we make use of the fact that the optimal *affine* subspace passes through the mean of the data. Hence, we construct the correlation matrix as

$$R_{ij} = \sum_{p=1}^{N} (x_i^{(p)} - \bar{x}_i)(x_j^{(p)} - \bar{x}_j)$$

where $\bar{x} = \frac{1}{N}\sum_{i=1}^{N} x^{(i)}$ is the mean of the data. The eigenvalues of R now provide us with information as to how close an approximation of the data is provided by a k-dimensional subspace; the goal is to choose k such that the fraction of the total energy in the subspace

$$\sum_{i=1}^{k} \lambda_i \Big/ \sum_{i=1}^{n} \lambda_i$$

is close to one, yet k is sufficiently small. Clearly this will not always be possible, with models which are better approximated by low-dimensional systems having relatively few large eigenvalues.

We now subtract the mean $\bar{x}$ from the dynamic equation, defining $e = x - \bar{x}$ to arrive at $\dot{e}(t) = f(e(t) + \bar{x})$. We would like to approximate $e(t)$ in this equation by $P^*Pe(t) \in S$.

Galerkin projection. The Galerkin projection has been used extensively to construct numerical solutions to partial differential equations. It can also be used to construct lower order *mathematical models* of a given dynamical system. The idea is to replace the given dynamics by an associated dynamics on a k-dimensional subspace $S \subset \mathbb{R}^n$ of the original state space, by projecting the vector field onto the tangent space of S. Using the coordinates defined previously, the resulting reduced order approximation is given by

$$\dot{y}(t) = Pf(P^*y(t) + \bar{x})$$

Clearly the subspace obtained by the Karhunen-Loève method depends upon the inner product on the Euclidean space in which the state space is embedded; physical meaning is an important indicator of the appropriate choice of inner product.

Computational application of this method requires only standard matrix computations, despite its application to nonlinear systems. Also the method has the feature that it separates the model reduction into two parts; that of finding a suitable subspace and that of performing the projection. This lends physical intuition to the procedure, and allows alternative projection schemes to be used.

We can expect such a procedure to work well for model reduction of the system within some given region of state space, and it is within such a specific region that data should be collected. In doing this, one must pay careful attention to symmetry, as in Glavaški [4] and references therein.

For control, a severe limitation is that the input-output behavior of the system is not taken into account. It is the generalization of the Karhunen-Loève reduction procedure to controlled systems that is the main focus of this paper, and we now proceed with this.

3. Model reduction of controlled systems

We now turn to the main problem addressed in this paper; that of model reducing nonlinear input-output systems of the form

$$\begin{aligned} \dot{x}(t) &= f(x(t), w(t)) \\ z(t) &= h(x(t)) \end{aligned} \tag{2}$$

Here $x(t) \in \mathbb{R}^n$ is the state of the system, $w(t) \in \mathbb{R}^p$, and $z(t) \in \mathbb{R}^q$. The function w is regarded as an input signal to the system, and the function z as an output signal. The goal of *model reduction* is to construct another nonlinear system of differential equations

$$\begin{aligned} \dot{q}(t) &= \hat{f}(q(t), w(t)) \\ z(t) &= \hat{h}(q(t)) \end{aligned}$$

where $q(t) \in \mathbb{R}^k$, and $k < n$, such that the *input-output behavior* of the two systems is similar, for states in some specific region of the the state space. That is, for a class of inputs w, we would like the outputs z of the two systems to be similar. Before proceeding further, we need some preliminary definitions from the theory of linear systems.

3.1. Linear systems

The system (2) is called *linear* if it has a representation

$$\begin{aligned} \dot{x}(t) &= Ax(t) + Bw(t) \\ z(t) &= Cx(t) \end{aligned} \tag{3}$$

where $x(t) \in \mathbb{R}^n$, and A, B, and C are matrices of appropriate dimension. This representation is known as the *state-space realization*, and the system is described by the triple (A, B, C). The linear system is called *stable* if the eigenvalues of A all have real part strictly negative.

Suppose the system in equation (3) is stable. Then, for $u \in L_2(-\infty, 0]$ and denoting $u_-(t) = u(-t)$, the state at time zero $x(0) = x_0$ is given by

$$x_0 = \int_0^\infty e^{As} B u_-(s)\, ds.$$

This defines the *controllability operator*, $\mathcal{C} : L_2[0,\infty) \to \mathbb{R}^n$ by $x_0 = \mathcal{C}u_-$. The following results can be found in [7].

Lemma 2. *Write $Y := \mathcal{C}\mathcal{C}^*$. Then Y is the smallest semipositive solution to the Lyapunov equation*

$$AY + YA^* + BB^* = 0. \tag{4}$$

The system is called controllable *if* $\operatorname{Im} \mathcal{C} = \mathbb{R}^n$, *in which case $Y > 0$ and equation (4) has a unique solution. The matrix Y is known as the* **controllability gramian.**

We can define a similar notion for the output. Denote the future output by $y_+(t) \in L_2[0,\infty)$. Then, define the *observability operator*, $\mathcal{O} : \mathbb{R}^n \to L_2[0,\infty)$ by $y_+(t) = \mathcal{O}x_0$, and hence $\mathcal{O}x_0 = Ce^{At}x_0$.

Lemma 3. *Write $X := \mathcal{O}^*\mathcal{O}$. Then X is the smallest semipositive solution to the Lyapunov equation*

$$A^*X + XA + C^*C = 0. \tag{5}$$

The system is said to be observable *if* $\ker \mathcal{O} = \{0\}$, *in which case $X > 0$ and equation (5) has a unique solution. The matrix X is known as the* **observability gramian.**

Both X and Y are $n \times n$ matrices, and they are given by the following integral formulae.

$$Y = \int_0^\infty e^{At} BB^* e^{A^*t}\, dt$$

$$X = \int_0^\infty e^{A^*t} C^* C e^{At}\, dt$$

3.2. Gramians and principal component analysis

The method of principal component analysis relies on the use of data to construct the correlation matrix, with the underlying assumption that this data is collected from 'typical' system trajectories. We make the assumption that the initial state of the system is zero, and parametrize the trajectories with respect to the system input u.

For theoretical purposes, we are not restricted to samples, and can simply construct the correlation matrix using the integral

$$R = \int_0^\infty (x(t) - \bar{x})(x(t) - \bar{x})^*\, dt \tag{6}$$

where $x(t)$ is the state of the system at time t, and $\bar{x}$ is the mean state. We now apply a Karhunen-Loève decomposition to the states of a nonlinear input-output system of the form given by equation (2), parametrized by the inputs.

Define the following sets:

$$\begin{aligned} \mathcal{T}^n &= \{T_1, \ldots, T_r;\ T_i \in \mathbb{R}^{n\times n},\ T_i^* T_i = I,\ i = 1, \ldots r\} \\ \mathcal{M} &= \{c_1, \ldots, c_s;\ c_i \in \mathbb{R},\ c_i > 0,\ i = 1, \ldots s\} \\ \mathcal{E}^n &= \{e_1, \ldots, e_n;\ \text{standard unit vectors in } \mathbb{R}^n\} \end{aligned}$$

Here $\mathcal{T}$ is an arbitrary set of r orthogonal matrices, and $\mathcal{M}$ is a set of s positive constants. Given a function $u \in \ell_\infty$, define the mean $\bar{u}$ by

$$\bar{u} := \lim_{T\to\infty} \frac{1}{T} \int_0^T u(t)\, dt$$

We make the standing assumptions that $x, z \in \ell_\infty$ and $z \in \ell_2$. For given initial conditions, these assumptions are satisfied for stable linear systems, and exponentially stable nonlinear systems, subject to a suitable definition of exponential stability.

Definition 4. *Let $\mathcal{T}^p$, $\mathcal{E}^p$ and $\mathcal{M}$ be given sets as described above. For the system (2), define the* ***empirical controllability gramian*** $\hat{Y}$ *by*

$$\hat{Y} = \sum_{l=1}^{r}\sum_{m=1}^{s}\sum_{i=1}^{p} \frac{1}{rsc_m^2}\int_0^\infty \Phi^{ilm}(t)\,dt$$

where $\Phi^{ilm}(t) \in \mathbb{R}$ is given by

$$\Phi^{ilm}(t) := (x^{ilm}(t) - \bar{x}^{ilm})(x^{ilm}(t) - \bar{x}^{ilm})^*$$

and $x^{ilm}(t)$ is the state of the system (2) corresponding to the impulsive input $w(t) = c_m T_l e_i \delta(t)$.

Lemma 5. *For any nonempty sets $\mathcal{T}^p$ and $\mathcal{M}$, the empirical controllability gramian $\hat{Y}$ of the stable linear system $\dot{x}(t) = Ax(t) + Bw(t)$ is equal to the usual controllability gramian Y.*

Proof. For the linear system,

$$\begin{aligned}\Phi^{ilm}(t) &= c_m^2(e^{At}BT_le_i)(e^{At}BT_le_i)^* \\ &= c_m^2 e^{At}BT_le_ie_i^*T_l^*B^*e^{A^*t}\end{aligned}$$

hence

$$\begin{aligned}\hat{Y} &= \int_0^\infty \sum_{l=1}^{r}\sum_{m=1}^{s}\sum_{i=1}^{p} \frac{1}{rs} e^{At}BT_le_ie_i^*T_l^*B^*e^{A^*t}\,dt \\ &= \int_0^\infty \sum_{m=1}^{s} \frac{1}{s} e^{At}BB^*e^{A^*t}\,dt = Y\end{aligned}$$

which is the desired result. ∎

A variant of this result was first shown by Moore [10]; for impulsive inputs distributed in the above sense on the unit ball in $\mathbb{R}^p$, the Karhunen-Loève decomposition of the states leads to a construction of the controllability gramian for linear systems.

The empirical controllability gramian is a computable generalization of this to nonlinear systems; it has the property that the eigenvectors of $\hat{Y}$ corresponding to non-zero eigenvalues span a subspace $O \subset \mathbb{R}^n$ which contains the set of states reachable using the chosen initial impulsive inputs.

However, for controlled systems, simply studying the input-state behavior is not enough. The next definition is the analogue of the previous one for the output behavior.

Definition 6. *Let $\mathcal{T}^n$, $\mathcal{E}^n$, and $\mathcal{M}$ be given sets as described above. For the system (2), define the* ***empirical observability gramian*** $\hat{X}$ *by*

$$\hat{X} = \sum_{l=1}^{r}\sum_{m=1}^{s} \frac{1}{rsc_m^2}\int_0^\infty T_l\Psi^{lm}(t)T_l^*\,dt$$

where $\Psi^{lm}(t) \in \mathbb{R}^{n\times n}$ is given by

$$\Psi^{lm}_{ij}(t) := (z^{ilm}(t) - \bar{z}^{ilm})^*(z^{jlm}(t) - \bar{z}^{jlm})$$

and $z^{ilm}(t)$ is the output of the system (2) corresponding to the initial condition $x_0 = c_m T_l e_i$.

Lemma 7. *For any nonempty sets $\mathcal{T}^n$ and $\mathcal{M}$, the empirical observability gramian $\hat{X}$ of the stable linear system $\dot{x}(t) = Ax(t) + Bw(t)$, $z(t) = Cx(t)$ is equal to the usual observability gramian X.*

Proof. For the linear system,

$$\begin{aligned}\Psi^{lm}_{ij}(t) &= c_m^2(Ce^{At}T_le_i)^*(Ce^{At}T_le_j) \\ &= c_m^2 e_i^*T_l^*e^{A^*t}C^*Ce^{At}T_le_j\end{aligned}$$

hence

$$\Psi^{lm}(t) = c_m^2 T_l^* e^{A^*t}C^*Ce^{At}T_l$$

and

$$\hat{X} = \sum_{m=1}^{s} \frac{1}{s}\int_0^\infty e^{A^*t}C^*Ce^{At}\,dt = X$$

which is the desired result. ∎

We now have the tools we need for empirical analysis of the input-output behavior of the nonlinear system. Rather than searching for exact controllability and observability submanifolds within the state space, our approach is to search for subspaces which approximate these manifolds. The advantage of this approach lies in the computation; all that is required is the solution of standard linear matrix eigenvalue problems.

Unlike in the autonomous case, we now have *two* important subspaces of the state space, and their corresponding eigenvalues. We can proceed in the same manner as for linear systems, and make use of the ideas of balanced realization theory to decide on which subspace to project. We therefore return to a description of linear systems theory.

3.3. Model reduction via balanced truncation

In this section we give a brief overview of the method of balanced truncation.

The *Hankel operator* of a linear system is the operator which maps (the reflection of) past inputs to future outputs, assuming the future input is zero. Denoting as before $u_-(t) = u(-t)$, the Hankel operator, Γ, is given by the map $\Gamma := \mathcal{O}\mathcal{C}$ so that $y_+ = \Gamma u_-$.

Since $x_0 = \mathcal{C}u_-$ and $y_+ = \mathcal{O}x_0$, it is immediate that the Hankel operator has at most rank n, equal to the state space dimension of the system. A singular value decomposition of this linear operator leads to n singular values, known as the *Hankel singular values*, which are independent of the particular state space realization of the system, and are a property of the system as an input-output map.

The Hankel singular values are the eigenvalues of $\Gamma\Gamma^*$. These are the eigenvalues of $\mathcal{O}\mathcal{C}\mathcal{C}^*\mathcal{O}^*$, which are the same as those of the $n \times n$ matrix $\mathcal{O}^*\mathcal{O}\mathcal{C}\mathcal{C}^* = XY$. The *balanced realization* gives a way to find which particular states correspond to which Hankel singular value.

Definition 8. *Suppose A is stable. The realization (A, B, C) is called* ***balanced*** *if the controllability gramian Y and the observability gramian X are equal and diagonal.*

We assume throughout the following that the system (A, B, C) is stable, controllable, and observable. In this case, there will always exist a nonsingular state transformation T such that the new realization (TAT^{-1}, TB, CT^{-1}) is balanced.

This leads to a method of model reduction known as *balanced truncation*, introduced by Moore [10] in the context of realization theory. The procedure is to truncate those states from the balanced realization corresponding to small Hankel singular values σ_i. If the states are ordered according to decreasing singular value, this is equivalent to applying a Galerkin projection to the balanced realization, where $P = \begin{bmatrix} I & 0 \end{bmatrix}$.

If (A, B, C) is balanced, the truncated system is given by

$$\dot{y}(t) = PAP^*y(t) + PBu(t)$$
$$z(t) = CP^*y(t)$$

The reduced-order model constructed via balanced truncation is independent of the original state space realization. A-priori error bounds in the induced 2-norm are known for the error between the original and the reduced system; these were first shown independently by Enns [3] and Glover [5].

3.4. Empirical balanced truncation

The empirical gramians give a precise method for deciding upon the importance of particular subspaces of the state space, with respect to the inputs and outputs of the system.

We propose to use these for model reduction in the same way as for linear systems; find a linear change of coordinates such that the empirical gramians are balanced, and perform a Galerkin projection onto the states corresponding to the largest eigenvalues of $\hat{X}\hat{Y}$.

Since for linear systems, the empirical gramians are exactly the usual gramians, this method is exactly balanced truncation when it is applied to a linear system. When applied to a nonlinear system, it requires only matrix computations, and results in a new nonlinear model.

Let T be the change of coordinates such that the system is balanced; that is $TYT^* = T^{-1*}XT^{-1} = \Sigma$, and let $P = \begin{bmatrix} I & 0 \end{bmatrix}$ be the $k \times n$ projection matrix. Applying the previous analysis leads to a reduced order model given by

$$\dot{y}(t) = PTf(T^{-1}P^*y(t), w(t))$$
$$z(t) = h(T^{-1}P^*y(t))$$

This is the final reduced order model. This *empirical balanced truncation* gives a reduced order model which takes into account the input-output behavior, and is directly computable from data.

4. Computation

We can directly apply Definitions 4 and 6 for construction of the empirical gramians from data, for nonlinear systems. The data may be taken either from simulation or from experiment. All that is needed is numerical approximation of the above integrals, using sampled-data from both $x(t)$ and $z(t)$. We can take advantage of knowledge of f to calculate $\dot{x}(t)$ in this computation to achieve higher order accuracy, and increase the allowable sample time.

To perform the required experiments, it is necessary to be able to set the initial state of the system, to measure the state at all times, and to apply approximately impulsive inputs. These conditions may be satisfied for certain mechanical systems, although in the case of fluid dynamics, for example, we are forced to rely on data from simulation.

We also need to choose the sets $\mathcal{T}$ and $\mathcal{M}$. A reasonable simple choice is $\mathcal{T} = \{I, -I\}$, since this corresponds to using both positive and negative inputs (or initial states) on each input separately. For distributed actuators, larger sets may be justified.

The sets $\mathcal{M}$ specify the size of inputs and states we are interested in. The choice is motivated by the magnitudes of the inputs and states actually seen in experiments.

As for autonomous systems, a crucial feature of the Karhunen-Loève procedure is that in general we should expect it to work well only in *limited regions of state space*. We do not expect to be able to approximate global behavior of the system well using a linear projection; however, it is perhaps more reasonable to expect to be able to construct low order models for systems within some given operating region.

A simple technique for balancing the empirical gramians $\hat{X}$ and $\hat{Y}$ is as follows. First, construct a Cholesky factorization of $\hat{Y}$ so that $\hat{Y} = ZZ^*$. Let $U\Sigma^2U^*$ be a singular value decomposition of $Z^*\hat{X}Z$, and let $T = \Sigma^{\frac{1}{2}}U^*Z^{-1}$. Then $T\hat{Y}T^* = \Sigma$, and $T^{-1*}\hat{X}T^{-1} = \Sigma$, as desired.

We can now change state coordinates of the nonlinear system, and truncate using the Galerkin projection. The rows of T give the modes of the system.

5. Mechanical links example

Not only are high-order systems hard to control, but it is difficult to develop intuition as to their behavior. One of the advantages of the procedure developed here is that it can be viewed as a selection of appropriate nonlinear mode shapes on which to project the dynamics. These mode shapes often are physically meaningful, and in this section we give examples for a simple, mildly nonlinear mechanical system.

The system considered here consists of five rigid rods, connected via torsional springs and dampers in the plane. The lowest rod is pinned to ground with a torsional spring, so that the system has a stable equilibrium in the upright vertical position.

The system has a single input, a torque about the lowest pin joint, and a single output, the horizontal displacement of the end of the last rod from the vertical symmetry axis.

5.1. Linearized model

We first analyze the linearization of the system about its stable equilibrium; Figure 1 shows the first four of the ten mode shapes and the corresponding singular values of the balanced realization. The singular values have been normalized so that they sum to one.

5.2. Nonlinear Lagrangian model

Application of the above empirical model reduction procedure to the nonlinear Lagrangian model of this system leads to a set of corresponding modes. For a particular choices of the magnitude constants c_i, Figure 2 shows these modes; we can see that the first two mode shapes have split into three.

Figure 3 shows the mode shapes at an increased value for the magnitude constants; the mode shapes become significantly different from those of the linearization.

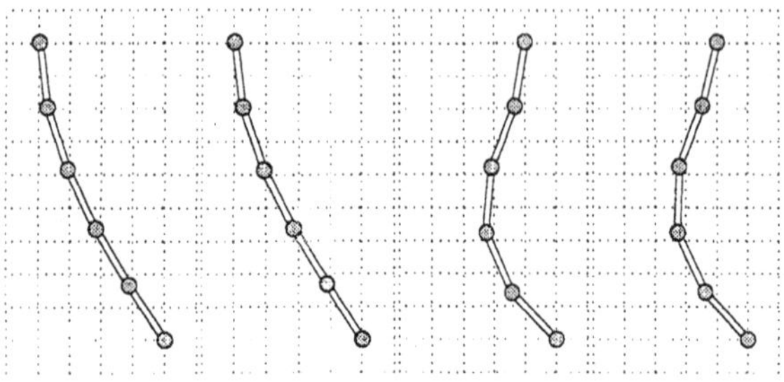

Figure 1: The balanced modes of the linearization, with the corresponding fraction of energy in each mode.

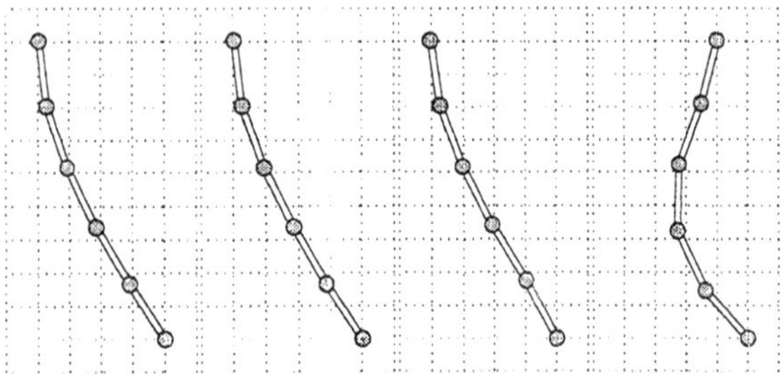

Figure 2: The empirical balanced modes of the nonlinear model, for small inputs.

6. Conclusions

In this paper we have presented new methods for model reduction of nonlinear systems. These methods combine the physical intuition of Karhunen-Loève techniques, as they have been applied to analysis of complex flows, with the control engineering methodology of balanced truncation.

The resulting method of model reduction is based on data. It requires only simple matrix computations, and may be applied directly to nonlinear systems. When applied to linear systems, the reduced order model constructed is exactly the usual balanced truncation of the system.

The choice of impulsive inputs, and initial states distributed uniformly on the unit sphere, are perhaps somewhat arbitrary methods of parametrizing trajectories for nonlinear systems. Many other methods might be suggested, and for control a very reasonable strategy would be to apply a Karhunen-Loève decomposition of the states seen during simulation of *closed-loop* behavior. There are several other possibilities that might be suggested; our focus here has been the construction of a computable extension of balanced truncation for nonlinear systems.

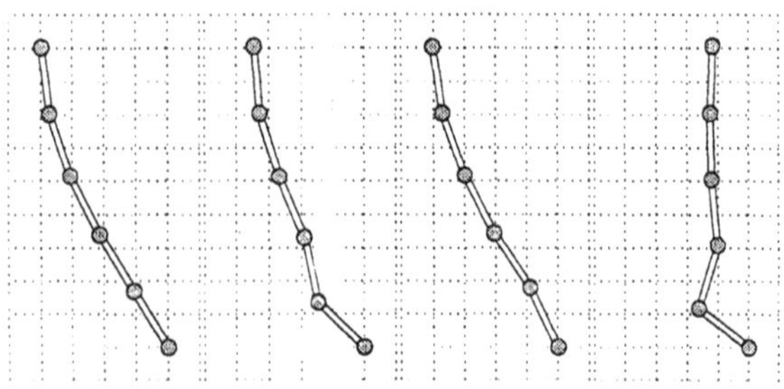

Figure 3: The empirical balanced modes of the nonlinear model, for large inputs.

For mechanical systems, a better approach is to perform the Karhunen-Loève expansion on the configuration space rather than the phase space, preserving the mechanical structure of the system in the reduction [8]. Similarly, taking account of symmetry [4] and other special structure present in the original system can be of great use, leading to reduced computational requirements and producing more accurate reduced-order models.

One advantage of the empirical procedure described in this paper is that often the simulation itself has been implemented using known heuristics; for example vortex methods are often used to simulate incompressible flow. Using data from such a simulation for model reduction, these heuristics can be incorporated into the control design process.

References

[1] G. Berkooz, P. Holmes, and J. L. Lumley. The proper orthogonal decomposition in the analysis of turbulent flows. *Annual Review of Fluid Mechanics*, 25:539–575, 1993.

[2] A. M. Bloch. Estimation, principal components, and Hamiltonian systems. *Systems and Control Letters*, 6:103–108, 1985.

[3] D. F. Enns. *Model reduction for control system design*. PhD thesis, Stanford university, 1984.

[4] S. Glavaški, J. E. Marsden, and R. M. Murray. Model reduction, centering, and the Karhunen-Loève expansion. *IEEE Conference on Decision and Control*, 1998.

[5] K. Glover. All optimal Hankel-norm approximations of linear multivariable systems and their L-infinity error bounds. *International Journal of Control*, 39:1115–1193, 1984.

[6] P. Holmes, J. L. Lumley, and G. Berkooz. *Turbulence, coherent structures, dynamical systems, and symmetry*. Cambridge University Press, 1996.

[7] T. Kailath. *Linear Systems*. Prentice-Hall, 1980.

[8] S. Lall and J. E. Marsden. Model reduction of mechanical systems. IGPP/NPACI Workshop on Accurate Simulation and Modeling of Physical Systems, 1998.

[9] J. L. Lumley. The structure of inhomogeneous turbulence. In A. M. Yaglom and V. I. Tatarski, editors, *Atmospheric turbulence and wave propagation*, pages 166–178. Nauka, Moscow, 1967.

[10] B. C. Moore. Principal component analysis in linear systems: controllability, observability, and model reduction. *IEEE Transactions on Automatic Control*, 26(1):17–32, February 1981.

[11] K. Pearson. On lines and planes of closest fit to points in space. *Philosophical magazine*, 2:609–629, 1901.

[12] J. M. A. Scherpen. H-infinity balancing for nonlinear systems. *International Journal of Robust and Nonlinear Control*, 6:645–668, 1996.

14th Triennial World Congress, Beijing, P.R. China

E-2c-19-6

A SERIES DESCRIBING THE EVOLUTION OF MECHANICAL CONTROL SYSTEMS

Francesco Bullo *

* *Coordinated Science Lab. and General Engineering*
University of Illinois at Urbana-Champaign
1308 W. Main St, Urbana, IL 61801, USA
Telephone: (217) 333-0656, Fax: (217) 244-1653
Email: bullo@uiuc.edu, *Url:* http://motion.csl.uiuc.edu

Abstract: We compute a series describing the evolution of a mechanical system starting at rest and subject to a time-varying external force. This generalizes various previous results and lays the foundation for the design of motion control algorithms for a large class of autonomous vehicles, robotic manipulators and locomotion devices.

Keywords: mechanical systems, motion planning, nonlinear controllability

1. INTRODUCTION

Underactuated mechanical systems provide a challenging research area of increasing interest in both application and theory. In this paper, we examine an important class of mechanical control systems, including autonomous vehicles and robotic locomotion devices.

A rich literature is available on the motion planning problem for kinematic systems, that is systems without drift. Numerous approaches include chained systems (Murray and Sastry, 1993), systems on Lie groups (Leonard and Krishnaprasad, 1995) and (Kolmanovsky and McClamroch, 1996), and the general solutions proposed in (Lafferriere and Sussmann, 1991). The enabling step in these works is the characterization of the evolution of the control system. In other words, the basic contributions of these papers is the computation of a "input history to final displacement" map. In its most general formulation this problems is solved by the Chen-Fliess-Sussmann series in (Sussmann, 1986) and by the logarithmic series in (Agračhev and Gamkrelidze, 1978). Motion control algorithms are then designed by inverting this "inputs to displacement" map.

Unfortunately, this body of literature is of limited applicability to a large class of mechanical systems that present dynamics. Recently, some progress in this direction has been obtained in (Bullo and Leonard, 1997) and (Bullo *et al.*, 1997). Under the assumption of small amplitude forcing, the authors compute the initial terms of a Taylor series describing the forced evolution. This result is related to the controllability analysis in (Lewis and Murray, 1997).

The main contribution of the paper is a series that describes the evolution of a forced mechanical system starting from rest. Mechanical systems are presented as second order systems on a configuration manifold. Instead of a series on the full phase space ($2n$ dimensional), the evolution is described as a flow on the configuration space Q. The treatment relies on some "chronological calculus" tools, see (Agračhev and Gamkrelidze, 1978), as opposed to the previous work based on the perturbation method. This improvement leads to numerous conceptual and computational advantages, one of which is the series this paper presents.

2. TOOLS FROM CHRONOLOGICAL CALCULUS

We present some basic result useful to describe composition and perturbation of flows of vector fields on manifolds. They are borrowed from the so-called "chronological calculus", as introduced in (Agračhev and Gamkrelidze, 1978), and employed in (Lafferriere and Sussmann, 1991; Kawski and Sussmann, 1997).

We begin by reviewing some notation:

(i) The order of composition of functions is $f(\phi(x)) = (f \circ \phi)(x)$.

(ii) x is a point on the manifold M, $f(x), g(x)$ are vector fields, and $[f, g](x) = \mathrm{ad}_f\, g(x)$ is their Lie bracket.

(iii) A non-autonomous differential equation on M is written as

$$\dot{x}(t) = f(x,t)$$
$$x(0) = x_0,$$

and the solution is denoted by

$$x(t) = \Phi^f_{0,t}(x_0).$$

(iv) If g is a vector field and ϕ is a diffeomorphism on M, the pull-back $\phi^* g$ is a vector field defined by

$$(\phi^* g)(x) \triangleq \left(T_x \phi^{-1} \circ g \circ \phi\right)(x).$$

2.1 *Variation of constants formula*

Consider the non-stationary differential equation

$$\dot{x}(t) = f(x,t) + g(x) \qquad (1)$$
$$x(0) = x_0,$$

where $f(x,t)$ and g are analytic vector fields. It is instructive to regard g as a perturbation to the vector field f and describe the flow of the previous differential equation in terms of a nominal and perturbed flow. As depicted in Figure 1, we have the following relationship between flows of vector fields (i.e., solutions of differential equations):

$$\Phi^{f+g}_{0,t} = \Phi^f_{0,t} \circ \Phi^{(\Phi^f_{0,t})^* g}_{0,t}. \qquad (2)$$

We formally state this result as follows.

Lemma 2.1. Let $\{x(t), t \in [0,T]\}$ be the solution to equation (1) and let $\{y(t), t \in [0,T]\}$ be the solution to

$$\dot{y}(t) = \left((\Phi^f_{0,t})^* g\right)(y) \qquad (3)$$
$$y(0) = x_0.$$

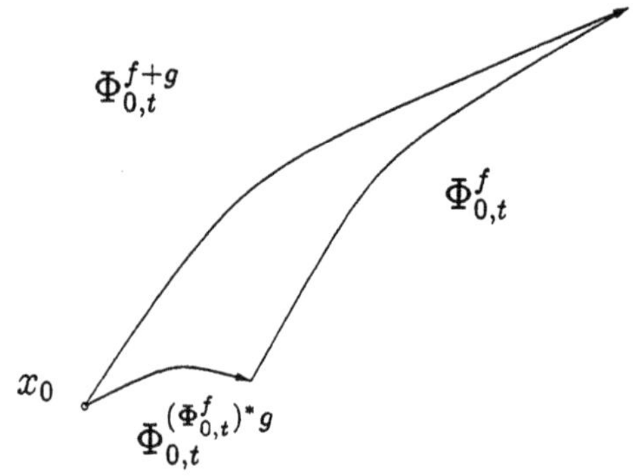

Fig. 1. The flow along $f + g$ is written as the composition of a flow along the pull-back system $(\Phi^f_{0,t})^* g$ and the flow along f.

Then it holds

$$x(T) = \Phi^f_{0,T}(y(T)).$$

2.2 *Formal expansions for the pull-back along a flow*

Motivated by equation (3) we investigate the pull-back of a vector field $g(x)$ along the flow of a time-varying vector field $f(x,t)$. It turns out that

$$\frac{d}{dt}\left(\Phi^f_{0,t}\right)^* g = \left(\Phi^f_{0,t}\right)^* [f(x,t), g(x)], \qquad (4)$$

where the Lie bracket between f and g is computed at t fixed. While this equality is a well-known fact for autonomous vector fields, it also holds for time-varying f. This statement is proved in (Agračhev and Gamkrelidze, 1978), see equation 3.3.

At fixed $x \in M$, we integrate equation (4) from time 0 to t to obtain

$$(\Phi^f_{0,t})^* g = g + \int_0^t \left((\Phi^f_{0,s})^* [f(s), g]\right) ds,$$

where we have dropped the argument x inside the integral in the interest of simplicity. We can iteratively apply the previous equality to obtain the formal expansion:

$$(\Phi^f_{0,t})^* g = g + \sum_{n=1}^{\infty} \int_0^t \cdots \int_0^{s_{n-1}} \left(\mathrm{ad}_{f(s_n)} \cdots \mathrm{ad}_{f(s_1)}\, g\right) ds_n \ldots ds_1. \qquad (5)$$

If the vector field f is time-independent, i.e., $f(x,t) = f(x)$, the previous formula reduces to the classic Campbell-Backer-Hausdorff formula:

$$(\Phi^f_{0,t})^* g = \sum_{n=0}^{\infty} \mathrm{ad}^n_f\, g \frac{t^n}{n!}.$$

3. MECHANICAL SYSTEMS AND HOMOGENEITY

In what follows we model mechanical systems on a configuration manifold Q. Because we are interested in local problems, we restrict our analysis to the case where $Q = \mathbb{R}^n$; we include some comments on the extension to the general manifold case. The fundamental feature of this setting is that we analyse *second order* differential equations, where the input is an acceleration (alternatively a force) and not a velocity.

Let $q = (q^1, \dots, q^n) \in \mathbb{R}^n$ be the configuration. We consider the following control system

$$\ddot{q}^i + \Gamma^i_{jk}(q)\dot{q}^j\dot{q}^k = \sum_{a=1}^{m} Y^i_a(q)u_a(t), \qquad (6)$$

where

(i) the summation convention is in place here and in what follows,
(ii) $\Gamma^k_{jk}(q)$ are n^3 analytic functions called the Christoffel symbols,
(iii) for $a = 1, \dots, m$, $Y_a(q)$ is an input vector fields and $Y^i_a(q,t)$ is its ith component with respect to the usual basis on $\mathbb{R}^n$. In what follows, we let

$$Y(q,t) = \sum_{a=1}^{m} Y_a(q)u_a(t).$$

Equation (6) describes a large class of mechanical systems with Lagrangian equal to kinetic energy, see (Lewis and Murray, 1997), with symmetries, see (Bullo *et al.*, 1997) and with nonholonomic constraints (Lewis, 1998; Lewis, 1997). For example, should the mechanical system be a robotic manipulator, then the Christoffel symbols can be easily computed via a well-known combination of partial derivatives of the inertia tensor.

3.1 *Lie algebraic structure*

We here review the Lie algebraic properties of the system (6). We start by rewriting the system in vector notation as

$$\frac{d}{dt}\begin{bmatrix} q \\ \dot{q} \end{bmatrix} = \begin{bmatrix} \dot{q} \\ -\Gamma(q,\dot{q}) \end{bmatrix} + \sum_{a=1}^{m} \begin{bmatrix} 0 \\ Y_a \end{bmatrix} u_a(t)$$

where $\Gamma(q,\dot{q})$ is the vector with components $\Gamma^i_{jk}(q)\dot{q}^j\dot{q}^k$. If we define $x = (q, \dot{q})$, and

$$Z_g = \begin{bmatrix} \dot{q} \\ -\Gamma(q,\dot{q}) \end{bmatrix} \quad \text{and} \quad Y^{\text{lift}}_a \triangleq \begin{bmatrix} 0 \\ Y_a \end{bmatrix},$$

then we can rewrite the system as

$$\dot{x} = Z_g(x) + \sum Y_a(x)^{\text{lift}} u_a.$$

Let $h_i(q,\dot{q})$ be the set of scalar functions on $\mathbb{R}^{2n}$, which are arbitrary functions of q and which are homogeneous polynomials in $\{\dot{q}^1, \dots, \dots q^n\}$ of degree i. Let $\mathcal{P}_i$ be the set of vector fields on $\mathbb{R}^{2n}$ which have the first n components in h_i and the second n components in h_{i+1}. It can be directly verified that

$$Z_g \in \mathcal{P}_1 \quad \text{and} \quad Y^{\text{lift}}_a \in \mathcal{P}_{-1}.$$

This set of vector fields has the useful properties:

(i) $\{[X,Y] \text{ s.t. } X \in \mathcal{P}_i, Y \in \mathcal{P}_j\} \subset \mathcal{P}_{i+j}$,
(ii) if $k \leq -2$, then $\mathcal{P}_k = \{0\}$,
(iii) if $k \geq 1$, then $X(q,0) = 0$ for all $X(q,\dot{q}) \in \mathcal{P}_k$.

On the base of these properties we investigate the Lie brackets between the drift Z_g and the inputs Y^{lift}_a. Iterating these brackets we have:

(1) $Z_g \in \mathcal{P}_1$ and $Y^{\text{lift}}_a \in \mathcal{P}_{-1}$.
(2) $[Z_g, Y^{\text{lift}}_a] \in \mathcal{P}_0$ and $[Y^{\text{lift}}_a, Y^{\text{lift}}_b] = 0$.
(3) $[Z_g, [Z_g, Y^{\text{lift}}_a]] \in \mathcal{P}_1$ and $[Y^{\text{lift}}_b, [Z_g, Y^{\text{lift}}_a]] \in \mathcal{P}_{-1}$,

and so forth, as the number of Lie brackets increases.

Remark 3.1. While the results in this section are presented in a coordinate dependent fashion, see also (Sontag and Sussmann, 1986), it is possible to turn them into geometric statements. Key concept is the notion of geometric homogeneity described in (Kawski, 1995) (mechanical systems are homogeneous with respect to the Liouville vector field).

3.2 *The symmetric product*

In this section we focus on the Lie bracket $[Y^{\text{lift}}_b, [Z_g, Y^{\text{lift}}_a]]$. Since this vector field belongs to $\mathcal{P}_{-1}$, there must exist a vector field on $\mathbb{R}^n$, which we denote $\langle Y_a : Y_b \rangle$, such that

$$\langle Y_a : Y_b \rangle^{\text{lift}} = [Y^{\text{lift}}_b, [Z_g, Y^{\text{lift}}_a]].$$

Such a vector field is called *symmetric product* between Y_b and Y_a and a direct computation shows that it satisfies

$$\langle Y_b : Y_a \rangle^i = \frac{\partial Y^i_a}{\partial q^j} Y^j_b + \frac{\partial Y^i_b}{\partial q^j} Y^j_a + \Gamma^i_{jk}\left(Y^j_a Y^k_b + Y^k_a Y^j_b \right).$$

For a more geometric definition of symmetric product we refer the reader to (Lewis and Murray, 1997). The adjective "symmetric" comes from the straightforward equality

$$\langle Y_a : Y_b \rangle = \langle Y_b : Y_a \rangle.$$

Finally, let $Y^{\text{lift}}(q,t) = \sum Y_a^{\text{lift}}(q)u_a(t)$ and, at fixed q, denote finite integrals with respect to time t as:

$$\overline{Y}(q,t) \triangleq \int_0^t Y(q,s)ds. \qquad (7)$$

We insist that q is fixed during the integration with respect to t. Via an integration by parts and thanks to the symmetry of the symmetric product, one can verify that

$$\int_0^t \int_0^{s_1} [Y^{\text{lift}}(s_2), [Y^{\text{lift}}(s_1), Z_g]]ds_2 ds_1 = -\frac{1}{2}\langle \overline{Y}(t) : \overline{Y}(t)\rangle^{\text{lift}}, \qquad (8)$$

where the $(q,\dot{q})$ dependency is dropped for simplicity.

3.3 *Solutions of ordinary differential equations with polynomial vector fields*

In this section we compute solutions to a few differential equations defined by the polynomial vector fields introduced above. In particular we can make significant simplifications in the following two cases:

(i) Let $Y(q,t)$ be a vector field on $\mathbb{R}^n$, and consider the differential equation on $x = (q,v) \in T\mathbb{R}^n$:

$$\dot{x} = Y^{\text{lift}}(q,t).$$

In coordinates the previous equation reads $\dot{q} = 0$ and $\dot{v} = Y(q,t)$. If $(q_0, v_0) \in T_{q_0}\mathbb{R}^n$ are the initial conditions we compute

$$\Phi_{0,t}^{Y^{\text{lift}}}\left(\begin{bmatrix} q_0 \\ v_0 \end{bmatrix}\right) = \begin{bmatrix} q_0 \\ v_0 + \overline{Y}(q_0,t) \end{bmatrix}.$$

(ii) For $m > 0$, consider the differential equation on $x = (q,v) \in T\mathbb{R}^n$

$$\dot{x} = X_0(q,v,t) + X_m(q,v,t) \qquad (9)$$
$$x(0) = (q_0, 0),$$

where the X_0 and X_m belong respectively to $\mathcal{P}_0$ and $\mathcal{P}_m$. We write X_0 as

$$X_0 = \begin{bmatrix} X_{0,1}(q,t) \\ X_{0,2}(q,v,t) \end{bmatrix}$$

where $X_{0,2}$ depends linearly on v. Since every component of the vector field X_m is at lest linear in v and since the initial velocity $v(0)$ is assumed zero, $v(t)$ remains zero for all time t. Accordingly, the differential equation (9) on $T\mathbb{R}^n$ reduces to a differential equation on $\mathbb{R}^n$. In short:

$$q(t) = \Phi_{0,t}^{X_{0,1}}(q_0)$$
$$v(t) = 0.$$

4. A SERIES FOR MECHANICAL SYSTEMS

We would like to describe the evolution of system (6) when starting from rest. We plan to do this by writing the total flow into a flow in the velocity variables and a flow in the configuration variables.

Theorem 4.1. Let $\{\gamma(t), t \in [0,T]\}$ denote the solution to the differential equation

$$\ddot{\gamma}^i + \Gamma_{jk}^i(\gamma)\dot{\gamma}^j\dot{\gamma}^k = Y^i(\gamma,t)$$

with initial conditions $\gamma(0) = q_0$ and $\dot{\gamma}(0) = 0$. Let the functions $\Gamma_{jk}^i(q)$ and the vector field $Y(q,t)$ be uniformly integrable and bounded analytic in a neighborhood of q_0. At fixed q, let $V_1(q,t) = \int_0^t Y(q,s)ds$ and

$$V_{k+1}(q,t) = -\int_0^t \Big\langle V_k(q,s) : \sum_{m=1}^{k-1} V_m(q,s) + \frac{1}{2}V_k(q,s)\Big\rangle ds.$$

There exists a sufficiently small T_c such that the series $\sum_{k=1}^{\infty} V_k(q,t)$ converges absolutely and uniformly in t and q for all $t \in [0,T_c]$ and for all q in a neighborhood of q_0. Over the same interval the solution $\gamma(t)$ satisfies

$$\dot{\gamma}(t) = \sum_{k=1}^{+\infty} V_k(\gamma(t),t). \qquad (10)$$

For a detailed proof of convergence of the series we refer to a forthcoming publication. In what follows, we present a formal derivation of the series expansion.

PROOF. Let k be a positive integer, $x_k \in \mathbb{R}^{2n}$, and let X_k and Y_k be time varying vector fields on $\mathbb{R}^n$. Consider the differential equation

$$\dot{x}_k = \left(Z_g + [X_k^{\text{lift}}, Z_g] + Y_k^{\text{lift}}\right)(x_k,t) \qquad (11)$$
$$x_k(0) = \begin{bmatrix} q_0 \\ 0 \end{bmatrix}.$$

We recover the mechanical system in equation (6) by setting $k = 1$, $X_1 = 0$, $Y_1 = Y(q,t)$, and accordingly $x(t) = x_1(t)$.

Using the variation of constants formula in equations (2.1) and (3), we set

$$x_k(t) = \Phi_{0,t}^{Y_k^{\text{lift}}}(x_{k+1}(t)) \qquad (12)$$

and

$$\dot{x}_{k+1} = \left(\left(\Phi_{0,t}^{Y_k^{\text{lift}}}\right)^{*}(Z_g + [X_k^{\text{lift}}, Z_g])\right)(x_{k+1}) \qquad (13)$$
$$x_{k+1}(0) = \begin{bmatrix} q_0 \\ 0 \end{bmatrix}.$$

where we compute the pull-back along the flow by means of the infinite series in equation (5). Remarkably, this series reduces to a finite sum. From the discussion in Section 3.1 on the Lie algebraic structure of the various vector fields, we have

$$\begin{aligned}\operatorname{ad}^m_{Y_k^{\text{lift}}} Z_g &= 0\\ \operatorname{ad}^m_{Y_k^{\text{lift}}}[X_k^{\text{lift}}, Z_g] &= 0,\end{aligned}$$

for all $m \geq 3$. With a little book-keeping we can exploit these equalities and compute

$$\begin{aligned}&\left(\Phi_{0,t}^{Y_k^{\text{lift}}}\right)^* (Z_g+[X_k^{\text{lift}}, Z_g])\\ &= Z_g+[X_k^{\text{lift}}, Z_g]+\int_0^t [Y_k^{\text{lift}}(s), (Z_g + [X_k^{\text{lift}}, Z_g])]ds\\ &\quad+\int_0^t\int_0^{s_1} [Y_k^{\text{lift}}(s_2), [Y_k^{\text{lift}}(s_1), Z_g]]ds_2 ds_1\\ &= Z_g + [X_k^{\text{lift}} + \overline{Y}_k^{\text{lift}}, Z_g] + [\overline{Y}_k^{\text{lift}}(s), [X_k^{\text{lift}}, Z_g]]\\ &\quad+\int_0^t\int_0^{s_1} [Y_k^{\text{lift}}(s_2), [Y_k^{\text{lift}}(s_1), Z_g]]ds_2 ds_1\\ &= Z_g + [X_k^{\text{lift}} + \overline{Y}_k^{\text{lift}}, Z_g] - \left\langle \overline{Y}_k^{\text{lift}} : X_k^{\text{lift}}\right\rangle\\ &\qquad -\frac{1}{2}\left\langle \overline{Y}_k^{\text{lift}} : \overline{Y}_k^{\text{lift}}\right\rangle,\end{aligned}$$

where we have used equation (8). Therefore, the differential equation describing the evolution of $x_{k+1}(t)$ is of the same form as equation (11), where

$$\begin{aligned}X_{k+1} &= X_k + \overline{Y}_k\\ Y_{k+1} &= -\left\langle \overline{Y}_k \ : \ X_k + \tfrac{1}{2}\overline{Y}_k\right\rangle.\end{aligned}$$

We easily compute

$$X_k = \sum_{m=1}^{k-1} \overline{Y}_m$$

and set

$$Y_{k+1} = -\left\langle \overline{Y}_k \ : \ \sum_{m=1}^{k-1} \overline{Y}_m + \tfrac{1}{2}\overline{Y}_k\right\rangle.$$

According to the the iteration in the statement of the theorem, we have proven that $V_k = \overline{Y}_k$.

In summary, the iteration procedure proves that, for any $k \geq 2$, the solution to the original mechanical system $x(t) = x_1(t)$ satisfies

$$x(t) = \left(\Phi_{0,t}^{Y_1^{\text{lift}}} \circ \Phi_{0,t}^{Y_2^{\text{lift}}} \circ \ldots \circ \Phi_{0,t}^{Y_{k-1}^{\text{lift}}}\right)(x_k(t)),$$

where $\{x_k(t), t \in [0,T]\}$ is the solution to equation (11). We immediately simplify this result as follows. Since the Lie bracket

$$[Y_i^{\text{lift}}(x, s_1), Y_j^{\text{lift}}(x, s_2)] = 0$$

for all i, j and for all s_1, s_2, the flows of Y_i^{lift} and Y_j^{lift} commute. Accordingly, we write

$$x(t) = \Phi_{0,t}^{\sum_{m=1}^{k-1} Y_m^{\text{lift}}}(x_k(t)).$$

Since the previous equality holds for all k, we next investigate the quantities $\sum_{m=1}^k Y_m(q,t)$ and $\{x_k(t), t \in [0,T]\}$ in the limit $k \to +\infty$.

Since the input $Y_1(q,t) = Y(q,t)$ is assumed analytic and bounded, one can prove the existence of a constant M and positive values c_k such that

$$\|\overline{Y}_k(q,t)\| \leq c_k t^{2k-1} \quad \text{and} \quad c_{k+1} \leq M c_k,$$

where $\|\cdot\|$ is an appropriate semi-norm. In the interest of brevity, we refer to a forthcoming publication for the proof of these statements.

In short, the series $\sum_{m=1}^\infty Y_m(q,t)$ converges absolutely and uniformly in q and t, for q in a neighborhood of q_0 and for sufficiently small t. We denote

$$Y_\infty(q,t) = \sum_{m=1}^\infty Y_m(q,t).$$

Since the vector field Y_∞^{lift} belongs to $\mathcal{P}_{-1}$ we can explicitly integrate its flow, see Section 3.3. From the initial condition $[q_0^T, 0^T]^T \in T_{q_0}\mathbb{R}^n$, we have

$$\Phi_{0,t}^{Y_\infty^{\text{lift}}}\left(\begin{bmatrix} q_0 \\ 0\end{bmatrix}\right) = \begin{bmatrix} q_0 \\ \overline{Y}_\infty(q_0,t)\end{bmatrix}. \tag{14}$$

Next, we investigate the limit as $k \to \infty$ of the solution $\{x_k(t), t \in [0,T]\}$ to the differential equation (11). The previous arguments lead to uniform convergence in q for small time t of the limits:

$$\lim_{k\to\infty} Y_k(q,t) = 0 \qquad \text{and} \qquad \lim_{k\to\infty} X_k(q,t) = \overline{Y}_\infty.$$

The limit $x_\infty(t) = \lim_{k\to\infty} x_k(t)$ satisfies

$$\begin{aligned}\dot{x}_\infty &= \left(Z_g + [\overline{Y}_\infty^{\text{lift}}, Z_g]\right)(x_\infty)\\ x_\infty(0) &= \begin{bmatrix} q_0 \\ 0\end{bmatrix}.\end{aligned}$$

This differential equation can be integrated as discussed in Section 3.3, since the initial velocity is zero and the vector field belongs to $\mathcal{P}_1 + \mathcal{P}_0$. From the initial condition $[q_0^T, 0^T]^T$, we have

$$x_\infty(t) = \begin{bmatrix} q_\infty(t) \\ \dot{q}_\infty(t)\end{bmatrix} = \begin{bmatrix} \Phi_{0,t}^{\overline{Y}_\infty}(q_0) \\ 0\end{bmatrix}. \tag{15}$$

The final result in the theorem follows from combining the two flows in equations (14) and (15).

Notice that equation (10) is well-defined, since at fixed q, the integration is performed with respect to the time variable. Using the abbreviated notation introduced in equation (7), a few terms of the iteration are computed as

$$V_1 = \overline{Y}$$
$$V_2 = -\frac{1}{2}\overline{\langle \overline{Y} : \overline{Y} \rangle}$$
$$V_3 = \frac{1}{2}\overline{\left\langle \overline{\langle \overline{Y} : \overline{Y} \rangle} : \overline{Y} \right\rangle} - \frac{1}{8}\overline{\left\langle \overline{\langle \overline{Y} : \overline{Y} \rangle} : \overline{\langle \overline{Y} : \overline{Y} \rangle} \right\rangle},$$

so that we can write

$$\dot{\gamma}(t) = \overline{Y}(\gamma, t) - \frac{1}{2}\overline{\langle \overline{Y} : \overline{Y} \rangle}(\gamma, t) + \frac{1}{2}\overline{\left\langle \overline{\langle \overline{Y} : \overline{Y} \rangle} : \overline{Y} \right\rangle}(\gamma, t) + O(t^7).$$

4.1 *Expansions under small amplitude forcing*

The series in equation (10) converges under the assumption of small final time T and bounded total acceleration $Y(q,t)$. Here we derive a series that converges under the assumption of bounded final time T and small acceleration $Y(q,t)$.

Let ϵ be a small positive constant. Motivated by the treatment in (Bullo *et al.*, 1997), consider a total acceleration dependent on ϵ and of the form:

$$Y(q,t,\epsilon) = \epsilon X(q,t).$$

Equation (10) is equivalent to

$$\dot{\gamma}(t) = \epsilon \overline{X}(\gamma, t) - \frac{\epsilon^2}{2}\overline{\langle \overline{X} : \overline{X} \rangle}(\gamma, t) + \frac{\epsilon^3}{2}\overline{\left\langle \overline{\langle \overline{X} : \overline{X} \rangle} : \overline{X} \right\rangle}(\gamma, t) + O(\epsilon^4). \quad (16)$$

This expression generalizes the results presented in Proposition 4.1 in (Bullo *et al.*, 1997), since in that context the treatment was restricted to manifold with Lie group structure and invariant vector fields. Notice that this series now converges under two different set of assumptions: either ϵ is bounded and t goes to zero (see main theorem above), or t is bounded and ϵ goes to zero (see statement in (Bullo *et al.*, 1997)).

Acknowledgments The author thanks Jim Radford for helpful discussions. This research was initially supported by the National Science Foundation under grant CMS-9502224.

5. REFERENCES

Agrac̆hev, A. A. and R. V. Gamkrelidze (1978). The exponential representation of flows and the chronological calculus. *Math. USSR Sbornik* **35**(6), 727–785.

Bullo, F. and N. E. Leonard (1997). Motion control for underactuated mechanical systems on Lie groups. In: *European Control Conference.* Brussels, Belgium.

Bullo, F., N. E. Leonard and A. D. Lewis (1997). Controllability and motion algorithms for underactuated Lagrangian systems on Lie groups. To appear, *IEEE Transactions on Automatic Control.* Also Technical Report CDS 97–013, California Institute of Technology.

Kawski, M. (1995). Geometric homogeneity and applications to stabilization. In: *IFAC World Conference.* San Francisco, CA.

Kawski, M. and H. J. Sussmann (1997). Noncommutative power series and formal Lie-algebraic techniques in nonlinear control theory. In: *Operators, Systems, and Linear Algebra* (U. Helmke, D. Pratzel-Wolters and E. Zerz, Eds.). pp. 111–128. Teubner. Stuttgart, Germany.

Kolmanovsky, I. and N. H. McClamroch (1996). Stabilizing feedback laws for internally actuated multibody systems in-space. *Nonlinear Analysis, Theory, Methods & Applications* **26**(9), 1461–1479.

Lafferriere, G. and H. J. Sussmann (1991). Motion planning for controllable systems without drift. In: *IEEE Conf. on Robotics and Automation.* Sacramento, CA. pp. 1148–1153.

Leonard, N. E. and P. S. Krishnaprasad (1995). Motion control of drift-free, left-invariant systems on Lie groups. *IEEE Transactions on Automatic Control* **40**(9), 1539–1554.

Lewis, A. D. (1997). Simple mechanical control systems with constraints. Submitted to the *IEEE Transactions on Automatic Control.*

Lewis, A. D. (1998). Affine connections and distributions with applications to nonholonomic mechanics. *Reports on Mathematical Physics* **42**(1/2), 135–164.

Lewis, A. D. and R. M. Murray (1997). Controllability of simple mechanical control systems. *SIAM Journal of Control and Optimization* **35**(3), 766–790.

Murray, R. M. and S. S. Sastry (1993). Nonholonomic motion planning: Steering using sinusoids. *IEEE Transactions on Automatic Control* **5**(38), 700–726.

Sontag, E. D. and H. J. Sussmann (1986). Time-optimal control of manipulators. In: *IEEE Conf. on Robotics and Automation.* San Francisco, CA. pp. 1692–1697.

Sussmann, H. J. (1986). A product expansion of the Chen series. In: *Theory and Applications of Nonlinear Control Systems* (C. I. Byrnes and A. Lindquist, Eds.). pp. 323–335. Elsevier. Oxford, UK.

AVERAGING APPROACH TO FEEDBACK CONTROL OF BIFURCATION PHENOMENA

Mark A. Pinsky

Dept. of Mathematics, University of Nevada, Reno
Reno, Nevada 89557, USA e-mail: pinsky@unr.edu

Abstract: In many cases unstable behavior of nonlinear systems can be better described as bifurcations occurring due to variation of actual or artificial systems parameters which attracts significant attention to a problem of robust and efficient control of bifurcation phenomena. This paper describes a method for robust control of local bifurcations and correspondent stability conditions. Control design is proposed as a three-step procedure which includes transformation of uncontrolled system to its averaged form, design of a control as function of systems unstable slow (averaged) variables and recovering of the control as function of initial measurable variables. This procedure directly controls the rate of decay of unstable systems modes and leaves uncontrolled their stable and oscillatory components, which decreases delay and better matches robustness and efficiency criteria. *Copyright© 1999 IFAC*

Keywords: Nonlinear Systems, Averaging Control.

INTRODUCTION

Feedback linearization approaches combined with nonlinear change of variables have become a prevailing trend in nonlinear control theory in the recent decades. In fact, these techniques fail for systems with critical linearization primarily concerned with revealing non-linear stability transitions (bifurcations). The problem of feedback stabilization of critical nonlinear systems has also attracted significant attention in the past few years (see review paper (Sontag, 1990) and references therein and more recent publications (Wang and Abed, 1995) and (Yabuno, 1997).

A method for control stabilization of local bifurcations using Poincare's normal forms was addressed in (Pinsky and Essary, 1994; Pinsky and Essary 1995). In this paper we establish sufficient conditions implying stability of local bifurcations in nonlinear systems with polynomial nonlinearities. These conditions structure the design of feedback control stabilizing the equilibrium solution and can be used for estimation of stability basin as well.

1. THE NORMAL FORMS METHOD AND BIFURCATION ANALYSIS

In many cases a model governing local transition behavior of a normal operating regime of a system can be written in the following form:

$$\dot{x} = A(\delta)x + \varepsilon f(x,\delta,\varepsilon), \qquad (1)$$

where $x \in R^n, f$ is a power series in x, and $f(0,\delta,\varepsilon) = 0, A$ is a square diagonal matrix of order n, $diag\, A = \{\lambda_1 ... \lambda_n\}$, $\delta \in R^m$ is a parameter vector and $0 < \varepsilon$ is a small parameter linked with the neighborhood size. Local stability analysis describes stability transition of a static solution $x=0$ which occurs when

eigenvalues of A cross the imaginary axis due to alteration of δ. Poincaré has drawn attention to the fact that analysis of nonlinear stability transition of the trivial solution: $x=0$ of (1) is coupled with revealing the corresponding bifurcation phenomenon of branching of this static solution. The global bifurcation analysis is rather complex and has been completed only in relatively simple cases. Fortunately, just local analysis of bifurcation phenomena often provides a clear picture on the development of transient behavior, which can be used in design of control terminating the instabilities.
Let us briefly describe the Poincaré normal forms method that attempts to simplify (1) with the aid of a close-to-identity change of variables

$$x = y + \varepsilon r(y,\varepsilon) \qquad (2)$$

Assume now that $f = f_{ms} = x^m e_s$, where $x^m = x_1^{m_1} \ldots x_n^{m_n}, m_i, i = 1,\ldots,n$ are positive integers, $m_1 + \ldots + m_n \geq 2, f_{ms} = const, e_s$ is a unit vector of the s-axis (namely, f is a vector with a single s-th nonzero component). Then (2) annihilates f in (1) if r satisfies the so-called homological equation (Wiggins, 1990):

$$L_A r = f(x) \qquad (3)$$

where $L_A r = (\partial r/\partial x)Ax - Ar$ is a Lie bracket and $\partial r/\partial x$ is the Jacobian of r. For the chosen form of f, (3) has a particular solution $r = r_{ms} x^m e_s$, where $r_{ms} = f_{ms}/((m;\lambda) - \lambda_s), (m;\lambda) = m_1\lambda_1 + \ldots + m_n\lambda_n, \lambda = \{\lambda_1,\ldots,\lambda_n\}$ is a vector formed from the eigenvalues of A. In fact, monomials conforming to the Poincaré resonance condition

$$\lambda_s = (m;\lambda) \qquad (4)$$

persist in (1) and are termed the resonance monomials. The Poincaré method successively eliminates all nonresonance monomials beginning with monomials of the lowest degree, in order to obtain the normal form which consists only of resonant monomials and turns out to be a significantly reduced equation.

2. GENERALIZED NORMAL FORMS METHOD

Unfortunately, lack of computability complicates adoption of the standard normal forms method in solving many applied problems. We describe here a more general approach to finding normal forms that is more efficient in computations and is applicable to a broader class of systems than the known techniques.
It is remarkable that in the context of the following algorithm the nonlinear resonance condition is shaped in a form congruent with an analogous relation from the theory of linear forced oscillations. Although it is easy to verify that the Poincaré resonance monomials are captured by the new condition, it is important to notice that this approach may also be used if this condition is inapplicable, which occurs, for example, if f is nonpolynomial. Separation of slow and fast variables becomes rather routine for the normal form equations and is presented in a closed form in this section.

Suppose that in the new variables equation (1) takes the form:

$$\dot{y} = Ay + \varepsilon B(y,\delta,\varepsilon) \qquad (5)$$

From (1), (2) and (5) follows the governing equation:

$$Ar - r_y Ay - r_t = -f(y+\varepsilon r) + B + \varepsilon r_y B \qquad (6)$$

where r_y is the Jacobian with respect to y and r_t is a partial derivative of r with respect to t. We consider (6) as a system of partial differential equations with respect to r. We seek a particular solution of (6) satisfying an additional condition:

$$\|f\| \to 0 \Rightarrow \|r\| \to 0$$

where $\|\cdot\|$ is a certain norm of a vector valued function. In other words, we are looking for a particular solution such that the norm of r tends to zero if the norm of f tends to zero as well. For small ε one may try to approximate the solution of (6) by the following successive approximations:

$$Ar_1 - r_{1,y}Ay - r_{1,t} = -f(y) \qquad (7)$$

$$\dot{r}_k - r_{k,y}\, Ay - r_{k,t} = -f(y+\varepsilon r_{k-1,y}) + B_k + \varepsilon r_{k-1,y}\, B_k,$$
$$k > 1$$

Note that (7) represents a set of linear partial differential equations with respect to r_k, $k \geq 1$. Let us write a characteristic equation for (7):

$$\dot{r}_1 = Ar_1 + f(y,\delta) - B_1 \qquad (8)$$

$$\dot{r}_k = Ar_k + f(y+\varepsilon r_{k-1},\delta) - \varepsilon r_{k-1,y}B_{k-1} - B_k, \quad k > 1 \qquad (9)$$

$$\dot{y} = Ay \qquad (10)$$

From (10) we see that $y = e^{At}c$, where $c=\{c_1, \ldots, c_n\}$ is a constant vector. The time-dependent terms in (8) and (9) should be split up into two distinct groups, namely, resonance and nonresonance terms.

Definition: $F(t)$ is called a resonance function if it can be represented in the following form:

$$F(t)=e^{At}N(c) \qquad (11)$$

where $N(c)\in R^n$ is a vector dependent upon a constant vector c, otherwise F is called a nonresonance function. In order to annihilate nonresonance terms we set: $B_k=e^{At}N_k(c)$ (recall that $c=e^{-At}y$). So, one is able to write:

$$B_k(y,\delta,\varepsilon)=e^{At}N_k(e^{-At}y,\delta,\varepsilon) \qquad (12)$$

The equations (11) and (12) naturally express a nonlinear resonance condition in a form adopted from the theory of linear forced oscillations. Although (12) looks distinct from (5), it turns out to be identical to the Poincaré condition in the case when $f_s(x)=f_{ms}x^m e_s, f_{ms}=const.$ Recalling that A is a diagonal matrix, and $f_s(e^{At}c)=f_{ms}e^{(m;\lambda)t}c^m e_s$ we see that f_s admits (13) if $(m;\lambda)=\lambda_s$. Note also that in this case $B_s=\alpha_{ms}y^m e_s$.

Observe that (8)-(10) coincide with the set of linear ODEs given by simple successive approximations if one sets: $B_k=0, k>1$. As is known, the resonance terms, usually present in these recursive sequences, force these iterations to diverge on large time intervals. The problem with such successive approximations is that they do not distinguish between resonance and nonresonance terms. In spite of this, one can use the available B_k to annihilate resonance terms in (8) and (9) and place them into the normal form. Because the resonance terms are exceptional, the normalization yields significant reduction of the initial system. The normal forms admit a closed-form complementary reduction in both the so-called slow and especially in the amplitude-phase variables, which inherently occur in the presence of complex-conjugate eigenvalues. The imaginary parts of the eigenvalues correspond with the oscillatory components of solutions, which are naturally filtered out when the normal form equations are written in the amplitude-phase variables.

Let us write the normal form equations as follows:

$$\dot{y}=Ay+\varepsilon e^{-At}N_k(e^{-At}y,\delta,\varepsilon)+O(\varepsilon^{k+1}) \qquad (13)$$

Denote a slow variable $c(t)$ by the formula:

$$y(t)=e^{At}c(t)$$

Then one obtains:

$$\dot{c}=\varepsilon N_k(c,\delta,\varepsilon)+O(\varepsilon^{k+1}) \qquad (14)$$

As is known, a couple of complex eigenvalues in A (say λ_k and $\overline{\lambda}_k$) relate to a couple of complex conjugate functions c_k and $\overline{c}_k$ in (14). These complex conjugate equations can be written as real valued equations. Assuming now for simplicity that all eigenvalues are complex conjugate couples:

$$\lambda_k=\alpha_k+I\omega_k;\overline{\lambda}_k=\alpha_k-I\omega, k=1,...,n/2, I=\sqrt{-1}$$

we present related c_k and $\overline{c}_k$ in the following exponential form:

$$c_k(t)=a_k(t)e^{I\rho_k(t)}, \overline{c}_k(t)=a_k(t)e^{-I\overline{\rho}_k(t)}, \quad k=1,...,n/2 \qquad (15)$$

where a_k and r_k are real functions, known as the amplitude and phase variables. Let us consider the application of (a,r) variables to the normal form equation in the case when all eigenvalues of A are located close to the imaginary axis (i.e. that $\alpha_k=O(\varepsilon), k=1,...,n/2$) which is the main concern in local bifurcation theory relating a bifurcation phenomenon with the eigenvalues transition through the imaginary axis. With this assumption, we write $\varepsilon\alpha_k$ instead of α_k, where now $\alpha_k=O(1)$ and next, we present (1) in a modified form:

$$\dot{x}=(\mathrm{Im}\,A)x+\varepsilon F(x,\delta,\varepsilon) \qquad (16)$$

where matrix $\mathrm{Im}\,A=diag[I\omega_1,-I\omega_1,...,I\omega_{n/2},-I\omega_{n/2}]$, $\mathrm{Re}\,A=diag[\varepsilon\alpha_1,\varepsilon\alpha_2,...,\varepsilon\alpha_{n/2},\varepsilon\alpha_{n/2}]$ and $\varepsilon F=\varepsilon f+(\mathrm{Re}\,A)x$. The eigenvalues of modified matrix (*Im A*) admit natural resonance condition: $\omega_k+(-\omega_k)=0, k=1,...,n/2$. We assume for simplicity that $(\omega_1,...,\omega_{n/2})$ are integer – noncommensurable which implies that there are no other resonance conditions. The application of the generalized normal form method to (16) yields the normal form equations which can be written in the amplitude-phase variables (15) as follows:

$$\dot{a}=\varepsilon\alpha a+\varepsilon\,\mathrm{Re}(N_k(a,\varepsilon))+O(\varepsilon^{k+1}),$$
$$\dot{\rho}=-\varepsilon a^{-1}\,\mathrm{Im}(N_k(a,\varepsilon))+O(\varepsilon^{k+1}) \qquad (17)$$

where $a=\{a_1,...,a_{n/2}\}$ $\rho=\{\rho_1,...,\rho_{n/2}\}$, and $O(\varepsilon^{k+1})$ stands for discarded terms of order ε^{k+1}. The most notable feature of the above equations is that the amplitude and phases are slow variables and the first variables are decoupled from the last.

In the complementary case when all real $\lambda_k=O(\varepsilon)$, we may write equation (1) in the form:

$$\dot{x}=\varepsilon F(x,\delta,\varepsilon)$$

where $\varepsilon F = \varepsilon f + \varepsilon Ax$. Applied to that equation, the generalized normalization technique yields the real normal form equations that can be written as (14).
Remark : Described adjustment of close-to-critical eigenvalues should be done when they cross or closely approach the Imaginary axis due to δ variation. Ignoring these circumstances yields a normal form, which is not capable of capturing a local bifurcation behavior, implied by transition of eigenvalues through the imaginary axis. Notice that when critical eigenvalues consist of both a number of zero eigenvalues and a certain number of incommensurable critical complex conjugate couples, the normal form in (a, ρ) variables can be written as (17) if we formally identify the slow variables related to zero eigenvalues with additional components in the vector a.

3. ESTIMATION OF RESIDUAL TERMS

As is known, see (Wiggins, 1990 for further details), successive normalization often diverges in practically important cases. Indeed, truncated normalization could be a valuable source in bifurcation analysis if the residuals are properly estimated. Equating

$$B_k = f(y + \varepsilon r_{k-1}) - \varepsilon r_{k-1,y} B_{k-1}, k > 1$$

we get $r_{k-1}=r_k$, which means that successive normalization has been stopped on the k-1 step. In this case, B_k consist of the resonance and nonresonance terms as well. Accumulating these nonresonance terms in residuals, we write the truncated normal form equations in the amplitude-phase variables as follows:

$$\dot{a} = \varepsilon \alpha a + \varepsilon \operatorname{Re}(N_{k-1}(a,\varepsilon)) + \varepsilon^k E(a,\rho,\varepsilon),$$
$$\dot{\rho} = -a^{-1} \varepsilon \operatorname{Im}(N_{k-1}(a,\varepsilon)) + \varepsilon^k e(a,\rho,\varepsilon) \qquad (20)$$

where amplitudes $a \in R^{n_1}$ and phases $\rho \in R^{n_2}$, $(n_1 - n_2)$ is the number of zero eigenvalues in matrix a, and residuals of truncated normalization: E and e are periodic functions with respect to each ρ_k with incommensurable periods T_k:

$$E(a, \rho_k + T_k) = E(a, \rho_k),\ e(a, \rho_k + T_k) = e(a, \rho_k)$$

Note that (20) are accurate equations derived from (1) by the closed form changes of variables. However, residuals in these equations depend upon both a and ρ variables, which forces us to use in stability analysis their ρ -invariant upper bounds represented as follows:

$$E \le \max_{\rho_{1\in} \in [0;T_1]} \dots \max_{\rho_n \in [0;T_n]} E(a,\rho,\varepsilon) = E^0(a,\varepsilon)$$
$$e \le \max_{\rho_{1\in} \in [0;T_1]} \dots \max_{\rho_n \in [0;T_n]} e(a,\rho,\varepsilon) = e^0(a,\varepsilon) \qquad (21)$$

Using these bounds we can formulate asymptotic stability conditions for a steady state solution of (20) with the aid of a Lyapunov function addressed in amplitude variables. Indeed, there are valuable advantages in application of the Lyapunov approach to the averaged equations (20) rather than to the initial ones. The averaged equations govern the evolution of slow variables which envelop trajectories of the initial system and separates the information valuable for stability analysis from the routine ones. Thus, relatively simple Lyapunov functions defined in the amplitude variables provide sharp sufficient stability conditions which in many cases assume explicit physical interpretation.

4. RESONANCE FEEDBACK STABILIZATION

Assume now that control $u = \{u_1, \dots, u_n\}$ enters equation (1) in the following way

$$\dot{x} = A(\delta)x + \varepsilon f(x,\delta,\varepsilon) + u(x,\delta,\varepsilon) \qquad (22)$$

where f is as in(1) and u : $R^{n+m+1} \to R^n$ is continuous in x and polynomial in δ and ε , and $u(0,\delta,\varepsilon)=0$. In this case u describes a portion of control stabilizing the solution $x=0$ of (22) which otherwise became unstable when δ reaches the bifurcation value $\delta = \delta^0$. A complimentary portion of control (which may alter the location of a steady state solution) is assumed to be adjoining to f. Direct application of a Lyapunov based approach to equation (22) ignores the distinction between resonance and nonresonance terms and leads to inefficient design and over-control. The resonance feedback stabilization method takes advantage of the normal forms methodology reducing an controlled plant (22) (with $u=0$) to its normal form (20) which allows to design control $U(a,\delta)$ as a function of averaged amplitude variables. In this case $U(a,\delta)$ should obey a specific "resonance" form, determined by the structure of A critical eigenvalues, which permits unique mapping of U into the original state variables. Notice that the effect of bounded unmodeled dynamics on the resonance feedback stabilization may be accounted for by formally adjoining these bonds, which are priory mapped into the amplitude-phase variables, into the residual bounds.

To explain the design of resonance feedback control, we write a controlled plant's equations in the amplitude-phase variables as follows:

$$\dot{a} = \varepsilon G(a,\rho,\delta,\varepsilon) \qquad (23)$$
$$\dot{\rho} = \varepsilon g(a,\rho,\delta,\varepsilon) \qquad (22)$$

$$G(a,\rho,\delta,\varepsilon)=Q(a,\delta,\varepsilon)+\varepsilon E(a,\rho,\delta,\varepsilon)+\varepsilon U(a,\delta),$$
$$g(a,\rho,\delta,\varepsilon)=g(a,\delta,\varepsilon)+\varepsilon e(a,\rho,\delta,\varepsilon)$$

where Q and q are resonance terms, E and e are residuals, $E(0,\rho,\delta,\varepsilon)=e(0,\rho,\delta,\varepsilon)=0$ and $U(0,\delta)=0$ is control. We intent to control only the amplitude variables since due to (15) explicit evolution of the phase variables is insignificant in stability analysis. However, their implicit effect on the amplitude variables is accounted for by the bound of E. In other words, this feedback design approach controls the rate of perturbations decay and leaves uncontrolled the oscillatory components of decaying trajectories determining by phases evolution. This contributes to the efficiency of the described feedback design procedure.

Let us sketch briefly the procedure of resonance control design. Using (22), we write

$$|G(a,\rho,\delta,\varepsilon)| < G^0(a,\delta,\varepsilon)$$

where $G^0 = Q(a,\delta,\varepsilon)+E^0(a,\delta,\varepsilon)+\varepsilon U(a,\delta)$. Next, we engage the Lyapunov stability condition aided by a Lyapunov function L=L(a) defined in a neighborhood around the origin a=0:

$$\left.\frac{dL}{dt}\right|_{\dot a=\varepsilon G} \le \sum_{l=1}^{n_1}\frac{\partial L}{\partial a_i}G_l^0(a,\delta,\varepsilon)\le D, \mu \le |a| \le a^0, \mu > 0$$
$$\left.\frac{dL}{dt}\right|_{\dot a=\varepsilon G} \le \sum_{l=1}^{n_1}\frac{\partial L}{\partial a_i}G_l^0(a,\delta,\varepsilon)\le 0, |a| \le \mu \qquad (24)$$

where $D \supset R > 0$ assignees the degree of relative stability, vector a^0 denotes a rectangular stability basin in a coordinates and μ is a small positive value. Note that, in the last relation, equality is achieved only for $a=0$. We summarize the above discussion in the following statement.

Theorem 1. Let Q,q,E and e are continuous and continuously differentiable functions of their variables, $U(a,\delta)$ is continuous in a and δ and its derivative $\frac{\partial U(a,\delta)}{\partial a}$ may be discontinuous only at $a=0$. Assume that $U(a,\delta)$ is chosen such that (24) is intact. Than solution $a=0$ is asymptotically stable and every solution emanating from a region $|a| < a^0$ asymptotically converge to zero.

Although a variety of Lyapunov functions can be adapted in feedback design, we show that the most simple one $L=\sum_{l=1}^{n_1} a_l^2$ yields a desirable controller. First, consider a case where all eigenvalues of diagonal matrix A are distinct from zero complex conjugate pairs located close to imaginary axis. Assume that both imaginary parts of eigenvalues are integer-noncommensurable and (22) is written in the form (16) where A eigenvalues turn out to be pure imaginary. In this case $a_i \ge 0, i=1,...,n/2$ and (24) can be written in simplified but more conservative form:

$$G^0(a,\delta,\varepsilon)\le -d, \mu \le a \le a^0 \qquad (25)$$
$$G^0(a,\delta,\varepsilon)\le 0, 0 \le a \le \mu$$

where $d=\{d_1,...d_{n_1}\}$ is a vector with positive components $d_i > 0, i=1,...,n$ and inner product of d and a: $(a,d) > D$. Note that the last inequality does not determine d_i values uniquely and can be further adapted in the control design. The last conditions can be written in the form:

$$\dot a \le -d, \mu \le a \le a^0$$
$$\dot a \le 0, 0 \le a < \eta$$

vividly showing that the resonance feedback controls least rates (d_i) of exponential decay of the amplitude variables a_i of unstable modes which is a principal concern in various engineering applications.

Asymptotic stability of $a=0$ solution of (23) implies asymptotic stability of $x=0$ solution of initial system (22) if $U(a,\delta)$ obeys the form determining by critically stable/unstable eigenvalues of A. This feedback can be uniquely mapped into y coordinates where it becomes a function of resonance monomial: $V=V(y,\delta)$. Subsequent mapping of $V=V(y,\delta)$ into original x variables is provided by inverse to (2) transformation: $y=\rho(x), \rho: R^n \to R^n, \rho(y+\varepsilon r(y))=y$ which exists in a certain open neighborhood γ_y of $y=0$ where (2) is a homomorphism. Let γ_x be image of γ_y yielding by (2). Moreover, stability basin Ω of (22) solution $x=0$ is determined as $\Omega=\Omega_x \cap \gamma_x$ where Ω_x is image in x coordinates of stability basin $\gamma_a : |a| \le a^0$ of (23) solution $a=0$. In this paper we focus on the case when all eigenvalues of $A: \lambda=\{\lambda_1,...,\lambda_{n/2},\bar\lambda_1,...,\bar\lambda_{n/2}\}$ and correspondent eigenvectors $y=\{y_1,...,y_{n/2},\bar y_1,...,\bar y_{n/2}\}$ turn out to be critical integer-incommensurable complex conjugate couples which allows us to choose

$$U(a,\delta)=\{U_1(\alpha,\delta),...,U_{n/2}(a,\delta)\} \text{ as}$$
$$U_k(a,\delta)=a_k f_k(a_1^2,...,a_{n/2}^2,\delta), k=1,...,n$$

where $U_k(a,\delta)$ is continuous in a and δ, and f_k may be discontinuous at $a_k=0$. This control is uniquely represented in the coordinates of A eigenvectors as:

$$V_k(y)=y_k f_k(y_1\bar{y}_1,\ldots,y_n\bar{y}_n) \quad (26)$$

and can be subsequently mapped into x coordinates. Now we formulate the stabilizability conditions in the following

Theorem 2. Let (22) represents dynamics of controlled system and equations (23), where $U(a,\delta)=0$, represent normalized to a certain order in ε uncontrolled equation (22) with $u(x,\delta,\varepsilon)=0$. Assume that all A eigenvalues are integer-incommensurable complex conjugate pairs and control $U(a,\delta)$ admits form (26) and chosen such that (25) intact. Define control in (22) as $u(x,\rho,\varepsilon)=V(\rho(x),\delta,\varepsilon)$, $x\subset\gamma_x$. Then x=0 solution of (23) is asymptotically stable and every solution from neighborhood Ω converge to zero.

5. CONTROL OF THE HOPF BIFURCATION

Now we apply both the normal forms analysis and the resonance feedback stabilization to controlling instabilities in the Van der-Pol equations:

$$\dot{x}_1 = x_2 + u_1(x_1,x_2),$$

$$\dot{x}_2 = -x_1 + \varepsilon(x_2 - x_2^3) + u_2(x_1,x_2) \quad (27)$$

where $u=\{u_1,u_2\}$ is a control and $u_i(0,0)=0,\ i=1,2$. Consider first a transient behavior occurring in the uncontrolled equations (u=0) which are known as a simple model revealing the Hopf bifurcation. The eigenvalues of the unperturbed system (27) (with $\varepsilon=0$) are $\omega_1 = I, \omega_2 = -I$. These eigenvalues satisfy the Poincaré resonance condition which follows from the identity: $\omega_1+\omega_2=0$, and the normal form takes the form (5). Assuming $\varepsilon=0.1$, we deduce the first order approximation to the normal form for (27) with $u=0$ and write these equations in (a,ρ) coordinates as:

$$\dot{a} = 0.05a - 0.15a^3 + \varepsilon^2 E(a,\rho), \qquad \dot{\rho} = \varepsilon^2 e(a,\rho)$$

where residuals $E(a,\rho)$ and $e(a,\rho)$ are obtained in the second iteration as previously mentioned.
The trivial solution $a=0$ is unstable in this case, but the system possesses a stable limit circle. Now we are able to engage feedback $U(a)$ and the residual bound:

$$E(a,\rho) < E^0 = 0.00928a^3 + 0.0346a^5 + 0.00253a^7 + 0.0002a^9$$

in the control design. Combining the resonance, residual, and feedback terms we get:

$$G_i^0 = 0.05a - 0.141a^3 + 0.0346a^5 + 0.00253a^7 + 0.0002a^9 + U_i(a)$$

Because in this case $a \geq 0$, the positive terms in G_i^0 contribute to instability. There are various ways to choose $U_i(a)$ in order to satisfy (25). We choose smooth feedback in the form:

$$U(a) = ka - 0.0350a^5 - 0.00260a^7 - 0.000210a^9$$

(where k is an adjustable parameter) which annihilates the small destabilizing (positive) nonlinear terms in G_i^0 and preserves the main nonlinear term ($-0.141a^3$) which contributes to stability.
We choose feedback in the form:

$$U(a) = U_1(a) = -2.05a - 0.035a^5 - 0.00260a^7 - 0.00021a^9$$

Mapping this function back to the original variables introduces control in measurable initial coordinates as follows:

$$u_{11}(x_1,x_2) = -2.05x_1 - 0.0022x_1^5 - 0.0044x_1^3x_2^2 - 0.0022x_1x_2^4$$

$$u_{12}(x_1,x_2) = -2.05x_2 - 0.0022x_2^5 - 0.0044x_1^2x_2^3 - 0.0022x_1^4x_2$$

where insignificant components in feedback are discarded for simplicity. Numerical simulation show that this control intact global asymptotic stability of equilibrium solution.

REFERENCES

Pinsky M.A. and B. Essary (1994). Analysis and resonance stabilization of bifurcation phenomena in aircraft flight. *AIAA J. of Guidance*, **17**, pp. 591-598.

Pinsky M.A. and B. Essary (1995). Resonance control of aircraft instabilities by smooth and continuous feedback. *AIAA J. of Guidance*, **18**, pp. 1298-1304.

Sontag, E.D. (1990). Feedback stabilization of nonlinear systems. In: *Robust Control of Linear Systems and Nonlinear Control* (M.A. Kaashoek, J.H. van Shorpen, and A.C.M. Ran. (ED)), **4**, pp. 61-81. Birkhauser, Boston.

Wang H. and E. Abed.(1995) Bifurcation control of a chaotic system. *Automatica J.IFAC*, **31**, pp. 1213-1226.

Wiggins, S. (1990). *Introduction to Applied Nonlinear Dynamical Systems and Chaos.* Springer-Verlag, New York.

Yabuno H. (1997). Bifurcation control of parametrically excited duffing system by a combined linear-plus-nonlinear feedback control. *Nonlinear Dynam.*, **12**, pp.263-274.

14th Triennial World Congress, Beijing, P.R. China

ON THE FEASIBILITY OF FINITE DIFFERENCES HOMOTOPY-BASED ALGORITHM FOR NONLINEAR OPTIMAL CONTROL PROBLEMS

Mazen Alamir *

* *Laboratoire d'Automatique de Grenoble, B.P. 46, 38402, Saint Martin d'Hères, France*
mazen.alamir @inpg.fr

Abstract: In this paper, we examine the feasibility of continuation methods when used to solve the set of nonlinear equations obtained by discretizing finite horizon optimal control problems. We prove that, under some uniform boundness conditions on the optimal solutions, under classical convergence conditions on the discretization scheme and a boundness condition on the horizon length, there is a minimal number of grid points (depending on an a-priori specified maximal size of initial condition and horizon length) such that the resulting set of nonlinear equations can be imbedded in a homotopy that admits a feasible continuation algorithm.

Keywords: Nonlinear Optimal Control, Continuation Methods, Finite Differences.

1 Introduction

In this paper, we investigate the possibility of using continuation methods to solve nonlinear optimal control problems with a penalty on the terminal state. The key feature that makes continuation methods attractive compared to many other iterative techniques is that it overcomes the local convergence nature of Newton-like iterative processes (Watson, 1979; Zhang and Han, 1993). For this reason, continuation approaches have been used to solve a very wide range of nonlinear problems including solution of general nonlinear equations in Banach spaces (Meyer, 1968), nonlinear two-points boundary value Problems (Deuflhard *et al.*, 1976; Watson, 1979; Yong and Xianrui, 1995) and the solution of nonlinear fluid dynamic equations (Liao, 1997). It is well known that one way to solve nonlinear optimal control problems is to solve a related set of nonlinear equations that may be obtained by means of discretization strategies like Finite elements, Finite differences or Orthogonal collocation (Polak, 1973; Reddien, 1979; Schwartz and Polak, 1996; Vlassenbroek and Dooren, 1988; Dontchev, 1996).
To solve the resulting set of nonlinear equations, Newton-like strategies are generally used together with the assumption that the gradient is everywhere nonsingular. The problem is that even if this assumption holds everywhere except on some set of measure 0, the trajectory generated by the corresponding iterations may be obliged to pass through the singular set for a large set of initial guesses. Continuation methods are precisely known to avoid such situations by subdividing the original global problem into several successive locally well posed regular problems. Generally, the key point is to prove feasibility of such a subdivision, namely that the original problem can be related to a trivial one by a finite number of subproblems (Avila, 1974).
In this paper, we use the very simple Euler finite differences in order to generate the set of nonlinear equations related to the optimal control problem. The consistency and the convergence of such discrete schemes have been extensively investigated (see (Dontchev and Hager, 1993; Dontchev, 1996) and the references therein) and will not be considered here. Instead, we suppose that suitable

conditions hold for the successive discrete approximations to converge to the solution of the continuous problem (see assumption 1). Our aim is then to investigate the feasibility of homotopy like algorithms when used to solve the set of nonlinear equations resulting from the supposed stable and consistent discretization scheme.
The paper is organized as follows. First, we recall some basic facts concerning continuation methods in section 2. Section 3 presents problem formulation and basic results.

2 Recalls on continuation methods

Let $F : \Omega \subset \mathcal{X} \to \mathcal{X}$ be a mapping on the linear normed space $\mathcal{X}$ and consider the problem of solving the equation:

$$F(X) = 0 \tag{1}$$

In the continuation framework, this problem is first imbedded into a family of problems of the form:

$$H(\lambda, X) = 0 \tag{2}$$

where $\lambda \in [0,1] \equiv J$ is a parameter. The imbedding (2) is chosen such that at $\lambda = 0$, (2) admits a known solution $X(0)$, while at $\lambda = 1$, the solution $X(1)$ of (2) also solves (1). For example, H might have the following form:

$$H(\lambda, X) = F(X) + (\lambda - 1)F(X_0) \tag{3}$$

When solving (2) locally to proceed from one λ to another, it is often convenient to rewrite it in order to obtain a fixed-point problem. This can be done by choosing $G(\lambda, X)$ such that

$$\left\{H(\lambda, \bar{X}(\lambda)) = 0\right\} \iff \left\{\bar{X}(\lambda) = G(\lambda, \bar{X}(\lambda))\right\} \tag{4}$$

and such that $G(\lambda, .)$ meets the requirement of fixed point theorems therefore enabling solutions of $H(\lambda, X) = 0$ to be found by means of the iterative process

$$X^{k+1} = G(\lambda, X^k) \tag{5}$$

This is the reason why a basic notion in this context is the domain of attraction of an iterative process that we recall hereafter

definition 1. [Domain of attraction of an iterative process]
Let $G : \Omega \subset \Re^q \to \Re^q$ be a given mapping. A non empty set $\Omega_0 \subset \Omega$ is a domain of attraction of the iterative process:

$$X^{k+1} = G(X^k) \quad , \quad k = 0, 1 \ldots \tag{6}$$

with respect to the point X^* if for any $X^0 \in \Omega_0$ we have $\{X^k\} \subset \Omega$ and $\lim_{i\to\infty} X^k = X^*$. If this is the case, X^* is called a point of attraction of (6) ♠

The feasibility of a continuation approach can be explained as follows.

definition 2. [feasibility of a continuation method]
Let us take a continuous path $\left\{(\lambda, \bar{X}(\lambda))\right\}_{\lambda\in[0,1]}$ defined as above.
If one can find a subdivision

$$0 = \lambda_1 < \lambda_2 < \ldots < \lambda_N = 1 \tag{7}$$

of $J = [0,1]$ such that the following condition holds:
There is some positive real $r > 0$ and a **continuous** mapping $G(\lambda, X)$ satisfying (4) such that:

- For all $i = 1, \ldots, N$, the **closed** ball $B_i(r) = B(\bar{X}(\lambda_i), r)$ (centered at $\bar{X}(\lambda_i)$ and of radius r) is a domain of attraction of the iteration process:
$$X^{k+1} = G(\lambda_i, X^k) \quad , \quad k = 0, 1, \ldots \tag{8}$$
with respect to $\bar{X}(\lambda_i)$.
- For all $i = 1, \ldots, N-1$, $\bar{X}(\lambda_i) \in int\Big(B_{i+1}(r)\Big)$,

where $int\Big(B_{i+1}(r)\Big)$ designate the interior of $B_{i+1}(r)$
Then we say that the subdivision (7) and the mapping $G(.,.)$ define a feasible continuation process to solve $H(1, X) = F(X) = 0$ ♠

The justification of the last definition is based on the following lemma (Avila, 1974)

lemma 1.
Under the assumptions of definition 2, for all $i = 1 \ldots N-1$, there is finite sequence of integers j_i such that the iterative process:

$$X_i^{k+1} = G(\lambda_i, X_i^k) \quad ; \quad X_i^0 \in B_i(r) \tag{9}$$

leads to a final point $X_i^{j_i}$ such that $X_i^{j_i} \in B_{i+1}(r)$. ♠

Therefore, starting from any $X_i^0 \in B_i(r)$ and executing the $(N-1)$ groups of iterations defined by (9) for $i = 1 \ldots N-1$, we obtain a point $X_{N-1}^{j_{N-1}} =: X_N^0 \in B_N(r)$, that is, in the domain of attraction of the solution $\bar{X}(1)$ we are searching for. For this last group of iterations ($i = N$ in (9)), we obtain

$$\lim_{k\to\infty} X_N^k = \bar{X}(1) \quad \text{solution of } F(X) = 0 \tag{10}$$

Therefore, in order to prove the feasibility of continuation schemes when solving a set of nonlinear equations (1), we must prove that the assumptions of definition 2 hold with some subdivision ($0 = \lambda_1 < \lambda_2 < \dots \lambda_N = 1$) and some continuous function $G(.,.)$ satisfying (4) used to define the iterative process (9). The following theorem is a trivial way to do so

Theorem 1. [Trivial Sufficient Conditions for feasibility]
Let us define a continuous path $\left\{(\lambda, \bar{X}(\lambda))\right\}_{\lambda\in[0,1]}$ as above. If there is a fixed $r > 0$ (independent of λ) and a **continuous map** G satisfying (4) such that for all $\lambda \in [0,1]$, $B(\bar{X}(\lambda), r)$ is a domain of attraction of the iterative process

$$X^{k+1} = G(\lambda, X^k) \tag{11}$$

then there is a finite subdivision $0 = \lambda_1 < \lambda_2 < \dots < \lambda_N = 1$ satisfying the assumptions of definition 2. ♠

It is important to note that the key feature in the assumptions of theorem 1 is the existence of a minimal size ($r > 0$) that holds for all the domains of attraction of the λ-dependent family of iterative processes (11). That is, this minimal size holds uniformly over all $\lambda \in [0,1]$. For this to be possible, some technical regularity requirements are needed. The following theorem due to (Avila, 1974) proposes a set of such sufficient conditions.

Theorem 2. [Technical Sufficient Conditions for Feasibility]

Consider the mapping $H : [0,1] \times D \to \Re^q$ where $D \in \Re^q$ is an open set. Assume that H has a continuous partial derivative H_x which for every $(\lambda_0, X_0) \in [0,1] \times D$ and some $\delta(\lambda_0, X_0)$ satisfies

$$\| H_x(\lambda, X) - H_x(\lambda, Y) \| \leq W(\lambda_0, X_0) \| X - Y \| \tag{12}$$

for all $|\lambda - \lambda_0| < \delta$ and all $X, Y \in B(X_0, \delta)$.
Assume furthermore that there is a continuous function $\bar{X} : [0,1] \to D$ satisfying $H(\lambda, \bar{X}(\lambda)) = 0$. If $H_x(\lambda, \bar{X}(\lambda))^{-1}$ exists and is bounded for each $\lambda \in [0,1]$ then the map G defined by

$$G(\lambda, X) := X - H_x(\lambda, X)^{-1} H(\lambda, X) \tag{13}$$

together with the continuous path $\left\{(\lambda, \bar{x}(\lambda))\right\}_{\lambda\in[0,1]}$ meet the requirements of theorem 1. ♠

3 Problem statement and discrete approximation

3.1 Problem Statement

We consider the following optimal control problem. Minimize the performance index

$$J(u) = \rho \| x(T) \|^2 + \frac{1}{2} \int_0^T \left[x^T(t) Q x(t) + u^T(t) R u(t) \right] dt \tag{14}$$

subject to the system equation

$$\dot{x}(t) = f(x(t)) + g(x(t)) u(t) \; ; \; x(0) = x_0 \in \Re^n \tag{15}$$

where $u(t) \in \Re^m$ is the control vector, $Q \in \Re^{n\times n}$ is positive nonnegative while $R \in \Re^{m\times m}$ is positive definite.
It is well known from optimal control theory that the solution of the above problem can be obtained by solving the two boundary values problem given by

$$\dot{x}(t) = f(x(t)) + g(x(t)) u(t) \tag{16}$$

$$\dot{p}(t) = -\left[\frac{\partial f}{\partial x}(x(t)) + \sum_{i=1}^{m} u_i \frac{\partial g_i}{\partial x}(x(t)) \right] p(t) - Q x(t) \tag{17}$$

$$u(t) = -R^{-1} g^T(x) p(t) \tag{18}$$

together with the boundary conditions

$$x(0) = x_0 \quad ; \quad p(T) = 2\rho x(T) \tag{19}$$

where $g(x) =: \left[g_1(x), \dots, g_m(x) \right] \in \Re^{n\times m}$.
In order to simplify the expressions, we shall use the following compact notations

$$\dot{x}(t) = S_1(x(t), p(t)) \tag{20}$$

$$\dot{p}(t) = S_2(x(t), p(t)) \tag{21}$$

$$x(0) = x_0 \quad ; \quad p(T) = 2\rho x(T) \tag{22}$$

3.2 The discrete approximation

In order to derive the set of nonlinear equations, we shall use finite difference discretization defined on a uniform grid given by:

$$t_1 = 0 \quad ; \quad t_{k+1} = t_k + h$$
$$h = \frac{T}{N_G - 1} \quad ; \quad N_G \geq 3 \tag{23}$$

such that $t_{N_G} = T$, where N_G is a given integer. Note that our result holds for any non uniform but still regular partition of $J = [0,1]$ in which the maximal step length tends to zero.
Denoting by $x^k = x(t_k) \in \Re^n$ and $p^k = p(t_k) \in \Re^n$, the value of the state vector and the adjoint state vector respectively at the grid points, the discretized version of (20)-(22) is given by

$$x^{k+1} - x^k = hS_1(x^k, p^k) \qquad k = 1 \dots N_G - 1 \quad (24)$$

$$p^{k+1} - p^k = hS_2(x^k, p^k) \qquad k = 1 \dots N_G - 1 \quad (25)$$

$$x^1 = x_0 \quad ; \quad p^{N_G} = 2\rho x^{N_G} \quad (26)$$

Now, define the vector $X \in \Re^{(2N_G-2)n}$ of unknowns by

$$X := \left[x^2, \dots, x^{N_G}, p^1, \dots, p^{N_G-1}\right]^T \quad (27)$$

Equations (24)-(26) can readily be written in the following compact form

$$A(\rho)X = b(x_0) + hS(x_0, X) \quad (28)$$

We shall use the following straightforward result:

lemma 2.
The set of equations (28) can be written in the following form

$$X = \Phi(\rho)x_0 + hM(\rho)S(x_0, X) \quad (29)$$

where $\Phi(\rho)$ and $M(\rho)$ are given by

$$\Phi(\rho) = (\, I_n \quad I_n \quad \dots \quad I_n \quad \rho I_n \quad \dots \quad \rho I_n \,)^T \quad ; \quad M(\rho) =$$

$$\left(\begin{array}{ccccc|ccccc}
I_n & 0_n & 0_n & \dots & 0_n & 0_n & 0_n & 0_n & \dots & 0_n \\
I_n & I_n & 0_n & \dots & 0_n & 0_n & 0_n & 0_n & \dots & 0_n \\
\vdots & \vdots & \vdots & \dots & \vdots & \vdots & \vdots & \dots & \vdots & \vdots \\
I_n & I_n & \dots & I_n & I_n & 0_n & 0_n & 0_n & \dots & 0_n \\
\hline
\rho I_n & \rho I_n & \rho I_n & \dots & \rho I_n & I_n & I_n & I_n & \dots & I_n \\
\rho I_n & \rho I_n & \rho I_n & \dots & \rho I_n & 0_n & I_n & I_n & \dots & I_n \\
\vdots & \vdots & \vdots & \dots & \vdots & \vdots & \vdots & \dots & \vdots & \vdots \\
\rho I_n & \rho I_n & \rho I_n & \dots & \rho I_n & 0_n & 0_n & 0_n & \dots & I_n
\end{array}\right) \quad (30)$$

♠

4 Feasibility of the homotopy based algorithm

We recall that using the finite difference discretization scheme, the continuous stationarity conditions (20)-(22) have been replaced by their finite dimensional discrete version (24)-(26) that can be reduced to (29).
When discretization of optimal control problems is concerned, two aspects can be separately addressed. The first is to examine the conditions under which the solutions to the discrete problem (29) -if any- converges to that of the original problem (20)-(22) when the maximum step size h tends to 0. Some answers to this question can be found in (Dontchev and Hager, 1993; Dontchev, 1996; Schwartz and Polak, 1996) and the references therein. In this paper, we assume that the conditions leading to convergence hold (see point 3 of assumption 1 hereafter). The second aspect is to examine numerical algorithms that enable efficient solving of the highly nonlinear set of equations (29). This includes examining how "*global*" is the convergence of such algorithms. This is the aim of our work as long as continuation methods are considered.

Following the continuation approach explained in section 2, let us consider the homotopy

$$H^{x_0}(\lambda, X) := X - \Phi(\rho)(\lambda x_0) - h.\lambda.M(\rho)S(\lambda x_0, X) \quad (31)$$

Note that $H^{x_0}(0,0) = 0$. Note also that $H^{x_0}(1, X) = 0$ iff (29) is satisfied. Hence, $H^{x_0}(\lambda, X)$ defines a family of sets of nonlinear equations into which (29) can be imbedded in order to apply homotopy like iterations as explained in section 2.

In order to establish our results, we need the following assumptions:

Assumption 1. [Necessary conditions]

Let $T_{max} > 0$ be a maximal horizon length over which optimal control problems (20)-(22) are to be considered.

(1) f and h in (15) are twice continuously differentiable,
(2) For all $r > 0$ there is a closed ball $B(0, \alpha(r)) \subset \Re^n$ such that for all $x_0 \in B(0, r)$, if $(x^*(., x_0), p^*(., x_0))$ is an optimal solution of the continuous optimal control problem (20)-(22) defined for some $T \leq T_{max}$, then we have the following boundness condition for all $t \in [0, T]$:

$$\left(x^*(t, x_0), p^*(t, x_0)\right) \in B(0, \alpha(r)) \times B(0, \alpha(r)) \quad (32)$$

(3) For all $r > 0$, and all $\epsilon > 0$, There is a sufficiently high integer N_G such that for all $h \leq \dfrac{T_{max}}{N_G - 1}$ and for all $x_0 \in B(0, r)$, if

$$\bar{X}(x_0) := \left[\bar{x}^2(x_0), \dots, \bar{x}^{N_G}(x_0), \bar{p}^1(x_0), \dots, \bar{p}^{N_G-1}(x_0)\right]^T$$

is a solution of (29) with initial condition x_0, then there is an optimal solution

$$\left(x^*(., x_0), p^*(., x_0)\right)$$

of (20)-(22) with $T = (N_G - 1)h$ and initial condition x_0 such that

$$\sup_{k=1\dots N_G} \| \bar{x}^k(x_0) - x^*(t_k, x_0) \| +$$

$$\sup_{k=1\dots N_G} \| \bar{p}^k(x_0) - p^*(t_k, x_0) \| < \epsilon \quad (33)$$

where the t_k's are defined by (23).

♠

Condition 2 of the above assumption states that all optimal solutions of continuous problems (20)-(22) for bounded initial conditions and bounded horizon lengths are "*uniformly*" bounded.
Condition 3 states that any solution of the set of nonlinear equations (29) parameterized by x_0 can

be viewed as the solution of the discrete version of some continuous optimal control problem with initial condition x_0 and horizon length $T = (N_G - 1)h$, and hence is "*close*" to a solution of this continuous problem. This condition is somehow a convergence condition on the corresponding finite difference discretization scheme.
We shall first prove that under the above conditions, there is a continuous path of solutions $\left(\lambda, \bar{X}(\lambda)\right)_{\lambda \in [0,1]}$ to $H^{x_0}(\lambda, x) = 0$

Proposition 1. [**Existence of a continuous path**]
Under assumption 1, for all $r > 0$, there is a maximal horizon length T_0 $(0 < T_0 \le T_{max})$ such that, for sufficiently high N_G, all continuous optimal control problems (20)-(22) defined for $T \le T_0$ and initial condition $x_0 \in B(0, r)$ lead to a set of nonlinear equations that admits a continuous path of solutions. Namely:
there is a continuous map $\bar{X} : [0,1] \to \Re^{(2N_G-2)n}$ such that for all $\lambda \in [0,1]$, we have

$$H\left(\lambda, \bar{X}(\lambda)\right) = 0 \tag{34}$$

Furthermore,

$$T_0 \ge \min(T_{max}, \frac{1}{8n^2 \max(1,\rho) K(\| x_0 \|)}) \tag{35}$$

where $K(\| x_0 \|)$ is given by (40) ♠

Sketch of the proof
Existence
We shall first prove that there is a compact set $\mathcal{B}_{x_0}$ that contains all solutions $\bar{X}$ of $H^{x_0}(\lambda, X) = 0$ with $\lambda \in [0,1]$ whenever $(N_G - 1)h \le T_{max}$.
Indeed, let $\bar{X}$ be a solution of $H^{x_0}(\lambda, X) = 0$ for some $\lambda \in [0,1]$, then $\bar{X}$ is a solution of (29) with initial condition $\lambda x_0 \in B(0, \| x_0 \|)$ and horizon length $\lambda(N_G - 1)h \le T_{max}$. Therefore, combining (32) of 2) and and (33) of 3) of assumption 1, we can easily prove that the set

$$\mathcal{B}_{x_0} := B\left(0, \alpha(\| x_0 \|) + \epsilon\right) \times \dots \times B\left(0, \alpha(\| x_0 \|) + \epsilon\right) \tag{36}$$

clearly meets the above requirement.
Define the set $\mathcal{I} \subset [0,1]$ by

$$\mathcal{I} := \left\{\lambda \in [0,1] \text{ s.t } H^{x_0}(\lambda, X) = 0 \text{ has a solution}\right\} \tag{37}$$

Clearly, it is sufficient to prove that $\mathcal{I}$ is non empty and that $\sup \mathcal{I} = 1$. First, note that $\mathcal{I}$ is non empty because it contains 0 (Indeed $H^{x_0}(0,0) = 0$). Being bounded, this set admits an upper bound $\bar{\lambda} \in [0,1]$. Suppose that $\bar{\lambda} < 1$, we shall prove that this leads to a contradiction.

To do so, note first that by continuity of H w.r.t its arguments, $\mathcal{I}$ is a closed set and therefore $\bar{\lambda} \in \mathcal{I}$. Let $\bar{X}(\bar{\lambda})$ be the corresponding solution, namely $H^{x_0}(\bar{\lambda}, \bar{X}(\bar{\lambda})) = 0$. If we can prove that under the assumptions of the proposition, $\frac{\partial H^{x_0}}{\partial X}(\bar{\lambda}, \bar{X}(\bar{\lambda}))$ is full rank, we can then apply the implicit function theorem to prove that there is convenient $d\lambda > 0$ and dX such that

$$H^{x_0}(\bar{\lambda} + d\lambda, \bar{X}(\bar{\lambda}) + dX) = 0$$

which contradicts the definition (maximality) of $\bar{\lambda}$.
Therefore, all we have to prove is that under the assumptions of the proposition, $\frac{\partial H^{x_0}}{\partial X}(\bar{\lambda}, \bar{X}(\bar{\lambda}))$ is full rank. We have by definition of H^{x_0} and using straightforward notations:

$$\frac{\partial H^{x_0}}{\partial X}(\bar{\lambda}, \bar{X}(\bar{\lambda})) = I_{(2N_G-2)n} - h\bar{\lambda}\Delta(\bar{\lambda}, x_0, \rho) \tag{38}$$

Using the special structure of $M(\rho)$ and $\Delta(\bar{\lambda}, x_0, \rho)$. According to the first point in the present proof, given the special structure (30) of $M(\rho)$ and $\frac{\partial S}{\partial X}(\bar{\lambda} x_0, \bar{X}(\bar{\lambda}))$, it is easy to prove the following inequality for all i, j

$$|\Delta_{ij}(\bar{\lambda}, x_0, \rho)| \le 2n \max(1, \rho) K(\| x_0 \|) \tag{39}$$

where

$$K(\| x_0 \|) := \max_{i=1,2 \ j=1 \dots n} \left(\sup_{(x,p) \in \mathcal{B}_{x_0}} \| \frac{\partial S_i}{\partial x_j}(x,p) \|, \sup_{(x,p) \in \mathcal{B}_{x_0}} \| \frac{\partial S_i}{\partial p_j}(x,p) \| \right) \tag{40}$$

therefore,

$$\| h\bar{\lambda}\Delta(\bar{\lambda}, x_0, \rho) \| \le 4n^2 \max(1, \rho) T K(\| x_0 \|) \tag{41}$$

Now, using the "*perturbation lemma*" (see (taylor, 1958)) it comes that if we take T_0 such that:

$$T_0 \le \min(T_{max}, \frac{1}{8n^2 \max(1,\rho) K(\| x_0 \|)}) \tag{42}$$

then for all $T \le T_0$, $\frac{\partial H^{x_0}}{\partial X}(\bar{\lambda}, \bar{X}(\bar{\lambda}))$ is full rank, and this ends the proof of the first part.

Continuity

Let $\lambda_1, \lambda_2 \in [0,1] \times [0,1]$ and the corresponding pair of solutions $(X_1, X_2) := (\bar{X}(\lambda_1), \bar{X}(\lambda_2)) \in \mathcal{B}_{x_0} \times \mathcal{B}_{x_0}$. We have by definition

$$X_2 - X_1 = (\lambda_2 - \lambda_1)\Phi(\rho)x_0 + h\lambda_2 M(\rho)\left[S(\lambda_2 x_0, X_2) - S(\lambda_1 x_0, X_1)\right] + h\left[\lambda_2 - \lambda_1\right] M(\rho) S(\lambda_1 x_0, X_1) \tag{43}$$

Using compactness arguments together with (41) one can find positive constants $\beta_1 > 0$, $\beta_2 > 0$, $\beta_3 > 0$, and $\beta_4 > 0$ such that (43) implies

$$\| X_2 - X_1 \| \leq \beta_1 |\lambda_2 - \lambda_1| + h\Big[\beta_2 \| X_2(\lambda_2) - X_1(\lambda_1) \| + \beta_3 |\lambda_2 - \lambda_1| + \beta_4 |\lambda_2 - \lambda_1|\Big] \quad (44)$$

with β_2 satisfying

$$\beta_2 \leq 4n^2 \max(1, \rho)(N_G - 1)K(\| x_0 \|) \quad (45)$$

and taking the limit when $|\lambda_2 - \lambda_1|$ tends to 0

$$\lim_{\lambda_1 \to \lambda_2} \| X_2(\lambda_2) - X_1(\lambda_1) \| \leq h\beta_2 \| X_2(\lambda_2) - X_1(\lambda_1) \| \quad (46)$$

if $T \leq T_0$ with T_0 satisfying (42) one has $h\beta_2 \leq 1/2$ and hence (46) implies

$$\lim_{\lambda_1 \to \lambda_2} \| X_2(\lambda_2) - X_1(\lambda_1) \| = 0$$

which proves the continuity of the path and hence ends the proof of proposition 1. ◇

Now, we have all we need to state the main result of this paper

Theorem 3. [Feasibility of homotopy-based algorithms in solving (29)]

Under the conditions 2) and 3) of assumption 1, if f and g in (15) are C^3, then for all $r > 0$ there is a maximal horizon length T_0 $(0 < T_0 \leq T_{max})$ such that, for sufficiently high N_G, all continuous optimal control problems (20)-(22) defined for $T \leq T_0$ and initial condition $x_0 \in B(0, r)$ lead to a set of nonlinear equations (29) for which the homotopy (31) together with

$$G^{x_0}(\lambda, X) := X - H_x^{x_0}(\lambda, X)^{-1} H^{x_0}(\lambda, X) \quad (47)$$

meet the requirements of theorem (2) and hence (47) can be used to define a feasible converging homotopy-based algorithm to solve (29). Furthermore a lower bound on T_0 is given by (35). ♠

Proof

All we have to prove is that under the assumptions of the theorem, those of theorem 2 hold.

Since f and g are supposed to be C^3, H^{x_0} naturally satisfies (12) and all the more so, it has a continuous partial derivative $H_x^{x_0}$. The existence of the continuous function $\bar{X}(\lambda)$ comes from proposition 1 while the existence and the boundness of $H_x^{x_0}(\lambda, \bar{X}(\lambda))^{-1}$ have already been established during the proof of proposition 1 and using the perturbation lemma. In particular, it has been shown that under the constraint $T \leq T_0$, one has

$$\| H_x^{x_0}(\lambda, \bar{X}(\lambda))^{-1} \| \leq 2$$

Therefore, theorem 2 applies and the conclusions of the present theorem follow. ♠

5 REFERENCES

Avila, J. H. (1974). The feasibility of continuation methods for nonlinear systems. *Siam J. Numer. Anal.* **11, No 1**, 102–122.

Deuflhard, P., H. J. Pesch and P. Rentrop (1976). A modified continuation method for numerical solution of nonlinear two-point boundary value problems by shooting techniques. *Numerische Mathematik* **26**, 327–343.

Dontchev, A. L. (1996). An a priori estimate for discrete approximations in nonlinear optimal control. *Siam J. Control and Optimization* **34, No 4**, 1315–1328.

Dontchev, A. L. and W. W. Hager (1993). Lipschitz stability in nonlinear control and optimization. *Siam J. Control and Optimization* **31**, 569–603.

Liao, S. J. (1997). High-order bem formulation for strongly nonlinear problems governed by quite general nonlinear differential operators. part 2 : Some 2d examples. *Int. J. for Numerical Methods in Fluids* **24, No 9**, 863–873.

Meyer, G. H. (1968). On solving nonlinear equations with one-parameter operator imbedding. *Siam J. Numerical Analysis* **5, No 4**, 739–752.

Polak, E. (1973). An historical survey of computational methods in optimal control. *Siam Review* **15, No 2**, 553–584.

Reddien, G. W. (1979). Collocation at gauss points as a discretization in optimal control. *Siam J. Control and Optimization* **17, No 2**, 298–306.

Schwartz, A. and E. Polak (1996). Consistent approximations for optimal control problems based on runge-kutta integration. *Siam J. Control and Optimization* **34, No 4**, 1235–1269.

taylor, A. (1958). *Introduction to Functional Analysis.* John Wiley, New York.

Vlassenbroek, J. and R. Van Dooren (1988). A chebychev technique for solving nonlinear optimal control problems. *IEEE Transactions on Automatic Control* **33, No 4**, 333–340.

Watson, L. T. (1979). An algorithme that is globally convergent with probability one for a class of nonlinear two-point boundary value problems. *Siam J. Numerical Analysis* **16, No 3**, 394–401.

Yong, L. and L. Xianrui (1995). Continuation theorems for boundary value problems. *Journal of Mathematical Analysis and Applications* **190**, 32–49.

Zhang, L. Q. and G. Q. Han (1993). Optimal homotopy methods for solving nonlinear systems. *Numerische Mathematik* **65**, 523–538.

CONSTRAINED NONLINEAR CONTROL DESIGN VIA ELLIPSOIDAL TECHNIQUES

Jie Chen[1], Zhihao Gong[1], Sándor M Veres[2]

Department of Automatic Control, Beijing Institute of Technology, P.O.Box327, Beijing 100081, P. R. China. Email: Chenjie@bit.edu.cn[1]
School of Electronic & Electrical Eng., The University of Birmingham, Edgbaston, B15 2TT, UK[2]

Abstract: The aim of this paper is to introduce new methods of nonlinear robust controller design under state and input constraints using ellipsoidal techniques. The controller stabilises and controls the plant to a final invariant set within certain time under bounded state-disturbances. The advantage of the ellipsoidal techniques is that they allow the computation of a controller in state space of relatively high dimension (≥20). Several simulation examples are shown to illustrate the implementation aspects. *Copyright © 1999 IFAC*

Keywords: Nonlinear systems, Robustness, Optimal control, Constraints, Geometric properties, Ellipsoidal methods.

1. INTRODUCTION

In most process control applications there is a need for controllers which are able to take into account constraints both on input and on state variables. In addition, the presence of model and state disturbances underline the importance of robust nonlinear control methods.

Nonlinear control under constraints has been the topic of some recent investigations, such as(Gilbert, et al., 1991; MayneMayne & Schroeder , 1994;J. Chen, S.M.Veres,1997) , both in theory and practice. Gilbert & Tan (1991,1994)defined the concept of maximal output admissible set and employed it to construct controllers. Mayne & Schroeder developed value functions for robust, time-optimal problems, and employed level sets to construct nonlinear controllers(MayneMayne & Schroeder ,1994a,b, 1997). Ellipsoidal techniques for uncertain systems have been developed by Kurzhanski & Valyi(1992).

Our goal in this paper is to develop an effective design method and numerical algorithm of robust constrained nonlinear control using ellipsoidal techniques.The results of this paper are organised in following way. Section 2 outlines the principles of the robust nonlinear control solution for a linear system under state and input constraints. Computation of robust control invariant sets using ellipsoids, the relevant ellipsoid approximations and guaranteed strategies are presented in Section 3. Simulated examples and the associated numerical optimisation problem are discussed in order to show the effectiveness of the method in Section 4.

2. THE CONSTRAINED CONTROL PROBLEM

2.1 Construction of The Controller

Consider a continuous time system of the form

$$\dot{x}(t) = A_c x(t) + B_c u(t, x(t)) + w(t), \qquad (1)$$

where $x(t) \in \Re^n$, $u(t,x(t)) \in \Re^m$ and $w(t) \in \Re^n$ A_c and B_c are $n{\times}n$ and $n{\times}m$ matrices with entries in the set of real numbers $\Re$. The input, state and disturbance are subject to the following constraints:

$u(t,x(t)) \in W$, $x(t) \in E$, $w(t) \in W_c$, where $\Omega \subset \Re^m$ is convex and compact, $E \subset \Re^n$ is convex and closed, and W_c is a convex compact set; each set containing the origin in its interior.

Assume that in the continuous time case the following hold:

(i) The pair (A_c, B_c) is controllable.
(ii) The state x is accessible.
(iii) Ω is a polytope in $\Re^m$, $0 \in int(\Omega)$ such that $u(k) \in \Omega,\ k \geq 0$.
(iv) E is a polyhedron in $\Re^n$, $0 \in \text{int}(E)$ such that $x(k) \in E,\ k \geq 0$.

Since the state perturbation w is present, the state cannot be steered to the origin and kept there. The best that can be hoped for is that the controller steers any initial state to a terminal set $\mathcal{X}_\infty$, (a control invariant set, see Definition 1 below), and maintains the state in this set once it is reached.

Model (1) can be approximated by the discrete time model

$$x[k+1] = A_d x[k] + B_d u[k, x[k]] + w[k]. \qquad (2)$$

It is assumed that sampling period T_s is such that A_d and B_d are controllable. $w(k) \in W$. W is a polytope in $\Re^n$, $0 \in \text{int}(W)$ .The initial state is $x[k_0]=x_0$

Definition 1 (Gilbert & Tan, 1991) The set $X \subset E$ is said to be robust control invariant for the discrete time system (2), if for any $x \in X$ there is a $u[k, x[k]] \in \Omega$ such that

$$A_d x[k] + B_d u[k, x[k]] + w[k] \subset X. \qquad (3)$$

If X is a robust control invariant set for (2), then there is a control law $u[k, x[k]] : x \to \Omega$, such that

$$A_d x[k] + B_d u[k, x[k]] + w[k] \subset X \qquad (4)$$

for all $x[k] \in X$, all $k \geq 0$, and every admissible disturbance sequence. For this system our conditions ensure the existence of a piece-wise constant value function. Level sets $\{X_k\}$ of the value function $f(x) = \inf_u \sup_w \{k \mid x(k) \in \mathcal{X}_\infty\}$, for this problem are generated by the following algorithm which is altered from Mayne and Schroeder(1994b). For later use, the geometric difference of two sets A and B is defined by: $A \succ B := \{x \mid x + B \subseteq A\}$.

Lemma 1 The sum of two polytopes, the intersection of two polytopes, and the intersection of a polytope and a polyhedron, are all polytopes. If **A** is a matrix, and X is a polytope, then AX is a polytope.

Theorem 1 (Mayne and Schroeder, 1994) Let the terminal set $\mathcal{X}_\infty$ ($X_o = \mathcal{X}_\infty$) be a robust control invariant polytope. Then

(i) the above algorithm is well defined and generates a non-decreasing sequence X_k of robust control invariant polytopes for the system (A_d, B_d, Ω, W, E);
(ii) for all k, all $x \in X_k$, there exists a $u \in \Omega$ such that $A_d x + B_d u[k, x] + w[k] \subset X_{k-1}$;
(iii) there exists a control law which satisfies the state and control constraints and which robustly steers the discrete time system from any initial state in X_k to the set $\mathcal{X}_\infty$ within k steps, subsequently maintaining the state in $\mathcal{X}_\infty$.

It is a direct consequence of Theorem 1 that Algorithm 1(Mayne & Schroeder,1994b) yields the level sets of the value function and the value function itself.

2.2 Detemination of Terminal Set $\mathcal{X}_\infty$

Consider a discrete time system with a linear model of the form

$$x[k+1] = A_d x[k] + B_d u[k, x[k]] + w[k]$$
$$u[k, x[k]] = Kx[k] \qquad (5)$$

With a suitable choice of matrices C_d and D_d and a set Y, all state and input constraints can be summarised by a single set inclusion:(Gilbert, et al., 1991,1994)

$$C_d x[k] + D_d u[k, x[k]] \in Y \qquad (6)$$

Let now a triple $A_d \in \Re^{n \times n}$, $C_d \in \Re^{p \times n}$, $Y \subset \Re^p$ be given. We are going to determine if the trajectory of the discrete-time system $x[k+1] = A_d x[k] + w[k]$, $y[k] = C_d x[k]$ satisfies the output constraint $y[k] \in Y$, for all $k \in Z^+$, where Z^+ is the set of nonnegative integers. Problem (5), (6) fits this format by making the assignments $A_d + B_d K \to A_d$, $C_d + D_d K \to C_d$, and $Y \to Y$.
Assume that in the discrete time case the following hold:

(v) The set Y is convex, compact, bounded and $0 \in \text{int}(Y)$.
(vi) C_d, A_d are observable, and A_d is Lyapunov stable.
(vii) The rank of $A_d + B_d K$ is the same as the dimension of the state space.
(viii) $A_d + B_d K$ is asymptotically stable.

For discrete time systems the terminal set can be defined by

$$\chi_\infty = \{x \in \Re^n : C_d A_d^k x \in Y \quad \forall k \in Z^+\} \qquad (7)$$

where Y is given by the inequalities:

$$Y = \{y \in \Re^p : \ f_i(y) \le 0, \ i = 1, \dots, s\}. \qquad (8)$$

Some properties of $\mathcal{X}_\infty$ are summarised in the following lemma.

Lemma 2 Let $\mathcal{X}_\infty$ be a terminal set of the form (7) for the system (A_d, C_d, Y). Suppose assumptions (vi) to (viii) are satisfied. Then
(i) each of the following properties of $\mathbf{Y}$: closure, convexity, symmetry and boundedness are inherited by $\mathcal{X}_\infty$.
(ii) $0 \in \text{int}(\mathcal{X}_\infty)$,
(iii) $\chi_\infty(A_d, C_d, aY) = a\chi_\infty(A_d, C_d, Y)$ for all $a \in \Re$.

Let

$\chi_k = \{x \in \Re^n : C_d A_d^t x \in Y \quad \forall t \le k\}, \quad k \in Z^+$.

Obviously, the set $\mathcal{X}_k$ satisfies the following condition: $\mathcal{X}_\infty \subset \mathcal{X}_{k_2} \subset \mathcal{X}_{k_1}$, $\forall k_1, k_2 \in Z^+$ such that $k_1 \le k_2$. $\mathcal{X}_\infty$ is finitely determined if for some $k \in Z^+$, $\mathcal{X}_\infty = \mathcal{X}_k$.

Theorem 2 Suppose the functions $f_i : \Re^p \to \Re$, i=1,2,…,s, are continuous, $f_i(0) \le 0$ i=1,2,…,s and assumption (vi) is satisfied. Then there are functions $g_i : \Re^n \to \Re$,

$g_i(x) = \sup\{f_i(C_d A_d^k x) : k \in Z^+\}$, i=1,2,…,s,

such that $\mathcal{X}_\infty = \{x \in \Re^n : g_i(x) \le 0, \ i = 1,\dots,s\}$.

If Y is given by (7) then $\mathcal{X}_\infty$ is finitely determined by

$$\mathcal{X}_\infty = \{x \in \Re^n : f_i(C_d A_d^k x) \le 0, \ i \in \{1,\dots,s\}, \ k \in \{0,\dots,k^*\}\} \qquad (9)$$

where k' is the smallest element in Z^+ such that $\mathcal{X}_\infty = \mathcal{X}_{k^*}$. It may happen that not all of the s inequalities in (9) are active in the sense that there exists a $x \in \chi_\infty$ such that $f_i(C_d A_d^k x) = 0$. The active inequalities have a special structure which is given by the following theorem.

Theorem 3 (Gilbert & Tan, 1991) Suppose $\mathcal{X}_\infty$ is given by (9). Then there exists a non-empty set of integers $S^* \subset \{1,\dots,s\}$ and indexes k_i^*, $i \in S^*$, such that
(i) $k^* = \max\{k_i^* : i \in S^*\}$;
(ii) $\chi_\infty = \{x \in \Re^n : f_i(C_d A_d^k x) \le 0, \ i \in S^*, \ k \in \{0,\dots,k_i^*\}\}$;
(iii) for all $i \in S^*$ and $k \in \{0,\dots,k_i^*\}$ there exists and $x \in \mathcal{X}_\infty$ such that $f_i(C_d A_d^k x) = 0$.

If $\mathbf{Y}$ is defined by (8) and the hypotheses of Lemma 2 are satisfied, then $\mathcal{X}_\infty$ can be calculated by Algorithm 2 from Gilbert and Tan(1991,1994).

2.3 Determination of A Controller

For discrete time system of form $x[k+1] = A_d x[k] + B_d u[k, x[k]]$, the control $u[k, x[k]]$ is chosen such that $A_d x[k] + B_d u[k, x[k]] \in E$.
Let $P := B_d^{-1}(E - A_d x[k])$, so the control must satisfies

$$u[k, x[k]] \in P. \qquad (10)$$

Furthermore, if the current state x[k] is in X_k, the state x[k] may be steered into next state X_{k-1} by an admissible control, whatever is the disturbance, i.e the next state must be in $X'_{k-1} = X_{k-1} \succ W$. It causes another condition that $u[k, x[k]]$ has to satisfy: $A_d x[k] + B_d u[k, x[k]] \in X'_{k-1}$. Let $Q := B_d^{-1}(X'_{k-1} - A_d x[k])$, so the control must satisfies

$$u[k, x[k]] \in Q, \qquad (11)$$

For control constraints as mentioned above, the control $u[k, x[k]]$ must obviously be inside the control constraint Ω, i.e

$$u[k, x[k]] \in \Omega \qquad (12)$$

From (10),(11),(12), therefore the control input has to satisfy

$$u[k, x[k]] \in R = P \cap Q \cap \Omega. \qquad (13)$$

3. ELLIPSOID APPROXIMATIONS

The information encoded in the sets $\{X_k\}$ defined in Section 2, i.e. the level sets of the value function for system (2), is sufficient to construct a controller. Since computation with polytopes $\{X_k\}$ can become too complicated for practical purposes, this section introduces ellipsoidal approximations.

3.1 Problem of Ellipsoid Optimisation

Algorithm 1 computes some polytopes starting from the terminal set $X_0 = \mathcal{X}_\infty$. It is assumed that all the constraints can be formulated with polytopes. Those polytopes that are computed will be ordered as to find which is the first one where the current state lies.

Then the control input $u[k, x[k]]$ is computed and kept constant during the next sampling period T_s. Most of the calculations are geometrical in nature, and all the functions can be computed using the Geometric Bounding Toolbox (GBT) (S.M.Veres,1996). For dimensions greater than 5, the algorithms of computing the sum of two polytopes, intersection of some polytopes and convex-hull set of a set of points are sometimes prohibitively slow on today's processors and it also demands large amounts of RAM. Therefore optimisation of the above algorithm using ellipsoidal techniques is presented in this section. This optimisation replaces all the polytopes by approximate ellipsoids. Computations with ellipsoids are faster than computations with polytopes, and ellipsoids require less memory space. Ellipsoidal approximations are able to deal with high dimensions of state space and with the disturbance, thereby keeping the controller robust. The principle of the algorithm is to replay all the polytope operations by ellipsoids so that it is ensured that the basic properties of the controller will remain intact. Therefore, the issue is how to organise a scheme of ellipsoidal approximations to perform these operations.

Set X is no longer initialised with the terminal set X_0 ($X_0 = \mathcal{X}_\infty$) but with a maximal ellipsoid $\overline{X}$ inscribed in X_0. This ellipsoid will be our terminal set where the state can be maintained by applying a controller that keeps the state in $\overline{X}_0'$. First, the controller determines in which ellipsoid the state lies by taking all the ellipsoids into account one by one.

Algorithm (Ellipsoid algorithm)

Given: polytope ($\boldsymbol{X}_0' = \mathcal{X}_\infty$) from Algorithm 2;

Step 1: compute maximal volume ellipsoid $\overline{\boldsymbol{X}}_0$ inscribed $\overline{\boldsymbol{X}}_0'$;

Step 2: let $k = 0$;

*Step 3: compute the minimal volume ellipsoid $\overline{W}$ cover **W**;*

Step 4: compute the tight inner bounding ellipsoid: $\overline{\boldsymbol{X}}_k' = \overline{\boldsymbol{X}}_k \succ \overline{W}$;

*Step 5: compute the maximal volume ellipsoids $\overline{\Omega}$ and $\overline{\boldsymbol{E}}$ inscribed in polytopes Ω and **E**;*

Step 6: compute the set:

$$\overline{\boldsymbol{X}}_{k+1} = \boldsymbol{A}_d^{-1}\overline{\boldsymbol{X}}_k' - \boldsymbol{A}_d^{-1}\boldsymbol{B}_d\overline{W};$$

Step 7: let k=k+1;

*Step 8: if all vertices of **E** are inside the $\overline{\boldsymbol{X}}_k$, go to step 9; else, go to step 4;*

Step 9: let k=0, and current state $\boldsymbol{x}[k] = x_0$ (initial state);

Step 10: compute which ellipsoid $\overline{\boldsymbol{X}}_{k'}$ the current state x[k] lies inside;

Step 11: if $\overline{\boldsymbol{X}}_{k'} = \overline{\boldsymbol{X}}_0$, stop; else, go to step 12.

Step 12: compute set $\boldsymbol{P} = \boldsymbol{B}_d^{-1}(\overline{\boldsymbol{E}} - \boldsymbol{A}_d\boldsymbol{x}[k])$;

Step 13: compute set $\boldsymbol{Q} = \boldsymbol{B}_d^{-1}(\overline{\boldsymbol{X}}_{k'-1}' - \boldsymbol{A}_d\boldsymbol{x}[k])$;

Step 14: compute set $\overline{\boldsymbol{R}} = \boldsymbol{P} \cap \boldsymbol{Q} \cap \overline{\Omega}$;

Step 15: compute the centre value of $\overline{R}$ as the single-valued control $\boldsymbol{u}[k, x[k]]$;

Step 16: compute $\mathbf{x}[k+1]$ by the control $\boldsymbol{u}[k, x[k]]$ is applied to the plant during a time T_s;

Step 17: let k=k+1, go to step10.

3.2 Ellipsoidal Techniques

Ellipsoids are given in the form $\mathcal{E} = \mathcal{E}(a, \mathcal{K}) = \{x: (x-a)^T\mathcal{K}(x-a) \le 1,\ x \in \Re^n\}$, where $\mathcal{K} = \{k_{ij}\}_{i,j=1}^n$ is a symmetric positive definite $n \times n$ dimensional matrix and a is the centre of the ellipsoid. Suppose that two ellipsoids $\mathcal{E}_1 = \mathcal{E}(a_1, \mathcal{K}_1)$, and $\mathcal{E}_2 = \mathcal{E}(a_2, \mathcal{K}_2)$ are given. The sum of them is not an ellipsoid in general, and their geometrical difference is not either. Let

$$\mathcal{K}^{1,2}(\pi) = (\pi_1 + \pi_2)(\pi_1^{-1}\mathcal{K}_1 + \pi_2^{-1}\mathcal{K}_2)$$

$$\mathcal{K}_+^{1,2}[S] = S^{-1}[(S\mathcal{K}_1S)^{1/2} + (S\mathcal{K}_2S)^{1/2}]^2 S^{-1},$$

$$\mathcal{K}_-^{1,2}[S] = S^{-1}[(S\mathcal{K}_1S)^{1/2} - (S\mathcal{K}_2S)^{1/2}]^2 S^{-1}., \quad (14)$$

where: $\pi = (\pi_1, \pi_2) \in \prod_-$

$$\prod_+ = \{(\pi_1, \pi_2) \in \Re^2 \mid \pi_1/\pi_2 \in [\lambda_{min}^{1/2}, \lambda_{max}^{1/2}]\},$$

$$\prod_- = \{(\pi_1, -\pi_2) \mid (\pi_1, \pi_2) \in \prod_+, \pi_1/\pi_2 < \lambda_{\min}\}, \quad (15)$$

where λ_{min}, λ_{max} are the smallest and the largest solutions to the equation $det(\mathcal{K}_1 - \lambda\mathcal{K}_2) = 0$, and S is any positive definite and symmetric matrices.

Theorem 4: (Kurzhanski & Valyi., 1991,1992)
(1) The following equalities are true:

$$\overline{\cup \{\mathcal{E}(a_1 + a_2, \mathcal{K}_+^{1,2}[S])\}} = \mathcal{E}(a_1, \mathcal{K}_1) + \mathcal{E}(a_2, \mathcal{K}_2)$$
$$= \cap\{\mathcal{E}(a_1 + a_2, \mathcal{K}^{1,2}(\pi)) \mid \pi \in \prod_+\}, \quad (16)$$

where $\overline{\mathcal{A}}$ stands for the closure of set $\mathcal{A}$.
(2) Under the condition $\mathcal{E}(a_1, \mathcal{K}_1) \succ \mathcal{E}(a_2, \mathcal{K}_2) \neq 0$, the following equations are true:

$$\overline{\cup \{\mathcal{E}(a_1 - a_2, \mathcal{K}^{1,2}(\pi)) \mid \pi \in \prod_-\}}$$
$$\subset \mathcal{E}(a_1, \mathcal{K}_1) \succ \mathcal{E}(a_2, \mathcal{K}_2) \subset \cap\{\mathcal{E}(a_1 - a_2, \mathcal{K}_+^{1,2}[S]\} \quad (17)$$

where S is any positive definite and symmetric matrices.

Further operations are needed to calculate maximal volume internal ellipsoid and intersection of ellipsoids.

(i) *Computation of the maximal volume ellipsoid inscribed in a polyhedron.*

Let X is a convex polyhedron given by a finite set of linear inequalities $X=\{x\!: b^T x \leq 1,\ i=1,...,m\}$. The problem of calculating the maximal volume ellipsoid inscribed in a polyhedron X can be transformed into the following mathematical programming problem

$$\min\{f_0^*(z) : f_i^*(z) \leq 0,\ i=1,...,m,\ z \in E\}$$

where: $K^{-1}=Q^*=\{q_{ij}\}_{i,j=1}^n$,

$$f_0^*(z) = -\ln\ \det Q_i^*$$

$$f_i^*(z) = \sqrt{b_i^T Q^* b_i} + b_i^T a - 1, \quad (18)$$

$E = \{z := \{\{q_{ij}\}_{i,j=1}^n, \{a_i\}_{i=1}^n\},\!: Q^*$ is positive definite$\}$. Two approaches will be considered to solve this problem(J. Chen, S. M. Veres, 1997).

(ii) *Computation of the intersection of ellipsoids*

Given ellipsoids $\mathcal{E}(a_1, \mathcal{K}_1)$, and $\mathcal{E}(a_2, \mathcal{K}_2)$, their intersection $\check{\mathcal{E}} = \mathcal{E}(a_1, \mathcal{K}_1) \cap \mathcal{E}(a_2, \mathcal{K}_2)$ could be approximated from a

$$\check{\mathcal{E}} = \sum_{i=1}^{2} \mathcal{E}(M_i a_i, M_i \mathcal{K}_i M_i^T) \quad \text{with } M_1 + M_2 = I,$$

and M_1, M_2 are $n \times n$ matrices and T stands for the transpose.

GBT can compute intersection of two ellipsoids, further research is planned about determination of $M_i$in case of several ellipsoids.

3.3 Guaranteed Strategies

This section presents how to find internal ellipsoids in the X_k sets. Part of this solution is to find ellipsoids $\overline{X}_k$, in which for any x there is a control input $u[k, x[k]] \in \Omega$ such that

$$A_d x[k] + B_d u[k, x[k]] \in X'_{k-1}. \quad (19)$$

Assuming

$$P := B_d^{-1}(E - A_d x[k]), \quad (20)$$

$Q := B_d^{-1}(X'_{k-1} - A_d x[k])$ are ellipsoids and $\overline{\Omega}$ is the maximal volume ellipsoids inscribed in polytopes Ω, the minimal volume ellipsoid $\overline{R} = \varepsilon(a_{\bar{u}}, k_{\bar{u}}) = \min\{R_i \mid R_i = P \cap Q \cap \overline{\Omega}\}$ may be found from the family of ellipsoids which is defined by a linear combination of inequalities defining the individual ellipsoids. Instead of the set-valued control, a single-valued control $u[k, x[k]]$ is selected, i.e. $u[k, x[k]] = a_{\bar{u}}$.

Theorem 5 Assume that for k=0, 1 ,... , N, the ellipsoids $\overline{E}_k$ and $\overline{X}_{k+1}$ are constructed so that,

(i) $\overline{E}_k \subset B_d u[k, x[k]] - \overline{X}_k$;

(ii) $\overline{X}_{k+1} \subset A_d^{-1} \overline{E}_k \cap \overline{E}$; hold.

Then $\overline{X}_N \subset X_N$, i.e. $\overline{X}_N$ is a set of states from which the set X_0 can be reached within N time periods.

4. NUMERICAL EXAMPLES

The experiment of inverted pendulum with cart consistes of a cart, a pendulum and an obstruction. As shown in Fig.4, the pendulum can rotate in the vertical plane by low torque u_2 and the cart can move by force u_1. The obstruction is fixed on top. This is multi-input and multi-output nonlinear system. The goal is that the inverted pendulum should avoid the obstruction andregain balance when it moves from position S_1 to position S_2 by using the above nonlinear control.

The state and input are subject to the following constraints. $x \in \Omega = P_1 \succ P_2$, is nonconvex polytope, where $P_1 = [-5\ -2\ -2\ -5] \times [5\ \ 2\ \ 2\ \ 5] \in \Re^4$, (where A is 4×4 matrix and b is 4×1 vector) is restriction which the inverted pendulum must avoid due to the construction. P_1 and P_2 are both convex polytopes .

$$u \in E = [-7\ -7\ -7\ -7] \times [7\ \ 7\ \ 7\ \ 7] \in \Re^4; \quad (21)$$

$$w \in W = [-0.1\ -0.1\ -0.1\ -0.1] \times [0.1\ \ 0.1\ \ 0.1\ \ 0.1] \in \Re^4; \quad (22)$$

Using the constrained nonlinear control, the results of simulations are presented below. Table 1 show the outputs of the system and state trajectory with initial state = $[4.95, 0, 0, 0]^T$, sampling period $T_s = 0.05$, and without white noise disturbance. As shown in

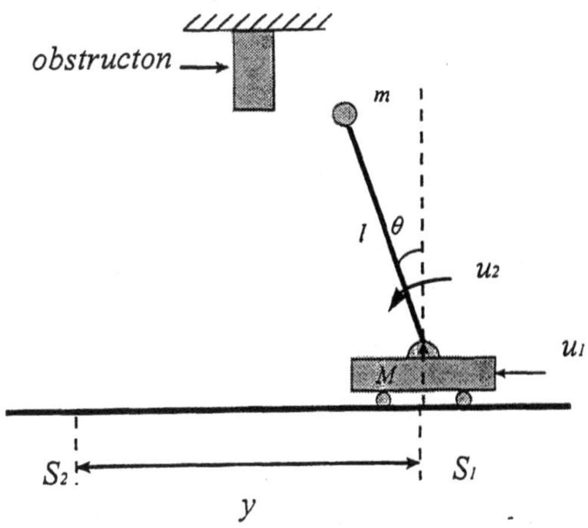

Fig. 4: Inverted pendulum

Table 1 :Control result I

Outputs	x_1	x_2	x_3	x_4
0	4.9500	0	0	0
1	4.7406	-0.1260	-0.0149	-0.5978
2	4.5250	-0.2576	-0.0508	-0.8492
3	4.3032	-0.3818	-0.0908	-0.7591
4	4.0852	-0.1174	-0.1585	-1.9674
5	3.8634	-0.5428	-0.2334	-1.0462
6	3.6281	-0.6727	-0.2695	-0.4078
7	3.3874	-0.7684	-0.2748	0.1950
8	3.1457	-0.7306	-0.2652	0.1889
9	2.9062	-0.6910	-0.2465	0.5666
10	2.6700	-0.6138	-0.2131	0.7772
11	2.4426	-0.3471	-0.1719	0.8813
12	2.2303	-0.0254	-0.1367	0.5344

Table 2: Control result II

Outputs	x_1	x_2	x_3	x_4
0	4.9500	0	0	0
1	4.8105	-0.4314	-0.0312	-0.0081
2	4.5960	-0.4758	-0.0938	-0.2882
3	4.2416	-0.5342	-0.1347	-0.2350
4	4.0352	0.4618	-0.2235	-3.8404
5	3.8284	-0.6348	-0.2989	-0.9264
6	3.6655	-0.7847	-0.2475	-0.0269
7	3.4094	-0.7327	-0.2896	0.0362
8	3.2195	-0.6187	-0.3629	0.1430
9	2.8994	-0.4488	-0.4303	0.0798
10	2.6514	-0.4413	-0.3485	0.1137
11	2.4713	-0.3433	-0.3233	0.1381
12	2.3153	-0.1584	-0.2300	0.1975
13	2.0196	0.1052	-0.0803	0.0096

Table 2, the state is robustly steered into the terminal polytope from initial state in 14 steps with disturbance.

5 CONCLUSION

The paper first outlined the design of robust nonlinear controllers under state and input constraints. Then ellipsoidal techniques have been presented. The computations using optimised ellipsoidal techniques are faster than using non-optimised ones. Ellipsoidal algorithms are able to perform calculation in a high-dimensional state-space. The controller can deal with the disturbance even with the ellipsoid approximation the controller is still robust. The computations were illustrated using the Geometric Bounding Toolbox, Version 5.2.

REFERENCES

Gilbert, E.G. and K.T.Tan(1991). Linear Systems with State and Control Constraints: the Theory and Application of Maximal Output Admissible Sets. *IEEE trans. on Auto. Control*, **AC-36**, 1008-1020.

Gilbert, E.G., I.Kolmanovsky and K.T.Tan(1994). Nonlinear control of Discrete-time Linear systems with State and Control Constraints: A Reference Governor with Global Convergence Properties. *Proceedings of 33rd. CDC*, 144-149.

J. Chen, S.M.Veres(1997). *Constrained Nonlinear Control Design via Ellipsoidal Techniques*. The University of Birmingham.

Mayne, D.Q. & W.R.Schroeder(1994a). Nonlinear Control of Constrained Dynamic System. *Int. J. of Control*, **60**, 1035-1043.

Mayne, D.Q. & W.R.Schroeder(1994b). Nonlinear Control of Constrained Linear Systems: regulation and Tracking. *Proceedings of 33rd. CDC*, 2370-2375.

Mayne, D.Q. & W.R.Schroeder(1997). Robust Time-Optimal Control of Constrained Linear Systems.

S.M.Veres *et al*(1996). *The Geometric Bounding Toolbox.*, Version 5.1 (licensed by the University of Birmingham, School of E&E Eng. WWW-reference: *http://www.eee.bham.ac.uk/gbt*).

Valyi, I.(1992). Ellipsoidal Techniques for Dynamic Systems. *Thesis for Obtaining the Candidate's Degree*, Budapest.

ON A POLYNOMIAL INTERPOLATION METHOD FOR THE EXACT LINEARIZATION PROBLEM

Eurice de Souza *,[1] **Edilberto P. Teixeira** **

* *Departamento de Engenharia Elétrica, Universidade de Brasília, FAX 0055(061)274.6651, ZIP 70910-070, Brasília, DF, BRAZIL, e-mail:eurice@ene.unb.br*
** *Departamento de Engenharia Elétrica, Universidade Federal de Uberlândia, Uberlândia, MG, BRAZIL*

Abstract: A polynomial interpolation method is described for the evaluation of a diffeomorphic transformation and state-feedback for the approximate linearization of a nonlinear system. The method can be applied for the states lying inside a parallelepiped. In the case of an involutive system, the linearization error is sufficiently small whenever the error on the evaluation of the partial derivatives is negligible. An example is shown to display the performance of the method. *Copyright © 1999 IFAC*

Keywords: Linearization Feedback Linearization Nonlinear Systems Polynomials Interpolation Algorithms Polynomial Methods.

1. INTRODUCTION

Exact linearization has been widely used to grant a smooth control of the trajectories of a nonlinear system and for robotic systems it has provided a satisfactory performance. The method has been well studied in the literature (Isidori 1989) and (Slotine and Li 1991) for scalar as well as for MIMO systems. The method is subdivided into:

- Input-Output Based Linearization Method;
- State-Feedback Based Linearization Method;

For the input-output linearization of an affine SISO system, the relative degree, that is defined as the number of time derivatives of the system output not depending on the system input, assumes an important role. For a system to be fully input-output linearizable, the relative degree, r, must be equal to the number of states, n. Similar results exist for MIMO systems. When $r < n$, two subsystems result, one linearized subsystem and one residual subsystem, known in the literature as *zero dynamics*. The zero dynamics has similar properties to those of a linear system described by a transfer function. The linearized system is stable whenever the residual subsystem behaves as a system with a transfer function with all the poles in the complex open left half plane. Around the equilibrium, the zero dynamics can be approximated by a linear system whose poles are equal to the zeros of the plant transfer function. Unfortunately, some nonlinear systems behave as nonminimal-phase systems, having a linear approximation with zeros on the right half complex plane. Thus, in general, the zero dynamics is unstable. Although several methods have been proposed to stabilize the residual subsystem (Devasia *et al.* 1996) and (Isidori and Byrnes 1990), stability is only guaranted around the equilibrium. For the state-feedback linearization problem, the following difficulties arise:

- the states must be measurable;
- some severe conditions should be satisfied for the problem to have an exact solution
- even with consistency, a difficult set of partial differential equations (PDE) need to be solved.

In this paper the two latter problems are considered. The system dynamical equation is sup-

[1] Partially supported by Universidade de Brasília. During 1996 and 1997, supported by FAPEMIG, the Minas Gerais State research support foundation

posed to be affine with respect to the input. Using the discretization and vectorization of the state coordinates, constrained to a parallelepiped, a method is introduced for the solution of the exact linearization problem. For the vectorized coordinates, Kronecker products are used to represent the Lie derivatives. So, the set of PDE of difficult solution is transformed into a system of algebraic equations, in general inconsistent. For an involutive system, the solution lies sufficiently close to the exact one, whenever the error of the partial derivatives approximation is sufficiently small.

2. PROBLEM FORMULATION

Consider a dynamical system described by the following differential equation

$$\dot{x} = f(x) + g(x)u, \tag{1}$$

where $x \in \Re^n$ is the state vector defined on the region X, supposed to be a parallelepiped. f and g are vector fields whose coordinates are supposed analytic functions of the state vectors and u is the system input. Without loss of generality, it is supposed that the point $x = 0$ is an equilibrium point of (1). Also consider the problem of finding a diffeomorphic transformation $z = \Theta(x)$ in the region X and the control

$$u = \alpha(x) + \beta(x)v, \tag{2}$$

where $\alpha(x)$ and $\beta(x)$ are smooth functions and v is the new input, such that, in the new coordinates, the system can be described by the linear differential equation

$$\dot{z} = Az + Bv, \tag{3}$$

where (A, B) is a controllable pair and A and B have appropriate dimensions.

Let $z = \{z^{(1)}, z^{(2)}, \ldots, z^{(n)}\}$ be the set of coordinates for the linearized system. Evaluating the time derivatives for the new state vector, one gets

$$\partial\Theta/\partial x \times (f(x) + g(x)\alpha(x)) = Az, \tag{4}$$

$$\partial\Theta/\partial x \times (g(x)\beta(x)) = B. \tag{5}$$

Basically, PDE's (4, 5) need to be solved to provide a change of coordinates plus the control action represented by the functions $\alpha(x)$ and $\beta(x)$.

A brief review of the main results of (Isidori 1989) is presented to establish the basis for the adequate interpretation of the results related to the poynomial solution. First, the derivative of a scalar function along a vector field is defined.

Definition 1. Let $f(.)$ be a vector field and ϕ a real function. The derivative of ϕ along f is defined as

$$L_f\phi := (\partial\phi/\partial x)\, f \tag{6}$$

Also, the recursive Lie derivative is defined by

$$L_f^k\phi := L_f(L_f^{k-1}\phi) \tag{7}$$

Consider a locally controllable system satisfying the involutivity conditions (Frobenius Theorem). In this case, the Brunowsky canonical form can be obtained by a nonlinear coordinate change combined to a state feedback. Hence, the standard procedure consists on supposing the matrices A and B in such canonical form. Thus, using the definition above, one can write the PDE's (4, 5)

$$L_g\phi = 0,\ L_gL_f\phi = 0,\ \ldots,\ L_gL_f^{n-2}\phi = 0, \tag{8}$$

$$L_gL_f^{n-1}\phi \neq 0. \tag{9}$$

The consistency of (8) depends on the involutivity of a certain distribution. The condition (9) is to prevent the nullity of the control channel.

2.1 *Consistency and Nonunicity*

Before attempting to search for the solution for the exact linearization problem, the important topic is considered about the nonunicity of the solution of (8, 9) (when it exists). In the sequell, a polynomial solution is searched in the form below. Also, all the restrictive equations for the problem are established only in terms of the Lie derivatives in the directions f and g. Hence, the consistency (or quasi-consistency) conditions should be established directly in terms of the polynomial analogue of equations (8, 9).

It seems that the matter of nonunicity has been explored more with the aim of getting the most simple solution (Isidori 1989) and (Slotine and Li 1991). In a recent paper (deSouza *et al.* 1998) established the generic solution for the complete linearization problem and showed how this can be used to exploit the possibilities beyond the pole assignment on the linearized diffeomorphic copy of the system. Here, the main result of (deSouza *et al.* 1998) is stated.

Theorem 1. Let a smooth function $z^{(1)} = \phi(x)$ be a first coordinate solution for the exact linearization of an affine single input system for x belonging to a neighborhood U of x^0. Then any solution can be described in the form $z^{(1)} = \theta(\phi(x))$, where $\theta(.)$ is a smooth function satisfying the nontriviality condition $\frac{\partial\theta}{\partial\phi} \neq 0$ in the region U.

According to the preceding results, if ϕ is a particular first coordinate solution, then the whole set of solutions can be generated by means of a generic function $\psi(x) = \theta(\phi(x))$, where θ is an arbitrary function such that $\frac{\partial\theta}{\partial\phi}$ does not vanish. Notice that the solution set for such a problem, when it exists, is nonunique. Furthermore, even in the case of a polynomial solution, there is an infinite set of polynomial solutions providing good approximations for the same problem. Next, polynomial functions and their respective degrees are defined.

2.2 *Existence of Polynomial Solutions*

Now, the special case is considered where (8, 9) has a polynomial solution. Considering that a polynomial does not provide an exact solution for the problem, a polynomial pseudosolution, as defined below, could yield an acceptable approximate solution, This is because the new coordinates are analytic functions of the state variables whenever the coordinate functions of f and g are analytic functions of the state vector x. First, a definition is given for a multivariable polynomial function.

Definition 2. Let $x \in \Re^n$. $\phi(x)$ is said to be a polynomial function of degree $a = \{a_1, \ldots, a_n\}$ if and only if it can be written in one of the forms

$$\phi(x_1, \ldots, x_{i-1}, x_i, x_{i+1}, \ldots, x_n) = \sum_{k=0}^{a_i} \alpha_k(x_1, \ldots, x_{i-1}, x_{i+1}, \ldots, x_n) \times x_i^k, \quad (10)$$

for $i = 1, \ldots, n$, *i.e.*, $\phi(x)$ is a polynomial of degree a_i with respect to the variable x_i.

Here is a generic polynomial function for $a = \{2, 3\}$:

$$\begin{aligned}\phi(x_1, x_2) = {} & \left(\alpha_{00} + \alpha_{10}x_2 + \alpha_{20}x_2^2 + \alpha_{30}x_2^3\right) + \\ & \left(\alpha_{01} + \alpha_{11}x_2 + \alpha_{21}x_2^2 + \alpha_{31}x_2^3\right) x_1 + \\ & \left(\alpha_{02} + \alpha_{12}x_2 + \alpha_{22}x_2^2 + \alpha_{32}x_2^3\right) x_1^2\end{aligned}$$

Now, exact polynomial solution is defined.

Definition 3. The PDE's (4, 5) admit an exact polynomial solution if for some $a = \{a_1, \ldots, a_n\}$ the equations (8, 9) are consistent.

3. VECTORIZATION OF THE PARALLELEPIPED

In the following, vectorization of the parallelepiped coordinates as well as Kronecker products are used to derive matrix vector relations for the numerical evaluation of the linearizing feedback with the associate diffeomorphic transformation. First, some *ad hoc* notations are introduced.

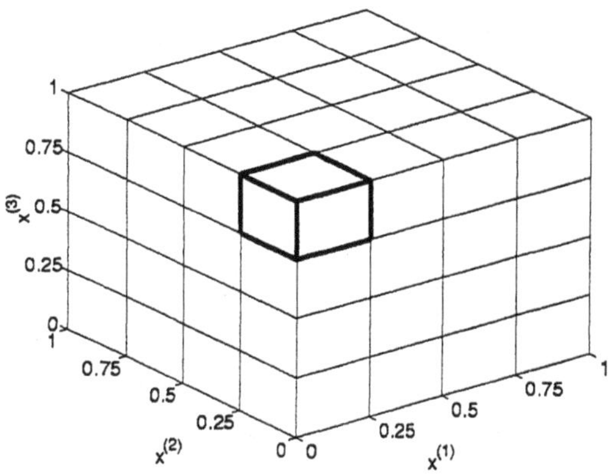

Fig. 1. Cellular Partition of the States

Let $x^{(j)}$ be the $j-th$ coordinate of the state vector,

$$x = \left[x^{(1)}\ x^{(2)}\ \ldots\ x^{(n)}\right]^T \quad (11)$$

then

$$\omega := \omega(x) = \omega\left(x^{(1)},\ x^{(2)},\ \ldots,\ x^{(n)}\right) \quad (12)$$

is a function defined on the system coordinates $x^{(j)}$, $j = 1, \ldots, n$. Consider that each coordinate $x^{(j)}$ has been discretized in m_k points. So, the following vector associated to the samples taken along the $x^{(j)}$ coordinate can be represented by

$$x_s^{(j)} := \left(x_{i_j}^{(j)}\right) = \left[x_1^{(j)},\ x_2^{(j)},\ \ldots,\ x_{m_j}^{(j)}\right) \quad (13)$$

Discretizing the coordinates (13), the function in (12) can be described by

$$\omega := \omega\left(i_1, \ldots, i_n\right) = \omega\left(x_{i_1}^{(1)}, \ldots, x_{i_n}^{(n)}\right) \quad (14)$$

$$i_1 = 1,\ \ldots,\ m_1,\ \ \ldots,\ i_n = 1,\ \ldots,\ m_n;$$

As shown in Fig. 1, the discretization of the state coordinates partitions the state parallelepiped into $m = (m_1 - 1) \times \ldots \times (m_n - 1)$ hypercube cells. Also, a typical hypercube cell is shown.

For the discretization of the function ω, the numbering of the vertices of the hypercube cells is used. So, the following vector coordinates are introduced for the discretized function $\hat{\omega}$,

$$\hat{\omega} := \hat{\omega}\left(l\left(i_1,\ \ldots,\ i_n\right)\right), \quad (15)$$

where the $l-$coordinates associated to the vector $\hat{\omega}$ can be expressed in terms of the coordinates $(i_1, \ldots, i_n)$, according to the one-to-one relation

$$l = i_1 + m_1 i_2 + \ldots + (m_1 \times \ldots \times m_{n-1}) i_n \quad (16)$$

3.1 *Evaluatin of the Lie Derivatives*

For the ease of visualization, set the dimension of the subspace X equal to 3. Also, consider that each coordinate has been discretized into 3 levels. Using $l-$coordinates in (16), the discrete function $\hat{\omega}$ in (15) can be represented by

$$\hat{\omega} = col(\hat{\omega}_{111}, \hat{\omega}_{112}, \hat{\omega}_{113}, \vdots \hat{\omega}_{121}, \hat{\omega}_{122}, \hat{\omega}_{123}, |\hat{\omega}_{131}, \hat{\omega}_{132}, \hat{\omega}_{133}, |\hat{\omega}_{211}, \ldots, \hat{\omega}_{233}, |\hat{\omega}_{311}, \ldots, \hat{\omega}_{333}). \tag{17}$$

Let D_i be the $m_i \times m_i$ matrix representation for the derivatives along the $x_{i_k}^{(i)}$ coordinates obtained through an appropriate interpolation scheme. Applying the derivative operator to the vectorized version of the function ω in (15), there results:

$$\frac{\partial \omega}{\partial x^{(1)}} := [[\hat{\omega}_{111}\hat{\omega}_{112}\hat{\omega}_{113}]\, D_1^T\, [\hat{\omega}_{121}\hat{\omega}_{122}\hat{\omega}_{123}]\, D_1^T \, [\hat{\omega}_{131}\hat{\omega}_{132}\hat{\omega}_{133}]\, D_1^T \ldots [\hat{\omega}_{131}\hat{\omega}_{132}\hat{\omega}_{133}]\, D_1^T]^T \tag{18}$$

The $l-$subscript indicates that the derivatives are evaluated at the $l-$coordinates. Let $D^{(k)}$ represent the partial derivative with respect to $x^{(k)}$, taken on the $l-$coordinates. Then, using Kronecker products, (18) can be written.

$$\frac{\partial \omega}{\partial x^{(1)}} := D^{(1)} \times \hat{\omega} = I_3 \otimes I_2 \otimes D_1 \times \hat{\omega} \tag{19}$$

where $I_k \in \Re^{m_k \times m_k}$ are identity matrices. Similar reasoning for other coordinates and, next, for arbitrary dimensions, yields

$$D^{(k)} = I_n \otimes \ldots I_{k+1} \otimes D_k \otimes I_{k-1} \ldots \otimes I_1. \tag{20}$$

Now, the discrete representation for the Lie derivative of ω along the vector field $\xi(x)$, $(\frac{\partial \omega}{\partial x}\xi(x))$, can be obtained. Expanding the directional derivative in the form

$$\left(\frac{\partial \omega}{\partial x}\xi(x)\right) = \sum_{s=1}^{n} \xi_s(x) \frac{\partial \omega}{\partial x^{(s)}},$$

and using approximations for the partial derivatives,

$$\left(\frac{\partial \omega}{\partial x}\xi(x)\right)\bigg|_l := \Xi\hat{\omega} = \sum_{s=1}^{n} \left[diag(\xi_s) D^{(s)}\right] \hat{\omega}. \tag{21}$$

Next, using (21) to approximate the Lie derivatives, error vetorized equations are deduced.

3.2 *Vectorizing the Linearization Error*

Let $A = (a_{ij}), B = (b_i), i, j = 1, \ldots, n$ be the state space representation with respect to a new set of (linearizing) coordinates, z, and let $u = \alpha(x) + \beta(x)v$, as in (2). The continuous linearization error can be written for $i = 1, \ldots, n$:

$$\epsilon^{(i)} = \frac{\partial z^{(i)}}{\partial x}(f + g\alpha + g\beta)\, v - \sum_{j=1}^{n} a_{ij} z^{(j)} - b_i v, \tag{22}$$

whose discrete version can be written as

$$\hat{\epsilon}^{(i)} = \sum_{s=1}^{n} diag\left(\hat{f}_s\right) D^{(s)} \hat{z}^{(i)} + \sum_{s=1}^{n} diag\,(\hat{\alpha} \odot \hat{g}_s)\, D^{(s)} \hat{z}^{(i)} - \sum_{j=1}^{n} a_{ij} \hat{z}^{(j)} + \sum_{s=1}^{n} diag\left(\hat{v} \odot \hat{\beta} \odot \hat{g}_s\right) D^{(s)} \hat{z}^{(i)} - b_i \hat{v}, \tag{23}$$

where " $\odot$ " is used to denote the Hadamard coordinate-by-coordinate product and $\hat{\alpha}$ and $\hat{\beta}$ are the discrete versions of the functions α and β:

$$\hat{\alpha} = [\hat{\alpha}_1 \ldots \hat{\alpha}_m]^T, \ \hat{\beta} = \left[\hat{\beta}_1 \ldots \hat{\beta}_m\right]^T. \tag{24}$$
$$m = m_1 \times m_2 \times \ldots \times m_n.$$

From the properties of Hadamard products, one gets

$$\hat{\epsilon}^{(i)} = (\Phi + diag(\hat{\alpha})\Gamma)\, \hat{z}^{(i)} - \sum_{j=1}^{n} a_{ij} \hat{z}^{(j)} + diag\,(\hat{v})\, diag(\hat{\beta})\Gamma \hat{z}^{(i)} - b_i \hat{v}, \tag{25}$$

where Φ and Γ are the discrete approximations for the Lie derivatives along f and g:

$$\Phi = \sum_{s=1}^{n} diag(\hat{f}_s) D^{(s)}, \tag{26}$$

$$\Gamma = \sum_{s=1}^{n} diag(\hat{g}_s) D^{(s)}, \tag{27}$$

Separating the $\hat{v}$ terms in (25) for $i = 1, \ldots, n$, there follow two equivalent error equations:

$$E_z = (\Phi + diag\,(\hat{\alpha})\,\Gamma)\, \hat{Z} - \hat{z}A^T \tag{28}$$

$$E_\beta = diag\left(\hat{\beta}\right) \Gamma \hat{Z} - B^T \otimes d_m, \tag{29}$$

where d_m and $\hat{Z}$ are the $m \times 1$ and $m \times n$ matrices defined below

$$d_m = [1\ 1\ \ldots\ 1]; \tag{30}$$

$$\hat{Z} = \left[\hat{z}^{(1)}\ \hat{z}^{(2)}\ \ldots\ \hat{z}^{n}\right] \tag{31}$$

4. SOLVING THE ERROR EQUATIONS

In order to formulate an efficient algorithm to solve the couple of equations (28, 29), first, the linearized system is considered in the Brunowsky canonical form,

$$[A \mid B] = \begin{bmatrix} 0 & I_{n-1} & \vdots & 0 \\ 0 & 0 & \vdots & 1 \end{bmatrix} \quad (32)$$

where the zeros represent blocks of zeros of appropriate dimensions. Then, from (29), the following conditions result

$$\Gamma \hat{z}^{(i)} = 0, \; i = 1, \ldots, n-1, \quad (33)$$

$$\hat{\beta} \odot \left(\Gamma \hat{z}^{(n)} \right) = d_m. \quad (34)$$

Considering the conditions (33), equations (28) can be written as

$$\Phi \hat{z}^{(i)} - \hat{z}^{(i+1)} = 0, \; i = 1, \ldots, n-1, \quad (35)$$

$$\Phi \hat{z}^{(n)} + \hat{\alpha} \odot \left(\Gamma \hat{z}^{(n)} \right) = 0. \quad (36)$$

Eliminating the vectors $\hat{z}^{(2)}, \ldots, \hat{z}^{(n)}$, from (35, 36), first one gets

$$M \hat{z}^{(1)} = 0, \quad (37)$$

$$M = \left[(\Gamma)^T \; (\Gamma\Phi)^T \; \ldots \; (\Gamma\Phi^{n-2})^T \right]^T. \quad (38)$$

Now, the vectors associated to the other coordinates and the linearizing vectors follow

$$\hat{z}^{(i+1)} = \Phi \hat{z}^{(i)}, \; i = 1, \ldots, n-1, \quad (39)$$

$$\hat{\alpha} = -\Phi^{(n)} \hat{z}^{(1)} / \left(\Gamma \Phi^{n-1} \hat{z}^{(1)} \right), \quad (40)$$

$$\hat{\beta} = d_m / \left(\Gamma \Phi^{n-1} \hat{z}^{(1)} \right), \quad (41)$$

the fraction sign being used here to denote coordinate-by-coordinate division.

Next, a proposition concerning exact polynomial solution for single input systems is presented

Proposition 1. Let L be the columns formed with the solution space of (37). For the existence of an exact polynomial solution of order $a = \{a_1, \ldots, a_n\}$ for the problem of state feedback exact linearization for the single input nonlinear dynamical system described by (1), the following conditions are necessary:

(1) the homogeneous equation (37) is consistent;
(2) there exists an element ζ^*, such that

$$\Gamma \Phi^{n-1} L \zeta^* > 0. \quad (42)$$

Proof: The demonstration of the first part is obvious and is omitted. For the second part, first assume $z^{(1)} = L\zeta^*$. Now, writing the latter identity in (34) in the form $\hat{\beta} \odot \left(\Gamma \Phi^{n-1} \hat{z}^{(1)} \right) = d_m$, in view of the continuity of the solution (polynomial solution, in the case), if, for any pair of adjacent samples, the input ponderation gain $\Gamma \Phi^{n-1} \hat{z}^{(1)}$ assumes distinct signals, using the Intermediate Value Theorem (applicable to continuous functions), in some intermediary point, the input ponderation will vanish.

NOTES:

- There is no loss of generality assuming the positive signal for $\Gamma \Phi^{n-1} \hat{z}^{(1)}$. Clearly, if $\hat{z}^{(1)} = L\zeta$ is a particular solution, then $\hat{z}^{(1)} = -L\zeta$ also is.
- The intermediate value theorem does not assure that, in the case of nonalternance of the signal in the interval between two adjacent samples, the solution cannot assume the trivial value in the interval.
- In the case of the problem (4, 5) admitting an exact polynomial solution, clearly there should be a vector ζ, such that $\Gamma \Phi^{n-1} L \zeta > 0$. This is stated by the theorem that follows.

Theorem 2. If the problem (4, 5) admits an exact polynomial solution,then there exists at least one vector ζ satisfying the nontriviality condition (42).

Proof: The proof is by contradiction, Suppose that (4, 5) admits a polynomial solution but there does not exist a vector ζ, such that the nontriviality condition (42) holds. But it contradicts the hypothesis of the theorem. The proof is complete.

4.1 *System Stabilization*

After the transformation of the system to the Brunowsky canonical form, an additional control action is necessary to stabilize the closed loop system. For this aim, the new control action $\hat{v}$ assumes the form $\hat{v} = \hat{Z}F + \hat{w}$, where $F \in \Re^n$ is a feedback gain vector and w is a new reference signal. Reasoning in terms of the linear model,

$$\dot{z} = Az + Bv, \quad v = F^T z + w. \quad (43)$$

The vector F can be fixed to shift the poles of $(A + BF^T)$ to a stable locus in the complex plane. The procedure can be implemented by the algorithm that follows.

Algorithm 1.

(1) For the affine single input system sampled as in (15, 16), build the matrices Φ, Γ and M as in (26, 27, 38). Solve the homogeneous system to express the solution as $\hat{z}^{(1)} = L\zeta$.

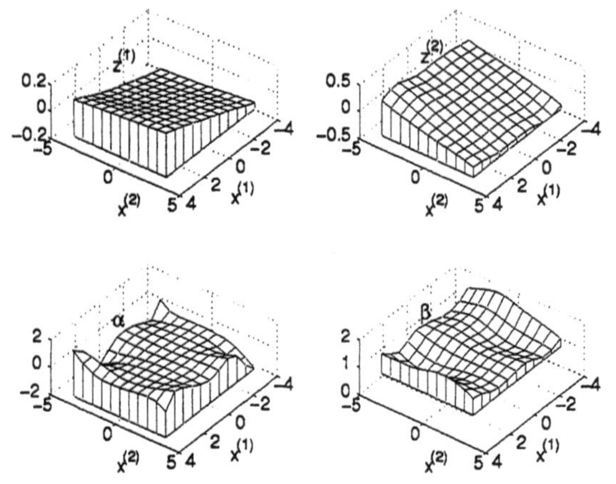

Fig. 2. Linearizing coordinates and functions.

(2) Find a solution of definite signal, *i.e.*, evaluate ζ^*, such that $\Gamma\Phi^{n-1}L\zeta^* > 0$.
(3) Take the sampled z−coordinates as $\hat{z}^{(1)} = L\zeta^*$, $\hat{z}^{(i+1)} = \Phi\hat{z}^{(i)}, i = 1, \ldots, n-1$.
(4) Evaluate the linearizing vectors adapting (40) in order to include the linearizing stabilizing actions, that is

$$\hat{\alpha} = -\Phi^{(n)}\hat{z}^{(1)} / \left(\Gamma\Phi^{n-1}\hat{z}^{(1)}\right) + \hat{Z}F,$$

4.1.1. *Example 1:* Consider the system described by the following dynamical equation

$$\dot{x} = \begin{bmatrix} x_2 + \frac{1}{5}x_1x_2 \\ -2x_1 \end{bmatrix} + \begin{bmatrix} -10 \\ 3 + \cos\sqrt{x_1^2 + x_2^2} \end{bmatrix} u \quad (44)$$

Using the expressions for f_1, f_2, g_1, g_2 given in (44), matrices Φ, Γ and M were built. With the tolerance used (10^{-5}), one solution space of dimension equal to 6 satisfying the consistency conditions was obtained. One adequate combination was taken to satisfy the nontriviality condition, yielding $\hat{z}^{(1)}$, $\hat{z}^{(2)}$, $\hat{\alpha}$ and $\hat{\beta}$ depicted in Fig. 2.

The closed loop poles were selected to stabilize the system and the system was simulated for a set of representative initial conditions. For each position x in the state space, the values of the linearizing feedback coordinates α and β were built using polynomial interpolation on the discrete values on the vertices of the hypercube cells stored in the vectors $\hat{\alpha}$ and $\hat{\beta}$. Simulation results can be summarized in the $x_1 \times x_2$ plots shown in Fig. 3.

Suppose the linearizing transformation and feedback are exact inside the parallelepiped. Once the trajectory escapes the interpolation region, calculation of the interpolation coefficients based on the dynamical behavior inside the parallelepiped may be used to approximate the system behavior on the outer region. However, as the trajectory escapes too far away from the interpolation region, stability may be lost. So, some zones of the interpolation region may sometimes be unstable.

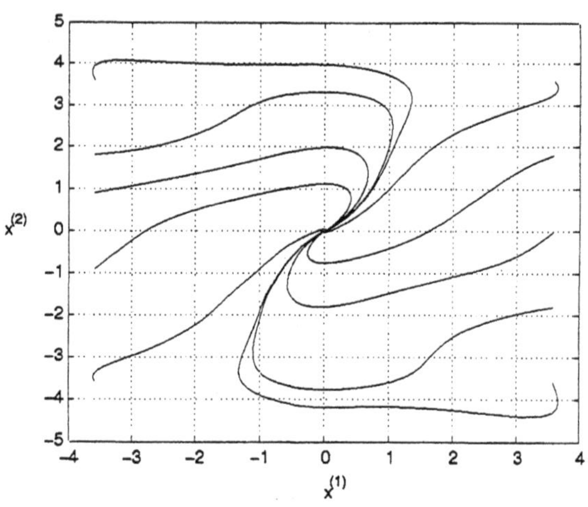

Fig. 3. Closed-loop system $x_1 \times x_2$ trajectories.

These arguments justify the importance of the choice of the linearized system poles to limit the overshoot of the trajectories next to the edges of the interpolation region. In the practice, with this procedure, the stability region may be enlarged.

5. CONCLUSION

A numerical method is presented for the exact linearization problem. The vector fields $f(x)$ and $g(x)$ are supposed to be analytic in terms of the state vector x and only the values of the vector fields on the vertices of the hypercube cells of the sample parallelepiped are required. No partial differential equation solution is required. Whenever the problem is consistent, an exact solution is obtained except for negligible rounding and interpolation errors. Simulation results point toward the need for an appropriate choice of the closed loop poles in order to prevent the state trajectory to escape the interpolation region.

6. REFERENCES

deSouza, E., E. P. Teixeira, J. M. Fernandes and H. R. Azevedo (1998). On the generic solution for the affine single input exact state feedback linearization problem. *Annals of the XII Brazilian Automatic Control Conference (XII CBA)* **IV**, 1415–1420.

Devasia, S., D. Chen and B. Paden (1996). Nonlinear inversion-based output tracking. *IEEE Trans. Automat. Contr.* **41**, 930–942.

Isidori, A. (1989). *Nonlinear Control Systems – An Introduction.* Springer Verlag. New York.

Isidori, A. and C. I. Byrnes (1990). Output regulation of nonlinear systems. *IEEE Trans. Automat. Contr.* **35**, 131–140.

Slotine, J.J. and W. Li (1991). *Applied Nonlinear Control.* Prentice Hall. New York.

14th Triennial World Congress, Beijing, P.R. China

E-2c-20-5

LINEARIZATION OF CONTROL SYSTEMS AND THE SWING-OUT

C. Yalçın Kaya *

* *School of Mathematics, University of South Australia, Mawson Lakes, S.A. 5095, Australia.* *yalcin.kaya@unisa.edu.au*

Abstract: Linearly approximating the nonlinear system so as to analyse it and design a control law, and then applying this control law to the original system has been a fairly common practice. The swing-out, as a measure of nonlinearity of a dynamical system, is incorporated in understanding the relationship between the small-time reachable sets of the nonlinear control system and its linear approximation in the plane. Linear approximation about nonequilibrium points are considered. Further facts are derived for a good linear approximation. Behaviour of the real system under a bang–bang control law particularly designed using the linear approximation of the system is examined. *Copyright ©1999 IFAC*

Keywords: Nonlinear systems, Swing-out, Small-time reachable sets, Bang–bang control.

1. INTRODUCTION

Linearizing the nonlinear system so as to analyse it and design a control law, and then applying this control law to the original system has been a fairly common practice. In doing so, variations in the dynamics are assumed to be small, so that the approximation would be valid. Sontag (1990) states the linearization dogma, "*Designs based on linearizations work locally for the original system*," and stresses that as it is not a theorem, it cannot be proved (like all other dogmas). However, he also points out that this dogma should be made more rigorous where possible. A compehensive survey on linearization, including *exact* linearization by means of state feedback, is given by Claude (1986). However we will focus only on the approximate linearization obtained through a first order Taylor series expansion.

The swing-out, which is a measure of nonlinearity of a dynamical system was previously introduced by Kaya & Noakes (1995). Given the linear approximation of a system about some operating point, it describes the rate the trajectory of the nonlinear system separates from the trajectory of its linear approximation at the point of approximation. Using the swing-out Kaya & Noakes (1995) examine the relationship between the small-time reachable sets of the nonlinear control system and its linear approximation in $\mathbb{R}^2$. In this work we continue this investigation by considering the swing-out in control systems in more detail. Further facts are derived for a good linear approximation. The linear approximation is considered to be carried out about a nonequilibrium point. Linearization about an equilibrium point is the subject of a future investigation.

A special emphasis is given to bang-bang control, given the fact that the bang-arcs define the small-time reachable sets for the systems in linear-analytic form. Although not directly used here, Hájek (1991) gives a theorem about local bang–bang optimal control with one switching, which should also be regarded as a motivation for considering this type of control regime. Behaviour of the real system under a bang–bang control particularly designed using the linear approximation of the system is examined.

2. THE SWING-OUT IN UNFORCED DYNAMICAL SYSTEMS

In this section, the formalism given in Kaya & Noakes (1995) for the swing-out in uncontrolled systems is outlined before furthering an investigation on the swing-out of control systems in Section 3.

Consider the unforced nonlinear dynamical system

$$\dot{x}(t) = X(x(t)) \qquad (1)$$

where $x(t) \in \mathbb{R}^n$, and X is a C^∞ vector field defined on $\mathbb{R}^n$. In subsequent references to this system it will often be convenient to omit reference to the dependence on t. Suppose the right-hand-side of the differential equation is given componentwise as the vector $X(x) = [a_1(x), \ldots, a_n(x)]^T$. Here $a_i : \mathbb{R}^n \to \mathbb{R}$. Then the vector field X can suitably be expressed as

$$X = \sum_{i=1}^{n} a_i(x) \left(\frac{\partial}{\partial x_i}\right)_x$$

Next we give an affine approximation to system (1) about some point $x(0) = x_0$ as

$$\dot{x}(t) = X_L(x(t)) \qquad (2)$$

where the affine vector field X_L is defined to be

$$X_L = \sum_{i=1}^{n} \left(a_i(x_0) + \frac{\partial a_i}{\partial x}(x_0)\,(x(t) - x_0)\right) \left(\frac{\partial}{\partial x_i}\right)_{x_0}$$

Note that when the vector field X_L is evaluated at x, the system (2) becomes

$$\dot{x} = X(x_0) + A(x - x_0) \qquad (3)$$

where A is an $n \times n$ matrix whose entries are given by $a_{ij} = (\partial a_i/\partial x_j)(x_0)$. When x_0 is an equilibrium point, i.e. $X(x_0) = \mathbf{0}$, (3) reduces to a linear system by allowing the transformation $z = x - x_0$ to get $\dot{z} = Az$, $z(0) = \mathbf{0}$. In this study system (3) will also be referred to as being *linearized at point* x_0 despite the constant term $X(x_0)$ which makes the system an *affine* one instead of linear.

Let Γ_{X,x_0} be the *integral curve* of X starting from x_0 for positive times. In other words, Γ_{X,x_0} is the solution curve of the dynamical system (1) with initial condition $x(0) = x_0$. Let $\Phi_{X,t} : V \to \mathbb{R}^n$, $V \subset \mathbb{R}^n$, be the *flow* of X such that $x_0 \mapsto \Gamma_{X,x_0}(t)$, that is to say, $\Phi_{X,t}(x_0) = \Gamma_{X,x_0}(t)$, $\Gamma_{X,x_0}(0) = x_0$.

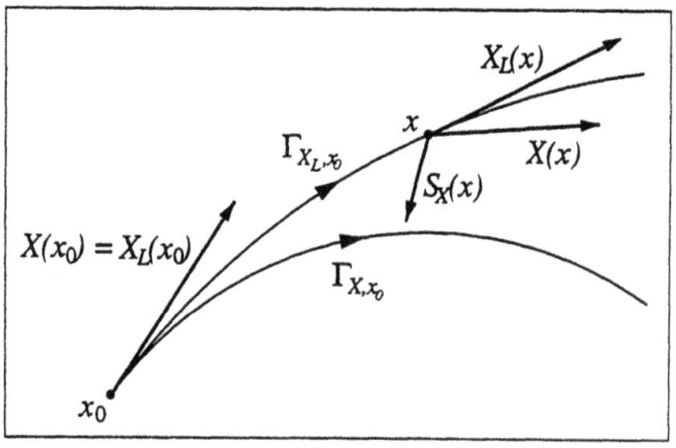

Fig. 1. An illustration of the swing-out S_X.

The *swing-out* S_X of X along X_L is a vector field given by

$$S_X = X - X_L \qquad (4)$$

The swing-out vector $S_X(x)$, as illustrated in Figure 1, is the rate of 'escape' of Γ_{X,x_0} from Γ_{X_L,x_0} at point x on Γ_{X_L,x_0}. It tells one how fast Γ_{X,x_0} deviates from Γ_{X_L,x_0} and therefore is a measure of nonlinearity. Note that $S_X(x_0) = 0$, or $X_L(x_0) = X(x_0)$, as expected.

An expansion involving the swing-out vector $S_X(x)$ can be given about x_0 by means of the Campbell-Baker-Hausdorff formula (Isidori 1989) as follows.

$$\begin{aligned}(\Phi_{X_L,-t})_*\; S_X(\Phi_{X_L,t}(x_0)) = {} & S_X(x_0) + t\, L_{X_L} S_X(x_0) \\ & + \frac{t^2}{2} L^2_{X_L} S_X(x_0) + O(t^3) \end{aligned} \qquad (5)$$

In this equation, the mapping $(\Phi_{X_L,-t})_*$ is the Jacobian of the flow Φ_{X_L} and it maps a vector in the tangent space at $x = \Phi_{X_L,t}(x_0)$ to another vector in the tangent space at x_0. It has the property that $(\Phi_{X_L,0})_* = \mathbf{1}$ and

$$(\Phi_{X_L,t})_*(\Phi_{X_L,-t})_* = \mathbf{1} \qquad (6)$$

where $\mathbf{1}$ is the identity map. In equation (5), $L_{X_L} S_X$, the Lie derivative of S_X with respect to X_L, can be written as the Lie bracket of X_L and S_X, i.e.

$$L_{X_L} S_X = [X_L, S_X]$$

where

$$[X_L, S_X](x) = \frac{\partial S_X}{\partial x}(x)\, X_L(x) - \frac{\partial X_L}{\partial x}(x)\, S_X(x) \qquad (7)$$

In (5), $L^2_{X_L} S_X = L_{X_L}(L_{X_L} S_X)$ and it is similarly expressed as

$$L^2_{X_L} S_X = [X_L, [X_L, S_X]]$$

Further, $[X_L, S_X] = [X_L, X - X_L] = [X_L, X]$ and $[X_L, [X_L, S_X]] = [X_L, [X_L, X]]$, using (4) and (7). Note that $[X_L, S_X](x_0) = \mathbf{0}$, since $X(x_0) = X_L(x_0)$ and $(\partial X/\partial x)(x_0) = (\partial X_L/\partial x)(x_0)$. Then (5) can be re-written by also using (6) as

$$S_X(x) = \frac{t^2}{2}(\Phi_{X_L,t})_* [X_L, [X_L, X]](x_0) + O(t^3) \quad (8)$$

Now, consider the following Taylor series expansion

$$\begin{aligned}(\Phi_{X_L,t})_* & [X_L, [X_L, X]](x_0) \\ &= (\Phi_{X_L,0})_* [X_L, [X_L, X]](x_0) + O(t) \\ &= [X_L, [X_L, X]](x_0) + O(t)\end{aligned}$$

Then equation (8) becomes

$$S_X(x) = \frac{t^2}{2}[X_L, [X_L, X]](x_0) + O(t^3) \quad (9)$$

The Lie bracket $[X_L, [X_L, X]]$ is actually the second time derivative of S_X along X_L to zeroth order. We shall denote it by $S_X^{''}$, i.e.

$$S_X^{''} = [X_L, [X_L, X]] \quad (10)$$

It is not difficult to verify that, in matrix form,

$$S_X^{''}(x_0) = \begin{bmatrix} X^T(x_0)\dfrac{\partial^2 a_1}{\partial x^2}(x_0)X(x_0) \\ \vdots \\ X^T(x_0)\dfrac{\partial^2 a_n}{\partial x^2}(x_0)X(x_0) \end{bmatrix} \quad (11)$$

3. THE SWING-OUT IN CONTROL SYSTEMS

The so-called linear-analytic control system with bounded single input is given by

$$\dot{x}(t) = f(x(t)) + u(t)\, g(x(t)), \quad |u(t)| \leq 1 \quad (12)$$

where $x(t) \in \mathbb{R}^n$, f and g are smooth vector fields in $\mathbb{R}^n$, and the control input function $u : \mathbb{R} \to \mathcal{U}$, $\mathcal{U}$ the input function space.

The following definitions are adopted from Krener & Schättler (1989). A *trajectory* of the system (12) corresponding to a control $u(\cdot)$ is a continuous curve $x(\cdot)$ satisfying $\dot{x}(t) = f(x(t)) + u(t)g(x(t))$ for almost all t. A point x_1 is *reachable* from a point x_0 within time T if and only if, for some admissible control, there exists a trajectory $x(\cdot)$ defined on an interval $[0, t]$, $t \leq T$, such that $x(0) = x_0$ and $x(t) = x_1$. The set of all such points x_1 is denoted by $\mathcal{R}(x_0, \leq T)$; $\mathcal{R}(x_0, T)$ denotes the set of points that are reachable exactly at time T. We shall denote $\mathcal{R}(x_0, \leq T)$ by $\mathcal{R}(x_0)$ for the sake of a simpler notation. The reachable set $\mathcal{R}(x_0)$ will be said to be *small-time reachable set* when T is sufficiently small for the convenience of the rest of this work. The small-time reachable set for the affine system (or system linearized at x_0) will be denoted by $\mathcal{R}_L(x_0)$.

The swing-out expression given in the previous section needs to be modified for a control system. If one considers system (12), the vector field X is expressed as

$$X = \sum_{i=1}^{n}(f_i(x) + u\, g_i(x))\left(\frac{\partial}{\partial x_i}\right)_x$$

and the vector field X_L of the system linearized about x_0 and $u_0 = 0$ is given by

$$X_L = \sum_{i=1}^{n} \left(f_i(x_0) + A(x - x_0) + b\,u\right)\left(\frac{\partial}{\partial x_i}\right)_{x_0}$$

where f_i and g_i are the coordinate functions of f and g, respectively, A an $n \times n$ matrix with entries $a_{ij} = \partial f_i/\partial x_j(x_0)$, and b a column vector in $\mathbb{R}^n$ with $b_i = g(x_0)$.

Consider the system (12) in $\mathbb{R}^2$, i.e., given

$$\dot{x} = f(x) + u\, g(x), \qquad |u| \leq 1 \quad (13)$$

$x \in \mathbb{R}^2$, f and g are smooth vector fields in $\mathbb{R}^2$, and the control input function $u : \mathbb{R} \to \mathcal{U}$, $\mathcal{U}$ the input function space.

Let $X_+ = f + g$ and $X_- = f - g$ be the *bang-bang vector fields*, corresponding to the control input values $u = +1$ and $u = -1$, respectively. It is assumed that X_+ and X_- are linearly independent at x_0. As pointed out in Krener and Schättler (1989), the boundary of the small-time reachable set $\mathcal{R}(x_0)$ consists of the integral curves Γ_{X_+,x_0} and Γ_{X_-,x_0}, where X_+ and X_- are linearly independent at x_0. So, $\mathcal{R}(x_0)$ itself is the union of this boundary and the open sector between Γ_{X_+,x_0} and Γ_{X_-,x_0}. A point x_1 in $\mathcal{R}(x_0)$ can be reached by switching from X_+ to X_- at some point x_s along Γ_{X_+,x_0}. Small-time reachability is depicted in Figure 2. Note that one can also reach the same point by using first X_- and then X_+.

Suppose that system (13) is linearized about $x(0) = x_0$ and $u(0) = 0$ to give

$$\dot{x} = f(x_0) + A(x - x_0) + bu\,, \quad |u| \leq 1 \quad (14)$$

The bang–bang vector fields of the linearized system are denoted by X_{L+} and X_{L-} corresponding to the respective control values $u = +1$ and $u = -1$. They are given by

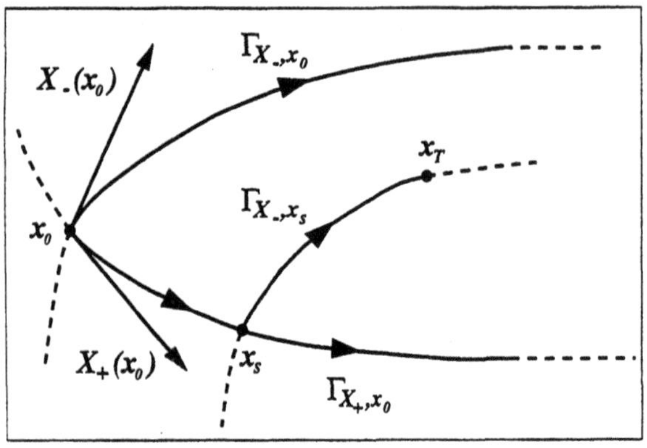

Fig. 2. Small-time reachability in $\mathbb{R}^2$.

$$X_{L+}(x) = f(x_0) + A(x - x_0) + b$$
$$X_{L-}(x) = f(x_0) + A(x - x_0) - b$$

Let S_+ be the swing-out vector of X_+ along X_{L+}, and S_- be the swing-out vector of X_- along X_{L-}. Define the *swing-out function* σ_+ of X_+ along X_{L+} by

$$\sigma_+(x) = \frac{\langle X_{L+}^{\perp}(x), S_+^{(r)}(x_0)\rangle}{\langle X_{L+}^{\perp}, X_{L-}\rangle(x)} \tag{15}$$

where $\langle\cdot,\cdot\rangle$ denotes the Euclidean inner product. The vector $X_{L+}^{\perp}$ is orthogonal to X_{L+}, i.e. $\langle X_{L+}^{\perp}, X_{L+}\rangle = 0$. Note that x lies in the integral curve of X_{L+}, i.e. $x(t) = \Gamma_{X_-,x_0}(t)$. Furthermore, $S_+^{(r)}$ denotes the rth time-derivative of the swing-out vector S_+ along the integral curve of X_{L+}, such that r is the smallest integer satisfying $S_+^{(r)}(x_0) \neq \mathbf{0}$. This choice is because of the fact that, in small time, $S_+(x)$ can be approximated by the first nonzero term of its Taylor series expansion. For the first and second derivatives the usual notation $S' = S^{(1)}$ and $S'' = S^{(2)}$ will be adopted. Recall that $S_+'(x_0) = [X_{L+}, X_+](x_0)$, and $S_+''(x_0) = [X_{L+}, [X_{L+}, X_+]](x_0)$. In this work it is assumed that $S_+''(x_0) \neq \mathbf{0}$ when $S_+'(x_0) = \mathbf{0}$.

The swing-out vector S_+ at x_0 is said to be *good* if $\sigma_+(x_0) < 0$. Otherwise it is said to be *bad*.

Similarly, the swing-out function σ_- of X_- along X_{L-} is defined by

$$\sigma_-(x) = \frac{\langle X_{L-}^{\perp}(x), S_-^{(r)}(x_0)\rangle}{\langle X_{L-}^{\perp}, X_{L+}\rangle(x)} \tag{16}$$

The relevant definitions and assumptions are similar to those made for σ_+ above. For the sake of re-emphasis, note that it is assumed that $S_-''(x_0) \neq \mathbf{0}$ when $S_-'(x_0) = \mathbf{0}$. The swing-out S_- at x_0 is said to be *good* if $\sigma_-(x_0) < 0$. Otherwise it is said to be *bad*.

The following theorem is from Kaya & Noakes (1995).

Theorem 3.1.

(a) If $S_+(x_0)$ and $S_-(x_0)$ are both good then $\mathcal{R}_L(x_0) \subset \mathcal{R}(x_0)$.
(b) If $S_+(x_0)$ and $S_-(x_0)$ are both bad then $\mathcal{R}(x_0) \subset \mathcal{R}_L(x_0)$.
(c) If either of $S_+(x_0)$ and $S_-(x_0)$ is bad then neither $\mathcal{R}(x_0) \subset \mathcal{R}_L(x_0)$ nor $\mathcal{R}_L(x_0) \subset \mathcal{R}(x_0)$.

Remark 3.1. If both of the swing-outs are good, then in small time any point that can be reached by the linearized system can also be reached by the nonlinear (or true) system. If, however, any of the swing-outs is bad, then there will be points which are reachable by the linearized system, but not reachable by the original nonlinear one.

When f and g are linearly independent, there exists a bang–bang control with only one switching to reach from x_0 to some $x_T \in Int\mathcal{R}(x_0)$. Suppose that this bang–bang control has been designed as a feedback control law by using the linearized system. So if the system were really linear (or affine, because of the term $X(x_0)$), first moving along the integral curve Γ_{X_{L+},x_0} and then along Γ_{X_{L-},x_c} would take the system to x_T. Here the curve Γ_{X_{L-},x_c} overlaps with the curve Γ_{-X_{L-},x_T}, the latter being called a *switching curve* for the particular problem. The other switching curve is Γ_{-X_{L+},x_T}, and it is used when one starts with X_{L-}, instead of X_{L+}. With this bang–bang scheme, the real system would move along Γ_{X_+,x_0} first, instead of Γ_{X_{L+},x_0}, and then along Γ_{X_-,x_s}, instead of Γ_{X_{L-},x_s}, after Γ_{X_+,x_0} intercepted Γ_{-X_{L-},x_T} at some point x_s. Since $\Gamma_{X_-,x_s} \neq \Gamma_{X_{L-},x_s}$ in general, the system would not get to the desired state x_T. Instead, it would travel until it possibly intercepts the other switching curve Γ_{-X_{L+},x_T} at some point x_r, and then continue travelling along Γ_{X_+,x_r} until it intercepts again the switching curve Γ_{-X_{L-},x_T}.

The following theorem, given in Kaya & Noakes (1995), points to certain facts when the above-mentioned control law is applied to the true system.

Theorem 3.2. Suppose that the target point $x_T \in Int\mathcal{R}(x_0)$. With the control law detailed above,

(a) If $S_+(x_0)$ and $S_-(x_0)$ are both bad, then x_T is not accessible,
(b) If $S_+(x_0)$ and $S_-(x_0)$ are both good, then the trajectory gets closer to x_T with each switching,
(c) If either of $S_+(x_0)$ and $S_-(x_0)$ is bad, then x_T is not accessible.

Remark 3.2. If both of the swing-outs are good, then in small time not only any point that can be reached by the linearized system can also be reached by the true system, but furthermore the bang–bang control law as described above would remain robust, albeit there will have to be more than one switching, and most possibly chattering.

A direct calculation yields

$$S'_+(x_0) = \frac{\partial g}{\partial x}(x_0)\,(f(x_0) + g(x_0)) \quad (17)$$

If $S'_+(x_0) = \mathbf{0}$, then

$$S''_+(x_0) = \left\{ \left(\frac{\partial^2 f}{\partial x^2}(x_0) + \frac{\partial^2 g}{\partial x^2}(x_0) \right) (f(x_0) + g(x_0)) + \frac{\partial g}{\partial x}(x_0) \frac{\partial f}{\partial x}(x_0) \right\} (f(x_0) + g(x_0)) \quad (18)$$

where the fact $S'_+(x_0) = \mathbf{0}$ has also been used. If the partial derivatives $\partial g/\partial x$ and $\partial f/\partial x$ commute, then the expression (18) in simplifies further by re-using the fact, $S'_+(x_0) = \mathbf{0}$:

$$S''_+(x_0) = \left(\frac{\partial^2 f}{\partial x^2}(x_0)(f(x_0) + g(x_0)) \right) \times (f(x_0) + g(x_0)) \quad (19)$$

This special case occurs especially when g is a constant vector field. In the rest of the paper, when $S'_+(x_0) = \mathbf{0}$, it is assumed that $\partial g/\partial x$ and $\partial f/\partial x$ commute at x_0, or more specifically the vector field g is constant.

The companions of equations (17) and (19) are given as

$$S'_-(x_0) = -\frac{\partial g}{\partial x}(x_0)\,(f(x_0) - g(x_0)) \quad (20)$$

and

$$S''_-(x_0) = \left(\frac{\partial^2 f}{\partial x^2}(x_0)(f(x_0) - g(x_0)) \right) \times (f(x_0) - g(x_0)) \quad (21)$$

Now suppose $r = 1$, i.e. $S'_+(x_0) \neq \mathbf{0}$. For the sake of computational simplicity, let $X^{\perp}_{L+}(x) = E^T X_{L+}(x)$, where the 2×2 "rotation" matrix E can either be chosen to be $\begin{bmatrix} 0 & 1 \\ -1 & 0 \end{bmatrix}$ or $\begin{bmatrix} 0 & -1 \\ 1 & 0 \end{bmatrix}$. When necessary substitutions are made, equation (15) becomes

$$\sigma_+(x) = \left\langle f + g + A(x - x_0),\, E\frac{\partial g}{\partial x}(f + g) \right\rangle \Big/ \langle f + g + A(x - x_0),\, E(f - g + A(x - x_0)) \rangle \quad (22)$$

where f, g, and $\partial g/\partial x$ are evaluated at x_0, but are not shown being so for a simpler appearance.

The above expression can further be simplified to get

$$\sigma_+(x) = \left\langle f + g + A(x - x_0),\, E\frac{\partial g}{\partial x}(f + g) \right\rangle \Big/ 2\left\langle g,\, E(f + A(x - x_0)) \right\rangle \quad (23)$$

We are mainly interested in the signs of the swing-out functions. So when f and g are linearly independent at x_0, it is convenient to consider the swing-out function σ_+ at x_0; namely,

$$\sigma_+(x_0) = \frac{\left\langle f + g,\, E\frac{\partial g}{\partial x}(f + g) \right\rangle}{2\,\langle g,\, Ef \rangle} \quad (24)$$

For completeness, we express below the swing-out function σ_- at x_0, by simply replacing gs by $-g$s.

$$\sigma_-(x_0) = \frac{\left\langle f - g,\, E\frac{\partial g}{\partial x}(f - g) \right\rangle}{2\,\langle g,\, Ef \rangle} \quad (25)$$

Theorem 3.3. Suppose f and g are linearly independent at x_0.

(a) If the matrix $E\frac{\partial g}{\partial x}$ is either positive- or negative-definite, then the signs of $\sigma_+(x_0)$ and $\sigma_-(x_0)$ are the same, i.e. either both $S_+(x_0)$ and $S_-(x_0)$ are good or both bad.

(b) Both $S_+(x_0)$ and $S_-(x_0)$ are good if either
(i) $E\frac{\partial g}{\partial x}$ is positive-definite and $\langle g, Ef \rangle < 0$, or
(ii) $E\frac{\partial g}{\partial x}$ is negative-definite and $\langle g, Ef \rangle > 0$.

Remark 3.3. The condition given in Theorem 3.3(b) guarantees that $\mathcal{R}_L(x_0) \subset \mathcal{R}(x_0)$ by Theorem 3.1. It also assures by Theorem 3.2 that the bang-bang control law explained before would remain robust for the true system.

In the case when $S'_+(x_0) = \mathbf{0}$, one has to use $S''_+(x_0)$ instead in the swing-out function expressions. In other words, The matrix $E(\partial g/\partial x)$ should be replaced by $E(\partial^2 f/\partial x^2)(f + g)$, namely

$$\sigma_+(x_0) = \frac{\left\langle f + g,\, \left(E\frac{\partial^2 f}{\partial x^2}(f + g)\right)(f + g) \right\rangle}{2\langle g,\, Ef \rangle} \quad (26)$$

and

$$\sigma_-(x_0) = -\frac{\left\langle f - g,\, \left(E\frac{\partial^2 f}{\partial x^2}(f - g)\right)(f - g) \right\rangle}{2\langle g,\, Ef \rangle} \quad (27)$$

Note however the negative sign in (27). Theorem 3.4 is then given below as the alternative to Theorem 3.3 for the case when $S'_+(x_0) = \mathbf{0}$ but $S''_+(x_0) \neq \mathbf{0}$.

Theorem 3.4. Suppose f and g are linearly independent at x_0, and $S'_+(x_0) = \mathbf{0}$. If the matrices $E\,\partial^2 f/\partial x^2 (f+g)$ and $E\,\partial^2 f/\partial x^2 (f-g)$ are either both positive-definite or both negative-definite, then the signs of $\sigma_+(x_0)$ and $\sigma_-(x_0)$ are opposite, and as a result either $S_+(x_0)$ or $S_-(x_0)$ is bad.

Remark 3.4. If f and g are linearly independent at x_0, $S'_+(x_0) = \mathbf{0}$, and the matrices $E\,\partial^2 f/\partial x^2 (f+g)$ and $E\,\partial^2 f/\partial x^2 (f-g)$ are either both positive-definite or both negative-definite, then $\mathcal{R}_L(x_0) \not\subset \mathcal{R}(x_0)$ by Theorem 3.1 and 3.4. Furthermore, under a previously described bang–bang control designed for the linearized system the true system would not converge to the terminal state by Theorem 3.2.

Example 3.1. Consider the system

$$\begin{aligned}\dot{x} &= f(x) + g(x)\,u \\ &= \begin{bmatrix} x_2^3 \\ x_1^3 \end{bmatrix} + \begin{bmatrix} 0 \\ 1 \end{bmatrix} u\,, \qquad |u| \leq 1 \end{aligned} \tag{28}$$

Since $S'_+(x_0) = \mathbf{0}$, it is required to consider the matrices $H_+ = E\,\partial^2 f/\partial x^2 (f+g)$ and $H_- = E\,\partial^2 f/\partial x^2 (f-g)$ at point x_0 about which the system is linearized.

$$H_+ = 6\,x_2^2 \begin{bmatrix} x_1\,x_2 & 0 \\ 0 & -(x_1^3+1) \end{bmatrix}$$

and

$$H_- = 6\,x_2^2 \begin{bmatrix} x_1\,x_2 & 0 \\ 0 & -(x_1^3-1) \end{bmatrix}.$$

The matrices H_+ and H_- are both positive-definite when $x_1 < -1$ and $x_2 < 0$; furthermore they are both negative-definite when $x_1 < -1$ and $x_2 > 0$. So, by Theorem 3.4, either $S_+(x_0)$ or $S_-(x_0)$ is bad, and as a result the linear approximation is not good, when $x_1 < -1$ and $x_2 \neq 0$. In other words, for x_0 with components $x_1 < -1$ and $x_2 \neq 0$, $\mathcal{R}_L(x_0) \not\subset \mathcal{R}(x_0)$. Even if one chooses some point x_T to be lying both in $\mathcal{R}_L(x_0)$ and $\mathcal{R}(x_0)$, the previously described bang–bang control law would not take the system from x_0 to x_T. The test given by Theorem 3.4 is inconclusive when $x_1 > -1$. In this case one needs to obtain and analyse the expressions for the swing-out functions σ_+ and σ_-. □

It may not always be possible to make use of the matrices H_+ and H_- in the way prescribed in Theorem 3.4, as the next example illustrates.

Example 3.2. Consider the system

$$\begin{aligned}\dot{x} &= f(x) + g(x)\,u \\ &= \begin{bmatrix} x_2^3 \\ \sin x_1 \end{bmatrix} + \begin{bmatrix} 0 \\ 1 \end{bmatrix} u\,, \qquad |u| \leq 1 \end{aligned} \tag{29}$$

One has $S'_+(x_0) = \mathbf{0}$. So it is necessary to check the matrices H_+ and H_- as in the previous example.

$$H_+ = \begin{bmatrix} -x_2^3 \sin x_1 & 0 \\ 0 & -6\,x_2(\sin x_1 + 1) \end{bmatrix}$$

and

$$H_- = \begin{bmatrix} -x_2^3 \sin x_1 & 0 \\ 0 & -6\,x_2(\sin x_1 - 1) \end{bmatrix}.$$

Calculations show that H_+ and H_- are nowhere both positive-definite or both negative-definite. So the test given by Theorem 3.4 is inconclusive. However, the swing-out functions σ_+ and σ_- can be calculated to be

$$\begin{aligned}\sigma_+ &= \frac{1}{2\,x_2^2}(x_2^8 \sin x_1 + 6\,(\sin x_1 + 1)^3) \\ \sigma_- &= -\frac{1}{2\,x_2^2}(x_2^8 \sin x_1 + 6\,(\sin x_1 - 1)^3)\end{aligned}$$

Then a few manipulations and close look reveal that σ_+ and σ_- are nowhere both negative. This means that the linear approximation of the system is nowhere good. □

Linearization about an equilibrium point and the associated swing-out is the subject of a future investigation.

REFERENCES

Claude, D. (1986). Everything you always wanted to know about linearization. In: *Algebraic and Geometric Methods in Nonlinear Control Theory (Fliess, M., & Harzewinkel, M., eds)*, Reidel, Dordrecht-Boston, Mass., pp. 181-226.

Isidori, A. (1989). *Nonlinear Control Systems.* Springer-Verlag, Berlin.

Hájek, O. (1991). *Control Theory in the Plane.* Springer-Verlag, Berlin.

Kaya, C.Y., & Noakes, J.L. (1995). Linearized control systems and small-time reachable sets. *IMA Journal of Mathematical Control and Information*, **12**, pp. 207-217.

Krener, A.J., & Schättler, H. (1989). The structure of small-time reachable sets in low dimensions. *SIAM J. Control and Optimization*, **27**(1), pp. 120-147.

Sontag, E. (1990). *Mathematical Control Theory.* Springer-Verlag, New York.

14th Triennial World Congress, Beijing, P.R. China

STABILITY ANALYSIS OF A CLASS OF SWITCHED SYSTEMS[1]

Ren Yanpin Zhang Zuo Wu Qiufeng

Automation Department of Tsinghua University
Beijing, P. R. China, 100084
e-mail: yanpren@263.net

Abstract: The stability analysis problem of a class of linear switched systems is investigated in this paper. Comparing with the extended Lyapunov function method currently used, a new approach to stability analysis of such systems, which is based on vector projection theory, is presented. As the results, several conditions for Lyapunov stability of linear switched systems are listed followed by concise proof. Finally, illustrative examples and simulation results are provided to show the efficacy of this new approach. *Copyright © 1999 IFAC*

Keywords: Hybrid systems, Switching systems, Lyapunov stability, Periodic motion, Equilibrium

1. INTRODUCTION

Stability is of great importance for dynamic systems. Switched system, as a specific class of hybrid dynamic systems, is defined as such a system that switches among several subsystems autonomously or under certain control commands issued by some higher processes (Branicky, 1994). The dynamics of each subsystem can usually be described by differential or difference equations. The stability of such kind of switched systems presents quite different characteristics as that of each subsystem. Formal stability analysis of such switched system has already gained great attentions worldwide (Branicky, 1994; Petterson and Lennartson, 1997; Wang, et al., 1996; Lemmon, 1996; Peleties and Decarlo, 1991; Li, et al., 1997; Ye, et al., 1996; Sen, 1997). And the extended Lyapunov function method is mostly used (Branicky, 1994; Petterson and Lennartson, 1997; Wang, et al., 1996; Peleties and Decarlo, 1991; Li, et al., 1997; Ye, et al., 1996; Sen, 1997). According to this method, if one can find appropriate multiple Lyapunov functions for different sub regions of the state space, the global stability of the switched system is guaranteed. Although theoretically perfect, when applied to general switched systems, this method will be confronted with, mainly, two problems. Firstly, it is difficult to carry out the state space partition in order that a Lyapunov function can be found for each obtained sub-region. Secondly, after the partition being made, to couple an appropriate Lyapunov function for each subspace is also a hard work. To conquer these, many researchers are searching for new methods. For example, in Petterson and Lennartson (1997), a linear matrix inequality (LMI) method was proposed for effective computation of quadratic Lyapunov functions.

Firstly motivated by the work of Branicky (1994) and Petterson and Lennartson (1997), the stability issue of a class of linear switched systems is investigated in this paper. Comparing with the existing multiple Lyapunov functions method, a new approach based on vector projection theory is presented. As the results, several stability conditions have been achieved. From the following detailed analysis and proof, it can be seen that, when applied to linear switched systems, this new approach is advantageous over the multiple Lyapunov functions method. Our approach is more effective for numerical computation and on the other hand, avoids the uncertainty in state space partition and Lyapunov functions construction.

[1]This work is supported by Chinese NSF (69674024).

Comparing with the work of Lemmon (1996), the subsystem considered here is more general.

2. FORMAL DESCRIPTION MODEL

A formal description model of general switched system is given by

$$\begin{cases} \dot{x} = f_{i_{\tau_k}}(x,t), \tau_k \le t < \tau_{k+1}, k \in Z^+ \\ i_{\tau_{k+1}} = \phi(i_{\tau_k}, x_{\tau_{k+1}}) \end{cases} \tag{1}$$

In this model, $x \in R^n$ is the continuous state of system; $t \in R^+$ stands for the global clock; $i_t \in Q$ is the discrete state of system where $Q = \{1,2\cdots q\}$ is a finite set containing the possible running modes of the switched system, $q \ge 2$; $f_{i_t}: R^n \times R^+ \to R^n$ and $\phi: R^n \times Q \to Q$ represents, respectively, continuous and discrete state transition dynamics. Additionally, $\tau_k, k \ge 1$ denotes the kth mode switching time instant with the natural assumption that $\tau_0 = 0$ and $i_{\tau_{k+1}} \ne i_{\tau_k}$. From (1) it is obvious that i_t is piecewise constant and right continuous ($\lim_{\tau \to t^+} i_\tau = i_{t^+}$).

The special class of linear switched systems considered throughout this paper is given as follows

$$\begin{cases} \dot{x}(t) = A_{i_{\tau_k}} x(t), \tau_k \le t < \tau_{k+1}, k \in Z^+ \\ S_{jk} = \left\{ x \middle| x \in \overline{OP_{jk}}, j,k \in Q, j \ne k, x \in R^n \right\} \end{cases} \tag{2}$$

where S_{jk} stands for the switching set from mode j to k. In this model, switching sets are radial lines starting from the original point O and passing through the points P_{jk}. And obviously one can conclude from (2) that the original point O is the common equilibrium of all subsystems.

3. MAIN RESULTS

Stability analysis of linear switched systems has been carried out and the main results are given in this section.

Theorem 1: Consider the linear switched system described by (2). If $\forall x \in \Omega_i$, $x^T A_i x \le 0$, $i \in Q$, $\Omega_i = \{x | \dot{x} = A_i x, x \in R^n, \tau_k \le t < \tau_{k+1}, k \ge 1\}$ holds, then the equilibrium $x_e = [0\cdots 0]^T$ is stable in the sense of Lyapunov.

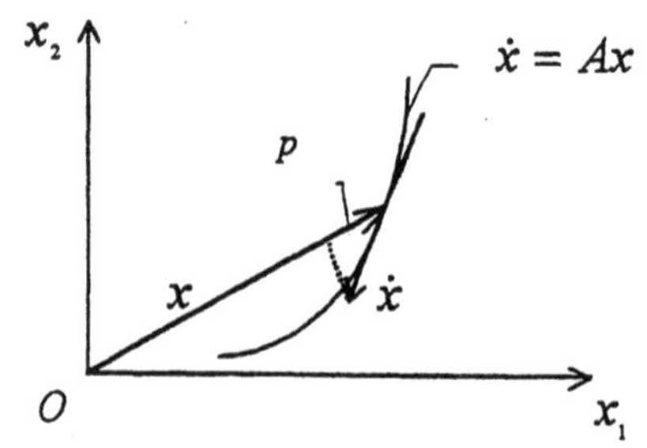

Fig. 1. Illustration of vector projection.

Proof: From Fig. 1, it can be easily seen that the projection of the vector $\dot{x}$ on x is

$$p = \frac{x^T \dot{x}}{\|x\|} = \frac{x^T A x}{\|x\|}$$

Hence, for the linear switched system, we have

$$\|x(t)\| = \|x(0)\| + \int_0^{\tau_1} \frac{x^T A_{i_{\tau_0}} x}{\|x\|} dt + \int_0^{\tau_2 - \tau_1} \frac{x^T A_{i_{\tau_1}} x}{\|x\|} dt + \cdots + \int_0^{t-\tau_n} \frac{x^T A_{i_{\tau_n}} x}{\|x\|} dt \tag{3}$$

where $\tau_{k+1} - \tau_k, k \ge 1$ is the total running time of subsystem i_{τ_k} between the kth and $(k+1)$th mode switching and no switching occur on $(\tau_n, t]$.

From the condition, we have

$$x^T A_{i_{\tau_k}} x \le 0, k \ge 1 \tag{4}$$

Substitution (4) to (3) gives

$$\|x(t)\| \le \|x(0)\| + \int_0^{\tau_1} \frac{x^T A_{i_{\tau_0}} x}{\|x\|} dt = \|x(0)\| + r_1 .$$

where r_1 is a finite real number.
From above, it follows that

$$\forall \varepsilon > 0, \exists \delta > 0, st: \|x(0)\| < \delta \Rightarrow \|x(t)\| < \varepsilon, t \ge 0.$$

which implies that the equilibrium $x_e = [0\,0\cdots 0]^T$ is stable in the sense of Lyapunov. ■

Remark: Theorem 1 implies that stability of linear switched system is ensured if, at all instants, the state trajectory keeps moving towards the equilibrium.

Corollary 1: Consider the linear switched system as in Theorem 1, the equilibrium $x_e = [0\,0\cdots 0]^T$ is stable in the sense of Lyapunov if and only if

$$\sum_{k=0}^{\infty} \int_0^{\tau_{k+1}-\tau_k} \frac{x^T A_{i_{\tau_k}} x}{\|x\|} dt \le W$$

holds where W is a nonnegative finite real constant.
Proof: Corollary 1 can be easily derived from the proof procedure of Theorem 1.

Corollary 1 gives a sufficient and necessary Lyapunov stability condition for linear switched system. However, the computation burden brought is considerable. The following discussion will show that when applied to a class of periodic linear switched system, corollary 1 becomes feasible.

Definition 1: Consider the mode switching sequence $I=\{i_{\tau_1}, i_{\tau_2}, \cdots i_{\tau_k} \cdots\}$ of the linear switched system above, if $\exists M>1$, such that $\forall k \geq 1, i_{\tau_{k+M}}=i_{\tau_k}$, then the system is called a periodic linear switched system with the period M.

Lemma 1: For any linear switched systems, if given

$$x_0 \in S_{jp},\ x_0' \in S_{jp},\ x_0, x_0' \neq 0,$$

$$x_0 \xrightarrow{t_1} x_{t_1},\ x_0' \xrightarrow{t_2} x_{t_2},\ x_{t_1}, x_{t_2} \neq 0,$$

under the assumption that there is no occurrence of switching on both $(0,t_1)$ and $(0,t_2)$, it follows that $x_{t_1} \in S_{pr}, x_{t_2} \in S_{pr}$ hold simultaneously if and only if $t_1=t_2$.

Proof: First we prove that $t_1=t_2$ is the sufficient condition.

According to the specific property of S_{jp} and the given conditions, we have

$$x_0' = \lambda x_0 (\lambda \in R^+) \tag{5}$$

$$x_{t_1} = e^{A_p t_1} x_0,\ x_{t_2} = e^{A_p t_2} x_0' \tag{6}$$

If $t_1=t_2$, in view of (5) and (6), it is clear that

$$x_{t_2} = e^{A_p t_2} x_0' = e^{A_p t_1} \lambda x_0 = \lambda x_{t_1}$$

which implies that $x_{t_1} \in S_{pr} \Leftrightarrow x_{t_2} \in S_{pr}$.

Then we prove that $t_1=t_2$ is also the necessary condition.

Without loss of generality, assume that $t_2>t_1$ and $x_0' \xrightarrow{t_1} x_{t_1}'$. As proved above, it follows that $x_{t_1} \in S_{pr} \Leftrightarrow x_{t_1}' \in S_{pr}$. Since $t_1 \in (0,t_2)$, it conflicts with the original assumption that there is no occurrence of switching on $(0,t_2)$.

This completes the proof. ∎

As for two-dimensional system, the intention of Lemma 1 is shown in Fig. 2.

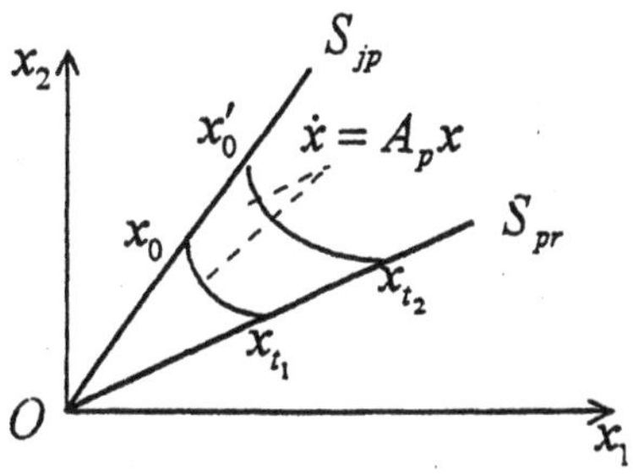

Fig. 2. Illustration of Lemma 1.

Theorem 2: A periodic linear switched system is asymptotically stable in the sense of Lyapunov if

$$\rho(T)<1$$

where

$$T=\prod_{k=M}^{1} e^{A_{i_{\tau_k}}(\tau_{k+1}-\tau_k)}$$

and $\rho(T)$ denotes the spectrum radius of matrix T.

Proof: Without loss of generality, assume that $i_{\tau_k}=k \bmod M, k \geq 1$. As a result of the inherent periodic property,

$$x_{\tau_k} \in S_{jp} \Rightarrow x_{\tau_{k+M}} \in S_{jp}$$

and

$$x_{\tau_{k+1}} \in S_{pr} \Rightarrow x_{\tau_{k+M+1}} \in S_{pr}$$

where

$$j=(k-1)\mathrm{mod}M$$
$$p=k \bmod M$$

and

$$r=(k+1)\mathrm{mod}M.$$

By Lemma 1, we have

$$\tau_{k+M+1}-\tau_{k+M}=\tau_{k+1}-\tau_k, k \geq 1 \tag{7}$$

Starting from the continuous state x_{τ_1}, after one period, the continuous state becomes

$$x_{\tau_{M+1}} = e^{A_M(\tau_{M+1}-\tau_M)} e^{A_{M-1}(\tau_M-\tau_{M-1})} \cdots e^{A_k(\tau_{k+1}-\tau_k)} \cdots e^{A_1(\tau_2-\tau_1)} x_{\tau_1}$$

$$= \prod_{k=M}^{1} e^{A_{i_{\tau_k}}(\tau_{k+1}-\tau_k)} x_{\tau_1} = T x_{\tau_1}, T \in R^{n \times n} \tag{8}$$

Similarly, after two periods, the continuous state is

$$x_{\tau_{2M+1}} = e^{A_M(\tau_{2M+1}-\tau_{2M})} e^{A_{M-1}(\tau_{2M}-\tau_{2M-1})} \cdots e^{A_k(\tau_{k+M+1}-\tau_{k+M})} \cdots e^{A_1(\tau_{M+2}-\tau_{M+1})} x_{\tau_{M+1}}$$

$$= \prod_{k=M}^{1} e^{A_{i_{\tau_k}}(\tau_{k+M+1}-\tau_{k+M})} x_{\tau_{M+1}} \tag{9}$$

Substitute (7) and (8) to (9), we have

$$x_{\tau_{2M+1}} = \prod_{k=M}^{1} e^{A_{i_{\tau_k}}(\tau_{k+1}-\tau_k)} x_{\tau_{M+1}} = T \cdot T x_{\tau_1} = T^2 x_{\tau_1}$$

Repeat the above procedures, we get the continuous states after $l(l \geq 2)$ periods

$$x_{\tau_{lM+1}} = T^l x_{\tau_1}$$

Thus,

$$\rho(T) < 1 \Rightarrow \lim_{l\to\infty} T^l = 0 \Rightarrow \lim_{l\to\infty} T^l x_{\tau_1} = 0$$
$$\Rightarrow \lim_{t\to\infty} x_t = 0 .$$

■

Corollary 2: Consider the periodic linear switched system in definition 1, the equilibrium $x_e = [00\cdots0]^T$ is stable in the sense of Lyapunov if and only if

$$\sum_{k=1}^{M} \int_0^{\tau_{k+1}-\tau_k} \frac{x^T A_{i_{\tau_k}} x}{\|x\|} dt \leq 0 \qquad (10)$$

Proof: Under the assumption $x_{\tau_1} \in S_{jp}$, we have $x_{\tau_{lM+1}} \in S_{jp}, l \geq 1$.

According to the property of S_{jp}, it follows that

$$\begin{cases} x_{M+1} = \lambda_1 x_{\tau_1} \\ x_{2M+1} = \lambda_2 x_{\tau_{M+1}} = \lambda_2 \lambda_1 x_{\tau_1} \\ \quad\vdots \\ x_{lM+1} = \lambda_l x_{\tau_{(l-1)M+1}} = \lambda_l \lambda_{l-1} \cdots \lambda_2 \lambda_1 x_{\tau_1} \\ \quad\vdots \end{cases} \qquad (11)$$

where $\lambda_l, l \geq 1$ are positive real numbers. Moreover, in view of the proof of Theorem 2, we have

$$\begin{cases} x_{\tau_{M+1}} = T x_{\tau_1} = \lambda_1 x_{\tau_1} \\ x_{\tau_{2M+1}} = T^2 x_{\tau_1} = \lambda_1^{\,2} x_{\tau_1} \\ \quad\vdots \\ x_{\tau_{lM+1}} = T^l x_{\tau_1} = \lambda_1^{\,l} x_{\tau_1} \\ \quad\vdots \end{cases} \qquad (12)$$

Comparing (11) with (12) turns out that $\lambda_l \equiv \lambda_1$, $l \geq 1$. Thus, we have

$$\begin{cases} \|x_{\tau_{M+1}}\| = \lambda_1 \|x_{\tau_1}\| \\ \|x_{\tau_{2M+1}}\| = \lambda_1 \|x_{\tau_{M+1}}\| \\ \quad\vdots \\ \|x_{\tau_{(l+1)M+1}}\| = \lambda_1 \|x_{\tau_{lM+1}}\| \\ \quad\vdots \end{cases}$$

which implies that

$$\|x_{\tau_{(l+1)M+1}}\| - \|x_{\tau_{lM+1}}\| \to \begin{cases} > 0, \lambda_1 > 1 \\ \leq 0, \lambda_1 \leq 1 \end{cases}, \forall l \geq 0 .$$

Since

$$\|x_{\tau_{(l+1)M+1}}\| - \|x_{\tau_{lM+1}}\| =$$
$$\sum_{k=1}^{M} \int_0^{\tau_{k+lM+1}-\tau_{k+lM}} \frac{x^T A_{i_{\tau_k}} x}{\|x\|} dt, \forall l \geq 0$$

Consequently, we can conclude that

$$\sum_{k=1}^{M} \int_0^{\tau_{k+1}-\tau_k} \frac{x^T A_{i_{\tau_k}} x}{\|x\|} dt \leq 0 \Leftrightarrow$$
$$\sum_{k=0}^{\infty} \int_0^{\tau_{k+1}-\tau_k} \frac{x^T A_{i_{\tau_k}} x}{\|x\|} dt \leq W$$

where W is a finite positive real number as mentioned in Corollary 1.

By Corollary 1, it turns out that (10) is the sufficient and necessary condition of Lyapunov stability.

Remark: Comparing with Corollary 1, when using Corollary 2 in stability analysis for periodic linear switched system, we only need to integrate the vector projection value on one period.

4. EXAMPLES AND DISCUSSION

In this section several examples are introduced and the stability analysis is shown along with simulation results.

Example 1 (Branicky, 1994): Consider the stability of a system switching between two linear subsystems periodically $\dot{x}(t) = A_i x(t), i = 1,2$. Given

$$A_1 = \begin{bmatrix} -0.1 & 1 \\ -10 & -0.1 \end{bmatrix}, A_2 = \begin{bmatrix} -0.1 & 10 \\ -1 & -0.1 \end{bmatrix}.$$

and

$$S_{12} = \{x | x_2 = 0\}, S_{21} = \{x | x_1 = 0\}.$$

Solution: In this example,

$$\Omega_1 = \{x | x_1 x_2 \geq 0\}, \Omega_2 = \{x | x_1 x_2 \leq 0\}$$

and

$$x^T A_1 x = -0.1(x_1^{\,2} + x_2^{\,2}) - 9x_1 x_2$$
$$x^T A_2 x = -0.1(x_1^{\,2} + x_2^{\,2}) + 9x_1 x_2$$

It is obvious that when $x \in \Omega_1, x^T A_1 x \leq 0$ and when $x \in \Omega_2$, $x^T A_2 x \leq 0$. By Theorem 1, we can conclude that this switched system is stable. Starting from $x_0 = [1 \quad 1]^T$ and subsystem 1, the continuous state trajectory is shown in Fig. 3.

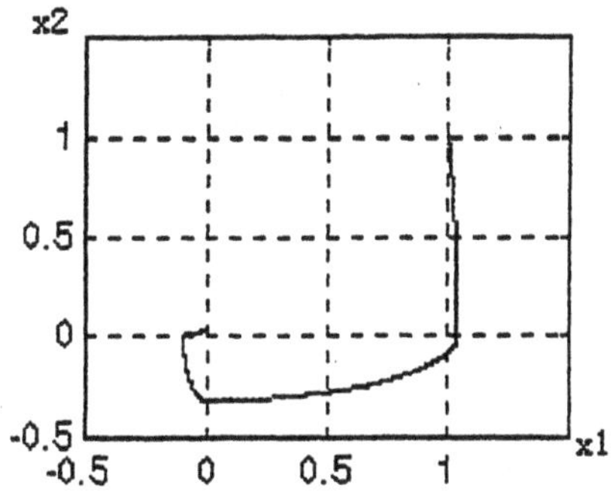

Fig. 3. The simulation results of example 1.

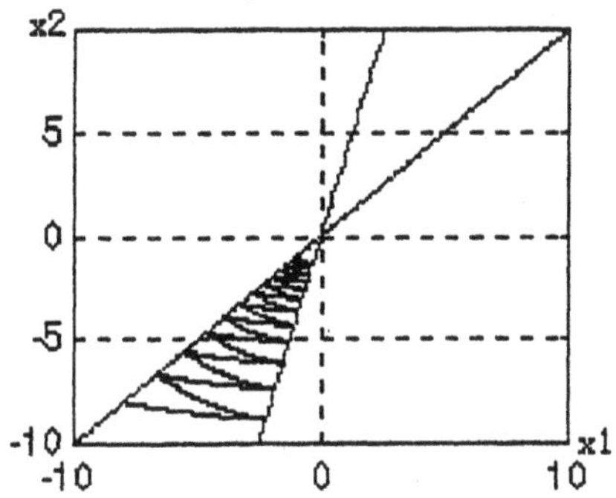

Fig. 4. The simulation results of example 2.

Example 2: Again we consider the stability of a periodic linear switched system $\dot{x}(t) = A_i x(t), i = 1,2$. Given

$$A_1 = \begin{bmatrix} 0.1 & -5 \\ 1 & 0.1 \end{bmatrix}, A_2 = \begin{bmatrix} 0.3 & 5 \\ -5 & 0.3 \end{bmatrix}$$

and

$$S_{12} = \{x | x_2 = 4x_1\}, S_{21} = \{x | x_2 = x_1\}.$$

Solution: Select a starting point $x_0 = [-8 \ -8]^T$, $x_0 \in S_{21}$. In this situation,

$$M = 2$$

and

$$\sum_{k=1}^{M} \int_0^{\tau_{k+1}-\tau_k} \frac{x^T A_{i_{\tau_k}} x}{\|x\|} dt =$$

$$\int_0^{\tau_2-\tau_1} \frac{x^T A_{i_{\tau_1}} x}{\|x\|} dt + \int_0^{\tau_3-\tau_2} \frac{x^T A_{i_{\tau_2}} x}{\|x\|} dt$$

Further, through a numerical integration procedure, the result turns out to be -1.91. According to Corollary 2, the system given is stable, which coincide with the simulation results shown in Fig. 4.

Remark: In this example, although the two subsystems are both unstable, the switched system turns out to be stable.

5. CONCLUSIONS

A new approach for stability analysis of a class of linear switched systems is demonstrated in this paper. This approach generalizes the work of Lemmon (1996) Moreover, comparing with the existing multiple Lyapunov function method, this approach avoids the uncertainty in state space partition and Lyapunov function construction.

There is still a far way to go for stability analysis of switched systems. In future research, our approach is expected to be extended so that it can act well in more general situations. On the other hand, the problem of searching stable switching rules for given subsystems will be investigated.

REFERENCES

Branicky, M.S. (1994). Stability of switched and hybrid systems. In: *Proc. of 33rd CDC*, pp. 3498-3503.

Lemmon, M.D. (1996). Stability of periodically switched systems of first order integrators. *Technical Report*, University of Notre Dame.

Li, Z.G., B.S. Cheong and X.H. Xu (1997). Stability of hybrid dynamic systems. *International Journal of Systems Science*, **28**, 837-846.

Peleties, P., R. Decarlo (1991). Asymptotic stability of m-switched systems using Lyapunov-like functions. In: *Proc. of American Control Conference*, pp. 1679-1684. Boston.

Petterson, S., B. Lennartson (1997). LMI for stability and robustness of hybrid systems. In: *Proc. of American Control Conference*, pp. 1714-1718.

Sen, M.D.L. (1997). On the hyperstability of a class of hybrid systems. *International Journal of System Science*, **28**, 925-934.

Wang, H.O., K. Tanaka and M. F. Griffin (1996). An approach to fuzzy control of nonlinear systems: stability and design issues. *IEEE Transactions on Fuzzy Systems*, **4**, 14-23.

Ye, H., A.N. Michel and L. Hou (1996). Stability analysis of discontinuous dynamical systems with applications. In: *Proc. of 13th IFAC*, pp. 461-466.

14th Triennial World Congress, Beijing, P.R. China

DYNAMIC COMPENSATOR DESIGN METHOD FOR A CLASS OF TIME-DELAY CONTROL SYSTEMS WITH SATURATING ACTUATORS

Jong-Koo Park * **and Chong-Ho Choi** **

* *School of Electrical & Computer Eng., Sungkyunkwan Univ., Suwon 440-746, Korea, pjk@yurim.skku.ac.kr*
** *School of Electrical Eng., ASRI and ERC-ACI by KOSEF, Seoul National Univ., Korea, chchoi@csl.snu.ac.kr*

Abstract: A dynamic compensator design method for a class of time-delay control systems with saturating actuators is proposed. Under the assumption that a controller was designed for a time-delay control system based on existing design techniques which shows desirable nominal performance, an additional compensator is introduced to provide smooth performance degradation under control input saturation. By taking the difference of the controller states in the absence and presence of saturating actuators as a performance index, a dynamical compensator which minimizes it is derived. The resulting dynamical compensator is expressed explicitly in the plant and controller parameters. The proposed method not only provides smooth performance degradation, but also guarantees the total stability of the overall systems. *Copyright © 1999 IFAC*

Keywords: Dynamic Compensator, Saturation, Time-Delay, Anti-Reset Windup, Generalized Delay Controller, Smith Predictor

1. INTRODUCTION

Saturations and time-delays are often observed together in control systems. To deal with both problems effectively, an appropriate design method is required. Up to now, only a few methods were reported to overcome the problems simultaneously. Chen *et al.* (1988) studied the problem of the stabilization of the saturating time-delay system with state feedback and sampled-state feedback, and they derived several sufficient conditions to ensure the stability of the system expressed in terms of norm inequalities. Chou *et al.* (1989) exploited the sufficient condition to stabilize a linear uncertain time-delay system containing saturating actuators by using a dynamical output feedback controller. The problem of robust stabilization of uncertain time-delay systems containing a saturating actuator was addressed by Niculescu *et al.* (1996). They proposed upper bounds for the delay which guarantee the stability of the closed-loop system by solving algebraic Riccati equations. Oucheriah (1996) considered a method to synthesize a globally stabilizing linear state feedback controller by means of an asymptotic observer for time-delay systems.

In this paper, we attempt a novel approach utilizing the two-step design method which is common in ARW techniques. By applying the concept of the ARW strategy to time-delay systems with saturating actuators, we propose a method to mitigate the adverse effects due to control input saturation. We assume that a controller has already been designed which shows satisfactory behaviors for a control system with known time-delays in the absence of saturating actuators. The controller may even be a generalized delay controller such as the Smith predictor. The objective is to provide an additional dynamical compensator which will exhibit smooth performance degradation of the closed loop system under saturation. In pursuit of this objective, we reformulate this problem into a state-tracking problem since outputs are combinations of internal states. The design parameters of the dynamic compensator are determined explicitly based on an objective function of controller states. Hence, in the two-step design procedure, once a linear controller is designed *a priori*, the resulting dynamical compensator is determined immediately.

2. Time-Delay Control Systems with Saturating Actuators

Most time-delay control systems with saturating actuators can be represented by Fig. 1, where τ_1 and τ_2 denote the control input delay and the output measurement delay, respectively. The plant and controller are expressed by the following minimal state-space representations

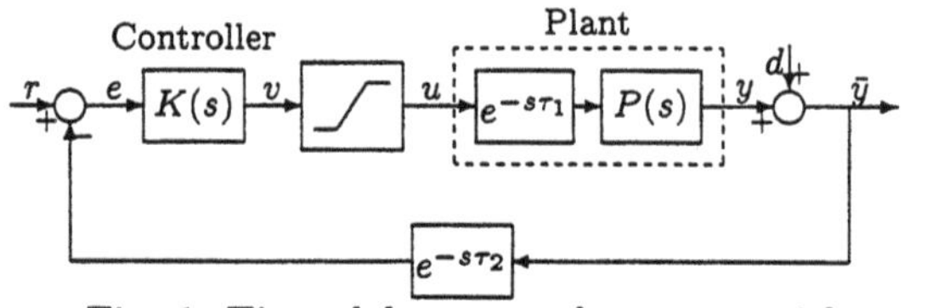

Fig. 1. Time-delay control systems with saturating actuators.

1) The dynamics of the plant $e^{-s\tau_1}P(s)$:

$$\dot{x}_p(t) = Ax_p(t) + Bu(t-\tau_1), \quad (1)$$
$$y(t) = Cx_p(t) + Du(t-\tau_1), \quad (2)$$
$$u(t) = sat(v(t)). \quad (3)$$

2) The dynamics of the controller $K(s)$:

$$\dot{x}_c(t) = Fx_c(t) + Ge(t), \quad (4)$$
$$v(t) = Hx_c(t) + Le(t), \quad (5)$$
$$e(t) = r(t) - y(t-\tau_2) - d(t-\tau_2). \quad (6)$$

The observer type approach is regarded as a standard configuration for saturating systems. Based on this approach, the dynamical compensation method for saturating control systems was presented in Park and Choi (1995, 1997). In this paper, the results will be extended to time-delay control systems with saturating actuators.

A compensation scheme utilizing the observer type technique is shown in Fig. 2. Here a block $\mathcal{N}$ is added to take the directionality of multiple control signals into consideration (Park and Choi, 1995, 1997). A simple solution for this problem is defined by

$$\bar{u}(t) := \mathcal{N}(v(t)) = \begin{cases} \alpha(t)v(t) & \text{if } v(t) \in S_{v(t)} \\ v(t) & \text{otherwise,} \end{cases}$$

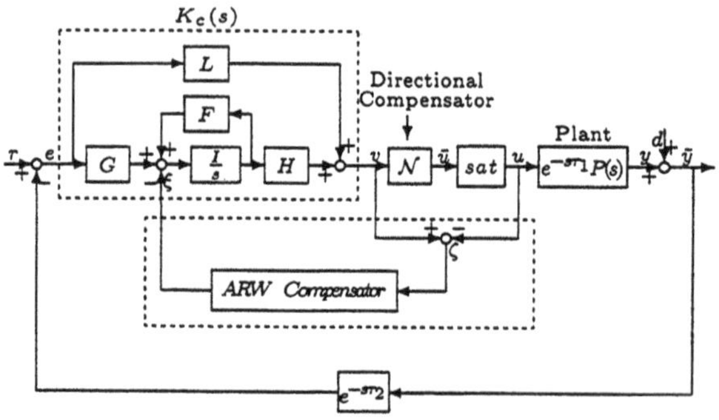

Fig. 2. The observer type compensation of time-delay control systems with saturating actuators.

where $S_{v(t)}$ is a set in which at least one element of $v(t)$ exceeds the saturation levels and

$$\alpha(t) := \left(\max_j \left(\frac{v_j(t)}{sat(v_j(t))}\right)\right)^{-1}.$$

3) The dynamics of the compensated controller $K_c(s)$:

$$\dot{x}_c(t) = Fx_c(t) + Ge(t) - \xi(t), \quad (7)$$
$$v(t) = Hx_c(t) + Le(t), \quad (8)$$
$$e(t) = r(t) - y(t-\tau_2) - d(t-\tau_2). \quad (9)$$

where $\xi(t) \subset \Re^q$ is the compensation signal. Its dynamics is given by the following.

4) The dynamics of the compensator $M(s)$:

$$\dot{x}_d(t) = \Phi x_d(t) + \Gamma\zeta(t), \quad (10)$$
$$\xi(t) = \Sigma x_d(t) + \Lambda\zeta(t), \quad (11)$$
$$\zeta(t) = v(t) - u(t). \quad (12)$$

where the matrices Φ, Γ, Σ and Λ are design parameters that should be appropriately determined for the overall system to behave well under saturation.

3. Derivation of the Dynamic Compensator

Let $x_c(t)$ and $\tilde{x}_c(t)$ be the controller states in the absence and presence of saturating actuators. Thus it is reasonable to regard the difference between $x_c(t)$ and $\tilde{x}_c(t)$ as an index of performance. The performance index is selected as

$$\min_{M(s)} J_1 = \int_0^\infty \|x_c(t) - \tilde{x}_c(t)\|^2 dt, \quad (13)$$

where $\|\cdot\|$ is the *Euclidean norm*.

For solving (13), we write the closed-loop dynamics of the plant, the compensated controller, and the dynamical compensator in the frequency domain depending on the absence or presence of saturating actuators. Here we provide notations to be used throughout the paper: The expression $\varrho(s)$, where s, $s = \sigma + j\omega$, is a complex variable, denotes the Laplace transform of $\varrho(t)$ for some variable ϱ. The symbol "$\tilde{}$" means that the corresponding variable operates in the presence of saturating actuators.

1) The dynamics of the closed-loop system in the absence of saturating actuators:

From (1)-(3), (7), (9) and $u(t) = v(t)$ and by the Laplace transform, we obtain the following equation.

$$\begin{bmatrix} sx_c(s) \\ sx_p(s) \end{bmatrix} = A_l(s,\tau_1,\tau_2) \begin{bmatrix} x_c(s) \\ x_p(s) \end{bmatrix} \tag{14}$$
$$+ B_l(s,\tau_1,\tau_2) \left[r(s) - e^{-s\tau_2}d(s)\right] + \begin{bmatrix} x_c(0) \\ x_p(0) \end{bmatrix},$$

where

$$A_l\,(s,\tau_1,\tau_2) := \begin{bmatrix} P & Q \\ R & S \end{bmatrix}, \; B_l(s,\tau_1,\tau_2) := \begin{bmatrix} T \\ U \end{bmatrix}, \tag{15}$$
$$P := F - e^{-s(\tau_1+\tau_2)}GD(I + e^{-s(\tau_1+\tau_2)}LD)^{-1}H,$$
$$Q := -e^{-s\tau_2}G[I - e^{-s(\tau_1+\tau_2)}D(I+e^{-s(\tau_1+\tau_2)}LD)^{-1}L]C,$$
$$R := e^{-s\tau_1}B(I + e^{-s(\tau_1+\tau_2)}LD)^{-1}H,$$
$$S := A - e^{-s(\tau_1+\tau_2)}B(I + e^{-s(\tau_1+\tau_2)}LD)^{-1}LC,$$
$$T := G[I - e^{-s(\tau_1+\tau_2)}D(I + e^{-s(\tau_1+\tau_2)}LD)^{-1}L],$$
$$U := e^{-s\tau_1}B(I + e^{-s(\tau_1+\tau_2)}LD)^{-1}L,$$

and $(x_c^T(0),\ x_p^T(0))^T$ is an initial state.

2) The dynamics of the closed-loop system in the presence of saturating actuators:

From (1)-(3), (7)-(12) and by the Laplace transform, we get

$$\begin{bmatrix} s\tilde{x}_c(s) \\ s\tilde{x}_d(s) \\ s\tilde{x}_p(s) \end{bmatrix} = A_{sat}(s,\tau_1,\tau_2) \begin{bmatrix} \tilde{x}_c(s) \\ \tilde{x}_d(s) \\ \tilde{x}_p(s) \end{bmatrix} + \begin{bmatrix} \tilde{x}_c(0) \\ \tilde{x}_d(0) \\ \tilde{x}_p(0) \end{bmatrix} \tag{16}$$

$$+ \begin{bmatrix} (G - \Lambda L)\left(r(s) - e^{-s\tau_2}d(s)\right) \\ + \left[\Lambda - e^{-s(\tau_1+\tau_2)}(G - \Lambda L)D\right]\widehat{sat(\tilde{v})} \\ \Gamma L\left(r(s) - e^{-s\tau_2}d(s)\right) \\ -\Gamma(I + e^{-s(\tau_1+\tau_2)}LD)\widehat{sat(\tilde{v})} \\ e^{-s\tau_1}B\widehat{sat(\tilde{v})} \end{bmatrix},$$

where

$$A_{sat}(s,\tau_1,\tau_2) := \begin{bmatrix} F - \Lambda H & -\Sigma & -e^{-s\tau_2}(G - \Lambda L)C \\ \Gamma H & \Phi & -e^{-s\tau_2}\Gamma LC \\ 0 & 0 & A \end{bmatrix} \tag{17}$$

and $(\tilde{x}_c^T(0),\ \tilde{x}_d^T(0),\ \tilde{x}_p^T(0))^T$ is an initial state. The expression $\widehat{sat(\tilde{v})}$ denotes the Laplace transform of $sat(\tilde{v})$ ($= \tilde{u}(t)$). When the function $sat(\cdot)$ operates in the linear region and the internal states of the dynamical compensator is set to be zero, (16) becomes (14).

Let $\mathcal{M}(t)$ be the impulse response matrix from $\zeta(t)$ to $\xi(t)$ in (10) and (11). That is,

$$\mathcal{M}(t) := \mathcal{L}^{-1}\{M(s)\} = \mathcal{L}^{-1}\{\Lambda + \Sigma(sI - \Phi)^{-1}\Gamma\}. \tag{18}$$

Then (7) can be rewritten as

$$\dot{x}_c(t) = Fx_c(t) + Ge(t) - \mathcal{M}(t) * (v(t) - u(t)), \tag{19}$$

where "$*$" denotes the convolution operator.

Thus from (1)-(3), (8), (9), (19), and the Laplace transform, another expression of the closed-loop system in the presence of saturating actuators is

$$\begin{bmatrix} s\tilde{x}_c(s) \\ s\tilde{x}_p(s) \end{bmatrix} = A_{arw}(s,\tau_1,\tau_2) \begin{bmatrix} \tilde{x}_c(s) \\ \tilde{x}_p(s) \end{bmatrix} + \begin{bmatrix} \tilde{x}_c(0) \\ \tilde{x}_p(0) \end{bmatrix} \tag{20}$$

$$+ \begin{bmatrix} (G - M(s)L)\left[r(s) - e^{-s\tau_2}d(s)\right] \\ + \left[M(s) - e^{-s(\tau_1+\tau_2)}(G - M(s)L)D\right]\widehat{sat(\tilde{v})} \\ e^{-s\tau_1}B\widehat{sat(\tilde{v})} \end{bmatrix},$$

where

$$A_{arw}(s,\tau_1,\tau_2) \tag{21}$$
$$:= \begin{bmatrix} F - M(s)H & -e^{-s\tau_2}[G - M(s)L]C \\ 0 & A \end{bmatrix}.$$

If the plant is open-loop unstable, the closed-loop system with saturating actuators can not be globally stabilized. Hence, we will only consider open-loop stable plants. Furthermore, for time-delay control systems in the absence of saturating actuators, it is assumed that the parameters of the linear controller, i.e., F, G, H and L, have been determined *a priori* by some design technique in consideration of the performance as well as the stability of the closed loop system. We make the following assumptions.

- (A1) The plant $P(s)$ is open-loop stable, i.e., the system matrix A of (1) has stable eigenvalues.
- (A2) In the absence of saturating actuators, the controller $K(s)$ was designed *a priori* taking the time-delay of the system into consideration so that the control system is stable and the nominal performance is acceptable.
- (A3) The system matrix of the closed-loop plant in the absence of saturating actuators, $S = A - e^{-s(\tau_1+\tau_2)}B(I + e^{-s(\tau_1+\tau_2)}LD)^{-1}LC$, has stable eigenvalues.

Theorem 1 : Consider the dynamical ARW scheme shown in Fig. 2. When Assumptions (A1)-(A3) are satisfied, the dynamical compensator $M^*(s)$ which optimizes the performance index (13) is uniquely determined by

$$
\begin{aligned}
&M^*(s) \qquad\qquad (22)\\
&= e^{-s(\tau_1+\tau_2)}GD(I + e^{-s(\tau_1+\tau_2)}LD)^{-1}\\
&\quad + e^{-s(\tau_1+\tau_2)}G[I - e^{-s(\tau_1+\tau_2)}D(I + e^{-s(\tau_1+\tau_2)}LD)^{-1}L]C\\
&\quad \times \left\{sI - [A - e^{-s(\tau_1+\tau_2)}B(I + e^{-s(\tau_1+\tau_2)}LD)^{-1}LC]\right\}^{-1}\\
&\quad \times B(I + e^{-s(\tau_1+\tau_2)}LD)^{-1}.\\
&= e^{-s(\tau_1+\tau_2)}GP(s)[I + e^{-s(\tau_1+\tau_2)}LP(s)]^{-1}.
\end{aligned}
$$

That is,

$$\Phi^* = A - e^{-s(\tau_1+\tau_2)}B(I + e^{-s(\tau_1+\tau_2)}LD)^{-1}LC, \quad (23)$$

$$\Gamma^* = e^{-s(\tau_1+\tau_2)}B(I + e^{-s(\tau_1+\tau_2)}LD)^{-1}, \quad (24)$$

$$\Sigma^* = G[I - e^{-s(\tau_1+\tau_2)}D(I + e^{-s(\tau_1+\tau_2)}LD)^{-1}L]C, \quad (25)$$

$$\Lambda^* = e^{-s(\tau_1+\tau_2)}GD(I + e^{-s(\tau_1+\tau_2)}LD)^{-1}. \quad (26)$$

The parameters of the dynamical compensator Φ^*, Γ^*, Σ^* and Λ^* presented in Theorem 1 are not directly realizable. To utilize the proposed dynamical compensator, we should know how to realize it. Substituting (23)-(26) into the Laplace transform equations of (10) and (11) (assume that the initial state is at rest), we obtain

$$sx_d(s) = Ax_d(s) + B\eta(s), \quad (27)$$

$$\xi(s) = GCx_d(s) + GD\eta(s), \quad (28)$$

where

$$
\begin{aligned}
\eta(s) :=& -e^{-s(\tau_1+\tau_2)}(I + e^{-s(\tau_1+\tau_2)}LD)^{-1}LCx_d(s) \quad (29)\\
&+ e^{-s(\tau_1+\tau_2)}(I + e^{-s(\tau_1+\tau_2)}LD)^{-1}\zeta(s).
\end{aligned}
$$

After multiplying $(I + e^{-s(\tau_1+\tau_2)}LD)$ to both sides of (29), we can rearrange it into

$$\eta(s) = e^{-s(\tau_1+\tau_2)}[-LCx_d(s) - LD\eta(s) + \zeta(s)]. \quad (30)$$

The time domain equation equivalent of (30) is

$$
\begin{aligned}
\eta(t) =& -LCx_d(t-\tau_1-\tau_2) - LD\eta(t-\tau_1-\tau_2) \quad (31)\\
&+ \zeta(t-\tau_1-\tau_2).
\end{aligned}
$$

Now, we can rewrite Theorem 1 as the following corollary based on the above derivation, and Fig. 3 depicts the results.

Corollary 1 : Consider the dynamical ARW scheme shown in Fig. 2. When Assumptions (A1)-(A3) are satisfied, the state-space realizable form of the dynamical compensator $M^*(s)$ which optimizes the performance index (13) is uniquely determined by

$$\dot{x}_d(t) = Ax_d(t) + B\eta(t), \quad (32)$$

$$\xi(t) = G\vartheta(t), \quad (33)$$

where

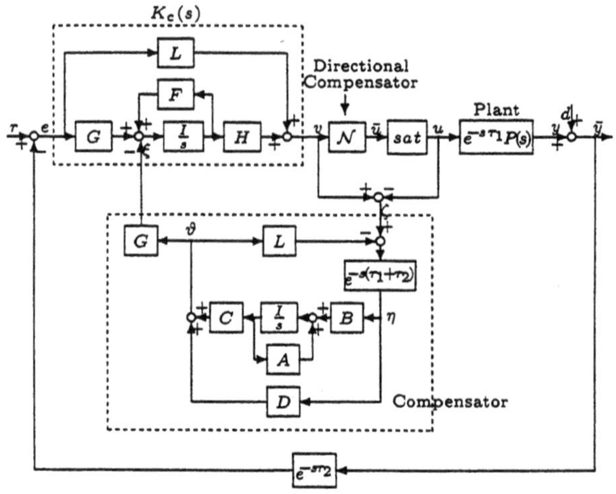

Fig. 3. Saturating control systems with time-delay adopting the dynamical anti-reset windup structure.

$$\eta(t) := \zeta(t-\tau_1-\tau_2) - L\vartheta(t-\tau_1-\tau_2), \quad (34)$$

$$\vartheta(t) := Cx_d(t) + D\eta(t). \quad (35)$$

Even though controller states can be maintained at unsaturated controller states via the dynamical compensation method, this may not be the best strategy with regard to performance when actuators escape from saturation. The remedy for this is to reset the states of the dynamical compensator to zero $(\tau_1 + \tau_2)$ seconds after whenever the actuators escape from saturation.

4. Stability of the Resulting Systems

The state representations of unsaturated and saturated closed-loop systems are described in (14) and (16), respectively. These two equations do not represent different systems, but express the same system dynamics depending on the absence or presence of saturating actuators. We first consider the ***total stability*** (Park and Choi, 1995) of the saturating systems when the proposed ARW method is adopted. Totally stable systems have the property that, for any bounded input, all the state variables as well as the output are bounded. We can state the following theorem.

Theorem 2 : Consider the time-delay control systems with saturating actuators and the ARW structure which is depicted in Fig. 3. If the following conditions are satisfied, then the overall system is ***totally stable.***

1. The system in the absence of saturating actuators satisfies Assumptions (A1)-(A3).
2. The dynamical ARW compensator is designed as in (32)-(35).

An asymptotic stability based on nonlinear stability analysis techniques can be addressed when exogenous inputs are not presented. However, it is not possible for the saturating systems to be

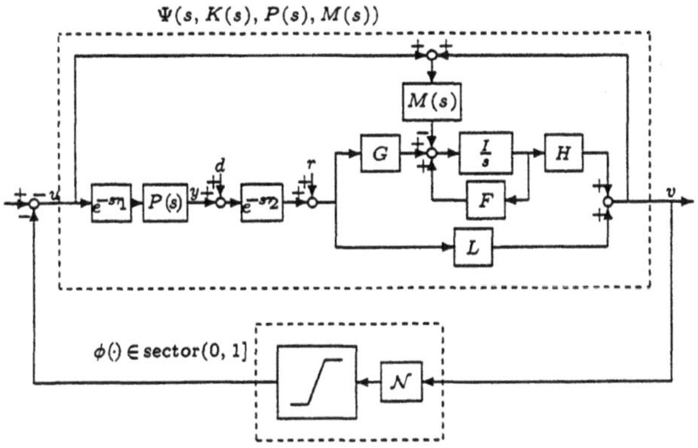

Fig. 4. Time-delay control systems with saturating actuators rearranged for stability analysis.

settled within the linear region of saturation for arbitrarily large but finite exogenous inputs. The reason is that one cannot expect a system with bounded controls to follow arbitrarily large reference inputs or to reject arbitrarily large disturbances. The system shown in Fig. 2 can be rearranged as Fig. 4.

By using the small-gain theorem, an asymptotic stability can be addressed. Let $\Psi(s, K(s), P(s), M(s))$ be the transfer function of the upper dashed box. Then

$$\begin{aligned}&\Psi(s, K(s), P(s), M(s)) \\ &= [I + H(sI - F)^{-1}M(s)]^{-1} \\ &\quad\times[e^{-s(\tau_1+\tau_2)}K(s)P(s) - H(sI - F)^{-1}M(s)].\end{aligned} \tag{36}$$

Theorem 3 : Assume that a system with saturating actuators satisfies Assumptions (A1)-(A3) and the dynamical compensator is designed as (22). Then the overall system is ***asymptotically stable*** in the absence of exogenous inputs $r(t)$ and $d(t)$ if the following condition is satisfied.

$$\|e^{-s(\tau_1+\tau_2)}LP(s)\|_\infty < 1 \tag{37}$$

Although the condition is conservative, Theorem 3 guarantees that the saturating system will settle within the linear region of saturation in the absence of exogenous inputs $r(t)$ and $d(t)$. Note, however, that Theorem 2 guarantees the total stability of the system even in the presence of them.

5. Extension to Generalized Delay Controllers

A generalized controller form of the delay compensator schemes is depicted in Fig. 5. If $K_1(s) = (1 + C(s)P(s))^{-1}C(s)$, $K_2(s) = P(s)$ and $K_3(s) = I$ with $\tau_1 = \tau_3$, and $\tau_2 = 0$, the controller corresponds to the Smith predictor or the Smith predictor in IMC form (Morari and Zafiriou, 1989),

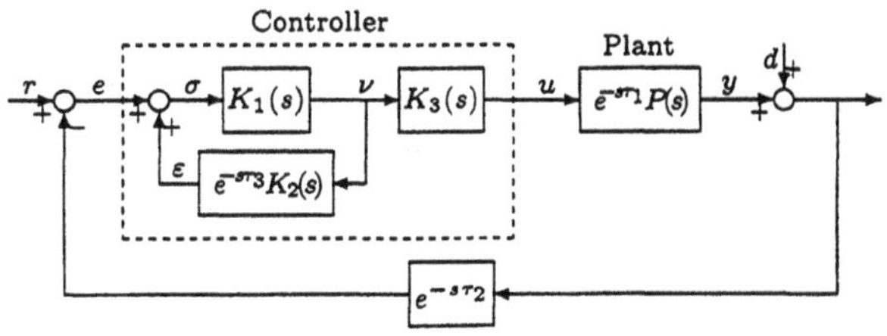

Fig. 5. Time-delay control systems with generalized delay controllers.

where $C(s)$ is a precompensator designed for the delay-free plant to give a desired response. When $K_1(s) = (I + \beta(s))^{-1}$, $K_2(s) = \beta(s)$ and $K_3(s) = K_d(s)$ with $\tau_1 = \tau_3$ and $\tau_2 = 0$, the controller is equivalent to the delay compensator scheme by Agamennoni *et al.* (1995), where $K_d(s)$ is a nominal controller for the delay-free plant and $\beta(s)$ is introduced to account for the time-delay. Most delay compensator schemes can be incorporated into this configuration.

The dynamical ARW scheme utilizing the observer type technique is shown in Fig. 6, where $K_1(s, \xi_1)$, $K_2(s, \xi_2)$ and $K_3(s, \xi_3)$ represent compensated controllers. Let the dynamics of the plant be the same as (1)-(3) and compensated controllers $K_1(s, \xi_1)$, $K_2(s, \xi_2)$ and $K_3(s, \xi_3)$ be expressed by the following state-space representations.

1) The dynamics of the compensated controller $K_1(s, \xi_1)$:

$$\dot{x}_{c1}(t) = F_1x_{c1}(t) + G_1\sigma(t) - \xi_1(t), \tag{38}$$
$$\nu(t) = H_1x_{c1}(t) + L_1\sigma(t), \tag{39}$$
$$\sigma(t) = r(t) + \varepsilon(t) - y(t - \tau_2) - d(t - \tau_2). \tag{40}$$

2) The dynamics of the compensated controller $e^{-s\tau_3}K_2(s, \xi_2)$:

$$\dot{x}_{c2}(t) = F_2x_{c2}(t) + G_2\nu(t - \tau_3) - \xi_2(t), \tag{41}$$
$$\varepsilon(t) = H_2x_{c2}(t) + L_2\nu(t - \tau_3). \tag{42}$$

3) The dynamics of the compensated controller $K_3(s, \xi_3)$:

$$\dot{x}_{c3}(t) = F_3x_{c3}(t) + G_3\nu(t) - \xi_3(t), \tag{43}$$
$$v(t) = H_3x_{c3}(t) + L_3\nu(t), \tag{44}$$

where $\xi(t) := (\xi_1^T(t), \xi_2^T(t), \xi_3^T(t))^T$ is the compensation signal and its dynamics are given in (10)-(12).

By utilizing the results in Section 3, we can state the following theorem (Park and Choi, 1998), and Fig. 7 depicts the proposed dynamical compensator $M^*(s)$.

Theorem 4 : Consider the dynamical ARW scheme shown in Fig. 6. When Assumptions (A1)-(A3) are

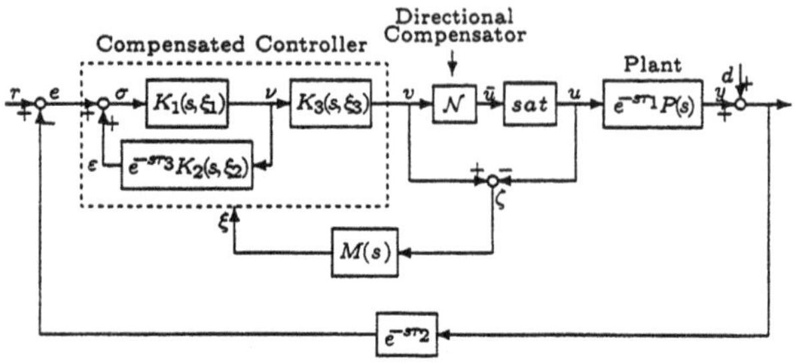

Fig. 6. Dynamical anti-reset windup method for time-delay control systems with saturating actuators (for generalized delay controllers).

satisfied, the state-space realizable form of the dynamical compensator $M^*(s)$ which optimizes the performance index (13) is uniquely determined by

$$\dot{x}_d(t) = Ax_d(t) + B\eta(t), \quad (45)$$
$$\xi_1(t) = G_1\rho(t), \quad (46)$$
$$\xi_2(t) = G_2L_1\rho(t-\tau_3), \quad (47)$$
$$\xi_3(t) = G_3L_1\rho(t), \quad (48)$$

where

$$\eta(t) := \zeta(t-\tau_1-\tau_2) - L_3L_1\rho(t-\tau_1-\tau_2), \quad (49)$$
$$\rho(t) := \vartheta(t) + \varphi(t), \quad (50)$$
$$\vartheta(t) := Cx_d(t) + D\eta(t), \quad (51)$$
$$\varphi(t) := L_2L_1\left[\vartheta(t-\tau_3) + \varphi(t-\tau_3)\right]. \quad (52)$$

The stability results in Section 4 is still effective in the proposed dynamical ARW method for the generalized delay controllers. More specifically, Theorem 2 is valid except for condition 2 that should be replaced with 'The dynamical ARW compensator is designed as in (45)-(52).' And Theorem 3 can be rephrased as follows.

Theorem 5 : Assume that the system with saturating actuators satisfies Assumptions (A1)-(A3) and the dynamical compensator is designed as in (45)-(52). Then the overall system is *asymptotically stable* in the absence of exogenous inputs $r(t)$ and $d(t)$ if the following condition is satisfied

$$\Big\| e^{-s(\tau_1+\tau_2)}L_3\left[I + e^{-s\tau_3}L_1 \right. \quad (53)$$
$$\left. \times \left(I - e^{-s\tau_3}L_2L_1\right)^{-1}L_2\right]L_1P(s)\Big\|_\infty < 1.$$

6. Conclusions

This paper proposed a dynamical anti-reset windup scheme which results in effective compensation for time-delay control systems with saturating actuators. The parameters of the dynamical compensator are derived explicitly based on the evolution of controller states in the absence or presence of saturating actuators. The proposed ARW scheme

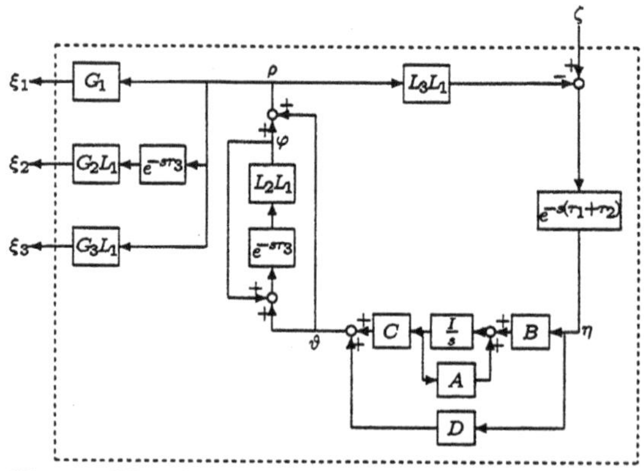

Fig. 7. The dynamical ARW compensator for the generalized delay controllers.

not only provides smooth performance degradation, but also guarantees the total stability of the resulting system. Once a controller is designed *a priori* for a known plant dynamics with time-delay, the dynamical compensator is determined immediately. The predesigned controller may be a generalized delay controller.

References

Agamennoni, O., J. L. Figueroa, J. A. Romagnoli and A. Palazoğlu (1995). Robust controller design for a class of uncertain multivariable systems. In *Proc. American Control Conf.*, Seattle, WA, pp. 3344-3347.

Chen, B.-S., S.-S. Wang and H.-C. Lu (1988). Stabilization of time-delay systems containing saturating actuators. *Int. J. Control*, **47**, 867-881.

Chou, J.-H., I.-R. Horng and B.-S. Chen (1989). Dynamical feedback compensator for uncertain time-delay systems containing saturating actuator. *Int. J. Control*, **49**, 961-968.

Landau, I. D. (1995). Robust digital control of systems with time delay (The smith predictor revisited). *Int. J. Control*, **62**, 325-347.

Morari, M. and E. Zafiriou (1989). ***Robust process control.*** Prentice-Hall, Englewood Cliffs, NJ.

Niculescu, S.-I., J.-M. Dion and L. Dugard (1996). Robust stabilization for uncertain time-delay systems containing saturating actuators. *IEEE Trans. Autom. Control*, **41**, 742-747.

Oucheriah, S. (1996). Global stabilization of a class of linear continuous time-delay systems with saturating controls. *IEEE Trans. Circuits and Systems*, **43**, 1012-1015.

Park, J.-K. and C.-H. Choi (1995). Dynamic compensation method for multivariable control systems with saturating actuators. *IEEE Trans. Autom. Control*, **40**, 1635-1640.

Park, J.-K. and C.-H. Choi (1997). Dynamical anti-reset windup method for discrete-time saturating systems. *Automatica*, **33**, 1055-1072.

Park, J.-K. and C.-H. Choi (1998). Dynamical anti-reset windup method for time-delay control systems with saturating actuators. *Submitted for publication.*

E-2c-21-2
14th Triennial World Congress, Beijing, P.R. China

A CONTROL LYAPUNOV FUNCTION BASED RECEDING HORIZON METHODOLOGY FOR INPUT CONSTRAINED NONLINEAR SYSTEMS

J. A. Primbs * M. Giannelli **

* *California Institute of Technology, Pasadena, CA, USA*
jprimbs@cds.caltech.edu
** *Dipartimento di Sistemi e Informatica, University of Florence,*
Via di S. Marta 3, 50139, Florence, Italy, moni@dsi.dsi.unifi.it

Abstract: Control Lyapunov Functions (CLFs) are used in conjunction with Receding Horizon Control (RHC) to develop a new class of control schemes for input-constrained nonlinear systems. First we develop a constrained version of pointwise min-norm controllers (Freeman and Kokotović, 1996). These controllers are then shown to be an extreme (zero horizon) case of a more general class of constrained receding horizon problems. The resulting controllers represent an extension of those presented in (Primbs *et al.*, 1998), and inherit a number of desirable properties. A simple example illustrates the use of this methodology.

Keywords: Nonlinear Predictive Control, Control Lyapunov Function, Input Saturation.

1. INTRODUCTION

Input saturation constraints plague nearly all physically based systems. For this reason, systems subject to input saturations have been a topic of interest for years. For linear plants, the theory is now well developed (see (Sussmann *et al.*, 1994) and references therein.) More recently, focus has shifted toward techniques for constrained nonlinear systems.

Two points of view have generally been employed. The first treats the problem from a stability viewpoint and requires a Control Lyapunov Function (CLF) (Kristić *et al.*, 1995). This approach consists of first deriving a CLF for the constrained system, and then determining a constrained input law consistent with the CLF. Research addressing the problem of determining a CLF for a constrained nonlinear system has become active lately. While similar to the unconstrained case in that no general systematic procedure exists for the derivation of a constrained CLF, procedures have emerged to construct CLFs for special classes of constrained systems. Results may be found in (Mazenc and Praly, 1996) for systems in the so called "forwarding" form and in (Freeman and Praly, 1998) for systems in the "backstepping" form. Extensions of Sontag's formula in the unconstrained case have even been developed in the constrained case (Lin and Sontag, 1991).

The second point of view exploits on-line computation to provide aggressive control actions that seek to optimize the performance objective. It forms the basis for *Model Predictive Control* (MPC), (Garcìa *et al.*, 1989), whose success has been well established. This success is due to the ease of direct incorporation of constraints in the on-line Receding Horizon (RH) computations. At the same time, the use of on-line computation has complicated the analysis of stability and performance properties of these schemes. Guaranteed stability has remained a concerned for some time. It is known that end constraints may be used for stability, but can lead to feasibility problems and complicate the on-line optimization (Mayne and Michalska, 1990).

At the intersection of the control Lyapunov and RH approaches to control is the methodology proposed in (Primbs *et al.*, 1998). It was shown that the CLF and RHC approaches are complementary, and can be placed in a single framework where the strengths of each scheme can be exploited, at the same time overcoming the drawbacks of each methodology individually.

In this paper, the results presented in (Primbs *et al.*, 1998) are extended to handle nonlinear systems subject to input constraints. In this way, the constraint handling capabilities of RHC are leveraged as well, leading to a unified approach to the control of constrained systems that incorporates both on-line processing opportunities and advances in CLF methods.

The paper is organized as follows. Section 2 motivates the approach by introducing the constrained nonlinear optimal control problem. Section 3 briefly reviews concepts from unconstrained nonlinear control design. Section 4 introduces a new pointwise min-norm scheme for constrained nonlinear systems. In Section 4.1 the pointwise min-norm scheme is extended to a receding horizon problem. Implementation issues are briefly discussed before providing an example that demonstrates the new methodology. Finally, concluding remarks summarize.

2. CONSTRAINED NONLINEAR OPTIMAL CONTROL AND PROBLEM STATEMENT

The reals will be denoted by $\mathbf{R}$, with $\mathbf{R}_+$ indicating the set of nonnegative real numbers. We will consider nonlinear systems of the form:

$$\dot{x} = f(x) + g(x)u, \qquad f(0) = 0 \tag{1}$$

with $x \in \mathbf{R}^n$ denoting the state and $f(x), g(x) \in \mathbf{C}^1$. The input will be constrained to lie in a specified set:

$$u \in \Omega_u \subset \mathbf{R}^m$$

where it is assumed that Ω_u contains a neighborhood of the origin. The basic motivation derives from the constrained infinite horizon nonlinear optimal control problem, stated as follows:

$$\begin{aligned} &\inf_{u(\cdot)} \int_0^\infty (q(x)) + u^T u)dt \\ &s.t.\ \dot{x} = f(x) + g(x)u \\ &\qquad u(\cdot) \in \Omega_u \end{aligned} \tag{2}$$

with $q(x) \in \mathbf{C}^1$, positive semi-definite. This is the standard nonlinear regulator problem, with the desired solution being a state feedback controller $u^* = k(x)$.

It is well known that the problem (2) is in general prohibitively difficult, even when there are no constraints. In the constrained case, issues of feasibility complicate the problem even further, leading to questions of the mere existence of a stabilizing controller. However, in special cases, procedures have emerged to provide suboptimal controllers for the problem (2).

As previously mentioned, the control Lyapunov function and receding horizon methodologies provide two possible approaches to the problem in (2). The most visible use of the receding horizon approach is so-called Model Predictive Control (MPC), which approximately solves the optimization problem (2) through on-line computation. While MPC has always been primarily motivated by constraints, it is only recently that methodologies for the derivation of CLFs have broached the subject of constraints (Mazenc and Praly, 1996; Janković *et al.*, 1996; Freeman and Praly, 1998). As this subject matures it will become increasingly useful to develop a theory analogous to that developed in (Primbs *et al.*, 1998), but for the constrained optimal control problem. Our aim is to extend those results to the case with saturated inputs.

3. MOTIVATION FROM THE UNCONSTRAINED PROBLEM

In this section, key results obtained for the unconstrained control problem are briefly summarized.

3.1 *Control Lyapunov Functions (CLFs)*

Consider the system (1). A CLF (Kristić *et al.*, 1995) is a $\mathbf{C}^1$, proper, positive definite function $V : \mathbf{R}^n \to \mathbf{R}_+$ such that

$$\inf_{u \in \mathbf{R}^m} \{V_x f + V_x g u\} < 0 \qquad \forall x \neq 0 \tag{3}$$

where $V_x := \frac{\partial V}{\partial x}$. A CLF may be used to design a stabilizing control law by posing the following optimization (Freeman and Kokotović, 1995):

Pointwise Min-Norm (PMN)

$$\begin{aligned} &minimize \qquad u^T(t)u(t) \\ &subject\ to\ V_x f + V_x g u(t) \leq -\sigma(x(t)) \end{aligned} \tag{4}$$

where σ is some positive definite real valued function. As pointed out in (Freeman and Primbs, 1996), a special choice of $\sigma(x)$ can be used

$$\sigma_s = \sqrt{(V_x f)^2 + q(x) V_x g g^T V_x^T} \tag{5}$$

with $q(x) \in \mathbf{C}^1$, positive semi-definite. In this case the resulting control action is a slight variation on Sontag's formula (Sontag, 1989):

$$u_s = \begin{cases} -\left[\dfrac{V_x f + \sigma_s}{V_x g g^T V_x^T}\right] g^T V_x^T & V_x g \neq 0_m \\ 0_m & V_x g = 0_m \end{cases} \quad (6)$$

This formula is well known for its continuity properties (Sontag, 1989).

3.2 *Receding Horizon Control (RHC) and CLFs*

Let V be a CLF, and u_σ and x_σ denote the control and state trajectories, respectively, obtained by solving the Pointwise Min-Norm (PMN) problem (4) with parameter $\sigma(x)$. Consider the following receding horizon optimal control problem:

RHC+CLF

$$min \int_t^{t+T} (q(x(\tau)) + u^T(\tau)u(\tau))d\tau \quad (7)$$

$$s.\ t.\ \dot{x}(\tau) = f(x(\tau)) + g(x(\tau))u(\tau) \quad (8)$$

$$V_x f + V_x g u(\tau) \leq -\epsilon(\tau)\sigma(x(\tau)) \quad (9)$$

$$V(x(t+T)) \leq V(x_\sigma(t+T)) \quad (10)$$

where $\tau \in [t, t+T]$ and $\epsilon(\tau)$ is a design parameter, and is chosen so that the implemented control actions from the receding horizon scheme satisfy $\dot{V} < 0$. Details of implementation choices can be found in Section 5. Corresponding to the use of splines to parameterize possible control trajectories, we may assume that this optimization is solved over piecewise $\mathbf{C}^1$ functions and that a minimizing solution exists (although such a minimum will not be required for stability.)

This optimization is solved on-line and implemented in a receding horizon fashion: the current control at time t and state $x(t)$ is obtained by determining on-line the optimal control $\hat{u}$ over the interval $[t, t+T]$, and then setting the control action to the initial optimizing solution $\hat{u}(t)$. These computations are repeated at each time step, updating the initial condition with the current state and resolving the above optimization.

This control scheme yields a number of appealing theoretical and implementation properties. Details can be found in (Primbs *et al.*, 1998).

With recent breakthroughs in techniques for deriving CLFs for systems under constraints (Freeman and Praly, 1998; Mazenc and Praly, 1996; Janković *et al.*, 1996), it is imperative that such a framework be extended to handle constrained systems as well.

4. A STABILIZING BOUNDED FEEDBACK CONTROL LAW

Suppose that a *constrained* CLF V is given for (1) such that

$$\inf_{u \in \Omega_u} \{V_x f + V_x g u\} < 0 \qquad \forall x \neq 0 \quad (11)$$

As in the unconstrained case, first a pointwise min-norm problem is introduced based on a CLF approach. Later, this approach is extended to a corresponding RHC problem.

As a direct extension of the unconstrained PMN problem (4), consider the following constrained formulation:

Constrained Pointwise Min-Norm (CPMN)

$$min \qquad u^T(t)u(t) \quad (12)$$

$$s.\ t.\ V_x f + V_x g u(t)) \leq -\hat{\sigma}(x(t)) \quad (13)$$

$$u(t) \in \Omega_u \quad (14)$$

with $\hat{\sigma}(x(t)) > 0$. In the unconstrained case, it is easy to see that the design parameter σ is only required to satisfy $V_x f \leq -\sigma$ when $V_x g = 0$ to be an admissible choice. However, the situation is now more complicated. Since the input is bounded, the stability constraint (13) may make the problem infeasible for an arbitrary choice of σ. To signify that σ must be chosen with this in mind, it is denoted by $\hat{\sigma}$ in the constrained problem. In order to avoid infeasibility, $\hat{\sigma}$ must be properly chosen. This can be accomplished by solving the following optimization problem:

Softened Pointwise Min-Norm (SPMN)

$$\min_{u,\mu} \qquad u^T(t)u(t) + \rho\mu^2 \quad (15)$$

$$s.\ t.\ V_x f + V_x g u(t) \leq -\sigma(x(t)) + \mu \quad (16)$$

$$\mu \geq 0 \quad (17)$$

$$-\sigma(x(t)) + \mu \leq 0 \quad (18)$$

$$u(t) \in \Omega_u \quad (19)$$

and setting

$$\hat{\sigma}(x(t)) = \sigma(x(t)) - \mu \quad (20)$$

with $\rho > 0$ a design knob.

Remark 1 - The softened pointwise min-norm problem (SPMN) (15)-(19) can be viewed as a pointwise min-norm problem in which the objective function contains the penalty term $\rho\mu^2$. In fact the stability constraint (14) in the standard pointwise min-norm problem may lead to infeasibility, and thus this term is used to "soften" this constraint. Hence, one may view σ as being the desired parameter for the pointwise min-norm problem, but due to the constraint (19) it must be compromised to $\hat{\sigma} = \sigma - \mu$. The parameter ρ measures one's averseness to deviations from the desired σ. For any value of ρ, the SPMN problem is feasible due to the condition (11).

Remark 2 - Even when condition (11) is not known to be satisfied, i.e. one is not sure whether the CLF V is valid for the constrained system, the above scheme is a reasonable approach to the design of a constrained control law. By removing the constraint (18) (which ensures that $\dot{V}$ is negative), and using a large value of ρ, the above problem will select u in the constraint set Ω_u that makes $\dot{V}$ as negative as possible whenever σ is not feasible for the CPMN problem (12)–(14). In this sense, the optimization will select a stabilizing control law in the bounded set Ω_u if such a law is possible.

There is the following important connection between the pointwise min-norm problems CPMN and SPMN:

Lemma 4.1. Let $(u^*(t), \mu^*(t))$ be the optimal solution of the problem SPMN for any given state $x(t)$. Then $u^*(t)$ is also the optimal solution of CPMN with $\hat{\sigma}(x(t)) = \sigma_s(x(t)) - \mu^*(t)$.

The proof is straightforward.

Remark 3 - The importance of Lemma 4.1 lies in the fact that we do not need to solve two optimization problems; i.e. first SPMN to obtain the optimal μ and thus compute $\hat{\sigma}$, and then CPMN to determine the optimal control input u using $\hat{\sigma}$. This allows us to always refer to the CPMN problem, even though the SPMN problem is effectively solved in order to obtain a feasible solution.

By determining a feasible $\hat{\sigma}$ for the CPMN problem, this allows the approach in (Primbs *et al.*, 1998) to be used to extend the CPMN controller to its natural RH extension.

4.1 *Receding horizon extensions*

Let $u_{\hat{\sigma}}$ and $x_{\hat{\sigma}}$ denote the control and state trajectories, respectively, obtained by solving the CPMN problem with parameter $\hat{\sigma}(x)$. Consider the following receding horizon optimal control problem:

<u>**Constrained RHC+CLF (CRHC+CLF)**</u>

$$min \int_t^{t+T} (q(x(\tau)) + u^T(\tau)u(\tau))d\tau \qquad (21)$$

$$s.\ t.\ \dot{x}(\tau) = f(x(\tau)) + g(x(\tau))u(\tau) \qquad (22)$$

$$V_x f + V_x g u(\tau)) \leq -\epsilon(\tau)\hat{\sigma}(x(\tau) \qquad (23)$$

$$V(x(t+T)) \leq V(x_{\hat{\sigma}}(t+T)) \qquad (24)$$

$$u(\cdot) \in \Omega_u \qquad (25)$$

where $\tau \in [t, t+T]$ and $\epsilon(\tau)$ is chosen as in the unconstrained case. This optimization is solved on-line and implemented in a receding horizon fashion.

The constraint (23) guarantees the stability of this scheme. On the other hand, the constraint (24) is oriented towards performance. It is obtained by first simulating the the control from the solution to the constrained pointwise min-norm problem (CPMN) for time T, which results in a state trajectory that ends at $x_{\hat{\sigma}}(t+T)$ and then evaluating the CLF at this point $V(x_{\hat{\sigma}}(t+T))$. In terms of level curves, the constraint (23) requires that the final state of all potential sequences lie inside the level curve of V that passes through $x_{\hat{\sigma}}(t+T)$. In the context of traditional receding horizon implementations, this constraint obviates the need for a terminal weight.

Theoretically, this scheme is the natural RH extension of the CPMN scheme. That is, the limit, as $T \to 0$, of the optimization problem CRHC+CLF is exactly the CPMN problem. The proof is analogous to the one in (Primbs *et al.*, 1998) for the unconstrained case. Thus the receding horizon optimization problem reduces to the pointwise min-norm scheme:

$$\begin{aligned} & min\ u^T u \\ & s.\ t.\ V_x f(x(t)) + V_x g(x(t))u(t) \\ & \qquad \leq -\hat{\sigma}(x(t)) \\ & \qquad u \in \Omega_u \end{aligned}$$

As in the unconstrained case, this scheme is particularly amenable to implementation due to the following properties:

- The optimization may be preempted without loss of stability. Since the constraint (23) ensures that V will be a Lyapunov function for the closed loop, any control that satisfies this constraint will be stabilizing. In particular, if the optimization cannot be completed, one may implement the current best solution and proceed without any loss of stability. Thus there is no requirement of a global optimum to the non-convex optimization CRHC+CLF to guarantee stability.
- Since the stability constraint (23) is independent of the objective function, the horizon may be varied on-line without compromising stability.
- An initial feasible trajectory is provided by the solution of the CPMN problem.

For more details of the properties of CLF based receding horizon schemes, see (Primbs *et al.*, 1998).

5. IMPLEMENTATION

The receding horizon strategy, as defined in CRHC+CLF (21)-(25), is not directly implementable because it requires the exact open loop optimal control for the nonlinear system over the interval $[t, t+T]$ to be recomputed at every current value of the state (i.e. infinitely often). An implementable strategy should update the optimal open loop control only discretely in time, at the beginning of each time interval of length T_s (usually $T_s \ll T$). Due to the fact that the receding horizon controller is only updated every T_s seconds, one should require the following of the function $\epsilon(\tau)$ in the CRHC+CLF scheme:

(1) $\epsilon(\tau) \leq 1$ for all $\tau \in [t, t+T]$.
(2) $\epsilon(\tau) > 0$ for all $\tau \in [t, t+T_s]$.

The first is to guarantee that the pointwise min-norm controller will be feasible for the receding horizon optimal control problem CRHC+CLF. The second ensures that V will be a Lyapunov function for the closed loop system since it requires $\dot{V}$ to be negative for all control actions that may be implemented. Beyond these mild requirements, $\epsilon(\tau)$ is a free design parameter.

While it is true that the optimization CRHC+CLF is only solved at discrete sampling times, it is also true that in general it is prohibitively difficult to find an exact global solution to the optimization. In general these optimization are solved over a finite number of decision parameters. This is done either by discretizing the control actions in time, or by choosing a finite set of basis functions and optimizing over their coefficients. Furthermore, note that this approach will only result in an approximate solution and in general no global optimizing solution will be found. Nevertheless, the proposed scheme does not require the exact solution, but rather only requires satisfaction of the constraint (23) to guarantee stability.

6. EXAMPLE

Consider the following dynamics:

$$\dot{x}_1 = -\frac{(1+0.1x_1^2)}{(1+0.1x_2^2)}x_2 + (1+2x_1^2)u_1 \quad (26)$$

$$\dot{x}_2 = \frac{20(1+0.1x_1^2)}{(1+0.1x_2^2)}x_1 + u_2 \quad (27)$$

with performance objective:

$$\inf_u \int_0^\infty (x_1^2 + 5x_2^2 + u_1^2 + u_2^2)dt$$

The inputs to the system are constrained by:

$$|u_1| \leq 1, \qquad |u_2| \leq 1$$

For this system, a constrained CLF is given by:

$$V = 10x_1^2 + 0.5x_2^2$$

A CPMN controller and its RH extension are both tested. In the CPMN scheme we selected the parameters $\rho = 1 \times 10^6$ and $\sigma(x)$ corresponding to Sontag's unconstrained formula (5). For the CRHC+CLF problem a sampling time of $T_s = 0.05$ was chosen, with a horizon of $T = 0.15$ seconds and $\epsilon(\tau) = 1 \times 10^{-4}$ over the interval $[t, t+T_s]$ and minus infinity otherwise. Simulation results from the initial condition $[1, 1]$ are shown in the figures. A summary of the cost achieved by each controller is given in Table 1.

Table 1

Controller	Cost
CPMN	27.1
CRHC+CLF (T=0.15)	22.9

Table 1. Cost of CPMN vs. CRHC+CLF controller

The control and state trajectories of the CPMN and CRHC+CLF controllers are contrasted in Fig. 1 and Fig. 2, respectively.

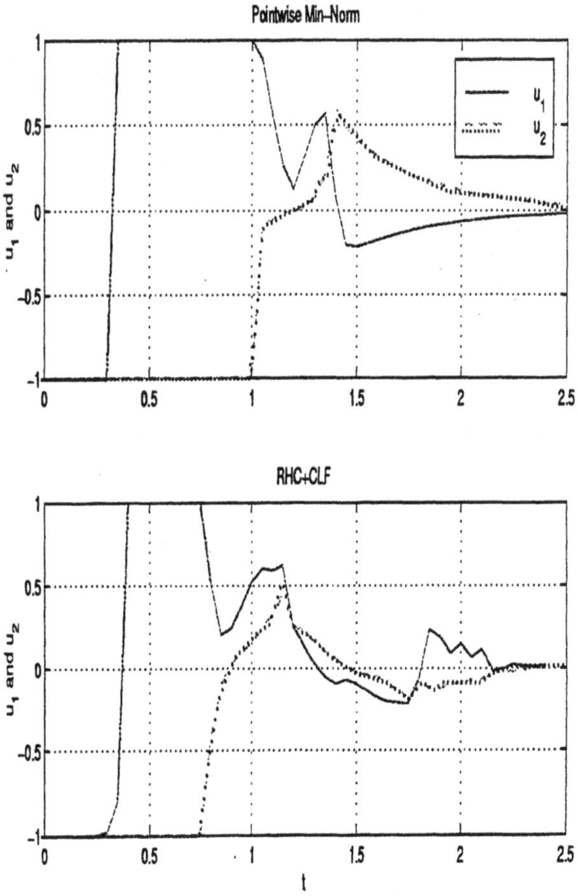

Fig. 1. Comparison of control trajectories from the CPMN and CRHC+CLF controller.

The results of this example follow the general trend of those given in (Primbs *et al.*, 1998) for unconstrained systems. While a PMN controller typically displays reasonable performance, especially under proper tuning of the parameters, the addition of a receding horizon often leads to significant improvements, even with the application of relatively short horizons.

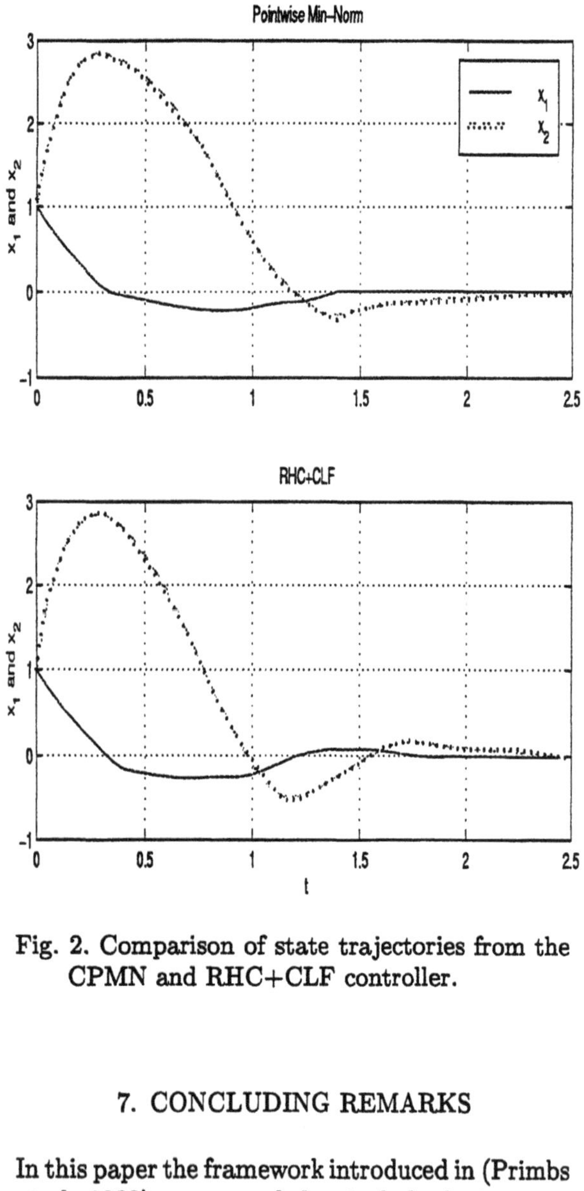

Fig. 2. Comparison of state trajectories from the CPMN and RHC+CLF controller.

7. CONCLUDING REMARKS

In this paper the framework introduced in (Primbs *et al.*, 1998) was extended to include the incorporation of control constraints. This first involved the development of a constrained pointwise min-norm control scheme. This scheme is based on a modification of the unconstrained pointwise min-norm scheme, and as well as providing a controller for the constrained system it generates the appropriate parameters required to establish a receding horizon extension. As in the unconstrained case, the resulting receding horizon extension inherits the stability properties of the pointwise min-norm controller in addition to providing a more flexible and implementable receding horizon scheme. As the theory for both control Lyapunov functions and receding horizon control improves, the required framework has been provided to take full advantage of those new contributions.

8. REFERENCES

Freeman, R. and J. A. Primbs (1996). Control lyapunov functions: New ideas from an old source. In: *Proc. of the 35th IEEE Conference on Decision and Control.* Kobe, Japan. pp. 3926–3931.

Freeman, R. and L. Praly (1998). Integrator backstepping for bounded controls and control rates. *IEEE Trans. Aut. Contr.* **43**, 258–262.

Freeman, R. and P. Kokotović (1995). Optimal nonlinear controllers for feedback linearizable systems. In: *Proc. of the American Control Conference '95.* Seattle, Washington, USA. pp. 2722–2726.

Freeman, R. and P. Kokotović (1996). Inverse optimality and robust stabilization. *Siam. J. Contr. Optimiz.* **34**, 1365–1391.

Garcìa, C. E., D. M. Prett and M. Morari (1989). Model predictive control: Theory and practice - a survey. *Automatica* **25**, 335–348.

Janković, M., R. Sepulchre and P. Kokotović (1996). Constructive lyapunov stabilization of nonlinear cascade systems. *IEEE Trans. Aut. Contr.* **41**, 1723–1735.

Kristić, M., I. Kanellakopoulos and P. Kokotović (1995). *Nonlinear and Adaptive Control Design.* John Wiley & Sons. New York.

Lin, Y. and E. D. Sontag (1991). A universal for stabilization with bounded controls. *System and Control Letters* **16**, 393–397.

Mayne, D.Q. and H. Michalska (1990). Receding horizon control of nonlinear systems. *IEEE Trans. Aut. Contr.* **35**, 814–824.

Mazenc, F. and L. Praly (1996). Adding integrations, saturated controls, and stabilization of feedforward systems. *IEEE Trans. Aut. Contr.* **41**, 1559–1578.

Primbs, J. A., V. Nevistić and J. C. Doyle (1998). On receding horizon extension and control lyapunov functions. In: *Proc. of the American Control Conference '98.* Philadelphia, Pennsylvania, USA.

Sontag, E. D. (1989). A 'universal' construction of artstein's theorem on nonlinear stabilization. *System and Control Letters* **13**, 117–123.

Sussmann, H.J., E.D. Sontag and Y. Yang (1994). A general result on the stabilization of linear systems using bounded control. *IEEE Trans. Aut. Contr.* **39**, 2411–2424.

A NONLINEAR PID CONTROLLER WITH APPLICATIONS

Yu-Chu Tian[1]

Institute of Industrial Process Control, Zhejiang University
Hangzhou 310027, P. R. China

Moses O. Tadé and Jinyu Tang

School of Chemical Engineering, Curtin University of Technology
GPO Box U 1987, Perth, WA 6845, Australia

Abstract: PID controllers have been the most commonly used industrial controllers. A linear PID controller is usually adequate for a nominal process. However, it often behaves poorly for a process with changes or uncertainties in operating conditions or environmental parameters. A nonlinear PID controller is introduced in this paper to alleviate this limitation. The controller gain and integral time are automatically adjusted according to the control error. Two simple nonlinear functions are proposed for the controller gain and integral time respectively. The proposed nonlinear controller is applied to two processes, a linear robotic process and a nonlinear thermoplastic injection molding process. Application examples show that the proposed nonlinear PID controller outperforms the linear PID controller in set-point tracking and load disturbance rejection. *Copyright © 1999 IFAC*

Keywords: PID controllers, control design, nonlinear control, control algorithms, control applications

1. INTRODUCTION

During the 1930s proportional-integral-derivative (PID) controllers became commercially available. Although there are numerous control algorithms in control theory and engineering, PID controllers have been the most popular and the most commonly used industrial controllers. This is primarily due to the simplicity and good performance of PID controllers. The controller principle, dynamics, design and tuning are discussed in numerous references.

A linear PID controller is described by the following relation

$$u(t) = K_p\left[e(t) + \frac{1}{T_i}\int_0^t e(t)dt + T_d\frac{de(t)}{dt}\right], \quad (1)$$

where u represents control action; K_p, T_i, and T_d are controller gain, integral time, and derivative time, respectively; e denotes the error between the process set-point and the actual output.

A PID controller consists of three terms representing three different control modes, i.e., proportional, integral, and derivative control actions. The proportional (P) control provides a simple correction proportional to control error to the manipulated variable. The integral (I) control integrates the control error over time to eliminate possible offset. The derivative (D) control considers the rate of

[1] Current address: School of Chemical Engineering, Curtin University of Technology, GPO Box U 1987, Perth, WA 6845, Australia. Telephone: +61-8-9266 3776. Fax: +61-8-9266 3554. E-mail: tiany@che.curtin.edu.au.

changes in the control error and takes an anticipatory action to correct the manipulated variable. For PID control, only three parameters are needed to be tuned. The three control modes of P, I, and D behave with some features of intelligence.

Compared with many model-based control algorithms, PID control is model-free and process-based. No process model is required to implement a PID control system. The controller parameters can be tuned by trial and error. This is an important advantage for process control where accurate process models are usually unavailable especially for complex processes such as reactor and distillation column processes.

Although linear PID control is usually adequate for a nominal process, it behaves poorly for a process with changes in operating conditions or environmental parameters. For example, a positive or negative deviation of process gain from its nominal value at which the PID controller is tuned results in oscillatory or sluggish process response. Requirements of fast process response and small overshoot are usually conflicting. In addition, a large control error leads to an aggressive integral control action that may result in undesirable reset windup.

Efforts have been made to overcome the above drawbacks. Cheung and Luyben (1980) investigated the performance of PI control with nonlinear gain. Nonlinear integral control action was considered by Ghreichi and Farison (1986). Recently, Seraji (1998) reported a class of nonlinear PID controllers, which are comprised of a sector-bounded nonlinear gain in cascade with a linear PID controller. An interesting nonlinear PI compensator was proposed by Shahruz and Schwartz (1997).

This paper introduces a new nonlinear PID (NPID) control algorithm to enhance linear PID control. The structure of the linear PID controller is retained in the NPID controller, but the controller gain and integral time are designed to be nonlinear functions of the control error. The proposed NPID controller is applied to two processes, a robotic process and a thermoplastic injection molding process.

2. A NONLINEAR PID CONTROLLER

The proposed NPID controller has the structure of the standard linear PID controller of equation (1). The controller gain and integral time are, however, not fixed. They are designed to be nonlinear functions of the control error e. Similar to equation (1), the NPID control algorithm is described by

$$u(t) = K_p(e)\left[e(t) + \frac{1}{T_i(e)}\int_0^t e(t)dt + T_d\frac{de(t)}{dt}\right]. \quad (2)$$

The nonlinear controller gain, $K_p(e)$, and integral time, $T_i(e)$, are respectively discussed below.

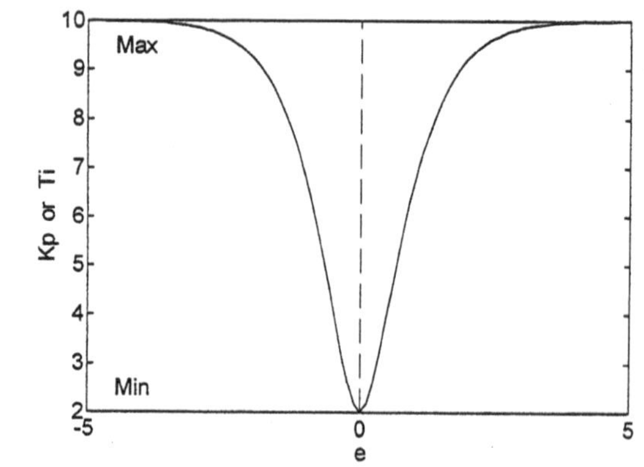

Figure 1: Nonlinear controller gain and integral time.

As discussed previously, requirements of fast process response and small overshoot are usually conflicting. A large controller gain is a benefit for fast reduction of a large control error, while it may lead to a large overshoot or an oscillatory response. A conservative controller gain is favorable for system stability, but it results in a sluggish response. These properties imply that the controller gain is expected to be aggressive for large control error or conservative for small control error. Therefore, we propose a three-parameter nonlinear controller gain as a function of the control error e

$$K_p = K_{\max} - (K_{\max} - K_{\min})(1 + a_p\,|e|)\exp(-a_p\,|e|), \quad (3)$$

where K_{max}, K_{min}, and a_p are positive parameters, $K_{max} > K_{min}$. It is clear that the maximum and minimum values of the controller gain are K_{max} and K_{min} respectively, i.e., the controller gain is bounded by $K_{max} \geq K_p \geq K_{min}$. Figure 1 gives a plot of the above nonlinear controller gain, K_p, as a function of e. It is seen that the controller gain, K_p, increases with the increase of $|e|$ and decreases with the decrease of $|e|$. K_p takes its maximum value, K_{max}, when $|e|$ tends to infinity and takes its minimum value, K_{min}, when e vanishes. To obtain an improved performance, K_{max} and K_{min} may be respectively selected to be larger and smaller than the nominal gain value of a linear PID controller. It is noted that the slope of K_p is determined by parameter a_p.

Similarly, the integral control action is expected to be conservative for large control error and aggressive for small control error. Hence, to improve integral control performance, a three-parameter nonlinear integral time is proposed as a function of the control error e

$$T_i = T_{\max} - (T_{\max} - T_{\min})(1 + a_i\,|e|)\exp(-a_i\,|e|), \quad (4)$$

where T_{max}, T_{min}, and a_i are positive parameters, $T_{max} > T_{min}$. The maximum and minimum values of the integral time are T_{max} and T_{min} respectively, i.e., the integral time is bounded by $T_{max} \geq T_i \geq T_{min}$. Figure 1 shows a plot of the above nonlinear integral time, T_i, as a function of e. It is clearly seen that the integral time, T_i, increases with the increase of $|e|$ and decreases with the decrease of $|e|$. This implies that the integral control becomes conservative or aggressive with the increase or decrease of $|e|$, respectively. T_i takes its maximum value, T_{max}, when $|e|$ tends to infinity and takes its minimum value, T_{min}, when e vanishes. T_{max} and T_{min} may be chosen to be larger and smaller than the nominal integral time of a linear PID controller. The slope of T_i is determined by parameter a_i.

It should be mentioned that the proposed NPID algorithm (2) has a fixed derivative time, T_d, although both controller gain, K_p, and integral time, T_i, are variable. This is based on the consideration that applications of PI control are more common than PID control in industry because process measurements are usually noisy.

To illustrate the superior performance of the proposed NPID controller, two applications are discussed in the next section.

3. APPLICATIONS

3.1 Robotic Process

The first application of the proposed NPID control is for a linear, second-order, robotic process discussed by Seraji (1998). The process is governed by

$$G(s) = \frac{K}{\omega^2 s^2 + 2\xi\omega s + 1}, K = 1, \omega = 0.2, \xi = 2. \qquad (5)$$

A unit step change in set-point is introduced at time $t = 0$ as an excitation of the process. Suppose that the process has the same load transfer function as equation (5). To test the load rejection ability of the NPID controller, a load step change of magnitude −0.5 is introduced at $t = 1$.

When a linear PI controller is applied, the tuned controller settings are $K_p = 5$ and $T_i = 0.2$, as discussed by Seraji (1998). A nonlinear PI controller is employed with a nonlinear gain of equation (3) and nonlinear integral time of equation (4). The controller settings are tuned to be $K_{max} = 15$, $K_{min} = 4$, $a_p = 1/0.6$, $T_{max} = 5$, $T_{min} = 0.4$, and $a_i = 1/0.6$. Figure 2 shows a comparison of the closed-loop responses of the standard linear PID controller (dash-dotted line) and the NPID controller (solid line). The corresponding variations of the nonlinear controller gain, K_p, and integral time, T_i, are plotted in Figure 3. Compared with the fixed parameter settings of the linear PID controller, it is clear from Figure 3 that K_p and T_i of the NPID controller are automatically adjusted based on control error. At $t = 0$, $|e| = 1$ is large, $K_p = 11.84$ and $T_i = 3$ are also large. K_p and T_i then decrease with the decrease of absolute e. At $t = 0.3$, K_p and T_i have been closed to their minimum values of 4 and 0.4 respectively. These automatic adjustments of K_p and T_i result in a fast set-point tracking without oscillations, as shown in Figure 2. When a load appears at $t = 1$ that leads to a control error, K_p and T_i change automatically to reject the load rapidly. It is clearly seen from Figure 2 that the NPID controller outperforms the linear PID controller in both set-point tracking and load disturbance rejection.

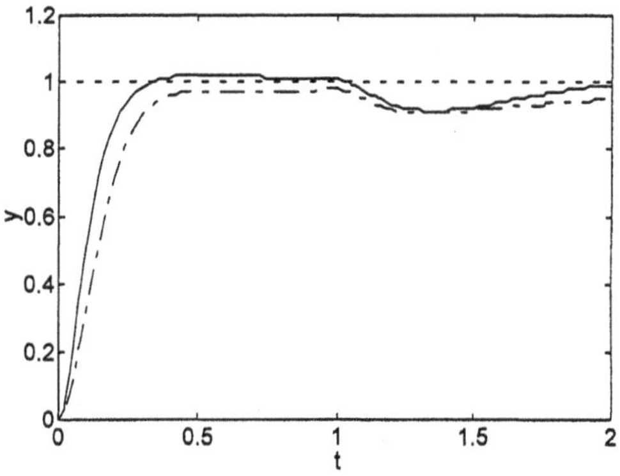

Figure 2: Closed-loop responses of the standard linear PID controller (dash-dotted line) and the NPID controller (solid line).

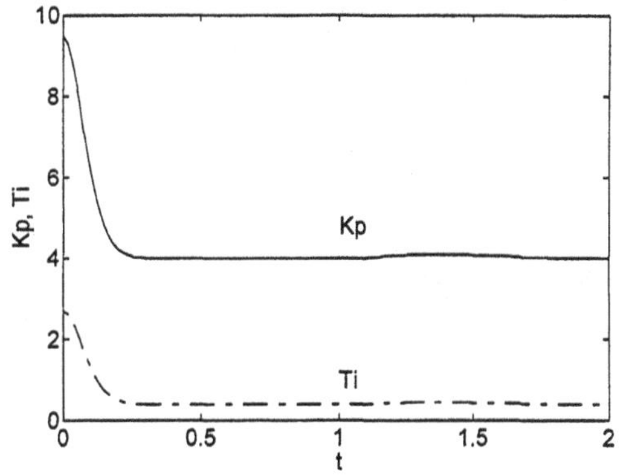

Figure 3: Plots of nonlinear controller gain K_p (solid line) and T_i (dash-dotted line) versus time t.

3.2 Thermoplastic Injection Molding Process

Now the application of the proposed NPID controller to a thermoplastic injection molding process is considered. Injection molding is an indispensable process in polymer processing industry. It produces a wide range of plastics economically and efficiently for industrial, agricultural, electronic, and household usage. A complete description of the injection molding process can be found in a number of books, say Rubin (1972). Injection molding is a typical cyclic process. Each cycle consists of three separate phases: filing, packing and holding, and cooling. The injection velocity control in the filling phase is considered here. The injection begins with a screw axial movement with a velocity profile and a continuous increase of the cavity pressure and ends with the cavity being completely filled.

The dynamic behaviors of injection molding are determined by a complex interaction among material properties, mold geometry and a number of process variables such as injection velocity, cavity pressure, nozzle pressure, hydraulic pressure, a group of barrel temperatures, nozzle temperature, etc. Among these variables, the injection velocity plays a critical role in injection molding quality. It indicates the rate of the polymer melt flowing to the mold cavity and significantly affects the cavity pressure and consequently the products quality such as shear rate and stress, shrinkage, impact strength, and surface quality. Productivity is also dependent on the velocity setting.

Efforts have been made for setting a proper velocity profile, say Yao, Bhattacharya, Kosior, *et al* (1994), Chao and Maul (1989). But keeping the velocity tracking a desired profile is not easy due to the complex effects of many factors on the velocity dynamics. Open-loop control is common at present for velocity control. However, it cannot give high control performance due to the complexity of the process dynamics and inevitable disturbances. Furthermore, an open-loop control algorithm developed for one injection molding machine and/or one mold cannot be applied to another machine and/or another mold due to the dynamics difference for different machines and molds. Even for the same machine and mold, the open-loop control performance deteriorates when operation conditions change. Optimization of the velocity profile is, thus, senseless unless the velocity can track the desired curve accurately in different production environments. Closed-loop control of the velocity is, therefore, vital for improving the velocity control performance and the products quality.

There have been a few reports on closed-loop control of the injection velocity. Experimental controls have been demonstrated by Zhang, Leonard and Speight (1996) with adaptive control and Tsoi and Gao (1998) with fuzzy logic. The design of these control schemes are, however, complicated for industrial implementation. The NPID controller proposed in this paper is adopted for closed-loop control of the injection velocity.

The injection velocity process behaves with nonlinearity and pure time delay, and it is time varying. A detailed analysis of injection velocity dynamics can be found in Tian and Gao (1998). A first-order plus delay process model with a nonlinear gain is adopted to approximate the injection velocity process

$$G(s) = \frac{K}{Ts+1}e^{-ds}. \quad (6)$$

The input and output variables of the above process transfer function is servo-valve opening (%) and injection velocity (m/s) respectively. For a cup mold, HDPE material, and the barrel temperature of 200°C, the model parameters are identified to be $K = 0.0006 \sim 0.0011 \text{m}\cdot\text{s}^{-1}/\%$ depending on excitations of the process, $T \approx 0.007$s, and $d \approx 0.015$s. It is also noted that there is a large pure time delay of about 0.08s at the beginning of injection.

Injection velocity control is a typical tracking problem. The desired injection velocity is determined according to mold, material, barrel temperature, etc. It is defined in injection velocity profile. Figure 4 shows a typical velocity profile, where v denotes injection velocity. Because the velocity profile is available before any control action is imposed on the process, set-point changes can thus be shifted forward by a period of process delay time (0.015s in this example) to compensate this delay and speed up the closed-loop response. On the other hand, the initial step change of the velocity profile is transformed into a smaller step change plus a ramp change to overcome possible initial peak in the closed-loop response. The shifting and transformation of the velocity profile has been discussed in detail by Tian and Gao (1998). Corresponding to the desired velocity curve in Figure 4 in solid line, a transformed and shifted velocity curve is also depicted in Figure 4 in dash-dotted line.

When a linear PI controller is employed, the controller parameters should be tuned in the worst case to ensure the system stability. The derivative control action is not employed here due to intensive noises in the process. The ITAE criterion is adopted to tune the controller parameters, resulting in the PI controller settings of K_p = 265%·s/m and T_i = 0.01035s for set-point tracking. These settings of the PI controller has been employed by Tian and Gao (1998) for the velocity control with a novel double-controller scheme. With the velocity profile of Figure 4, a typical closed-loop system response is shown in Figure 5 in dash-dotted line (Tian and Gao, 1998).

The proposed NPID control of equations (2) through (4) is now applied to the injection velocity control. As discussed above, the derivative control action is not employed, implying that $T_d = 0$. The controller parameters are tuned to be $K_{max} = 320$, $K_{min} = 255$, $a_p = 250$, $T_{max} = 0.0115$, $T_{min} = 0.0086$, and $a_i = 250$. It

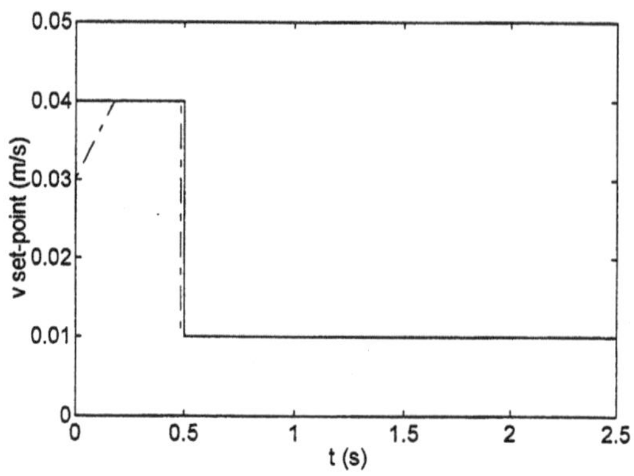

Figure 4: A typical injection velocity profile.

is clear that both K_p and T_i increase with the increase of control error e. Conversely, both K_p and T_i decrease as e is reduced. Thus, a large e leads to an aggressive proportional control and a conservative integral control. A small e results in a conservative proportional control and an aggressive integral control. A typical closed-loop response of the nonlinear PI controller is shown in Figure 5 in solid line. It is clear from this figure that the nonlinear PI control behaves with a faster set-point tracking than the linear PI control. Figures 6 and 7 depict the corresponding variations of K_p and T_i versus t.

Extensive simulations and experiments of injection velocity control have shown that the proposed NPID controller outperforms the linear PID controller.

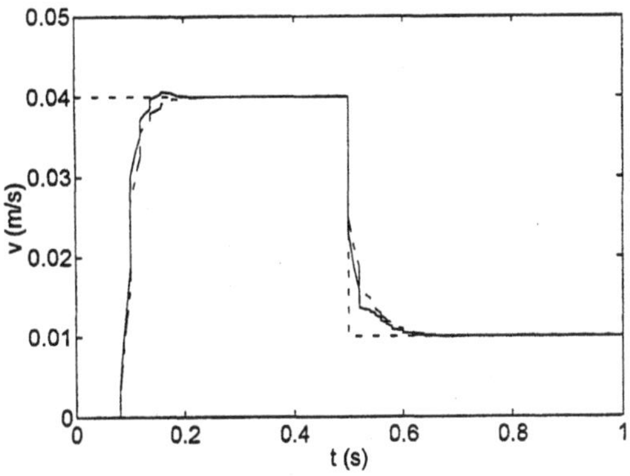

Figure 5: Closed-loop responses of the NPID controller (solid line) and the linear PID controller (dash-dotted line).

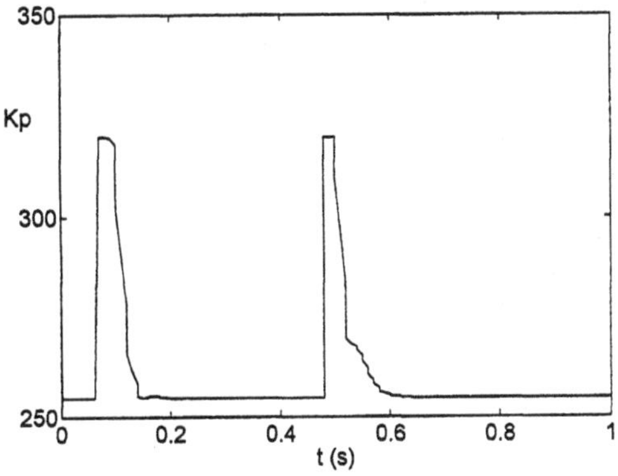

Figure 6: Nonlinear controller gain K_p versus t.

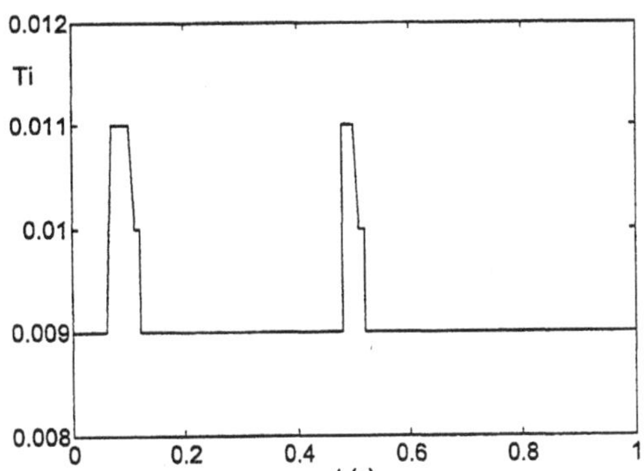

Figure 7: Nonlinear controller integral time T_i.

4. CONCLUSIONS

A NPID controller has been introduced with high performance. The controller has the structure of the standard linear PID controller, while its gain and integral time are designed to be nonlinear functions of control error. Two simple nonlinear functions have been proposed for the nonlinear controller gain and integral time, which increase with the increase of absolute control error and decrease with the decrease of absolute control error. The proposed NPID controller have been applied to two processes, a second-order, linear, robotic process and a first-order plus delay, nonlinear, thermoplastic injection molding process. The applications have shown that the proposed NPID controller is superior to the standard linear PID controller. The proposed NPID controller is, therefore, a promising tool for industrial process control.

REFERENCES

Chao O.-Y. and Maul G. P. (1989). Optimal injection velocity profiling. *Int J of Prod Res*, **27(11)**, 1917-1934

Cheung T.-F. and Luyben W. L. (1980). Nonlinear and nonconventional liquid level controllers. *Ind & Eng Chem Foundation*, **19**, 93-98.

Ghreichi G. Y. and Farison J. B. (1986). Sampled-data PI controller with nonlinear integrator. *Proc of IECON"86*, Milwaukee, WI, 451-456.

Rubin I. I. (1972). Injection molding theory and practice. John Wiely & Sons.

Seraji H. (1998). A new class of nonlinear PID controllers with robotic applications. *J of Robotic Syst*, **15(3)**, 161-181.

Shahruz S. M. and Schwartz A. L. (1997). Nonlinear PI compensators that achieve high performance. *J of Dynamic Syst, Measurement & Control*, **119**, 105-110.

Tian Y.-C. and Gao F. (1998). Injection velocity control of thermoplastic injection molding via a double-controller scheme. Submitted to *Ind & Eng Chem Res*.

Tsoi H. P. and F. Gao. (1998). Control of injection velocity using a fuzzy logic rule based controller for thermoplastic injection molding. *Poly Eng Sci*, in print.

Yao P. Y. V., Bhattacharya S. N., Kosior E.I., *et al.* (1994). Control of modling defects using programmed injection velocity. *Materials Forum*, **17(3)**, 247-250.

Zhang C. Y., Leonard J. and Speight R. G. (1996). Adaptive controller performance used for injection velocity control during filling phase. in *Proc of Ann Tech Conf – ANTEC*, **1**, Soc of Plastics Engineers, Brookfield, CT, USA, 593-597.

VARIABLE STRUCTURE CONTROL OF LINEAR DELAYED LARGE-SCALE SYSTEMS *

Zhang Xinzheng

Guangdong Univ. of Technology, Guangzhou 510090, P.R.China. Email: lucy@gdut.edu.cn

Liu Yongqing

Department of Automation, South China University of Technology, Guangzhou 510641, P.R.Chima

Abstract: This paper presents the design of decontralized variable structure controllers in the linear large scale control systems with time-delays, including some design methods of control systems under certain given conditions as well as local decentralized controllers and the multilayer decentralized variable structure controllers by means of the concept and method of approaching rates. Copyright © 1999 IFAC

Keywords: Variable structure control; time-delays large scale interconnected system; local decentralized controller; multilayer decentralized controller

1. INTRODUCTION

In engineering and social practice, propagation of any signal and information is always to exist time delays, Due to the existence of some perturbation factor, some loops temporary interruption or get on again, this kinks of control system with delays will take place some change in structure. Therefore, there are problems of the structure and variable structure control (VSC) not only in the retarded interconnected control systems but also in the large scale interconnected control systems with delays, namely, the control system bear relation to $x(t-\tau)$ of the state variable $x(t)$. We are again in need of studying multilevel hierarchy multi-delays interconnected control system in practice, now, that is appearing from structure and interconnected stabilization problem of the linear continuous interconnected control systems with delays, its mathematical model is always differential equation with time delay variables. The problems of the interconnected control system are naturally in infinitive dimension space, so the structure and (VSC) for interconnected linear constant interconnected control system with the time-delays becomes more difficult and more complicated. In this paper, based on Lyapunov stability theory, the concept and method of approaching rates the problem is solved.

The variable structure control (VSC) concepts and theorical foundation have been generalized to the control design of systems with delays[1-4]. Outstanding advantage of delay VSC is tthat the sliding mode has invariance (or completely robustness) for the parameter perturbation and the external disrurbance of the system, which is the

*The project was supported by the science fundof the chinese nature science(69874005). This work is supported by Natural Science Foundation of Guangdong Province and Shandong Province(970235)

problem that robust control needs to solve; its major shortcoming is that delay is produced by inertia in the switching equipment, the control round-trip switching with finite frequency in sliding process, therefore oscillation is caused by delays. This leads to study vsc (of linear large scale control system with time-delays) but as far as the authors know, there have not been any other papers dealing with structure and VSC of interconnected control systems with delays.

Today, the technology of decentralized VSC design of the multi-delay large scale interconnected control system has been generalized [5], especially structure of the time-delay linear large scale interconnected control systems under some given conditions. In practice, to simplify the design problems of control law, decrease the expenditure of communication and improve the motion properties of controlled systems, it is necessary to design locally decentralized VS controllers of the large scale control systems with time-delay in order to satisfy the different needs. The approach presented in this paper can be extended to other cases of the multi-delay nonlinear large scale control systems with time-delay input control loop.

2. PROBLEM FORMULATION

Considering a linear large scale interconnected control systems with time-delays $\sum$ compcsed of N subsystems $\sum_i$, $i \in N = \{1, 2, \cdots, N\}$, each of which is described by

$$\begin{aligned}\textstyle\sum_i : \dot{x}_i = A_i x_i + B_i^0 u_i + \sum_{r=1}^{N_1} \overline{A}_{ij}^{(r)} x_j(t) + \\ \sum_{s=1}^{N_2} \overline{A}_{ij}^{(s)} x_j(t - \tau_{1ij}^{(s)}) = G_{1i} \quad (1) \\ x_i(t) = \phi_i(t), \quad u_i(t) = \psi_i(t), \\ -\tau \le t \le 0, \quad i = 1, \cdots, N\end{aligned}$$

where $x_i \in R^{n_i}$ and $u_i \in R^{m_i}$ are respectively the state vector and the input control vector of i-th subsystems; $x = (x_1^{\top}, \cdots, x_N^{\top})^{\top} \in R^n$ and $u = (u_1^{\top}, \cdots, u_N^{\top})^{\top} \in R^m$ are the state vector and the input control vector of the combination large scale control systems; $\phi_i(t) \in C$, $\psi_i(t) \in C$ $(i = 1, \cdots, N)$ are initial vector function and initial control vector function of i-th subsystems; $\overline{A}_{ij}^{(r)} = A_{ij}^{(r)} E_{1ij}^{(r)}$, $\overline{A}_{ij}^{(s)} = A_{ij}^{(s)} E_{2ij}^{(s)}$, are all $n_i \times n_i$ coefficient matrix; B_i^0, are $n_i \times m_i$ control matrix; m, m_i, n, m, M and N are all positive integers, and $n = n_1 + \cdots + n_N$, $m = m_1 + \cdots + m_N$.

Definition 1 We call the fundamental mutual connection matrix for corresponding systems (1) to the matrix $\overline{E}_1^{(r)}(t) = (\overline{e}_{1ij}^{(r)}(t))_{n_1 \times n_1}$, $\overline{E}_2^{(s)}(t) = (\overline{e}_{2ij}^{(s)}(t))_{n_1 \times n_1}$,

If

$$\begin{cases} \overline{e}_{1ij}^{(r)}(t) = \begin{cases} 1, & x_j(t) \text{ is in } G_{1i}, \\ 0, & x_j(t) \text{ isn't in } G_{1i}; \end{cases} \\ \overline{e}_{2ij}^{(s)}(t) = \begin{cases} 1, & x_j(t - \tau_{1ij}^{(s)}) \text{ is in } G_{1i}, \\ 0, & x_j(t - \tau_{1ij}^{(s)}) \text{ isn't in } G_{1i}; \end{cases} \end{cases}$$

Definition 2 We call that matrices $E_1^{(r)}(t) = (e_{1ij}^{(r)}(t))_{n_1 \times n_1}$, $E_2^{(s)}(t) = (e_{2ij}^{(s)}(t))_{n_1 \times n_1}$, are mutual connection matrices to be produced by corresponding fundamental mutual connection matrices $\overline{E}_1^{(r)}$, $\overline{E}_2^{(s)}$, if $\overline{e}_{\beta ij}^{(\alpha)} = 0$, $e_{\beta ij}^{(\alpha)} = 0, (\alpha = r, s, d;\ \beta = 1, 2, 3)$. and if $\overline{e}_{\beta ij}^{(\alpha)} = 1$ then either $e_{\beta ij}^{(\alpha)} = 0$ or $e_{\beta ij}^{(\alpha)} = 1$

denote that $E_1^{(r)} \in \overline{E}_1^{(r)}$, $E_2^{(s)} \in \overline{E}_2^{(s)}$.

For i-th $(i = 1, \cdots, N)$ subsystems of Eq. (1), suppose that the following conditions hold.

A1) Each state x_i can be locally accessible.
A2) A_i, $A_{ij}^{(r)}$, $A_{ij}^{(s)}$ and B_i^0, $E_{1ij}^{(r)}$, $E_{2ij}^{(s)}$, $(i = 1, \cdots, N)$ are all constant matrix;
A3) B_i^0 is a matrix with the column rank filled;
A4) (A_i, B_i^0) is a pair of controllable matrix;
A5) $E_1^{(r)} A_{ij}^{(r)} \in Im(B_i^0)$, $E_2^{(s)} A_{ij}^{(s)} \in Im(B_i^0)$.

We are to design a series of decentralized VSC laws $u_i(t)$ $(i = 1, \cdots, N)$ so as to make the equilibrium state be global asymptotically stable for the closed-loop time-delay linear large scale system.

3. DESIGN OF THE SWITCHING FUNCTION

Suppose that the switching functions have the following linear forms

$$S_i(t) = C_i X_i(t), \ i = 1, \cdots, N \quad (2)$$

The switching surfaces are given by

$$C_i X_i(t) = 0, \ i = 1, \cdots, N \quad (3)$$

where $S_i \in R^{m_i}$; $C_i \in R^{m_i \times n_i}$; is to be determined. First of all, choose proper transform matrix $C_i (i = 1, \cdots, N)$ so that the equation of the sliding mode is asymptotically stable. In view of A2)-A4), we can always carry out a series of local state transformations.

$$y_i(t) = T_i x_i(t), \ i = 1, \cdots, N \quad (4)$$

we can choose T_i so that

$$y = T_i x = \begin{bmatrix} T_i^{(1)} \\ T_i^{(2)} \end{bmatrix} \quad x = \begin{bmatrix} y_i^{(1)} \\ y_i^{(2)} \end{bmatrix} \quad (5)$$

where $T_i^{(1)} \in R^{(n_i - m_i) \times n_i}$, $T_i^{(2)} \in R^{m_i \times n_i}$, Defining

$$T_i\overline{A}_{ij}^{(r)}T^{-1} = \begin{bmatrix} \hat{A}_{ir}^{11} & \hat{A}_{ir}^{12} \\ \hat{A}_{ir}^{21} & \hat{A}_{ir}^{22} \end{bmatrix},$$

$$T_i\overline{A}_{ij}^{(s)}T^{-1} = \begin{bmatrix} \hat{A}_{is}^{11} & \hat{A}_{is}^{12} \\ \hat{A}_{is}^{21} & \hat{A}_{is}^{22} \end{bmatrix}, \quad T_iB_{ij}^{0} = \begin{bmatrix} 0 \\ B_i \end{bmatrix};$$

$$T_i\overline{B}_{ij}^{(d)} = \begin{bmatrix} 0 \\ \hat{B}_{ij}^{(s)} \end{bmatrix};$$

$$T_iA_iT_i^{-1} = \begin{bmatrix} A_i^{11} & A_i^{12} \\ A_i^{21} & A_i^{22} \end{bmatrix} \quad (i = 1, \cdots, N) \tag{6}$$

In the local state linear transformations (4), the control system (1) is changed into the following form

$$\dot{y}_i^{(1)}(t) = A_i^{11}y_i^{(1)}(t) + A_i^{12}y_i^{(2)}(t) \tag{7}$$

$$\begin{aligned} \dot{y}_i^{(2)}(t) = A_i^{21}y_i^{(1)}(t) + A_i^{22}y_i^{(2)}(t) + B_iu_i^2 \\ + \sum_{r=1}^{N_1}[\hat{A}_{ir}^{21}y_r^{(1)}(t) + \hat{A}_{ir}^{22}y_s^{(2)}(t)] \\ + \sum_{s=1}^{N_2}\hat{A}_{is}^{21}y_s^{(1)}(t-\tau_{ij}^{(s)}) + \hat{A}_{is}^{22}y_s^{(2)}(t-\tau_{ij}^{(s)})] \\ (i = 1, \cdots, N) \end{aligned} \tag{8}$$

$$\begin{aligned} x_i(t) = T_i\phi_i(t),\ u_i(t) = T_i\psi_i(t), \\ -\tau \le t \le 0, \quad (i = 1, \cdots, N) \end{aligned} \tag{9}$$

where $y_i^{(1)}(t) \in R^{n_i - m_i}$, $y_i^{(2)}(t) \in R^{m_i}$, $y_i = ((y_i^{(1)})^{\top}, (y_i^{(2)})^{\top})^{\top}$, and the matrices A_i^{11}, A_i^{12}, A_i^{21},A_i^{22}, and B_i all have corresponding dimensions. The equation of the switching surface becomes

$$C_i^{(1)}y_i^{(1)}(t) + C_i^{(2)}y_i^{(2)}(t) = 0, \quad i = 1, \cdots, N \tag{10}$$

where $C_iT_i^1 = (C_i^{(1)}|C_i^{(2)})$.

Lemma 1[6] If (A_i, B_i) is a pair of controllable matrices, then (A_i^{11}, A_i^{12}) is also a pair of controllable matrices.

Obviously, $C_i^{(2)}$ can be taken as $m_i \times m_i$ dimension invertible matrix. To make the calculation simple and convenient, we take

$$C_i^{(2)} = B_i^{-1} \tag{11}$$

From differential equations (7),(10) and (11), the sliding mode equation of i-th subsystem is as follows,

$$\dot{y}_i^{(1)}(t) = (A_i^{11} - A_i^{12}B_iC_i^{(1)})y_i^{(1)}(t) \tag{12}$$

From Lemma 1, we know that (A_i^{11}, A_i^{12}) is a pair of controllable matrices. Because of invertibility of B_i, we can properly choose the matrix $C_i^{(1)}$ which makes the sliding mode (12) of i-th subsystem be asymptotically stable, approach to its equilibrium at any needed tate or have arbitrary previously expected poles, and $C_i = (C_i^{(1)}|C_i^{(2)})$. The determination of $C_i^{(1)}$ and $C_i^{(2)}$ makes the equation of the switching surface and the switching function of the sliding mode be asymptotically stable.

4. DASIGN OF VSC LAW OF TIME-DELAYS LINEAR LARGE SCALE INTERCONNECTED CONTROL SYSTEM

By using the concept and method of approaching rate established by W.B.Gao in [7], we provide two decentralized variable structure control schemes of Eq.(1) linear large scale interconnected control system with time-delays.

4.1 Design of the Local Decentralized VSC Law

For interconnected item suppose that the following inequalities of norm hold.

$$\begin{aligned} ||S_i(t)||^{\frac{1}{2}}||B_i^{-1}\sum_{r=1}^{N_1}[\hat{A}_{ir}^{21}y_r^{(1)}(t) + \hat{A}_{ir}^{22}y_s^{(2)}(t)] \\ + \sum_{s=1}^{N_2}\hat{A}_{is}^{21}y_s^{(1)}(t-\tau_{ij}^{(s)}) + \hat{A}_{is}^{22}y_s^{(2)}(t-\tau_{ij}^{(s)})] \\ \le \sum_{j=1}^{N}\rho_{ij}||S_j(t)||^{\frac{1}{2}}|||y_j(t)||, \quad (i = 1, \cdots, N) \end{aligned} \tag{13}$$

Choose the VSC law in order that the attainable condition $S^{\top}\dot{S} < 0$ is satisfied. That is, the state orbits from any point of the state space can all get to the switching surface in finite time, and can eventually slip to the origin on the switching surface. Therefore, the global asymptotic stability of the close-loop system is achieved. For this purpose, we can easily obtain the following formula.

$$\begin{aligned} S_i(t)|_{(1)} = C_i^{(1)}\dot{y}_i^{(1)}(t) + C_i^{(2)}\dot{y}_i^{(2)}(t) = \\ C_i^{(1)}(A_i^{11}y_i^{(1)}(t) + A_i^{12}y_i^{(2)}(t)) + \\ B_i^{-1}(A_i^{21}y_i^{(1)}(t) + A_i^{22}y_i^{(2)}(t) + B_iU_i) + \\ B_i^{-1}\sum_{r=1}^{N_1}[\hat{A}_{ir}^{21}y_r^{(1)}(t) + \hat{A}_{ir}^{22}y_s^{(2)}(t)] \\ + \sum_{s=1}^{N_2}\hat{A}_{is}^{21}y_s^{(1)}(t-\tau_{ij}^{(s)}) + \hat{A}_{is}^{22}y_s^{(2)}(t-\tau_{ij}^{(s)})] \\ i = 1, \cdots, N \end{aligned} \tag{14}$$

where $y = (y_1^{\top}, \cdots, y_N^{\top})^{\top}$. Let the locally decentralized variable control law of the i-th subsystem $(i = 1, \cdots, N)$ be the following form.

$$\begin{aligned} u_i(t) = -[C_i^{(1)}A_i^{11} + B_i^{-1}A_i^{21}]y_i^{(1)}(t) \\ -[C_i^{(1)}A_i^{12} + B_i^{-1}A_i^{22}]y_i^{(2)}(t) \\ -[g_i + h_i]||y_i(t)||^2]\frac{S_i(t)}{||S_i(t)||} \end{aligned} \tag{15}$$

where $(i = 1, \cdots, N)$, g_i and all proper positytve

constants. We have

$$S^{\top}\dot{S}=\sum_{i=1}^{N}S^{\top}\dot{S}_i=-\sum_{i=1}^{N}S_i^{\top}\{B_i^{-1}\sum_{r=1}^{N_1}[\hat{A}_{ir}^{21}y_r^{(1)}(t)+$$
$$\hat{A}_{ir}^{22}y_s^{(2)}(t)]+\sum_{s=1}^{N_2}\hat{A}_{is}^{21}y_s^{(1)}(t-\tau_{ij}^{(s)})+$$
$$\hat{A}_{is}^{22}y_s^{(2)}(t-\tau_{ij}^{(s)})]-[g_i+h_i||y_i(t)||^2]\frac{S_i(t)}{||S_i(t)||}\}\leq$$
$$-\sum_{i=1}^{N}|[g_i+h_i||y_i(t)||^2]||S_i(t)||-$$
$$\sum_{i=1}^{N}\rho_{ij}||y_i||||S_i(t)||^{1/2}||S_j(t)||^{1/2}|=-F^{\top}QF \tag{16}$$

where $F = (F_1^{\top}, F_1^{\top}||y||)^{\top}$, $F_i = (||S_1(t)||^{1/2},\cdots,||S_N(t)||^{1/2})^{\top}$, $||y|| = (||y_1(t)||,\cdots,||y_N(t)||)^{\top}$, $Q=\begin{bmatrix} G & -P/2 \\ -P/2 & H \end{bmatrix}$, $G=\text{diag}\,(g_1,\cdots,g_N)$, $H=\text{diag}\,(h_1,\cdots,h_N)$, $P=(\rho_{ij})_{N\times N}$.

Therefore, we can obtain the following theorem.

Theorem 1 The sliding mode equation (12) of i-th subsystem of Eq.(1) is the asymptotically stable if and only if all chosen $g_i > 0$ and $h_i > 0 \ (i = 1,\cdots,N)$ are great enough so that Q is a diagonal dominant matrix.

4.2 Design of Multilayer Decentralized VSC Law

From Section 4.1, we known that the local decentralized VSC law is not always realized in all cases. So we suppose that the decentralized VSC law $u_i(t) \ (i=1,\cdots,N)$ of the i-th subsystem has the following multilayer form

$$u_i(t)=u_i^l(t)+u_i^g(t) \tag{17}$$

where u_i^l and u_i^g are respectively the local control component and the global control component of the i-th subsystem, u_i^l only uses the state information up to the i-th subsystem, while u_i^g can use the information of all subsystems. Therefore (17) is called the multi-order decentralized control law.

From formula (14), if we let

$$\begin{aligned}u_i^l(t)=-\{&C_i^{(1)}[A_i^{11}y_i^{(1)}(t)+A_i^{12}y_i^{(2)}(t)]\\&+B_i^{-1}[A_i^{21}y_i^{(1)}(t)+A_i^{22}y_i^{(2)}(t)\}\\&-g_iS_i(t)-h_isgn(S_i(t))\end{aligned} \tag{18}$$

and

$$u_i^g(t)=-B_i^{-1}[\sum_{r=1}^{N_1}[\hat{A}_{ir}^{21}y_r^{(1)}(t)+\hat{A}_{ir}^{22}y_s^{(2)}(t)$$
$$+\sum_{s=1}^{N_2}\hat{A}_{is}^{21}y_s^{(1)}(t-\tau_{ij}^{(s)})+\hat{A}_{is}^{22}y_s^{(2)}(t-\tau_{ij}^{(s)})] \tag{19}$$

where g_i, h_i are all proper positive numbers, thus the approaching condition

$$S_i^{\top}\dot{S}_i=-g_i||S_i(t)||^2-h_iS_i^{\top}(t)sgn(S_i(t))\leq 0 \tag{20}$$

being satisfied.

5. DESIGN OF VS CLOSED-LOOP SYSTEMS WITHOUT MATCHING CONDITIONS

If the matcing condition A5) is not satisfied but the other assumed conditions in Section 2 hold, then the main design work of the decentralized VSC closed-loop system is to design the switching function and the switching surface equations, the design of decentralized VSC law being the same as Section 4.

When the matching condition A5) in Section 4 does not hold, the interconnected terms of the i-th subsystem have the following form

$$\sum_{s=1,j\neq 1}^{N}A_{1is}x_s(t)+\sum_{j=1}^{M}\sum_{s=1}^{N}\Phi_{1is}x_s(t-\tau) \tag{21}$$

Letting

$$T_iA_{1is}T_i^1=\begin{bmatrix}A_{1is}^{11} & A_{1is}^{12}\\ A_{1is}^{21} & A_{1is}^{22}\end{bmatrix},$$
$$T_i\Phi_{1is}T_i^{-1}=\begin{bmatrix}\Phi_{1is}^{11} & \Phi_{1is}^{12}\\ \Phi_{1is}^{21} & \Phi_{1is}^{22}\end{bmatrix}$$

The siiding made equation of i-th subsystem, with the same from of switching surface equation, becomes

$$\begin{aligned}\dot{y}_i(t)=&A_{1ii}^{11}y_i^{(1)}(t)+A_{1ii}^{12}y_i^{(2)}(t)+\\&\sum_{s=1,s\neq i}^{N}A_{1is}^{11}y_s^{(1)}(t)+A_{1is}^{12}y_s^{(2)}(t)+\\&\sum_{j=1}^{M}(\Phi_{1is}^{11}y_s^{(1)}(t-\tau_{js})+\Phi_{1is}^{12}y_s^{(2)}(t-\tau_{js})\end{aligned} \tag{22}$$

Furthermore, letting $Y_r = ((Y_1^{(r)})^{\top},\cdots,Y_N^{(r)})^{\top})^{\top}$, $C_r = \text{diag}(C_1^{(r)},\cdots,C_N^{(r)})$, $r=1,2,;$ $B=\text{diag}\,(B_1,\cdots,B_N)$, on the switching surface we have the following formula

$$Y_2=-C_2^{-1}C_1Y_1 \tag{23}$$

Taking C_2 as

$$C_2=B^{-1} \tag{24}$$

and introducing certain transformation, we obtain the vector form of sliding mode equation

$$\begin{aligned}Y_1^{(1)}(t)=&(A_1-A_2BC_1)Y_1(t)+\\&A_3Y_1(t)+\sum_{j=1}^{M}P_jY_1(t-\tau_j)\end{aligned} \tag{25}$$

where

$$A_1 = diag(A_{1ll}^{11}, \cdots, A_{1NN}^{11}),$$
$$A_2 = diag(A_{1ll}^{12}, \cdots, A_{1NN}^{12}),$$
$$A_3 = diag(\sum_{s=1,s\neq i}^{N} (A_{1ls}^{11} - A_{1ls}^{12} B_1 C_1^{(1)}), \cdots, \sum_{s=1,s\neq i}^{N} (A_{1Ns}^{11} - A_{1Ns}^{12} B_N C_N^{(1)})),$$

$$P_j = diag(\sum_{s=1,s\neq i}^{N} (\Phi_{1ls}^{11} - \Phi_{1ls}^{12} B_1 C_1^{(1)}), \cdots, \sum_{s=1,s\neq i}^{N} (\Phi_{1Ns}^{11} - \Phi_{1Ns}^{12} B_N C_N^{(1)})),$$

$$(j = 1, \cdots, M)$$

We can determine the stability of the sliding mode equation (25) in the light of the criteria of small-delay or delay-independent stability or asymptotic stability. Now that (A_1, A_2) is a pair of controllable matrices, we can always proper choose the matrix C to determine the boundary of the delay τ so as to make the sliding mode has various stabilities.

The method presented in this paper can be easily expanded to the other multi-delay nonlinear large scale control systems with multi-delay input control vectors.

6. CONCLUSION

This paper dealt with the decentralized VSC problems linear large scale interconnected control systems (1) with time-delays under different conditions, and offered the design methods of local and multilayer decentralized VS controllers of these systems by making clever use of the property of diagonal dominant matrices as well as the concept and method of approaching rate established by W.B.Gao in [7], thus ensuring the global asymptotic stability of these large scale VS close-loop control systems.

REFERENCES

Jafarove E.M.Analysis and synthesis of multidimensional VSS with delays in sliding modes. Preprints of 11-th IFAC World Congress on Automatic Control in the Service of mankind, Tallinn, U.S.S.R, 1990,6:46-49

Yuan F.S. and Liu Y.Q. On the variable struceure control of N-dimension invariant systems with single time-delay and multi input, Advances in Modelling Simulation, 1992,31(3):57-64

Hu Y.M, Zhou QJ. Variable struceure control of control systems with delay, Acte Automatica Sinica, 1991 (in Chinese), 17(5):587-591

Gao C C, Liu Y Q. Stabilization and synthesis of two order linear constant neutral variable structure control systems. Kongzhi Lilun Yu Yingyong, To Appear, 1996(in Chinese)

Gao C C, Liu Y Q. On the design of decentralized variable structure control of neutral type nonlinear large scale control system with delays, Proceedings of IEEE'96 International Conference on Systems, Man and Cybernation, 1996, Beijing.

Matthews G P, Decentralized variable structure control for class of multiinput multioutput nonlinear systems, Proc 24th Annual IEEE Conf on Decision and Control, Fort Lauderriale, Foride, 1985:1719-1724

Gao W B, The basic theory of the variable structure control. China Press of Sciences and Technology, Bejjing, China, 1990(in Chinese):22-24

Gao Cunchen, Liu Yongqing. Variable structure control of multi-delays neutral nonlinear large scale control systems with time-delay input control, Journal of South China University of Technology, 1996,24(5):102-108

14th Triennial World Congress, Beijing, P.R. China

E-2c-21-6

EFFECTS OF MAGNITUDE SATURATION IN CONTROL OF BIFURCATIONS

Yong Wang and Richard M. Murray

Division of Engineering and Applied Science
California Institute of Technology
Pasadena, CA 91125, USA

Abstract: Motivated by problems such as active control of rotating stall in compression systems, an analysis of the effects of controller magnitude saturation in feedback stabilization of steady-state bifurcations is performed. In particular the region of attraction to the stabilized bifurcated equilibria is solved for feedback controllers with magnitude saturation limits using the technique of center manifold reduction and bifurcation analysis. It has been shown that the stability boundary is the saturation envelope formed by the unstable equilibria for the closed loop system when the controllers saturate. The framework allows the design of feedback control laws to achieve desirable size of region of attraction when the noise is modeled as a closed set of initial conditions in the phase space. Modular the phase of the limit cycles, the qualitative behavior in the Hopf bifurcation case is the same as some cases in the steady-state bifurcations. *Copyright © 1999 IFAC*

Keywords: Bifurcation, center manifold, magnitude saturation, region of attraction, saturation envelope

1. INTRODUCTION

Feedback stabilization is one of the most important issues in control theory. Although controllability is a mathematically generic concept for linear time invariant systems, linearly uncontrollable dynamical systems are encountered in practice. For example, the three state Moore Greitzer model describing rotating stall and surge behavior in compression systems with down stream bleed valve actuators is linearly uncontrollable around the nominal equilibria denoting steady axisymmetric flow. Typically for an axial compression system the Hopf bifurcation to rotating stall is subcritical and there is a hysteresis associated with it (see Figure 1(a)). The stall inception is catastrophic. However, control laws can be designed to change the bifurcation characteristic from subcritical to supercritical (see [7]), whereby the hysteresis associated with the subcritical bifurcation is eliminated and the stall inception is progressive (see Figure 1(b)). But in applications, the magnitude and rate saturation of the actuators can seriously limit the performance of the controllers (see [9,10,13]). Due to the magnitude saturation of the actuators there is a hysteresis for the closed loop system and the region of attraction to the stabilized stall equilibria achieved by the controllers could be very small (see Figure 1(c)). The goal in this paper is to generalize the techniques developed in [9,10] to nonlinear dynamical systems with steady-state or Hopf bifurcations.

Much research has been done on feedback stabilization of nonlinear systems that possess linearly uncontrollable modes (see [1–4,6,8,14]). Under certain nondegeneracy conditions, necessary and sufficient conditions have been obtained in [12] for the existence of sufficiently smooth state feedbacks such that the bifurcation for the closed loop system is supercritical, provided that it is a single-input system with simple steady-state or Hopf bi-

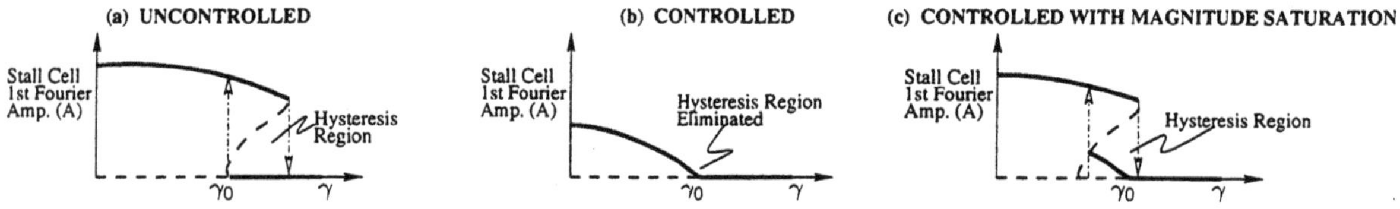

Fig. 1. Control of rotating stall in compression system using bleed valve actuators.

furcations for which the critical modes are linearly uncontrollable. Also in [12], explicit constructions of the stabilizing state feedbacks are given. The contribution of this paper is to estimate the region of attraction to the stabilized equilibria for the controllers designed in [9] subject to magnitude saturation limits.

2. REVIEW OF PREVIOUS RESULTS

Consider the following single-input system

$$\dot{y} = f_\mu(y, u), \tag{1}$$

where $y \in \mathbb{R}^{n+1} (n \geq 1)$ is the state variable, $\mu \in \mathbb{R}$ is a bifurcation parameter, and $u \in \mathbb{R}$ is the control input. We assume that the vector field is sufficiently smooth and the system has a steady-state solution $y_0(\mu)$. $\lambda(\mu)$ is a simple real eigenvalue of the linearization around $y_0(\mu)$ and the eigenspace associated with $\lambda(\mu)$ is the only linearly uncontrollable mode. Under these assumptions, we use Taylor series expansion followed by a linear transformation to decouple the uncontrollable mode from the controllable modes. The resulting standard form is given by

$$\begin{aligned}
\dot{x} &= d\mu x + q_{11}x^2 + q_{12}\tilde{x}x + q_{13}xu + \tilde{x}^T q_{22}\tilde{x} \\
&\quad + q_{23}\tilde{x}u + q_{33}u^2 + c_{111}x^3 + c_{112}\tilde{x}x^2 \\
&\quad + c_{113}x^2u + \tilde{x}^T c_{122}\tilde{x}x + c_{123}\tilde{x}ux + c_{133}xu^2 \\
&\quad + c_{222}(\tilde{x},\tilde{x},\tilde{x}) + \tilde{x}^T c_{223}\tilde{x}u + c_{233}\tilde{x}u^2 \\
&\quad + c_{333}u^3 + \cdots, && (2) \\
\dot{\tilde{x}} &= A\tilde{x} + Bu + \tilde{q}_{11}x^2 + \tilde{q}_{12}\tilde{x}x + \tilde{q}_{13}xu \\
&\quad + \tilde{q}_{22}(\tilde{x},\tilde{x}) + \tilde{q}_{23}\tilde{x}u + \tilde{q}_{33}u^2 + \cdots, && (3)
\end{aligned}$$

where $x \in \mathbb{R}$, $\tilde{x} \in \mathbb{R}^n$, $A \in \mathbb{R}^{n\times n}$, and $B \in \mathbb{R}^{n\times 1}$, and (A, B) is in controller canonical form. The state x represents the uncontrollable mode and the vector $\tilde{x}$ represents the controllable modes.

Let $e_n = [\,0 \cdots 0\ 1\,]^T \in \mathbb{R}^n$. In the following, we denote $q_{12}^n = q_{12}e_n$, $q_{22}^{nn} = e_n^T q_{22} e_n$, etc. Define

$$\begin{aligned}
\Upsilon_1 &= q_{12}^n + q_{13}a_n, \\
\Upsilon_2 &= q_{22}^{nn} + q_{23}^n a_n + q_{33}a_n^2, \\
\Pi_1^S &= [\,0\ q_{12}^1 \cdots q_{12}^{n-1}\,] \\
&\quad + q_{13}[\,1\ a_1 \cdots a_{n-1}\,], \\
\Pi_2^S &= 2[\,0\ q_{22}^{n1} \cdots q_{22}^{n,n-1}\,] \\
&\quad + a_n[\,0\ q_{23}^1 \cdots q_{23}^{n-1}\,] \\
&\quad + (q_{23}^n + 2q_{33}a_n)[\,1\ a_1 \cdots a_{n-1}\,], \\
\alpha_0 &= c_{111} - \Pi_1^S\tilde{q}_{11}, && (4) \\
\alpha_1 &= c_{112}^n + c_{113}a_n - \Pi_1^S(\tilde{q}_{12}^n + \tilde{q}_{13}a_n) \\
&\quad - \Pi_2^S\tilde{q}_{11}, && (5) \\
\alpha_2 &= c_{122}^{nn} + c_{123}^n a_n + c_{133}a_n^2 \\
&\quad - \Pi_1^S(\tilde{q}_{22}^{nn} + \tilde{q}_{23}^n a_n + \tilde{q}_{33}a_n^2) \\
&\quad - \Pi_2^S(\tilde{q}_{12}^n + \tilde{q}_{13}a_n), && (6) \\
\alpha_3 &= c_{222}^{nnn} + c_{223}^{nn}a_n + c_{233}^n a_n^2 + c_{333}a_n^3 \\
&\quad - \Pi_2^S(\tilde{q}_{22}^{nn} + \tilde{q}_{23}^n a_n + \tilde{q}_{33}a_n^2), && (7) \\
P(\zeta) &= \alpha_0 + \alpha_1\zeta + \alpha_2\zeta^2 + \alpha_3\zeta^3. && (8)
\end{aligned}$$

where $p(s) = \det(sI - A) = s^n + a_1 s^{n-1} + \cdots + a_{n-1}s + a_n$. Also define $\Delta_\Upsilon = \Upsilon_1^2 - 4q_{11}\Upsilon_2$, $\Delta_\alpha = \alpha_1^2 - 4\alpha_2\alpha_0$, and $\rho = -\frac{\Upsilon_1}{2\Upsilon_2}$.

Under some nondegeneracy conditions (see [12]), there exists a sufficiently smooth feedback

$$\begin{aligned}
u = F(\tilde{x}, x) &= K_1\tilde{x} + K_2x + K_3x^2 + K_4\tilde{x}x \\
&\quad + \tilde{x}^T K_5\tilde{x} + \mathcal{O}\left(\{x,\tilde{x}\}^3\right), && (9)
\end{aligned}$$

such that $(0, 0)$ is asymptotically stable if and only if one of the conditions in Table 1 is satisfied. The construction of the stabilizing feedbacks is given in Table 2, where $K = \frac{K_2}{\bar{a}_n}$, and

$$\begin{aligned}
\check{\alpha}_0 &= \frac{\Upsilon_1}{\bar{a}_n}[\,1\ \bar{a}_1 \cdots \bar{a}_{n-1}\,]\tilde{Q}(K), \\
\hat{\alpha}_0 &= \frac{1}{\bar{a}_n}\sqrt{\Delta_\Upsilon}[\,1\ \bar{a}_1 \cdots \bar{a}_{n-1}\,]\tilde{Q}(K), \\
\tilde{Q}(K) &= \tilde{q}_{11} + (\tilde{q}_{12}^n + \tilde{q}_{13}a_n)K \\
&\quad + (\tilde{q}_{22}^{nn} + \tilde{q}_{23}^n a_n + \tilde{q}_{33}a_n^2)K^2, && (10)
\end{aligned}$$

where $\det[sI - (A + BK_1)] = s^n + \bar{a}_1 s^{n-1} + \cdots + \bar{a}_{n-1}s + \bar{a}_n$. The construction procedure consists of the following two steps: First, choose K_1 such that $A + BK_1$ is Hurwitz, then let $K_4 = 0$, $K_5 = 0$, and choose K_2, K_3 as shown in Table 2.

3. EFFECTS OF MAGNITUDE LIMITS IN STEADY-STATE BIFURCATIONS

We consider the system (1) with state feedback $u = \alpha(y)$ constrained by controller magnitude

	q_{11}	Υ_1	Υ_2	Δ_Υ	α_3	α_2	α_1	α_0	Δ_α	$P(\rho)$
SS-1	$=0$							<0		
SS-2	$=0$	$\neq 0$						≥ 0		
SS-3	$=0$	$=0$	$=0$		$\neq 0$			≥ 0		
SS-4	$=0$	$=0$	$=0$		$=0$	<0		≥ 0		
SS-5	$=0$	$=0$	$=0$		$=0$	$=0$	$\neq 0$	≥ 0		
SS-6	$=0$	$=0$	$=0$		$=0$	>0		≥ 0	>0	
SS-7	$\neq 0$	$\neq 0$	$=0$							
SS-8	$\neq 0$		$\neq 0$	>0						
SS-9	$\neq 0$		$\neq 0$	$=0$						<0

Table 1. Different cases when steady-state bifurcations are stabilizable.

	K	K_3
SS-1	0	$-[1\ \bar{a}_1\ \cdots\ \bar{a}_{n-1}]\tilde{q}_{11}$
SS-2	0	$\alpha_0 + \frac{\Upsilon_1}{\bar{a}_n}\left(K_3 + [1\ \bar{a}_1\ \cdots\ \bar{a}_{n-1}]\tilde{q}_{11}\right) < 0$
SS-3	$P(K) < 0$	0
SS-4	$P(K) < 0$	0
SS-5	$P(K) < 0$	0
SS-6	$-\frac{\alpha_1}{2\alpha_2}$	0
SS-7	$-\frac{q_{11}}{\Upsilon_1}$	$\Upsilon_1 \frac{K_3}{\bar{a}_n} + \tilde{\alpha}_0 + P(K) < 0$
SS-8	$\frac{1}{2\Upsilon_2}\left(-\Upsilon_1 + \sqrt{\Delta_\Upsilon}\right)$	$\frac{K_3}{\bar{a}_n}\sqrt{\Delta_\Upsilon} + \hat{\alpha}_0 + P(K) < 0$
SS-9	$-\frac{\Upsilon_1}{2\Upsilon_2}$	0

Table 2. Construction of stabilizing feedback for steady-state bifurcations.

limits:

$$u = \sigma(\alpha(y)) := \begin{cases} u_{mag}, & \text{if } \alpha(y) \geq u_{mag}, \\ \alpha(y), & \text{if } |\alpha(y)| < u_{mag}, \\ -u_{mag}, & \text{if } \alpha(y) \leq -u_{mag}. \end{cases}$$

By the same procedure in the previous section, we transform the system into a standard form (2) and (3) with $u = \sigma(F(\tilde{x}, x))$, where $F(\tilde{x}, x)$ is given by (9). For the simplicity of discussion, we assume the controllable subsystem is stable, i.e., A is Hurwitz for all values of μ. From the construction procedure in the above section it is sufficient to only consider the state feedback on the linearly uncontrollable mode x, i.e., $K_1 = 0$, $\bar{a}_j = a_j$ $(j = 1, \cdots, n)$, and $u = \sigma(F(x))$.

The goal in this section is to analyze the qualitative bifurcation behavior of the closed loop system under the constraints of controller magnitude saturation. We approach this problem in the following way: first, we derive the dynamics on the center manifold for the closed loop system without controller magnitude saturation, then we add saturation limits to the dynamics on the center manifold. By doing so, we could estimate the region of attraction for the stabilized equilibria. The details of the center manifold reduction can be found in [11]. We analyze the effects of magnitude limits case by case. For **SS-1**, the bifurcation for the uncontrolled system is already supercritical. For **SS-2**, the dynamics on the center manifold is given by

$$\dot{x} = d\mu x + b_1 x u + b_2 x^3,$$
$$u = \sigma\left(K_3 x^2\right)$$

where $b_1 = \frac{\Upsilon_1}{a_n}$, and

$$b_2 = c_{111} + \frac{1}{a_n}\left(q_{12}^n \left[\, 1\ a_1\ \cdots\ a_{n-1} \,\right] - a_n \left[\, 0\ q_{12}^1\ \cdots\ q_{12}^{n-1} \,\right]\right)\tilde{q}_{11}.$$

Without loss of generality we assume $d < 0$ and $b_1 < 0$, then the bifurcation for the open loop system is a subcritical pitchfork bifurcation since $b_2 > 0$ (see Figure 2(a)). If $K_3 > 0$ is small, then the bifurcation for closed loop system is given in Figure 2(a). The bifurcation for the closed loop system is still a subcritical bifurcation, but the region of attraction to the nominal equilibria $x = 0$ is enlarged. The enlarged region is given by the shaded area in Figure 2(a). If we increase K_3, then the bifurcation for the closed loop system is a supercritical pitchfork bifurcation (see Figure 2(b)). Also, the region enlarged from control is given by the shaded region. The intersection between the stabilized equilibria and the saturation envelope signifies the operating limit, i.e., any trajectory from a nonzero initial condition to the left of it will go to infinity. As the gain K_3 goes to infinity, the saturation envelope is a closed curve shown in Figure 2(b). The equation of the saturation envelope is given by

$$\mu = \frac{1}{d}\left(|b_1| u_{mag} + b_2 x^2\right).$$

It is clear that any trajectory from a nonzero initial conditions will go to infinity if the system operates at $\mu < \mu^* := \frac{|b_1|}{d} u_{mag}$. In other words, the region between the saturation envelope and the unstable bifurcated equilibria of the uncontrolled system is the maximal augmented region of attraction due to the magnitude saturation limit.

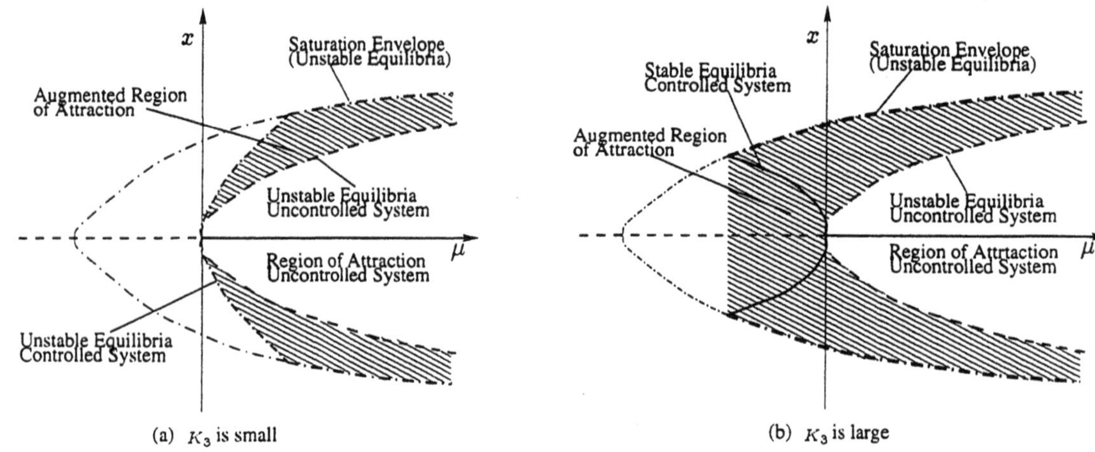

Fig. 2. Bifurcation diagrams of **SS-2** for two different gains.

For **SS-3**, the dynamics on the center manifold is given by

$$\dot{x} = d\mu x + \alpha_0 x^3 + \alpha_1 x^2 u + \alpha_2 x u^2 + \alpha_3 u^3,$$
$$u = \sigma(Kx)$$

where α_j $(j = 0, \cdots, 3)$ are given by (4), (5), (6), and (7). The saturation envelope is given by

$$\mu = -\frac{1}{d}\left(\alpha_0 x^2 \pm \alpha_1 u_{mag} x + \alpha_2 u_{mag}^2 \pm \frac{\alpha_3 u_{mag}^3}{x}\right).$$

The bifurcation diagrams could be characterized by two cases and they are given in Figure 3(a) and Figure 3(b). The interpretations to Figure 3(a) and Figure 3(b) for the region of attraction are the same as the previous case. In Figure 3(a) one branch of the saturation envelope has at most one point at which $\frac{d\mu}{dx} = 0$, whereas in Figure 3(b) there are two points on the saturation envelope at which $\frac{d\mu}{dx} = 0$. The solid line parts on the saturation envelope in Figure 3(b) are stable equilibria. If $|K|$ is sufficiently large, this stable equilibrium branch attracts the initial conditions between the upper and lower branch of the unstable equilibria on the saturation envelope. The difference between the saturation envelopes of **SS-2** and **SS-3** is that the saturation envelope of **SS-2** is closed but the saturation envelope of **SS-3** is open and goes off along the μ-axis. This implies that if the noise (i.e., initial conditions) is infinitely small, then **SS-3** will give larger region of stable operating range.

For **SS-4**, the dynamics on the center manifold is given by

$$\dot{x} = d\mu x + \alpha_0 x^3 + \alpha_1 x^2 u + \alpha_2 x u^2,$$
$$u = \sigma(Kx)$$

where α_j $(j = 0, 1, 2)$ are given by (4), (5), and (6). The saturation envelope is given by

$$\mu = -\frac{1}{d}\left(\alpha_0 x^2 \pm \alpha_1 u_{mag} x + \alpha_2 u_{mag}^2\right).$$

The bifurcation diagram is given by Figure 3(c). The saturation envelope is a closed curve. μ_1 is the maximum of operating range that can be achieved, i.e., any trajectory from a nonzero initial condition of x diverges to infinity. Also the stable equilibria for $\mu \in [\mu_1, \mu_2]$ are determined by the controller saturation limit, so increasing gain will not make them arbitrarily closed to the nominal equilibria $x = 0$.

For **SS-5**, the dynamics on the center manifold is given by

$$\dot{x} = d\mu x + \alpha_0 x^3 + \alpha_1 x^2 u + \alpha_2 x u^2,$$
$$u = \sigma(Kx),$$

where α_j $(j = 0, 1, 2)$ are given by (4), (5), and (6). The saturation envelope is given by

$$\mu = -\frac{1}{d}\left(\alpha_0 x^2 \pm \alpha_1 u_{mag} x + \alpha_2 u_{mag}^2\right).$$

The bifurcation diagram is given by Figure 3(d). In this case, we have $\mu_2 = 0$, so it is impossible to bring the stable equilibria arbitrarily close to the nominal equilibria for $\mu \in [\mu_1, 0)$ since the stable equilibria are determined by the controller magnitude saturation limit.

For **SS-6**, the dynamics on the center manifold is given by

$$\dot{x} = d\mu x + \alpha_0 x^3 + \alpha_1 x^2 u,$$
$$u = \sigma(Kx),$$

where α_0 and α_1 are given by (4) and (5), respectively. The saturation envelope is given by

$$\mu = -\frac{1}{d}\left(\alpha_0 x^2 \pm \alpha_1 u_{mag} x\right).$$

The bifurcation diagram is given by Figure 3(e). In this case, we have $\mu_1 < 0 < \mu_2$, so the bifurcation for the closed loop system is subcritical if the gain is very large, but the region of attraction to the stable part of the saturation envelope is larger than that to the nominal equilibria for the open loop system.

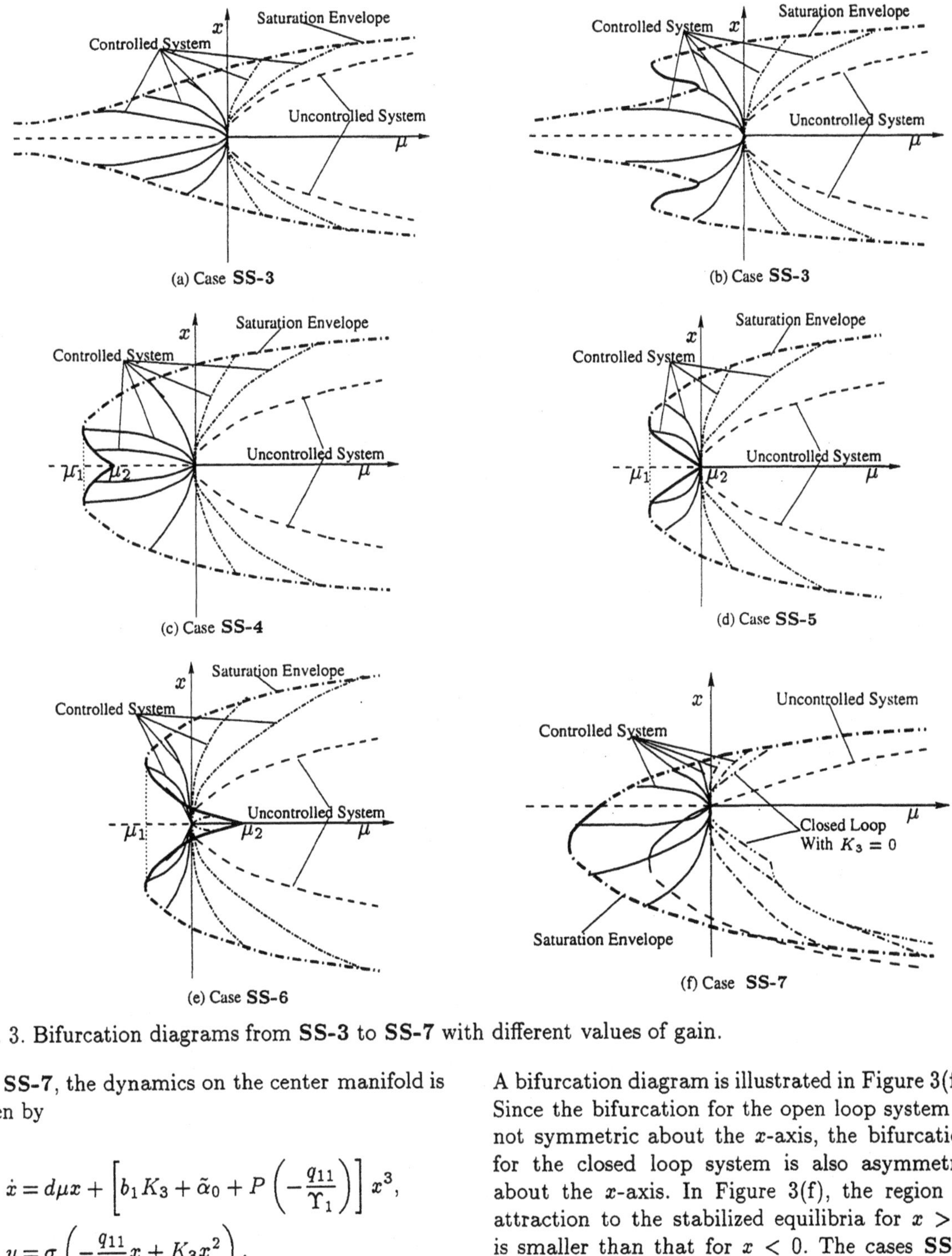

(a) Case **SS-3** (b) Case **SS-3** (c) Case **SS-4** (d) Case **SS-5** (e) Case **SS-6** (f) Case **SS-7**

Fig. 3. Bifurcation diagrams from **SS-3** to **SS-7** with different values of gain.

For **SS-7**, the dynamics on the center manifold is given by

$$\dot{x} = d\mu x + \left[b_1 K_3 + \tilde{\alpha}_0 + P\left(-\frac{q_{11}}{\Upsilon_1}\right)\right] x^3,$$
$$u = \sigma\left(-\frac{q_{11}}{\Upsilon_1}x + K_3 x^2\right),$$

where

$$\tilde{\alpha}_0 = \frac{\Upsilon_1}{a_n}\left[1\ a_1\ \cdots\ a_{n-1}\right]\tilde{Q}\left(-\frac{q_{11}}{\Upsilon_1}\right),$$

$P(\cdot)$ is given by (8), and $\tilde{Q}(\cdot)$ is given by (10). The saturation envelope is given by

$$\mu = -\frac{\Upsilon_1}{da_n}\left(\pm u_{mag} + \frac{q_{11}}{\Upsilon_1}x\right) - \frac{1}{d}\left[\tilde{\alpha}_0 + P\left(-\frac{q_{11}}{\Upsilon_1}\right)\right]x^2,$$
$$\pm u_{mag} = -\frac{q_{11}}{\Upsilon_1}x + K_3 x^2.$$

A bifurcation diagram is illustrated in Figure 3(f). Since the bifurcation for the open loop system is not symmetric about the x-axis, the bifurcation for the closed loop system is also asymmetric about the x-axis. In Figure 3(f), the region of attraction to the stabilized equilibria for $x > 0$ is smaller than that for $x < 0$. The cases **SS-8** and **SS-9** can be similarly analyzed.

The above is a complete discussion of the effects of magnitude saturation in feedback stabilization of steady state bifurcations. The most important feature is that the boundary of the region of attraction to the stabilized equilibria for all the gains is the saturation envelope formed by the unstable equilibria for the closed loop system when the controllers saturate. We remark here that for the cases when there does not exist a sufficiently smooth feedback such that the bifurcation for the closed loop system is supercritical, we can design feedbacks to enlarge the region of attraction. As in the stabilizable cases discussed above, the boundary for the enlarged region of attraction is formed by the saturation envelope.

Similar to the steady-state bifurcation, we have classified the stabilizability for Hopf bifurcations in [12]. We have also given the construction of the stabilizing feedback controllers in [12]. We remark here that the techniques for analyzing the region of attraction to the stabilized equilibria in steady-state bifurcations here in this paper apply to the Hopf bifurcations. The only difference is that in the Hopf bifurcation case, we have to take into account the normal form transformation in addition to the center manifold reduction. The qualitative behavior of the feedback controllers with magnitude saturation for the Hopf bifurcation case is the same as the steady-state bifurcations discussed in this paper. The bifurcation and the saturation envelope for the Hopf bifurcations are the same as **SS-2** to **SS-6** except that the equilibria in the bifurcation diagram are replaced by limit cycles. Due to space limitation, we will not discuss the Hopf bifurcation case.

4. CONCLUSIONS AND FUTURE WORK

In this paper we have analyzed the effects of magnitude saturation limits in control of steady-state bifurcations by estimating the region of attraction to the stabilized equilibria. The boundary of the region of attraction which is the saturation envelope which is formed by the unstable equilibria when the controllers saturate. The significance of the analysis is that it helps understand the qualitative behavior of the magnitude limits in control of bifurcations. A direct application of the theory developed here is in active control of rotating stall, where the effects of magnitude saturation for the bleed valve actuator are the same as the case **SS-2** (see [9,10]), and the effects of magnitude limits for the pulsed air injection control are the same as the case **SS-3**.

We have also addressed that the effects of controller magnitude saturation for the Hopf bifurcation case are qualitatively the same as some cases in the steady-state bifurcation, except that the saturation enveloped is formed by the unstable limit cycles when the controllers saturate. For future work we need to address the effects of bandwidth and rate limits in addition to the magnitude limits.

ACKNOWLEDGMENT We would like to thank Professors J. E. Marsden and J. D. Paduano for many useful suggestions. We would also like to thank S. Yeung for a number of helpful discussions.

5. REFERENCES

[1] Abed, E. H., and J.-H. Fu (1986). Local Feedback Stabilization and Bifurcation Control, I. Hopf Bifurcation, *System and Control Letters*, No. 7, pp. 11-17.

[2] Abed, E. H., and J.-H Fu (1987). Local Feedback Stabilization and Bifurcation Control, II. Stationary Bifurcation, *System and Control Letters*, No. 7, pp. 467-473.

[3] Aeyels, D. (1985). Stabilization of a Class of Nonlinear Systems by a Smooth Feedback, *System and Control Letters*, No. 5, pp. 289-294.

[4] Fu, J.-H., and E. H. Abed (1993). Linear Feedback Stabilization of Nonlinear Systems with an Uncontrollable Critical Mode, *Automatica*, Vol. 29, No. 4, pp. 999-1010.

[5] Guckenheimer, J., and P. Holmes (1983). *Nonlinear Oscillations, Dynamical Systems, and Bifurcation of Vector Fields*. Spring-Verlag.

[6] Kang, W. (1998). Bifurcation and Normal Form of Nonlinear Control Systems, Part I and Part II, *SIAM Journal on Control and Optimization*, Vol. 36, No. 1, pp. 193-232.

[7] Liaw, D. C., and E. H. Abed (1996) Control of Compressor Stall Inception: A Bifurcation-Theoretic Approach, *Automatica*, Vol. 32, No. 1, pp. 109-115.

[8] Mehra, R. K. (1977). Catastrophe Theory, Nonlinear System Identification and Bifurcation Control, *Proc. of the Joint Automatic Control Conference*, pp. 823-831.

[9] Wang, Y., and R. M. Murray (1997). Effects of Noise, Magnitude Saturation and Rate Limits on Rotating Stall Control, *Proc. IEEE Conference on Decision and Control*, pp. 4682-4688.

[10] Wang, Y., and R. M. Murray (1998). Effects of the Shape of Compressor Characteristics on Actuator Requirements for Rotating Stall Control, *Proc. 1998 American Control Conference*, pp. 1996-2001.

[11] Wang, Y., and R. M. Murray (1998). Feedback Stabilization of Steady-State and Hopf Bifurcations, *CDS Report 98-005, Caltech.*

[12] Wang, Y., and R. M. Murray (1998). Feedback Stabilization of Steady-State and Hopf Bifurcations, *Proc. IEEE Conference on Decision and Control*, pp. 2431-2437.

[13] Yeung, S., and R. M. Murray (1997). Reduction of Bleed Valve Rate Requirements for Control of Rotating Stall Using Continuous Air Injection, *Proc. IEEE Conference on Control Applications*, pp. 683-690.

[14] Zaslavsky, B. (1996). Feedback Stabilization of Connected Nonlinear Oscillators with Uncontrollable Linearization, *Systems and Control Letters*, No. 27, pp. 181-185.

AUTHOR INDEX

KEYWORD INDEX

www.ingramcontent.com/pod-product-compliance
Ingram Content Group UK Ltd.
Pitfield, Milton Keynes, MK11 3LW, UK
UKHW052301280926
15471UKWH00006B/1041

9 780080 432168